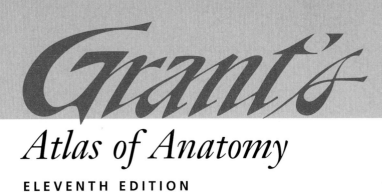

Grant's
Atlas of Anatomy
ELEVENTH EDITION

1 THORAX 1

2 ABDOMEN 91

3 PELVIS AND PERINEUM 183

4 BACK 273

5 LOWER LIMB 337

6 UPPER LIMB 457

7 HEAD 587

8 NECK 721

9 CRANIAL NERVES 793

INDEX 827

Grant's
Atlas of Anatomy
ELEVENTH EDITION

ANNE M.R. AGUR, B.Sc. (OT), M.Sc., Ph.D.

Associate Professor in Division of Anatomy, Department of Surgery

Department of Physical Therapy and Department of Occupational Therapy

Division of Biomedical Communications, Department of Surgery

Institute of Medical Science

Faculty of Medicine

University of Toronto

Toronto, Ontario, Canada

ARTHUR F. DALLEY II, Ph.D.

Professor of Cell and Developmental Biology

Department of Cell and Developmental Biology

Vanderbilt University

School of Medicine

Nashville, Tennessee, U.S.A.

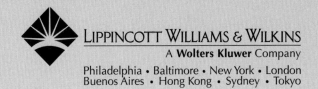

LIPPINCOTT WILLIAMS & WILKINS
A **Wolters Kluwer** Company

Philadelphia • Baltimore • New York • London
Buenos Aires • Hong Kong • Sydney • Tokyo

Editor: Betty Sun
Managing Editors: Beth Goldner, Amy Oravec, Crystal Taylor
Marketing Manager: Joseph Schott
Production Editor: Jennifer Glazer
Production Coordinator: Wayne Hubbel
Compositor: Maryland Composition
Printer: R.R. Donnelley

351 West Camden Street
Baltimore, Maryland 21201-2436 USA

530 Walnut Street
Philadelphia, Pennsylvania 19106-3621 USA

Printed in the United States of America

By J.C.B. Grant:
First Edition, 1943
Second Edition, 1947
Third Edition, 1951
Fourth Edition, 1956
Fifth Edition, 1962
Sixth Edition, 1972
By J.E. Anderson:
Seventh Edition, 1978
Eighth Edition, 1983
By A.M.R. Agur:
Ninth Edition, 1991
By A.M.R. Agur and M.J. Lee:
Tenth Edition, 1999

Library of Congress Cataloging-in-Publication Data is available. (Softcover: 0-7817-4255-2, hardcover: 0-7817-4256-0)

To purchase additional copies of this book, call our customer service department at **(800) 638-3030** or fax orders to **(301) 824-7390**. International customers should call **(301) 714-2324**.

Visit Lippincott Williams & Wilkins on the Internet: http://www.lww.com. Lippincott Williams & Wilkins customer service representatives are available from 8:30 am to 6:00 pm, EST, Monday through Friday.

05 06 07 08
3 4 5 6 7 8 9 10

To my husband Enno and my children Erik and Kristina
for their support and encouragement
and
To my mother Valentina Vilde and late father Raimond Vilde
(A.M.R.A.)

To Muriel
My bride, best friend, counselor, and mother of our boys;
To my family
Tristan, Lana and Elijah Gray,
Denver and Skyler
With great appreciation for their support, humor and patience;
And in loving memory of my nephew and devoted fan
Tymon Jeremy Downs, B.S.N., Nurse Anesthetist-in-training
R.I.P.
(A.F.D.)

DR. JOHN CHARLES BOILEAU GRANT
1886—1973

Dr. J.C.B. Grant in his office, McMurrich Building, University of Toronto, 1946. Through his textbooks, Dr. Grant made an indelible impression on the teaching of anatomy throughout the world.

by **Dr. Carlton G. Smith**, M.D., Ph.D. (1905–2003)

The life of J.C. Boileau Grant has been likened to the course of the seventh cranial nerve as it passes out of the skull: complicated, but purposeful.[1] He was born in the parish of Lasswade in Edinburgh, Scotland, on February 6, 1886. Dr. Grant studied medicine at the University of Edinburgh from 1903 to 1908. Here, his skill as a dissector in the laboratory of the renowned anatomist, Dr. Daniel John Cunningham (1850–1909), earned him a number of awards.

Following graduation, Dr. Grant was appointed the resident house officer at the Infirmary in Whitehaven, Cumberland. From 1909 to 1911, Dr. Grant demonstrated anatomy in the University of Edinburgh, followed by two years at the University of Durham, at Newcastle-on-Tyne in England, in the laboratory of Professor Robert Howden, editor of *Gray's Anatomy*.

With the outbreak of World War I in 1914, Dr. Grant joined the Royal Army Medical Corps and served with distinction. He was mentioned in dispatches in September 1916, received the Military Cross in September 1917 for "conspicuous gallantry and devotion to duty during attack," and received a bar to the Military Cross in August 1918.[1] In October 1919, released from the Royal Army, he accepted the position of Professor of Anatomy at the University of Manitoba in Winnipeg, Canada. With the front-line medical practitioner in mind, he endeavored to "bring up a generation of surgeons who knew exactly what they were doing once an operation had begun."[1] Devoted to research and learning, Dr. Grant took interest in other projects, such as performing anthropometric studies of Indian tribes in northern Manitoba during the 1920s. In Winnipeg, Dr. Grant met Catriona Christie, whom he married in 1922.

Dr. Grant was known for his reliance on logic, analysis, and deduction as opposed to rote memory. While at the University of Manitoba, Dr. Grant began writing *A Method of Anatomy*, *Descriptive and Deductive*, which was published in 1937 (by Williams & Wilkins Co.).

In 1930, Dr. Grant accepted the position of Chair of Anatomy at the University of Toronto. He stressed the value of a "clean" dissection, with the structures well defined. This required the delicate touch of a sharp scalpel, and students soon learned that a dull tool was an anathema. Instructive dissections were made available in the Anatomy Museum, a means of student review on which Dr. Grant placed a high priority. Many of these illustrations have been included in *Grant's Atlas of Anatomy*. The first edition of the *Atlas*, published in 1943 (by Williams & Wilkins Co.), was the first anatomical atlas to be published in North America. *Grant's Dissector* preceded the *Atlas* in 1940.

Dr. Grant remained at the University of Toronto until his retirement in 1956. At that time, he became Curator of the Anatomy Museum in the University. He also served as Visiting Professor of Anatomy at the University of California at Los Angeles, where he taught for 10 years.

Dr. Grant died in 1973 of cancer. Through his teaching method, still presented in the Grant's textbooks, Dr. Grant's life interest—human anatomy—lives on. In their eulogy, colleagues and friends Ross MacKenzie and J. S. Thompson said: "Dr. Grant's knowledge of anatomical fact was encyclopedic, and he enjoyed nothing better than sharing his knowledge with others, whether they were junior students or senior staff. While somewhat strict as a teacher, his quiet wit and boundless humanity never failed to impress. He was, in the very finest sense, a scholar and a gentleman."[1]

Reference

1. Robinson C. Canadian Medical Lives. J.C. Boileau Grant: Anatomist Extraordinary. Markham, Ontario, Canada: Associated Medical Services Inc./Fithzenry & Whiteside, 1993.

PREFACE

Preparing a new edition of an atlas, even a classic with as long and rich a history as *Grant's Atlas of Anatomy*, requires intensive research and creativity. It is not enough to rely on a solid reputation. Medical and Health Sciences education, and the role of anatomy instruction and application within it, are continually changing with the evolution of educational models and developments in the administration of health care. To complement these, new and improved methods of presenting content are being enabled by rapid advances in publishing technology.

Vision

The revision process began by evaluating the existing content of the *Atlas*, and then soliciting and analyzing feedback from medical students and faculty members from around the world. Many of these ideas are incorporated to provide you with this, the Eleventh Edition of *Grant's Atlas of Anatomy*. You will find the *Atlas* to be:

- A companion in the dissecting room, guiding you through each step of exploring and exposing structures;
- A portable "anatomical museum," useful in determining relationships with superficial structures removed; reviewing regional anatomy when away from the dissection lab, or when cadavers are not available; and preparing for examinations;
- A lifetime companion as a reference book for all fields of health care.

Key Features

Organization and Layout. Ease-of-use has always been the goal in determining the organization and layout of the *Atlas*. In this edition sequential illustrations are resized to maintain scale, so that relevant relationships can be better appreciated throughout the series, and the need for reorientation is minimized. All figures were considered in terms of the preceding and subsequent figures, with special attention given to figures on facing pages, using the two-page spread to greatest advantage. The layout is uniform in the Eleventh Edition with all legends placed at the bottom of the page, improving the use of white space. Legends are also reduced in length and updated. The observations and comments that accompany the illustrations draw attention to salient points and significant structures that might otherwise escape notice. Their purpose is to augment the illustrations without providing exhaustive description. Readability, clarity, and practicality were emphasized in the preparation of this edition. In addition, the number of illustrations per page is kept to a minimum to highlight relevant structures. Although the basic organization by body region is maintained, every figure within every chapter was scrutinized to ensure logical and effective placement. More than 100 new figures were added to complete dissection sequences or to clarify existing material. Also, additional tables, such as overviews of muscles, nerves, and blood vessels, are included.

Classic Dissection Illustrations. A unique feature of *Grant's Atlas of Anatomy* is that rather than providing an idealized view of human anatomy, the classic illustrations are drawn from actual dissections and readily relate to students' own preparations in the lab. Many of the specimens depicted in the *Atlas* are displayed in the Grant Museum of Anatomy at the University of Toronto. Several key dissection illustrations from the Grant's collection that are found to be particularly instructive are reintroduced in the Eleventh Edition, and many of the figures are updated with a more vibrant color palette. Re-colorization of the figures will continue in future editions. The accuracy, simplicity, and beauty of these illustrations have always been an acclaimed feature of *Grant's Atlas*. New illustrations in the classic tradition are added to most chapters to complete dissection sequences, e.g., superficial and deep dissections of the muscles of the anterior aspect of arm in Chapter 6.

Schematic Illustrations. Full-color schematic illustrations supplement many dissection figures to clarify anatomical concepts, emphasize the relationships of structures, or provide an overview of the body region being studied. Many new schematic illustrations are added to this edition. All labeled illustrations were reviewed to ensure that they conform to Dr. Grant's desire to "keep it simple"—extraneous labels were deleted and some labels were added to identify key structures and make the illustrations as useful as possible to students. In addition, many new, simple orientation drawings are included to facilitate understanding of the viewpoint and context of dissected areas.

Surface Anatomy. New surface anatomy images are a highlight of the Eleventh Edition. The new color photographs provide clearly demarcated surface features that form a basis for the sequential dissection illustrations that follow and a basis for physical examination.

Diagnostic Images. Because medical imaging has taken on increased importance in the diagnosis and treatment of injuries and illnesses, this edition features approximately 250 clinically relevant magnetic resonance images (MRIs), computed tomography (CT) scans, ultrasound scans, radiographs, and corresponding orientation drawings. These images are included within each chapter and in an overview of sectional imaging at the end of each chapter. We believe that we achieved our goal of safeguarding the historical strengths of *Grant's Atlas of Anatomy*, while bringing the content up to date, thus enhancing the value and usefulness of the *Atlas* for both students and practitioners.

Anne M.R. Agur
Arthur F. Dalley II

ACKNOWLEDGMENTS

Starting with the first edition of this atlas published in 1943, many people have given generously of their talents and expertise and we acknowledge their participation with heartfelt gratitude. Most of the original carbon dust halftones on which this book is based were created by **Dorothy Foster Chubb**, a pupil of Max Brödel and one of Canada's first professionally trained medical artists. She was later joined by **Nancy Joy**, who is Professor Emeritus in the Division of Biomedical Communications, Department of Surgery, University of Toronto. Mrs. Chubb was mainly responsible for the artwork of the first two editions and the sixth edition; Miss Joy for those in between. In subsequent editions, additional line and half-tone illustrations by **Elizabeth Blackstock**, **Elia Hopper Ross**, and **Marguerite Drummond** were added.

Much credit is also due to **Charles E. Storton** for his role in the preparation of the majority of the original dissections and preliminary photographic work. We also wish to acknowledge the work of Dr. **James Anderson**, a pupil of Dr. Grant, under whose stewardship the seventh and eighth editions were published.

The following individuals also provided invaluable contributions to the previous editions of the *Atlas*, and are gratefully acknowledged: C.A. Armstrong, P.G. Ashmore, D. Baker, D.A. Barr, J.V. Basmajian, S. Bensley, D. Bilbey, J. Bottos, W. Boyd, J. Callagan, H.A. Cates, S.A. Crooks, M. Dickie, J.W.A. Duckworth, F.B. Fallis, J.B. Francis, J.S. Fraser, P. George, R.K. George, M.G. Gray, B.L. Guyatt, C.W. Hill, W.J. Horsey, B.S. Jaden, M.J. Lee, G.F. Lewis, I.B. MacDonald, D.L. MacIntosh, R.G. MacKenzie, S. Mader, K.O. McCuaig, D. Mazierski, W.R. Mitchell, K. Nancekivell, A.J.A. Noronha, S. O'Sullivan, W. Pallie, W.M. Paul, D. Rini, C. Sandone, C.H. Sawyer, A.I. Scott, J.S. Simpkins, J.S. Simpson, C.G. Smith, I.M. Thompson, J.S. Thompson, N.A. Watters, R.W. Wilson, B. Vallecoccia, and K. Yu.

Eleventh Edition

We are indebted to our colleagues and former professors for their encouragement—especially Dr. **Keith L. Moore** for his expert advice and to Drs. **Daniel O. Graney**, **Douglas J. Gould**, and **Jerzy St. Gielecki** for their invaluable input.

We extend our gratitude to the medical artists, **Caitlin Duckwall** of Dragonfly Media Group and **Valerie Oxorn**, who contributed new and modified illustrations, and to **Anne Rayner** of Vanderbilt University Medical Center's Medical Art Group for creating the new surface anatomy photographs.

Special thanks go to everyone at **Lippincott Williams & Wilkins**—especially Crystal Taylor and Beth Goldner, Senior Managing Editors; Amy Oravec, Managing Editor; Betty Sun, Executive Editor; Jennifer Glazer, Senior Production Editor; Wayne Hubbel, Senior Production Coordinator; and Jonathan Dimes, Associate Director, Art Management. All of your efforts and expertise are much appreciated.

We would like to acknowledge the reviewers who provided expert advice on the development of this and previous editions of the *Atlas*. Finally, we would like to thank the hundreds of instructors and students who have over the years communicated via the publisher and directly with the editors their suggestions for how this *Atlas* might be improved. We hope that readers will find many of these suggestions incorporated into the Eleventh Edition and will continue to provide their valuable input.

Anne M.R. Agur
Arthur F. Dalley II

CONTENTS

Dr. John Charles Boileau Grant **vi**

Preface **vii**

Acknowledgments **viii**

List of Tables **xi**

Table and Figure Credits **xii**

1 THORAX 1
Pectoral Region 2
Breast 4
Bony Thorax and Joints 10
Thoracic Wall 17
Thoracic Contents 25
Pleural Cavity 28
Mediastinum 29
Lungs and Pleura 30
Bronchi and Bronchopulmonary Segments 36
Innervation and Lymphatic Drainage of Lungs 42
External Heart 44
Coronary Vessels 52
Internal Heart and Valves 56
Conducting System of Heart 62
Pericardial Sac 63
Superior Mediastinum and Great Vessels 64
Diaphragm 71
Posterior Thorax 72
Scans: Axial, Sagittal, Coronal 82

2 ABDOMEN 91
Overview 92
Anterolateral Abdominal Wall 94
Inguinal Region 100
Testes 106
Peritoneum and Peritoneal Cavities 112
Stomach 120
Pancreas, Duodenum, and Bile Duct 125
Intestines 132
Liver and Gallbladder 142
Portal Venous System 152
Retroperitoneal Viscera 154
Posterior Abdominal Wall and Diaphragm 162
Aorta and Autonomic Nerves 167
Lymphatic Drainage 172
Review 175
Scans: Sagittal, Axial, Coronal 176

3 PELVIS AND PERINEUM 183
Bony Pelvis and Ligaments 184
Floor and Walls of Pelvis 192
Overview of Male Pelvis 196
Rectum 198
Urinary Bladder, Prostate, and Seminal Vesicles 204
Vessels of Male Pelvis 212
Lymphatic Drainage and Innervation of Male Pelvis 218
Lumbosacral Plexus 220
Overview of Female Pelvis 224
Uterus, Ovaries, and Vagina 226
Vessels of Female Pelvis 234
Lymphatic Drainage and Innervation of Female Pelvis 236
Overview of Perineum 242
Male Perineum 244
Female Perineum 256
Scans: Male 264
Scans: Female 268

4 BACK 273
Vertebral Column and Overview of Vertebrae 274
Cervical Vertebrae 282
Thoracic Vertebrae 286
Lumbar Vertebrae 288
Sacrum and Sacroiliac Joint 291
Anomalies of Vertebrae 296
Ligaments of Vertebral Column and Intervertebral Discs 298
Vertebral Venous Plexuses 305
Muscles of Back 306
Suboccipital Region 316
Craniovertebral Joints 320
Spinal Cord and its Environment 322
Spinal Nerves and Segmental Innervation 330
Imaging of Vertebrae and Spinal Cord 334

5 LOWER LIMB 337
Skeleton of Lower Limb 338
Cutaneous Innervation of Lower Limb 340
Venous and Lymphatic Drainage of Lower Limb 342
Fascia and Fascial Compartments of Lower Limb 345
Arteries and Nerves of Lower Limb 346
Inguinal Region and Femoral Triangle 350
Anterior and Medial Thigh 356
Lateral Thigh 365
Gluteal Region and Posterior Thigh 366

Hip Joint 376
Knee Region 384
Knee Joint 390
Anterior and Lateral Leg and Dorsum of Foot 404
Posterior Leg 415
Sole of Foot 424
Ankle and Foot Joints 432
Bony Anomalies 449
Sectional Anatomy of Thigh and Leg 450
Sectional Anatomy and Imaging: Thigh and Leg 454

6 UPPER LIMB 457
Skeleton of Upper Limb 458
Cutaneous Innervation of Upper Limb 462
Venous and Lymphatic Drainage of Upper Limb 464
Fascia and Fascial Compartments of Upper Limb 468
Arteries and Nerves of Upper Limb 470
Pectoral Region 474
Axilla, Axillary Vessels, and Brachial Plexus 481
Scapular Region and Superficial Back 492
Arm and Rotator Cuff 498
Joints of Pectoral Girdle and Shoulder 508
Elbow Region 518
Elbow Joint 524
Anterior Aspect of Forearm 530
Anterior Aspect of Wrist and Palm of Hand 538
Posterior Aspect of Forearm 554
Posterior Aspect of Wrist and Dorsum of Hand 558
Lateral Aspect of Wrist and Hand 564
Medial Aspect of Wrist and Hand 567
Bones and Joints of Wrist and Hand 568
Function of Hand: Grips and Pinches 576
Sectional Anatomy and Imaging: Arm and Forearm 578

7 HEAD 587
Cranium 588
Face and Scalp 600
Circulation and Innervation of Cranial Cavity 606
Meninges, Meningeal Spaces 610
Cranial Base and Cranial Nerves 614
Blood Supply of Brain 622
Overview of Brain and Ventricles 626
Telencephalon (Cerebrum) and Diencephalon 629
Brainstem and Cerebellum 635
Anatomical Sections of Brain 638
Orbit and Eyeball 640

Parotid Region 652
Temporal Region and Infratemporal Fossa 654
Temporomandibular Joint 662
Tongue 666
Palate 672
Teeth 675
Nose, Paranasal Sinuses, and Pterygopalatine Fossa 680
Ear 696
Lymphatic Drainage of Head 710
Imaging of Head: Coronal, Axial, Sagittal Scans 711

8 NECK 721
Cervical Fascia and Regions of Neck 722
Lateral Cervical Region (Posterior Triangle) 730
Veins and Arteries of Neck 734
Anterior Cervical Region (Anterior Triangle) 738
Root of Neck 756
Prevertebral Region 761
External Cranial Base 764
Pharynx 768
Isthmus of Fauces and Tonsils 772
Larynx 778
Sectional Anatomy and Imaging: Neck 786

9 CRANIAL NERVES 793
Overview of Cranial Nerves 794
Cranial Nerve Nuclei 798
Cranial Nerve I: Olfactory 800
Cranial Nerve II: Optic 801
Cranial Nerves III, IV, and VI: Oculomotor, Trochlear, and Abducent 803
Cranial Nerve V: Trigeminal 806
Cranial Nerve VII: Facial 812
Cranial Nerve VIII: Vestibulocochlear 814
Cranial Nerve IX: Glossopharyngeal 816
Cranial Nerve X: Vagus 818
Cranial Nerve XI: Spinal Accessory 820
Cranial Nerve XII: Hypoglossal 821
Summary of Autonomic Ganglia of Head 822
Summary of Cranial Nerve Lesions 823
Sectional Imaging of Cranial Nerves: Transverse and Coronal Scans 824

INDEX 827

LIST OF TABLES

1 THORAX
1.1 Muscles of Thoracic Wall **21**
1.2 Muscles of Respiration **24**
1.3 Surface Markings of Pleura and Lungs **31**

2 ABDOMEN
2.1 Principal Muscles of Anterolateral Abdominal Wall **96**
2.2 Boundaries of the Inguinal Canal **102**
2.3 Parts and Relationships of Duodenum **125**
2.4 Schema of Terminology for Subdivisions of the Liver **146**
2.5 Principal Muscles of Posterior Abdominal Wall **164**
2.6 Nerves of Lumbar Plexus **165**
2.7 Referred Pain **169**

3 PELVIS AND PERINEUM
3.1 Differences between Male and Female Pelves **190**
3.2 Muscles of Pelvic Walls and Floor **192**
3.3 Arteries of Male Pelvis **213**
3.4 Nerves of Sacral and Coccygeal Plexuses **223**
3.5 Arteries of Female Pelvis **235**
3.6 Muscles of Perineum **246**

4 BACK
4.1 Typical Cervical Vertebrae (C3–C7) **284**
4.2 Thoracic Vertebrae **286**
4.3 Lumbar Vertebrae **289**
4.4 Intrinsic (Deep) Back Muscles **315**
4.5 Muscles of the Atlanto-occipital and Atlantoaxial Joints **318**

5 LOWER LIMB
5.1 Motor Nerves of Lower Limb **349**
5.2 Muscles of Anterior Thigh **358**
5.3 Muscles of Medial Thigh **360**
5.4 Muscles of Gluteal Region **368**
5.5 Muscles of Posterior Thigh (Hamstring) **369**
5.6 Nerves of Gluteal Region **374**
5.7 Bursae Around Knee **397**
5.8 Muscles of the Anterior Compartment of the Leg **405**
5.9 Muscles of the Lateral Compartment of the Leg **411**
5.10 Muscles of the Posterior Compartment of the Leg **414**
5.11 Arterial Supply of the Leg and Foot **423**
5.12 Muscles in Sole of Foot—First Layer **428**
5.13 Muscles in Sole of Foot—Second Layer **429**
5.14 Muscles in Sole of Foot—Third Layer **430**
5.15 Muscles in Sole of Foot—Fourth Layer **431**
5.16 Joints of Foot **442**

6 UPPER LIMB
6.1 Anterior Thoracoappendicular Muscles **479**
6.2 Arteries of the Proximal Upper Limb **484**
6.3 Branches of Brachial Plexus **487**
6.4 Superficial Back Muscles **496**
6.5 Rotator Cuff Muscles **497**
6.6 Arm Muscles **500**
6.7 Arteries of Forearm **530**
6.8 Muscles on the Anterior Surface of the Forearm **533**
6.9 Muscles of Hand **545**
6.10 Arteries of Hand **553**
6.11 Muscles on the Posterior Surface of the Forearm **555**

7 HEAD
7.1 Main Muscles of Facial Expression **603**
7.2 Nerves of Face and Scalp **605**
7.3 Arteries of Face and Scalp **606**
7.4 Veins of Face **607**
7.5 Arterial Supply to the Brain **625**
7.6 Arteries of Orbit **647**
7.7 Muscles of Orbit **648**
7.8 Actions of Muscles of the Orbit **649**
7.9 Muscles Acting on the Temporomandibular Joint **662**
7.10 Movements of the Temporomandibular Joint **663**
7.11 Muscles of Tongue **667**
7.12 Muscles of Soft Palate **674**
7.13 Primary and Secondary Dentition **679**

8 NECK
8.1 Platysma **722**
8.2 Cervical Triangles and Contents **724**
8.3 Sternocleidomastoid and Trapezius **725**
8.4 Prevertebral Muscles **729**
8.5 Suprahyoid and Infrahyoid Muscles **740**
8.6 Prevertebral Muscles **760**
8.7 Muscles of Pharynx **767**
8.8 Muscles of Larynx **784**

9 CRANIAL NERVES
9.1 Summary of Cranial Nerves **797**
9.2 Olfactory Nerve (CN I) **800**
9.3 Optic Nerve (CN II) **801**
9.4 Oculomotor (CN III), Trochlear (CN IV), and Abducent (CN VI) Nerves **804**
9.5 Trigeminal Nerve (CN V) **806**
9.6 Branches of Ophthalmic Nerve (CN V^1) **807**
9.7 Branches of Maxillary Nerve (CN V^2) **808**
9.8 Branches of Mandibular Nerve (CN V^3) **810**
9.9 Facial Nerve (CN VII), Including Motor Root and Nervus Intermedius **812**
9.10 Vestibulocochlear Nerve (CN VIII) **814**
9.11 Glossopharyngeal Nerve (CN IX) **816**
9.12 Vagus Nerve (CN X) **819**
9.13 Spinal Accessory Nerve (CN XI) **820**
9.14 Hypoglossal Nerve (CN XII) **821**
9.15 Autonomic Ganglia of the Head **822**
9.16 Summary of Cranial Nerves **823**

TABLE AND FIGURE CREDITS

Tables

Tables that appear in this atlas are reproduced or modified from: Moore KL, Dalley AF. Clinically Oriented Anatomy. 4th Ed. Philadelphia: Lippincott Williams & Wilkins, 1999 and/or Moore KL, Agur AMR. Essential Clinical Anatomy. 2nd Ed. Philadelphia: Lippincott Williams & Wilkins, 2002 – *except Tables 1.2, 1.3, and 2.2.*

Figures

All sources are published by Lippincott Williams & Wilkins, unless otherwise indicated.

CHAPTER 1

1.5A,B	Courtesy of K. Bukhanov, University of Toronto, Canada.
1.5C	Dean D, Herbener TE. Cross-Sectional Human Anatomy, 2000:25. Orientation drawing from Moore KL, Dalley AF. Clinically Oriented Anatomy. 4th Ed., 1999 (Fig 1.18A).
1.23	Courtesy of M.A. Haider, University of Toronto, Canada.
1.33A	Courtesy of D.E. Sanders, University of Toronto, Canada.
1.33B	Courtesy of S. Herman, University of Toronto, Canada.
1.33C	Courtesy of E.L. Lansdown, University of Toronto, Canada.
1.36	Courtesy of J. Heslin, Toronto, Canada.
1.47B,D	Courtesy of I. Morrow, University of Manitoba, Canada.
1.48B	Courtesy of J. Heslin, Toronto, Canada.
1.64B	Courtesy of E.L. Lansdown, University of Toronto, Canada.
1.70C	Moore KL, Dalley AF. Clinically Oriented Anatomy. 4th Ed., 1999 (Fig. 1.62).
1.79A-F	MRIs courtesy of M.A. Haider, University of Toronto, Canada; orientation drawing from Moore KL, Agur AMR. Essential Clinical Anatomy, 1995:56 (Fig. 2.12A).
1.80A,B	MRIs courtesy of M.A. Haider, University of Toronto, Canada; orientation drawing from Moore KL, Agur AMR. Essential Clinical Anatomy, 1995:71 (Fig. 2.16).
1.81A-D	MRIs courtesy of M.A. Haider, University of Toronto, Canada.
1.82	Dean D, Herbener TE. Cross-Sectional Human Anatomy, 2000:6 (Fig. 2.7).
1.83A-F	Madden ME. Introduction to Sectional Anatomy, 2001:61,63,65,67,71,73 (Figs. 2.37B,2.38B,2.39B,2.40B,2.42B,2.43B)

CHAPTER 2

2.29	Moore KL, Dalley AF. Clinically Oriented Anatomy. 4th Ed., 1999 (Fig. 2.46).
2.30	Courtesy of J. Heslin, Toronto, Canada.
2.31A,C,D	Courtesy of E.L. Lansdown, University of Toronto, Canada.
2.31B	Courtesy of J. Heslin, Toronto, Canada.
2.32	Moore KL, Dalley AF. Clinically Oriented Anatomy. 4th Ed., 1999:239 (Table 2.7).
2.37C	Courtesy of G.B. Haber, University of Toronto, Canada.
2.41A	Courtesy of C.S. Ho, University of Toronto, Canada.
2.41B	Courtesy of E.L. Lansdown, University of Toronto, Canada.
2.44A	Courtesy of E.L. Lansdown, University of Toronto, Canada.
2.44B	Courtesy of J. Heslin, Toronto, Canada.
2.46	Courtesy of K. Sniderman, University of Toronto, Canada.
2.52B	Courtesy of A.M. Arenson, University of Toronto, Canada.
2.53A-E	Moore KL, Dalley AF. Clinically Oriented Anatomy. 4th Ed., 1999 (Fig. 2.53).
2.56A,B	Courtesy of J. Heslin, Toronto, Canada.
2.58A,B	Courtesy of G.B. Haber, University of Toronto, Canada.

2.61B	Courtesy of G.B. Haber, University of Toronto, Canada.
2.66B	Courtesy of M. Asch, University of Toronto, Canada.
2.66C	Courtesy of E.L. Lansdown, University of Toronto, Canada.
2.68B(right)	Courtesy of M. Asch, University of Toronto, Canada.
2.77A,B	Moore KL, Dalley AF. Clinically Oriented Anatomy. 4th Ed., 1999:234.
2.83A-F	MRIs courtesy of M.A. Haider, University of Toronto, Canada.
2.83B,H	National Library of Medicine's™ Visible Human Project®, Visible Man images: 1499,1625. (Appeared in Dean D, Herbener TE. Cross-Sectional Human Anatomy, 2000:28,50.)
2.84A-D	MRIs courtesy of M.A. Haider, University of Toronto, Canada.
2.85A-D	MRIs courtesy of M.A. Haider, University of Toronto, Canada.
2.86A-C	Courtesy of A.M. Arenson, University of Toronto, Canada.
2.86E,G	Dean D, Herbener TE. Cross-Sectional Human Anatomy, 2000:45,53.
2.86D,E	Courtesy of J. Lai, University of Toronto, Canada.

CHAPTER 3

3.1B,C	Moore KL, Dalley AF. Clinically Oriented Anatomy. 4th Ed., 1999:508,507 (Figs. 5.5, 5.4B).
3.2B	Moore KL, Dalley AF. Clinically Oriented Anatomy. 4th Ed., 1999:333 (Fig. 3.1C).
3.10B,C	From Moore KL, Dalley AF. Clinically Oriented Anatomy. 4th Ed., 1999:386 (Fig. 3.30A,B).
3.11	Modified from Moore KL, Dalley AF. Clinically Oriented Anatomy. 4th Ed., 1999:386 (Fig. 3.31B).
3.14D	Modified from Moore KL, Dalley AF. Clinically Oriented Anatomy. 4th Ed., 1999:361 (Fig. 3.16C).
3.18A-C	Ultrasounds courtesy of A. Toi, University of Toronto, Canada.
3.20A	Uflacker R. Atlas of Vascular Anatomy: An Angiographic Approach, 1997:611.
3.23B	Modified from Moore KL, Agur AF. Clinically Oriented Anatomy. 4th Ed., 1995:166 (Fig. 4.12B).
3.25A	Moore KL, Agur AMR. Essential Clinical Anatomy. 2nd Ed., 2002:263 (Fig. 4.30).
3.33A	Courtesy of A.M. Arenson, University of Toronto, Canada.
3.33B	Moore KL, Dalley AF. Clinically Oriented Anatomy. 4th Ed., 1999:419 (Fig. 3.52A).
3.33C	Moore KL, Agur AMR. Essential Clinical Anatomy. 2nd Ed., 2002:245 (Fig. 4.21A,B).
3.33D	Courtesy of E.L. Lansdown, University of Toronto, Canada.
3.36B	Modified from Moore KL, Dalley AF. Clinically Oriented Anatomy. 4th Ed., 1999:160 (Fig. 4.9B).
3.37B	Modified from Moore KL, Dalley AF. Clinically Oriented Anatomy. 4th Ed., 1999:170 (Fig. 4.15B).
3.42A-E	Moore KL, Dalley AF. Clinically Oriented Anatomy. 4th Ed., 1999.
3.47C	Moore KL, Agur AMR. Essential Clinical Anatomy. 2nd Ed., 2002:263 (Fig. 4.30).
3.49B	Moore KL, Dalley AF. Clinically Oriented Anatomy. 4th Ed., 1999:407 (Figs. 3.42B).
3.50A	Moore KL, Dalley AF. Clinically Oriented Anatomy. 4th Ed., 1999:410 (Figs. 3.45).
3.51	Orientation drawing from Moore KL, Dalley AF. Clinically Oriented Anatomy. 4th Ed., 1999:407 (Figs. 3.42B).
3.60A-E	Courtesy of M.A. Haider, University of Toronto, Canada.
3.61A-D	Courtesy of M.A. Haider, University of Toronto, Canada.
3.62A-E	MRIs courtesy of M.A. Haider, University of Toronto, Canada.
3.63A,B	Courtesy of M.A. Haider, University of Toronto, Canada.
3.64A,B	Ultrasounds courtesy of A.M. Arenson, University of Toronto, Canada.
Table 3.1	Orientation drawings from Moore KL, Dalley AF. Clinically Oriented Anatomy. 4th Ed., 1999:336 (Table 3.1A,B).

CHAPTER 4

4.1B	Courtesy of D. Salonen, University of Toronto, Canada.
4.7B,D	Courtesy of Drs. E. Becker and P. Bobechke, University of Toronto, Canada.
4.8F	Courtesy of Drs. E. Becker and P. Bobechke, University of Toronto, Canada.
4.9C,D	Courtesy of J. Heslin, Toronto, Canada.
4.13D	Courtesy of E. Becker, University of Toronto, Canada.
4.14	Courtesy of E. Becker, University of Toronto, Canada.
4.15A	Courtesy of E. Becker, University of Toronto, Canada.
4.19A,B	Courtesy of E. Becker, University of Toronto, Canada.
4.21B,C	Courtesy of E. Becker, University of Toronto, Canada.
4.23B	From Moore KL, Agur AMR. Essential Clinical Anatomy, 1995:200 (bottom).
4.46C	From Moore KL, Agur AMR. Essential Clinical Anatomy, 1995:27 (Fig. 1.15A).
4.52A,B,E	Courtesy of D. Armstrong, University of Toronto, Canada.
4.52C,D,F-M	Courtesy of D. Salonen, University of Toronto, Canada.
Table 4.1	Left illustration from Moore KL, Agur AMR. Essential Clinical Anatomy, 1995:205 (Fig. 5.3).
Table 4.2	Bottom illustration (anterior view) from Moore KL, Agur AMR. Essential Clinical Anatomy, 1995:425.

CHAPTER 5

5.1A,B	Overlays from Moore KL, Dalley AF. Clinically Oriented Anatomy. 4th Ed., 1999:504,505 (Figs. 5.1,5.2A).
5.2B,D	Courtesy of P. Babyn, University of Toronto, Canada.
5.4A-D	Moore KL, Dalley AF. Clinically Oriented Anatomy. 4th Ed., 1999:529 (Fig. 5.12). A and B are based on Fender FA. Foerster's scheme of the dermatomes. Arch Neurol Psychiatry 1939;41:699. C and D are based on Keegan JJ, Garrett FD. The segmental distribution of the cutaneous nerves in the limbs of man. Anat Rec 1948;102:409.
5.6A	Moore KL, Dalley AF. Clinically Oriented Anatomy. 4th Ed., 1999:525 (Fig. 5.10D).
5.6B	Roche Lexikon Medizin. 4th Ed. Munich: Urban & Schwarzenberg, 1998. (Appeared in Moore KL, Dalley AF. Clinically Oriented Anatomy. 4th Ed., 1999:527.)
5.7A,B	Moore KL, Dalley AF. Clinically Oriented Anatomy. 4th Ed., 1999:528 (Fig. 5.11).
5.8A-D	Moore KL, Dalley AF. Clinically Oriented Anatomy. 4th Ed., 1999:523 (Fig. 5.9).
5.12B	Courtesy of E.L. Lansdown, University of Toronto, Canada.
5.13B	Moore KL, Dalley AF. Clinically Oriented Anatomy. 4th Ed., 1999:544 (Fig. 5.18).
5.24B	Moore KL, Agur AMR. Essential Clinical Anatomy. 2nd Ed., 2002:351.
5.27, 5.29	Orientation drawings from Moore KL, Agur AMR. Essential Clinical Anatomy, 1995:10 (Fig. 1.3).
5.33B	Courtesy of D. Salonen, University of Toronto, Canada.
5.35C	From Moore KL, Agur AMR. Essential Clinical Anatomy, 1995:268 (top).
5.48A	Moore KL, Dalley AF. Clinically Oriented Anatomy. 4th Ed., 1999:626 (Fig. 5.61).
5.49A,B	Courtesy of P. Bobechko, University of Toronto, Canada.
5.49C	Courtesy of D. Salonen, University of Toronto, Canada.
5.50B,C	Courtesy of D. Salonen, University of Toronto, Canada.
5.50, 5.52	Orientation drawings from Basmajian JV, Slonecker CE. Grant's Method of Anatomy: A Clinical Problem-Solving Approach. 11th Ed. Baltimore: Williams & Wilkins, 1989:xix,xviii.
5.51	Courtesy of P. Bobechko, University of Toronto, Canada.
5.52B,C	Courtesy of D. Salonen, University of Toronto, Canada.
5.65A	Courtesy of D. K. Sniderman, University of Toronto, Canada.
5.68D, 5.69B	Courtesy of E. Becker, University of Toronto, Canada.
5.72B	Courtesy of P. Bobechko, University of Toronto, Canada.
5.73B	Courtesy of W. Kucharczyk, University of Toronto, Canada.
5.74B	Courtesy of W. Kucharczyk, University of Toronto, Canada.
5.88C	Dean D, Herbener TE. Cross-Sectional Human Anatomy, 2000:95.
5.89C	Dean D, Herbener TE. Cross-Sectional Human Anatomy, 2000:105.
5.90A,B	Courtesy of D. Salonen, University of Toronto, Canada; orientation drawing from Moore KL, Agur AMR. Essential Clinical Anatomy, 1995:246 (Fig. 6.11C).
5.90C	Courtesy of D. Salonen, University of Toronto, Canada.
5.91A-D	Courtesy of D. Salonen, University of Toronto, Canada.
Table 5.2	Illustrations from Moore KL, Agur AMR. Essential Clinical Anatomy. 2nd Ed., 2002:324 (Fig. 6.9B,C,E).
Table 5.4	Illustrations from Moore KL, Agur AMR. Essential Clinical Anatomy. 2nd Ed., 2002:348 (Fig. B-D).
Table 5.10	Left illustration from Moore KL, Agur AMR. Essential Clinical Anatomy. 2nd Ed., 2002:366 (Fig. 6.23B); middle and right illustrations from Moore KL, Agur AMR. Essential Clinical Anatomy, 1995:257 (Figs. 6.16C,D).
Table 5.11	Illustrations from Moore KL, Dalley AF. Clinically Oriented Anatomy. 4th Ed., 1999:583.
Table 5.13	Right illustration modified from Moore KL, Dalley AF. Clinically Oriented Anatomy. 4th Ed., 1999:597 (Fig. 5.43B).

CHAPTER 6

6.3C	Courtesy of D. Armstrong, University of Toronto, Canada.
6.5A,B	Based on Fender FA. Foerster's scheme of the dermatomes. Arch Neurol Psychiatry 1939;41:688. (Appeared in Moore KL, Dalley AF. Clinically Oriented Anatomy. 4th Ed., 1999:682,683.)
6.5C,D	Based on Keegan JJ, Garrett FD. The segmental distribution of the cutaneous nerves in the limbs of man. Anat Rec 1948;102:409.
6.7	Moore KL, Dalley AF. Clinically Oriented Anatomy. 4th Ed., 1999:686 (Fig. 6.13).
6.9	Moore KL, Dalley AF. Clinically Oriented Anatomy. 4th Ed., 1999:680 (Fig. 6.9).
6.10	Moore KL, Dalley AF. Clinically Oriented Anatomy. 4th Ed., 1999:681 (Fig. 6.10).
6.11B	Moore KL, Agur AMR. Essential Clinical Anatomy, 1995:288 (Fig. 7.3A).
6.11D	Moore KL, Dalley AF. Clinically Oriented Anatomy. 4th Ed., 1999:773.
6.17A-C	Modified from Moore KL, Agur AMR. Essential Clinical Anatomy. 2nd Ed., 2002:423 (Fig. 7.10A-C).
6.20B	Moore KL, Dalley AF. Clinically Oriented Anatomy. 4th Ed., 1999:700 (Fig. 6.23A).
6.21B,C	Moore KL, Dalley AF. Clinically Oriented Anatomy. 4th Ed., 1999:700,701 (Figs. 6.23B,6.24A).
6.22A	Courtesy of D. Armstrong, University of Toronto, Canada.
6.39B	Moore KL, Dalley AF. Clinically Oriented Anatomy. 4th Ed., 1999:787 (Fig. 6.62C).
6.43B	Moore KL, Dalley AF. Clinically Oriented Anatomy. 4th Ed., 1999:791 (Fig. 6.65A).
6.44B	Moore KL, Dalley AF. Clinically Oriented Anatomy. 4th Ed., 1999:786 (Fig. 6.62A,B).
6.45B	Moore KL, Dalley AF. Clinically Oriented Anatomy. 4th Ed., 1999:790 (Fig. 6.64A).
6.47	Courtesy of E. Becker, University of Toronto, Canada.
6.48A,B	Courtesy of D. Salonen, University of Toronto, Canada.
6.50A,B,D,E,	Moore KL, Agur AMR. Essential Clinical Anatomy, 1995.
6.52C	Courtesy of E. Becker, University of Toronto, Canada.
6.53A,B	Radiographs courtesy of J. Heslin, Toronto, Canada; orientation drawings from Moore KL, Agur AMR. Essential Clinical Anatomy, 1995:338,339.
6.54B	Courtesy of D. Salonen, University of Toronto, Canada.
6.55B	Courtesy of E. Becker, University of Toronto, Canada.
6.58A	Courtesy of K. Sniderman, University of Toronto, Canada.
6.66	Redrawn from Moore KL, Agur AMR. Essential Clinical Anatomy, 1995:325.
6.67A,B	Moore KL, Dalley AF. Clinically Oriented Anatomy. 4th Ed., 1999:765 (Fig. 6.54A,B).
6.71B	Moore KL, Dalley AF. Clinically Oriented Anatomy. 4th Ed., 1999:772 (Fig. 6.57B).
6.72C	Moore KL, Dalley AF. Clinically Oriented Anatomy. 4th Ed., 1999:764 (Fig. 6.53B).
6.75A	Courtesy of D. Armstrong, University of Toronto, Canada.
6.81F	Courtesy of E. Becker, University of Toronto, Canada.
6.84A,B	Courtesy of E. Becker, University of Toronto, Canada.
6.85B	Courtesy of D. Armstrong, University of Toronto, Canada.
6.92B-D	Courtesy of D. Salonen, University of Toronto, Canada.
9.93B-D	Courtesy of D. Salonen, University of Toronto, Canada; orientation drawing from Moore KL, Dalley AF. Clinically Oriented Anatomy. 4th Ed., 1999:800 (Fig. 6.72 left).

6.94A-C	Courtesy of D. Salonen, University of Toronto, Canada.
6.96A-D	Courtesy of R. Leekam, University of Toronto and West End Diagnostic Imaging, Canada
Table 6.2	Illustration from Moore KL, Dalley AF. Clinically Oriented Anatomy. 4th Ed., 1999:702.
Table 6.3	Illustration from Moore KL, Dalley AF. Clinically Oriented Anatomy. 4th Ed., 1999:710.
Table 6.6	Illustrations from Moore KL, Agur AMR. Essential Clinical Anatomy. 2nd Ed., 2002:444,445 (Figs. 7.17A,B;7.18A).
Table 6.4	Illustration from Moore KL, Agur AMR. Essential Clinical Anatomy, 1995:298 (top).
Table 6.9	Illustration B from Moore KL, Agur AMR. Essential Clinical Anatomy, 1995:330 (left); unlettered illustrations on left from Moore KL, Dalley AF. Clinically Oriented Anatomy. 4th Ed., 1999:769.
Table 6.10	Illustrations modified from Moore KL, Dalley AF. Clinically Oriented Anatomy. 4th Ed., 1999:773.

CHAPTER 7

7.5A,B	Courtesy of E. Becker, University of Toronto, Canada.
7.6B,E,F	Courtesy of D. Armstrong, University of Toronto, Canada.
7.7B,C	Moore KL, Dalley AF. Clinically Oriented Anatomy. 4th Ed., 1999:856 (Fig. 7.8B,C).
7.11A	Moore KL, Agur AMR. Essential Clinical Anatomy, 1995:356 (top).
7.12C	Moore KL, Dalley AF. Clinically Oriented Anatomy. 4th Ed., 1999:839.
7.13B	Moore KL, Dalley AF. Clinically Oriented Anatomy. 4th Ed., 1999:876,873 (Figs. 7.16B,7.14).
7.13C-E	Moore KL, Agur AMR. Essential Clinical Anatomy, 1995:362,356.
7.15B	Moore KL, Dalley AF. Clinically Oriented Anatomy. 4th Ed., 1999:884 (Fig. 7.23B).
7.16B	Moore KL, Dalley AF. Clinically Oriented Anatomy. 4th Ed., 1999:881 (Fig. 7.21B).
7.22B	Moore KL, Agur AMR. Essential Clinical Anatomy, 1995:360 (Figs. 8.10B)
7.23B	Moore KL, Dalley AF. Clinically Oriented Anatomy. 4th Ed., 1999:893 (Fig. 7.27).
7.25A-C	Courtesy of D. Armstrong, University of Toronto, Canada.
7.26B,C,E,F	Moore KL, Agur AMR. Essential Clinical Anatomy. 2nd Ed., 2002:525 (Fig. 8.12A,D,B,C).
7.28-7.32	(*Except 7.30B*) Colorized from photographs provided courtesy of C. G. Smith, which appeared in Smith CG. Serial Dissections of the Human Brain. Baltimore: Urban & Schwarzenberg, Inc and Toronto: Gage Publishing Ltd., 1981.
7.33A-D	Colorized from photographs provided courtesy of C. G. Smith, which appeared in Smith CG. Serial Dissections of the Human Brain. Baltimore: Urban & Schwarzenberg, Inc and Toronto: Gage Publishing Ltd., 1981.
7.34-7.35	(*Except 7.34C, 7.35B*) Colorized from photographs provided courtesy of C.G. Smith, which appeared in Smith CG. Serial Dissections of the Human Brain. Baltimore: Urban & Schwarzenberg, Inc and Toronto: Gage Publishing Ltd., 1981.
7.34C	Colorized illustration provided courtesy of C. G. Smith, which appeared in Smith CG. Basic Neuroanatomy, 2nd Ed. Toronto: University of Toronto Press, 1971.
7.38C	Courtesy of W. Kucharczyk, University of Toronto, Canada.
7.39C	Courtesy of W. Kucharczyk, University of Toronto, Canada.
7.40B	Moore KL, Dalley AF. Clinically Oriented Anatomy. 4th Ed., 1999:909 (Fig. 7.36B).
7.41A,B	Moore KL, Dalley AF. Clinically Oriented Anatomy. 4th Ed., 1999:904 (Fig. 7.34A,B).
7.42A	Courtesy of J.R. Buncic, University of Toronto, Canada.
7.43, 7.46	Orientation drawings from Moore KL, Agur AMR. Essential Clinical Anatomy, 1995:382 (Table 8.8).
7.44C,E	Moore KL, Dalley AF. Clinically Oriented Anatomy. 4th Ed., 1999:917 (Fig. 7.40).
7.51B,C	Illustrations from Moore KL, Agur AMR. Essential Clinical Anatomy. 2nd Ed., 2002:552 (Fig. 8.29A,B); CTs and MRIs from Langland OE, Langlais RP, Preece JW. Principles of Dental Imaging, 2002:278 (Figs. 11.32A,B;11.33A,B).
7.52C,D	Moore KL, Dalley AF. Clinically Oriented Anatomy. 4th Ed.,

	1999:946 (Fig. 7.60A,B).
7.56B	Moore KL, Dalley AF. Clinically Oriented Anatomy. 4th Ed., 1999:945 (Fig. 7.58).
7.58A	Langland OE, Langlais RP, Preece JW. Principles of Dental Imaging, 2002:334 (Fig. 14.1).
7.58D	Courtesy of M.J. Pharoah, University of Toronto, Canada.
7.59D	Moore KL, Dalley AF. Clinically Oriented Anatomy. 4th Ed., 1999:929 (Fig. 7.48A).
7.60B,C	Woelfel JB, Scheid RC. Dental Anatomy: Its Relevance to Dentistry. 6th Ed., 2002:86,46 (Figs. 3.5,1.29).
7.63	Orientation drawing modified from Moore KL, Dalley AF. Clinically Oriented Anatomy. 4th Ed., 1999:954 (Fig. 7.66).
7.64	Orientation drawing modified from Moore KL, Dalley AF. Clinically Oriented Anatomy. 4th Ed., 1999:958 (Fig. 7.69).
7.70B	Moore KL, Dalley AF. Clinically Oriented Anatomy. 4th Ed., 1999:952 (Fig. 7.64C).
7.71B	Courtesy of E. Becker, University of Toronto, Canada.
7.74B	Courtesy of D. Armstrong, University of Toronto, Canada.
7.76B	Courtesy of W. Kucharczyk, University of Toronto, Canada.
7.80D	Courtesy of Welch Allen, Inc. Skaneateles Falls, NY. (Appeared in Moore KL, Dalley AF. Clinically Oriented Anatomy. 4th Ed., 1999:966.)
7.86B,C,D	Moore KL, Dalley AF. Clinically Oriented Anatomy. 4th Ed., 1999:973 (Fig. 7.80B,C,D).
7.87A,B	Moore KL, Dalley AF. Clinically Oriented Anatomy. 4th Ed., 1999:975,974 (Figs. 7.82,7.81).
7.88C	Moore KL, Agur AMR. Essential Clinical Anatomy, 1995:352 (Fig. 8.6B).
7.89A-C	Courtesy of W. Kucharczyk, University of Toronto, Canada.
7.90A,B	Courtesy of W. Kucharczyk, University of Toronto, Canada.
7.91A-C	MRIs courtesy of D. Armstrong, University of Toronto, Canada; orientation drawing from Moore KL, Dalley AF. Clinically Oriented Anatomy. 4th Ed., 1999:888 (Fig. 7.25, superior view).
7.92A-E	MRIs courtesy of D. Armstrong, University of Toronto, Canada; orientation drawing from Moore KL, Dalley AF. Clinically Oriented Anatomy. 4th Ed., 1999:888 (Fig. 7.25B).
7.93A-F	MRIs courtesy of D. Armstrong, University of Toronto, Canada; orientation drawing from Moore KL, Dalley AF. Clinically Oriented Anatomy. 4th Ed., 1999:888 (Fig. 7.25B).
7.94A-F	MRIs courtesy of D. Armstrong, University of Toronto, Canada; orientation drawing from Moore KL, Dalley AF. Clinically Oriented Anatomy. 4th Ed., 1999:888 (Fig. 7.25A).
Table 7.1	Left illustration from Moore KL, Agur AMR. Essential Clinical Anatomy. 2nd Ed., 2002:507; right illustration from Moore KL, Agur AMR. Essential Clinical Anatomy, 1995:351.
Table 7.2	Illustrations from Moore KL, Dalley AF. Clinically Oriented Anatomy. 4th Ed., 1999:860.
Table 7.3	Illustrations from Moore KL, Dalley AF. Clinically Oriented Anatomy. 4th Ed., 1999:866.
Table 7.4	Illustration from Moore KL, Dalley AF. Clinically Oriented Anatomy. 4th Ed., 1999:867.
Table 7.7	Bottom illustration from Moore KL, Dalley AF. Clinically Oriented Anatomy. 4th Ed., 1999:910 (B).
Table 7.8	Bottom illustration from Moore KL, Dalley AF. Clinically Oriented Anatomy. 4th Ed., 1999:910 (A).
Table 7.9	Illustrations from Clay JH, Pounds DM. Basic Clinical Massage Therapy: Integrating Anatomy and Treatment, 2002:76,74,79 (Figs. 3.17,3.15,3.19).
Table 7.11	Illustrations from Moore KL, Dalley AF. Clinically Oriented Anatomy. 4th Ed., 1999:942.
Table 7.12	Top and bottom right illustrations from Moore KL, Agur AMR. Essential Clinical Anatomy. 2nd Ed., 2002:559 (Fig. 8.35A,B); bottom left illustration from Clay JH, Pounds DM. Basic Clinical Massage Therapy: Integrating Anatomy and Treatment, 2002:80 (Fig. 3.22).

CHAPTER 8

8.1A-C	Moore KL, Dalley AF. Clinically Oriented Anatomy. 4th Ed., 1999:999 (Fig. 8.4).
8.2A-C	Moore KL, Agur AMR. Essential Clinical Anatomy. 2nd Ed., 2002:280.
8.5A	Moore KL, Agur AMR. Essential Clinical Anatomy. 2nd Ed., 2002:611 (Fig. 9.7).

8.14B	Moore KL, Dalley AF. Clinically Oriented Anatomy, 4th Ed., 1999:945 (Fig. 7.58).
8.16D,F	Moore KL, Agur AMR. Essential Clinical Anatomy. 2nd Ed., 2002:620 (Figs. 9.13B,C).
8.17C	Moore KL, Dalley AF. Clinically Oriented Anatomy. 4th Ed., 1999:1031 (Fig. 8.21B).
8.19A	Moore KL, Agur AMR. Essential Clinical Anatomy. 2nd Ed., 2002:619 (Fig. 9.12).
8.19B	Courtesy of M. Keller, University of Toronto, Canada.
8.19C	Courtesy of W. Kucharczyk, University of Toronto, Canada.
8.23B	Moore KL, Agur AMR. Essential Clinical Anatomy. 2nd Ed., 2002:202 (Fig. 3.45B).
8.30B	From Biebgott B. The Anatomical Basis of Dentistry. Philadelphia: WB Saunders Co. 1982.
8.32A,C	Orientation drawings from Moore KL, Agur AMR. Essential Clinical Anatomy, 1995:440 (Figs. 9.10).
8.39B,C	Courtesy of W. Kucharczyk, University of Toronto, Canada.
8.40A	Courtesy of D. Salonen, University of Toronto, Canada.
8.40B	Courtesy of W. Kucharczyk, University of Toronto, Canada.
8.41B	Courtesy of D. Salonen, University of Toronto, Canada.
8.42A-C	Courtesy of D. Salonen, University of Toronto, Canada; orientation drawing from Moore KL, Agur AMR. Essential Clinical Anatomy, 1995.
Table 8.2	Illustrations from Moore KL, Dalley AF. Clinically Oriented Anatomy. 4th Ed., 1999:1004.
Table 8.4	Bottom illustrations on left page from Clay JH, Pounds DM. Basic Clinical Massage Therapy: Integrating Anatomy and Treatment, 2002:92 (Fig. 3.40); top left illustration on right page from Moore KL, Agur AMR. Essential Clinical Anatomy, 1995:413; top right illustration on right page from Moore KL, Agur AMR. Essential Clinical Anatomy. 2nd Ed., 2002:614.

Table 8.5	Left illustration from Clay JH, Pounds DM. Basic Clinical Massage Therapy: Integrating Anatomy and Treatment, 2002:86 (Fig. 3.31); right illustrations from Moore KL, Agur AMR. Essential Clinical Anatomy, 1995:418 (top).

CHAPTER 9

9.6A-F	Courtesy of W. Kucharczyk, University of Toronto, Canada.
9.7A-C	Courtesy of W. Kucharczyk, University of Toronto, Canada.
Table 9.1	Illustration from Moore KL, Dalley AF. Clinically Oriented Anatomy. 4th Ed., 1999:1084.
Table 9.2	Top illustration from Moore KL, Agur AMR. Essential Clinical Anatomy, 1995:449 (Fig. 10.2B); bottom illustration from Moore KL, Dalley AF. Clinically Oriented Anatomy. 4th Ed., 1999:1089 (Fig. 9.3B).
Table 9.3	Illustration from Moore KL, Dalley AF. Clinically Oriented Anatomy. 4th Ed., 1999:1091 (Fig. 9.4A,B).
Table 9.7	Bottom illustration on right page from Moore KL, Agur AMR. Essential Clinical Anatomy, 1995:396 (Fig. 8.22B)
Table 9.9	Bottom right illustration from Moore KL, Agur AMR. Essential Clinical Anatomy, 1995:456 (Fig. 10.6B).
Table 9.10	Top left illustration from Moore KL, Dalley AF. Clinically Oriented Anatomy. 4th Ed., 1999; bottom left illustration from Moore KL, Dalley AF. Clinically Oriented Anatomy. 4th Ed., 1999:975 (Fig. 7.82).
Table 9.12	Right illustration from Moore KL, Agur AMR. Essential Clinical Anatomy, 1995:459 (Table 10.2).
Table 9.13	Illustrations from Moore KL, Dalley AF. Clinically Oriented Anatomy. 4th Ed., 1999:1101 (Table 9.4).
Table 9.15	Illustration from Moore KL, Dalley AF. Clinically Oriented Anatomy. 4th Ed., 1999:1101.

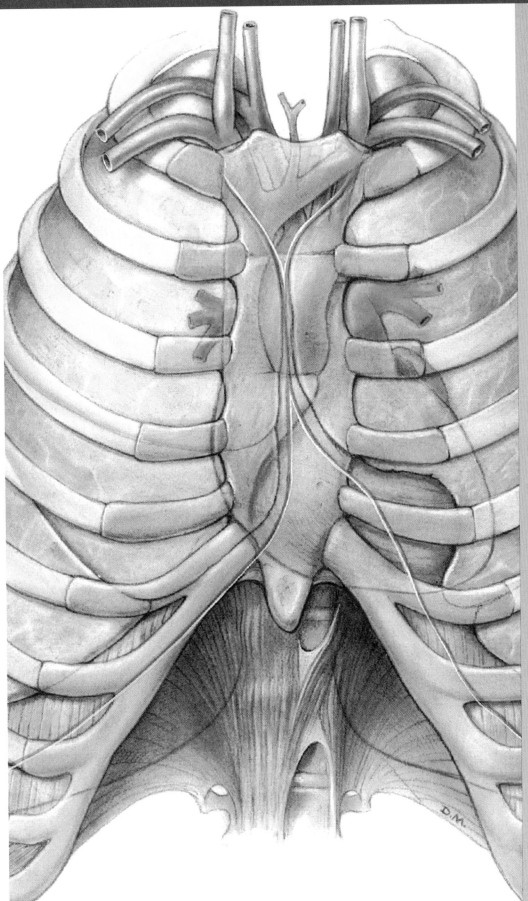

- Pectoral Region **2**
- Breast **4**
- Bony Thorax and Joints **10**
- Thoracic Wall **17**
- Thoracic Contents **25**
- Pleural Cavity **28**
- Mediastinum **29**
- Lungs and Pleura **30**
- Bronchi and Bronchopulmonary Segments **36**
- Innervation and Lymphatic Drainage of Lungs **42**
- External Heart **44**
- Coronary Vessels **52**
- Internal Heart and Valves **56**
- Conducting System of Heart **62**
- Pericardial Sac **63**
- Superior Mediastinum and Great Vessels **64**
- Diaphragm **71**
- Posterior Thorax **72**
- Scans: Axial, Sagittal, Coronal **82**

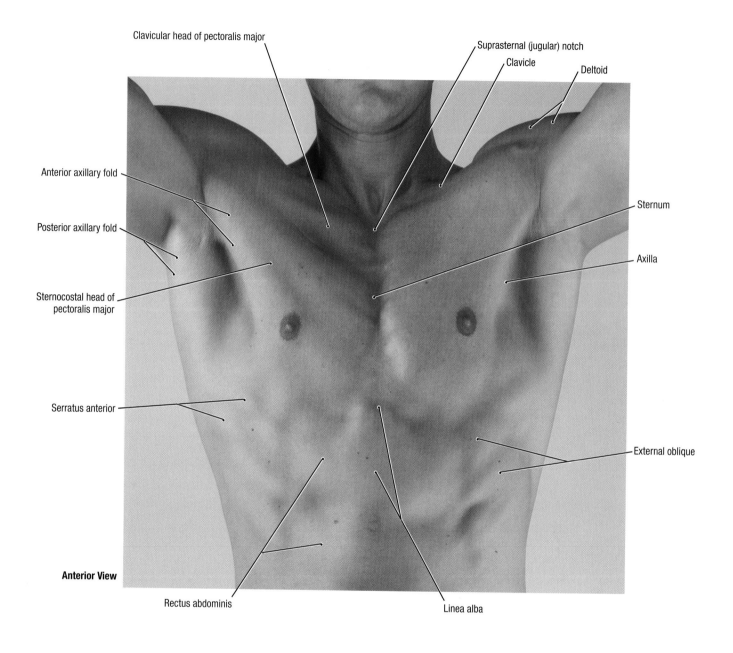

Clavicular head of pectoralis major

Suprasternal (jugular) notch

Clavicle

Deltoid

Anterior axillary fold

Posterior axillary fold

Sternocostal head of pectoralis major

Sternum

Axilla

Serratus anterior

External oblique

Anterior View

Rectus abdominis

Linea alba

1.1 Surface anatomy of the male pectoral region

- The subject is adducting the shoulders against resistance to demonstrate the pectoralis major muscle.
- The pectoralis major muscle has two parts, the sternocostal and clavicular heads.
- The anterior axillary fold is formed by the inferior border of the sternocostal head of pectoralis major muscle.
- The deltopectoral triangle (infraclavicular fossa) is bounded by the clavicle superiorly, the deltoid muscle laterally and the clavicular head of pectoralis major muscle medially.

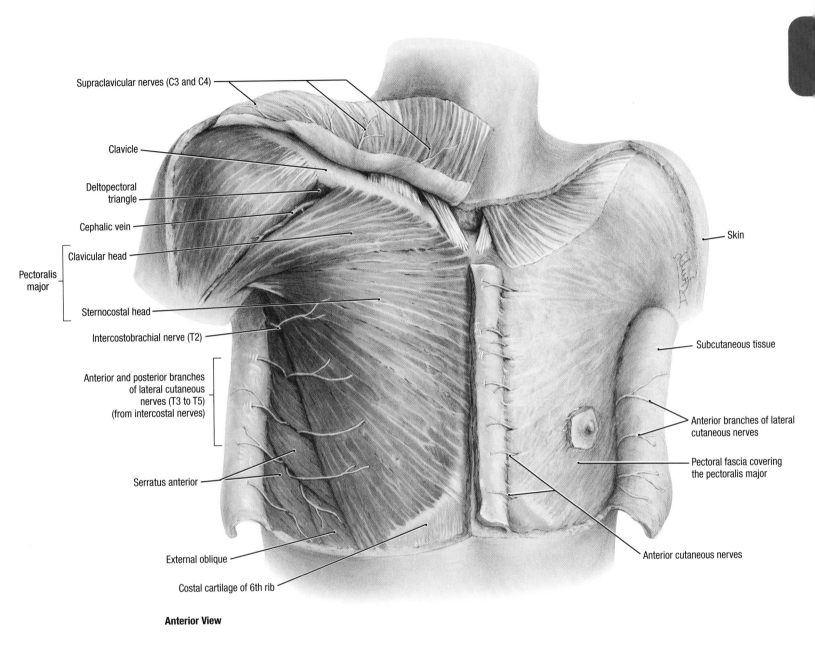

Supraclavicular nerves (C3 and C4)

Clavicle

Deltopectoral
triangle

Cephalic vein

Clavicular head

Pectoralis
major

Sternocostal head

Intercostobrachial nerve (T2)

Anterior and posterior branches
of lateral cutaneous
nerves (T3 to T5)
(from intercostal nerves)

Serratus anterior

External oblique

Costal cartilage of 6th rib

Skin

Subcutaneous tissue

Anterior branches of lateral
cutaneous nerves

Pectoral fascia covering
the pectoralis major

Anterior cutaneous nerves

Anterior View

1.2   Superficial dissection, male pectoral region

- The platysma muscle, which descends to the 2nd or 3rd rib, is cut short on the right side; together with the supraclavicular nerves, it is reflected on the left side.
- The thin pectoral fascia covers the pectoralis major.
- The clavicle lies deep to the subcutaneous tissue and the platysma muscle.
- The cephalic vein passes deeply in the deltopectoral triangle to join the axillary vein.
- Supraclavicular (C3 and C4) and upper thoracic nerves (T2 to T6) supply cutaneous innervation to the pectoral region.

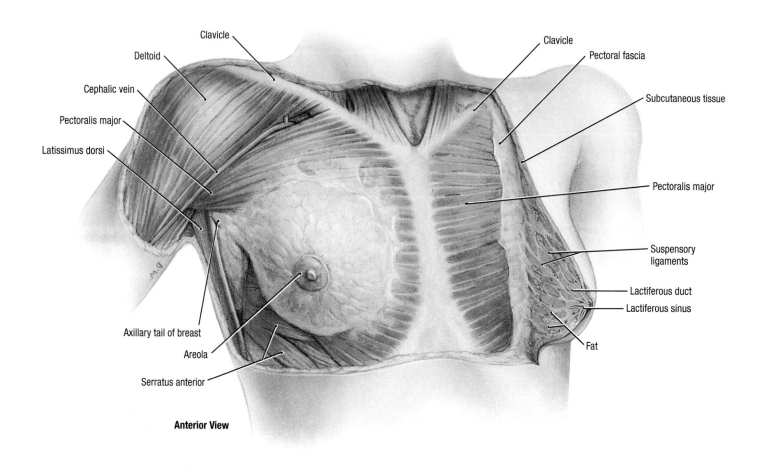

Anterior View

Clavicle

Deltoid

Cephalic vein

Pectoralis major

Latissimus dorsi

Axillary tail of breast

Areola

Serratus anterior

Clavicle

Pectoral fascia

Subcutaneous tissue

Pectoralis major

Suspensory ligaments

Lactiferous duct

Lactiferous sinus

Fat

1.3   **Superficial dissection, female pectoral region**

- On the specimen's right side the skin is removed, on the left side the breast is sagittally sectioned.
- The breast extends from the 2nd to the 6th ribs, and the axillary tail projects into the axilla.
- The region of loose connective tissue between the pectoral fascia and the deep surface of the breast, the retromammary bursa, permits the breast to move on the deep fascia.

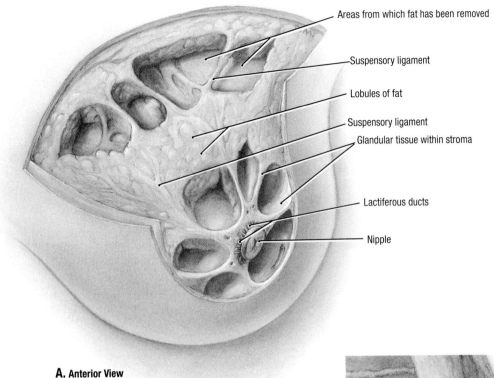

Areas from which fat has been removed

Suspensory ligament

Lobules of fat

Suspensory ligament

Glandular tissue within stroma

Lactiferous ducts

Nipple

A. Anterior View

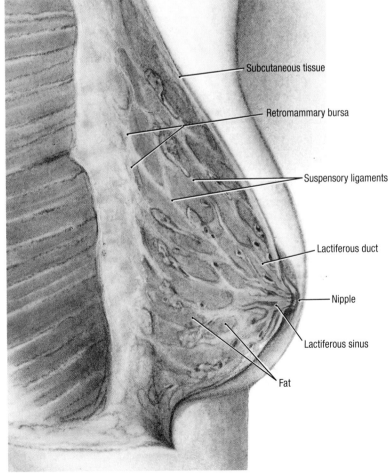

Subcutaneous tissue

Retromammary bursa

Suspensory ligaments

Lactiferous duct

Nipple

Lactiferous sinus

Fat

B. Sagittal Section

1.4   Female mammary gland

A. The breast consists primarily of fat compartmentalized between connective and glandular tissue septa. The lactiferous ducts (usually 15 to 20 in number) open on the nipple; the glandular tissue lies within a dense (fibro-) areolar stroma, from which suspensory ligaments extend to the deeper layers of the skin. Collections of superficial fat were scooped out from between the septa. **B.** Sagittal section through breast.

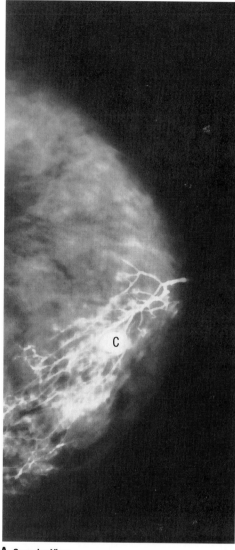

A. Superior View

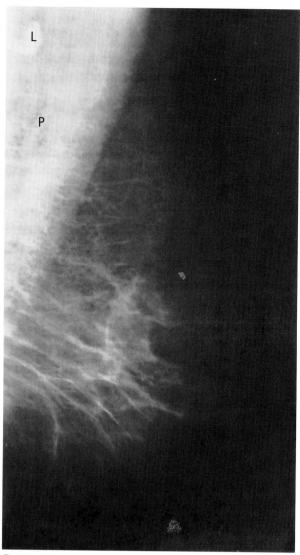

B. Lateral View

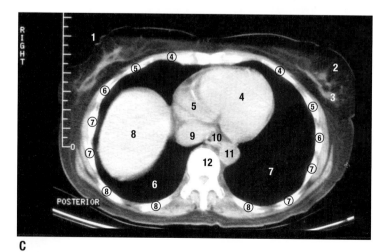

C

1. Nipple
2. Lactiferous ducts
3. Suspensory ligaments
4. Left ventricle
5. Right atrium
6. Right lung
7. Left lung
8. Liver
9. Inferior vena cava
10. Esophagus
11. Descending aorta
12. T9 vertebra

Circled numbers: corresponding rib numbers

1.5   **Imaging of breast**

A. Galactogram. Contrast has been injected into a lactiferous duct, outlining the branching pattern of its tributaries. Note the presence of a ductal cyst (*C*). **B.** Normal mammogram. Observe the connective tissue network of the breast. The stroma is radiopaque and changes with age and during lactation. Pectoralis major muscle (*P*) and an axillary lymph node (*L*) can also be seen. **C.** Axial computed tomographic (CT) scan through the female breast (T5 level).

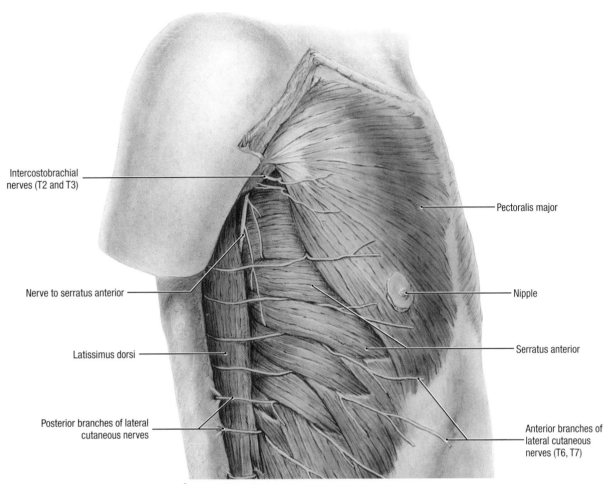

Intercostobrachial nerves (T2 and T3)

Nerve to serratus anterior

Latissimus dorsi

Posterior branches of lateral cutaneous nerves

Pectoralis major

Nipple

Serratus anterior

Anterior branches of lateral cutaneous nerves (T6, T7)

A. Anterolateral View

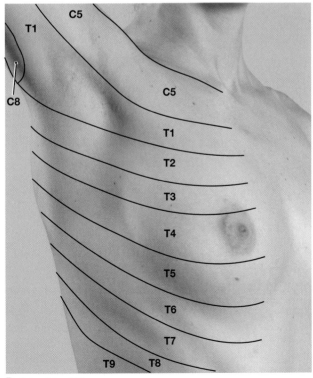

C5

T1

C8

C5

T1

T2

T3

T4

T5

T6

T7

T9   T8

B. Anterolateral View

1.6   **Bed of breast**

A. This superficial dissection shows the posterior and anterior branches of the lateral cutaneous nerves. **B.** Dermatomes.

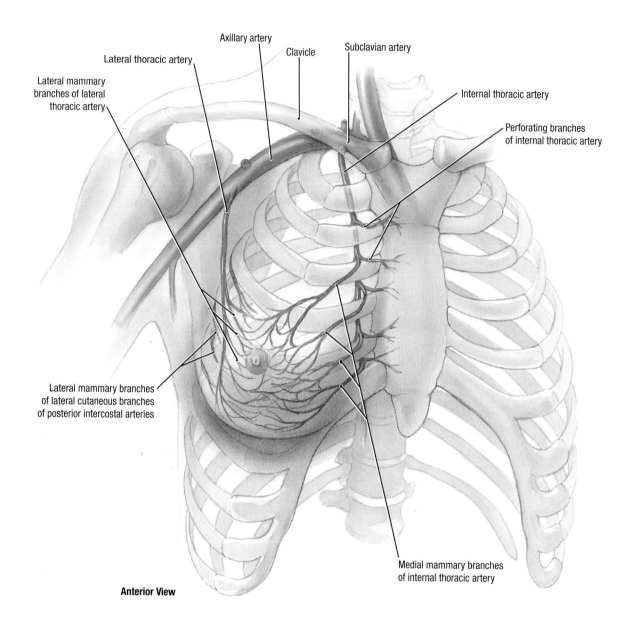

Lateral thoracic artery

Axillary artery

Clavicle

Subclavian artery

Lateral mammary branches of lateral thoracic artery

Internal thoracic artery

Perforating branches of internal thoracic artery

Lateral mammary branches of lateral cutaneous branches of posterior intercostal arteries

Medial mammary branches of internal thoracic artery

Anterior View

1.7   Arterial supply of the breast

Arteries enter the breast from its superomedial and superolateral aspects; vessels also penetrate the deep surface of the breast. The blood supply is from the medial mammary branches of the internal thoracic artery, lateral mammary branches from the lateral thoracic artery, and lateral mammary branches of lateral cutaneous branches of the posterior intercostal arteries. The arteries branch profusely and anastomose with each other.

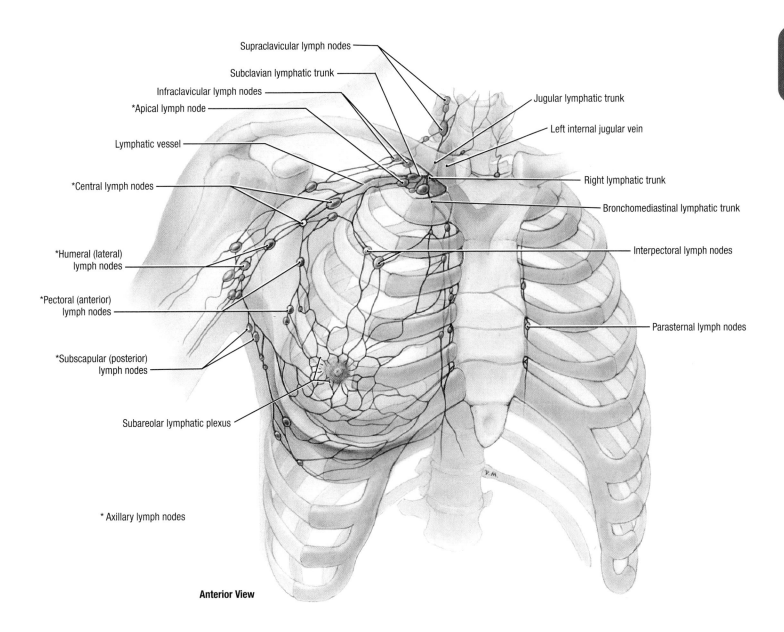

Anterior View

1.8 Lymphatic drainage of breast

Lymph drained from the upper limb and breast passes through nodes arranged irregularly in groups: (a) pectoral, along the inferior border of the pectoralis minor muscle; (b) subscapular, along the subscapular artery and veins; (c) humeral, along the distal part of the axillary vein; (d) central, at the base of the axilla embedded in axillary fat; and (e) apical, along the axillary vein between the clavicle and the pectoralis minor muscle. Most of the breast drains to the subclavian lymph trunk, which joins the venous system at the junction of the subclavian and internal jugular veins. The medial part of the breast drains to the parasternal nodes, which are located along the internal thoracic vessels.

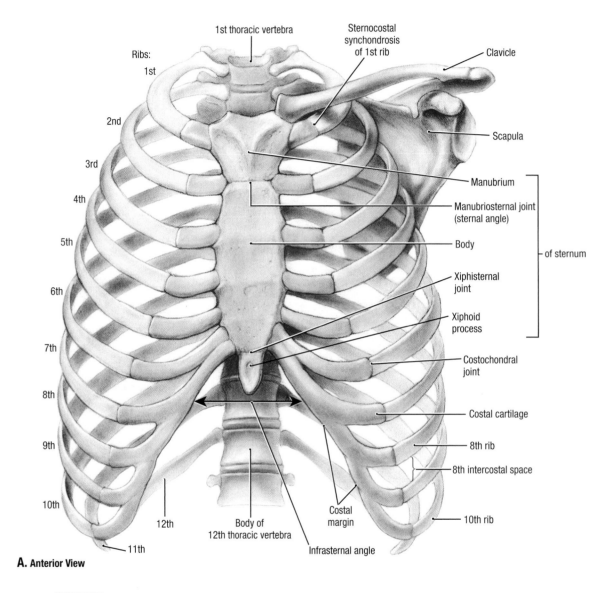

A. Anterior View

1.9 Bony thorax

- The skeleton of the thorax consists of 12 thoracic vertebrae, 12 pairs of ribs and costal cartilages, and the sternum.
- Anteriorly, forming the costal margin, the superior seven costal cartilages articulate with the sternum; the 8th, 9th, and 10th cartilages articulate with the cartilage above; the 11th and 12th are "floating" ribs, i.e., their cartilages do not articulate anteriorly.
- The clavicle lies over the anterosuperior aspect of the 1st rib, making it difficult to palpate.
- The 2nd rib is easy to locate because its costal cartilage articulates with the sternum at the sternal angle, located at the junction of the manubrium and body of the sternum.
- The 3rd to 10th ribs can be palpated in sequence inferolaterally from the 2nd rib; the fused costal cartilages of the 7th to 10th ribs form the costal arch (margin), and the tips of the 11th and 12th ribs can be palpated posterolaterally.

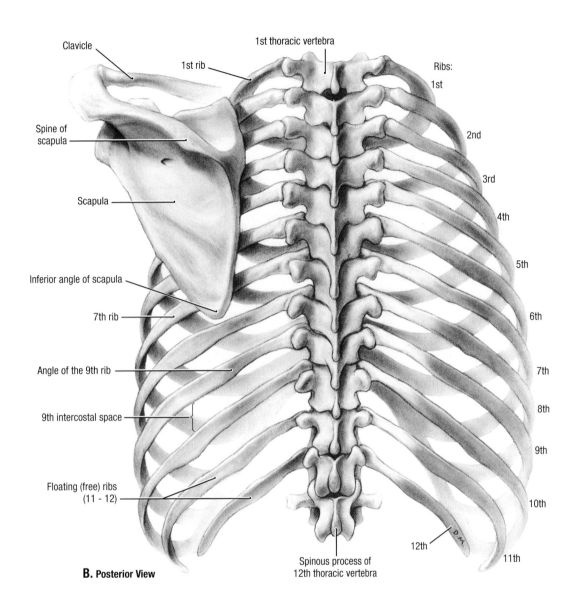

B. Posterior View

Labels on the figure:
Clavicle
1st rib
1st thoracic vertebra
Ribs:
1st
2nd
3rd
4th
5th
6th
7th
8th
9th
10th
11th
12th
Spine of scapula
Scapula
Inferior angle of scapula
7th rib
Angle of the 9th rib
9th intercostal space
Floating (free) ribs (11 - 12)
Spinous process of 12th thoracic vertebra

1.9   Bony thorax (continued)

- The superior thoracic aperture (thoracic inlet) is the doorway between the thoracic cavity and the neck region; it is bounded by the 1st thoracic vertebra, the 1st ribs and their cartilages, and the manubrium of the sternum.
- Each rib articulates posteriorly with the vertebral column.
- Posteriorly, all ribs angle inferiorly; anteriorly, the 3rd to 10th costal cartilages angle superiorly.
- The scapula is suspended from the clavicle and crosses the 2nd to 7th ribs.

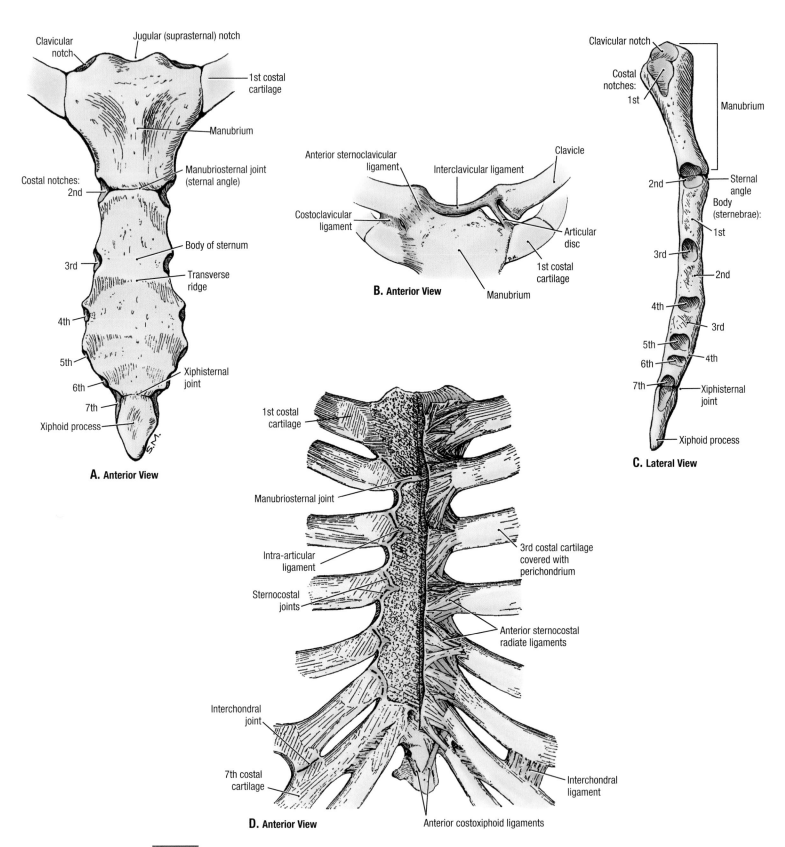

Clavicular notch — Clavicular notch

Jugular (suprasternal) notch

1st costal cartilage

Manubrium

Manubriosternal joint (sternal angle)

Costal notches: 2nd

Body of sternum

3rd

Transverse ridge

4th

5th

6th

7th

Xiphisternal joint

Xiphoid process

S.M.

A. Anterior View

Anterior sternoclavicular ligament

Interclavicular ligament

Clavicle

Costoclavicular ligament

Articular disc

1st costal cartilage

Manubrium

B. Anterior View

Clavicular notch

Costal notches: 1st

Manubrium

2nd

Sternal angle

Body (sternebrae): 1st

3rd

2nd

4th

3rd

5th

4th

6th

7th

Xiphisternal joint

Xiphoid process

C. Lateral View

1st costal cartilage

Manubriosternal joint

Intra-articular ligament

Sternocostal joints

3rd costal cartilage covered with perichondrium

Anterior sternocostal radiate ligaments

Interchondral joint

7th costal cartilage

Interchondral ligament

Anterior costoxiphoid ligaments

D. Anterior View

1.10   **Sternum and associated joints**

A. Parts of the anterior aspect of the sternum. **B.** Sternoclavicular joint. **C.** Features of the lateral aspect of the sternum. **D.** Sternocostal and interchondral joints. On the right side of the specimen, the cortex of the sternum and the external surface of the costal cartilages have been shaved away.

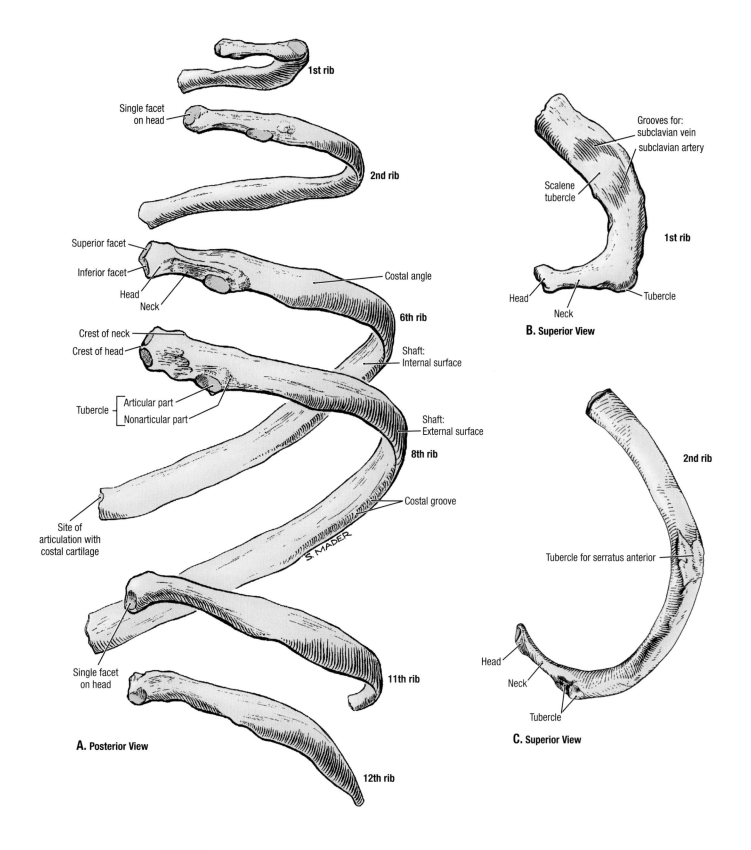

1.11 **Ribs**

A. "Typical" (6th and 8th) and "atypical" (1st and 2nd, 11th and 12th) ribs. **B.** First rib.
C. Second rib.

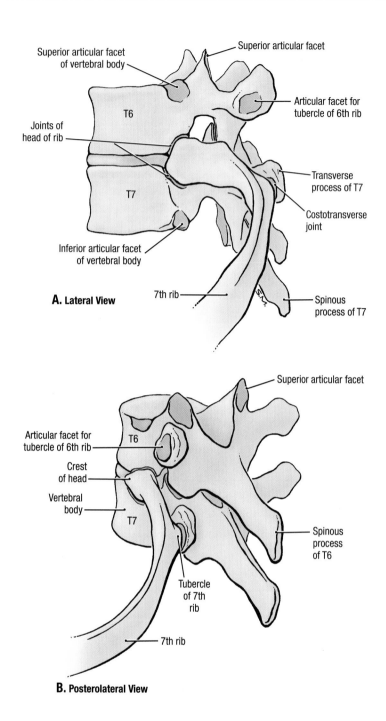

A. Lateral View

B. Posterolateral View

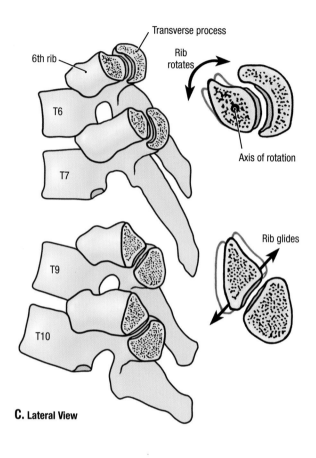

C. Lateral View

1.12 Costovertebral articulations

- The costovertebral articulations include the articulation of the head of the rib with two adjacent vertebral bodies and the tubercle of the rib with the transverse process of a vertebra.
- There are two articular facets on the head of the rib: a larger, inferior one for articulation with the vertebral body of its own number, and a smaller, superior facet for the vertebral body above.
- The crest of the head of the rib separates the two articular facets.

- The smooth articular part of the tubercle of the rib articulates with the transverse process of the same numbered vertebra at the costotransverse joint.
- In **C**: At the 1st to 7th costotransverse joints, the ribs rotate, increasing the anteroposterior diameter of the thorax; at the 8th, 9th, and 10th, they glide, increasing the transverse diameter of the upper abdomen.

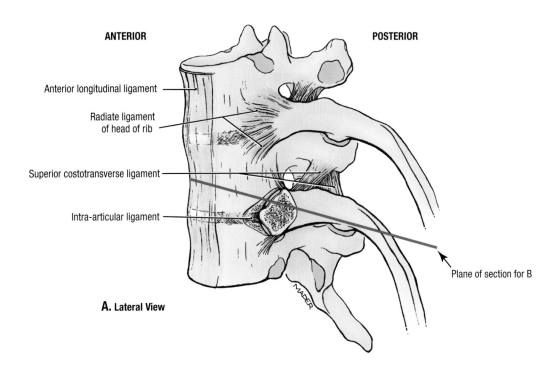

ANTERIOR   POSTERIOR

Anterior longitudinal ligament

Radiate ligament
of head of rib

Superior costotransverse ligament

Intra-articular ligament

Plane of section for B

A. Lateral View

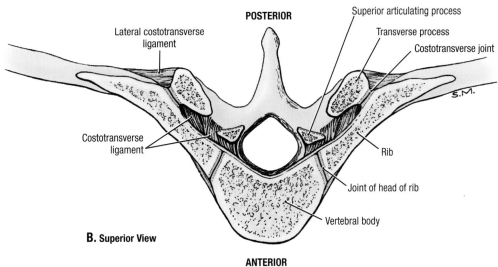

POSTERIOR   Superior articulating process

Lateral costotransverse   Transverse process
ligament
                                            Costotransverse joint

Costotransverse
ligament

Rib

Joint of head of rib

Vertebral body

B. Superior View

ANTERIOR

1.13 Ligaments of costovertebral articulations

A:

- The radiate ligament joins the head of the rib to two vertebral bodies and the interposed intervertebral disc.
- The superior costotransverse ligament joins the crest of the neck of the rib to the transverse process above.
- The intraarticular ligament joins the crest of the head of the rib to the intervertebral disc.

B:

- The vertebral body, transverse processes, superior articulating processes, and posterior elements of the articulating ribs have

been transversely sectioned to visualize the joint surfaces and ligaments.

- The costotransverse ligament joins the posterior aspect of the neck of the rib to the adjacent transverse process.
- The lateral costotransverse ligament joins the nonarticulating part of the tubercle of the rib to the tip (apex) of the transverse process.
- The articular surfaces (blue) of the synovial plane costovertebral joints.

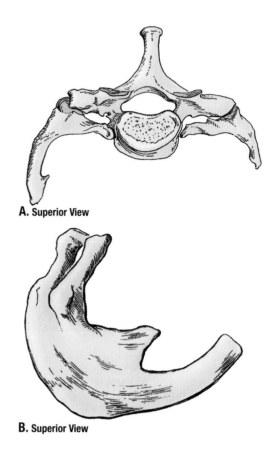

A. Superior View

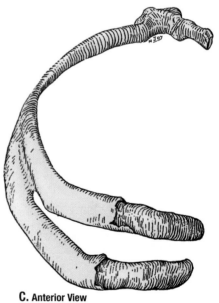

B. Superior View

C. Anterior View

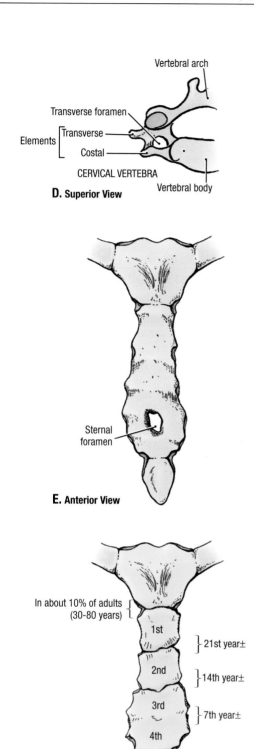

Vertebral arch

Transverse foramen

Elements ⎡ Transverse
        ⎣ Costal

CERVICAL VERTEBRA

Vertebral body

D. Superior View

Sternal foramen

E. Anterior View

In about 10% of adults (30-80 years)

1st

2nd

3rd

4th

21st year±

14th year±

7th year±

Commonly after middle life

F. Anterior View

1.14 Rib and sternum anomalies

A. Cervical ribs. This is an enlarged costal element of the 7th cervical vertebra. (Compare with diagrammatic cervical vertebra in **D.**) Cervical ribs can be unilateral or bilateral, and large and palpable or detectable only radiologically. It can be asymptomatic or, through pressure on the most inferior root of the brachial plexus, can produce sensory and motor changes over the distribution of the ulnar nerve. **B.** Bifid rib. The superior component of this 3rd rib is supernumerary and articulated with the lateral aspect of the 1st sternebra. The inferior component articulated at the junction of the 1st and 2nd sternebrae. **C.** Bicipital rib. In this specimen, there has been partial fusion of the first two thoracic ribs. **E.** Sternal foramen. **F.** Ossification of sternum.

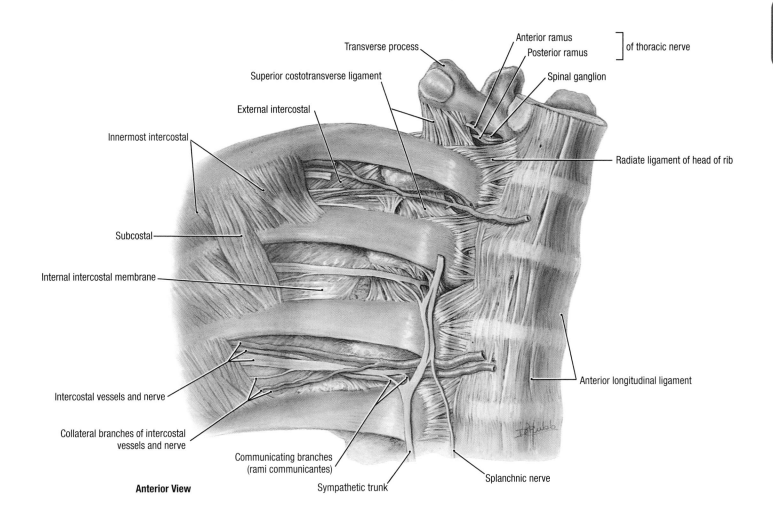

Anterior View

1.15   Vertebral ends of internal aspect of intercostal spaces

- Portions of the innermost intercostal muscle that bridge two intercostal spaces are called subcostal muscles.
- The internal intercostal membrane, in the middle space, is continuous medially with the superior costotransverse ligament.
- Note the order of the structures in the most inferior space: intercostal vein, artery, and nerve; note also their collateral branches.
- The anterior ramus crosses anterior to the superior costotransverse ligament; the posterior primary ramus is posterior to it.
- The intercostal nerves attach to the sympathetic trunk by rami communicantes; the splanchnic nerve is a visceral branch of the trunk.

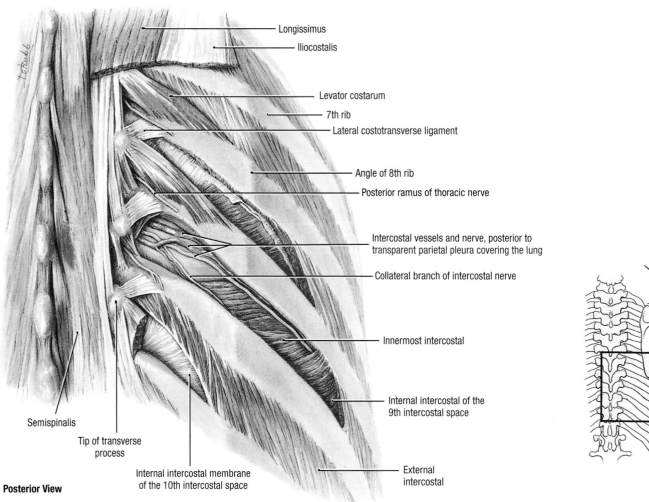

Longissimus

Iliocostalis

Levator costarum

7th rib

Lateral costotransverse ligament

Angle of 8th rib

Posterior ramus of thoracic nerve

Intercostal vessels and nerve, posterior to transparent parietal pleura covering the lung

Collateral branch of intercostal nerve

Innermost intercostal

Internal intercostal of the 9th intercostal space

Semispinalis

Tip of transverse process

Internal intercostal membrane of the 10th intercostal space

External intercostal

Posterior View

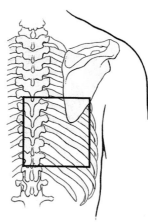

1.16   Vertebral ends of external aspect of inferior intercostal spaces

- The iliocostalis and longissimus muscles have been removed, exposing the levator costarum muscle. Of the five intercostal spaces shown, the superior two (6th and 7th) are intact. In the 8th and 10th spaces, varying portions of the external intercostal muscle have been removed to reveal the underlying internal intercostal membrane, which is continuous with the internal intercostal muscle. In the 9th space, the levator costarum muscle has been removed to show the intercostal vessels and nerve.
- The intercostal vessels and nerve appear medially between the superior costotransverse ligament and the transparent parietal pleura covering the lung; they disappear laterally between the internal and innermost intercostal muscles.
- The intercostal nerve is the most inferior of the neurovascular trio and the least sheltered in the intercostal groove; a collateral branch arises near the angle of the rib.

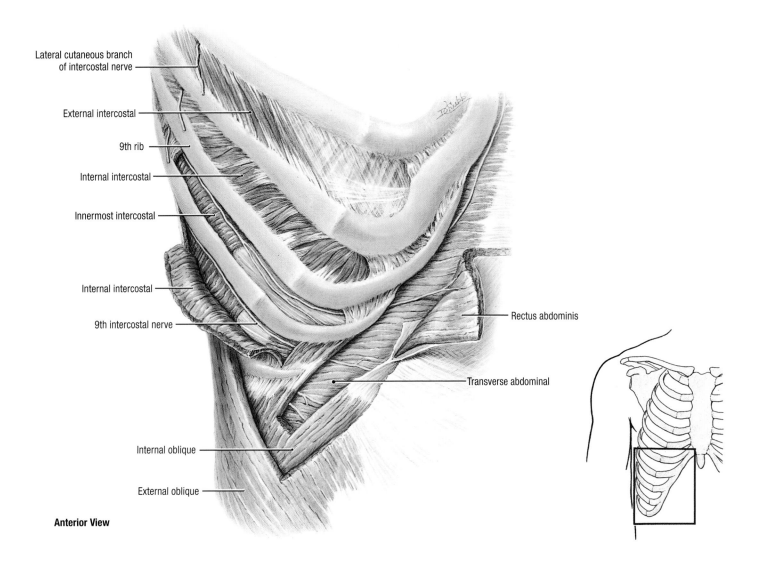

Anterior View

1.17 Anterior ends of inferior intercostal spaces

- The fibers of the external intercostal and external oblique muscles run inferomedially.
- The internal intercostal and internal oblique muscles are in continuity at the ends of the 9th, 10th, and 11th intercostal spaces.
- The intercostal nerves lie deep to the internal intercostal muscle but superficial to the innermost intercostal muscle; anteriorly, these nerves lie superficial to the transverse thoracic or transverse abdominal muscles.
- Intercostal nerves run parallel to the ribs and costal cartilages; on reaching the abdominal wall, nerves T7 and T8 continue superiorly, T9 continues nearly horizontally, and T10 continues inferomedially toward the umbilicus. These nerves provide cutaneous innervation in overlapping segmental bands.

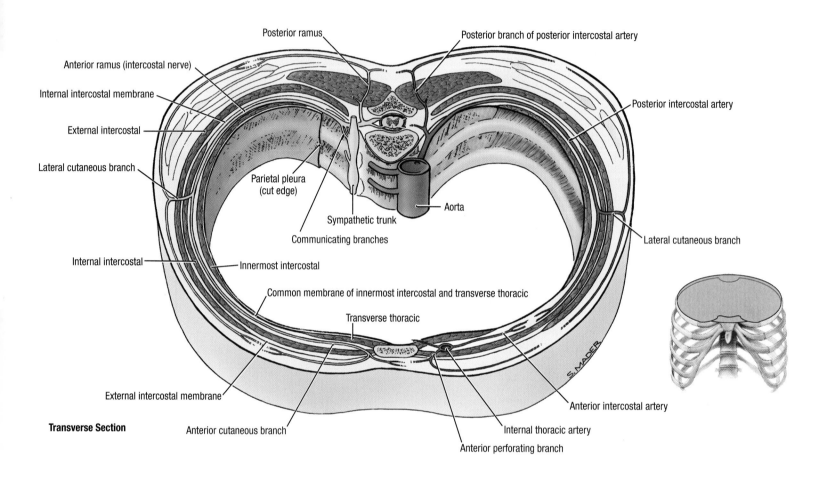

Posterior ramus

Posterior branch of posterior intercostal artery

Anterior ramus (intercostal nerve)

Internal intercostal membrane

External intercostal

Lateral cutaneous branch

Posterior intercostal artery

Parietal pleura
(cut edge)

Sympathetic trunk

Communicating branches

Aorta

Lateral cutaneous branch

Internal intercostal

Innermost intercostal

Common membrane of innermost intercostal and transverse thoracic

Transverse thoracic

External intercostal membrane

Transverse Section

Anterior cutaneous branch

Anterior perforating branch

Internal thoracic artery

Anterior intercostal artery

1.18   Contents of intercostal space, transverse section

- The diagram is simplified by showing nerves on the right and arteries on the left.
- The three musculomembranous layers are the external intercostals muscle and membrane, internal intercostals muscle and membrane, and the innermost intercostals, transverse thoracic muscle, and the membrane connecting them.
- The intercostal nerves are the anterior rami of spinal nerves T1 to T11; the anterior ramus of T12 is the subcostal nerve.
- Posterior intercostal arteries are branches of the aorta (the superior two spaces are supplied from the superior intercostal branch of the costocervical trunk); the anterior intercostals arteries are branches of the internal thoracic artery or its branch, the musculophrenic artery.
- The posterior rami innervate the deep back muscles and skin adjacent to the vertebral column.

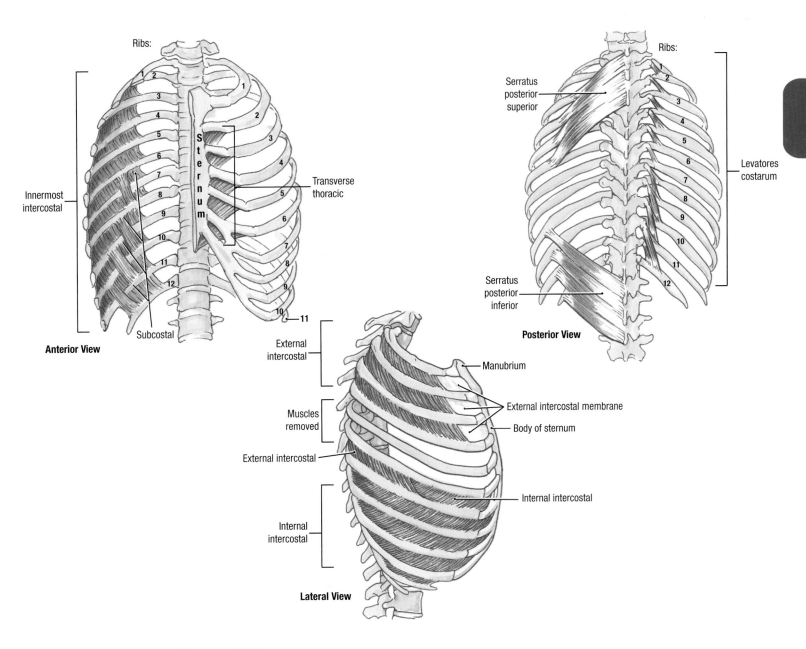

Anterior View

Ribs:

Innermost intercostal

Sternum

Transverse thoracic

Subcostal

Posterior View

Ribs:

Serratus posterior superior

Serratus posterior inferior

Levatores costarum

Lateral View

External intercostal

Muscles removed

External intercostal

Internal intercostal

Internal intercostal

Manubrium

External intercostal membrane

Body of sternum

Internal intercostal

TABLE 1.1   MUSCLES OF THORACIC WALL

MUSCLE	SUPERIOR ATTACHMENT	INFERIOR ATTACHMENT	INNERVATION	ACTION[a]
External intercostal	Inferior border of ribs	Superior border of ribs below	Intercostal nerve	Elevate ribs
Internal intercostal	Inferior border of ribs	Superior border of ribs below	Intercostal nerve	Depress ribs
Innermost intercostal	Inferior border of ribs	Superior border of ribs below	Intercostal nerve	Probably elevate ribs
Transverse thoracic	Posterior surface of lower sternum	Internal surface of costal cartilages 2–6	Intercostal nerve	Depress ribs
Subcostal	Internal surface of lower ribs near their angles	Superior borders of 2nd or 3rd ribs below	Intercostal nerve	Elevate ribs
Levatores costarum	Transverse processes of T7–T11	Subjacent ribs between tubercle and angle	Posterior rami of C8–T11 nerves	Elevate ribs
Serratus posterior superior	Nuchal ligament, spinous processes of C7 to T3 vertebrae	Superior borders of 2nd to 4th ribs	Second to fifth intercostal nerves	Elevate ribs
Serratus posterior inferior	Spinous processes of T11 to L2 vertebrae	Inferior borders of 8th to 12th ribs near their angles	Anterior rami of T9–T12 nerves	Depress ribs

[a]All intercostal muscles keep intercostal spaces rigid, thereby preventing them from bulging out during expiration and from being drawn in during inspiration. Role of individual intercostal muscles and accessory muscles of respiration in moving the ribs is difficult to interpret despite many electromyograpic studies.

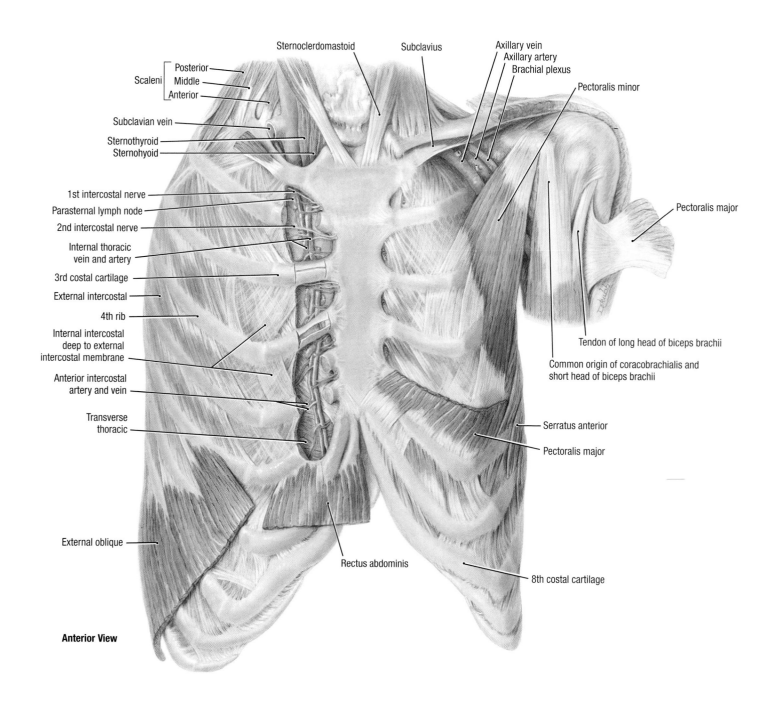

Scaleni
- Posterior
- Middle
- Anterior

Sternoclerdomastoid
Subclavius
Axillary vein
Axillary artery
Brachial plexus
Pectoralis minor

Subclavian vein
Sternothyroid
Sternohyoid

Pectoralis major

1st intercostal nerve
Parasternal lymph node
2nd intercostal nerve
Internal thoracic vein and artery
3rd costal cartilage
External intercostal
4th rib
Internal intercostal deep to external intercostal membrane
Anterior intercostal artery and vein
Transverse thoracic

Tendon of long head of biceps brachii
Common origin of coracobrachialis and short head of biceps brachii

Serratus anterior
Pectoralis major

External oblique

Rectus abdominis

8th costal cartilage

Anterior View

1.19 **External aspect of thoracic wall**

- H-shaped cuts were made through the perichondrium of the 3rd and 4th cartilages to shell out segments of cartilage.
- The internal thoracic (internal mammary) vessels run inferiorly deep to the costal cartilages and just lateral to the edge of the sternum, providing intercostal branches.
- The parasternal lymph nodes (green) receive lymphatic vessels from the anterior parts of intercostal spaces, the costal pleura and diaphragm, and the medial part of the breast.
- The subclavian vessels are "sandwiched" between the 1st rib and clavicle and are "padded" by the subclavius.

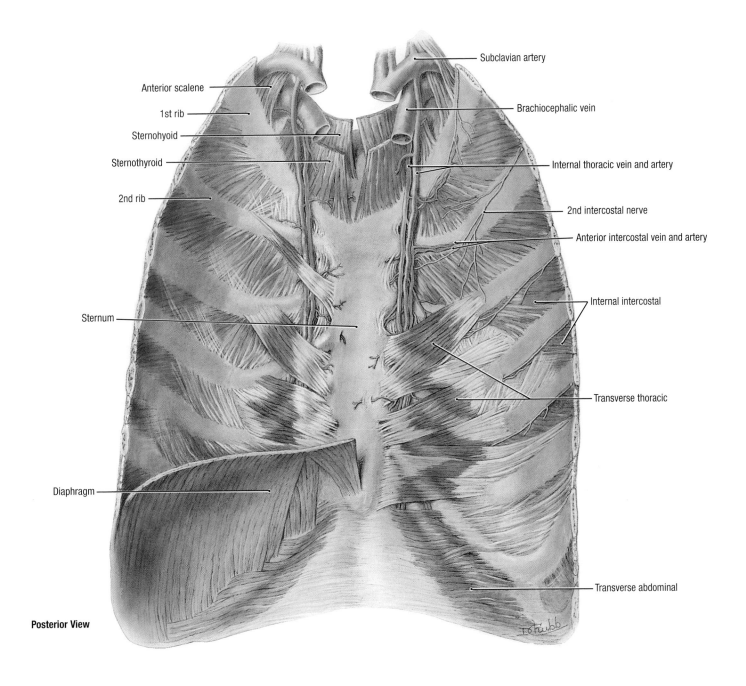

Anterior scalene

1st rib

Sternohyoid

Sternothyroid

2nd rib

Sternum

Diaphragm

Posterior View

Subclavian artery

Brachiocephalic vein

Internal thoracic vein and artery

2nd intercostal nerve

Anterior intercostal vein and artery

Internal intercostal

Transverse thoracic

Transverse abdominal

1.20   Internal aspect of anterior thoracic wall

- The inferior portions of the internal thoracic vessels are covered posteriorly by the transverse thoracic muscle; the superior portions are in contact with parietal pleura (removed).
- The transverse thoracic is continuous with the transverse abdominal; these form the innermost layer of the three flat muscles of the thoracoabdominal wall.
- The internal thoracic (internal mammary) artery arises from the subclavian artery and is accompanied by two communicating veins (venae comitantes) up to the 2nd costal cartilage in this specimen and, superior to this, by the single internal thoracic vein, which drains into the brachiocephalic vein.

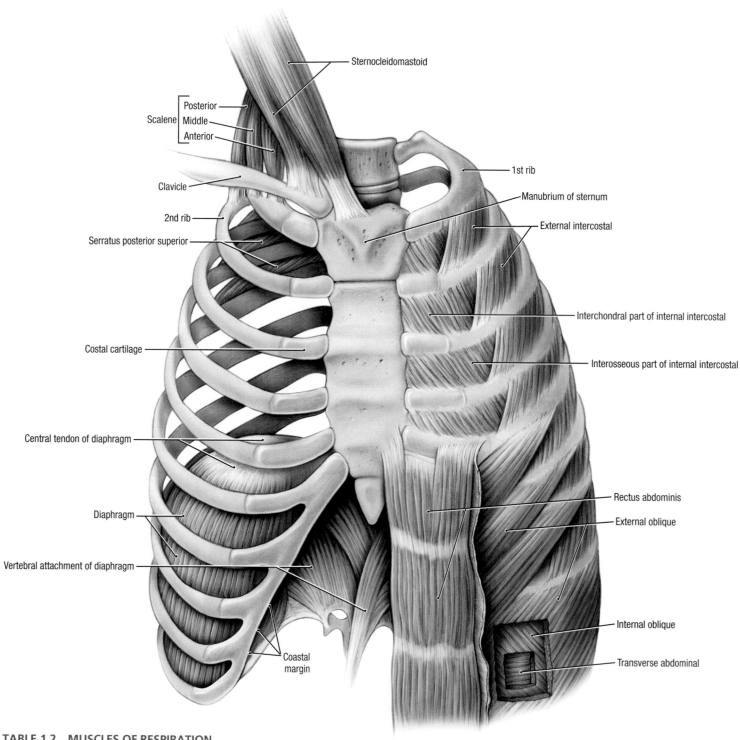

Sternocleidomastoid

Scalene — Posterior / Middle / Anterior

Clavicle

2nd rib

Serratus posterior superior

Costal cartilage

Central tendon of diaphragm

Diaphragm

Vertebral attachment of diaphragm

Coastal margin

1st rib

Manubrium of sternum

External intercostal

Interchondral part of internal intercostal

Interosseous part of internal intercostal

Rectus abdominis

External oblique

Internal oblique

Transverse abdominal

TABLE 1.2   MUSCLES OF RESPIRATION

		INSPIRATION	EXPIRATON
Normal (Quiet)	Major	Diaphragm (active contraction)	Passive (elastic) recoil of lungs and thoracic cage
	Minor	*Tonic contraction* of external intercostals and interchondral portion of internal intercostals to resist negative pressure	*Tonic contraction* of muscles of anterolateral abdominal walls (rectus abdominis, external and internal obliques, transverse abdominal) to antagonize diaphragm by maintaining intraabdominal pressure
Active (Forced)		In addition to the above, *active contraction* of	In addition to the above, *active contraction* of
		Sternocleidomastoid, superior trapezius, pectoralis minor, and scalenes, to elevate and fix upper rib cage	Muscles of anterolateral abdominal wall (antagonizing diaphragm by increasing intraabdominal pressure bracket and by pulling inferiorly on and fixing inferior costal margin):
		External intercostals, interchondral portion of internal intercostals, subcostals, levator costarum, and serratus posterior superior[a] to elevate ribs	● rectus abdominis ● external and internal obliques ● transverse abdominal
			Internal intercostal (interosseous part) and serratus posterior inferior[a] to depress ribs

[a]Recent studies indicate the serratus posterior superior and inferior muscles may serve primarily as organs of proprioception rather than motion.

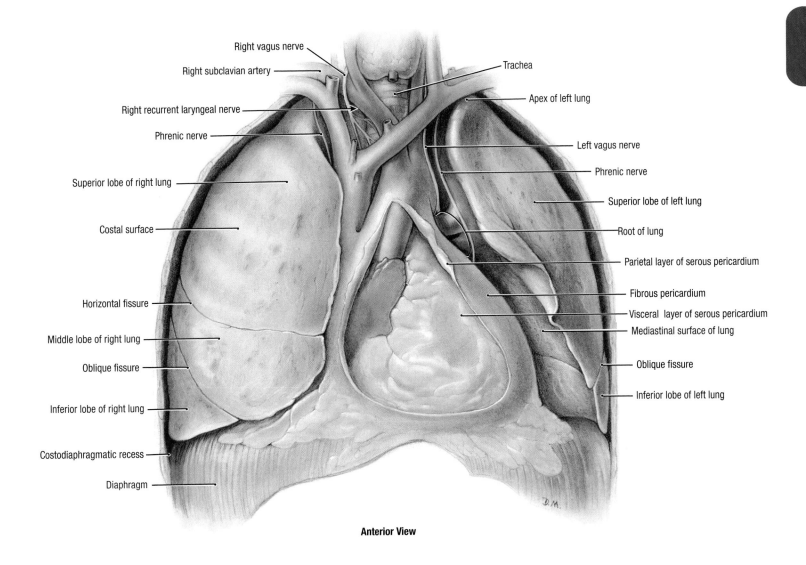

Right vagus nerve

Right subclavian artery

Trachea

Right recurrent laryngeal nerve

Apex of left lung

Phrenic nerve

Left vagus nerve

Superior lobe of right lung

Phrenic nerve

Costal surface

Superior lobe of left lung

Root of lung

Parietal layer of serous pericardium

Horizontal fissure

Fibrous pericardium

Middle lobe of right lung

Visceral layer of serous pericardium

Oblique fissure

Mediastinal surface of lung

Inferior lobe of right lung

Oblique fissure

Inferior lobe of left lung

Costodiaphragmatic recess

Diaphragm

Anterior View

1.21 Thoracic contents in situ

- The fibrous pericardium, lined by the parietal layer of serous pericardium, is removed anteriorly to expose the heart and great vessels.
- The right lung has three lobes; the superior lobe is separated from the middle lobe by the horizontal fissure, and the middle lobe is separated from the inferior lobe by the oblique fissure; the left lung has two lobes, superior and inferior, separated by the oblique fissure.
- The anterior border of the left lung is reflected laterally to visualize the phrenic nerve passing anterior to the root of the lung and the vagus nerve lying anterior to the arch of the aorta and then passing posterior to the root of the lung.
- As the right vagus nerve passes anterior to the right subclavian artery, it gives rise to the recurrent branch and then divides to contribute fibers to the esophageal, cardiac, and pulmonary plexuses.

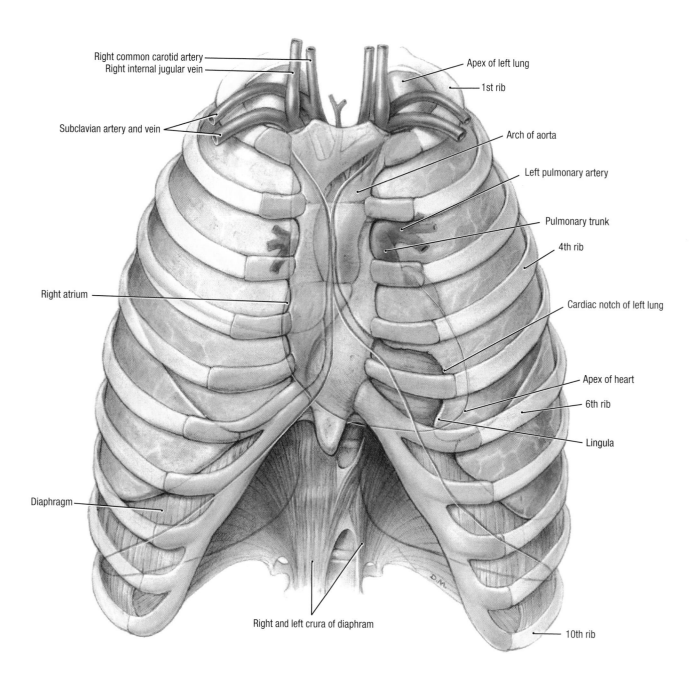

Right common carotid artery
Right internal jugular vein

Subclavian artery and vein

Right atrium

Diaphragm

Right and left crura of diaphram

Apex of left lung
1st rib
Arch of aorta
Left pulmonary artery
Pulmonary trunk
4th rib
Cardiac notch of left lung
Apex of heart
6th rib
Lingula
10th rib

1.22 Topography of the lungs and mediastinum

- The apex of the lungs is at the level of the neck of the 1st rib, and the inferior border of the lungs is at the 6th rib in the left midclavicular line and the 8th rib at the lateral aspect of the bony thorax at the midaxillary line.
- Note the cardiac notch of the left lung and the deviation of the parietal pleura away from the median plane toward the left side in the region of the notch.
- Note the inferior reflection of parietal pleura at the 8th costochondral junction in the midclavicular line, at the 10th rib in

the midaxillary line, and at the level of the neck of the 12th rib on each side of the vertebral column.
- The apex of the heart is in the 5th intercostal space at the left midclavicular line.
- The right atrium forms the right border of the heart and extends just beyond the lateral margin of the sternum.
- The great vessels lie superior and posterior to the heart; their branches pass through the superior thoracic aperture.

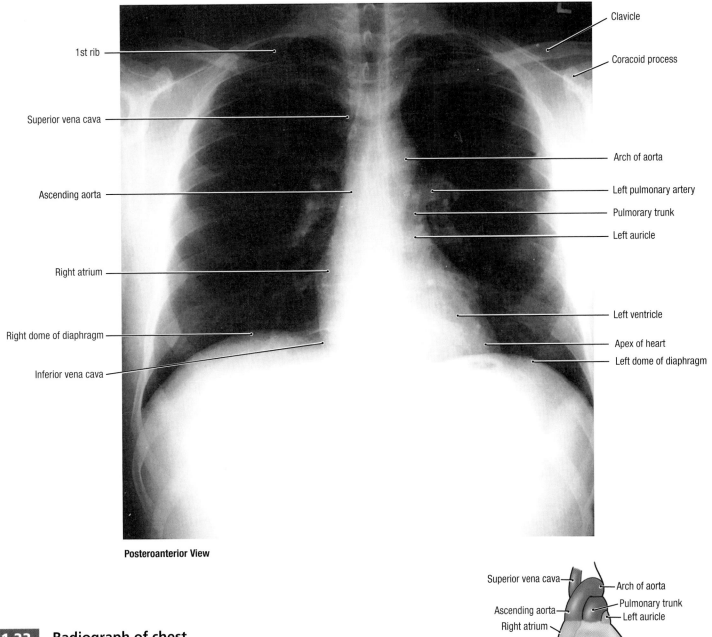

Posteroanterior View

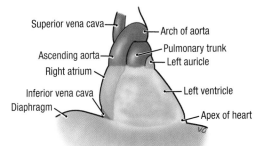

1.23   **Radiograph of chest**

- The right dome of the diaphragm is higher than the left dome due primarily to the large underlying liver.
- The convex right mediastinal border of the heart is formed by the right atrium; above this, the superior vena cava and ascending aorta produce less convex borders.
- The left mediastinal border of the cardiac silhouette is formed by the arch of the aorta, pulmonary trunk, left auricle (normally not prominent), and left ventricle.
- Follow the 1st rib from its articulation with the body of the 1st thoracic vertebra (T1) to where it curves laterally and then medially to cross the clavicle.

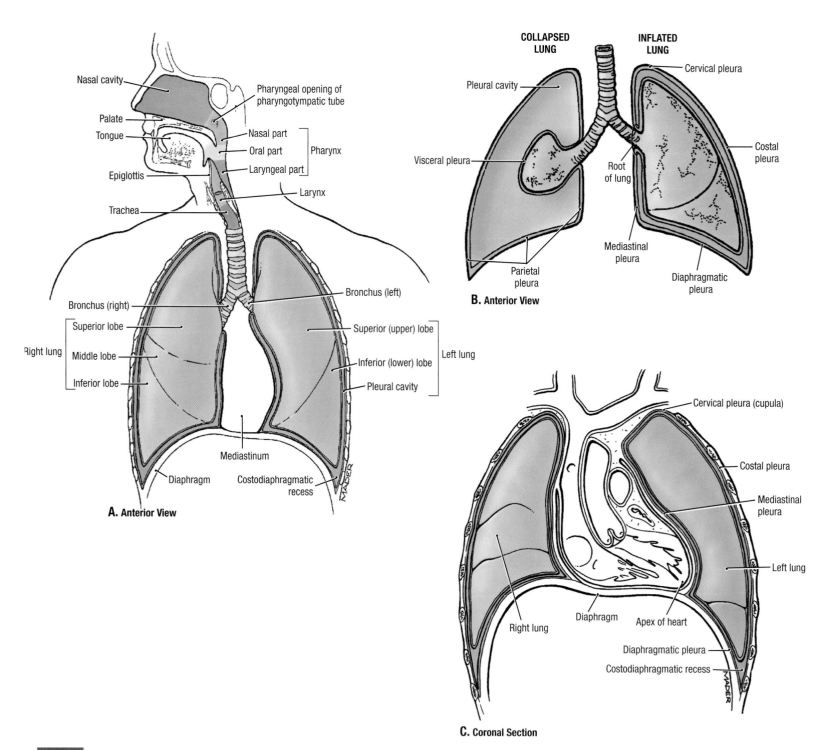

A. Anterior View

B. Anterior View

C. Coronal Section

1.24 Respiratory system

A. Overview. **B.** Pleural cavity and pleura. **C.** Coronal section through heart and lungs.

- The lungs invaginate a continuous membranous pleural sac; the visceral (pulmonary) pleura covers the lungs, and the parietal pleura lines the thoracic cavity; the visceral and parietal pleurae are continuous around the root of the lung.
- The pleural cavity is a potential space between the visceral and parietal pleurae that contains a thin layer of fluid. When the lung collapses (as in B), the pleural cavity becomes a "real" space and may contain air, blood, etc.
- The parietal pleura can be divided regionally into the costal, diaphragmatic, mediastinal, and cervical pleura; note the costodiaphragmatic recess.

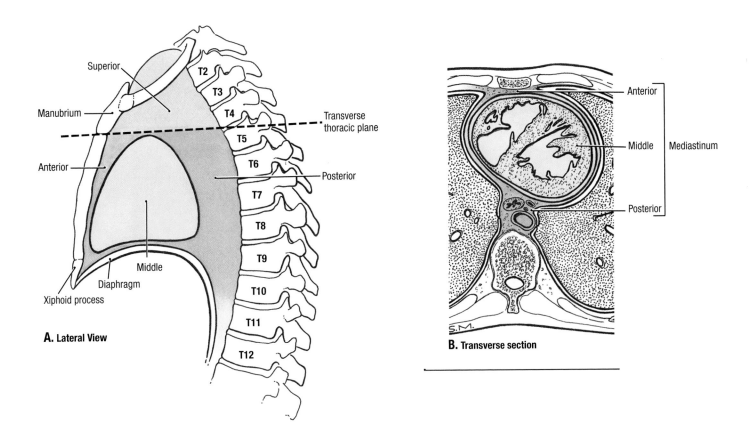

A. Lateral View

B. Transverse section

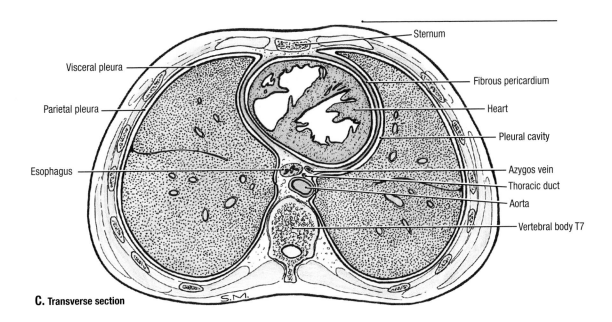

C. Transverse section

1.25   **Mediastinum**

A. Subdivisions of the mediastinum. Transverse sections through mediastinum (**B**) and thorax (**C**). The mediastinum is located between the right and left pleural sacs and consists of four parts: the superior, anterior, middle, and posterior medistinum.

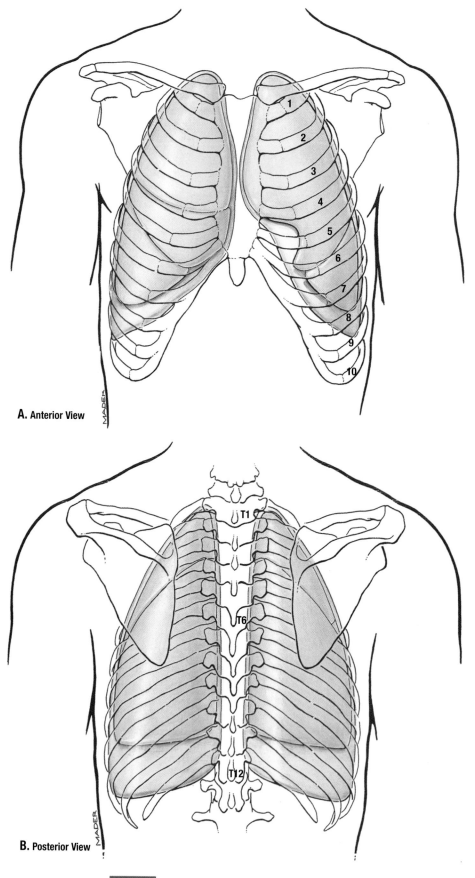

A. Anterior View

B. Posterior View

1.26 **Extent of pleura and lungs**

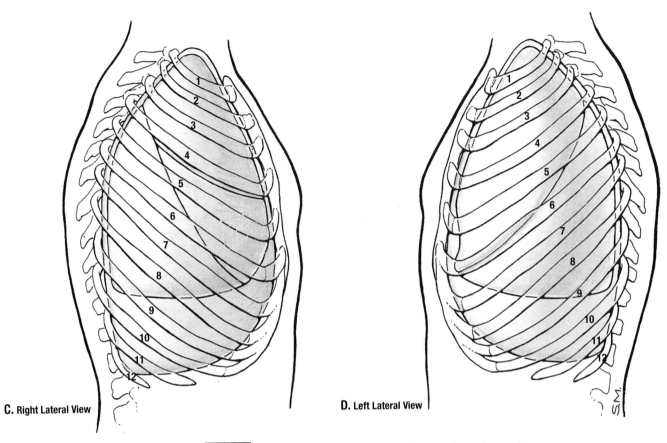

C. Right Lateral View   **D.** Left Lateral View

1.26   **Extent of pleura and lungs (continued)**

TABLE 1.3
SURFACE MARKINGS OF PLEURA

LEVEL	LEFT PLEURA	RIGHT PLEURA
Apex	About 1½ inches superior to middle of clavicle	About 1½ inches superior to middle of clavicle
4th costal cartilage	Midline (anteriorly)	Midline (anteriorly)
6th costal cartilage	Lateral margin of sternum	Midline (anteriorly)
8th costal cartilage	Midclavicular line	Midclavicular line
10th rib	Midaxillary line	Midaxillary line
11th rib	Line of inferior angle of scapula	Line of inferior angle of scapula
12th rib	Lateral border of erector spinae to T12 spinous process (slightly lower level than right pleura)	Lateral border of erector spinae to T12 spinous process

SURFACE MARKINGS OF LUNGS

LEVEL	LEFT LUNG	RIGHT LUNG
Apex	About 1½ inches superior to middle of clavicle	About 1½ inches superior to middle of clavicle
2nd costal cartilage	Midline (anteriorly)	Midline (anteriorly)
4th costal cartilage	Lateral margin of sternum	Lateral margin of sternum
6th costal cartilage	Follows 4th costal cartilage, turns inferiorly to 6th costal cartilage in the midclavicular line (cardiac notch)	Midclavicular line
8th rib	Midaxillary line	Midaxillary line
10th rib	Line of inferior angle of scapula to T10 spinous process	Line of inferior angle of scapula to T10 spinous process

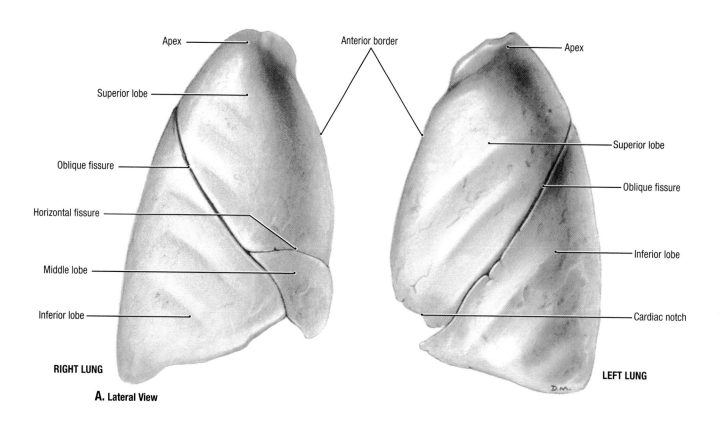

Apex — Anterior border — Apex

Superior lobe

Oblique fissure

Horizontal fissure

Middle lobe

Inferior lobe

Superior lobe

Oblique fissure

Inferior lobe

Cardiac notch

RIGHT LUNG

LEFT LUNG

A. Lateral View

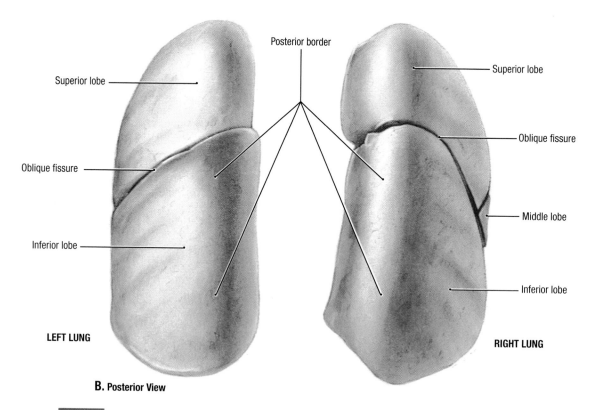

Posterior border

Superior lobe

Oblique fissure

Inferior lobe

Superior lobe

Oblique fissure

Middle lobe

Inferior lobe

LEFT LUNG

RIGHT LUNG

B. Posterior View

1.27 **Lungs**

The right lung usually has three lobes, and the left lung, two lobes. The oblique and horizontal fissures of the right lung, and the oblique fissure of the left lung may be incomplete or absent in some specimens.

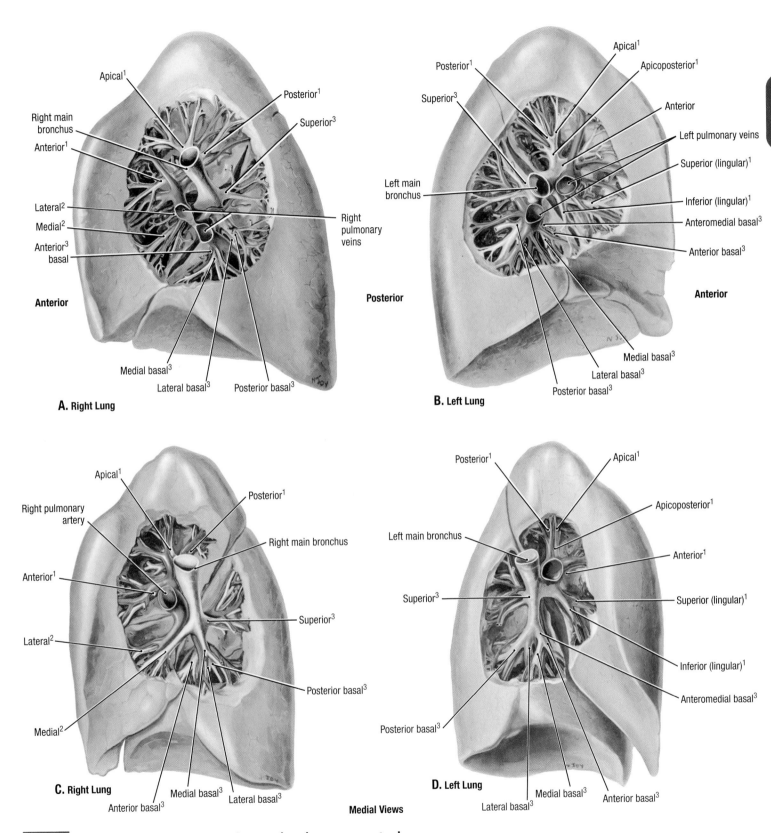

1.28   **Bronchi, pulmonary veins and pulmonary arteries**

A and **C.** Right lungs. **B** and **D.** Left lungs. Superscripts indicate segmental bronchi to [1]the superior lobe, [2]middle lobe, and [3]inferior lobe. The pulmonary veins (*pink*; oxygenated blood) and pulmonary arteries (*purple*; deoxygenated blood) of fresh lungs were filled with latex, the bronchi (*gray*) were inflated with air. The tissues surrounding the bronchi and veins were moistened and cut away.

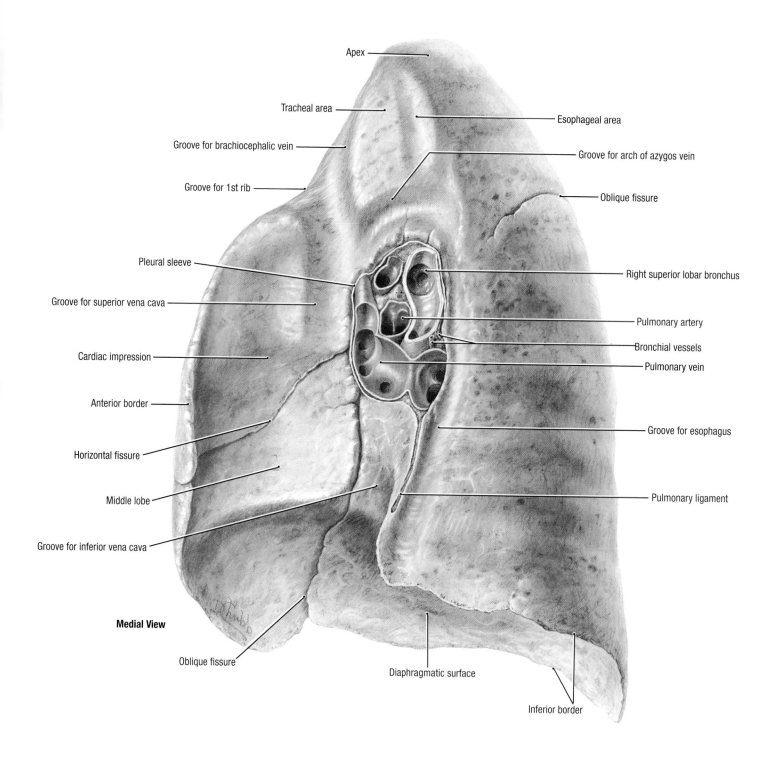

Apex

Tracheal area

Esophageal area

Groove for brachiocephalic vein

Groove for arch of azygos vein

Groove for 1st rib

Oblique fissure

Pleural sleeve

Right superior lobar bronchus

Groove for superior vena cava

Pulmonary artery

Cardiac impression

Bronchial vessels

Pulmonary vein

Anterior border

Groove for esophagus

Horizontal fissure

Middle lobe

Pulmonary ligament

Groove for inferior vena cava

Medial View

Oblique fissure

Diaphragmatic surface

Inferior border

1.29 Mediastinal (medial) surface of right lung

The embalmed lung shows impressions of the structures with which it comes into contact clearly demarcated as surface features; the base is contoured by the domes of the diaphragm; the costal surface bears the impressions of the ribs; distended vessels leave their mark, but nerves do not. The oblique fissure is incomplete here.

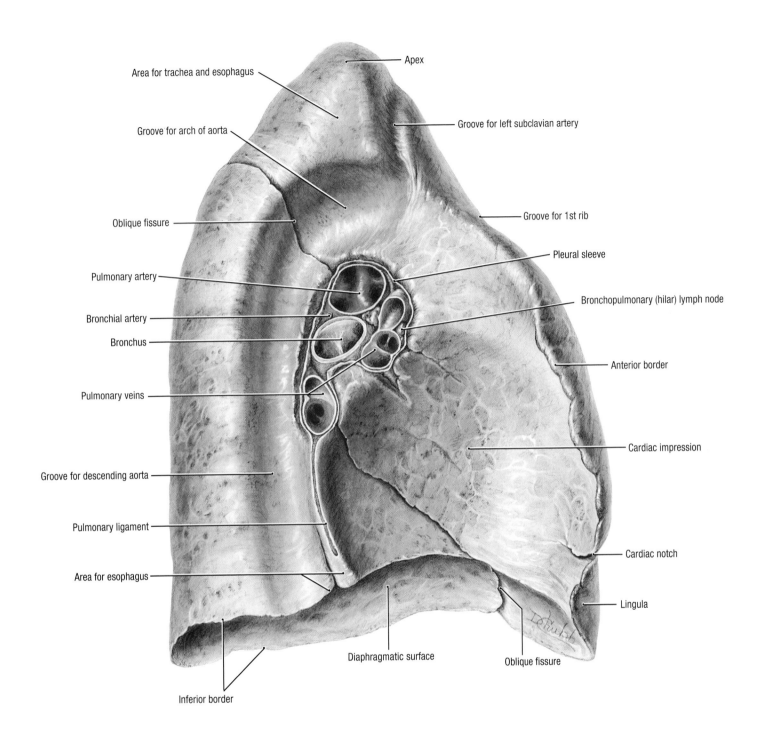

Apex

Area for trachea and esophagus

Groove for left subclavian artery

Groove for arch of aorta

Oblique fissure

Groove for 1st rib

Pleural sleeve

Pulmonary artery

Bronchopulmonary (hilar) lymph node

Bronchial artery

Bronchus

Anterior border

Pulmonary veins

Cardiac impression

Groove for descending aorta

Pulmonary ligament

Cardiac notch

Area for esophagus

Lingula

Inferior border

Diaphragmatic surface

Oblique fissure

1.30   Mediastinal (medial) surface of left lung

Note the site of contact with the esophagus, between the descending aorta and the inferior
end of the pulmonary ligament. In the right and left roots, the artery is superior, the
bronchus is posterior, one vein is anterior, and the other is inferior; in the right root, the
bronchus to the superior lobe (also called the *eparterial bronchus*) is the most superior structure.

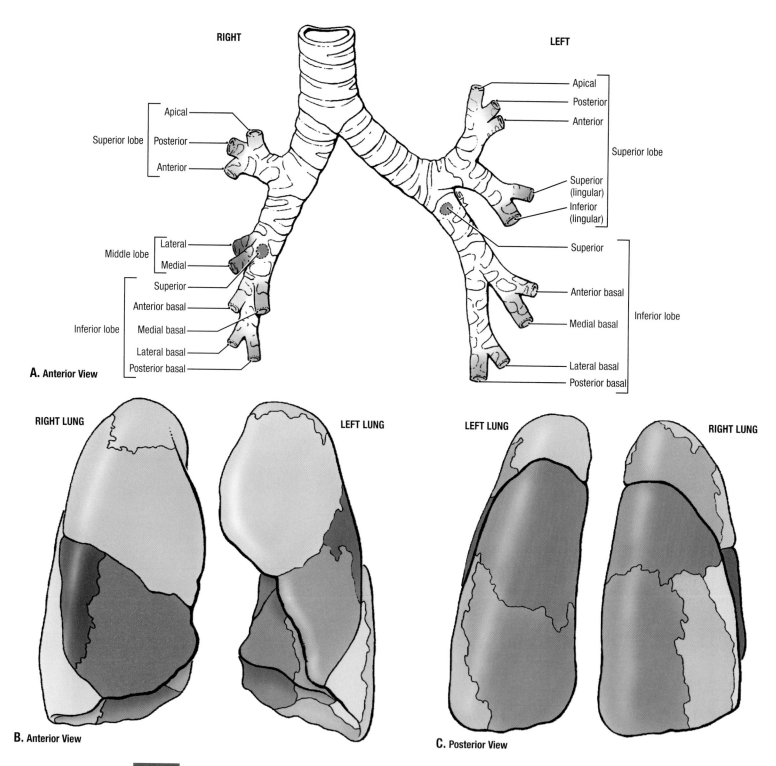

A. Anterior View

RIGHT

Superior lobe
- Apical
- Posterior
- Anterior

Middle lobe
- Lateral
- Medial

Inferior lobe
- Superior
- Anterior basal
- Medial basal
- Lateral basal
- Posterior basal

LEFT

Superior lobe
- Apical
- Posterior
- Anterior
- Superior (lingular)
- Inferior (lingular)

Inferior lobe
- Superior
- Anterior basal
- Medial basal
- Lateral basal
- Posterior basal

B. Anterior View

RIGHT LUNG   LEFT LUNG

C. Posterior View

LEFT LUNG   RIGHT LUNG

1.31   **Segmental bronchi and bronchopulmonary segments**

A. There are 10 tertiary or segmental bronchi on the right, and 8 on the left. Note that on the left, the apical and posterior bronchi arise from a single stem, as do the anterior basal and medial basal. **B** to **F.** A bronchopulmonary segment consists of a tertiary bronchus, pulmonary vein and artery, and the portion of lung they serve. These structures are surgically separable to allow segmental resection of the lung. To prepare these specimens, the tertiary bronchi of fresh lungs were isolated within the hilus and injected with latex of various colors. Minor variations in the branching of the bronchi result in variations in the surface patterns.

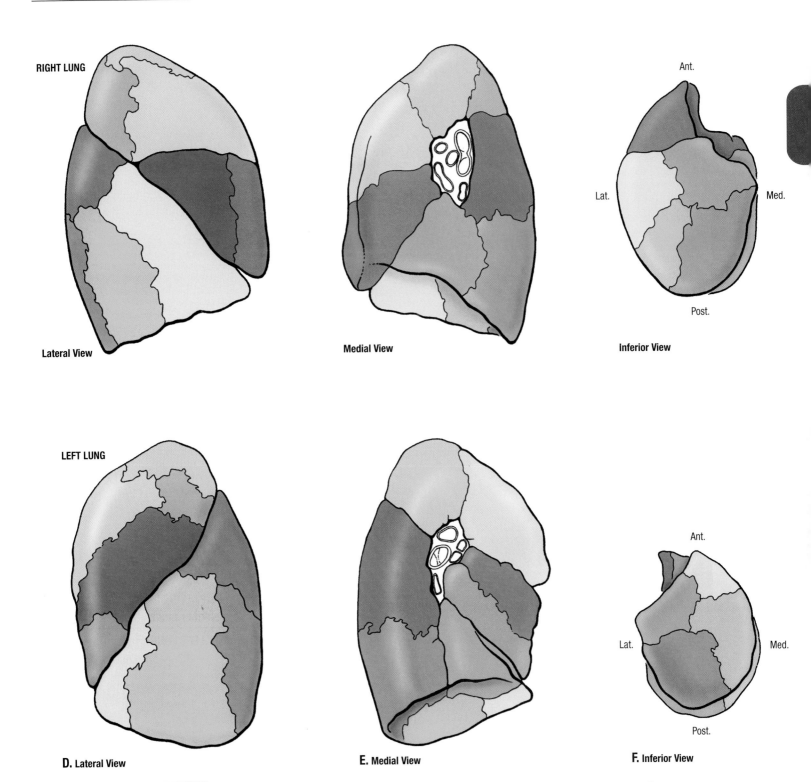

RIGHT LUNG

Lateral View

Medial View

Ant.

Lat.   Med.

Post.

Inferior View

LEFT LUNG

D. Lateral View

E. Medial View

Ant.

Lat.   Med.

Post.

F. Inferior View

1.31   **Segmental bronchi and bronchopulmonary segments** *(continued)*

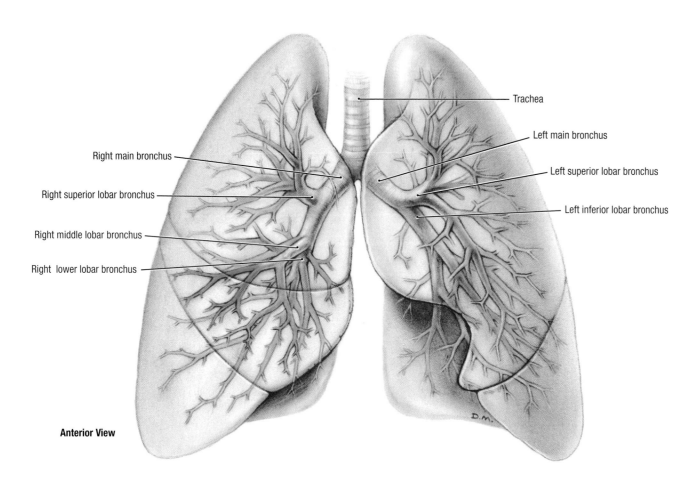

Trachea

Left main bronchus

Left superior lobar bronchus

Left inferior lobar bronchus

Right main bronchus

Right superior lobar bronchus

Right middle lobar bronchus

Right lower lobar bronchus

Anterior View

1.32 Trachea and bronchi in situ

- The segmental (tertiary) bronchi are color coded.
- The trachea bifurcates into right and left main (primary) bronchi; the right main bronchus is shorter, wider, and more vertical than the left. Therefore, it is more likely that foreign objects will become lodged in the right main bronchus.
- The right main bronchus gives off the right superior lobe bronchus (eparterial bronchus) before entering the hilum (hilus) of the lung; after entering the hilum, the right middle and inferior lobar bronchi branch off.
- The left main bronchus divides into the left superior and left inferior lobar bronchi; the lobar bronchi further divide into segmental (tertiary) bronchi.

Segmental bronchi:

RIGHT LUNG	LEFT LUNG
Superior Lobe	**Superior Lobe**

RIGHT LUNG

Superior Lobe
- ▢ Apical
- ▢ Posterior
- ▢ Anterior

Middle Lobe
- ▢ Lateral
- ▢ Medial

Inferior Lobe
- ▢ Superior
- ▢ Anterior basal
- ▢ Medial basal
- ▢ Lateral basal
- ▢ Posterior basal

LEFT LUNG

Superior Lobe
- ▢ Apical ⎤ Apicoposterior
- ▢ Posterior ⎦
- ▢ Anterior
- ▢ Superior ⎤ Lingular
- ▢ Inferior ⎦

Inferior Lobe
- ▢ Superior
- ▢ Anterior basal ⎤ Anteromedial basal
- ▢ Medial basal ⎦
- ▢ Lateral basal
- ▢ Posterior basal

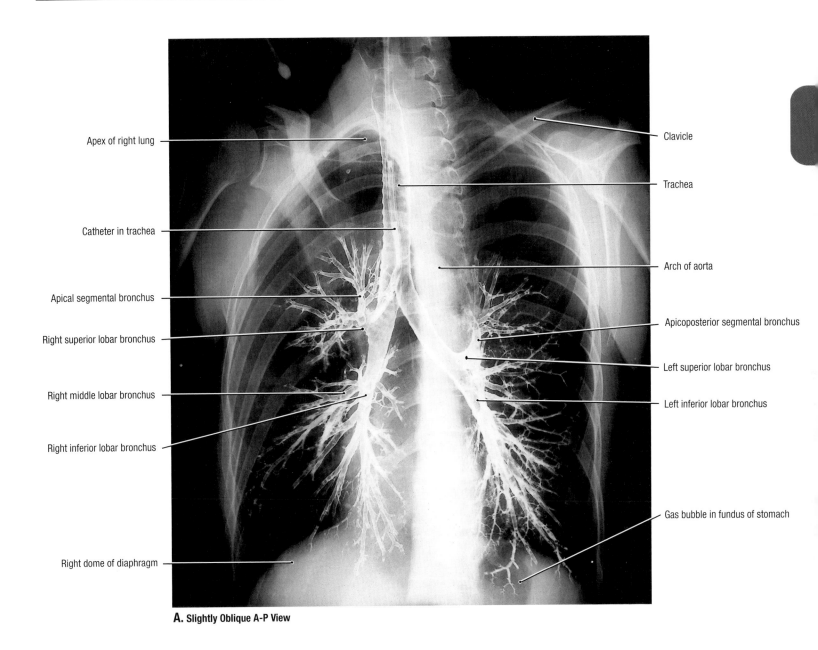

Apex of right lung

Catheter in trachea

Apical segmental bronchus

Right superior lobar bronchus

Right middle lobar bronchus

Right inferior lobar bronchus

Right dome of diaphragm

Clavicle

Trachea

Arch of aorta

Apicoposterior segmental bronchus

Left superior lobar bronchus

Left inferior lobar bronchus

Gas bubble in fundus of stomach

A. Slightly Oblique A-P View

1.33   **Bronchograms**

A. Bronchogram of tracheobronchial tree. *(Continued on next page)*

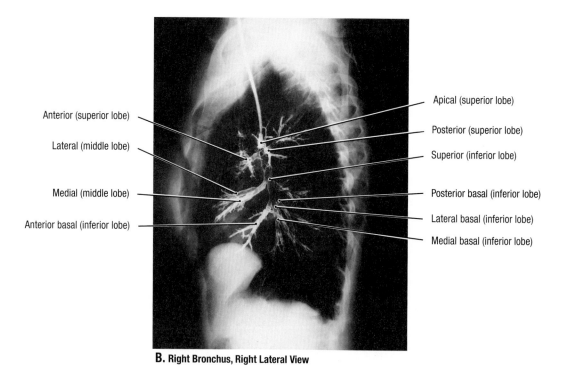

Anterior (superior lobe)

Lateral (middle lobe)

Medial (middle lobe)

Anterior basal (inferior lobe)

Apical (superior lobe)

Posterior (superior lobe)

Superior (inferior lobe)

Posterior basal (inferior lobe)

Lateral basal (inferior lobe)

Medial basal (inferior lobe)

B. Right Bronchus, Right Lateral View

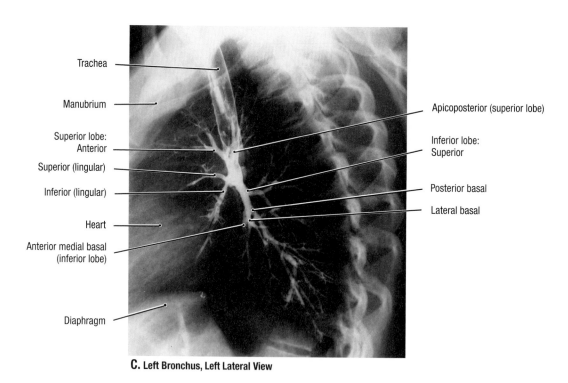

Trachea

Manubrium

Superior lobe:
Anterior

Superior (lingular)

Inferior (lingular)

Heart

Anterior medial basal
(inferior lobe)

Diaphragm

Apicoposterior (superior lobe)

Inferior lobe:
Superior

Posterior basal

Lateral basal

C. Left Bronchus, Left Lateral View

1.33 Bronchograms (continued)

B. Right lateral bronchogram, showing segmental bronchi. **C.** Left lateral bronchogram, showing segmental bronchi.

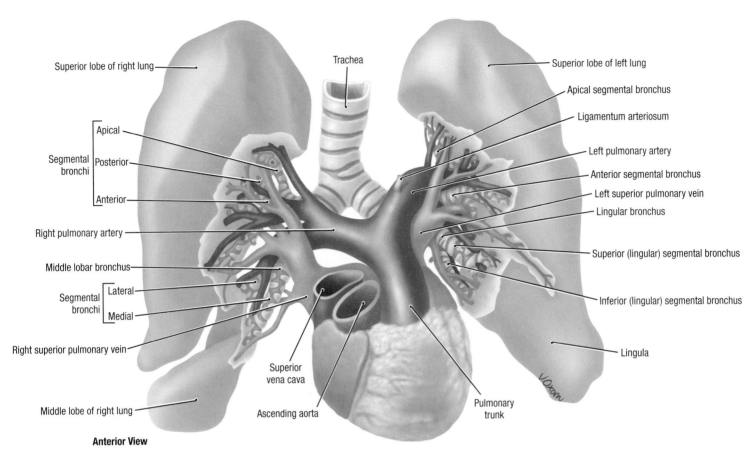

Anterior View

1.34   **Pulmonary artery, lungs retracted**

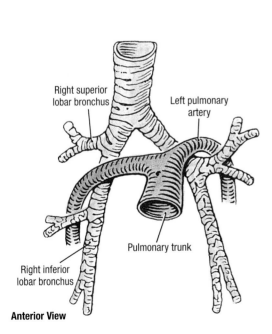

Anterior View

1.35   **Relationship of bronchi and pulmonary arteries**

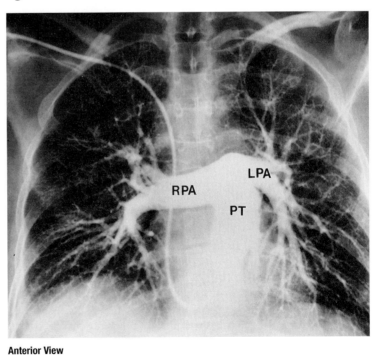

Anterior View

1.36   **Pulmonary angiogram**

Note the catheter located in the right ventricle and pulmonary trunk (PT); the pulmonary trunk dividing into a longer right pulmonary artery (RPA) that passes under the aortic arch and shorter left pulmonary artery (LPA); The branches of the right and left pulmonary arteries follow the corresponding segmental bronchi.

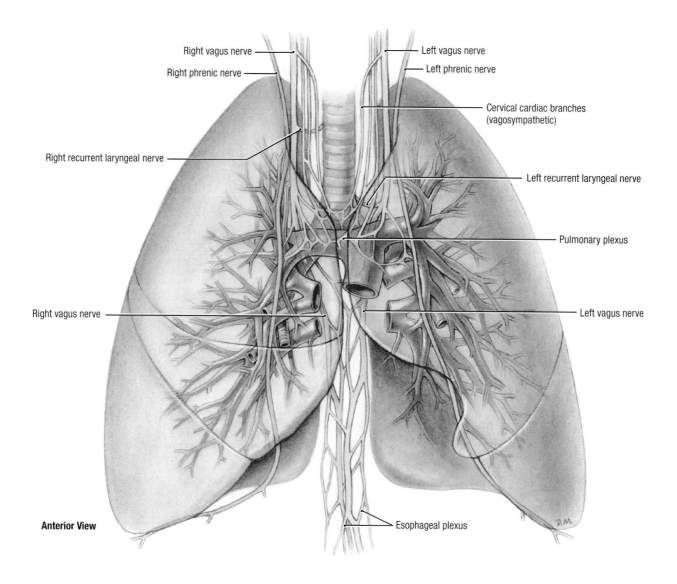

Right vagus nerve

Right phrenic nerve

Right recurrent laryngeal nerve

Right vagus nerve

Anterior View

Left vagus nerve

Left phrenic nerve

Cervical cardiac branches
(vagosympathetic)

Left recurrent laryngeal nerve

Pulmonary plexus

Left vagus nerve

Esophageal plexus

1.37　Innervation of lungs

- The pulmonary plexuses, located anterior and posterior to the roots of the lungs, receive sympathetic contributions from the right and left sympathetic trunks (2nd to 5th thoracic ganglia, not shown) and parasympathetic contributions from the right and left vagus nerves; cell bodies of postsynaptic parasympathetic neurons are in the pulmonary plexuses and along the branches of the pulmonary tree.

- The right and left vagus nerves continue inferiorly from the posterior pulmonary plexus to contribute fibers to the esophageal plexus.

- Branches from the pulmonary plexuses continue along the bronchi and pulmonary vasculature into the lungs.

- The phrenic nerves pass anterior to the root of the lung on their way to the diaphragm.

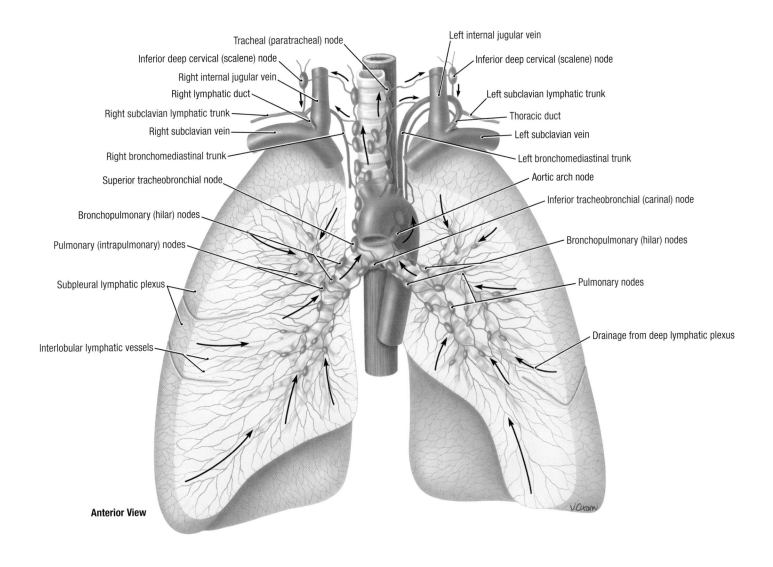

Tracheal (paratracheal) node
Inferior deep cervical (scalene) node
Right internal jugular vein
Right lymphatic duct
Right subclavian lymphatic trunk
Right subclavian vein
Right bronchomediastinal trunk
Superior tracheobronchial node
Bronchopulmonary (hilar) nodes
Pulmonary (intrapulmonary) nodes
Subpleural lymphatic plexus
Interlobular lymphatic vessels

Left internal jugular vein
Inferior deep cervical (scalene) node
Left subclavian lymphatic trunk
Thoracic duct
Left subclavian vein
Left bronchomediastinal trunk
Aortic arch node
Inferior tracheobronchial (carinal) node
Bronchopulmonary (hilar) nodes
Pulmonary nodes
Drainage from deep lymphatic plexus

Anterior View

1.38 Lymphatic drainage of lungs

- Lymphatic vessels originate in the subpleural (superficial) and deep lymphatic plexuses.
- The subpleural lymphatic plexus, is superficial, lying deep to the visceral pleura and drains lymph from the surface of the lung to the bronchopulmonary (hilar) nodes.
- The deep lymphatic plexus is in the lung and follows the bronchi and pulmonary vessels to the pulmonary and then bronchopulmonary nodes located at the root of the lung.
- All lymph from the lungs enters the inferior (carinal) and superior tracheobronchial nodes and then continues to the right and left bronchomediastinal trunks to drain into the venous system via the right lymphatic and thoracic ducts; lymph from the left inferior lobe passes largely to the right side;
- Lymph from the parietal pleura drains into lymph nodes of the thoracic wall (Fig. 1.71).

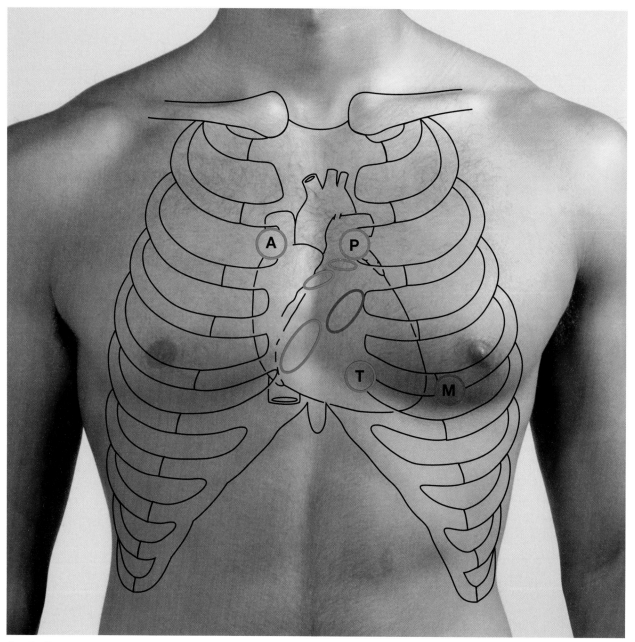

Anterior View

1.39   **Surface markings of the heart, heart valves, and their auscultation areas**

- The location of each heart valve in situ is indicated by a colored oval and the area of auscultation of the valve is indicated as a circle of the same color containing the first letter of the valve name: the tricuspid valve (T) is green, the mitral valve (M) is purple, the pulmonary valve (P) is pink and the aortic valve (A) is blue.
- The auscultation areas are sites where the sounds of each of the heart's valves can be heard most distinctly through a stethoscope.
- The aortic (A) and pulmonary (P) auscultation areas are in the 2nd intercostal space to the right and left of the sternal border; the tricuspid area (T) is near the left sternal border in the 5th or 6th intercostal space; the mitral valve (M) is heard best near the apex of the heart in the 5th intercostal space in the midclavicular line.

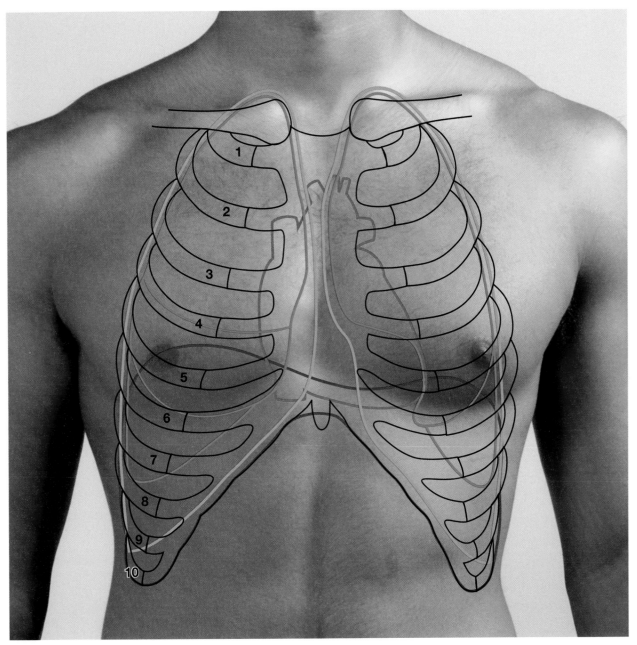

Anterior View

1.40   **Surface markings of the heart, lungs, and diaphragm**

- Outlined are the heart (red), lungs (green), parietal pleura (blue), and diaphragm (purple).
- The superior border of the heart is represented by a slightly oblique line joining the 3rd costal cartilages; the convex right side of the heart projects lateral to the sternum and inferiorly lying at the 6th or 7th costochondral junction; the inferior border of the heart lying superior to the central tendon of the diaphragm and sloping slightly inferiorly to the apex at the 5th interspace at the midclavicular line.
- The right dome of the diaphragm is higher than the left because of the large size of the liver inferior to the dome; during expiration the right dome reaches as high as the 5th rib and the left dome ascends to the 5th intercostal space.
- The left pleural cavity is smaller than the right because of the projection of the heart to the left side.

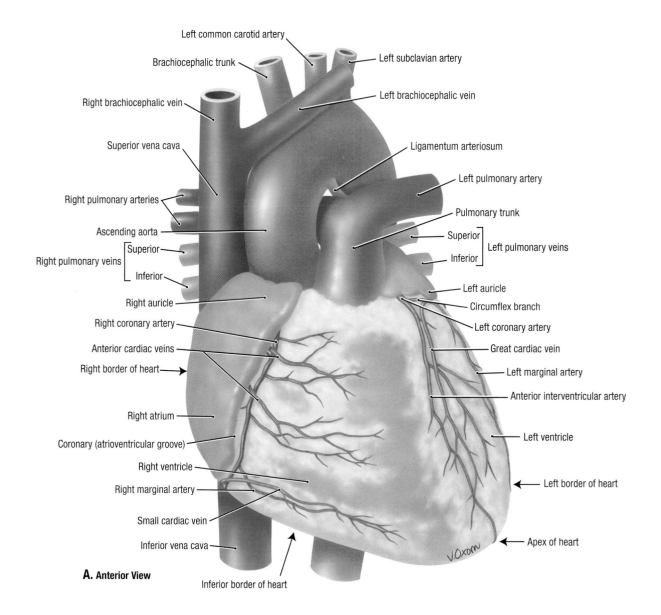

Left common carotid artery

Brachiocephalic trunk

Left subclavian artery

Right brachiocephalic vein

Left brachiocephalic vein

Superior vena cava

Ligamentum arteriosum

Left pulmonary artery

Right pulmonary arteries

Pulmonary trunk

Ascending aorta

Superior ⎤
        ⎥ Left pulmonary veins
Inferior ⎦

Superior ⎤
        ⎥ Right pulmonary veins
Inferior ⎦

Right auricle

Left auricle

Circumflex branch

Right coronary artery

Left coronary artery

Anterior cardiac veins

Great cardiac vein

Right border of heart

Left marginal artery

Anterior interventricular artery

Right atrium

Coronary (atrioventricular groove)

Left ventricle

Right ventricle

Left border of heart

Right marginal artery

Small cardiac vein

Inferior vena cava

Apex of heart

A. Anterior View

Inferior border of heart

1.41   Heart and great vessels

A:

- The right border of the heart, formed by the right atrium, is slightly convex and almost in line with the superior vena cava.
- The inferior border is formed primarily by the right ventricle and a small part of the left ventricle.
- The left border is formed primarily by the left ventricle and a small portion of the left auricle.
- The pulmonary artery bifurcates inferior to the arch of the aorta into a right (that passes under the arch) and left pulmonary artery.
- The ligamentum arteriosum passes from the origin of the left pulmonary artery to the arch of the aorta.

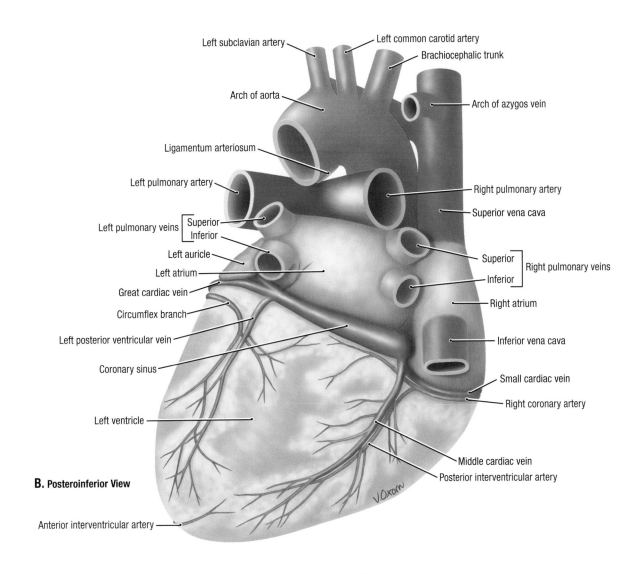

B. Posteroinferior View

Labels (clockwise from top):
- Left subclavian artery
- Left common carotid artery
- Brachiocephalic trunk
- Arch of aorta
- Arch of azygos vein
- Ligamentum arteriosum
- Left pulmonary artery
- Right pulmonary artery
- Superior vena cava
- Left pulmonary veins { Superior / Inferior }
- Left auricle
- Superior
- Right pulmonary veins { Superior / Inferior }
- Left atrium
- Great cardiac vein
- Right atrium
- Circumflex branch
- Left posterior ventricular vein
- Inferior vena cava
- Coronary sinus
- Small cardiac vein
- Right coronary artery
- Left ventricle
- Middle cardiac vein
- Posterior interventricular artery
- Anterior interventricular artery

V.Oxorn

1.41 **Heart and great vessels (continued)**

B:
- Most of the left atrium and left ventricle are visible in this posteroinferior view.
- The right and left pulmonary veins open into the left atrium.
- The right and left pulmonary arteries are just superior and parallel to the pulmonary veins.
- The arch of the aorta is arched in two planes: superiorly and to the left.
- The azygos vein arches over the right pulmonary vessels (and bronchus).

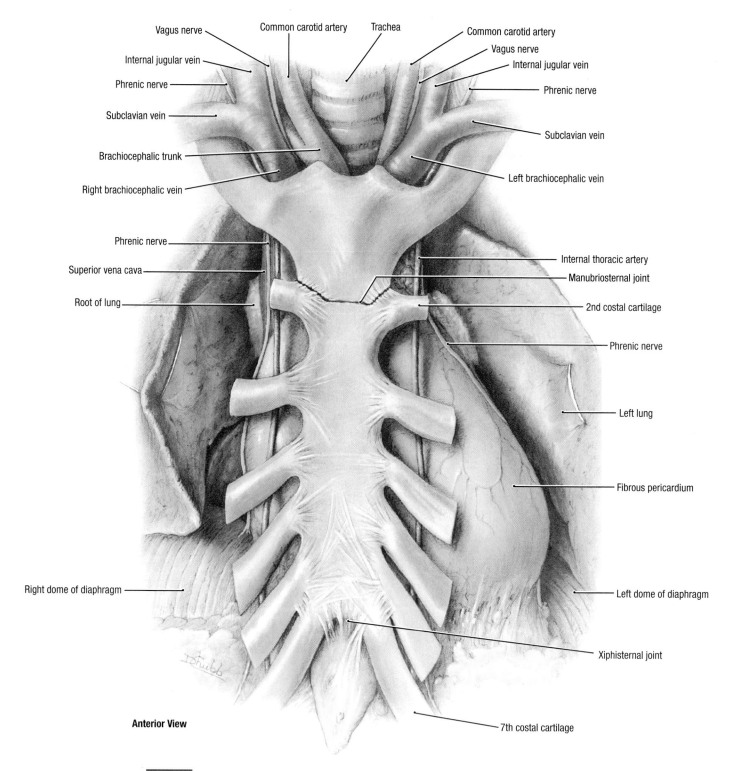

Vagus nerve — Common carotid artery — Trachea — Common carotid artery — Vagus nerve

Internal jugular vein — Internal jugular vein

Phrenic nerve — Phrenic nerve

Subclavian vein — Subclavian vein

Brachiocephalic trunk — Left brachiocephalic vein

Right brachiocephalic vein

Phrenic nerve — Internal thoracic artery

Superior vena cava — Manubriosternal joint

Root of lung — 2nd costal cartilage

Phrenic nerve

Left lung

Fibrous pericardium

Right dome of diaphragm — Left dome of diaphragm

Xiphisternal joint

Anterior View

7th costal cartilage

1.42   Pericardial sac in relation to sternum

- The pericardial sac lies posterior to the body of the sternum, extending from just superior to the sternal angle to the level of the xiphisternal joint; approximately two thirds lies to the left of the median plane.
- The heart lies between the sternum and the anterior mediastinum anteriorly and the vertebral column and the posterior mediastinum posteriorly; in cardiac compression, the sternum is depressed 4 to 5 cm, forcing blood out of the heart and into the great vessels.
- Internal thoracic arteries arise from the subclavian arteries and descend posterior to the costal cartilages, running lateral to the sternum and anterior to the pleura.

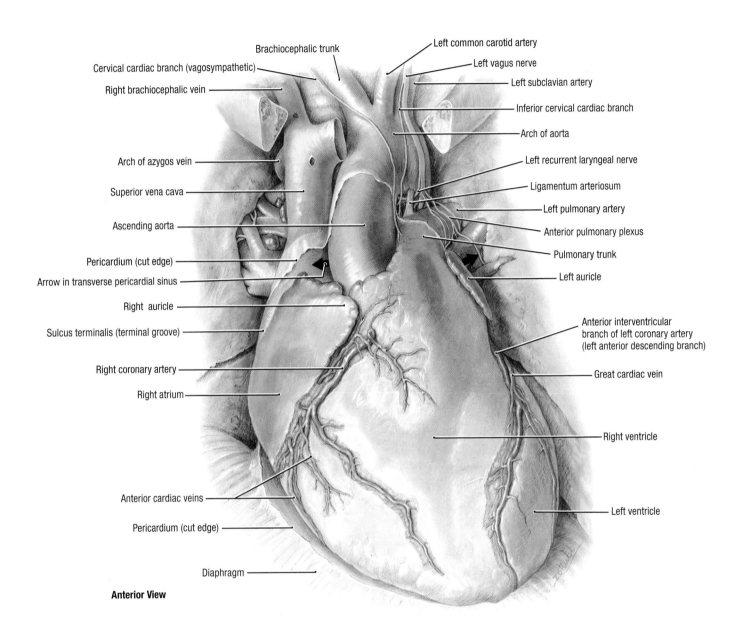

Brachiocephalic trunk
Cervical cardiac branch (vagosympathetic)
Right brachiocephalic vein
Arch of azygos vein
Superior vena cava
Ascending aorta
Pericardium (cut edge)
Arrow in transverse pericardial sinus
Right auricle
Sulcus terminalis (terminal groove)
Right coronary artery
Right atrium
Anterior cardiac veins
Pericardium (cut edge)
Diaphragm

Left common carotid artery
Left vagus nerve
Left subclavian artery
Inferior cervical cardiac branch
Arch of aorta
Left recurrent laryngeal nerve
Ligamentum arteriosum
Left pulmonary artery
Anterior pulmonary plexus
Pulmonary trunk
Left auricle
Anterior interventricular branch of left coronary artery (left anterior descending branch)
Great cardiac vein
Right ventricle
Left ventricle

Anterior View

1.43 Sternocostal (anterior) surface of heart and great vessels in situ

- The entire right auricle and much of the right atrium are visible anteriorly, but only a small portion of the left auricle is visible; the auricles, like a closing claw, grasp the pulmonary artery and ascending aorta from a posterior approach.
- The ligamentum arteriosum passes from the origin of the left pulmonary artery to the arch of the aorta.
- The right coronary artery courses in the anterior atrioventricular groove, and the anterior interventricular branch of left coronary artery (anterior descending branch) courses in the anterior interventricular groove.
- The left vagus nerve passes lateral to the aortic arch and then posterior to the root of the lung; the recurrent laryngeal nerve passes inferior to the aortic arch posterior to the ligamentum arteriosum.
- The great cardiac vein ascends with the anterior interventricular branch of the left coronary artery to drain into the coronary sinus posteriorly.

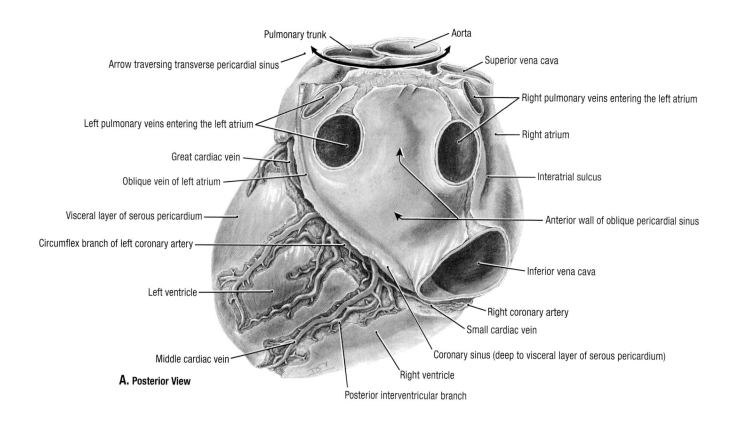

Pulmonary trunk

Aorta

Arrow traversing transverse pericardial sinus

Superior vena cava

Right pulmonary veins entering the left atrium

Left pulmonary veins entering the left atrium

Right atrium

Great cardiac vein

Oblique vein of left atrium

Interatrial sulcus

Visceral layer of serous pericardium

Anterior wall of oblique pericardial sinus

Circumflex branch of left coronary artery

Inferior vena cava

Left ventricle

Right coronary artery

Small cardiac vein

Coronary sinus (deep to visceral layer of serous pericardium)

Middle cardiac vein

A. Posterior View

Right ventricle

Posterior interventricular branch

1.44 Heart and pericardial sac

- This heart (A) was removed from the interior of the pericardial sac (B).
- The entire base, or posterior surface, and part of the diaphragmatic or inferior surface of the heart are in view.
- The superior vena cava and larger inferior vena cava join the superior and inferior aspects of the right atrium.
- The left atrium forms the greater part of the base (posterior surface) of the heart.
- The left coronary artery in this specimen is dominant, since it supplies the posterior interventricular branch.
- Most branches of cardiac veins cross branches of the coronary arteries superficially.

- The visceral layer of serous pericardium (epicardium) covers the surface of the heart and reflects onto the great vessels; from around the great vessels, the serous pericardium reflects to line the internal aspect of the fibrous pericardium as the parietal layer of serous pericardium.
- Note the cut edges of the reflections of serous pericardia around the arterial vessels (the pulmonary trunk and aorta) and venous vessels (the superior and inferior venae cavae and the pulmonary veins).

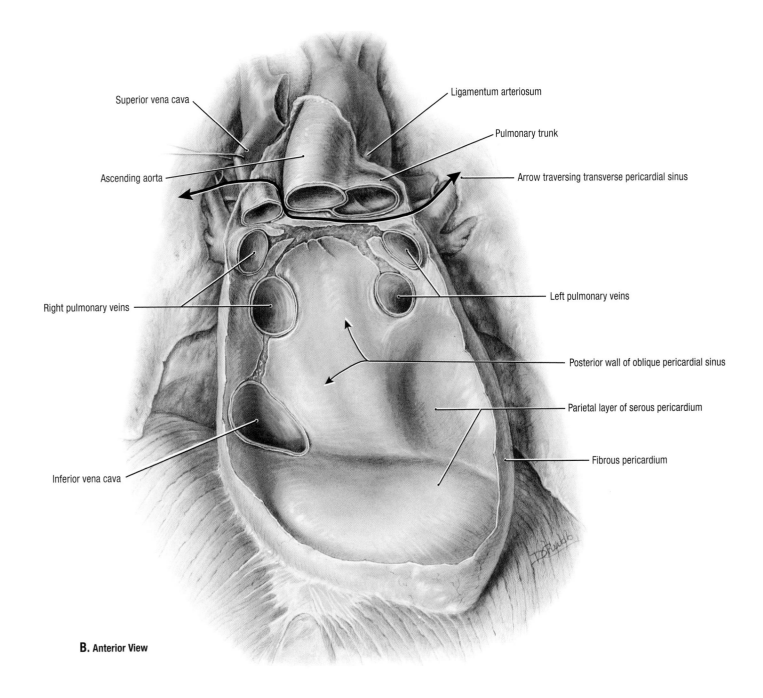

B. Anterior View

1.44   Heart and pericardial sac *(continued)*

- Interior of pericardial sac. Eight vessels were severed to excise the heart: two caval veins (superior and inferior venae cavae), four pulmonary veins, and two pulmonary arteries.
- The oblique sinus is bounded anteriorly by the visceral layer of serous pericardium covering the left atrium (**A**), posteriorly by the parietal layer of serous pericardium lining the fibrous pericardium, and superiorly and laterally by the reflection of serous pericardium around the four pulmonary veins and the superior and inferior venae cavae (**B**).

- The transverse sinus is bounded anteriorly by the serous pericardium covering the posterior aspect of the pulmonary trunk and aorta, and posteriorly by the visceral pericardium covering the atria (**A**).
- The apex of the pericardial sac is near the junction of the ascending aorta and arch of the aorta.
- The superior vena cava is partly inside and partly outside the pericardium, and the ligamentum arteriosum is entirely outside.

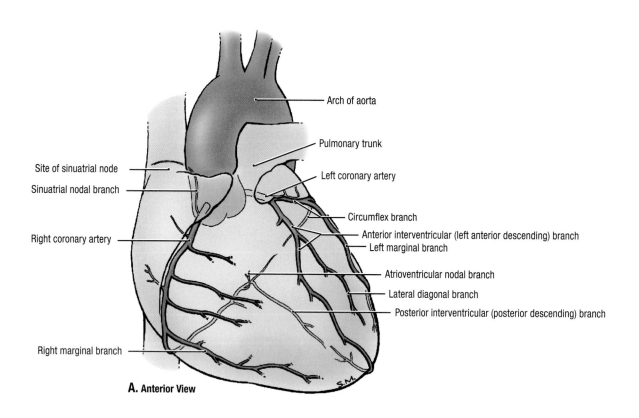

Arch of aorta

Pulmonary trunk

Site of sinuatrial node

Sinuatrial nodal branch

Left coronary artery

Circumflex branch

Anterior interventricular (left anterior descending) branch

Left marginal branch

Right coronary artery

Atrioventricular nodal branch

Lateral diagonal branch

Posterior interventricular (posterior descending) branch

Right marginal branch

A. Anterior View

1.45   Coronary arteries

- The right coronary artery travels in the coronary groove (sulcus) to reach the posterior surface of the heart, where it anastomoses with the circumflex branch of the left coronary artery. Early in its course, it gives off the sinuatrial (SA) nodal artery that supplies the right atrium and SA node; major branches are a marginal branch supplying much of the anterior wall of the right ventricle, an atrioventricular (AV) nodal artery given off near the posterior border of the interventricular septum, and a posterior interventricular artery in the interventricular groove that anastomoses with the anterior interventricular artery, a branch of the left coronary artery.
- The left coronary artery divides into a circumflex branch that passes posteriorly to anastomose with the right coronary on the posterior aspect of the heart and an anterior descending branch in the interventricular groove; the origin of the SA nodal artery is variable and may be a branch of the left coronary artery.
- The interventricular septum receives its blood supply from septal branches of the two descending branches: typically the anterior two thirds from the left coronary, and the posterior one third from the right.

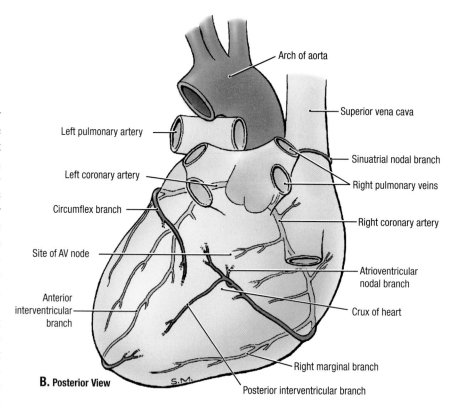

Arch of aorta

Superior vena cava

Left pulmonary artery

Sinuatrial nodal branch

Left coronary artery

Right pulmonary veins

Circumflex branch

Right coronary artery

Site of AV node

Atrioventricular nodal branch

Anterior interventricular branch

Crux of heart

Right marginal branch

B. Posterior View

Posterior interventricular branch

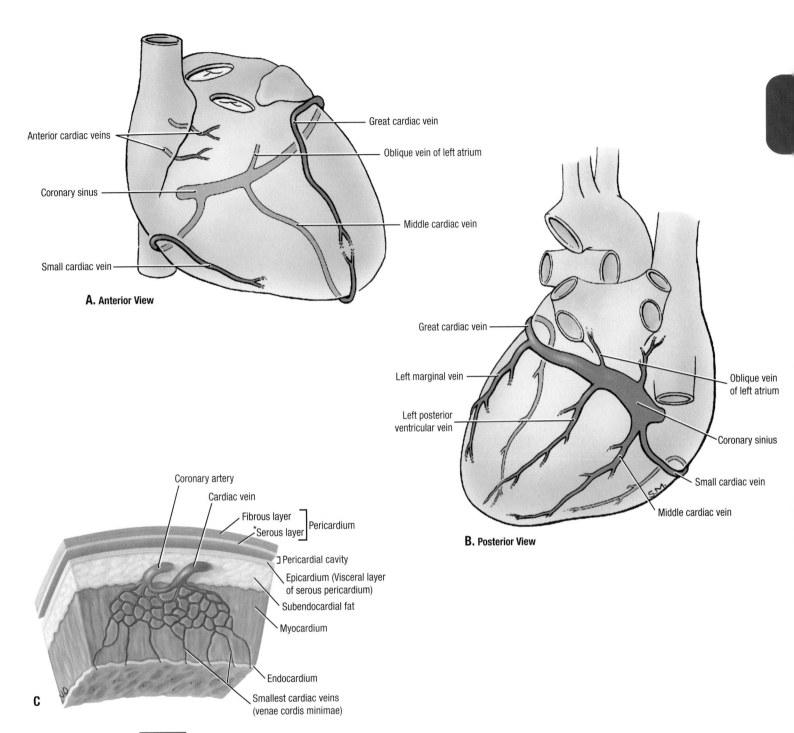

A. Anterior View

Anterior cardiac veins

Coronary sinus

Small cardiac vein

Great cardiac vein

Oblique vein of left atrium

Middle cardiac vein

B. Posterior View

Great cardiac vein

Left marginal vein

Left posterior ventricular vein

Oblique vein of left atrium

Coronary sinius

Small cardiac vein

Middle cardiac vein

Coronary artery

Cardiac vein

Fibrous layer

*Serous layer

Pericardium

Pericardial cavity

Epicardium (Visceral layer of serous pericardium)

Subendocardial fat

Myocardium

Endocardium

Smallest cardiac veins (venae cordis minimae)

C

1.46   **Cardiac veins**

The coronary sinus is the major venous drainage vessel of the heart; it is located posteriorly in the atrioventricular (coronary) groove and drains into the right atrium. The great, middle, and small cardiac veins, the oblique vein of the left atrium, and the posterior vein of the left ventricle are the principal vessels draining into the coronary sinus; the anterior cardiac veins drain directly into the right atrium. The smallest cardiac veins (venae cordis minimae) drain the myocardium directly into the atria and ventricles **(C)**. In C the asterisk (*) indicates the parietal layer of serous pericardium. The cardiac veins accompany the coronary arteries and their branches, e.g., the great cardiac vein and the anterior interventricular artery, middle cardiac vein, and posterior interventricular artery.

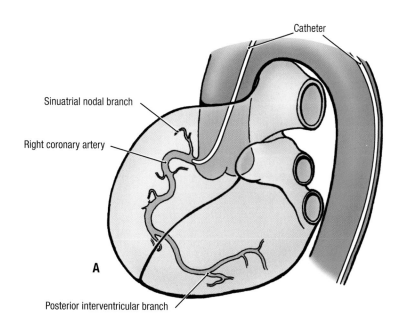

Catheter

Sinuatrial nodal branch

Right coronary artery

Posterior interventricular branch

A

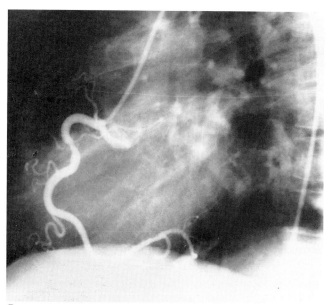

B. Left Anterior Oblique View

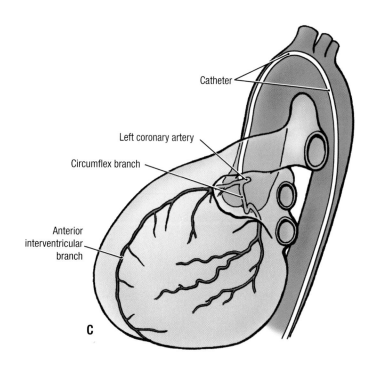

Catheter

Left coronary artery

Circumflex branch

Anterior interventricular branch

C

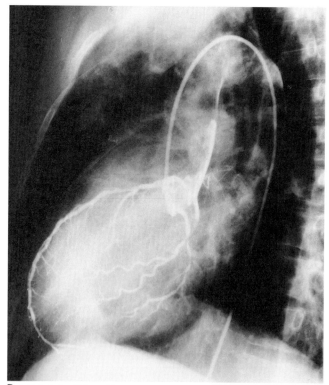

D. Left Anterior Oblique View

1.47 **Coronary arteriograms with orientation drawings**

Right (**A** and **B**) and left (**C** and **D**) coronary arteriograms.

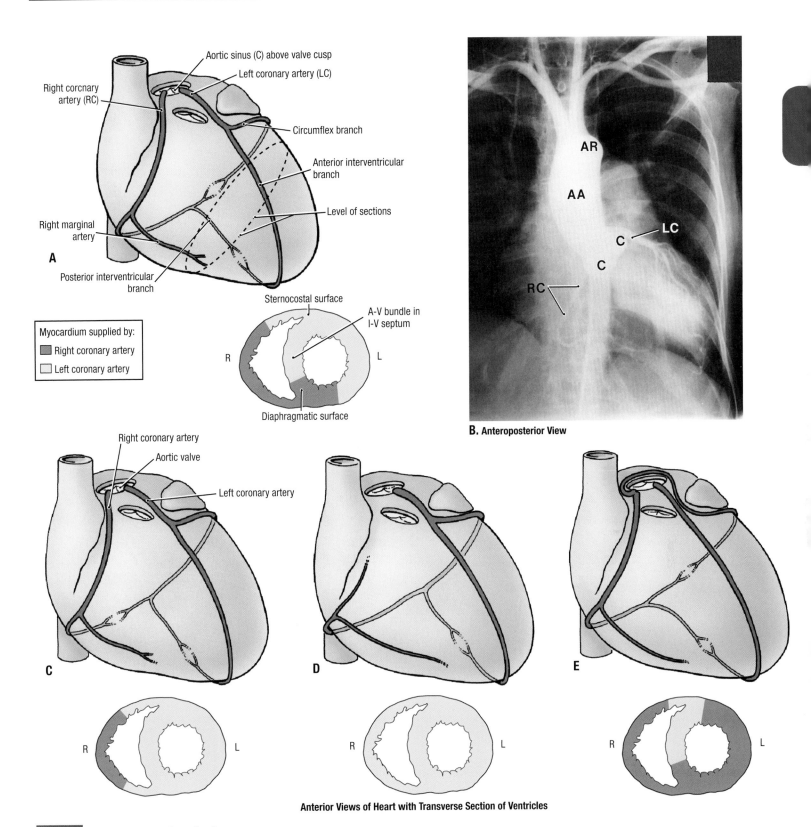

B. Anteroposterior View

Anterior Views of Heart with Transverse Section of Ventricles

1.48   Coronary circulation

A. In most cases, the right and left coronary arteries share equally in the blood supply to the heart. The dotted line indicates the plane of the cross-section demonstrating the parts of the myocardium supplied by the right and left coronary arteries. Atrioventricular bundle (*A-V bundle*), Interventricular septum (*I-V septum*) **B.** Aortic angiogram. Observe arch of aorta (*AR*), ascending aorta (*AA*), cusp of aortic valve (*C*), left coronary artery (*LC*), and right coronary artery (*RC*). **C.** Dominant left coronary artery. In approximately 15% of hearts, the left coronary artery is dominant in that the posterior interventricular branch comes off the circumflex branch. **D.** Single coronary artery. **E.** Circumflex branch emerging from the right coronary sinus.

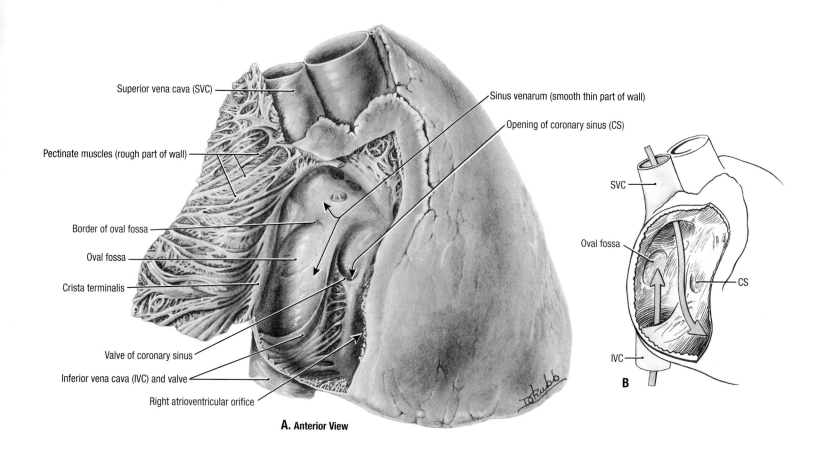

Superior vena cava (SVC)

Pectinate muscles (rough part of wall)

Border of oval fossa

Oval fossa

Crista terminalis

Valve of coronary sinus

Inferior vena cava (IVC) and valve

Right atrioventricular orifice

Sinus venarum (smooth thin part of wall)

Opening of coronary sinus (CS)

SVC

Oval fossa

CS

IVC

B

A. Anterior View

1.49   **Right atrium**

A. Interior of right atrium. **B.** Blood flow into atrium from the superior and inferior vena cavae.

- The smooth part of the atrial wall is formed by the absorption of the right horn of the sinus venosus, and the rough part is formed from the primitive atrium.
- Crista terminalis, the valve of the inferior vena cava, and the valve of the coronary sinus separate the smooth part from the rough part.
- The pectinate muscle passes anteriorly from the crista terminalis like teeth from the back of a comb; the crista underlies the sulcus terminalis (not shown), a groove visible externally on the posterolateral surface of the right atrium between the superior and inferior venae cavae.

- The superior and inferior venae cavae and the coronary sinus open onto the smooth part of the right atrium; the anterior cardiac veins and venae cordis minimae (not shown) also open into the atrium.
- The right atrioventricular, or tricuspid, orifice, is situated at the anterior aspect of the atrium.
- In B, the inflow from the superior vena cava is directed toward the tricuspid orifice, whereas blood from the inferior vena cava is directed toward the fossa ovalis.

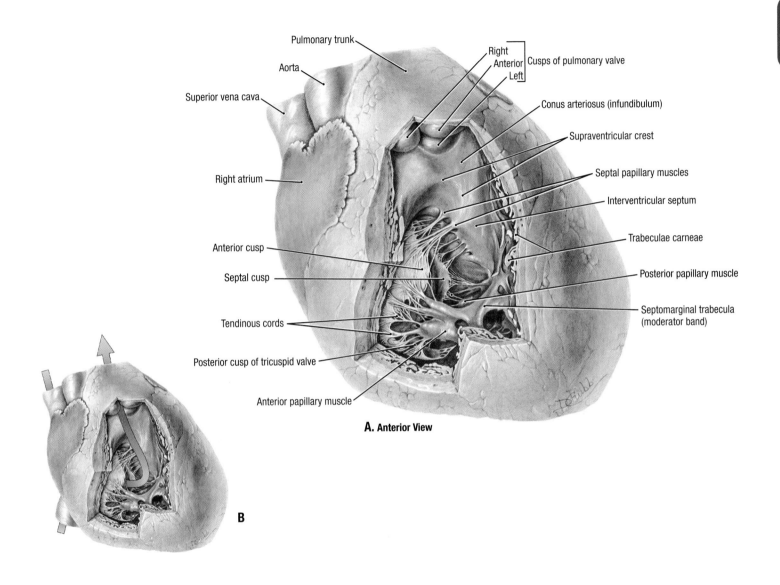

A. Anterior View

1.50   **Right ventricle**

A. Interior of right ventricle. **B.** Blood flow through right heart.

- The entrance to this chamber, the right atrioventricular or tricuspid orifice, is situated posteriorly; the exit, the orifice of the pulmonary trunk, is superior.
- The outflow portion of the chamber inferior to the pulmonary orifice (conus arteriosus or infundibulum) has a smooth, funnel-shaped wall; the remainder of the ventricle is rough with fleshy trabeculae.

- There are three types of trabeculae: mere ridges, bridges attached only at each end, and fingerlike projections called papillary muscles. The anterior papillary muscle rises from the anterior wall, the posterior (not labeled) from the posterior wall, and a series of small septal papillae from the septal wall.
- The septomarginal trabecula, here thick, extends from the septum to the base of the anterior papillary muscle.

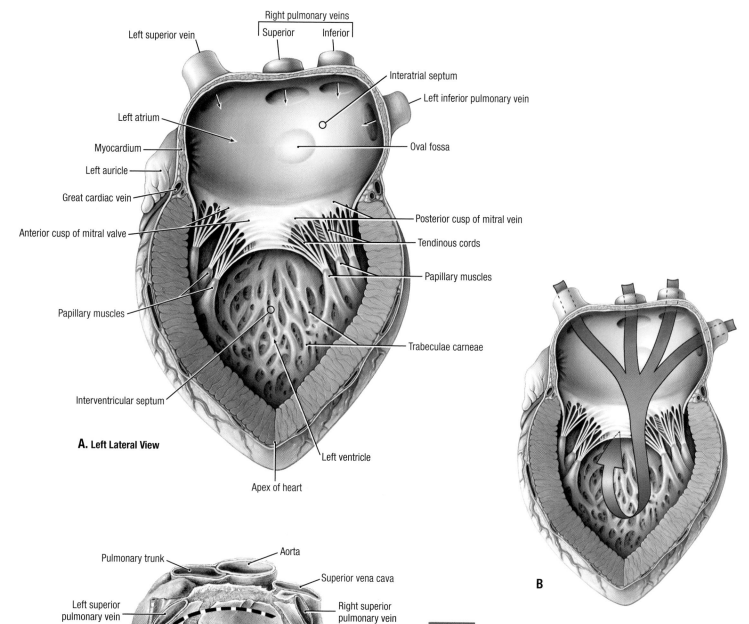

Right pulmonary veins
Superior   Inferior

Left superior vein

Interatrial septum

Left inferior pulmonary vein

Left atrium

Myocardium

Oval fossa

Left auricle

Great cardiac vein

Posterior cusp of mitral vein

Anterior cusp of mitral valve

Tendinous cords

Papillary muscles

Papillary muscles

Trabeculae carneae

Interventricular septum

Left ventricle

Apex of heart

A. Left Lateral View

B

Pulmonary trunk

Aorta

Superior vena cava

Left superior
pulmonary vein

Right superior
pulmonary vein

Left inferior
pulmonary vein

Right inferior
pulmonary vein

Line of dissection
in A and B

Inferior vena cava

1.51 **Left atrium and left ventricle.**

A. Interior of left heart. **B.** Blood flow into left heart.

- A diagonal cut was made from the base of the heart to the apex, passing between the superior and inferior pulmonary veins and through the posterior cusp of the mitral valve, followed by retraction (spreading) of the left heart wall on each side of the incision.
- The entrances (pulmonary veins) to the left atrium are posterior, and the exit (left atrioventricular or mitral orifice) is anterior.
- The left side of the valve of the foramen ovale is also seen on the left side of the interatrial septum, although the left side is not usually as distinct as the right side is within the right atrium,. The floor of the fossa is the remnant of the fetal septum primum; the crescent-shaped ridge (limbus) partially surrounding the fossa is the remnant of the septum secundum
- Except for that of the auricle, the atrial wall is smooth.

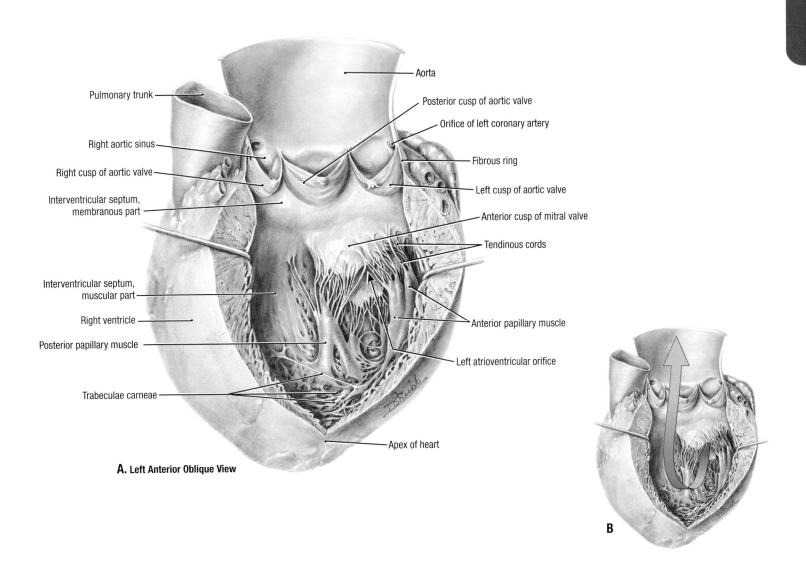

Pulmonary trunk

Right aortic sinus

Right cusp of aortic valve

Interventricular septum, membranous part

Interventricular septum, muscular part

Right ventricle

Posterior papillary muscle

Trabeculae carneae

Aorta

Posterior cusp of aortic valve

Orifice of left coronary artery

Fibrous ring

Left cusp of aortic valve

Anterior cusp of mitral valve

Tendinous cords

Anterior papillary muscle

Left atrioventricular orifice

Apex of heart

A. Left Anterior Oblique View

B

1.52   **Left ventricle**

A. Interior of left ventricle. **B.** Blood flow through the left ventricle.
- The chamber has a conical shape.
- The entrance (left atrioventricular, bicuspid, or mitral orifice) is situated posteriorly, and the exit (aortic orifice) is superior.
- The left ventricular wall is thin and muscular near the apex, thick and muscular above, and thin and fibrous (nonelastic) at the aortic orifice.

- Trabeculae carneae form ridges, bridges, and papillary muscles, as in the right ventricle.
- Two large papillary muscles, the anterior from the anterior wall and the posterior from the posterior wall, control the adjacent halves of two cusps of the mitral valve with chordae tendineae.
- The anterior cusp of the mitral valve intervenes between the inlet (mitral orifice) and the outlet (aortic orifice).

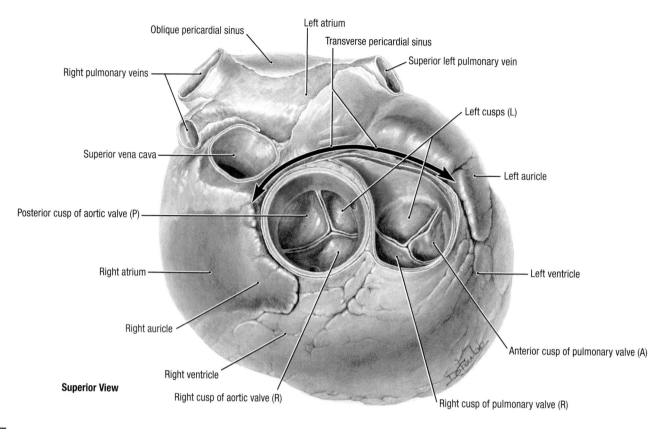

Oblique pericardial sinus

Left atrium

Transverse pericardial sinus

Right pulmonary veins

Superior left pulmonary vein

Left cusps (L)

Superior vena cava

Left auricle

Posterior cusp of aortic valve (P)

Left ventricle

Right atrium

Left ventricle

Right auricle

Anterior cusp of pulmonary valve (A)

Superior View

Right ventricle

Right cusp of aortic valve (R)

Right cusp of pulmonary valve (R)

1.53 **Excised heart**

- The ventricles are positioned anteriorly, the atria posteriorly.
- The roots of the aorta and pulmonary artery, which conduct blood from the ventricles, are placed anterior to the atria and their incoming blood vessels (the superior vena cava and pulmonary veins).
- The aorta and pulmonary artery are enclosed within a common tube of serous pericardium and partly embraced by the auricles of the atria.

- The transverse pericardial sinus curves posterior to the enclosed stems of the aorta and pulmonary trunk and anterior to the superior vena cava and upper limits of the atria.
- The three cusps of the aortic and pulmonary valves—and the names of the cusps—have a developmental origin, as explained in Fig. 1.54.

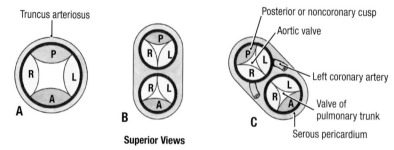

Truncus arteriosus

Posterior or noncoronary cusp

Aortic valve

Left coronary artery

Valve of pulmonary trunk

Serous pericardium

A **B** **C**

Superior Views

1.54 **Pulmonary and aortic valve names**

The names of these cusps have a developmental origin: the truncus arteriosus with four cusps **(A)** splits to form two valves, each with three cusps **(B)**. The heart undergoes partial rotation to the left on its axis, resulting in the arrangement of cusps shown in **C**.

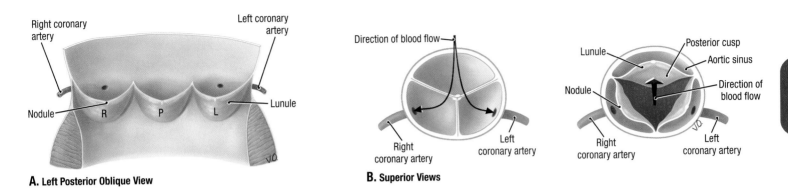

A. Left Posterior Oblique View

B. Superior Views

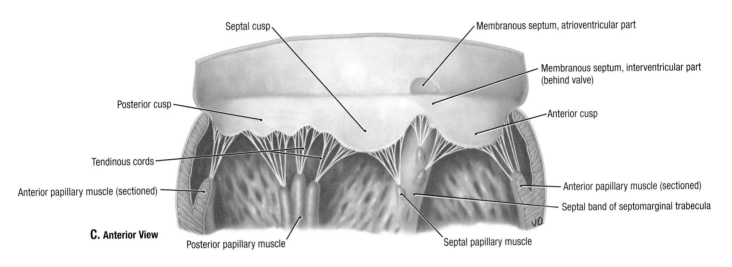

C. Anterior View

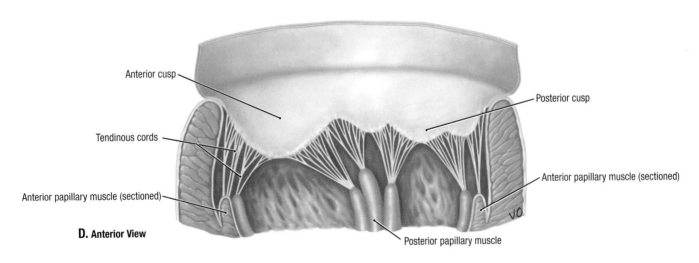

D. Anterior View

1.55   Valves of the heart

A and **B.** Semilunar valves. **C** and **D.** Atrioventricular valves

In **(A)**, as in Figure 1.52A, the anulus of the aortic valve has been incised between the right and left cusps and spread open. Each cusp of the semilunar valves bears a nodule in the midpoint of its free edge, flanked by thin connective tissue areas (lunules). When the ventricles relax to fill (diastole), backflow of blood from aortic recoil or pulmonary resistance fills the sinus (space between cusp and dilated part of the aortic or pulmonary wall), causing the nodules and lunules meet centrally, closing the valve **(B)**. The orifices of the coronary arteries in the right and left aortic sinuses may be covered by the cusp when outflowing blood forces the valve open during ventricular contraction (systole); filling of the arteries occurs during diastole (when ventricular walls are relaxed) as backflow "inflates" the cusps to close the valve. Tendinous cords pass from the tips of the papillary muscles to the free margins and ventricular surfaces of the cusps of the tricuspid **(C)** and mitral **(D)** valves. Each papillary muscle or muscle group controls the adjacent sides of two cusps, resisting valve prolapse during systole.

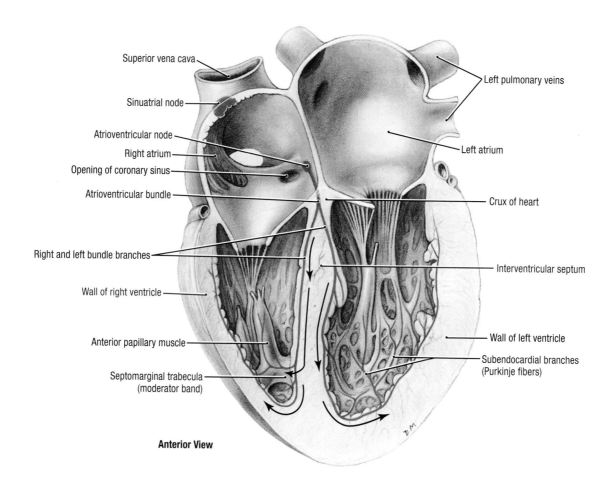

Superior vena cava

Sinuatrial node

Atrioventricular node

Right atrium

Opening of coronary sinus

Atrioventricular bundle

Right and left bundle branches

Wall of right ventricle

Anterior papillary muscle

Septomarginal trabecula (moderator band)

Left pulmonary veins

Left atrium

Crux of heart

Interventricular septum

Wall of left ventricle

Subendocardial branches (Purkinje fibers)

Anterior View

1.56 **Conduction system of heart, coronal section**

- The sinuatrial (SA) node in the wall of the right atrium near the superior end of the sulcus terminalis extends over the anterior aspect of the opening of the superior vena cava. The SA node is the "pacemaker" of the heart because it initiates muscle contraction and determines the heart rate. It is supplied by the sinuatrial nodal artery, usually a branch of the right coronary artery (see Fig. 1.45A–B), but it may be a branch of the left.

- Contraction spreads through the atrial wall (myogenic induction) until it reaches the atrioventricular (AV) node in the interatrial septum just superior to the opening of the coronary sinus. The AV node is supplied by the atrioventricular nodal artery, usually arising from the right coronary artery posteriorly at the inferior margin of the interatrial septum.

- The AV bundle, usually supplied by the right coronary artery, passes from the AV node in the membranous part of the interventricular septum, dividing into right and left bundle branches on either side of the muscular part of the interventricular septum.

- The right bundle branch travels inferiorly in the interventricular septum to the anterior wall of the ventricle, then through the septomarginal trabecula to the anterior papillary muscle; excitation spreads throughout the right ventricular wall through a network of subendocardial branches from the right bundle (Purkinje fibers).

- The left bundle branch lies beneath the endocardium on the left side of the interventricular septum and branches to enter the anterior and posterior papillary muscles and the wall of the left ventricle; further branching into a plexus of subendocardial branches (Purkinje fibers) allows the impulses to be conveyed throughout the left ventricular wall. The bundle branches are usually supplied by the left coronary, except the posterior limb of the left bundle branch, which is supplied by both coronary arteries.

- Damage to the cardiac conduction system (often by compromised blood supply as in coronary artery disease) leads to disturbances of muscle contraction. Damage to the AV node results in "heart block" because the atrial excitation wave does not reach the ventricles, which begin to contract independently at their own, slower rate. Damage to one of the branches results in "bundle branch block," in which excitation goes down the unaffected branch to cause systole of that ventricle; the impulse then spreads to the other ventricle, producing later, asynchronous contraction.

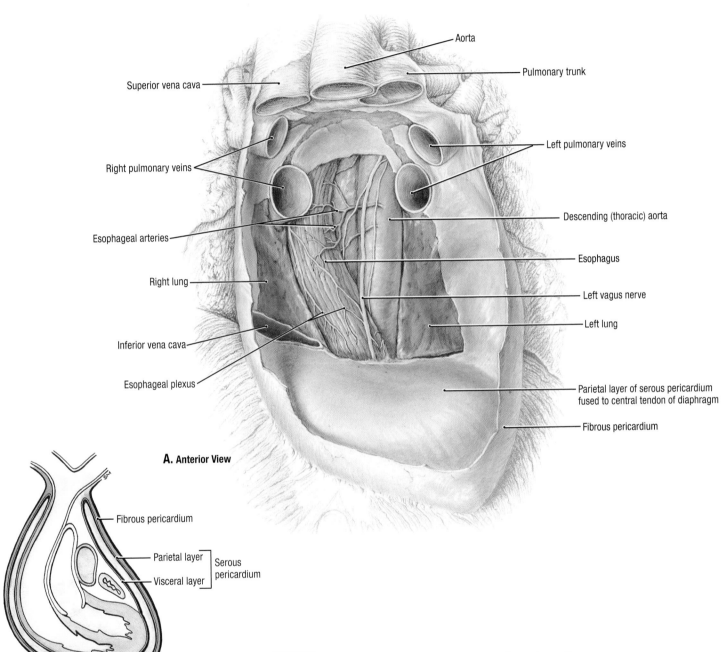

A. Anterior View

B. Diagrammatic Coronal Section

Fibrous pericardium

Parietal layer ⎫ Serous
Visceral layer ⎭ pericardium

Labels in figure A:
Aorta
Pulmonary trunk
Superior vena cava
Left pulmonary veins
Right pulmonary veins
Descending (thoracic) aorta
Esophageal arteries
Esophagus
Right lung
Left vagus nerve
Inferior vena cava
Left lung
Esophageal plexus
Parietal layer of serous pericardium fused to central tendon of diaphragm
Fibrous pericardium

1.57   Relationships of heart and pericardium

A. Posterior relationships. **B.** Structure of pericardial sac. In **A,** the fibrous and parietal layers of serous pericardium have been removed from posterior and lateral to the oblique sinus. The esophagus in this specimen is deflected to the right; it usually lies in contact with the aorta. **B** demonstrates that the fibrous pericardium is lined by a double-layered membranous sac, the serous pericardium. The outer parietal layer lines the fibrous pericardium and is continuous with the inner visceral layer, or epicardium, that covers the heart and great vessels. A thin film of fluid between the visceral and parietal layers allows the heart to move within the sac.

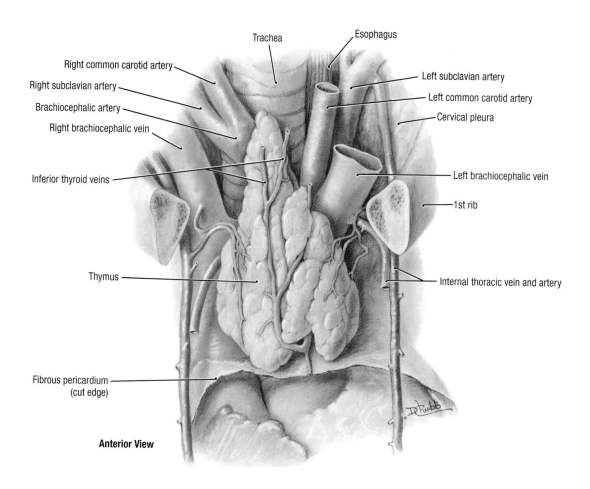

Trachea

Esophagus

Right common carotid artery

Right subclavian artery

Brachiocephalic artery

Right brachiocephalic vein

Inferior thyroid veins

Left subclavian artery

Left common carotid artery

Cervical pleura

Left brachiocephalic vein

1st rib

Thymus

Internal thoracic vein and artery

Fibrous pericardium
(cut edge)

Anterior View

1.58 **Superior mediastinum I: superficial dissection**

The sternum and ribs have been excised and the pleurae removed. It is unusual in an adult to see such a discrete thymus, which is impressive during puberty but subsequently regresses and is largely replaced by fat and fibrous tissue.

- The thymus lies in the superior mediastinum, overlapping the pericardial sac inferiorly and extending superiorly into the neck (in this specimen, farther than usual).
- A longitudinal fissure divides the thymus into two asymmetrical lobes, a larger right and smaller left; these two developmentally separate parts are easily separated from each other by blunt dissection.
- Thymic arteries arise from the internal thoracic arteries; thymic veins drain to the brachiocephalic and internal thoracic veins, and communicate superiorly with the inferior thyroid veins.

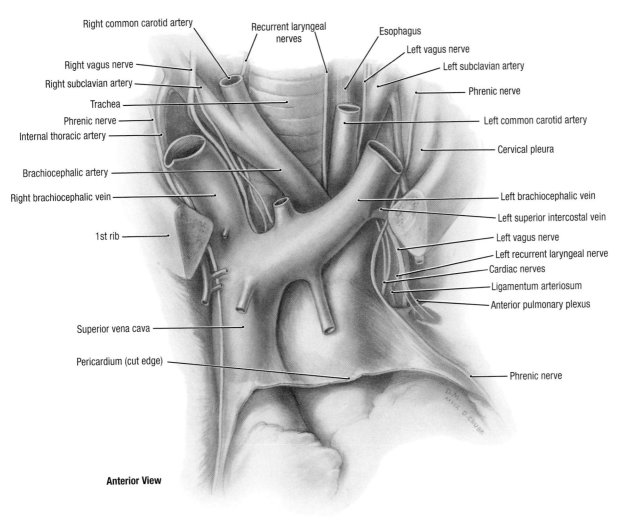

Right common carotid artery — Recurrent laryngeal nerves — Esophagus — Left vagus nerve — Left subclavian artery — Phrenic nerve — Left common carotid artery — Cervical pleura — Left brachiocephalic vein — Left superior intercostal vein — Left vagus nerve — Left recurrent laryngeal nerve — Cardiac nerves — Ligamentum arteriosum — Anterior pulmonary plexus — Phrenic nerve

Right vagus nerve — Right subclavian artery — Trachea — Phrenic nerve — Internal thoracic artery — Brachiocephalic artery — Right brachiocephalic vein — 1st rib — Superior vena cava — Pericardium (cut edge)

Anterior View

| 1.59 | **Superior mediastinum II: root of neck** |

- The thymus gland has been removed.
- The great veins lie anterior to the great arteries.
- The arch of the aorta lies mostly in a sagittal plane.
- The ligamentum arteriosum is external to the pericardial sac.

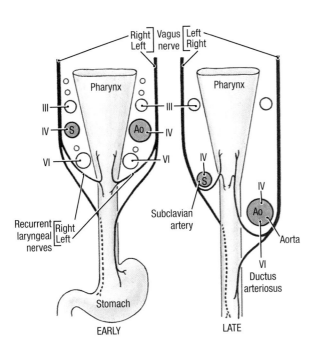

Right Left — Vagus nerve — Left Right — Pharynx — III — IV — S — VI — Recurrent laryngeal nerves — Right Left — Stomach — **EARLY**

Pharynx — III — Ao — IV — VI — Subclavian artery — IV — S — Ao — IV — Aorta — VI — Ductus arteriosus — **LATE**

| 1.60 | **Great vessels and nerves** |

Scheme to explain the asymmetrical courses of the right and left recurrent laryngeal nerves. III, IV, and VI are embryonic aortic arches. Arch VI disappears on the right, leaving the right recurrent laryngeal nerve to pass under arch IV, which becomes the right subclavian artery. Arch VI becomes part of the ductus arteriosus on the left side, and arch IV "descends" to become the arch of the aorta; thus the left recurrent laryngeal nerve is pulled into the thorax.

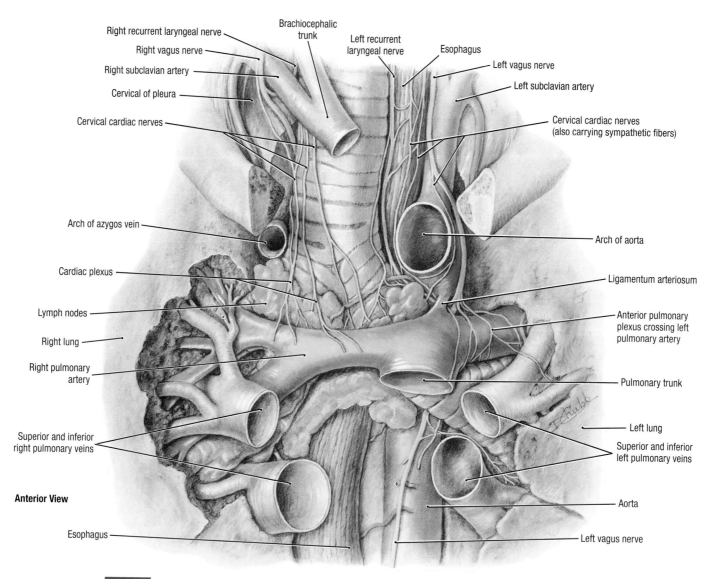

1.61 **Superior mediastinum III: cardiac plexus and pulmonary arteries**

The cardiac plexus is primarily related to the great vessels (ascending aorta and superior vena cava) and heart, which have been removed, leaving the cardiac plexus on the anterior aspect of the tracheal bifurcation—a secondary relationship.

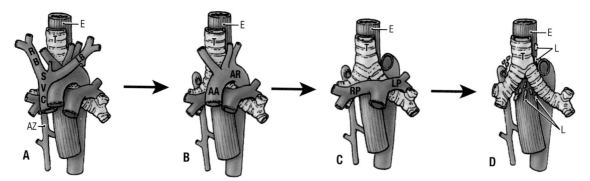

1.62 **Relations of great vessels and trachea**

Observe, from superficial to deep: **(A)** the right *(RB)* and left *(LB)* brachiocephalic veins form the superior vena cava *(SVC)*, and receive the arch of the azygos vein *(AZ)* posteriorly; **(B)** the ascending aorta *(AA)* and arch of the aorta *(AR)* arch over the right pulmonary artery and left main bronchus; **(C)** the pulmonary arteries *(RP* and *LP)*; and **(D)** the tracheobronchial lymph nodes *(L)* at the tracheal bifurcation *(T)*

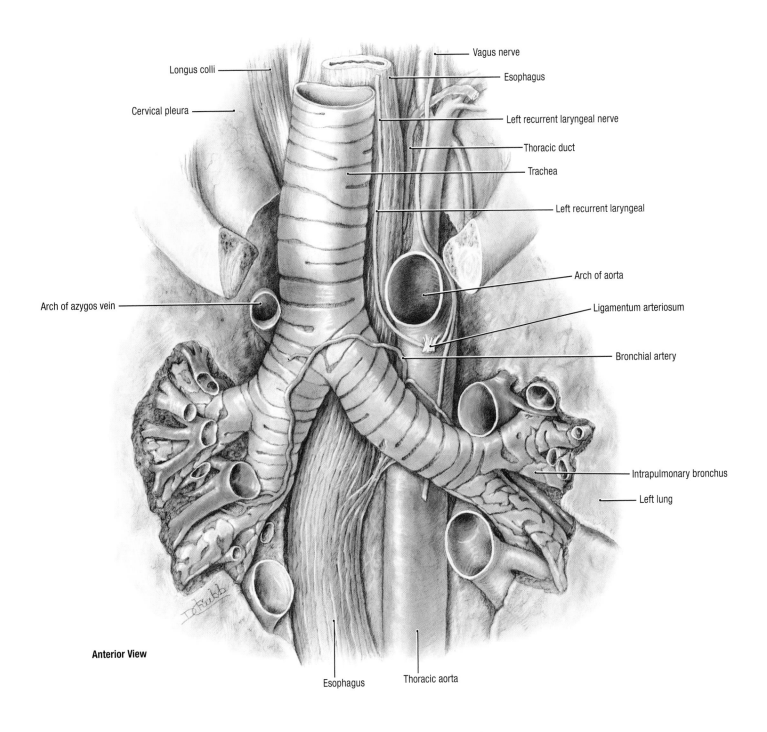

Longus colli

Cervical pleura

Arch of azygos vein

Vagus nerve

Esophagus

Left recurrent laryngeal nerve

Thoracic duct

Trachea

Left recurrent laryngeal

Arch of aorta

Ligamentum arteriosum

Bronchial artery

Intrapulmonary bronchus

Left lung

Anterior View

Esophagus   Thoracic aorta

1.63   **Superior mediastinum IV: tracheal bifurcation and bronchi**

- Note the four parallel structures: the trachea, esophagus, left recurrent laryngeal nerve, and thoracic duct. The esophagus bulges to the left of the trachea, the recurrent nerve lies in the angle between the trachea and esophagus, and the duct is at the left side of the esophagus.
- The arch of the aorta passes posterior to the left of these four structures as it arches over the left main bronchus; the arch of the azygos vein passes anterior to their left as it arches over the right main bronchus.

- The right main bronchus is (1) more vertical, (2) of greater caliber and (3) shorter than the left main bronchus.
- The U-shaped tracheal "rings" commonly bifurcate; a wedge-like cartilage (carina) occurs at the bifurcation.
- The bronchial arteries (more commonly on the posterior aspect of the bronchi) primarily supply the structures of the root of the lung rather than the lung tissue (parenchyma).

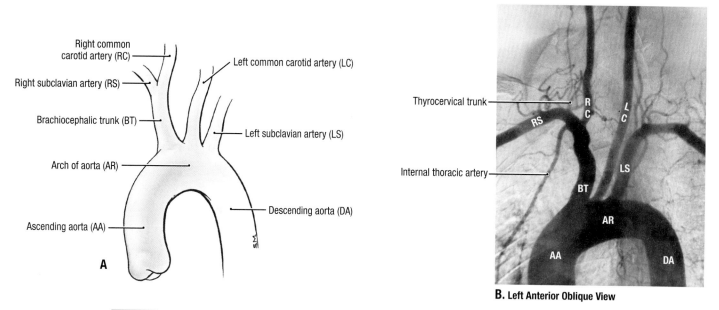

B. Left Anterior Oblique View

1.64   **Branches of aortic arch**

A. Aortic arch. **B.** Aortic angiogram. Observe the ascending aorta *(AA)*, the arch of the aorta *(AR)*, the descending aorta *(DA)*, the brachiocephalic *(BT)* trunk (artery) branching into the right subclavian *(RS)* and right common carotid *(RC)* arteries, and the left subclavian *(LS)* and left common carotid *(LC)* arteries arising directly from the aorta.

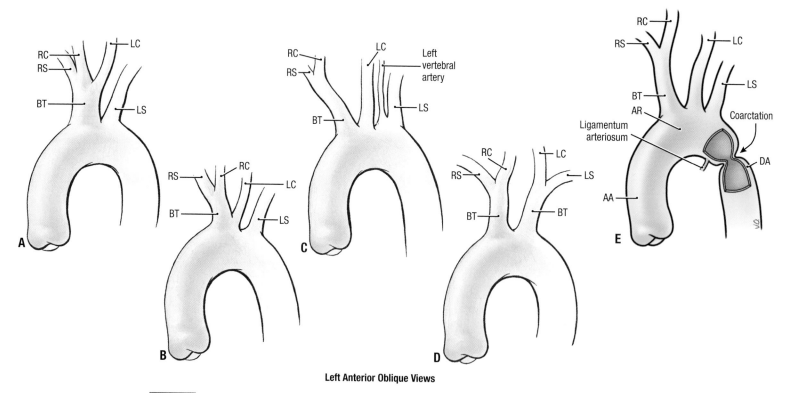

Left Anterior Oblique Views

1.65   **Variations in origins of branches of aortic arch**

The most common pattern (65%) is shown in Figure 1.64. Less common variations include **(A and B)** left common carotid artery originating from the brachiocephalic trunk (27%); **(C)** each of the four arteries originating independently from the arch of the aorta (2.5%); **(D)** right and left brachiocephalic trunks originating from the arch of the aorta (1.2%).

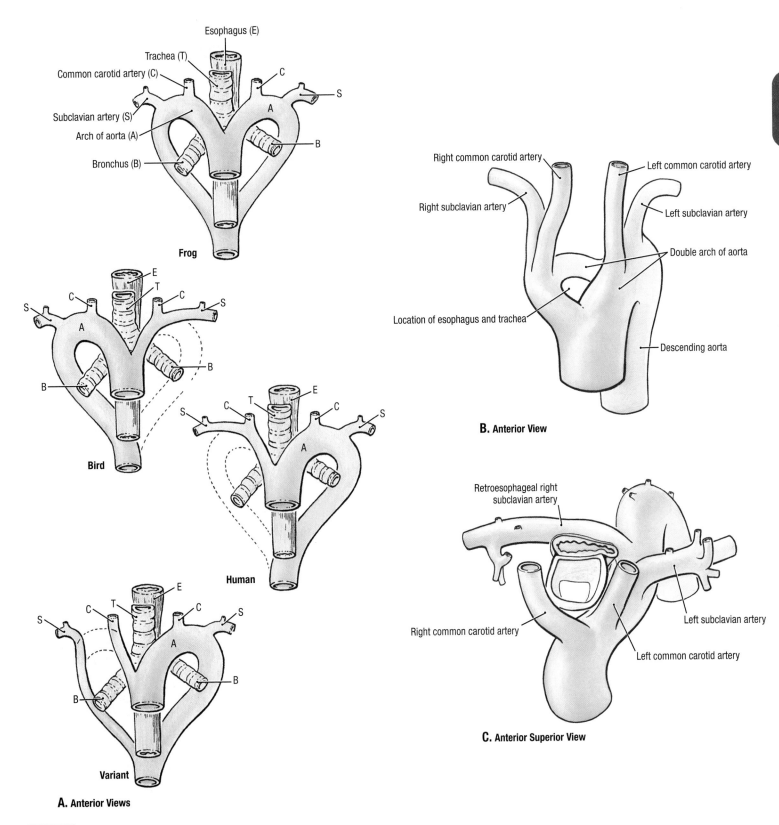

Esophagus (E)

Trachea (T)

Common carotid artery (C)

Subclavian artery (S)

Arch of aorta (A)

Bronchus (B)

Frog

Bird

Human

Variant

A. Anterior Views

Right common carotid artery

Right subclavian artery

Location of esophagus and trachea

Left common carotid artery

Left subclavian artery

Double arch of aorta

Descending aorta

B. Anterior View

Retroesophageal right subclavian artery

Right common carotid artery

Left subclavian artery

Left common carotid artery

C. Anterior Superior View

1.66   Arch of the aorta

A. Comparative anatomy. The double aortic arch of the frog; the right aortic arch of the bird; the left aortic arch of the mammal, including man, and a variant. **B.** Double aortic arch. The right and left aortic arches persist completely, as in the frog. In this rare condition, the esophagus and trachea pass through the so-formed "aortic ring." **C.** Retroesophageal right subclavian artery. The artery arises as the last branch of the arch of the aorta, passing posterior to the esophagus and trachea. The right recurrent laryngeal nerve takes a direct course to the larynx.

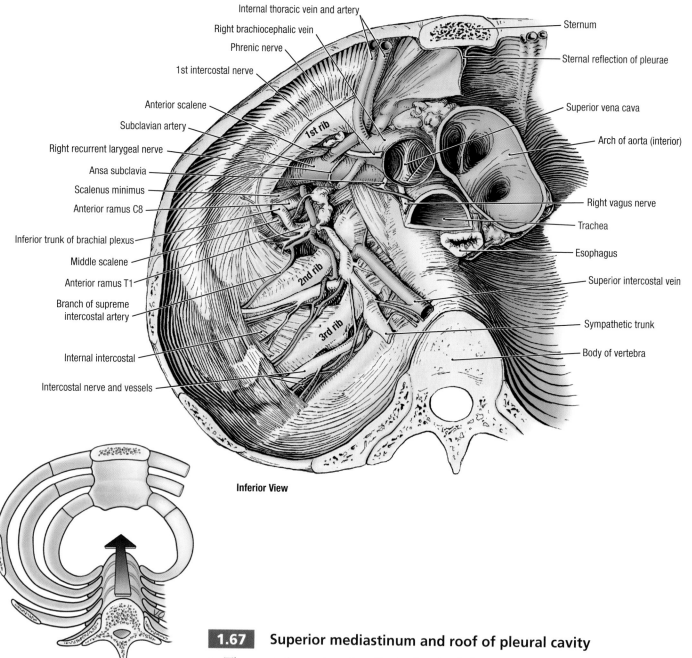

Internal thoracic vein and artery

Right brachiocephalic vein

Phrenic nerve

1st intercostal nerve

Anterior scalene

Subclavian artery

Right recurrent larygeal nerve

Ansa subclavia

Scalenus minimus

Anterior ramus C8

Inferior trunk of brachial plexus

Middle scalene

Anterior ramus T1

Branch of supreme intercostal artery

Internal intercostal

Intercostal nerve and vessels

1st rib

2nd rib

3rd rib

Sternum

Sternal reflection of pleurae

Superior vena cava

Arch of aorta (interior)

Right vagus nerve

Trachea

Esophagus

Superior intercostal vein

Sympathetic trunk

Body of vertebra

Inferior View

1.67   Superior mediastinum and roof of pleural cavity

- The cervical, costal, and mediastinal parietal pleura and endothoracic fascia have been removed to demonstrate structures traversing the superior thoracic aperture.
- The first part of the subclavian artery disappears as it crosses the first rib anterior to the anterior scalene muscle.
- The ansa subclavian from the sympathetic trunk and right recurrent laryngeal nerve from the vagus are seen looping inferior to the subclavian artery.
- The anterior rami of C8 and T1 merge to form the inferior trunk of the brachial plexus, which crosses the first rib posterior to the anterior scalene muscle.

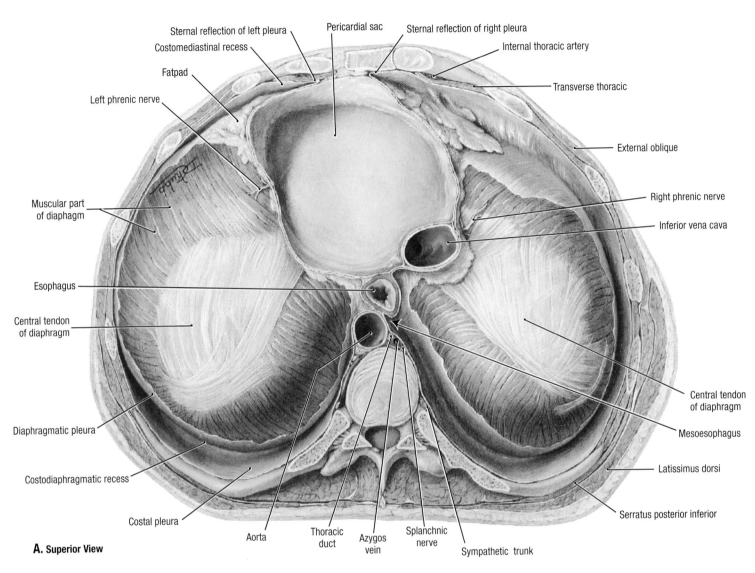

Sternal reflection of left pleura
Costomediastinal recess
Fatpad
Left phrenic nerve
Pericardial sac
Sternal reflection of right pleura
Internal thoracic artery
Transverse thoracic
External oblique
Muscular part of diaphagm
Right phrenic nerve
Inferior vena cava
Esophagus
Central tendon of diaphragm
Central tendon of diaphragm
Mesoesophagus
Diaphragmatic pleura
Latissimus dorsi
Costodiaphragmatic recess
Serratus posterior inferior
Costal pleura
Aorta
Thoracic duct
Azygos vein
Splanchnic nerve
Sympathetic trunk

A. Superior View

1.68 Diaphragm and pericardial sac

A. The diaphragmatic pleura is mostly removed. The pericardial sac is situated on the anterior half of the diaphragm; one third is to the right of the median plane, and two thirds to the left. The most caudal point is anterior and to the left, corresponding to the apex of the heart. Note also that anterior to the pericardium, the sternal reflection of the left pleural sac approaches but fails to meet that of the right sac in the median plane; and on reaching the vertebral column, the costal pleura becomes the mediastinal pleura. **B.** Between the inferior part of the esophagus and the aorta, the right and left layers of mediastinal pleura form a dorsal mesoesophagus.

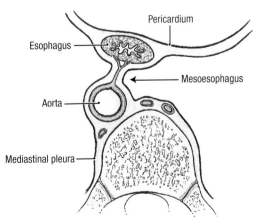

Pericardium
Esophagus
Mesoesophagus
Aorta
Mediastinal pleura

B. Superior View

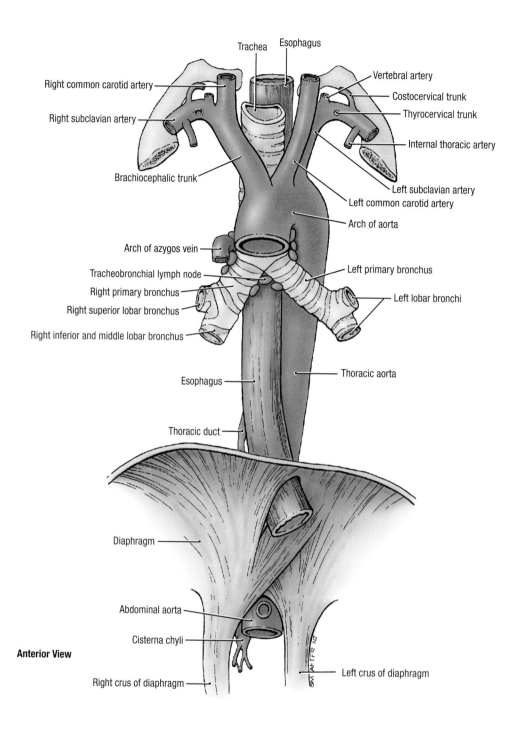

Trachea   Esophagus

Right common carotid artery

Right subclavian artery

Brachiocephalic trunk

Vertebral artery

Costocervical trunk

Thyrocervical trunk

Internal thoracic artery

Left subclavian artery

Left common carotid artery

Arch of aorta

Arch of azygos vein

Tracheobronchial lymph node

Right primary bronchus

Right superior lobar bronchus

Right inferior and middle lobar bronchus

Left primary bronchus

Left lobar bronchi

Esophagus

Thoracic aorta

Thoracic duct

Diaphragm

Abdominal aorta

Cisterna chyli

Anterior View

Right crus of diaphragm

Left crus of diaphragm

1.69   Esophagus, trachea, and aorta

- The anterior relations of the thoracic part of the esophagus from superior to inferior are the trachea (from origin at cricoid cartilage to bifurcation), right and left bronchi, inferior tracheobronchial lymph nodes, pericardium (not shown) and, finally, the diaphragm.
- The arch of the aorta passes posterior to the left of these four structures as it arches over the left main bronchus; the arch of the azygos vein passes anterior to their left as it arches over the right main bronchus.

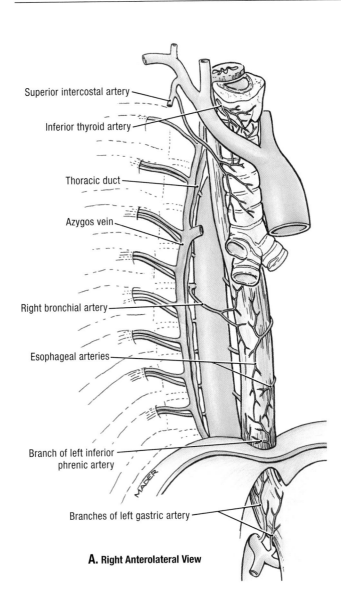

Superior intercostal artery

Inferior thyroid artery

Thoracic duct

Azygos vein

Right bronchial artery

Esophageal arteries

Branch of left inferior phrenic artery

Branches of left gastric artery

A. Right Anterolateral View

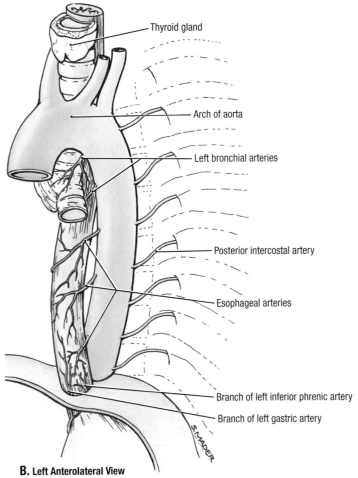

Thyroid gland

Arch of aorta

Left bronchial arteries

Posterior intercostal artery

Esophageal arteries

Branch of left inferior phrenic artery

Branch of left gastric artery

B. Left Anterolateral View

1.70   Arterial supply to trachea and esophagus

A and **B.** The continuous anastomotic chain of arteries on the esophagus are formed (a) by branches of the right and left inferior thyroid and right superior intercostal arteries superiorly, (b) by the unpaired median aortic (bronchial and esophageal) branches, and (c) by branches of the left gastric and left inferior phrenic arteries inferiorly. The right bronchial artery usually arises from the superior left bronchial or 3rd right posterior intercostal artery (here the 5th) or from the aorta directly. The unpaired median aortic branches also supply the trachea and bronchi. **C.** Branches of the thoracic aorta. The inferiormost superior phrenic branches, which supply the posterior aspect of the diaphragm, are not shown.

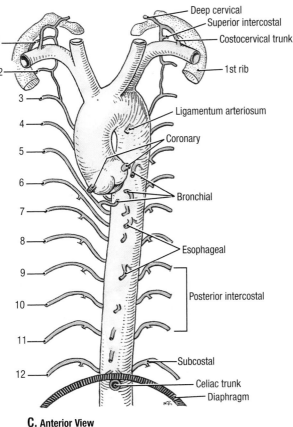

Deep cervical

Superior intercostal

Costocervical trunk

1st rib

Ligamentum arteriosum

Coronary

Bronchial

Esophageal

Posterior intercostal

Subcostal

Celiac trunk

Diaphragm

C. Anterior View

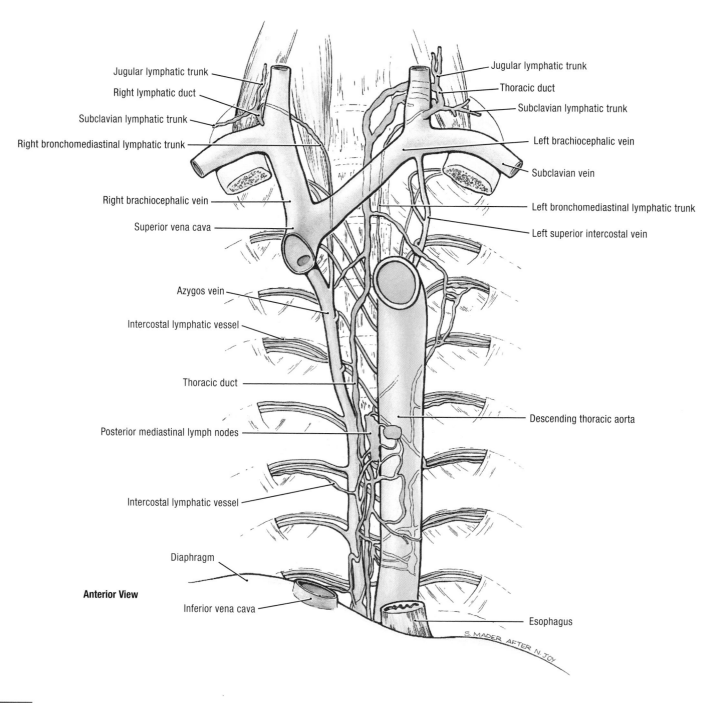

Jugular lymphatic trunk

Right lymphatic duct

Subclavian lymphatic trunk

Right bronchomediastinal lymphatic trunk

Right brachiocephalic vein

Superior vena cava

Azygos vein

Intercostal lymphatic vessel

Thoracic duct

Posterior mediastinal lymph nodes

Intercostal lymphatic vessel

Diaphragm

Anterior View

Inferior vena cava

Jugular lymphatic trunk

Thoracic duct

Subclavian lymphatic trunk

Left brachiocephalic vein

Subclavian vein

Left bronchomediastinal lymphatic trunk

Left superior intercostal vein

Descending thoracic aorta

Esophagus

S. MADER AFTER N. JOY

1.71   Thoracic duct

- The descending aorta is pulled slightly to the left, and the azygos vein slightly to the right.
- The thoracic duct (a) originates from the chyle cistern at the T12 vertebral level, (b) ascends on the vertebral column between the azygos vein and the descending aorta, (c) passes to the left at the junction of the posterior and superior mediastina, and continues its ascent to the neck, where (d) it arches laterally to enter the venous system near or at the angle of union of the left internal jugular and subclavian veins (left venous angle).
- The thoracic duct is commonly (a) plexiform (resembling a net-

work) in the posterior mediastinum and (b) splits and reunites in the neck.
- The duct receives branches from the intercostal spaces on both sides through several collecting trunks, as well as branches from posterior mediastinal structures.
- The termination of the thoracic duct typically receives the jugular, subclavian, and bronchomediastinal trunks.
- The right lymph duct is very short and formed by the union of the right jugular, subclavian, and bronchomediastinal trunks.

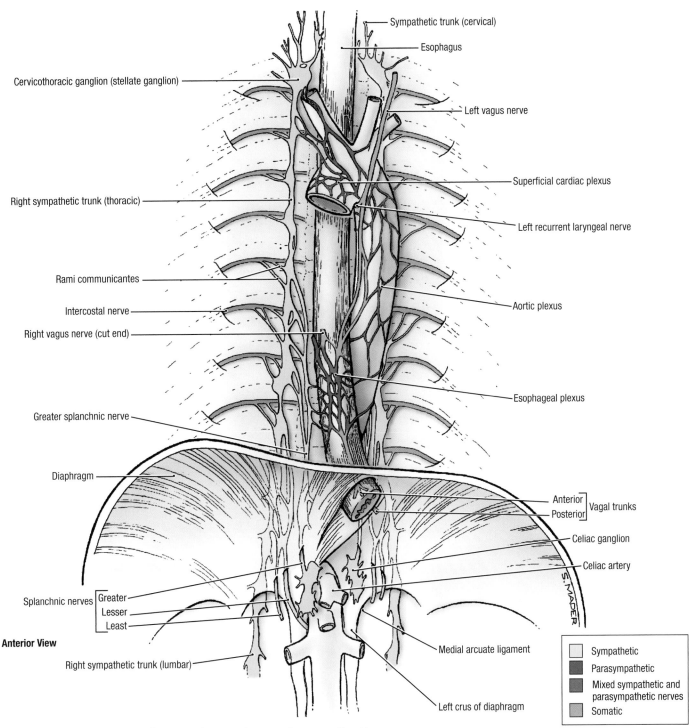

Sympathetic trunk (cervical)

Esophagus

Cervicothoracic ganglion (stellate ganglion)

Left vagus nerve

Superficial cardiac plexus

Right sympathetic trunk (thoracic)

Left recurrent laryngeal nerve

Rami communicantes

Aortic plexus

Intercostal nerve

Right vagus nerve (cut end)

Esophageal plexus

Greater splanchnic nerve

Diaphragm

Anterior ⎫ Vagal trunks
Posterior ⎭

Celiac ganglion

Celiac artery

Splanchnic nerves ⎧ Greater
　　　　　　　　⎨ Lesser
　　　　　　　　⎩ Least

Medial arcuate ligament

Anterior View

Right sympathetic trunk (lumbar)

Left crus of diaphragm

S. MADER

	Sympathetic
	Parasympathetic
	Mixed sympathetic and parasympathetic nerves
	Somatic

1.72　Autonomic nerves of posterior and superior mediastina

- The anterior rami of spinal nerves T1 to L2 or 3 are each connected to the sympathetic trunk by two rami communicantes: white rami communicans conduct presynaptic sympathetic fibers to the sympathetic trunks from the anterior rami, and gray rami communicans conduct postsynaptic sympathetic fibers from the sympathetic ganglia to the anterior rami for sympathetic innervation of the body wall.
- For the sympathetic innervation of thoracic viscera, the presynaptic fibers synapse with postsynaptic cell bodies in the upper thoracic sympathetic ganglia, and the postsynaptic fibers are distributed to thoracic viscera by thoracic splanchnic nerves

(not shown) via the thoracic autonomic plexuses (e.g., cardiac, esophageal).
- Presynaptic fibers traverse the lower thoracic sympathetic ganglia to pass to abdominal ganglia (e.g., celiac) by the greater, lesser, and least splanchnic nerves, where they will synapse with postsynaptic neurons for innervation of abdominal viscera.
- The left vagus and recurrent laryngeal nerves contribute parasympathetic fibers to the pulmonary, cardiac, and esophageal plexuses and the cut end of the right vagus nerve, which also contributes fibers to these plexuses.

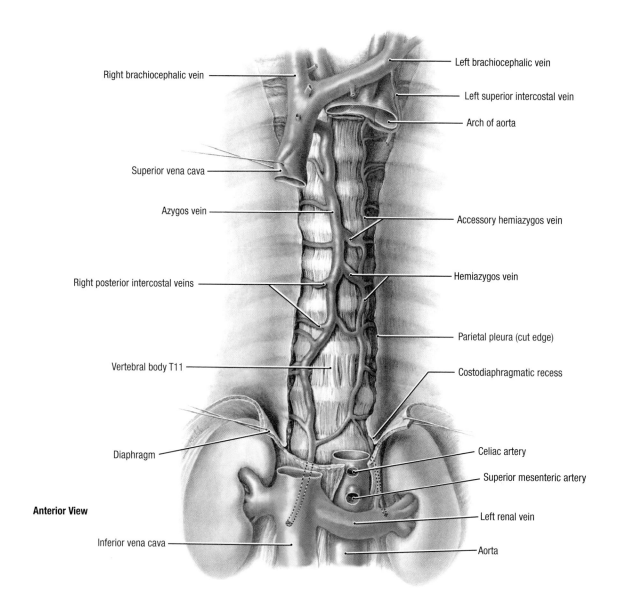

Right brachiocephalic vein

Left brachiocephalic vein

Left superior intercostal vein

Arch of aorta

Superior vena cava

Azygos vein

Accessory hemiazygos vein

Hemiazygos vein

Right posterior intercostal veins

Parietal pleura (cut edge)

Vertebral body T11

Costodiaphragmatic recess

Diaphragm

Celiac artery

Superior mesenteric artery

Anterior View

Left renal vein

Inferior vena cava

Aorta

1.73 Azygos system of veins in the thorax

The paired and approximately symmetrical longitudinal azygos and hemiazygos veins, immediately anterior to the vertebral column, receive the respective right and all but the upper left posterior intercostal veins. The azygos vein on the right side communicates inferiorly with the inferior vena cava, and the hemiazygos vein, on the left side, communicates inferiorly with the left renal vein. In this specimen, the hemiazygos, accessory hemiazygos, and left superior intercostal veins are continuous, but commonly they are discontinuous. The hemiazygos vein crosses the vertebral column at approximately T9, and the accessory hemiazygos vein crosses at T8 to enter the azygos vein. In this specimen, there are four cross-connecting channels between the azygos and hemiazygos systems. The azygos vein arches superior to the root of the right lung at T4 to drain into the superior vena cava.

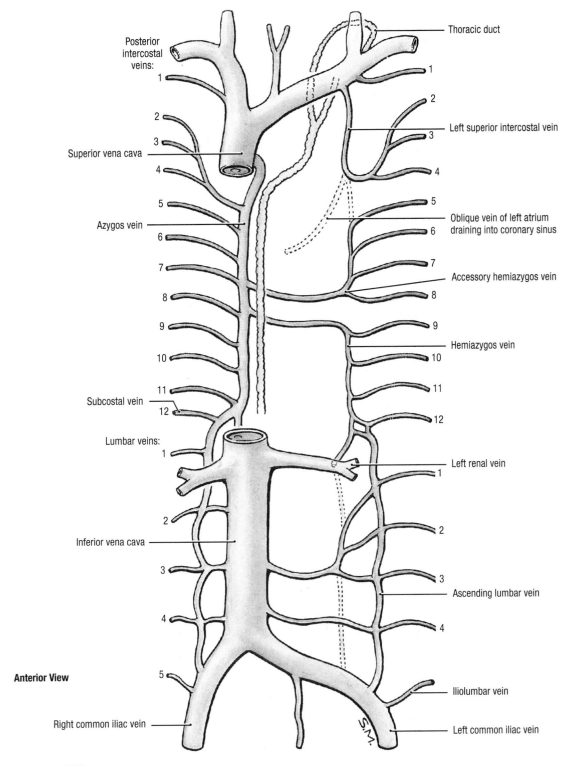

Posterior intercostal veins:

Thoracic duct

Superior vena cava

Left superior intercostal vein

Azygos vein

Oblique vein of left atrium draining into coronary sinus

Accessory hemiazygos vein

Hemiazygos vein

Subcostal vein

Lumbar veins:

Left renal vein

Inferior vena cava

Ascending lumbar vein

Anterior View

Iliolumbar vein

Right common iliac vein

Left common iliac vein

1.74   Azygos, hemiazygos, posterior intercostal, and lumbar veins

This system is of great importance in continued venous return to the heart if there is an obstruction of the venae cavae. The ascending lumbar veins connect the common iliac veins to the lumbar veins and join the subcostal veins to become the lateral roots of the azygos and hemiazygos veins; the medial roots of the azygos and hemiazygos veins are usually from the inferior vena cava and left renal vein, if present. Typically the upper four left posterior intercostal veins drain into the left brachiocephalic vein, directly and via the left superior intercostal veins.

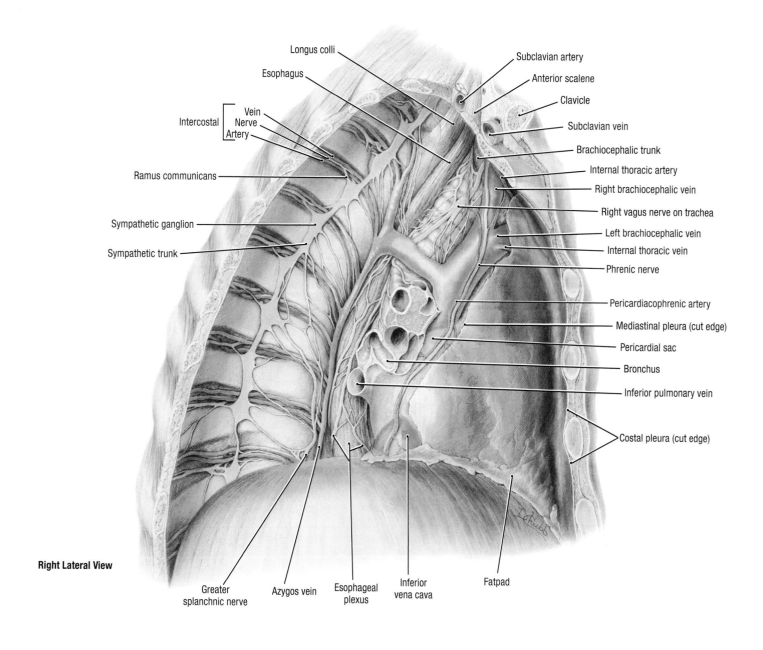

Longus colli

Esophagus

Intercostal — Vein / Nerve / Artery

Ramus communicans

Sympathetic ganglion

Sympathetic trunk

Subclavian artery

Anterior scalene

Clavicle

Subclavian vein

Brachiocephalic trunk

Internal thoracic artery

Right brachiocephalic vein

Right vagus nerve on trachea

Left brachiocephalic vein

Internal thoracic vein

Phrenic nerve

Pericardiacophrenic artery

Mediastinal pleura (cut edge)

Pericardial sac

Bronchus

Inferior pulmonary vein

Costal pleura (cut edge)

Right Lateral View

Greater splanchnic nerve

Azygos vein

Esophageal plexus

Inferior vena cava

Fatpad

1.75 Mediastinum, right side

- The costal and mediastinal pleurae have mostly been removed, exposing the underlying structures. Compare with the mediastinal surface of the right lung in Figure 1.29.
- The right side of the mediastinum is the "blue side," dominated by the arch of the azygos vein and the superior vena cava.
- Both the trachea and the esophagus are visible from the right side.
- The right vagus nerve descends on the medial surface of the trachea, passes medial to the arch of the azygos vein, posterior to the root of the lung, and then enters the esophageal plexus.
- The right phrenic nerve passes anterior to the root of the lung lateral to both venae cavae.

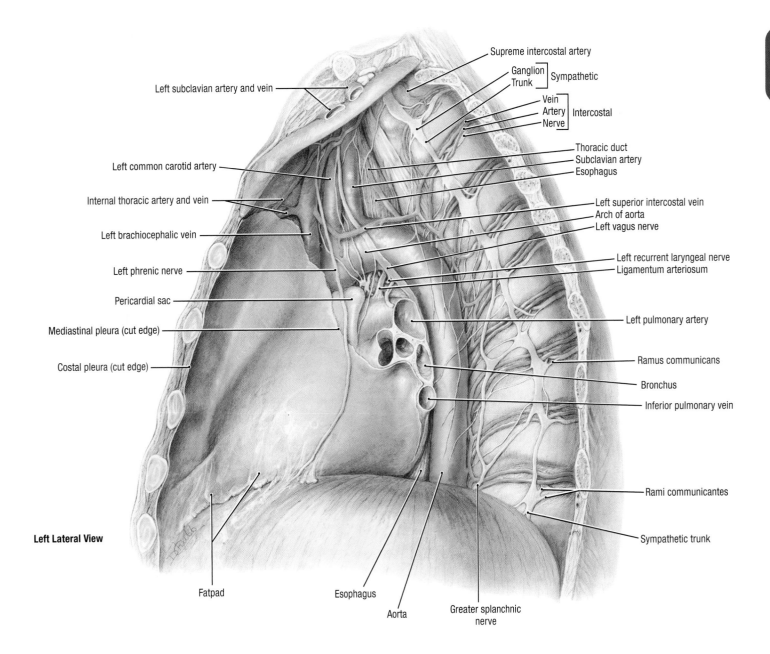

Left Lateral View

1.76   Mediastinum, left side

- Compare with the mediastinal surface of the left lung in Figure 1.30.
- The left side of the mediastinum is the "red side," dominated by the arch and descending portion of the aorta, the left common carotid and subclavian arteries; the latter obscure the trachea from view.
- The thoracic duct can be seen on the left side of the esophagus.
- The left vagus nerve passes posterior to the root of the lung, sending its recurrent laryngeal branch around the ligamentum arteriosum inferior, then medial to the aortic arch.
- The phrenic nerve passes anterior to the root of the lung and penetrates the diaphragm more anteriorly than on the right side.

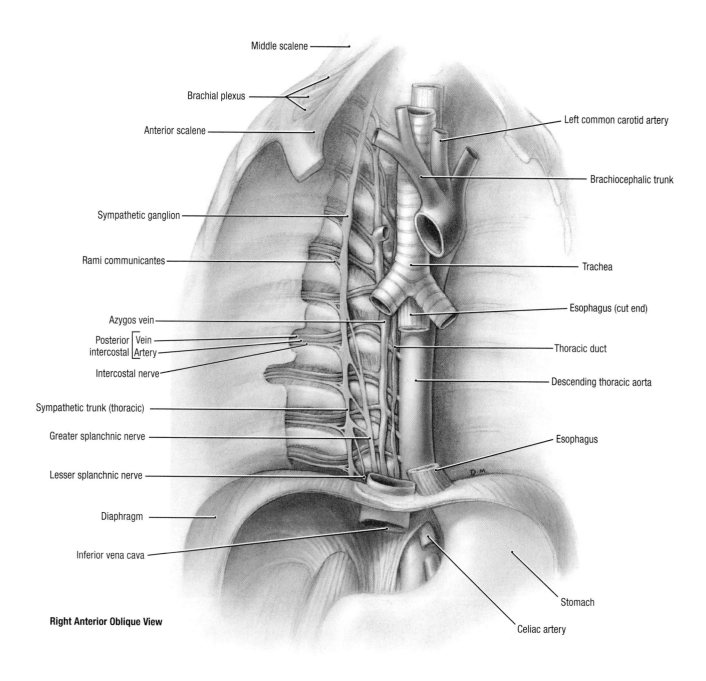

Middle scalene

Brachial plexus

Anterior scalene

Left common carotid artery

Brachiocephalic trunk

Sympathetic ganglion

Rami communicantes

Trachea

Azygos vein

Esophagus (cut end)

Posterior [Vein
intercostal [Artery

Thoracic duct

Intercostal nerve

Descending thoracic aorta

Sympathetic trunk (thoracic)

Greater splanchnic nerve

Esophagus

Lesser splanchnic nerve

Diaphragm

Inferior vena cava

Stomach

Right Anterior Oblique View

Celiac artery

1.77 **Posterior mediastinum**

- In this specimen, the parietal pleura is intact on the left side and partially removed on the right side. A portion of the esophagus, between the bifurcation of the trachea and the diaphragm, is also removed.
- The thoracic sympathetic trunk is connected to each intercostal nerve by rami communicantes.
- The greater splanchnic nerve is formed by fibers from the 5th to 10th thoracic ganglia, and the lesser splanchnic nerve receives fibers from the 10th and 11th thoracic ganglia. Both nerves contain presynaptic and visceral afferent fibers.
- The azygos vein ascends anterior to the intercostal vessels and to the right of the thoracic duct and aorta.

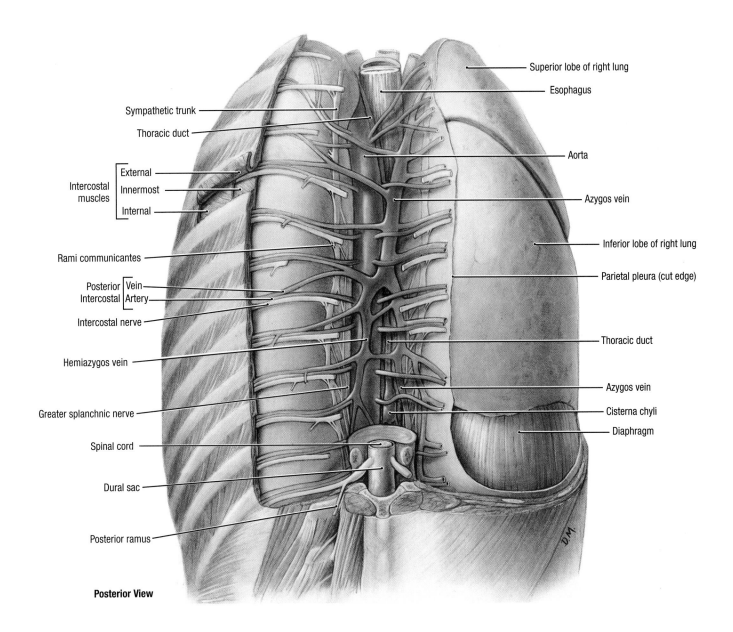

Superior lobe of right lung

Esophagus

Sympathetic trunk

Thoracic duct

Aorta

Intercostal muscles — External / Innermost / Internal

Azygos vein

Rami communicantes

Inferior lobe of right lung

Posterior Intercostal — Vein / Artery

Parietal pleura (cut edge)

Intercostal nerve

Hemiazygos vein

Thoracic duct

Greater splanchnic nerve

Azygos vein

Cisterna chyli

Spinal cord

Diaphragm

Dural sac

Posterior ramus

Posterior View

1.78   Mediastinum and lungs

- The thoracic vertebral column and thoracic cage are removed on the right. On the left, the ribs and intercostal musculature are removed posteriorly as far laterally as the angles of the ribs. The parietal pleura is intact on the left side but partially removed on the right to reveal the visceral pleura covering the right lung.
- The azygos vein is on the right side, and the hemiazygos vein is on the left, crossing the midline (usually at T9, but higher in this specimen) to join the azygos vein. The accessory hemiazygos vein is absent in this specimen; instead, three most superior posterior intercostal veins drain directly into the azygos vein.

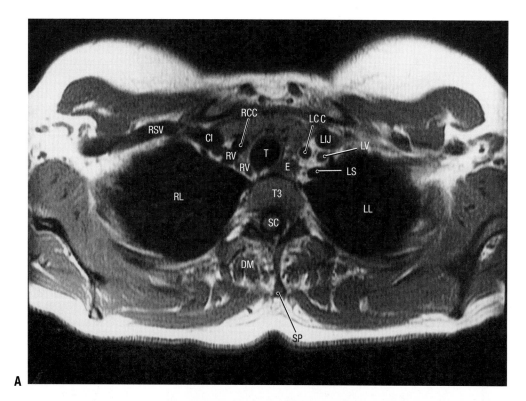

CI	Confluence of internal jugular vein
DM	Deep back muscles
E	Esophagus
LCC	Left common carotid artery
LIJ	Left internal jugular vein
LL	Left lung
LS	Left subclavian artery
LV	Left vertebral artery
M	Manubrium
PM	Pectoralis major
RBC	Right brachiocephalic vein
RCC	Right common carotid artery
RL	Right lung
RSV	Right subclavian vein
RV	Right vertebral artery
SC	Spinal cord
SP	Spinous process
ST	Sternoclavicular joint
T	Trachea
TH	Thymus
T3	Vertebral body (T3)
T4	Vertebral body (T4)

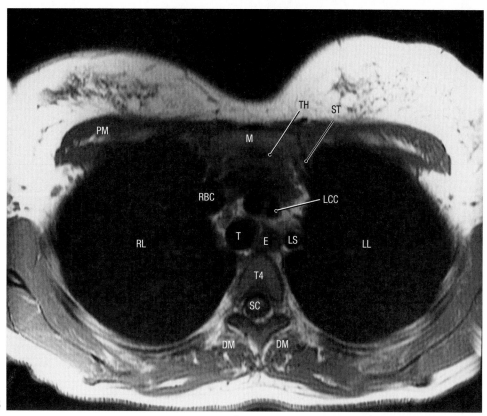

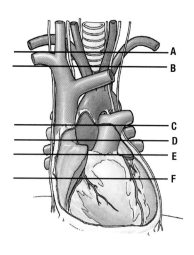

1.79 Transverse, or horizontal (axial), MRIs of the thorax (A through F)

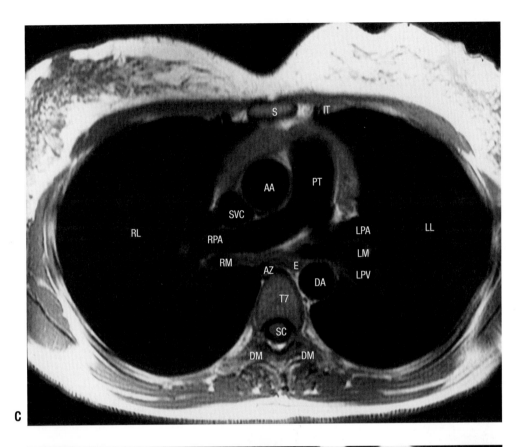

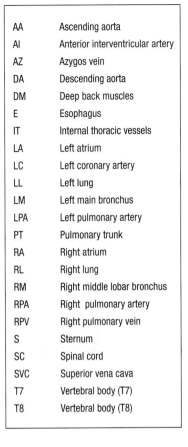

AA	Ascending aorta
AI	Anterior interventricular artery
AZ	Azygos vein
DA	Descending aorta
DM	Deep back muscles
E	Esophagus
IT	Internal thoracic vessels
LA	Left atrium
LC	Left coronary artery
LL	Left lung
LM	Left main bronchus
LPA	Left pulmonary artery
PT	Pulmonary trunk
RA	Right atrium
RL	Right lung
RM	Right middle lobar bronchus
RPA	Right pulmonary artery
RPV	Right pulmonary vein
S	Sternum
SC	Spinal cord
SVC	Superior vena cava
T7	Vertebral body (T7)
T8	Vertebral body (T8)

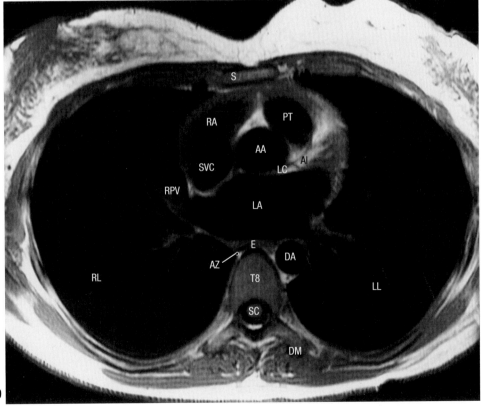

1.79   Transverse, or horizontal (axial), MRIs of the thorax *(continued)*

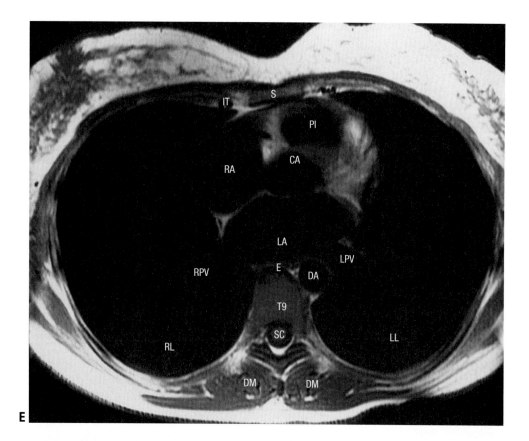

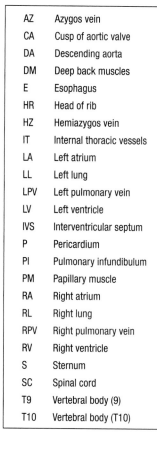

AZ	Azygos vein
CA	Cusp of aortic valve
DA	Descending aorta
DM	Deep back muscles
E	Esophagus
HR	Head of rib
HZ	Hemiazygos vein
IT	Internal thoracic vessels
LA	Left atrium
LL	Left lung
LPV	Left pulmonary vein
LV	Left ventricle
IVS	Interventricular septum
P	Pericardium
PI	Pulmonary infundibulum
PM	Papillary muscle
RA	Right atrium
RL	Right lung
RPV	Right pulmonary vein
RV	Right ventricle
S	Sternum
SC	Spinal cord
T9	Vertebral body (9)
T10	Vertebral body (T10)

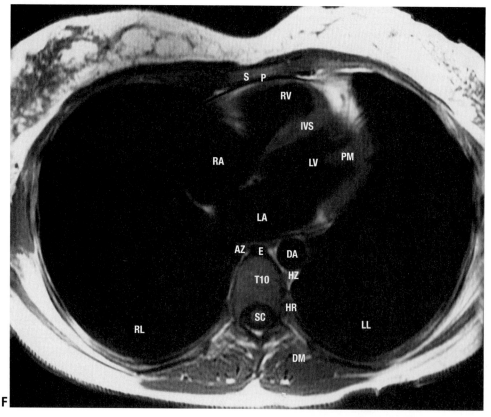

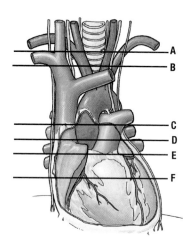

1.79   Transverse, or horizontal (axial), MRIs of the thorax *(continued)*

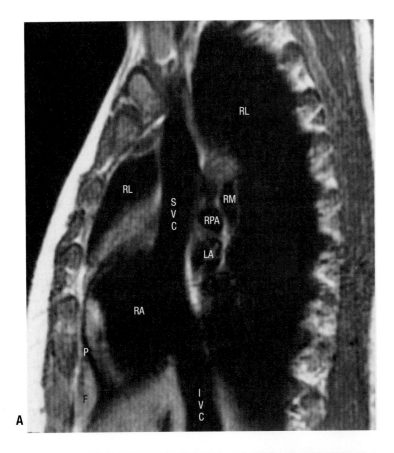

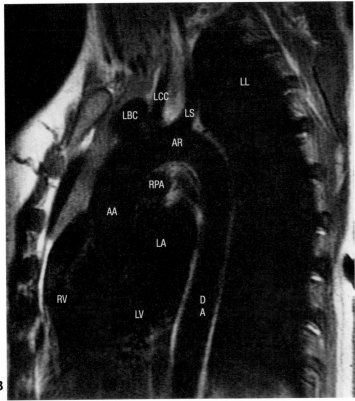

AR	Arch of aorta
AA	Ascending aorta
DA	Descending aorta
F	Fat
IVC	Inferior vena cava
LA	Left atrium
LBC	Left brachiocephalic vein
LCC	Left common carotid artery
LL	Left lung
LS	Left subclavian artery
LV	Left ventricle
P	Pericardium
RA	Right atrium
RL	Right lung
RM	Right main bronchus
RPA	Right pulmonary artery
RV	Right ventricle
SVC	Superior vena cava

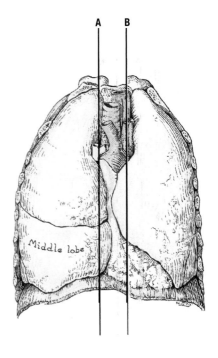

1.80   **Sagittal MRIs of the thorax**

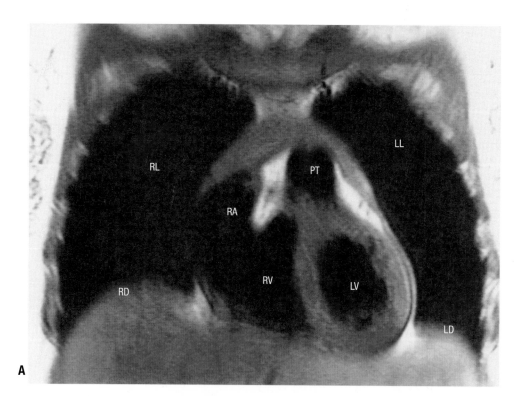

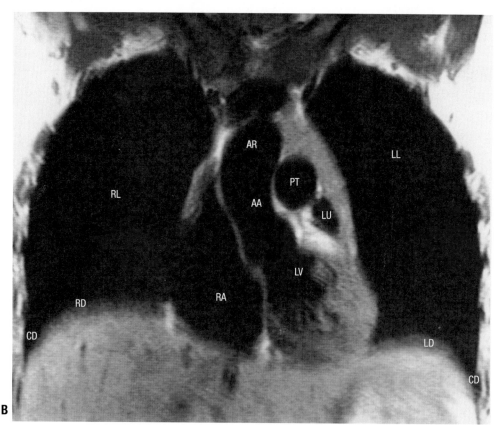

AA	Ascending aorta
AR	Arch of aorta
CD	Costodiaphragmatic recess
LD	Left dome of diaphragm
LL	Left lung
LU	Left auricle
LV	Left ventricle
PT	Pulmonary trunk
RA	Right atrium
RD	Right dome of diaphragm
RL	Right lung
RV	Right ventricle

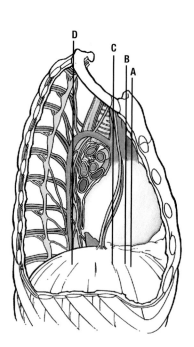

1.81   **Coronal MRIs of the thorax (A through D)**

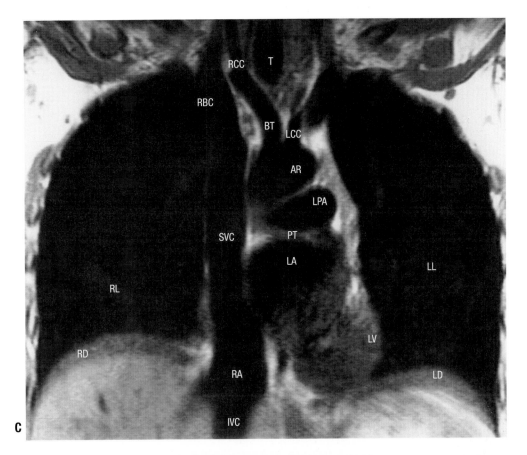

AR	Arch of aorta
AZ	Azygos vein
BT	Brachiocephalic trunk (artery)
DA	Descending aorta
IVC	Inferior vena cava
LV	Left ventricle
LA	Left atrium
LCC	Left common carotid artery
LD	Left dome of diaphragm
LL	Left lung
LPA	Left pulmonary artery
PT	Pulmonary trunk
RA	Right atrium
RBC	Right brachiocephalic vein
RCC	Right common carotid artery
RD	Right dome of diaphragm
RL	Right lung
SVC	Superior vena cava
T	Trachea

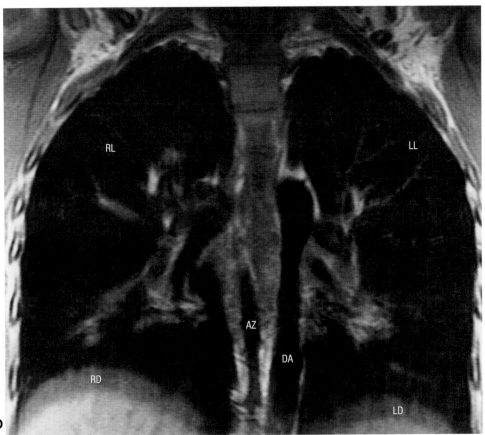

1.81   **Coronal MRIs of the thorax** *(continued)*

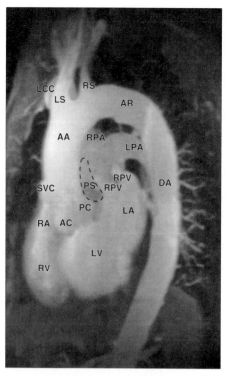

Oblique Sagittal View

1.82 MR angiogram of heart and great vessels

AA	Ascending aorta	LV	Left ventricle
AC	Anterior cusp of aortic valve	PC	Posterior cusp of aortic valve
AR	Arch of aorta	PS	Transverse pericardial
AZ	Azygos vein		sinus
DA	Descending aorta	PT	Pulmonary trunk
DLPA	Descending branch of left pulmonary	RA	Right atrium
	artery	RPA	Right pulmonary artery
E	Esophagus	RPV	Right pulmonary vein
IS	Interventricular septum	RS	Right subclavian artery
LA	Left atrium	RV	Right ventricle
LCC	Left common carotid	SLPV	Left superior pulmonary vein
	artery	SRPV	Right superior pulmonary vein
LPA	Left pulmonary artery	SVC	Superior vena cava
LS	Left subclavian artery		

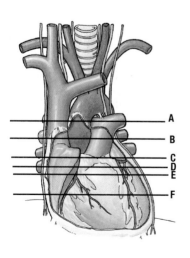

1.83 Transverse or horizontal (axial) CTs of the thorax (A through F)

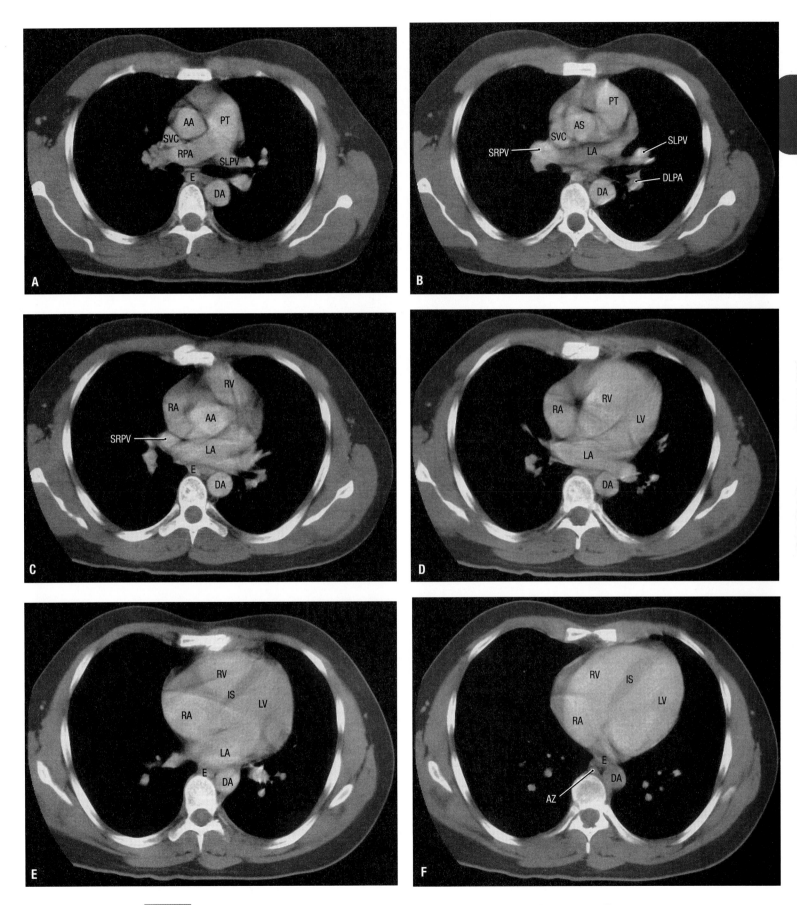

1.83   Transverse or horizontal (axial) CTs of the thorax *(continued)*

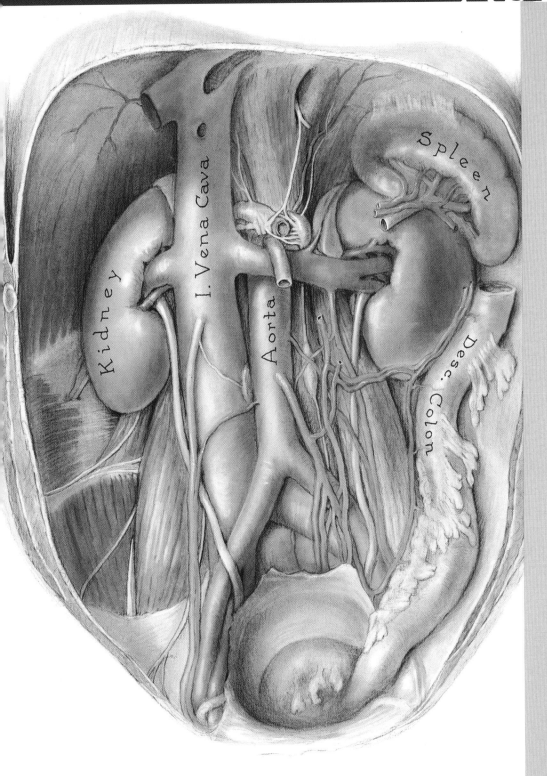

- Overview 92
- Anterolateral Abdominal Wall 94
- Inguinal Region 100
- Testes 106
- Peritoneum and Peritoneal Cavities 112
- Stomach 120
- Pancreas, Duodenum, and Bile Duct 125
- Intestines 132
- Liver and Gallbladder 142
- Portal Venous System 152
- Retroperitoneal Viscera 154
- Posterior Abdominal Wall and Diaphragm 162
- Aorta and Autonomic Nerves 167
- Lymphatic Drainage 172
- Review 175
- Scans: Sagittal, Axial, Coronal 176

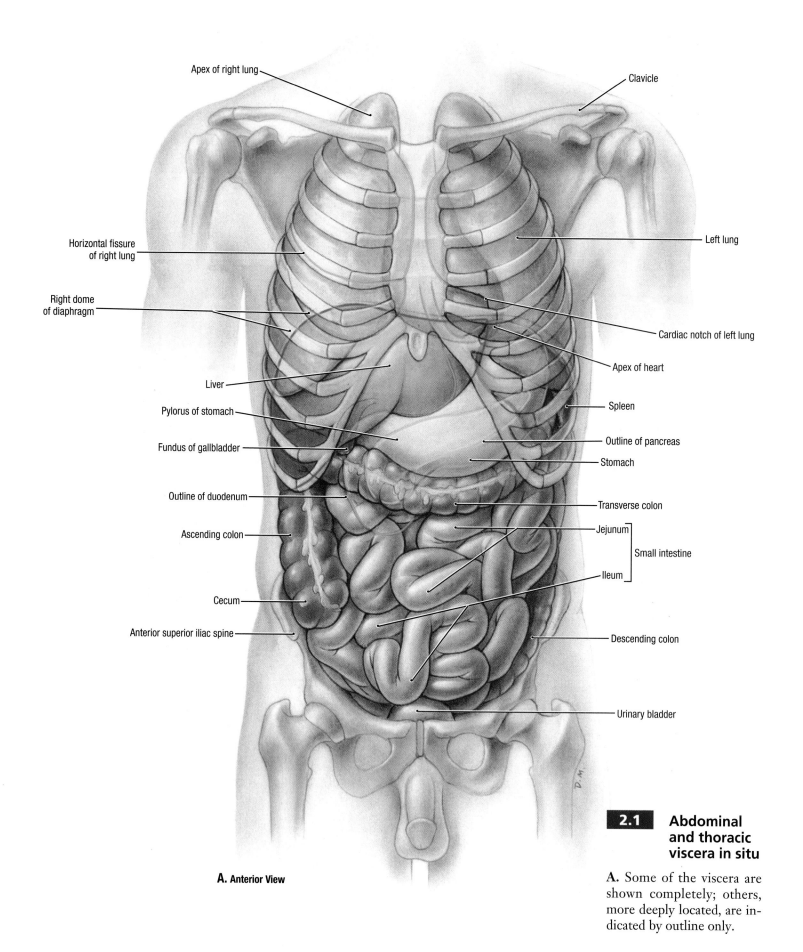

Apex of right lung

Clavicle

Horizontal fissure of right lung

Left lung

Right dome of diaphragm

Cardiac notch of left lung

Apex of heart

Liver

Spleen

Pylorus of stomach

Outline of pancreas

Fundus of gallbladder

Stomach

Outline of duodenum

Transverse colon

Ascending colon

Jejunum
Small intestine

Ileum

Cecum

Anterior superior iliac spine

Descending colon

Urinary bladder

A. Anterior View

2.1 **Abdominal and thoracic viscera in situ**

A. Some of the viscera are shown completely; others, more deeply located, are indicated by outline only.

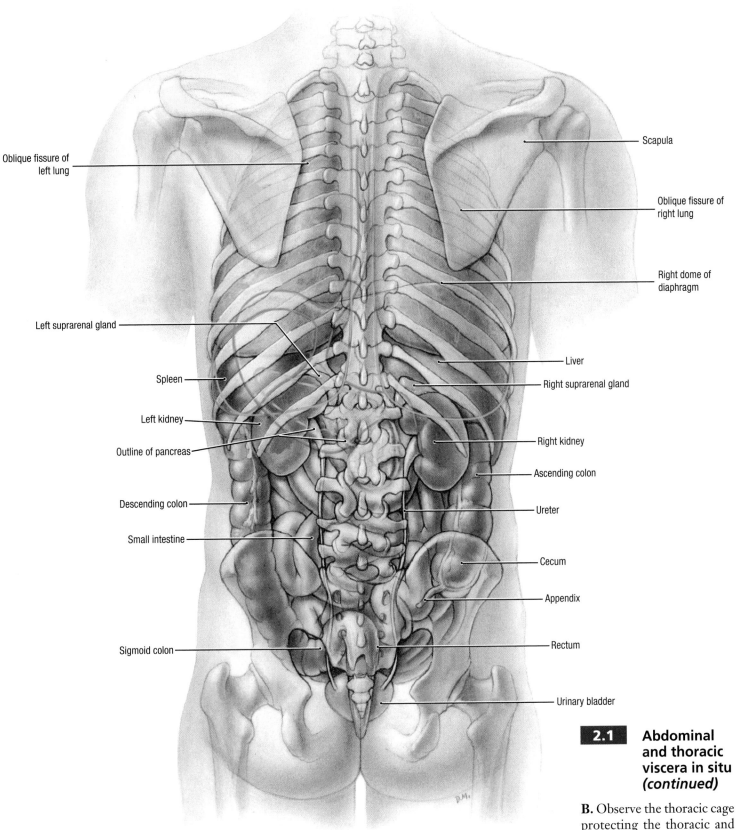

Oblique fissure of
left lung

Scapula

Oblique fissure of
right lung

Right dome of
diaphragm

Left suprarenal gland

Liver

Spleen

Right suprarenal gland

Left kidney

Right kidney

Outline of pancreas

Ascending colon

Descending colon

Ureter

Small intestine

Cecum

Appendix

Sigmoid colon

Rectum

Urinary bladder

B. Posterior View

2.1 **Abdominal
and thoracic
viscera in situ
(continued)**

B. Observe the thoracic cage
protecting the thoracic and
upper abdominal viscera,
and the pelvis supporting
and protecting the lower abdominal viscera.

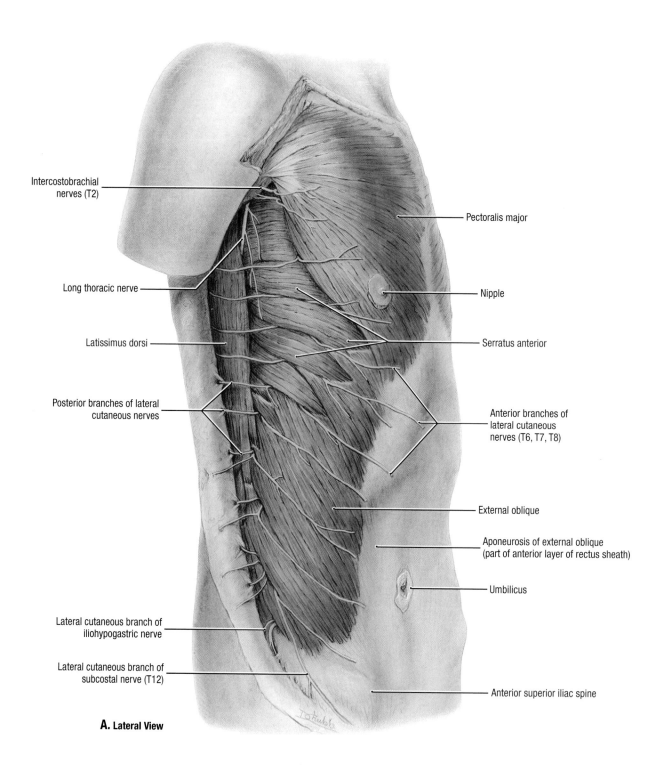

Intercostobrachial nerves (T2)

Long thoracic nerve

Latissimus dorsi

Posterior branches of lateral cutaneous nerves

Lateral cutaneous branch of iliohypogastric nerve

Lateral cutaneous branch of subcostal nerve (T12)

Pectoralis major

Nipple

Serratus anterior

Anterior branches of lateral cutaneous nerves (T6, T7, T8)

External oblique

Aponeurosis of external oblique (part of anterior layer of rectus sheath)

Umbilicus

Anterior superior iliac spine

A. Lateral View

2.2 Trunk

A. This superficial dissection shows the external oblique muscle and the anterior and posterior branches of the lateral cutaneous nerves. Observe the costal attachments of the serratus anterior muscle interdigitating with those of the external oblique muscle.

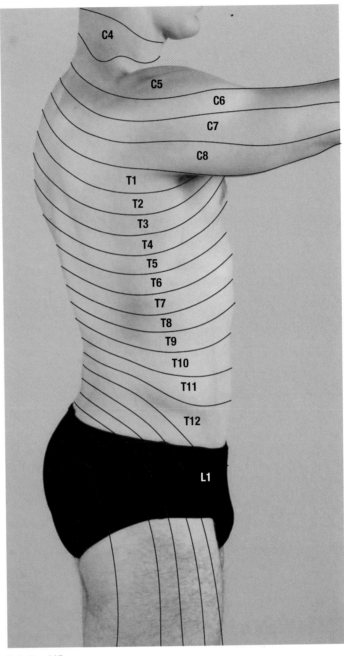

B. Lateral View

2.2   Trunk

B. Dermatomes. The thoracoabdominal (T7–T11) nerves run between the external and internal oblique muscles to supply sensory innervation to the overlying skin. The subcostal nerve (T12) runs along the inferior border of the 12th rib to supply the skin over the anterior superior iliac spine and hip. The iliohypogastric nerve (L1) innervates the skin over the iliac crest and hypogastric region and the ilioinguinal nerve (L1), the skin of the medial aspect of the thigh, the scrotum or labium majus and mons pubis.

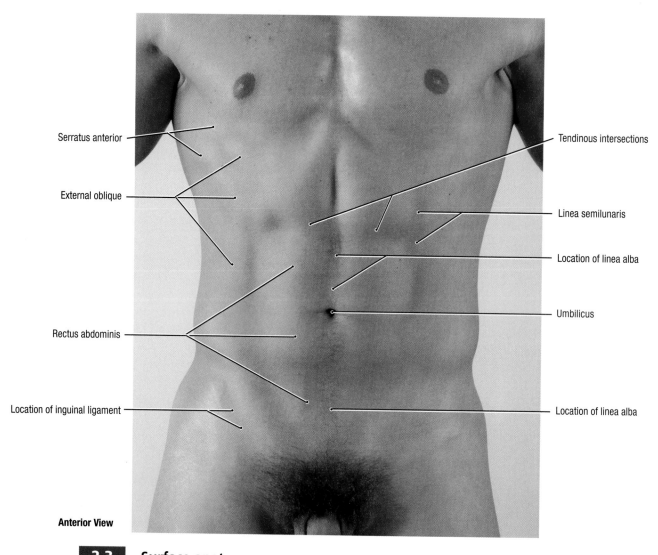

2.3 **Surface anatomy**

Serratus anterior

External oblique

Rectus abdominis

Location of inguinal ligament

Tendinous intersections

Linea semilunaris

Location of linea alba

Umbilicus

Location of linea alba

Anterior View

TABLE 2.1 PRINCIPAL MUSCLES OF ANTEROLATERAL ABDOMINAL WALL

MUSCLE	ORIGIN	INSERTION	INNERVATION	ACTION(S)
External oblique	External surfaces of 5th to 12th ribs	Linea alba pubic tubercle, and anterior half of iliac crest	Inferior six thoracic nerves and subcostal nerve	Compress and support abdominal visceral flex and rotate trunk
Internal oblique	Thoracolumbar fascia, anterior two-thirds of iliac crest, and lateral half of inguinal ligament	Inferior borders of 10th–12th ribs, linea alba, and pubis via conjoint tendon	Anterior rami of inferior six thoracic and 1st lumbar nerves	
Transverse abdominal	Internal surfaces of 7th to 12th costal cartilages, thoracolumber fascia, iliac crest, and lateral third of inguinal ligament	Linea alba with aponeurosis of internal oblique, pubic crest, and pecten pubis via conjoint tendon		Compress and support abdominal visceral
Rectus abdominis	Pubic symphysis and pubic crest	Xiphoid process and 5th to 7th costal cartilages	Anterior rami of inferior six thoracic nerves	Flexes trunk and compresses abdominal visceral

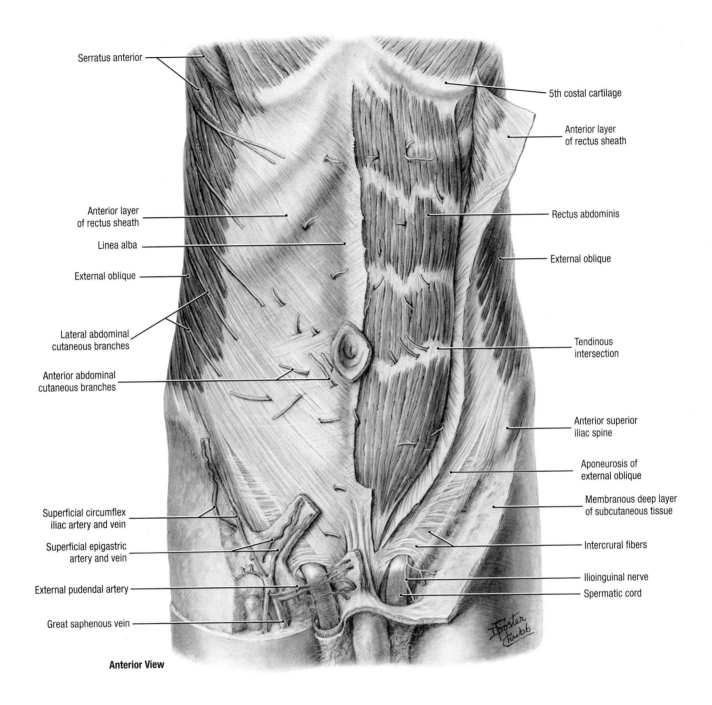

Serratus anterior

5th costal cartilage

Anterior layer
of rectus sheath

Anterior layer
of rectus sheath

Rectus abdominis

Linea alba

External oblique

External oblique

Lateral abdominal
cutaneous branches

Tendinous
intersection

Anterior abdominal
cutaneous branches

Anterior superior
iliac spine

Aponeurosis of
external oblique

Membranous deep layer
of subcutaneous tissue

Superficial circumflex
iliac artery and vein

Superficial epigastric
artery and vein

Intercrural fibers

External pudendal artery

Ilioinguinal nerve

Spermatic cord

Great saphenous vein

Anterior View

2.4   **Anterior abdominal wall, superficial dissection**

The anterior layer of the rectus sheath is reflected on the left side
of the specimen.

- The anterior cutaneous nerves of the T8 to T12 spinal nerves
  pierce the rectus abdominis muscle and anterior layer of its
  sheath; spinal nerve T10 supplies the region of the umbilicus.
- The three superficial inguinal branches of the femoral artery
  (superficial circumflex iliac artery, superficial epigastric artery,
  and external pudendal artery) and the great (long) saphenous
  vein lie in the fatty superficial layer of subcutaneous tissue.

- The superficial inguinal ring is an oblique, triangular opening
  of the inguinal canal, 2 to 3 cm long; its central point is superior
  to the pubic tubercle. The spermatic cord and ilioinguinal
  nerve pass through the superficial inguinal ring.
- The linea alba, usually indicated superficially by a vertical skin
  groove in the midline, is a subcutaneous fibrous band extending
  from the sternum to the symphysis pubis.

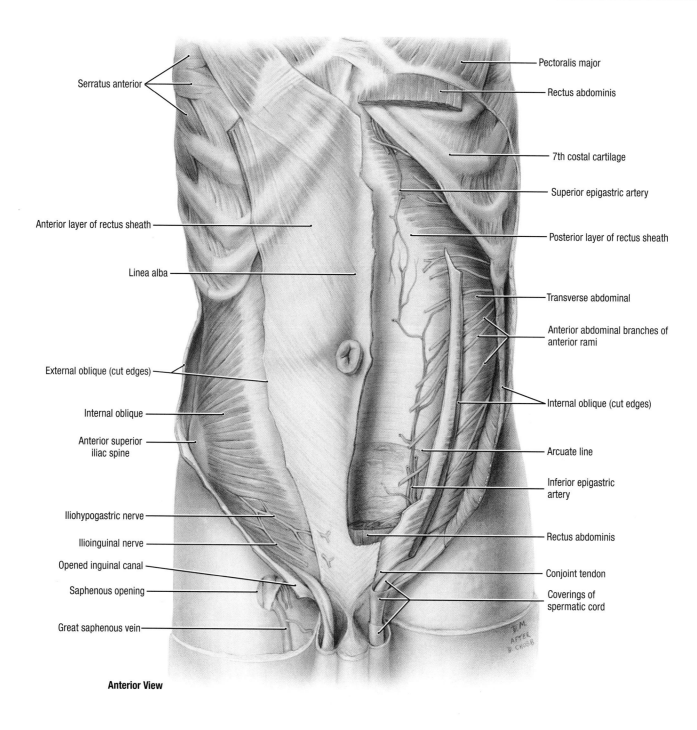

Serratus anterior

Pectoralis major

Rectus abdominis

7th costal cartilage

Superior epigastric artery

Anterior layer of rectus sheath

Posterior layer of rectus sheath

Linea alba

Transverse abdominal

Anterior abdominal branches of anterior rami

External oblique (cut edges)

Internal oblique (cut edges)

Internal oblique

Anterior superior iliac spine

Arcuate line

Inferior epigastric artery

Iliohypogastric nerve

Rectus abdominis

Ilioinguinal nerve

Opened inguinal canal

Conjoint tendon

Saphenous opening

Coverings of spermatic cord

Great saphenous vein

Anterior View

2.5   Anterior abdominal wall, deep dissection

On the right side of the specimen, most of the external oblique muscle is excised. On the left, the rectus abdominis muscle is excised, and the internal oblique muscle is divided.

- The fibers of the internal oblique muscle run horizontally at the level of the anterior superior iliac spine, obliquely upward superior to this level, and obliquely downward inferior to it.
- The arcuate line is at the level of the anterior superior iliac spine.
- The anastomosis between the superior and inferior epigastric arteries indirectly unites the subclavian artery of the upper limb to the external iliac arteries of the lower limb. The anastomosis can become functionally patent in response to slowly developing occlusion of the aorta.

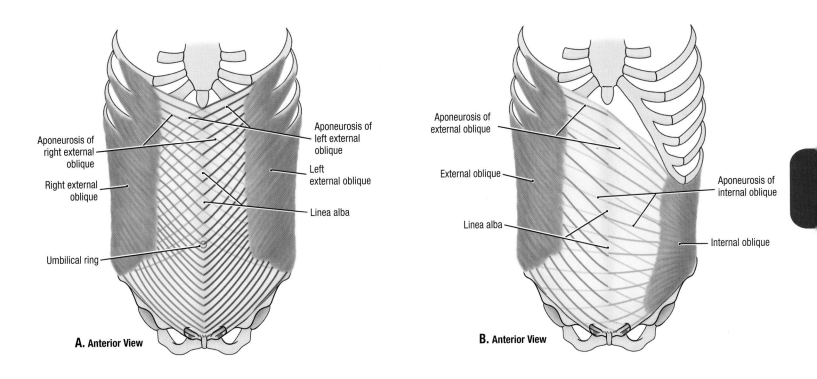

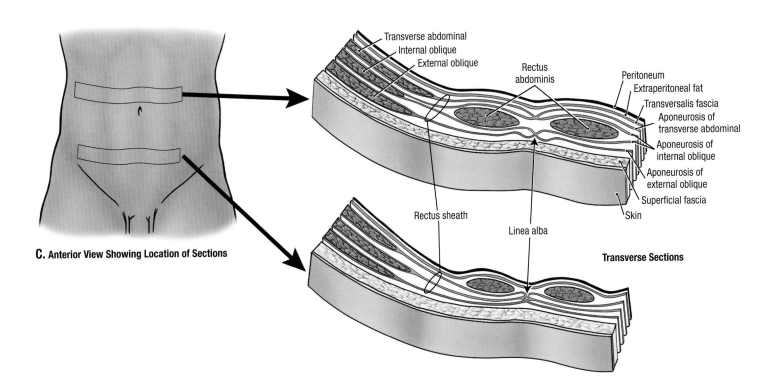

2.6   Structure of the anterolateral abdominal wall

A. Intramuscular interdigitation of superficial and deep fibers within the aponeuroses of the right and left external obliques. **B.** Intermuscular interdigitation of fibers between the aponeuroses of the contralateral external and internal obliques. **C.** Structure of the rectus sheath: transverse sections superior and inferior to the arcuate line.

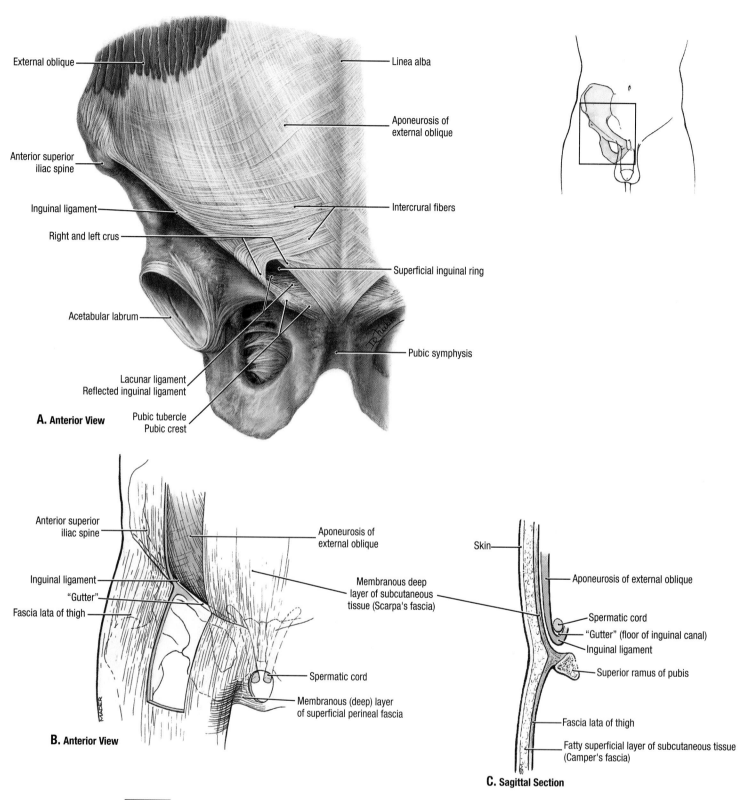

External oblique

Linea alba

Aponeurosis of external oblique

Anterior superior iliac spine

Inguinal ligament

Intercrural fibers

Right and left crus

Superficial inguinal ring

Acetabular labrum

Pubic symphysis

Lacunar ligament
Reflected inguinal ligament

Pubic tubercle
Pubic crest

A. Anterior View

Anterior superior iliac spine

Aponeurosis of external oblique

Inguinal ligament

Membranous deep layer of subcutaneous tissue (Scarpa's fascia)

"Gutter"

Fascia lata of thigh

Spermatic cord

Membranous (deep) layer of superficial perineal fascia

B. Anterior View

Skin

Aponeurosis of external oblique

Spermatic cord

"Gutter" (floor of inguinal canal)

Inguinal ligament

Superior ramus of pubis

Fascia lata of thigh

Fatty superficial layer of subcutaneous tissue (Camper's fascia)

C. Sagittal Section

2.7 **Inguinal region of male—I**

A. Aponeurosis of external oblique and inguinal ligament. **B** and **C.** Membranous deep layer of superficial fascia. The membranous deep layer of fascia blends with the fascia lata of the thigh, a finger breadth inferior to the inguinal ligament. The fascia follows along the penis and spermatic cord to the scrotum and continues posteriorly to blend with the deep layer of superficial fascia of the perineum. The floor of the inguinal canal is formed by the superior surface of the in-curving inguinal ligament, which forms a shallow trough or "gutter."

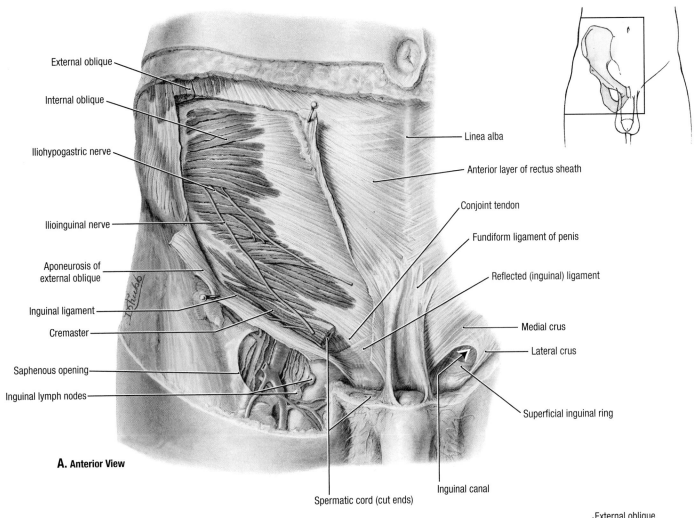

External oblique

Internal oblique

Iliohypogastric nerve

Ilioinguinal nerve

Aponeurosis of
external oblique

Inguinal ligament

Cremaster

Saphenous opening

Inguinal lymph nodes

Linea alba

Anterior layer of rectus sheath

Conjoint tendon

Fundiform ligament of penis

Reflected (inguinal) ligament

Medial crus

Lateral crus

Superficial inguinal ring

A. Anterior View

Spermatic cord (cut ends)

Inguinal canal

2.8   Inguinal region of male—II

A. Internal oblique and cremaster. Part of the aponeurosis of the external oblique muscle is cut away, and the spermatic cord is cut short. **B.** Schematic illustration.

- The fleshy fibers of the internal oblique muscle run horizontally at the level of the anterior superior iliac spine, pass superomedially from the iliac crest, and arch inferomedially from the inguinal ligament.
- The cremaster muscle covers the spermatic cord.
- The reflected (inguinal) ligament is formed by aponeurotic fibers of the external oblique muscle and lies anterior to the conjoint tendon. The conjoint tendon is formed by the fusion of the aponeurosis of the internal oblique and transverse abdominal muscles.

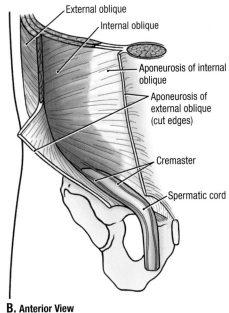

External oblique

Internal oblique

Aponeurosis of internal oblique

Aponeurosis of external oblique (cut edges)

Cremaster

Spermatic cord

B. Anterior View

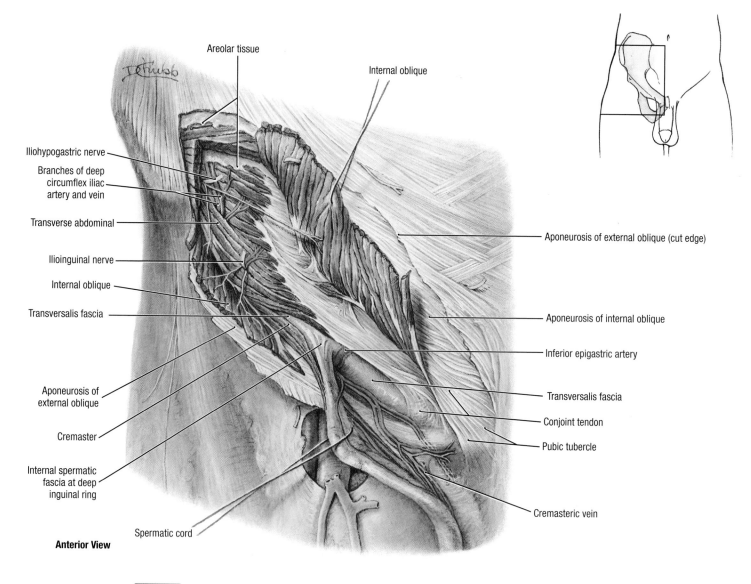

Areolar tissue

Internal oblique

Iliohypogastric nerve

Branches of deep circumflex iliac artery and vein

Transverse abdominal

Ilioinguinal nerve

Internal oblique

Transversalis fascia

Aponeurosis of external oblique

Cremaster

Internal spermatic fascia at deep inguinal ring

Spermatic cord

Aponeurosis of external oblique (cut edge)

Aponeurosis of internal oblique

Inferior epigastric artery

Transversalis fascia

Conjoint tendon

Pubic tubercle

Cremasteric vein

Anterior View

2.9 Inguinal region of male—III

The internal oblique muscle is reflected, and the spermatic cord is retracted.
- The internal oblique portion of the conjoint tendon is attached to the pubic crest, and the transverse portion runs parallel to the pectineal line.
- The iliohypogastric and ilioinguinal nerves supply the internal oblique and transverse abdominal muscles.
- The transversalis fascia is evaginated to form the tubular internal spermatic fascia. The mouth of the tube, called the *deep inguinal ring*, is situated lateral to the inferior epigastric vessels.

TABLE 2.2 BOUNDARIES OF THE INGUINAL CANAL

	LATERAL THIRD	MIDDLE THIRD	MEDIAL THIRD
Posterior wall	Transversalis fascia Deep inguinal ring	Transversalis fascia	Tranversalis fascia Conjoint tendon
Anterior wall	Aponeurosis of external oblique Internal oblique	Aponeurosis of external oblique	Aponeurosis of external oblique Superficial inguinal ring
Roof	Arching fibers of internal oblique and transverse abdominal muscles		
Floor	Inguinal ligament	Inguinal ligament	Inguinal ligament Lacunar ligament

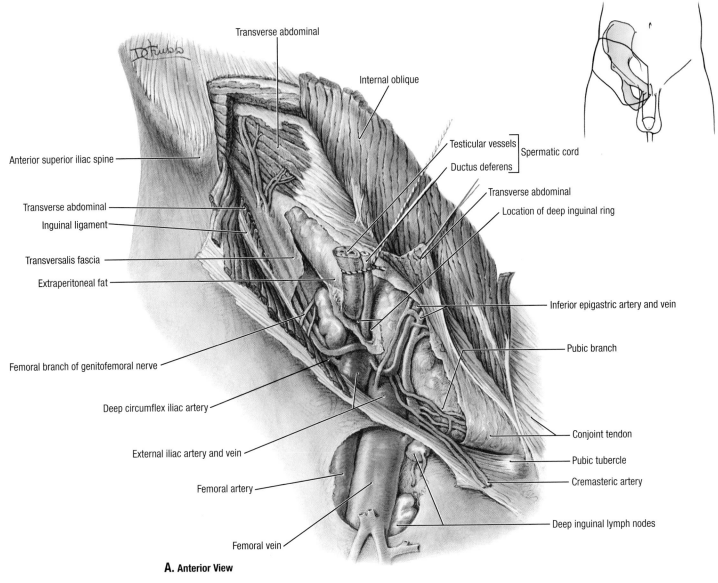

Transverse abdominal

Internal oblique

Testicular vessels ⎤
Ductus deferens ⎦ Spermatic cord

Transverse abdominal

Location of deep inguinal ring

Anterior superior iliac spine

Transverse abdominal

Inguinal ligament

Transversalis fascia

Extraperitoneal fat

Inferior epigastric artery and vein

Femoral branch of genitofemoral nerve

Pubic branch

Deep circumflex iliac artery

Conjoint tendon

External iliac artery and vein

Pubic tubercle

Cremasteric artery

Femoral artery

Deep inguinal lymph nodes

Femoral vein

A. Anterior View

2.10   Inguinal region of male—IV

A. Transverse abdominal muscle and transversalis fascia. The inguinal part of the transverse abdominal muscle and transversalis fascia is partially cut away, the spermatic cord is excised, and the ductus deferens is retracted. **B.** Schematic illustration.

- The deep inguinal ring is located a finger breadth superior to the inguinal ligament at the midpoint between the anterior superior iliac spine and pubic tubercle.
- Note that the testicular vessels and ductus deferens part company at the deep inguinal ring.
- Observe the proximity of the external iliac artery and vein to the inguinal canal.
- The external iliac artery has two branches, the deep circumflex iliac and inferior epigastric arteries. Note also the cremasteric and pubic branches of the latter.

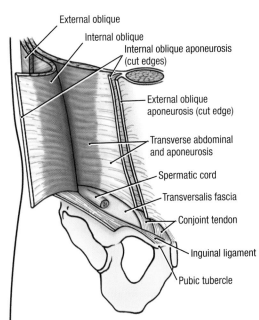

External oblique

Internal oblique

Internal oblique aponeurosis (cut edges)

External oblique aponeurosis (cut edge)

Transverse abdominal and aponeurosis

Spermatic cord

Transversalis fascia

Conjoint tendon

Inguinal ligament

Pubic tubercle

B. Anterior View

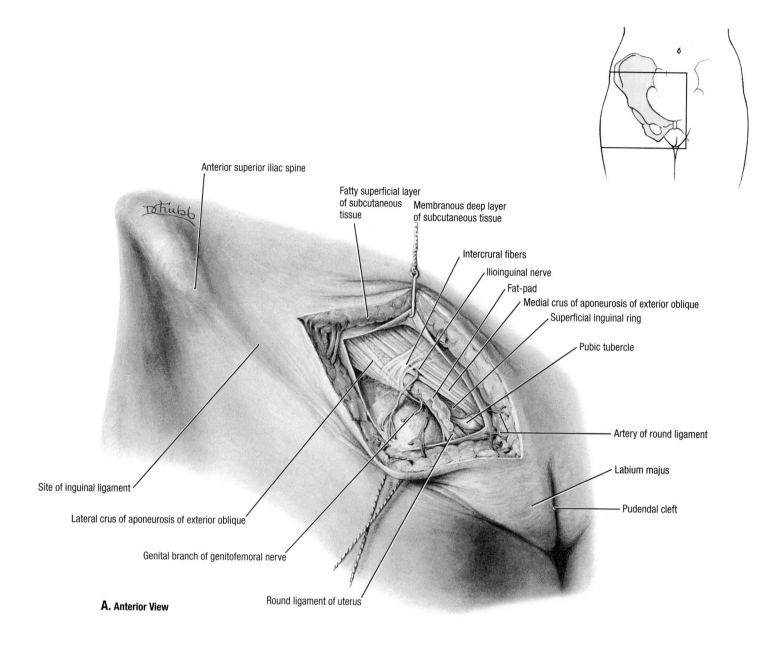

Anterior superior iliac spine

Fatty superficial layer of subcutaneous tissue

Membranous deep layer of subcutaneous tissue

Intercrural fibers

Ilioinguinal nerve

Fat-pad

Medial crus of aponeurosis of exterior oblique

Superficial inguinal ring

Pubic tubercle

Artery of round ligament

Labium majus

Pudendal cleft

Site of inguinal ligament

Lateral crus of aponeurosis of exterior oblique

Genital branch of genitofemoral nerve

Round ligament of uterus

A. Anterior View

2.11   Inguinal canal of female

Progressive dissections of the female inguinal canal **(A–D).**

- Note in **A,** the small superficial inguinal ring. Its crura are prevented from spreading by the intercrural fibers. Passing through the superficial inguinal ring are the round ligament of the uterus, a closely applied fat-pad, the genital branch of the genitofemoral nerve, and the artery of the round ligament of the uterus.
- The cremaster muscle does not extend beyond the superficial inguinal ring **(B).**
- The round ligament breaks up into strands as it leaves the inguinal canal and approaches the labium majus **(C).**
- The external iliac artery and vein are close to the inguinal canal **(D).**

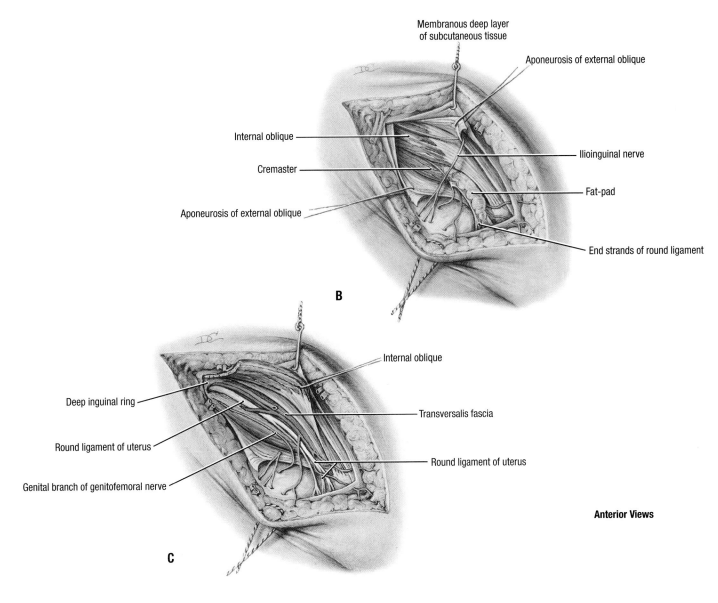

Membranous deep layer of subcutaneous tissue

Aponeurosis of external oblique

Internal oblique

Ilioinguinal nerve

Cremaster

Fat-pad

Aponeurosis of external oblique

End strands of round ligament

B

Internal oblique

Deep inguinal ring

Transversalis fascia

Round ligament of uterus

Round ligament of uterus

Genital branch of genitofemoral nerve

Anterior Views

C

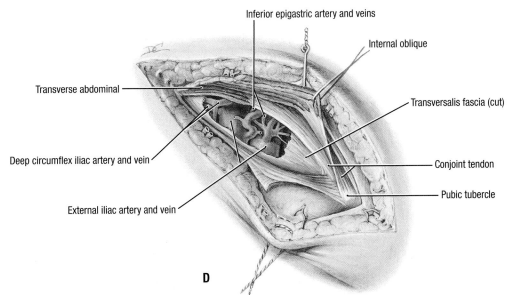

Inferior epigastric artery and veins

Internal oblique

Transverse abdominal

Transversalis fascia (cut)

Deep circumflex iliac artery and vein

Conjoint tendon

Pubic tubercle

External iliac artery and vein

D

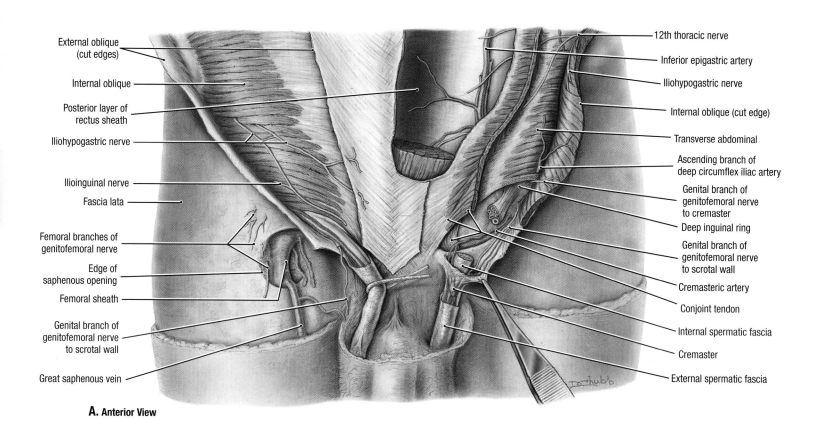

External oblique (cut edges)

Internal oblique

Posterior layer of rectus sheath

Iliohypogastric nerve

Ilioinguinal nerve

Fascia lata

Femoral branches of genitofemoral nerve

Edge of saphenous opening

Femoral sheath

Genital branch of genitofemoral nerve to scrotal wall

Great saphenous vein

12th thoracic nerve

Inferior epigastric artery

Iliohypogastric nerve

Internal oblique (cut edge)

Transverse abdominal

Ascending branch of deep circumflex iliac artery

Genital branch of genitofemoral nerve to cremaster

Deep inguinal ring

Genital branch of genitofemoral nerve to scrotal wall

Cremasteric artery

Conjoint tendon

Internal spermatic fascia

Cremaster

External spermatic fascia

A. Anterior View

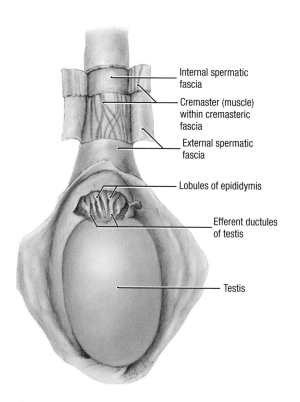

Internal spermatic fascia

Cremaster (muscle) within cremasteric fascia

External spermatic fascia

Lobules of epididymis

Efferent ductules of testis

Testis

B. Lateral View

2.12 Anterior abdominal wall, deep dissection

A. *Right:* The external oblique and its aponeurosis are cut away to reveal the iliohypogastric and ilioinguinal nerves. The spermatic cord is pulled medially to reveal the genital branch of the genitofemoral nerve. *Left:* The internal oblique is divided and reflected, and a section of the spermatic cord is excised, revealing fibers of the internal oblique and transverse abdominal, arching medially to form the aponeurotic conjoint tendon. **B.** Coverings of the spermatic cord include the internal spermatic fascia, derived from the transversalis fascia; the cremaster muscle and fascia, from the internal oblique and transverse abdominal; and the external spermatic fascia, from the external oblique aponeurosis.

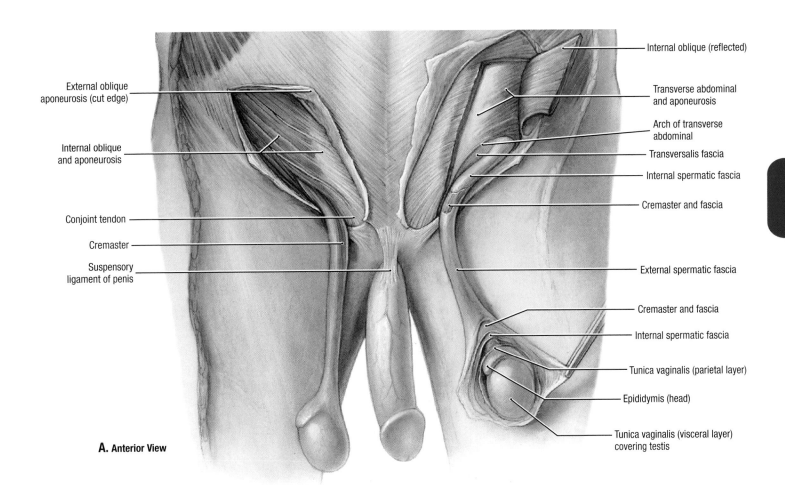

External oblique
aponeurosis (cut edge)

Internal oblique
and aponeurosis

Conjoint tendon

Cremaster

Suspensory
ligament of penis

Internal oblique (reflected)

Transverse abdominal
and aponeurosis

Arch of transverse
abdominal

Transversalis fascia

Internal spermatic fascia

Cremaster and fascia

External spermatic fascia

Cremaster and fascia

Internal spermatic fascia

Tunica vaginalis (parietal layer)

Epididymis (head)

Tunica vaginalis (visceral layer)
covering testis

A. Anterior View

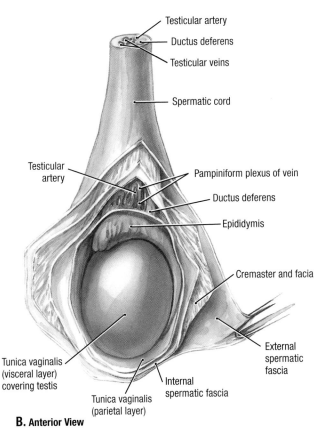

Testicular artery

Ductus deferens

Testicular veins

Spermatic cord

Testicular
artery

Pampiniform plexus of vein

Ductus deferens

Epididymis

Cremaster and facia

External
spermatic
fascia

Tunica vaginalis
(visceral layer)
covering testis

Internal
spermatic fascia

Tunica vaginalis
(parietal layer)

B. Anterior View

2.13   **Inguinal canal, spermatic cord, and testis**

A. Inguinal canal. The aponeurosis of the external oblique muscle is incised and reflected, revealing the deeper internal oblique muscle. A window has been made by cutting and reflecting the internal oblique muscle laterally to expose the deeper transverse abdominal muscle and aponeurosis. **B.** Coverings of the spermatic cord and testis. All of the layers covering the testis have been cut open sequentially: the external spermatic fascia, the cremaster muscle and fascia, the internal spermatic fascia, and the visceral and parietal layers of the tunica vaginalis of the testis.

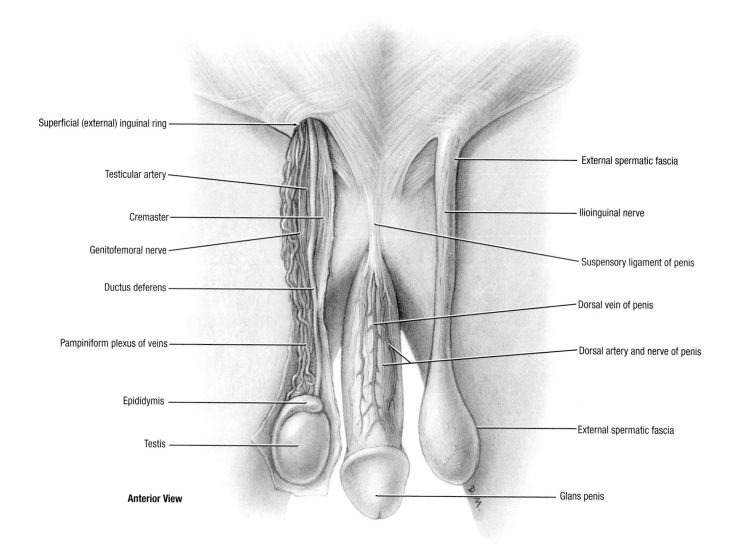

Superficial (external) inguinal ring

Testicular artery

Cremaster

Genitofemoral nerve

Ductus deferens

Pampiniform plexus of veins

Epididymis

Testis

Anterior View

External spermatic fascia

Ilioinguinal nerve

Suspensory ligament of penis

Dorsal vein of penis

Dorsal artery and nerve of penis

External spermatic fascia

Glans penis

2.14 Vessels and nerves of the penis and contents of spermatic cord

The superficial fascia covering the penis is removed to expose the median deep dorsal vein and the bilateral dorsal arteries and nerves of the penis. On the specimen's right, the coverings of the spermatic cord and testis are reflected, and the contents of the cords are separated. The testicular artery has been dissected away from the pampiniform plexus of veins. Lymphatic vessels and autonomic nerve fibers (not shown) are also present. On the specimen's left, the spermatic cord passes through the external inguinal ring and picks up a covering of external spermatic fascia from the margins of the external inguinal ring. The ilioinguinal nerve supplies the skin at the base of the penis and the anterior aspect of the scrotum. The cremasteric vessels supply the coverings of the cord and cremaster muscle.

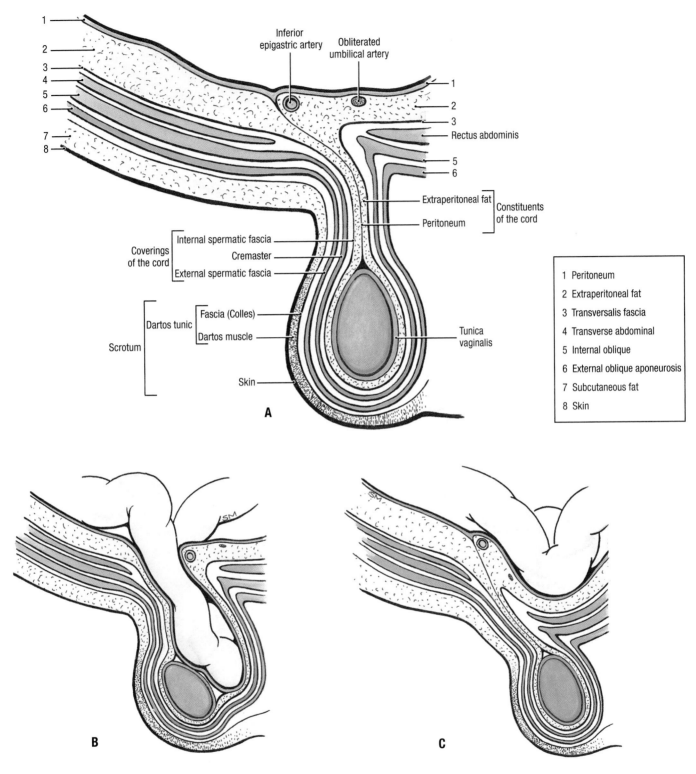

1 Peritoneum
2 Extraperitoneal fat
3 Transversalis fascia
4 Transverse abdominal
5 Internal oblique
6 External oblique aponeurosis
7 Subcutaneous fat
8 Skin

2.15 Coverings of spermatic cord and testes, and inguinal hernias

A. Schematic illustration. The scrotum and testis are assumed to have been raised to the level of the superficial inguinal ring. **B.** Indirect inguinal hernia. **C.** Direct inguinal hernia.

- The tunica vaginalis is derived from peritoneum.
- The dartos tunic consists of membranous superficial (Colles) fascia that is continuous with the deep membranous layer of subcutaneous tissue of the abdomen but lacks fat, replacing it with smooth muscle fibers of dartos muscle.
- The deep inguinal ring is lateral to the inferior epigastric artery.

Indirect inguinal hernias pass through this ring, with the sac following the course of the spermatic cord **(B).** Direct inguinal hernias bulge through the abdominal wall, medial to the inferior epigastric artery **(C).**

- Cremasteric fascia is derived from the fascia of internal oblique (superficial and deep surfaces), cremaster muscle from internal oblique arising from the inguinal ligament.

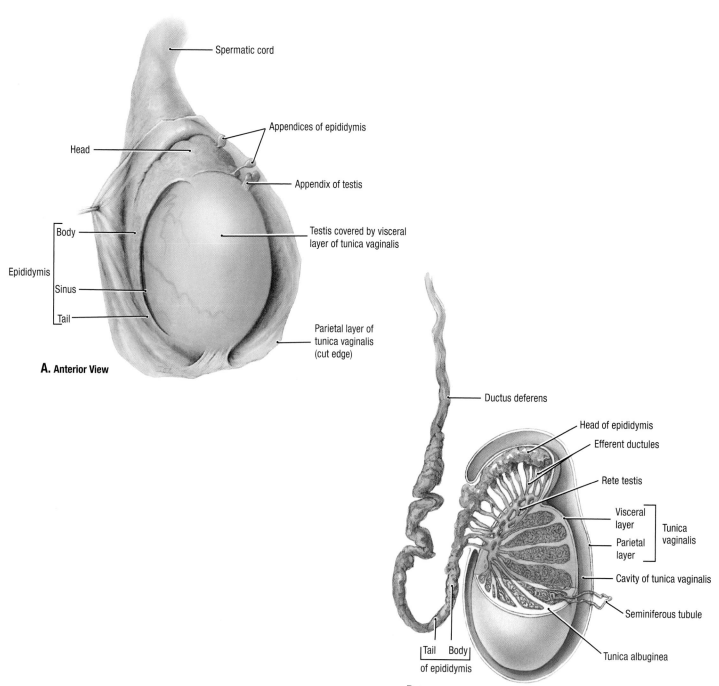

A. Anterior View

Spermatic cord

Appendices of epididymis

Head

Appendix of testis

Body

Testis covered by visceral
layer of tunica vaginalis

Epididymis

Sinus

Tail

Parietal layer of
tunica vaginalis
(cut edge)

Ductus deferens

Head of epididymis

Efferent ductules

Rete testis

Visceral
layer   Tunica
               vaginalis
Parietal
layer

Cavity of tunica vaginalis

Seminiferous tubule

Tail   Body
of epididymis

Tunica albuginea

**B. Longitinal Section of Tunica Vaginalis;
Testis Sectioned in Sagittal and Transverse Planes**

2.16   Testis and epididymis

A. Testis. The tunica vaginalis has been incised longitudinally to expose its cavity, surrounding the testis anteriorly and laterally and extending between the testis and epididymis at the sinus of the epididymis. Note the epididymis, lying posterolateral to the testis. It indicates to which side a testis belongs, for it is on the right side of the right testis and on the left side of the left testis. **B.** Structure of the epididymis and testis. Note the pyramidal compartments of the seminiferous tubules, shown semidiagrammatically; each of the 250 compartments contains two or three hairlike seminiferous tubules that join in the mediastinum testis to form a rete.

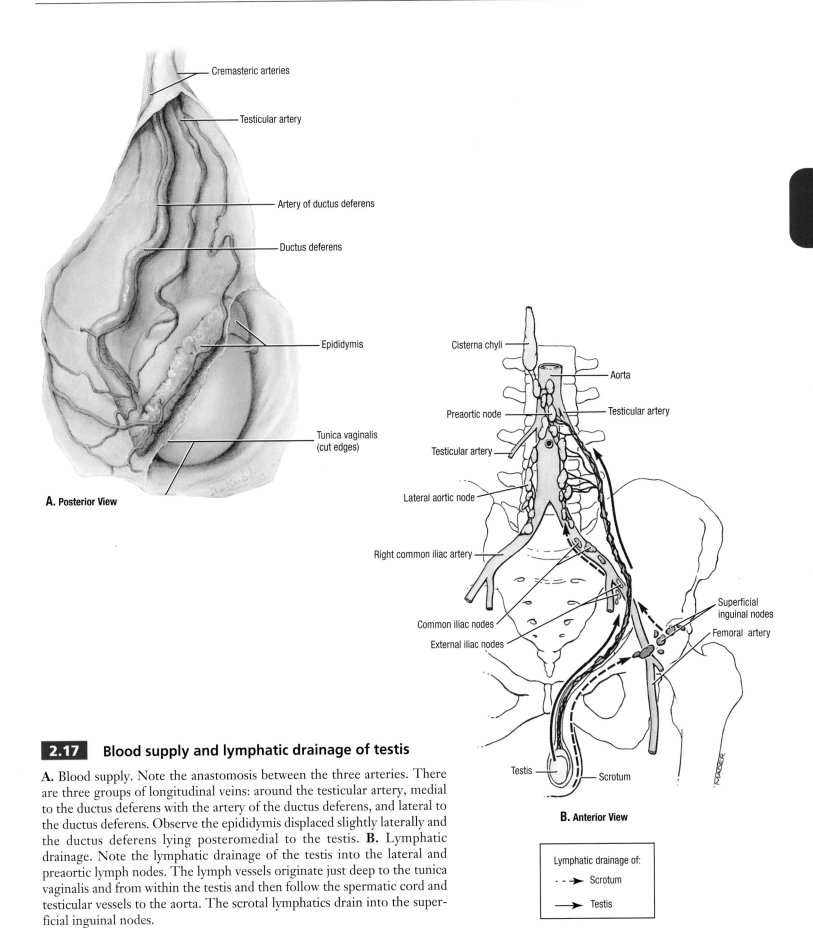

Cremasteric arteries

Testicular artery

Artery of ductus deferens

Ductus deferens

Epididymis

Tunica vaginalis
(cut edges)

A. Posterior View

Cisterna chyli

Aorta

Preaortic node

Testicular artery

Testicular artery

Lateral aortic node

Right common iliac artery

Superficial
inguinal nodes

Femoral artery

Common iliac nodes

External iliac nodes

Testis

Scrotum

B. Anterior View

2.17   **Blood supply and lymphatic drainage of testis**

A. Blood supply. Note the anastomosis between the three arteries. There are three groups of longitudinal veins: around the testicular artery, medial to the ductus deferens with the artery of the ductus deferens, and lateral to the ductus deferens. Observe the epididymis displaced slightly laterally and the ductus deferens lying posteromedial to the testis. **B.** Lymphatic drainage. Note the lymphatic drainage of the testis into the lateral and preaortic lymph nodes. The lymph vessels originate just deep to the tunica vaginalis and from within the testis and then follow the spermatic cord and testicular vessels to the aorta. The scrotal lymphatics drain into the superficial inguinal nodes.

Lymphatic drainage of:

- - ► Scrotum

——► Testis

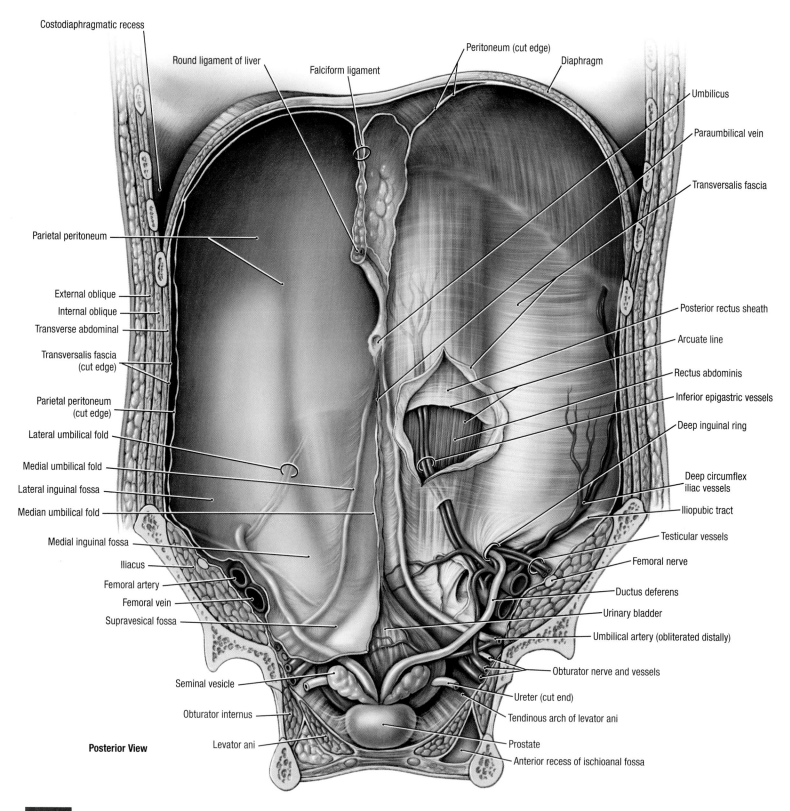

2.18 **Posterior aspect of the anterolateral abdominal wall**

Note the attachment of the falciform ligament and round ligament of the liver to the abdominal wall. The median umbilical fold extends from the apex of the urinary bladder to the umbilicus and covers the median umbilical ligament (the remnant of the urachus). The two medial umbilical folds cover the medial umbilical ligaments (the remnants of the occluded fetal umbilical arteries).

Two lateral umbilical folds cover the inferior epigastric vessels. Note also the supravesical fossae between the median and medial umbilical folds, the medial inguinal fossae (inguinal triangles) between the medial and lateral umbilical folds, and the lateral inguinal fossae lateral to the lateral umbilical folds.

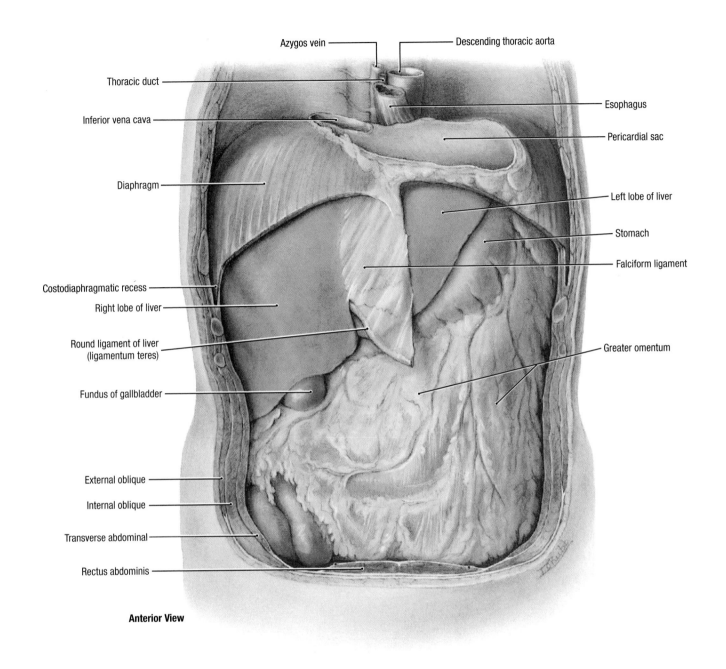

Azygos vein — Descending thoracic aorta

Thoracic duct —

Inferior vena cava —

— Esophagus

— Pericardial sac

Diaphragm —

— Left lobe of liver

— Stomach

— Falciform ligament

Costodiaphragmatic recess —

Right lobe of liver —

Round ligament of liver
(ligamentum teres) —

— Greater omentum

Fundus of gallbladder —

External oblique —

Internal oblique —

Transverse abdominal —

Rectus abdominis —

Anterior View

2.19 **Abdominal contents, undisturbed**

The falciform ligament, with the round ligament of the liver (ligamentum teres) in its free edge, is severed at its attachment to the abdominal wall and diaphragm in the median plane. Its attachment to the liver is to the right of the median plane; it resists displacement of the liver to the right.

- The gallbladder projects inferior to the sharp, inferior border of the liver.
- The internal oblique muscle is the thickest of the three flat abdominal muscles.
- The costodiaphragmatic recesses (right and left) of the pleural cavities extend between the diaphragm and superior abdominal viscera and the body wall.
- Two thirds of the pericardial sac lies to the left of the median plane; its apex, the lowest and leftmost point, overlies the stomach.

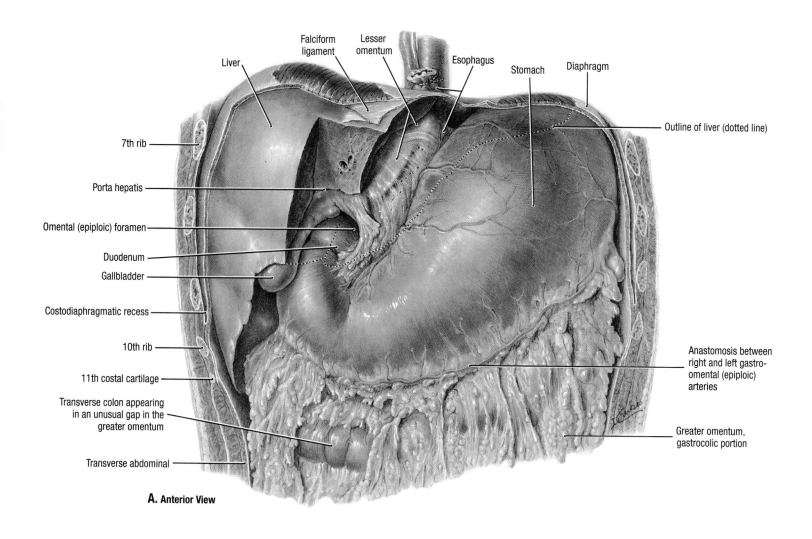

Liver — Falciform ligament — Lesser omentum — Esophagus — Stomach — Diaphragm

7th rib —

Porta hepatis —

Omental (epiploic) foramen —

Duodenum —

Gallbladder —

Costodiaphragmatic recess —

10th rib —

11th costal cartilage —

Transverse colon appearing in an unusual gap in the greater omentum —

Transverse abdominal —

Outline of liver (dotted line)

Anastomosis between right and left gastro-omental (epiploic) arteries

Greater omentum, gastrocolic portion

A. Anterior View

2.20   **Stomach and omenta**

A. Lesser and greater omenta. The stomach is inflated with air, and the left part of the liver is cut away, but the outline of the liver is demarcated with a *dotted line*. The gallbladder, followed superiorly, leads to the free margin of the lesser omentum and serves as a guide to the epiploic foremen, which lies posterior to that free margin. The lesser omentum, thickened at its free margin but thin elsewhere, has many perforations, and the caudate lobe of the liver is visible through it. **B.** Lesser omentum. Two sagittal cuts have been made through the liver: one at the fissure for the ligamentum venosum and the other at the right limit of the porta hepatis. These two cuts have been joined by a coronal cut. Note that the lesser omentum may be regarded as the "mesentery" of the bile passages because they occupy its free edge.

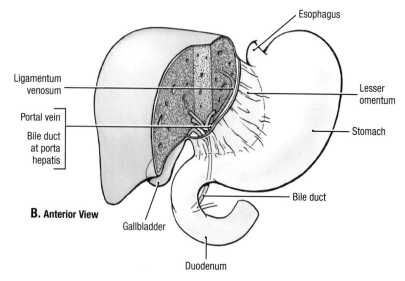

Esophagus

Ligamentum venosum

Portal vein

Bile duct at porta hepatis

Lesser omentum

Stomach

Bile duct

B. Anterior View

Gallbladder

Duodenum

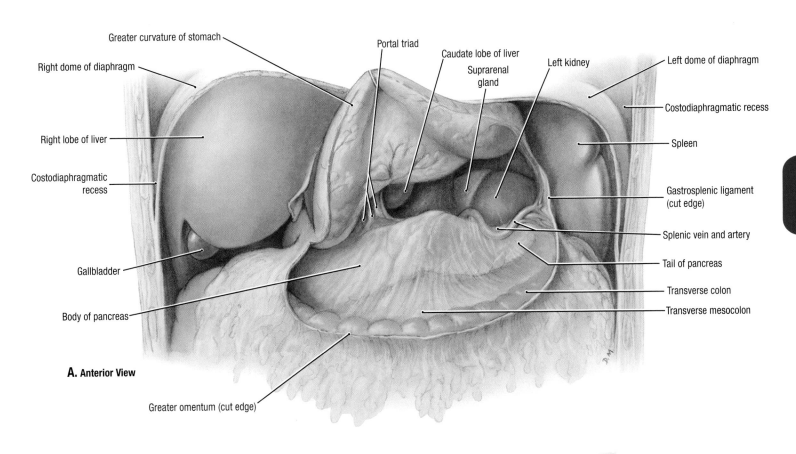

Greater curvature of stomach

Right dome of diaphragm

Right lobe of liver

Costodiaphragmatic recess

Gallbladder

Body of pancreas

Greater omentum (cut edge)

Portal triad

Caudate lobe of liver

Suprarenal gland

Left kidney

Left dome of diaphragm

Costodiaphragmatic recess

Spleen

Gastrosplenic ligament (cut edge)

Splenic vein and artery

Tail of pancreas

Transverse colon

Transverse mesocolon

A. Anterior View

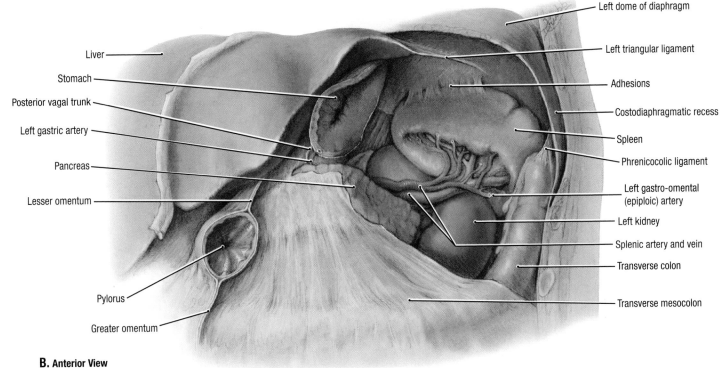

Liver

Stomach

Posterior vagal trunk

Left gastric artery

Pancreas

Lesser omentum

Pylorus

Greater omentum

Left dome of diaphragm

Left triangular ligament

Adhesions

Costodiaphragmatic recess

Spleen

Phrenicocolic ligament

Left gastro-omental (epiploic) artery

Left kidney

Splenic artery and vein

Transverse colon

Transverse mesocolon

B. Anterior View

2.21   **Posterior relationships of omental bursa**

A. Opened omental bursa. The greater omentum and gastro-splenic ligament have been cut along the greater curvature of the stomach, and the stomach is reflected superiorly. **B.** Stomach bed. The stomach is excised. The peritoneum of the omental bursa cov-ering the stomach bed is largely removed, as is the peritoneum of the greater sac covering the inferior part of the kidney and pan-creas. The pancreas is unusually short, and the adhesions binding the spleen to diaphragm are pathological, but not unusual.

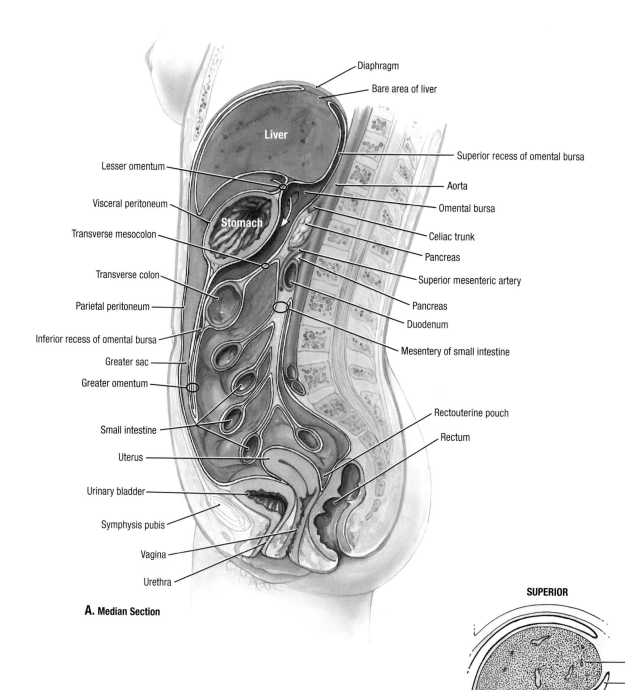

A. Median Section

Diaphragm

Bare area of liver

Liver

Superior recess of omental bursa

Lesser omentum

Aorta

Visceral peritoneum

Omental bursa

Stomach

Celiac trunk

Transverse mesocolon

Pancreas

Transverse colon

Superior mesenteric artery

Parietal peritoneum

Pancreas

Inferior recess of omental bursa

Duodenum

Greater sac

Mesentery of small intestine

Greater omentum

Small intestine

Rectouterine pouch

Uterus

Rectum

Urinary bladder

Symphysis pubis

Vagina

Urethra

SUPERIOR

Liver

Superior recess of omental bursa

Lesser omentum

Omental (epiploic) foramen

ANTERIOR

POSTERIOR

Pancreas

Stomach

Omental bursa (lesser sac)

Transverse mesocolon

Transverse colon

Inferior recess of omental bursa (fused)

Greater omentum

B. Median Section

INFERIOR

2.22   Peritoneal cavity

A. Greater sac and omental bursa (lesser sac). The peritoneal cavity consists of the greater sac and omental bursa. The superior recess of the omental bursa is between the liver and the posterior attachment of the diaphragm. The inferior recess of the omental bursa is between the two double layers of the greater omentum. In the adult, the inferior recess usually only extends inferiorly as far as the transverse colon because of fusion of the two double peritoneal layers at birth. **B.** Omental bursa, schematic sagittal section. In **A** and **B,** the *arrow* passes from the greater sac through the omental foramen into the omental bursa (lesser sac).

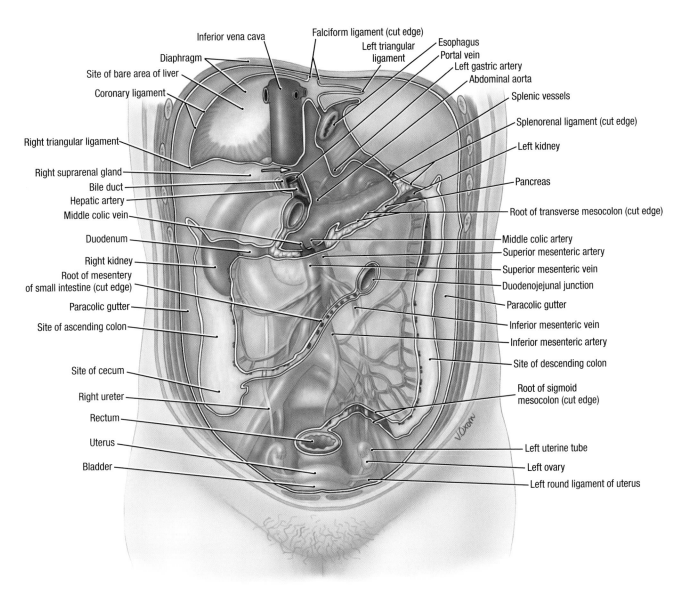

A. Anterior View

2.23   Posterior wall of the peritoneal cavity

A. Roots of the peritoneal reflections. **B.** Omental bursa, schematic transverse section. The gastrosplenic and splenorenal ligaments tether the spleen in place between the stomach and the kidney; the ligaments form a pedicle (stalk), through which blood vessels run to and from the hilum of the spleen. These ligaments are double layers of peritoneum that form the left boundary of the omental bursa (lesser sac); the inner layer consists of peritoneum lining the omental bursa, and the outer layer consists of peritoneum lining the peritoneal cavity (greater sac).

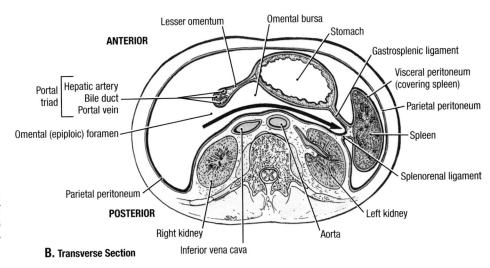

B. Transverse Section

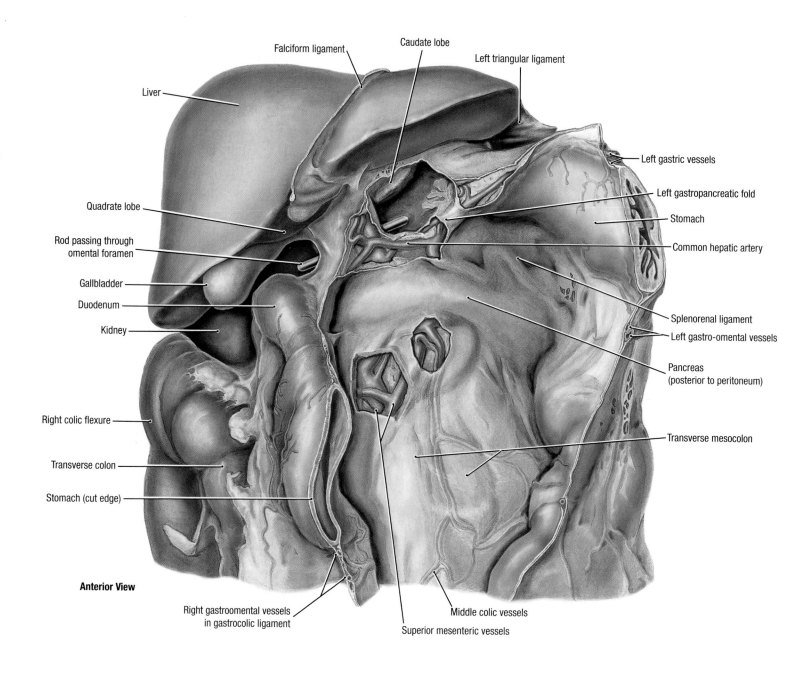

Falciform ligament

Caudate lobe

Left triangular ligament

Liver

Left gastric vessels

Quadrate lobe

Left gastropancreatic fold

Stomach

Rod passing through omental foramen

Common hepatic artery

Gallbladder

Duodenum

Kidney

Splenorenal ligament

Left gastro-omental vessels

Pancreas (posterior to peritoneum)

Right colic flexure

Transverse mesocolon

Transverse colon

Stomach (cut edge)

Anterior View

Right gastroomental vessels in gastrocolic ligament

Middle colic vessels

Superior mesenteric vessels

2.24 Omental bursa, opened

The anterior wall of the bursa, consisting of the stomach with its two omenta and the vessels along its curvatures, has been sectioned sagittally; the two parts have been turned to the left and right, the body of the stomach on the left side, and the pyloric part and the first part of the duodenum on the right. Note that the *rod* passes through the omental foramen into the omental bursa. The right kidney forms the posterior wall of the hepatorenal pouch, and the pancreas lies somewhat horizontally on the posterior wall of the bursa.

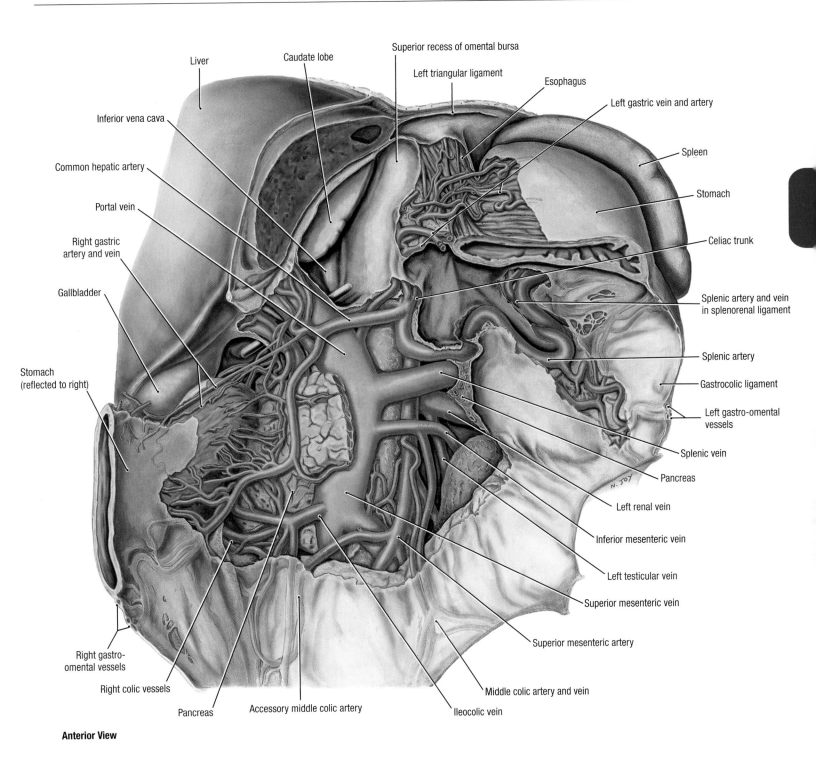

Liver

Caudate lobe

Superior recess of omental bursa

Left triangular ligament

Esophagus

Left gastric vein and artery

Inferior vena cava

Spleen

Common hepatic artery

Stomach

Portal vein

Celiac trunk

Right gastric
artery and vein

Splenic artery and vein
in splenorenal ligament

Gallbladder

Splenic artery

Gastrocolic ligament

Stomach
(reflected to right)

Left gastro-omental
vessels

Splenic vein

Pancreas

Left renal vein

Inferior mesenteric vein

Left testicular vein

Superior mesenteric vein

Right gastro-
omental vessels

Superior mesenteric artery

Right colic vessels

Middle colic artery and vein

Pancreas

Accessory middle colic artery

Ileocolic vein

Anterior View

2.25   **Posterior wall of omental bursa**

The peritoneum of the posterior wall has been mostly removed, and a section of the pancreas has been excised. The *rod* passes through the omental foramen.

- The esophageal branches of the left gastric vessels are applied to the esophagus.
- Note the celiac trunk, giving rise to the left gastric artery that arches superiorly, the splenic artery that runs tortuously to the left, and the common hepatic artery that runs to the right, passing anterior to the portal vein.
- The portal vein is formed posterior to the neck of the pancreas by the union of the superior mesenteric and splenic veins, with the inferior mesenteric vein joining at or near the angle of union.

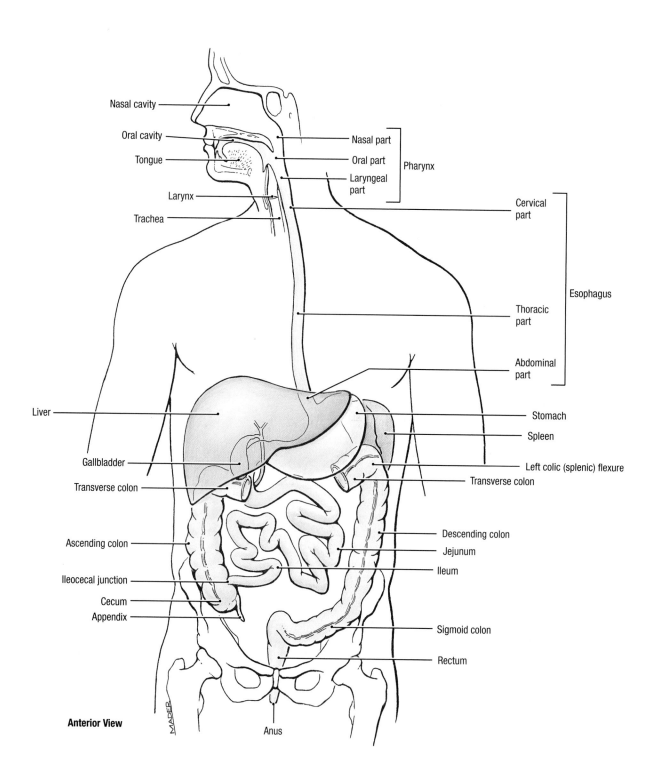

Nasal cavity

Oral cavity

Tongue

Larynx

Trachea

Nasal part

Oral part

Laryngeal part

Pharynx

Cervical part

Esophagus

Thoracic part

Abdominal part

Liver

Gallbladder

Transverse colon

Ascending colon

Ileocecal junction

Cecum

Appendix

Stomach

Spleen

Left colic (splenic) flexure

Transverse colon

Descending colon

Jejunum

Ileum

Sigmoid colon

Rectum

Anterior View

Anus

2.26 **Digestive system**

The head is sagittally sectioned and turned laterally. The digestive system extends from the lips to the anus and consists of the oral cavity, pharynx, esophagus, stomach, small intestine (duodenum, jejunum, ileum), and large intestine (appendix; cecum; ascending, transverse, descending, and sigmoid colon; rectum; and anal canal). Associated organs include the liver, gallbladder, and pancreas.

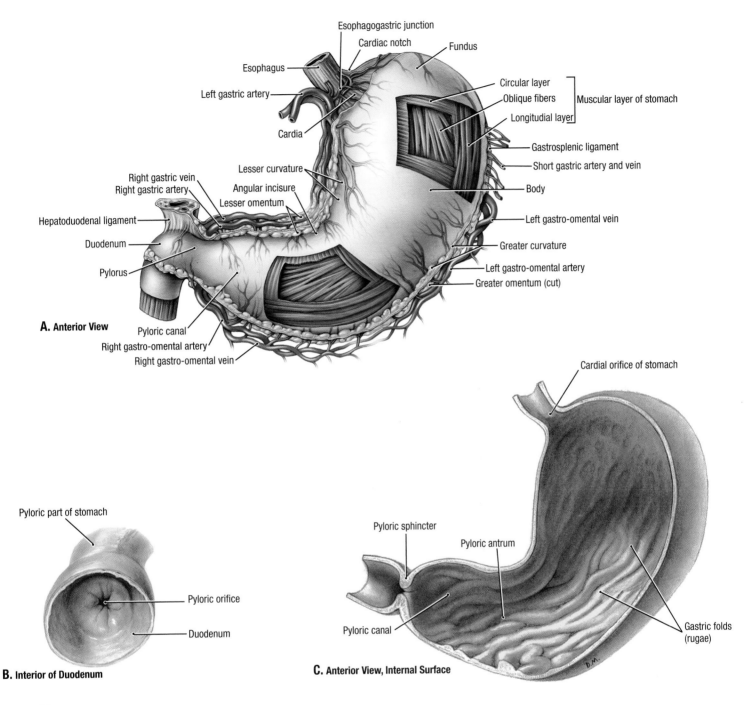

A. Anterior View

B. Interior of Duodenum

C. Anterior View, Internal Surface

2.27   Stomach

A. External surface. **B.** Pylorus, viewed from the duodenum. **C.** Internal surface (mucous membrane), anterior wall removed.

- The esophagogastric junction is usually located to the left of the midline, at the 11th thoracic vertebra, and the pylorus is usually located at the L1 vertebral level to the right of the midline at the transpyloric plane; the angular incisure separates the body from the pyloric region of the stomach **(A).**
- The lesser omentum attaches to the lesser curvature of the stomach and contains the right and left gastric vessels. The continuous greater omentum and gastrosplenic ligament attach to the greater curvature of the stomach. The right and left gastroomental vessels follow the greater curvature **(A).**

- The muscular layer of the stomach consists of longitudinal and circular layers and oblique fibers **(A).**
- The pylorus projects into the 1st (superior) part of the duodenum. The first 4 cm of the duodenum has no circular folds (plicae circulares), but the mucous membrane may be ridged **(B).**
- When the stomach is contracted the gastric mucosa is thrown into longitudinal ridges called *gastric folds (rugae)*, best seen along the greater curvature and pylorus; the pyloric sphincter is a ring of circular muscle at the junction of the stomach and duodenum **(C).**

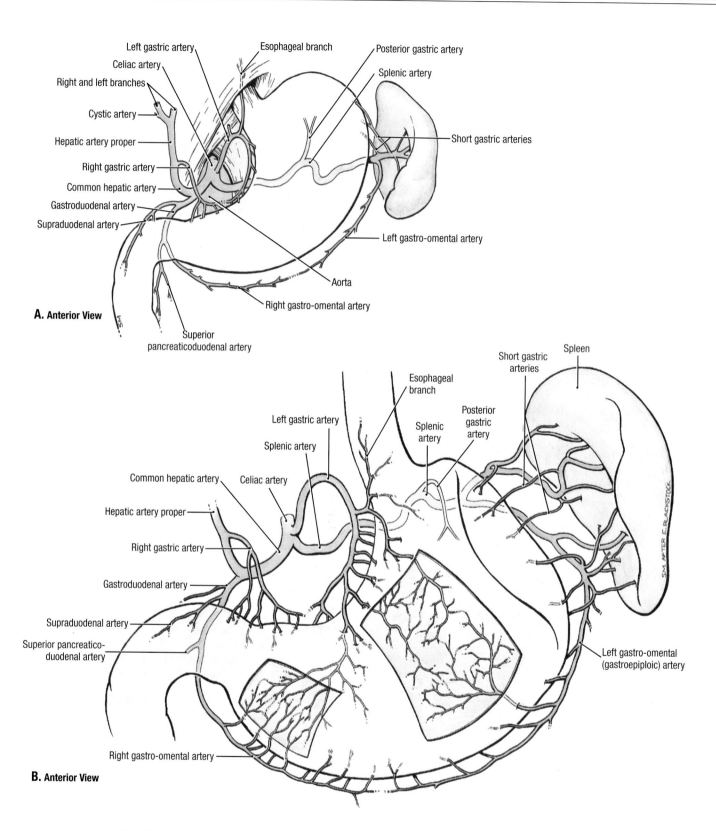

A. Anterior View

Left gastric artery
Celiac artery
Right and left branches
Cystic artery
Hepatic artery proper
Right gastric artery
Common hepatic artery
Gastroduodenal artery
Supraduodenal artery
Esophageal branch
Posterior gastric artery
Splenic artery
Short gastric arteries
Left gastro-omental artery
Aorta
Right gastro-omental artery
Superior pancreaticoduodenal artery

B. Anterior View

Common hepatic artery
Hepatic artery proper
Right gastric artery
Gastroduodenal artery
Supraduodenal artery
Superior pancreatico-duodenal artery
Right gastro-omental artery
Celiac artery
Left gastric artery
Splenic artery
Esophageal branch
Splenic artery
Posterior gastric artery
Short gastric arteries
Spleen
Left gastro-omental (gastroepiploic) artery

2.28 Celiac artery

A. Branches of the celiac artery. The celiac artery is a branch of the abdominal aorta, arising from immediately inferior to the aortic hiatus of the diaphragm. The artery is usually 1 to 2 cm long and divides into the left gastric, common hepatic, and splenic arteries. **B.** Arteries of stomach and spleen. The serous and muscular coats are removed from two areas of the stomach, revealing the anastomotic networks in the submucous coat.

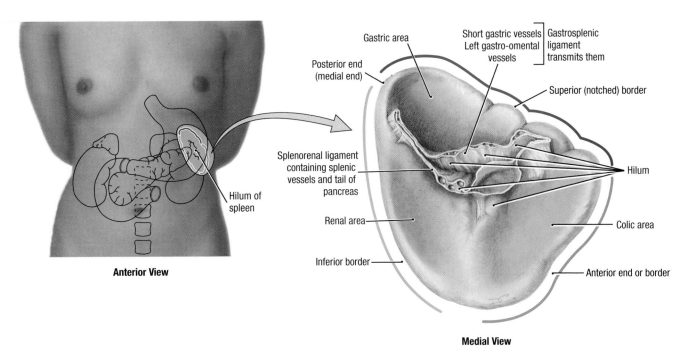

Anterior View

Gastric area

Posterior end (medial end)

Short gastric vessels · Gastrosplenic
Left gastro-omental } ligament
vessels · transmits them

Superior (notched) border

Splenorenal ligament
containing splenic
vessels and tail of
pancreas

Hilum

Renal area

Colic area

Inferior border

Anterior end or border

Medial View

2.29   Visceral surface of the spleen

Left: The surface anatomy of the spleen. *Right:* Note the impressions of structures in contact with the spleen.

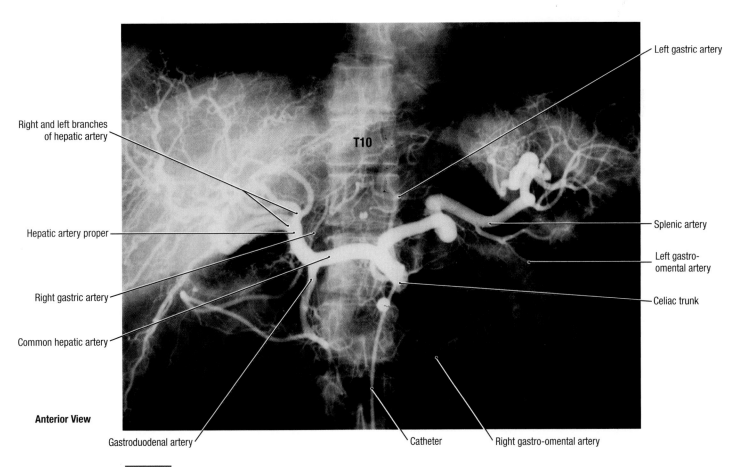

Left gastric artery

Right and left branches
of hepatic artery

T10

Hepatic artery proper

Splenic artery

Right gastric artery

Left gastro-
omental artery

Common hepatic artery

Celiac trunk

Anterior View

Gastroduodenal artery

Catheter

Right gastro-omental artery

2.30   Celiac arteriogram

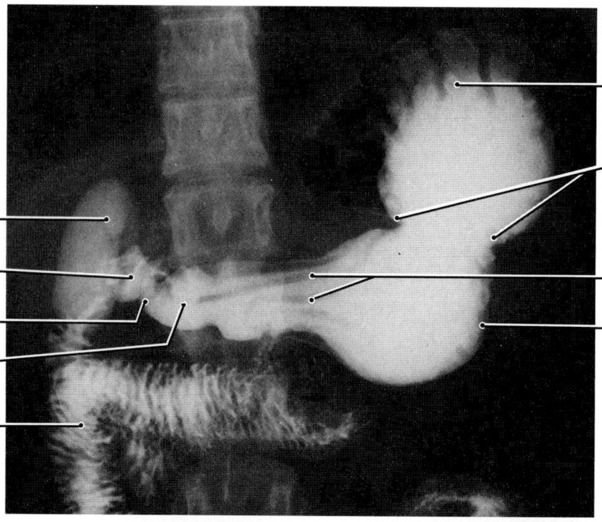

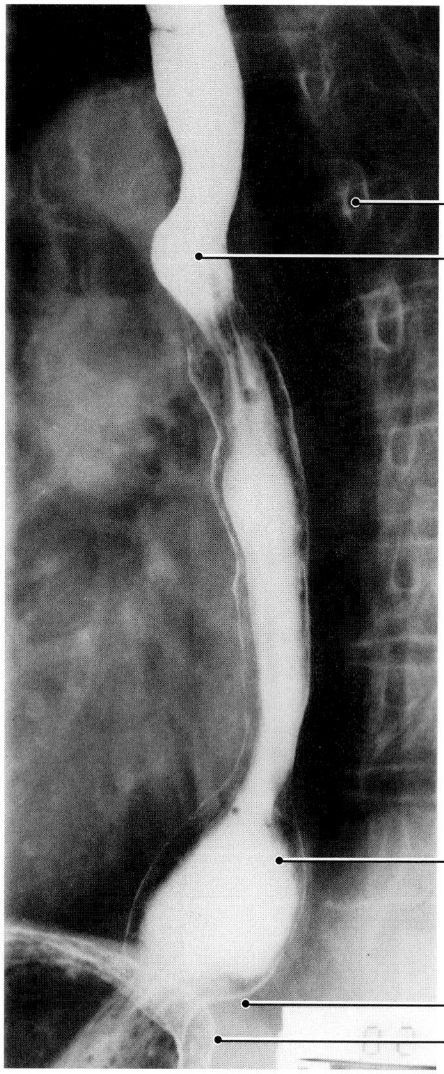

A. Lateral View

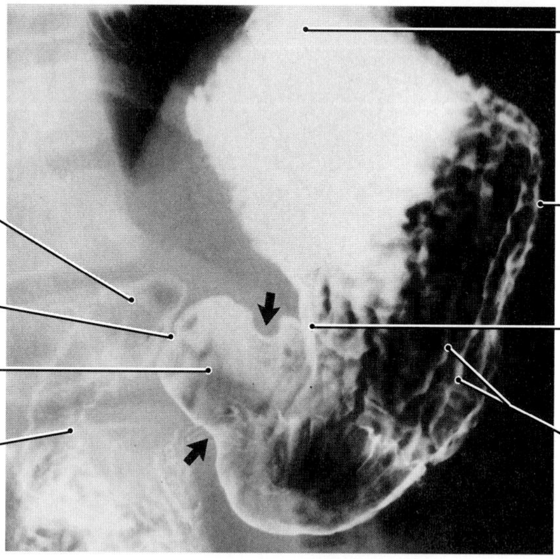

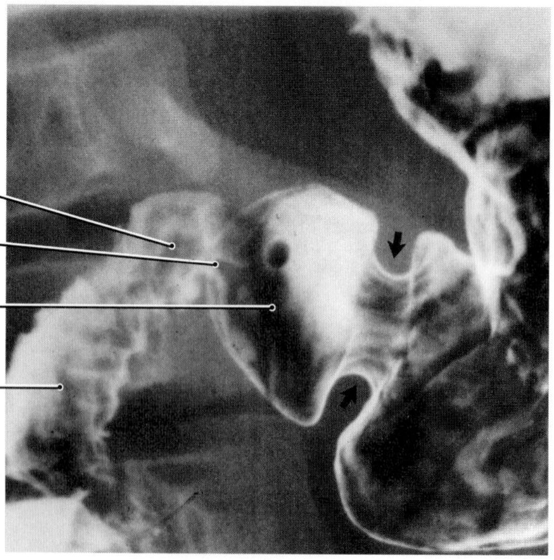

B

- Fundus of stomach
- Peristaltic wave
- Gastric folds
- Greater curvature
- Gallbladder
- Duodenal cap
- Pylorus
- Pyloric antrum
- Jejunum

Transverse process

Esophagus

Phrenic ampulla (seen only radiologically)

Diaphragm

Stomach

C

- Fundus
- Greater curvature
- Angular incisure
- Gastric folds
- Duodenal cap
- Pylorus
- Pyloric antrum
- Duodenum

D

- Duodenal cap
- Pylorus
- Pyloric antrum
- Duodenum

Anterior Views (B–D)

2.31 Radiographs of esophagus, stomach, small intestine (barium swallow)

A. Esophagus. The esophageal (phrenic) ampulla is the distensible portion of the esophagus seen only radiologically. **B.** Stomach, small intestine, and gallbladder. Note additional contrast in gallbladder. **C.** Stomach and small intestine. **D.** Pyloric antrum and duodenal cap.

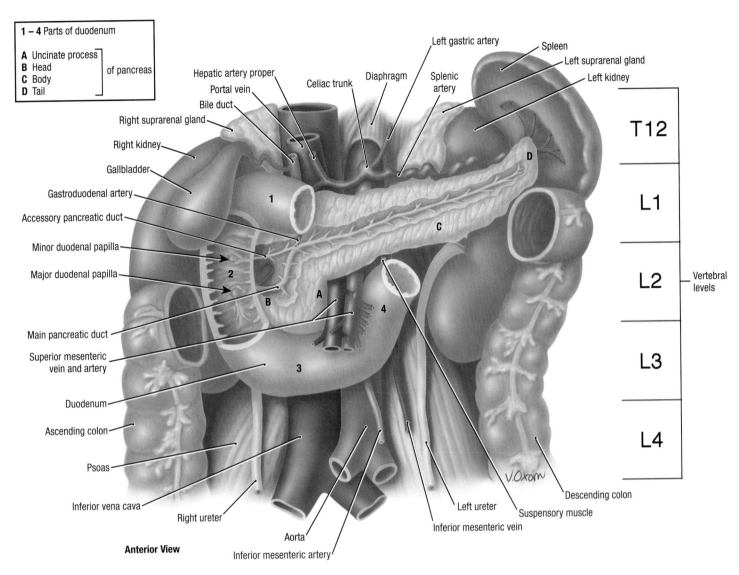

1 – 4 Parts of duodenum

A Uncinate process
B Head
C Body
D Tail
  of pancreas

Left gastric artery
Spleen
Left suprarenal gland
Left kidney

Hepatic artery proper
Celiac trunk
Diaphragm
Splenic artery
Portal vein
Bile duct
Right suprarenal gland
Right kidney
Gallbladder
Gastroduodenal artery
Accessory pancreatic duct
Minor duodenal papilla
Major duodenal papilla
Main pancreatic duct
Superior mesenteric vein and artery
Duodenum
Ascending colon
Psoas
Inferior vena cava
Right ureter
Aorta
Inferior mesenteric artery
Left ureter
Inferior mesenteric vein
Suspensory muscle
Descending colon

T12
L1
L2
L3
L4

Vertebral levels

V.Oxorn

Anterior View

2.32   **Parts and relationships of pancreas and duodenum**

TABLE 2.3   PARTS AND RELATIONSHIPS OF DUODENUM

PART OF DUODENUM	ANTERIOR	POSTERIOR	MEDIAL	SUPERIOR	INFERIOR	VERTEBRAL LEVEL
Superior (1st part)	Peritoneum Gallbladder Quadrate lobe	Bile duct Gastroduodenal artery Portal vein IVC		Neck of gallbladder	Neck of pancreas	Anterolateral to L1 vertebra
Descending (2nd part)	Transverse colon Transverse mesocolon Coils of small intestine	Hilum of right kidney Renal vessels Ureter Psoas major	Head of pancreas Pancreatic duct Bile duct			Right of L2–L3 vertebrae
Horizontal (3rd part)	SMA SMV Coils of small intestine	Right psoas major IVC Aorta Right ureter		Head and uncinate process of pancreas Superior mesenteric vessels		Anterior to L3 vertebra
Ascending (4th part)	Beginning of root of mesentery Coils of jejunum	Left psoas major Left margin of aorta	Head of pancreas	Body of pancreas		Left of L3 vertebra

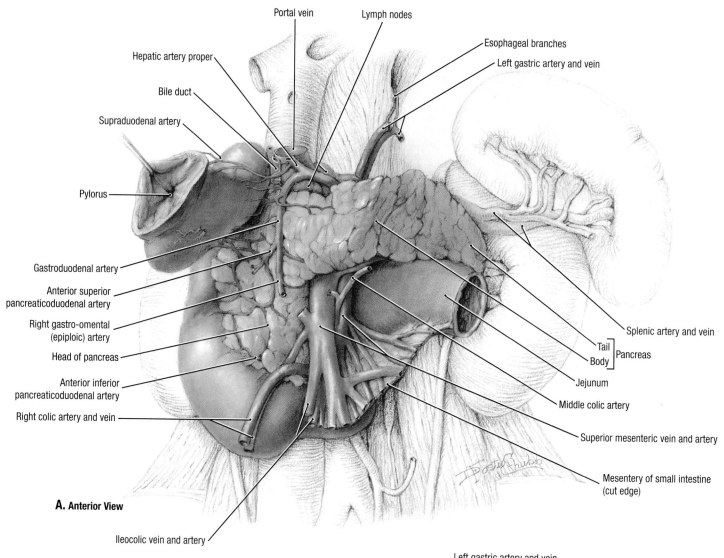

Portal vein
Lymph nodes
Hepatic artery proper
Esophageal branches
Bile duct
Left gastric artery and vein
Supraduodenal artery
Pylorus
Gastroduodenal artery
Anterior superior
pancreaticoduodenal artery
Right gastro-omental
(epiploic) artery
Head of pancreas
Splenic artery and vein
Tail
Body   } Pancreas
Anterior inferior
pancreaticoduodenal artery
Jejunum
Right colic artery and vein
Middle colic artery
Superior mesenteric vein and artery
Mesentery of small intestine
(cut edge)

A. Anterior View

Ileocolic vein and artery

2.33   Duodenum and pancreas

A. Duodenum and pancreas in situ. The duodenum is molded
around the head of the pancreas; its 1st, or superior, part (retracted)
passes posteriorly, superiorly, and to the right. The remaining parts
(2nd, 3rd, and 4th) overlap by the pancreas; near the junction of its
3rd and 4th parts, the duodenum is crossed by the superior mesenteric vessels, which descend anterior to the uncinate process as it
enters the root of the mesentery. The tail of the pancreas, here
short, usually abuts on the spleen. **B.** Posterior surface of duodenum, pancreas, and bile duct. Only the end of the 1st part of the
duodenum is in view; the bile duct descends in a fissure (opened up)
in the posterior part of the head of the pancreas.

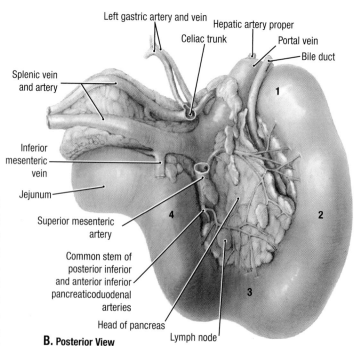

Left gastric artery and vein
Hepatic artery proper
Celiac trunk
Portal vein
Bile duct
Splenic vein
and artery
1
Inferior
mesenteric
vein
Jejunum
Superior mesenteric
artery
4
2
Common stem of
posterior inferior
and anterior inferior
pancreaticoduodenal
arteries
3
Head of pancreas
Lymph node

B. Posterior View

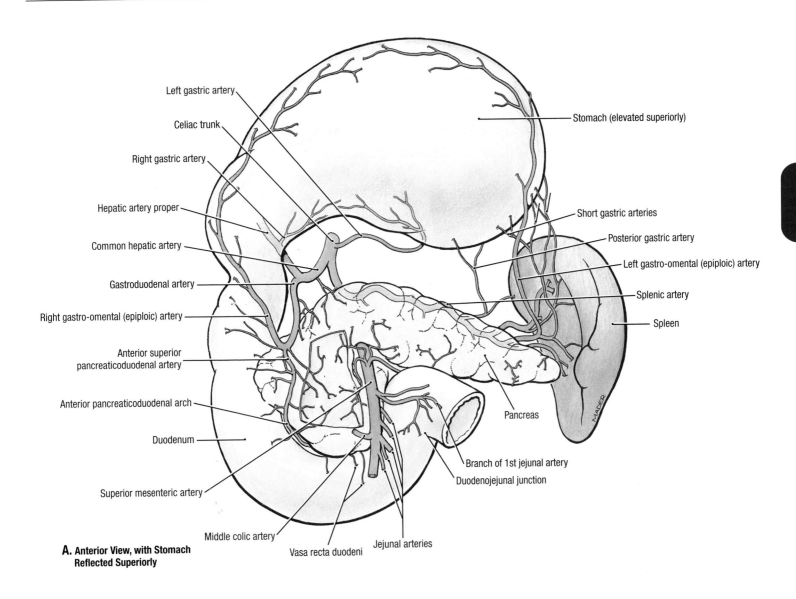

Left gastric artery

Celiac trunk

Right gastric artery

Hepatic artery proper

Common hepatic artery

Gastroduodenal artery

Right gastro-omental (epiploic) artery

Anterior superior
pancreaticoduodenal artery

Anterior pancreaticoduodenal arch

Duodenum

Superior mesenteric artery

Middle colic artery

Vasa recta duodeni

Jejunal arteries

Stomach (elevated superiorly)

Short gastric arteries

Posterior gastric artery

Left gastro-omental (epiploic) artery

Splenic artery

Spleen

Pancreas

Branch of 1st jejunal artery

Duodenojejunal junction

**A. Anterior View, with Stomach
Reflected Superiorly**

2.34   **Blood supply to the pancreas,
duodenum, and spleen**

A. Celiac trunk and superior mesenteric artery. **B.** Pancreatic and pancreaticoduodenal arteries.

- The anterior superior pancreaticoduodenal branch of the gastroduodenal artery and the anterior inferior pancreaticoduodenal branch of the superior mesenteric artery form the anterior pancreaticoduodenal arch anterior to the head of the pancreas. The posterior superior and posterior inferior branches of the same two arteries form the posterior pancreaticoduodenal arch posterior to the pancreas. The anterior and posterior inferior arteries arise from a common stem.
- From each arch, vasa recta duodeni pass to the anterior and posterior surfaces of the 2nd, 3rd, and 4th parts of the duodenum.
- Arteries supplying the pancreas are derived from the common hepatic artery, gastroduodenal artery, pancreaticoduodenal arches, splenic artery, and superior mesenteric artery.

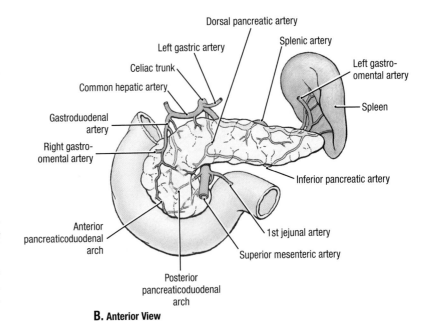

Dorsal pancreatic artery

Splenic artery

Left gastric artery

Celiac trunk

Common hepatic artery

Gastroduodenal artery

Right gastro-omental artery

Left gastro-omental artery

Spleen

Inferior pancreatic artery

1st jejunal artery

Superior mesenteric artery

Anterior pancreaticoduodenal arch

Posterior pancreaticoduodenal arch

B. Anterior View

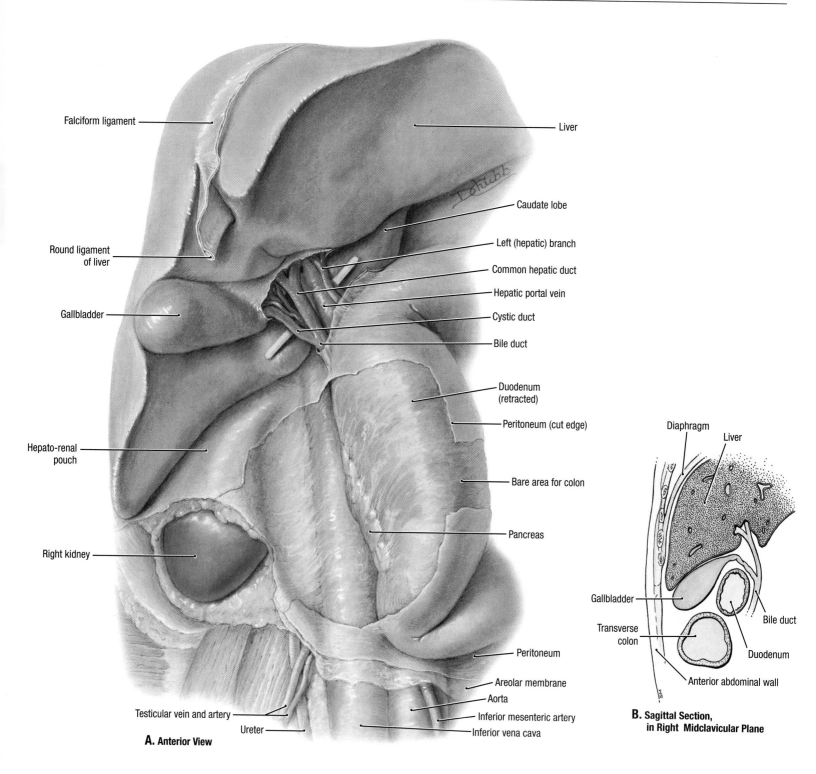

Falciform ligament

Liver

Caudate lobe

Left (hepatic) branch

Common hepatic duct

Round ligament of liver

Hepatic portal vein

Cystic duct

Gallbladder

Bile duct

Duodenum (retracted)

Peritoneum (cut edge)

Hepato-renal pouch

Bare area for colon

Pancreas

Right kidney

Peritoneum

Areolar membrane

Aorta

Testicular vein and artery

Inferior mesenteric artery

Ureter

Inferior vena cava

A. Anterior View

Diaphragm

Liver

Gallbladder

Transverse colon

Bile duct

Duodenum

Anterior abdominal wall

B. Sagittal Section, in Right Midclavicular Plane

2.35   Exposure of the portal triad—I

A. Dissection. The portal triad typically consists of the portal vein (posteriorly), the hepatic artery proper (ascending from the left), and the bile passages (descending to the right). In this specimen, the hepatic artery proper is replaced by a left hepatic artery, arising directly from the common hepatic artery, and a right hepatic artery, arising from the superior mesenteric artery (a common variation). A *rod* is passed through the omental (epiploic) foramen. The lesser omentum and transverse colon are removed, and the peritoneum is cut along the right border of the duodenum; this part of the duodenum is retracted anteriorly. The space opened up reveals two smooth areolar membranes applied to each other; one membrane covers the posterior aspect of the second part of the duodenum and the head of the pancreas, and the other covers the aorta, inferior vena cava, renal vessels, and perirenal fat. **B.** Relations of gallbladder, schematic sagittal section. Note the gallbladder contacts the visceral surface of the liver, the transverse colon, and the superior part of the duodenum.

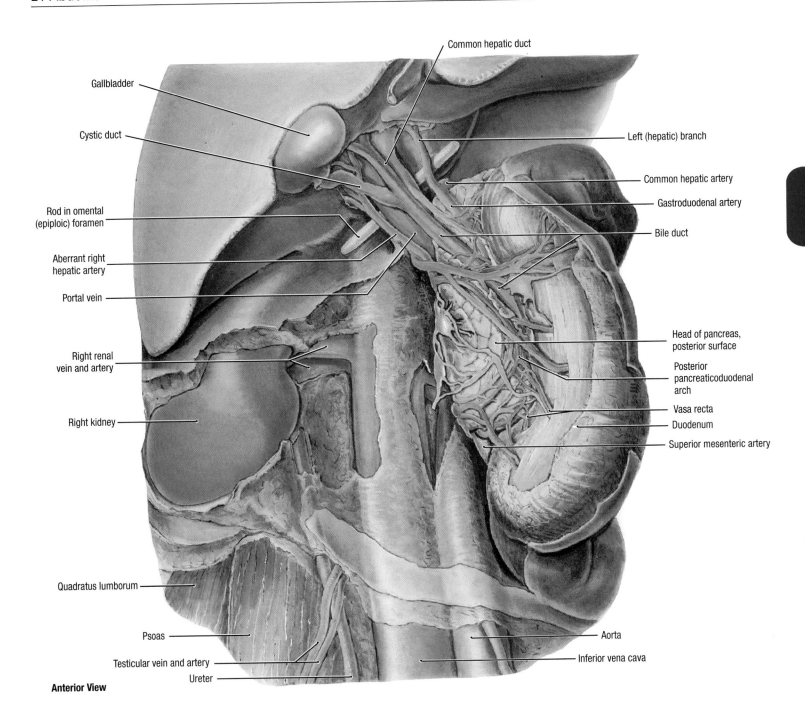

Common hepatic duct

Gallbladder

Cystic duct

Left (hepatic) branch

Common hepatic artery

Gastroduodenal artery

Rod in omental
(epiploic) foramen

Bile duct

Aberrant right
hepatic artery

Portal vein

Head of pancreas,
posterior surface

Right renal
vein and artery

Posterior
pancreaticoduodenal
arch

Vasa recta

Right kidney

Duodenum

Superior mesenteric artery

Quadratus lumborum

Psoas

Aorta

Testicular vein and artery

Inferior vena cava

Ureter

Anterior View

2.36   Exposure of the portal triad—II

In this dissection, the duodenum is retracted anteriorly and to the left, taking the head of the pancreas with it. In effect, the omental foramen has been enlarged inferiorly. The areolar membrane covering the pancreas and duodenum is largely removed, and that covering the great vessels is partly removed. Of the two posterior pancreaticoduodenal arteries that form the posterior arch, the inferior arises from the superior mesenteric artery and, in this specimen, the superior from the right hepatic artery; usually the superior arises from the gastroduodenal artery. The posterior superior pancreaticoduodenal vein drains into the portal vein. Note the close relationship of the inferior vena cava to the portal vein; they are separated by the omental foramen. A portacaval shunt to divert the portal circulation into the caval system may be performed by an end-to-side anastomosis at this site.

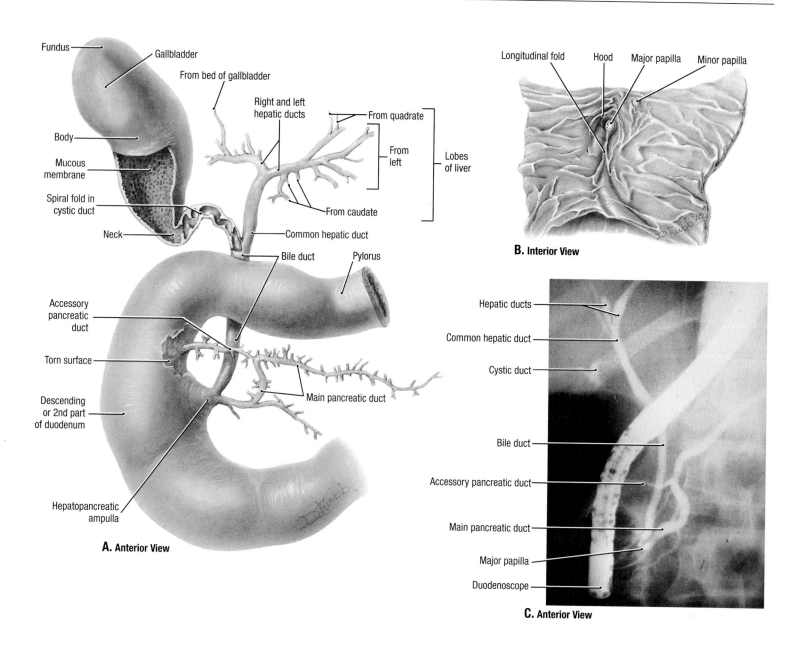

A. Anterior View

Fundus
Gallbladder
From bed of gallbladder
Right and left hepatic ducts
From quadrate
From left
Lobes of liver
Body
Mucous membrane
Spiral fold in cystic duct
From caudate
Neck
Common hepatic duct
Bile duct
Pylorus
Accessory pancreatic duct
Torn surface
Main pancreatic duct
Descending or 2nd part of duodenum
Hepatopancreatic ampulla

B. Interior View

Longitudinal fold
Hood
Major papilla
Minor papilla

C. Anterior View

Hepatic ducts
Common hepatic duct
Cystic duct
Bile duct
Accessory pancreatic duct
Main pancreatic duct
Major papilla
Duodenoscope

2.37 Bile and pancreatic ducts

A. Extrahepatic bile passages and pancreatic ducts. The mucous membrane of the gallbladder has a low, honeycomb surface, whereas the cystic duct is sinuous, with its mucous membrane forming a spiral fold (spiral valve). The right and left hepatic ducts collect bile from the liver; the common hepatic duct unites with the cystic duct just superior to the duodenum to form the bile duct. After descending posterior to the 1st part of the duodenum, the bile duct and the accessory pancreatic duct are joined by the main pancreatic duct; these open on the duodenal papilla. **B.** Interior of the second part of the duodenum. Observe the larger duodenal papilla, projecting into the duodenum approximately 9 cm from the pylorus. A small duodenal papilla of the accessory pancreatic duct lies just anterosuperior to it. **C.** Endoscopic retrograde cholangiography and pancreatography (ERCP) of the bile and pancreatic ducts.

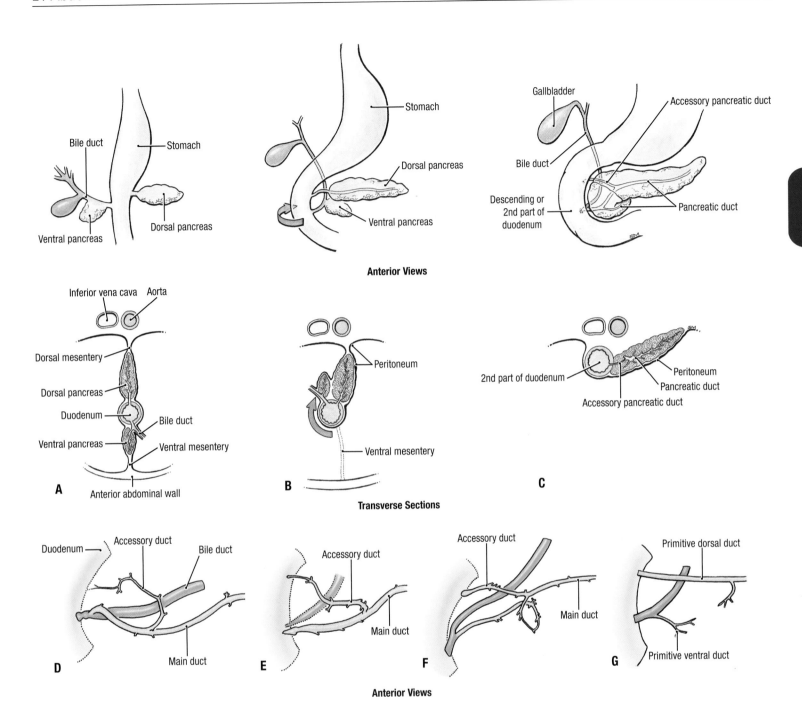

2.38   Development and variability of the pancreatic ducts

A–C. Anterior views *(top)* and transverse sections *(bottom)* of the stages in the development of the pancreas. **A.** The small, primitive ventral bud arises in common with the bile duct, and a larger, primitive dorsal bud arises independently from the duodenum. **B.** The 2nd, or descending, part of the duodenum rotates on its long axis, which brings the ventral bud and bile duct posterior to the dorsal bud. **C.** A connecting segment unites the dorsal duct to the ventral duct, whereupon the duodenal end of the dorsal duct atrophies, and the direction of flow within it is reversed. **D–G.** Common variations of the pancreatic duct. **D.** An accessory duct that has lost its connection with the duodenum. **E.** An accessory duct that is large enough to relieve an obstructed main duct. **F.** An accessory duct that could probably substitute for the main duct. **G.** A persisting primitive dorsal duct unconnected to the primitive ventral duct.

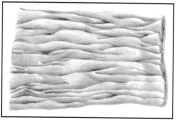

A. Proximal Jejunum

B. Proximal Ileum

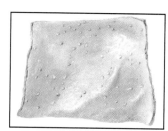

C. Distal Ileum

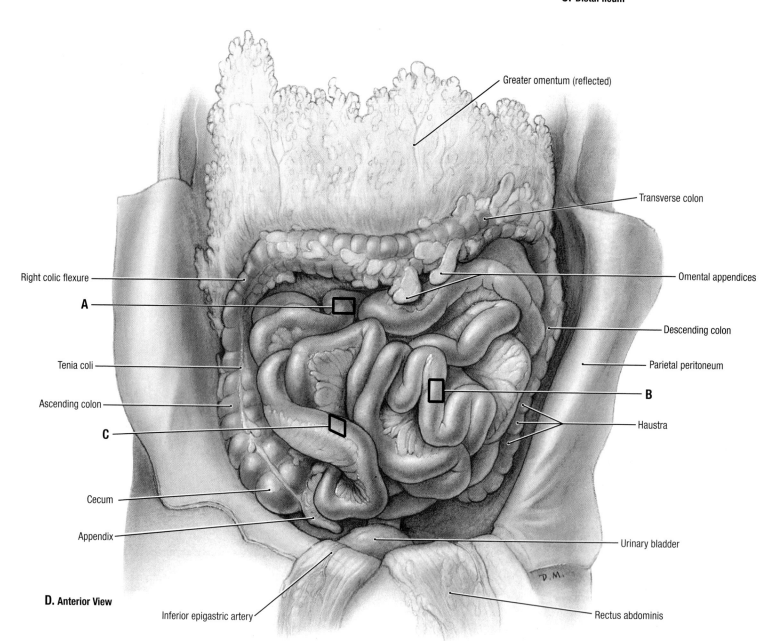

Greater omentum (reflected)

Transverse colon

Right colic flexure

A

Omental appendices

Descending colon

Tenia coli

Parietal peritoneum

Ascending colon

B

C

Haustra

Cecum

Appendix

Urinary bladder

D. Anterior View

Inferior epigastric artery

Rectus abdominis

2.39   **Intestines in situ, interior of small intestine**

A. Proximal jejunum. The circular folds (plicae circulares) are tall, closely packed, and commonly branched. **B.** Proximal ileum. The circular folds are low and becoming sparse. The caliber of the gut is reduced, and the wall is thinner. **C.** Distal ileum. Circular folds are absent, and solitary lymph nodules stud the wall. **D.** Intestines in situ, greater omentum reflected. The ileum is reflected to expose the appendix in the lower right quadrant. The appendix usu-ally lies posterior to the cecum (retrocecal) or, as in this case, projects over the pelvic brim. Note the extensive coiling of the jejunum and ileum of the small intestine (together approximately 6 m in length). Also observe the distinguishing features of the large intes-tine: its position around the small intestine; the teniae coli, or lon-gitudinal muscle bands; the sacculations, or haustra; and the omental appendices.

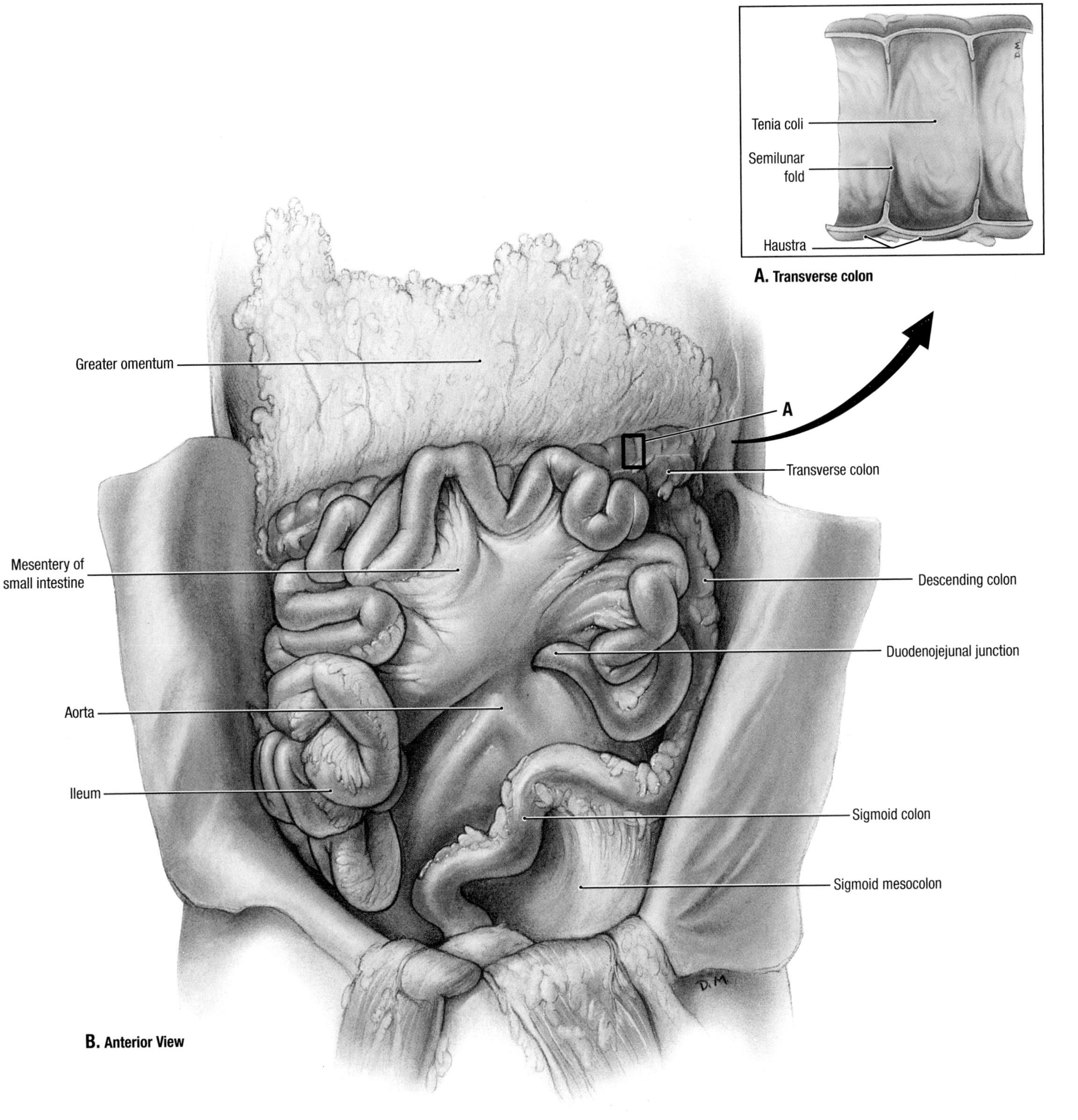

A. Transverse colon

Tenia coli
Semilunar fold
Haustra

Greater omentum

A

Transverse colon

Mesentery of small intestine

Descending colon

Duodenojejunal junction

Aorta

Ileum

Sigmoid colon

Sigmoid mesocolon

B. Anterior View

2.40 Sigmoid mesocolon and mesentry of small intestine, interior of transverse colon

A. Transverse colon. The semilunar folds and teniae coli form prominent features on the smooth-surfaced wall. **B**. Sigmoid mesocolon and mesentry of the small intestine.

- The duodenojejunal junction is situated to the left of the median plane.
- The mesentery of the small intestine fans out extensively from its short root to accommodate the length of the jejunum and ileum.

- The descending colon, the narrowest part of the large intestine, spans from the left colic flexure to the pelvic brim, where it is continuous with the sigmoid colon.
- The descending colon is retroperitoneal, but the sigmoid colon has a mesentery, the sigmoid mesocolon; the sigmoid colon is continuous with the rectum at the point at which the sigmoid mesocolon ends.

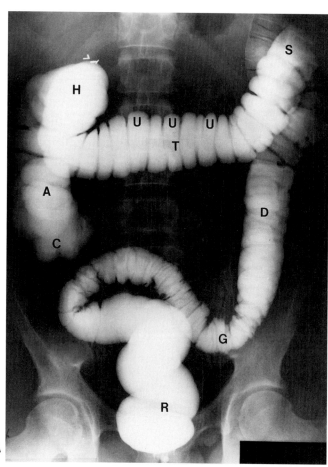

Posterioanterior Radiographs

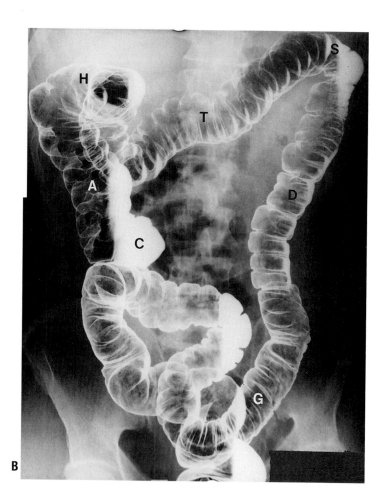

2.41 **Barium enema of the colon**

A. Single-contrast study. A barium enema has filled the colon. Observe the relative levels of the hepatic and splenic flexures. *C*, cecum; *A*, ascending colon; *H*, hepatic (right colic) flexure; *T*, transverse colon; *S*, splenic (left colic) flexure; *D*, descending colon; *G*, sigmoid colon; *R*, rectum; *U*, haustra. **B.** Double-contrast study. Barium can be seen coating the walls of the colon, which is distended with air, providing a vivid view of the mucosal relief and haustra.

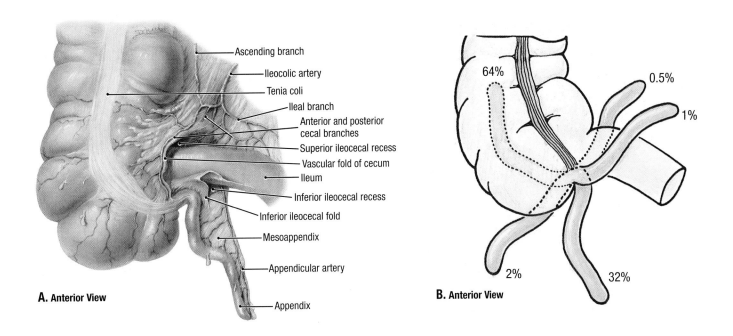

A. Anterior View

- Ascending branch
- Ileocolic artery
- Tenia coli
- Ileal branch
- Anterior and posterior cecal branches
- Superior ileocecal recess
- Vascular fold of cecum
- Ileum
- Inferior ileocecal recess
- Inferior ileocecal fold
- Mesoappendix
- Appendicular artery
- Appendix

B. Anterior View

64%   0.5%   1%   2%   32%

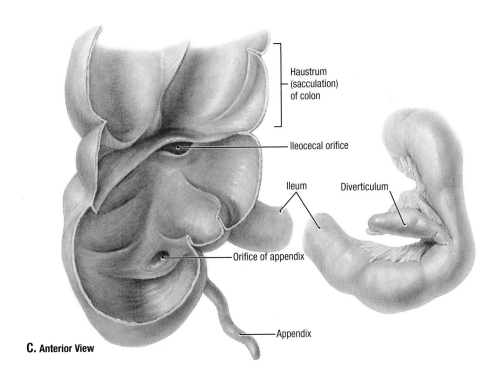

C. Anterior View

- Haustrum (sacculation) of colon
- Ileocecal orifice
- Ileum
- Diverticulum
- Orifice of appendix
- Appendix

2.42 Ileocecal region and appendix

A. Blood supply. The appendix in one free border of the mesoappendix, with the appendicular artery in the other. The inferior ileocecal fold is bloodless, whereas the superior ileocecal fold is called the *vascular fold of the cecum*. **B.** The approximate incidence of various locations of the appendix. The appendix can be long or short, and it can occupy any position consistent with its length. **C.** Interior of a dried cecum and ileal diverticulum (of Meckel). This cecum was filled with air until dry, opened, and varnished. Ileal diverticulum is a congenital anomaly that occurs in 1 to 2% of persons. It is a pouchlike remnant (3–6 cm long) of the proximal part of the yolk stalk, typically within 50 cm of the ileocecal junction. It sometimes becomes inflamed and produces pain that may mimic that produced by appendicitis.

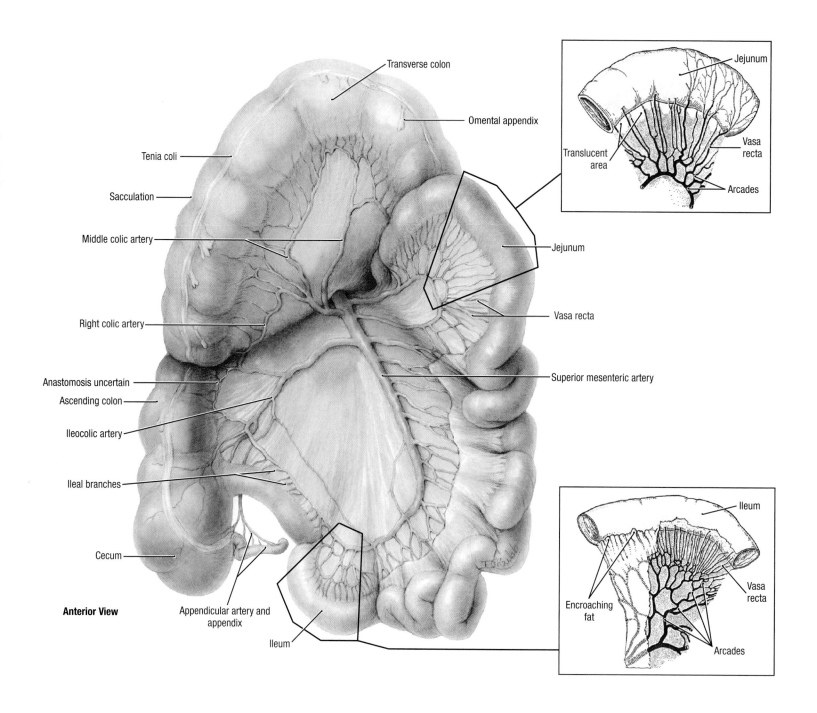

Transverse colon

Omental appendix

Tenia coli

Sacculation

Middle colic artery

Right colic artery

Anastomosis uncertain

Ascending colon

Ileocolic artery

Ileal branches

Cecum

Anterior View

Appendicular artery and appendix

Ileum

Jejunum

Vasa recta

Superior mesenteric artery

Jejunum

Translucent area

Vasa recta

Arcades

Ileum

Encroaching fat

Vasa recta

Arcades

2.43 **Superior mesenteric artery, jejunum, and ileum**

The peritoneum is partially stripped off.

- The superior mesenteric artery ends by anastomosing with one of its own branches, the ileal branch of the ileocolic artery.
- On the *inset*, drawings of jejunum and ileum compare the diameter, thickness of wall, number of arterial arcades, long or short vasa recta, presence of translucent (fat free) areas at the mesenteric border, and fat encroaching on the wall of the gut between the jejunum and ileum.

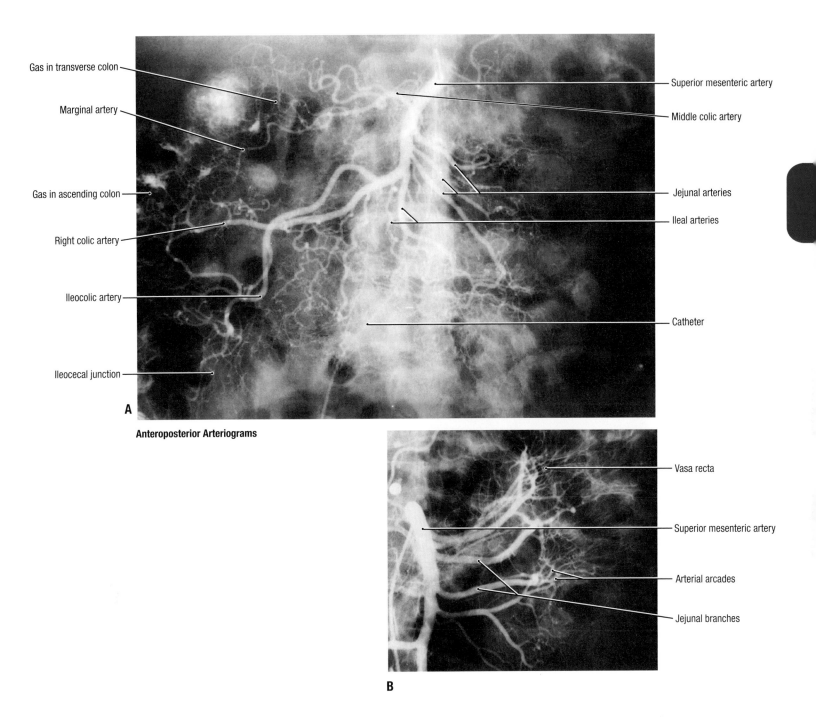

Gas in transverse colon

Marginal artery

Gas in ascending colon

Right colic artery

Ileocolic artery

Ileocecal junction

A

Superior mesenteric artery

Middle colic artery

Jejunal arteries

Ileal arteries

Catheter

Anteroposterior Arteriograms

Vasa recta

Superior mesenteric artery

Arterial arcades

Jejunal branches

B

2.44 **Superior mesenteric arteriograms**

A. Branches of superior mesenteric artery. Consult Figure 2.43 to identify the branches. **B.** Enlargement to show the jejunal branches, arterial arcades, and the vasa recta.

• The branches of the superior mesenteric artery include, from its left side, 12 or more jejunal and ileal branches that anastomose to form arcades from which vasta recta pass to the small intestine and, from its right side, the middle colic, ileocolic, and commonly (but not here) an independent right colic artery that anastomose to form a marginal artery from which vasa recta pass to the large intestine. The two inferior pancreaticoduodenal arteries arise from the main artery, either directly or in conjunction with the first jejunal branch.

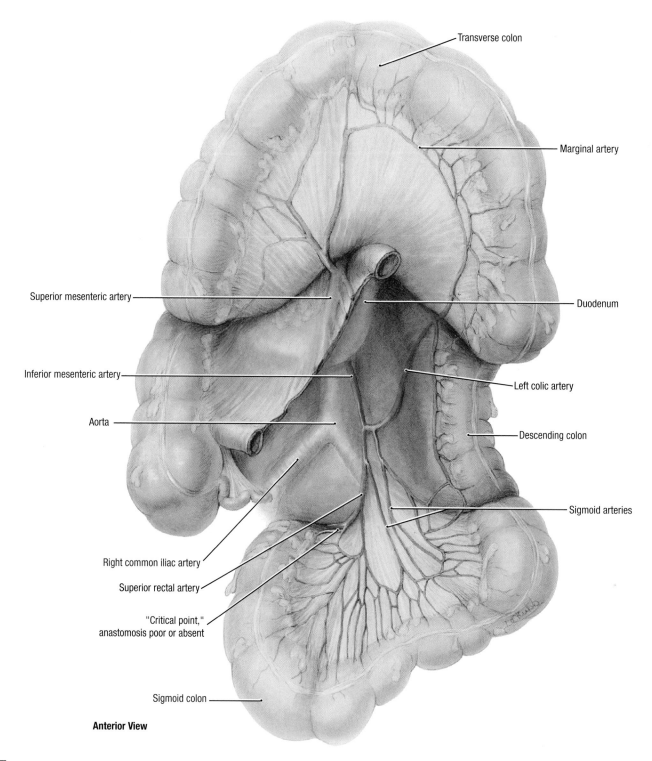

Transverse colon

Marginal artery

Superior mesenteric artery

Duodenum

Inferior mesenteric artery

Left colic artery

Aorta

Descending colon

Sigmoid arteries

Right common iliac artery

Superior rectal artery

"Critical point,"
anastomosis poor or absent

Sigmoid colon

Anterior View

2.45 Inferior mesenteric artery

The mesentery of the small intestine has been cut at its root and discarded with the jejunum and ileum.

- The inferior mesenteric artery arises posterior to the 4th part of the duodenum, about 4 cm superior to the bifurcation of the aorta; on crossing the left common iliac artery, it becomes the superior rectal artery.
- The branches of the inferior mesenteric artery include the left

colic artery and several sigmoid arteries; the inferior two sigmoid arteries branch from the superior rectal artery.
- The point at which the last artery to the colon branches from the superior rectal artery is known as the "critical point"; this branch has poor or no anastomotic connections with the superior rectal artery.

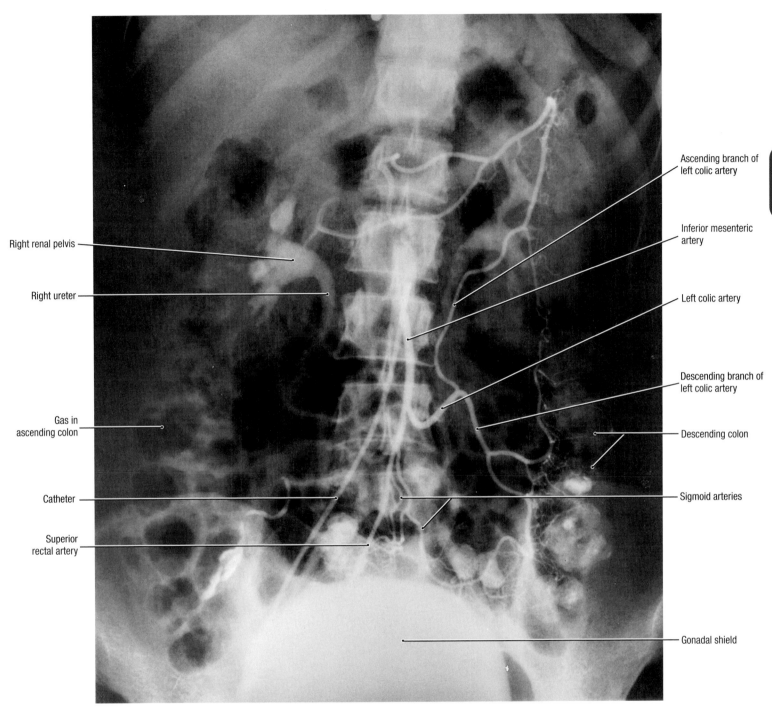

Right renal pelvis

Right ureter

Gas in
ascending colon

Catheter

Superior
rectal artery

Ascending branch of
left colic artery

Inferior mesenteric
artery

Left colic artery

Descending branch of
left colic artery

Descending colon

Sigmoid arteries

Gonadal shield

Posteroanterior arterigram

2.46   Inferior mesenteric arteriogram

- The left colic artery courses to the left toward the descending colon and splits into ascending and descending branches.
- The sigmoid arteries, two to four in number, supply the sigmoid colon.
- The superior rectal artery, which is the continuation of the inferior mesenteric artery, supplies the rectum; the superior rectal anastomoses with branches of the middle and inferior rectal arteries (from the internal iliac artery).

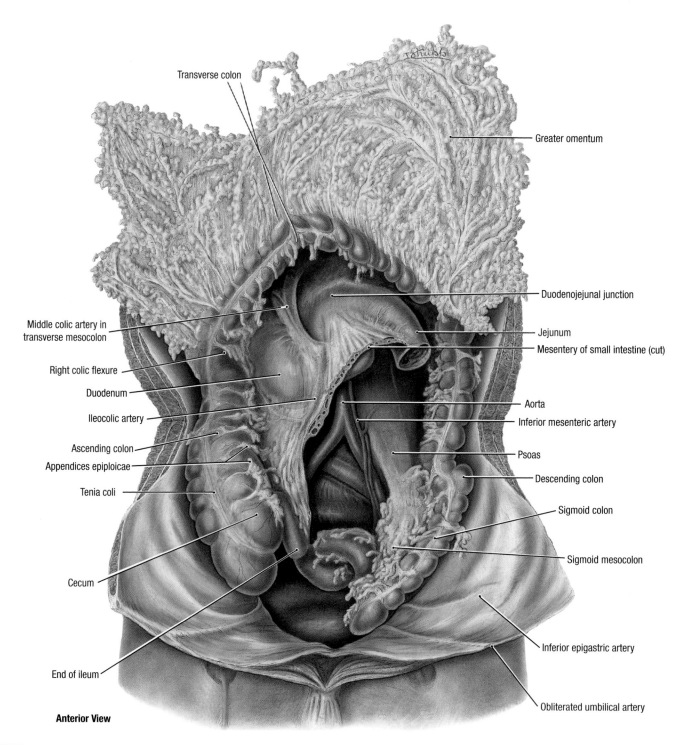

Transverse colon

Greater omentum

Duodenojejunal junction

Middle colic artery in transverse mesocolon

Jejunum

Mesentery of small intestine (cut)

Right colic flexure

Duodenum

Ileocolic artery

Aorta

Inferior mesenteric artery

Ascending colon

Psoas

Appendices epiploicae

Descending colon

Tenia coli

Sigmoid colon

Sigmoid mesocolon

Cecum

Inferior epigastric artery

End of ileum

Obliterated umbilical artery

Anterior View

2.47 Peritoneum of posterior abdominal cavity

The greater omentum is retracted superiorly, along with the transverse colon and transverse mesocolon. The appendix had been surgically removed. This dissection is continued in Figure 2.48.

- The duodenojejunal junction is situated to the left of the median plane and directly inferior to the root of the transverse mesocolon.
- The first few centimeters of the jejunum descend inferiorly and to the left, anterior to the left kidney. The last few centimeters of the ileum ascend superiorly and to the right, out of the pelvic cavity.

- The root of the mesentery of the small intestine, approximately 15 to 20 cm in length, extends between the duodenojejunal junction and ileocecal junction. The mesentery fans extensively to accommodate the length of jejunum and ileum.
- The large intestine forms $3\frac{1}{2}$ sides of a square "picture frame" around the jejunum and ileum. On the right are the cecum and ascending colon, superior is the transverse colon, on the left is the descending and sigmoid colon, inferiorly is the sigmoid colon.

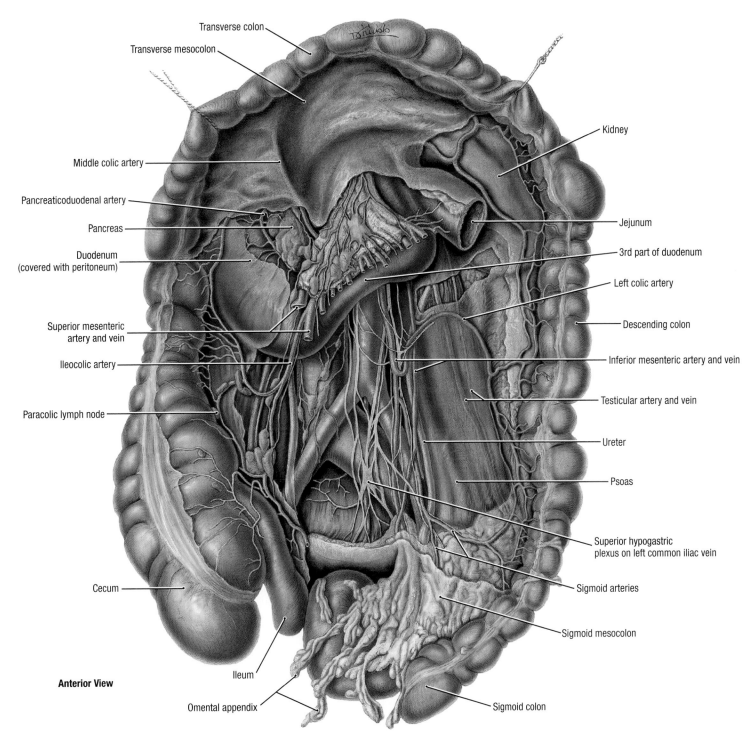

Transverse colon

Transverse mesocolon

Kidney

Middle colic artery

Pancreaticoduodenal artery

Pancreas

Jejunum

Duodenum
(covered with peritoneum)

3rd part of duodenum

Left colic artery

Descending colon

Superior mesenteric
artery and vein

Inferior mesenteric artery and vein

Ileocolic artery

Testicular artery and vein

Paracolic lymph node

Ureter

Psoas

Superior hypogastric
plexus on left common iliac vein

Cecum

Sigmoid arteries

Ileum

Sigmoid mesocolon

Anterior View

Omental appendix

Sigmoid colon

2.48   Posterior abdominal cavity with peritoneum removed

The jejunal and ileal branches (cut) pass from the left side of the superior mesenteric artery. The right colic artery here is a branch of the ileocolic artery. This is the same specimen as in Figure 2.46.

- The duodenum is large in diameter before crossing the superior mesenteric vessels and narrow afterward.
- On the right side, there are lymph nodes on the colon, paracolic nodes beside the colon, and nodes along the ileocolic artery, which drain into nodes ventral to the pancreas.

- The intestines and intestinal vessels lie on a resectable plane anterior to that of the testicular vessels; these in turn lie anterior to the plane of the kidney, its vessels, and the ureter.
- The superior hypogastric plexus lies within the bifurcation of the aorta and ventral to the left common iliac vein, the body of the 5th lumbar vertebra, and the 5th intervertebral disc.

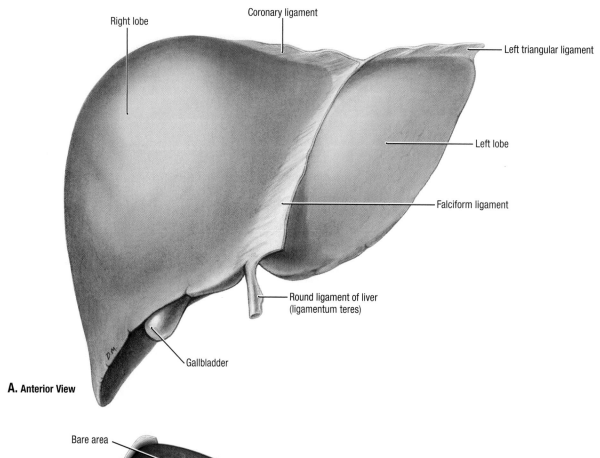

A. Anterior View

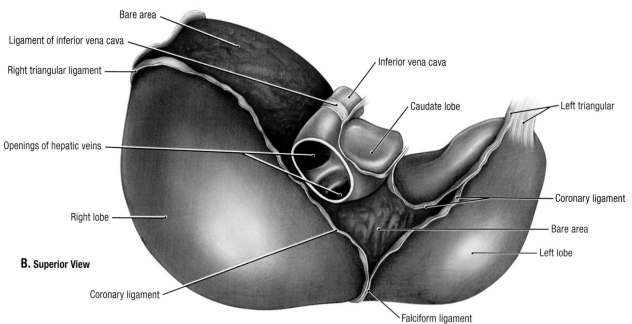

B. Superior View

2.49 **Diaphragmatic (anterior and superior) surface of liver**

A. Isolated specimen. The falciform ligament has been severed close to its attachment to the diaphragm and anterior abdominal wall. The fundus of the gallbladder is visible at the inferior border of the liver. Note the right and left lobes of the liver and the falciform ligament. The round ligament of the liver (ligamentum teres) is the obliterated umbilical vein that carried oxygenated blood from the placenta to the liver before birth. It lies within the free edge of the falciform ligament. **B.** Liver removed. The two layers of peritoneum that form the falciform ligament separate over the superior aspect of the liver to form the anterior layer of the coronary ligament on the right and the left triangular ligament on the left.

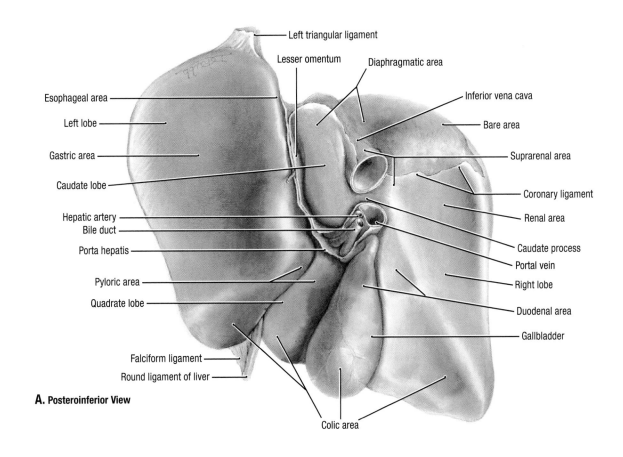

Left triangular ligament

Lesser omentum

Diaphragmatic area

Esophageal area

Inferior vena cava

Left lobe

Bare area

Gastric area

Suprarenal area

Caudate lobe

Coronary ligament

Hepatic artery

Renal area

Bile duct

Porta hepatis

Caudate process

Portal vein

Pyloric area

Right lobe

Quadrate lobe

Duodenal area

Gallbladder

Falciform ligament

Round ligament of liver

A. Posteroinferior View

Colic area

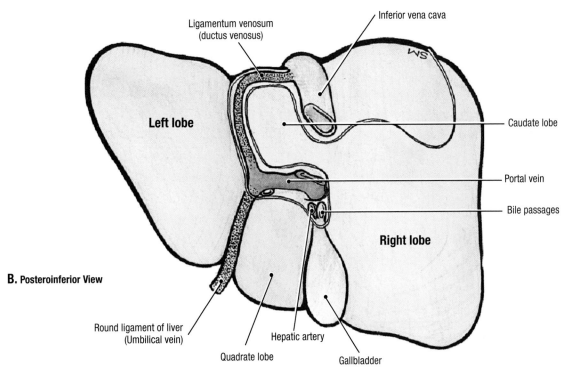

Ligamentum venosum
(ductus venosus)

Inferior vena cava

Left lobe

Caudate lobe

Portal vein

Bile passages

Right lobe

B. Posteroinferior View

Round ligament of liver
(Umbilical vein)

Hepatic artery

Quadrate lobe

Gallbladder

2.50   Visceral (posteroinferior) surface of liver

A. Isolated specimen. Note the caudate lobe is separated from the quadrate lobe by the porta hepatis and joined to the right lobe by the caudate process. **B.** Round ligament of liver and ligamentum venosum. The round ligament of liver includes the obliterated remains of the umbilical vein that carried well-oxygenated blood from the placenta to the fetus. The ligamentum venosum is the fibrous remnant of the fetal ductus venosus that shunted blood from the umbilical vein to the inferior vena cava, short circuiting the liver.

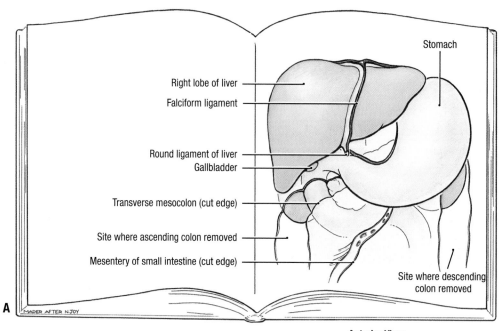

Anterior View

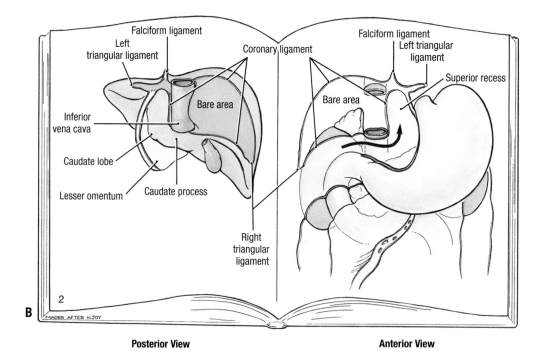

Posterior View   **Anterior View**

2.51   Liver and its posterior relations, schematic illustration

A. Liver in situ. The jejunum, ileum, and the ascending, transverse, and descending colon
have been removed. **B.** The liver is drawn schematically on a page in a book, so that as the
page is turned, the liver is reflected to the right to reveal its posterior surface, and on the facing page, the posterior relations that compose the bed of the liver are viewed. The *arrow*
traverses the site of the omental foramen. The bare area is triangular, and the coronary ligament that surrounds it is three-sided; its left side, or base, is between the inferior vena cava
and caudate lobe, and its apex is at the right triangular ligament, where the superior and inferior layers of the coronary ligament meet.

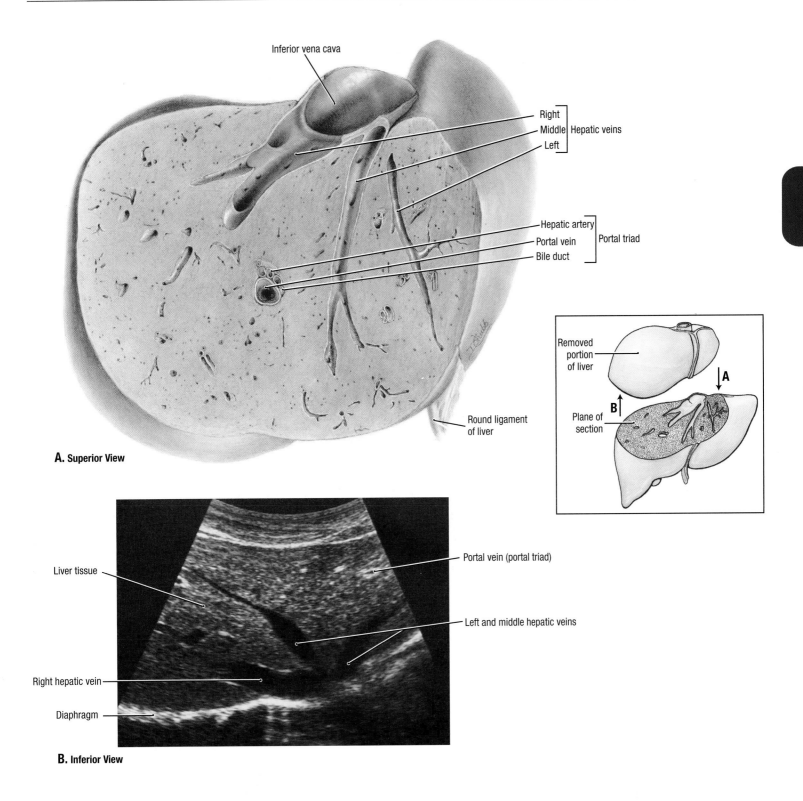

A. Superior View

B. Inferior View

2.52   Hepatic veins

A. Approximately horizontal section of liver with the posterior aspect at top of page. Note the multiple perivascular fibrous capsules sectioned throughout the cut surface, each containing a portal triad (the portal vein, hepatic artery, bile ductules) plus lymph vessels. Interdigitating with these are branches of the three main hepatic veins (right, middle, and left), which, unaccompanied and lacking capsules, converge on the inferior vena cava. **B.** Ultrasound scan. The transducer was placed under the costal margin, and directed posteriorly producing an inverted image corresponding to **A.**

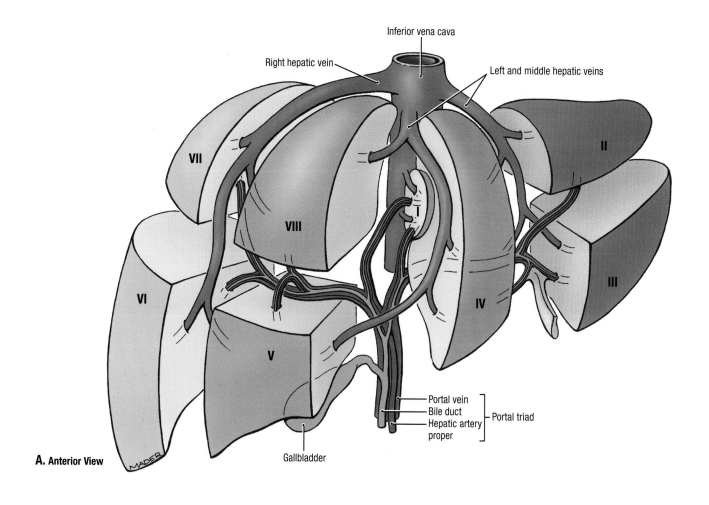

A. Anterior View

Inferior vena cava

Right hepatic vein

Left and middle hepatic veins

VII

VIII

II

VI

IV

III

V

Portal vein
Bile duct — Portal triad
Hepatic artery proper

Gallbladder

TABLE 2.4   SCHEMA OF TERMINOLOGY FOR SUBDIVISIONS OF THE LIVER

ANATOMICAL TERM	RIGHT LOBE		LEFT LOBE		CAUDATE LOBE	
	Right (part of) liver [Right portal lobe*]		Left (part of) liver [Left portal lobe⁺]		Posterior (part of) liver	
Functional/ surgical term**	Right lateral division division	Right medial division	Left medial division	Left lateral division	[Right caudate lobe*]	[Left caudate lobe⁺]
	Posterior lateral segment **Segment VII** [Posterior superior area]	Posterior medial segment **Segment VIII** [Anterior superior area]	[Medial superior area] Left medial segment **Segment IV** [Medial inferior area = quadrate lobe]	Lateral segment **Segment II** [Lateral superior area]	Posterior segment **Segment I**	
	Right anterior lateral segment **Segment VI** [Posterior inferior area]	Anterior medial segment **Segment V** [Anterior inferior area]		Left anterior lateral segment **Segment III** [Lateral inferior area]		

**The labels in the table and figure above reflect the new Terminologia Anatomica: *International Anatomical Terminology*. Previous terminology is in brackets.
*⁺Under the schema of the previous terminology, the caudate lobe was divided into right and left halves, and
*the right half of the caudate lobe was considered a subdivision of the right portal lobe;
⁺the left half of the caudate lobe was considered a subdivision of the left portal lobe.

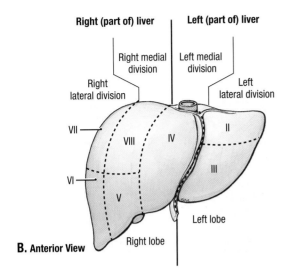

B. Anterior View

Right (part of) liver — Left (part of) liver

Right medial division — Left medial division

Right lateral division — Left lateral division

VII · VIII · IV · II · VI · V · III

Left lobe

Right lobe

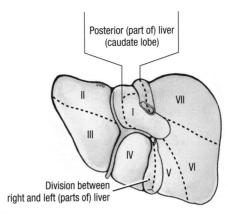

C. Posteroinferior View

Posterior (part of) liver (caudate lobe)

II · VII · III · I · IV · V · VI

Division between right and left (parts of) liver

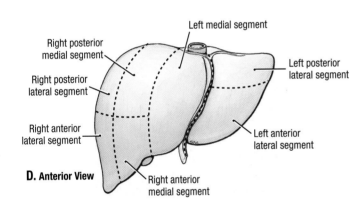

D. Anterior View

Right posterior medial segment

Right posterior lateral segment

Right anterior lateral segment

Right anterior medial segment

Left medial segment

Left posterior lateral segment

Left anterior lateral segment

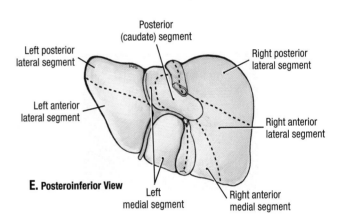

E. Posteroinferior View

Left posterior lateral segment

Left anterior lateral segment

Posterior (caudate) segment

Right posterior lateral segment

Right anterior lateral segment

Left medial segment

Right anterior medial segment

2.53   Segments of the liver

A. Schematic illustration of the segmental and vascular anatomy of the liver. The liver is divisible functionally into almost equal halves. The right and left parts of the liver are served by the right and left portal veins, hepatic arteries, and bile passages. The two parts of the liver are demarcated by a plane passing through the gallbladder fossa and the fossa for the inferior vena cava. The right and left parts of the liver are subdivided into segments. Each segment is supplied by a branch of the hepatic artery, bile duct, and portal vein. The hepatic veins interdigitate between the structures of the portal triad and are intersegmental in that they drain adjacent segments. Each of the segments can be numerically identified as in **A, B,** and **C** or named as in **D** and **E.** Since the right and left hepatic arteries and ducts and branches of the right and left portal veins do not communicate, it is possible to perform hepatic lobectomies (removal of the right or left part of the liver) and segmentectomies.

A. **Anterior View**

Cystic artery

Gallbladder

Gallbladder

Round ligament of liver
(obliterated umbilical vein)

Cystic artery
(superficial branch)

Cystic duct

Right branch

Common
hepatic duct

Right hepatic branch

Left hepatic duct

Left branch

Left hepatic branch

Ligamentum venosum
(obliterated ductus
venosus)

Inferior vena cava

Bile duct

Hepatic portal vein

Hepatic artery proper

B. **Inferior View**

Cystic veins

Fossa for gallbladder

to Liver

to Left portal vein

Common hepatic duct

Right gastric vein

Anterior cystic vein

Posterior cystic vein

Cystic duct

Bile duct

Posterior superior
pancreaticoduodenal vein

C. **Inferior View with
Gallbladder Retracted**

2.54 Porta hepatis and gallbladder

A. Intrahepatic courses of the hepatic portal vein (blue), hepatic artery (red), and bile passages (green). Note the segmental distribution of these vessels. The branches of the cystic artery are not intrahepatic, but lie posterior to the liver on the surface of the gallbladder and cystic duct. **B.** Porta hepatis and cystic artery. The inferior vascular attachments of the liver are severed, and the inferior border of the liver is raised to demonstrate the visceral surface (as in the orientation drawing). The cystic artery arises from the right hepatic artery and divides into superficial and deep branches that arborize on the respective surfaces of the gallbladder. **C.** Venous drainage of gallbladder and cystic ducts.

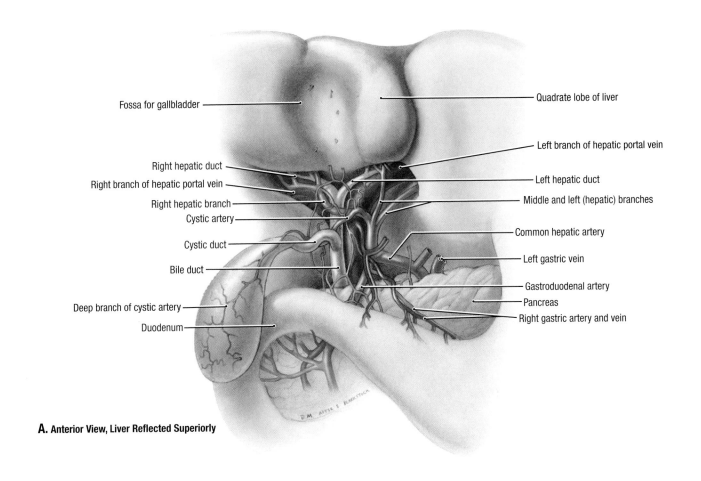

Fossa for gallbladder

Quadrate lobe of liver

Left branch of hepatic portal vein

Right hepatic duct

Right branch of hepatic portal vein

Left hepatic duct

Right hepatic branch

Middle and left (hepatic) branches

Cystic artery

Common hepatic artery

Cystic duct

Left gastric vein

Bile duct

Gastroduodenal artery

Pancreas

Deep branch of cystic artery

Right gastric artery and vein

Duodenum

A. Anterior View, Liver Reflected Superiorly

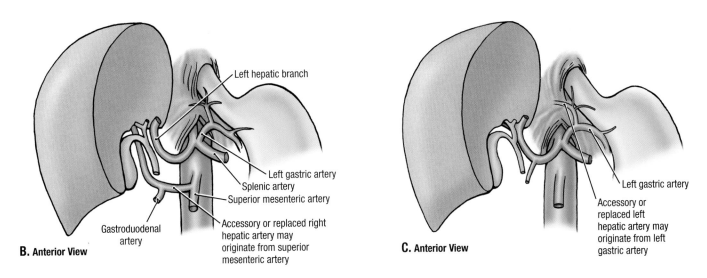

Left hepatic branch

Left gastric artery

Splenic artery

Superior mesenteric artery

Gastroduodenal artery

Accessory or replaced right hepatic artery may originate from superior mesenteric artery

B. Anterior View

Left gastric artery

Accessory or replaced left hepatic artery may originate from left gastric artery

C. Anterior View

2.55   Vessels in porta hepatis

A. Hepatic and cystic vessels. The liver is reflected superiorly. The gallbladder, freed from its bed, or fossa, has remained nearly in its anatomical position, pulled slightly to the right. The deep branch of the cystic artery on the deep, or attached, surface of the gallbladder anastomoses with branches of the superficial branch of the cystic artery and sends twigs into the bed of the gallbladder. Veins (not all shown) accompany most arteries. **B.** Aberrant (accessory or replaced) right hepatic artery. **C.** Aberrant left hepatic artery.

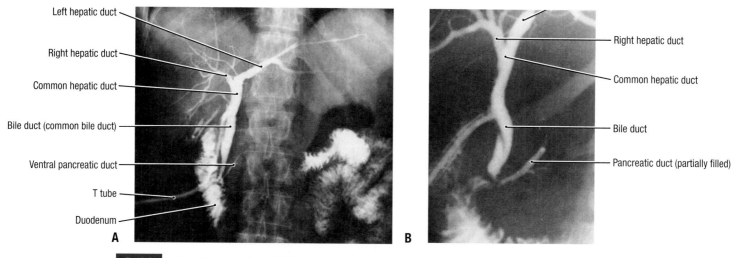

2.56 Radiographs of biliary passages

After a cholecystectomy (removal of the gallbladder), contrast medium was injected with a T tube inserted into the bile passages. The biliary passages are visualized in the superior abdomen in **A** and are more localized in **B**.

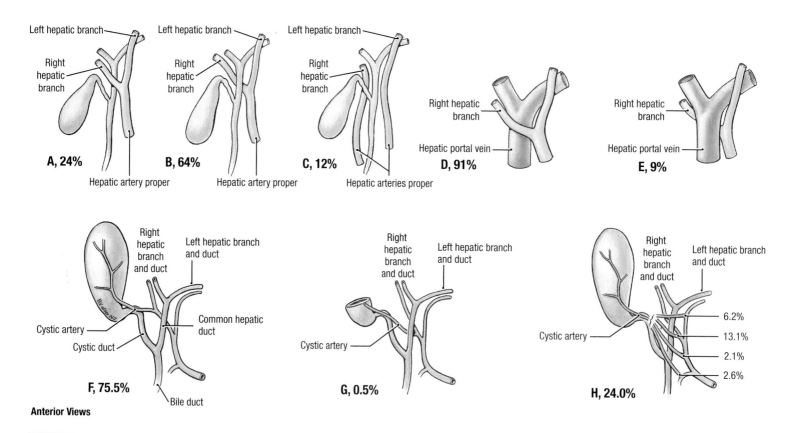

Anterior Views

2.57 Variations in hepatic and cystic arteries

In a study of 165 cadavers, five patterns were observed. **A.** Right hepatic artery crossing anterior to bile passages, 24%. **B.** Right hepatic artery crossing posterior to bile passages, 64%. **C.** Aberrant artery arising from the superior mesenteric artery, 12%. The artery crossed anterior (**D**) to the portal vein in 91%, and posterior (**E**) in 9%. The cystic artery usually arises from the right hepatic artery in the angle between the common hepatic duct and cystic duct, without crossing the common hepatic duct (**F and G**). However, when it arises on the left of the bile passages, it almost always crosses anterior to the passages (**H**).

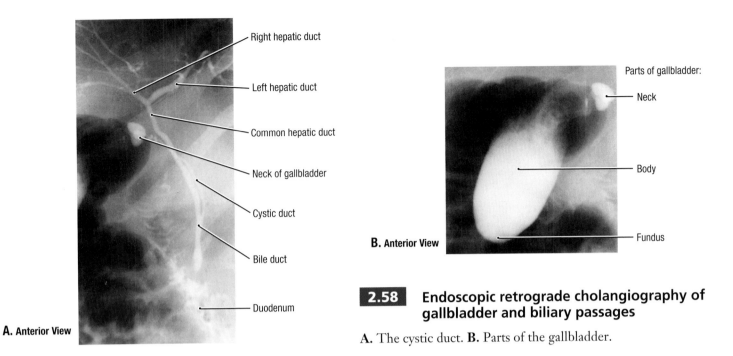

A. Anterior View

Right hepatic duct

Left hepatic duct

Common hepatic duct

Neck of gallbladder

Cystic duct

Bile duct

Duodenum

Parts of gallbladder:

Neck

Body

Fundus

B. Anterior View

2.58 Endoscopic retrograde cholangiography of gallbladder and biliary passages

A. The cystic duct. **B.** Parts of the gallbladder.

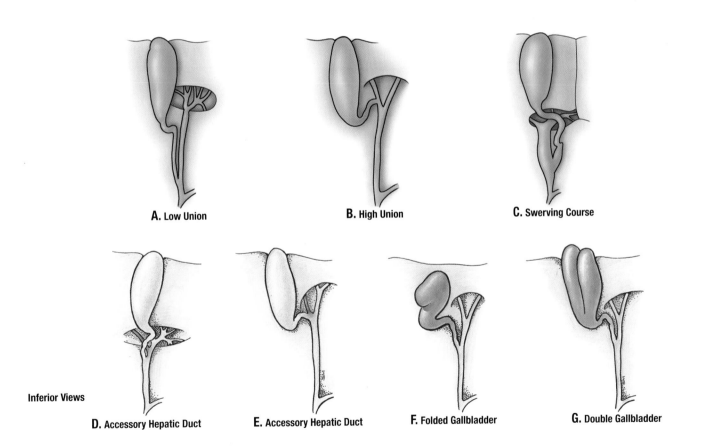

A. Low Union

B. High Union

C. Swerving Course

Inferior Views

D. Accessory Hepatic Duct

E. Accessory Hepatic Duct

F. Folded Gallbladder

G. Double Gallbladder

2.59 Variations of cystic and hepatic ducts and gallbladder

The cystic duct usually lies on the right side of the common hepatic duct, joining it just above the 1st part of the duodenum, but this varies as in **A–C.** Of 95 gallbladders and bile passages studied, 7 had accessory ducts. Of these, 4 joined the common hepatic duct near the cystic duct **(D),** 2 joined the cystic duct **(E),** and 1 was an anastomosing duct connecting the cystic with the common hepatic duct. **F.** Folded gallbladder. **G.** Double gallbladder.

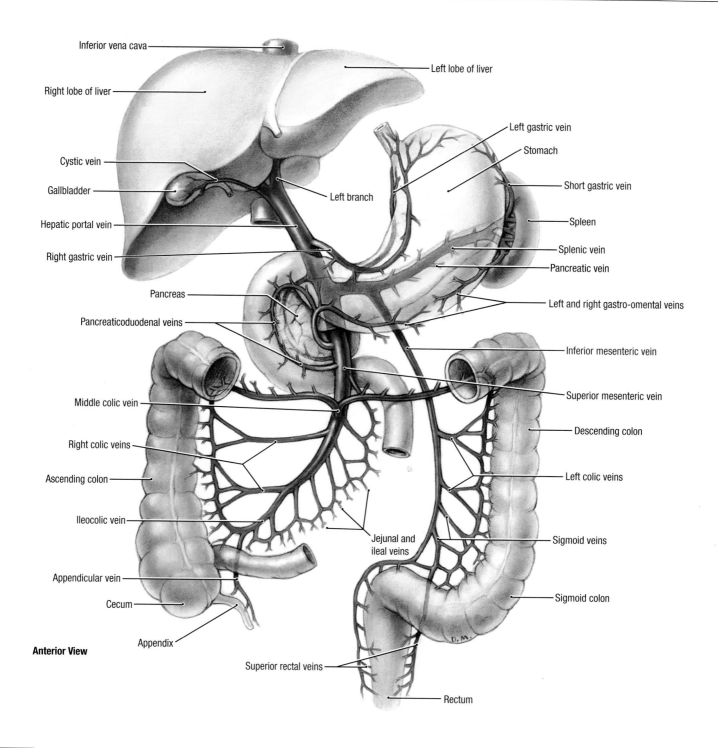

Inferior vena cava
Left lobe of liver
Right lobe of liver
Cystic vein
Gallbladder
Hepatic portal vein
Right gastric vein
Left branch
Left gastric vein
Stomach
Short gastric vein
Spleen
Splenic vein
Pancreatic vein
Pancreas
Pancreaticoduodenal veins
Left and right gastro-omental veins
Inferior mesenteric vein
Superior mesenteric vein
Middle colic vein
Right colic veins
Ascending colon
Ileocolic vein
Descending colon
Left colic veins
Jejunal and ileal veins
Sigmoid veins
Appendicular vein
Cecum
Sigmoid colon
Appendix
Superior rectal veins
Rectum

Anterior View

| 2.60 | **Portal venous system** |

- The portal vein drains venous blood from the gastrointestinal tract, spleen, pancreas, and gallbladder to the sinusoids of the liver; from here, the blood is conveyed to the systemic venous system by the hepatic veins that drain directly to the inferior vena cava.
- The portal vein forms posterior to the neck of the pancreas by the union of the superior mesenteric and splenic veins, with the inferior mesenteric vein joining at or near the angle of union.
- The splenic vein drains blood from the inferior mesenteric, left gastroomental (epiploic), short gastric, and pancreatic veins.
- The right gastroomental, pancreaticoduodenal, jejunal, ileal,

right, and middle colic veins drain into the superior mesenteric vein.
- The inferior mesenteric vein commences in the rectal plexus as the superior rectal vein and, after crossing the common iliac vessels, becomes the inferior mesenteric vein; branches include the sigmoid and left colic veins.
- The portal vein divides into right and left branches at the porta hepatis. The left branch carries mainly, but not exclusively, blood from the inferior mesenteric, gastric, and splenic veins, and the right branch carries blood mainly from the superior mesenteric vein.

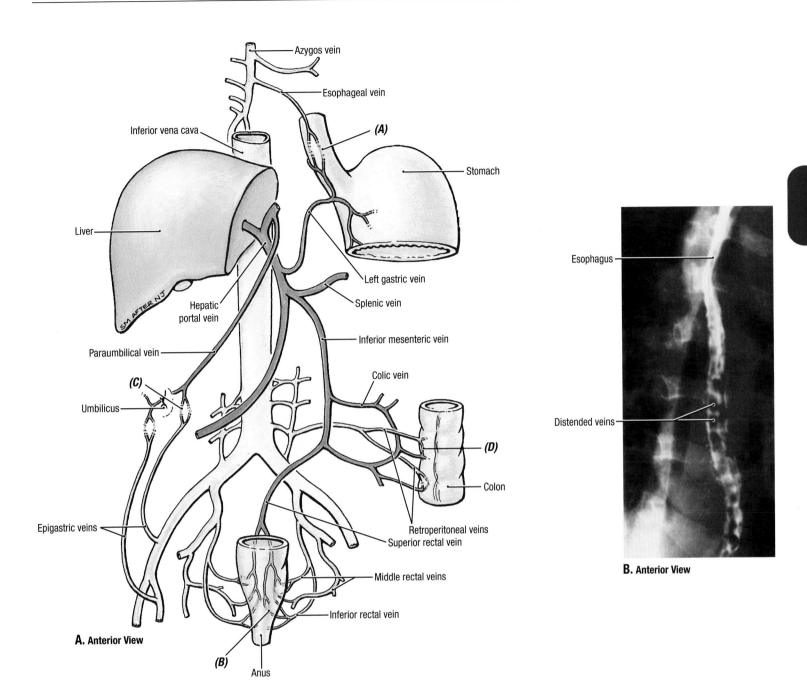

A. Anterior View

B. Anterior View

2.61   Portacaval system

A. Portacaval system. In this diagram, portal tributaries are *dark blue*, and systemic tributaries and communicating veins are *light blue*. In portal hypertension (as in hepatic cirrhosis), the portal blood cannot pass freely through the liver, and the portocaval anastomoses become engorged, dilated, or varicose; as a consequence, these veins may rupture. The sites of the portocaval anastomosis shown are between *(A)* esophageal veins draining into the azygos vein (systemic) and left gastric vein (portal), which when dilated are esophageal varices, also shown in radiograph **B**; *(B)* the inferior and middle rectal veins, draining into the inferior vena cava (systemic) and the superior rectal vein continuing as the inferior mesenteric vein (portal) (hemorrhoids include these dilated vessels); *(C)* paraumbilical veins (portal) and small epigastric veins of the anterior abdominal wall (systemic), which when varicose form "caput medusae" (so named because of the resemblance of the radiating veins to the serpents on the head of Medusa, a character in Greek mythology); and *(D)* twigs of colic veins (portal) anastomosing with systemic retroperitoneal veins. **B.** Esophageal varices. Note the distended veins of the esophagus.

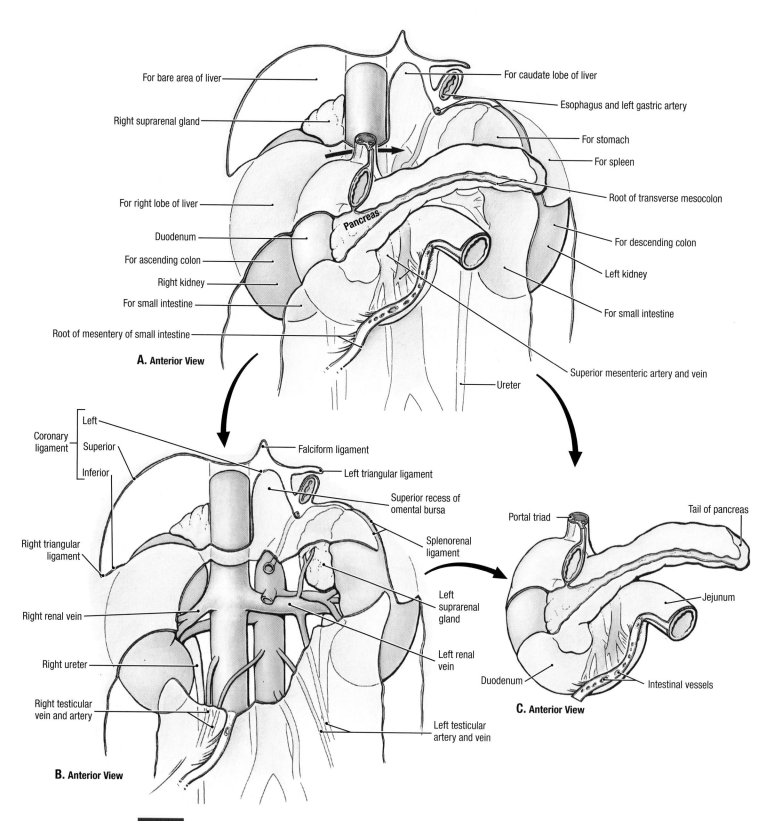

A. Anterior View

For bare area of liver

Right suprarenal gland

For right lobe of liver

Duodenum

For ascending colon

Right kidney

For small intestine

Root of mesentery of small intestine

For caudate lobe of liver

Esophagus and left gastric artery

For stomach

For spleen

Root of transverse mesocolon

For descending colon

Left kidney

For small intestine

Pancreas

Superior mesenteric artery and vein

Ureter

B. Anterior View

Coronary ligament
- Left
- Superior
- Inferior

Right triangular ligament

Right renal vein

Right ureter

Right testicular vein and artery

Falciform ligament

Left triangular ligament

Superior recess of omental bursa

Splenorenal ligament

Left suprarenal gland

Left renal vein

Left testicular artery and vein

C. Anterior View

Portal triad

Tail of pancreas

Jejunum

Duodenum

Intestinal vessels

2.62 **Posterior abdominal viscera and their anterior relations**

Note the peritoneal coverings (*yellow*). **A.** Duodenum and pancreas in situ. Note the line of attachment of the root of the transverse mesocolon to the body and tail of the pancreas; the viscera contacting specific regions are indicated by the term *for*. The omental (epiploic) foramen is traversed by an *arrow*. **B.** After removal of duodenum and pancreas. The three parts of the coronary ligament are attached to the diaphragm, except where the inferior vena cava, suprarenal gland, and kidney intervene. **C.** Pancreas and duodenum removed from **A.**

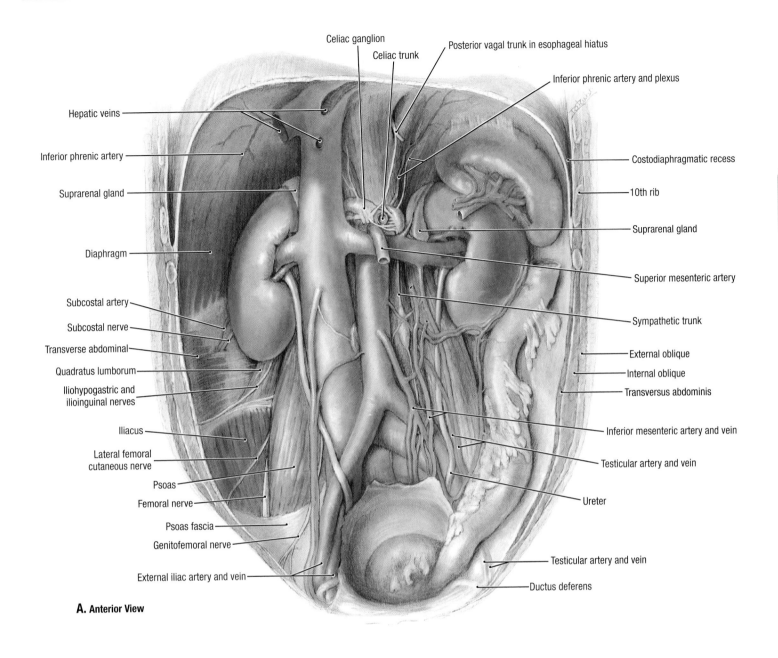

Celiac ganglion
Celiac trunk
Posterior vagal trunk in esophageal hiatus
Inferior phrenic artery and plexus

Hepatic veins

Inferior phrenic artery

Suprarenal gland

Diaphragm

Subcostal artery
Subcostal nerve
Transverse abdominal
Quadratus lumborum
Iliohypogastric and
ilioinguinal nerves

Iliacus

Lateral femoral
cutaneous nerve
Psoas
Femoral nerve
Psoas fascia
Genitofemoral nerve

External iliac artery and vein

Costodiaphragmatic recess
10th rib

Suprarenal gland

Superior mesenteric artery

Sympathetic trunk

External oblique
Internal oblique
Transversus abdominis

Inferior mesenteric artery and vein

Testicular artery and vein

Ureter

Testicular artery and vein
Ductus deferens

A. Anterior View

2.63   Viscera and vessels of posterior abdominal wall

A. Great vessels, kidneys, and suprarenal glands. **B.** Relationships of left renal vein and third part of duodenum to aorta and superior mesenteric artery.

- The abdominal aorta is shorter and smaller in caliber than the inferior vena cava.
- The inferior mesenteric artery arises about 4 cm superior to the aortic bifurcation and crosses the left common iliac vessels to become the superior rectal artery.
- The left renal vein drains the left testis, left suprarenal gland, and left kidney; the renal arteries are posterior to the renal veins.
- The ureter crosses the external iliac artery just beyond the common iliac bifurcation; the blood supply to the ureter comes from three main sources: the renal artery superiorly, a vesical artery inferiorly, and either the common iliac artery or the aorta.
- The testicular vessels cross anterior to the ureter and join the ductus deferens at the deep inguinal ring.
- In **B**, the left renal vein and duodenum pass between the aorta posteriorly and the superior mesenteric artery, anteriorly; they may be compressed like nuts in a nutcracker.

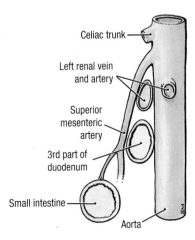

Celiac trunk

Left renal vein
and artery

Superior
mesenteric
artery

3rd part of
duodenum

Small intestine

Aorta

B. Lateral View (from left)

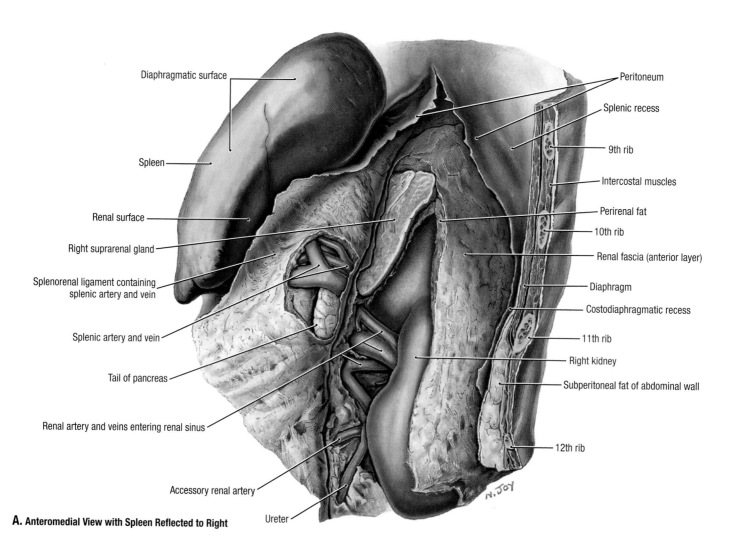

A. Anteromedial View with Spleen Reflected to Right

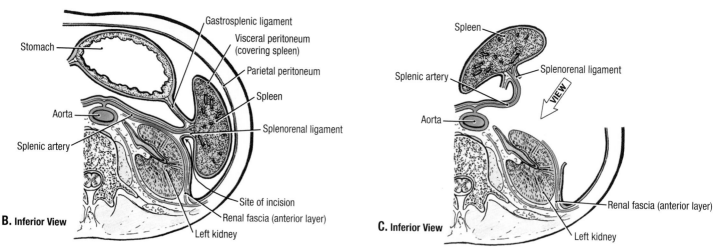

B. Inferior View

C. Inferior View

2.64 **Exposure of the left kidney and suprarenal gland**

A. Dissection. **B.** Schematic section with spleen and splenorenal ligament intact. **C.** Procedure used in **A** to expose the kidney. The spleen and splenorenal ligament are reflected anteriorly, with the splenic vessels and tail of the pancreas. Part of the fatty capsule of the kidney is removed. Note the proximity of the splenic vein and left renal vein, enabling a splenorenal shunt to be established surgically to relieve portal hypertension.

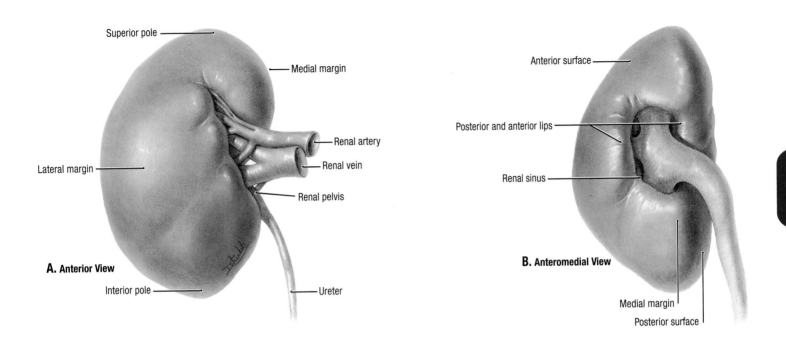

A. Anterior View

- Superior pole
- Medial margin
- Renal artery
- Renal vein
- Lateral margin
- Renal pelvis
- Interior pole
- Ureter

B. Anteromedial View

- Anterior surface
- Posterior and anterior lips
- Renal sinus
- Medial margin
- Posterior surface

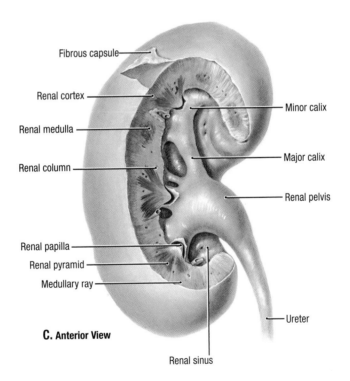

C. Anterior View

- Fibrous capsule
- Renal cortex
- Renal medulla
- Renal column
- Renal papilla
- Renal pyramid
- Medullary ray
- Minor calix
- Major calix
- Renal pelvis
- Ureter
- Renal sinus

2.65   Structure of kidney

A. External features. The superior pole of the kidney is usually wider than the inferior pole and closer to the median plane. Approximately 25% of kidneys may have a 2nd, 3rd, and even 4th accessory renal artery branching from the aorta. These multiple vessels enter through the renal sinus or at the superior or inferior pole. **B**. Renal sinus. The renal sinus is a vertical "pocket" on the medial side of the kidney. Tucked into the pocket are the renal pelvis and renal vessels. **C**. Renal calices. The anterior wall of the renal sinus has been cut away to expose the renal pelvis and the calices.

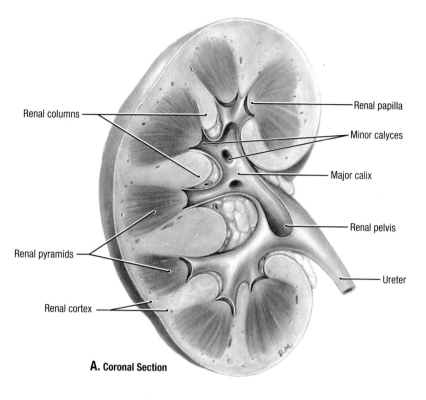

Renal columns

Renal papilla

Minor calyces

Major calix

Renal pelvis

Renal pyramids

Ureter

Renal cortex

A. Coronal Section

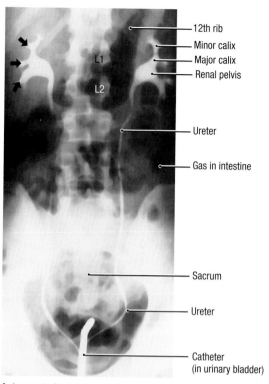

12th rib

Minor calix

Major calix

Renal pelvis

L1

L2

Ureter

Gas in intestine

Sacrum

Ureter

Catheter (in urinary bladder)

B. Anteroposterior Pyelogram

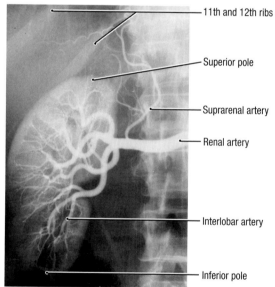

11th and 12th ribs

Superior pole

Suprarenal artery

Renal artery

Interlobar artery

Inferior pole

C. Anteroposterior Arteriogram

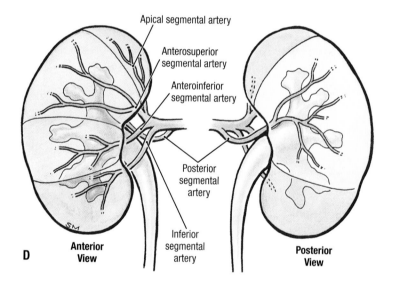

Apical segmental artery

Anterosuperior segmental artery

Anteroinferior segmental artery

Posterior segmental artery

Inferior segmental artery

D **Anterior View**

Posterior View

2.66 Kidney

A. Internal features. **B.** Pyelogram. Radiopaque material occupies the cavities that normally conduct urine. Note the papillae (indicated with *arrows*) bulging into the minor calices, which empty into a major calyx that opens, in turn, into the renal pelvis drained by the ureter. **C.** Renal arteriogram. **D.** Segmental arteries. Typically, the renal artery divides into five arteries, each supplying a segment of the kidney; only the apical and inferior arteries supply the whole thickness of the kidney.

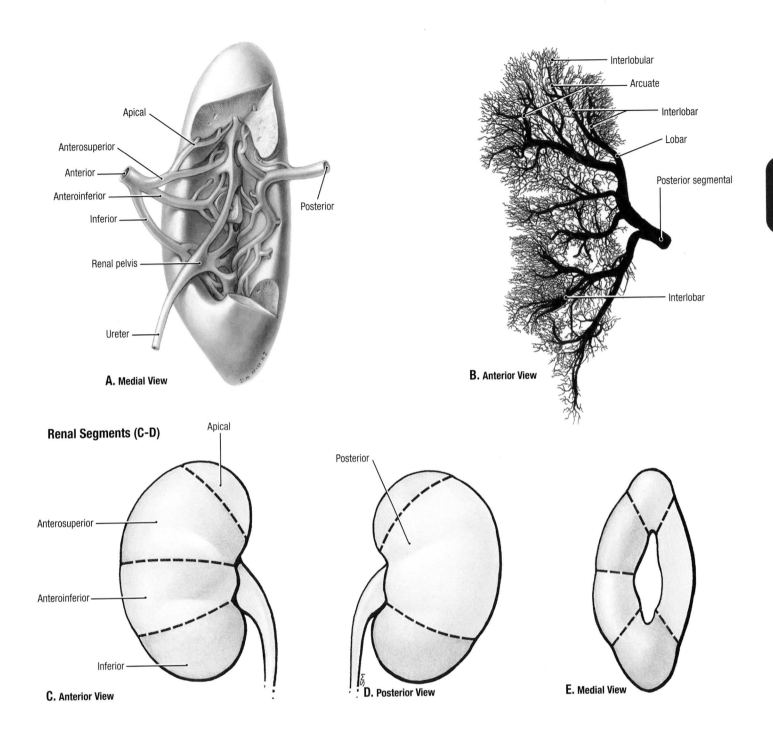

A. Medial View

B. Anterior View

Renal Segments (C-D)

C. Anterior View

D. Posterior View

E. Medial View

2.67   Segments of the kidneys

A. Dissection of segmental arteries at the renal sinus. **B.** Corrosion cast of posterior segmental artery of the kidney. Near the junction of the medulla and cortex, arcuate arteries arise at right angles to the parent stem. Segmental arteries do not anastomose; they are end arteries. **C–E.** Segments of the kidneys. The kidney has five segments: apical, anterosuperior (upper anterior), anteroinferior (middle anterior), inferior (lower), and posterior.

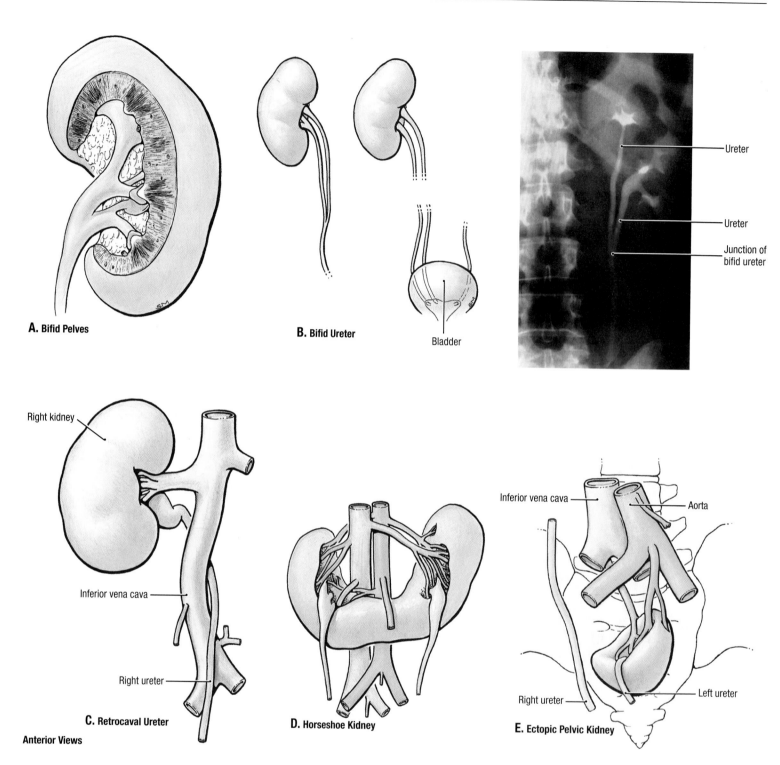

A. Bifid Pelves

B. Bifid Ureter

Bladder

Ureter

Ureter

Junction of bifid ureter

Right kidney

Inferior vena cava

Right ureter

C. Retrocaval Ureter

Anterior Views

D. Horseshoe Kidney

Inferior vena cava

Aorta

Right ureter

Left ureter

E. Ectopic Pelvic Kidney

2.68   Anomalies of kidney and ureter

A. Bifid pelves. The pelves are almost replaced by two long major calices, which extend outside the sinus. **B.** Duplicated, or bifid, ureters. These can be unilateral or bilateral, and complete or incomplete. **C.** Retrocaval ureter. The ureter courses posterior and then anterior to the inferior vena cava. **D.** Horseshoe kidney. The right and left kidneys are fused in the midline. **E.** Ectopic pelvic kidney. Pelvic kidneys have no fatty capsule and can be unilateral or bilateral. During childbirth, they may cause obstruction and suffer injury.

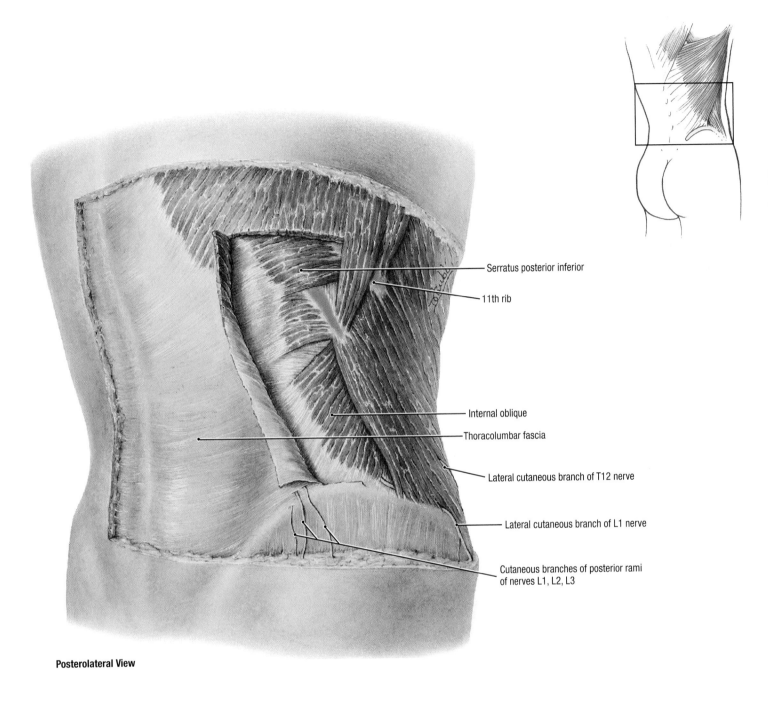

Serratus posterior inferior

11th rib

Internal oblique

Thoracolumbar fascia

Lateral cutaneous branch of T12 nerve

Lateral cutaneous branch of L1 nerve

Cutaneous branches of posterior rami of nerves L1, L2, L3

Posterolateral View

2.69 **Posterior abdominal wall—I**

The latissimus dorsi is partially reflected.
- The external oblique muscle has an oblique, free posterior border that extends from the tip of the 12th rib to the midpoint of the iliac crest.
- The internal oblique muscle extends posteriorly beyond the border of the external oblique muscle.

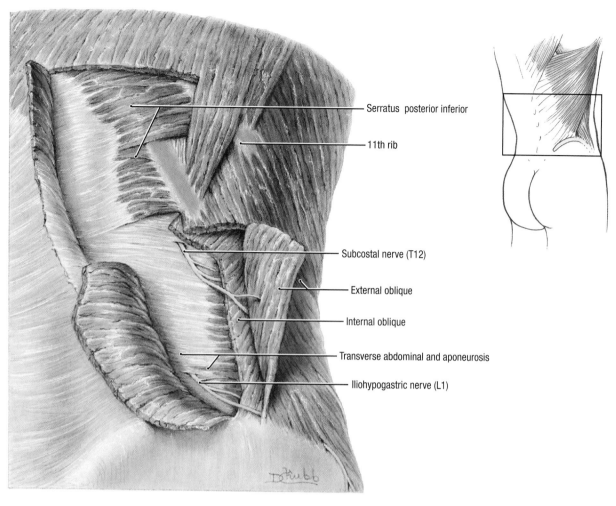

Serratus posterior inferior

11th rib

Subcostal nerve (T12)

External oblique

Internal oblique

Transverse abdominal and aponeurosis

Iliohypogastric nerve (L1)

Posterolateral View

2.70 Posterior abdominal wall—II

The external oblique muscle is incised and reflected laterally, and the internal oblique muscle is incised and reflected medially; the transverse abdominal muscle and its posterior aponeurosis are exposed where pierced by the subcostal (T12) and iliohypogastric (L1) nerves. These nerves give off motor twigs and lateral cutaneous branches and continue anteriorly between the internal oblique and transverse abdominal muscles.

2.71 Posterior abdominal wall—III and renal fascia

A. Dissection. The posterior aponeurosis of the transverse abdominal is divided between the subcostal and iliohypogastric nerves and lateral to the oblique lateral border of the quadratus lumborum muscle; the retroperitoneal fat surrounding the kidney is exposed. The renal fascia is within this fat; the portion of fat internal to the renal fascia is termed *fatty renal capsule* (perirenal fat), and the fat immediately external is *pararenal fat*. **B.** Renal fascia and fat, schematic transverse section.

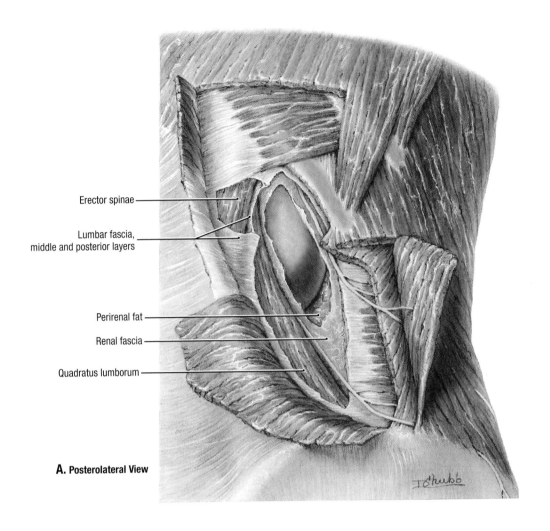

Erector spinae

Lumbar fascia,
middle and posterior layers

Perirenal fat

Renal fascia

Quadratus lumborum

A. Posterolateral View

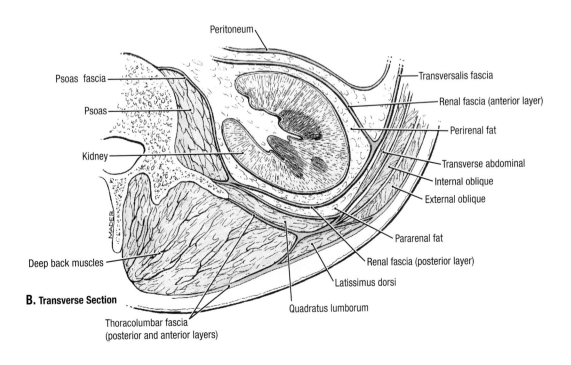

Peritoneum

Psoas fascia

Psoas

Kidney

Deep back muscles

Transversalis fascia

Renal fascia (anterior layer)

Perirenal fat

Transverse abdominal

Internal oblique

External oblique

Pararenal fat

Renal fascia (posterior layer)

Latissimus dorsi

Quadratus lumborum

Thoracolumbar fascia
(posterior and anterior layers)

B. Transverse Section

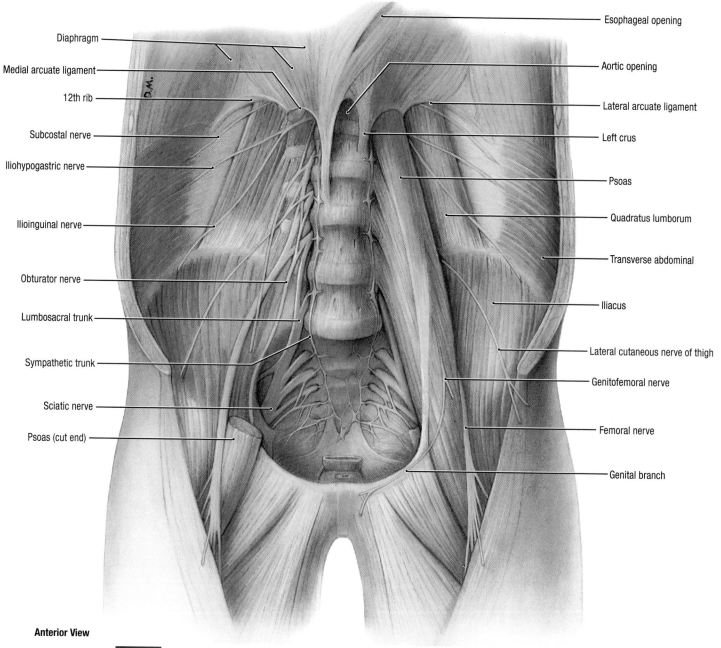

Diaphragm

Medial arcuate ligament

12th rib

Subcostal nerve

Iliohypogastric nerve

Ilioinguinal nerve

Obturator nerve

Lumbosacral trunk

Sympathetic trunk

Sciatic nerve

Psoas (cut end)

Esophageal opening

Aortic opening

Lateral arcuate ligament

Left crus

Psoas

Quadratus lumborum

Transverse abdominal

Iliacus

Lateral cutaneous nerve of thigh

Genitofemoral nerve

Femoral nerve

Genital branch

Anterior View

2.72 **Lumbar plexus and vertebral attachment of diaphragm**

TABLE 2.5 PRINCIPAL MUSCLES OF POSTERIOR ABDOMINAL WALL

MUSCLE	SUPERIOR ATTACHMENTS	INFERIOR ATTACHMENT(S)	INNERVATION	ACTIONS
Psoas major[a]	Transverse processes of lumbar vertebrae; sides of bodies of T12–L5 vertebrae and intervening invertebral discs	By a strong tendon to lesser trochanter of femur	Lumbar plexus via anterior branches of L2–L4 nerves	Acting inferiorly with iliacus, it flexes thigh at hip; acting superiorly, it flexes vertebral column laterally; it is used to balance the trunk; when sitting it acts inferiorly with iliacus to flex trunk
Iliacus[a]	Superior two thirds of iliac fossa, ala of sacrum; and anterior sacroiliac ligaments	Lesser trochanter of femur and body inferior to it, and to psoas major tendon	Femoral nerve (L2–L4)	Flexes thigh and stabilizes hip joint; acts with psoas major
Quadratus lumborum	Medial half of inferior border of 12th rib and tips of lumbar transverse processes	Iliolumbar ligament and internal lip of iliac crest	Anterior branches of T12 and L1–L4 nerves	Extends and laterally flexed vertebral column; fixes 12th rib during inspiration

[a]Psoas major and iliacus muscles are often described together as the iliopsoas muscle when flexion of the thigh is discussed. The iliopsoas is the chief flexor of the thigh, and when thigh is fixed, it is a strong flexor of the trunk (e.g., during situps).

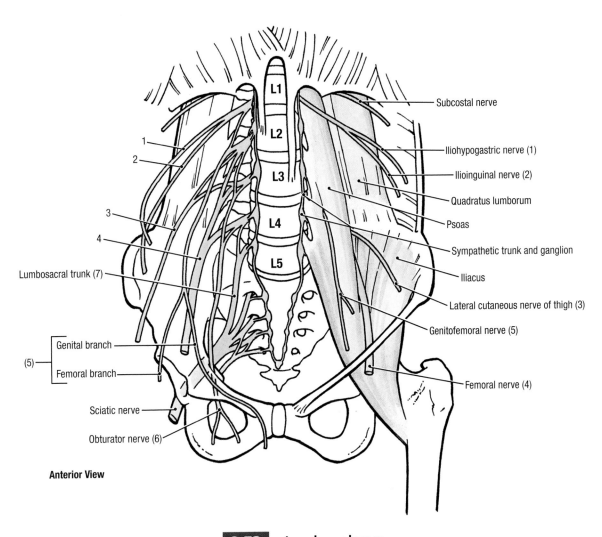

2.73   Lumbar plexus

Labels on figure:

L1, L2, L3, L4, L5

Subcostal nerve
Iliohypogastric nerve (1)
Ilioinguinal nerve (2)
Quadratus lumborum
Psoas
Sympathetic trunk and ganglion
Iliacus
Lateral cutaneous nerve of thigh (3)
Genitofemoral nerve (5)
Femoral nerve (4)

1
2
3
4
Lumbosacral trunk (7)
(5)
Genital branch
Femoral branch
Sciatic nerve
Obturator nerve (6)

Anterior View

TABLE 2.6   NERVES OF LUMBAR PLEXUS

1 and 2. The ilioinguinal and iliohypogastric nerves (L1) arise from the anterior ramus of L1 and enter the abdomen posterior to the medial arcuate ligaments and pass inferolaterally, anterior to the quadratus lumborum muscle; they pierce the transversus abdominis muscle near the anterior superior iliac spine and pass through the internal and external oblique muscles to supply the skin of the suprapubic and inguinal regions

3. The lateral femoral cutaneous nerve (L2, L3) runs inferolaterally on the iliacus muscle and enters the thigh posterior to the inguinal ligament, just medial to the anterior superior iliac spine; it supplies the skin on the anterolateral surface of the thigh

4. The femoral nerve (L2–L4) emerges from the lateral border of the psoas and innervates the iliacus muscle and the extensor muscles of the knee

5. The genitofemoral nerve (L1, L2) pierces the anterior surface of the psoas major muscle and runs inferiorly on it deep to the psoas fascia; it divides lateral to the common and external iliac arteries into femoral and genital branches

6. The obturator nerve (L2–L4) emerges from the medial border of the psoas to supply the adductor muscles of the thigh

7. The lumbosacral trunk (L4, L5) passes over the ala (wing) of the sacrum and decends into the pelvis to take part in the formation of the sacral plexus along with the anterior rami of S1–S4 nerves

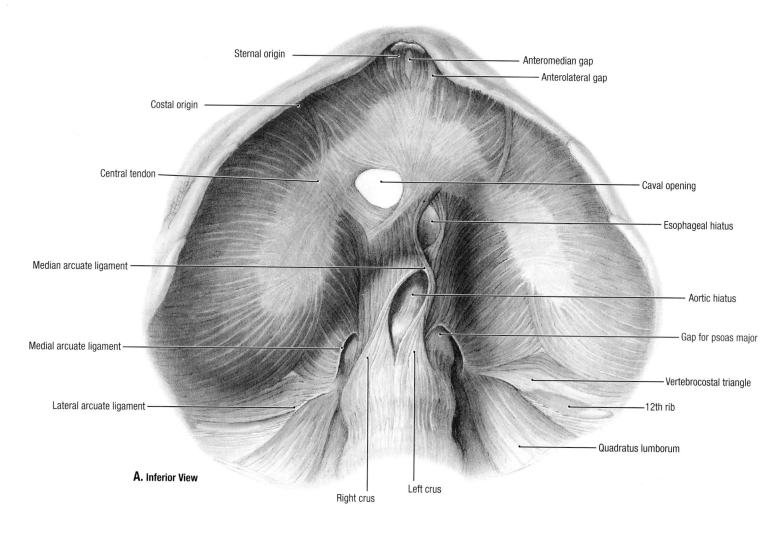

A. Inferior View

Sternal origin

Anteromedian gap

Anterolateral gap

Costal origin

Central tendon

Caval opening

Esophageal hiatus

Median arcuate ligament

Aortic hiatus

Gap for psoas major

Medial arcuate ligament

Vertebrocostal triangle

12th rib

Lateral arcuate ligament

Quadratus lumborum

Right crus

Left crus

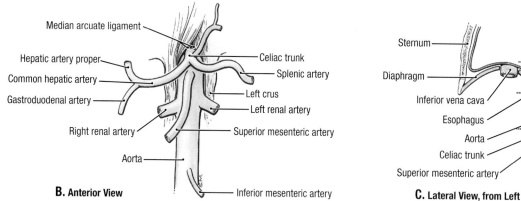

B. Anterior View

Median arcuate ligament

Hepatic artery proper

Common hepatic artery

Gastroduodenal artery

Right renal artery

Aorta

Celiac trunk

Splenic artery

Left crus

Left renal artery

Superior mesenteric artery

Inferior mesenteric artery

C. Lateral View, from Left

Sternum

Diaphragm

Inferior vena cava

Esophagus

Aorta

Celiac trunk

Superior mesenteric artery

T8

T10

T12

2.74 Diaphragm

A. Dissection. The clover-shaped central tendon is the aponeurotic insertion of the muscle. The diaphragm in this specimen fails to arise from the left lateral arcuate ligament, leaving a potential opening, the vertebrocostal triangle, through which abdominal contents may be herniated into the thoracic cavity. **B.** Median arcuate ligament and branches of the aorta. **C.** Openings of the diaphragm. There are three major openings through which major structures pass from the thorax into the abdomen: the caval opening for the inferior vena cava, most anterior, at the T8 vertebral level to the right of the midline; the esophageal hiatus, intermediate, at T10 level and to the left; and the aortic hiatus, which allows the aorta to pass posterior to the vertebral attachment of the diaphragm in the midline at T12.

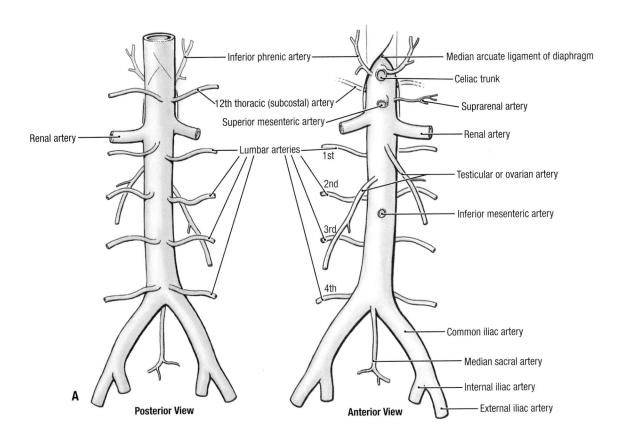

Inferior phrenic artery
Median arcuate ligament of diaphragm
Celiac trunk
12th thoracic (subcostal) artery
Suprarenal artery
Superior mesenteric artery
Renal artery
Renal artery
Lumbar arteries
1st
2nd
Testicular or ovarian artery
3rd
Inferior mesenteric artery
4th
Common iliac artery
Median sacral artery
Internal iliac artery
External iliac artery

A   **Posterior View**   **Anterior View**

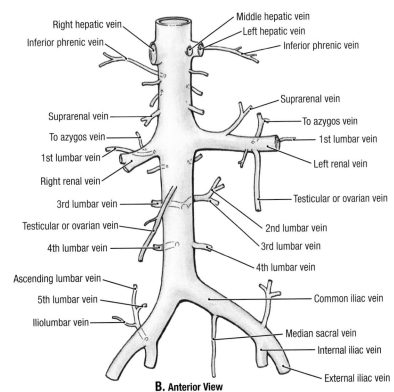

Right hepatic vein
Middle hepatic vein
Left hepatic vein
Inferior phrenic vein
Inferior phrenic vein
Suprarenal vein
Suprarenal vein
To azygos vein
To azygos vein
1st lumbar vein
1st lumbar vein
Left renal vein
Right renal vein
3rd lumbar vein
Testicular or ovarian vein
Testicular or ovarian vein
2nd lumbar vein
4th lumbar vein
3rd lumbar vein
4th lumbar vein
Ascending lumbar vein
5th lumbar vein
Common iliac vein
Iliolumbar vein
Median sacral vein
Internal iliac vein
External iliac vein

B. Anterior View

2.75   **Abdominal aorta and inferior vena cava and their branches**

A. Branches of abdominal aorta. **B.** Tributaries of the inferior vena cava. The tributaries of the inferior vena cava **(B)** correspond to branches of the aorta **(A).**

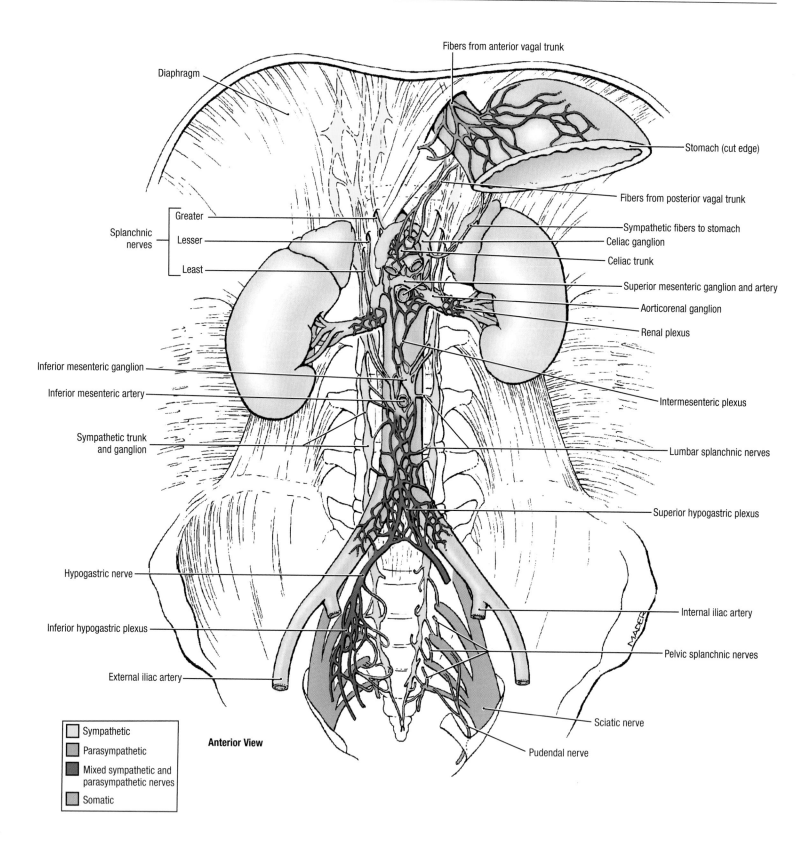

Diaphragm

Fibers from anterior vagal trunk

Stomach (cut edge)

Fibers from posterior vagal trunk

Splanchnic nerves
- Greater
- Lesser
- Least

Sympathetic fibers to stomach

Celiac ganglion

Celiac trunk

Superior mesenteric ganglion and artery

Aorticorenal ganglion

Renal plexus

Inferior mesenteric ganglion

Inferior mesenteric artery

Intermesenteric plexus

Sympathetic trunk and ganglion

Lumbar splanchnic nerves

Superior hypogastric plexus

Hypogastric nerve

Internal iliac artery

Inferior hypogastric plexus

Pelvic splanchnic nerves

External iliac artery

Sciatic nerve

Pudendal nerve

Anterior View

- ☐ Sympathetic
- ☐ Parasympathetic
- ■ Mixed sympathetic and parasympathetic nerves
- ☐ Somatic

2.76 **Autonomic nerve supply of the abdomen**

Abdominopelvic nerve plexuses and ganglia.

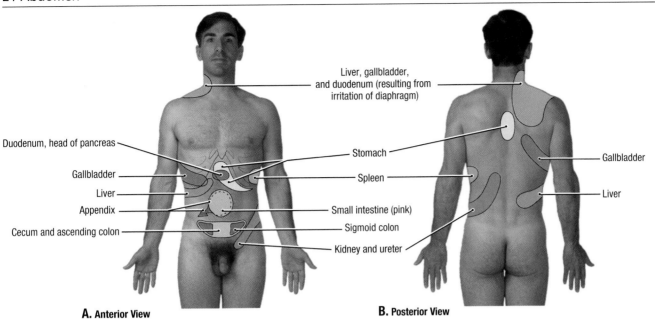

A. Anterior View

Liver, gallbladder, and duodenum (resulting from irritation of diaphragm)

Duodenum, head of pancreas

Gallbladder

Liver

Appendix

Cecum and ascending colon

Stomach

Spleen

Small intestine (pink)

Sigmoid colon

Kidney and ureter

B. Posterior View

Gallbladder

Liver

2.77 Surface projections of visceral pain

Pain arising from a viscus (organ) varies from dull to severe but is poorly localized. It radiates to the part of the body supplied by somatic sensory fibers associated with the same spinal ganglion and segment of the spinal cord that receive visceral sensory (autonomic) fibers from the viscus concerned. The pain is interpreted by the brain as though the irritation occurred in the area of skin supplied by the posterior roots of the affected segments. This is called *visceral referred pain*.

TABLE 2.7   REFERRED PAIN

ORGAN	NERVE SUPPLY	SPINAL CORD	REFERRED SITE AND CLINICAL EXAMPLE
Stomach	Anterior and posterior vagal trunks. Presynaptic sympathetic fibers reach celiac and other ganglia through greater splanchnic nerves	T6–T9 or T10	Epigastric and left hypochondriac regions (e.g., gastric peptic ulcer)
Duodenum	Vagus nerves; presynaptic sympathetic fibers reach celiac and superior mesenteric ganglia through greater splanchnic nerves	T5–T9 or T10	Epigastric region (e.g., duodenal peptic ulcer) peptic ulcer) Right shoulder if ulcer perforates
Pancreatic head	Vagus and thoracic splanchnic nerves	T8–T9	Inferior part of epigastric region (e.g., pancreatitis)
Small intestine jejunum and ileum	Posterior vagal trunks; presynaptic sympathetic fibers reach celiac ganglion through greater splanchnic nerves	T5–T9	Periumbilical region (e.g., acute intestinal obstruction)
Colon	Vagus nerves; presynaptic sympathetic fibers reach celiac, superior mesenteric, and inferior mesenteric ganglia through greater splanchnic nerves	T10–T12 (proximal colon)	Hypogastric region (e.g., ulcerative colitis)
	Parasympathetic supply to distal colon is derived from pelvic splanchnic nerves through hypogastric nerves and inferior hypogastric plexus	L1–L3 (distal colon)	Left lower quadrant (e.g., sigmoiditis)
Spleen	Celiac plexus, especially from greater splanchnic nerve	T6–T8	Left hypochondriac region (e.g., splenic infarct)
Appendix	Sympathetic and parasympathetic nerves from superior mesenteric plexus; afferent nerve fibers accompany sympathetic nerves to T10 segment of spinal cord	T10	Periumbilical region and later to right lower quadrant (e.g., appendicitis)
Gallbladder and liver	Nerves are derived from celiac plexus (sympathetic), vagus nerve (parasympathetic), and right phrenic nerve (sensory)	T6–T9	Epigastric region and later to right hypochondriac region; may cause pain on posterior thoracic wall or right shoulder owing to diaphragmatic irritation
Kidneys/ureters	Nerves arise from the renal plexus and consist of sympathetic, parasympathetic, and visceral afferent fibers from thoracic and lumbar splanchnics and the vagus nerve	T11–T12	Small of back, flank (lumbar quadrant), extending to groin (inguinal region) and genitals (e.g., renal or ureteric calculi)

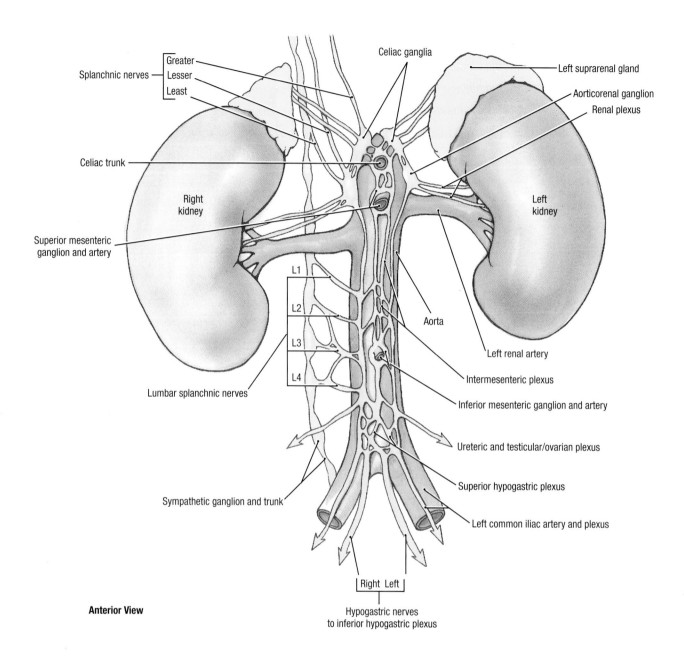

Greater
Lesser — Splanchnic nerves
Least

Celiac ganglia

Left suprarenal gland

Aorticorenal ganglion

Renal plexus

Celiac trunk

Right kidney

Left kidney

Superior mesenteric ganglion and artery

L1

L2

L3

L4

Aorta

Left renal artery

Intermesenteric plexus

Inferior mesenteric ganglion and artery

Lumbar splanchnic nerves

Ureteric and testicular/ovarian plexus

Superior hypogastric plexus

Sympathetic ganglion and trunk

Left common iliac artery and plexus

Right Left

Anterior View

Hypogastric nerves
to inferior hypogastric plexus

2.78 Autonomic nerve supply to the abdomen and pelvis

- Simplified diagram of abdominal nerve plexuses and ganglia. Sympathetic and parasympathetic nerves mingle in the tangle of nerve plexuses anterior to the aorta; both types of fibers reach their destinations by "piggybacking" on the branches of the abdominal aorta. This network is variable and difficult to dissect.
- The sympathetic splanchnic nerves synapse in preaortic ganglia, e.g., greater splanchnic nerve in the celiac ganglion, lesser splanchnic nerve in the renal plexus (aorticorenal ganglion), and lumbar splanchnics in the ganglia of the intermesenteric and superior hypogastric plexuses.
- The sympathetic trunk lies on the vertebral bodies and descends along the vertebral attachment of the psoas major; the trunk is slender where it enters the abdomen, its ganglia are ill defined, and approximately six lumbar splanchnic nerves leave it anteromedially;
- Parasympathetic fibers through the posterior and anterior vagal trunks (formerly right and left vagus nerves) are distributed to the foregut and midgut. Pelvic splanchnic nerves from branches of the anterior primary rami of sacral spinal nerves 2, 3, and 4 supply parasympathetic fibers to the hindgut and pelvic viscera.

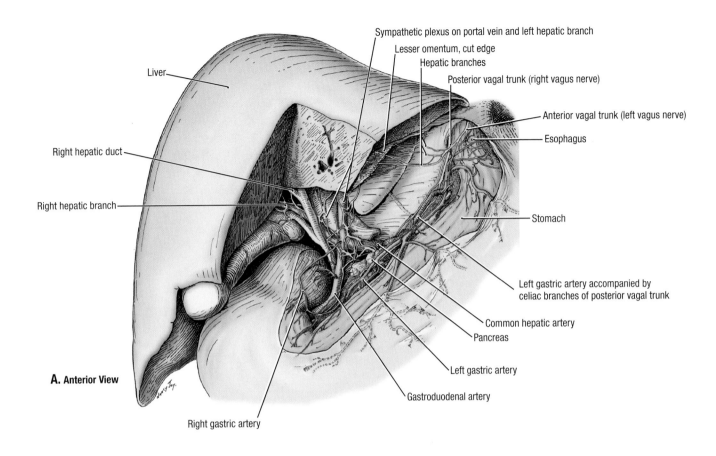

A. Anterior View

- Liver
- Right hepatic duct
- Right hepatic branch
- Sympathetic plexus on portal vein and left hepatic branch
- Lesser omentum, cut edge
- Hepatic branches
- Posterior vagal trunk (right vagus nerve)
- Anterior vagal trunk (left vagus nerve)
- Esophagus
- Stomach
- Left gastric artery accompanied by celiac branches of posterior vagal trunk
- Common hepatic artery
- Pancreas
- Left gastric artery
- Gastroduodenal artery
- Right gastric artery

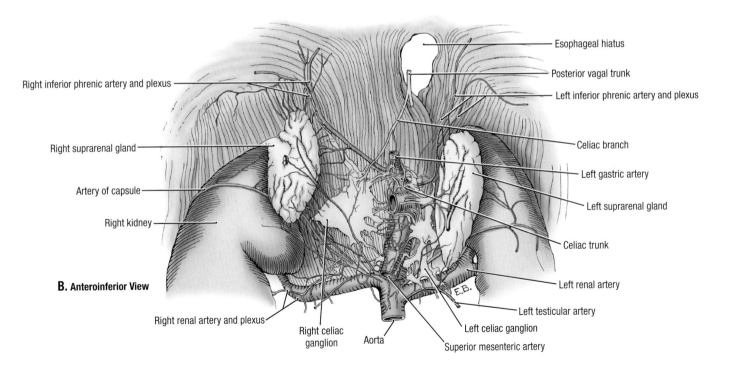

B. Anteroinferior View

- Right inferior phrenic artery and plexus
- Right suprarenal gland
- Artery of capsule
- Right kidney
- Esophageal hiatus
- Posterior vagal trunk
- Left inferior phrenic artery and plexus
- Celiac branch
- Left gastric artery
- Left suprarenal gland
- Celiac trunk
- Left renal artery
- Left testicular artery
- Left celiac ganglion
- Superior mesenteric artery
- Aorta
- Right celiac ganglion
- Right renal artery and plexus

2.79   **Vagus nerves in abdomen**

A. Anterior and posterior vagal trunks. **B.** Celiac plexus and ganglia and suprarenal glands.

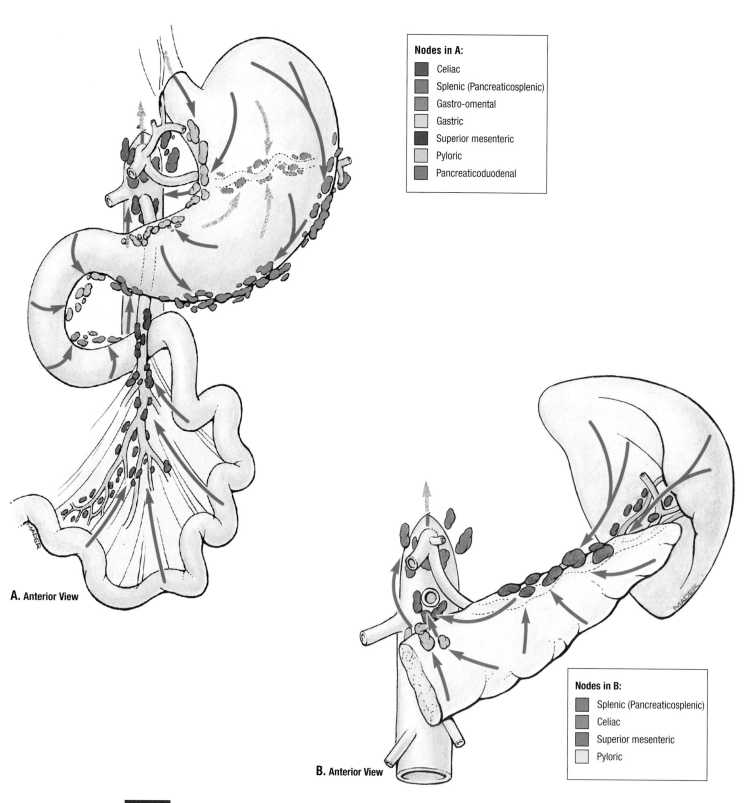

Nodes in A:
■ Celiac
■ Splenic (Pancreaticosplenic)
■ Gastro-omental
■ Gastric
■ Superior mesenteric
■ Pyloric
■ Pancreaticoduodenal

A. Anterior View

Nodes in B:
■ Splenic (Pancreaticosplenic)
■ Celiac
■ Superior mesenteric
□ Pyloric

B. Anterior View

2.80 **Lymphatic drainage**

A. Stomach and small intestine. **B.** Spleen and pancreas. **C.** Large intestine. **D.** Liver and kidney. The *arrows* indicate the direction of lymph flow; each group of lymph nodes is color coded. Lymph from the abdominal nodes drains into the cisterna chyli, origin of the inferior end of the thoracic duct. The thoracic duct receives all lymph that forms inferior to the diaphragm and left upper quadrant (thorax and left upper limb) and empties into the junction of the left subclavian and left internal jugular veins.

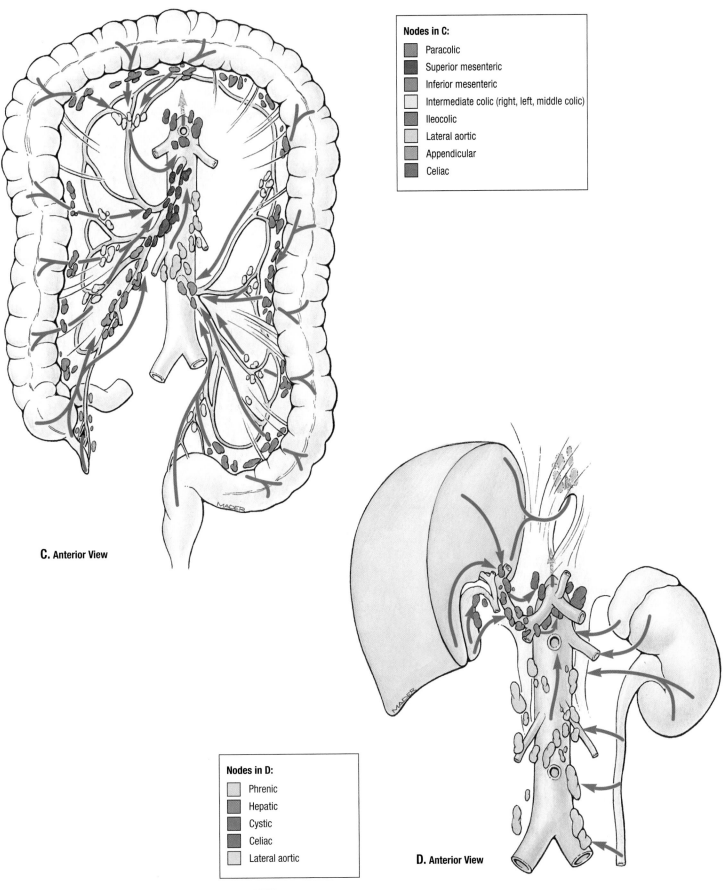

Nodes in C:

- Paracolic
- Superior mesenteric
- Inferior mesenteric
- Intermediate colic (right, left, middle colic)
- Ileocolic
- Lateral aortic
- Appendicular
- Celiac

C. Anterior View

Nodes in D:

- Phrenic
- Hepatic
- Cystic
- Celiac
- Lateral aortic

D. Anterior View

2.80   **Lymphatic drainage** *(continued)*

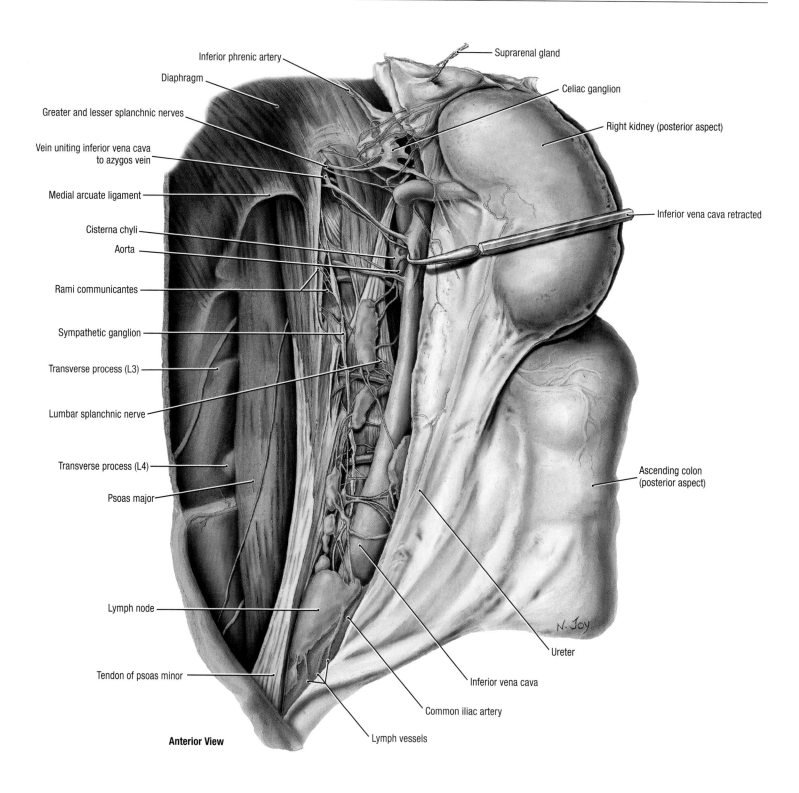

Inferior phrenic artery

Diaphragm

Greater and lesser splanchnic nerves

Vein uniting inferior vena cava to azygos vein

Medial arcuate ligament

Cisterna chyli

Aorta

Rami communicantes

Sympathetic ganglion

Transverse process (L3)

Lumbar splanchnic nerve

Transverse process (L4)

Psoas major

Lymph node

Tendon of psoas minor

Suprarenal gland

Celiac ganglion

Right kidney (posterior aspect)

Inferior vena cava retracted

Ascending colon (posterior aspect)

Ureter

Inferior vena cava

Common iliac artery

Lymph vessels

Anterior View

2.81 **Lumbar lymph nodes, sympathetic trunk, nerves, and ganglia**

The right suprarenal gland, kidney, ureter, and colon are reflected to the left; the inferior vena cava is pulled medially, and the third and fourth lumbar veins are removed. In this specimen, both splanchnic nerves, the sympathetic trunk, and a communicating vein pass through an unusually wide cleft in the right crus. Recall that the splanchnic nerves convey preganglionic fibers arising from the cell bodies in the (thoracolumbar) sympathetic trunk. The greater splanchnic nerve is from thoracic ganglia 5 to 9, and the lesser from thoracic ganglia 10 to 11.

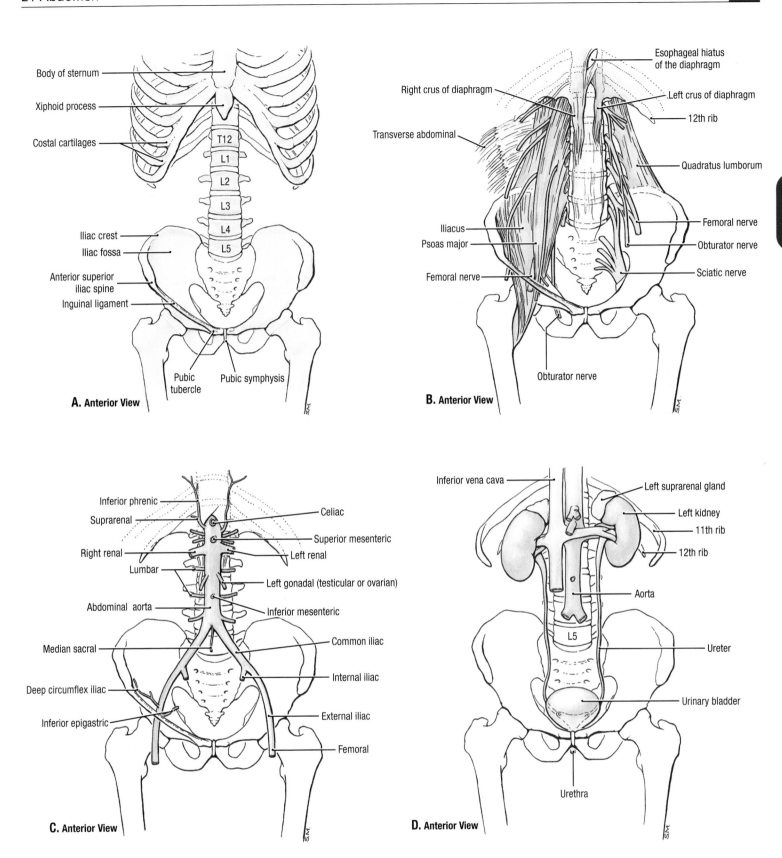

2.82 **Posterior abdominal wall, overview**

A. Skeleton. **B.** Musculature and lumbosacral plexus. **C.** Abdominal aorta. **D.** Urinary system.

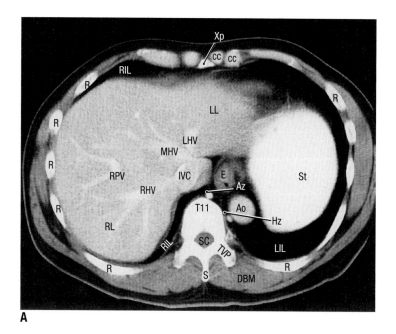

A

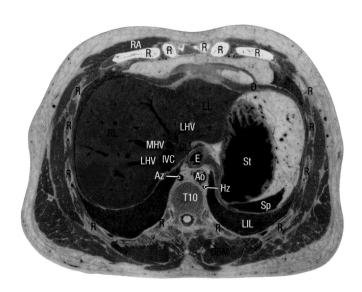

B

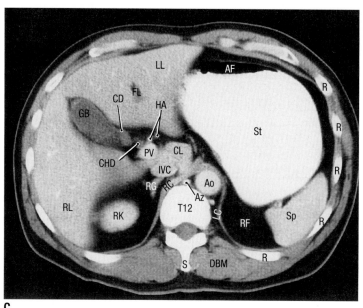

C

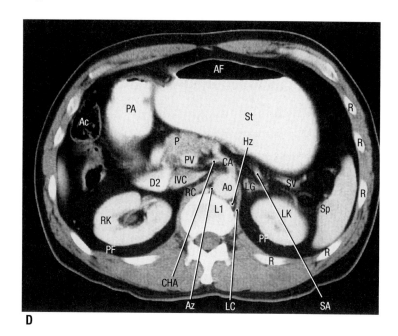

D

Ac	Ascending colon	DBM	Deep back muscles	LC	Left crus of diaphragm
AF	Air-fluid level of stomach	Dc	Descending colon	LG	Left suprarenal gland
Ao	Aorta	D2	2nd part of duodenum	LHV	Left hepatic vein
Az	Azygos vein	D3	3rd part of duodenum	LIL	Left inferior lobe of lung
CA	Celiac artery	E	Esophagus	LK	Left kidney
cc	Costal cartilage	FL	Falciform ligament	LL	Left lobe of liver
CD	Cystic duct	GB	Gallbladder	LRV	Left renal vein
CHA	Common hepatic artery	HA	Hepatic artery	LU	Left ureter
CHD	Common hepatic duct	Hz	Hemiazygos vein	MHV	Middle hepatic vein
CL	Caudate lobe of liver	IMV	Inferior mesenteric vein split	P	Pancreas
D	Diaphragm	IVC	Inferior vena cava	PA	Pyloric antrum of stomach

2.83   **Transverse or horizontal (axial) MRIs of the abdomen**

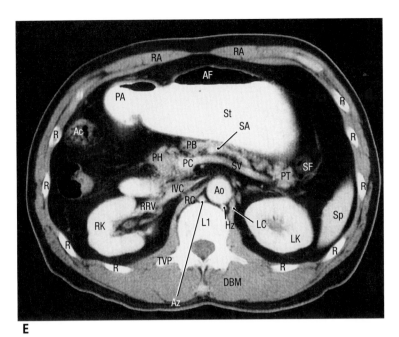

E

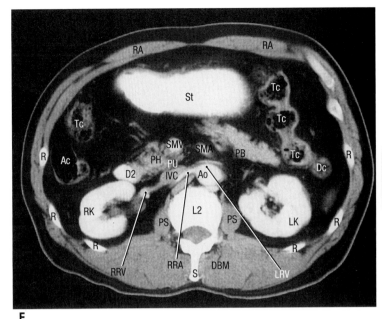

F

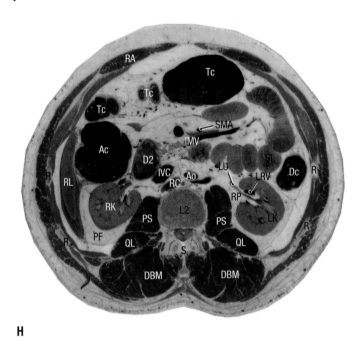

G

H

PB	Body of pancreas	R	Rib	RP	Renal pelvis	SI	Small intestine
PC	Portal confluence	RA	Rectus abdominis	RPV	Portal vein	SMA	Superior mesenteric artery
PF	Perirenal fat	RC	Right crus of diaphragm	RRA	Right renal artery	SMV	Superior mesenteric vein
PH	Head of pancreas	RF	Retroperitoneal fat	RRV	Right renal vein	Sp	Spleen
PS	Psoas muscle	RG	Right suprarenal gland	RU	Right ureter	St	Stomach
PT	Tail of pancreas	RHV	Right hepatic vein	S	Spinous process	SV	Splenic vein
PU	Uncinate process of pancreas	RIL	Right inferior lobe of lung	SA	Splenic artery	Tc	Transverse colon
PV(R)	Branches of right portal vein	RK	Right kidney	SC	Spinal cord	TVP	Transverse process
QL	Quadratus lumborum	RL	Right lobe of liver	SF	Splenic flexure	Xp	Xiphoid process

2.83   Transverse or horizontal (axial) MRIs of the abdomen *(continued)*

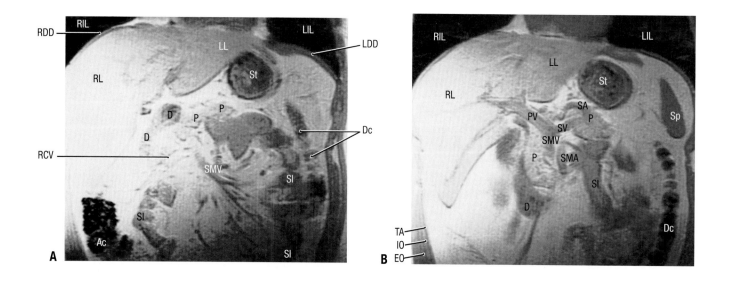

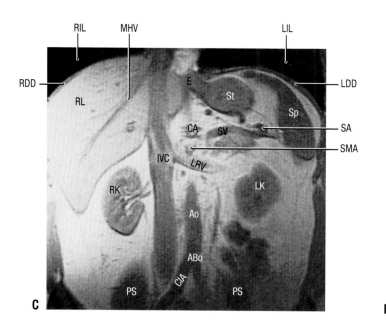

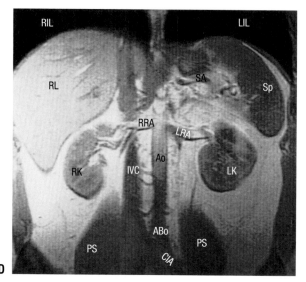

ABo	Aortic bifurcation	LIL	Left lung (inferior lobe)	RK	Right kidney
Ac	Ascending colon	LK	Left kidney	RL	Right lobe of liver
Ao	Aorta	LL	Left lobe of liver	RRA	Right renal artery
CA	Celiac artery	LRA	Left renal artery	SA	Splenic artery
CIA	Common iliac artery	LRV	Left renal vein	SI	Small intestine
D	Duodenum	MHV	Middle hepatic vein	SMA	Superior mesenteric artery
Dc	Descending colon	P	Pancreas	SMV	Superior mesenteric vein
E	Esophagus	PV	Portal vein	Sp	Spleen
EO	External oblique	PS	Psoas	St	Stomach
IO	Internal oblique	RCV	Right colic vein	SV	Splenic vein
IVC	Inferior vena cava	RDD	Right dome of diaphragm	TA	Transversus abdominis
LDD	Left dome of diaphragm	RIL	Right lung (inferior lobe)		

2.84   Coronal MRIs of the abdomen

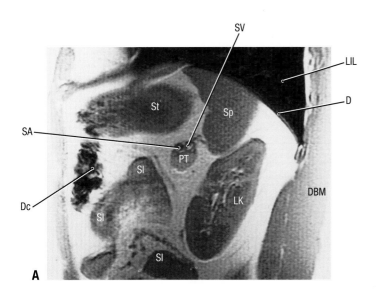

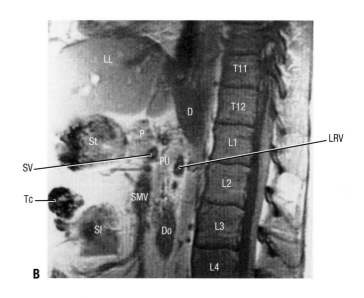

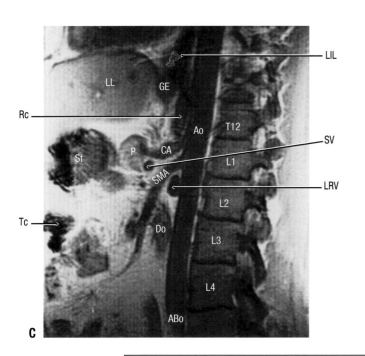

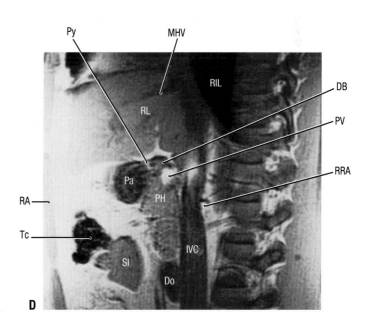

Ao	Aorta	LL	Left lobe of liver	RC	Right crus
ABo	Bifurcation of aorta	LRV	Left renal vein	RIL	Inferior lobe of right lung
CA	Celiac artery	MHV	Middle hepatic vein	RL	Right lobe of right liver
D	Diaphragm	P	Pancreas	RRA	Right renal artery
DB	Bulb of duodenum	Pa	Pyloric antrum	SA	Splenic artery
Dc	Descending colon	PC	Portal confluence	SI	Small intestine
Do	Duodenum	PH	Head of pancreas	SMA	Superior mesenteric artery
DBM	Deep back muscles	PT	Tail of pancreas	SMV	Superior mesenteric vein
GE	Gastroesophageal junction	PV	Portal vein	Sp	Spleen
IVC	Inferior vena cava	PU	Uncinate process of pancreas	St	Stomach
LIL	Inferior lobe of left lung	Py	Pylorus of stomach	SV	Splenic vein
LK	Left kidney	RA	Rectus abdominus	Tc	Transverse colon

2.85 Sagittal MRIs of the abdomen

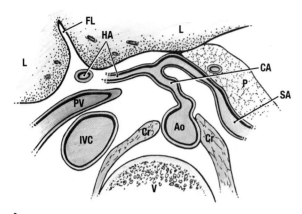
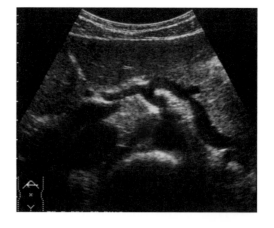
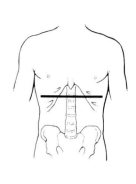

A. Transverse Section, Inferior View

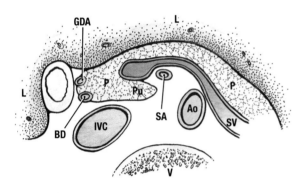
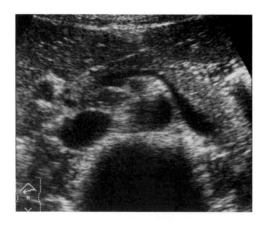
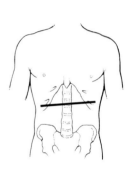

B. Transverse Section, Inferior View

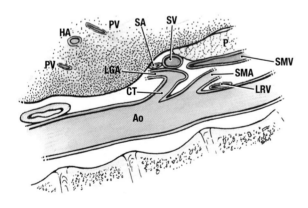
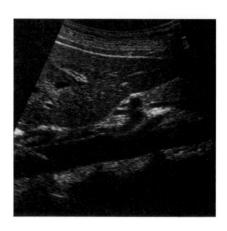
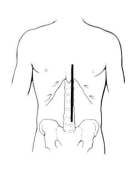

C. Median Section, Right Lateral View

2.86 Ultrasound scans and MR angiogram of the abdomen

A. Transverse ultrasound scan through celiac trunk. **B.** Transverse ultrasound scan through pancreas. **C** and **D.** Sagittal ultrasound scans through the aorta, celiac trunk, and superior mesenteric artery. (**D.** With Doppler.) **E.** MR angiogram of abdominal aorta and branches. **F.** Transverse ultrasound scan at hilum of left kidney with the left renal artery and vein. (with Doppler) **G.** Sagittal ultrasound scan of the right kidney. *Note:* Doppler ultrasonography is used to view moving blood within blood vessels and to display blood flow in color, superimposed on the two-dimensional image.

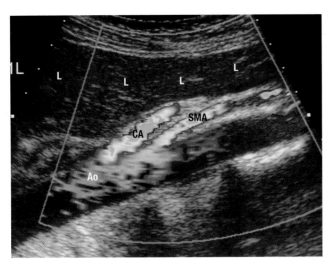

D. Median Section, Right Lateral View

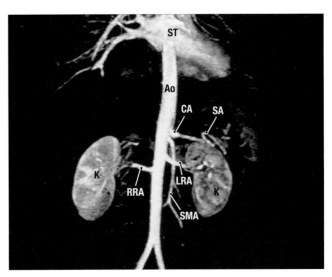

E. Anterior View

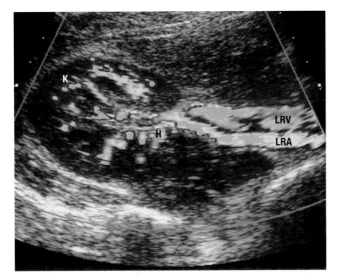

F. Transverse Section

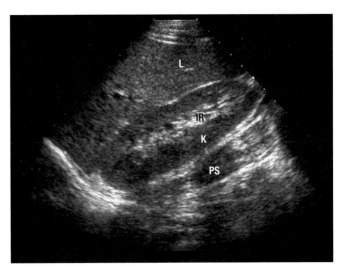

G. Sagittal Section, Right Lateral View

Ao	Aorta	IR	Intrarenal fat	PV	Portal vein
BD	Bile duct	IVC	Inferior vena cava	PVC	Portal venous confluence
CA	Celiac artery	K	Cortex of kidney	RRA	Right renal artery
Cr	Crus of diaphragm	L	Liver	SA	Splenic artery
D	Duodenum	LGA	Left gastric artery	SMA	Superior mesenteric artery
FL	Falciform ligament	LRA	Left renal artery	SMV	Superior mesenteric vein
GDA	Gastroduodenal artery	LRV	Left renal vein	ST	Stomach
GE	Gastroesophageal junction	P	Pancreas	SV	Splenic vein
H	Hilum of kidney	PS	Psoas	V	Vertebra
HA	Hepatic artery	Pu	Uncinate process of pancreas		

2.86   Ultrasound scans and MR angiogram of the abdomen *(continued)*

CHAPTER 3

PELVIS AND PERINEUM

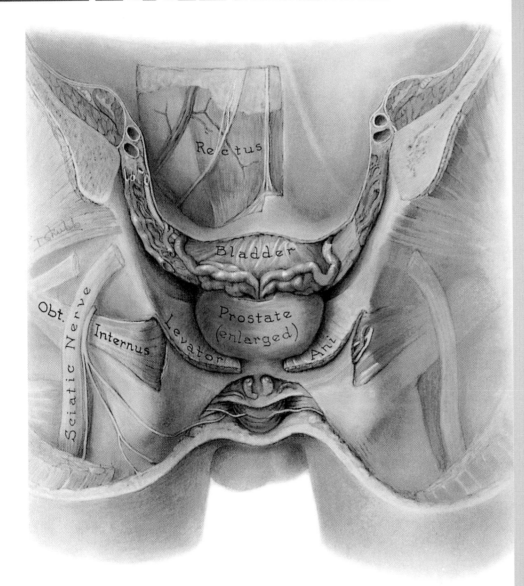

- Bony Pelvis and Ligaments **184**
- Floor and Walls of Pelvis **192**
- Overview of Male Pelvis **196**
- Rectum **198**
- Urinary Bladder, Prostate, and Seminal Vesicles **204**
- Vessels of Male Pelvis **212**
- Lymphatic Drainage and Innervation of Male Pelvis **218**
- Lumbrosacral Plexus **220**
- Overview of Female Pelvis **224**
- Uterus, Ovaries, and Vagina **226**
- Vessels of Female Pelvis **234**
- Lymphatic Drainage and Innervation of Female Pelvis **236**
- Overview of Perineum **242**
- Male Perineum **244**
- Female Perineum **256**
- Scans: Male **264**
- Scans: Female **268**

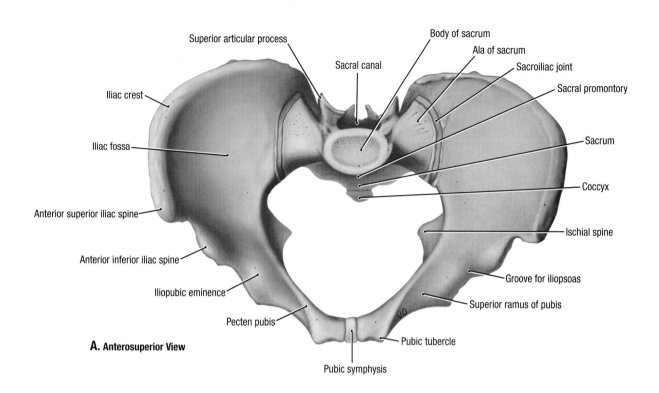

Superior articular process

Sacral canal

Body of sacrum

Ala of sacrum

Sacroiliac joint

Sacral promontory

Iliac crest

Iliac fossa

Sacrum

Anterior superior iliac spine

Coccyx

Anterior inferior iliac spine

Ischial spine

Iliopubic eminence

Groove for iliopsoas

Superior ramus of pubis

Pecten pubis

Pubic tubercle

A. Anterosuperior View

Pubic symphysis

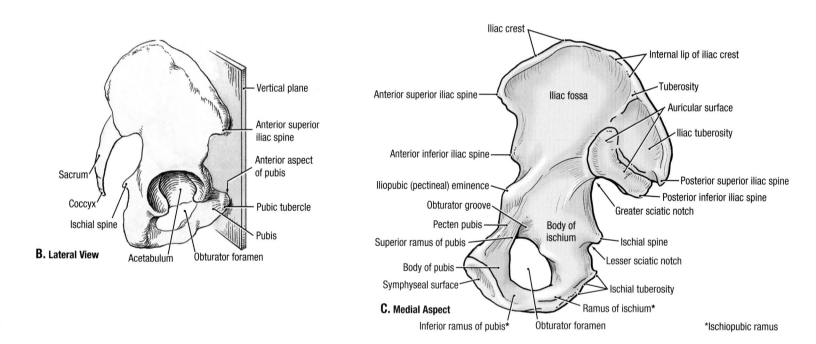

Iliac crest

Internal lip of iliac crest

Vertical plane

Anterior superior iliac spine

Iliac fossa

Tuberosity

Auricular surface

Anterior superior
iliac spine

Iliac tuberosity

Anterior aspect
of pubis

Anterior inferior iliac spine

Sacrum

Iliopubic (pectineal) eminence

Posterior superior iliac spine

Pubic tubercle

Coccyx

Obturator groove

Posterior inferior iliac spine

Ischial spine

Pubis

Pecten pubis

Body of
ischium

Greater sciatic notch

Superior ramus of pubis

Ischial spine

B. Lateral View

Acetabulum

Obturator foramen

Body of pubis

Lesser sciatic notch

Symphyseal surface

Ischial tuberosity

C. Medial Aspect

Ramus of ischium*

Inferior ramus of pubis*

Obturator foramen

*Ischiopubic ramus

3.1 **Pelvis, anatomical position**

A. Bony pelvis. **B.** Placement of hip bone in anatomical position. Note that in the anatomical position, the anterior superior iliac spine and the anterior aspect of the pubis lie in the
same vertical plane. **C.** Features of hip bone.

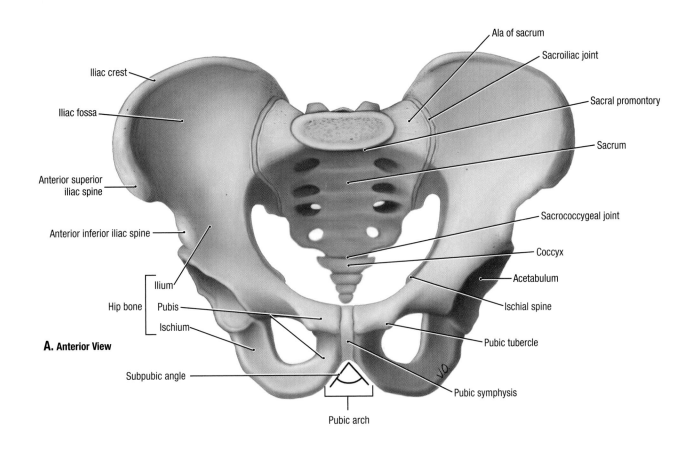

A. Anterior View

Iliac crest

Iliac fossa

Anterior superior iliac spine

Anterior inferior iliac spine

Ilium

Hip bone — Pubis

Ischium

Subpubic angle

Pubic arch

Ala of sacrum

Sacroiliac joint

Sacral promontory

Sacrum

Sacrococcygeal joint

Coccyx

Acetabulum

Ischial spine

Pubic tubercle

Pubic symphysis

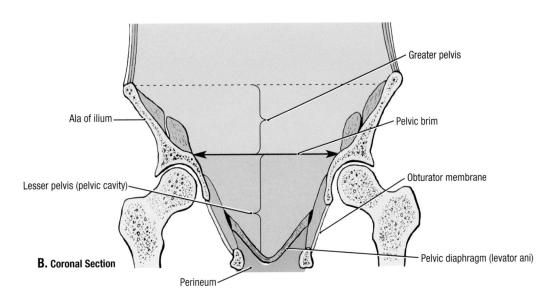

B. Coronal Section

Ala of ilium

Lesser pelvis (pelvic cavity)

Perineum

Greater pelvis

Pelvic brim

Obturator membrane

Pelvic diaphragm (levator ani)

3.2   Bones and divisions of pelvis

A. Bones of pelvis. The three bones composing the pelvis are the pubis, ischium, and ilium.
B. Lesser and greater pelvis, schematic illustration. The plane of the pelvic brim *(double-headed arrow)* separates the greater pelvis (part of the abdominal cavity) from the lesser pelvis (pelvic cavity). The pelvic diaphragm is formed mainly by the levator ani.

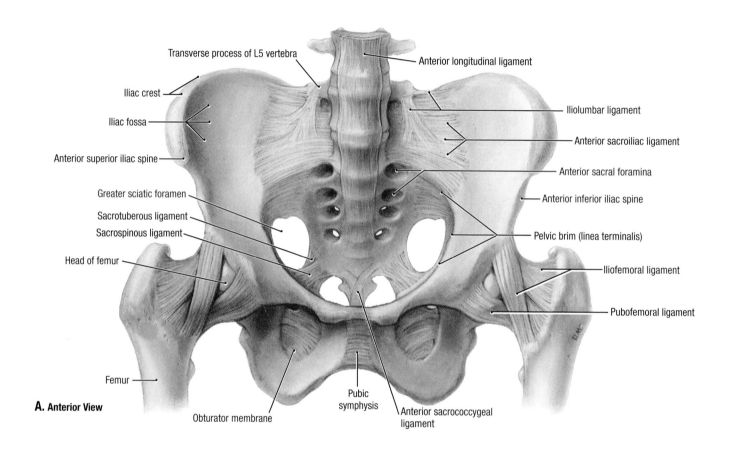

Transverse process of L5 vertebra

Anterior longitudinal ligament

Iliac crest

Iliac fossa

Anterior superior iliac spine

Greater sciatic foramen

Sacrotuberous ligament

Sacrospinous ligament

Head of femur

Femur

A. Anterior View

Obturator membrane

Pubic symphysis

Anterior sacrococcygeal ligament

Iliolumbar ligament

Anterior sacroiliac ligament

Anterior sacral foramina

Anterior inferior iliac spine

Pelvic brim (linea terminalis)

Iliofemoral ligament

Pubofemoral ligament

3.3 **Pelvis and pelvic ligaments**

A. Ligaments of pelvis, anterior aspect of pelvis. **B.** Male pelvis in situ. **C.** Female pelvis in situ.

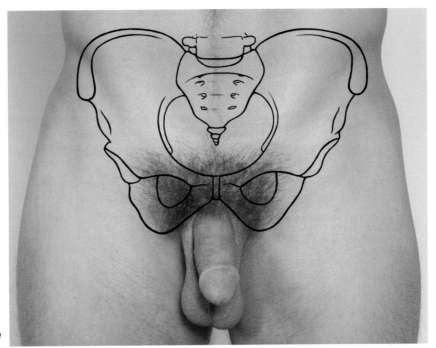

B. Anterior View

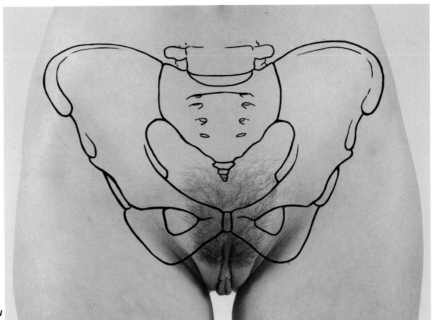

C. Anterior View

3.3   Pelvis and pelvic ligaments (*continued*)

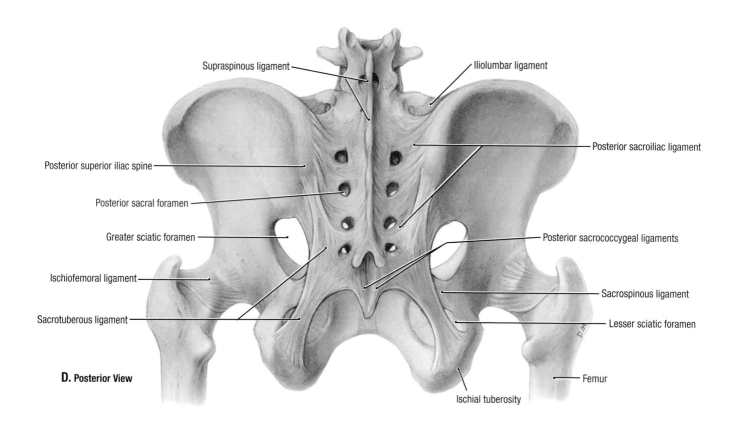

Supraspinous ligament

Iliolumbar ligament

Posterior superior iliac spine

Posterior sacral foramen

Greater sciatic foramen

Ischiofemoral ligament

Sacrotuberous ligament

Posterior sacroiliac ligament

Posterior sacrococcygeal ligaments

Sacrospinous ligament

Lesser sciatic foramen

Femur

Ischial tuberosity

D. Posterior View

3.3 **Pelvis and pelvic ligaments** *(continued)*
D. Ligaments of pelvis, posterior aspect of pelvis. **E.** Male pelvis in situ. **F.** Female pelvis in situ.

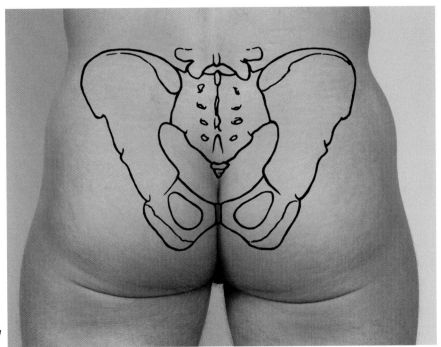

E. Posterior View

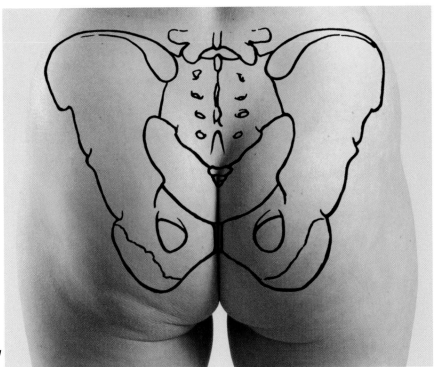

F. Posterior View

3.3   Pelvis and pelvic ligaments *(continued)*

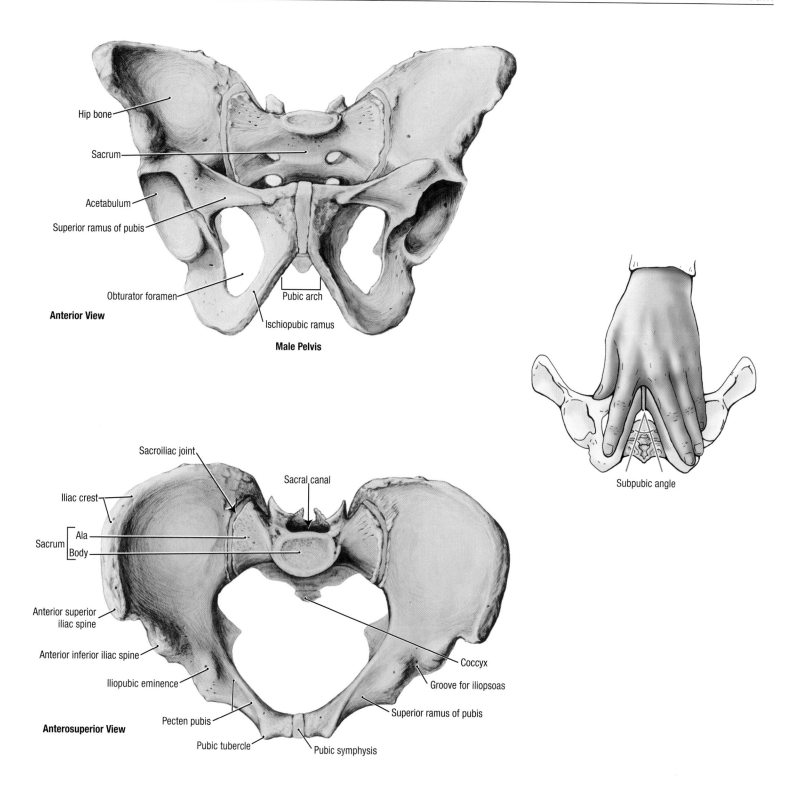

Hip bone

Sacrum

Acetabulum

Superior ramus of pubis

Obturator foramen

Anterior View

Pubic arch

Ischiopubic ramus

Male Pelvis

Subpubic angle

Sacroiliac joint

Sacral canal

Iliac crest

Sacrum — Ala / Body

Anterior superior iliac spine

Anterior inferior iliac spine

Iliopubic eminence

Pecten pubis

Anterosuperior View

Pubic tubercle

Pubic symphysis

Coccyx

Groove for iliopsoas

Superior ramus of pubis

TABLE 3.1   DIFFERENCES BETWEEN MALE AND FEMALE PELVES

BONY PELVIS	MALE	FEMALE
General structure	Thick and heavy	Thin and light
Greater pelvis (pelvis major)	Deep	Shallow
Lesser pelvis (pelvis minor)	Narrow and deep	Wide and shallow
Pelvic inlet (superior pelvic aperture)	Heart-shaped	Oval or rounded

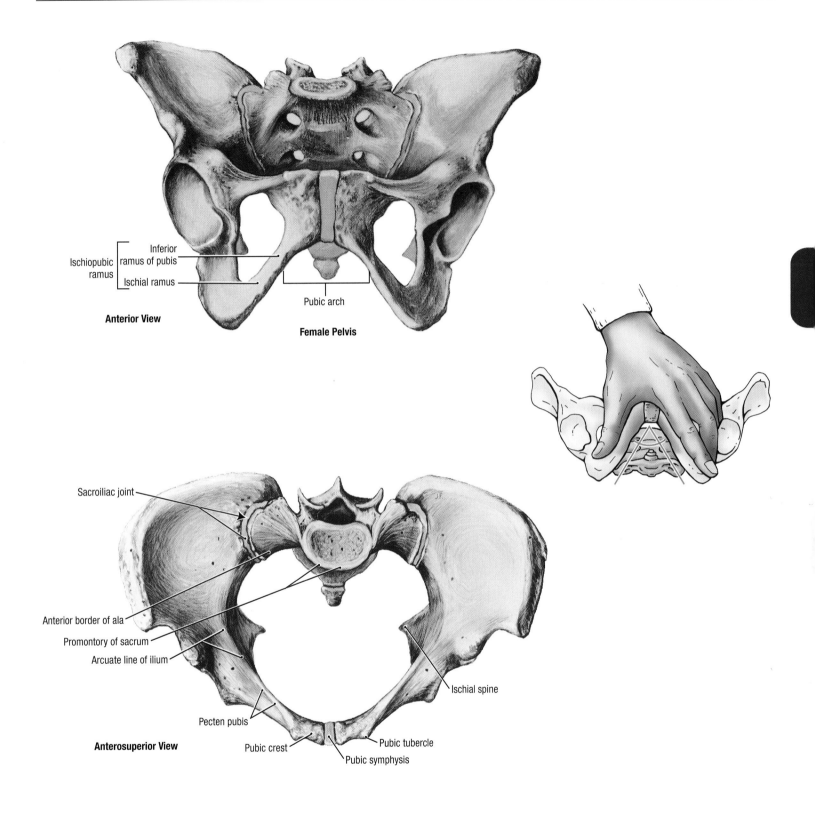

Ischiopubic ramus {
Inferior ramus of pubis
Ischial ramus

Pubic arch

Anterior View

Female Pelvis

Sacroiliac joint

Anterior border of ala

Promontory of sacrum

Arcuate line of ilium

Ischial spine

Pecten pubis

Anterosuperior View

Pubic crest

Pubic tubercle

Pubic symphysis

TABLE 3.1   DIFFERENCES BETWEEN MALE AND FEMALE PELVES (continued)

BONY PELVIS	MALE	FEMALE
Pelvis outlet (inferior pelvic aperture)	Comparatively small	Comparatively large
Pubic arch and subpubic angle	Narrow	Wide
Obturator foramen	Round	Oval
Acetabulum	Large	Small

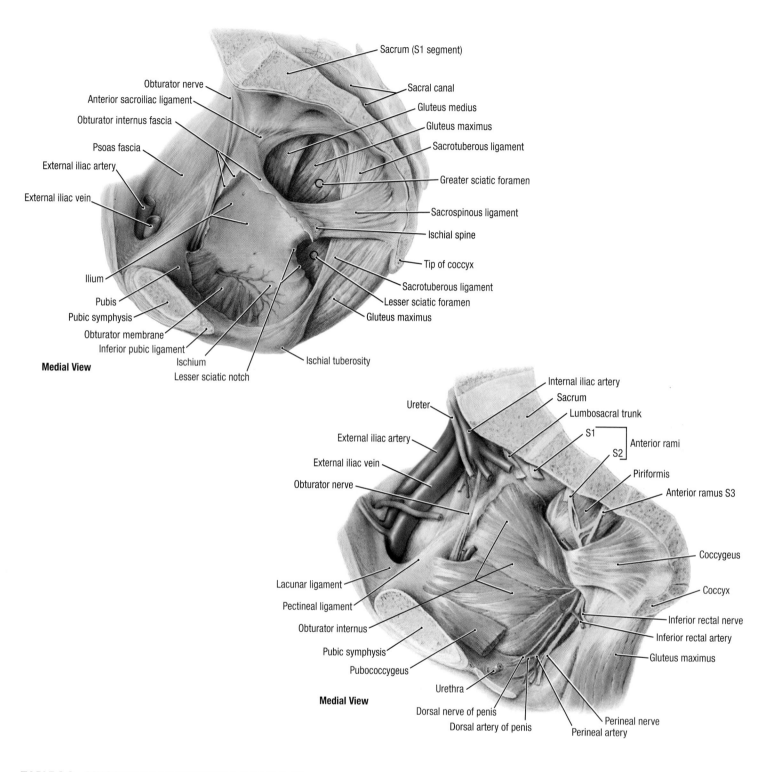

Medial View

Medial View

TABLE 3.2 MUSCLES OF PELVIC WALLS AND FLOOR

MUSCLE	PROXIMAL ATTACHMENT	DISTAL ATTACHMENT	INNERVATION	MAIN ACTION
Obturator internus	Pelvic surfaces of ilium and ischium; obturator membrane	Greater trochanter of femur	Nerve to obturator internus (L5, S1, and S2)	Rotates thigh laterally; assists in holding head of femur in acetabulum
Piriformis	Pelvic surface of second to fourth sacral segments: superior margin of greater sciatic notch, and sacrotuberous ligament		Anterior rami of S1 and S2	Rotates thigh laterally; abducts thigh; assists in holding head of femur in acetabulum

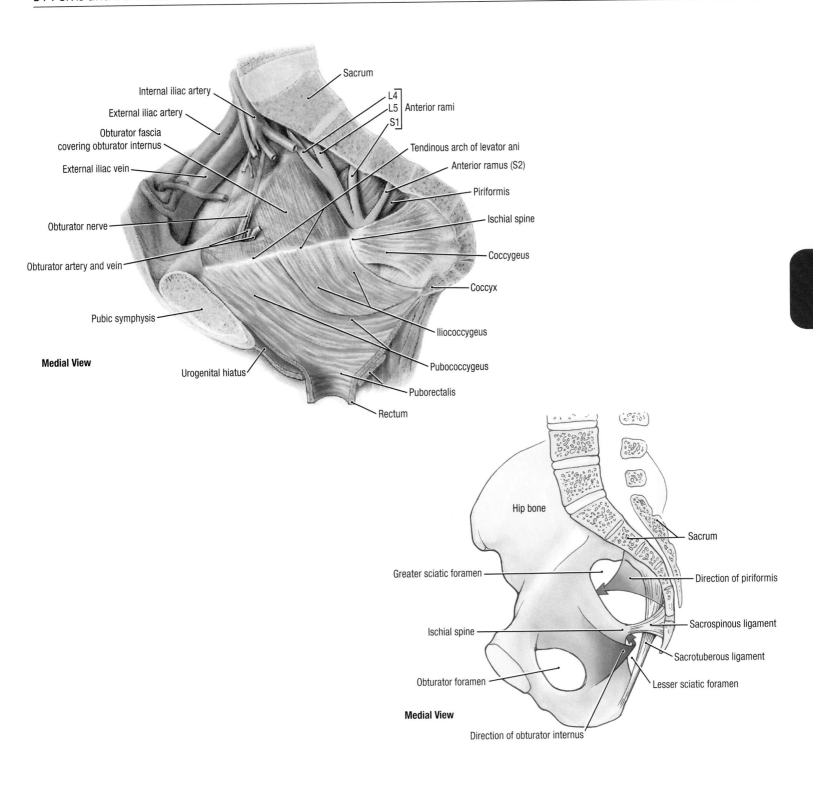

Medial View

- Sacrum
- Internal iliac artery
- External iliac artery
- Obturator fascia covering obturator internus
- External iliac vein
- Obturator nerve
- Obturator artery and vein
- Pubic symphysis
- Urogenital hiatus
- L4, L5, S1 Anterior rami
- Tendinous arch of levator ani
- Anterior ramus (S2)
- Piriformis
- Ischial spine
- Coccygeus
- Coccyx
- Iliococcygeus
- Pubococcygeus
- Puborectalis
- Rectum

Medial View

- Hip bone
- Sacrum
- Greater sciatic foramen
- Direction of piriformis
- Ischial spine
- Sacrospinous ligament
- Sacrotuberous ligament
- Obturator foramen
- Lesser sciatic foramen
- Direction of obturator internus

TABLE 3.2   MUSCLES OF PELVIC WALLS AND FLOOR *(continued)*

MUSCLE	PROXIMAL ATTACHMENT	DISTAL ATTACHMENT	INNERVATION	MAIN ACTION
Levator ani (pubococcygeus, puborectalis, and iliococcygeus)	Body of pubis, tendinous arch of obturator fascia, and ischial spine	Perineal body, coccyx, anococcygeal ligament, walls of prostate or vagina, rectum, and anal canal	Branches of S4 and pudendal	Form pelvic diaphragm; help to support the pelvic viscera and resist increases in intraabdominal pressure
Coccygeus (ischiococcygeus)	Ischial spine	Inferior end of sacrum	Branches of S4 and S5	

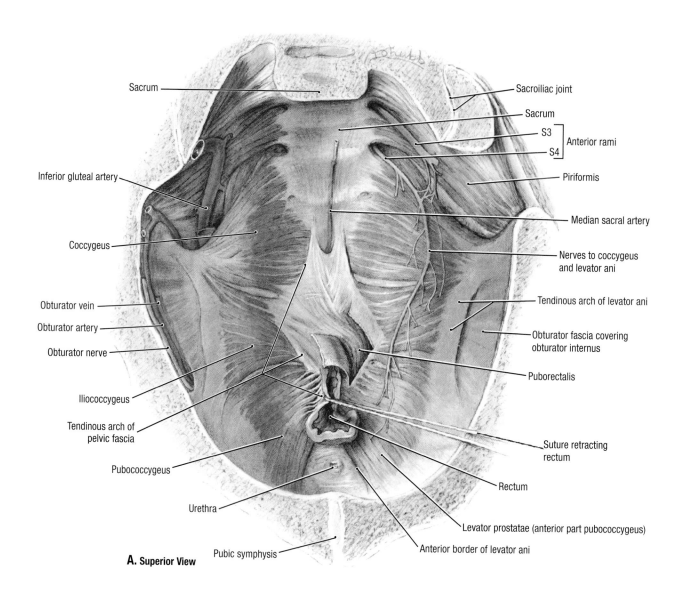

Sacrum

Sacroiliac joint

Sacrum

S3

S4

Anterior rami

Inferior gluteal artery

Piriformis

Coccygeus

Median sacral artery

Nerves to coccygeus and levator ani

Obturator vein

Tendinous arch of levator ani

Obturator artery

Obturator fascia covering obturator internus

Obturator nerve

Puborectalis

Iliococcygeus

Tendinous arch of pelvic fascia

Suture retracting rectum

Pubococcygeus

Rectum

Urethra

Levator prostatae (anterior part pubococcygeus)

Pubic symphysis

Anterior border of levator ani

A. Superior View

3.4 Floor of male pelvis, pelvic diaphragm

A. Dissection. The pelvic viscera are removed, and the bony pelvis has been cut in the transverse plane to show the levator ani and coccygeus muscles. **B.** Schematic illustration.

- The pubococcygeus muscle arises mainly from the pubic bone, the iliococcygeus muscle from the tendinous arch, and the coccygeus muscle from the ischial spine.
- In the male, the anterior part of the pubococcygeus muscle that lies adjacent to the prostate is the levator prostatae.
- Although not part of the pelvic diaphragm, the piriformis assists in closure of the pelvic outlet, largely occluding the greater sciatic foramen.

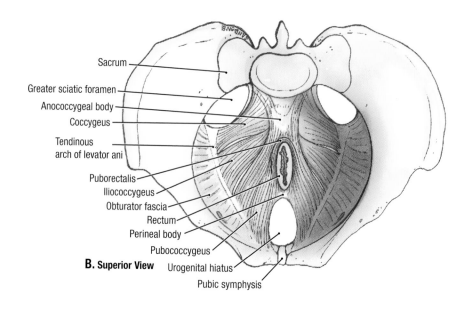

Sacrum

Greater sciatic foramen

Anococcygeal body

Coccygeus

Tendinous arch of levator ani

Puborectalis

Iliococcygeus

Obturator fascia

Rectum

Perineal body

Pubococcygeus

B. Superior View Urogenital hiatus

Pubic symphysis

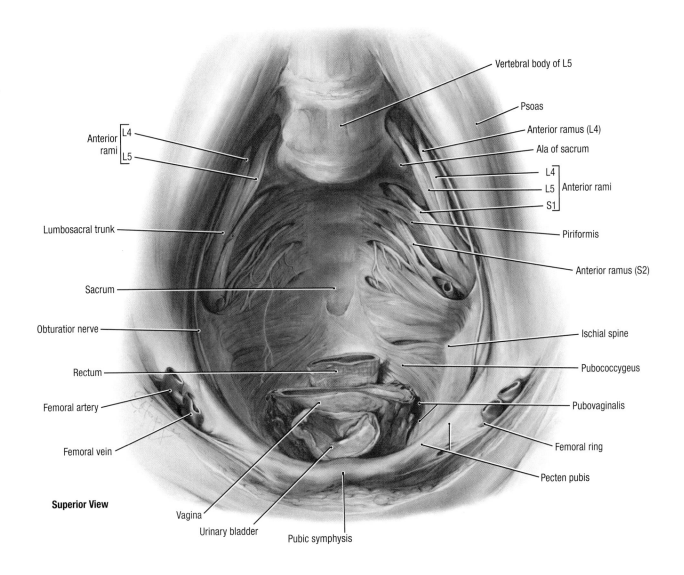

Superior View

3.5   Floor of female pelvis

The pelvic viscera are removed to reveal the levator ani and coccygeus muscles.

- Note the relative positions of the bladder, vagina, and rectum as they penetrate the pelvic floor.
- Branches of S3 and S4 nerves supply the levator ani and coccygeus muscles; the pudendal nerve, through its perineal branch, also supplies the levator ani muscle (see Table 3.2).
- The obturator nerve runs along the lateral wall of the pelvis and enters the thigh by passing through the obturator foramen.
- The anterior rami of L4–S4 are part of the sacral plexus, almost all of which exits the pelvis via the greater sciatic foramen with the piriformis.

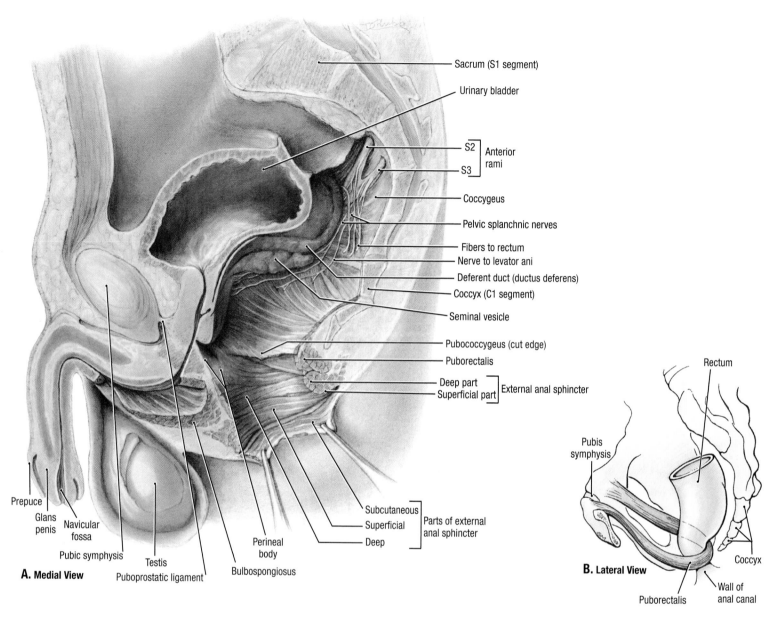

A. Medial View

Prepuce
Glans penis
Navicular fossa
Pubic symphysis
Testis
Puboprostatic ligament
Perineal body
Bulbospongiosus
Subcutaneous
Superficial
Deep
Parts of external anal sphincter

Sacrum (S1 segment)
Urinary bladder
S2 / S3 Anterior rami
Coccygeus
Pelvic splanchnic nerves
Fibers to rectum
Nerve to levator ani
Deferent duct (ductus deferens)
Coccyx (C1 segment)
Seminal vesicle
Pubococcygeus (cut edge)
Puborectalis
Deep part / Superficial part External anal sphincter

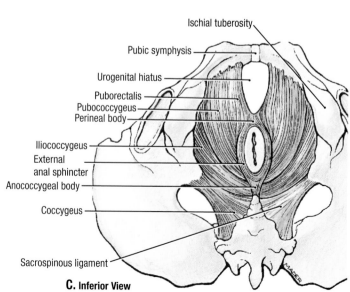

B. Lateral View

Rectum
Pubis symphysis
Puborectalis
Coccyx
Wall of anal canal

Ischial tuberosity
Pubic symphysis
Urogenital hiatus
Puborectalis
Pubococcygeus
Perineal body
Iliococcygeus
External anal sphincter
Anococcygeal body
Coccygeus
Sacrospinous ligament

C. Inferior View

3.6 Levator ani

A. Levator ani, median section. The rectum, anal canal, and bulb of the penis are removed from the specimen in Figure 3.7A. The subcutaneous fibers of the external anal sphincter are reflected with forceps. The pubococcygeus muscle is cut to remove the anal canal, to which it is, in part, attached. **B.** Puborectalis. The innermost part of the pubococcygeus muscle, the puborectalis, forms a U-shaped muscular "sling" around the anorectal junction, which maintains the anorectal flexure.

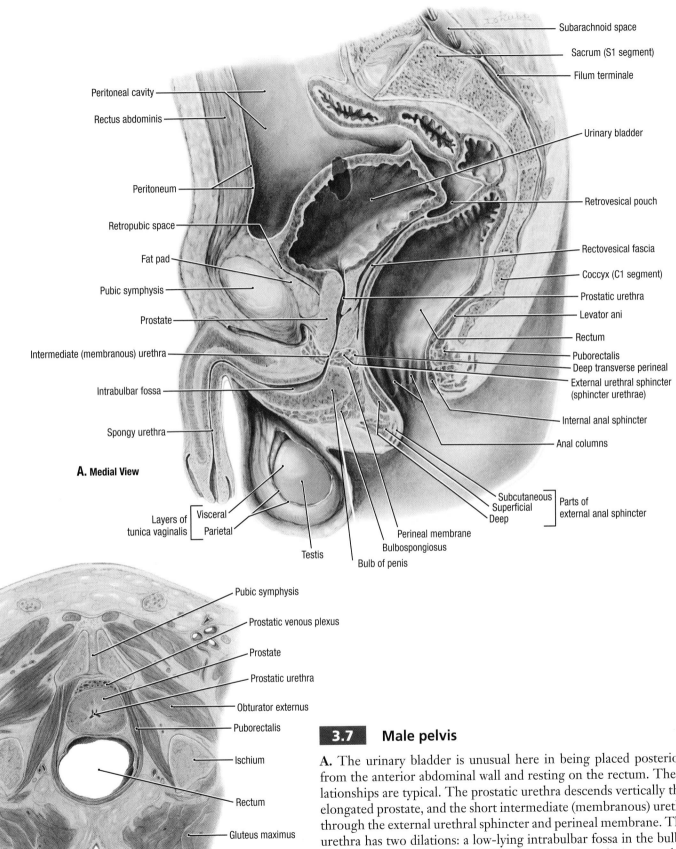

Peritoneal cavity

Rectus abdominis

Peritoneum

Retropubic space

Fat pad

Pubic symphysis

Prostate

Intermediate (membranous) urethra

Intrabulbar fossa

Spongy urethra

A. Medial View

Layers of tunica vaginalis { Visceral | Parietal }

Testis

Bulb of penis

Bulbospongiosus

Perineal membrane

Subcutaneous | Parts of
Superficial | external anal sphincter
Deep

Subarachnoid space

Sacrum (S1 segment)

Filum terminale

Urinary bladder

Retrovesical pouch

Rectovesical fascia

Coccyx (C1 segment)

Prostatic urethra

Levator ani

Rectum

Puborectalis

Deep transverse perineal

External urethral sphincter (sphincter urethrae)

Internal anal sphincter

Anal columns

Pubic symphysis

Prostatic venous plexus

Prostate

Prostatic urethra

Obturator externus

Puborectalis

Ischium

Rectum

Gluteus maximus

B. Transverse Section

3.7 Male pelvis

A. The urinary bladder is unusual here in being placed posteriorly, away from the anterior abdominal wall and resting on the rectum. The other relationships are typical. The prostatic urethra descends vertically through an elongated prostate, and the short intermediate (membranous) urethra passes through the external urethral sphincter and perineal membrane. The spongy urethra has two dilations: a low-lying intrabulbar fossa in the bulb, and the navicular fossa in the glans penis. The tunica vaginalis is opened to expose the testis. **B.** Section through the prostate demonstrating the puborectalis (puborectal sling) encircling the rectum, pulling it anteriorly against the prostate.

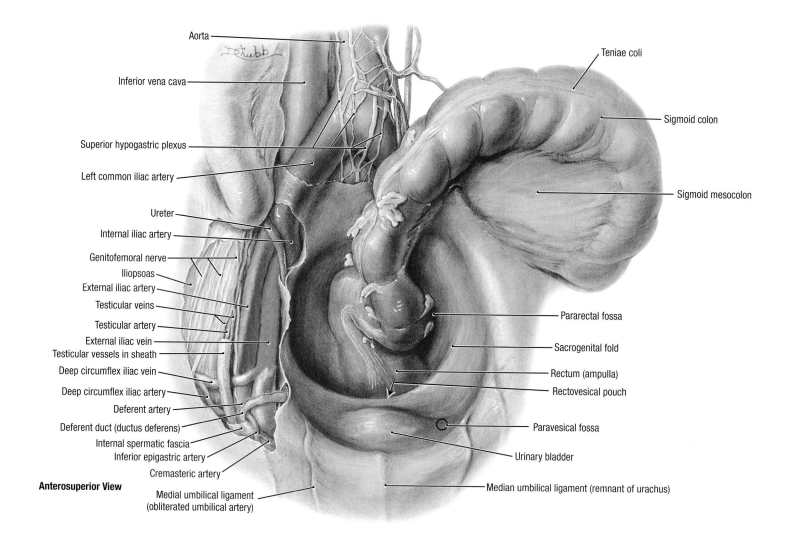

Aorta

Inferior vena cava

Superior hypogastric plexus

Left common iliac artery

Ureter

Internal iliac artery

Genitofemoral nerve

Iliopsoas

External iliac artery

Testicular veins

Testicular artery

External iliac vein

Testicular vessels in sheath

Deep circumflex iliac vein

Deep circumflex iliac artery

Deferent artery

Deferent duct (ductus deferens)

Internal spermatic fascia

Inferior epigastric artery

Cremasteric artery

Anterosuperior View

Medial umbilical ligament
(obliterated umbilical artery)

Teniae coli

Sigmoid colon

Sigmoid mesocolon

Pararectal fossa

Sacrogenital fold

Rectum (ampulla)

Rectovesical pouch

Paravesical fossa

Urinary bladder

Median umbilical ligament (remnant of urachus)

3.8   Male pelvis and surrounding structures

- The sigmoid colon begins at the left pelvic brim and becomes the rectum anterior to the third sacral segment in the midline.
- The attachment of the mesentery of the sigmoid colon, the sigmoid mesocolon, to the pelvic wall has an inverted V shape.
- The superior hypogastric plexus lies inferior to the bifurcation of the aorta and anterior to the left common iliac vein.
- The ureter adheres to the external aspect of the peritoneum, crosses the external iliac vessels, and descends anterior to the internal iliac artery. The deferent duct and its artery also adhere to the peritoneum, cross the external iliac vessels, and then hook around the inferior epigastric artery to join the other components of the spermatic cord.
- The extraperitoneal genitofemoral nerve lies on the psoas fascia; the femoral branches become cutaneous in the thigh, and the genital branch supplies the cremaster muscle and then becomes cutaneous in the anterior perineum.

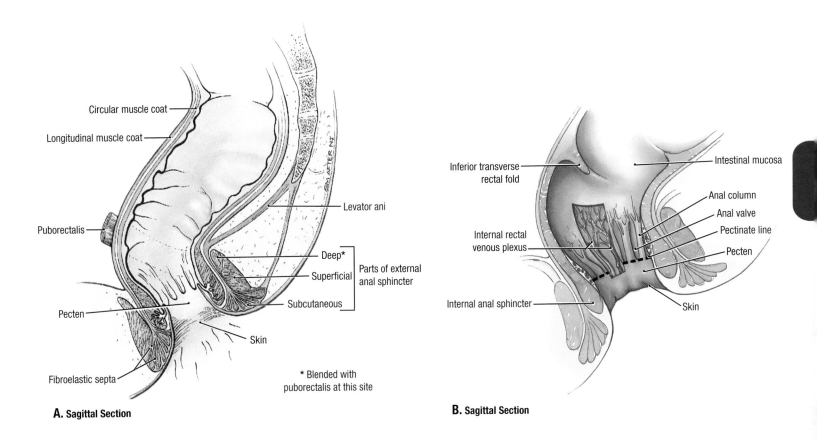

A. Sagittal Section

* Blended with
puborectalis at this site

B. Sagittal Section

3.9   Anal sphincters and anal canal

A. External and internal anal sphincters. **B.** Features of the anal canal.

- The internal anal sphincter is a thickening of the inner, circular, muscular coat of the anal canal.
- There are three continuous zones of the external anal sphincter: deep, superficial, and subcutaneous; the deep part intermingles with the puborectalis muscle posteriorly.
- The longitudinal muscle layer of the rectum separates the internal and external anal sphincters and terminates in the subcutaneous tissue and skin around the anus.
- The anal columns are 5 to 10 vertical folds of mucosa separated by anal valves; they contain portions of the rectal venous plexus.
- The pecten is a smooth area of hairless stratified epithelium that lies between the anal valves superiorly and the inferior border of the internal anal sphincter inferiorly.
- The pectinate line is an irregular line at the base of the anal valves where the intestinal mucosa is continuous with the pecten; this indicates the junction of the superior part of the anal canal (derived from hindgut) and the inferior part of the anal canal (derived from the proctodeum). Innervation is visceral proximal to the line and somatic distally; lymphatic drainage is to the pararectal nodes proximally and to the superficial inguinal nodes distally.

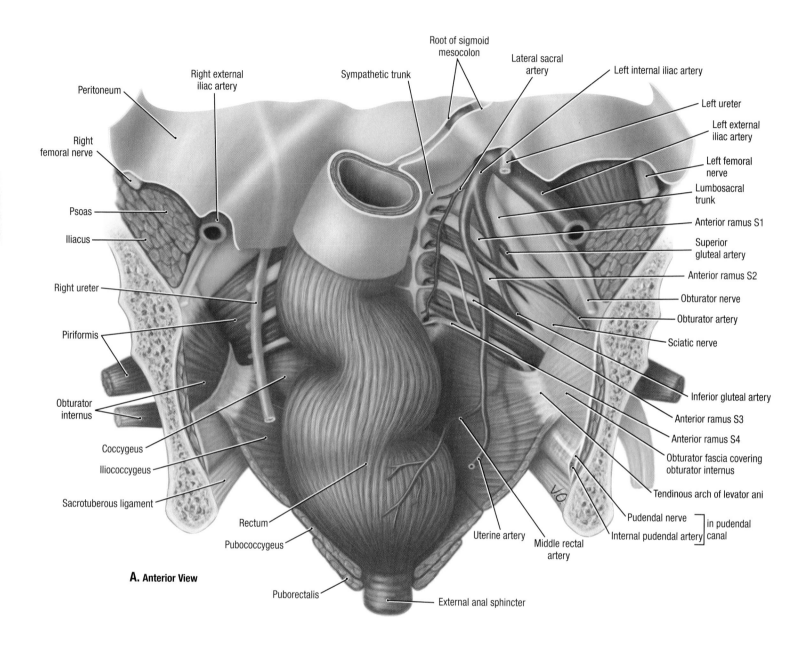

Right external
iliac artery

Peritoneum

Root of sigmoid
mesocolon

Sympathetic trunk

Lateral sacral
artery

Left internal iliac artery

Left ureter

Left external
iliac artery

Right
femoral nerve

Left femoral
nerve

Lumbosacral
trunk

Psoas

Iliacus

Anterior ramus S1

Superior
gluteal artery

Right ureter

Anterior ramus S2

Obturator nerve

Piriformis

Obturator artery

Sciatic nerve

Obturator
internus

Inferior gluteal artery

Anterior ramus S3

Coccygeus

Anterior ramus S4

Iliococcygeus

Obturator fascia covering
obturator internus

Sacrotuberous ligament

Tendinous arch of levator ani

Rectum

Pubococcygeus

Uterine artery

Middle rectal
artery

Pudendal nerve ⎤ in pudendal
Internal pudendal artery ⎦ canal

Puborectalis

External anal sphincter

A. Anterior View

3.10   **Rectum and anal canal**

A. Dissection of rectum. **B.** Arterial supply. **C.** Venous drainage.
In **B:**

- The branches of the right and left divisions of the superior rectal artery obliquely encir-
cle the rectum.
- The middle rectal arteries (branches of the internal iliac arteries) are usually small; in this
specimen, the right artery is small, but the left one is large. The inferior rectal arteries
branch from the internal pudendal arteries.

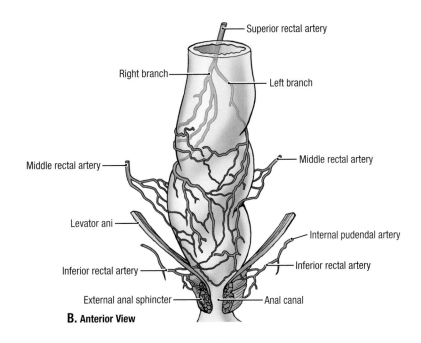

B. Anterior View

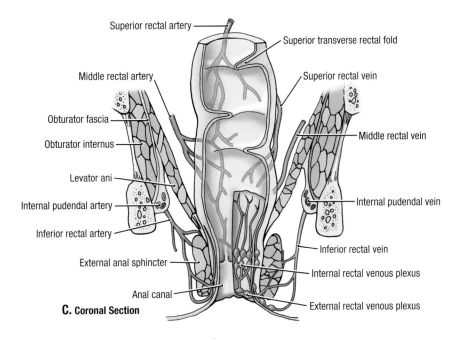

C. Coronal Section

3.10 **Rectum and anal canal (continued)**

In **C**:
- The superior, middle, and inferior rectal veins drain the rectum and anal canal; there are anastomoses between the plexuses formed by all three veins.
- The rectal venous plexus surrounds the distal rectum and anal canal and consists of an internal rectal plexus deep to the epithelium of the anal canal and an external rectal plexus external to the muscular coats of the wall of the anal canal.
- The superior rectal vein drains into the portal system, and the middle and inferior veins drain into the systemic system; thus, this is an important area of portacaval anastomosis. The normal varicosities of the internal rectal plexus are included in internal hemorrhoids.

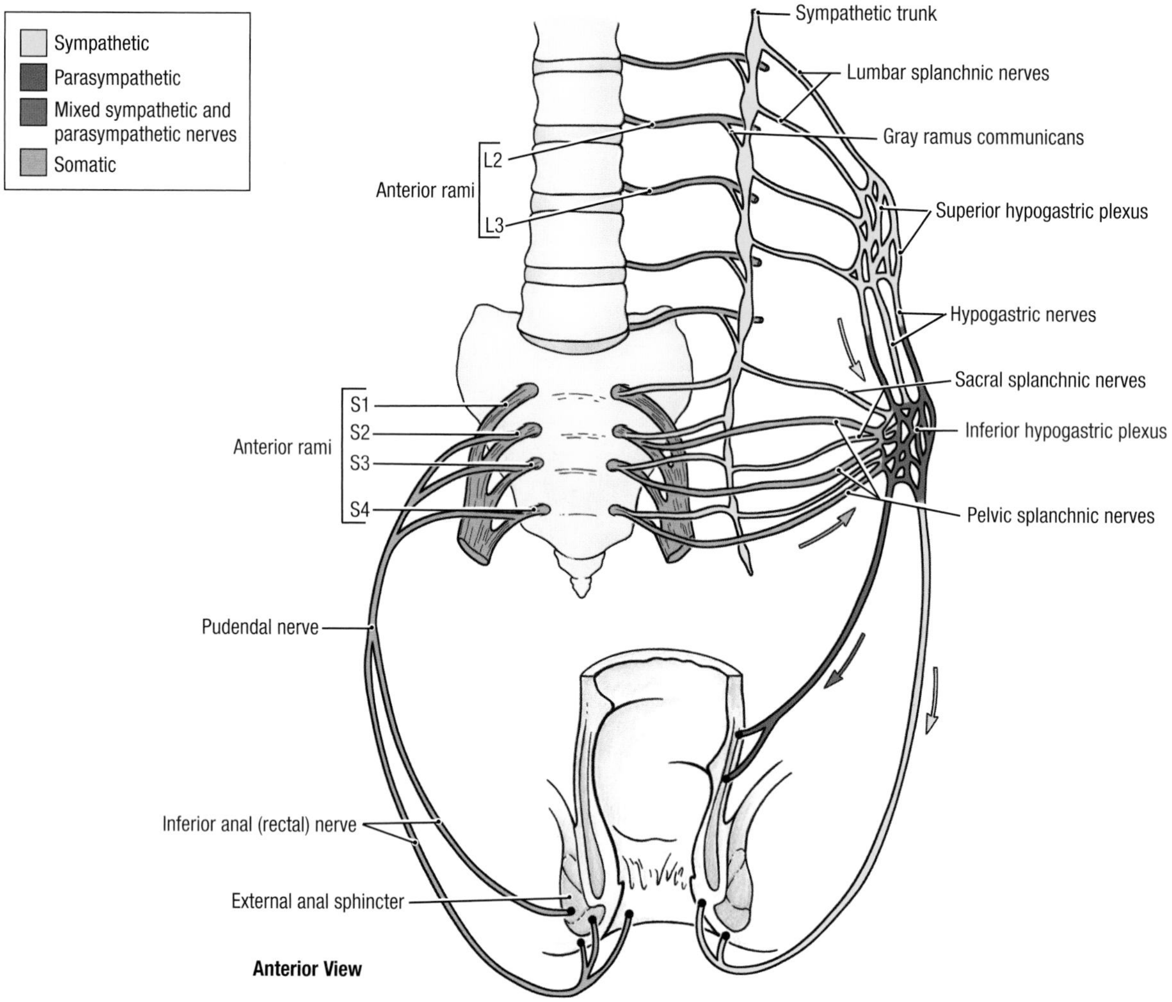

Anterior View

3.11 Innervation of rectum and anal canal

- On the right side of the figure: The voluntary external anal sphincter and perianal skin are innervated by the inferior anal (rectal) nerve, a branch of the pudendal nerve (S2, S3, and S4).
- On the left: The autonomic innervation is from the parasympathetic pelvic splanchnic nerves (S2, S3, and S4), sympathetic hypogastric nerves (from the superior hypogastric plexus), and visceral (sacral) splanchnic nerves (from the 2nd and 3rd sacral sympathetic ganglia). The sympathetic and parasympathetic fibers become intermingled in the inferior hypogastric plexus, located on the lateral wall of the rectum. The fibers are conveyed from the plexus to the wall of the rectum and the involuntary internal anal sphincter.

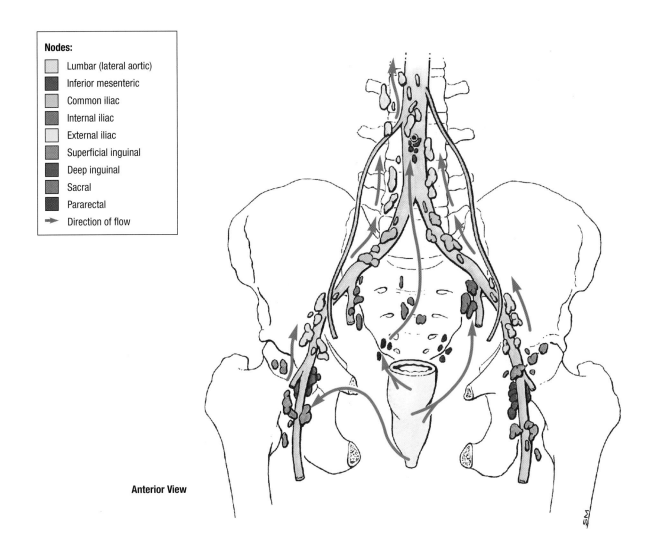

Nodes:

- Lumbar (lateral aortic)
- Inferior mesenteric
- Common iliac
- Internal iliac
- External iliac
- Superficial inguinal
- Deep inguinal
- Sacral
- Pararectal
- ➤ Direction of flow

Anterior View

3.12   Lymphatic drainage of rectum

- The lymphatic vessels from the superior part of the rectum ascend along the superior rectal vessels to the pararectal and sacral lymph nodes. Lymph then passes to abdominal lymph nodes in the inferior part of the mesentry of the sigmoid colon and from them to the inferior mesenteric and lateral aortic (lumbar) lymph nodes.
- Superior to the pectinate line, the lymphatic vessels ascend with the middle rectal arteries and drain into the internal iliac nodes.
- Inferior to the pectinate line the lymphatic vessels drain into the superficial inguinal lymph nodes.

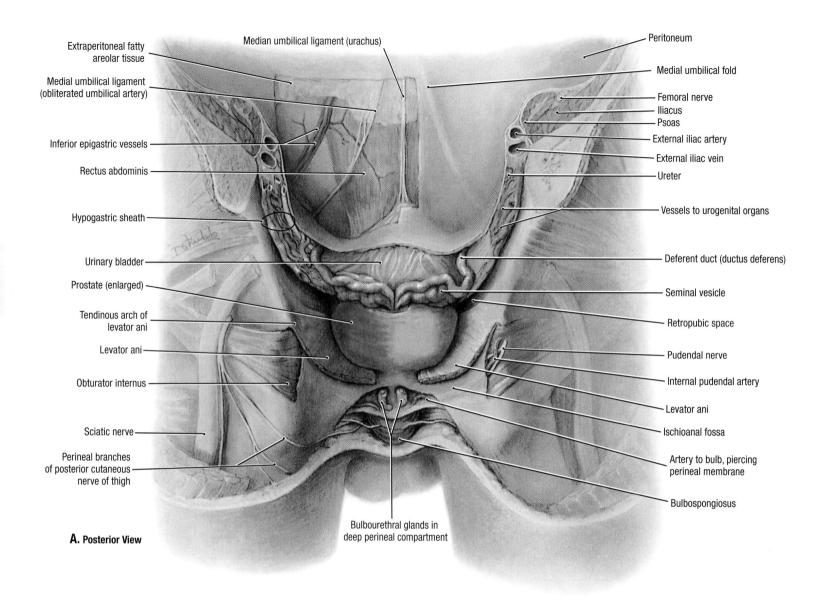

Extraperitoneal fatty areolar tissue

Medial umbilical ligament (obliterated umbilical artery)

Inferior epigastric vessels

Rectus abdominis

Hypogastric sheath

Urinary bladder

Prostate (enlarged)

Tendinous arch of levator ani

Levator ani

Obturator internus

Sciatic nerve

Perineal branches of posterior cutaneous nerve of thigh

Median umbilical ligament (urachus)

Peritoneum

Medial umbilical fold

Femoral nerve

Iliacus

Psoas

External iliac artery

External iliac vein

Ureter

Vessels to urogenital organs

Deferent duct (ductus deferens)

Seminal vesicle

Retropubic space

Pudendal nerve

Internal pudendal artery

Levator ani

Ischioanal fossa

Artery to bulb, piercing perineal membrane

Bulbospongiosus

Bulbourethral glands in deep perineal compartment

A. Posterior View

3.13   Posterior approach to anterior pelvis and perineum

A. Dissection anterior to rectum (removed), view of anterior portion from behind. **B.** Umbilical ligaments, posterior surface of inferior part of anterior abdominal wall and anterior pelvic viscera. **C.** Schematic coronal section through the anterior pelvis (plane of urinary bladder and prostate).

- In **A** and **B,** the inferior epigastric artery and communicating veins enter the rectus sheath to form the lateral umbilical ligament. The medial umbilical ligament is formed by the obliterated umbilical artery, and the median umbilical ligament is formed by the urachus.
- In **A,** the femoral nerve lies between the psoas and iliacus muscles outside the psoas fascia, which is attached to the pelvic brim; the external iliac artery and vein lie within the psoas fascia.
- The deferent duct and ureter are subperitoneal; near the bladder, the ureter accompanies a "leash" of vesical vessels enclosed in the hypogastric sheath.
- The levator ani muscle and its fascial coverings separate the retropubic space (pelvis) from the ischioanal fossa (perineum).
- The bulbourethral glands and the artery to the bulb lie superior to the perineal membrane.

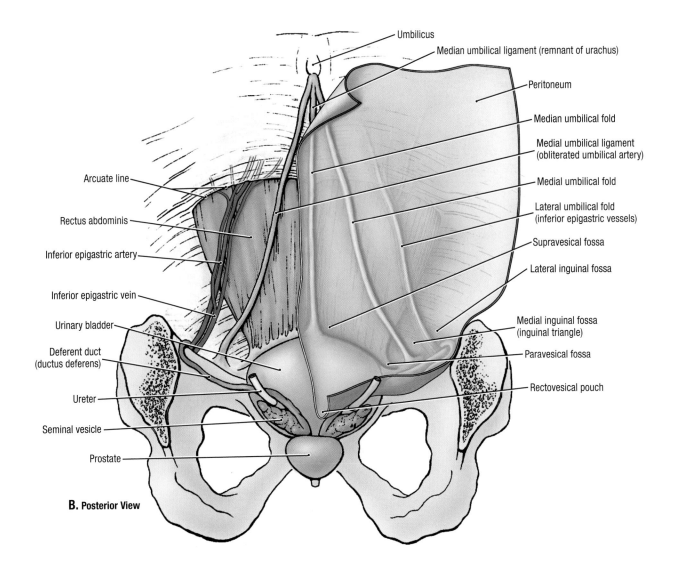

Umbilicus

Median umbilical ligament (remnant of urachus)

Peritoneum

Median umbilical fold

Medial umbilical ligament (obliterated umbilical artery)

Medial umbilical fold

Lateral umbilical fold (inferior epigastric vessels)

Supravesical fossa

Lateral inguinal fossa

Medial inguinal fossa (inguinal triangle)

Paravesical fossa

Rectovesical pouch

Arcuate line

Rectus abdominis

Inferior epigastric artery

Inferior epigastric vein

Urinary bladder

Deferent duct (ductus deferens)

Ureter

Seminal vesicle

Prostate

B. Posterior View

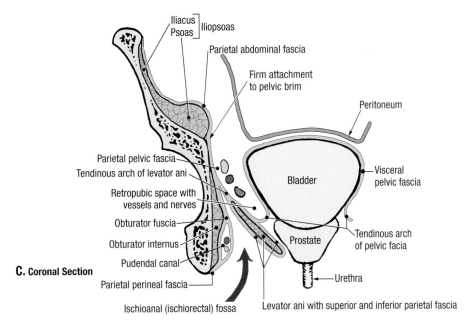

Iliacus ⎤
Psoas ⎦ Iliopsoas

Parietal abdominal fascia

Firm attachment to pelvic brim

Peritoneum

Parietal pelvic fascia

Tendinous arch of levator ani

Retropubic space with vessels and nerves

Obturator fuscia

Obturator internus

Pudendal canal

Parietal perineal fascia

Ischioanal (ischiorectal) fossa

Bladder

Visceral pelvic fascia

Prostate

Tendinous arch of pelvic facia

Urethra

Levator ani with superior and inferior parietal fascia

C. Coronal Section

3.13   **Posterior approach to anterior pelvis and perineum (*continued*)**

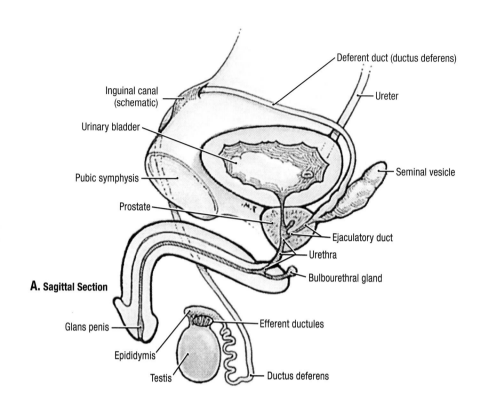

A. Sagittal Section

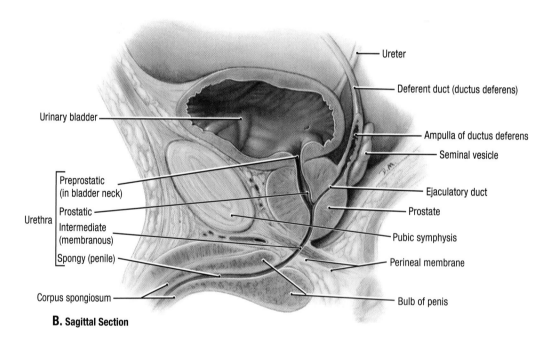

B. Sagittal Section

3.14   **Male pelvis**

A. Overview of urogenital system. **B.** Bladder, prostate, and deferent duct. The ejaculatory duct (approximately 2 cm in length) is formed by the union of the deferent duct and duct of the seminal vesicle; it passes anteriorly and inferiorly through the substance of the prostate to enter the prostatic urethra on the seminal colliculus.

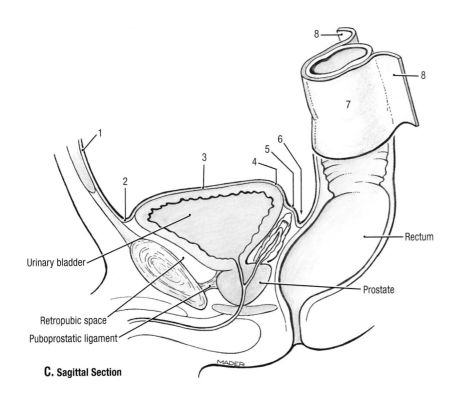

Male

Peritoneum passes:
- From the anterior abdominal wall *(1)*
- Superior to the pubic bone *(2)*
- On the superior surface of the urinary bladder *(3)*
- 2 cm inferiorly on the posterior surface of the urinary bladder *(4)*
- On the superior ends of the seminal glands *(5)*
- Posteriorly to line the rectovesical pouch *(6)*
- To cover the rectum *(7)*
- Posteriorly to become the sigmoid mesocolon *(8)*

Rectum

Prostate

Urinary bladder

Retropubic space

Puboprostatic ligament

C. Sagittal Section

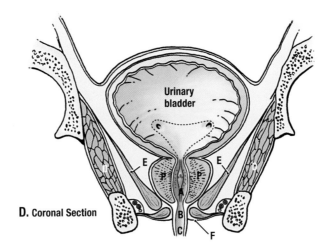

D. Coronal Section

3.14   **Male pelvis** *(continued)*

C. Peritoneum covering male pelvic organs. **D.** Coronal section through urinary bladder and prostate. Observe the parts of the urethra: prostatic *(A)*, intermediate (membranous) *(B)*, and spongy *(C)*, as well as the obturator internus *(D)* and levator ani *(E)* muscles, the perineal membrane *(F)*, and the prostate *(P)*.

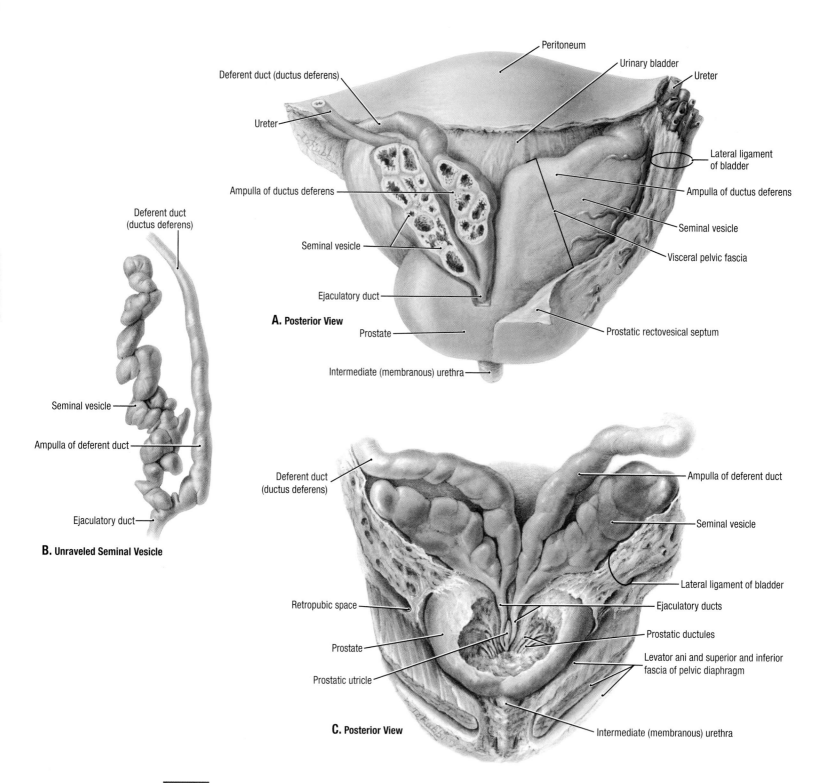

A. Posterior View

B. Unraveled Seminal Vesicle

C. Posterior View

3.15 Seminal vesicles and prostate

A. Bladder, deferent ducts, seminal vesicles, and prostate. The left seminal vesicle and ampulla of the deferent duct are dissected and opened; part of the prostate is cut away to expose the ejaculatory duct. **B.** Seminal vesicle, unraveled. The vesicle is a tortuous tube with numerous dilatations. The ampulla of the deferent duct (vas deferens) has similar dilatations. **C.** Prostate, dissected posteriorly. The ejaculatory duct (approximately 2 cm in length) is formed by the union of the deferent duct and the duct of the seminal vesicle; it passes anteriorly and inferiorly through the substance of the prostate to enter the prostatic urethra on the seminal colliculus. The prostatic utricle lies between the ends of the two ejaculatory ducts. The prostatic ductules mostly open onto the prostatic sinus.

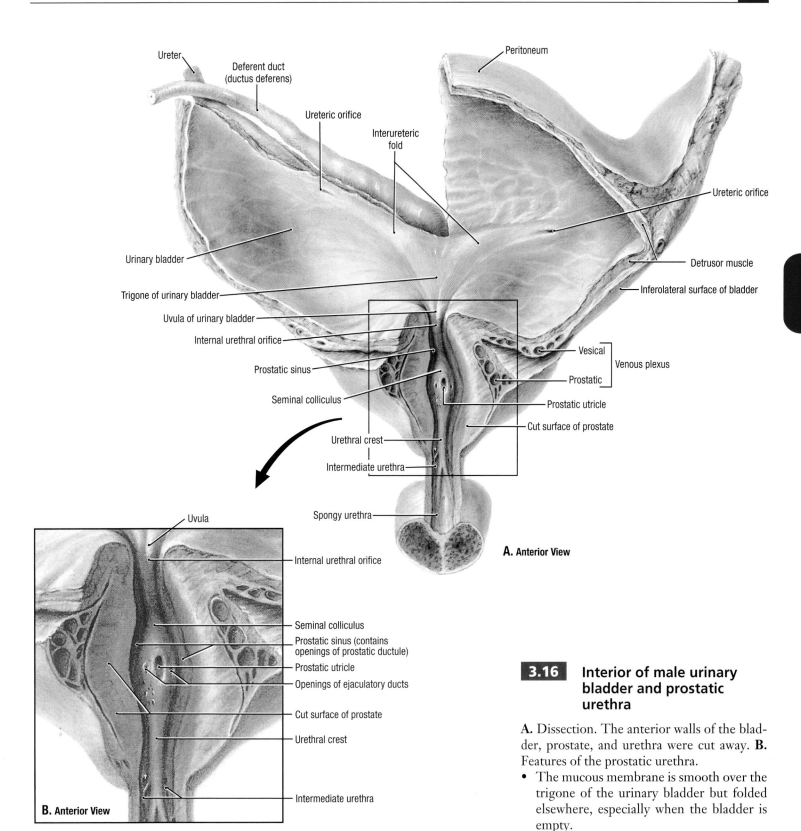

Ureter
Deferent duct (ductus deferens)
Ureteric orifice
Interureteric fold
Peritoneum
Ureteric orifice
Urinary bladder
Detrusor muscle
Inferolateral surface of bladder
Trigone of urinary bladder
Uvula of urinary bladder
Internal urethral orifice
Prostatic sinus
Vesical
Prostatic
Venous plexus
Seminal colliculus
Prostatic utricle
Cut surface of prostate
Urethral crest
Intermediate urethra
Spongy urethra

A. Anterior View

Uvula
Internal urethral orifice
Seminal colliculus
Prostatic sinus (contains openings of prostatic ductule)
Prostatic utricle
Openings of ejaculatory ducts
Cut surface of prostate
Urethral crest
Intermediate urethra

B. Anterior View

3.16 Interior of male urinary bladder and prostatic urethra

A. Dissection. The anterior walls of the bladder, prostate, and urethra were cut away. **B.** Features of the prostatic urethra.

- The mucous membrane is smooth over the trigone of the urinary bladder but folded elsewhere, especially when the bladder is empty.
- The opening of the prostatic utricle is in the seminal colliculus on the urethral crest; there is an orifice of an ejaculatory duct on each side of the prostatic utricle. The prostatic fascia encloses a venous plexus.
- In this specimen, the urethral crest extends farther superiorly and bifurcates more inferiorly than usual.

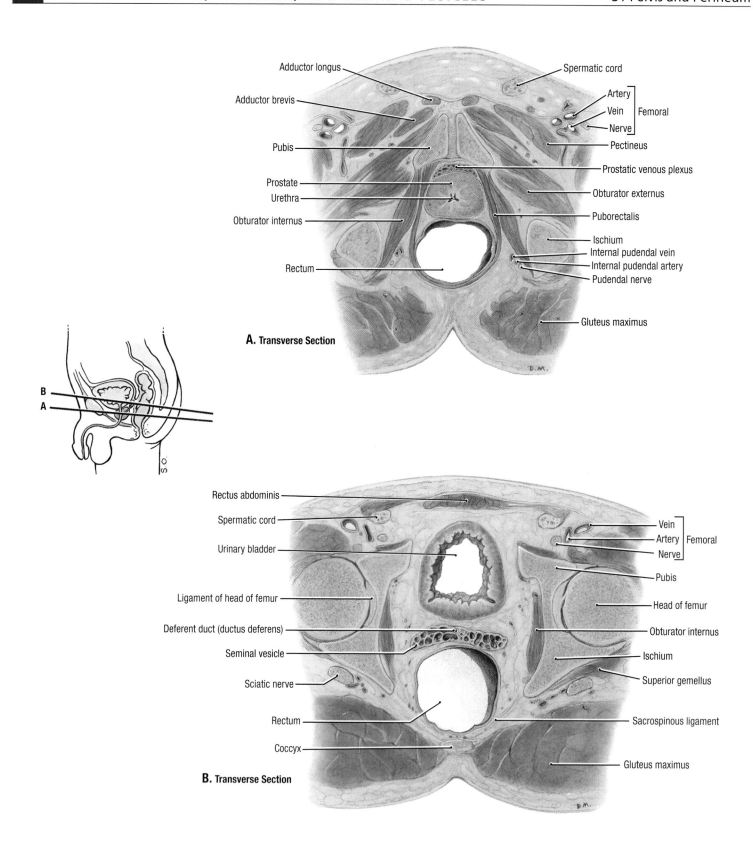

Adductor longus
Adductor brevis
Pubis
Prostate
Urethra
Obturator internus
Rectum

Spermatic cord
Artery
Vein — Femoral
Nerve
Pectineus
Prostatic venous plexus
Obturator externus
Puborectalis
Ischium
Internal pudendal vein
Internal pudendal artery
Pudendal nerve
Gluteus maximus

A. Transverse Section

Rectus abdominis
Spermatic cord
Urinary bladder
Ligament of head of femur
Deferent duct (ductus deferens)
Seminal vesicle
Sciatic nerve
Rectum
Coccyx

Vein
Artery — Femoral
Nerve
Pubis
Head of femur
Obturator internus
Ischium
Superior gemellus
Sacrospinous ligament
Gluteus maximus

B. Transverse Section

3.17   Male pelvis, transverse sections

A. Transverse section through prostate and puborectalis. **B.** Transverse section through urinary bladder and seminal vesicles.

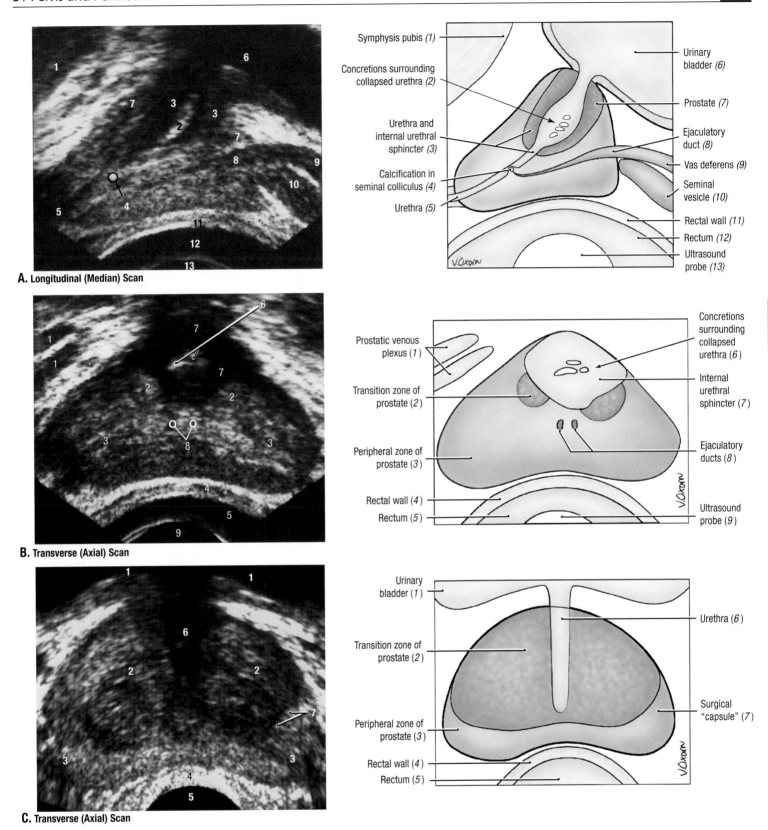

A. Longitudinal (Median) Scan

B. Transverse (Axial) Scan

C. Transverse (Axial) Scan

3.18   **Ultrasound scans of male pelvis**

A. In this longitudinal (transrectal) ultrasound scan, the probe was inserted into the rectum to scan the anteriorly located prostate. The ducts of the glands in the peripheral zone open into the prostatic sinuses, whereas the ducts of the glands in the central (internal) zone open into the prostatic sinuses and onto the seminal colliculus. The large peripheral zone is the common site for carcinomas. The central zone is the site of enlargement in cases of benign prostate hypertrophy. **B** and **C.** Transverse (axial) transrectal ultrasound scans. **B.** Normal prostate of young male. **C.** Benign prostatic hyperplasia. In **C**, note the enlarged transition zone. The transition zone of the prostate normally starts becoming hyperplastic after age 30. The *numbers in parentheses* correspond to labels on the ultrasound scan.

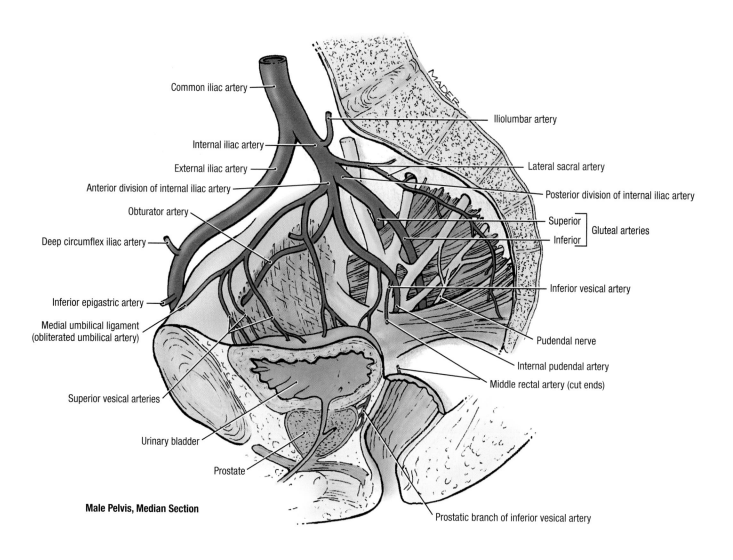

Common iliac artery

Internal iliac artery

External iliac artery

Anterior division of internal iliac artery

Obturator artery

Deep circumflex iliac artery

Inferior epigastric artery

Medial umbilical ligament
(obliterated umbilical artery)

Superior vesical arteries

Urinary bladder

Prostate

Iliolumbar artery

Lateral sacral artery

Posterior division of internal iliac artery

Superior ⎤
        ⎥ Gluteal arteries
Inferior ⎦

Inferior vesical artery

Pudendal nerve

Internal pudendal artery

Middle rectal artery (cut ends)

Prostatic branch of inferior vesical artery

Male Pelvis, Median Section

TABLE 3.3   ARTERIES OF MALE PELVIS[a]

ARTERY	ORIGIN	COURSE	DISTRIBUTION
Testicular (gonadal)	Abdominal aorta	Descends retroperitoneally; testicular artery passes into deep inguinal ring	Testis
Internal iliac	Common iliac artery	Passes over brim of pelvis to reach pelvic cavity	Main blood supply to pelvic organs, gluteal muscles, and perineum
Anterior division of internal iliac artery	Internal iliac artery	Passes anteriorly and divides into visceral branches and obturator artery	Pelvic viscera and muscles in medial compartment of thigh
Umbilical	Anterior division of internal iliac artery	Short pelvic course and ends as superior vesical artery in females	Ductus deferens
Obturator	Anterior division of internal iliac artery	Runs anteroinferiorly on lateral pelvic wall	Pelvic muscles, nutrient artery to ilium, and head of femur
Superior vesical	Remnant of proximal part of umbilical artery	Passes to superior aspect of urinary bladder	Superior aspect of urinary bladder
Artery to ductus deferens	Inferior vesical artery	Runs retroperitoneally to ductus deferens	Ductus deferens
Inferior vesical	Anterior division of internal iliac artery	Passes retroperitoneally to inferior portion of urinary bladder in males	Urinary bladder, seminal vesicle and prostate
Middle rectal	Anterior division of internal iliac artery	Descends in pelvis to rectum	Seminal vesicle, prostate, and rectum
Internal pudendal	Anterior division of internal iliac artery	Leaves pelvis through greater sciatic foramen and enters perineum by passing through lesser sciatic foramen	Piriformis, coccygeus, levator ani, and gluteal muscles
Posterior division of internal iliac artery	Internal iliac artery	Passes posteriorly and gives rise to parietal branches	Pelvic wall and gluteal region
Iliolumbar	Posterior division of internal iliac artery	Ascends anterior to sacroiliac joint and posterior to common iliac vessels and psoas major	Iliacus, psoas major, quadratus lumborum muscles, and cauda equina in vertebral canal
Lateral sacral (superior and inferior)	Posterior division of internal iliac artery	Run on superficial aspect of piriformis	Piriformis and vertebral canal

[a]See figure on opposite page.

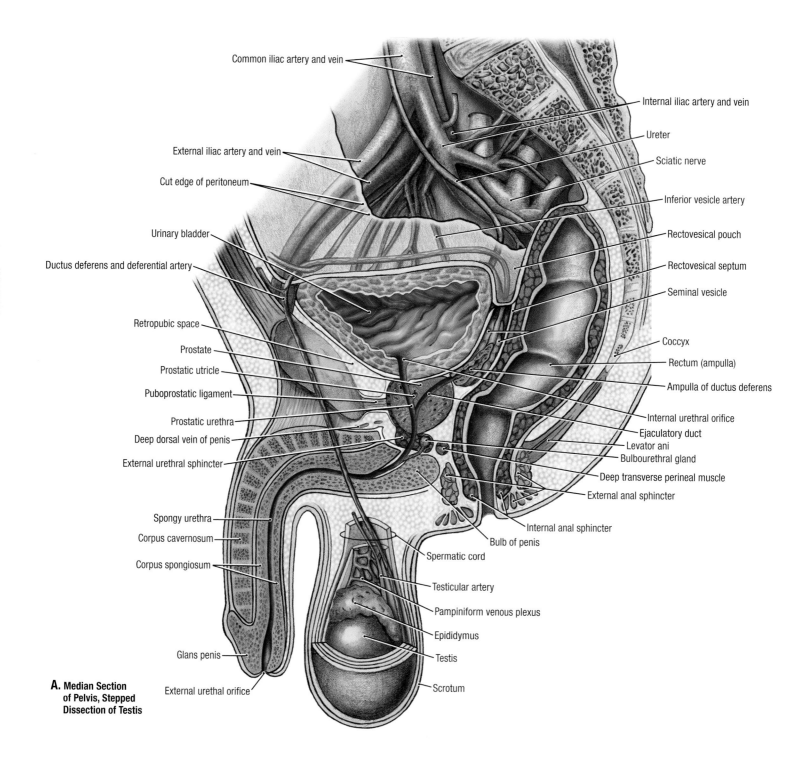

Common iliac artery and vein

External iliac artery and vein

Cut edge of peritoneum

Urinary bladder

Ductus deferens and deferential artery

Retropubic space

Prostate

Prostatic utricle

Puboprostatic ligament

Prostatic urethra

Deep dorsal vein of penis

External urethral sphincter

Spongy urethra

Corpus cavernosum

Corpus spongiosum

Glans penis

External urethal orifice

Internal iliac artery and vein

Ureter

Sciatic nerve

Inferior vesicle artery

Rectovesical pouch

Rectovesical septum

Seminal vesicle

Coccyx

Rectum (ampulla)

Ampulla of ductus deferens

Internal urethral orifice

Ejaculatory duct

Levator ani

Bulbourethral gland

Deep transverse perineal muscle

External anal sphincter

Internal anal sphincter

Bulb of penis

Spermatic cord

Testicular artery

Pampiniform venous plexus

Epididymus

Testis

Scrotum

A. Median Section of Pelvis, Stepped Dissection of Testis

3.19 **Male pelvic organs and lateral pelvic wall**

A. Male pelvis and perineum.

- Most of the pelvic viscera are subperitoneal, embedded in a matrix of fatty endopelvic fascia.
- The genital tract is demonstrated in its entirely; it merges with the urinary tract in the prostatic urethra.
- The deferential artery is arising as a branch of the inferior epigastric artery in this specimen.
- The ampullae of the deferent ducts and seminal vesicles occupy the rectovesical septum between the bladder and rectum.

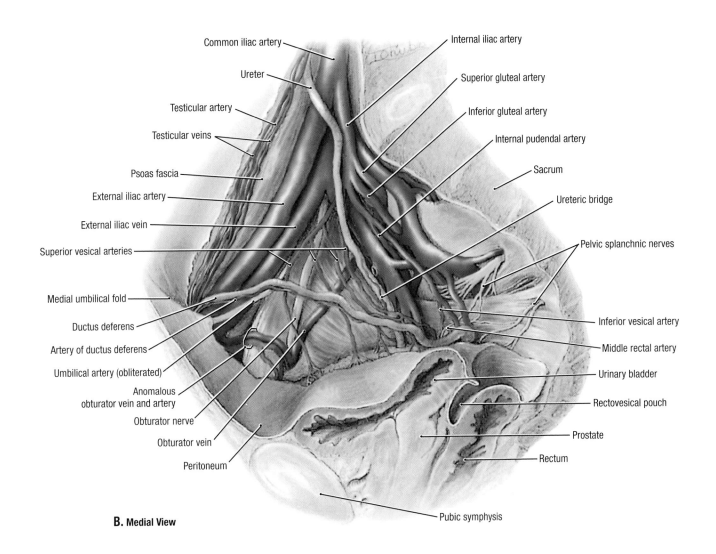

Common iliac artery

Ureter

Testicular artery

Testicular veins

Psoas fascia

External iliac artery

External iliac vein

Superior vesical arteries

Medial umbilical fold

Ductus deferens

Artery of ductus deferens

Umbilical artery (obliterated)

Anomalous
obturator vein and artery

Obturator nerve

Obturator vein

Peritoneum

Internal iliac artery

Superior gluteal artery

Inferior gluteal artery

Internal pudendal artery

Sacrum

Ureteric bridge

Pelvic splanchnic nerves

Inferior vesical artery

Middle rectal artery

Urinary bladder

Rectovesical pouch

Prostate

Rectum

Pubic symphysis

B. Medial View

3.19 **Male pelvic organs and lateral pelvic wall *(continued)***

B. Lateral wall of pelvis.
- The ureter and deferent duct take a subperitoneal course across the external iliac vessels, umbilical artery, and obturator nerve and vessels.
- The ureter crosses the external iliac artery at its origin (common iliac bifurcation), and the ductus deferens crosses the external iliac artery at its termination (deep inguinal ring).
- In this specimen, an anomalous (replaced) obturator artery branches from the inferior epigastric artery; there are normal and anomalous (accessory) obturator veins.

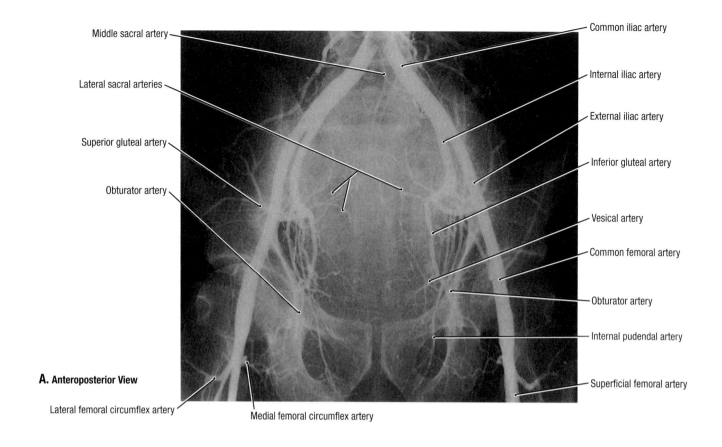

Middle sacral artery

Lateral sacral arteries

Superior gluteal artery

Obturator artery

A. Anteroposterior View

Lateral femoral circumflex artery

Medial femoral circumflex artery

Common iliac artery

Internal iliac artery

External iliac artery

Inferior gluteal artery

Vesical artery

Common femoral artery

Obturator artery

Internal pudendal artery

Superficial femoral artery

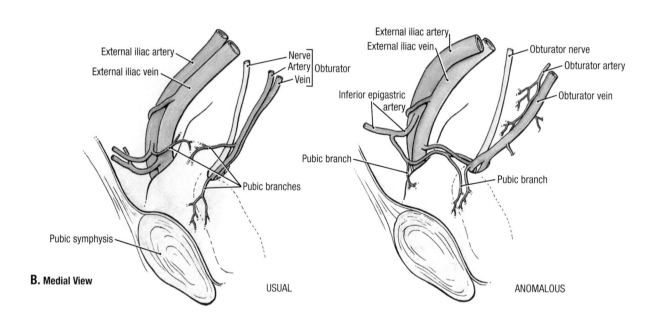

External iliac artery

External iliac vein

Nerve
Artery } Obturator
Vein

Pubic branches

Pubic symphysis

B. Medial View

USUAL

External iliac artery

External iliac vein

Inferior epigastric artery

Obturator nerve

Obturator artery

Obturator vein

Pubic branch

Pubic branch

ANOMALOUS

3.20   Pelvic angiography and usual and anomalous obturator arteries

A. Pelvic angiogram. **B.** Usual and anomalous obturator arteries. The pubic branch of the obturator artery usually anastomoses posterior to the body of the pubis with the pubic branch of the inferior epigastric artery. In the anomalous case illustrated here, the obturator artery (replaced) arises from the inferior epigastric artery via the pubic anastomosis. Dr. Grant reported that in a study of 283 limbs, the obturator artery arose from the internal iliac in 70%, from the inferior epigastric in 25.4%, and nearly equally from both in 4.6%.

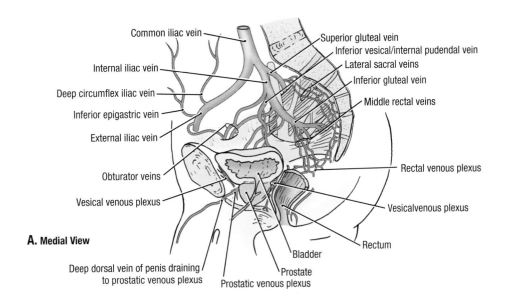

A. Medial View

Common iliac vein
Internal iliac vein
Deep circumflex iliac vein
Inferior epigastric vein
External iliac vein
Obturator veins
Vesical venous plexus
Deep dorsal vein of penis draining to prostatic venous plexus
Prostatic venous plexus
Prostate
Bladder
Rectum
Vesicalvenous plexus
Rectal venous plexus
Middle rectal veins
Inferior gluteal vein
Lateral sacral veins
Inferior vesical/internal pudendal vein
Superior gluteal vein

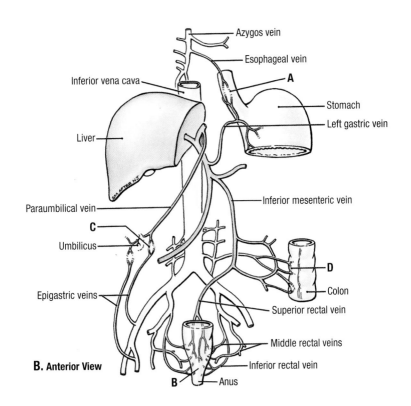

B. Anterior View

Azygos vein
Esophageal vein
A
Stomach
Inferior vena cava
Left gastric vein
Liver
Paraumbilical vein
Inferior mesenteric vein
C
Umbilicus
D
Epigastric veins
Colon
Superior rectal vein
Middle rectal veins
Inferior rectal vein
B
Anus

3.21   **Pelvic veins of male pelvis**

A. Pelvic veins and venous plexuses. **B.** Portal–systemic anastomoses. These communications provide collateral circulation in case of obstruction in the liver or portal vein. In this figure, the portal tributaries are *darker blue*, and systemic tributaries are *lighter blue*. *A–D* indicate sites of portal systemic anastomosis. *A*, between esophageal veins; *B*, between rectal veins; *C*, paraumbilical veins (portal) anastomosing with small epigastric veins of the anterior abdominal wall; *D*, twigs of colic veins (portal) anastomosing with retroperitoneal veins.

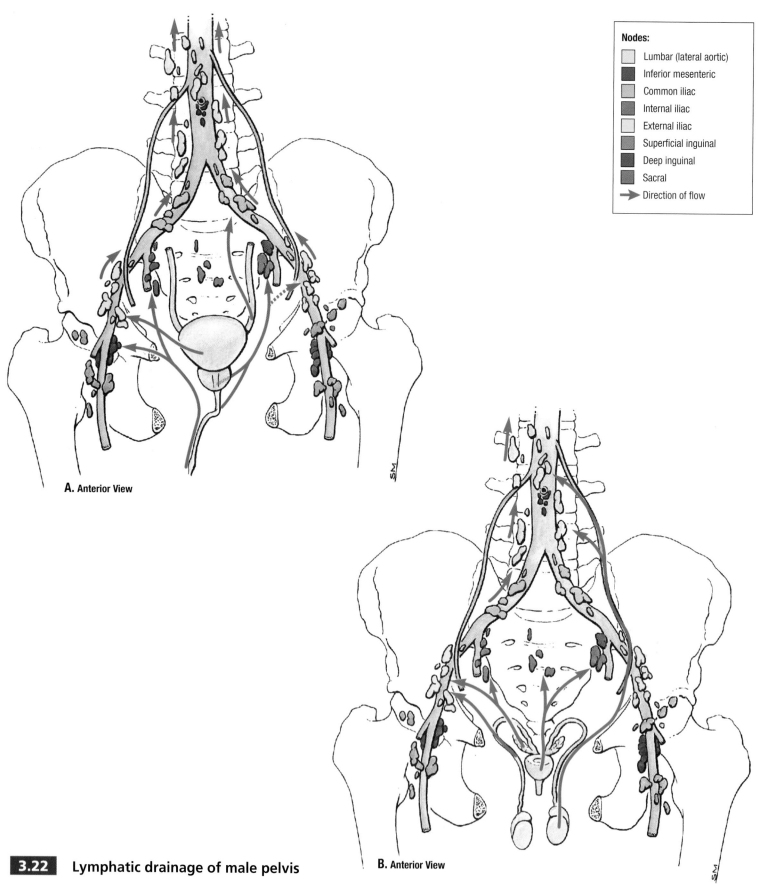

A. Anterior View

B. Anterior View

3.22 **Lymphatic drainage of male pelvis**

A. Lymphatic drainage of the ureters, urinary bladder, prostate, and urethra. **B.** Lymphatic drainage of the testis, deferent duct, prostate, and seminal vesicles.

Nodes:
- Lumbar (lateral aortic)
- Inferior mesenteric
- Common iliac
- Internal iliac
- External iliac
- Superficial inguinal
- Deep inguinal
- Sacral
- → Direction of flow

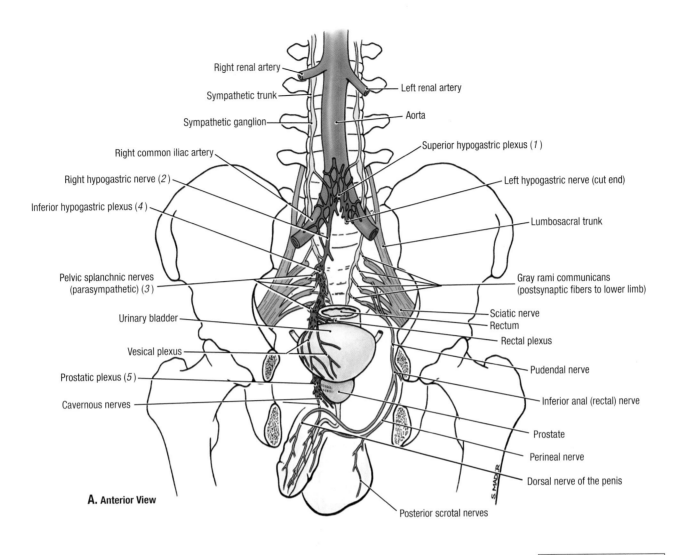

A. Anterior View

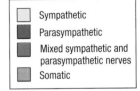

3.23   Innervation of male pelvis and genitalia

A. Autonomic and somatic innervation. **B.** Autonomic innervation of testis, deferent duct, prostate, and seminal vesicles.

- The autonomic pelvic plexuses include the inferior hypogastric, vesical, middle rectal, and prostatic. The pelvic plexuses consist of sympathetic and parasympathetic fibers. The right and left hypogastric nerves join the superior hypogastric plexus to the inferior hypogastric plexus.
- The right and left inferior hypogastric plexuses continue as the vesical plexus, supplying the urinary bladder, seminal vesicles, and deferent duct; as the middle rectal plexus, supplying the rectum; and as the prostatic plexus, supplying the prostate, seminal vesicles, corpus spongiosum, corpora cavernosum, and urethra. Vasoconstrictor fibers innervate the male genital organs and play an important role in ejaculation.
- The parasympathetic pelvic splanchnic nerves arise as visceral branches of the anterior rami of S2, S3, and S4 and supply motor fibers to the descending colon, sigmoid colon, rectum, and pelvic organs and vasodilator fibers to the erectile tissue of the penis.
- The sympathetic sacral splanchnic nerves arise from the sympathetic trunk and join the inferior hypogastric plexuses. The pudendal nerve, a branch of the sacral plexus, arises from the anterior rami of S2, S3, and S4; this nerve is part of the voluntary somatic nervous system and innervates the perineal region.

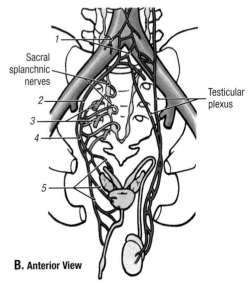

B. Anterior View

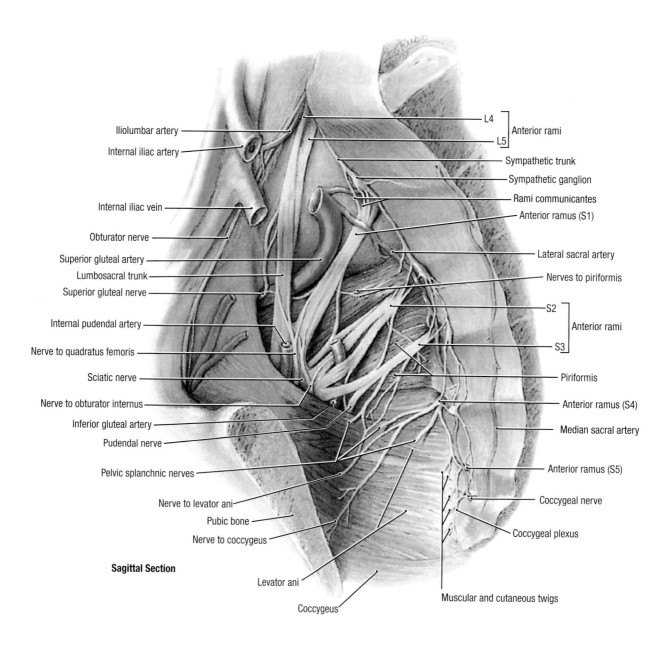

Iliolumbar artery

Internal iliac artery

Internal iliac vein

Obturator nerve

Superior gluteal artery

Lumbosacral trunk

Superior gluteal nerve

Internal pudendal artery

Nerve to quadratus femoris

Sciatic nerve

Nerve to obturator internus

Inferior gluteal artery

Pudendal nerve

Pelvic splanchnic nerves

Nerve to levator ani

Pubic bone

Nerve to coccygeus

Sagittal Section

Levator ani

Coccygeus

L4
L5 } Anterior rami

Sympathetic trunk

Sympathetic ganglion

Rami communicantes

Anterior ramus (S1)

Lateral sacral artery

Nerves to piriformis

S2
S3 } Anterior rami

Piriformis

Anterior ramus (S4)

Median sacral artery

Anterior ramus (S5)

Coccygeal nerve

Coccygeal plexus

Muscular and cutaneous twigs

3.24 Sacral and coccygeal nerve plexuses

- The sympathetic trunk or its ganglia send gray rami communicantes to each sacral nerve and to the coccygeal nerve.
- The anterior ramus from L4 joins that of L5 to form the lumbosacral trunk.
- The anterior rami of S1 and S2 supply the piriformis muscle; S3 and S4 supply the coccygeus and levator ani muscles.
- The sciatic nerve arises from anterior rami of L4, L5, S1, S2, and S3; the pudendal nerve from S2, S3, and S4; and the coccygeal plexus from S4, S5, and coccygeal segments.
- The iliolumbar artery accompanies the L5 nerve, and the branches of the lateral sacral artery accompany the sacral nerves. The superior gluteal artery often passes posteriorly between L5 and S1.

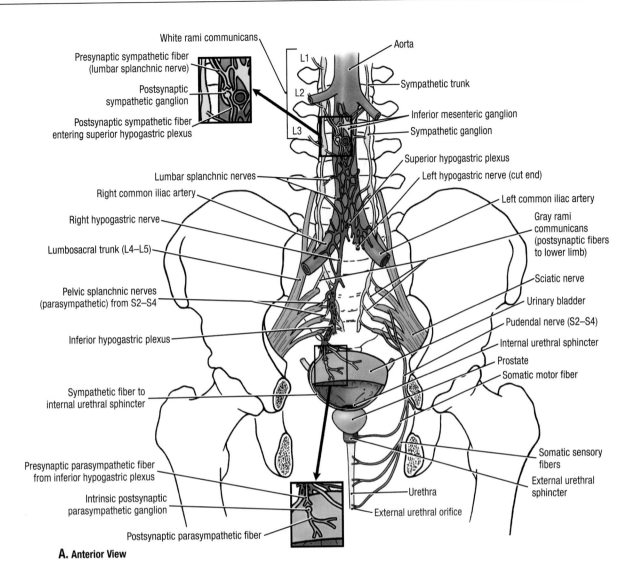

White rami communicans
Presynaptic sympathetic fiber (lumbar splanchnic nerve)
Postsynaptic sympathetic ganglion
Postsynaptic sympathetic fiber entering superior hypogastric plexus
Lumbar splanchnic nerves
Right common iliac artery
Right hypogastric nerve
Lumbosacral trunk (L4–L5)
Pelvic splanchnic nerves (parasympathetic) from S2–S4
Inferior hypogastric plexus
Sympathetic fiber to internal urethral sphincter
Presynaptic parasympathetic fiber from inferior hypogastric plexus
Intrinsic postsynaptic parasympathetic ganglion
Postsynaptic parasympathetic fiber

Aorta
Sympathetic trunk
Inferior mesenteric ganglion
Sympathetic ganglion
Superior hypogastric plexus
Left hypogastric nerve (cut end)
Left common iliac artery
Gray rami communicans (postsynaptic fibers to lower limb)
Sciatic nerve
Urinary bladder
Pudendal nerve (S2–S4)
Internal urethral sphincter
Prostate
Somatic motor fiber
Somatic sensory fibers
External urethral sphincter
Urethra
External urethral orifice

L1
L2
L3

A. Anterior View

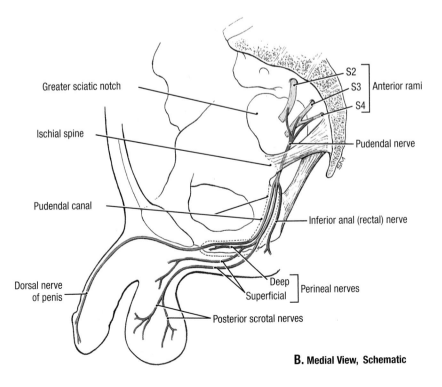

Greater sciatic notch
Ischial spine
Pudendal canal
Dorsal nerve of penis
Posterior scrotal nerves
Deep / Superficial } Perineal nerves
Inferior anal (rectal) nerve
Pudendal nerve
S2
S3 } Anterior rami
S4

B. Medial View, Schematic

3.25   **Innervation of male perineum, bladder, and urethra**

A. Nerve supply to the bladder and urethra. **B.** Distribution of the pudendal nerve.

- The pelvic splanchnic nerves (parasympathetic) are motor nerves to the detrusor muscle of the bladder; they stimulate emptying of the bladder, dilation of the blood vessels, and erection of the penis and are inhibitory to the internal urethral sphincter of males.
- Sympathetic innervation through the superior hypogastric plexus (T12, L1, L2, L3) is motor to ureteric musculature, trigonal muscle, and the muscles of the urethral crest; it also supplies the muscle of the epididymis, deferent duct, seminal vesicle and prostate. The plexus stimulates ejaculation of seminal fluid into the urethra.
- The pudendal nerve is motor to the external urethral sphincter and perianal muscles and sensory to the glans penis, urethra, and most of the skin of the perineum.

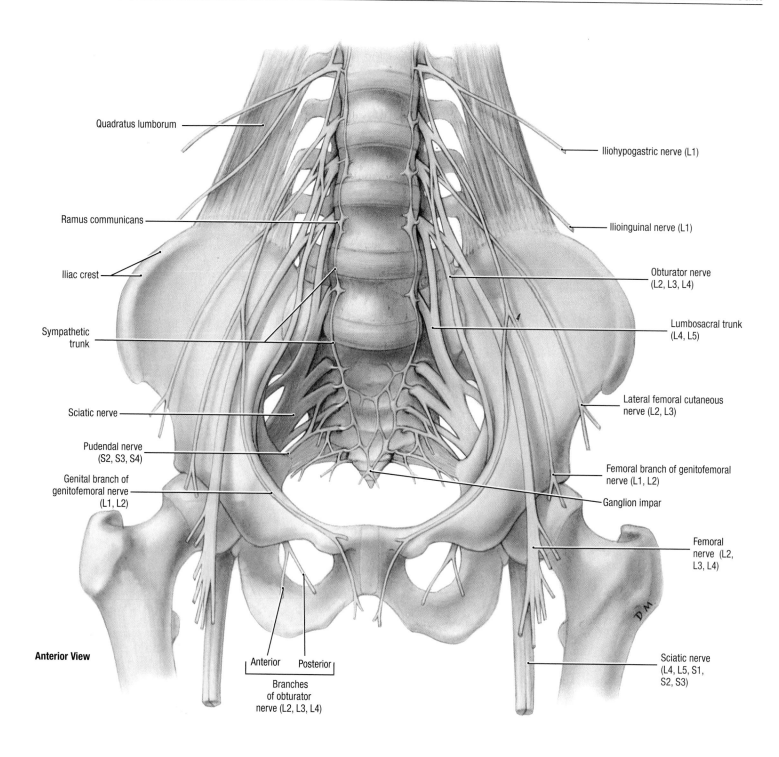

Quadratus lumborum

Ramus communicans

Iliac crest

Sympathetic trunk

Sciatic nerve

Pudendal nerve (S2, S3, S4)

Genital branch of genitofemoral nerve (L1, L2)

Iliohypogastric nerve (L1)

Ilioinguinal nerve (L1)

Obturator nerve (L2, L3, L4)

Lumbosacral trunk (L4, L5)

Lateral femoral cutaneous nerve (L2, L3)

Femoral branch of genitofemoral nerve (L1, L2)

Ganglion impar

Femoral nerve (L2, L3, L4)

Sciatic nerve (L4, L5, S1, S2, S3)

Anterior View

Anterior   Posterior

Branches of obturator nerve (L2, L3, L4)

3.26 **Overview of lumbosacral plexus**

- The lumbosacral trunk provides continuity between the lumbar and sacral plexuses.
- The anterior rami of (T12), L1, L2, L3, and L4 form the lumbar plexus, and the anterior rami of L4, L5, S1, S2, and S3 form the sacral plexus.
- The sciatic nerve passes laterally through the greater sciatic foramen.
- The femoral nerve terminates in the femoral triangle by splitting into many branches just distal to the inguinal ligament.
- The obturator nerve passes through the obturator foramen with the obturator artery and vein to supply the medial aspect of the thigh.

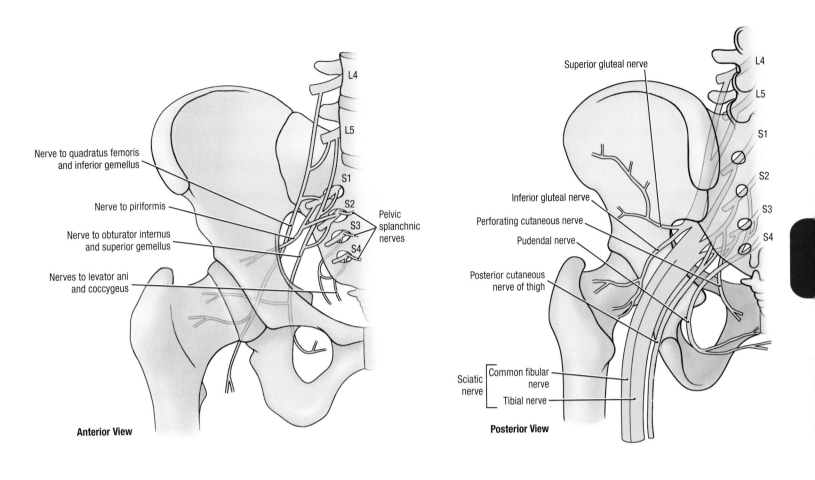

Nerve to quadratus femoris
and inferior gemellus

Nerve to piriformis

Nerve to obturator internus
and superior gemellus

Nerves to levator ani
and coccygeus

L4

L5

S1

S2

S3

S4

Pelvic
splanchnic
nerves

Anterior View

Superior gluteal nerve

Inferior gluteal nerve

Perforating cutaneous nerve

Pudendal nerve

Posterior cutaneous
nerve of thigh

Sciatic
nerve

Common fibular
nerve

Tibial nerve

L4

L5

S1

S2

S3

S4

Posterior View

TABLE 3.4   NERVES OF SACRAL AND COCCYGEAL PLEXUSES

NERVE	ORIGIN	DISTRIBUTION
Sciatic	L4, L5, S1, S2, S3	Articular branches to hip joint and muscular branches to flexors of knee in thigh and all muscles in leg and foot
Superior gluteal	L4, L5, S1	Gluteus medius and gluteus minimus muscles
Inferior gluteal	L5, S1, S2	Gluteus maximus
Nerve to piriformis	S1, S2	Piriformis muscle
Nerve to quadratus femoris and inferior gemellus	L4, L5, S1	Quadratus femoris and inferior gemellus muscles
Nerve to obturator internus and superior gemellus	L5, S1, S2	Obturator internus and superior gemellus muscles
Pudendal	S2, S3, S4	Structures in perineum: sensory to genitalia; muscular branches to perineal muscles, external urethral sphincter, and external anal sphincter
Nerves to levator ani and coccygeus	S3, S4	Levator ani and coccygeus muscles
Posterior femoral cutaneous	S2, S3	Cutaneous branches to buttock and uppermost medial and posterior surfaces of thigh
Perforating cutaneous	S2, S3	Cutaneous branches to medial part of buttock
Pelvic splanchnic	S2, S3, S4	Pelvic viscera via inferior hypogastric and pelvic plexuses

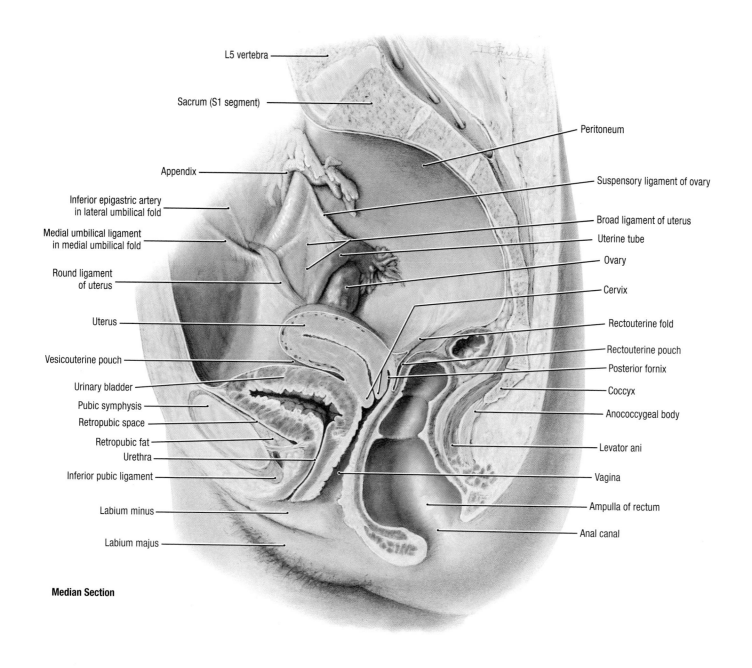

L5 vertebra

Sacrum (S1 segment)

Peritoneum

Appendix

Suspensory ligament of ovary

Inferior epigastric artery
in lateral umbilical fold

Broad ligament of uterus

Medial umbilical ligament
in medial umbilical fold

Uterine tube

Ovary

Round ligament
of uterus

Cervix

Uterus

Rectouterine fold

Vesicouterine pouch

Rectouterine pouch

Urinary bladder

Posterior fornix

Pubic symphysis

Coccyx

Retropubic space

Anococcygeal body

Retropubic fat

Urethra

Levator ani

Inferior pubic ligament

Vagina

Labium minus

Ampulla of rectum

Labium majus

Anal canal

Median Section

3.27   Female pelvis

- The uterine tube and the ovary commonly lie against the lateral wall of the pelvis.
- The uterus is bent on itself (anteflexed) at the junction of its body and the cervix; the cervix, opening on the anterior wall of the vagina, has a short, round, anterior lip and a long, thin, posterior lip.
- The urethra, the vagina, and the rectum are parallel to one another and to the pelvic brim; the uterus is nearly at right angles to these structures when the bladder is empty.

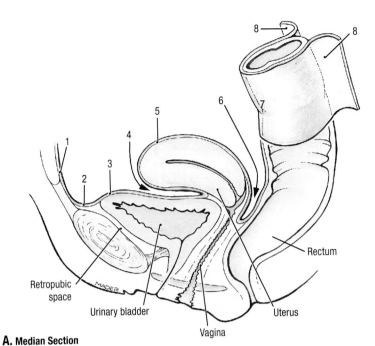

Female:
Peritoneum passes:
- From the anterior abdominal wall *(1)*
- Superior to the pubic bone *(2)*
- On the superior surface of the urinary bladder *(3)*
- From the bladder to the uterus, forming the vesicouterine pouch *(4)*
- On the fundus and body of the uterus, posterior formix, and all of the vagina *(5)*
- Between the rectum and uterus, forming the rectouterine pouch *(6)*
- On the anterior and lateral sides of the rectum *(7)*
- Posteriorly to become the sigmoid mesocolon *(8)*

A. Median Section

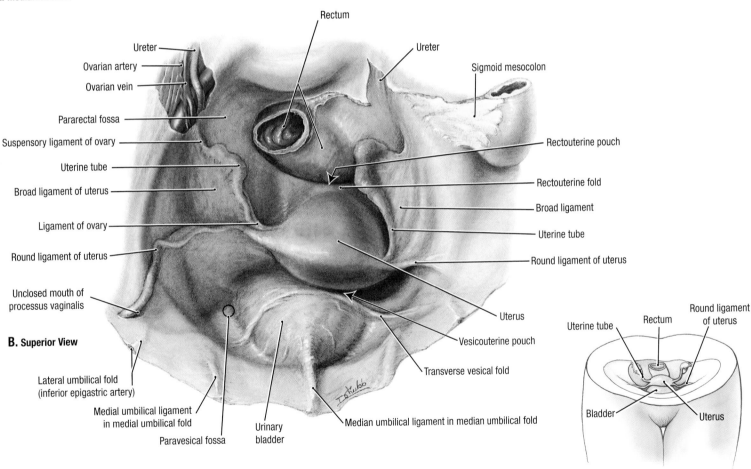

B. Superior View

3.28   Peritoneum of female pelvis

A. Peritoneum covering female pelvic organs. **B.** True pelvis with peritoneum intact, viewed from above. The uterus is usually asymmetrically placed; in this specimen, it leans to the left. The round ligament of the female takes the same subperitoneal course as the deferent duct of the male. The free edge of the medial four fifths of the broad ligament contains the uterine tube, and that of the lateral one fifth, the ovarian vessels in the suspensory ligament of the ovary.

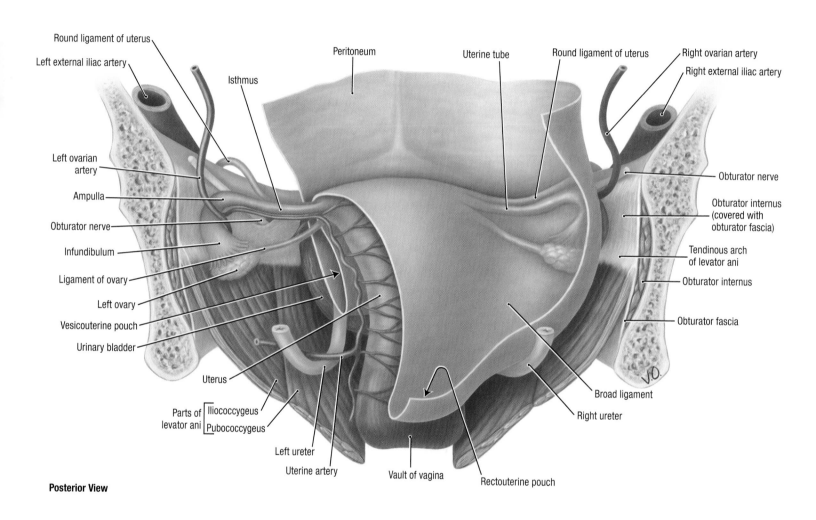

Round ligament of uterus

Left external iliac artery

Isthmus

Peritoneum

Uterine tube

Round ligament of uterus

Right ovarian artery

Right external iliac artery

Left ovarian artery

Ampulla

Obturator nerve

Infundibulum

Ligament of ovary

Left ovary

Vesicouterine pouch

Urinary bladder

Uterus

Parts of levator ani { Iliococcygeus / Pubococcygeus

Left ureter

Uterine artery

Vault of vagina

Rectouterine pouch

Obturator nerve

Obturator internus (covered with obturator fascia)

Tendinous arch of levator ani

Obturator internus

Obturator fascia

Broad ligament

Right ureter

Posterior View

3.29   **Uterus in situ, vesicouterine and rectouterine pouch**

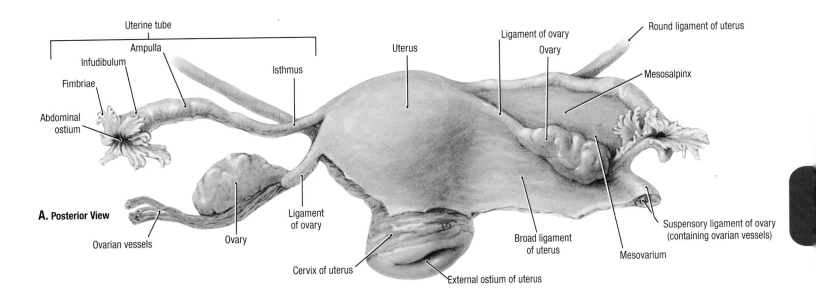

A. Posterior View

Uterine tube
Ampulla
Infudibulum
Fimbriae
Abdominal ostium
Ovarian vessels
Ovary
Ligament of ovary
Isthmus
Cervix of uterus
External ostium of uterus
Uterus
Ligament of ovary
Ovary
Round ligament of uterus
Mesosalpinx
Broad ligament of uterus
Mesovarium
Suspensory ligament of ovary (containing ovarian vessels)

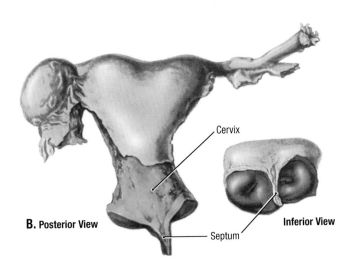

B. Posterior View

Cervix
Septum

Inferior View

3.30 Uterus (normal and bipartite) and its adnexa

A. Uterus and adnexa, removed from cadaver. On the specimen's left, the broad ligament of the uterus is removed; on the specimen's right, the mesentery of the uterus and uterine tube is called the *broad ligament*. The ovary is attached (1) to the broad ligament by a mesentery of its own, called the *mesovarium*; (2) to the uterus by the ligament of the ovary; and (3) near the pelvic brim, by the suspensory ligament of the ovary containing the ovarian vessels. The part of the broad ligament superior to the level of the mesovarium is called the *mesosalpinx*. **B.** Bipartite uterus. Note the absence of the fundus of the uterus and in its place a median depression, the remains of a median vaginal septum, and a cervix with two openings.

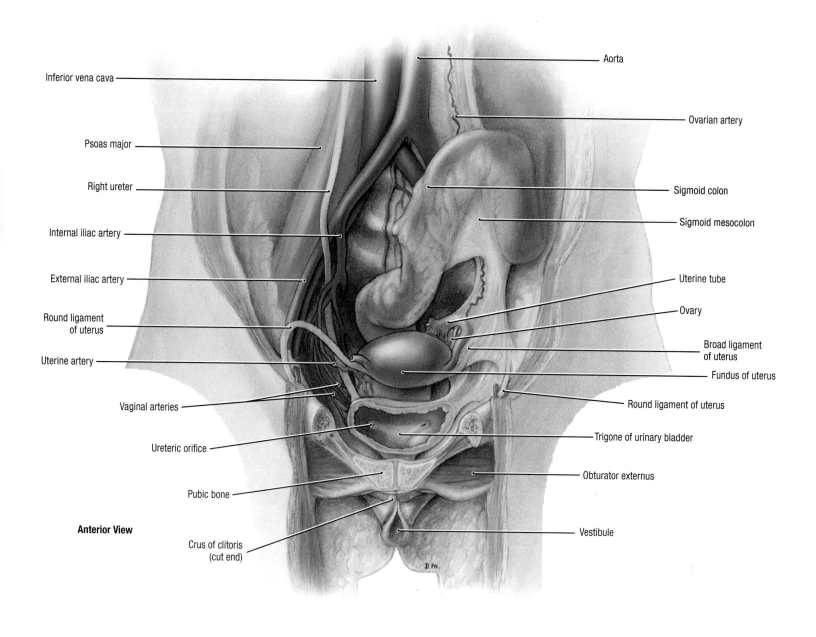

Inferior vena cava

Psoas major

Right ureter

Internal iliac artery

External iliac artery

Round ligament of uterus

Uterine artery

Vaginal arteries

Ureteric orifice

Pubic bone

Anterior View

Crus of clitoris (cut end)

Aorta

Ovarian artery

Sigmoid colon

Sigmoid mesocolon

Uterine tube

Ovary

Broad ligament of uterus

Fundus of uterus

Round ligament of uterus

Trigone of urinary bladder

Obturator externus

Vestibule

3.31   Female genital organs

- Part of the pubic bones, the anterior aspect of the bladder, and, on the specimen's right side, the uterine tube, ovary, broad ligament, and peritoneum covering the lateral wall of the pelvis have been removed.
- The uterine artery is located in the base of the broad ligament and runs superiorly along the lateral margin of the uterus.
- The vaginal arteries, branching from the uterine artery, supply the cervix and anterior surface of the vagina; the vaginal arteries, arising from the internal iliac artery, supply the posterior surface of the vagina.
- The right ureter crosses the external iliac artery at the bifurcation of the common iliac artery; note the close proximity of the right ureter to the cervix of the uterus and lateral fornix of the vagina, where it is crossed by the uterine artery.

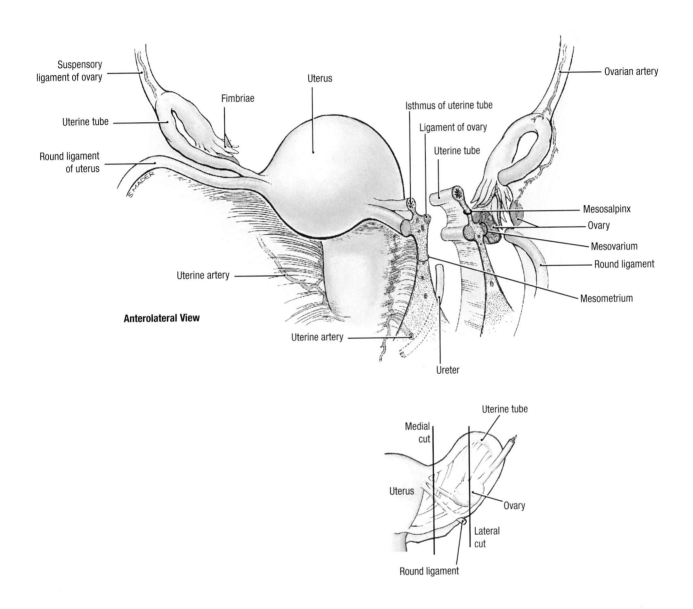

Anterolateral View

3.32   Uterus and broad ligament

Two paramedian sections show "mesenteries" with the prefix *meso-*. *Salpinx* is Greek for trumpet or tube, *metro* for uterus. The mesentery of the uterus and uterine tube is called *the broad ligament*. The major part of the broad ligament, the mesometrium, is attached to the uterus. The ovary is attached: to the broad ligament by a mesentery of its own, called the *mesovarium;* to the uterus by the ligament of the ovary; and near the pelvic brim, by the suspensory ligament of the ovary containing the ovarian vessels. The part of the broad ligament superior to the level of the mesovarium is called the *mesosalpinx.*

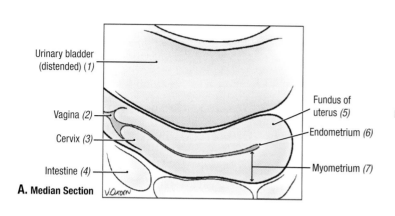

Urinary bladder (distended) (1)

Vagina (2)

Cervix (3)

Intestine (4)

A. Median Section

Fundus of uterus (5)

Endometrium (6)

Myometrium (7)

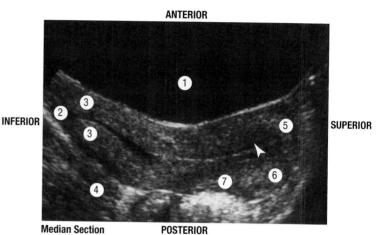

ANTERIOR

INFERIOR

SUPERIOR

Median Section

POSTERIOR

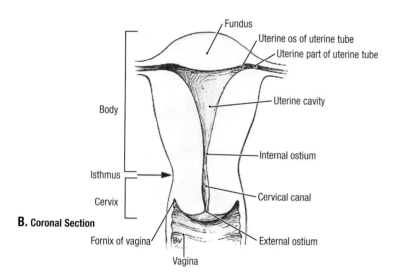

Fundus

Uterine os of uterine tube

Uterine part of uterine tube

Uterine cavity

Body

Internal ostium

Isthmus

Cervix

Cervical canal

Fornix of vagina

External ostium

Vagina

B. Coronal Section

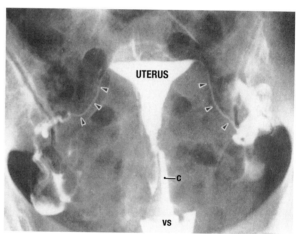

UTERUS

C

VS

Hysterosalpingogram

3.33 Imaging of uterus and ovaries

A. Sagittal ultrasound scan and orientation drawing (*numbers in parentheses* correspond to labels on the ultrasound scan). **B.** Hysterosalpingogram. Radiopaque contrast medium was injected via a cannula (*c*) into the uterus through the external ostium; *vs*, vaginal speculum in vagina. The triangular uterine cavity is clearly outlined. Contrast medium has traveled through the uterine tubes (*arrowheads*) to the infundibulum and leaked into the peritoneal cavity (pararectal fossae) on both sides.

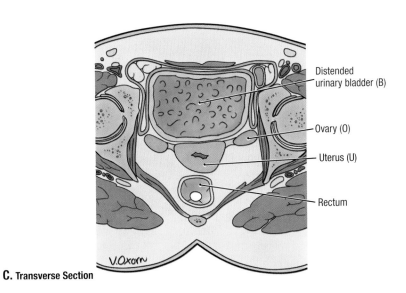

Distended urinary bladder (B)

Ovary (O)

Uterus (U)

Rectum

C. Transverse Section

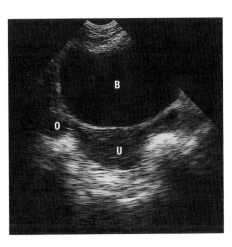

Transverse (Axial) Scan

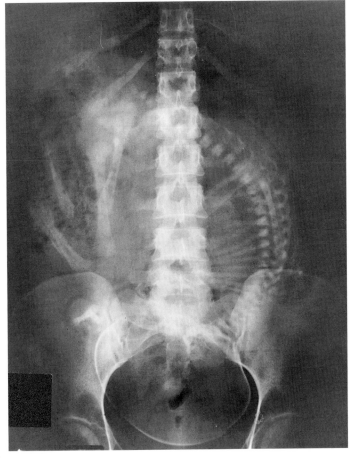

D. Anteroposterior View

3.33   **Imaging of uterus and ovaries (*continued*)**

C. Transverse ultrasound scan through the urinary bladder, uterus, and ovaries. **D.** Radiograph of fetus.

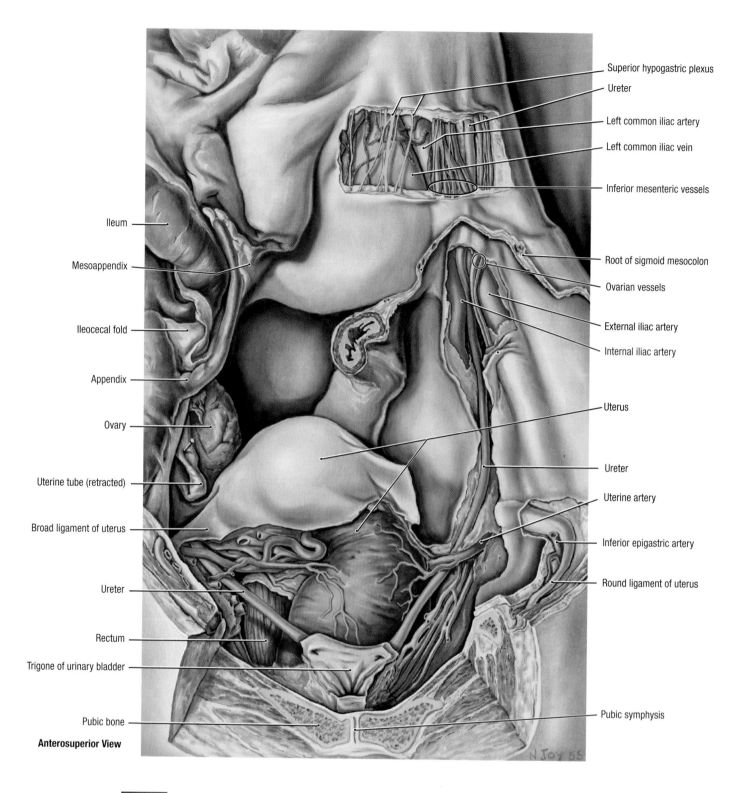

Ileum

Mesoappendix

Ileocecal fold

Appendix

Ovary

Uterine tube (retracted)

Broad ligament of uterus

Ureter

Rectum

Trigone of urinary bladder

Pubic bone

Anterosuperior View

Superior hypogastric plexus

Ureter

Left common iliac artery

Left common iliac vein

Inferior mesenteric vessels

Root of sigmoid mesocolon

Ovarian vessels

External iliac artery

Internal iliac artery

Uterus

Ureter

Uterine artery

Inferior epigastric artery

Round ligament of uterus

Pubic symphysis

3.34 **Female ureter in pelvis**

- Most of the pubic symphysis and most of the bladder (except the trigone of its floor) have been removed.
- The left ureter is crossed by the ovarian vessels and nerves; the apex of the inverted V-shaped root of the sigmoid mesocolon is situated anterior to the left ureter.
- The left ureter crosses the external iliac artery at the bifurcation of the common iliac artery and then descends anterior to the internal iliac artery; its course is subperitoneal from where it enters the pelvis to where it passes deep to the broad ligament and is crossed by the uterine artery.

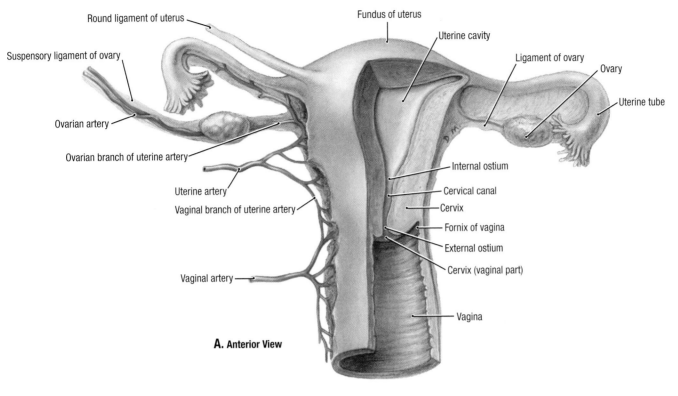

Round ligament of uterus
Suspensory ligament of ovary
Ovarian artery
Ovarian branch of uterine artery
Uterine artery
Vaginal branch of uterine artery
Vaginal artery

Fundus of uterus
Uterine cavity
Ligament of ovary
Ovary
Uterine tube
Internal ostium
Cervical canal
Cervix
Fornix of vagina
External ostium
Cervix (vaginal part)
Vagina

A. Anterior View

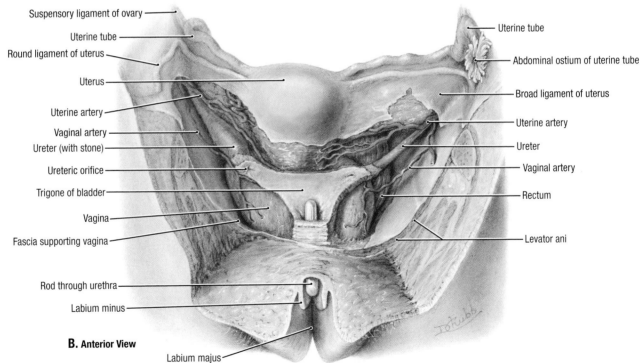

Suspensory ligament of ovary
Uterine tube
Round ligament of uterus
Uterus
Uterine artery
Vaginal artery
Ureter (with stone)
Ureteric orifice
Trigone of bladder
Vagina
Fascia supporting vagina
Rod through urethra
Labium minus

Uterine tube
Abdominal ostium of uterine tube
Broad ligament of uterus
Uterine artery
Ureter
Vaginal artery
Rectum
Levator ani
Labium majus

B. Anterior View

3.35 **Uterus and its adnexa**

A. Blood supply. On the specimen's left side, part of the uterine wall with the round ligament and the vaginal wall have been cut away to expose the cervix, uterine cavity, and thick muscular wall of the uterus, the myometrium. On the specimen's right side, the ovarian artery (from the aorta) and uterine artery (from the internal iliac) supply the ovary, uterine tube, and uterus and anastomose in the broad ligament along the lateral aspect of the uterus. The uterine artery sends a uterine branch to supply the uterine body and fundus and a vaginal branch to supply the cervix and vagina. **B.** Uterus and broad ligament. The pubic bones and the bladder, trigone excepted, are removed, as in Figure 3.34.

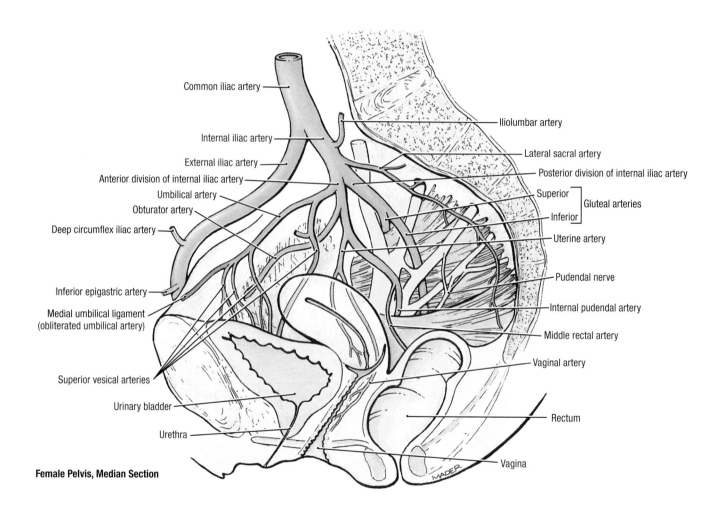

Common iliac artery

Internal iliac artery

External iliac artery

Anterior division of internal iliac artery

Umbilical artery

Obturator artery

Deep circumflex iliac artery

Inferior epigastric artery

Medial umbilical ligament
(obliterated umbilical artery)

Superior vesical arteries

Urinary bladder

Urethra

Iliolumbar artery

Lateral sacral artery

Posterior division of internal iliac artery

Superior ⎫
          ⎬ Gluteal arteries
Inferior ⎭

Uterine artery

Pudendal nerve

Internal pudendal artery

Middle rectal artery

Vaginal artery

Rectum

Vagina

Female Pelvis, Median Section

TABLE 3.5   ARTERIES OF FEMALE PELVIS

ARTERY	ORIGIN	COURSE	DISTRIBUTION
Ovarian	Abdominal aorta	Descends retroperitoneally; testicular artery passes through inguinal canal into scrotum; ovarian artery crosses pelvic brim, coursing medially in suspensory ligament to ovary	Abdominal and/or pelvic ureter, ovary, and (ampullary end of) uterine tube
Superior rectal	Continuation of inferior mesenteric artery	Crosses left common iliac vessels and descends into the pelvis between the layers of the sigmoid mesocolon	Upper part of rectum; anastomoses with middle and inferior rectal arteries
Median sacral	Posterior aspect of abdominal aorta	Descends in median line over L4 and L5 vertebrae and the sacrum and coccyx	Lower lumbar vertebrae, sacrum, and coccyx
Internal iliac	Common iliac	Passes over pelvic brim to reach pelvic cavity	Main blood supply to pelvic organs, gluteal muscles, and perineum
Anterior division of internal iliac	Internal iliac	Passes anteriorly and divides into visceral branches and obturator artery	Pelvic viscera and muscles in medial compartment of thigh
Umbilical	Anterior division of internal iliac	Obliterates becoming medical umbilical ligament after running a short pelvic course during which it gives rise to superior vesical arteries	Superior aspect of urinary bladder
Superior vesical	Patent (proximal) part of umbilical	Usually multiple, these arteries pass to the superior aspect of the urinary bladder	Superior aspect of urinary bladder, pelvic portion of ureter
Obturator	Anterior division of internal iliac	Runs anteroinferiorly on lateral pelvic wall to exit pelvis via obturator canal	Pelvic muscles, nutrient artery to ilium, head of femur, muscles of medial compartment of thigh
Uterine	Anterior division of internal iliac	Runs medially in base of broad ligament superior to cardinal ligament, crossing superior to ureter, to sides of uterus	Uterus, ligaments of uterus, uterine tube, and vagina
Vaginal	Uterine artery	Arises lateral to ureter and descends inferior to it to lateral aspect of vagina	Vagina; branches to inferior part of urinary bladder and termination of ureter
Internal pudendal	Anterior division of internal iliac	Leaves pelvis through greater sciatic foramen and enters perineum (ischioanal fossa) by passing through lesser sciatic foramen	Main artery to perineum, including muscles and skin of anal and urogenital triangles; erectile bodies
Middle rectal	Anterior division of internal iliac	Descends in pelvis to lower part of rectum	Lower part of rectum
Inferior gluteal	Anterior division of internal iliac	Exits pelvis via greater sciatic foramen, passing inferior to piriformis	Pelvic diaphragm (coccygeus and levator ani), piriformis, quadratus femoris, uppermost hamstrings, gluteus maximus, sciatic nerve
Posterior division of internal iliac	Internal iliac	Passes posteriorly and gives rise to parietal branches	Pelvic wall and gluteal region
Iliolumbar	Posterior division of internal iliac	Ascends anterior to sacroiliac joint and posterior to common iliac vessels and psoas major	Psoas major, iliacus and quadratus lumborum muscles, cauda equina in vertebral canal
Lateral sacral (superior and inferior)	Posterior division of internal iliac	Runs on anteromedial aspect of piriformis to send branches into pelvic sacral foramina	Piriformis, structures in sacral canal, erector spinae and overlying skin
Superior gluteal	Posterior division of internal iliac	Exits pelvis via greater sciatic foramen, passing superior to piriformis	Piriformis, all 3 gluteal muscles, tensor fascia lata

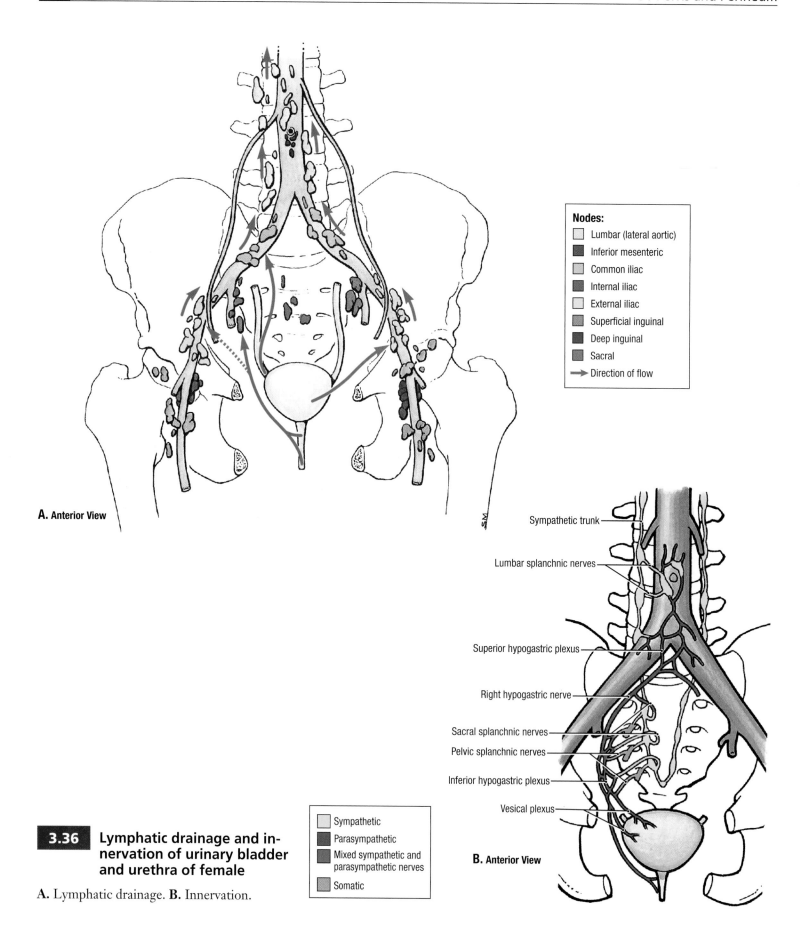

Nodes:
- ☐ Lumbar (lateral aortic)
- ■ Inferior mesenteric
- ■ Common iliac
- ■ Internal iliac
- ☐ External iliac
- ■ Superficial inguinal
- ■ Deep inguinal
- ■ Sacral
- → Direction of flow

A. Anterior View

Sympathetic trunk

Lumbar splanchnic nerves

Superior hypogastric plexus

Right hypogastric nerve

Sacral splanchnic nerves

Pelvic splanchnic nerves

Inferior hypogastric plexus

Vesical plexus

B. Anterior View

3.36 **Lymphatic drainage and innervation of urinary bladder and urethra of female**

A. Lymphatic drainage. B. Innervation.

- ☐ Sympathetic
- ■ Parasympathetic
- ■ Mixed sympathetic and parasympathetic nerves
- ☐ Somatic

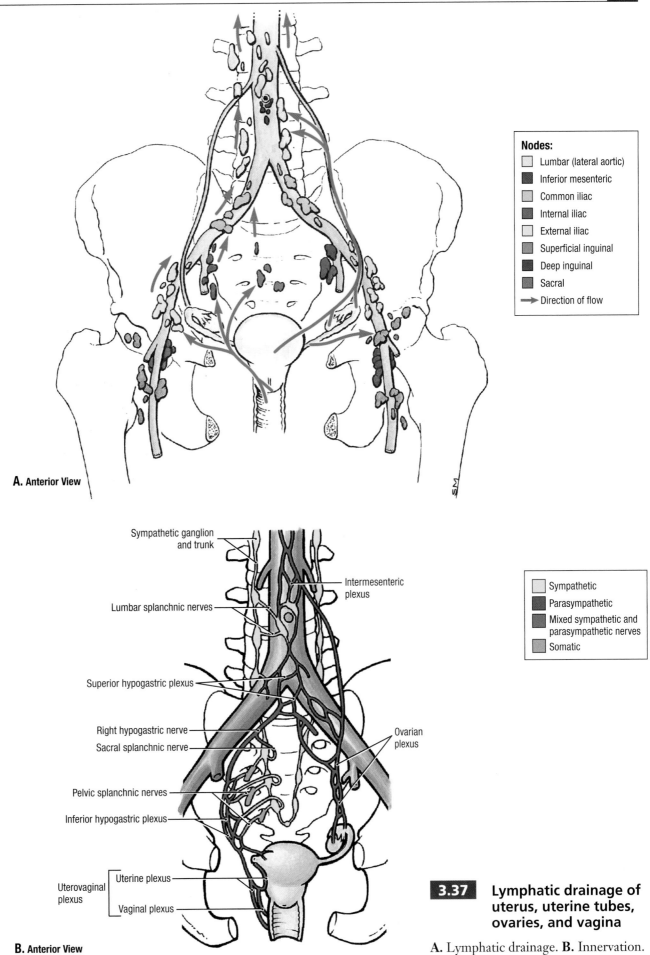

Nodes:
- Lumbar (lateral aortic)
- Inferior mesenteric
- Common iliac
- Internal iliac
- External iliac
- Superficial inguinal
- Deep inguinal
- Sacral
- → Direction of flow

A. Anterior View

Sympathetic ganglion and trunk

Intermesenteric plexus

Lumbar splanchnic nerves

Superior hypogastric plexus

Right hypogastric nerve

Sacral splanchnic nerve

Ovarian plexus

Pelvic splanchnic nerves

Inferior hypogastric plexus

Uterovaginal plexus — Uterine plexus

Vaginal plexus

- Sympathetic
- Parasympathetic
- Mixed sympathetic and parasympathetic nerves
- Somatic

B. Anterior View

3.37 Lymphatic drainage of uterus, uterine tubes, ovaries, and vagina

A. Lymphatic drainage. **B.** Innervation.

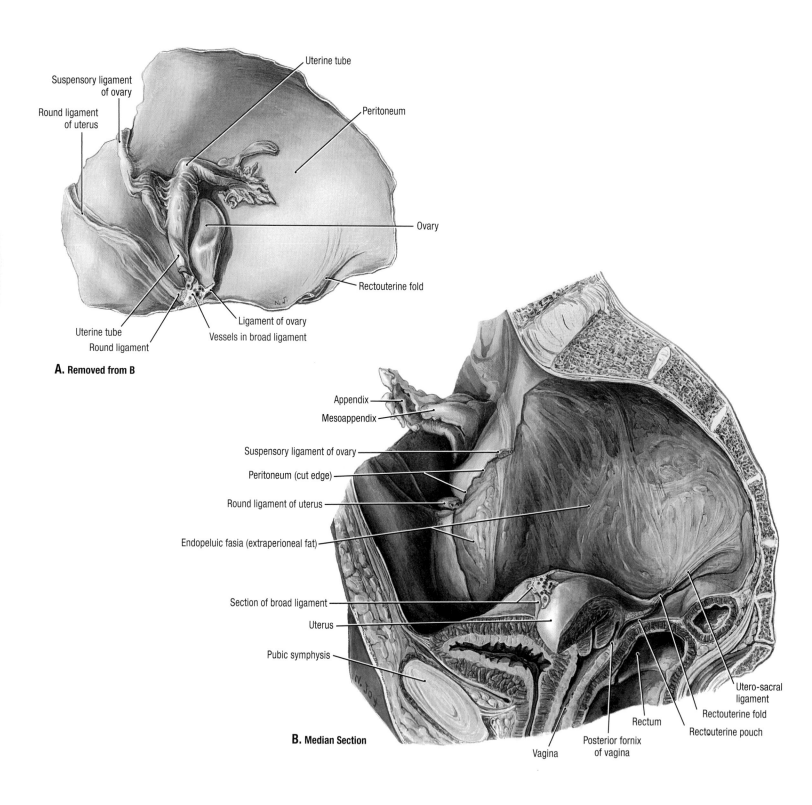

A. Removed from B

- Suspensory ligament of ovary
- Round ligament of uterus
- Uterine tube
- Peritoneum
- Ovary
- Rectouterine fold
- Ligament of ovary
- Vessels in broad ligament
- Uterine tube
- Round ligament

B. Median Section

- Appendix
- Mesoappendix
- Suspensory ligament of ovary
- Peritoneum (cut edge)
- Round ligament of uterus
- Endopeluic fasia (extraperioneal fat)
- Section of broad ligament
- Uterus
- Pubic symphysis
- Vagina
- Posterior fornix of vagina
- Rectum
- Utero-sacral ligament
- Rectouterine fold
- Rectouterine pouch

3.38   Broad ligament and its related structures

Broad ligament and its related structures (**A**) and the pelvis, otherwise intact, from which these have been removed (**B**). The round ligament and suspensory ligament of the ovary have been cut at the pelvic brim. The broad ligament, the ligament of the ovary, and branches of uterine vessels have been cut close to the uterus. In **B**, note the exposed sub-peritoneal fatty-areolar tissue (endopelvic fascia).

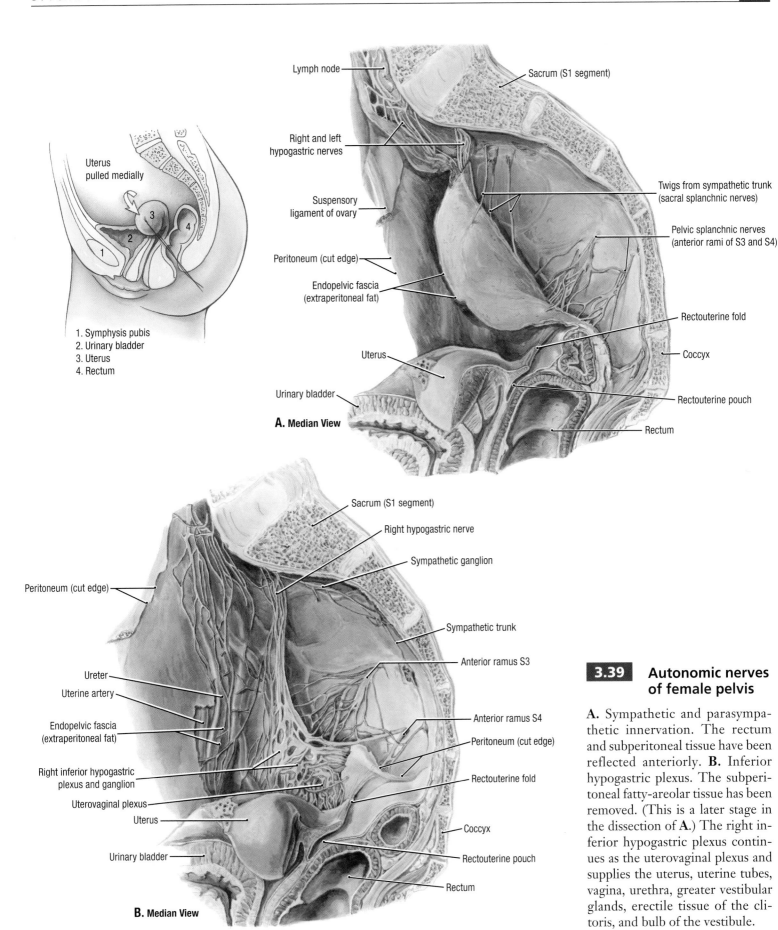

Uterus
pulled medially

1. Symphysis pubis
2. Urinary bladder
3. Uterus
4. Rectum

Lymph node

Right and left
hypogastric nerves

Suspensory
ligament of ovary

Peritoneum (cut edge)

Endopelvic fascia
(extraperitoneal fat)

Uterus

Urinary bladder

A. Median View

Sacrum (S1 segment)

Twigs from sympathetic trunk
(sacral splanchnic nerves)

Pelvic splanchnic nerves
(anterior rami of S3 and S4)

Rectouterine fold

Coccyx

Rectouterine pouch

Rectum

Peritoneum (cut edge)

Ureter

Uterine artery

Endopelvic fascia
(extraperitoneal fat)

Right inferior hypogastric
plexus and ganglion

Uterovaginal plexus

Uterus

Urinary bladder

B. Median View

Sacrum (S1 segment)

Right hypogastric nerve

Sympathetic ganglion

Sympathetic trunk

Anterior ramus S3

Anterior ramus S4

Peritoneum (cut edge)

Rectouterine fold

Coccyx

Rectouterine pouch

Rectum

3.39   Autonomic nerves of female pelvis

A. Sympathetic and parasympathetic innervation. The rectum and subperitoneal tissue have been reflected anteriorly. **B.** Inferior hypogastric plexus. The subperitoneal fatty-areolar tissue has been removed. (This is a later stage in the dissection of **A.**) The right inferior hypogastric plexus continues as the uterovaginal plexus and supplies the uterus, uterine tubes, vagina, urethra, greater vestibular glands, erectile tissue of the clitoris, and bulb of the vestibule.

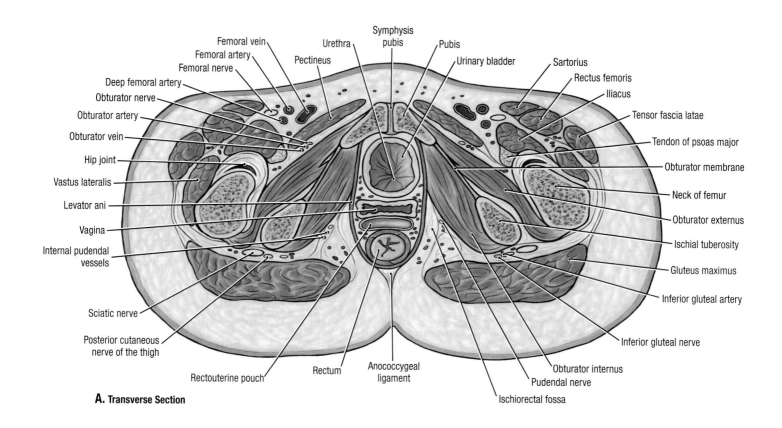

A. Transverse Section

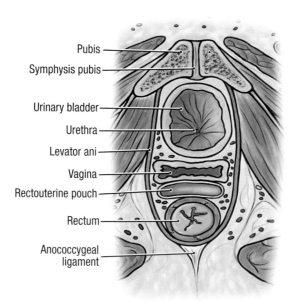

B. Transverse Section

3.40   Transverse section through the female pelvis

A. Transverse section through the ischial tuberosities. **B.** Enlargement of central part of section including the bladder, vagina, rectum, and rectouterine pouch.

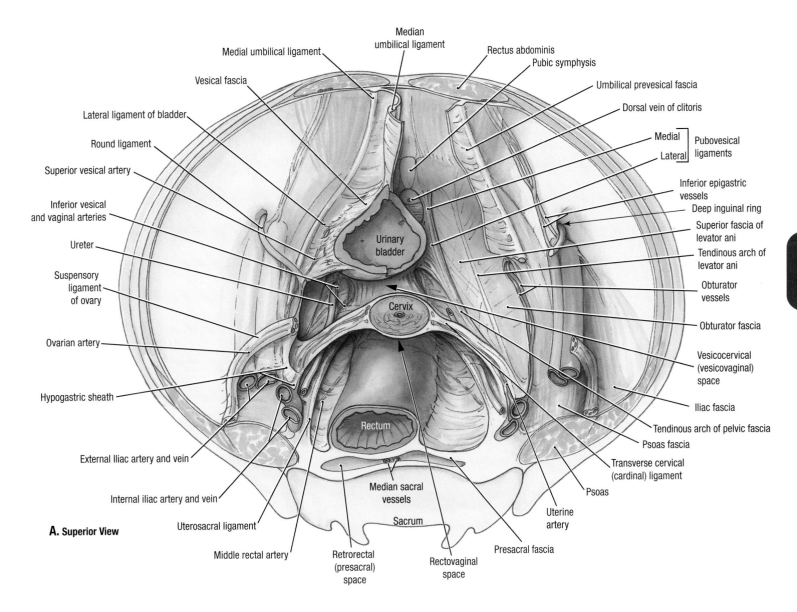

A. Superior View

3.41 Pelvic fascia and the supporting mechanism of the cervix and upper vagina

A. Pelvic viscera and endopelvic fascia. **B.** Schematic illustration of fascial ligaments and spaces.

- Note the parietal pelvic fascia covering the obturator internus and levator ani muscles and the visceral pelvic fascia surrounding the pelvic organs. These membranous fasciae are continuous where the organs penetrate the pelvic floor, forming a tendinous arch of pelvic fascia bilaterally.
- The subperitoneal endopelvic fascia lies between, and is continuous with, both visceral and parietal layers of pelvic fascia. The loose, areolar portions of the endopelvic fascia have been removed; the fibrous, condensed portions remain. Note the condensation of this fascia into the hypogastric sheath, containing the vessels to the pelvic viscera, the ureters, and (in the male) the deferent duct.
- Observe the ligamentous extensions of the hypogastric sheath: the lateral ligament of the urinary bladder, the transverse cervical ligament at the base of the broad ligament, and a less prominent lamina posteriorly containing the middle rectal vessels.

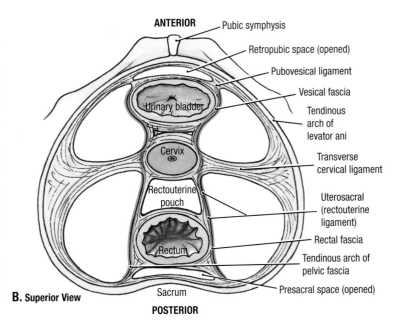

B. Superior View

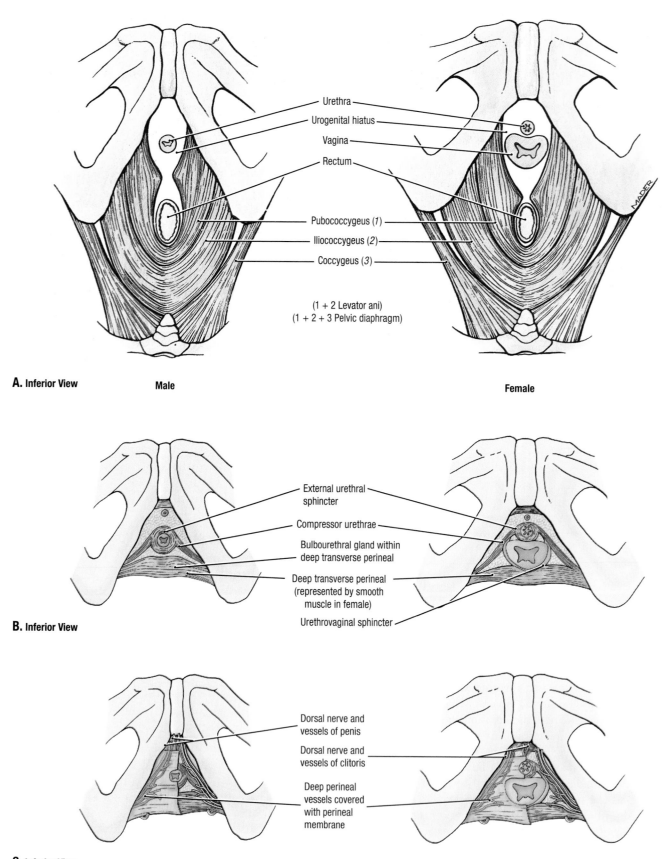

Urethra
Urogenital hiatus
Vagina
Rectum
Pubococcygeus (*1*)
Iliococcygeus (*2*)
Coccygeus (*3*)

(1 + 2 Levator ani)
(1 + 2 + 3 Pelvic diaphragm)

A. Inferior View   **Male**   **Female**

External urethral sphincter
Compressor urethrae
Bulbourethral gland within deep transverse perineal
Deep transverse perineal (represented by smooth muscle in female)
Urethrovaginal sphincter

B. Inferior View

Dorsal nerve and vessels of penis
Dorsal nerve and vessels of clitoris
Deep perineal vessels covered with perineal membrane

C. Inferior View

3.42   **Male and female perineum**

A. Levator ani. **B.** External urethral sphincter, urethrovaginal sphincter, and compressor urethrae. **C.** Perineal membrane and vessels.

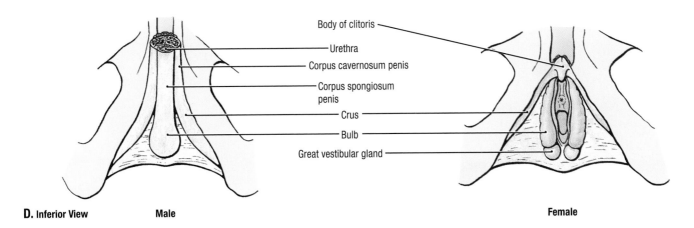

Body of clitoris

Urethra

Corpus cavernosum penis

Corpus spongiosum penis

Crus

Bulb

Great vestibular gland

D. Inferior View   Male   Female

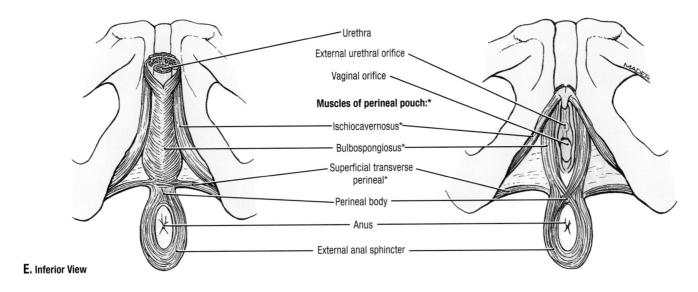

Urethra

External urethral orifice

Vaginal orifice

Muscles of perineal pouch:*

Ischiocavernosus*

Bulbospongiosus*

Superficial transverse perineal*

Perineal body

Anus

External anal sphincter

E. Inferior View

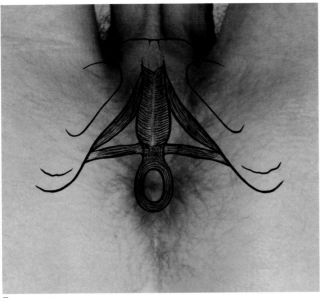

F. Inferior View

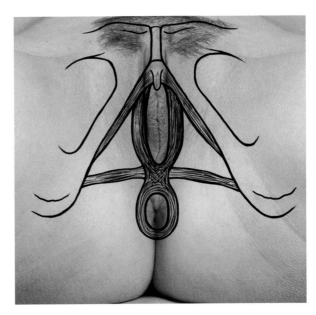

3.42   Male and female perineum

D. Crura and bulb of penis and clitoris. **E.** Muscles of superficial perineal compartment. **F.** Structures of the perineum in situ. These illustrations show the layers of the perineum built up from deep to superficial.

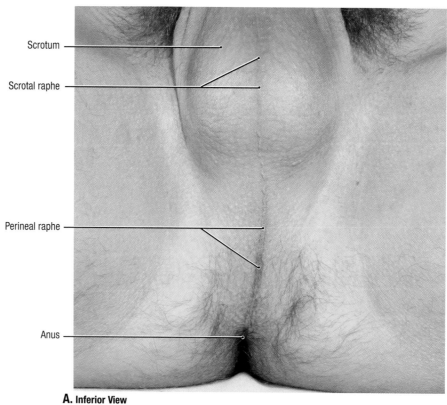

Scrotum

Scrotal raphe

Perineal raphe

Anus

A. Inferior View

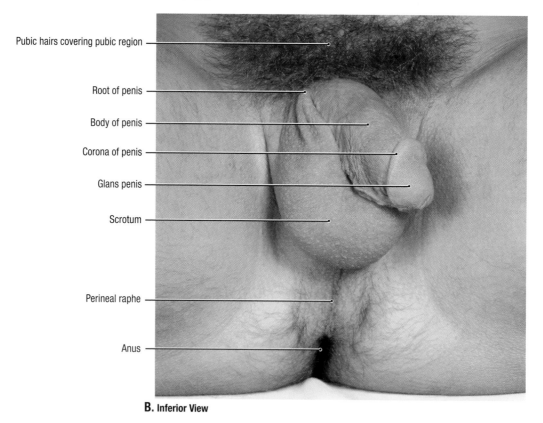

Pubic hairs covering pubic region

Root of penis

Body of penis

Corona of penis

Glans penis

Scrotum

Perineal raphe

Anus

B. Inferior View

3.43 **Surface anatomy of male perineum**

A. Scrotum and anal region. **B.** Penis, scrotum, and anal region.

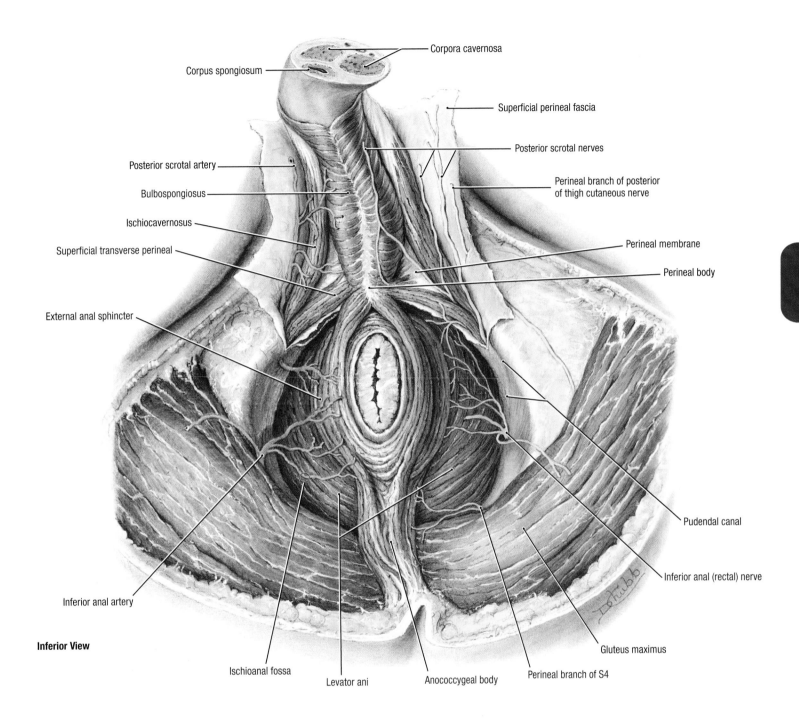

Corpora cavernosa

Corpus spongiosum

Posterior scrotal artery

Bulbospongiosus

Ischiocavernosus

Superficial transverse perineal

External anal sphincter

Inferior anal artery

Inferior View

Ischioanal fossa

Levator ani

Anococcygeal body

Perineal branch of S4

Gluteus maximus

Inferior anal (rectal) nerve

Pudendal canal

Perineal body

Perineal membrane

Perineal branch of posterior of thigh cutaneous nerve

Posterior scrotal nerves

Superficial perineal fascia

3.44   Dissection of Male Perineum—I

Superficial dissection.

- The anal canal is surrounded by the external anal sphincter; there is an ischioanal (ischiorectal) fossa on each side.
- The superficial fibers of the external anal sphincter anchor the anus anteriorly to the perineal body and posteriorly, via the anococcygeal body (ligament), to the coccyx and skin of the gluteal cleft.
- The ischioanal fossae are bound medially by the levator ani muscle, laterally by the obturator internus fascia, posteriorly by the gluteus maximus muscle overlying the sacrotuberous ligament, and anteriorly by the perineal membrane.
- The inferior anal nerve leaves the pudendal canal and, with the

perineal branch of S4, supplies the external anal sphincter; its cutaneous twigs have been removed. The branch hooking around the gluteus maximus muscle is commonly replaced by the perforating cutaneous nerve (from the sacral plexus).
- The superficial perineal fascia was incised in the midline, freed from its attachment to the base of the perineal membrane, and reflected.
- The cutaneous nerves course in the superficial perineal compartment.
- The perineal membrane is exposed between the three paired muscles of the superficial compartment.

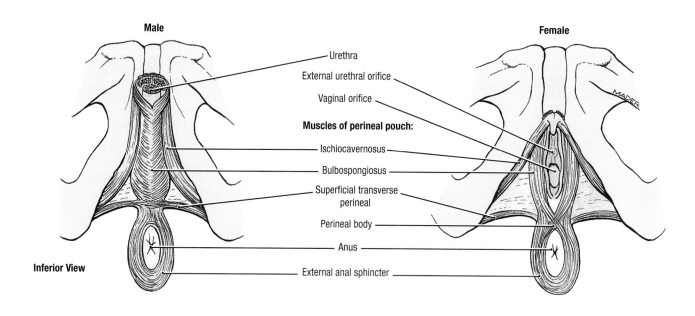

Male

Female

Urethra

External urethral orifice

Vaginal orifice

Muscles of perineal pouch:

Ischiocavernosus

Bulbospongiosus

Superficial transverse perineal

Perineal body

Anus

External anal sphincter

Inferior View

TABLE 3.6   MUSCLES OF PERINEUM

MUSCLE	ORIGIN	INSERTION	INNERVATION	ACTION(S)
External anal sphincter	Skin and fascia surrounding anus and coccyx via anococcygeal ligament	Perineal body	Inferior anal nerve	Closes anal canal; works with bulbospongiosus to support and fix perineal body
Bulbospongiosus	*Male*: Median raphe, ventral surface of bulb of penis, and perineal body *Female*: Perineal body	*Male*: Corpus spongiosum and cavernosa and fascia of bulb of penis nerve *Female*: Fascia of corpus cavernosa	Deep branch of perineal nerve, branch of pudendal	*Male*: Compresses bulb of urethra and assists in erection of penis *Female*: Reduces lumen of vagina and assists in erection of clitoris
Ischiocavernosus	Internal surface of ischiopubic ramus and ischial tuberosity (compressor urethrae portion only)	Crus of penis or clitoris		Maintains erection of penis or clitoris by compression of crura
Superficial transverse perineal		Perineal body		Supports perineal body
Deep transverse perineal		Median raphe, perineal body, and external anal sphincter		Fixes perineal body
External urethral sphincter		Surrounds urethra; in males, also ascends anterior aspect of prostate; in females also encloses vagina		*Female*: Compresses urethra and vagina *Male*: Compresses intermediate (membranous) urethra

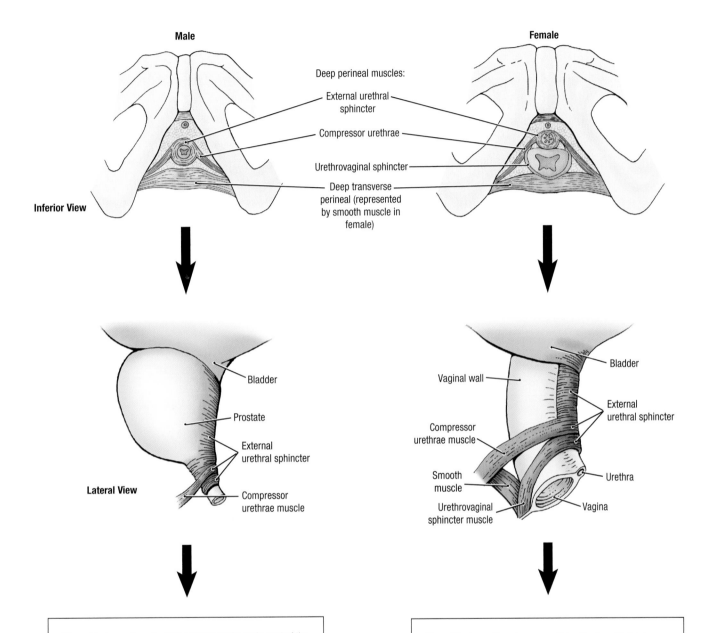

Male

Female

Deep perineal muscles:

External urethral sphincter

Compressor urethrae

Urethrovaginal sphincter

Deep transverse perineal (represented by smooth muscle in female)

Inferior View

Bladder

Prostate

External urethral sphincter

Compressor urethrae muscle

Lateral View

Bladder

Vaginal wall

External urethral sphincter

Compressor urethrae muscle

Smooth muscle

Urethra

Urethrovaginal sphincter muscle

Vagina

The sphincter urethrae is more tubelike. In the male, part of the muscle encircles the intermediate (membranous) part of the urethra, and part extends to the base of the bladder, investing the prostatic urethra anteriorly and anterolaterally. This muscle is called the *external urethral sphincter*.

The sphincter urethrae is a urogenital sphincter with three parts: (1) a superior part that extends to the base of the bladder, the external urethral sphincter muscle; (2) a part extending inferolaterally to the ischial ramus on each side, the compressor urethrae muscle; and (3) a bandlike part encircling the vagina and urethra, the urethrovaginal sphincter.

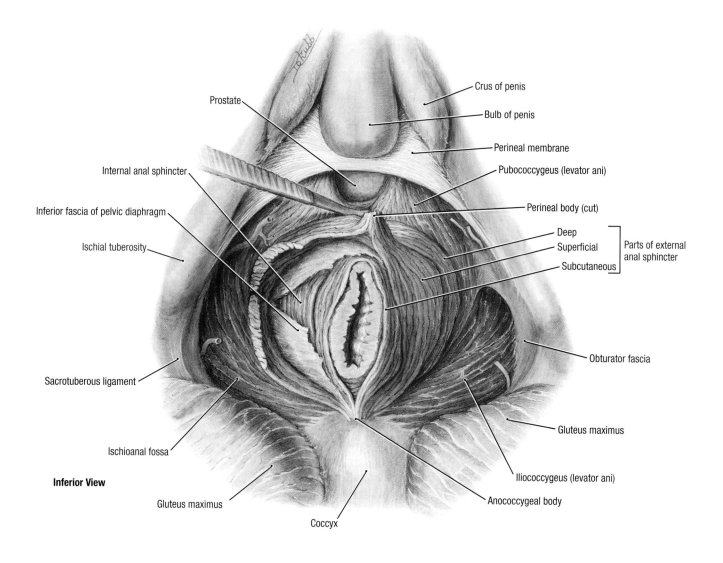

Prostate

Internal anal sphincter

Inferior fascia of pelvic diaphragm

Ischial tuberosity

Sacrotuberous ligament

Ischioanal fossa

Inferior View

Gluteus maximus

Coccyx

Crus of penis

Bulb of penis

Perineal membrane

Pubococcygeus (levator ani)

Perineal body (cut)

Deep
Superficial   Parts of external
               anal sphincter
Subcutaneous

Obturator fascia

Gluteus maximus

Iliococcygeus (levator ani)

Anococcygeal body

3.45   Dissection of male perineum—II

- The external anal sphincter has three continuous regions: subcutaneous, encircling the anal orifice; superficial, anchoring the anus in the median plane to the perineal body anteriorly and to the coccyx posteriorly; and deep, forming a wide, encircling band.
- On the left of the figure: The superficial and deep parts of the sphincter are reflected, and the underlying sheet, consisting of areolar tissue, levator ani fibers, and the outer, longitudinal, muscular coat, is cut to reveal the inner, circular, muscular coat, which is thickened to form the internal anal sphincter.

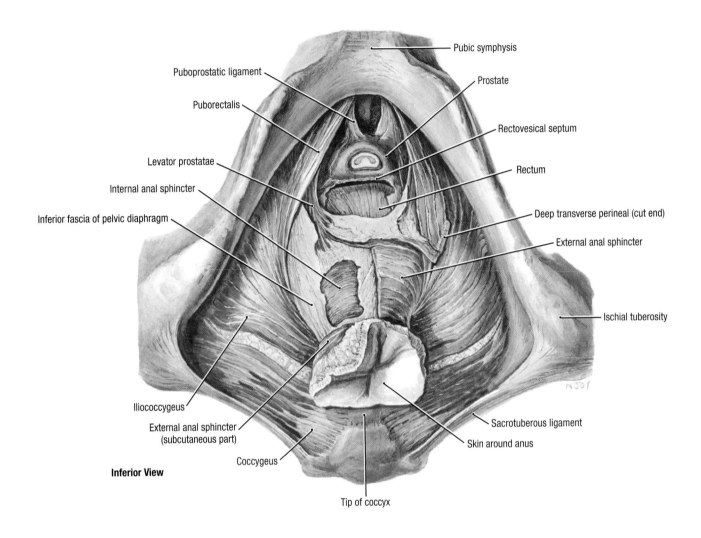

Pubic symphysis

Puboprostatic ligament

Prostate

Puborectalis

Rectovesical septum

Levator prostatae

Rectum

Internal anal sphincter

Inferior fascia of pelvic diaphragm

Deep transverse perineal (cut end)

External anal sphincter

Ischial tuberosity

Iliococcygeus

External anal sphincter
(subcutaneous part)

Sacrotuberous ligament

Skin around anus

Coccygeus

Inferior View

Tip of coccyx

3.46   **Dissection of male perineum—III**

The longitudinal muscle coat of the rectum and its fascia blend with the levator ani muscle and its fasciae to form a fibroelastic tube that descends between the external and internal anal sphincters. From this tube, septa pass (1) through the internal sphincter to the submucous coat, (2) through the external sphincter to the skin, and (3) as the anal intermuscular septum, inferior to the internal sphincter.

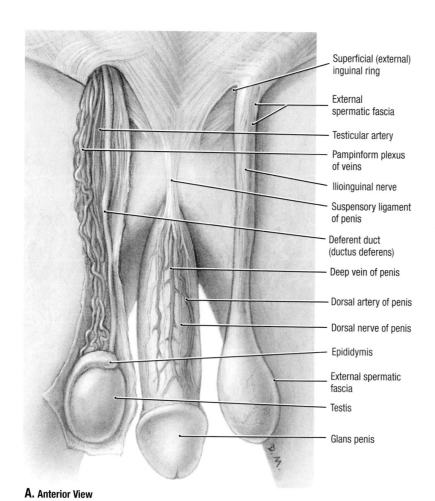

Superficial (external) inguinal ring
External spermatic fascia
Testicular artery
Pampiniform plexus of veins
Ilioinguinal nerve
Suspensory ligament of penis
Deferent duct (ductus deferens)
Deep vein of penis
Dorsal artery of penis
Dorsal nerve of penis
Epididymis
External spermatic fascia
Testis
Glans penis

A. Anterior View

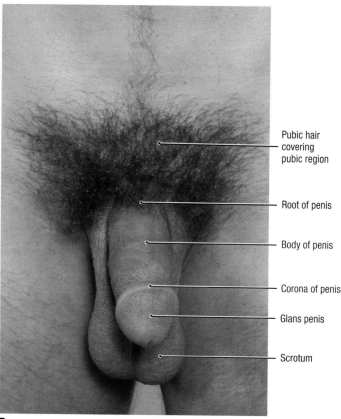

Pubic hair covering pubic region
Root of penis
Body of penis
Corona of penis
Glans penis
Scrotum

B. Anterior View

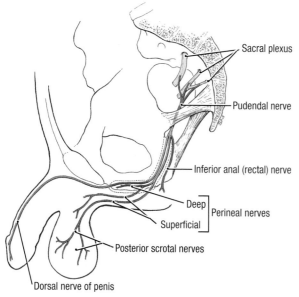

Sacral plexus
Pudendal nerve
Inferior anal (rectal) nerve
Deep
Superficial } Perineal nerves
Posterior scrotal nerves
Dorsal nerve of penis

C. Medial View, Schematic

3.47 **Spermatic cord and penis**

A. Vessels and nerves of penis and contents of spermatic cord. **B.** Surface anatomy. **C.** Course and distribution of pudendal nerve.
In **A**:
- The superficial and deep fasciae covering the penis are removed to expose the midline deep dorsal vein and the bilateral dorsal arteries and nerves of the penis. The triangular suspensory ligament of the penis attaches to the region of the pubic symphysis and blends with the deep fascia of the penis.
- On the specimen's left, the spermatic cord passes through the external inguinal ring and picks up a covering of external spermatic fascia from the margins of the superficial inguinal ring.
- On the specimen's right, the coverings of the spermatic cord and testis are incised and reflected, and the contents of the cord are separated. The spermatic cord contains the deferent duct (with its vessels), the testicular artery (dissected away from the surrounding pampiniform plexus of veins), the genital branch of the genitofemoral nerve, and fibers of the cremaster muscle. Lymphatic vessels and autonomic nerve fibers (not shown here) are also present.

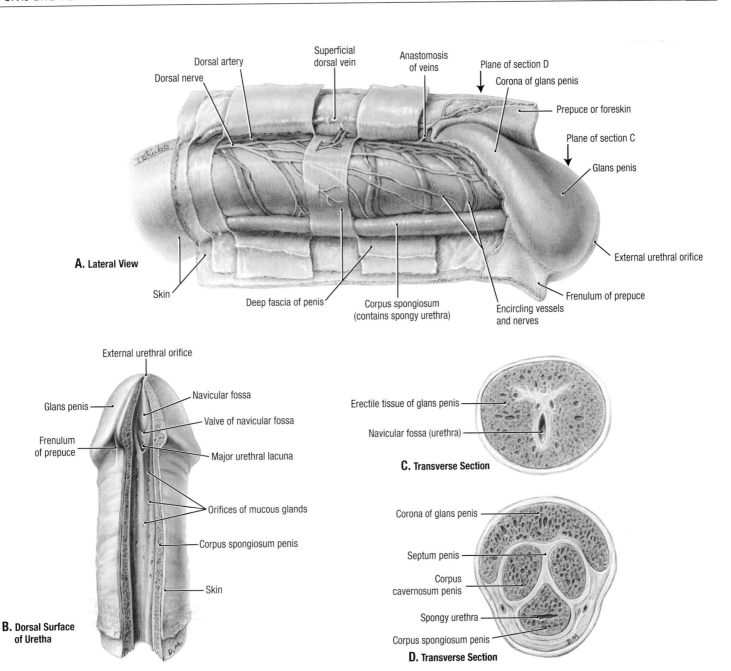

A. Lateral View

Dorsal nerve
Dorsal artery
Superficial dorsal vein
Anastomosis of veins
Plane of section D
Corona of glans penis
Prepuce or foreskin
Plane of section C
Glans penis
External urethral orifice
Frenulum of prepuce
Encircling vessels and nerves
Corpus spongiosum (contains spongy urethra)
Deep fascia of penis
Skin

B. Dorsal Surface of Uretha

External urethral orifice
Navicular fossa
Glans penis
Valve of navicular fossa
Frenulum of prepuce
Major urethral lacuna
Orifices of mucous glands
Corpus spongiosum penis
Skin

C. Transverse Section

Erectile tissue of glans penis
Navicular fossa (urethra)

D. Transverse Section

Corona of glans penis
Septum penis
Corpus cavernosum penis
Spongy urethra
Corpus spongiosum penis

3.48   Penis

A. Dissection. The skin, subcutaneous tissue, and deep fascia of the penis and the prepuce are reflected separately. **B.** Spongy urethra, interior. A longitudinal incision was made on the urethral surface of the penis and carried through the floor of the urethra, allowing a view of the dorsal surface of the interior of the urethra. **C.** Transverse section through the proximal part of the glans penis. **D.** Transverse section through the distal part of the glans penis. For **C** and **D**, see indication of level of section in **A.**

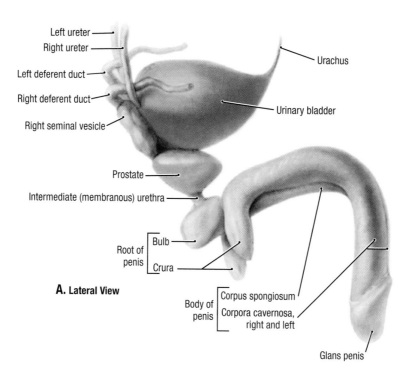

Left ureter

Right ureter

Left deferent duct

Right deferent duct

Right seminal vesicle

Urachus

Urinary bladder

Prostate

Intermediate (membranous) urethra

Root of penis — Bulb / Crura

Body of penis — Corpus spongiosum / Corpora cavernosa, right and left

Glans penis

A. Lateral View

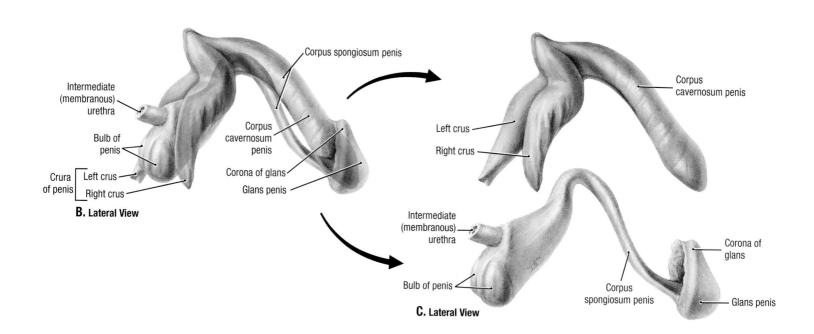

Corpus spongiosum penis

Intermediate (membranous) urethra

Corpus cavernosum penis

Bulb of penis

Corpus cavernosum penis

Left crus

Right crus

Crura of penis — Left crus / Right crus

Corona of glans

Glans penis

B. Lateral View

Intermediate (membranous) urethra

Corona of glans

Bulb of penis

Corpus spongiosum penis

Glans penis

C. Lateral View

3.49   Male urogenital system

A. Pelvic components of genital and urinary tracts and erectile bodies of perineum. **B.** Dissection of male erectile bodies (corpora cavernosa and corpus spongiosum). **C.** Corpus spongiosum and corpora cavernosa, separated. The corpora cavernosa is bent where the penis is suspended by the suspensory ligament of the penis from the pubic symphysis. The corpus spongiosum extends posteriorly as the bulb of the penis and terminates anteriorly as the glans.

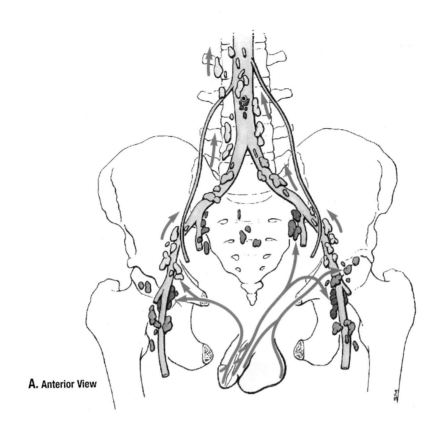

Nodes:
- Lumbar (lateral aortic)
- Inferior mesenteric
- Common iliac
- Internal iliac
- External iliac
- Superficial inguinal
- Deep inguinal
- Sacral
- → Direction of flow

A. Anterior View

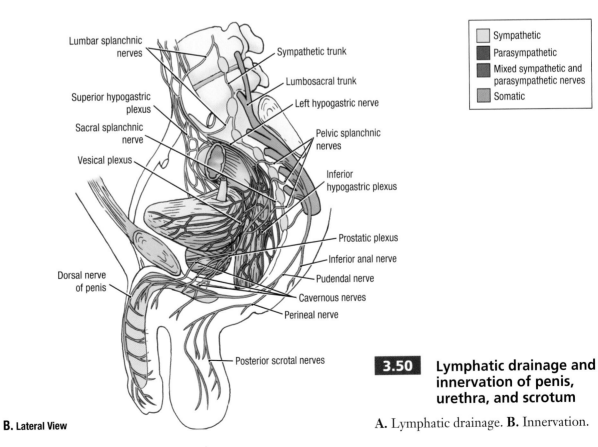

- Sympathetic
- Parasympathetic
- Mixed sympathetic and parasympathetic nerves
- Somatic

Lumbar splanchnic nerves

Superior hypogastric plexus

Sacral splanchnic nerve

Vesical plexus

Dorsal nerve of penis

Posterior scrotal nerves

Sympathetic trunk

Lumbosacral trunk

Left hypogastric nerve

Pelvic splanchnic nerves

Inferior hypogastric plexus

Prostatic plexus

Inferior anal nerve

Pudendal nerve

Cavernous nerves

Perineal nerve

B. Lateral View

3.50 **Lymphatic drainage and innervation of penis, urethra, and scrotum**

A. Lymphatic drainage. **B.** Innervation.

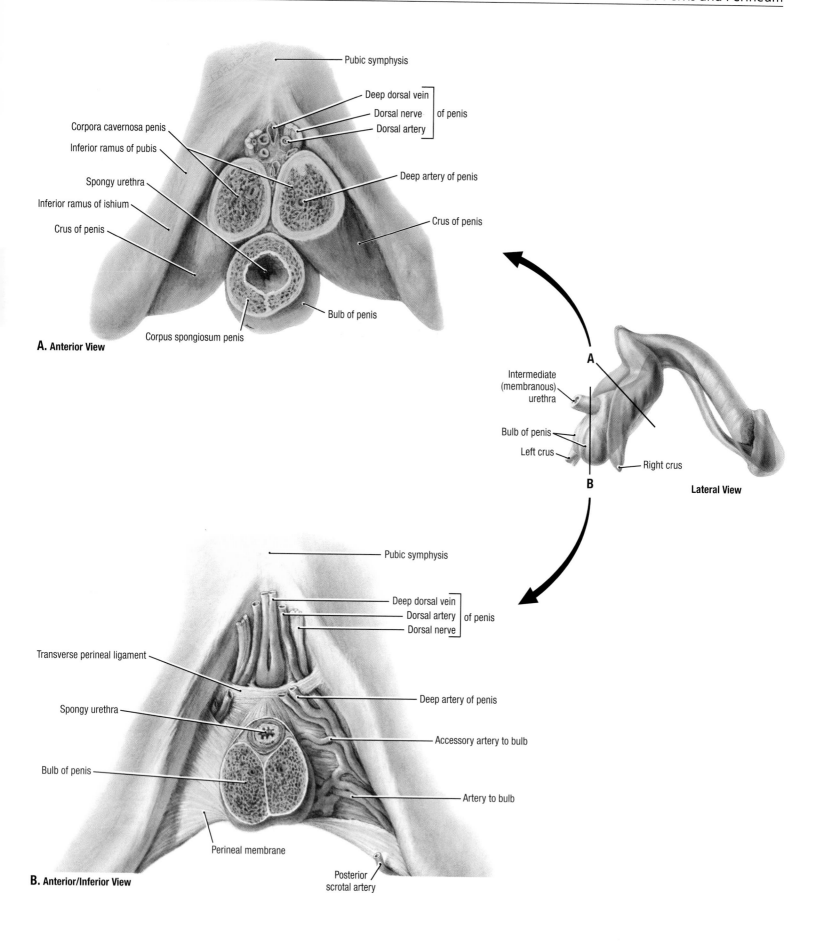

Pubic symphysis

Deep dorsal vein
Dorsal nerve ⎤ of penis
Dorsal artery

Corpora cavernosa penis

Inferior ramus of pubis

Deep artery of penis

Spongy urethra

Inferior ramus of ishium

Crus of penis

Crus of penis

Bulb of penis

Corpus spongiosum penis

A. Anterior View

Intermediate
(membranous)
urethra

Bulb of penis

Left crus

Right crus

Lateral View

Pubic symphysis

Deep dorsal vein
Dorsal artery ⎤ of penis
Dorsal nerve

Transverse perineal ligament

Deep artery of penis

Spongy urethra

Accessory artery to bulb

Bulb of penis

Artery to bulb

Perineal membrane

Posterior
scrotal artery

B. Anterior/Inferior View

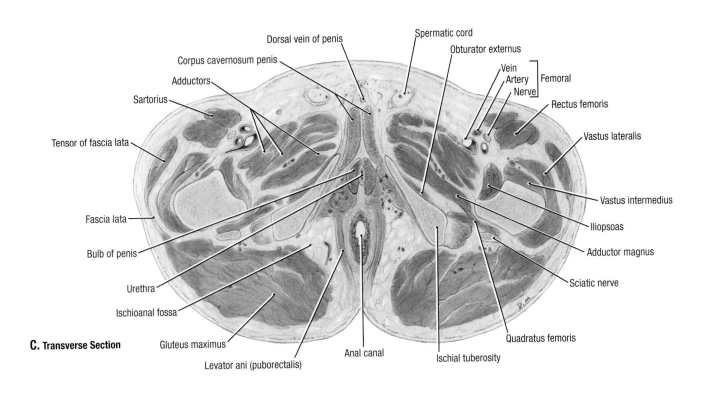

Dorsal vein of penis
Corpus cavernosum penis
Adductors
Sartorius
Tensor of fascia lata
Fascia lata
Bulb of penis
Urethra
Ischioanal fossa
Gluteus maximus
Levator ani (puborectalis)
Anal canal
Ischial tuberosity
Quadratus femoris
Sciatic nerve
Adductor magnus
Iliopsoas
Vastus intermedius
Vastus lateralis
Rectus femoris
Femoral { Vein / Artery / Nerve }
Obturator externus
Spermatic cord

C. Transverse Section

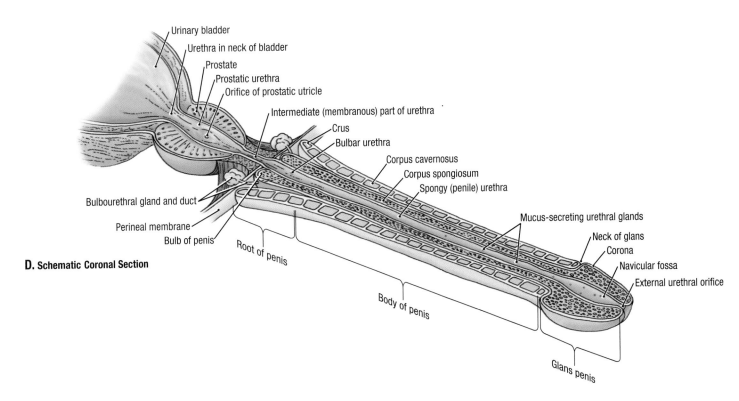

Urinary bladder
Urethra in neck of bladder
Prostate
Prostatic urethra
Orifice of prostatic utricle
Intermediate (membranous) part of urethra
Crus
Bulbar urethra
Corpus cavernosus
Corpus spongiosum
Spongy (penile) urethra
Mucus-secreting urethral glands
Neck of glans
Corona
Navicular fossa
External urethral orifice
Glans penis
Body of penis
Root of penis
Bulb of penis
Perineal membrane
Bulbourethral gland and duct

D. Schematic Coronal Section

3.51 Root of penis

A. The crura and bulb of penis have been sectioned transversely. The urethra is dilated within the bulb of the penis as the intrabulbar fossa. **B.** Transverse section through bulb of penis with crura removed. The bulb is cut posterior to the entry of the intermediate urethra. On the left side, the perineal membrane is partially removed, opening the deep perineal compartment. **C.** Transverse section of body passing through the bulb of the penis. **D.** Urethra and related structures.

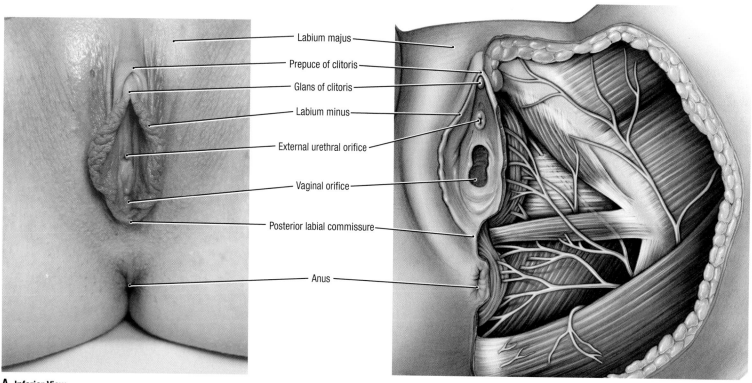

Labium majus
Prepuce of clitoris
Glans of clitoris
Labium minus
External urethral orifice
Vaginal orifice
Posterior labial commissure
Anus

A. Inferior View

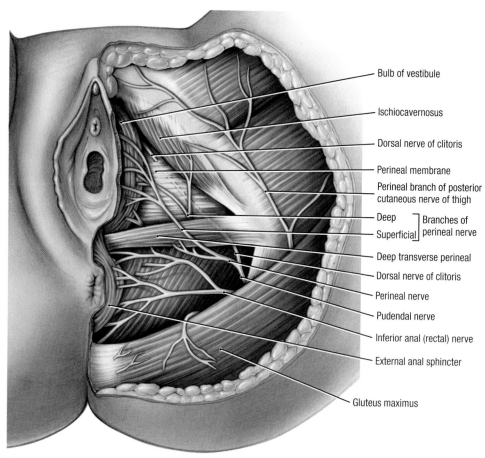

Bulb of vestibule

Ischiocavernosus

Dorsal nerve of clitoris

Perineal membrane

Perineal branch of posterior cutaneous nerve of thigh

Deep — Branches of
Superficial — perineal nerve

Deep transverse perineal

Dorsal nerve of clitoris

Perineal nerve

Pudendal nerve

Inferior anal (rectal) nerve

External anal sphincter

Gluteus maximus

B. Inferior View

3.52 **Surface anatomy and innervation of the female
perineum.**

A. Surface anatomy. **B.** Innervation.

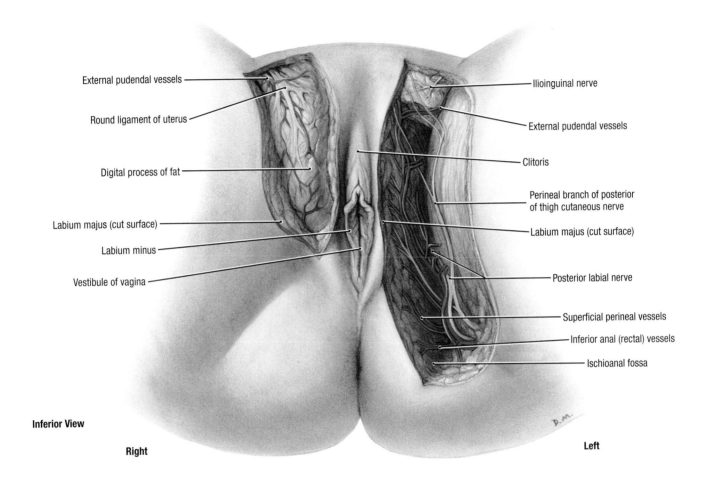

External pudendal vessels

Round ligament of uterus

Digital process of fat

Labium majus (cut surface)

Labium minus

Vestibule of vagina

Ilioinguinal nerve

External pudendal vessels

Clitoris

Perineal branch of posterior of thigh cutaneous nerve

Labium majus (cut surface)

Posterior labial nerve

Superficial perineal vessels

Inferior anal (rectal) vessels

Ischioanal fossa

Inferior View

Right   **Left**

3.53   Female perineum—I

Superficial dissection. On the right side of the specimen:
- A long digital process of fat lies deep to the subcutaneous fatty tissue and descends into the labium majus.
- The round ligament of the uterus ends as a branching band of fascia that spreads out superficial to the fatty digital process.

On the left side of the specimen:
- Most of the fatty digital process is removed.
- The posterior labial vessels and nerves (S2, S3) are joined by the perineal branch of the posterior femoral cutaneous nerve (S1, S2, S3) and run anteriorly to the mons pubis. At the mons pubis the vessels anastomose with the external pudendal vessels and the nerves overlap in supply with the ilioinguinal nerve (L1). The mons pubis is the rounded fatty prominence anterior to the pubic symphysis, pubic tubercle, and superior pubic rami.

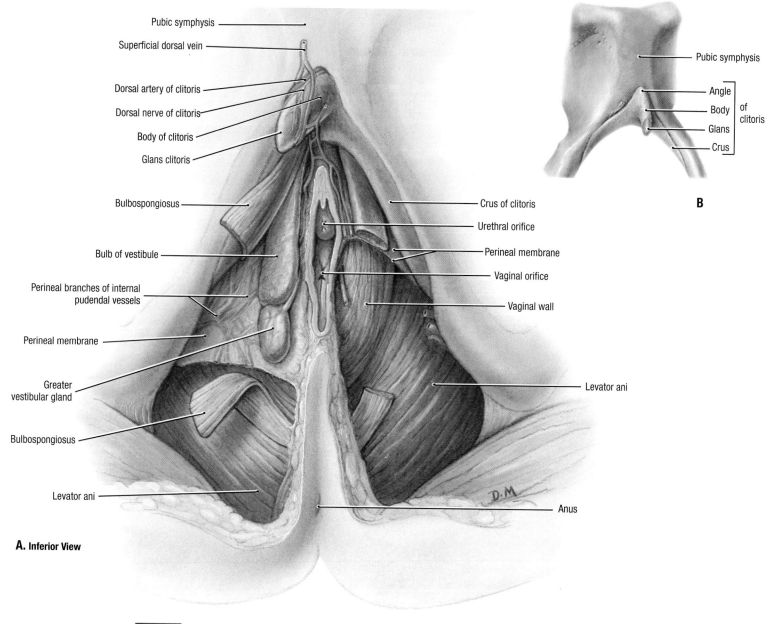

Pubic symphysis

Superficial dorsal vein

Dorsal artery of clitoris

Dorsal nerve of clitoris

Body of clitoris

Glans clitoris

Bulbospongiosus

Bulb of vestibule

Perineal branches of internal pudendal vessels

Perineal membrane

Greater vestibular gland

Bulbospongiosus

Levator ani

Crus of clitoris

Urethral orifice

Perineal membrane

Vaginal orifice

Vaginal wall

Levator ani

Anus

A. Inferior View

Pubic symphysis

Angle

Body

Glans

Crus

of clitoris

B

3.54   Female perineum—II

A. Deeper dissection. **B.** Clitoris.

In **A**:

- The bulbospongiosus muscle is reflected on the right side and mostly removed on the left side; the posterior portion of the bulb of the vestibule and greater vestibular gland have been removed on the left side.
- The glans clitoris is displaced to the right so that the distribution of the dorsal vessels and nerve of the clitoris can be seen.
- A homologue of the bulb of the penis, the bulb of the vestibule, is split by the vagina so that it appears as two masses of elongated erectile tissue that lie along the sides of the vaginal orifice; veins connect the bulbs of the vestibule to the glans of the clitoris.
- On the specimen's right side, the greater vestibular gland is situated at the posterior end of the bulb; both structures are covered by bulbospongiosus muscle.
- On the specimen's left side, the perineal membrane is cut away, thereby revealing the external aspect of the vaginal wall.

In **B**:

- The body of the clitoris is composed of two corpora cavernosa suspended by a suspensory ligament and capped by the glans.

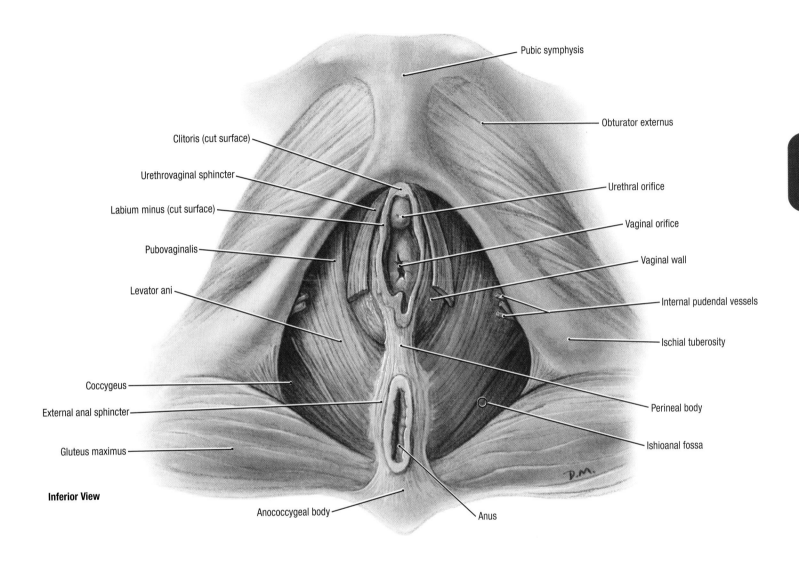

Inferior View

- Pubic symphysis
- Obturator externus
- Clitoris (cut surface)
- Urethrovaginal sphincter
- Urethral orifice
- Labium minus (cut surface)
- Vaginal orifice
- Pubovaginalis
- Vaginal wall
- Levator ani
- Internal pudendal vessels
- Ischial tuberosity
- Coccygeus
- External anal sphincter
- Perineal body
- Gluteus maximus
- Ishioanal fossa
- Anococcygeal body
- Anus

3.55   Female perineum—III

- The perineal membrane and deep transverse perineal muscle have been removed.
- The most anterior medial part of the levator ani (pubovaginalis) muscle meet posterior to the vaginal orifice.
- The urethrovaginal sphincter, part of the external urethral sphincter, rests on the urethra and straddles the vagina.
- The labia minora, cut short, bound the vestibule of the vagina.

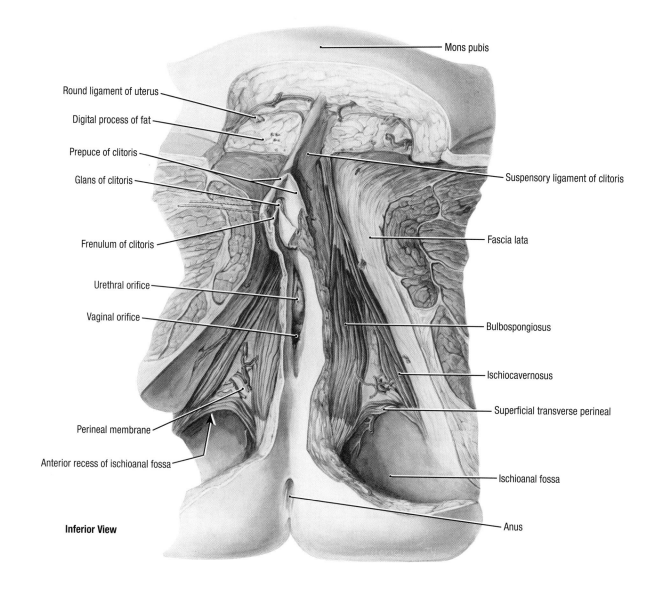

Mons pubis

Round ligament of uterus

Digital process of fat

Prepuce of clitoris

Glans of clitoris

Suspensory ligament of clitoris

Frenulum of clitoris

Fascia lata

Urethral orifice

Vaginal orifice

Bulbospongiosus

Ischiocavernosus

Superficial transverse perineal

Perineal membrane

Anterior recess of ischioanal fossa

Ischioanal fossa

Inferior View

Anus

3.56   Female perineum—IV

- Note the thickness of the subcutaneous fatty tissue of the mons pubis and the encapsulated digital process of fat deep to this. The suspensory ligament of the clitoris descends from the linea alba.
- Anteriorly, each labium minus forms two laminae or folds: the lateral laminae of the labia pass on each side of the glans clitoris and unite, forming a hood that partially or completely covers the glans, the prepuce (foreskin) of the clitoris. The medial laminae of the labia merge posterior to the glans, forming the frenulum of the clitoris.
- There are three muscles on each side: bulbospongiosus, ischiocavernosus, and superficial transverse perineal; the perineal membrane is revealed between them.
- The bulbospongiosus muscle overlies the bulb of the vestibule and the great vestibular gland. In the male, the muscles of the two sides are united by a median raphe; in the female, the orifice of the vagina separates the two.

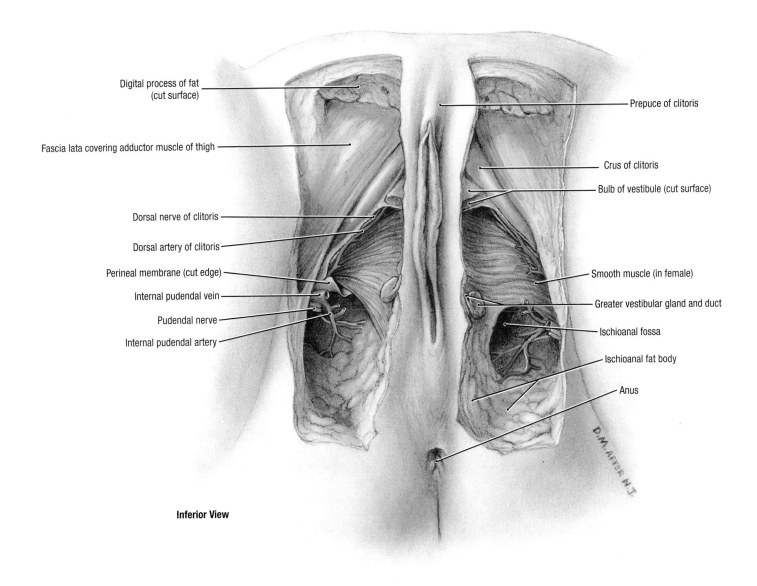

Digital process of fat (cut surface)

Fascia lata covering adductor muscle of thigh

Dorsal nerve of clitoris

Dorsal artery of clitoris

Perineal membrane (cut edge)

Internal pudendal vein

Pudendal nerve

Internal pudendal artery

Prepuce of clitoris

Crus of clitoris

Bulb of vestibule (cut surface)

Smooth muscle (in female)

Greater vestibular gland and duct

Ischioanal fossa

Ischioanal fat body

Anus

Inferior View

3.57   **Female perineum—V**

- The bulbs of the vestibule have been removed, except at their pubic ends, but the greater vestibular glands and ducts remain. The perineal membrane is removed, except for a marginal fringe.
- The dorsal nerve and artery to the clitoris run anteriorly and give twigs to the bulb and crus.
- The greater vestibular glands are located in the superficial perineal pouch.

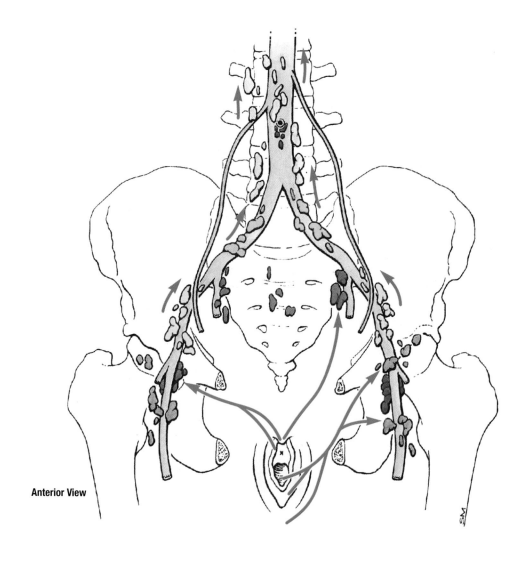

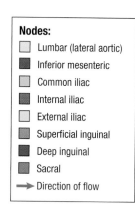

Nodes:

- Lumbar (lateral aortic)
- Inferior mesenteric
- Common iliac
- Internal iliac
- External iliac
- Superficial inguinal
- Deep inguinal
- Sacral
- → Direction of flow

Anterior View

3.58 Lymphatic drainage of the external female genitalia

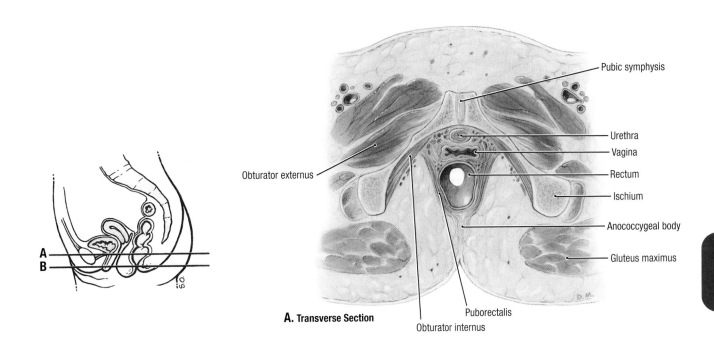

A. Transverse Section

- Pubic symphysis
- Urethra
- Vagina
- Rectum
- Ischium
- Anococcygeal body
- Gluteus maximus
- Obturator externus
- Puborectalis
- Obturator internus

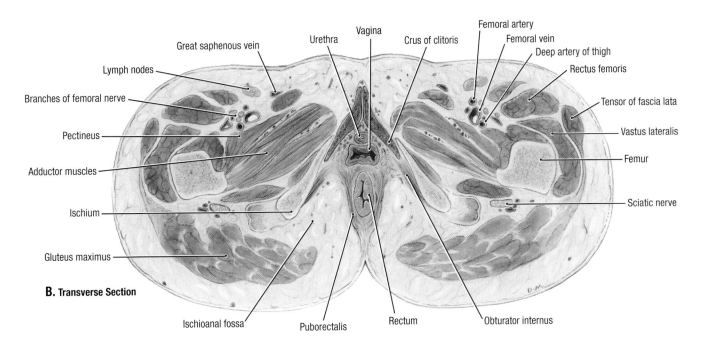

B. Transverse Section

- Great saphenous vein
- Urethra
- Vagina
- Crus of clitoris
- Femoral artery
- Femoral vein
- Deep artery of thigh
- Rectus femoris
- Tensor of fascia lata
- Vastus lateralis
- Femur
- Sciatic nerve
- Lymph nodes
- Branches of femoral nerve
- Pectineus
- Adductor muscles
- Ischium
- Gluteus maximus
- Ischioanal fossa
- Puborectalis
- Rectum
- Obturator internus

3.59   **Transverse sections of female perineum**

A. Section through vagina and urethra at base of urinary bladder. **B.** Section through vagina, urethra, and crura of clitoris.

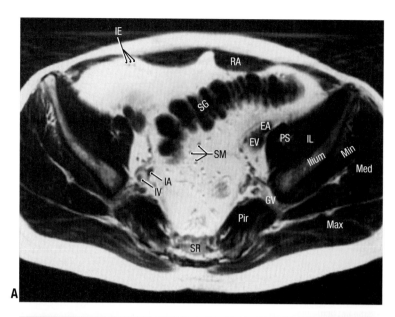

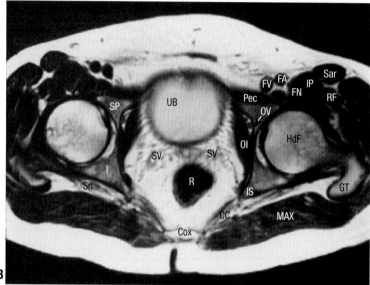

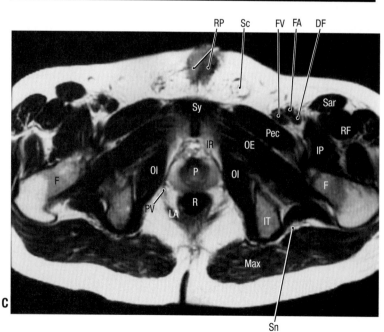

A	Anus
Ad	Adductor muscles
C	Conjoint ramus
Cav	Corpus cavernosum penis
CC	Coccygeus
Cox	Coccyx
Cr	Crus of penis
DF	Deep femoral artery
EA	External iliac artery
EV	External iliac vein
F	Femur
FA	Femoral artery
FN	Femoral nerve
FV	Femoral vein
GT	Greater trochanter
GV	Superior gluteal vein
HdF	Head of femur
I	Body of ischium
IA	Internal iliac artery
IAF	Ischioanal fossa
IE	Inferior epigastric vessels
IL	Iliacus
IP	Iliopsoas
IR	Inferior ramus of pubis
IS	Ischial spine
IT	Ischial tuberosity
IV	Internal iliac vein
LA	Levator ani
Max	Gluteus maximus

3.60 Transverse (axial) MRIs of the male pelvis and perineum

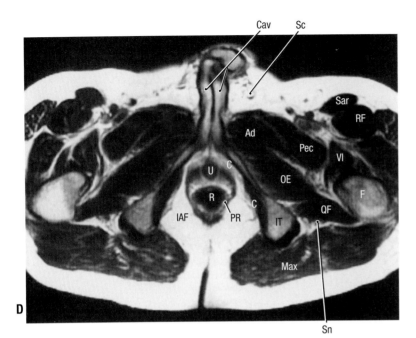

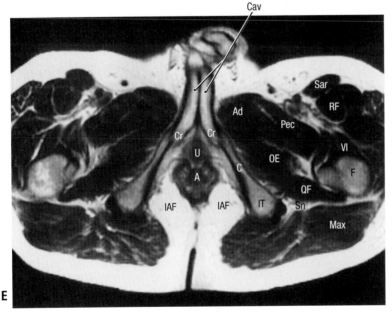

Med	Gluteus medius
Min	Gluteus minimus
OE	Obturator externus
OI	Obturator internus
OV	Obturator vessels and nerve
P	Prostate
Pec	Pectineus
Pir	Piriformis
PR	Puborectalis
PS	Psoas
PV	Pelvic vessels and nerves
QF	Quadratus femoris
R	Rectum
RA	Rectus abdominis
RF	Rectus femoris
RP	Root of penis
Sar	Sartorius
Sc	Spermatic cord
SG	Sigmoid colon
SM	Sigmoidal vessels in mesentery of sigmoid colon
Sn	Sciatic nerve
SP	Superior ramus of pubis
SR	Sacrum
SV	Seminal vesicle
Sy	Pubic symphysis
U	Urethra
UB	Urinary bladder
VI	Vastus intermedius

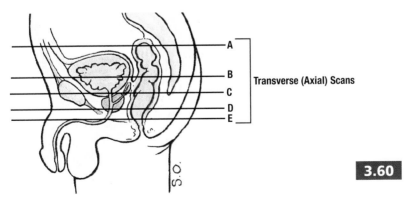

3.60 **Transverse (axial) MRIs of the male pelvis and perineum** *(continued)*

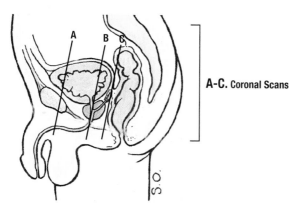

A–C. Coronal Scans

Key for A–C

A	Anus
Ad	Adductors
CA	Common iliac artery
Cav	Corpus cavernosum penis
Cs	Corpus spongiosum penis
CV	Common iliac vein
DC	Descending colon
EA	External iliac artery
EV	External iliac vein
FA	Femoral artery
FV	Femoral vein
HdF	Head of femur
IL	Iliacus
In	Intestine
IR	Inferior rectal nerve and vessels
LA	Levator ani
LS	Lumbosacral trunk
OE	Obturator externus
OI	Obturator internus
P	Prostate
Pec	Pectineus
PS	Psoas
Pu	Pubic bone
PV	Pelvic vessels and nerves
R	Rectum
Sac	Sacrum
SG	Sigmoid colon
SV	Seminal vesicle
Sy	Pubic symphysis
U	Urethra
UB	Urinary bladder

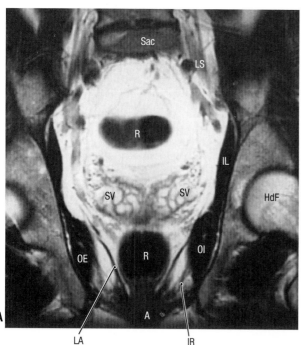

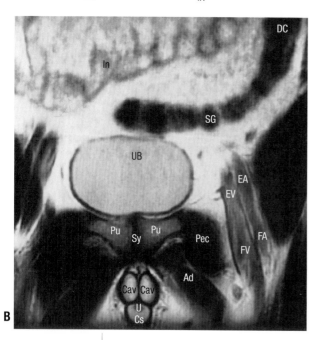

3.61 **Coronal and sagittal MRIs of the male pelvis and perineum**

A–C. Coronal scans. **D.** Sagittal scan.

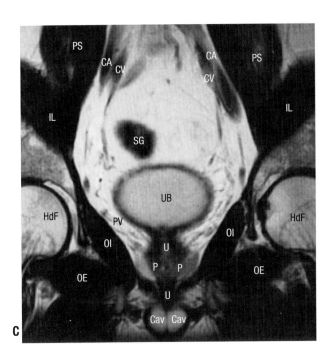

C

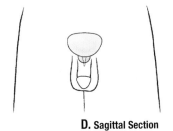

D. Sagittal Section

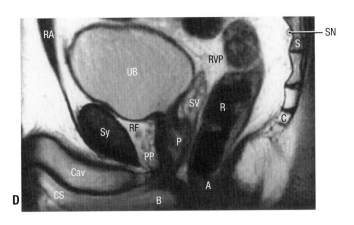

D

Key for D	
A	Anus
B	Bulb of penis
C	Coccyx
Cav	Corpus cavernosum penis
CS	Corpus spongiosum penis
P	Prostate
PP	Prostatic venous plexus
R	Rectum
RA	Rectus abdominis
RF	Retropubic fat
RVP	Rectovesical pouch
S	Sacrum
SN	Sacral nerves
SV	Seminal vesicle
Sy	Pubic symphysis
UB	Urinary bladder

3.61 **Coronal and sagittal MRIs of the male pelvis and perineum** *(continued)*

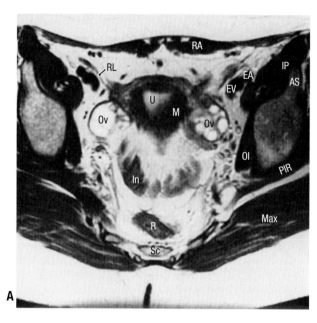

A

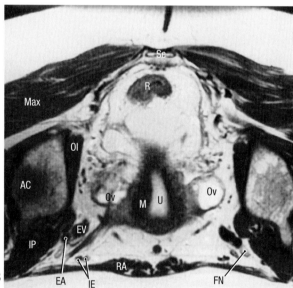

B

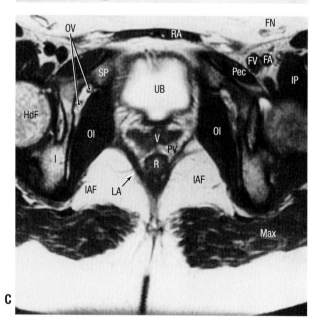

C

A	Anus
AC	Acetabulum
Ad	Adductor muscles
AS	Anterior superior iliac spine
C	Conjoint ramus
EA	External iliac artery
EV	External iliac vein
FA	Femoral artery
FN	Femoral nerve
FV	Femoral vein
HdF	Head of femur
I	Ilium
IAF	Ischioanal fossa
IE	Inferior epigastric vessels
In	Intestine
IP	Iliopsoas
IT	Ischial tuberosity
LA	Levator ani
LM	Labia majus
M	Myometrium
Max	Gluteus maximus
OE	Obturator externus
OI	Obturator internus
Ov	Ovary
OV	Obturator vessels
Pec	Pectineus
PIR	Piriformis
Pm	Perineal membrane
Pu	Pubic bone
PV	Perivaginal veins
QF	Quadratus femoris
R	Rectum
RA	Rectus abdominis
RL	Round ligament
Sc	Sacrum
SP	Superior ramus of pubis
Sy	Pubis symphysis
U	Uterus
UB	Urinary bladder
Ur	Urethra
V	Vagina
Ve	Vestibule

3.62 Transverse (axial) MRIs of the female pelvis and perineum

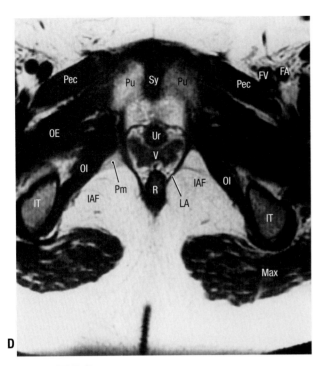

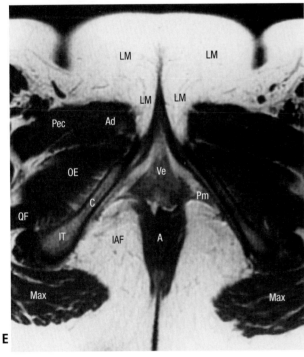

A	Anus
AC	Acetabulum
Ad	Adductor muscles
AS	Anterior superior iliac spine
C	Conjoint ramus
EA	External iliac artery
EV	External iliac vein
FA	Femoral artery
FN	Femoral nerve
FV	Femoral vein
HdF	Head of femur
I	Ilium
IAF	Ischioanal fossa
IE	Inferior epigastric vessels
In	Intestine
IP	Iliopsoas
IT	Ischial tuberosity
LA	Levator ani
LM	Labia majus
M	Myometrium
Max	Gluteus maximus
OE	Obturator externus
OI	Obturator internus
Ov	Ovary
OV	Obturator vessels
Pec	Pectineus
PIR	Piriformis
Pm	Perineal membrane
Pu	Pubic bone
PV	Perivaginal veins
QF	Quadratus femoris
R	Rectum
RA	Rectus abdominis
RL	Round ligament
Sc	Sacrum
SP	Superior ramus of pubis
Sy	Pubis symphysis
U	Uterus
UB	Urinary bladder
Ur	Urethra
V	Vagina
Ve	Vestibule

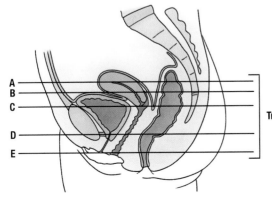

Transverse (Axial) Scans

3.62   **Transverse (axial) MRIs of the female pelvis and perineum** *(continued)*

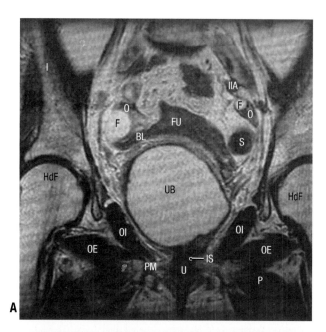

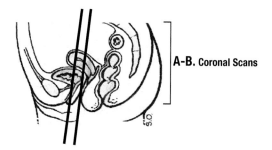

A-B. Coronal Scans

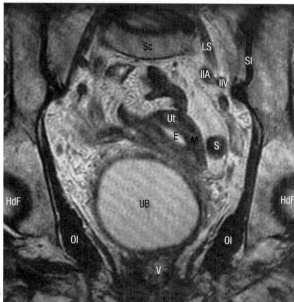

Key for A and B

BL	Broad ligament
E	Endometrium
F	Follicle
FU	Fundus of uterus
HdF	Head of femur
I	Ilium
IIA	Internal iliac artery
IIV	Internal iliac vein
IS	Internal urethral sphincter
LS	Lumbosacral trunk
M	Myometrium
O	Ovary
OE	Obturator externus
OI	Obturator internus
P	Pectineus
PM	Perineal membrane
S	Sigmoid colon
Sc	Sacrum
SI	Sacroiliac joint
U	Urethra
UB	Urinary bladder
Ut	Uterus
V	Vagina

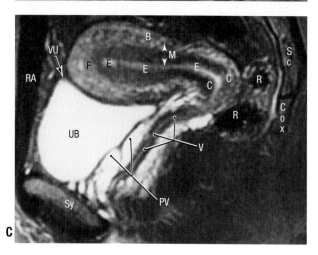

Key for C

B	Body of uterus
C	Cervix of uterus
Cox	Coccyx
E	Endometrium
F	Fundus of uterus
M	Myometrium
PV	Perivaginal veins
R	Rectum
RA	Rectus abdominis
Sc	Sacrum
Sy	Pubis symphysis
UB	Urinary bladder
V	Vagina
VU	Vesicouterine pouch

3.63 **Coronal and sagittal MRIs of the female pelvis and perineum**

A and **B.** Coronal scans. **C.** Sagittal scan.

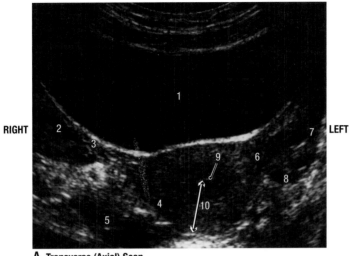

A. Transverse (Axial) Scan

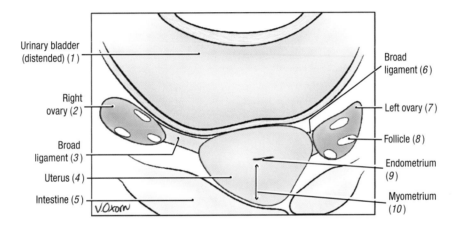

Urinary bladder (distended) (*1*)

Right ovary (*2*)

Broad ligament (*3*)

Uterus (*4*)

Intestine (*5*)

Broad ligament (*6*)

Left ovary (*7*)

Follicle (*8*)

Endometrium (*9*)

Myometrium (*10*)

V.Oxorn

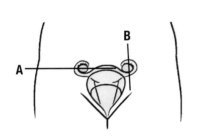

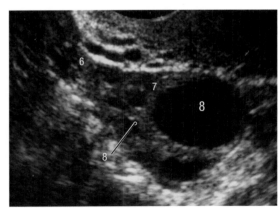

B. Sagittal Scan

3.64   **Ultrasound scans of female pelvis**

A. Transabdominal axial (transverse) scan through uterus and ovaries. **B.** Transvaginal sagittal scan of left ovary (*numbers in parentheses* correspond to labels on the ultrasound scans).

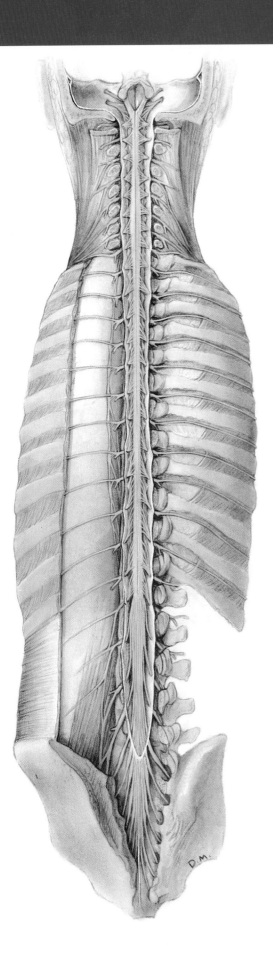

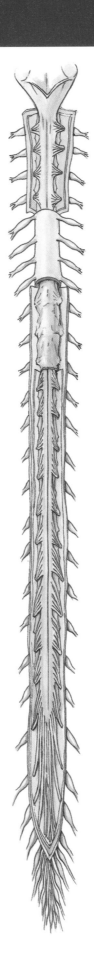

■ Vertebral Column and
  Overview of Vertebrae **274**
■ Cervical Vertebrae **282**
■ Thoracic Vertebrae **286**
■ Lumbar Vertebrae **288**
■ Sacrum and Sacroiliac
  Joint **291**
■ Anomalies of Vertebrae **296**
■ Ligaments of Vertebral
  Column and Intervertebral
  Discs **298**
■ Vertebral Venous Plexuses **305**
■ Muscles of the Back **306**
■ Suboccipital Region **316**
■ Craniovertebral Joints **320**
■ Spinal Cord and its
  Environment **322**
■ Spinal Nerves and
  Segmental Innervation **330**
■ Imaging of Vertebrae and
  Spinal Cord **334**

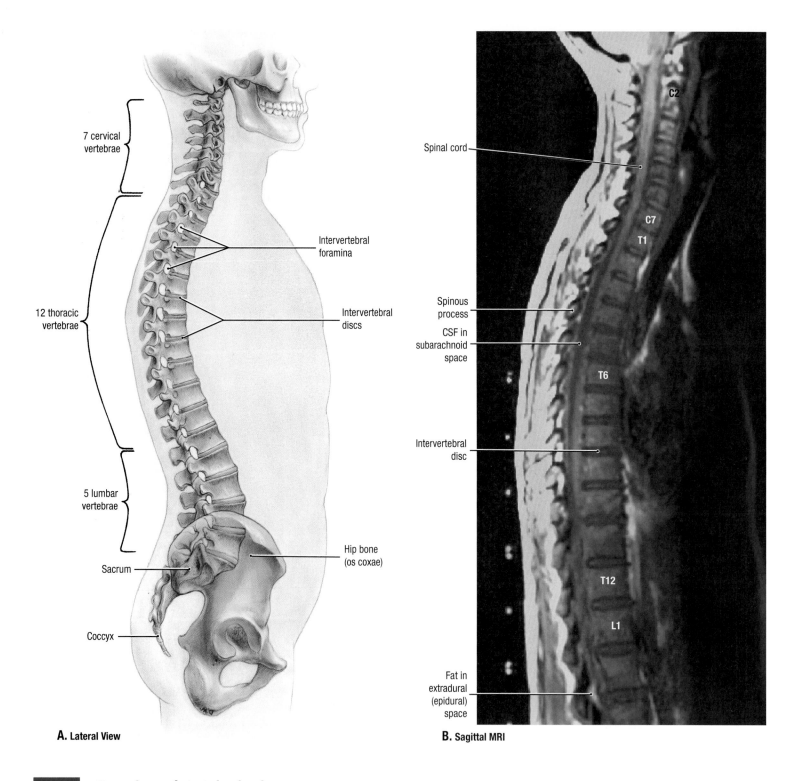

A. Lateral View

B. Sagittal MRI

4.1 Overview of vertebral column

A. Vertebral column showing articulation with skull and hip bone.
B. Sagittal MRI, lateral view.

- The vertebral column usually comprises 24 separate (presacral) vertebrae, 5 fused vertebrae in the sacrum and a variable number of fused and separate coccygeal vertebrae; of the 24 separate vertebrae, 12 support ribs (thoracic), 7 are in the neck (cervical), and 5 are in the lumbar region (lumbar).
- Vertebrae forming the posterior walls of the bony cavities (the thoracic vertebrae posterior to the thoracic cavity, and the

sacrum and coccyx posterior to the pelvic cavity) are concave anteriorly; elsewhere (in the cervical and lumbar regions) they are convex anteriorly.

- The spinal nerves exit the vertebral (spinal) canal via the intervertebral foramina. There are 8 cervical, 12 thoracic, 5 lumbar, 5 sacral, and 1 to 2 coccygeal spinal nerves.
- Note the size and shape of the vertebral bodies, the direction of the spinous processes, and the spinal cord in the vertebral canal (in **B**).

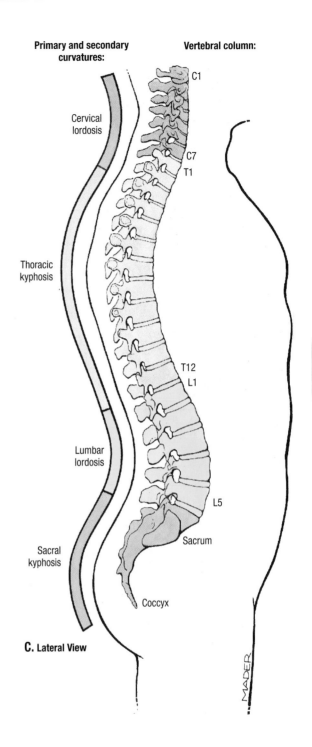

Primary and secondary curvatures:

Vertebral column:

Cervical lordosis

Thoracic kyphosis

Lumbar lordosis

Sacral kyphosis

C1

C7
T1

T12
L1

L5

Sacrum

Coccyx

C. Lateral View

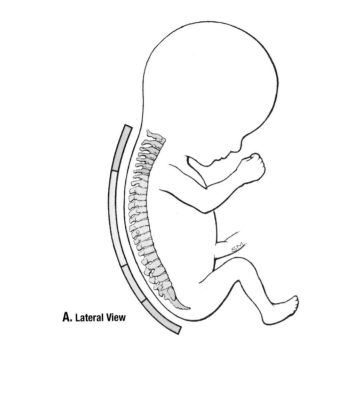

A. Lateral View

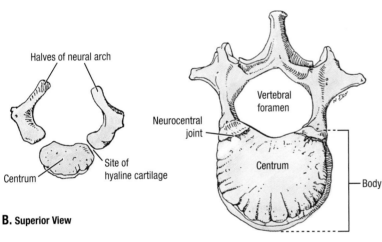

Halves of neural arch

Centrum

Site of hyaline cartilage

Neurocentral joint

Vertebral foramen

Centrum

Body

B. Superior View

4.2   Curvatures of vertebral column

Cervical vertebrae are *red*, thoracic vertebrae are *brown*, and lumbar vertebrae are *yellow*; the sacrum and coccyx are *orange*. **A.** Fetus. Note the C-shaped curvature of the fetal spine, which is concave anteriorly over its entire length. **B.** Development of the vertebrae. At birth, a vertebra consists of three bony parts (two halves of the neural arch and the centrum) united by hyaline cartilage. At age 2, the halves of each neural arch begin to fuse, proceeding from the lumbar to the cervical region; at approximately age 7, the arches begin to fuse to the centrum, proceeding from the cervical to lumbar regions. **C.** Adult. The four curvatures of the adult vertebral column include the cervical curve, which is convex

anteriorly and lies between vertebrae C1 and T2; the thoracic curve, which is concave anteriorly, between vertebrae T2 and T12; the lumbar curve, convex anteriorly and lying between T12 and the lumbosacral joint; and the sacrococcygeal curve, concave anteriorly and spanning from the lumbosacral joint to the tip of the coccyx. The anteriorly concave thoracic and sacrococcygeal curves are primary curves, and the anteriorly convex cervical and lumbar curves are secondary curves that develop after birth. The cervical curve develops when the child begins to hold the head up, and the lumbar curve develops when the child begins to walk.

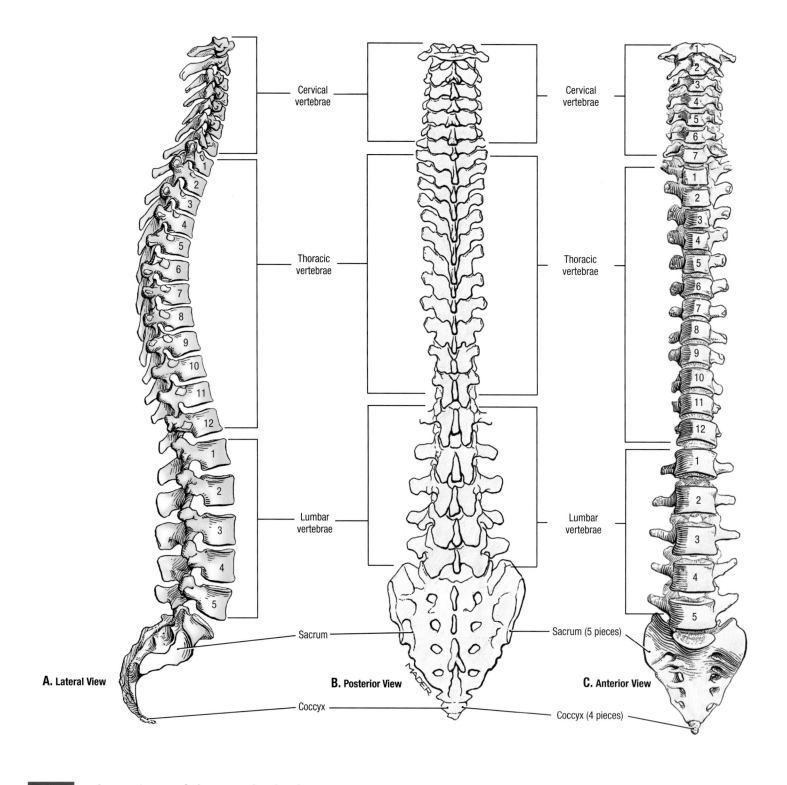

Cervical vertebrae

Thoracic vertebrae

Lumbar vertebrae

Sacrum

A. Lateral View

Coccyx

Cervical vertebrae

Thoracic vertebrae

Lumbar vertebrae

Sacrum (5 pieces)

B. Posterior View

Coccyx

Cervical vertebrae

Thoracic vertebrae

Lumbar vertebrae

Sacrum (5 pieces)

C. Anterior View

Coccyx (4 pieces)

MADER

4.3 **Three views of the vertebral column**

- Observe that there are 24 separate presacral vertebrae and that the 5 sacral vertebrae are fused to form the sacrum. The 4 coccygeal segments usually fuse late in life to form the coccyx;
- The superior surface of the body of a cervical vertebra ends on each side in an upturned, superior lip; hence, it is concave from side to side, and the inferior surface ends anteriorly in a down-turned, inferior lip. The superior and inferior surfaces of thoracic and lumbar bodies are flat.

- Transverse processes in the cervical region point laterally, inferiorly, and anteriorly, ending in two tubercles with a gutter between them. In the thoracic region, they point laterally, posteriorly, and superiorly, have a facet for the tubercle of a rib, and are stout. In the lumbar region, they point laterally, and are long and slender.
- Spinous processes are bifid in the cervical, spinelike in thoracic, and oblong in lumbar vertebrae.

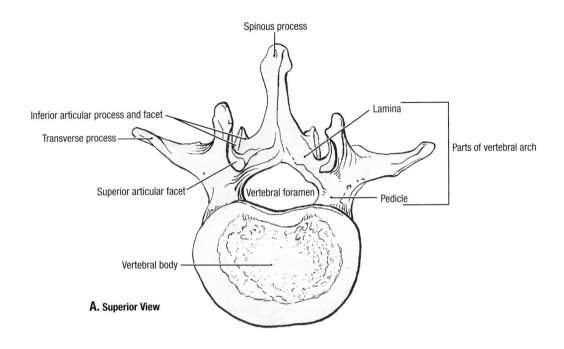

A. Superior View

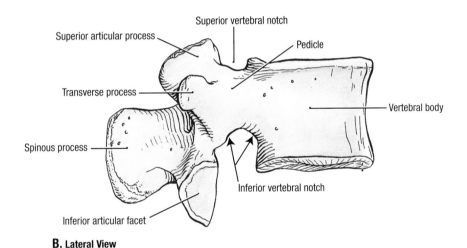

B. Lateral View

4.4   Typical vertebra

A typical vertebra (e.g., the 2nd lumbar vertebra) consists of the following parts:

- A vertebral body, situated anteriorly, functions to support weight.
- A vertebral arch, posterior to the body, which, with the body, encloses the vertebral foramen. Collectively, the vertebral foramina constitute the vertebral canal, in which the spinal cord lies. The function of a vertebral arch is to protect the spinal cord. The vertebral arch consists of two rounded pedicles, one on each side, which arise from the body, and two flat plates called *laminae* that unite posteriorly in the midline.
- Three processes, two transverse and one spinous, which provide attachment for muscles and are the levers that help move the vertebrae.
- Four articular processes, two superior and two inferior, each having an articular facet. The articular processes project superiorly and inferiorly from the vertebral arch and come into apposition with the articular facet of the corresponding processes of the vertebrae above and below. The direction of the articular facets determines the nature of the movement between adjacent vertebrae and prevents the vertebrae from slipping anteriorly.

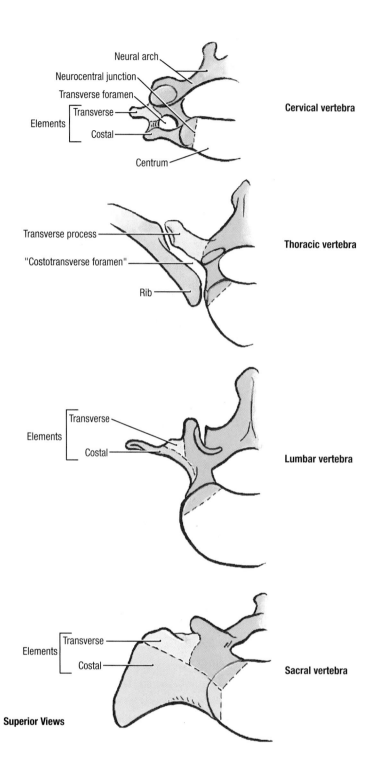

Neural arch
Neurocentral junction
Transverse foramen
Elements [Transverse
Costal
Centrum

Cervical vertebra

Transverse process
"Costotransverse foramen"
Rib

Thoracic vertebra

Elements [Transverse
Costal

Lumbar vertebra

Elements [Transverse
Costal

Sacral vertebra

Superior Views

4.5 Homologous parts of vertebrae

Note the centrum *(uncolored)*, the neural arch *(dark pink)* and its transverse element *(pink)*, and the rib, or costal element *(yellow)*. A rib is a free costal element in the thoracic region; in the cervical and lumbar regions, it is represented by the anterior part of a transverse process, and in the sacrum, by the anterior part of the lateral mass. The heads of the ribs (thoracic region) articulate with the sides of the vertebral bodies posterior to the neurocentral junctions.

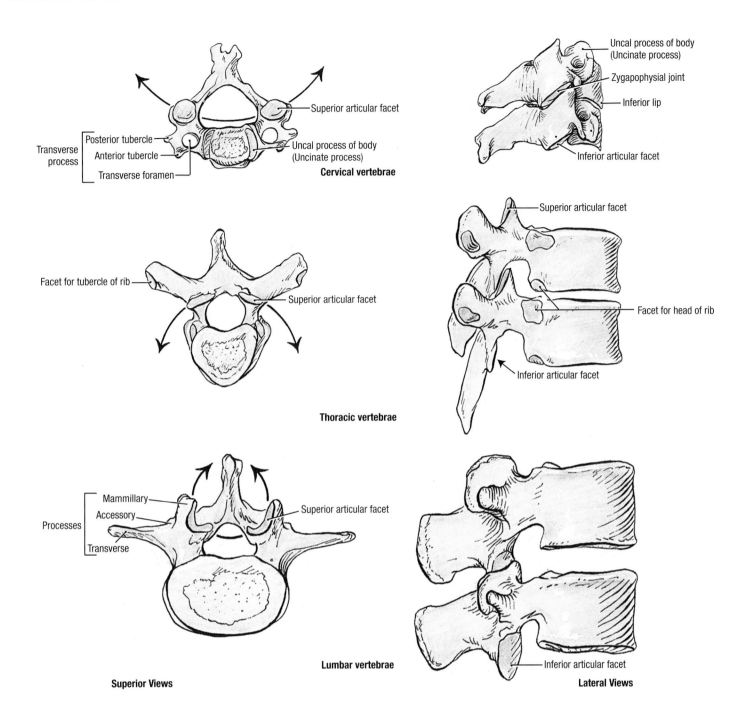

Cervical vertebrae

Thoracic vertebrae

Lumbar vertebrae

Superior Views   **Lateral Views**

4.6   Vertebral features and movements

Direction of movement is indicated by *arrows*.

- In the thoracic and lumbar regions, the superior articular facets lie posterior to the pedicles, and the inferior facets are anterior to the laminae. Superior articular facets in the cervical region face mainly superiorly, in the thoracic region, mainly posteriorly, and in the lumbar region, mainly medially. The change in direction is gradual from cervical to thoracic but abrupt from thoracic to lumbar.
- Although movements between adjacent vertebrae are relatively small, especially in the thoracic region, the summation of all the small movements produces a considerable range of movement of the vertebral column as a whole.

- Movements of the vertebral column are freer in the cervical and lumbar regions than in the thoracic region. Lateral bending is freest in the cervical and lumbar regions; flexion of the vertebral column is greatest in the cervical region; extension is most marked in the lumbar region, but the interlocking articular processes prevent rotation.
- The thoracic region is most stable due to the external support from the articulations of the ribs and costal cartilages with the sternum. The direction of the articular facets permit rotation but flexion, and lateral bending is severely restricted.

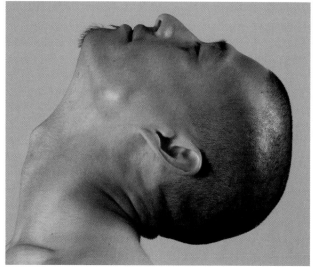

A. Lateral View

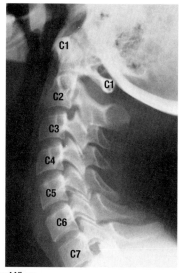

B. Lateral View

C. Lateral View

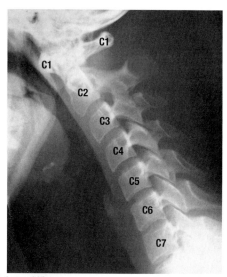

D. Lateral View

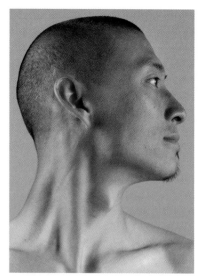

E. Anterior View

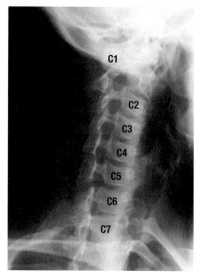

F. Oblique View

4.7 Surface anatomy with radiographic correlation of selected movements of the cervical spine

A. Extension of the neck. **B.** Radiograph of the extended cervical spine. **C.** Flexion of the neck. **D.** Radiograph of the flexed cervical spine. **E.** Head turned (rotated) to left. **F.** Radiograph of cervical spine rotated to left.

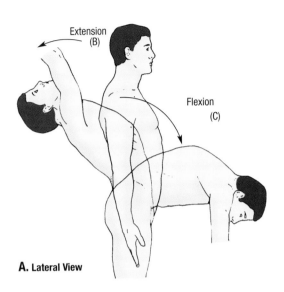

A. Lateral View

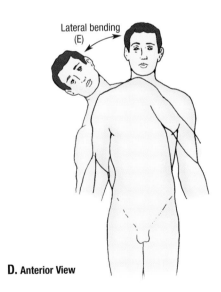

D. Anterior View

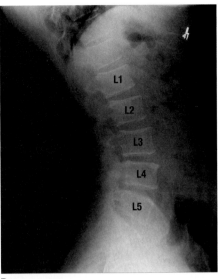

B. Lateral View

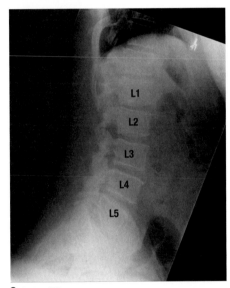

C. Lateral View

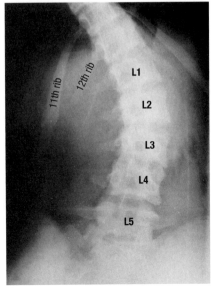

E. Anteroposterior View

4.8 **Surface anatomy with radiographic correlation of selected movements of the lumbar spine**

A. Flexion and extension of the trunk. **B.** Radiograph of the extended lumbar spine. **C.** Radiograph of the flexed lumbar spine. **D.** Lateral bending of the trunk. **E.** Radiograph of the lumbar spine during lateral bending.

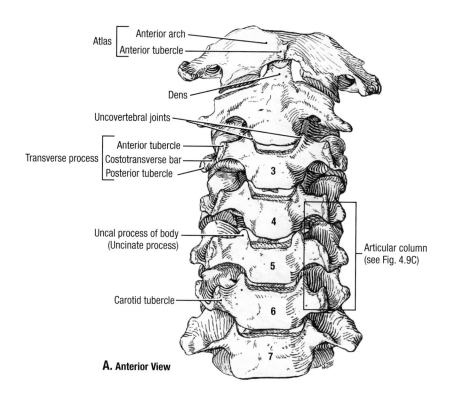

Atlas
Anterior arch
Anterior tubercle

Dens

Uncovertebral joints

Transverse process
Anterior tubercle
Costotransverse bar
Posterior tubercle

3

4

Uncal process of body
(Uncinate process)

5

Articular column
(see Fig. 4.9C)

Carotid tubercle

6

7

A. Anterior View

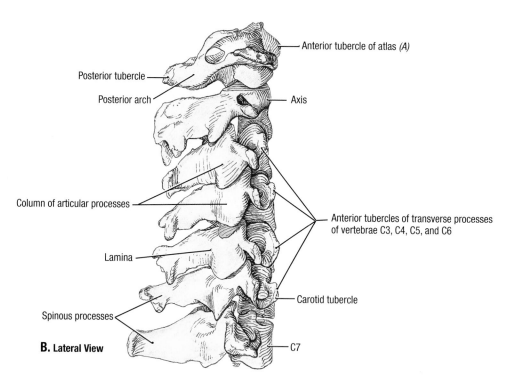

Anterior tubercle of atlas *(A)*

Posterior tubercle

Posterior arch

Axis

Column of articular processes

Anterior tubercles of transverse processes
of vertebrae C3, C4, C5, and C6

Lamina

Carotid tubercle

Spinous processes

C7

B. Lateral View

4.9   **Cervical spine**

A and **B.** Articulated cervical vertebrae. **C** and **D.** Radiographs of cervical vertebrae. Note in **A** the overlapping transverse processes *(boxed area)*. Note in **B** the laterally located column of articular processes. Note in **C** the margins of the *(black)* column of air in the trachea *(arrowheads)*. Note in **D** that the bodies of the 2nd to 7th cervical vertebrae are numbered and the anterior arch of the atlas *(A)* is in a plane anterior to the curved line joining the anterior borders of the vertebral bodies.

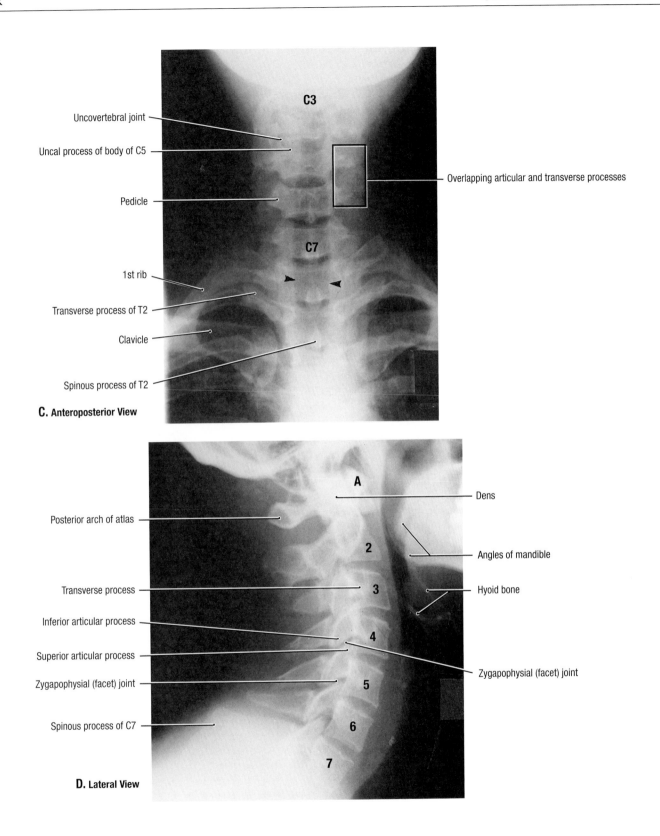

C. **Anteroposterior View**

Uncovertebral joint

Uncal process of body of C5

Pedicle

1st rib

Transverse process of T2

Clavicle

Spinous process of T2

C3

C7

Overlapping articular and transverse processes

D. **Lateral View**

Posterior arch of atlas

Transverse process

Inferior articular process

Superior articular process

Zygapophysial (facet) joint

Spinous process of C7

A

2

3

4

5

6

7

Dens

Angles of mandible

Hyoid bone

Zygapophysial (facet) joint

4.9   Cervical spine *(continued)*

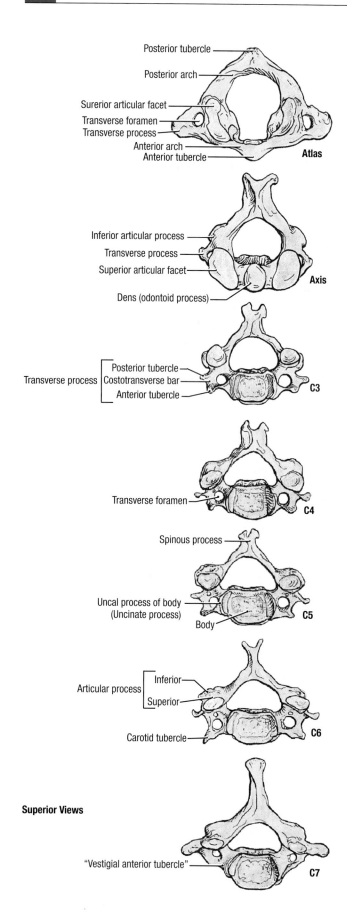

Posterior tubercle

Posterior arch

Surerior articular facet

Transverse foramen

Transverse process

Anterior arch

Anterior tubercle

Atlas

Inferior articular process

Transverse process

Superior articular facet

Dens (odontoid process)

Axis

Transverse process ⎡ Posterior tubercle
Costotransverse bar
Anterior tubercle ⎦

C3

Transverse foramen

C4

Spinous process

Uncal process of body
(Uncinate process)

Body

C5

Articular process ⎡ Inferior
Superior ⎦

Carotid tubercle

C6

Superior Views

"Vestigial anterior tubercle"

C7

TABLE 4.1   TYPICAL CERVICAL VERTEBRAE (C3–C7)[a]

PART	DISTINCTIVE CHARACTERISTICS
Body	Small and wider from side to side than anteroposteriorly; superior surface is concave and inferior surface is convex
Vertebral foramen	Large and triangular
Transverse processes	Transverse foramina (foramina transversaria); small or absent in C7; vertebral arteries and accompanying venous and sympathetic plexuses pass through foramina, except C7, which transmits only small accessory vertebral veins; anterior and posterior tubercles
Articular processes	Superior facets directed superoposteriorly; inferior facets directed inferoanteriorly; obliquely placed facets are most nearly horizontal in this region
Spinous process	Short (C3–C5) and bifid (C3–C5); process of C6 is long but that of C7 is longer (for this reason, C7 is called vertebra prominens)

[a]C1 and C2 vertebrae are atypical.

4.10   **Cervical vertebrae**

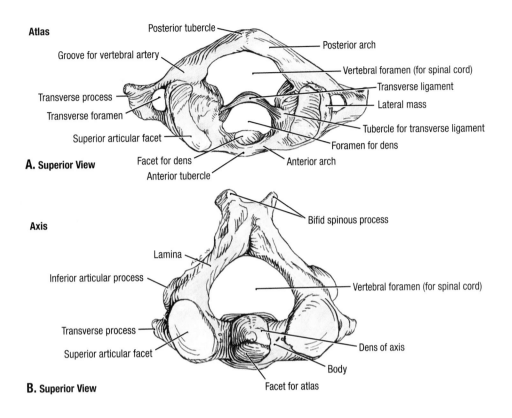

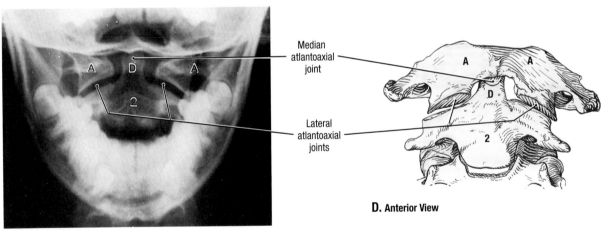

4.11   **Atlas and axis and the atlantoaxial joint**

A. Atlas. **B.** Axis. **C.** Radiograph taken through the open mouth. **D.** Articulated atlas and axis.

- Observe in **A** and **B** that the large vertebral foramen of the atlas is divided into two foramina by the transverse ligament; the spinal cord lies in the larger posterior foramen, and the dens of the axis fits tightly into the smaller anterior foramen. Articular facets on the dens of the axis articulate anteriorly with the anterior arch of the atlas and posteriorly with the transverse ligament.

- Observe in **C** and **D** that the atlantoaxial joint consists of two lateral and one medial articulation. The dens, or odontoid process *(D)*, projects superiorly from the body of the axis so that it lies between the lateral masses of the atlas *(A)* and articulates with the anterior arch of the atlas (medial articulation). The synovial joint between the lateral masses of C1 and the superior articular facets of C2 form the lateral articulations.

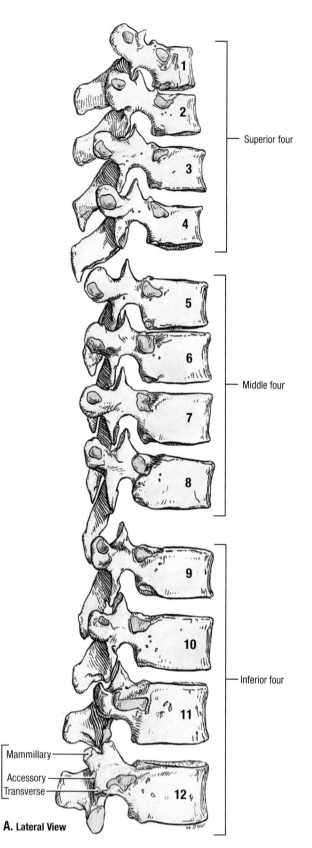

A. Lateral View

Superior four

Middle four

Inferior four

Processes
- Mammillary
- Accessory
- Transverse

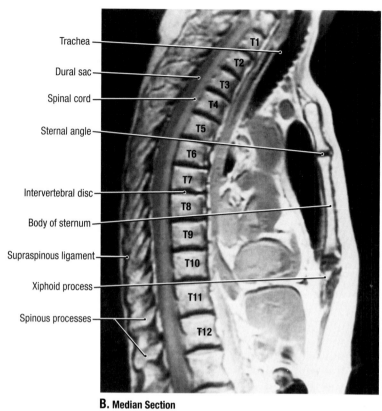

Trachea
Dural sac
Spinal cord
Sternal angle
Intervertebral disc
Body of sternum
Supraspinous ligament
Xiphoid process
Spinous processes

T1, T2, T3, T4, T5, T6, T7, T8, T9, T10, T11, T12

B. Median Section

TABLE 4.2   THORACIC VERTEBRAE

PART	DISTINCTIVE CHARACTERISTICS
Body	Heart-shaped; has one or two costal facets for articulation with head of rib
Vertebral foramen	Circular and smaller than those of cervical and lumbar vertebrae
Transverse processes	Long and strong and extend posterolaterally; length diminishes from T1–T12 (T1–T10 have transverse costal facets for articulation with tubercle of a rib)
Articular processes	Superior facets directed posteriorly and slightly laterally; inferior facets directed anteriorly and slightly medially; plane of facets lies on an arc centered about vertebral body
Spinous process	Long and slopes posteroinferiorly; tip extends to level of vertebral body below

4.12   **Thoracic vertebrae**

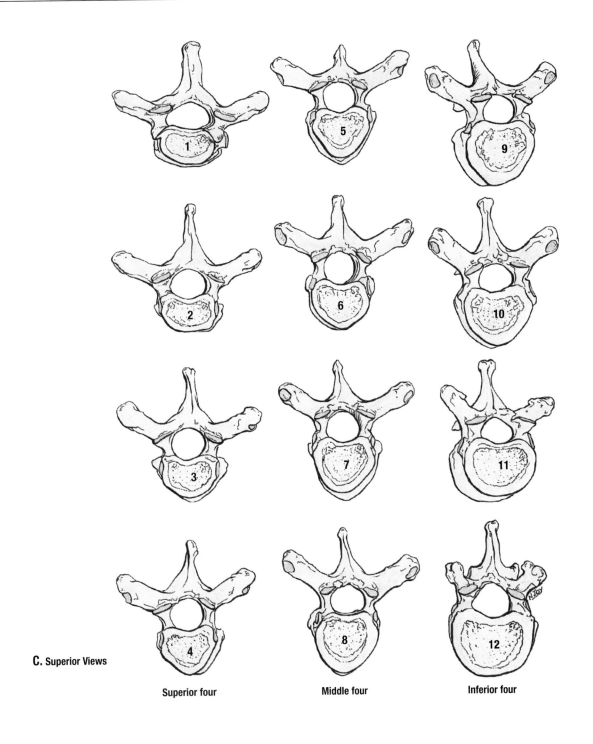

C. Superior Views

Superior four   Middle four   Inferior four

4.12   Thoracic vertebrae *(continued)*

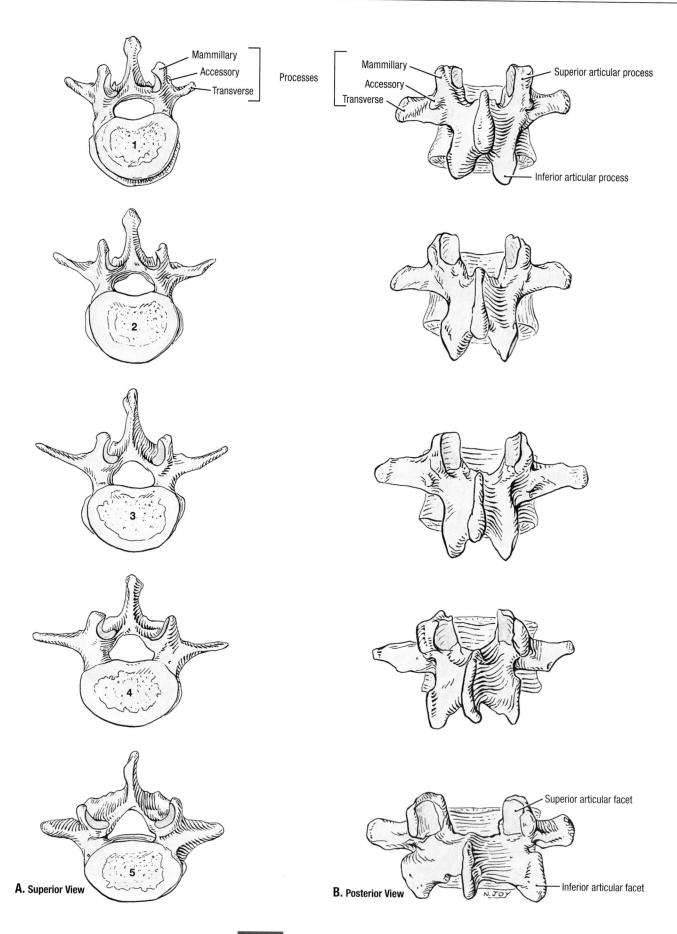

Mammillary
Accessory
Transverse
} Processes

Mammillary
Accessory
Transverse

Superior articular process

Inferior articular process

Superior articular facet

Inferior articular facet

A. Superior View

B. Posterior View

N. JOY

4.13 **Lumbar vertebrae**

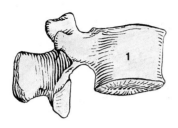

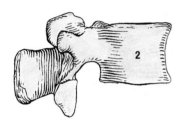

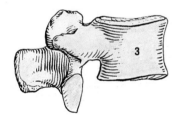

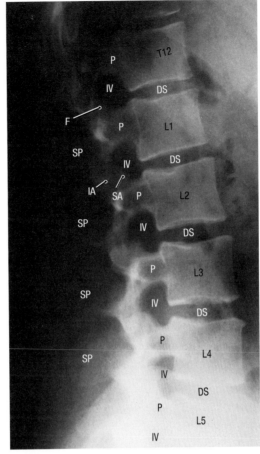

D. Lateral View

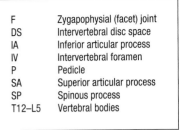

F	Zygapophysial (facet) joint
DS	Intervertebral disc space
IA	Inferior articular process
IV	Intervertebral foramen
P	Pedicle
SA	Superior articular process
SP	Spinous process
T12–L5	Vertebral bodies

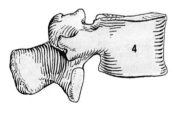

C. Lateral View

TABLE 4.3   LUMBAR VERTEBRAE

PART	DISTINCTIVE CHARACTERISTICS
Body	Massive; kidney-shaped when viewed superiorly
Vertebral foramen	Triangular; larger than in thoracic vertebrae and smaller than in cervical vertebrae
Transverse processes	Long and slender; accessory process on posterior surface of base of each process
Articular processes	Superior facets directed posteromedially (or medially); inferior facets directed anterolaterally (or laterally); mamillary process on posterior surface of each superior articular process
Spinous process	Short and sturdy; thick, broad, and hatchet-shaped

4.13   **Lumbar vertebrae *(continued)***

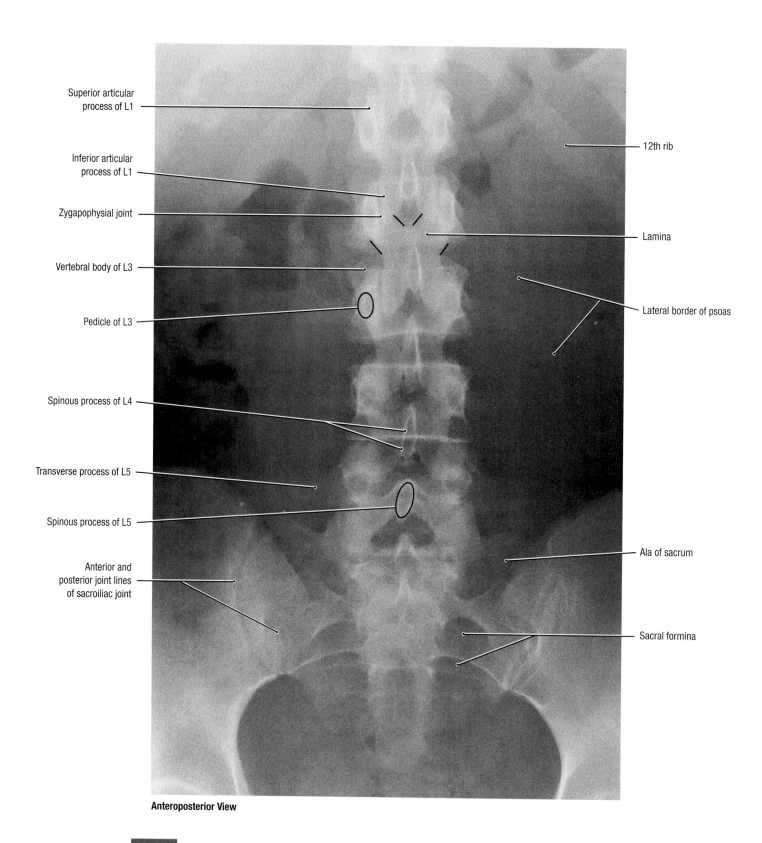

Superior articular process of L1

12th rib

Inferior articular process of L1

Zygapophysial joint

Lamina

Vertebral body of L3

Pedicle of L3

Lateral border of psoas

Spinous process of L4

Transverse process of L5

Spinous process of L5

Ala of sacrum

Anterior and posterior joint lines of sacroiliac joint

Sacral formina

Anteroposterior View

4.14 **Radiograph of inferior thoracic and lumbosacral spine**

Note the bodies and processes of the five lumbar vertebrae, the labeled spinous and transverse processes of L5, the sinuous sacroiliac joint, the lateral margin of the right and left psoas muscles, and the 12th rib.

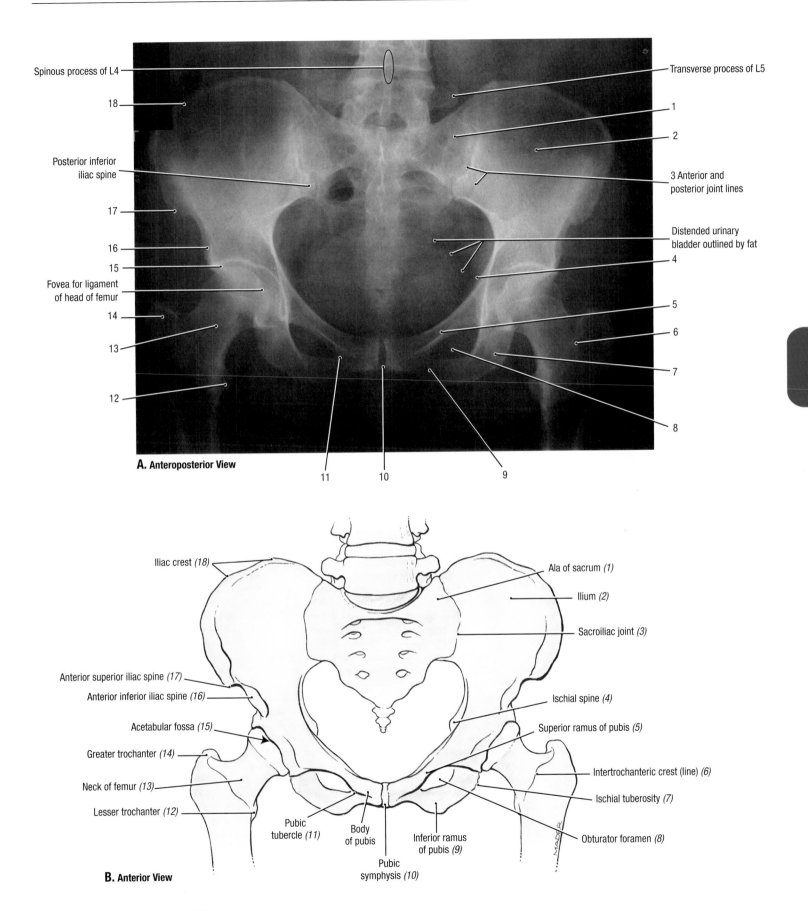

Spinous process of L4

Transverse process of L5

18

1

2

Posterior inferior
iliac spine

3 Anterior and
posterior joint lines

17

Distended urinary
bladder outlined by fat

16

4

15

Fovea for ligament
of head of femur

5

14

6

13

7

12

8

A. Anteroposterior View

11   10   9

Iliac crest (18)

Ala of sacrum (1)

Ilium (2)

Sacroiliac joint (3)

Anterior superior iliac spine (17)

Anterior inferior iliac spine (16)

Ischial spine (4)

Acetabular fossa (15)

Superior ramus of pubis (5)

Greater trochanter (14)

Neck of femur (13)

Intertrochanteric crest (line) (6)

Ischial tuberosity (7)

Lesser trochanter (12)

Pubic
tubercle (11)

Body
of pubis

Inferior ramus
of pubis (9)

Obturator foramen (8)

Pubic
symphysis (10)

B. Anterior View

4.15   **Pelvis**

A. Radiograph of pelvis. **B.** Bony pelvis with articulated femora.

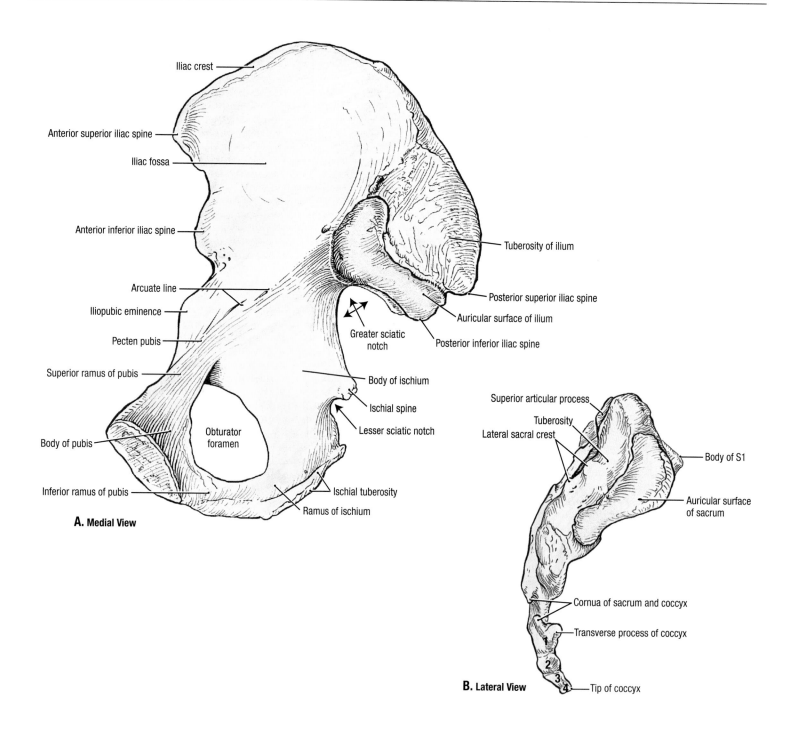

A. Medial View

- Iliac crest
- Anterior superior iliac spine
- Iliac fossa
- Anterior inferior iliac spine
- Arcuate line
- Iliopubic eminence
- Pecten pubis
- Superior ramus of pubis
- Body of pubis
- Inferior ramus of pubis
- Obturator foramen
- Ramus of ischium
- Ischial tuberosity
- Lesser sciatic notch
- Ischial spine
- Body of ischium
- Greater sciatic notch
- Posterior inferior iliac spine
- Auricular surface of ilium
- Posterior superior iliac spine
- Tuberosity of ilium

B. Lateral View

- Superior articular process
- Tuberosity
- Lateral sacral crest
- Body of S1
- Auricular surface of sacrum
- Cornua of sacrum and coccyx
- Transverse process of coccyx
- Tip of coccyx

4.16 **Hip bone, sacrum and coccyx**

A. Hip bone. **B.** Sacrum and coccyx.

- Each hip bone consists of three bones: ilium, ischium, and pubis. The ilium is the superior, larger part of the hip bone, forming the superior part of the acetabulum, the deep socket on the lateral aspect of the hip bone that articulates with the head of the femur. The ischium forms the posteroinferior part of the acetabulum and hip bone. The pubis forms the anterior part of the acetabulum and anteromedial part of the hip bone.
- Anterosuperiorly, the auricular, ear-shaped surface the sacrum articulates with the auricular surface of the ilium; posterosuperiorly, note the tuberosity of the sacrum and ilium for the attachment of the posterior sacroiliac and interosseous sacroiliac ligaments.

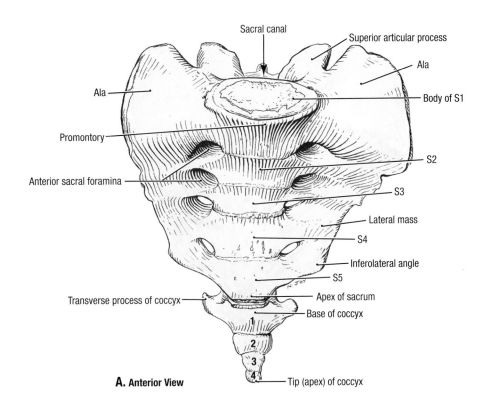

A. Anterior View

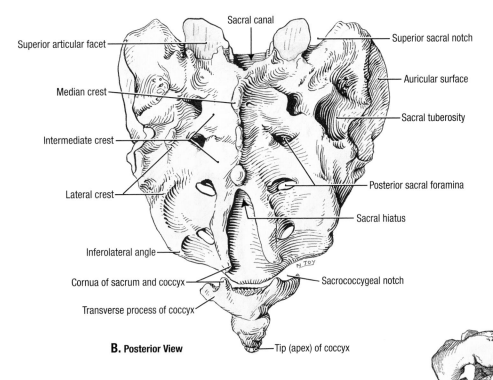

B. Posterior View

4.17   Sacrum and coccyx

A. Pelvic (anterior) surface. **B.** Posterior surface. **C.** Sacrum in youth.

- In **A** the five sacral bodies are demarcated in the mature sacrum by four transverse lines ending laterally in four pairs of anterior sacral foramina and that the coccyx has four pieces—the first having a pair of transverse processes and a pair of cornua (horns).
- The costal (lateral) elements begin to fuse around puberty. The bodies begin to fuse from inferior to superior at about the 17th to 18th year, with fusion usually completed by the 23rd year.

C. Anterior View

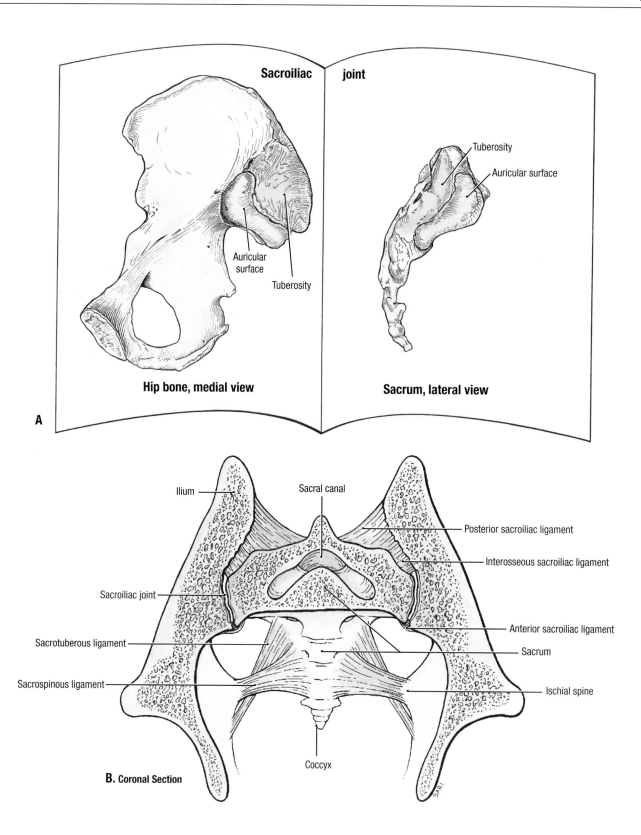

Sacroiliac joint

Tuberosity

Auricular surface

Auricular surface

Tuberosity

Hip bone, medial view

Sacrum, lateral view

A

Ilium

Sacral canal

Posterior sacroiliac ligament

Interosseous sacroiliac ligament

Sacroiliac joint

Anterior sacroiliac ligament

Sacrotuberous ligament

Sacrum

Sacrospinous ligament

Ischial spine

Coccyx

B. Coronal Section

| **4.18** | **Articular surfaces of sacroiliac joint and ligaments** |

A. Articular surfaces. Note the auricular surface (articular area, *blue*) of the sacrum and hip bone and the roughened areas superior and posterior to the auricular areas *(orange)* for the attachment of the interosseous sacroiliac ligament. **B.** Sacroiliac ligaments. Note the sacroiliac joints and the strong interosseous sacroiliac ligament that lies inferior and anterior to the posterior sacroiliac ligament and consists of short fibers connecting the sacral tuberosity to the tuberosity of the ilium. The sacrum is suspended from the ilia by the sacroiliac ligaments.

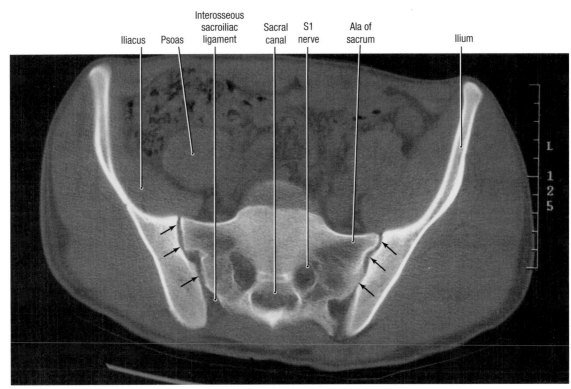

Iliacus   Psoas   Interosseous sacroiliac ligament   Sacral canal   S1 nerve   Ala of sacrum   Ilium

A. Transverse (axial) CT Scan

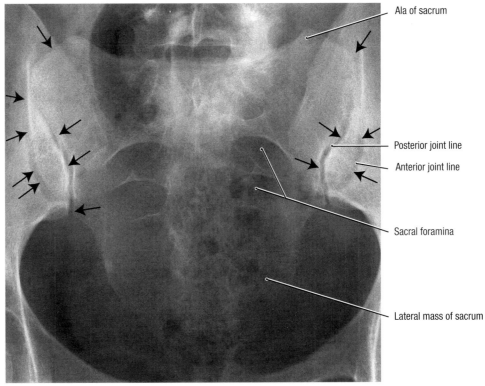

Ala of sacrum

Posterior joint line

Anterior joint line

Sacral foramina

Lateral mass of sacrum

B. Anteroposterior View

4.19   **Imaging of the sacroiliac joint**

A. CT scan. The sacroiliac joint is indicated by *arrows*. Note that the articular surfaces of the ilium and sacrum have irregular shapes that result in partial interlocking of the bones. The sacroiliac joint is oblique, with the anterior aspect of the joint situated lateral to the posterior aspect of the joint. **B.** Radiograph. Because of the oblique placement of the sacroiliac joints, the anterior and posterior joint lines appear separately.

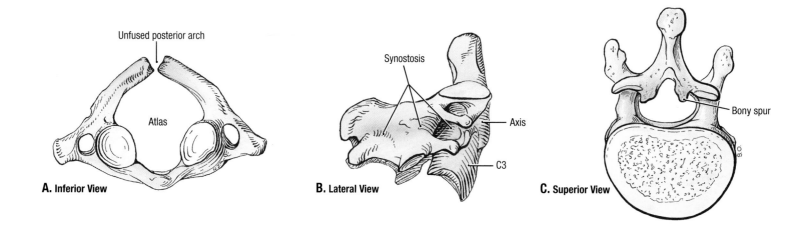

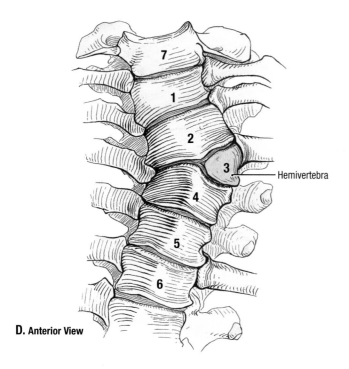

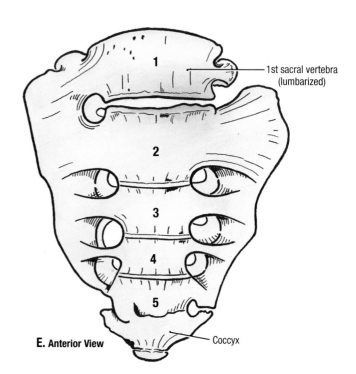

4.20 Anomalies of the vertebrae

A. Unfused posterior arch of the atlas. The centrum fused to the right and left halves of the neural arch, but the arch did not fuse in the midline posteriorly. **B.** Synostosis (fusion) of vertebrae C2 (axis) and C3. **C.** Bony spurs. Sharp bony spurs may grow from the laminae inferiorly into the ligamenta flava, thereby reducing the lengths of the functional portions of these elastic bands. When the vertebral column is flexed, the ligaments may be torn. **D.** Hemivertebra. The entire right half of T3 and the corresponding rib are absent. The left lamina and the spine are fused with those of T4, and the left intervertebral foramen is reduced in size. Observe the associated scoliosis (lateral curvature of the spine). **E.** Transitional lumbosacral vertebra. Here, the 1st sacral vertebra is partly free (lumbarized). Not uncommonly, the 5th lumbar vertebra is partly fused to the sacrum (sacralized).

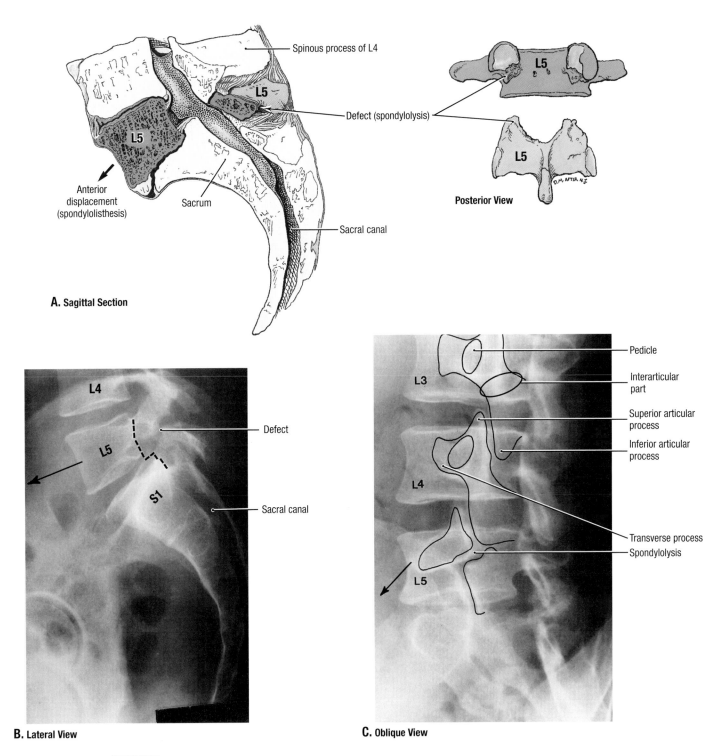

A. Sagittal Section

Posterior View

B. Lateral View

C. Oblique View

4.21   **Spondylolysis and spondylolisthesis**

A. Articulated and isolated spondylolytic L5 vertebra. The vertebra has an oblique defect (spondylolysis) through the interarticular part (pars interarticularis). The defect may be traumatic or congenital in origin. The interarticular part is the region of the lamina of a lumbar vertebra that is located between the superior and inferior articular processes. Also, the vertebral body of L5 has slipped anteriorly (spondylolisthesis). **B** and **C.** Radiographs. In **B,** the *dotted line* following the posterior vertebral margins of L5 and the sacrum shows the anterior displacement of L5 *(arrow)*. In **C,** note the superimposed outline of a dog: the head is the transverse process, the eye is the pedicle, and the ear is the superior articular process. The lucent cleft across the "neck" of the dog is the spondylolysis; the anterior displacement *(arrow)* is the spondylolisthesis.

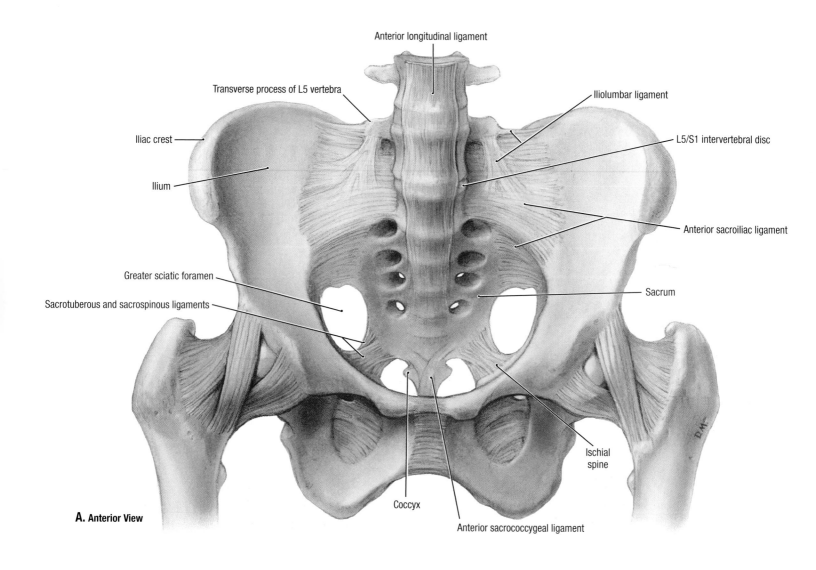

Anterior longitudinal ligament

Transverse process of L5 vertebra

Iliolumbar ligament

Iliac crest

L5/S1 intervertebral disc

Ilium

Anterior sacroiliac ligament

Greater sciatic foramen

Sacrum

Sacrotuberous and sacrospinous ligaments

Ischial spine

Coccyx

Anterior sacrococcygeal ligament

A. Anterior View

4.22 **Lumbar and pelvic ligaments**

- The iliolumbar ligaments unite the ilia and transverse processes of L5; the lumbosacral portion of the ligaments descend to the alae of the sacrum and blends with the anterior sacroiliac ligaments.
- The sacrococcygeal ligaments correspond to the anterior and posterior longitudinal ligaments of the other intervertebral joints.
- The anterior sacroiliac ligaments unite the anterior aspect of the joint.

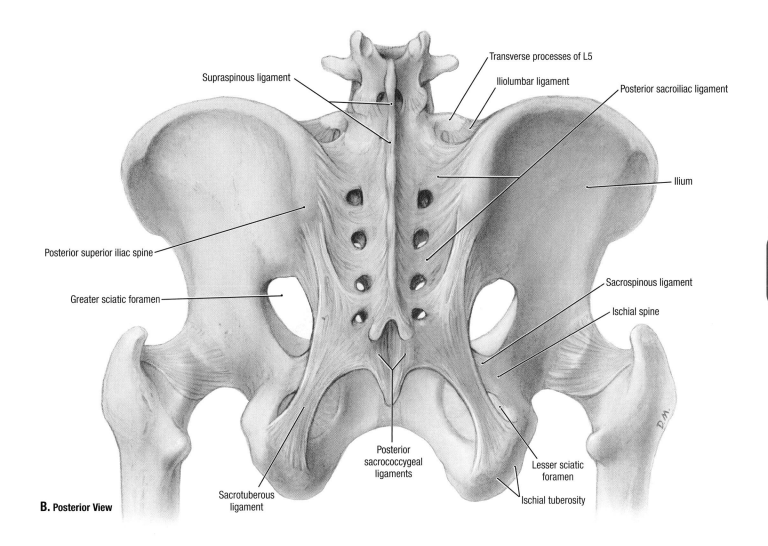

Transverse processes of L5

Iliolumbar ligament

Posterior sacroiliac ligament

Supraspinous ligament

Ilium

Posterior superior iliac spine

Sacrospinous ligament

Greater sciatic foramen

Ischial spine

Posterior sacrococcygeal ligaments

Lesser sciatic foramen

B. Posterior View

Sacrotuberous ligament

Ischial tuberosity

4.22   **Lumbar and pelvic ligaments (continued)**

- Note the posterior sacroiliac ligaments between the sacrum and tuberosity of the ilium. Short fibers pass between the first and second transverse tubercles of the sacrum and the iliac tuberosity, and long fibers pass between the posterior superior iliac spine and third and fourth transverse tubercles of the sacrum.
- The sacrotuberous ligaments attach the sacrum, ilium, and coccyx to the ischial tuberosity; the sacrospinous ligaments unite the sacrum and coccyx to the ischial spine. The sacrotuberous and sacrospinous ligaments convert the sciatic notches of the hip bones into greater and lesser sciatic foramina.

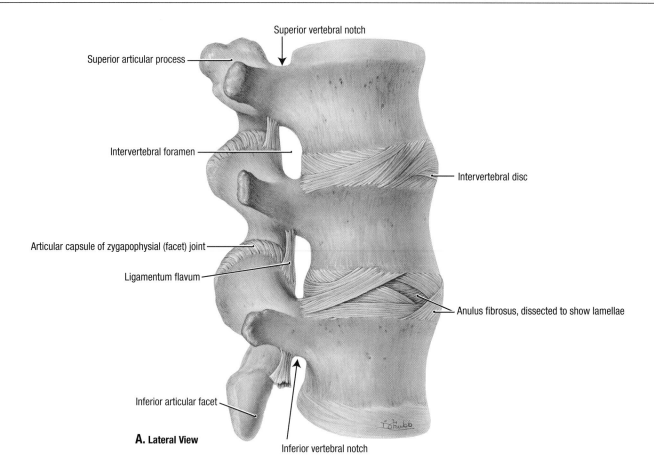

Superior vertebral notch

Superior articular process

Intervertebral foramen

Intervertebral disc

Articular capsule of zygapophysial (facet) joint

Ligamentum flavum

Anulus fibrosus, dissected to show lamellae

Inferior articular facet

A. Lateral View

Inferior vertebral notch

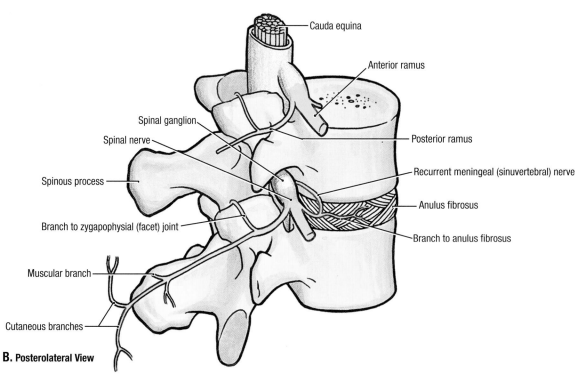

Cauda equina

Anterior ramus

Spinal ganglion

Spinal nerve

Posterior ramus

Spinous process

Recurrent meningeal (sinuvertebral) nerve

Anulus fibrosus

Branch to zygapophysial (facet) joint

Branch to anulus fibrosus

Muscular branch

Cutaneous branches

B. Posterolateral View

4.23   Intervertebral disc and ligaments

A. Anulus fibrosus and intervertebral foramen. Sections have been removed from the superficial layers of the inferior intervertebral disc to show the change in direction of the fibers in the concentric layers of the anulus fibrosus. **B.** Innervation of zygapophysial joint and intervertebral disc. The sensory innervation of the zygapophysial (facet) joint is from a medial branch of the posterior ramus.

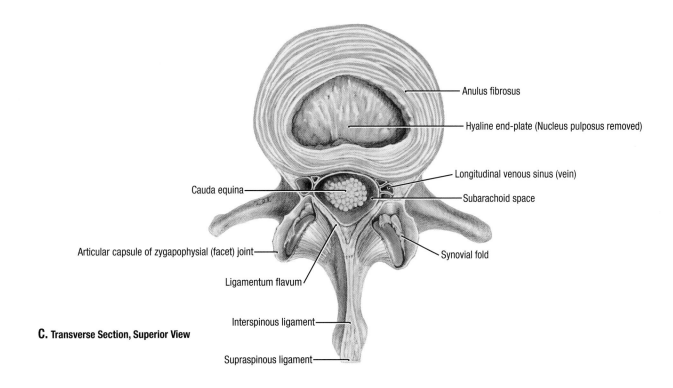

Anulus fibrosus

Hyaline end-plate (Nucleus pulposus removed)

Longitudinal venous sinus (vein)

Cauda equina

Subarachoid space

Articular capsule of zygapophysial (facet) joint

Synovial fold

Ligamentum flavum

Interspinous ligament

C. Transverse Section, Superior View

Supraspinous ligament

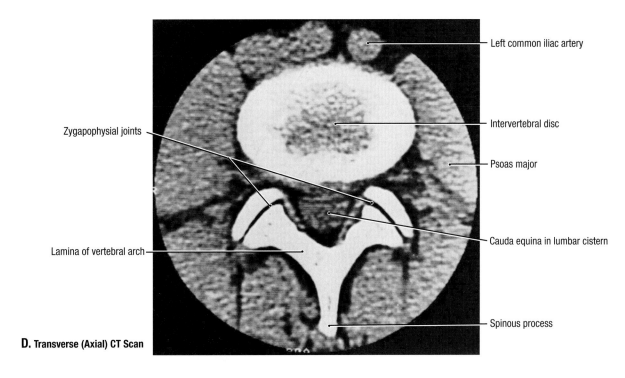

Left common iliac artery

Zygapophysial joints

Intervertebral disc

Psoas major

Lamina of vertebral arch

Cauda equina in lumbar cistern

Spinous process

D. Transverse (Axial) CT Scan

4.23   Intervertebral disc and ligaments (*continued*)

C. Transverse section. The nucleus pulposus has been removed, and the cartilaginous epiphyseal plate exposed. Note that the rings of the anulus fibrosus are fewer posteriorly, and consequently this portion of the annulus fibrosus is thinner. Note in **C** the continuity of the ligamentum flavum, the interspinous, and supraspinous ligaments, and the articular capsule of the zygapophysial (facet) joint. Observe also the longitudinal vertebral venous sinuses (veins) extending extradurally throughout the length of the vertebral canal and the cauda equina of the spinal cord lying within the subarachnoid space. **D.** CT image of L4/L5 intervertebral disc.

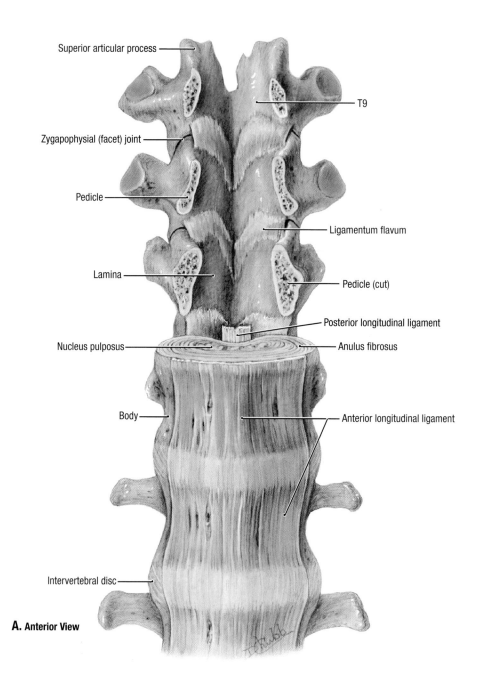

Superior articular process

Zygapophysial (facet) joint

Pedicle

Lamina

Nucleus pulposus

Body

Intervertebral disc

T9

Ligamentum flavum

Pedicle (cut)

Posterior longitudinal ligament

Anulus fibrosus

Anterior longitudinal ligament

A. Anterior View

4.24 Intervertebral discs: ligaments and movements

A. Anterior longitudinal ligament and ligamenta flava. The pedicles of T9 to T11 were sawed through, and the posterior aspect of the bodies are shown in **B.**

B. Posterior longitudinal ligament. **C.** Intervertebral disc during loading and movement.

- The anterior and posterior longitudinal ligaments are ligaments of the vertebral bodies; the ligamenta flava are ligaments of the vertebral arches.
- The anterior longitudinal ligament consists of broad, strong, fibrous bands that are attached to the intervertebral discs and vertebral bodies anteriorly and are perforated by the foramina for arteries and veins passing to and from the vertebral bodies.
- The ligamenta flava, composed of elastic fibers, extend between adjacent laminae; right and left ligaments converge in the median plane. They extend laterally to the articular processes, where they blend with the capsule of the zygapophysial joint.

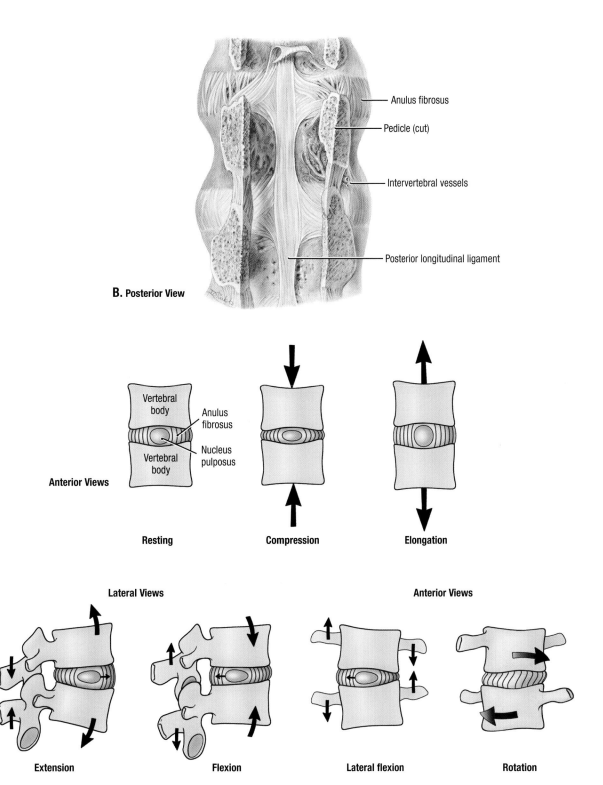

B. Posterior View

- Anulus fibrosus
- Pedicle (cut)
- Intervertebral vessels
- Posterior longitudinal ligament

Vertebral body — Anulus fibrosus — Nucleus pulposus — Vertebral body

Anterior Views

Resting   **Compression**   **Elongation**

Lateral Views   **Anterior Views**

C

Extension   **Flexion**   **Lateral flexion**   **Rotation**

4.24   Intervertebral discs: ligaments and movements *(continued)*

- The posterior longitudinal ligament is a narrow band passing from disc to disc, spanning the posterior surfaces of the vertebral bodies. Note the diamond shape of the ligament posterior to each disc, where it exchanges fibers with the anulus; the ligament extends to the sacrum inferiorly and becomes the strong tectorial membrane cranially (in **B**).
- The movement or loading of the intervertebral disc changes its shape and the position of the nucleus pulposus. Flexion and extension movements cause compression and elongation simultaneously.

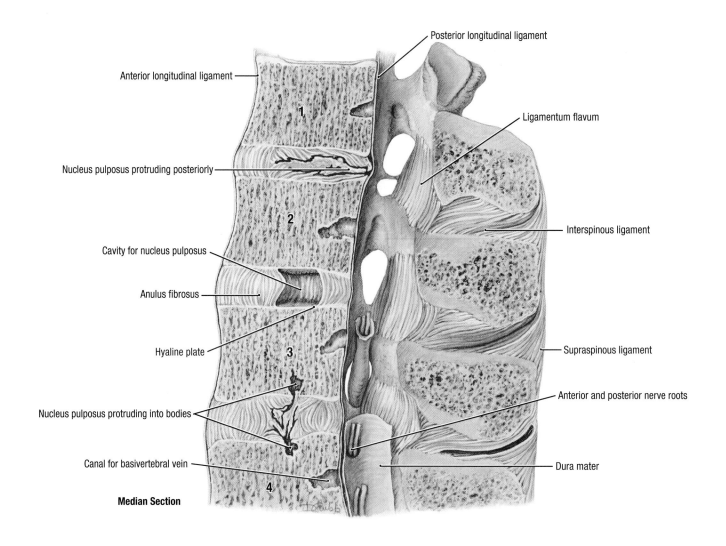

Anterior longitudinal ligament

Posterior longitudinal ligament

Nucleus pulposus protruding posteriorly

Ligamentum flavum

Cavity for nucleus pulposus

Interspinous ligament

Anulus fibrosus

Hyaline plate

Nucleus pulposus protruding into bodies

Supraspinous ligament

Anterior and posterior nerve roots

Canal for basivertebral vein

Dura mater

Median Section

4.25   **Lumbar region of vertebral column**

The nucleus pulposus of the normal disc between L2 and L3 has been removed from the en-closing anulus fibrosus.

- The ligamentum flavum extends from the superior border and adjacent part of the pos-terior aspect of one lamina to the inferior border and adjacent part of the anterior aspect of the lamina above and extends laterally to become continuous with the fibrous capsule of the zygapophysial joint.
- The obliquely placed interspinous ligament unites the superior and inferior borders of two adjacent spines.
- The supraspinous ligament extends as far inferiorly as L4 or L5.
- The bursa between L3 and L4 spines is presumably the result of habitual hyperextension, which brings the lumbar spines into contact.
- Two degenerative changes are demonstrated. The pulp of the disc between L1 and L2 has herniated posteriorly through the anulus; spinal nerves are vulnerable to the pressure of an extruded nucleus pulposus through a torn anulus fibrosus (the most common site of a disc lesion is between L5 and S1), and the pulp of the disc between L3 and L4 has her-niated through the cartilaginous epiphyseal plates into the bodies of the vertebrae supe-riorly and inferiorly.

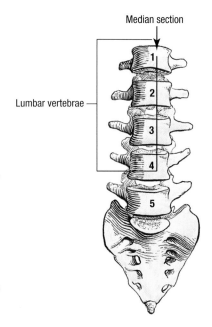

Median section

Lumbar vertebrae

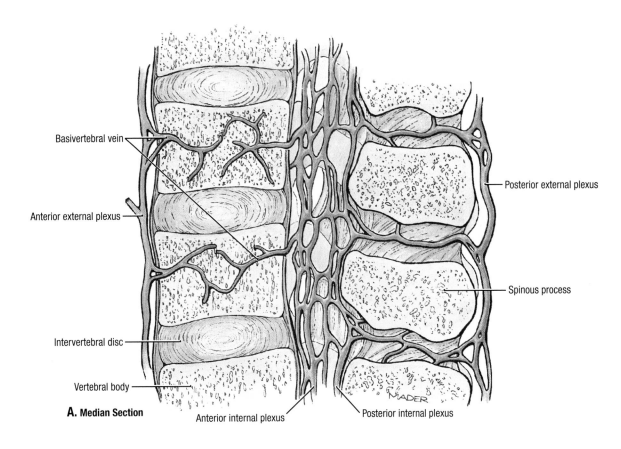

Basivertebral vein

Anterior external plexus

Intervertebral disc

Vertebral body

Posterior external plexus

Spinous process

A. Median Section

Anterior internal plexus

Posterior internal plexus

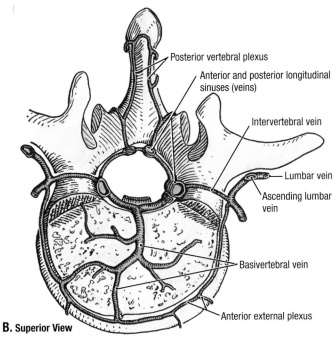

Posterior vertebral plexus

Anterior and posterior longitudinal sinuses (veins)

Intervertebral vein

Lumbar vein

Ascending lumbar vein

Basivertebral vein

Anterior external plexus

B. Superior View

4.26   Vertebral venous plexuses

A. Median section of lumbar spine. **B.** Superior view of lumbar vertebra with the vertebral body sectioned transversely.

- There are internal and external plexuses, communicating with each other and with both systemic veins and the portal system. Infection and tumors can spread from the systemic and portal areas (e.g., prostate, breast) to the vertebral venous system and lodge in the vertebrae, spinal cord, brain, or skull.

- The internal plexus: The vertebral canal contains a plexus of thin-walled, valveless veins that surround the dura mater. Anterior and posterior longitudinal channels (venous sinuses) can be discerned in this plexus. Cranially, the internal plexus communicates through the foramen magnum with the occipital and basilar sinuses; at each spinal segment, the plexus receives veins from the spinal cord and a basivertebral vein from the vertebral body. The plexus is drained by intervertebral veins that pass through the intervertebral and sacral foramina to the vertebral, intercostal, lumbar, and lateral sacral veins.

- The external plexus: Through the body of each vertebra come veins that form a small anterior external vertebral plexus, and through the ligamenta flava pass veins that form a well-marked posterior external vertebral plexus. In the cervical region, these plexuses communicate freely with the occipital and deep cervical veins, which receive blood from the sigmoid sinus and mastoid and condyloid emissary veins. In the thoracic, lumbar, and pelvic regions, the azygos (or hemiazygos), ascending lumbar, and lateral sacral veins, respectively, further link segment to segment.

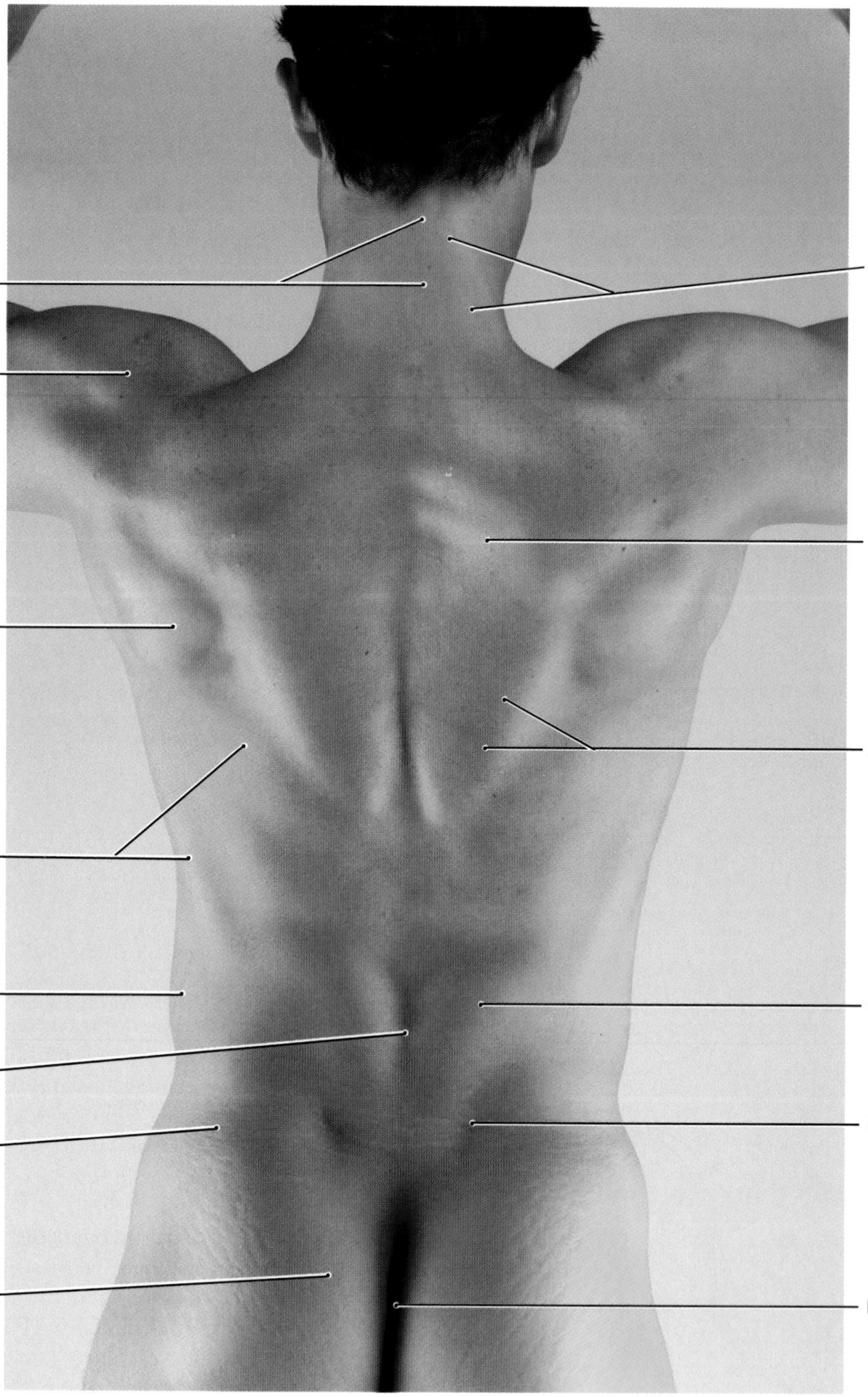

Site of nuchal ligament

Spinal (posterior) part of deltoid

Teres major

Latissimus dorsi

External oblique

Posterior median furrow

Gluteus medius

Gluteus maximus

Posterior View

Descending (superior) part of trapezius

Transverse (middle) part of trapezius

Ascending (inferior) part of trapezius

Erector spine

Site of posterior superior iliac spine

Intergluteal cleft

4.27 Surface anatomy of back

- The upper limbs are elevated, so the scapulae have moved laterally and rotated superiorly on the thoracic wall.
- The latissimus dorsi and teres major muscles form the posterior axillary fold.
- The trapezius muscle has three parts: descending, transverse, and ascending.
- Note the deep, midline furrow that separates the longitudinal bulges formed by the contracted erector spinae group of muscles;
- Dimples (depressions) indicate the site of the posterior superior iliac spines, which usually lie at the level of the sacroiliac joint.

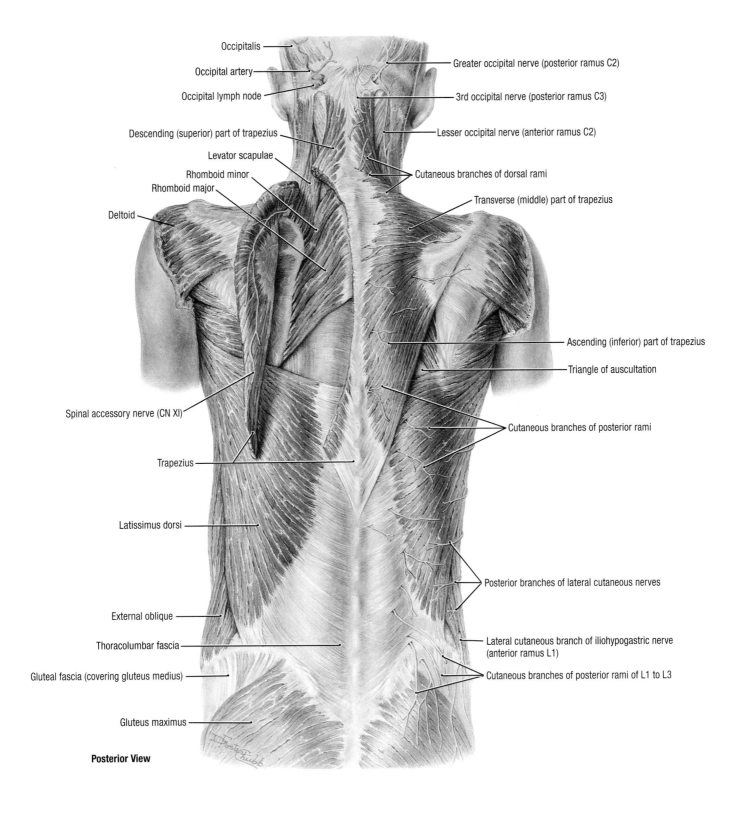

Occipitalis

Occipital artery

Occipital lymph node

Descending (superior) part of trapezius

Levator scapulae

Rhomboid minor

Rhomboid major

Deltoid

Spinal accessory nerve (CN XI)

Trapezius

Latissimus dorsi

External oblique

Thoracolumbar fascia

Gluteal fascia (covering gluteus medius)

Gluteus maximus

Posterior View

Greater occipital nerve (posterior ramus C2)

3rd occipital nerve (posterior ramus C3)

Lesser occipital nerve (anterior ramus C2)

Cutaneous branches of dorsal rami

Transverse (middle) part of trapezius

Ascending (inferior) part of trapezius

Triangle of auscultation

Cutaneous branches of posterior rami

Posterior branches of lateral cutaneous nerves

Lateral cutaneous branch of iliohypogastric nerve
(anterior ramus L1)

Cutaneous branches of posterior rami of L1 to L3

4.28   **Superficial muscles of back**

On the *left*, the trapezius muscle is reflected. Observe two layers: the trapezius and latissimus dorsi muscles, and the levator scapulae and rhomboids minor and major. These muscles help attach the upper limb to the trunk.

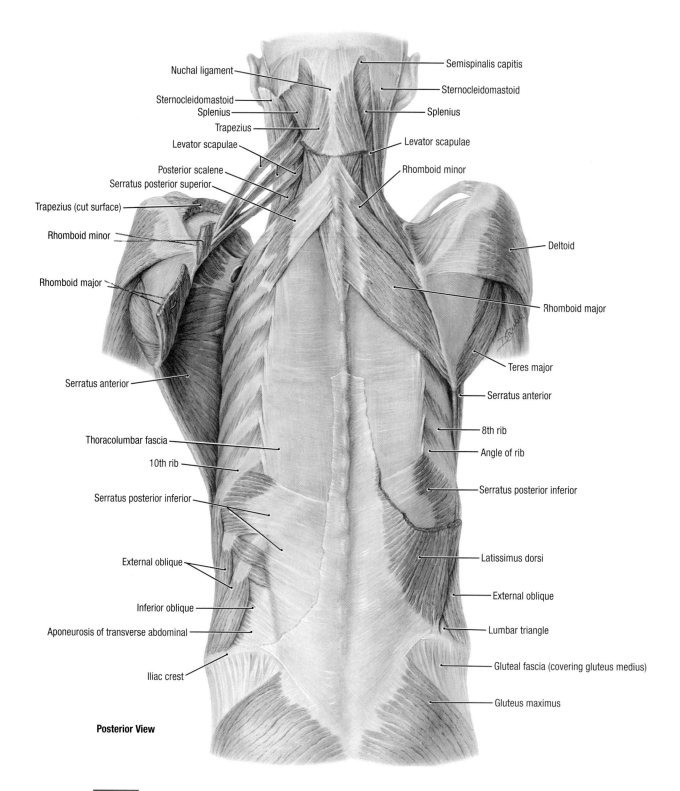

Nuchal ligament

Sternocleidomastoid
Splenius
Trapezius
Levator scapulae
Posterior scalene
Serratus posterior superior

Trapezius (cut surface)

Rhomboid minor

Rhomboid major

Serratus anterior

Thoracolumbar fascia

10th rib

Serratus posterior inferior

External oblique

Inferior oblique

Aponeurosis of transverse abdominal

Iliac crest

Semispinalis capitis
Sternocleidomastoid
Splenius
Levator scapulae
Rhomboid minor

Deltoid

Rhomboid major

Teres major
Serratus anterior

8th rib
Angle of rib

Serratus posterior inferior

Latissimus dorsi

External oblique

Lumbar triangle

Gluteal fascia (covering gluteus medius)

Gluteus maximus

Posterior View

4.29 **Intermediate muscles of back**

The trapezius and latissimus dorsi muscles are largely cut away on both sides. On the *left*, the rhomboid muscles have been severed, allowing the vertebral border of the scapula to be raised from the thoracic wall. The serratus posterior superior and inferior form the intermediate layer of muscles, passing from the vertebral spines to the ribs; the two muscles slope in opposite directions and are muscles of respiration. The thoracolumbar fascia extends laterally to the angles of the ribs, becoming thin superiorly and passing deep to the serratus posterior superior muscle. It gives attachment to the latissimus dorsi and serratus posterior inferior muscles.

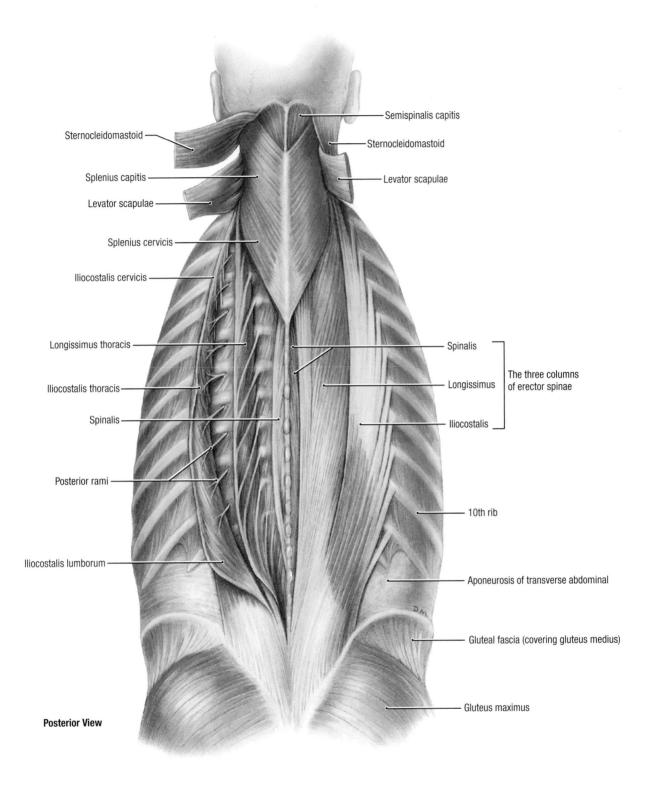

Sternocleidomastoid

Splenius capitis

Levator scapulae

Splenius cervicis

Iliocostalis cervicis

Longissimus thoracis

Iliocostalis thoracis

Spinalis

Posterior rami

Iliocostalis lumborum

Semispinalis capitis

Sternocleidomastoid

Levator scapulae

Spinalis

Longissimus

Iliocostalis

The three columns of erector spinae

10th rib

Aponeurosis of transverse abdominal

Gluteal fascia (covering gluteus medius)

Gluteus maximus

Posterior View

4.30　**Deep muscles of back: splenius and erector spinae**

On the *right* of the body, note the erector spinae muscles in situ, lying between the spinous processes medially and the angles of the ribs laterally and splitting into three longitudinal columns: iliocostalis laterally, longissimus in the middle, and spinalis medially. On the *left*, the longissimus muscle, the intermediate column, is pulled laterally to show the insertion into the transverse processes and ribs; not shown here are its extensions to the neck and head, longissimus cervicis and capitis.

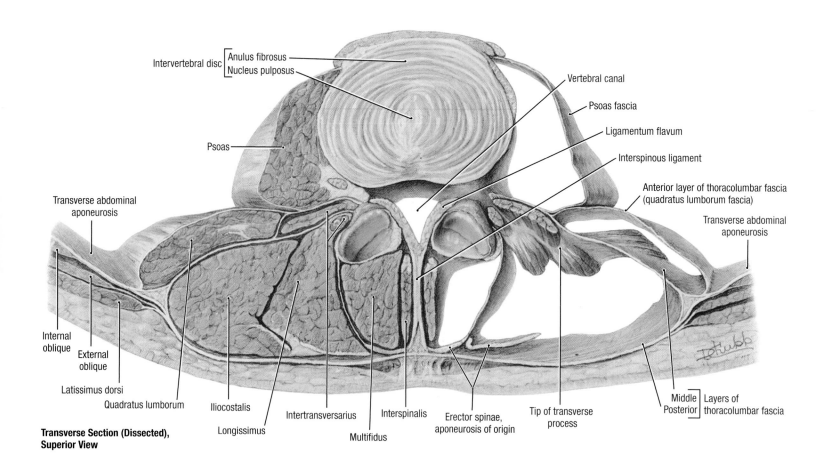

Intervertebral disc { Anulus fibrosus
Nucleus pulposus

Vertebral canal

Psoas fascia

Ligamentum flavum

Interspinous ligament

Anterior layer of thoracolumbar fascia
(quadratus lumborum fascia)

Transverse abdominal
aponeurosis

Psoas

Transverse abdominal
aponeurosis

Internal
oblique

External
oblique

Latissimus dorsi

Quadratus lumborum

Iliocostalis

Longissimus

Intertransversarius

Interspinalis

Multifidus

Erector spinae,
aponeurosis of origin

Tip of transverse
process

Middle | Layers of
Posterior | thoracolumbar fascia

**Transverse Section (Dissected),
Superior View**

4.31   Transverse section of back muscles and thoracolumbar fascia

Section is at the lumbar level. On the *left*, the muscles are seen in their fascial sheaths or compartments; on the *right*, the muscles have been removed from their sheaths. The deep back muscles extend from the pelvis to the cranium and are enclosed in fascia. This fascia attaches medially to the nuchal ligament, the tips of the spinous processes, the supraspinous ligament, and the median crest of the sacrum. The lateral attachment of the fascia is to the cervical and lumbar transverse processes and to the angles of the ribs. The thoracic and lumbar parts of the fascia are named *thoracolumbar fascia*. The aponeurosis of the transverse abdominal and posterior aponeurosis of internal oblique muscles split into two strong sheets, the middle and posterior layers of the thoracolumbar fascia. The anterior layer of thoracolumbar fascia is the deep fascia of the quadratus lumborum (quadratus lumborum fascia). The posterior layer of the thoracolumbar fascia provides proximal attachment for the latissimus dorsi muscle and, at a higher level, the serratus posterior inferior muscle.

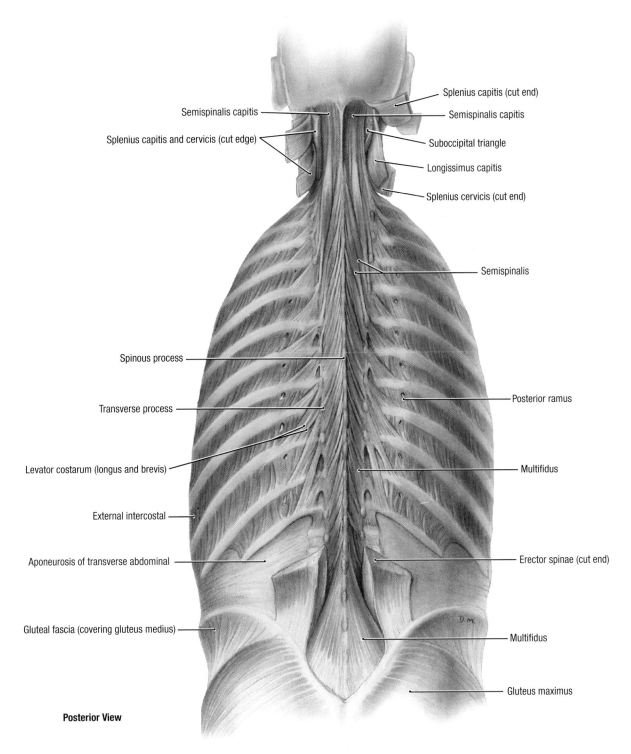

Semispinalis capitis

Splenius capitis and cervicis (cut edge)

Splenius capitis (cut end)

Semispinalis capitis

Suboccipital triangle

Longissimus capitis

Splenius cervicis (cut end)

Semispinalis

Spinous process

Transverse process

Posterior ramus

Levator costarum (longus and brevis)

Multifidus

External intercostal

Aponeurosis of transverse abdominal

Erector spinae (cut end)

Gluteal fascia (covering gluteus medius)

Multifidus

Gluteus maximus

Posterior View

4.32   **Deep muscles of back: semispinalis and multifidus**

- The semispinalis, multifidus, and rotatores muscles constitute the transversospinalis group of deep muscles. In general, their bundles pass obliquely in a superomedial direction, from transverse processes to spinous processes in successively deeper layers. The bundles of semispinalis span approximately five interspaces, those of multifidus approximately three, and those of rotatores, one or two (see Fig. 4.34).
- The semispinalis (thoracis, cervicis, and capitis) muscles extend from the lower thoracic region to the skull; the semispinalis capitis, a powerful extensor muscle, originates from the lower cervical and upper thoracic vertebrae and inserts into the occipital bone between the superior and inferior nuchal lines.
- The multifidus muscle extends from the sacrum to the spine of the axis, emerges from the lumbosacral region from the aponeurosis of the erector spinae, the sacrum, and mammillary processes of the lumbar vertebrae, and inserts into spinous processes approximately three segments higher.

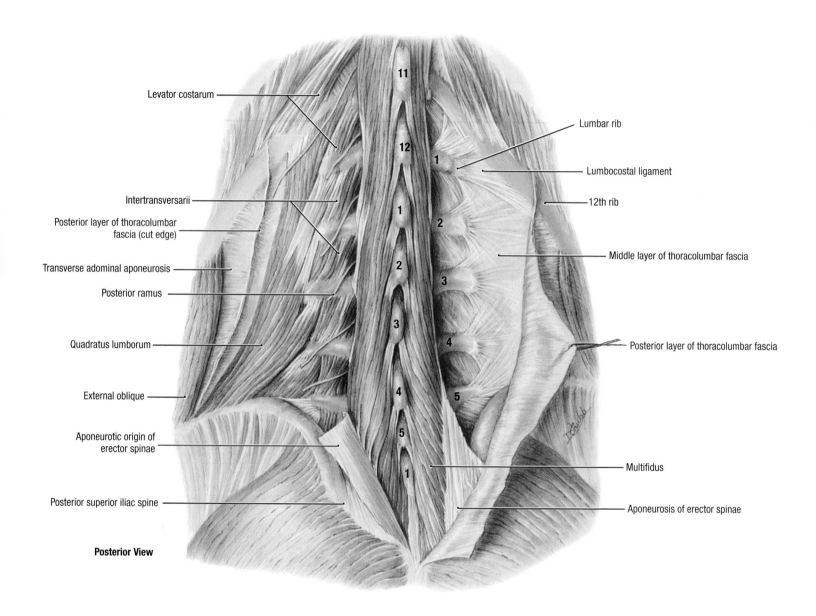

Levator costarum

Intertransversarii

Posterior layer of thoracolumbar
fascia (cut edge)

Transverse adominal aponeurosis

Posterior ramus

Quadratus lumborum

External oblique

Aponeurotic origin of
erector spinae

Posterior superior iliac spine

Posterior View

Lumbar rib

Lumbocostal ligament

12th rib

Middle layer of thoracolumbar fascia

Posterior layer of thoracolumbar fascia

Multifidus

Aponeurosis of erector spinae

4.33   Back: multifidus, quadratus lumborum, and lumbar fascia

Right: After removal of erector spinae at the L1 level, the middle layer of thoracolumbar fascia extends from the tip of each lumbar transverse process in a fan-shaped manner. Also note a short lumbar rib at the level of L1. *Left:* After removal of the posterior and middle layers of thoracolumbar fascia, the lateral border of the quadratus lumborum muscle is oblique, and the medial border is in continuity with the intertransversarii.

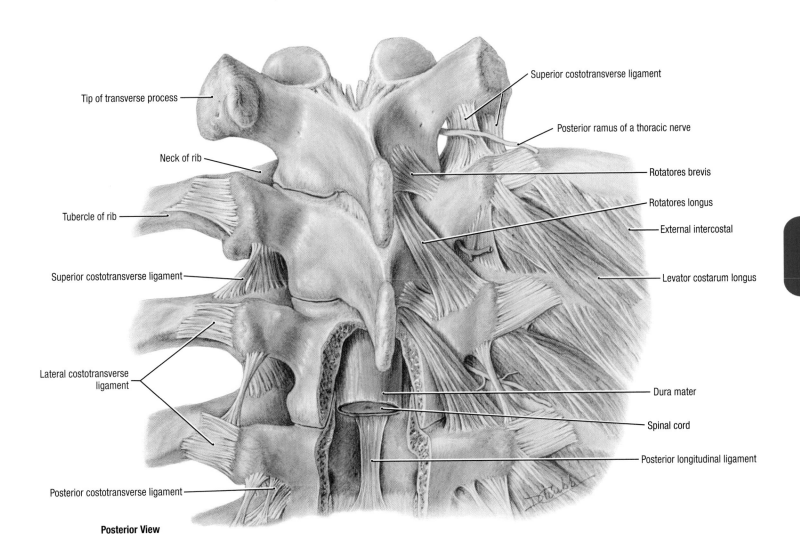

Tip of transverse process

Neck of rib

Tubercle of rib

Superior costotransverse ligament

Lateral costotransverse ligament

Posterior costotransverse ligament

Superior costotransverse ligament

Posterior ramus of a thoracic nerve

Rotatores brevis

Rotatores longus

External intercostal

Levator costarum longus

Dura mater

Spinal cord

Posterior longitudinal ligament

Posterior View

4.34   **Rotatores and costotransverse ligaments**

- Of the three layers of transversospinalis, or oblique muscles of the back (semispinalis, multifidus, rotatores), the rotatores are the deepest and shortest. They pass from the root of one transverse process superomedially to the junction of the transverse process and lamina of the vertebra above. Some (rotatores longi) span two vertebrae.
- The levatores costarum pass from the tip of one transverse process inferiorly to the rib below; some (levatores longi) span two ribs.
- The superior costotransverse ligament splits laterally into two sheets, between which lie the levator costae and external intercostal; the posterior ramus of a thoracic nerve passes posterior to this ligament.
- The lateral costotransverse ligament is strong and joins the tubercle of the rib to the tip of the transverse process.

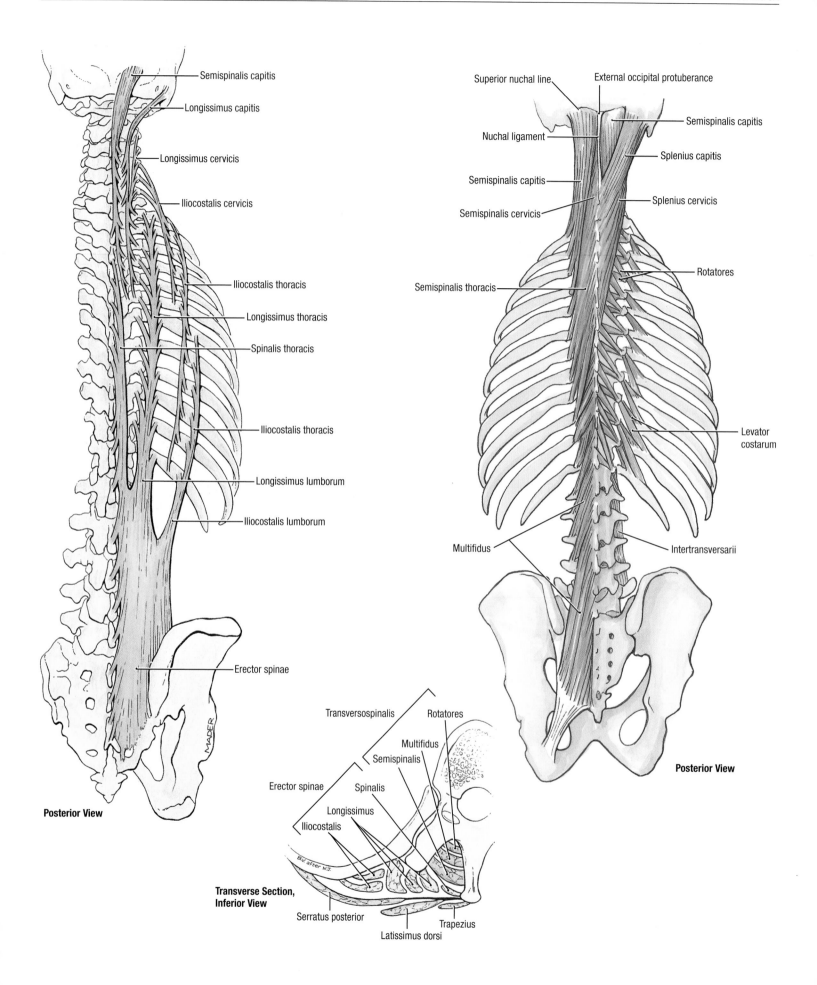

Semispinalis capitis

Longissimus capitis

Longissimus cervicis

Iliocostalis cervicis

Iliocostalis thoracis

Longissimus thoracis

Spinalis thoracis

Iliocostalis thoracis

Longissimus lumborum

Iliocostalis lumborum

Erector spinae

Posterior View

Superior nuchal line

External occipital protuberance

Semispinalis capitis

Nuchal ligament

Splenius capitis

Semispinalis capitis

Semispinalis cervicis

Splenius cervicis

Semispinalis thoracis

Rotatores

Levator costarum

Multifidus

Intertransversarii

Posterior View

Transversospinalis

Rotatores

Multifidus

Semispinalis

Erector spinae

Spinalis

Longissimus

Iliocostalis

Transverse Section, Inferior View

Serratus posterior

Trapezius

Latissimus dorsi

TABLE 4.4   INTRINSIC (DEEP) BACK MUSCLES[a]

MUSCLES	ORIGIN	INSERTION	NERVE SUPPLY[b]	MAIN ACTIONS
Superficial layer				
Splenius	Arises from nuchal ligament and spinous processes of C7–T3 or T4 vertebrae	*Splenius capitis:* fibers run superolaterally to mastoid process of temporal; bone and lateral third of superior nuchal line of occipital bone *Splenius cervicis:* posterior tubercles of transverse processes of C1–C3 or C4 vertebrae		*Acting alone,* they laterally bend to side of active muscles; *acting together,* they extend head and neck
Intermediate layer				
Erector spinae	Arises by a broad tendon from posterior part of iliac crest, posterior surface of sacrum, sacral and inferior lumbar spinous processes, and supraspinous ligament	*Iliocostalis:* lumborum, thoracis, and cervicis, fibers run superiorly to angles of lower ribs and cervical transverse processes *Longissimus:* thoracis, cervicis, and capitis, fibers run superiorly to ribs between tubercles and angles, to transverse processes in thoracic and cervical regions, and to mastoid process of temporal bone *Spinalis:* thoracis, cervicis, and capitis; fibers run superiorly to spinous processes in the upper thoracic region and to skull	Posterior rami of spinal nerves	*Acting bilaterally:* they extend vertebral column and head; as back is flexed they control movement by gradually lengthening their fibers; *acting unilaterally,* they laterally bend vertebral column
Deep layer				
Transversospinalis	Semispinalis arises from thoracic and cervical transverse processes	*Semispinalis:* thoracis, cervicis, and capitis; fibers run superomedially and attach to occipital bone and spinous processes in thoracic and cervical regions, spanning four to six segments		Extend head and thoracic and cervical regions of vertebral column and rotate them contralaterally
	Multifidus arises from sacrum and ilium, transverse processes of T1–L5, and articular processes of C4–C7	Fibers pass superomedially to spinous processes, spanning two to four segments		Stabilizes vertebrae during local movements of vertebral column
	Rotatores arise from transverse processes of vertebrae; best developed in thoracic region	Pass superomedially and attach to junction of lamina and transverse process of vertebra of origin or into spinous process above their origin, spanning one to two segments		Stabilize vertebrae and assist with local extension and rotary movements of vertebral column
Minor deep layer				
Interspinales	Superior surfaces of spinous processes of cervical and lumbar vertebrae	Inferior surfaces of spinous processes of vertebrae superior to vertebrae of origin	Posterior rami of spinal nerves	Aid in extension and rotation of vertebral column
Intertransversarii	Transverse processes of cervical and lumbar vertebrae	Transverse processes of adjacent vertebrae	Posterior and anterior rami of spinal nerves	Aid in lateral bending of vertebral column; acting bilaterally, they stabilize vertebral column
Levator costarum	Tips of transverse processes of C7 and T1–T11 vertebrae	Pass inferolaterally and insert on rib between its tubercle and angle	Posterior rami of C8–T11 spinal nerves[c]	Elevate ribs, assisting inspiration Assist with lateral bending of vertebral column

[a]See figures on opposite page.
[b]Most back muscles are innervated by posterior rami of spinal nerves, but a few are innervated by anterior rami. Intertransversarii of cervical region are supplied by anterior rami.
[c]Levator costarum were once said to be innervated by anterior rami, but investigators now agree that they are innervated by posterior rami.

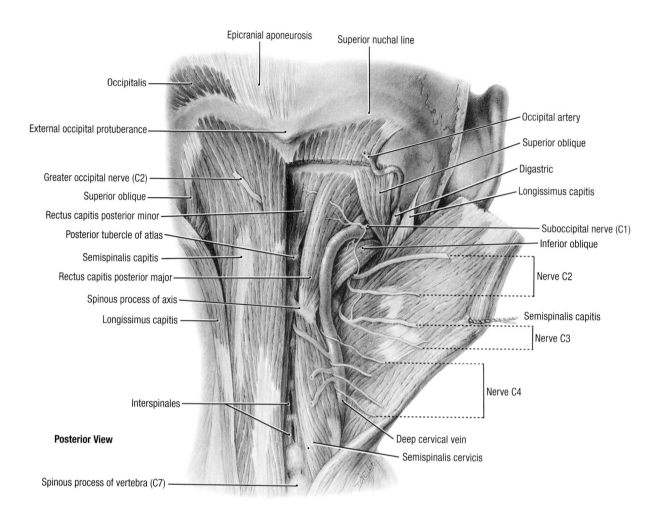

Epicranial aponeurosis

Superior nuchal line

Occipitalis

Occipital artery

External occipital protuberance

Superior oblique

Digastric

Greater occipital nerve (C2)

Longissimus capitis

Superior oblique

Rectus capitis posterior minor

Suboccipital nerve (C1)

Posterior tubercle of atlas

Inferior oblique

Semispinalis capitis

Rectus capitis posterior major

Nerve C2

Spinous process of axis

Semispinalis capitis

Longissimus capitis

Nerve C3

Nerve C4

Interspinales

Posterior View

Deep cervical vein

Semispinalis cervicis

Spinous process of vertebra (C7)

4.35 Suboccipital region—I

The trapezius, sternocleidomastoid, and splenius muscles are removed. The right semi-spinalis muscle is cut and turned laterally.

- The semispinalis capitis, the great extensor muscle of the head and neck, forms the posterior wall of the suboccipital region. It is pierced by the greater occipital nerve (posterior ramus of C2) and has free medial and lateral borders at this high level.
- The greater occipital nerve, when followed caudally, leads to the inferior border of the inferior oblique muscle, around which it turns. Following the inferior border of the inferior oblique muscle medially from the nerve leads to the spinous process of the axis; followed laterally, this leads to the transverse process of the atlas.
- Five muscles (all paired) are attached to the spinous process of the axis: inferior oblique, rectus capitis posterior major, semispinalis cervicis, multifidus, and interspinalis; the latter two are largely concealed by the semispinalis cervicis.
- The suboccipital triangle is bounded by three muscles: inferior oblique, superior oblique, and rectus capitis posterior major.
- The occipital veins along with the suboccipital nerve (posterior ramus of C1) emerge through the suboccipital triangle to join the deep cervical vein.

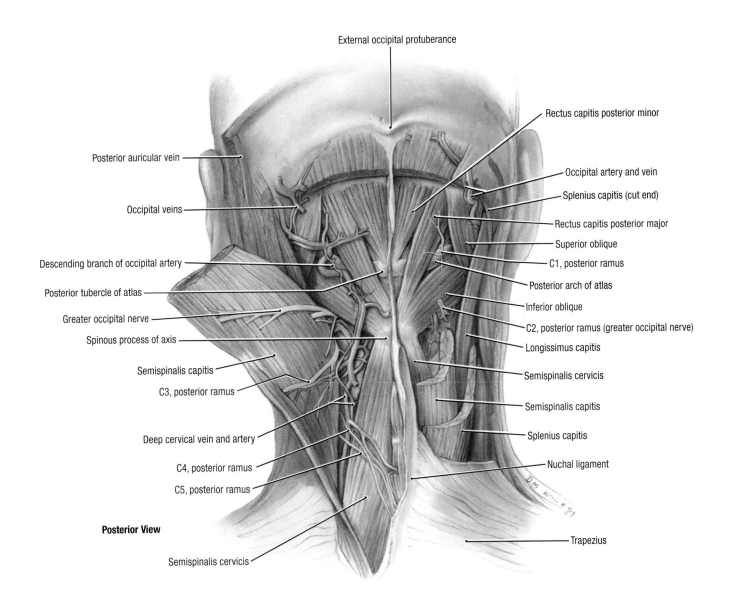

External occipital protuberance

Posterior auricular vein

Occipital veins

Descending branch of occipital artery

Posterior tubercle of atlas

Greater occipital nerve

Spinous process of axis

Semispinalis capitis

C3, posterior ramus

Deep cervical vein and artery

C4, posterior ramus

C5, posterior ramus

Posterior View

Semispinalis cervicis

Rectus capitis posterior minor

Occipital artery and vein

Splenius capitis (cut end)

Rectus capitis posterior major

Superior oblique

C1, posterior ramus

Posterior arch of atlas

Inferior oblique

C2, posterior ramus (greater occipital nerve)

Longissimus capitis

Semispinalis cervicis

Semispinalis capitis

Splenius capitis

Nuchal ligament

Trapezius

4.36   **Suboccipital region—II**

The semispinalis capitis is reflected on the *left* and removed on the *right* side of the body.

- The suboccipital region contains four pairs of structures: two straight muscles, the rectus capitis posterior major and minor; two oblique muscles, the superior oblique and inferior oblique; two nerves (posterior rami), C1 suboccipital (motor) and C2 greater occipital (sensory); and two arteries, the occipital and vertebral.
- The ligamentum nuchae, which represents the cervical part of the supraspinous ligament, is a median, thin, fibrous partition attached to the spinous processes of cervical vertebrae and the external occipital crest; its posterior border gives origin to the trapezius muscle and extends superiorly to the external occipital protuberance.
- The rectus capitis posterior minor muscle arises from the posterior tubercle of the atlas and thus lies on a deeper plane than the rectus capitis posterior major muscle, which arises from a spinous process.
- The suboccipital nerve (posterior ramus of C1) supplies the three muscles bounding the suboccipital triangle and also the rectus capitis minor muscle and communicates with the greater occipital nerve.
- The descending branch of the occipital artery anastomoses with the deep cervical artery, a branch of the subclavian.
- The posterior arch of the atlas forms the floor of the suboccipital triangle.

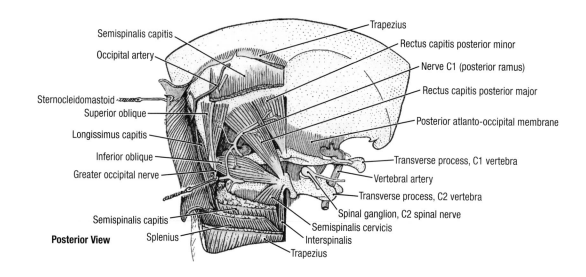

Posterior View

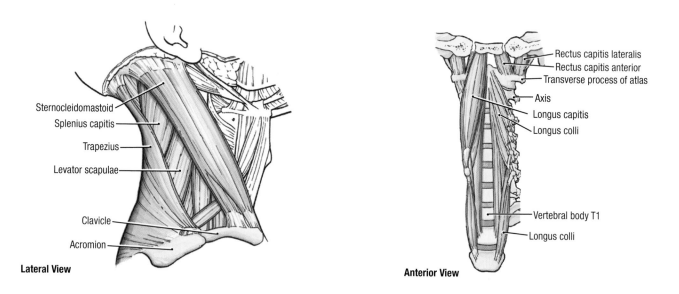

Lateral View

Anterior View

TABLE 4.5   MUSCLES OF THE ATLANTO-OCCIPITAL AND ATLANTOAXIAL JOINTS

MOVEMENTS OF ATLANTO-OCCIPITAL JOINTS

FLEXION	EXTENSION	LATERAL BENDING
Longus capitis Rectus capitis anterior Anterior fibers of sternocleidomastoid	Rectus capitis posterior major and minor Superior oblique Semispinalis capitis Splenius capitis Longissimus capitis Trapezius	Sternocleidomastoid Superior and inferior oblique Rectus capitis lateralis Longissimus capitis Splenius capitis

ROTATION OF ATLANTOAXIAL JOINTS[a]

IPSILATERAL[b]	CONTRALATERAL
Inferior oblique Rectus capitis posterior, major and minor Longissimus capitis Splenius capitis	Sternocleidomastoid Semispinalis capitis

[a]Rotation is the specialized movement at these joints. Movement of one joint involves the other.
[b]Same side to which head is rotated.

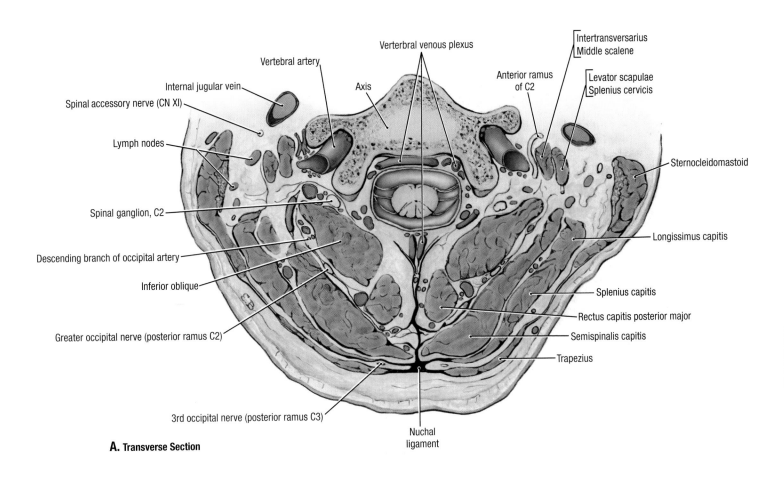

Vertebral venous plexus

Vertebral artery

Internal jugular vein

Spinal accessory nerve (CN XI)

Lymph nodes

Axis

Anterior ramus of C2

Intertransversarius Middle scalene

Levator scapulae Splenius cervicis

Spinal ganglion, C2

Descending branch of occipital artery

Inferior oblique

Greater occipital nerve (posterior ramus C2)

3rd occipital nerve (posterior ramus C3)

Nuchal ligament

Sternocleidomastoid

Longissimus capitis

Splenius capitis

Rectus capitis posterior major

Semispinalis capitis

Trapezius

A. Transverse Section

4.37   Nuchal region

A. Transverse section at the level of the axis. **B.** Vertebral artery.
- The trapezius, splenius, and semispinalis capitis form the roof of the suboccipital triangle.
- The anterior ramus of C2 passes anteriorly lateral to the vertebral artery and the posterior ramus, which is ascending posterior to the inferior oblique.
- In **B,** the vertebral artery arises from the subclavian artery and passes through the transverse foramina of the superior six cervical vertebrae.

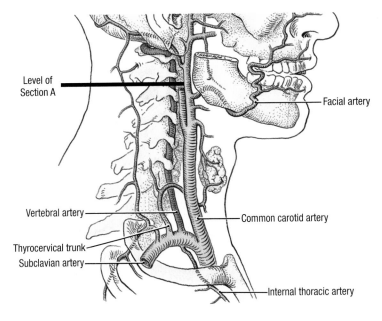

Level of Section A

Facial artery

Vertebral artery

Common carotid artery

Thyrocervical trunk

Subclavian artery

Internal thoracic artery

B. Lateral View

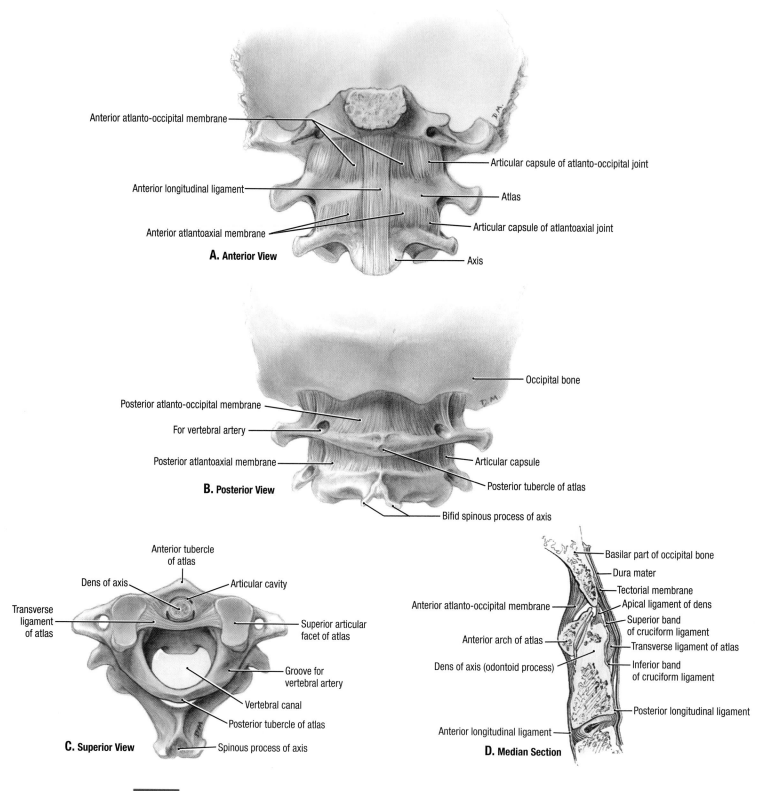

A. Anterior View

- Anterior atlanto-occipital membrane
- Anterior longitudinal ligament
- Anterior atlantoaxial membrane
- Articular capsule of atlanto-occipital joint
- Atlas
- Articular capsule of atlantoaxial joint
- Axis

B. Posterior View

- Posterior atlanto-occipital membrane
- For vertebral artery
- Posterior atlantoaxial membrane
- Occipital bone
- Articular capsule
- Posterior tubercle of atlas
- Bifid spinous process of axis

C. Superior View

- Anterior tubercle of atlas
- Dens of axis
- Articular cavity
- Transverse ligament of atlas
- Superior articular facet of atlas
- Groove for vertebral artery
- Vertebral canal
- Posterior tubercle of atlas
- Spinous process of axis

D. Median Section

- Basilar part of occipital bone
- Dura mater
- Tectorial membrane
- Apical ligament of dens
- Anterior atlanto-occipital membrane
- Superior band of cruciform ligament
- Anterior arch of atlas
- Transverse ligament of atlas
- Dens of axis (odontoid process)
- Inferior band of cruciform ligament
- Posterior longitudinal ligament
- Anterior longitudinal ligament

4.38   **Ligaments of atlanto-occipital and atlantoaxial joints**

A. Anterior atlantoaxial and atlanto-occipital membranes. The anterior longitudinal ligament blends in the midline with the anterior atlanto-occipital and anterior atlantoaxial membranes. **B.** Posterior atlantoaxial and atlanto-occipital membranes. **C and D.** Median atlantoaxial joint. In **D,** observe that the layers of tissue, from posterior to anterior, are dura mater, posterior longitudinal ligament continued superiorly as the tectorial membrane, transverse ligament of the atlas, inferior and superior bands of the cruciform ligament, and the apical ligament of dens, stretching from the apex of the dens to the occipital bone.

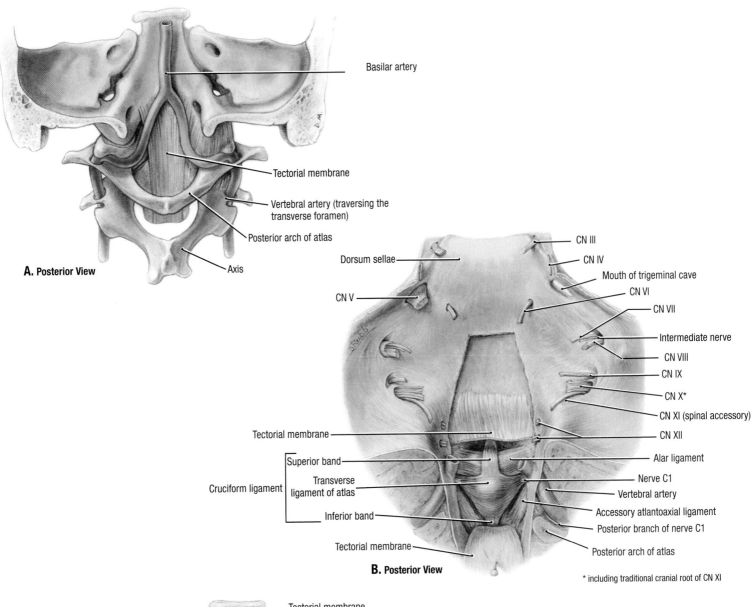

A. Posterior View

- Basilar artery
- Tectorial membrane
- Vertebral artery (traversing the transverse foramen)
- Posterior arch of atlas
- Axis

B. Posterior View

- Dorsum sellae
- CN V
- Tectorial membrane
- Cruciform ligament
  - Superior band
  - Transverse ligament of atlas
  - Inferior band
- Tectorial membrane
- CN III
- CN IV
- Mouth of trigeminal cave
- CN VI
- CN VII
- Intermediate nerve
- CN VIII
- CN IX
- CN X*
- CN XI (spinal accessory)
- CN XII
- Alar ligament
- Nerve C1
- Vertebral artery
- Accessory atlantoaxial ligament
- Posterior branch of nerve C1
- Posterior arch of atlas

* including traditional cranial root of CN XI

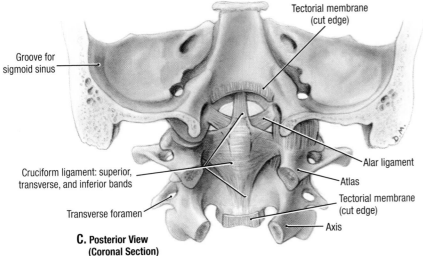

- Tectorial membrane (cut edge)
- Groove for sigmoid sinus
- Cruciform ligament: superior, transverse, and inferior bands
- Transverse foramen
- Alar ligament
- Atlas
- Tectorial membrane (cut edge)
- Axis

C. Posterior View (Coronal Section)

4.39 **Craniovertebral joints and vertebral artery**

A. Vertebral arteries. The vertebral arteries enter the skull through the foramen magnum and join to form the basilar artery. **B.** Ligaments and cranial nerves. **C.** Cruciform and alar ligaments.

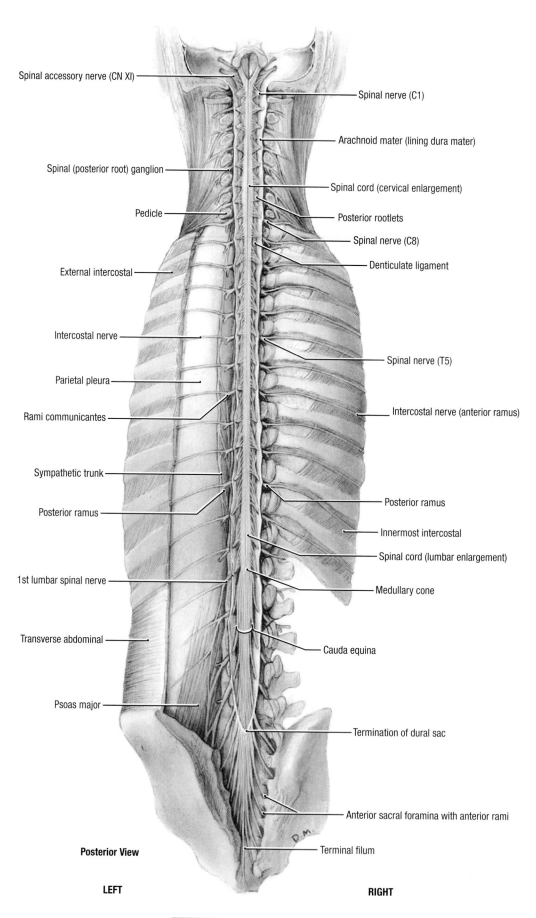

Spinal accessory nerve (CN XI)

Spinal nerve (C1)

Arachnoid mater (lining dura mater)

Spinal (posterior root) ganglion

Spinal cord (cervical enlargement)

Pedicle

Posterior rootlets

Spinal nerve (C8)

Denticulate ligament

External intercostal

Intercostal nerve

Spinal nerve (T5)

Parietal pleura

Rami communicantes

Intercostal nerve (anterior ramus)

Sympathetic trunk

Posterior ramus

Posterior ramus

Innermost intercostal

Spinal cord (lumbar enlargement)

1st lumbar spinal nerve

Medullary cone

Transverse abdominal

Cauda equina

Psoas major

Termination of dural sac

Anterior sacral foramina with anterior rami

Posterior View

Terminal filum

LEFT

RIGHT

4.40 **Spinal cord in situ**

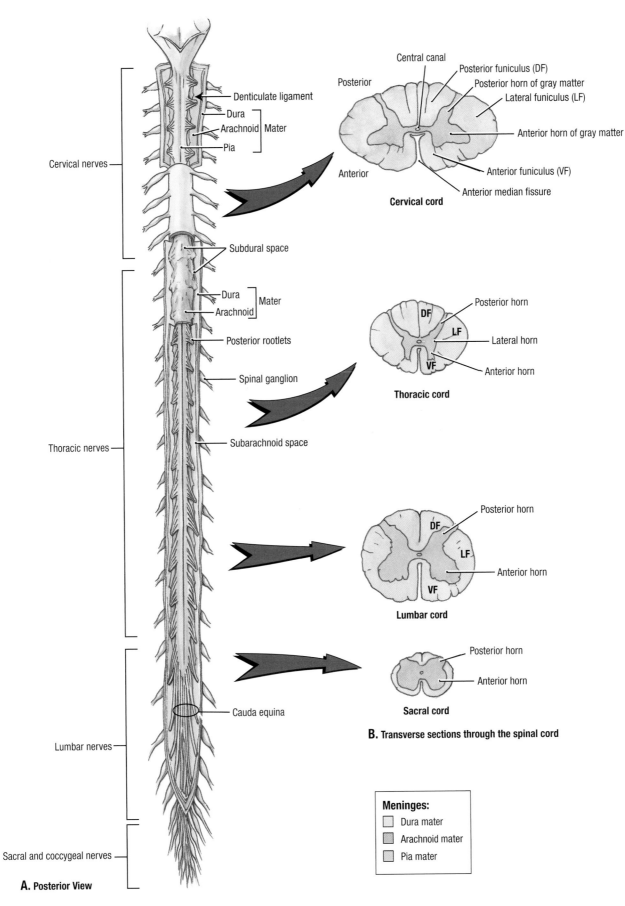

Denticulate ligament

Dura ⎱
Arachnoid ⎰ Mater

Pia

Cervical nerves

Subdural space

Dura ⎱ Mater
Arachnoid ⎰

Posterior rootlets

Spinal ganglion

Thoracic nerves

Subarachnoid space

Cauda equina

Lumbar nerves

Sacral and coccygeal nerves

A. Posterior View

Central canal

Posterior

Posterior funiculus (DF)

Posterior horn of gray matter

Lateral funiculus (LF)

Anterior horn of gray matter

Anterior funiculus (VF)

Anterior

Anterior median fissure

Cervical cord

Posterior horn

DF

LF   Lateral horn

VF   Anterior horn

Thoracic cord

Posterior horn

DF

LF

Anterior horn

VF

Lumbar cord

Posterior horn

Anterior horn

Sacral cord

B. Transverse sections through the spinal cord

Menenges:
☐ Dura mater
☐ Arachnoid mater
☐ Pia mater

4.41   **Isolated spinal cord and spinal nerve roots with coverings and regional sections**

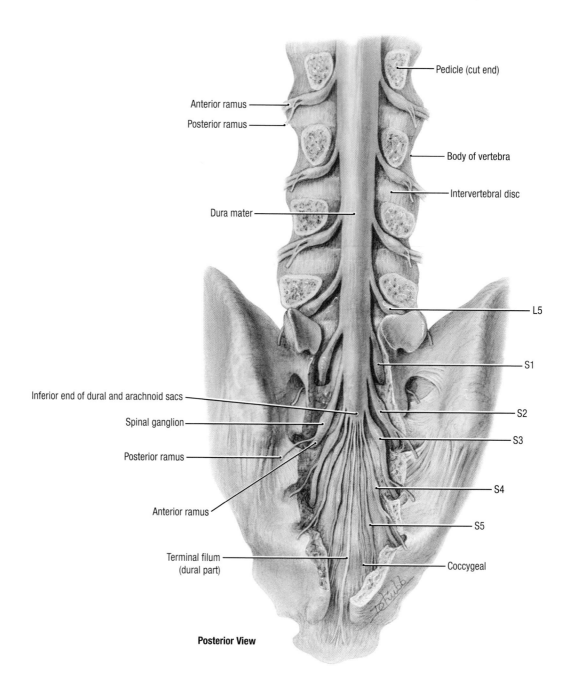

Anterior ramus

Posterior ramus

Dura mater

Inferior end of dural and arachnoid sacs

Spinal ganglion

Posterior ramus

Anterior ramus

Terminal filum
(dural part)

Pedicle (cut end)

Body of vertebra

Intervertebral disc

L5

S1

S2

S3

S4

S5

Coccygeal

Posterior View

4.42 **Inferior end of dural sac—I**

The posterior parts of the lumbar vertebrae and sacrum were removed.
- The inferior limit of the dural sac is at the level of the posterior superior iliac spine (body of 2nd sacral vertebra); the dura continues as the dural part of the terminal filum (coccygeal ligament).
- Note the lumbar spinal ganglia in the intervertebral foramina and the sacral spinal ganglia, somewhat asymmetrically placed within the sacral canal.
- The posterior rami are smaller than the anterior rami.

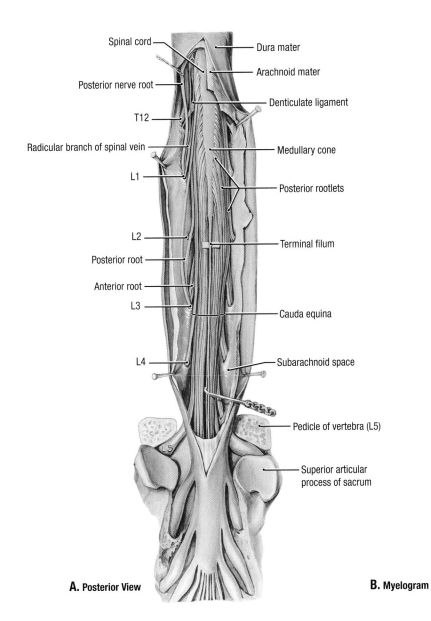

Spinal cord — Dura mater
Arachnoid mater
Posterior nerve root —
Denticulate ligament
T12 —
Radicular branch of spinal vein — Medullary cone
L1 —
Posterior rootlets
L2 — Terminal filum
Posterior root —
Anterior root —
L3 — Cauda equina
L4 — Subarachnoid space
Pedicle of vertebra (L5)
Superior articular process of sacrum

A. Posterior View

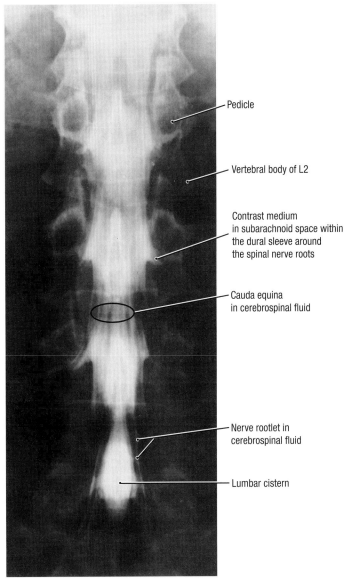

Pedicle
Vertebral body of L2
Contrast medium in subarachnoid space within the dural sleeve around the spinal nerve roots
Cauda equina in cerebrospinal fluid
Nerve rootlet in cerebrospinal fluid
Lumbar cistern

B. Myelogram

4.43   **Inferior end of dural sac—II**

A. Inferior end of dural sac, opened. **B.** Myelogram of the lumbar region of the vertebral column. Contrast medium was injected into the subarachnoid space. **C.** Termination of spinal cord, in situ, sagittal section.

- The medullary cone, or conical lower end of the spinal cord, continues as a glistening thread, the pial part of the terminal filum, which descends with the posterior and anterior nerve roots; these constitute the cauda equina.
- In the adult, the spinal cord usually ends at the level of the disc between L1 and L2.
- The subarachnoid space usually ends at the level of the disc between S1 and S2, but it can be more inferior.
- Variations: 95% of cords end within the limits of the bodies of L1 and L2, whereas 3% end posterior to the inferior half of T12 and 2% posterior to L3.

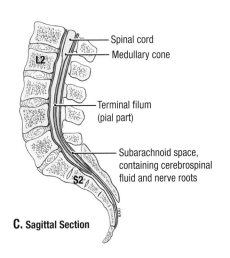

Spinal cord
Medullary cone
Terminal filum (pial part)
Subarachnoid space, containing cerebrospinal fluid and nerve roots

C. Sagittal Section

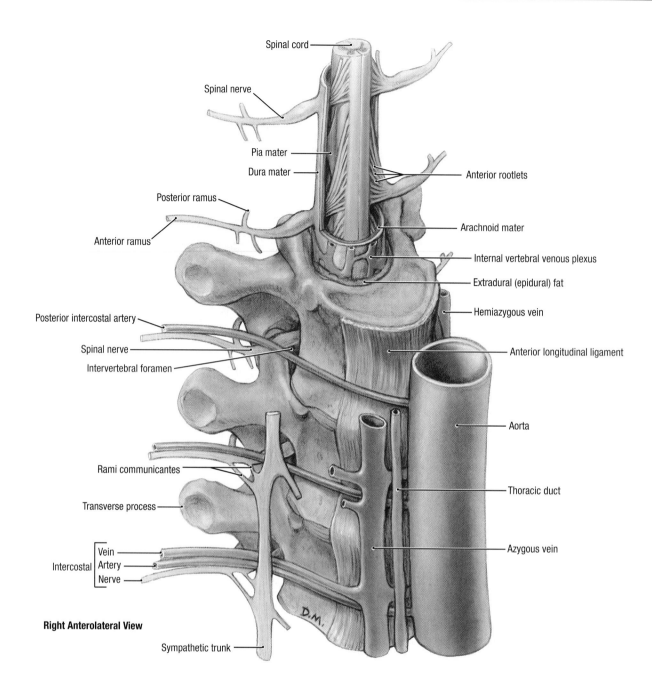

Spinal cord

Spinal nerve

Pia mater

Dura mater

Anterior rootlets

Posterior ramus

Anterior ramus

Arachnoid mater

Internal vertebral venous plexus

Extradural (epidural) fat

Hemiazygous vein

Posterior intercostal artery

Spinal nerve

Intervertebral foramen

Anterior longitudinal ligament

Aorta

Rami communicantes

Transverse process

Thoracic duct

Intercostal { Vein / Artery / Nerve }

Azygous vein

Right Anterolateral View

Sympathetic trunk

D.M.

4.44   Spinal cord and prevertebral structures

The vertebrae have been removed superiorly to expose the spinal cord and meninges.

- The aorta descends to the left of the midline, with the thoracic duct and azygos vein to its right.
- Typically, the azygos vein is on the right side of the vertebral bodies, and the hemiazygous vein is on the left.
- The thoracic sympathetic trunk and ganglia lie lateral to the thoracic vertebrae; the rami communicantes connect the sympathetic ganglia with the spinal nerve.
- A sleeve of dura mater surrounds the spinal nerves and blends with the sheath (epineurium) of the spinal nerve.
- The spinal cord is covered with pia mater and anchored laterally by the denticulate ligaments.
- The dura mater is separated from the walls of the vertebral canal by extradural (epidural) fat and the internal vertebral venous plexus.

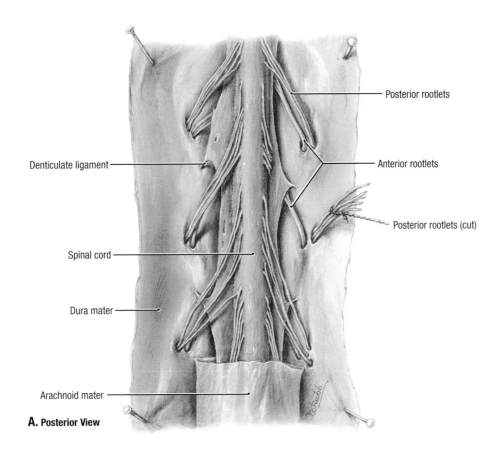

Posterior rootlets

Denticulate ligament

Anterior rootlets

Posterior rootlets (cut)

Spinal cord

Dura mater

Arachnoid mater

A. Posterior View

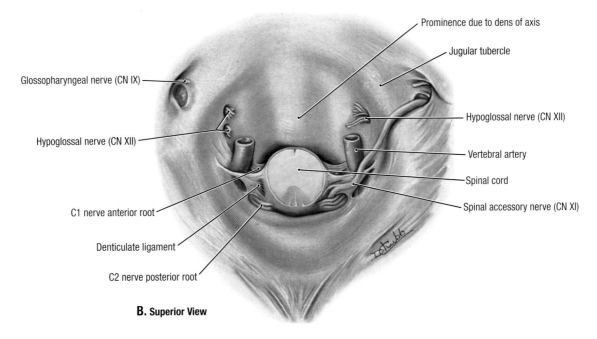

Prominence due to dens of axis

Jugular tubercle

Glossopharyngeal nerve (CN IX)

Hypoglossal nerve (CN XII)

Hypoglossal nerve (CN XII)

Vertebral artery

Spinal cord

C1 nerve anterior root

Spinal accessory nerve (CN XI)

Denticulate ligament

C2 nerve posterior root

B. Superior View

4.45   **Spinal cord and meninges**

A. Dural sac cut open. The denticulate ligament anchors the cord to the dural sac between successive nerve roots by means of strong, toothlike processes. The anterior nerve roots lie anterior to the denticulate ligament, and the posterior nerve roots lie posterior to the ligament. **B.** Structures of vertebral canal seen through foramen magnum. The spinal cord (or medulla), vertebral arteries, spinal accessory nerve (CN XI), and most superior tooth of the denticulate ligament pass through the foramen magnum within the meninges.

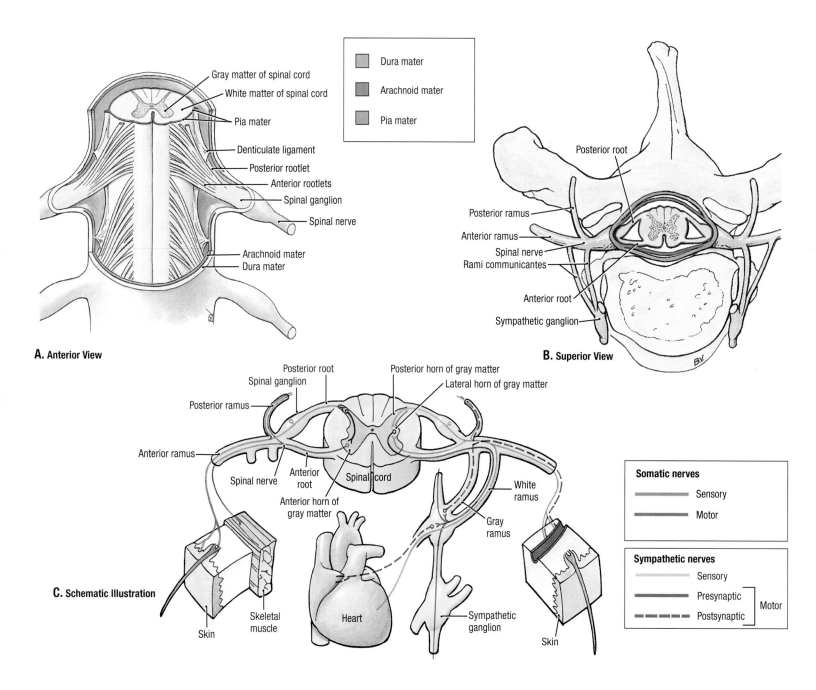

A. Anterior View

Gray matter of spinal cord
White matter of spinal cord
Pia mater
Denticulate ligament
Posterior rootlet
Anterior rootlets
Spinal ganglion
Spinal nerve
Arachnoid mater
Dura mater

Dura mater
Arachnoid mater
Pia mater

B. Superior View

Posterior root
Posterior ramus
Anterior ramus
Spinal nerve
Rami communicantes
Anterior root
Sympathetic ganglion

C. Schematic Illustration

Posterior root
Spinal ganglion
Posterior ramus
Anterior ramus
Spinal nerve
Anterior root
Anterior horn of gray matter
Posterior horn of gray matter
Lateral horn of gray matter
Spinal cord
White ramus
Gray ramus
Skin
Skeletal muscle
Heart
Sympathetic ganglion
Skin

Somatic nerves
Sensory
Motor

Sympathetic nerves
Sensory
Presynaptic — Motor
Postsynaptic

4.46 Formation of spinal nerves

A. Relationship to dural sac. **B.** In situ in the vertebral canal. **C.** Components of a typical spinal nerve.

- The dural sac consists of the dura mater and arachnoid mater. The pia mater covers the spinal cord and projects laterally as the denticulate ligaments, which separate the rows of posterior and anterior rootlets.
- Cerebrospinal fluid circulates between the pia and arachnoid in the subarachnoid space.
- On each side, two rows of rootlets attach to the cord; the posterior rootlets carry sensory information to the CNS, and the anterior rootlets convey motor information from the CNS to the periphery. Several rootlets combine to form posterior and anterior roots at each segment; posterior and anterior roots unite to form a spinal nerve.
- The enlarged portion of the posterior root, the spinal (dorsal root) ganglion, contains cell bodies of sensory neurons.

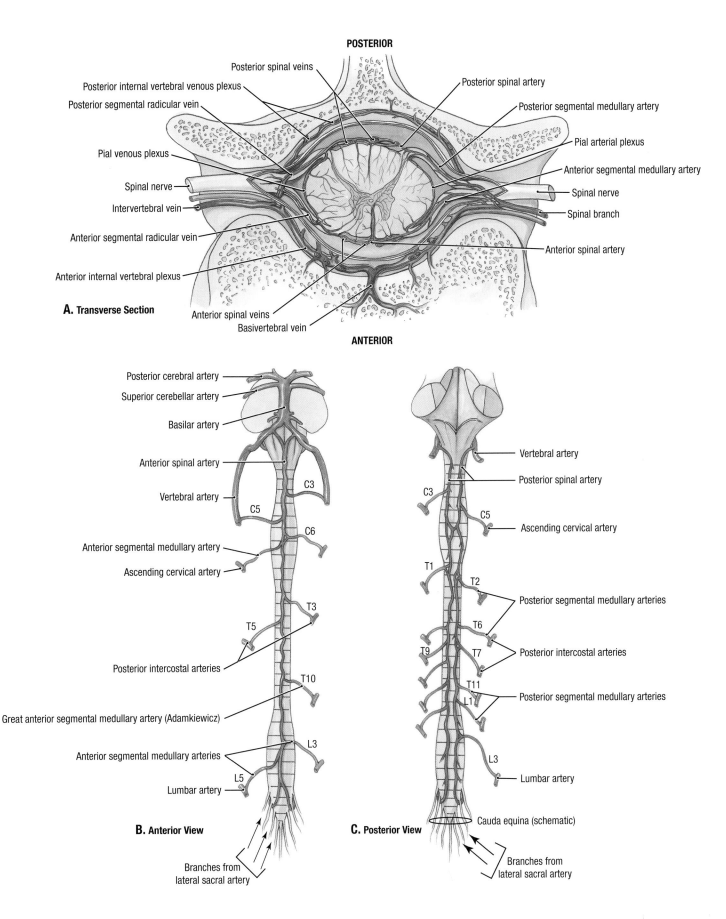

POSTERIOR

Posterior spinal veins
Posterior internal vertebral venous plexus
Posterior segmental radicular vein
Pial venous plexus
Spinal nerve
Intervertebral vein
Anterior segmental radicular vein
Anterior internal vertebral plexus

Posterior spinal artery
Posterior segmental medullary artery
Pial arterial plexus
Anterior segmental medullary artery
Spinal nerve
Spinal branch
Anterior spinal artery

A. Transverse Section

Anterior spinal veins
Basivertebral vein

ANTERIOR

Posterior cerebral artery
Superior cerebellar artery
Basilar artery
Anterior spinal artery
Vertebral artery
Anterior segmental medullary artery
Ascending cervical artery
Posterior intercostal arteries
Great anterior segmental medullary artery (Adamkiewicz)
Anterior segmental medullary arteries
Lumbar artery

C3
C5
C6
T3
T5
T10
L3
L5

B. Anterior View

Branches from lateral sacral artery

Vertebral artery
Posterior spinal artery
Ascending cervical artery
Posterior segmental medullary arteries
Posterior intercostal arteries
Posterior segmental medullary arteries
Lumbar artery
Cauda equina (schematic)

C3
C5
T1
T2
T6
T9
T7
T11
L1
L3

C. Posterior View

Branches from lateral sacral artery

4.47   **Blood vessels of spinal cord**

A. Arterial supply and venous drainage. **B** and **C.** Arterial supply. The anterior and posterior spinal arteries are reinforced at irregular intervals by anterior and posterior segmental medullary arteries; at levels where these arteries do not occur, radicular arteries supply the posterior and anterior roots, but they do not reinforce the spinal arteries. The great anterior segmental medullary artery ("Adamkiewicz artery") occurs on the left side in 65% of people. It reinforces the circulation to two thirds of the spinal cord.

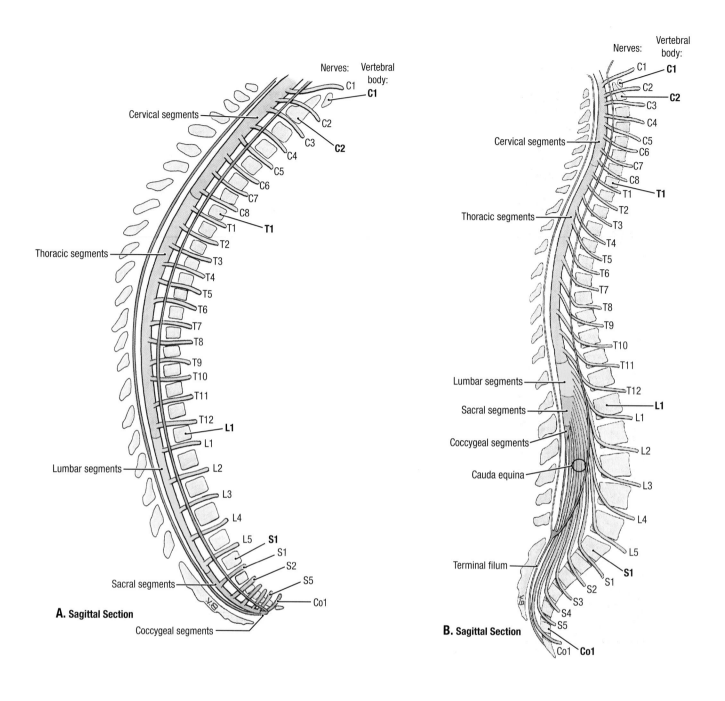

A. Sagittal Section

B. Sagittal Section

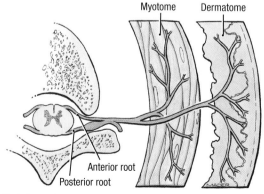

C. Schematic Transverse Section

4.48   **Subarachnoid space and segmental innervation**

A. Spinal cord at 12 weeks' gestation. **B.** Spinal cord of an adult. Early in development, the spinal cord and vertebral (spinal) canal are nearly equal in length. The canal grows longer, so spinal nerves have an increasingly longer course to reach the intervertebral foramen at the correct level for their exit. The spinal cord proper terminates at L2, and the remaining spinal nerves, seeking their intervertebral foramen of exit, form the cauda equina. **C.** Schematic diagram of segmental innervation: dermatome, and myotome.

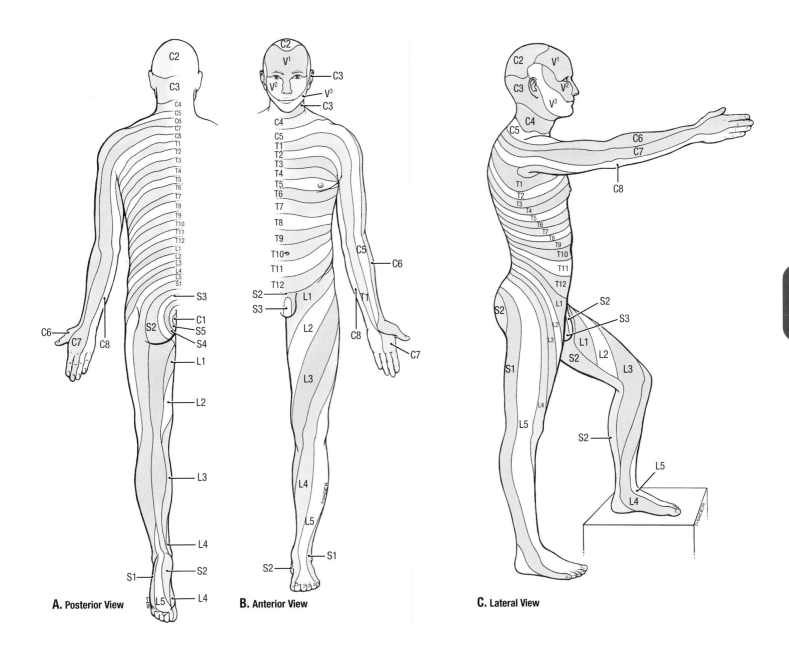

A. Posterior View   B. Anterior View   C. Lateral View

4.49   **Dermatomes**

A dermatome is an area of skin supplied by the posterior (sensory) root of a spinal nerve. In the neck and trunk, each segment is horizontally disposed, except C1, which has no sensory component. The dermatomes of the limbs from the 5th cervical to the 1st thoracic and from the 3rd lumbar to the 2nd sacral vertebrae extend as a series of bands from the midline of the trunk posteriorly into the limbs. Note that there is considerable overlapping of adjacent dermatomes (i.e., each segmental nerve overlaps the territories of its neighbors). As a result, no anesthesia occurs unless two or more consecutive posterior roots have lost their functions. (This dermatome map is based on J. J. Keegan and F. D. Garrett, 1948.)

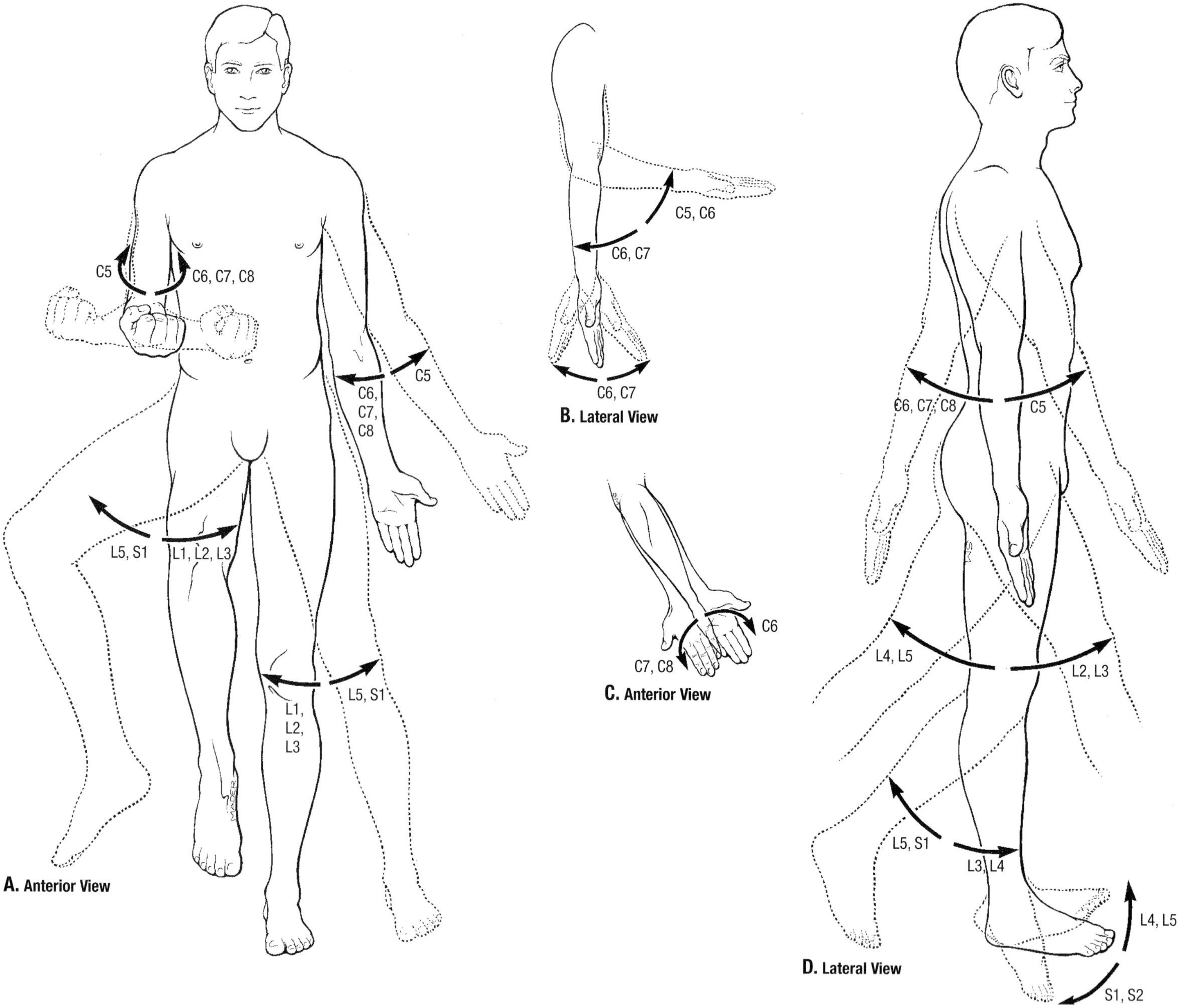

A. Anterior View

B. Lateral View

C. Anterior View

D. Lateral View

4.50 **Segmental innervation of limb movements (myotome function)**

A myotome is the segmental innervation of skeletal muscle by the anterior (motor) root(s) of the spinal nerve(s). When adjacent myotomes fuse during embryonic development, the resultant muscle can be innervated by one or both of its segmental spinal nerves. This diagram indicates the myotomes that produce the movements of the joints of the upper and lower extremities. **A.** Medial and lateral rotation of shoulder and hip, abduction and adduction of shoulder and hip. **B.** Flexion and extension of elbow and wrist. **C.** Pronation and supination of forearm. **D.** Flexion and extension of shoulder, hip, and knee, dorsiflexion and plantar flexion of ankle.

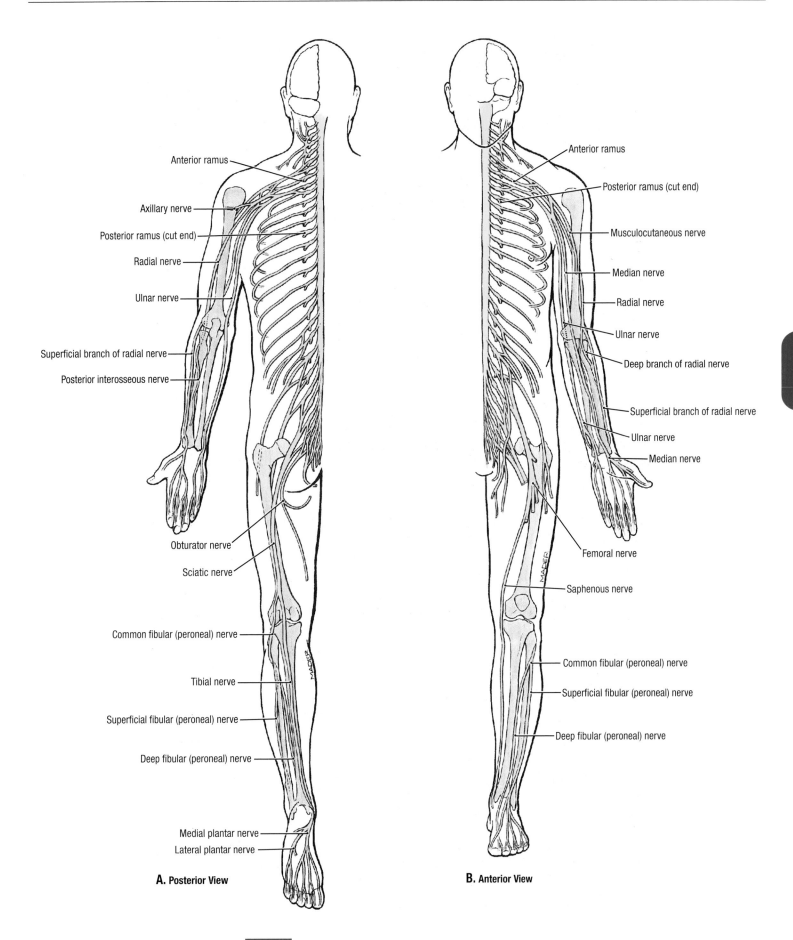

4.51   **Overview of somatic nervous system**

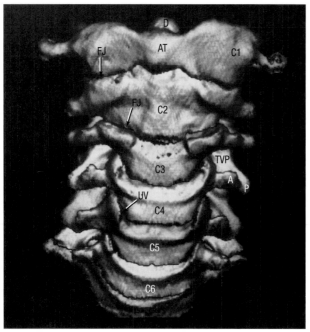

A. Anterior View

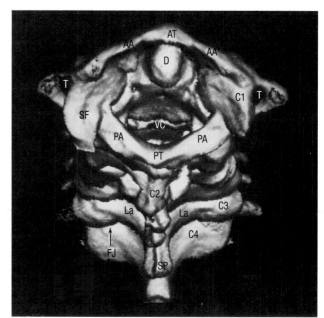

B. Posterior View

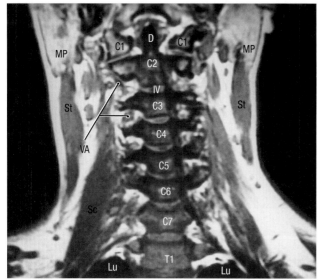

C. Coronal MRI

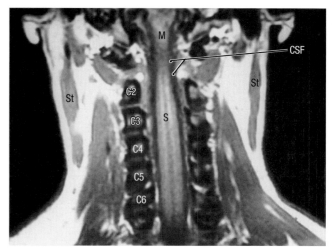

D. Coronal MRI

A	Anterior tubercle of transverse process	Lu	Lungs	SP	Spinous process
AA	Anterior arch of C1	M	Medulla oblongata	St	Sternocleidomastoid
AT	Anterior tubercle of C1	MP	Mastoid process	T	Transverse foramen
C1–T1	Vertebrae	P	Posterior tubercle of transverse process	TVP	Transverse process
CSF	Cerebrospinal fluid in subarachnoid space	PA	Posterior arch of C1	UV	Uncovertebral joint
D	Dens (odontoid) process of C2	PT	Posterior tubercle of C1	VA	Vertebral artery
FJ	Zygapophysial (facet) joint	S	Spinal cord	VC	Vertebral canal
IV	Intervertebral disc	Sc	Scalene muscles		
La	Lamina	SF	Superior articular facet of C1		

4.52 **Imaging of spine**

A and **B.** Three-dimensional computer-generated images of cervical spine. **C** and **D.** T1 coronal MRIs of cervical spine. **E.** Three-dimensional computer-generated image of the base of the skull and atlas. **F.** T1 sagittal MRI of the cervical and upper thoracic spine. **G** and **H.** T1 coronal MRIs of thoracic spine and spinal cord.

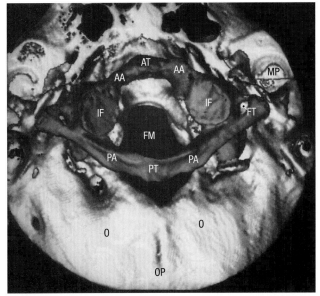

E. Posteroinferior View

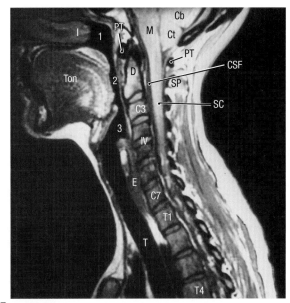

F. Sagittal MRI

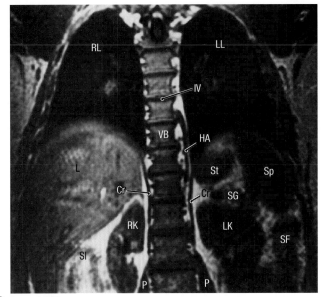

G. Coronal MRI

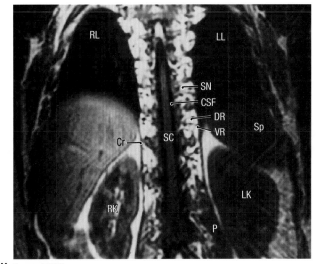

H. Coronal MRI

AA	Anterior arch of C1	IV	Intervertebral disc	SG	Suprarenal gland
AT	Anterior tubercle of C1	L	Liver	SI	Small intestine
C2-T4	Vertebral bodies	LK	Left kidney	SN	Spinal nerve
Cb	Cerebellum	LL	Left lung	SP	Spinous process
Cr	Crus of diaphragm	M	Medulla oblongata	Sp	Spleen
CSF	Cerebrospinal fluid in subarachnoid space	MP	Mastoid process	St	Stomach
Ct	Tonsil of cerebellum	O	Occipital bone of skull	Ton	Tongue
D	Dens (odontoid) process of C2	OP	External occipital protuberance	T	Trachea
DR	Posterior ramus	P	Psoas muscle	VB	Vertebral body
E	Esophagus	PA	Posterior arch of C1	VR	Anterior ramus
FM	Foramen magnum	PT	Posterior tubercle of C1	1	Nasopharynx
FT	Transverse foramen	RK	Right Kidney	2	Oropharynx
HA	Hemiazygos vein	RL	Right lung	3	Laryngopharynx
I	Inferior concha	SC	Spinal cord		
IF	Inferior articular facet of C1	SF	Splenic flexure		

4.52   **Imaging of spine (continued)**

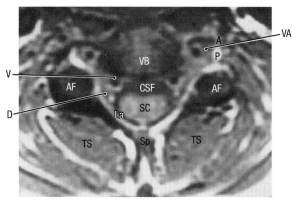

I. Transverse (axial) MRI

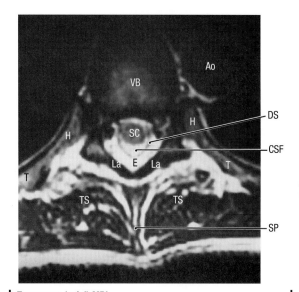

J. Transverse (axial) MRI

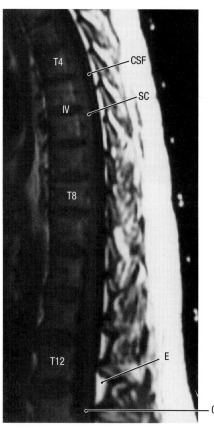

L. Sagittal MRI

A	Anterior tubercle
AF	Articular facet
Ao	Aorta
CE	Cauda equina in cerebrospinal fluid
CM	Medullary cone
CSF	Cerebrospinal fluid in subarachnoid space
D	Posterior rootlet
DS	Dural sac
E	Epidural (extradural) fat
FJ	Zygapophysial (facet) joint
H	Head of rib
IV	Intervertebral disc
La	Lamina
N	Spinal nerve in intervertebral foramen
P	Posterior tubercle
Pe	Pedicle
SC	Spinal cord
SP	Spinous process
T	Tubercle of rib
T4-L5	Vertebral bodies
TS	Transversospinalis muscles
V	Anterior rootlet
VA	Vertebral artery in transverse foramen
VB	Vertebral body

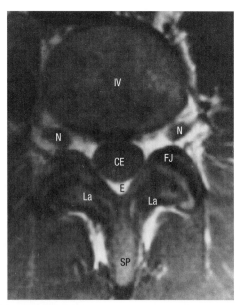

K. Transverse (axial) MRI

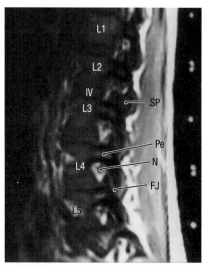

M. Sagittal MRI

4.52 **Imaging of spine** *(continued)*

I. T1 axial (transverse) MRI of cervical spine. **J.** T2 axial (transverse) MRI of thoracic spine. **K.** T1 axial (transverse) MRI of lumbar spine. **L.** T1 sagittal MRI of thoracic and upper lumbar spine. **M.** T1 sagittal MRI of lumbar spine.

CHAPTER 5

LOWER LIMB

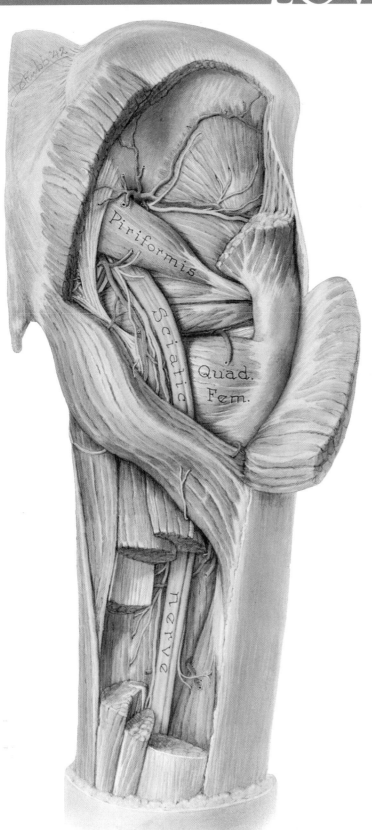

- Skeleton of Lower Limb **338**
- Cutaneous Innervation of Lower Limb **340**
- Venous and Lymphatic Drainage of Lower Limb **342**
- Fascia and Fascial Compartments of Lower Limb **345**
- Arteries and Nerves of Lower Limb **346**
- Inguinal Region and Femoral Triangle **350**
- Anterior and Medial Thigh **356**
- Lateral Thigh **365**
- Gluteal Region and Posterior Thigh **366**
- Hip Joint **376**
- Knee Region **384**
- Knee Joint **390**
- Anterior and Lateral Leg and Dorsum of Foot **404**
- Posterior Leg **415**
- Sole of Foot **424**
- Ankle and Foot Joints **432**
- Bony Anomalies **449**
- Sectional Anatomy of Thigh and Leg **450**
- Sectional Anatomy and Imaging: Thigh and Leg **454**

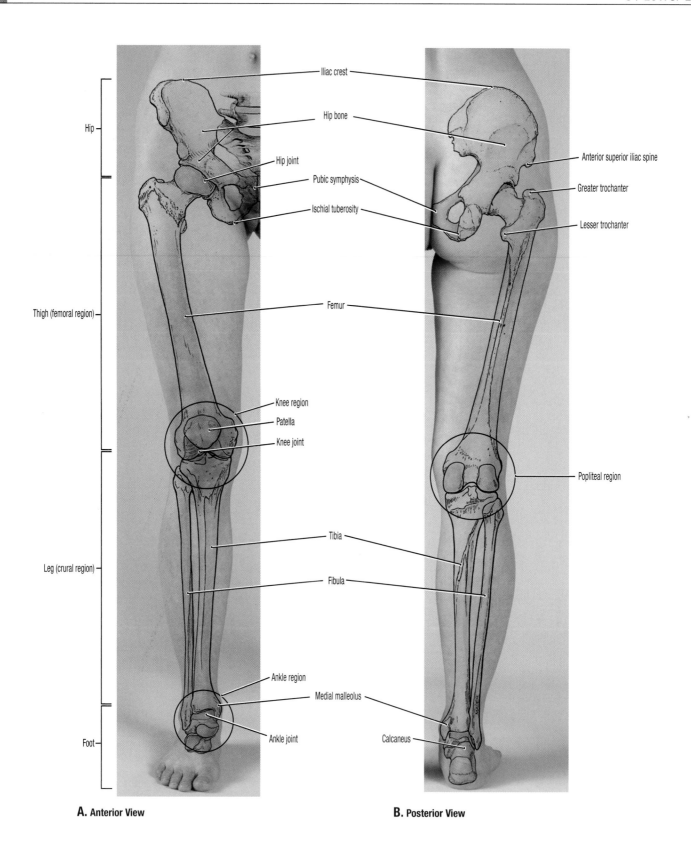

Iliac crest

Hip bone

Hip joint

Pubic symphysis

Ischial tuberosity

Anterior superior iliac spine

Greater trochanter

Lesser trochanter

Hip

Thigh (femoral region)

Femur

Knee region

Patella

Knee joint

Popliteal region

Tibia

Fibula

Leg (crural region)

Ankle region

Medial malleolus

Ankle joint

Calcaneus

Foot

A. Anterior View

B. Posterior View

5.1   **Regions and bones of lower limb**

The hip bones meet anteriorly at the symphysis pubis and articulate with the sacrum posteriorly. The femur articulates with the hip bone proximally and the tibia distally. The tibia
and fibula are the bones of the leg that join the foot at the ankle.

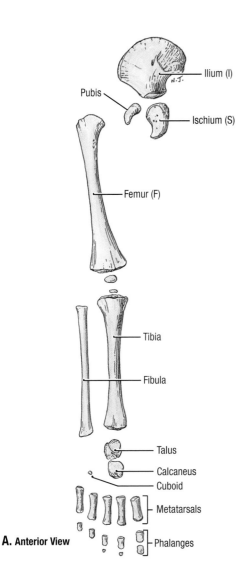

A. Anterior View

Pubis

Ilium (I)

Ischium (S)

Femur (F)

Tibia

Fibula

Talus

Calcaneus

Cuboid

Metatarsals

Phalanges

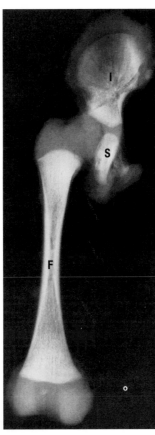

B. Anteroposterior View

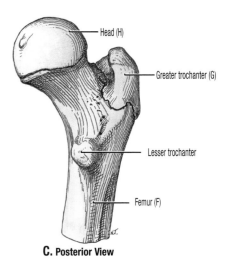

Head (H)

Greater trochanter (G)

Lesser trochanter

Femur (F)

C. Posterior View

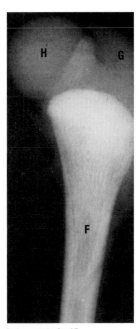

D. Anteroposterior View

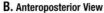

5.2   Lower limb development

A. Bones of lower limb at birth. The hip bone can be divided into three primary parts: ilium, ischium, and pubis. The diaphyses (bodies) of the long bones are well ossified. Some epiphyses (growth plates) and tarsal bones have begun to ossify, including the distal epiphysis of the femur, proximal epiphysis of the tibia, calcaneus, talus, and cuboid. **B** and **D.** Anteroposterior radiographs of postmortem specimens of newborns show the bony *(white)* and cartilaginous *(gray)* components of the femur and hip bone. **C.** Epiphyses at proximal end of femur. The epiphysis of the head of the femur begins to ossify during the 1st year, that of the greater trochanter before the 5th year, and that of the lesser trochanter before the 14th year. These usually fuse completely with the body (shaft) before the end of the 18th year.

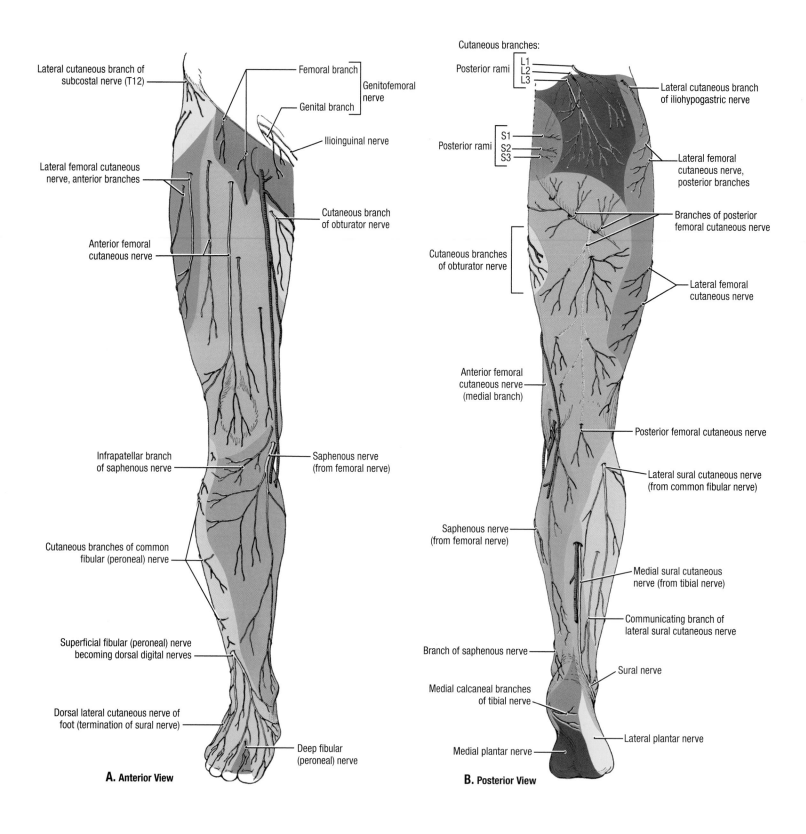

Lateral cutaneous branch of subcostal nerve (T12)

Femoral branch

Genital branch

Genitofemoral nerve

Ilioinguinal nerve

Lateral femoral cutaneous nerve, anterior branches

Cutaneous branch of obturator nerve

Anterior femoral cutaneous nerve

Infrapatellar branch of saphenous nerve

Saphenous nerve (from femoral nerve)

Cutaneous branches of common fibular (peroneal) nerve

Superficial fibular (peroneal) nerve becoming dorsal digital nerves

Dorsal lateral cutaneous nerve of foot (termination of sural nerve)

Deep fibular (peroneal) nerve

A. Anterior View

Cutaneous branches:

Posterior rami

L1
L2
L3

Lateral cutaneous branch of iliohypogastric nerve

Posterior rami

S1
S2
S3

Lateral femoral cutaneous nerve, posterior branches

Branches of posterior femoral cutaneous nerve

Cutaneous branches of obturator nerve

Lateral femoral cutaneous nerve

Anterior femoral cutaneous nerve (medial branch)

Posterior femoral cutaneous nerve

Lateral sural cutaneous nerve (from common fibular nerve)

Saphenous nerve (from femoral nerve)

Medial sural cutaneous nerve (from tibial nerve)

Branch of saphenous nerve

Communicating branch of lateral sural cutaneous nerve

Medial calcaneal branches of tibial nerve

Sural nerve

Medial plantar nerve

Lateral plantar nerve

B. Posterior View

5.3   Cutaneous nerves of lower limb

Cutaneous nerves in the subcutaneous tissue supply the skin of the lower limb. In **B,** the medial sural cutaneous nerve (*sural* is Latin for calf) is joined just proximal to the posterior aspect of the ankle by a communicating branch of the lateral sural cutaneous nerve to form the sural nerve. The level of the junction is variable and is low in this specimen.

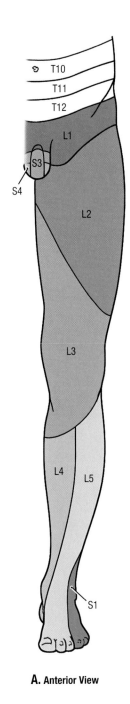

A. Anterior View

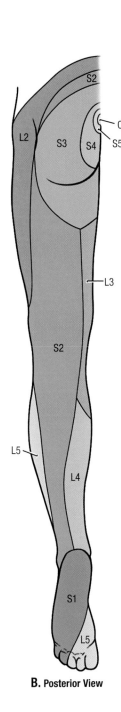

B. Posterior View

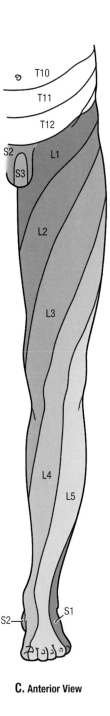

C. Anterior View

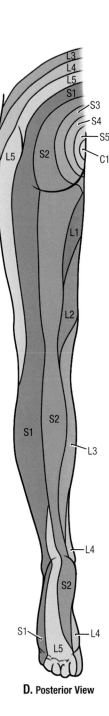

D. Posterior View

5.4 Dermatomes of lower limb

The area of skin supplied by cutaneous branches from a spinal nerve is called a *dermatome*. Adjacent dermatomes may overlap except at the axial line, which is the line of junction of dermatomes supplied from discontinuous spinal levels. Two alternate dermatome maps are commonly used. **A.** and **B.** are based on Fender FA. Foerster's scheme of the dermatomes. Arch Neurol Psychiatr 1939;41:688. **C.** and **D.** are based on Keegan JJ and Garrett FD. The segmental distribution of the cutaneous nerves in the limbs of man. Anat Rec 1948;102:409.

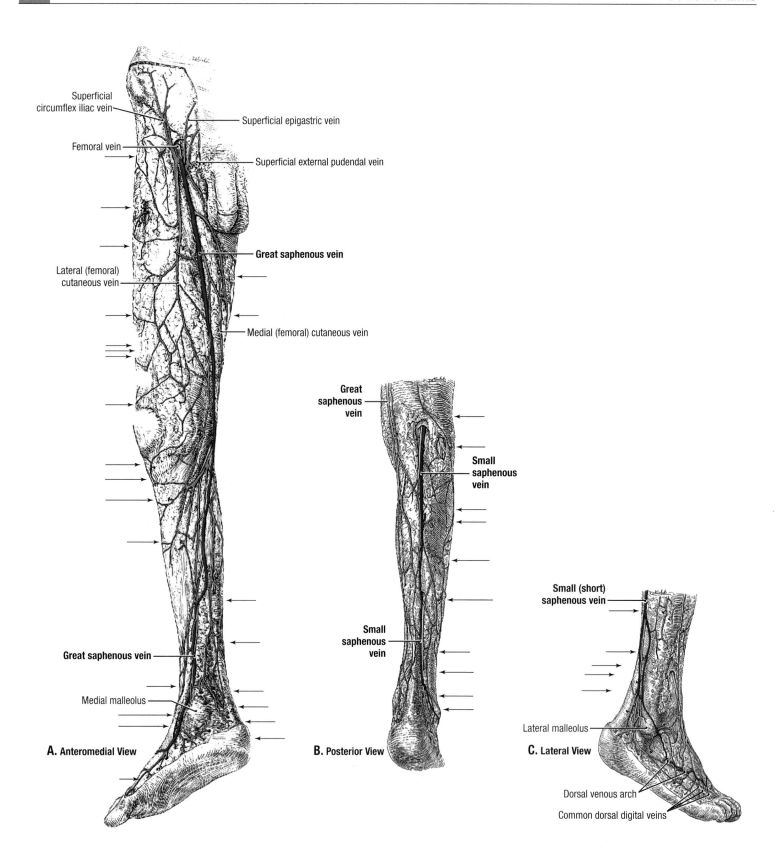

Superficial circumflex iliac vein

Superficial epigastric vein

Femoral vein

Superficial external pudendal vein

Great saphenous vein

Lateral (femoral) cutaneous vein

Medial (femoral) cutaneous vein

Great saphenous vein

Small saphenous vein

Great saphenous vein

Medial malleolus

Small saphenous vein

A. Anteromedial View

B. Posterior View

Small (short) saphenous vein

C. Lateral View

Lateral malleolus

Dorsal venous arch

Common dorsal digital veins

5.5   Superficial veins of lower limb

The *arrows* indicate where perforating veins penetrate the deep fascia. Blood is continuously shunted from these superficial veins in the subcutaneous tissue to deep veins via the perforating veins.

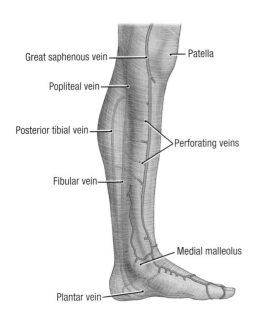

Great saphenous vein

Patella

Popliteal vein

Posterior tibial vein

Perforating veins

Fibular vein

Medial malleolus

Plantar vein

A. Medial View

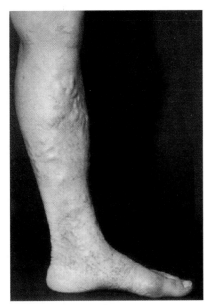

B. Medial View, Varicose Veins

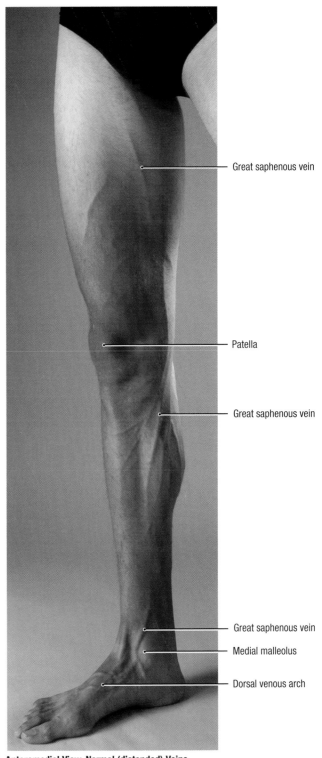

Great saphenous vein

Patella

Great saphenous vein

Great saphenous vein

Medial malleolus

Dorsal venous arch

**C. Anteromedial View, Normal (distended) Veins
(following exercise)**

5.6   Surface anatomy of the superficial veins of lower limb

A. Perforating veins, schematic diagram. The perforating veins drain the blood from the great saphenous vein to the posterior tibial and fibular veins. **B.** Varicose veins. Varicose veins form when deep fascia or the valves that usually prevent blood flow from the deep veins through the perforating veins to the superficial veins are incompetent. As a result, the superficial veins become enlarged, tortuous, and dilated. **C.** Normal veins, distended following exercise.

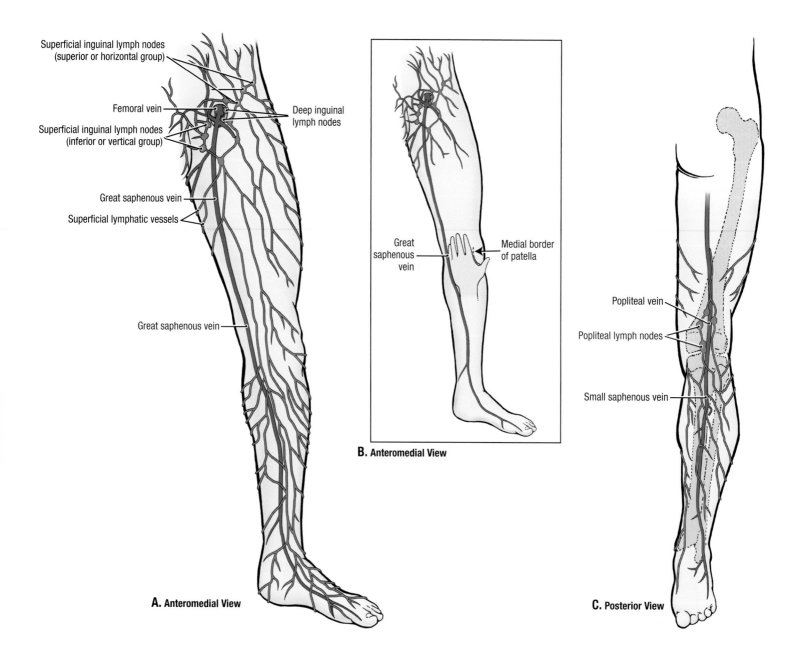

Superficial inguinal lymph nodes
(superior or horizontal group)

Femoral vein

Superficial inguinal lymph nodes
(inferior or vertical group)

Deep inguinal
lymph nodes

Great saphenous vein

Superficial lymphatic vessels

Great saphenous vein

A. Anteromedial View

Great
saphenous
vein

Medial border
of patella

B. Anteromedial View

Popliteal vein

Popliteal lymph nodes

Small saphenous vein

C. Posterior View

5.7   Superficial lymphatic drainage of lower limb

The superficial lymphatic vessels converge on and accompany the saphenous veins and their tributaries in the superficial fascia. The lymphatic vessels along the great saphenous vein drain into the superficial inguinal lymph nodes; those along the small saphenous vein drain into the popliteal lymph nodes. Lymph from the superficial inguinal nodes drains to the deep inguinal and external iliac nodes. Lymph from the popliteal nodes ascends through deep lymphatic vessels accompanying the deep blood vessels to the deep inguinal nodes. In **B,** Note that the great saphenous vein lies anterior to the medial malleolus and a hand's breadth posterior to the medial aspect of the patella.

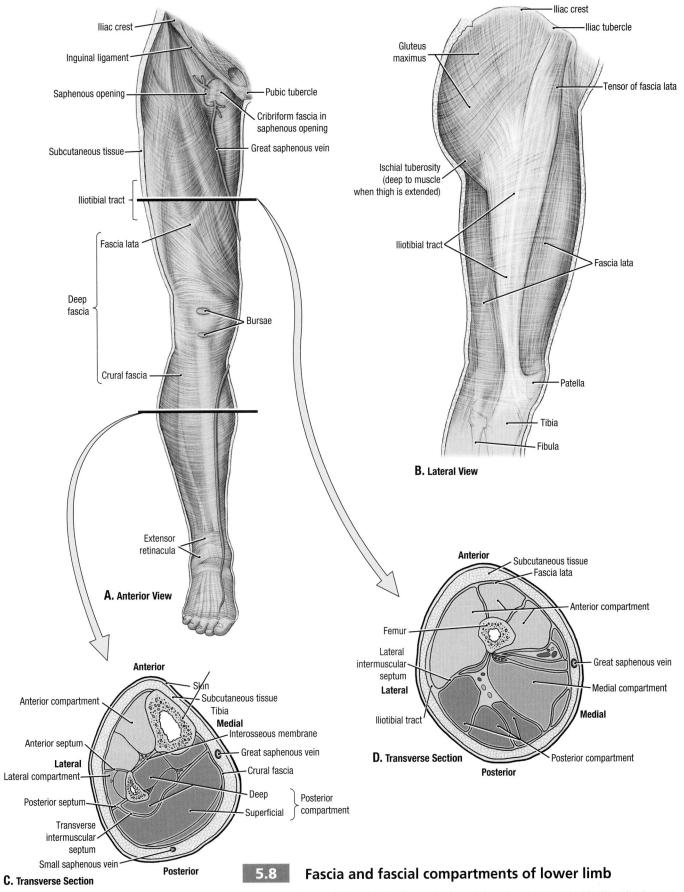

5.8 Fascia and fascial compartments of lower limb

A. Deep fascia of the thigh (fascia lata) and leg (crural fascia). **B.** Iliotibial tract. **C.** Fascial compartments of leg. **D.** Fascial compartments of thigh.

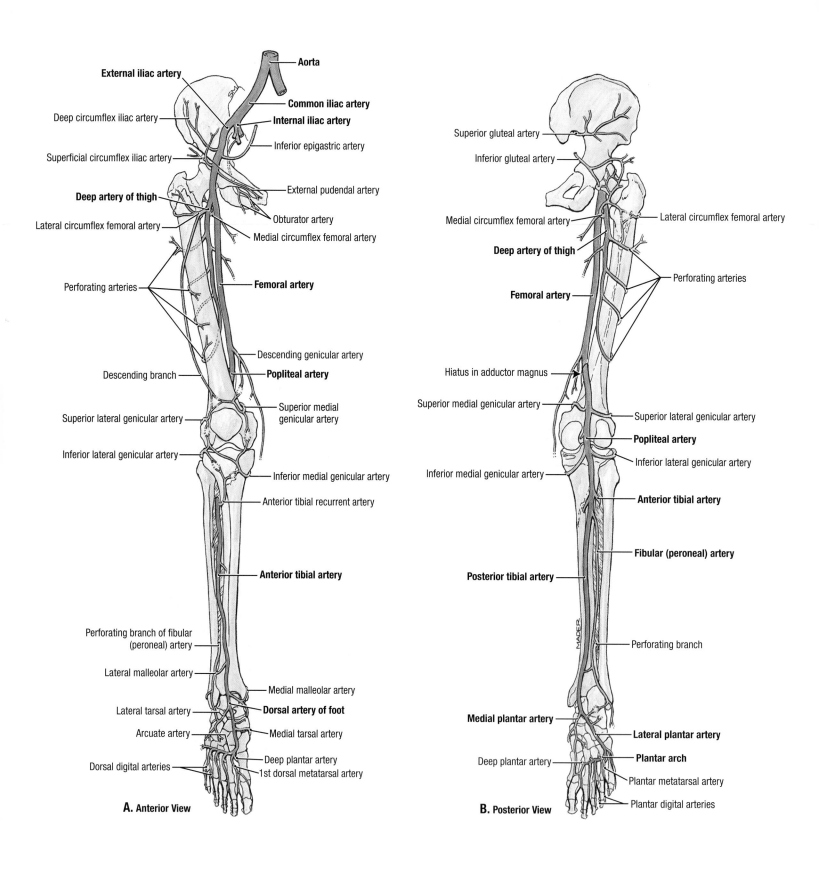

Aorta

External iliac artery

Deep circumflex iliac artery

Common iliac artery

Internal iliac artery

Inferior epigastric artery

Superficial circumflex iliac artery

External pudendal artery

Deep artery of thigh

Obturator artery

Lateral circumflex femoral artery

Medial circumflex femoral artery

Perforating arteries

Femoral artery

Descending genicular artery

Descending branch

Popliteal artery

Superior medial genicular artery

Superior lateral genicular artery

Inferior lateral genicular artery

Inferior medial genicular artery

Anterior tibial recurrent artery

Anterior tibial artery

Perforating branch of fibular (peroneal) artery

Lateral malleolar artery

Medial malleolar artery

Lateral tarsal artery

Dorsal artery of foot

Arcuate artery

Medial tarsal artery

Dorsal digital arteries

Deep plantar artery

1st dorsal metatarsal artery

A. Anterior View

Superior gluteal artery

Inferior gluteal artery

Medial circumflex femoral artery

Lateral circumflex femoral artery

Deep artery of thigh

Perforating arteries

Femoral artery

Hiatus in adductor magnus

Superior medial genicular artery

Superior lateral genicular artery

Popliteal artery

Inferior medial genicular artery

Inferior lateral genicular artery

Anterior tibial artery

Fibular (peroneal) artery

Posterior tibial artery

Perforating branch

Medial plantar artery

Lateral plantar artery

Deep plantar artery

Plantar arch

Plantar metatarsal artery

Plantar digital arteries

B. Posterior View

5.9 **Overview of arteries of lower limb**

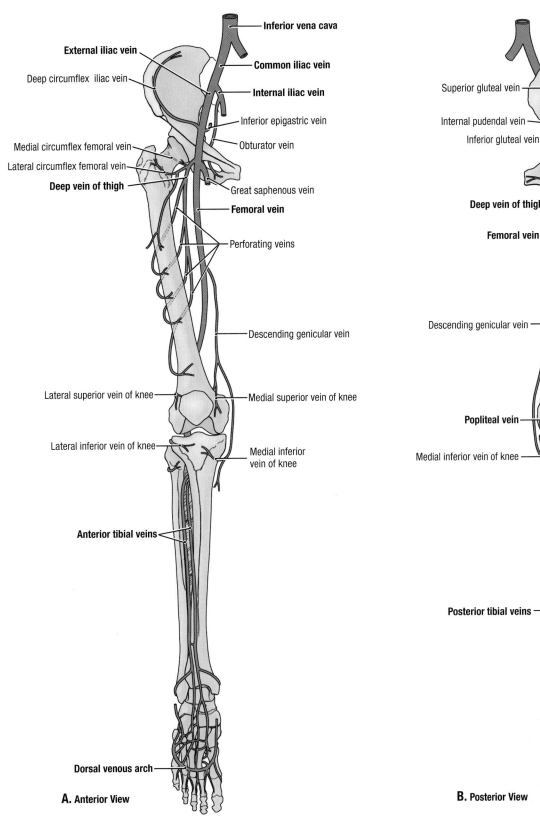
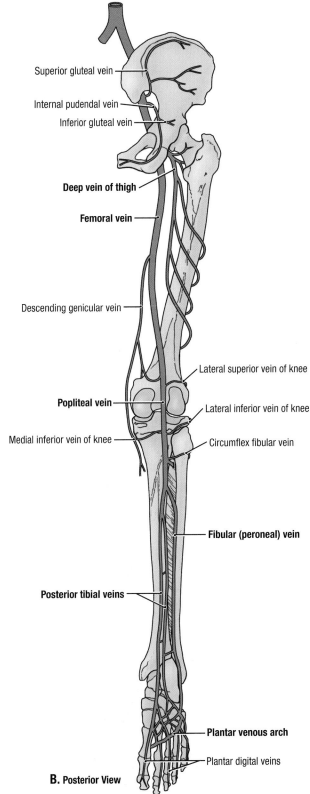

Inferior vena cava

External iliac vein

Deep circumflex iliac vein

Common iliac vein

Internal iliac vein

Inferior epigastric vein

Medial circumflex femoral vein

Obturator vein

Lateral circumflex femoral vein

Deep vein of thigh

Great saphenous vein

Femoral vein

Perforating veins

Descending genicular vein

Lateral superior vein of knee

Medial superior vein of knee

Lateral inferior vein of knee

Medial inferior
vein of knee

Anterior tibial veins

Dorsal venous arch

A. Anterior View

Superior gluteal vein

Internal pudendal vein

Inferior gluteal vein

Deep vein of thigh

Femoral vein

Descending genicular vein

Lateral superior vein of knee

Popliteal vein

Lateral inferior vein of knee

Medial inferior vein of knee

Circumflex fibular vein

Fibular (peroneal) vein

Posterior tibial veins

Plantar venous arch

Plantar digital veins

B. Posterior View

5.10   **Overview of deep veins of lower limb**

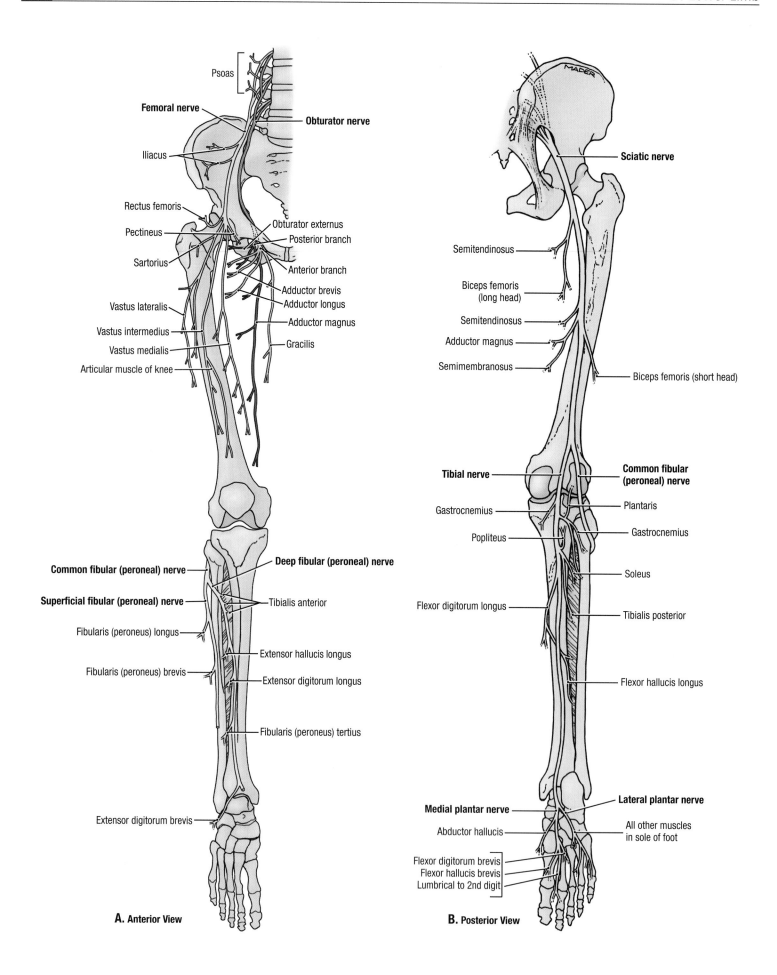

Psoas

Femoral nerve

Obturator nerve

Iliacus

Rectus femoris

Obturator externus

Pectineus

Posterior branch

Sartorius

Anterior branch

Adductor brevis

Adductor longus

Vastus lateralis

Adductor magnus

Vastus intermedius

Gracilis

Vastus medialis

Articular muscle of knee

Sciatic nerve

Semitendinosus

Biceps femoris (long head)

Semitendinosus

Adductor magnus

Semimembranosus

Biceps femoris (short head)

Common fibular (peroneal) nerve

Deep fibular (peroneal) nerve

Superficial fibular (peroneal) nerve

Tibialis anterior

Fibularis (peroneus) longus

Extensor hallucis longus

Fibularis (peroneus) brevis

Extensor digitorum longus

Fibularis (peroneus) tertius

Extensor digitorum brevis

Tibial nerve

Common fibular (peroneal) nerve

Gastrocnemius

Plantaris

Popliteus

Gastrocnemius

Soleus

Flexor digitorum longus

Tibialis posterior

Flexor hallucis longus

Medial plantar nerve

Lateral plantar nerve

Abductor hallucis

All other muscles in sole of foot

Flexor digitorum brevis
Flexor hallucis brevis
Lumbrical to 2nd digit

A. Anterior View

B. Posterior View

TABLE 5.1   MOTOR NERVES OF LOWER LIMB

NERVE	ORIGIN	COURSE	DISTRIBUTION IN LEG
Femoral	Lumbar plexus (L2–L4)	Passes deep to midpoint of inguinal ligament, lateral to femoral vessels, and divides into muscular and cutaneous branches	Supplies anterior thigh muscles, hip and knee joints, and skin on antero-medial side of thigh
Obturator	Lumbar plexus (L2–L4)	Enters thigh through obturator foramen and divides; its anterior branch descends between adductor longus and adductor brevis; its posterior branch descends between adductor brevis and adductor magnus	Anterior branch supplies adductor longus, adductor brevis, gracilis, and pectineus; posterior branch supplies obturator externus and adductor magnus
Sciatic	Sacral plexus (L4–S3)	Enters gluteal region through greater sciatic foramen inferior to piriformis, descends along posterior aspect of thigh, and divides proximal to knee into tibial and common fibular nerves	Innervates hamstrings by its tibial division, except for short head of biceps femoris, which is innervated by its common fibular division; provides articular branches to hip and knee joints
Tibial	Sciatic nerve	Forms as sciatic nerve bifurcates at apex of popliteal fossa; descends through popliteal fossa and lies on popliteus; runs inferiorly on the tibialis posterior with the posterior tibial vessels; terminates beneath the flexor retinaculum by dividing into the medial and lateral plantar nerves	Supplies posterior muscles of leg and knee joint
Common fibular	Sciatic nerve	Forms as sciatic bifurcates at apex of popliteal fossa and follows medial border of biceps femoris and its tendon; passes over posterior aspect of head of fibula and then winds around neck of fibula deep to fibularis longus, where it divides into deep and superficial fibular nerves	Supplies skin on lateral part of posterior aspect of leg via its branch, the lateral sural cutaneous nerve; also supplies knee joint via its articular branch
Superficial fibular	Common fibular nerve	Arises between fibularis longus and neck of fibula and descends in lateral compartment of the leg; pierces deep fascia at distal third of leg to become subcutaneous	Supplies fibularis longus and brevis and skin on distal third of anterior surface of leg and dorsum of foot
Deep fibular	Common fibular nerve	Arises between fibularis longus and neck of fibula; passes through extensor digitorum longus and descends on interosseous membrane; crosses distal end of tibia and enters dorsum of foot	Supplies anterior muscles of leg, dorsum of foot, and skin of first interdigital cleft; sends articular branches to joints it crosses

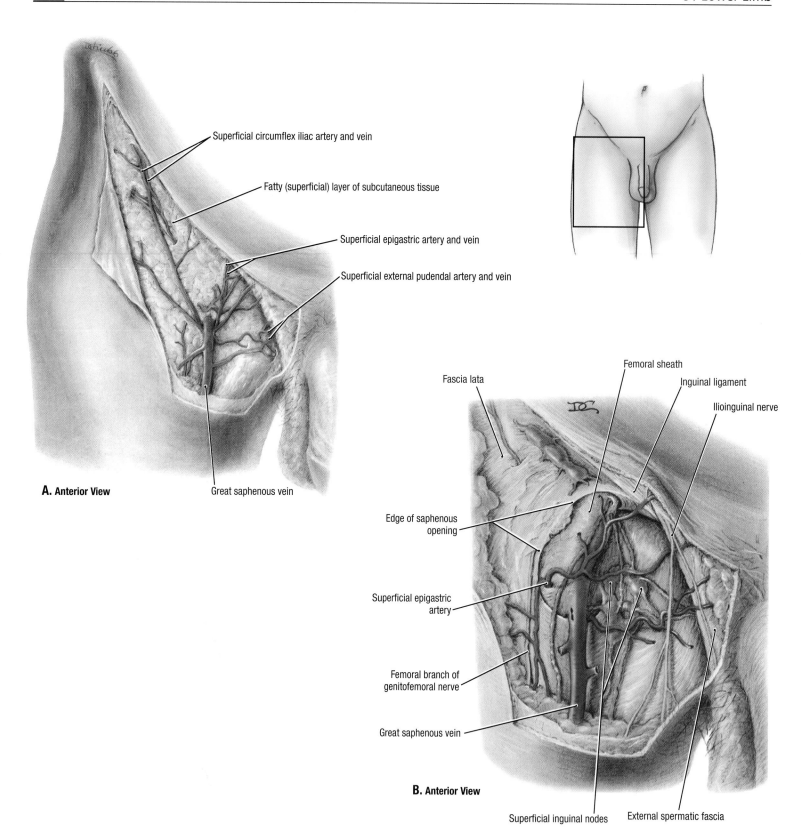

A. Anterior View

Superficial circumflex iliac artery and vein

Fatty (superficial) layer of subcutaneous tissue

Superficial epigastric artery and vein

Superficial external pudendal artery and vein

Great saphenous vein

Fascia lata

Femoral sheath

Inguinal ligament

Ilioinguinal nerve

Edge of saphenous opening

Superficial epigastric artery

Femoral branch of genitofemoral nerve

Great saphenous vein

B. Anterior View

Superficial inguinal nodes

External spermatic fascia

5.11   **Superficial inguinal vessels and saphenous opening**

A. Superficial inguinal vessels. The arteries are branches of the femoral artery, and the veins are tributaries of the great saphenous vein. **B.** Saphenous opening. Observe the oval shape of this opening in the fascia lata. The great saphenous vein passes through the opening to join the femoral vein that is located deeply in the femoral sheath. The opening has sharp superior, lateral, and inferior free margins.

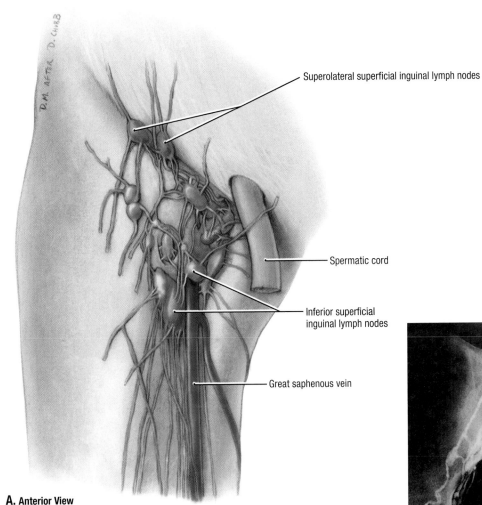

Superolateral superficial inguinal lymph nodes

Spermatic cord

Inferior superficial inguinal lymph nodes

Great saphenous vein

A. Anterior View

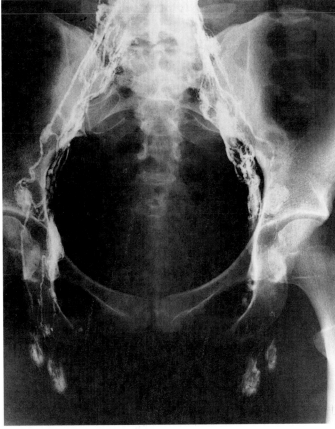

B. Anteroposterior View

5.12   Inguinal lymph nodes

A. Dissection. **B.** Lymphangiogram.

- Observe the arrangement of the nodes: a proximal chain parallel to the inguinal ligament (superolateral superficial inguinal nodes) and a distal chain on the sides of the great saphenous vein (inferior superficial inguinal nodes). Efferent vessels leave these nodes and pass deep to the inguinal ligament to enter the external iliac nodes. Some of the lymphatic vessels traverse the femoral canal, and others ascend alongside the femoral artery and vein, some inside the femoral sheath, and some outside it.
- Note the anastomosis between the lymph vessels.

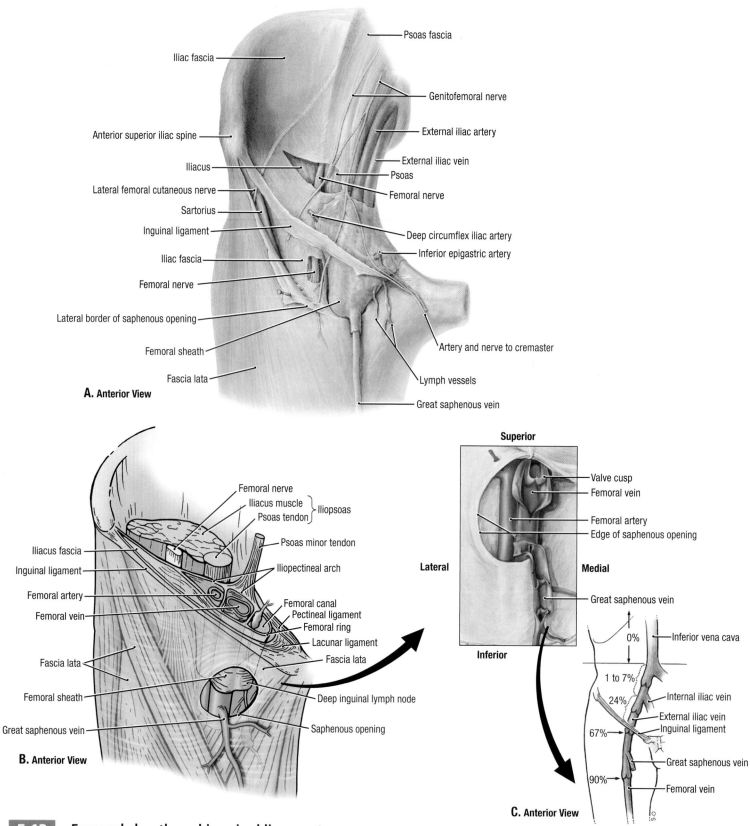

A. Anterior View

B. Anterior View

C. Anterior View

5.13   Femoral sheath and inguinal ligament

A. Dissection. The muscles of the abdominal wall are cut away from the superior border of the inguinal ligament, and the fascia lata is removed from the inferior border of the ligament. **B.** Schematic illustration. The femoral sheath contains the femoral artery, vein, and lymph vessels, but the femoral nerve, lying posterior to the iliacus

fascia, is outside the femoral sheath. **C.** Valves of the proximal part of femoral and great saphenous veins. The valve is usually composed of two cusps and permits blood flow toward the heart, but not in the reverse direction. Percentage incidence of valves between the proximal femoral vein and inferior vena cava is shown.

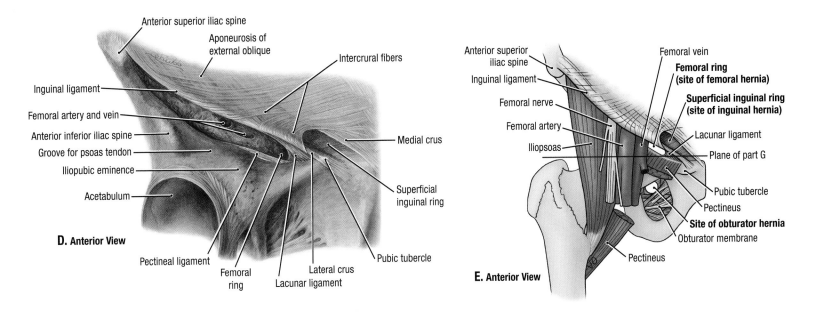

D. Anterior View

Anterior superior iliac spine
Aponeurosis of external oblique
Intercrural fibers
Inguinal ligament
Femoral artery and vein
Anterior inferior iliac spine
Groove for psoas tendon
Iliopubic eminence
Acetabulum
Medial crus
Superficial inguinal ring
Pubic tubercle
Pectineal ligament
Femoral ring
Lacunar ligament
Lateral crus

E. Anterior View

Anterior superior iliac spine
Inguinal ligament
Femoral nerve
Femoral artery
Iliopsoas
Femoral vein
Femoral ring (site of femoral hernia)
Superficial inguinal ring (site of inguinal hernia)
Lacunar ligament
Plane of part G
Pubic tubercle
Pectineus
Site of obturator hernia
Obturator membrane
Pectineus

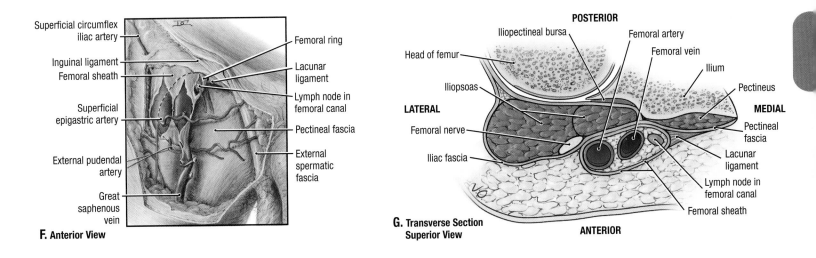

F. Anterior View

Superficial circumflex iliac artery
Inguinal ligament
Femoral sheath
Superficial epigastric artery
External pudendal artery
Great saphenous vein
Femoral ring
Lacunar ligament
Lymph node in femoral canal
Pectineal fascia
External spermatic fascia

G. Transverse Section Superior View

POSTERIOR
Iliopectineal bursa
Head of femur
Iliopsoas
Femoral nerve
Iliac fascia
LATERAL
Femoral artery
Femoral vein
Ilium
Pectineus
MEDIAL
Pectineal fascia
Lacunar ligament
Lymph node in femoral canal
Femoral sheath
ANTERIOR

5.13   Femoral sheath and inguinal ligament (continued)

D. Inguinal ligament and femoral ring. **E.** Three hernia sites: inguinal, femoral, and obturator. **F.** Femoral sheath and femoral ring. The lateral margin of the saphenous opening is cut away. **G.** Transverse section at the level of the hip joint. The plane of section is indicated in **E.**

In **D** and **E**:
- The inguinal ligament is formed by inferior, underturned fibers of the aponeurosis of the external oblique muscle.
- Intercrural fibers from the superolateral margin of the superficial inguinal ring prevent the crura of the ring from spreading anteriorly.
- The fibers of the inguinal ligament attach to the pectineal line as the lacunar and pectineal ligaments. The lacunar ligament forms the medial boundary of the femoral ring, through which a femoral hernia can descend.

- Observe the site of the femoral hernia through the femoral ring, of the obturator hernia through the obturator foramen, and of the inguinal hernias at the superficial inguinal ring.
- The femoral artery and vein lie anterior to the fascia covering the iliopsoas and pectineus muscles, and the femoral nerve lies posterior to the fascia.

In **F** and **G**:
- Observe the three main components of the femoral sheath: the lateral part for the femoral artery; the middle for the femoral vein; and the medial one, the femoral canal, for lymph nodes and lymphatic vessels.
- The proximal end of the femoral canal, called the *femoral ring*, is bounded medially by the lacunar ligament, anteriorly by the inguinal ligament, posteriorly by the pectineus muscle and its fascia, and laterally by the femoral vein.

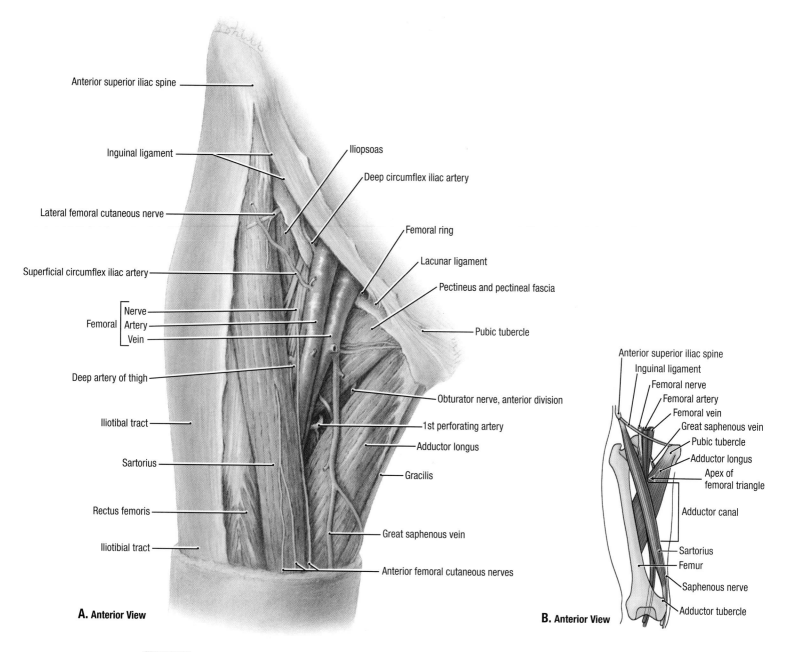

Anterior superior iliac spine

Inguinal ligament

Lateral femoral cutaneous nerve

Superficial circumflex iliac artery

Femoral — Nerve / Artery / Vein

Deep artery of thigh

Iliotibial tract

Sartorius

Rectus femoris

Iliotibial tract

A. Anterior View

Iliopsoas

Deep circumflex iliac artery

Femoral ring

Lacunar ligament

Pectineus and pectineal fascia

Pubic tubercle

Obturator nerve, anterior division

1st perforating artery

Adductor longus

Gracilis

Great saphenous vein

Anterior femoral cutaneous nerves

Anterior superior iliac spine

Inguinal ligament

Femoral nerve

Femoral artery

Femoral vein

Great saphenous vein

Pubic tubercle

Adductor longus

Apex of femoral triangle

Adductor canal

Sartorius

Femur

Saphenous nerve

Adductor tubercle

B. Anterior View

5.14 Femoral triangle

A. Dissection. **B.** Schematic illustration to show the course of the femoral artery.

- Note the boundaries of the triangle: the inguinal ligament, which extends from the anterior superior iliac spine to the pubic tubercle, forming the base; the medial border of the sartorius muscle, forming the lateral side; the medial border of the adductor longus muscle, forming the medial side (some authors regard the lateral border of the adductor longus muscle as the medial side of the triangle); and the point at which the two converging sides meet distally, forming the apex.
- The femoral artery enters the femoral triangle midway between the anterior superior iliac spine and pubic tubercle and exits where the medial border of the sartorius muscle crosses the lateral border of the adductor longus muscle. Pulsation of the femoral artery can be felt just distal to the inguinal ligament, midway between the anterior superior iliac spine and the pubic tubercle.

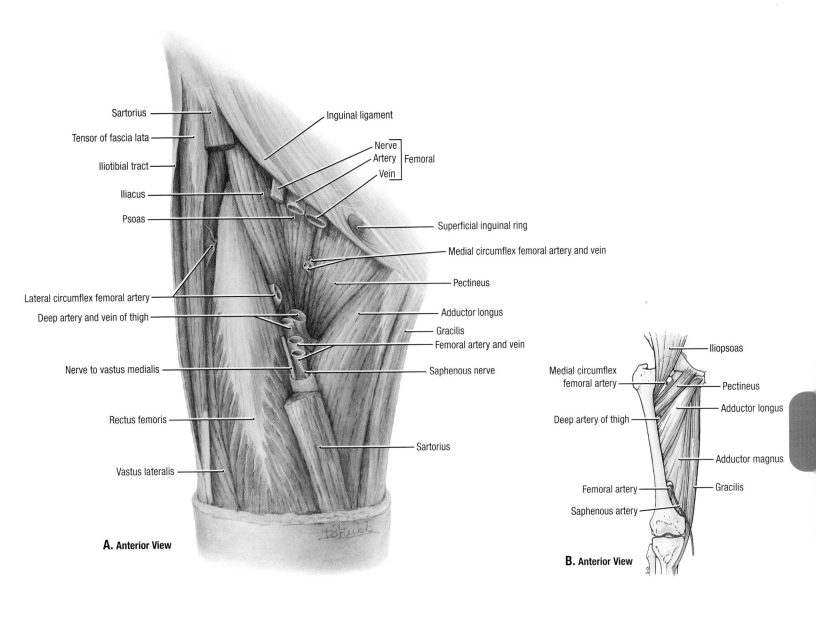

A. Anterior View

- Sartorius
- Tensor of fascia lata
- Iliotibial tract
- Iliacus
- Psoas
- Lateral circumflex femoral artery
- Deep artery and vein of thigh
- Nerve to vastus medialis
- Rectus femoris
- Vastus lateralis

- Inguinal ligament
- Nerve
- Artery — Femoral
- Vein
- Superficial inguinal ring
- Medial circumflex femoral artery and vein
- Pectineus
- Adductor longus
- Gracilis
- Femoral artery and vein
- Saphenous nerve
- Sartorius

B. Anterior View

- Medial circumflex femoral artery
- Deep artery of thigh
- Femoral artery
- Saphenous artery
- Iliopsoas
- Pectineus
- Adductor longus
- Adductor magnus
- Gracilis

5.15   Floor of femoral triangle

A. Dissection. Sections are removed from the sartorius muscle, femoral vessels, and femoral nerve. **B.** Schematic illustration.

- The floor of the femoral triangle is a "trough," with sloping lateral and medial walls; the trough is shallow at the base and deep at the apex.
- At the apex of the femoral triangle, four vessels (the femoral artery and vein and the deep artery and vein of the thigh) and two nerves (the nerve to vastus medialis and the saphenous nerve) pass into the adductor (subsartorial) canal, which is located deep to the sartorius muscle (Fig. 5.14B).

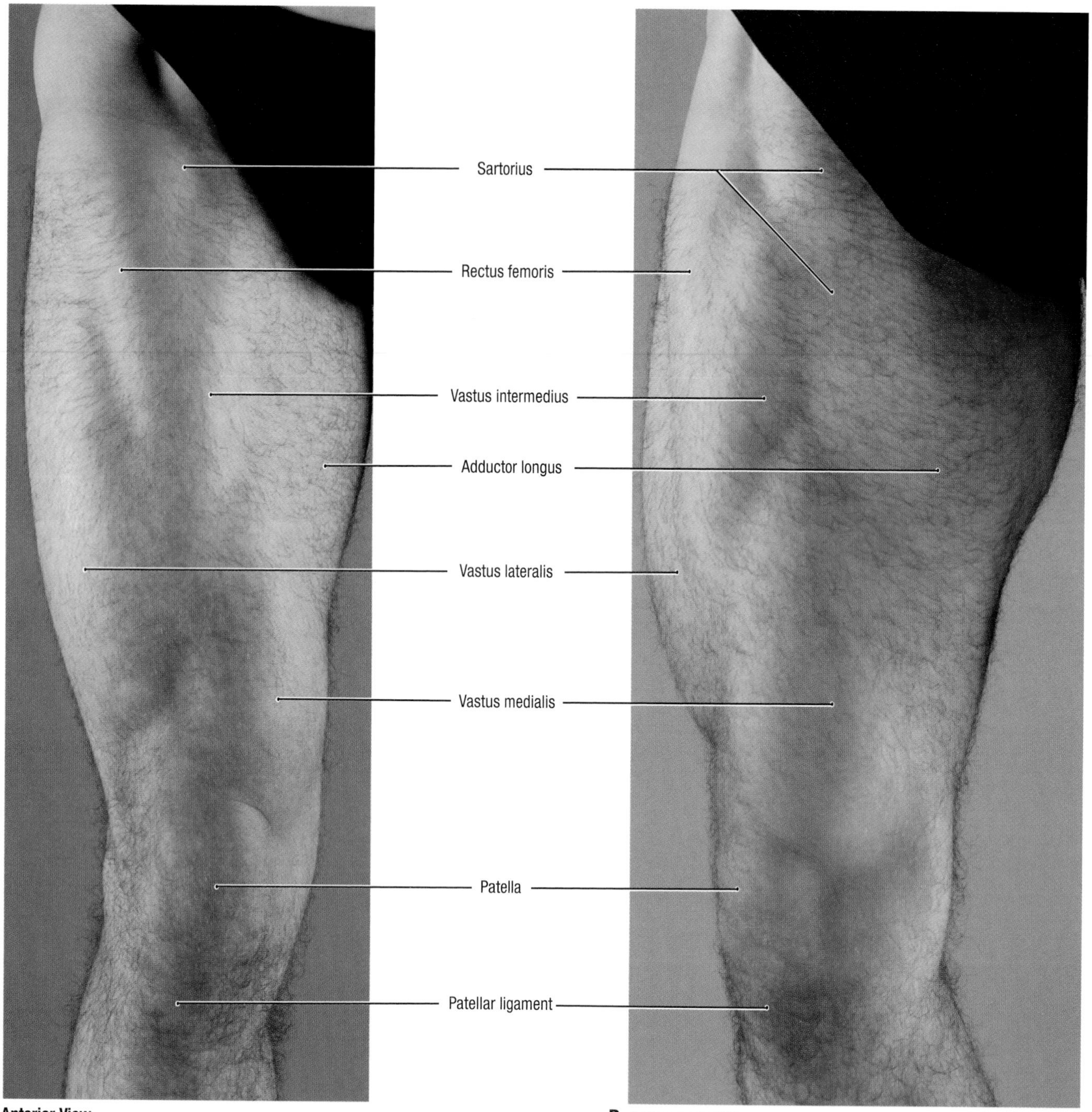

A. Anterior View

B. Anteromedial View

Sartorius

Rectus femoris

Vastus intermedius

Adductor longus

Vastus lateralis

Vastus medialis

Patella

Patellar ligament

5.16   **Surface anatomy of the anterior and medial aspects of the thigh**

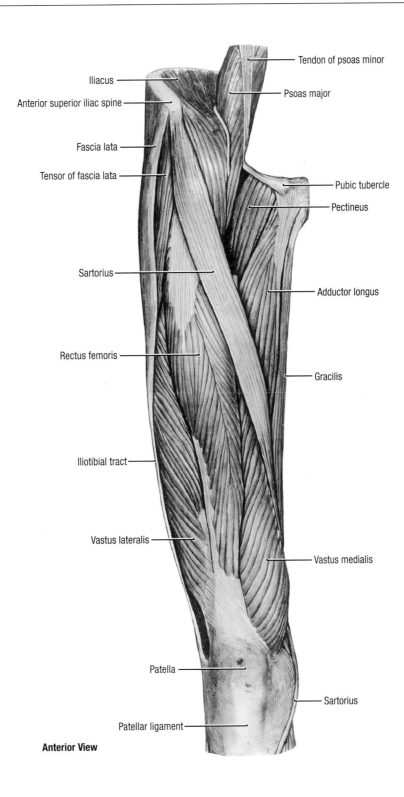

Iliacus

Anterior superior iliac spine

Fascia lata

Tensor of fascia lata

Sartorius

Rectus femoris

Iliotibial tract

Vastus lateralis

Patella

Patellar ligament

Tendon of psoas minor

Psoas major

Pubic tubercle

Pectineus

Adductor longus

Gracilis

Vastus medialis

Sartorius

Anterior View

5.17   Anterior and medial thigh muscles—I

The quadriceps (consisting of the vastus lateralis, medialis, and intermedius, and rectus femoris muscles) insert on the patella and, through the patellar ligament, to the tibial tuberosity. The vastus medialis muscle attaches to the base and proximal two thirds of the medial surface of the patella, and the vastus lateralis attaches mainly to the base of the patella and slightly onto the lateral surface.

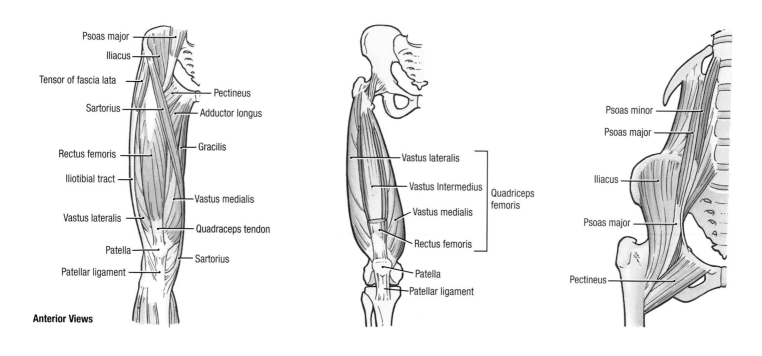

Anterior Views

TABLE 5.2 MUSCLES OF ANTERIOR THIGH

MUSCLE	PROXIMAL ATTACHMENT	DISTAL ATTACHMENT	INNERVATION[a]	MAIN ACTIONS
Iliopsoas				
Psoas major	Sides of T12–L5 vertebrae and discs between them; transverse processes of all lumbar vertebrae	Lesser trochanter of femur	Anterior rami of lumbar nerves (**L1, L2,** and L3)	Act jointly in flexing thigh at hip joint and in stabilizing this joint[b]
Iliacus	Iliac crest, iliac fossa, ala of sacrum and anterior sacroiliac ligaments	Tendon of psoas major, lesser trochanter, and femur distal to it	Femoral nerve (L2 and L3)	
Tensor of fascia lata	Anterior superior iliac spine and anterior part of iliac crest	Iliotibial tract that attaches to lateral condyle of tibia	Superior gluteal (L4 and L5)	Abducts, medially rotates, and flexes thigh; helps to keep knee extended; steadies trunk on thigh
Sartorius	Anterior superior iliac spine and superior part of notch inferior to it	Superior part of medial surface of tibia	Femoral nerve (L2 and L3)	Flexes, abducts, and laterally rotates thigh at hip joint; flexes leg at knee joint[c]
Quadiceps femoris				
Rectus femoris	Anterior inferior iliac spine and ilium superior to acetabulum	Base of patella and by patellar ligament to tibial tuberosity	Femoral nerve (L2, **L3,** and **L4**)	Extends leg at knee joint; rectus femoris also steadies hip joint and helps iliopsoas to flex thigh
Vastus lateralis	Greater trochanter and lateral lip of linea aspera of femur			
Vastus medialis	Intertrochanteric line and medial lip of linea aspera of femur			
Vastus intermedius	Anterior and lateral surfaces of body of femur			

[a]Numbers indicate spinal cord segmental innervation of nerves (e.g., L1, L2, and L3 indicate that nerves supplying psoas major are derived from first three lumbar segments of the spinal cord; boldface type [**L1, L2**] indicates main segmental innervation). Damage to one or more of these spinal cord segments or to motor nerve roots arising from these segments results in paralysis of the muscles concerned.

[b]Psoas major is also a postural muscle that helps control deviation of trunk and is active during standing.

[c]Four actions of sartorius (L. *sartor*, tailor) produce the once common cross-legged sitting position used by tailors—hence the name.

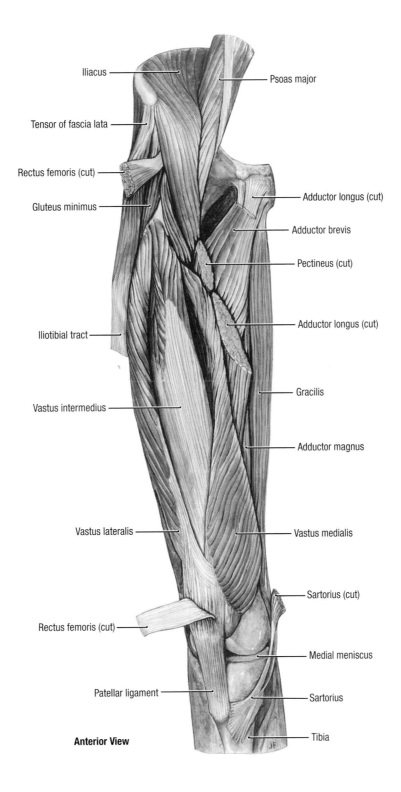

Iliacus

Psoas major

Tensor of fascia lata

Rectus femoris (cut)

Adductor longus (cut)

Gluteus minimus

Adductor brevis

Pectineus (cut)

Adductor longus (cut)

Iliotibial tract

Gracilis

Vastus intermedius

Adductor magnus

Vastus lateralis

Vastus medialis

Sartorius (cut)

Rectus femoris (cut)

Medial meniscus

Patellar ligament

Sartorius

Tibia

Anterior View

5.18   **Medial thigh muscles—II**

Deep dissection of muscles of the anterior and medial aspects of thigh. Note that the central portions of the muscle bellies of the sartorius, rectus femoris, pectineus, and adductor longus muscles have been removed.

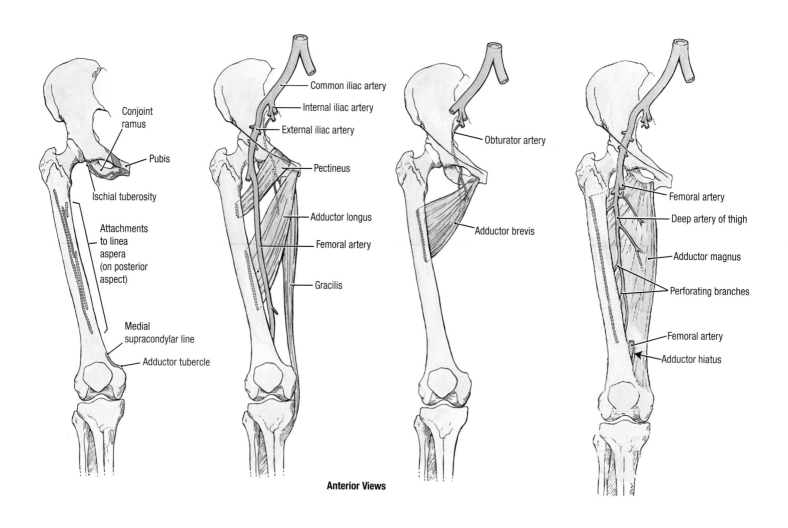

Anterior Views

TABLE 5.3 MUSCLES OF MEDIAL THIGH

MUSCLE	PROXIMAL ATTACHMENT	DISTAL ATTACHMENT[a]	INNERVATION[b]	MAIN ACTIONS
Pectineus	Superior ramus of pubis	Pectineal line of femur, just inferior to lesser trochanter	Femoral nerve (**L2** and L3) may receive a branch from obturator nerve	Adducts and flexes thigh; assists with medial rotation of thigh
Adductor longus	Body of pubis inferior to pubic crest	Middle third of linea aspera of femur	Obturator nerve, anterior branch (L2, **L3**, and L4)	Adducts thigh
Adductor brevis	Body and inferior ramus of pubis	Pectineal line and proximal part of linea aspera of femur	Obturator nerve (L2, **L3**, and L4)	Adducts thigh and, to some extent, flexes it
Adductor magnus	Inferior ramus of pubis, ramus of ischium (adductor part), and ischial tuberosity	Gluteal tuberosity, linea aspera, medial supracondylar line (adductor part), and adductor tubercle of femur (hamstring part)	*Adductor part:* obturator nerve (L2, **L3**, and **L4**) *Hamstring part:* tibial part of sciatic nerve (**L4**)	Adducts thigh; its adductor part also flexes thigh, and its hamstring part extends it
Gracilis	Body and inferior ramus of pubis	Superior part of medial surface of tibia	Obturator nerve (**L2** and L3)	Adducts thigh, flexes leg, and helps rotate it medially
Obturator externus	Margins of obturator foramen and obturator membrane	Trochanteric fossa of femur	Obturator nerve (L3 and **L4**)	Laterally rotates thigh; steadies head of femur in acetabulum

Collectively, the first five muscles listed are the adductors of the thigh, but their actions are more complex (e.g., they act as flexors of the hip joint during flexion of the knee joint and are active during walking).

[a]See Figure 5.20 for muscle attachments.

[b]See Table 5.1 for explanation of segmental innervation.

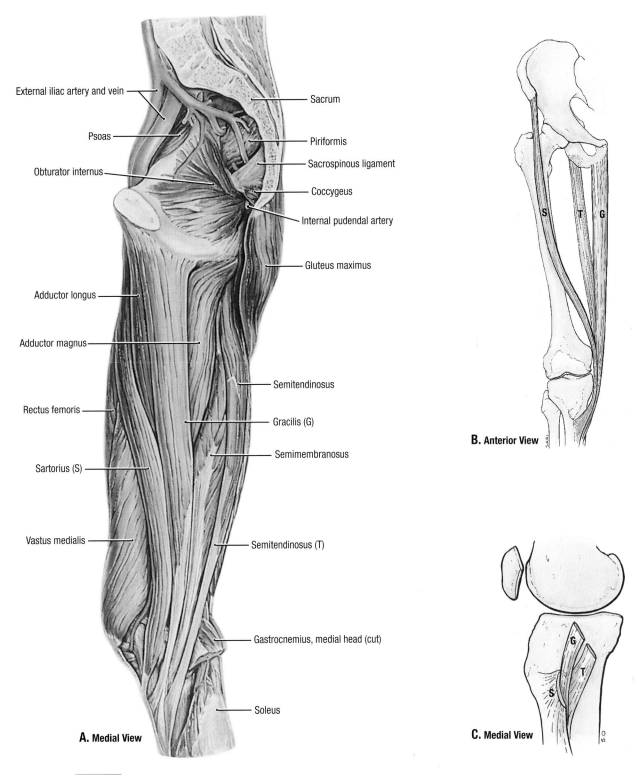

External iliac artery and vein

Psoas

Obturator internus

Adductor longus

Adductor magnus

Rectus femoris

Sartorius (S)

Vastus medialis

Sacrum

Piriformis

Sacrospinous ligament

Coccygeus

Internal pudendal artery

Gluteus maximus

Semitendinosus

Gracilis (G)

Semimembranosus

Semitendinosus (T)

Gastrocnemius, medial head (cut)

Soleus

A. Medial View

B. Anterior View

C. Medial View

5.19   Muscles of medial aspect of thigh

A. Dissection **B.** Muscular tripod. Relationships of sartorius *(S)*, gracilis *(G)*, and semitendinosus *(T)* muscles in the thigh. The muscles forming an inverted tripod arise from three different components of the hip bone, course within three different compartments, perform
three different functions, and are innervated by three different nerves yet share a common
distal attachment. **C.** Distal attachment of sartorius *(S)*, gracilis *(G)*, and semitendinosus *(T)*
muscles. All three tendons become thin and aponeurotic and are collectively referred to as
the *pes anserinus*. The aponeurotic fibers of the sartorius muscle curve posteriorly superior to
the insertion of the gracilis muscle.

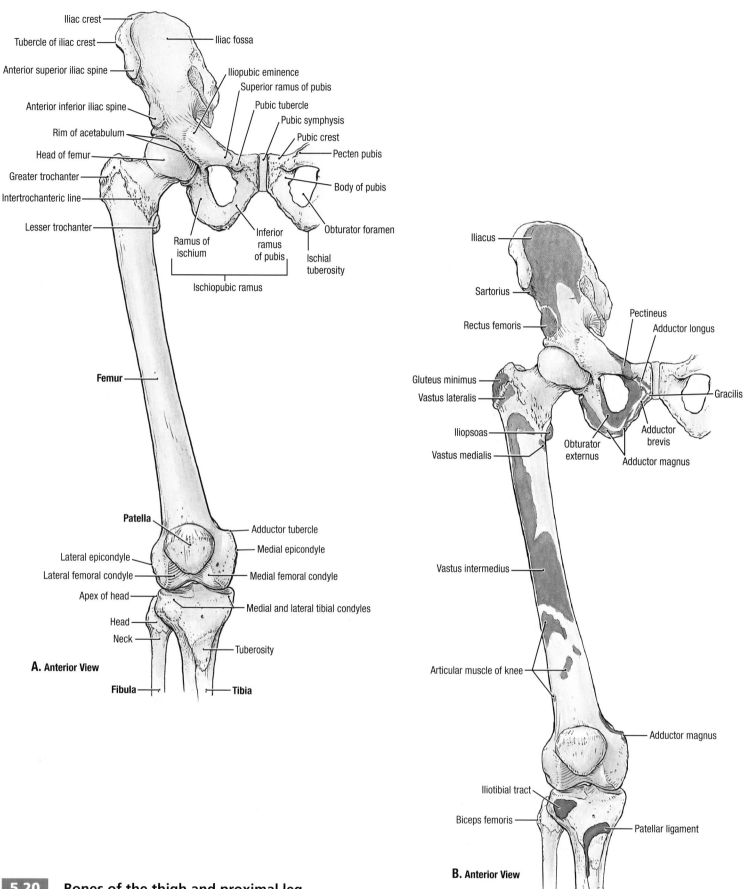

A. Anterior View

Iliac crest
Tubercle of iliac crest
Anterior superior iliac spine
Anterior inferior iliac spine
Rim of acetabulum
Head of femur
Greater trochanter
Intertrochanteric line
Lesser trochanter
Iliac fossa
Iliopubic eminence
Superior ramus of pubis
Pubic tubercle
Pubic symphysis
Pubic crest
Pecten pubis
Body of pubis
Obturator foramen
Ischial tuberosity
Ramus of ischium
Inferior ramus of pubis
Ischiopubic ramus
Femur
Patella
Adductor tubercle
Lateral epicondyle
Medial epicondyle
Lateral femoral condyle
Medial femoral condyle
Apex of head
Medial and lateral tibial condyles
Head
Neck
Tuberosity
Fibula
Tibia

B. Anterior View

Iliacus
Sartorius
Rectus femoris
Gluteus minimus
Vastus lateralis
Iliopsoas
Vastus medialis
Vastus intermedius
Articular muscle of knee
Pectineus
Adductor longus
Gracilis
Adductor brevis
Obturator externus
Adductor magnus
Adductor magnus
Iliotibial tract
Biceps femoris
Patellar ligament

5.20 **Bones of the thigh and proximal leg**

A. Bony features, anterior aspect. **B.** Muscle attachment sites, anterior aspect.
C. Bony features, posterior aspect. **D.** Muscle attachment sites, posterior aspect.

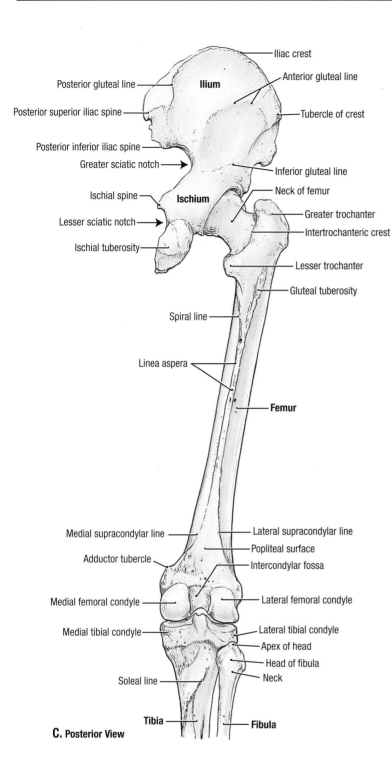

C. Posterior View

Iliac crest

Posterior gluteal line

Ilium

Anterior gluteal line

Posterior superior iliac spine

Tubercle of crest

Posterior inferior iliac spine

Greater sciatic notch

Inferior gluteal line

Ischial spine

Ischium

Neck of femur

Lesser sciatic notch

Greater trochanter

Ischial tuberosity

Intertrochanteric crest

Lesser trochanter

Gluteal tuberosity

Spiral line

Linea aspera

Femur

Medial supracondylar line

Lateral supracondylar line

Adductor tubercle

Popliteal surface

Intercondylar fossa

Medial femoral condyle

Lateral femoral condyle

Medial tibial condyle

Lateral tibial condyle

Apex of head

Head of fibula

Neck

Soleal line

Tibia

Fibula

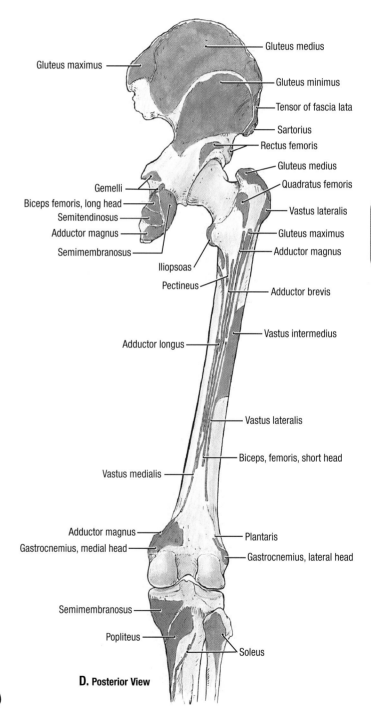

D. Posterior View

Gluteus maximus

Gluteus medius

Gluteus minimus

Tensor of fascia lata

Sartorius

Rectus femoris

Gluteus medius

Gemelli

Quadratus femoris

Biceps femoris, long head

Vastus lateralis

Semitendinosus

Gluteus maximus

Adductor magnus

Adductor magnus

Semimembranosus

Iliopsoas

Adductor brevis

Pectineus

Vastus intermedius

Adductor longus

Vastus lateralis

Biceps, femoris, short head

Vastus medialis

Adductor magnus

Plantaris

Gastrocnemius, medial head

Gastrocnemius, lateral head

Semimembranosus

Popliteus

Soleus

5.20   **Bones of the thigh and proximal leg** *(continued)*

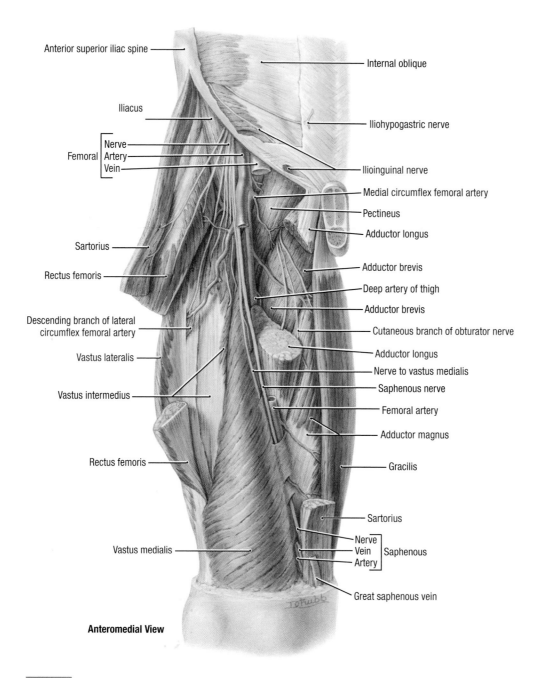

Anterior superior iliac spine

Iliacus

Femoral { Nerve / Artery / Vein

Sartorius

Rectus femoris

Descending branch of lateral
circumflex femoral artery

Vastus lateralis

Vastus intermedius

Rectus femoris

Vastus medialis

Internal oblique

Iliohypogastric nerve

Ilioinguinal nerve

Medial circumflex femoral artery

Pectineus

Adductor longus

Adductor brevis

Deep artery of thigh

Adductor brevis

Cutaneous branch of obturator nerve

Adductor longus

Nerve to vastus medialis

Saphenous nerve

Femoral artery

Adductor magnus

Gracilis

Sartorius

Saphenous { Nerve / Vein / Artery

Great saphenous vein

Anteromedial View

5.21 **Anteromedial aspect of thigh**

- The limb is rotated laterally.
- The femoral nerve breaks up into several nerves on entering the thigh.
- The femoral artery lies between two motor territories: that of the obturator nerve, which is medial, and that of the femoral nerve, which is lateral. No motor nerve crosses anterior to the femoral artery, but the twig to the pectineus muscle crosses posterior to the femoral artery.
- The nerve to the vastus medialis muscle and the saphenous nerve accompany the femoral artery into the adductor canal. The saphenous nerve and artery and their anastomotic accompanying vein emerge from the canal distally between the sartorius and gracilis muscles.
- The deep artery of the thigh arises approximately 4 cm inferior to the inguinal ligament, lies posterior to the femoral artery, and disappears posterior to the adductor longus muscle. It supplies the thigh through the medial and lateral circumflex femoral branches and the perforating arteries that pass through the adductor magnus muscle on their way to the posterior aspect of the thigh. Both the femoral and lateral femoral circumflex arteries have descending branches that contribute to the anastomoses around the knee.

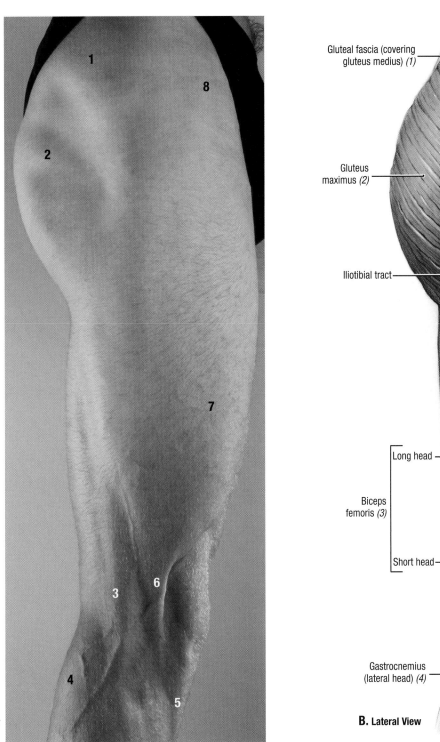

A. Lateral View

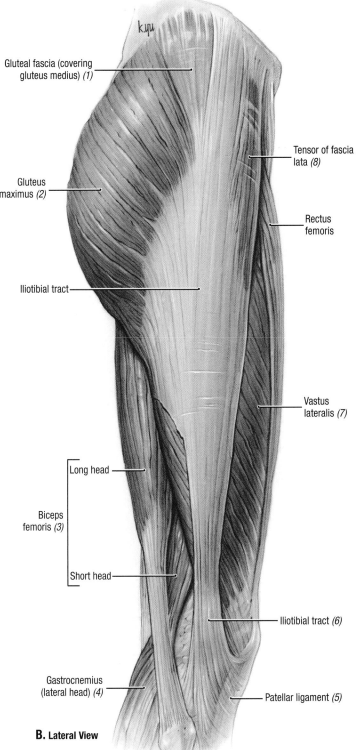

Gluteal fascia (covering gluteus medius) (1)

Gluteus maximus (2)

Iliotibial tract

Biceps femoris (3)

Long head

Short head

Gastrocnemius (lateral head) (4)

Tensor of fascia lata (8)

Rectus femoris

Vastus lateralis (7)

Iliotibial tract (6)

Patellar ligament (5)

B. Lateral View

5.22   **Lateral aspect of thigh**

A. Surface anatomy (*numbers* refer to structures in **B**). **B.** Dissection showing the iliotibial tract. The posterior edge of the iliotibial tract (a thickening of the fascia lata, which serves as a tendon for the gluteus maximus and tensor of fascia lata) attaches to the lateral condyle of the tibia and the biceps femoris tendon attaches on the head of the fibula.

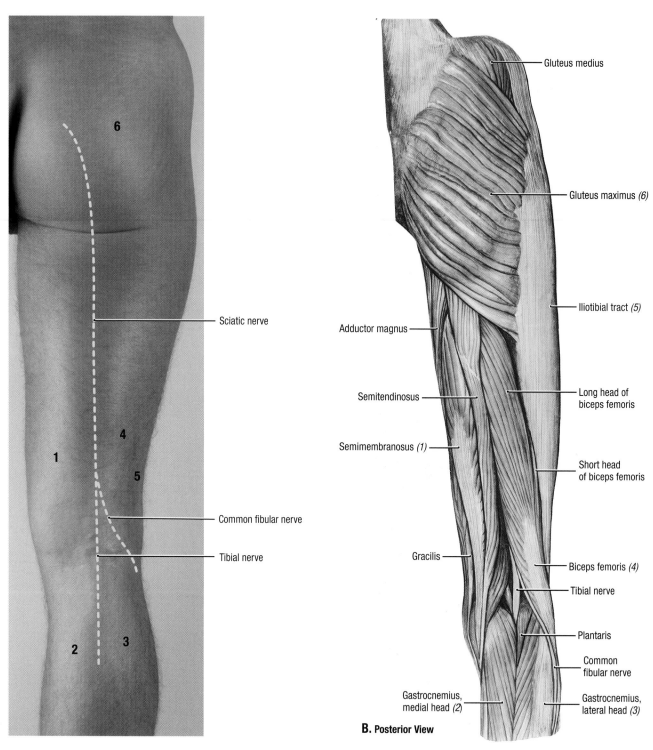

A. Posterior View

6

Sciatic nerve

1

4

5

Common fibular nerve

Tibial nerve

2

3

Gluteus medius

Gluteus maximus (6)

Iliotibial tract (5)

Adductor magnus

Semitendinosus

Long head of biceps femoris

Semimembranosus (1)

Short head of biceps femoris

Gracilis

Biceps femoris (4)

Tibial nerve

Plantaris

Common fibular nerve

Gastrocnemius, medial head (2)

Gastrocnemius, lateral head (3)

B. Posterior View

5.23 **Muscles of the gluteal region and posterior aspect of thigh—I**

A. Surface anatomy (*numbers* refer to structures in **B**). **B.** Superficial dissection of muscles of gluteal region and posterior thigh.

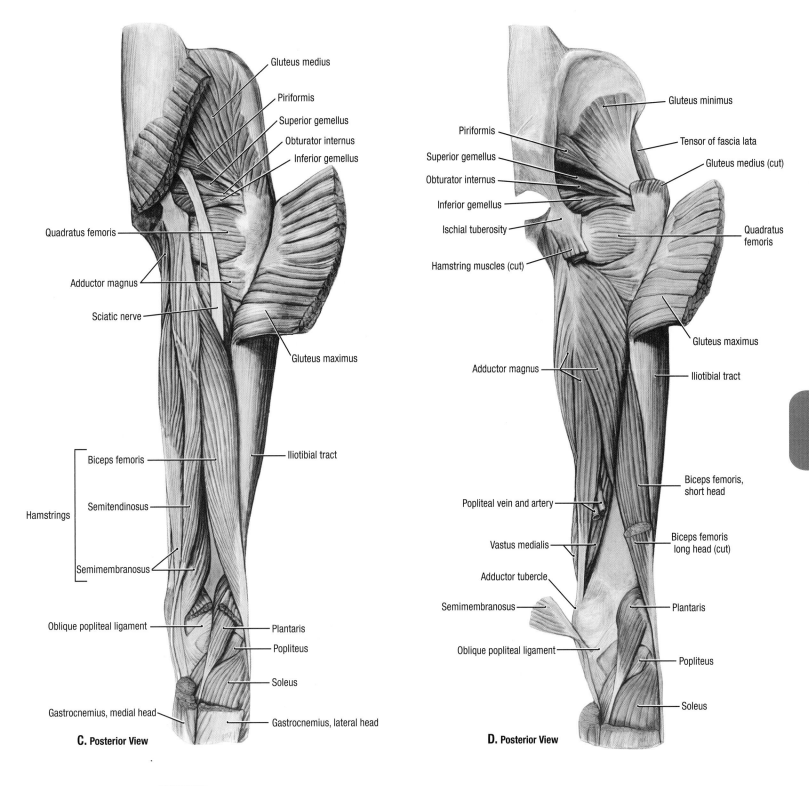

C. Posterior View

D. Posterior View

5.23 **Muscles of gluteal region and posterior aspect of thigh—II and III**

C. Muscles of gluteal region and posterior thigh with gluteus maximus reflected. **D.** Adductor magnus muscle. The adductor magnus is a large muscle with two parts: one belongs to the adductor group and the other to the hamstring group. The adductor part is innervated by the obturator nerve and originates from the inferior ramus of the ischium and pubis (conjoint ramus); it inserts on the linea aspera and medial supracondylar line of the femur (see Fig. 5.20). The hamstring part originates from the ischial tuberosity and inserts via a palpable tendon into the adductor tubercle; the hamstring part is innervated by the tibial portion of the sciatic nerve.

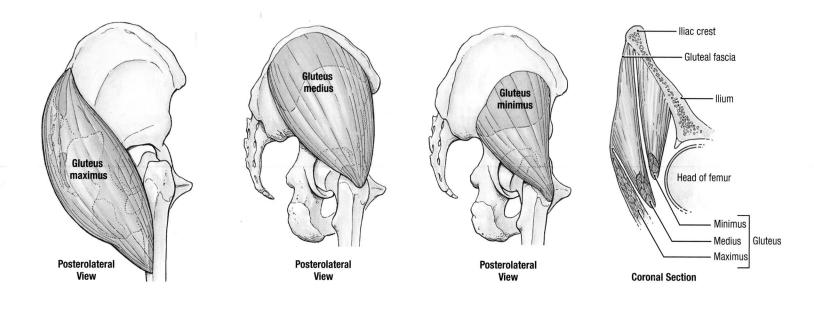

Posterolateral View · Posterolateral View · Posterolateral View · Coronal Section

Labels (Coronal Section): Iliac crest, Gluteal fascia, Ilium, Head of femur, Minimus / Medius / Maximus — Gluteus

Labels: Gluteus maximus, Gluteus medius, Gluteus minimus

TABLE 5.4 MUSCLES OF GLUTEAL REGION

MUSCLE	PROXIMAL ATTACHMENT[a]	DISTAL ATTACHMENT[a]	INNERVATION[b]	MAIN ACTIONS
Gluteus maximus	Ilium posterior to posterior gluteal line, dorsal surface of sacrum and coccyx, and sacrotuberous ligament	Most fibers end in iliotibial tract that inserts into lateral condyle of tibia; some fibers insert on gluteal tuberosity of femur	Inferior gluteal nerve (L5, **S1**, and **S2**)	Extends thigh and assists in its lateral rotation; steadies thigh and assists in raising trunk from flexed position
Gluteus medius	External surface of ilium between anterior and posterior gluteal lines	Lateral surface of greater trochanter of femur	Superior gluteal nerve (**L5** and S1)	Abduct and medially rotate thigh; steady pelvis on leg when opposite leg is raised
Gluteus minimus	External surface of ilium between anterior and inferior gluteal lines	Anterior surface of greater trochanter of femur		
Piriformis	Anterior surface of sacrum and sacrotuberous ligament	Superior border of greater trochanter of femur	Branches of anterior rami of S1 and S2	
Obturator internus	Pelvic surface of obturator membrane and surrounding bones	Medial surface of greater trochanter of femur[c]	Nerve to obturator internus (L5 and S1); *Superior gemellus:* same nerve supply as obturator internus; *Inferior gemellus:* same nerve supply as quadratus femoris	Laterally rotate extended thigh and abduct flexed thigh; steady femoral head in acetabulum
Gemelli, superior and inferior	Superior gemellus from ischial spine; inferior gemellus from ischial tuberosity			
Quadratus femoris	Lateral border of ischial tuberosity	Quadrate tubercle on intertrochanteric crest of femur and inferior to it	Nerve to quadratus femoris (L5 and S1)	Laterally rotates thigh,[d] steadies femoral head in acetabulum

[a]See Figure 5.20 for muscle attachments.

[b]See Table 5.1 for explanation of segmental innervation.

[c]Gemelli muscles blend with the tendon of obturator internus muscle as the tendon attaches to greater trochanter of femur.

[d]There are six lateral rotators of the thigh: piriformis, obturator internus, gemelli (superior and inferior), quadratus femoris, and obturator externus. These muscles also stabilize the hip joint.

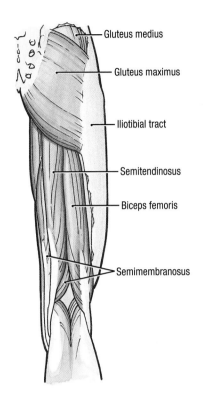

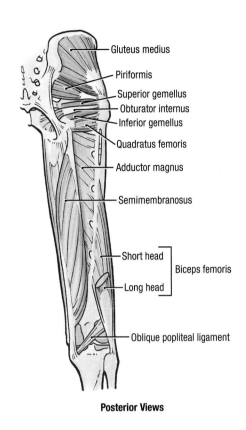

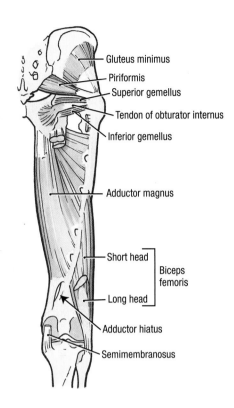

Posterior Views

TABLE 5.5   MUSCLES OF POSTERIOR THIGH (HAMSTRING)

MUSCLE[a]	PROXIMAL ATTACHMENT[a]	DISTAL ATTACHMENT[a]	INNERVATION[b]	MAIN ACTIONS
Semitendinosus	Ischial tuberosity	Medial surface of superior part of tibia	Tibial division of sciatic nerve (**L5**, **S1**, and S2)	Extend thigh; flex leg and rotate it medially; when thigh and leg are flexed, they can extend trunk
Semimembranosus		Posterior part of medial condyle of tibia; reflected attachment forms oblique popliteal ligament to lateral femoral condyle		
Biceps femoris	Long head: ischial tuberosity; Short head: linea aspera and lateral supracondylar line of femur	Lateral side of head of fibula; tendon is split at this site by fibular collateral ligament of knee	*Long head*: tibial division of sciatic nerve (L5, **S1**, and S2); *Short head*: common fibular (peroneal) division of sciatic nerve (L5, **S1**, and S2)	Flexes leg and rotates it laterally; extends thigh (e.g., when starting to walk)

[a]See Figure 5.20 for muscle attachments.

[b]See Table 5.1 for explanation of segmental innervation.

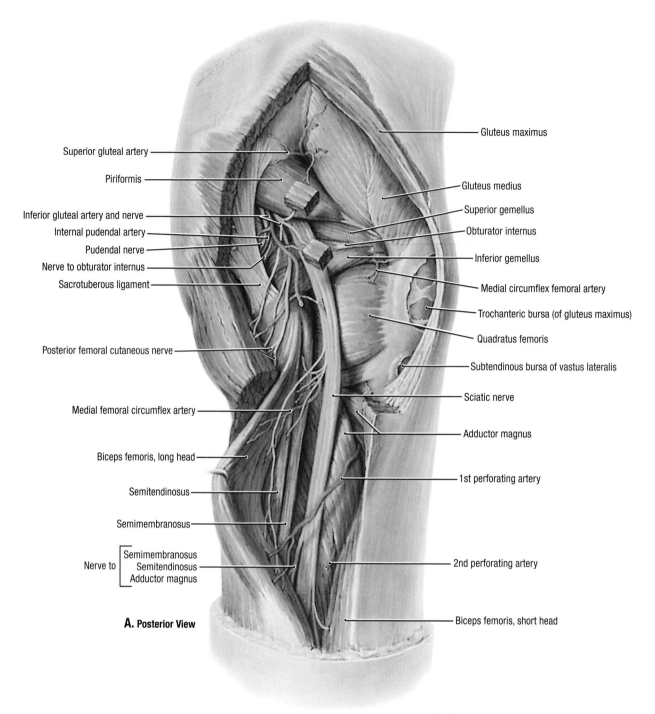

Superior gluteal artery

Piriformis

Inferior gluteal artery and nerve

Internal pudendal artery

Pudendal nerve

Nerve to obturator internus

Sacrotuberous ligament

Posterior femoral cutaneous nerve

Medial femoral circumflex artery

Biceps femoris, long head

Semitendinosus

Semimembranosus

Nerve to [Semimembranosus / Semitendinosus / Adductor magnus]

Gluteus maximus

Gluteus medius

Superior gemellus

Obturator internus

Inferior gemellus

Medial circumflex femoral artery

Trochanteric bursa (of gluteus maximus)

Quadratus femoris

Subtendinous bursa of vastus lateralis

Sciatic nerve

Adductor magnus

1st perforating artery

2nd perforating artery

Biceps femoris, short head

A. Posterior View

5.24 **Muscles of gluteal region and posterior aspect of thigh—IV**

A. Dissection. The gluteus maximus muscle is split superiorly and inferiorly, and the middle part is excised; two cubes remain to identify its nerve. **B.** Intragluteal injection. Injections can be made safely only into the superolateral part of the buttock, avoiding injury to the sciatic and gluteal nerves.

In **A:**

- Observe that the gluteus maximus is the only muscle to cover the greater trochanter; it is aponeurotic and has underlying bursae where it glides on the trochanter and the aponeurosis of the vastus lateralis muscle.
- The inferior gluteal nerve enters the gluteus maximus muscle by two chief branches near its center.
- Branches arise from the medial side of the sciatic nerve at variable levels to supply the hamstrings and part of the adductor magnus muscle; only the branch to the biceps muscle (short head) arises from its lateral side.

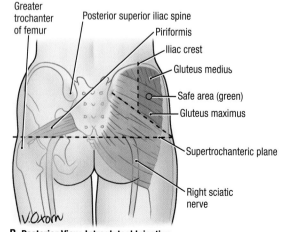

Greater trochanter of femur

Posterior superior iliac spine

Piriformis

Iliac crest

Gluteus medius

Safe area (green)

Gluteus maximus

Supertrochanteric plane

Right sciatic nerve

B. Posterior View, Intragluteal Injection

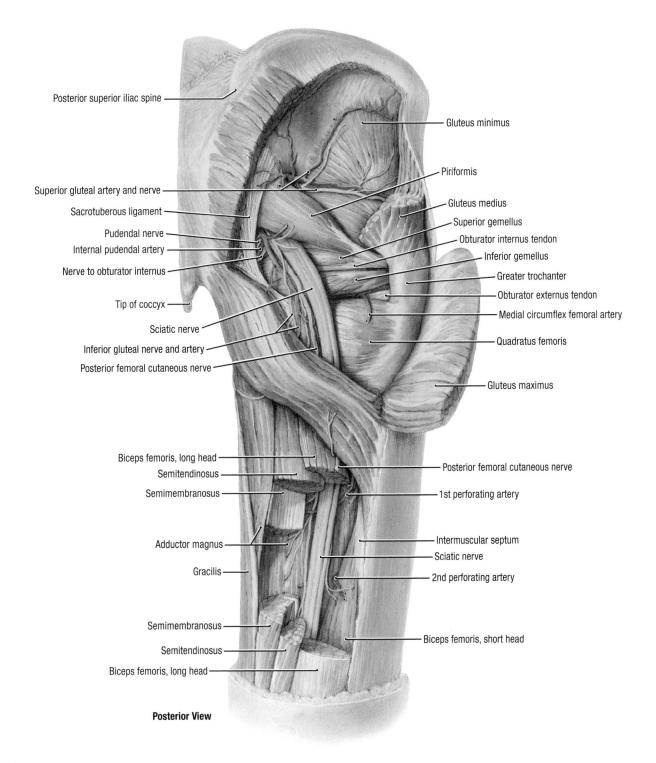

Posterior superior iliac spine

Superior gluteal artery and nerve

Sacrotuberous ligament

Pudendal nerve

Internal pudendal artery

Nerve to obturator internus

Tip of coccyx

Sciatic nerve

Inferior gluteal nerve and artery

Posterior femoral cutaneous nerve

Biceps femoris, long head

Semitendinosus

Semimembranosus

Adductor magnus

Gracilis

Semimembranosus

Semitendinosus

Biceps femoris, long head

Gluteus minimus

Piriformis

Gluteus medius

Superior gemellus

Obturator internus tendon

Inferior gemellus

Greater trochanter

Obturator externus tendon

Medial circumflex femoral artery

Quadratus femoris

Gluteus maximus

Posterior femoral cutaneous nerve

1st perforating artery

Intermuscular septum

Sciatic nerve

2nd perforating artery

Biceps femoris, short head

Posterior View

5.25   **Muscles of gluteal region and posterior aspect of thigh—V**

The proximal three quarters of the gluteus maximus muscle is reflected, and parts of the gluteus medius and the three hamstring muscles are excised.

- The superior gluteal vessels and nerves appear superior to the piriformis muscle; all other vessels and nerves appear inferior to it.
- The superior gluteal artery divides into superficial and deep branches; the superficial branch supplies the gluteus maximus muscle, and the deep branch divides into a superior branch that

anastomoses with arteries of the region and an inferior branch that supplies the gluteus medius and minimus muscles.

- The inferior gluteal artery supplies the buttock, proximal part of the thigh, and sciatic nerve.
- The gluteus maximus muscle consists of bundles of parallel fibers. It is rhomboidal, and the deep fascia covering it is thin.
- The gluteus medius muscle arises in part from the covering deep fascia, which is strong and thick.

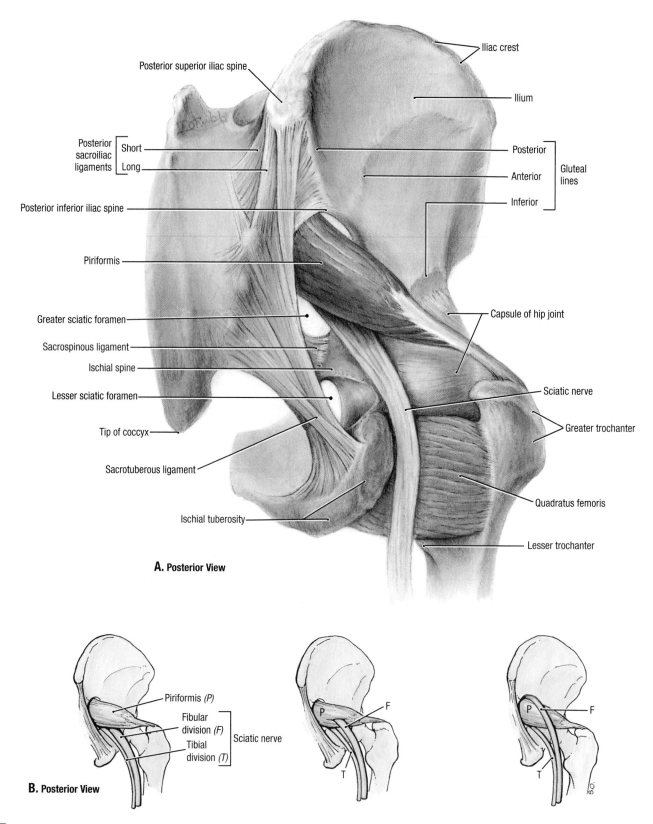

A. Posterior View

B. Posterior View

5.26 **Piriformis, sciatic nerve, and ligaments of gluteal region**

A. Dissection. Note that the tip of the coccyx lies superior to the level of the ischial tuberosity and inferior to that of the ischial spine and that the lateral border of the sciatic nerve lies midway between the lateral surface of the greater trochanter and the medial surface of the ischial tuberosity, provided the body is in anatomical position. **B.** Relationship of sciatic nerve to piriformis muscle. Of 640 limbs studied in Dr. Grant's laboratory, in 87%, the tibial and fibular (peroneal) divisions passed inferior to the piriformis (left); in 12.2%, the fibular (peroneal) division passed through the piriformis (center); and in 0.5% the fibular (peroneal) division passed superior to the piriformis (right).

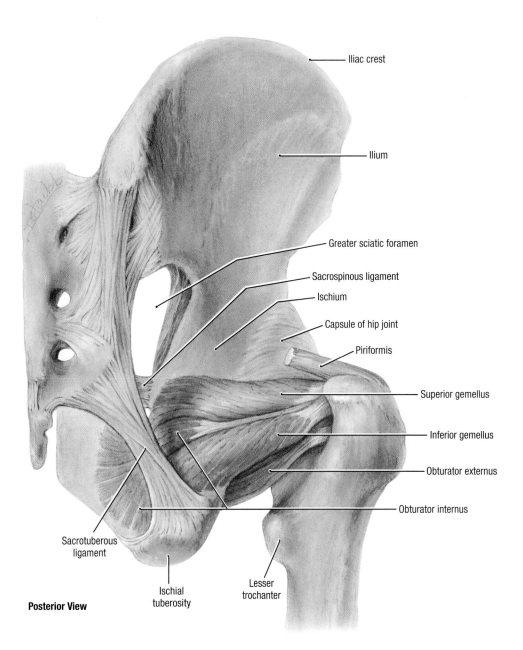

Iliac crest

Ilium

Greater sciatic foramen

Sacrospinous ligament

Ischium

Capsule of hip joint

Piriformis

Superior gemellus

Inferior gemellus

Obturator externus

Obturator internus

Sacrotuberous ligament

Ischial tuberosity

Lesser trochanter

Posterior View

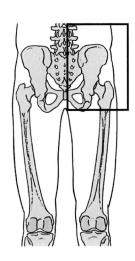

5.27   Obturator internus and externus muscles

- The obturator internus and gemelli muscles fill the gap between the piriformis muscle superiorly and the quadratus femoris muscle inferiorly.
- The obturator externus muscle passes obliquely, inferior to the neck of femur, to its distal attachment.
- The inferior aspect of the ischial tuberosity is on the level of the lesser trochanter.

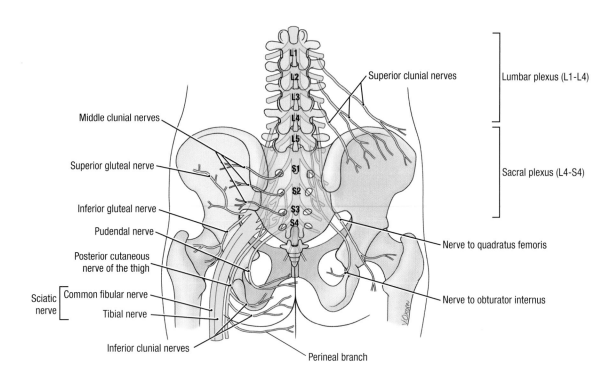

TABLE 5.6   NERVES OF GLUTEAL REGION

NERVE	ORIGIN	COURSE	DISTRIBUTION IN GLUTEAL REGION
Clunial (superior, middle, and inferior)	Superior: posterior rami of L1–L3 nerves Middle: posterior rami of S1–S3 nerves Inferior: posterior cutaneous nerve of thigh (anterior rami of S2–S3)	Superior nerves cross iliac crest; middle nerves exit through posterior sacral foramina and enter gluteal region; inferior nerves curve around inferior border of gluteus maximus	Supplies skin of buttock or gluteal region as far as greater trochanter
Sciatic	Sacral plexus (L4–S3)	Leaves pelvis through greater sciatic foramen inferior to piriformis and enters gluteal region	Supplies no muscles in gluteal region
Posterior cutaneous nerve of thigh	Sacral plexus (S1–S3)	Leaves pelvis through greater sciatic foramen inferior to piriformis, runs deep to gluteus maximus, and emerges from its inferior border	Supplies skin of buttock through inferior clunial branches and skin over posterior aspect of thigh and calf; lateral perineum, upper medial thigh via perineal branch
Superior gluteal	Anterior rami of L4–S1 nerves	Leaves pelvis through greater sciatic foramen superior to piriformis and runs between gluteus medius and minimus	Innervates gluteus medius, gluteus minimus, and tensor of fascia lata
Inferior gluteal	Anterior rami of L5–S2 nerves	Leaves pelvis through greater sciatic foramen inferior to piriformis and divides into several branches	Supplies gluteus maximus
Nerve to quadratus femoris	Anterior rami of L4, L5, and S1 nerves	Leaves pelvis through greater sciatic foramen deep to sciatic nerve	Innervates hip joint, inferior gemellus, and quadratus femoris
Pudendal	Anterior rami of S2–S4 nerves	Enters gluteal region through greater sciatic foramen inferior to piriformis; descends posterior to sacrospinous ligament; enters perineum through lesser sciatic foramen	Supplies most of the innervation to the perineum; supplies no structures in gluteal region
Nerve to obturator internus	Anterior rami of L5, S1, and S2 nerves	Enters gluteal region through greater sciatic foramen inferior to piriformis; descends posterior to ischial spine; enters lesser sciatic foramen and passes to obturator internus	Supplies superior gemellus and obturator internus

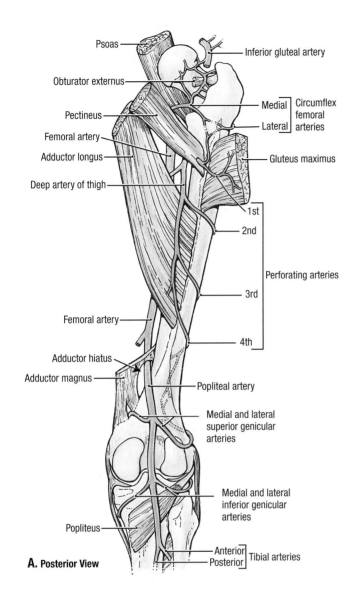

A. Posterior View

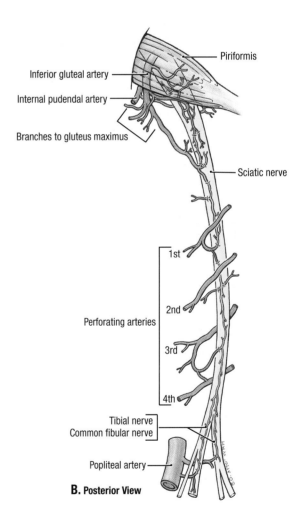

B. Posterior View

5.28 **Blood supply of posterior aspect of the thigh**

A. Deep artery of thigh and perforating branches. **B.** Blood supply of the sciatic nerve. Observe the continuous anastomotic chain of arteries.

Anterior superior iliac spine

Anterior inferior iliac spine

Rectus femoris

Iliofemoral ligament

Greater trochanter

Intertrochanteric line

Lesser trochanter

A. Anterior View

Acetabular labrum

Head of femur

Pectineus

Pectineal fascia

Pectineal ligament

Pubic tubercle

Anterior branch ⎤
⎥ Obturator nerve
Posterior branch ⎦

Obturator externus

Piriformis

Obturator internus and gemelli

Gluteus minimus

Vastus lateralis

Fovea (pit) for ligament of head

Iliofemoral ligament

Iliopsoas

B. Anterior View

5.29 **Hip joint**

A. Iliofemoral ligament. **B.** Muscle attachments of anterior aspect of the proximal femur.
In **A:**

- The head of the femur is exposed just medial to the iliofemoral ligament and faces superiorly, medially, and anteriorly. At the site of the subtendinous bursa of psoas, the capsule is weak or (as in this specimen) partially deficient, but it is guarded by the psoas tendon.
- The iliofemoral ligament, shaped like an inverted "Y." Superiorly it is attached deep to the rectus femoris muscle; the ligament becomes tight on medial rotation of the femur.
- The pectineus muscle is thin, and its fascia blends with the pectineal ligament.

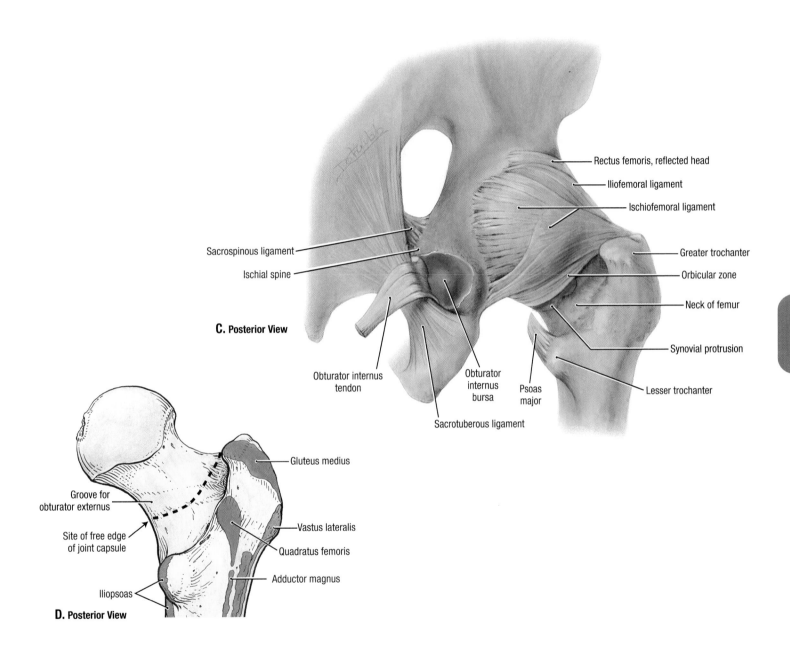

C. Posterior View

Rectus femoris, reflected head
Iliofemoral ligament
Ischiofemoral ligament
Greater trochanter
Orbicular zone
Neck of femur
Synovial protrusion
Lesser trochanter
Psoas major
Sacrotuberous ligament
Obturator internus bursa
Obturator internus tendon
Ischial spine
Sacrospinous ligament

D. Posterior View

Gluteus medius
Groove for obturator externus
Site of free edge of joint capsule
Iliopsoas
Vastus lateralis
Quadratus femoris
Adductor magnus

5.29   Hip joint *(continued)*

C. Ischiofemoral ligament. **D.** Muscle attachments on to the posterior aspect of proximal femur. In **C**:

- The fibers of the capsule spiral to become taut during extension and medial rotation of the femur.
- The synovial membrane protrudes inferior to the fibrous capsule and forms a bursa for the tendon of the obturator externus muscle. Note the large subtendinous bursa of the obturator internus at the lesser sciatic notch, where the tendon turns 90° to attach to the greater trochanter.

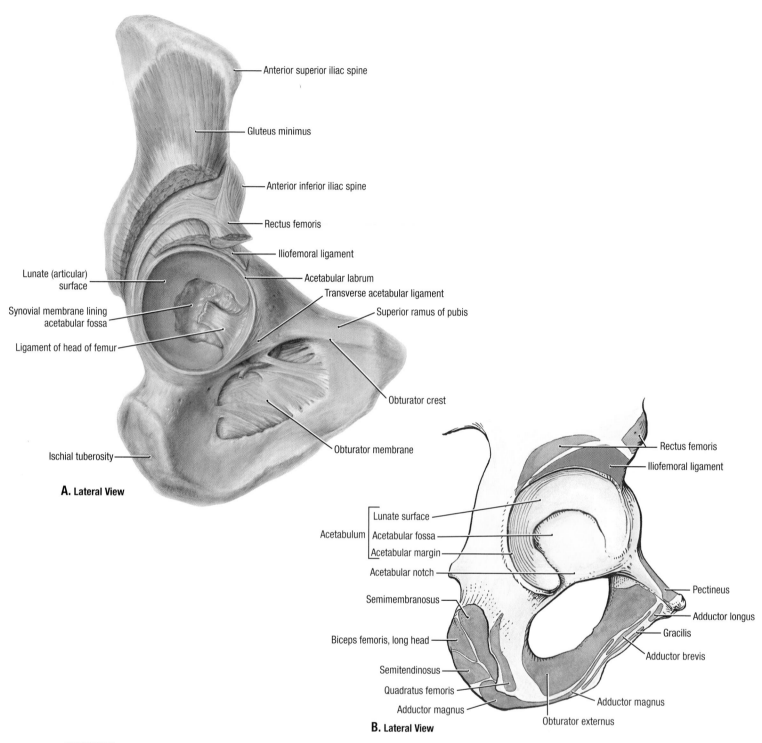

Anterior superior iliac spine

Gluteus minimus

Anterior inferior iliac spine

Rectus femoris

Iliofemoral ligament

Acetabular labrum

Transverse acetabular ligament

Lunate (articular) surface

Superior ramus of pubis

Synovial membrane lining acetabular fossa

Ligament of head of femur

Obturator crest

Ischial tuberosity

Obturator membrane

A. Lateral View

Rectus femoris

Iliofemoral ligament

Lunate surface

Acetabulum — Acetabular fossa

Acetabular margin

Acetabular notch

Pectineus

Semimembranosus

Adductor longus

Biceps femoris, long head

Gracilis

Adductor brevis

Semitendinosus

Quadratus femoris

Adductor magnus

Adductor magnus

Obturator externus

B. Lateral View

5.30   Acetabular region

A. Dissection of acetabulum. **B.** Muscle attachments of acetabular region.

In **A**:

- Note that the transverse acetabular ligament bridges the acetabular notch.
- The acetabular labrum is attached to the acetabular rim and transverse acetabular ligament and forms a complete ring around the head of the femur.
- The ligament of the head of the femur lies between the head of the femur and the acetabulum. These fibers are attached superiorly to the pit (fovea) on the head of the femur and inferiorly to the transverse acetabular ligament and the margins of the acetabular notch; the artery of the ligament of the head of the femur passes through the acetabular notch and into the ligament of the head of the femur.

A. Lateral View

Anterior gluteal line
Posterior gluteal line
Posterior superior iliac spine
Posterior inferior iliac spine
Greater sciatic notch
Ischial spine
Lesser sciatic notch
Body of ischium
Ischial tuberosity

Iliac crest
Anterior superior iliac spine
Inferior gluteal line
Anterior inferior iliac spine
Lunate surface
Acetabular fossa
Acetabular notch
Acetabulum
Pubic tubercle
Obturator foramen
Inferior ramus of pubis
Ramus of ischium

B. Lateral View

Site of triradiate cartilage
Ischium
Ilium
Pubis

5.31   Hip bone

A. Features of the lateral aspect. Note that in anatomical position, the anterior superior iliac spine and pubic tubercle are in the same coronal plane, and the ischial spine and superior end of the pubic symphysis are in the same horizontal plane; the internal aspect of the body of the pubis faces superiorly, and the acetabulum faces inferolaterally. **B.** Hip bone in youth. The three parts of the hip bone (ilium, ischium, and pubis) meet in the acetabulum at the triradiate synchondrosis. One or more primary centers of ossification appear in the triradiate cartilage at approximately the 12th year. Secondary centers of ossification appear along the length of the iliac crest, at the anterior inferior iliac spine, the ischial tuberosity, and the symphysis pubis at about puberty; fusion is usually complete by age 23.

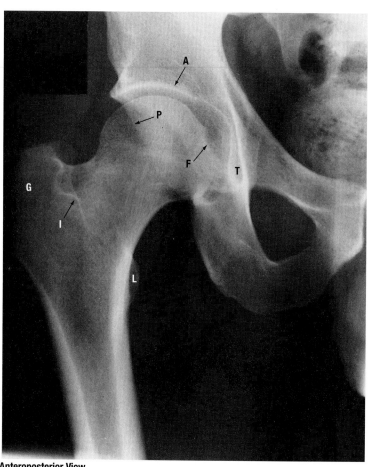

A. Anteroposterior View

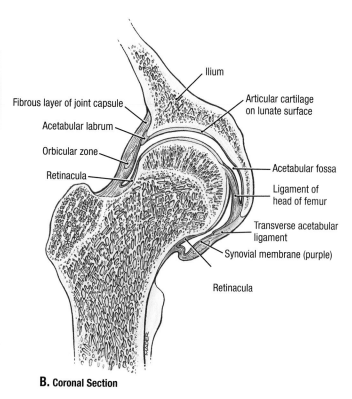

Ilium

Fibrous layer of joint capsule

Acetabular labrum

Orbicular zone

Retinacula

Articular cartilage on lunate surface

Acetabular fossa

Ligament of head of femur

Transverse acetabular ligament

Synovial membrane (purple)

Retinacula

B. Coronal Section

5.32 **Radiograph and coronal section of hip joint**

A. Radiograph. On the femur, note the greater (G) and lesser (L) trochanters, the intertrochanteric crest (I), and the pit or fovea (F) for the ligament of the head. On the pelvis, note the roof (A) and posterior rim (P) of the acetabulum and the "teardrop" appearance (T) caused by the superimposition of structures at the inferior margin of the acetabulum. **B.** Coronal section. Observe the bony trabeculae projecting into the head of the femur. The ligament of the head of the femur becomes taut during adduction of the hip joint, such as when crossing the legs.

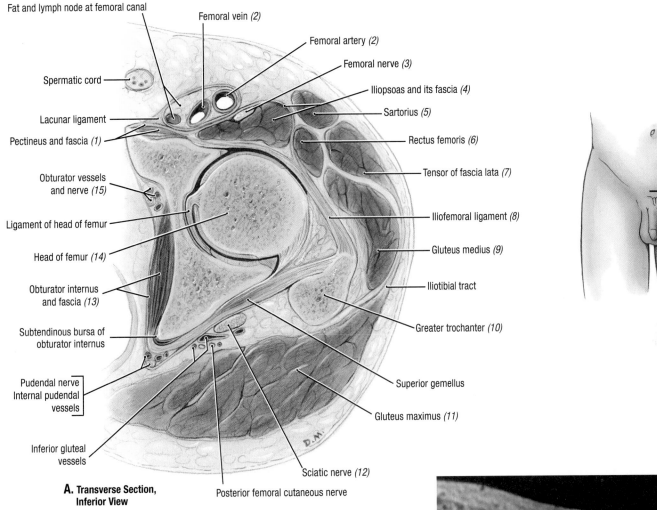

Fat and lymph node at femoral canal
Femoral vein (2)
Femoral artery (2)
Femoral nerve (3)
Spermatic cord
Iliopsoas and its fascia (4)
Lacunar ligament
Sartorius (5)
Pectineus and fascia (1)
Rectus femoris (6)
Obturator vessels and nerve (15)
Tensor of fascia lata (7)
Ligament of head of femur
Iliofemoral ligament (8)
Head of femur (14)
Gluteus medius (9)
Obturator internus and fascia (13)
Iliotibial tract
Subtendinous bursa of obturator internus
Greater trochanter (10)
Pudendal nerve
Internal pudendal vessels
Superior gemellus
Gluteus maximus (11)
Inferior gluteal vessels
Sciatic nerve (12)
Posterior femoral cutaneous nerve

A. Transverse Section, Inferior View

5.33 **Transverse section through thigh at level of hip joint**

A. Transverse section. **B.** MRI (*numbers* refer to structures in **A**).

In **A**:

- The fibrous capsule of the joint is thick where it forms the iliofemoral ligament and thin posterior to the subtendinous bursa of psoas and tendon.
- The femoral sheath, enclosing the femoral artery, vein, lymph node, lymph vessels, and fat, is free, except posteriorly where, between the psoas and pectineus muscles, it is attached to the capsule of the hip joint.
- The femoral vein is located at the interval between the psoas and pectineus muscles. The femoral nerve lies between the iliacus muscle and fascia.

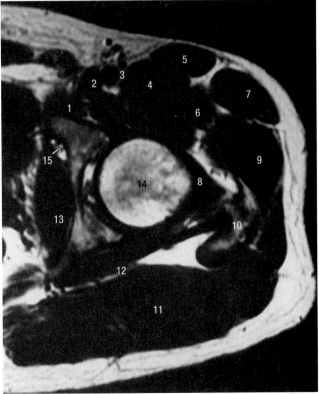

B. Transverse MRI

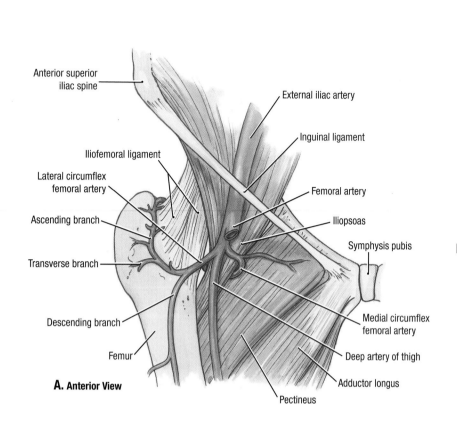

A. Anterior View

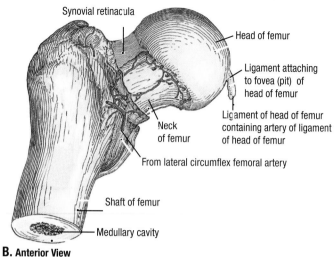

B. Anterior View

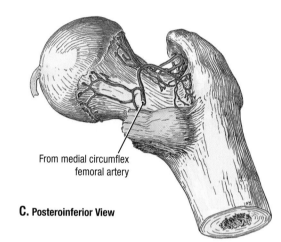

C. Posteroinferior View

5.34 Blood supply to head of femur

A. Medial and lateral circumflex femoral arteries in femoral triangle. **B.** Branches of lateral circumflex femoral artery. **C.** Branches of medial circumflex femoral artery.

- Branches of the medial and lateral circumflex femoral arteries ascend on the posterosuperior and posteroinferior parts of the neck of the femur. The vessels ascend in synovial retinacula—reflections of synovial membrane along the neck of the femur. The retinacula (in **B** and **C**) have been mostly removed; thus, the vessels can be clearly visualized.
- The branches of the medial and lateral circumflex femoral arteries perforate the bone just distal to the head of the femur where they anastomose with branches from the artery of the ligament of the head of the femur and with medullary branches located within the shaft of the femur.
- The ligament of the head of the femur usually contains the artery of the ligament of the head of the femur, a branch of the obturator artery. The artery enters the head of the femur only when the center of the ossification has extended to the pit (fovea) for the ligament of the head (12th to 14th year). When present, this anastomosis persists even in advanced age; however, in 20% of persons, it is never established.
- Fractures of the femoral neck often disrupt the blood supply to the head of the femur. The medial circumflex femoral artery supplies most of the blood to the head and neck of the femur and is often torn when the femoral neck is fractured. In some cases, the blood supplied by the artery of the ligament of the head may be the only blood received by the proximal fragment of the femoral head, which may be inadequate. If the blood vessels are ruptured, the fragment of bone may receive no blood and undergo aseptic necrosis.

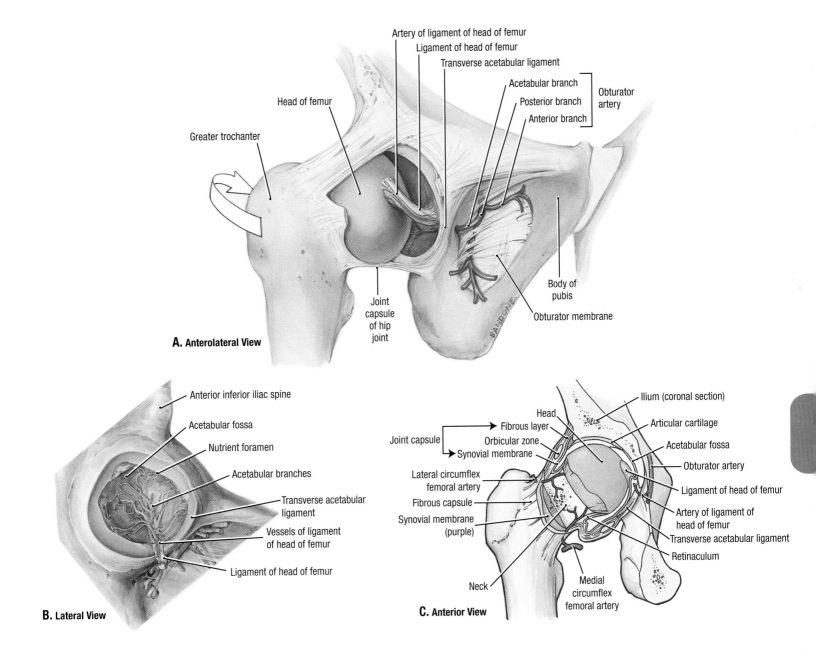

A. **Anterolateral View**

B. **Lateral View**

C. **Anterior View**

5.35   **Blood vessels of acetabular fossa and ligament of head of femur**

A. Obturator artery. The hip joint has been dislocated to reveal the ligament of the head of the femur. Note that the obturator artery divides into anterior and posterior branches, and the acetabular branch arises from the posterior branch. The artery of the ligament of the head of the femur is a branch of the acetabular artery and can be seen traveling in the ligament to the head of the femur. **B.** Acetabular artery and vein. The acetabular branches (artery and vein) pass through the acetabular foramen and enter the acetabular fossa, where they diverge in the fatty areolar tissue. The branches radiate to the margin of the fossa, where they enter nutrient foramina. **C.** Blood supply of the head and neck of the femur. A section of bone has been removed from the femoral neck.

SUPERIOR

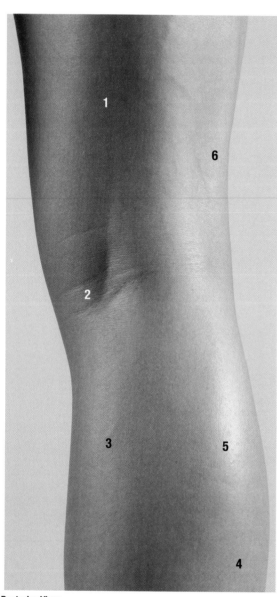

A. Posterior View

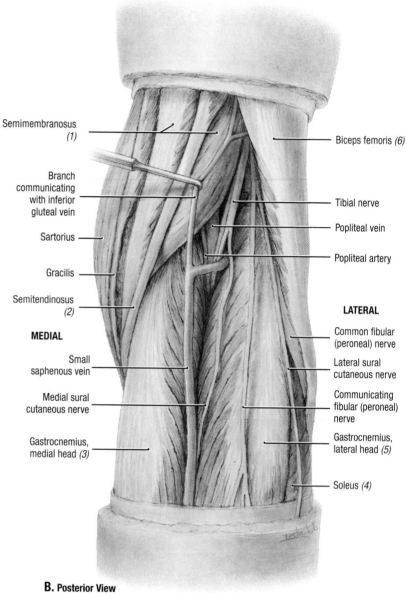

Semimembranosus *(1)*

Branch communicating with inferior gluteal vein

Sartorius

Gracilis

Semitendinosus *(2)*

MEDIAL

Small saphenous vein

Medial sural cutaneous nerve

Gastrocnemius, medial head *(3)*

Biceps femoris *(6)*

Tibial nerve

Popliteal vein

Popliteal artery

LATERAL

Common fibular (peroneal) nerve

Lateral sural cutaneous nerve

Communicating fibular (peroneal) nerve

Gastrocnemius, lateral head *(5)*

Soleus *(4)*

B. Posterior View

INFERIOR

5.36 **Popliteal fossa**

A. Surface anatomy (*numbers* refer to structures in **B**). **B.** Superficial dissection.

- The two heads of the gastrocnemius muscle are embraced on the medial side by the semi-membranosus muscle, which is overlaid by the semitendinosus muscle, and on the lateral side by the biceps femoris muscle.
- The small saphenous vein runs between the two heads of the gastrocnemius muscle. Deep to this vein is the medial sural cutaneous nerve, which, followed proximally, leads to the tibial nerve. The tibial nerve is superficial to the popliteal vein, which, in turn, is superficial to the popliteal artery.
- The common fibular (peroneal) nerve follows the posterior border of the biceps femoris muscle and, in this specimen, gives off two cutaneous branches.

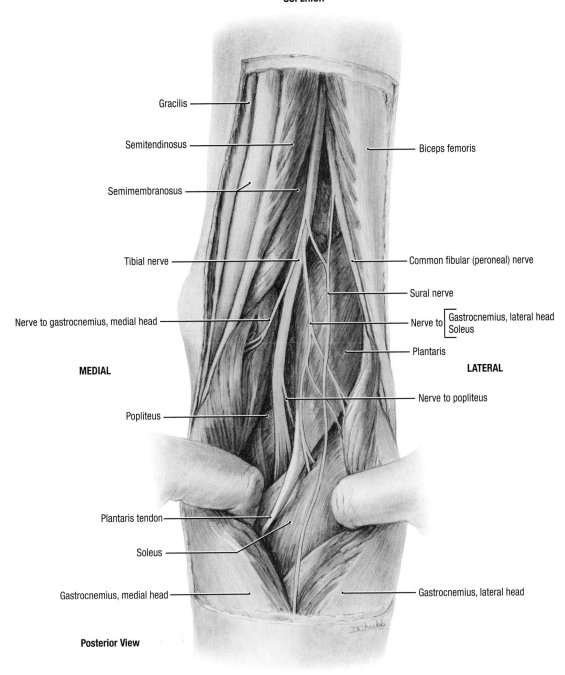

SUPERIOR

Gracilis

Semitendinosus

Biceps femoris

Semimembranosus

Tibial nerve

Common fibular (peroneal) nerve

Sural nerve

Nerve to — Gastrocnemius, lateral head / Soleus

Nerve to gastrocnemius, medial head

MEDIAL   Plantaris   LATERAL

Nerve to popliteus

Popliteus

Plantaris tendon

Soleus

Gastrocnemius, medial head

Gastrocnemius, lateral head

Posterior View

INFERIOR

5.37   Nerves of popliteal fossa

The two heads of the gastrocnemius muscle are separated.

- A cutaneous branch of the tibial nerve joins a cutaneous branch of the common fibular (peroneal) nerve to form the sural nerve. In this specimen, the junction is high; usually it is 5 to 8 cm proximal to the ankle.
- All motor branches in this region emerge from the tibial nerve, one branch from its medial side and the others from its lateral side; hence, it is safer to dissect on the medial side.

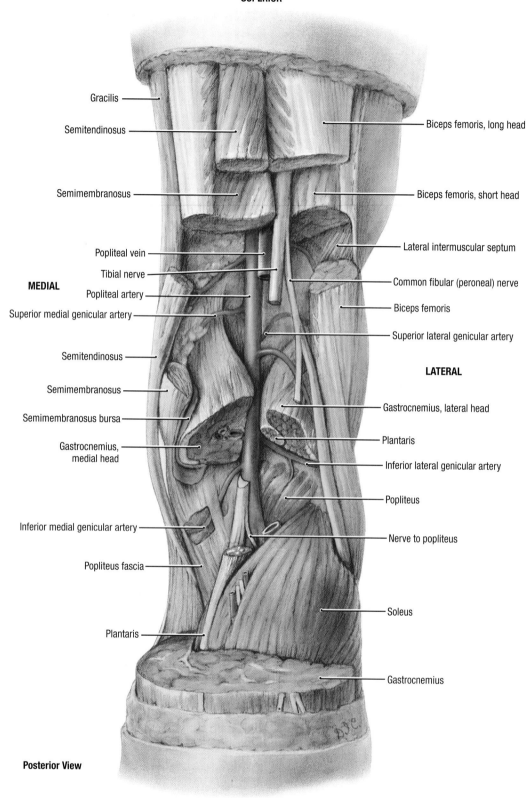

SUPERIOR

Gracilis

Semitendinosus

Semimembranosus

Popliteal vein

Tibial nerve

MEDIAL

Popliteal artery

Superior medial genicular artery

Semitendinosus

Semimembranosus

Semimembranosus bursa

Gastrocnemius, medial head

Inferior medial genicular artery

Popliteus fascia

Plantaris

Biceps femoris, long head

Biceps femoris, short head

Lateral intermuscular septum

Common fibular (peroneal) nerve

Biceps femoris

Superior lateral genicular artery

LATERAL

Gastrocnemius, lateral head

Plantaris

Inferior lateral genicular artery

Popliteus

Nerve to popliteus

Soleus

Gastrocnemius

Posterior View

INFERIOR

5.38 **Deep dissection of popliteal fossa**

The popliteal artery lies on the floor of the popliteal fossa. The floor is formed by the femur, capsule of the knee joint, and popliteus fascia. The popliteal artery gives off genicular branches that also lie on the floor of the fossa and ends by bifurcating into the anterior and posterior tibial arteries at the proximal border of the soleus muscle.

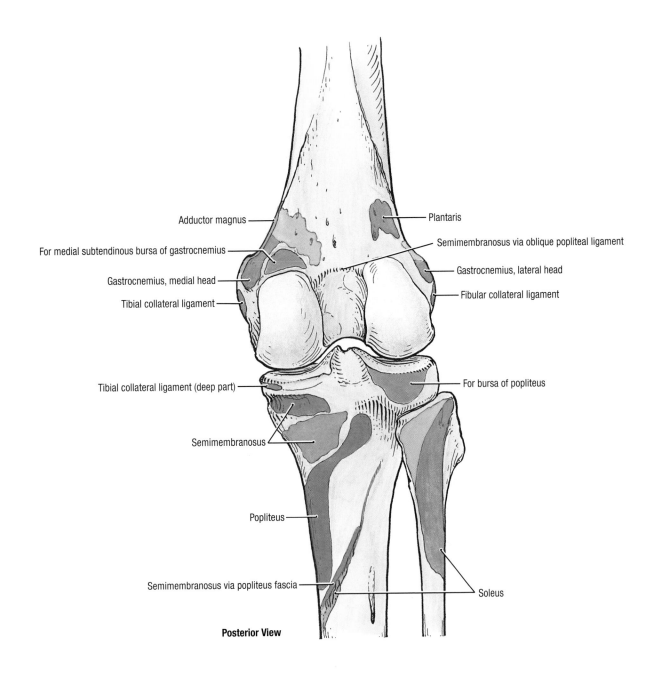

Adductor magnus

For medial subtendinous bursa of gastrocnemius

Gastrocnemius, medial head

Tibial collateral ligament

Plantaris

Semimembranosus via oblique popliteal ligament

Gastrocnemius, lateral head

Fibular collateral ligament

Tibial collateral ligament (deep part)

For bursa of popliteus

Semimembranosus

Popliteus

Semimembranosus via popliteus fascia

Soleus

Posterior View

5.39 **Attachments of muscles of popliteal region**

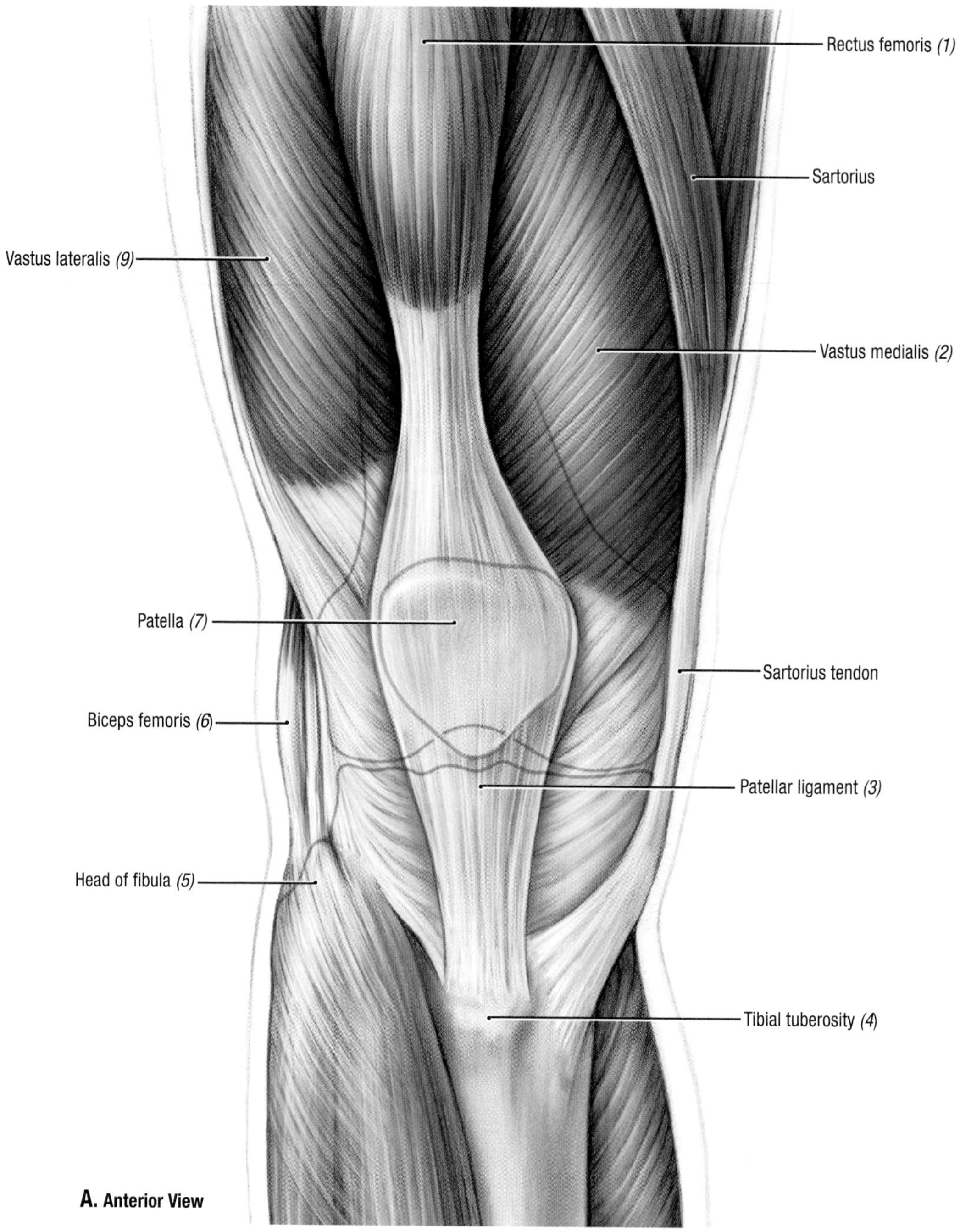

Rectus femoris *(1)*

Sartorius

Vastus lateralis *(9)*

Vastus medialis *(2)*

Patella *(7)*

Sartorius tendon

Biceps femoris *(6)*

Patellar ligament *(3)*

Head of fibula *(5)*

Tibial tuberosity *(4)*

A. Anterior View

5.40 **Anterior aspect of knee**

A. Distal thigh and knee regions.
Note that the tendons of the four parts of the quadriceps unite to form the quadriceps tendon, a broad band that attaches to the patella. The patellar ligament, a continuation of the quadriceps tendon, attaches the patella to the tibial tuberosity.

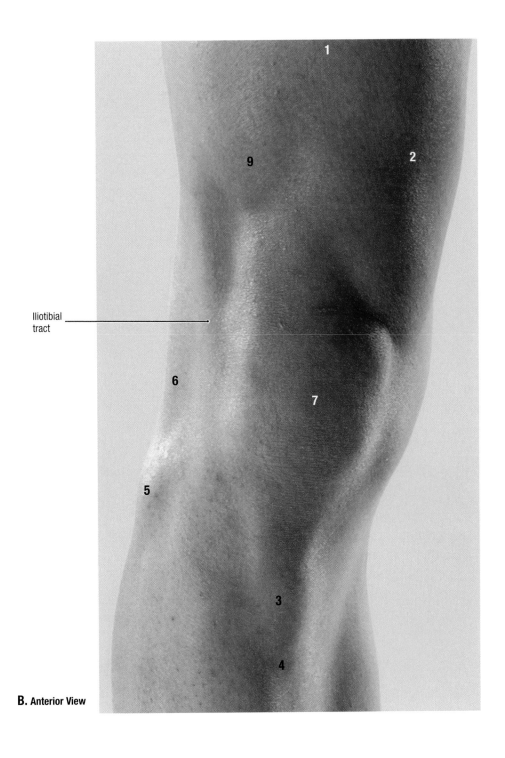

Iliotibial tract

B. Anterior View

5.40   **Anterior aspect of knee (continued)**

B. Surface anatomy (*numbers* refer to structures in **A**).

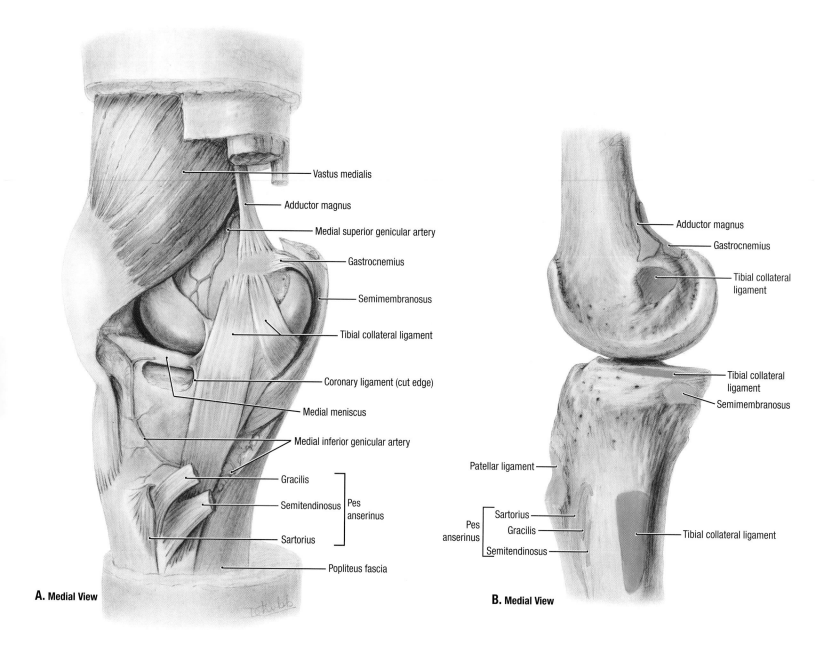

A. Medial View

- Vastus medialis
- Adductor magnus
- Medial superior genicular artery
- Gastrocnemius
- Semimembranosus
- Tibial collateral ligament
- Coronary ligament (cut edge)
- Medial meniscus
- Medial inferior genicular artery
- Gracilis ⎤
- Semitendinosus ⎬ Pes anserinus
- Sartorius ⎦
- Popliteus fascia

B. Medial View

- Adductor magnus
- Gastrocnemius
- Tibial collateral ligament
- Tibial collateral ligament
- Semimembranosus
- Patellar ligament
- Pes anserinus ⎡ Sartorius
- ⎢ Gracilis
- ⎣ Semitendinosus
- Tibial collateral ligament

5.41 **Medial aspect of knee**

A. Dissection. The bandlike part of the tibial collateral ligament attaches to the medial epi-condyle of the femur, bridges superficial to the insertion of the semimembranosus muscle, and crosses the medial inferior genicular artery. Distally, the ligament is crossed by the three tendons forming the pes anserinus (sartorius, gracilis, and semitendinosus). **B.** Bones, show-ing muscle and ligament attachments.

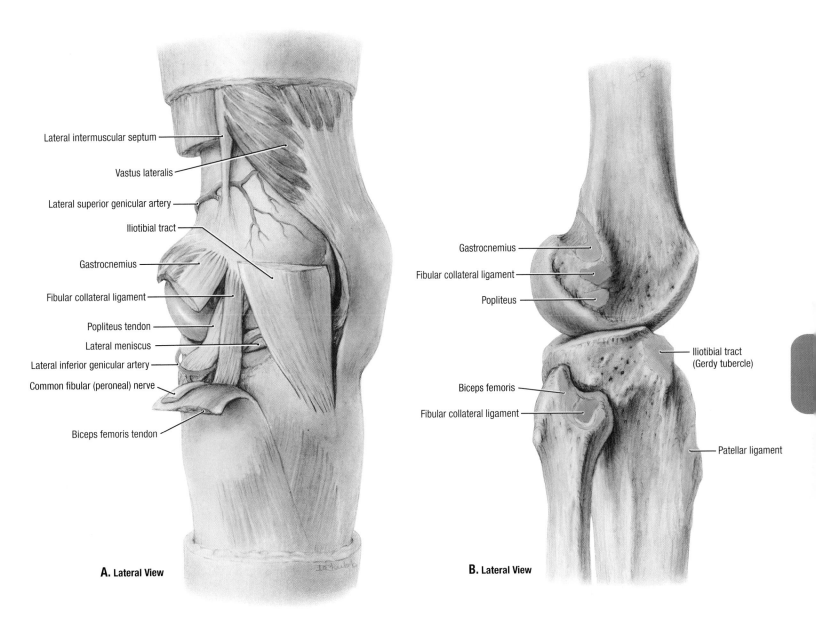

Lateral intermuscular septum

Vastus lateralis

Lateral superior genicular artery

Iliotibial tract

Gastrocnemius

Fibular collateral ligament

Popliteus tendon

Lateral meniscus

Lateral inferior genicular artery

Common fibular (peroneal) nerve

Biceps femoris tendon

A. Lateral View

Gastrocnemius

Fibular collateral ligament

Popliteus

Iliotibial tract
(Gerdy tubercle)

Biceps femoris

Fibular collateral ligament

Patellar ligament

B. Lateral View

5.42 **Lateral aspect of knee**

A. Dissection. **B.** Bones, showing muscle and ligament attachments.
Three structures arise from the lateral epicondyle and are uncovered by reflecting the biceps
muscle: the gastrocnemius muscle is posterosuperior; the popliteus muscle is anteroinferior;
and the fibular collateral ligament is in between, crossing superficial to the popliteus muscle. The lateral inferior genicular artery courses along the lateral meniscus.

SUPERIOR

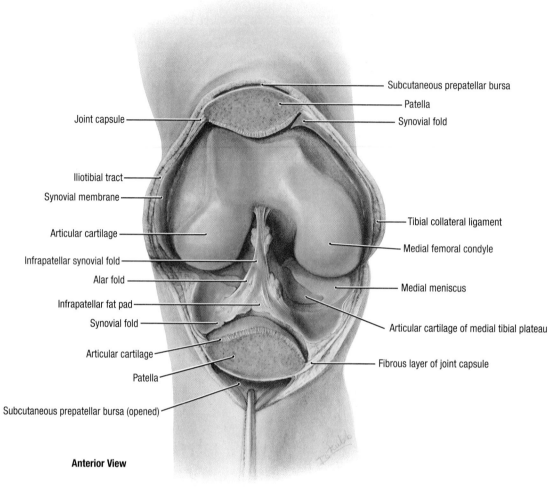

Subcutaneous prepatellar bursa

Patella

Joint capsule

Synovial fold

Iliotibial tract

Synovial membrane

Tibial collateral ligament

Articular cartilage

Medial femoral condyle

Infrapatellar synovial fold

Alar fold

Infrapatellar fat pad

Medial meniscus

Synovial fold

Articular cartilage

Articular cartilage of medial tibial plateau

Patella

Fibrous layer of joint capsule

Subcutaneous prepatellar bursa (opened)

Anterior View

INFERIOR

5.43 **Open knee joint**

The patella is sawed, the skin and joint capsule are cut, and the joint is flexed.
- The articular cartilage of the patella is not uniform in thickness.
- The infrapatellar synovial fold resembles a partially collapsed tent, with its apex attached to the intercondylar notch of the femur and its base inferior to the patella.
- A fracture of the patella would bring the prepatellar bursa into communication with the joint cavity.

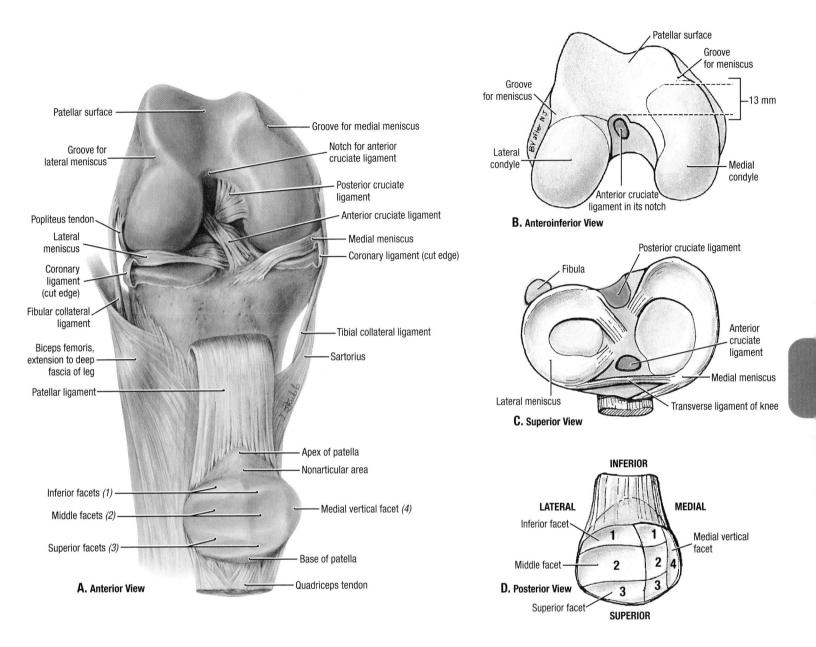

A. Anterior View

- Patellar surface
- Groove for lateral meniscus
- Popliteus tendon
- Lateral meniscus
- Coronary ligament (cut edge)
- Fibular collateral ligament
- Biceps femoris, extension to deep fascia of leg
- Patellar ligament
- Inferior facets (1)
- Middle facets (2)
- Superior facets (3)
- Groove for medial meniscus
- Notch for anterior cruciate ligament
- Posterior cruciate ligament
- Anterior cruciate ligament
- Medial meniscus
- Coronary ligament (cut edge)
- Tibial collateral ligament
- Sartorius
- Apex of patella
- Nonarticular area
- Medial vertical facet (4)
- Base of patella
- Quadriceps tendon

B. Anteroinferior View

- Patellar surface
- Groove for meniscus
- Groove for meniscus
- 13 mm
- Lateral condyle
- Medial condyle
- Anterior cruciate ligament in its notch

C. Superior View

- Posterior cruciate ligament
- Fibula
- Anterior cruciate ligament
- Medial meniscus
- Lateral meniscus
- Transverse ligament of knee

D. Posterior View

- INFERIOR
- LATERAL
- MEDIAL
- Inferior facet
- Medial vertical facet
- Middle facet
- Superior facet
- SUPERIOR

5.44 **Articular surfaces and ligaments of the knee joint**

A. Flexed knee joint with patella reflected. There are indentations on the sides of the femoral condyles at the junction of the patellar and tibial articular areas. The lateral tibial articular area is shorter than the medial one. The notch at the anterolateral part of the intercondylar notch is for the anterior cruciate ligament on full extension. **B.** Distal femur. **C.** Tibial plateaus. **D.** Articular surfaces of patella. The three paired facets (superior, middle, and inferior) on the posterior surface of the patella articulate with the patellar surface of the femur successively during (1) extension, (2) slight flexion, (3) flexion, and the most medial vertical facet on the patella (4) articulates during full flexion with the cresenteric facet on the medial margin of the intercondylar notch of the femur.

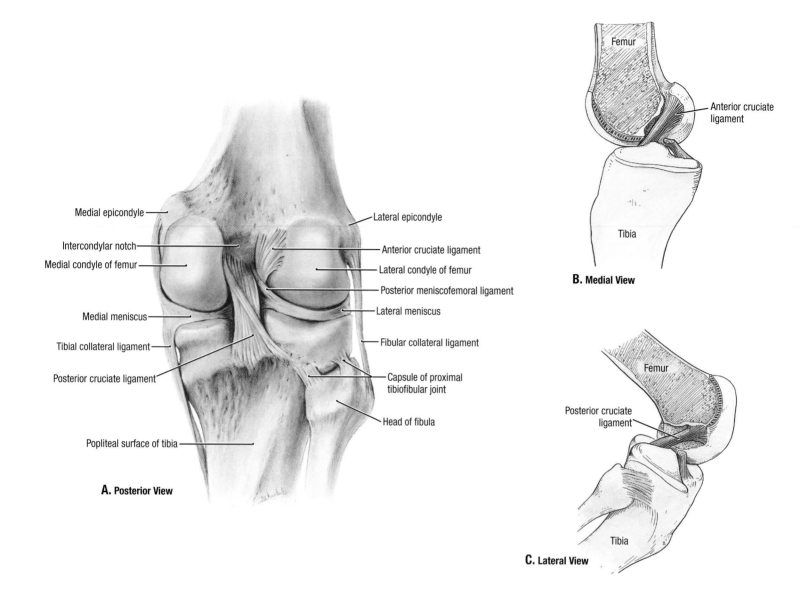

Medial epicondyle
Intercondylar notch
Medial condyle of femur
Medial meniscus
Tibial collateral ligament
Posterior cruciate ligament
Popliteal surface of tibia

A. Posterior View

Lateral epicondyle
Anterior cruciate ligament
Lateral condyle of femur
Posterior meniscofemoral ligament
Lateral meniscus
Fibular collateral ligament
Capsule of proximal tibiofibular joint
Head of fibula

Femur
Anterior cruciate ligament
Tibia

B. Medial View

Femur
Posterior cruciate ligament
Tibia

C. Lateral View

5.45   Ligaments of the knee joint

A. Posterior aspect of joint. The bandlike medial (tibial) collateral ligament is attached to the medial meniscus, and the cordlike lateral (fibular) collateral ligament is separated from the lateral meniscus by the width of the popliteus tendon (removed). The posterior cruciate ligament is joined by a cord from the lateral meniscus called the *posterior meniscofemoral ligament*. The posterior meniscofemoral ligament attaches to the medial condyle of the femur just posterior to the attachment of the posterior cruciate ligament. **B.** Anterior cruciate ligament. **C.** Posterior cruciate ligament. In each illustration, half the femur is sagittally sectioned and removed with the proximal part of the corresponding cruciate ligament. Note that the posterior cruciate ligament prevents the femur from sliding anteriorly on the tibia, particularly when the knee is flexed. The anterior cruciate ligament prevents the femur from sliding posteriorly on the tibia, preventing hyperextension of the knee, and limits medial rotation of the femur when the foot is on the ground (i.e., when the leg is fixed).

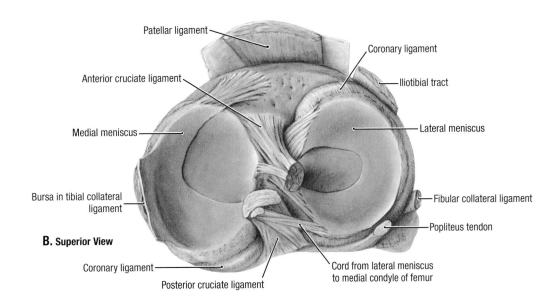

A. Superior View

Anterior intercondylar area

Medial intercondylar tubercle

Articular surface of medial condyle

Lateral intercondylar tubercle

Articular surface of lateral condyle

Posterior intercondylar area

B. Superior View

Patellar ligament

Coronary ligament

Anterior cruciate ligament

Iliotibial tract

Medial meniscus

Lateral meniscus

Bursa in tibial collateral ligament

Fibular collateral ligament

Popliteus tendon

Coronary ligament

Cord from lateral meniscus to medial condyle of femur

Posterior cruciate ligament

5.46   Cruciate ligaments and menisci

A. Attachments to tibia. **B.** Menisci in situ.

- The lateral tibial condyle is flatter, shorter from anterior to posterior, and more circular. The medial condyle is concave, longer from anterior to posterior, and more oval.
- The menisci conform to the shapes of the surfaces on which they rest. Because the horns of the lateral meniscus are attached close together and its coronary ligament is slack, this meniscus can slide anteriorly and posteriorly on the (flat) condyle; because the horns of the medial meniscus are attached further apart, its movements on the (concave) condyle are restricted.
- Note the bursa between the long and short parts of the medial (tibial) collateral ligament of the knee.

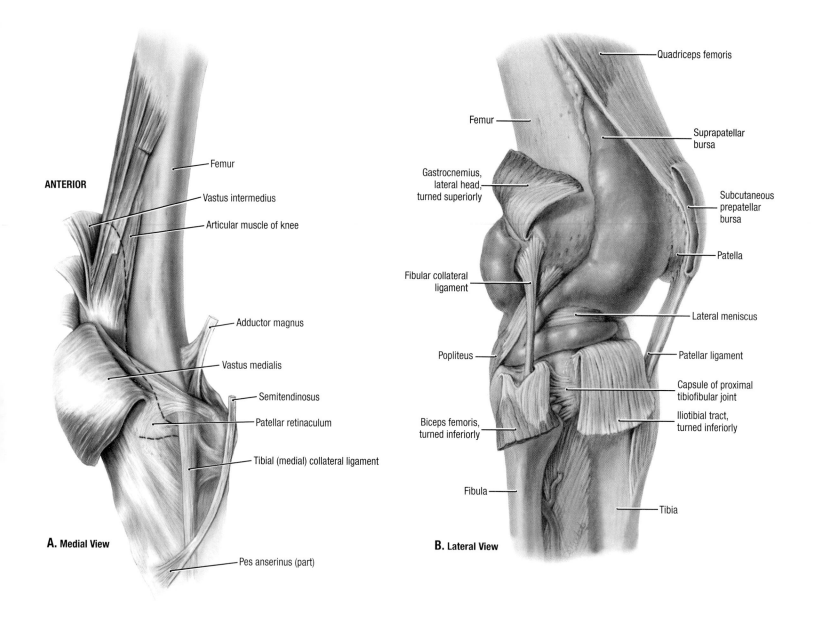

ANTERIOR

Femur

Vastus intermedius

Articular muscle of knee

Adductor magnus

Vastus medialis

Semitendinosus

Patellar retinaculum

Tibial (medial) collateral ligament

A. Medial View

Pes anserinus (part)

Quadriceps femoris

Femur

Gastrocnemius, lateral head, turned superiorly

Suprapatellar bursa

Subcutaneous prepatellar bursa

Fibular collateral ligament

Patella

Lateral meniscus

Popliteus

Patellar ligament

Capsule of proximal tibiofibular joint

Iliotibial tract, turned inferiorly

Biceps femoris, turned inferiorly

Fibula

Tibia

B. Lateral View

5.47   **Articular muscle of the knee and distended knee joint**

A. Articular muscle of knee. This muscle lies deep to the vastus intermedius muscle and consists of fibers arising from the anterior surface of the femur proximally and attaching into the synovial capsule of the knee joint distally. The articular muscle of the knee pulls the synovial capsule of the suprapatellar bursa (dotted line) superiorly during extension of the knee so that it will not be caught between the patella and femur within the knee joint. **B.** Medial aspect of knee. Latex was injected into the joint cavity and fixed with acetic acid. The distended synovial capsule was exposed and cleaned. The gastrocnemius muscle was reflected proximally, and the biceps femoris muscle and the iliotibial tract were reflected distally. The extent of the synovial capsule: superiorly, it rises superior to the patella, where it rests on a layer of fat that allows it to glide freely with movements of the joint; this superior part is called the *suprapatellar bursa;* posteriorly, it rises as high as the origin of the gastrocnemius muscle; laterally, it curves inferior to the lateral femoral epicondyle, where the popliteus tendon and fibular collateral ligament are attached; and inferiorly, it bulges inferior to the lateral meniscus, overlapping the tibia (the coronary ligament is removed to show this).

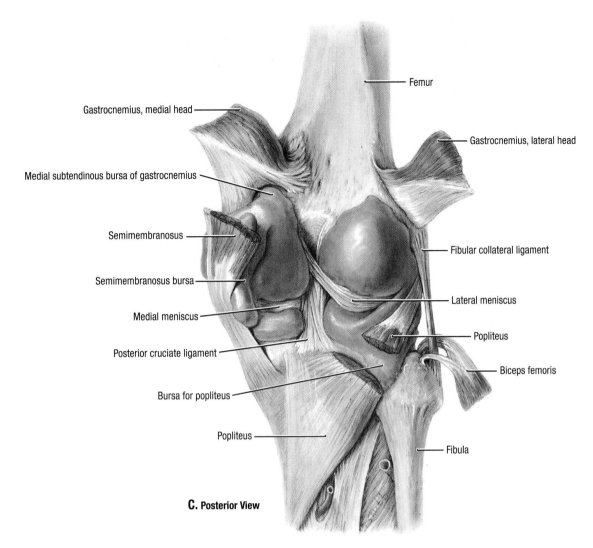

C. Posterior View

5.47   **Distended knee joint** *(continued)*

Both heads of the gastrocnemius muscle are reflected proximally, the biceps muscle is reflected distally, and a section is removed from the popliteus muscle. The posterior cruciate ligament is exposed posteriorly without opening the synovial capsule. Note that the origins of the gastrocnemius muscle limit the proximal extent of the synovial capsule.

TABLE 5.7   BURSAE AROUND KNEE

BURSAE	LOCATIONS	BURSAE	LOCATIONS
Suprapatellar	Between femur and tendon of quadriceps femoris	Semimembranosus	Located between medial head of gastrocnemius and semimembranosus tendon
Popliteus	Between tendon of popliteus and lateral condyle of tibia	Subcutaneous prepatellar	Lies between skin and anterior surface of patella
Anserine	Separates tendons of sartorius, gracilis, and semitendinosus from tibia and tibial collateral ligament	Subcutaneous infrapatellar	Located between skin and tibial tuberosity
Gastrocnemius	Lies deep to proximal attachment of tendon of medial head of gastrocnemius	Deep infrapatellar	Lies between patellar ligament and anterior surface of tibia

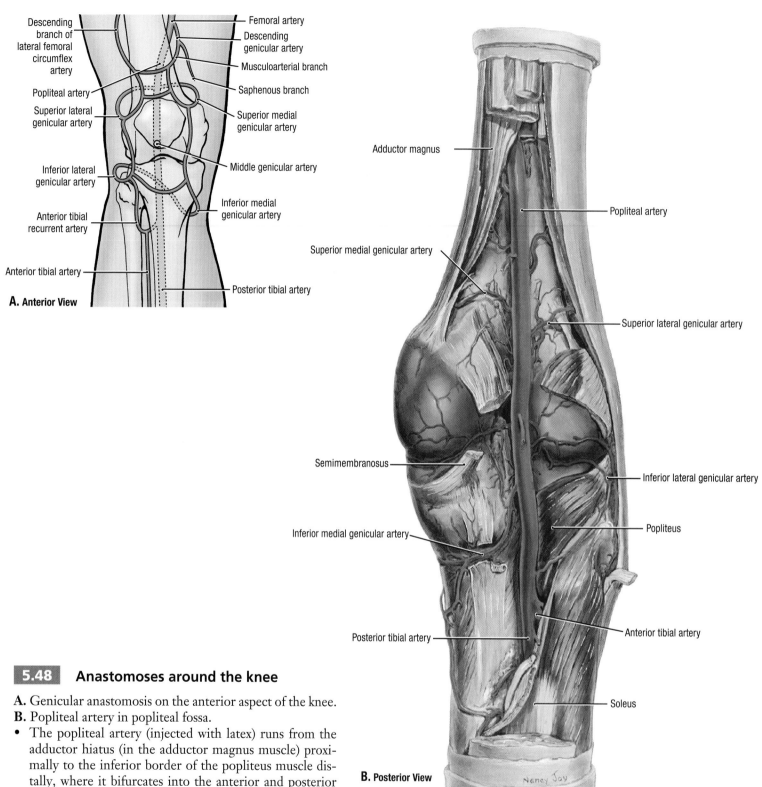

Descending branch of lateral femoral circumflex artery
Popliteal artery
Superior lateral genicular artery
Inferior lateral genicular artery
Anterior tibial recurrent artery
Anterior tibial artery
A. Anterior View

Femoral artery
Descending genicular artery
Musculoarterial branch
Saphenous branch
Superior medial genicular artery
Middle genicular artery
Inferior medial genicular artery
Posterior tibial artery

Adductor magnus
Superior medial genicular artery
Semimembranosus
Inferior medial genicular artery
Posterior tibial artery

Popliteal artery
Superior lateral genicular artery
Inferior lateral genicular artery
Popliteus
Anterior tibial artery
Soleus

Nancy Joy

B. Posterior View

5.48 **Anastomoses around the knee**

A. Genicular anastomosis on the anterior aspect of the knee.

B. Popliteal artery in popliteal fossa.

- The popliteal artery (injected with latex) runs from the adductor hiatus (in the adductor magnus muscle) proximally to the inferior border of the popliteus muscle distally, where it bifurcates into the anterior and posterior tibial arteries.

- The three anterior relations of the popliteal artery include the femur (fat intervening), the joint capsule of the knee; and the popliteus muscle (covered with popliteus fascia).

- The superior and inferior genicular branches of the popliteal artery course along the femur, with nothing intervening except the popliteus tendon.

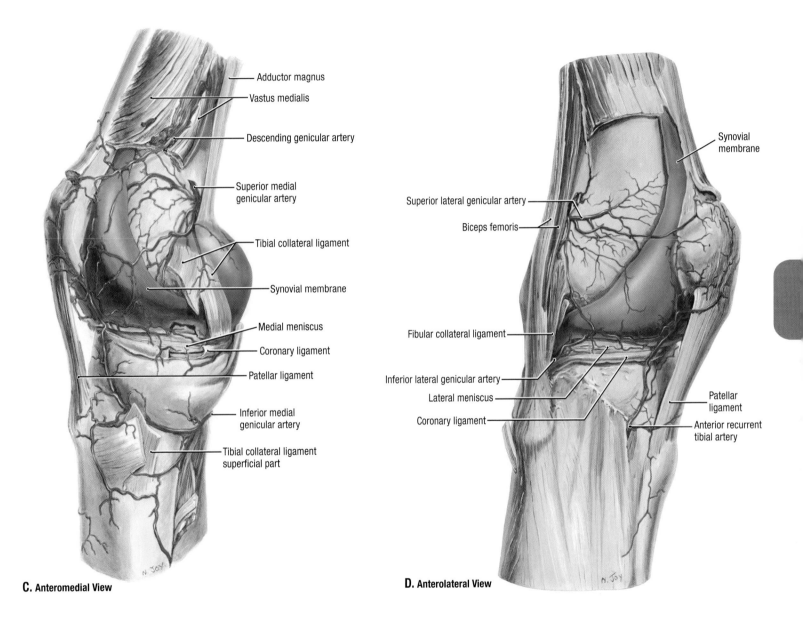

C. **Anteromedial View**

Adductor magnus
Vastus medialis
Descending genicular artery
Superior medial genicular artery
Tibial collateral ligament
Synovial membrane
Medial meniscus
Coronary ligament
Patellar ligament
Inferior medial genicular artery
Tibial collateral ligament superficial part

D. **Anterolateral View**

Synovial membrane
Superior lateral genicular artery
Biceps femoris
Fibular collateral ligament
Inferior lateral genicular artery
Lateral meniscus
Coronary ligament
Patellar ligament
Anterior recurrent tibial artery

5.48   Anastomoses around the knee *(continued)*

C. Medial aspect of the knee showing superior and inferior medial genicular arteries. **D.**
Lateral aspect of the knee showing superior and inferior lateral genicular arteries.

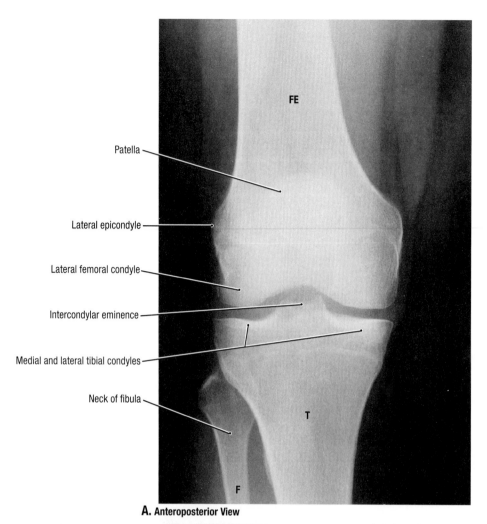

A. Anteroposterior View

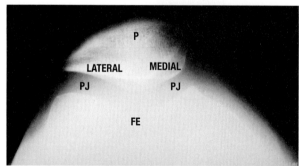

B. Skyline View

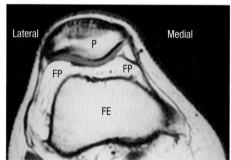

C. Transverse MRI

5.49   **Imaging of the knee and patellofemoral articulation**

A. Anteroposterior radiograph of knee. **B.** Radiograph of patella (knee joint flexed). *FE*, femur; *FP*, fat pad; *P*, patella; *PJ*, patellofemoral joint. **C.** Transverse MRI showing the patellofemoral joint.

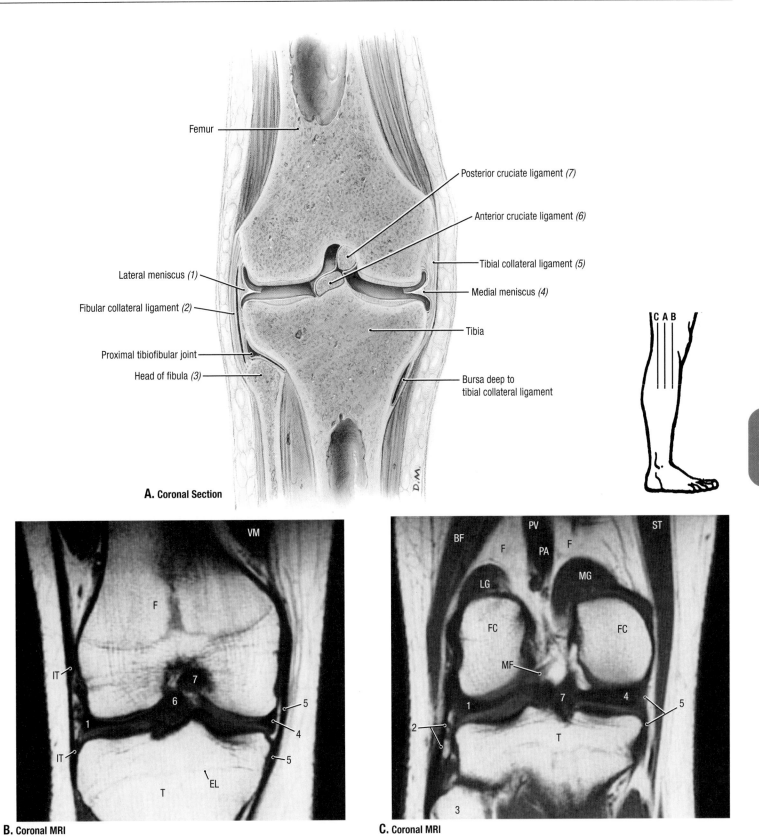

A. Coronal Section

B. Coronal MRI

C. Coronal MRI

5.50   Coronal section and MRIs of knee

A. Section through intercondylar notch of femur, tibia, and fibula.
B. MRI through intercondylar notch of femur and tibia. **C.** MRI
through femoral condyles tibia and fibula. *Numbers* in MRIs refer
to structures in **A.** *VM*, vastus medialis; *EL*, epiphyseal line; *IT*, iliotibial tract; *FC*, femoral condyle; *BF*, biceps femoris; *ST*, semitendinosus; *LG*, lateral head of gastrocnemius; *MG*, medial head of
gastrocnemius; *PV*, popliteal vein; *PA*, popliteal artery; *F*, fat in
popliteal fossa; *MF*, meniscofemoral ligament.

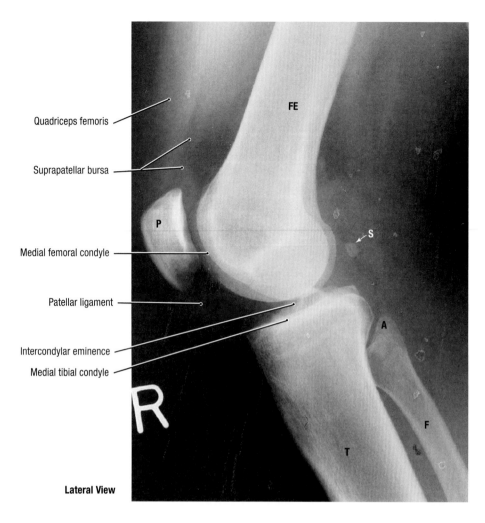

Quadriceps femoris

Suprapatellar bursa

Medial femoral condyle

Patellar ligament

Intercondylar eminence
Medial tibial condyle

Lateral View

| 5.51 | **Radiograph of knee** |

Lateral radiograph of flexed knee. *FE*, femur; *T*, tibia; *F*, fibula; *A*, apex of fibula; *S*, fabella; *P*, patella. The fabella is a sesamoid bone in the lateral head of gastrocnemius muscle.

| 5.52 | **Sagittal section and MRIs of knee** |

A (opposite page). Section through lateral aspect of intercondylar notch of femur. **B.** MRI through medial aspect of intercondylar notch of femur showing cruciate ligaments. **C.** MRI through medial femoral and tibial condyles. *Numbers* in MRIs refer to structures in **A.** *SM*, semimembranosus; *ST*, semitendinosus; *MG*, medial head of gastrocnemius; *VM*, vastus medialis; *PF*, prefemoral fat; *SF*, suprapatellar fat; *AM*, anterior horn of medial meniscus; *PM*, posterior horn of medial meniscus; *PV*, popliteal vessels.

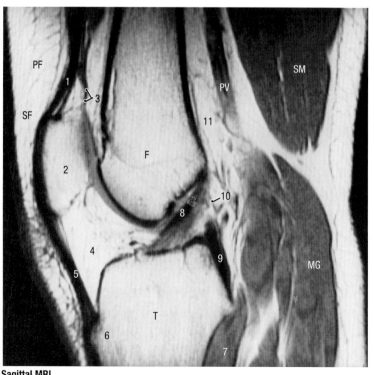

B. Sagittal MRI

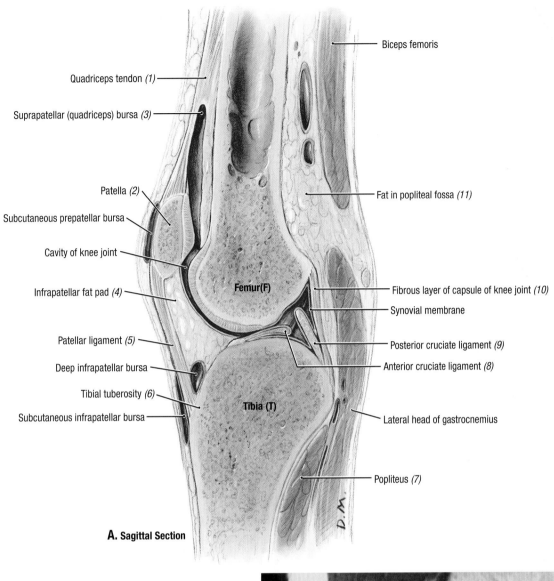

Biceps femoris

Quadriceps tendon *(1)*

Suprapatellar (quadriceps) bursa *(3)*

Patella *(2)*

Subcutaneous prepatellar bursa

Cavity of knee joint

Femur(F)

Infrapatellar fat pad *(4)*

Patellar ligament *(5)*

Deep infrapatellar bursa

Tibial tuberosity *(6)*

Subcutaneous infrapatellar bursa

Tibia (T)

Fat in popliteal fossa *(11)*

Fibrous layer of capsule of knee joint *(10)*

Synovial membrane

Posterior cruciate ligament *(9)*

Anterior cruciate ligament *(8)*

Lateral head of gastrocnemius

Popliteus *(7)*

A. Sagittal Section

A B C

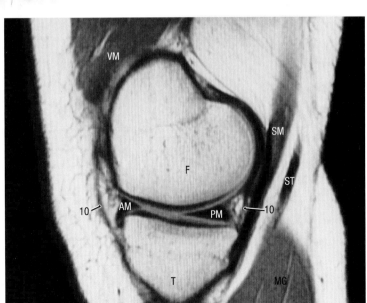

VM

SM

F

ST

10   AM   PM   10

T   MG

C. Sagittal MRI

5.52   **Sagittal section and MRIs of knee** *(continued)*

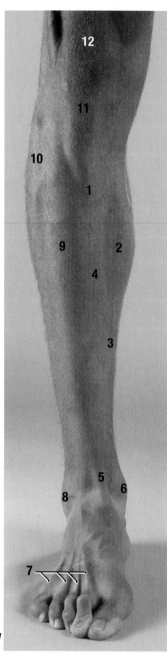

A. Anterior View

Patella *(12)*
Iliotibial tract
Patellar ligament *(11)*
Tibial tuberosity *(1)*
Gastrocnemius, medial head *(2)*
Fibularis longus *(10)*
Soleus *(3)*
Tibialis anterior *(9)*
Medial surface of tibia *(4)*
Extensor digitorum longus
Fibularis (Peroneus) brevis
Tendon of tibialis anterior *(5)*
Extensor digitorum longus
Extensor hallucis longus
Superior extensor retinaculum
Inferior extensor retinaculum
(8) Lateral malleolus
Medial malleolus *(6)*
Fibularis tertius muscle and tendon
(7) Tendons of extensor digitorum longus
Tendon of extensor hallucis longus
Extensor hallucis brevis
Extensor digitorum brevis

B. Anterior View

5.53 **Anterior leg—superficial**

A. Surface anatomy (*numbers* refer to structures labeled in **B**). **B.** Dissection.

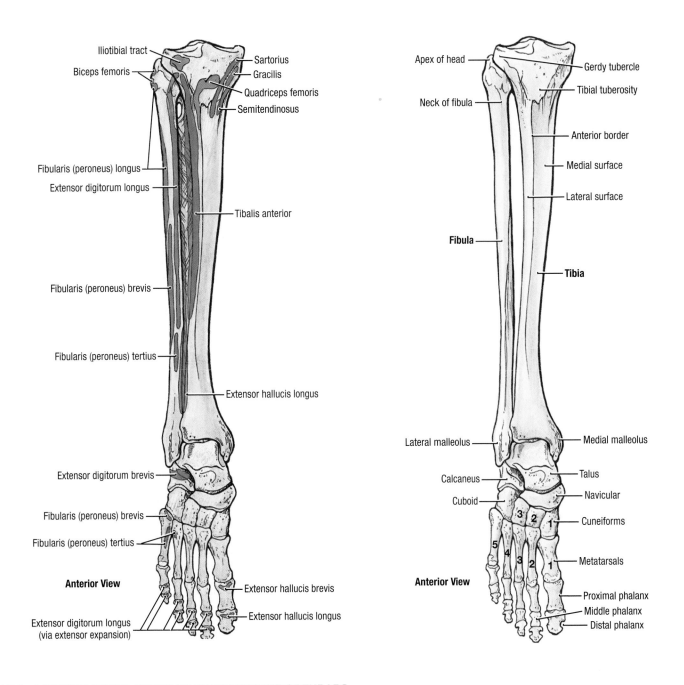

TABLE 5.8   MUSCLES OF THE ANTERIOR COMPARTMENT OF THE LEG

MUSCLE	PROXIMAL ATTACHMENT	DISTAL ATTACHMENT	INNERVATION[a]	MAIN ACTIONS
Tibialis anterior	Lateral condyle and superior half of lateral surface of tibia	Medial and inferior surfaces of medial cuneiform and base of first metatarsal	Deep fibular (peroneal) nerve (**L4** and L5)	Dorsiflexes ankle and inverts foot
Extensor hallucis longus	Middle part of anterior surface of fibula and interosseous membrane	Dorsal aspect of base of distal phalanx of great toe (hallux)	Deep fibular (peroneal) nerve (L5 and S1)	Extends great toe and dorsiflexes ankle
Extensor digitorum longus	Lateral condyle of tibia and superior three fourths and anterior surface of interosseous membrane	Middle and distal phalanges of lateral four digits		Extends lateral four digits and dorsiflexes ankle
Fibularis (peroneus) tertius	Inferior third of anterior surface of fibula and interosseus membrane	Dorsum of base of fifth metatarsal		Dorsiflexes ankle and aids in eversion of foot

[a]See Table 5.1 for explanation of segmental innervation.

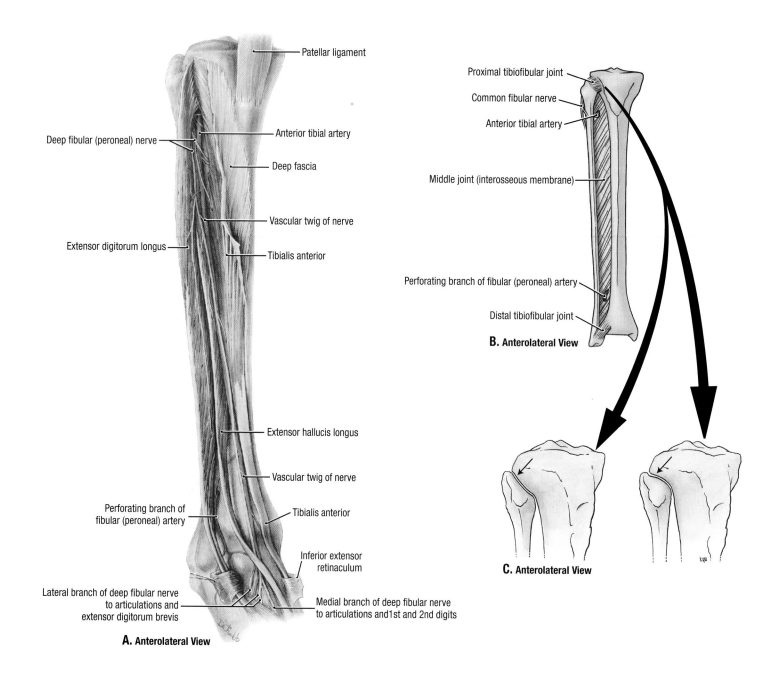

A. Anterolateral View

Patellar ligament

Deep fibular (peroneal) nerve

Anterior tibial artery

Deep fascia

Vascular twig of nerve

Extensor digitorum longus

Tibialis anterior

Extensor hallucis longus

Vascular twig of nerve

Perforating branch of fibular (peroneal) artery

Tibialis anterior

Inferior extensor retinaculum

Lateral branch of deep fibular nerve to articulations and extensor digitorum brevis

Medial branch of deep fibular nerve to articulations and1st and 2nd digits

Proximal tibiofibular joint

Common fibular nerve

Anterior tibial artery

Middle joint (interosseous membrane)

Perforating branch of fibular (peroneal) artery

Distal tibiofibular joint

B. Anterolateral View

C. Anterolateral View

5.54   Anterior leg—deep and tibiofibular joints

A. Deep dissection of the anterior compartment of the leg. The muscles are separated to display the anterior tibial artery and deep fibular nerve. **B.** Superior and inferior tibiofibular joints and interosseous membrane. **C.** Superior tibiofibular joint. Oblique form of joint surfaces is depicted on the *left* and the horizontal form on the *right*. The proximal tibiofibular joint *(arrows)* has important functions in the dissipation of torsional stress applied at the ankle and in the dissipation of lateral bending moment of the tibia. Generally, the greater the angle inclination, the smaller the surface area of the joint. Rotation at this joint occurs during dorsiflexion of the ankle.

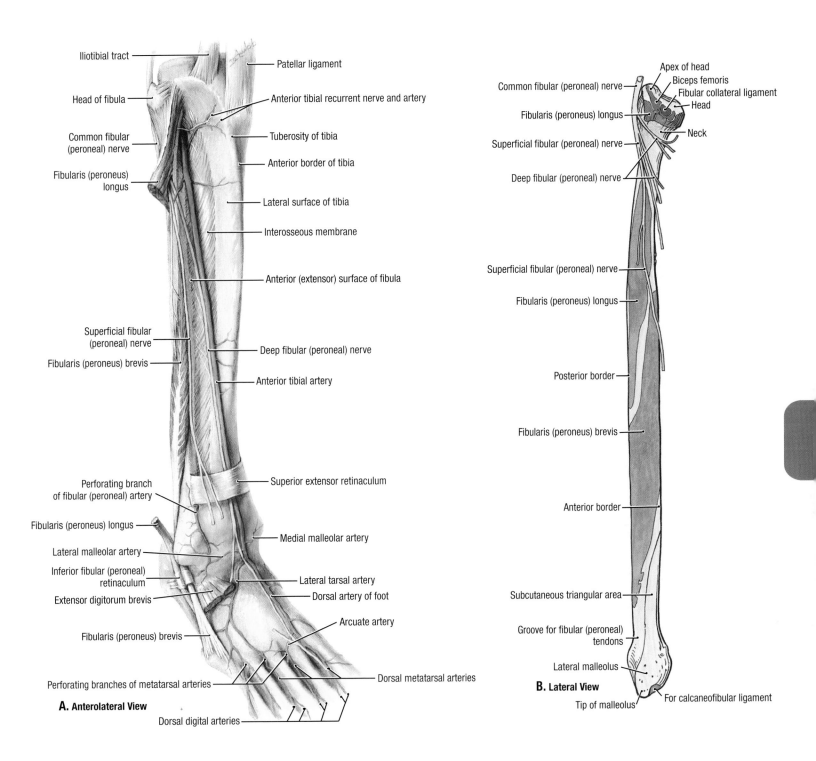

Iliotibial tract

Head of fibula

Common fibular (peroneal) nerve

Fibularis (peroneus) longus

Patellar ligament

Anterior tibial recurrent nerve and artery

Tuberosity of tibia

Anterior border of tibia

Lateral surface of tibia

Interosseous membrane

Anterior (extensor) surface of fibula

Superficial fibular (peroneal) nerve

Fibularis (peroneus) brevis

Deep fibular (peroneal) nerve

Anterior tibial artery

Perforating branch of fibular (peroneal) artery

Fibularis (peroneus) longus

Lateral malleolar artery

Inferior fibular (peroneal) retinaculum

Extensor digitorum brevis

Fibularis (peroneus) brevis

Perforating branches of metatarsal arteries

Superior extensor retinaculum

Medial malleolar artery

Lateral tarsal artery

Dorsal artery of foot

Arcuate artery

Dorsal metatarsal arteries

A. Anterolateral View

Dorsal digital arteries

Apex of head

Biceps femoris

Fibular collateral ligament

Head

Common fibular (peroneal) nerve

Fibularis (peroneus) longus

Superficial fibular (peroneal) nerve

Deep fibular (peroneal) nerve

Neck

Superficial fibular (peroneal) nerve

Fibularis (peroneus) longus

Posterior border

Fibularis (peroneus) brevis

Anterior border

Subcutaneous triangular area

Groove for fibular (peroneal) tendons

Lateral malleolus

B. Lateral View

Tip of malleolus

For calcaneofibular ligament

5.55   Arteries and nerves of anterior and lateral aspects of leg and dorsum of foot

A. Dissection, anterolateral view. The anterior crural muscles (muscles of the anterior compartment) were removed, and the fibularis (peroneus) longus muscle was excised. The common fibular (peroneal) nerve is exposed. It lies in contact with the posterior aspect of the head of the fibula, and its branches are applied directly to the neck and body of the fibula deep to the fibularis (peroneus) longus muscle. The anterior tibial artery enters the anterior compartment in contact with the medial side of the neck of the fibula.

The superficial fibular (peroneal) nerve follows the anterior border of the fibularis brevis muscle, which guides it to the surface to become cutaneous. **B.** Muscle attachments on the lateral aspect of the fibula. The relations of the common fibular nerve and branches are illustrated. The lateral surface of the fibula spirals slightly; thus, the proximal end is directed more laterally. The distal end is grooved and faces posteriorly, allowing the lateral malleolus to act as a pulley for the long and short fibularis (peroneal) tendons.

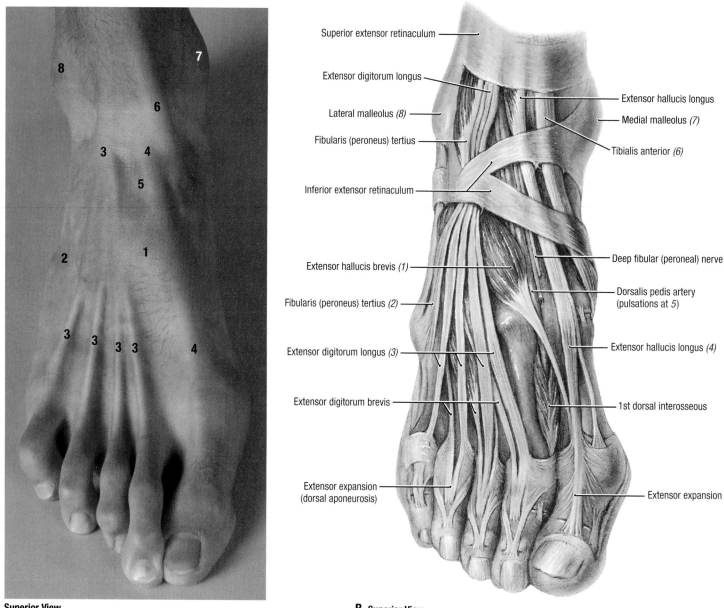

A. Superior View

B. Superior View

5.56 **Dorsum of foot**

A. Surface anatomy (*numbers* refer to structures labeled in **B**). **B.** Dissection. The dorsal vein of foot and deep fibular nerve are cut short.

- At the ankle, the dorsal artery of foot (dorsalis pedis artery) and deep fibular nerve lie midway between the malleoli.
- On the dorsum of the foot, the dorsal artery of foot is crossed by the extensor hallucis brevis muscle and disappears between the two heads of the first dorsal interosseous muscle.
- The inferior extensor retinaculum is "Y" shaped and crosses the tendons of the anterior compartment anteromedially.

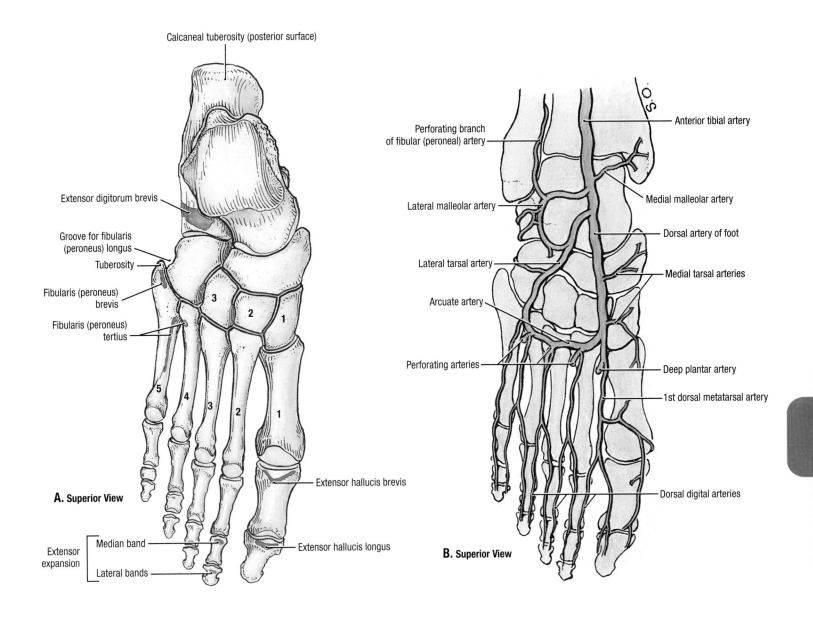

A. Superior View

Calcaneal tuberosity (posterior surface)

Extensor digitorum brevis

Groove for fibularis (peroneus) longus

Tuberosity

Fibularis (peroneus) brevis

Fibularis (peroneus) tertius

Extensor expansion
Median band
Lateral bands

Extensor hallucis brevis

Extensor hallucis longus

B. Superior View

Perforating branch of fibular (peroneal) artery

Lateral malleolar artery

Lateral tarsal artery

Arcuate artery

Perforating arteries

Anterior tibial artery

Medial malleolar artery

Dorsal artery of foot

Medial tarsal arteries

Deep plantar artery

1st dorsal metatarsal artery

Dorsal digital arteries

5.57　Bones and arteries of the dorsum of foot

A. Bones of foot. **B.** Arterial supply.

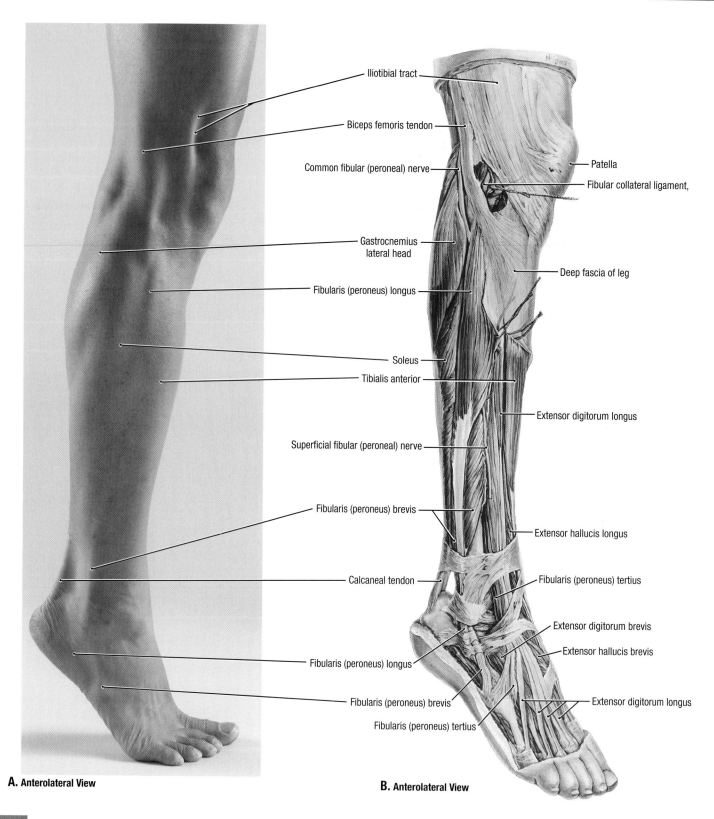

A. Anterolateral View

B. Anterolateral View

5.58 **Muscles of lateral aspect of the leg and foot**

A. Surface anatomy. **B.** Dissection

- The two fibular (peroneal) muscles both attach to two thirds of the fibula, the fibularis (peroneus) longus muscle to the proximal two thirds, and the fibularis (peroneus) brevis muscle to the distal two thirds. Where they overlap, the fibularis brevis muscle lies anteriorly.
- The fibularis (peroneus) longus muscle enters the foot by hook-

ing around the cuboid and traveling medially to the base of the first metatarsal and medial cuneiform.

- The common fibular (peroneal) nerve is in contact with the neck of the fibula deep to the fibularis longus muscle. Here it is vulnerable to injury with serious implications; because it supplies the extensor and everter muscle groups, loss of function results in foot-drop (inability to dorsiflex the ankle) and difficulty in everting the foot.

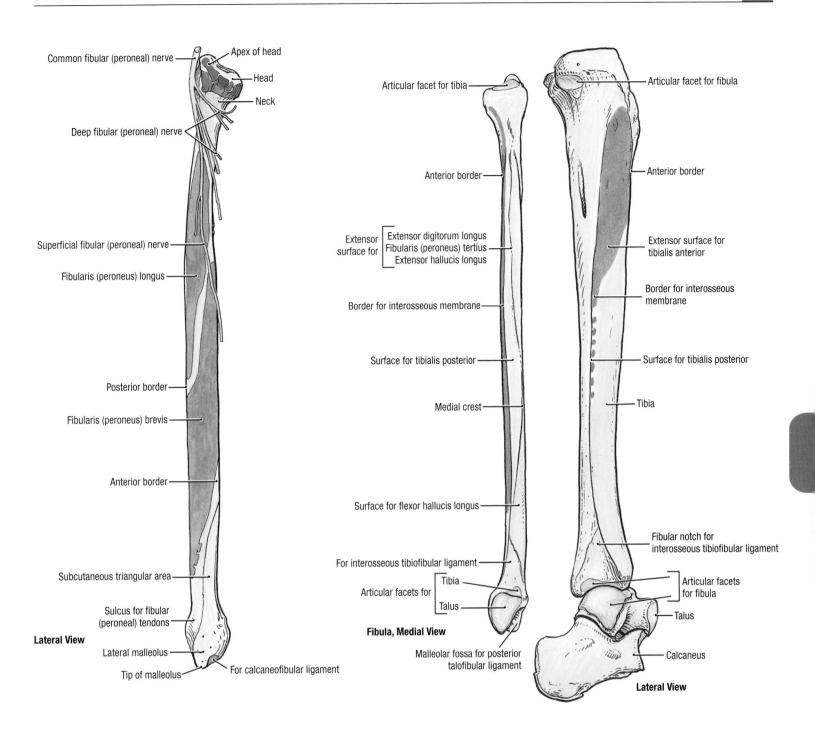

Common fibular (peroneal) nerve — Apex of head
Head
Neck
Deep fibular (peroneal) nerve

Superficial fibular (peroneal) nerve
Fibularis (peroneus) longus
Posterior border
Fibularis (peroneus) brevis
Anterior border
Subcutaneous triangular area
Sulcus for fibular (peroneal) tendons
Lateral View
Lateral malleolus
Tip of malleolus — For calcaneofibular ligament

Articular facet for tibia
Anterior border
Extensor surface for — Extensor digitorum longus / Fibularis (peroneus) tertius / Extensor hallucis longus
Border for interosseous membrane
Surface for tibialis posterior
Medial crest
Surface for flexor hallucis longus
For interosseous tibiofibular ligament
Articular facets for — Tibia / Talus
Fibula, Medial View
Malleolar fossa for posterior talofibular ligament

Articular facet for fibula
Anterior border
Extensor surface for tibialis anterior
Border for interosseous membrane
Surface for tibialis posterior
Tibia
Fibular notch for interosseous tibiofibular ligament
Articular facets for fibula
Talus
Calcaneus
Lateral View

TABLE 5.9   MUSCLES OF THE LATERAL COMPARTMENT OF THE LEG

MUSCLE	PROXIMAL ATTACHMENT	DISTAL ATTACHMENT	INNERVATION[a]	MAIN ACTIONS
Fibularis (peroneus) longus	Head and superior two thirds of lateral surface of fibula	Base of first metatarsal and medial cuneiform	Superficial fibular (peroneal) nerve (**L5**, **S1**, and **S2**)	Evert foot and weakly plantarflex ankle
Fibularis (peroneus) brevis	Inferior two thirds of lateral surface of fibula	Dorsal surface of tuberosity on lateral side of base of fifth metatarsal		

[a]See Table 5.1 for explanation of segmental innervation.

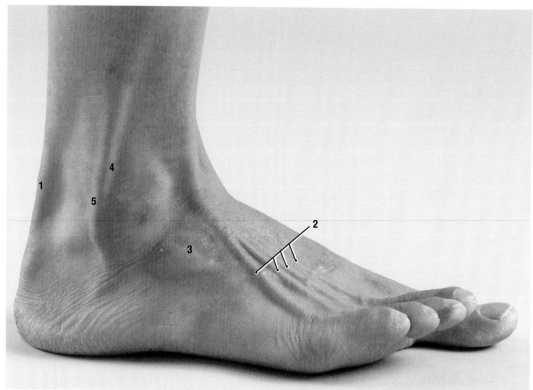

A. Lateral View

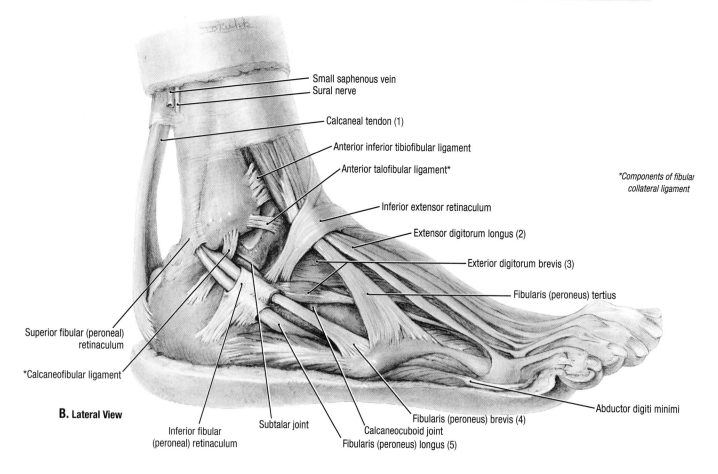

Small saphenous vein
Sural nerve

Calcaneal tendon (1)

Anterior inferior tibiofibular ligament

Anterior talofibular ligament*

*Components of fibular
collateral ligament

Inferior extensor retinaculum

Extensor digitorum longus (2)

Exterior digitorum brevis (3)

Fibularis (peroneus) tertius

Abductor digiti minimi

Superior fibular (peroneal)
retinaculum

*Calcaneofibular ligament

B. Lateral View

Inferior fibular
(peroneal) retinaculum

Subtalar joint

Calcaneocuboid joint

Fibularis (peroneus) longus (5)

Fibularis (peroneus) brevis (4)

5.59 Synovial sheaths and tendons at the ankle

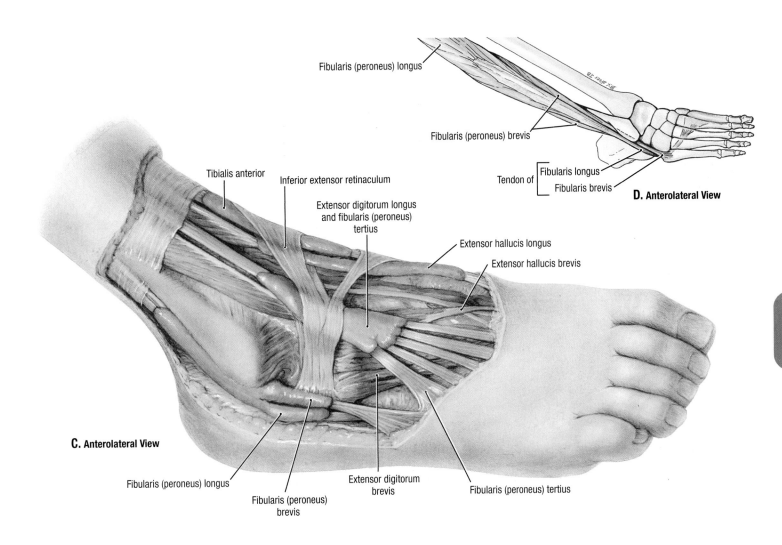

Fibularis (peroneus) longus

Fibularis (peroneus) brevis

Tibialis anterior

Inferior extensor retinaculum

Extensor digitorum longus and fibularis (peroneus) tertius

Tendon of

Fibularis longus

Fibularis brevis

D. Anterolateral View

Extensor hallucis longus

Extensor hallucis brevis

C. Anterolateral View

Fibularis (peroneus) longus

Fibularis (peroneus) brevis

Extensor digitorum brevis

Fibularis (peroneus) tertius

| 5.59 | **Synovial sheaths and tendons at the ankle (*continued*)** |

A. Surface anatomy (*numbers* refer to structures labeled in **B**). **B.** Tendons at the lateral aspect of the ankle. The ankle, subtalar, and calcaneocuboid joints are exposed to reveal their positions. **C.** Synovial sheaths of tendons on the anterolateral aspect of the ankle. The tendons of the fibularis (peroneus) longus and fibularis (peroneus) brevis muscles are enclosed in a common synovial sheath posterior to the lateral malleolus. This sheath splits into two, one for each tendon, posterior to the fibular (peroneal) trochlea. **D.** Schematic illustration of fibularis longus and brevis.

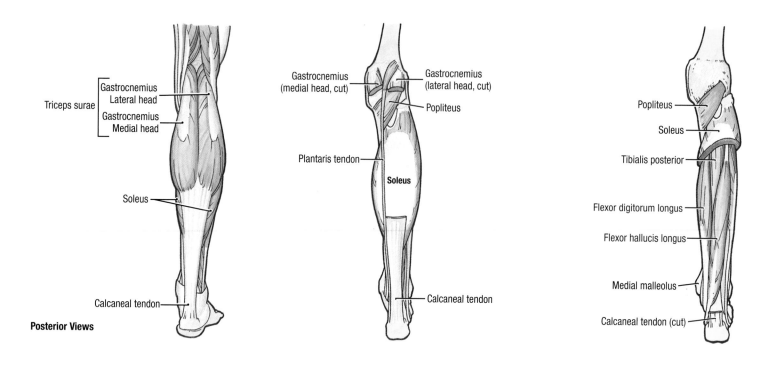

Posterior Views

TABLE 5.10 MUSCLES OF THE POSTERIOR COMPARTMENT OF THE LEG

MUSCLE	PROXIMAL ATTACHMENT	DISTAL ATTACHMENT	INNERVATION[a]	MAIN ACTIONS
Superficial muscles				
Gastrocnemius	Lateral head: lateral aspect of lateral condyle of femur; Medial head: popliteal surface of femur, superior to medial condyle	Posterior surface of calcaneus with calcaneal tendon (tendocalcaneus)	Tibial nerve (S1 and S2)	Plantarflexes ankle, raises heel during walking, and flexes leg at knee joint
Soleus	Posterior aspect of head of fibula, superior fourth of posterior surface of fibula, soleal line and medial border of tibia			Plantarflexes ankle and steadies leg on foot
Plantaris	Inferior end of lateral supracondylar line of femur and oblique popliteal ligament			Weakly assists gastrocnemius in plantarflexing ankle and flexing knee
Deep muscles				
Popliteus	Lateral surface of lateral condyle of femur and lateral meniscus	Posterior surface of tibia, superior to soleal line	Tibial nerve (**L4**, L5, and S1)	Weakly flexes knee and unlocks it
Flexor hallucis longus	Inferior two thirds of posterior surface of fibula and inferior part of interosseous membrane	Base of distal phalanx of great toe (hallux)	Tibial nerve (**S2** and S3)	Flexes great toe at all joints and plantarflexes ankle; supports medial longitudinal arch of foot
Flexor digitorum longus	Medial part of posterior surface of tibia inferior to soleal line, and by a broad tendon to fibula	Bases of distal phalanges of lateral four digits		Flexes lateral four digits and plantarflexes ankle; supports longitudinal arches of foot
Tibialis posterior	Interosseous membrane, posterior surface of tibia inferior to soleal line, and posterior surface of fibula	Tuberosity of navicular, cuneiform, and cuboid and bases of second, third, and fourth metatarsals	Tibial nerve (L4 and L5)	Plantarflexes ankle and inverts foot

[a]See Table 5.1 for explanation of segmental innervation.

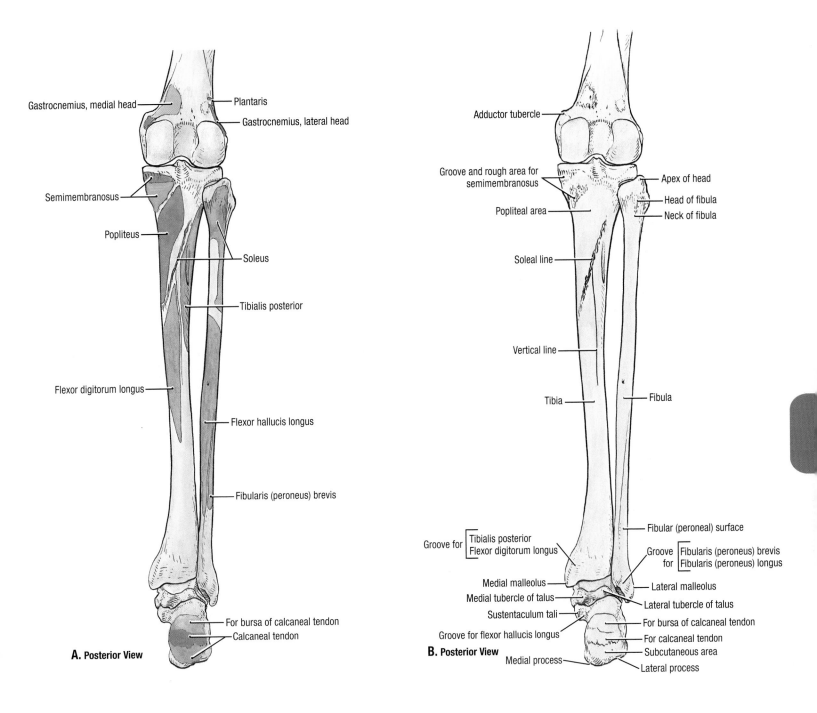

Gastrocnemius, medial head
Plantaris
Gastrocnemius, lateral head
Semimembranosus
Popliteus
Soleus
Tibialis posterior
Flexor digitorum longus
Flexor hallucis longus
Fibularis (peroneus) brevis
For bursa of calcaneal tendon
Calcaneal tendon

A. Posterior View

Adductor tubercle
Groove and rough area for semimembranosus
Apex of head
Head of fibula
Popliteal area
Neck of fibula
Soleal line
Vertical line
Tibia
Fibula
Fibular (peroneal) surface
Groove for { Tibialis posterior / Flexor digitorum longus }
Groove for { Fibularis (peroneus) brevis / Fibularis (peroneus) longus }
Medial malleolus
Lateral malleolus
Medial tubercle of talus
Lateral tubercle of talus
Sustentaculum tali
For bursa of calcaneal tendon
Groove for flexor hallucis longus
For calcaneal tendon
Medial process
Subcutaneous area
Lateral process

B. Posterior View

5.60   **Bones of the posterior aspect of the leg**

A. Muscle attachments. **B.** Features of bones.

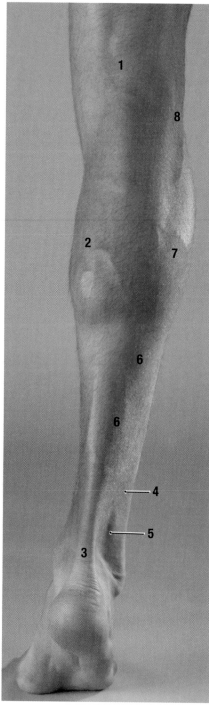

A. Posterior View

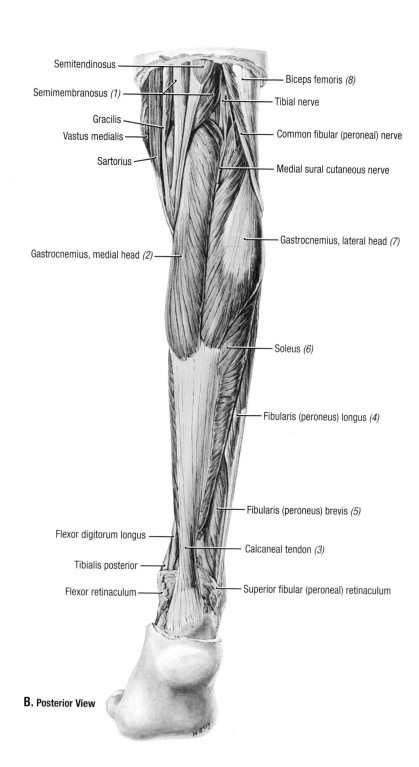

Semitendinosus

Semimembranosus *(1)*

Gracilis

Vastus medialis

Sartorius

Gastrocnemius, medial head *(2)*

Biceps femoris *(8)*

Tibial nerve

Common fibular (peroneal) nerve

Medial sural cutaneous nerve

Gastrocnemius, lateral head *(7)*

Soleus *(6)*

Fibularis (peroneus) longus *(4)*

Fibularis (peroneus) brevis *(5)*

Flexor digitorum longus

Tibialis posterior

Flexor retinaculum

Calcaneal tendon *(3)*

Superior fibular (peroneal) retinaculum

B. Posterior View

5.61   **Posterior leg, superficial compartment**

A. Surface anatomy. *Numbers* refer to labeled structures in **B. B.** Gastrocnemius muscle. The proximal attachments of this muscle lie deep to the distal ends of the hamstring muscles, (semitendinosus, semimembranosus, and biceps femoris). The two bellies of the gastrocnemius cover the distal popliteal fossa.

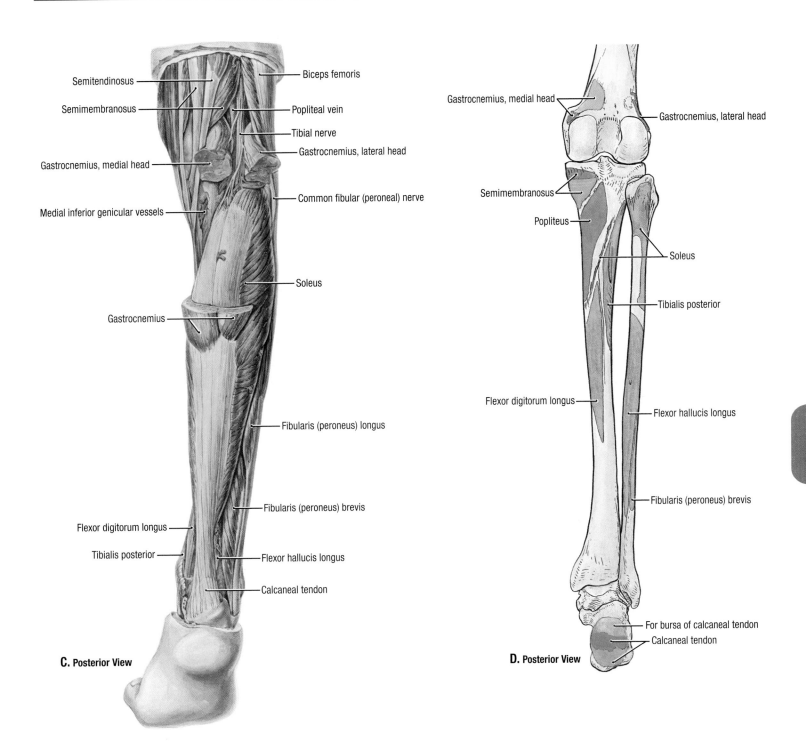

Semitendinosus
Semimembranosus
Gastrocnemius, medial head
Medial inferior genicular vessels
Gastrocnemius
Flexor digitorum longus
Tibialis posterior

Biceps femoris
Popliteal vein
Tibial nerve
Gastrocnemius, lateral head
Common fibular (peroneal) nerve
Soleus
Fibularis (peroneus) longus
Fibularis (peroneus) brevis
Flexor hallucis longus
Calcaneal tendon

C. Posterior View

Gastrocnemius, medial head
Semimembranosus
Popliteus
Flexor digitorum longus

Gastrocnemius, lateral head
Soleus
Tibialis posterior
Flexor hallucis longus
Fibularis (peroneus) brevis
For bursa of calcaneal tendon
Calcaneal tendon

D. Posterior View

5.61   **Posterior leg, superficial compartment** *(continued)*

C. Soleus muscle. The fleshy bellies of the gastrocnemius muscle are largely excised, and the proximal attachment of the soleus muscle is thereby exposed; the plantaris muscle is absent from this specimen. **D.** Bones of leg showing muscle attachments.

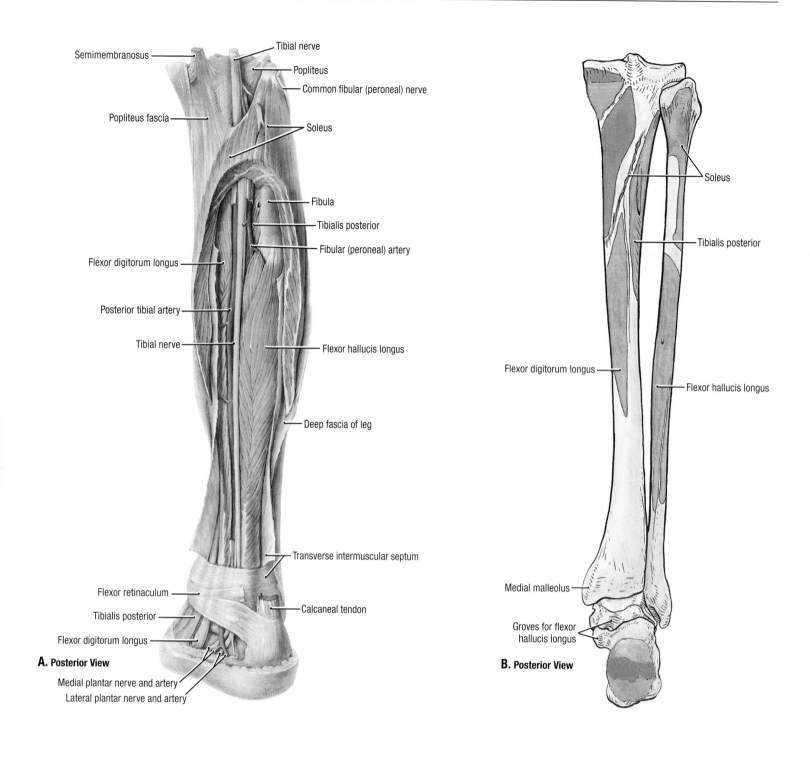

A. Posterior View

Semimembranosus
Tibial nerve
Popliteus
Common fibular (peroneal) nerve
Popliteus fascia
Soleus
Fibula
Tibialis posterior
Fibular (peroneal) artery
Flexor digitorum longus
Posterior tibial artery
Tibial nerve
Flexor hallucis longus
Deep fascia of leg
Transverse intermuscular septum
Flexor retinaculum
Tibialis posterior
Calcaneal tendon
Flexor digitorum longus
Medial plantar nerve and artery
Lateral plantar nerve and artery

B. Posterior View

Soleus
Tibialis posterior
Flexor digitorum longus
Flexor hallucis longus
Medial malleolus
Groves for flexor hallucis longus

5.62   Dissection of posterior leg, deep compartment

A. Superficial dissection. The calcaneal tendon (Achilles tendon) is cut, and the gastrocnemius muscle is removed and a horseshoe-shaped proximal part of the soleus muscle remains in place. Observe the bipennate structure of the large flexor hallucis longus and the smaller flexor digitorum longus muscles. The posterior tibial artery and the tibial nerve descend between these two muscles on a layer of fascia that covers the tibialis posterior. **B.** Bones of leg showing muscle attachments. Note that the tibial and fibular parts of the proximal part of the soleus forms an inverted "U." The proximal attachment of tibialis posterior is largely from the interosseous membrane, which has been removed.

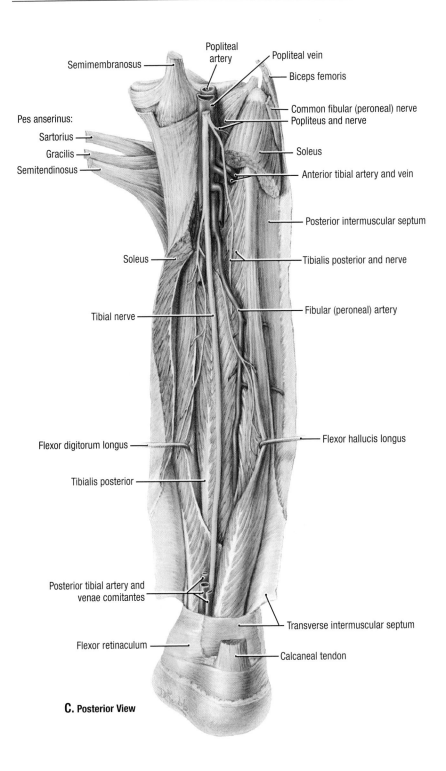

C. Posterior View

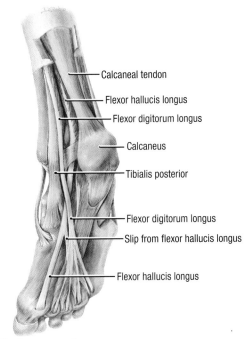

D. Anteromedial View

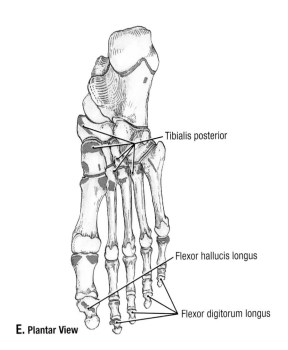

E. Plantar View

5.62   **Dissection of posterior leg, deep compartment** *(continued)*

C. Deeper dissection. The soleus muscle is largely cut away, the two long digital flexors are pulled apart, and the posterior tibial artery is partly excised. Note that the tibialis posterior lies deep to the two long digital flexors and that the fibular (peroneal) artery is overlapped by the flexor hallucis longus muscle. **D.** Crossing of muscles (tendons) of the deep compartment posterior to the medial malleolus and into the sole of the foot. **E.** Bones of foot showing muscle attachments.

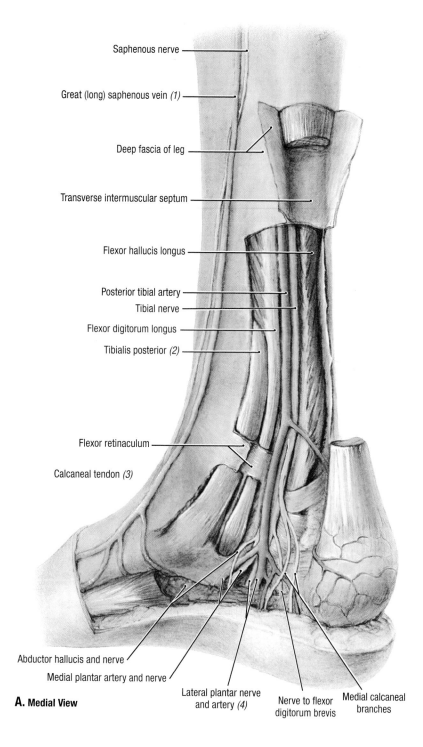

Saphenous nerve

Great (long) saphenous vein *(1)*

Deep fascia of leg

Transverse intermuscular septum

Flexor hallucis longus

Posterior tibial artery

Tibial nerve

Flexor digitorum longus

Tibialis posterior *(2)*

Flexor retinaculum

Calcaneal tendon *(3)*

Abductor hallucis and nerve

Medial plantar artery and nerve

Lateral plantar nerve and artery *(4)*

Nerve to flexor digitorum brevis

Medial calcaneal branches

A. Medial View

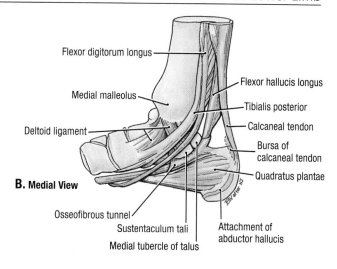

Flexor digitorum longus

Medial malleolus

Deltoid ligament

Flexor hallucis longus

Tibialis posterior

Calcaneal tendon

Bursa of calcaneal tendon

Quadratus plantae

Osseofibrous tunnel

Sustentaculum tali

Medial tubercle of talus

Attachment of abductor hallucis

B. Medial View

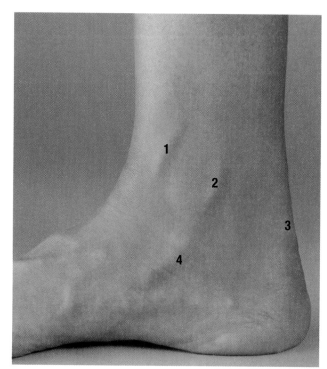

C. Medial View

5.63 Medial ankle region

A. Dissection. The calcaneal tendon and posterior part of the abductor hallucis was excised. **B.** Schematic illustration of the tendons passing posterior to medial malleolus. **C.** Surface anatomy (*numbers* refer to structures labeled in **A**).

- The posterior tibial artery and the tibial nerve lie between the flexor digitorum longus and flexor hallucis longus muscles and divide into medial and lateral plantar branches on the surface of the osseofibrous tunnel of the flexor hallucis longus muscle.

- The pulsations of the posterior tibial artery are palpated just posterior to the medial malleolus.
- The tibialis posterior and flexor digitorum longus tendons occupy separate osseofibrous tunnels posterior to the medial malleolus, which the tendons use as a pulley; only the flexor hallucis longus muscle uses the sustentaculum tali as a pulley.
- The medial and lateral plantar nerves lie within the fork of the medial and lateral plantar arteries.

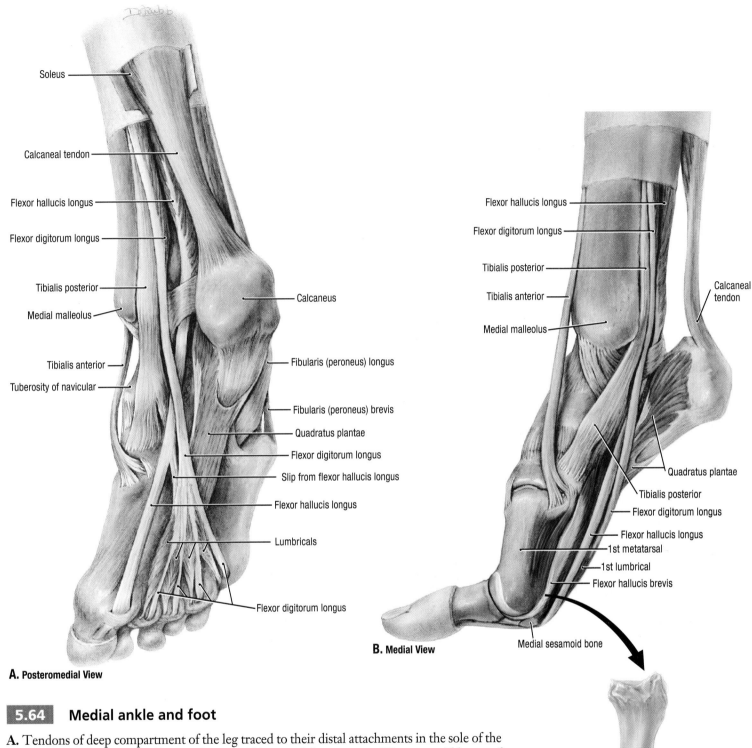

A. Posteromedial View

B. Medial View

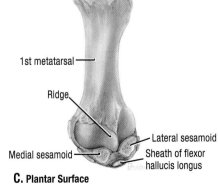

C. Plantar Surface

5.64   Medial ankle and foot

A. Tendons of deep compartment of the leg traced to their distal attachments in the sole of the foot. **B.** Foot raised as in walking and sesamoid bones of the great toe. The sesamoid bones of the great toe are bound together and located on each side of a bony ridge on the first metatarsal.

- The sesamoid bones are a "footstool" for the first metatarsal, giving it increased height; by inserting into the flexor digitorum longus muscle, the quadratus plantae muscle acts as a guy wire, modifying the oblique pull of the flexor tendons.
- The flexor hallucis longus muscle uses three pulleys: a groove on the posterior aspect of the distal end of the tibia, a groove on the posterior aspect of the talus, and a groove inferior to the sustentaculum tali.
- The flexor digitorum longus muscle crosses superficial to the tibialis posterior, posterior to the medial malleolus, and superficial to the flexor hallucis longus muscle at the tuberosity of the navicular bone.

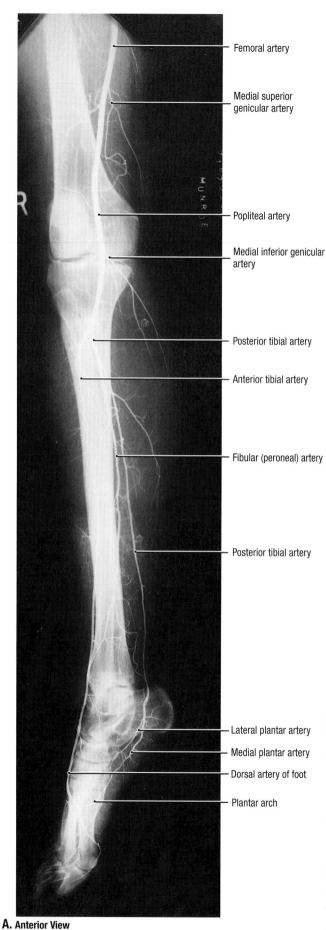

A. Anterior View

Femoral artery

Medial superior genicular artery

Popliteal artery

Medial inferior genicular artery

Posterior tibial artery

Anterior tibial artery

Fibular (peroneal) artery

Posterior tibial artery

Lateral plantar artery

Medial plantar artery

Dorsal artery of foot

Plantar arch

Anterior tibial artery

Dorsal artery of foot

Perforating branch of fibular (peroneal) artery

Calcaneal branch

B. Lateral View

Popliteal artery

Popliteal artery

Anterior tibial artery

Popliteus

Fibular (peroneal) artery

Anterior tibial artery

Posterior tibial artery

D. Posterior View

Flexor hallucis longus

Communicating branch

Perforating branch

Posterior tibial artery

Calcaneal branch

C. Posterior View

5.65 Popliteal arteriogram and arterial anomalies

A. Popliteal arteriogram. The femoral artery becomes the popliteal artery at the adductor hiatus. The anterior tibial artery supplies the anterior compartment of the leg and the ankle; it continues as the dorsal artery of the foot. The posterior tibial artery supplies the posterior compartment of the leg and terminates as the medial and lateral plantar arteries; its major branch is the fibular artery. It supplies the posterior and lateral compartments of the leg. **B.** Anomalous dorsal artery of the foot. The perforating branch of the fibular artery rarely continues as the dorsal artery of the foot, but when it does, the anterior tibial artery ends proximal to the ankle or is a slender vessel. **C.** Absence of posterior tibial artery. Compensatory enlargement of the fibular artery was found to occur in approximately 5% of limbs. **D.** High division of popliteal artery. Along with the anterior tibial artery descending anterior to the popliteus muscle; this anomaly was found to occur in approximately 2% of limbs.

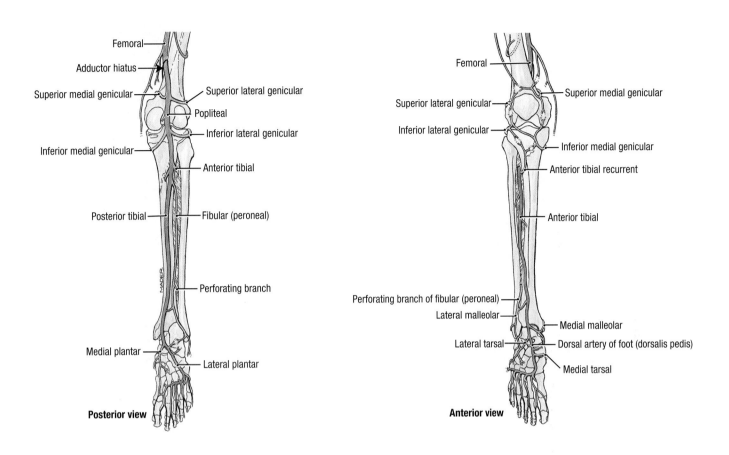

Posterior view

Femoral
Adductor hiatus
Superior medial genicular
Superior lateral genicular
Popliteal
Inferior lateral genicular
Inferior medial genicular
Anterior tibial
Posterior tibial
Fibular (peroneal)
Perforating branch
Medial plantar
Lateral plantar

Anterior view

Femoral
Superior medial genicular
Superior lateral genicular
Inferior lateral genicular
Inferior medial genicular
Anterior tibial recurrent
Anterior tibial
Perforating branch of fibular (peroneal)
Lateral malleolar
Medial malleolar
Lateral tarsal
Dorsal artery of foot (dorsalis pedis)
Medial tarsal

TABLE 5.11 ARTERIAL SUPPLY OF THE LEG AND FOOT

ARTERY	ORIGIN	COURSE	DISTRIBUTION IN LEG
Popliteal	Continuation of femoral artery at adductor hiatus in adductor magnus	Passes through popliteal fossa to leg; ends at lower border of popliteus muscle by dividing into anterior and posterior tibial arteries	Superior, middle, and inferior genicular arteries to both lateral and medial aspects of knee
Anterior tibial	Popliteal	Passes between tibia and fibula into anterior compartment through gap in superior part of interosseous membrane and descends on this membrane between tibialis anterior and extensor digitorum longus	Anterior compartment of leg
Dorsal artery of foot (dorsalis pedis)	Continuation of anterior tibial artery distal to inferior extensor retinaculum	Descends anteromedially to first interosseous space and divides into plantar and arcuate arteries	Muscles on dorsum of foot; pierces first dorsal interosseous muscle as deep plantar artery to contribute to formation of plantar arch
Posterior tibial	Popliteal	Passes through posterior compartment of leg and terminates distal to flexor retinaculum by dividing into medial and lateral plantar arteries	Posterior and lateral compartments of leg; circumflex fibular branch joins anastomoses around knee; nutrient artery passes to tibia
Fibular (peroneal)	Posterior tibial	Descends in posterior compartment adjacent to posterior intermuscular septum	Posterior compartment of leg: perforating branches supply lateral compartment of leg
Medial plantar	Posterior tibial	Passes distally in foot between abductor hallucis and flexor digitorum brevis	Supplies mainly muscles of great toe and skin on medial side of sole
Lateral plantar	Posterior tibial	Runs anterolaterally deep to abductor hallucis and flexor digitorum brevis and then arches medially to form deep plantar arch	Supplies remainder of sole of foot

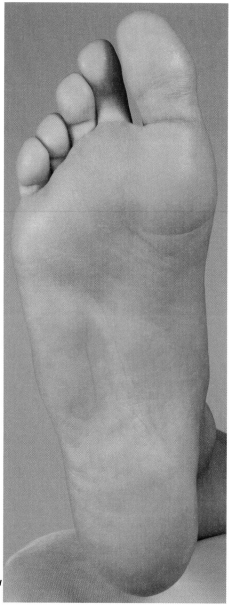

A. Plantar View

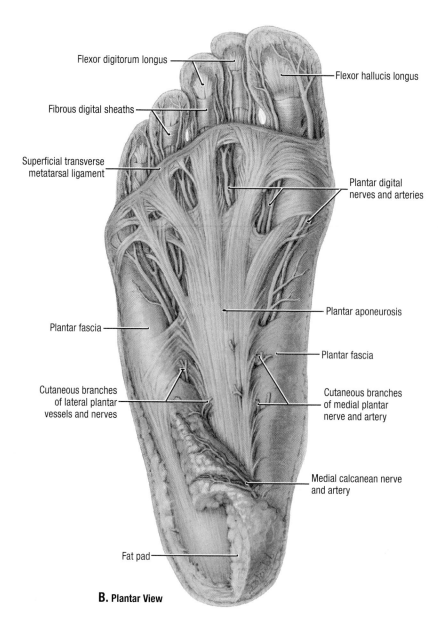

Flexor digitorum longus

Flexor hallucis longus

Fibrous digital sheaths

Superficial transverse
metatarsal ligament

Plantar digital
nerves and arteries

Plantar aponeurosis

Plantar fascia

Plantar fascia

Cutaneous branches
of lateral plantar
vessels and nerves

Cutaneous branches
of medial plantar
nerve and artery

Medial calcanean nerve
and artery

Fat pad

B. Plantar View

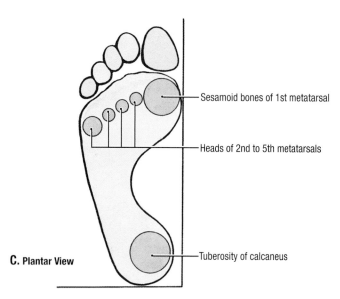

Sesamoid bones of 1st metatarsal

Heads of 2nd to 5th metatarsals

Tuberosity of calcaneus

C. Plantar View

5.66   **Sole of foot, superficial**

A. Surface anatomy. **B.** Plantar aponeurosis. **C.** Weight-bearing areas.
The weight of the body is transmitted to the talus from the tibia and
fibula. It is then transmitted to the tuberosity of the calcaneus, the
heads of the second to fifth metatarsals, and the sesamoid bones of the
first digit.

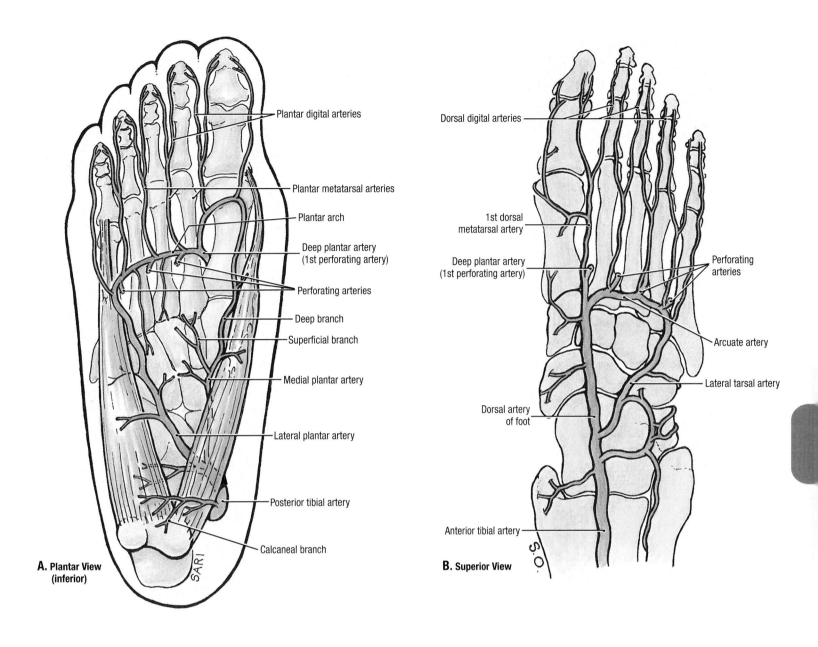

Plantar digital arteries

Plantar metatarsal arteries

Plantar arch

Deep plantar artery
(1st perforating artery)

Perforating arteries

Deep branch

Superficial branch

Medial plantar artery

Lateral plantar artery

Posterior tibial artery

Calcaneal branch

**A. Plantar View
(inferior)**

Dorsal digital arteries

1st dorsal
metatarsal artery

Deep plantar artery
(1st perforating artery)

Perforating
arteries

Arcuate artery

Lateral tarsal artery

Dorsal artery
of foot

Anterior tibial artery

B. Superior View

| 5.67 | **Arteries of foot** |

A. Arteries on plantar surface of foot. **B.** Arteries on dorsum of foot. The dorsal arterial arch
is formed by the dorsal artery of the foot, which also contributes to the plantar arch formed
by the medial and lateral plantar arteries.

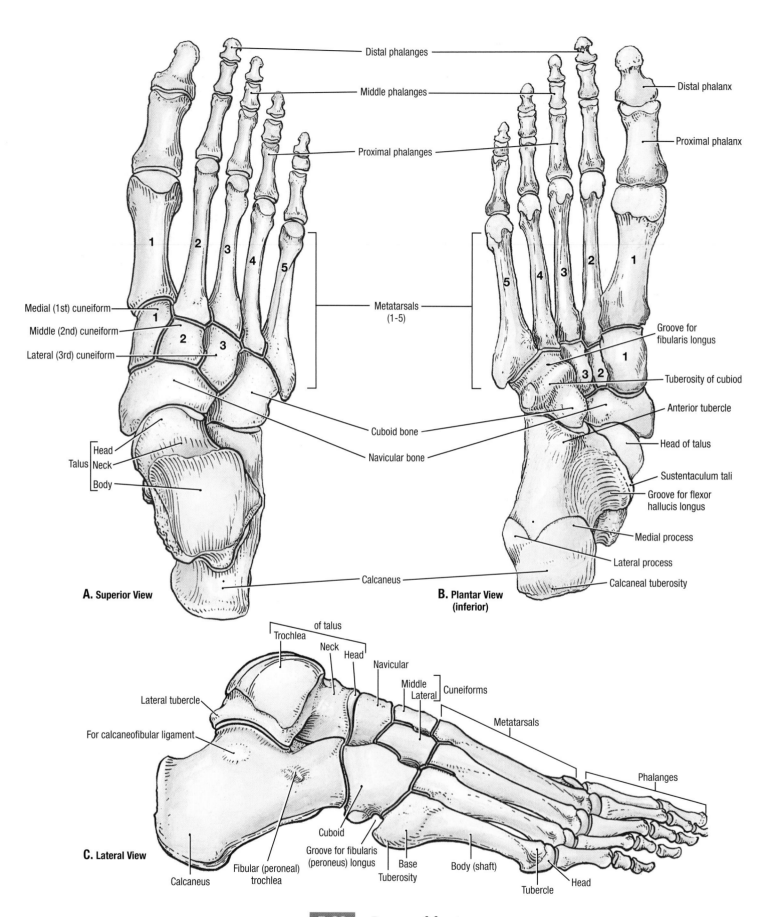

Distal phalanges

Middle phalanges

Proximal phalanges

Distal phalanx

Proximal phalanx

Medial (1st) cuneiform

Middle (2nd) cuneiform

Lateral (3rd) cuneiform

Metatarsals
(1-5)

Groove for
fibularis longus

Tuberosity of cubiod

Anterior tubercle

Head of talus

Sustentaculum tali

Groove for flexor
hallucis longus

Cuboid bone

Navicular bone

Head
Talus Neck
Body

Medial process

Lateral process

Calcaneus

Calcaneal tuberosity

A. Superior View

B. Plantar View
(inferior)

Trochlea
of talus
Neck
Head

Navicular

Middle
Lateral

Cuneiforms

Metatarsals

Lateral tubercle

For calcaneofibular ligament

Phalanges

Cuboid

Groove for fibularis
(peroneus) longus

Base

Tuberosity

Body (shaft)

C. Lateral View

Calcaneus

Fibular (peroneal)
trochlea

Tubercle

Head

5.68   Bones of foot

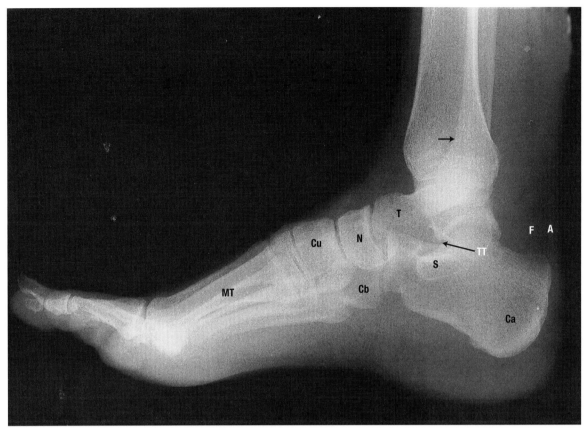

D. Medial View

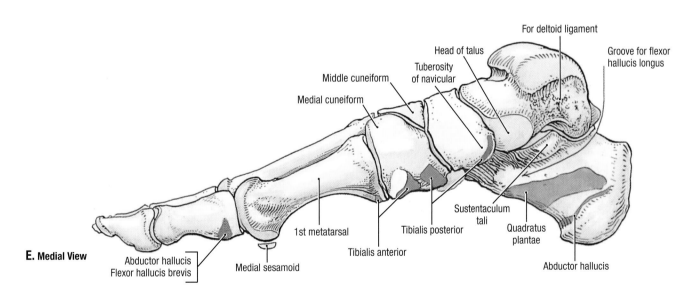

For deltoid ligament

Head of talus

Tuberosity of navicular

Middle cuneiform

Medial cuneiform

Groove for flexor hallucis longus

Sustentaculum tali

1st metatarsal

Tibialis posterior

Quadratus plantae

Tibialis anterior

E. Medial View

Abductor hallucis
Flexor hallucis brevis

Medial sesamoid

Abductor hallucis

5.68   **Bones of foot *(continued)***

A. Dorsal surface. **B.** Plantar surface. **C.** Lateral aspect. The lateral part of the longitudinal arch of the foot consists of the calcaneus, cuboid, and fourth and fifth metatarsals. **D.** Lateral radiograph of ankle and foot. *T*, talus; *Ca*, calcaneus; *S*, sustentaculum tali; *N*, navicular; *Cu*, cuneiforms; *Cb*, cuboid; *MT*, metatarsal; *TT*, tarsal tunnel; *A*, Achilles tendon; *F*, fat; *arrowhead*, superimposed tibia and fibula. **E.** Medial aspect. The medial part of the longitudinal arch of the foot consists of the calcaneus, talus, navicular, three cuneiforms, and first, second, and third metatarsals.

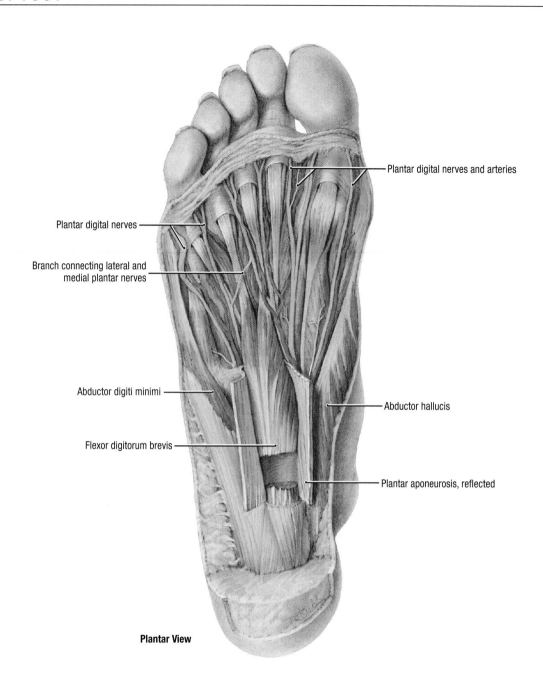

Plantar digital nerves and arteries

Plantar digital nerves

Branch connecting lateral and
medial plantar nerves

Abductor digiti minimi

Flexor digitorum brevis

Abductor hallucis

Plantar aponeurosis, reflected

Plantar View

TABLE 5.12   MUSCLES IN SOLE OF FOOT—FIRST LAYER

MUSCLE	PROXIMAL ATTACHMENT	DISTAL ATTACHMENT	INNERVATION	MAIN ACTIONS
Abductor hallucis	Medial process of tuberosity of calcaneus, flexor retinaculum, and plantar aponeurosis	Medial side of base of proximal phalanx of first digit	Medial plantar nerve (S2 and **S3**)	Abducts and flexes
Flexor digitorum brevis	Medial process of tuberosity of calcaneus, plantar aponeurosis, and intermuscular septa	Both sides of middle phalanges of lateral four digits		Flexes lateral four digits
Abductor digiti minimi	Medial and lateral processes of tuberosity of calcaneus, plantar aponeurosis, and intermuscular septa	Lateral side of base of proximal phalanx of fifth digit	Lateral plantar nerve (S2 and **S3**)	Abducts and flexes fifth digit

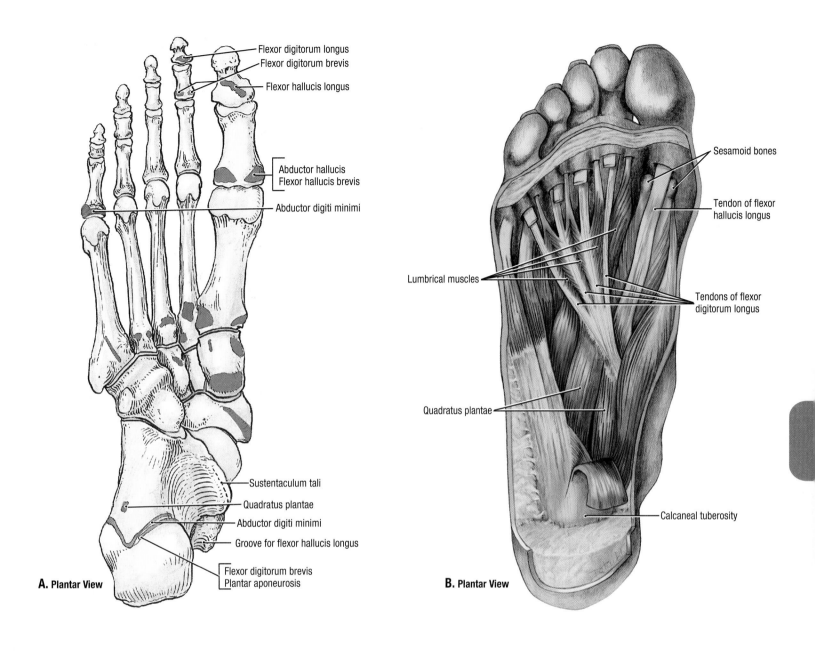

A. Plantar View

- Flexor digitorum longus
- Flexor digitorum brevis
- Flexor hallucis longus
- Abductor hallucis / Flexor hallucis brevis
- Abductor digiti minimi
- Sustentaculum tali
- Quadratus plantae
- Abductor digiti minimi
- Groove for flexor hallucis longus
- Flexor digitorum brevis / Plantar aponeurosis

B. Plantar View

- Sesamoid bones
- Tendon of flexor hallucis longus
- Tendons of flexor digitorum longus
- Lumbrical muscles
- Quadratus plantae
- Calcaneal tuberosity

TABLE 5.13   MUSCLES IN SOLE OF FOOT—SECOND LAYER

MUSCLE	PROXIMAL ATTACHMENT	DISTAL ATTACHMENT	INNERVATION	MAIN ACTIONS
Quadratus plantae	Medial surface and lateral margin of plantar surface of calcaneus	Posterolateral margin of tendon of flexor digitorum longus	Lateral plantar nerve (S2 and **S3**)	Assists flexor digitorum longus in flexing lateral four digits
Lumbricals	Tendons of flexor digitorum longus	Medial aspect of extensor expansion over lateral four digits	*Medial one:* medial plantar nerve (S2 and **S3**); *Lateral three:* lateral plantar nerve (S2 and **S3**)	Flex proximal phalanges and extend middle and distal phalanges of lateral four digits

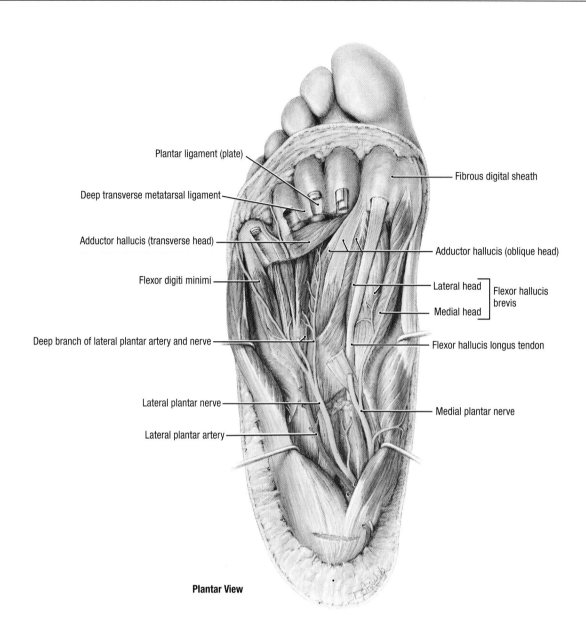

Plantar ligament (plate)

Deep transverse metatarsal ligament

Adductor hallucis (transverse head)

Flexor digiti minimi

Deep branch of lateral plantar artery and nerve

Lateral plantar nerve

Lateral plantar artery

Fibrous digital sheath

Adductor hallucis (oblique head)

Lateral head ⎤ Flexor hallucis
Medial head ⎦ brevis

Flexor hallucis longus tendon

Medial plantar nerve

Plantar View

TABLE 5.14 MUSCLES IN SOLE OF FOOT—THIRD LAYER

MUSCLE	PROXIMAL ATTACHMENT	DISTAL ATTACHMENT	INNERVATION	MAIN ACTIONS
Flexor hallucis brevis	Plantar surfaces of cuboid and lateral cuneiforms	Both sides of base of proximal phalanx of first digit	Medial plantar nerve (S2 and **S3**)	Flexes proximal phalanx of first digit
Adductor hallucis	*Oblique head:* bases of metatarsals 2–4; *Transverse head:* plantar ligaments of metatarsophalangeal joints	Tendons of both heads attach to lateral side of base of proximal phalanx of first digit	Deep branch of lateral plantar nerve (S2 and **S3**)	Adducts first digit; assists in maintaining transverse arch of foot
Flexor digiti minimi	Base of fifth metatarsal	Base of proximal phalanx of fifth digit	Superficial branch of lateral plantar nerve (S2 and **S3**)	Flexes proximal phalanx of fifth digit, thereby assisting with its flexion

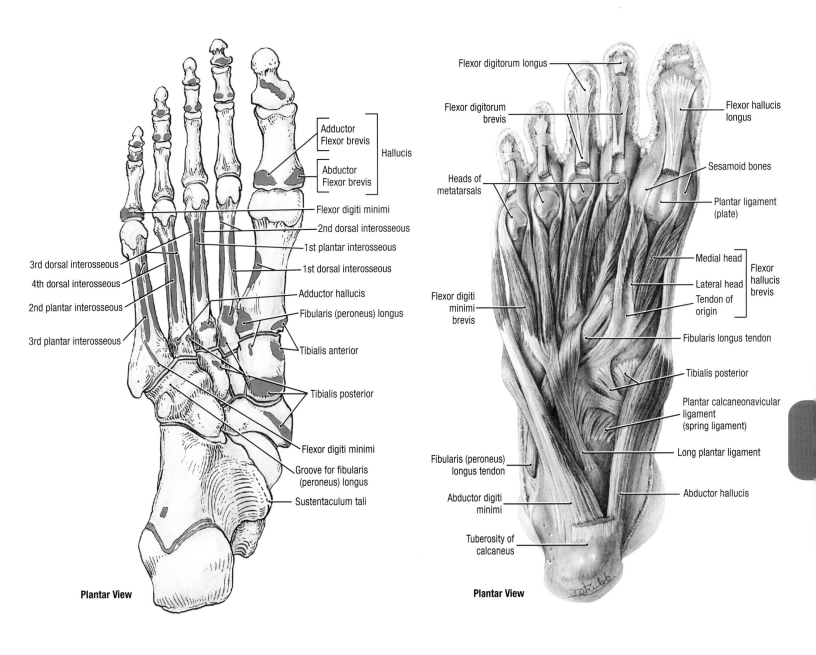

Plantar View

Adductor
Flexor brevis } Hallucis

Abductor
Flexor brevis

Flexor digiti minimi

2nd dorsal interosseous

1st plantar interosseous

1st dorsal interosseous

Adductor hallucis

Fibularis (peroneus) longus

Tibialis anterior

Tibialis posterior

Flexor digiti minimi

Groove for fibularis
(peroneus) longus

Sustentaculum tali

3rd dorsal interosseous

4th dorsal interosseous

2nd plantar interosseous

3rd plantar interosseous

Plantar View

Flexor digitorum longus

Flexor digitorum
brevis

Flexor hallucis
longus

Sesamoid bones

Plantar ligament
(plate)

Heads of
metatarsals

Medial head

Lateral head } Flexor
hallucis
brevis

Tendon of
origin

Fibularis longus tendon

Tibialis posterior

Plantar calcaneonavicular
ligament
(spring ligament)

Long plantar ligament

Abductor hallucis

Flexor digiti
minimi
brevis

Fibularis (peroneus)
longus tendon

Abductor digiti
minimi

Tuberosity of
calcaneus

TABLE 5.15   MUSCLES IN SOLE OF FOOT—FOURTH LAYER

MUSCLE	PROXIMAL ATTACHMENT	DISTAL ATTACHMENT	INNERVATION	MAIN ACTIONS
Plantar interossei (three muscles)	Bases and medial sides of metatarsals 3–5	Medial sides of bases of proximal phalanges of third to fifth digits	Lateral plantar nerve (S2 and S3)	Adduct digits (2–4) and flex metatarsophalangeal joints
Dorsal interossei (four muscles)	Adjacent sides of metatarsals 1–5	First: medial side of proximal phalanx of second digit. Second to fourth: lateral sides of second to fourth digits		Abduct digits (2–4) and flex metatarsophalangeal joints

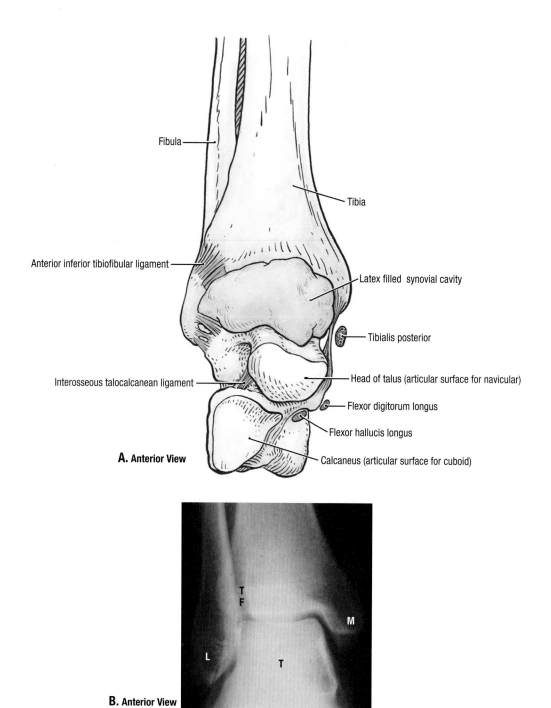

A. Anterior View

Fibula

Tibia

Anterior inferior tibiofibular ligament

Latex filled synovial cavity

Tibialis posterior

Head of talus (articular surface for navicular)

Interosseous talocalcanean ligament

Flexor digitorum longus

Flexor hallucis longus

Calcaneus (articular surface for cuboid)

B. Anterior View

5.69   Joint cavity of ankle joint

A. Ankle joint with synovial cavity distended with injected latex. **B.** Radiograph of anterior aspect of ankle.

- Observe the extension of the synovial membrane onto the neck of the talus.
- The anterior articular surfaces of the talus and calcaneus are each convex from side to side; thus the foot can be inverted and everted at the transverse tarsal joint.
- Note the relations of the tendons to the sustentaculum tali: the flexor hallucis longus inferior to it, flexor digitorum longus along its medial aspect, and tibialis posterior superior to it and in contact with the deltoid ligament. *L*, lateral malleolus; *M*, medial malleolus; *T*, talus; *TF*, inferior tibiofibular joint.

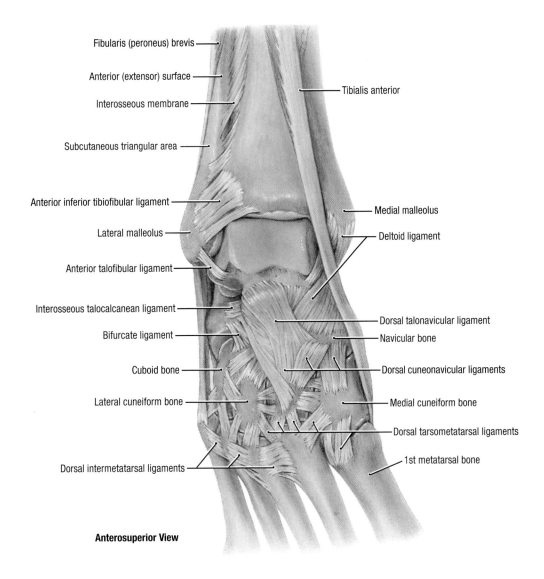

Fibularis (peroneus) brevis

Anterior (extensor) surface

Interosseous membrane

Subcutaneous triangular area

Anterior inferior tibiofibular ligament

Lateral malleolus

Anterior talofibular ligament

Interosseous talocalcanean ligament

Bifurcate ligament

Cuboid bone

Lateral cuneiform bone

Dorsal intermetatarsal ligaments

Tibialis anterior

Medial malleolus

Deltoid ligament

Dorsal talonavicular ligament

Navicular bone

Dorsal cuneonavicular ligaments

Medial cuneiform bone

Dorsal tarsometatarsal ligaments

1st metatarsal bone

Anterosuperior View

5.70   Ankle joint and ligaments of dorsum of foot

The ankle joint is plantar flexed, and its anterior capsular fibers are removed.
- The fibers of the interosseous membrane and ligaments uniting the fibula to the tibia are directed to resist the inferior pull of the muscles but allow the fibula to be forced superiorly.
- The anterior talofibular ligament is a weak band that is easily torn.

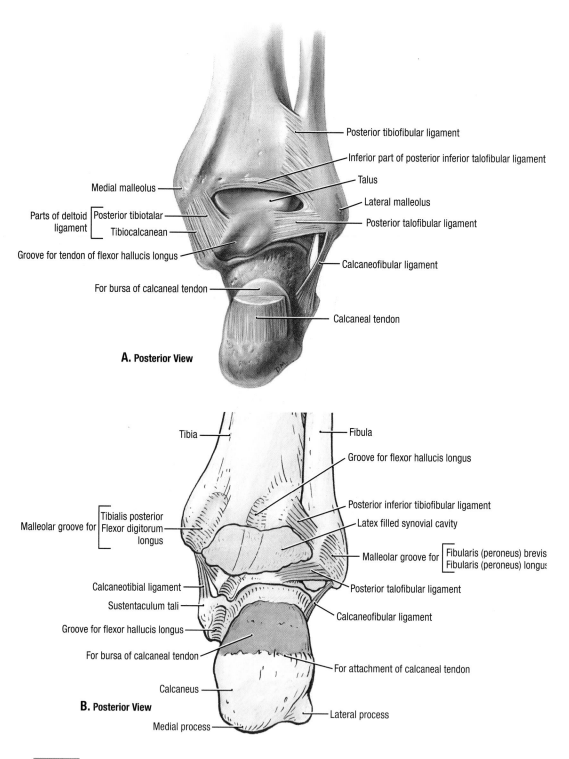

A. Posterior View

B. Posterior View

5.71 **Posterior aspect of the ankle joint**

A. Dissection. **B.** Distended ankle joint. Observe the grooves for the flexor hallucis longus muscle, which crosses the middle of the ankle joint posteriorly, the two tendons posterior to the medial malleolus, and the two tendons posterior to the lateral malleolus.

- The posterior aspect of the ankle joint is strengthened by the transversely oriented posterior tibiofibular and posterior talofibular ligaments.
- The calcaneofibular ligament stabilizes the joint laterally, and the posterior tibiotalar and tibiocalcanean parts of the deltoid ligament stabilize it medially.
- The groove for the flexor hallucis tendon is between the medial and lateral tubercles of the talus and continues inferior to the sustentaculum tali.

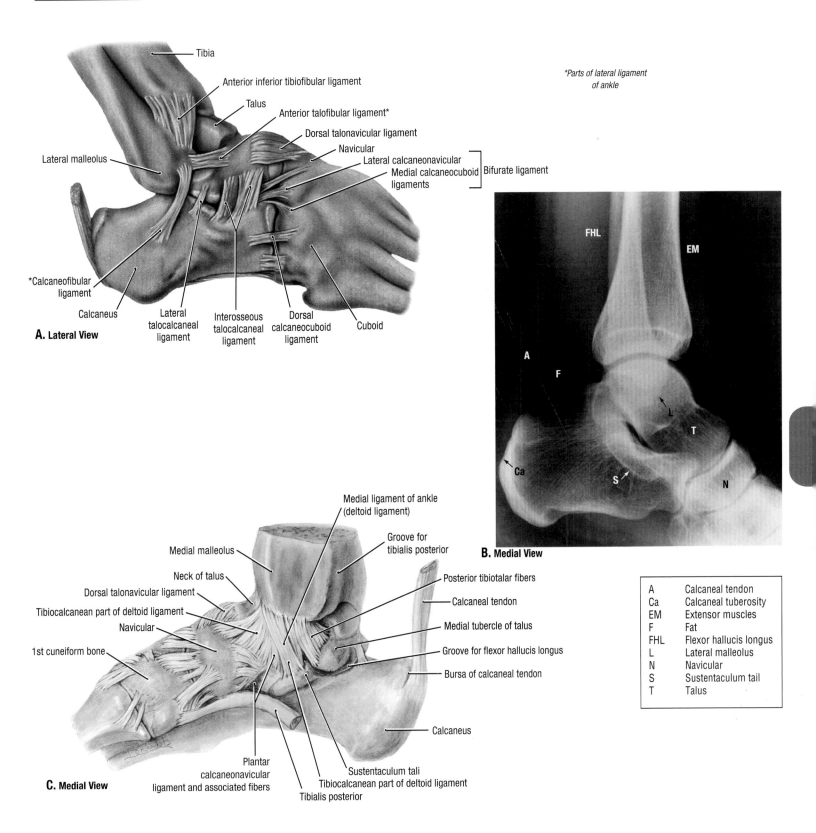

A. Lateral View

Tibia

Anterior inferior tibiofibular ligament

Talus

Anterior talofibular ligament*

Dorsal talonavicular ligament

Navicular

Lateral malleolus

Lateral calcaneonavicular

Medial calcaneocuboid ligaments } Bifurate ligament

*Parts of lateral ligament of ankle

*Calcaneofibular ligament

Calcaneus

Lateral talocalcaneal ligament

Interosseous talocalcaneal ligament

Dorsal calcaneocuboid ligament

Cuboid

B. Medial View

FHL

EM

A

F

L

T

Ca

S

N

A	Calcaneal tendon
Ca	Calcaneal tuberosity
EM	Extensor muscles
F	Fat
FHL	Flexor hallucis longus
L	Lateral malleolus
N	Navicular
S	Sustentaculum tail
T	Talus

C. Medial View

Medial ligament of ankle (deltoid ligament)

Groove for tibialis posterior

Medial malleolus

Neck of talus

Dorsal talonavicular ligament

Tibiocalcanean part of deltoid ligament

Navicular

1st cuneiform bone

Posterior tibiotalar fibers

Calcaneal tendon

Medial tubercle of talus

Groove for flexor hallucis longus

Bursa of calcaneal tendon

Calcaneus

Plantar calcaneonavicular ligament and associated fibers

Sustentaculum tali

Tibiocalcanean part of deltoid ligament

Tibialis posterior

5.72 Ligaments of the ankle

A. Lateral ligaments. The ankle is plantar flexed; thus part of the body of the talus is exposed, and the foot is inverted. **B.** Lateral radiograph. **C.** Medial ligaments. The tibialis posterior is displaced from its "bed" of the medial malleolus, deltoid ligament, and plantar calcaneo-navicular (spring) ligament. The deltoid ligament is attached superiorly to the medial malle-olus of the tibia and inferiorly to the talus, navicular, and calcaneus.

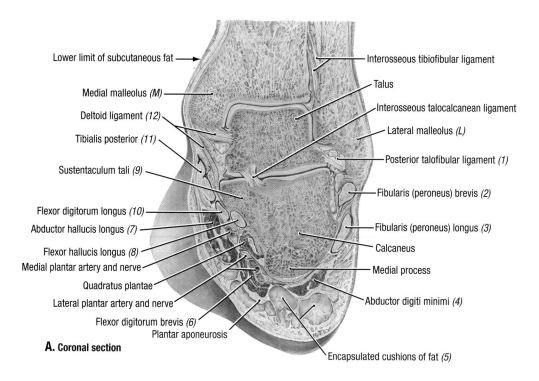

Lower limit of subcutaneous fat →

Medial malleolus *(M)*

Deltoid ligament *(12)*

Tibialis posterior *(11)*

Sustentaculum tali *(9)*

Flexor digitorum longus *(10)*
Abductor hallucis longus *(7)*

Flexor hallucis longus *(8)*
Medial plantar artery and nerve

Quadratus plantae
Lateral plantar artery and nerve

Flexor digitorum brevis *(6)*
Plantar aponeurosis

Interosseous tibiofibular ligament

Talus

Interosseous talocalcanean ligament

Lateral malleolus *(L)*

Posterior talofibular ligament *(1)*

Fibularis (peroneus) brevis *(2)*

Fibularis (peroneus) longus *(3)*

Calcaneus

Medial process

Abductor digiti minimi *(4)*

Encapsulated cushions of fat *(5)*

A. Coronal section

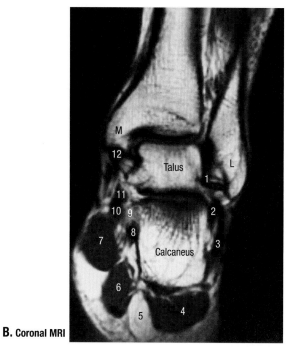

B. Coronal MRI

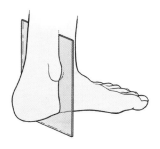

5.73   **Coronal section and MRI through the ankle**

A. Coronal section. **B.** Coronal MRI (*numbers* in **B** refer to structures labeled in **A**).

• The tibia rests on the talus, and the talus rests on the calcaneus; between the calcaneus and the skin are several encapsulated cushions of fat.

• The lateral malleolus descends much farther inferiorly than the medial malleolus.

• The interosseous band between the talus and calcaneus separates the subtalar, or posterior, talocalcanean joint from the talocalcaneonavicular joint.

• The sustentaculum tali acts as a pulley for the flexor hallucis longus muscle and gives attachment to the calcaneotibial band of the deltoid ligament.

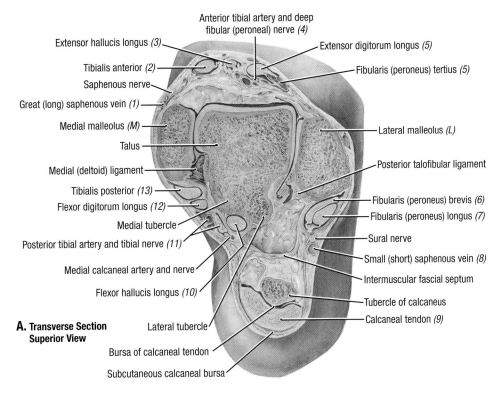

A. Transverse Section Superior View

Anterior tibial artery and deep fibular (peroneal) nerve *(4)*

Extensor hallucis longus *(3)*

Tibialis anterior *(2)*

Saphenous nerve

Great (long) saphenous vein *(1)*

Medial malleolus *(M)*

Talus

Medial (deltoid) ligament

Tibialis posterior *(13)*

Flexor digitorum longus *(12)*

Medial tubercle

Posterior tibial artery and tibial nerve *(11)*

Medial calcaneal artery and nerve

Flexor hallucis longus *(10)*

Lateral tubercle

Bursa of calcaneal tendon

Subcutaneous calcaneal bursa

Extensor digitorum longus *(5)*

Fibularis (peroneus) tertius *(5)*

Lateral malleolus *(L)*

Posterior talofibular ligament

Fibularis (peroneus) brevis *(6)*

Fibularis (peroneus) longus *(7)*

Sural nerve

Small (short) saphenous vein *(8)*

Intermuscular fascial septum

Tubercle of calcaneus

Calcaneal tendon *(9)*

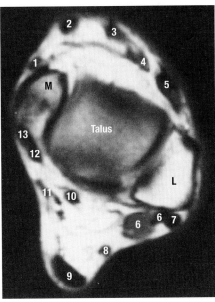

B. Transverse MRI

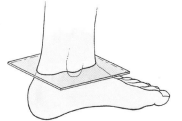

5.74   **Transverse section and MRI through the ankle**

A. Transverse section. **B.** Transverse MRI (*numbers* in **B** refer to structures labeled in **A**).

- The body of the talus is wedge shaped and positioned between the malleoli, which are bound to it by the deltoid and posterior talofibular ligaments.
- The flexor hallucis longus muscle lies within its osseofibrous sheath between the medial and lateral tubercles of the talus.
- There is a small, inconstant subcutaneous bursa superficial to the calcaneal tendon and a large, constant bursa deep to it (bursa of calcaneal tendon) that contains a long synovial fold.

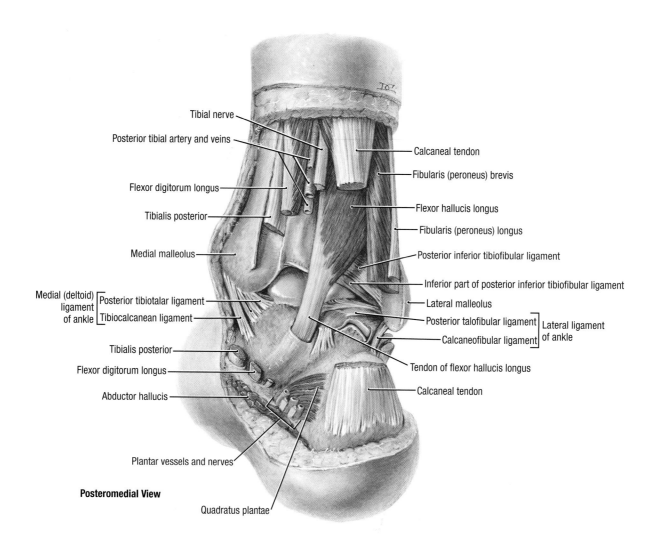

Tibial nerve

Posterior tibial artery and veins

Flexor digitorum longus

Tibialis posterior

Medial malleolus

Medial (deltoid) ligament of ankle ⌈ Posterior tibiotalar ligament ⌊ Tibiocalcanean ligament

Tibialis posterior

Flexor digitorum longus

Abductor hallucis

Plantar vessels and nerves

Posteromedial View

Quadratus plantae

Calcaneal tendon

Fibularis (peroneus) brevis

Flexor hallucis longus

Fibularis (peroneus) longus

Posterior inferior tibiofibular ligament

Inferior part of posterior inferior tibiofibular ligament

Lateral malleolus

Posterior talofibular ligament ⌉ Lateral ligament
Calcaneofibular ligament ⌋ of ankle

Tendon of flexor hallucis longus

Calcaneal tendon

5.75 Medial ankle

- The flexor hallucis longus muscle is midway between the medial and lateral malleoli; the tendons of the flexor digitorum and tibialis posterior are medial to it, and the tendons of the fibularis longus and brevis are lateral to it.
- The posterior tibial artery and the tibial nerve lie medial to the flexor hallucis longus muscle proximally and distally, after bifurcating posterolateral to it.
- The strongest parts of the ligaments of the ankle are those that prevent anterior displacement of the leg bones, namely, the posterior part of the deltoid (posterior tibiotalar), the posterior talofibular, the tibiocalcanean, and the calcaneofibular.

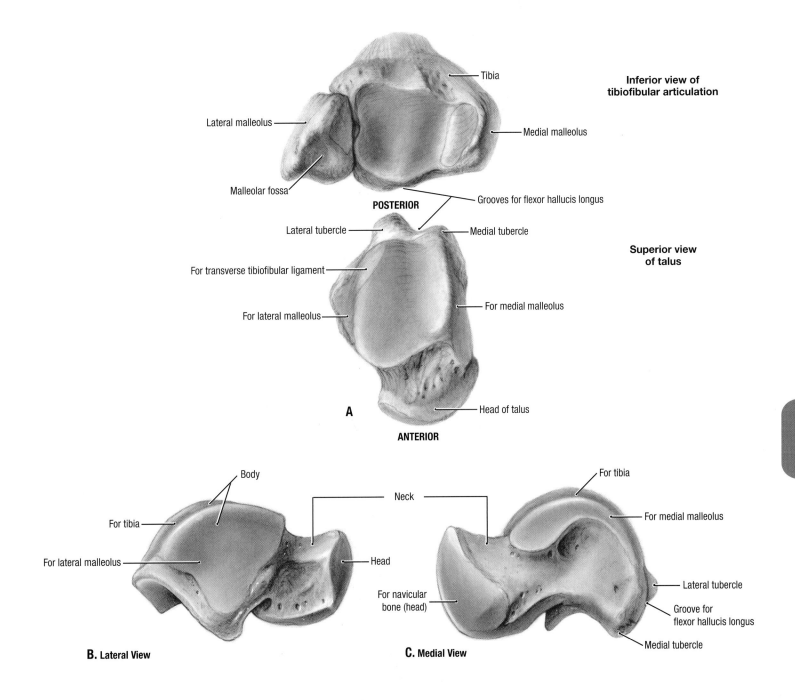

Inferior view of tibiofibular articulation

Tibia

Lateral malleolus

Medial malleolus

Malleolar fossa

Grooves for flexor hallucis longus

POSTERIOR

Lateral tubercle

Medial tubercle

Superior view of talus

For transverse tibiofibular ligament

For lateral malleolus

For medial malleolus

A

Head of talus

ANTERIOR

Body

Neck

For tibia

For tibia

For lateral malleolus

For medial malleolus

Head

For navicular bone (head)

Lateral tubercle

Groove for flexor hallucis longus

Medial tubercle

B. Lateral View

C. Medial View

5.76 Articular surfaces of the ankle joint

A. Superior aspect of talus separated from distal ends of tibia and fibula. The superior articular surface of the talus is broader anteriorly than posteriorly; hence the medial and lateral malleoli, which grasp the sides of the talus, tend to be forced apart in dorsiflexion. The fully dorsiflexed position is stable compared with the fully plantar flexed position. In plantar flexion, when the tibia and fibula articulate with the narrower posterior part of the superior articular surface of the talus, some side-to-side movement of the joint is allowed, accounting for the instability of the joint in this position. **B.** Lateral aspect of talus. Observe the lateral, triangular area for articulation with the lateral malleolus. **C.** Medial aspect of talus. Observe the comma-shaped area for articulation with the medial malleolus.

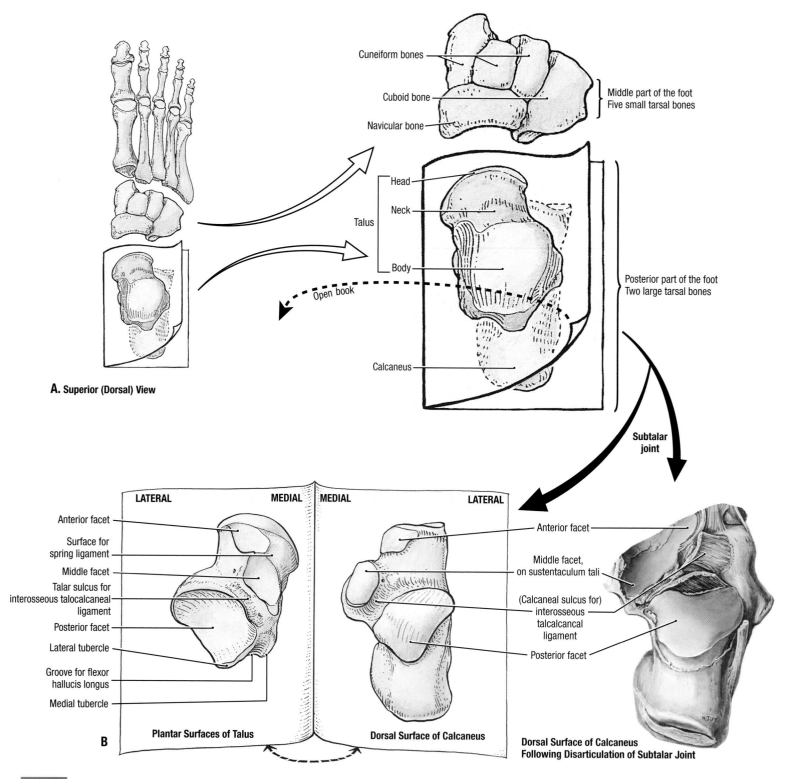

Cuneiform bones

Cuboid bone

Navicular bone

Middle part of the foot
Five small tarsal bones

Head

Neck

Talus

Body

Open book

Posterior part of the foot
Two large tarsal bones

Calcaneus

A. Superior (Dorsal) View

Subtalar
joint

LATERAL   MEDIAL   MEDIAL   LATERAL

Anterior facet

Surface for
spring ligament

Middle facet

Talar sulcus for
interosseous talocalcaneal
ligament

Posterior facet

Lateral tubercle

Groove for flexor
hallucis longus

Medial tubercle

Anterior facet

Middle facet,
on sustentaculum tali

(Calcaneal sulcus for)
interosseous
talcalcancal
ligament

Posterior facet

B   **Plantar Surfaces of Talus**   **Dorsal Surface of Calcaneus**   **Dorsal Surface of Calcaneus
Following Disarticulation of Subtalar Joint**

5.77   **Talocalcanean joint**

A. Bones of foot, dorsal view. **B.** Bony surfaces of talocalcanean joints. The plantar surface of the talus and dorsal surface of the cal-caneus are displayed as pages in a book.

- The joints of inversion and eversion are the subtalar (posterior talocalcanean) joint, talocalcaneonavicular joint, and transverse tarsal (combined calcaneocuboid and talonavicular) joint.
- The talus is part of the ankle joint, of the posterior and anterior talocalcanean joints, and of the talonavicular joint.
- The posterior and anterior talocalcanean joints are separated

from each other by the talar sulcus and calcaneal sulcus, which, when the talus and calcaneus are in articulation, become the tarsal sinus.

- The subtalar joint has its own synovial cavity, whereas the talonavicular and anterior talocalcanean joints share a common synovial cavity.
- The space between the navicular bone and the middle talar facet on the sustentaculum tali is bridged by the plantar calca-neonavicular ligament.

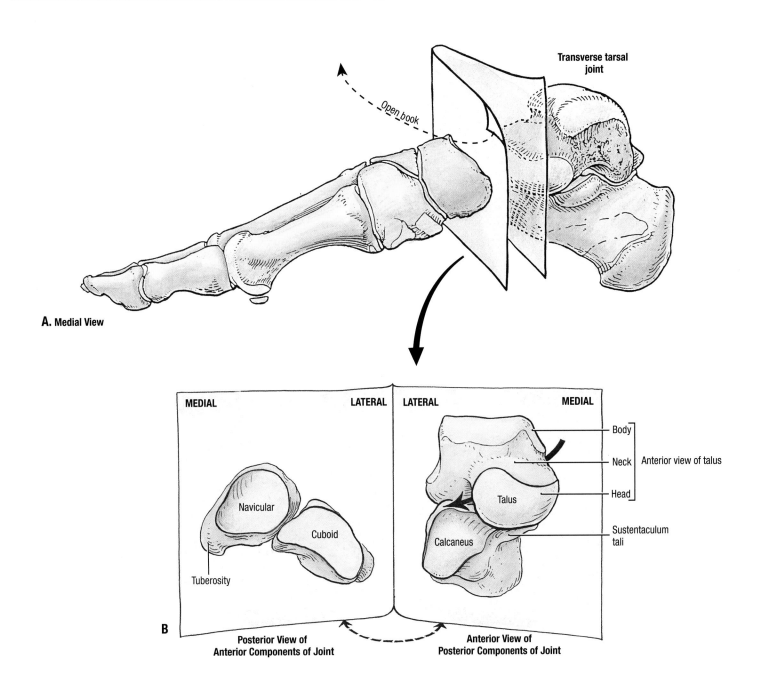

5.78   Transverse tarsal joint

A. Bones of foot, medial view. **B.** Bony surfaces of transverse tarsal joint. This compound joint includes the talonavicular and calcaneocuboid articulations. The posterior surfaces of the navicular and cuboid bones and the anterior surfaces of the talus and calcaneus are displayed as pages in a book. The *black arrow* traverses the tarsal sinus, in which the interosseous talocalcanean ligament is located.

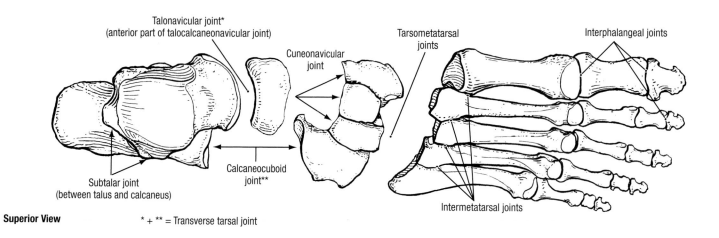

Talonavicular joint*
(anterior part of talocalcaneonavicular joint)

Cuneonavicular joint

Tarsometatarsal joints

Interphalangeal joints

Subtalar joint
(between talus and calcaneus)

Calcaneocuboid joint**

Intermetatarsal joints

Superior View   * + ** = Transverse tarsal joint

TABLE 5.16   JOINTS OF FOOT

JOINT	TYPE	ARTICULAR SURFACE	ARTICULAR CAPSULE	LIGAMENTS	MOVEMENTS
Subtalar	Plane type of synovial joint	Inferior surface of body of talus articulates with superior surface of calcaneus	Fibrous capsule is attached to margins of articular surfaces	Medial, lateral, and posterior talocalcaneal ligaments support capsule; interosseous talocalcaneal ligament binds bones together	Inversion and eversion of foot
Talocalcaneo-navicular	Synovial joint; talonavicular part is ball and socket type	Head of talus articulates with calcaneus and navicular bones	Fibrous capsule incompletely encloses joint	Plantar calcaneonavicular ("spring") ligament supports head of talus	Gliding and rotary movements are possible
Calcaneocuboid	Plane type of synovial joint	Anterior end of calcaneus articulates with posterior surface of cuboid	Fibrous capsule encloses joint	Dorsal calcaneocu-boid ligament, plantar calcaneocu-boid ligament, and long plantar ligament support fibrous capsule	Inversion and eversion of foot
Cuneonavicular	Plane type of synovial joint	Anterior navicular articulates with posterior surface of cuneiforms	Common fibrous capsule encloses joint	Dorsal and plantar ligaments	Little movement possible
Tarsometatarsal	Plane type of synovial joint	Anterior tarsal bones articulate with bases of metatarsal bones	Fibrous capsule encloses joint	Dorsal, plantar, and interosseous ligaments	Gliding or sliding
Intermetatarsal	Plane type of synovial joint	Bases of metatarsal bones articulate with each other	Fibrous capsule encloses each joint	Dorsal, plantar, and interosseous ligaments bind bones together	Little individual movement of bones possible
Metatarso-phalangeal	Condyloid type of synovial joint	Heads of metatarsal bones articulate with bases of proximal phalanges	Fibrous capsule encloses each joint	Collateral ligaments support capsule on each side; plantar ligament supports plantar part of capsule	Flexion, extension, and some abduction, adduction, and circumduction
Interphalangeal	Hinge type of synovial joint	Head of one phalanx articulates with base of one distal to it	Fibrous capsule encloses each joint	Collateral and plantar ligaments support joints	Flexion and extension

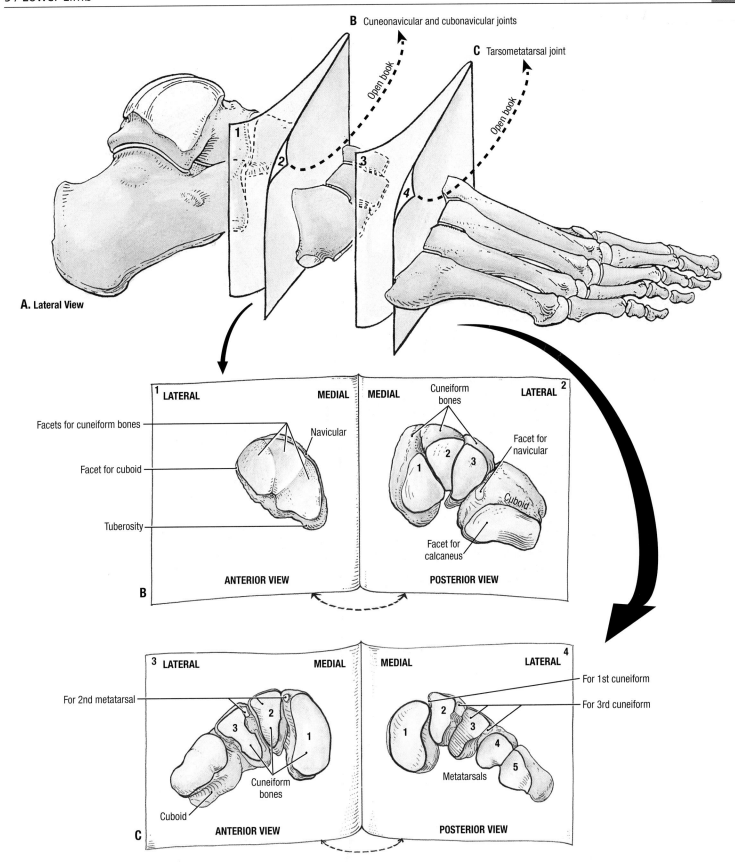

B Cuneonavicular and cubonavicular joints

C Tarsometatarsal joint

Open book

Open book

A. Lateral View

1 LATERAL — **MEDIAL** | **MEDIAL** — **LATERAL 2**

Facets for cuneiform bones
Facet for cuboid
Tuberosity
Navicular

Cuneiform bones
Facet for navicular
Cuboid
Facet for calcaneus

1 · 2 · 3

B — **ANTERIOR VIEW** — **POSTERIOR VIEW**

3 LATERAL — **MEDIAL** | **MEDIAL** — **LATERAL 4**

For 2nd metatarsal
Cuneiform bones
Cuboid

For 1st cuneiform
For 3rd cuneiform
Metatarsals

C — **ANTERIOR VIEW** — **POSTERIOR VIEW**

5.79   Cuneonavicular, cubonavicular, and tarsometatarsal joints

A. Bones of foot, lateral view. **B.** Bony surfaces of the cuneonavicular and cubonavicular joints. The anterior surface of the navicular bone, posterior surfaces of the three cuneiform bones, and medial and posterior surfaces of the cuboid bone are displayed as

pages in a book. **C.** Bony surfaces of the tarsometatarsal joints. The anterior surfaces of the cuboid and three cuneiform bones and the posterior surfaces of the bases of the five metatarsal bones are displayed as pages in a book.

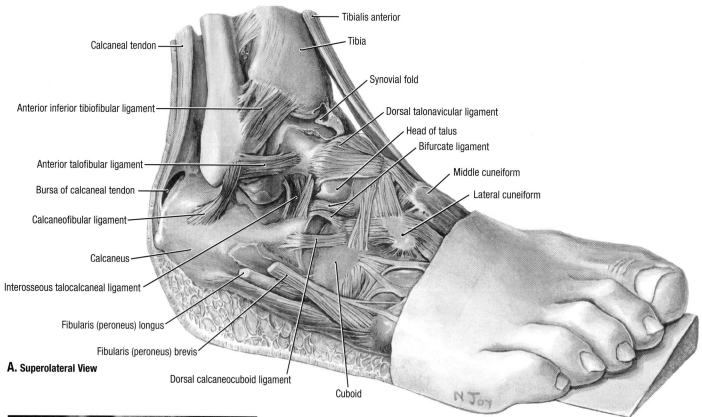

Tibialis anterior

Tibia

Calcaneal tendon

Synovial fold

Anterior inferior tibiofibular ligament

Dorsal talonavicular ligament

Head of talus

Bifurcate ligament

Anterior talofibular ligament

Middle cuneiform

Bursa of calcaneal tendon

Lateral cuneiform

Calcaneofibular ligament

Calcaneus

Interosseous talocalcaneal ligament

Fibularis (peroneus) longus

Fibularis (peroneus) brevis

A. Superolateral View

Dorsal calcaneocuboid ligament

Cuboid

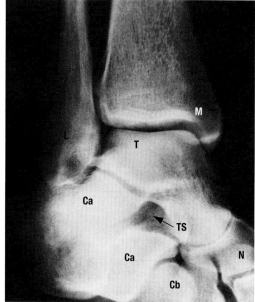

B. Lateral View

5.80   **Ankle and foot**

A. Dorsum of foot. The foot has been inverted to demonstrate articular surfaces and tightened ligaments. The exposed articular surfaces include the posterior talar facet of the calcaneus, the anterior surface of the calcaneus, and the head of the talus, all of which are palpable. Because inversion of the foot is commonly associated with plantar flexion of the ankle joint, the superior and lateral articular surfaces of the body of the talus are also commonly exposed. **B.** Lateral radiograph of ankle. *M*, medial malleolus; *L*, lateral malleolus; *T*, talus; *Ca*, calcaneus; *N*, navicular; *Cb*, cuboid; *TS*, tarsal sinus.

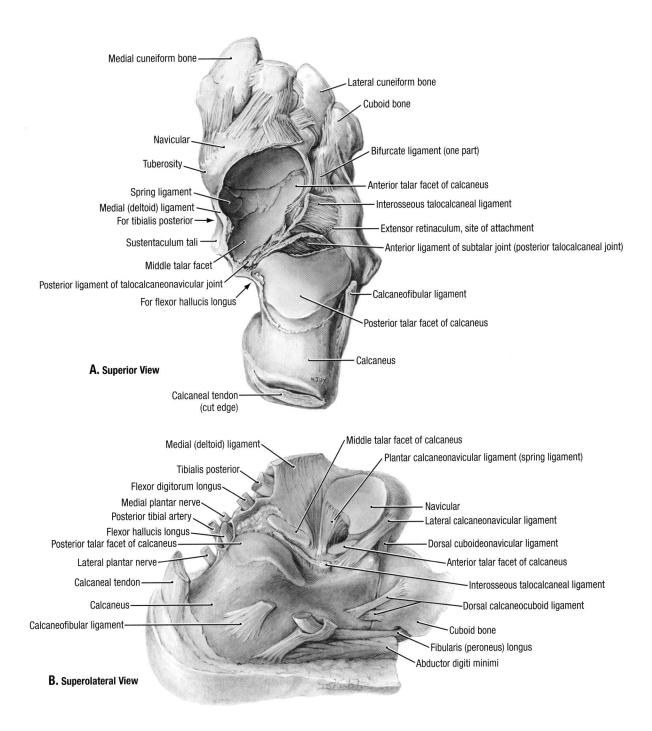

A. Superior View

- Medial cuneiform bone
- Lateral cuneiform bone
- Cuboid bone
- Navicular
- Bifurcate ligament (one part)
- Tuberosity
- Anterior talar facet of calcaneus
- Spring ligament
- Interosseous talocalcaneal ligament
- Medial (deltoid) ligament
- For tibialis posterior
- Extensor retinaculum, site of attachment
- Sustentaculum tali
- Anterior ligament of subtalar joint (posterior talocalcaneal joint)
- Middle talar facet
- Posterior ligament of talocalcaneonavicular joint
- Calcaneofibular ligament
- For flexor hallucis longus
- Posterior talar facet of calcaneus
- Calcaneus
- Calcaneal tendon (cut edge)

B. Superolateral View

- Medial (deltoid) ligament
- Middle talar facet of calcaneus
- Plantar calcaneonavicular ligament (spring ligament)
- Tibialis posterior
- Flexor digitorum longus
- Medial plantar nerve
- Navicular
- Posterior tibial artery
- Lateral calcaneonavicular ligament
- Flexor hallucis longus
- Dorsal cuboideonavicular ligament
- Posterior talar facet of calcaneus
- Anterior talar facet of calcaneus
- Lateral plantar nerve
- Interosseous talocalcaneal ligament
- Calcaneal tendon
- Dorsal calcaneocuboid ligament
- Calcaneus
- Cuboid bone
- Calcaneofibular ligament
- Fibularis (peroneus) longus
- Abductor digiti minimi

5.81 **Joints of inversion and eversion**

The joints of inversion and eversion are the subtalar (posterior talocalcanean) joint, talocalcaneonavicular joint, and transverse tarsal (combined calcaneocuboid and talonavicular) joint. **A.** Posterior and middle parts of foot with talus removed. **B.** Posterior part of foot with talus removed. The convex posterior talar facet is separated from the concave, middle, and anterior facets by the interosseous talocalcanean ligament within the tarsal sinus.

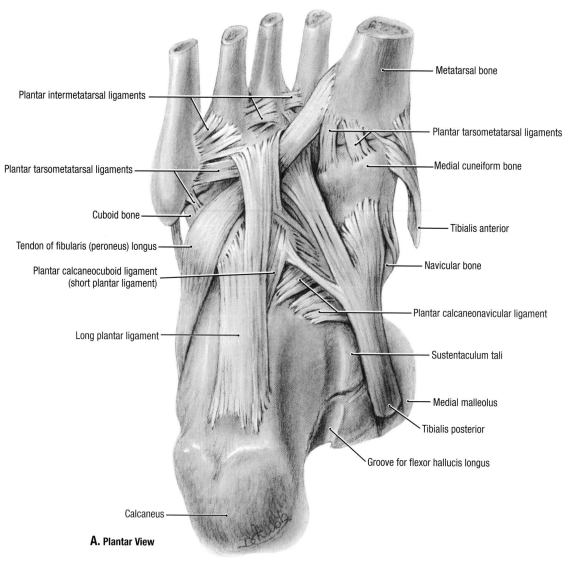

Plantar intermetatarsal ligaments

Plantar tarsometatarsal ligaments

Cuboid bone

Tendon of fibularis (peroneus) longus

Plantar calcaneocuboid ligament
(short plantar ligament)

Long plantar ligament

Calcaneus

A. Plantar View

Metatarsal bone

Plantar tarsometatarsal ligaments

Medial cuneiform bone

Tibialis anterior

Navicular bone

Plantar calcaneonavicular ligament

Sustentaculum tali

Medial malleolus

Tibialis posterior

Groove for flexor hallucis longus

5.82   Ligaments of sole of foot—I

A. Dissection of superficial ligaments. **B.** Bones lying deep to ligaments of **A.**

In **A:**

- The head of the talus is exposed between the sustentaculum tali of the calcaneus and the navicular.
- Note the insertions of three long tendons: fibularis (peroneus) longus, tibialis anterior, and tibialis posterior.
- The tendon of the fibularis (peroneus) longus muscle crosses the sole of the foot in the groove anterior to the ridge of the cuboid, is bridged by some fibers of the long plantar ligament, and inserts into the base of the first metatarsal.
- Observe the slips of the tibialis posterior tendon extending to the bones anterior to the transverse tarsal joint.

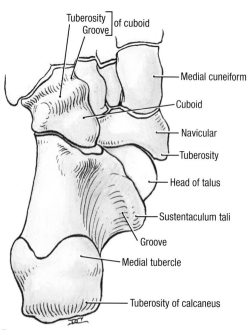

Tuberosity ⎤ of cuboid
Groove   ⎦

Medial cuneiform

Cuboid

Navicular

Tuberosity

Head of talus

Sustentaculum tali

Groove

Medial tubercle

Tuberosity of calcaneus

B. Plantar View

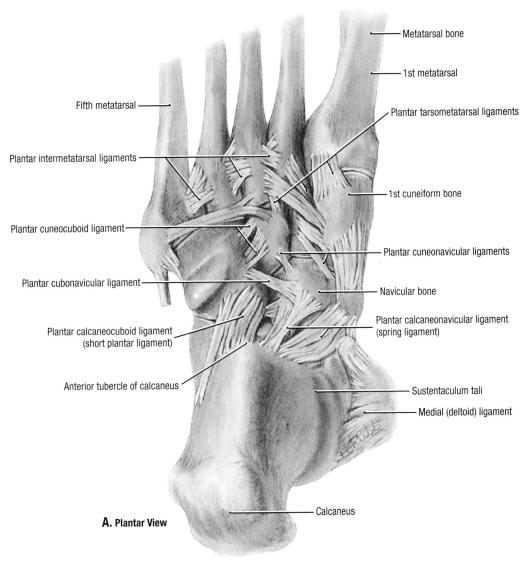

- Metatarsal bone
- 1st metatarsal
- Fifth metatarsal
- Plantar tarsometatarsal ligaments
- Plantar intermetatarsal ligaments
- 1st cuneiform bone
- Plantar cuneocuboid ligament
- Plantar cuneonavicular ligaments
- Plantar cubonavicular ligament
- Navicular bone
- Plantar calcaneocuboid ligament (short plantar ligament)
- Plantar calcaneonavicular ligament (spring ligament)
- Anterior tubercle of calcaneus
- Sustentaculum tali
- Medial (deltoid) ligament
- Calcaneus

A. Plantar View

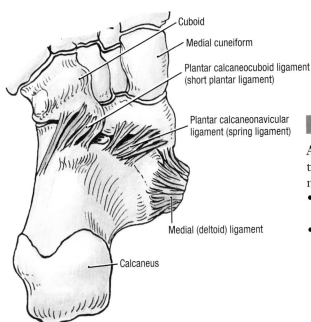

- Cuboid
- Medial cuneiform
- Plantar calcaneocuboid ligament (short plantar ligament)
- Plantar calcaneonavicular ligament (spring ligament)
- Medial (deltoid) ligament
- Calcaneus

B. Plantar View

5.83   **Ligaments of sole of foot—II**

A. Dissection of the deep ligaments. **B.** Support for head of talus. The head of the talus is supported by the plantar calcaneonavicular ligament (spring ligament) and the tendon of the tibialis posterior.

- The plantar calcaneocuboid (short plantar) and plantar calcaneonavicular (spring) ligaments are ligaments of the transverse tarsal joint.
- The ligaments of the anterior foot diverge laterally and posteriorly from each side of the long axis of the third metatarsal and third cuneiform; hence a posterior thrust received by the first metatarsal, as when rising on the big toe while in walking, is transmitted directly to the navicular and talus by the first cuneiform and indirectly by the second metatarsal, second cuneiform, third metatarsal, and third cuneiform.
- A posterior thrust received by the fourth and fifth metatarsals is transmitted directly to the cuboid and calcaneus.

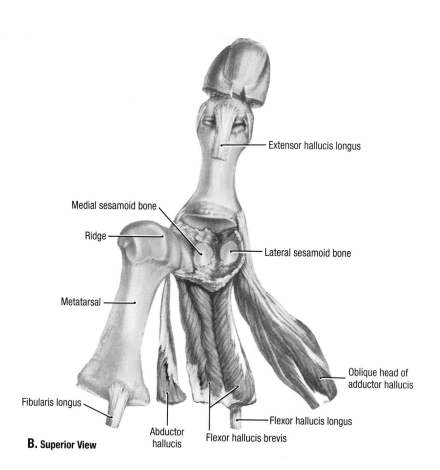

A. Plantar Surface of 1st Metatarsal

Sheath of flexor hallucis longus

Lateral sesamoid bone

Medial sesamoid bone

B. Superior View

Extensor hallucis longus

Medial sesamoid bone

Ridge

Lateral sesamoid bone

Metatarsal

Oblique head of adductor hallucis

Fibularis longus

Abductor hallucis

Flexor hallucis brevis

Flexor hallucis longus

5.84   Metatarsophalangeal joint of the great toe

A. First metatarsal and sesamoid bones. The sesamoid bones of the great toe (hallux) are bound together and located on each side of a bony ridge on the first metatarsal. **B.** Dissection. The joint capsule of the first metatarsophalangeal joint has been cut dorsally, and the first metatarsal reflected medially to expose the sesamoid bones.

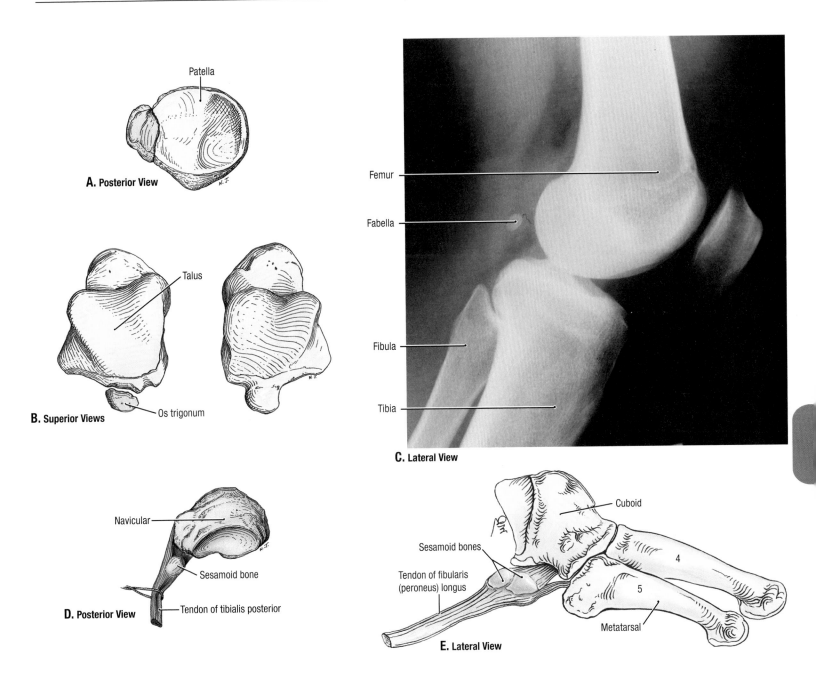

A. Posterior View

B. Superior Views

- Patella
- Talus
- Os trigonum

C. Lateral View

- Femur
- Fabella
- Fibula
- Tibia

D. Posterior View

- Navicular
- Sesamoid bone
- Tendon of tibialis posterior

E. Lateral View

- Cuboid
- Sesamoid bones
- Tendon of fibularis (peroneus) longus
- 4
- 5
- Metatarsal

5.85 Bony anomalies

A. Bipartite patella. Occasionally, the superolateral angle of the patella ossifies independently and remains discrete. **B.** Os trigonum. The lateral (posterior) tubercle of the talus has a separate center of ossification that appears from the ages of 7 to 13 years; when this fails to fuse with the body of the talus, as in the left bone of this pair, it is called an *os trigonum*. It was found in 7.7% of 558 adult feet; 22 were paired, and 21 were unpaired. **C.** Fabella. A sesamoid bone (fabella) in the lateral head of the gastrocnemius muscle was present in 21.6% of 116 limbs. **D.** Sesamoid bone in the tendon of tibialis posterior. A sesamoid bone was found in 23% of 348 adults. **E.** Sesamoid bone in the tendon of fibularis (peroneus) longus. A sesamoid bone was found in 26% of 92 feet. In this specimen, it is bipartite, and the fibularis (peroneus) longus muscle has an additional attachment to the 5th metatarsal bone.

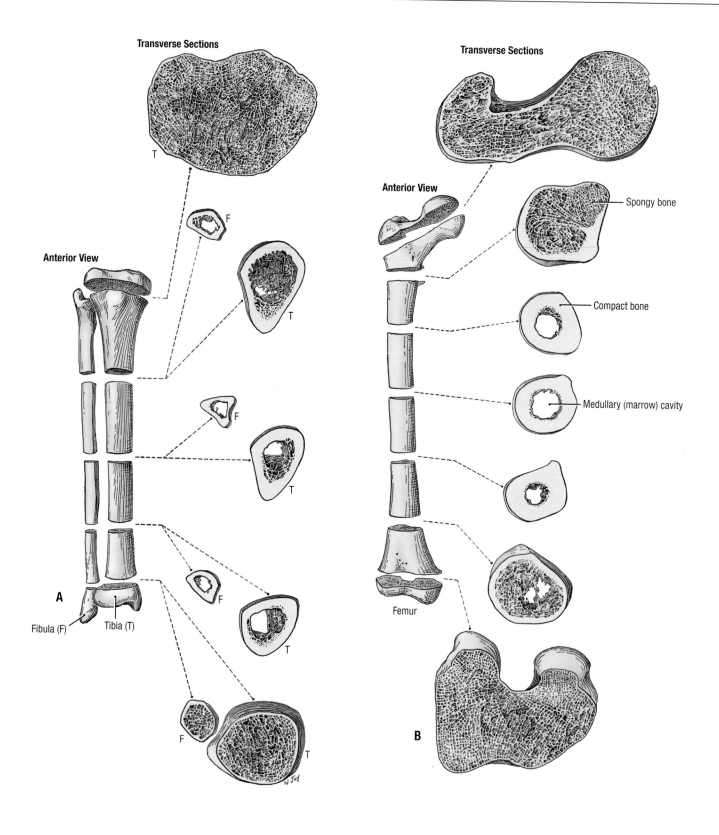

Transverse Sections

Transverse Sections

Anterior View

Anterior View

Spongy bone

Compact bone

Medullary (marrow) cavity

Fibula (F)   Tibia (T)

Femur

5.86   **Transverse sections of femur, tibia, and fibula**

A. Tibia and fibula. **B.** Femur. These sections demonstrate the differences in thickness between the compact and spongy (cancellous) bone throughout the length of the bones. Also the width of the medullary (marrow) cavity can be compared at various levels.

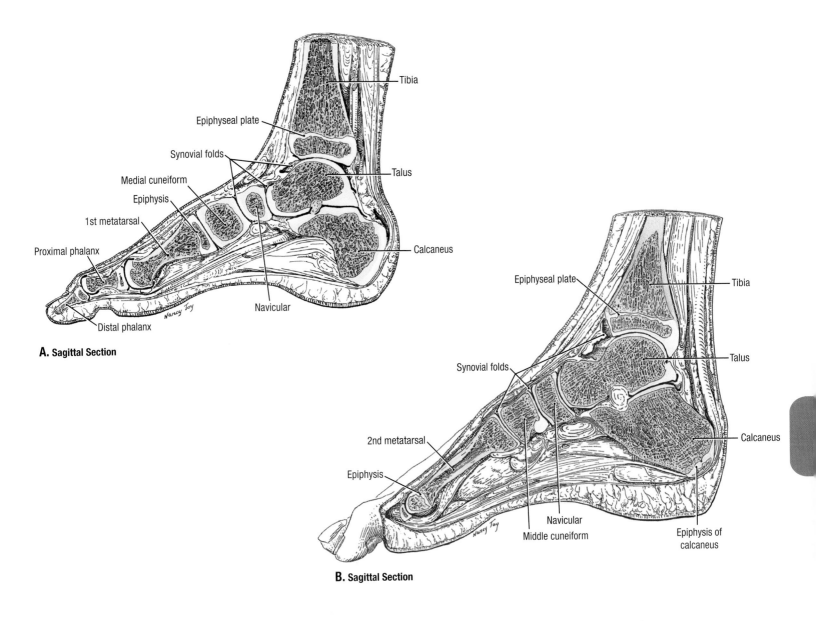

A. Sagittal Section

B. Sagittal Section

5.87   Foot development—sagittal sections

A. Foot of child age 4. **B.** Foot of child age 10.

- In the foot of the younger child **(A)**, epiphyses of long bones (tibia, metatarsals, and phalanges) ossify like short bones, with the ossification centers being enveloped in cartilage. Ossification has already extended to the surface of the larger tarsal bones.

- In the foot of the older child **(B)**, ossification has spread to the dorsal and plantar surfaces of all tarsal bones in view, and cartilage persists on the articular surfaces only.

- The traction epiphysis of the calcaneus for the calcaneal tendon and plantar aponeurosis begins to ossify from the ages of 6 to 10 years.

- The first metatarsal bone is similar to a phalanx in that its epiphysis is at the base instead of the head, as in the second and other metatarsal bones.

- The tuberosity of the calcaneus and the sesamoid bones of the first and the heads of the second to fifth metatarsals (here the second) support the longitudinal arch of the foot; the medial part of the longitudinal arch is higher and more mobile than the lateral.

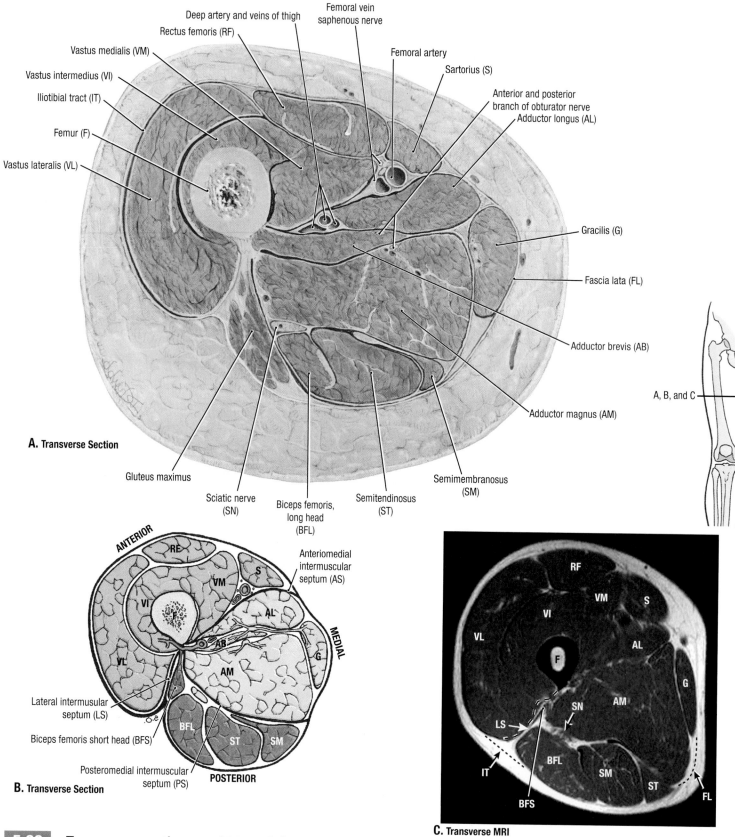

A. Transverse Section

Deep artery and veins of thigh
Rectus femoris (RF)
Vastus medialis (VM)
Vastus intermedius (VI)
Iliotibial tract (IT)
Femur (F)
Vastus lateralis (VL)
Femoral vein
saphenous nerve
Femoral artery
Sartorius (S)
Anterior and posterior
branch of obturator nerve
Adductor longus (AL)
Gracilis (G)
Fascia lata (FL)
Adductor brevis (AB)
Adductor magnus (AM)
Gluteus maximus
Sciatic nerve (SN)
Biceps femoris, long head (BFL)
Semitendinosus (ST)
Semimembranosus (SM)

A, B, and C

B. Transverse Section

ANTERIOR
RF
VM
S
VI
AL
F
AB
G
VL
AM
MEDIAL
Anteriomedial intermuscular septum (AS)
Lateral intermusular septum (LS)
Biceps femoris short head (BFS)
Posteromedial intermuscular septum (PS)
BFL
ST
SM
POSTERIOR

C. Transverse MRI

RF
VM
S
VI
AL
VL
F
G
SN
AM
LS
BFL
IT
SM
BFS
ST
FL

5.88   Transverse sections and MRI of thigh

A. Dissection. **B.** Schematic overview of compartments of thigh.
C. T1 transverse (axial) MRIs of thigh (*numbers* in **C** refer to structures labeled in **A** and **B**).
The thigh has three compartments, each with its own nerve supply and primary functional groups. The anterior group is supplied by the femoral nerve and functions to extend the knee, the medial group is supplied by the obturator nerve and functions to adduct the hip, and the posterior group is supplied by the sciatic nerve and functions to flex the knee.

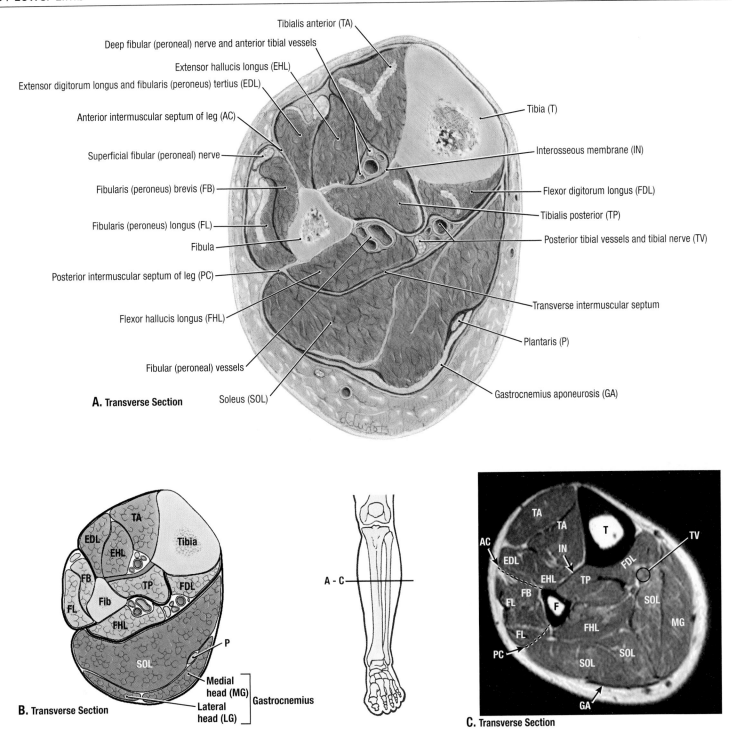

A. Transverse Section

- Tibialis anterior (TA)
- Deep fibular (peroneal) nerve and anterior tibial vessels
- Extensor hallucis longus (EHL)
- Extensor digitorum longus and fibularis (peroneus) tertius (EDL)
- Anterior intermuscular septum of leg (AC)
- Superficial fibular (peroneal) nerve
- Fibularis (peroneus) brevis (FB)
- Fibularis (peroneus) longus (FL)
- Fibula
- Posterior intermuscular septum of leg (PC)
- Flexor hallucis longus (FHL)
- Fibular (peroneal) vessels
- Soleus (SOL)
- Tibia (T)
- Interosseous membrane (IN)
- Flexor digitorum longus (FDL)
- Tibialis posterior (TP)
- Posterior tibial vessels and tibial nerve (TV)
- Transverse intermuscular septum
- Plantaris (P)
- Gastrocnemius aponeurosis (GA)

B. Transverse Section

TA, EDL, EHL, FB, FL, Fib, FHL, Tibia, TP, FDL, P, SOL, Medial head (MG), Lateral head (LG) Gastrocnemius

A - C

C. Transverse Section

TA, TA, AC, EDL, EHL, FB, FL, FL, PC, IN, TP, T, TV, FDL, SOL, MG, FHL, SOL, SOL, GA, F

5.89 Transverse sections and MRI of leg

A. Dissection. **B.** Schematic overview of compartments of leg. **C.** T1 transverse (axial) MRIs of leg (*numbers* in **C** refer to structures labeled in **A** and **B**).

The anterior compartment (*brown*) is bounded by the tibia, interosseous membrane, fibular, anterior intermuscular septum, and crural fascia. It contains the anterior tibial vessels, deep fibular nerve, and tibialis anterior, extensor hallucis longus, and extensor digitorum muscles. The unyielding walls of this compartment can lead to necrosis of the muscles if pressure increases in the compartment following injury or ischemia. The lateral compartment (*light brown*) is bounded by the fibula, anterior and posterior inter-muscular septa, and the crural fascia. It contains the superficial fibular (peroneal) nerve and fibularis longus and brevis muscles. The posterior compartment (*light and dark green*) is bounded by the tibia, interosseous membrane, fibula, posterior intermuscular septum, and crural fascia. This compartment is subdivided by two coronal septa into three subcompartments. The deepest subcompartment contains the tibialis posterior; the intermediate contains the flexor hallucis longus, flexor digitorum longus, and posterior tibial vessels and tibial nerve; and the most superficial contains the soleus, gastrocnemius, and plantaris muscles.

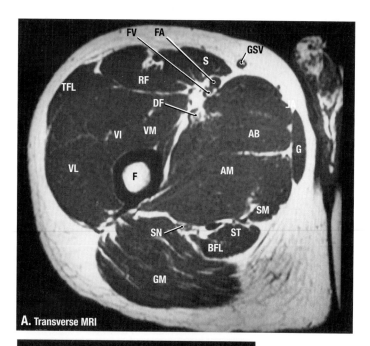

A. Transverse MRI

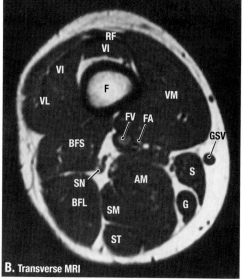

B. Transverse MRI

AB	Adductor brevis
AL	Adductor longus
AM	Adductor magnus
BF	Biceps femoris
BFL	Long head of biceps femoris
BFS	Short head of biceps femoris
DF	Deep vessels of thigh
F	Femur
FA	Femoral artery
FV	Femoral vein
G	Gracilis
GM	Gluteus maximus
GSV	Great saphenous vein
H	Head of femur
OE	Obturator externus
RF	Rectus femoris
S	Sartorius
SM	Semimembranosus
SN	Sciatic nerve
ST	Semitendinosus
TFL	Tensor of fascia lata
UB	Urinary bladder
VI	Vastus intermedius
VL	Vastus lateralis
VM	Vastus medialis

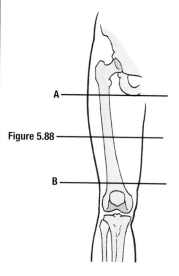

Figure 5.88

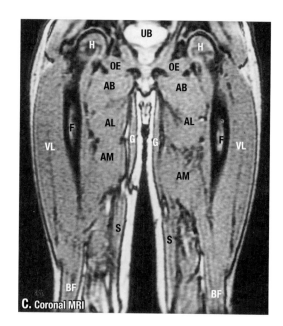

C. Coronal MRI

5.90   T1 transverse and coronal MRIs of the thigh

A. Proximal. **B.** Distal (transverse [axial] MRIs). **C.** Coronal MRI.

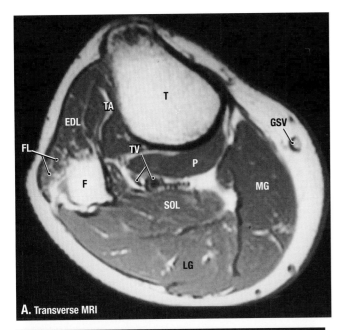

A. Transverse MRI

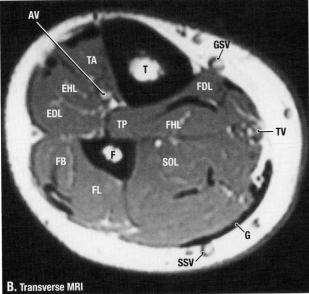

B. Transverse MRI

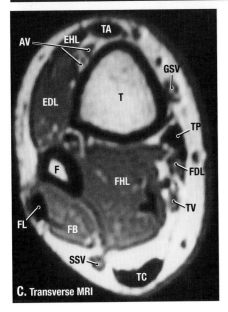

C. Transverse MRI

AV	Anterior tibial vessels and deep fibular nerve
EDL	Extensor digitorum longus
EHL	Extensor hallucis longus
F	Fibula
FB	Fibularis brevis
FDL	Flexor digitorum longus
FHL	Flexor hallucis longus
FL	Fibularis longus
G	Gracilis
GM	Gluteus maximus
GSV	Great saphenous vein
HF	Head of fibula
LG	Lateral head of gastrocnemius
MG	Medial head of gastrocnemius
MM	Medial malleolus
P	Popliteus
SOL	Soleus
SSV	Small saphenous vein
T	Tibia
TA	Tibialis anterior
Ta	Talus
TC	Calcaneal tendon
TP	Tibialis posterior
TV	Tibial nerve and posterior tibial vessels

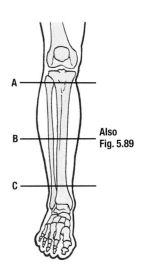

Also Fig. 5.89

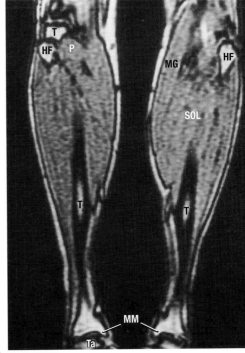

D. Coronal MRI

5.91   **T1 transverse and coronal MRIs of the leg**

A. Proximal. **B.** Middle. **C.** Distal. (Transverse [axial] MRIs). **D.** Coronal MRI.

CHAPTER 6

UPPER LIMB

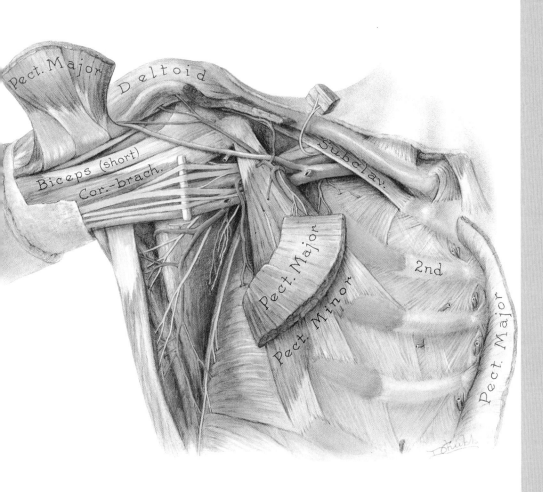

- Skeleton of Upper Limb **458**
- Cutaneous Innervation of Upper Limb **462**
- Venous and Lymphatic Drainage of Upper Limb **464**
- Fascia and Fascial Compartments of Upper Limb **468**
- Arteries and Nerves of Upper Limb **470**
- Pectoral Region **474**
- Axilla, Axillary Vessels, and Brachial Plexus **481**
- Scapular Region and Superficial Back **492**
- Arm and Rotator Cuff **498**
- Joints of Pectoral Girdle and Shoulder **508**
- Elbow Region **518**
- Elbow Joint **524**
- Anterior Aspect of Forearm **530**
- Anterior Aspect of Wrist and Palm of Hand **538**
- Posterior Aspect of Forearm **554**
- Posterior Aspect of Wrist and Dorsum of Hand **558**
- Lateral Aspect of Wrist and Hand **564**
- Medial Aspect of Wrist and Hand **567**
- Bones and Joints of Wrist and Hand **568**
- Function of Hand: Grips and Pinches **576**
- Sectional Anatomy and Imaging: Arm and Forearm **578**

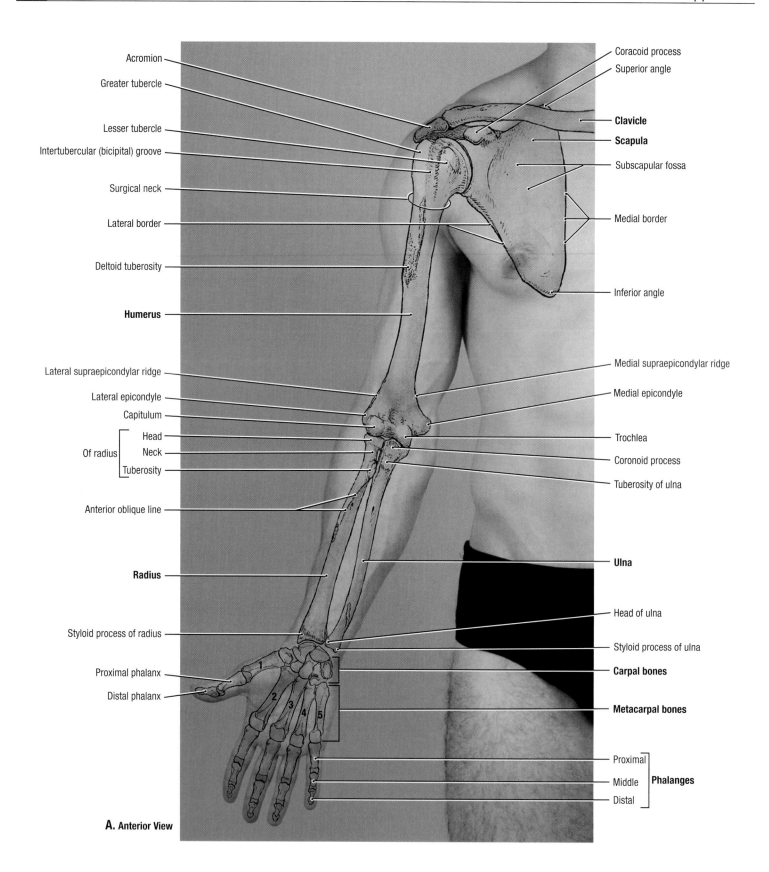

Acromion

Greater tubercle

Lesser tubercle

Intertubercular (bicipital) groove

Surgical neck

Lateral border

Deltoid tuberosity

Humerus

Lateral supraepicondylar ridge

Lateral epicondyle

Capitulum

Of radius { Head

Neck

Tuberosity

Anterior oblique line

Radius

Styloid process of radius

Proximal phalanx

Distal phalanx

Coracoid process

Superior angle

Clavicle

Scapula

Subscapular fossa

Medial border

Inferior angle

Medial supraepicondylar ridge

Medial epicondyle

Trochlea

Coronoid process

Tuberosity of ulna

Ulna

Head of ulna

Styloid process of ulna

Carpal bones

Metacarpal bones

Proximal ⎤

Middle   ⎬ **Phalanges**

Distal   ⎦

A. Anterior View

6.1   Regions and bones of upper limb

The joints divide the upper limb into four main regions: the shoulder, arm, forearm, and hand. The pectoral (shoulder) girdle is an incomplete ring of bones formed by the scapulae and clavicles and is joined medially to the manubrium of the sternum.

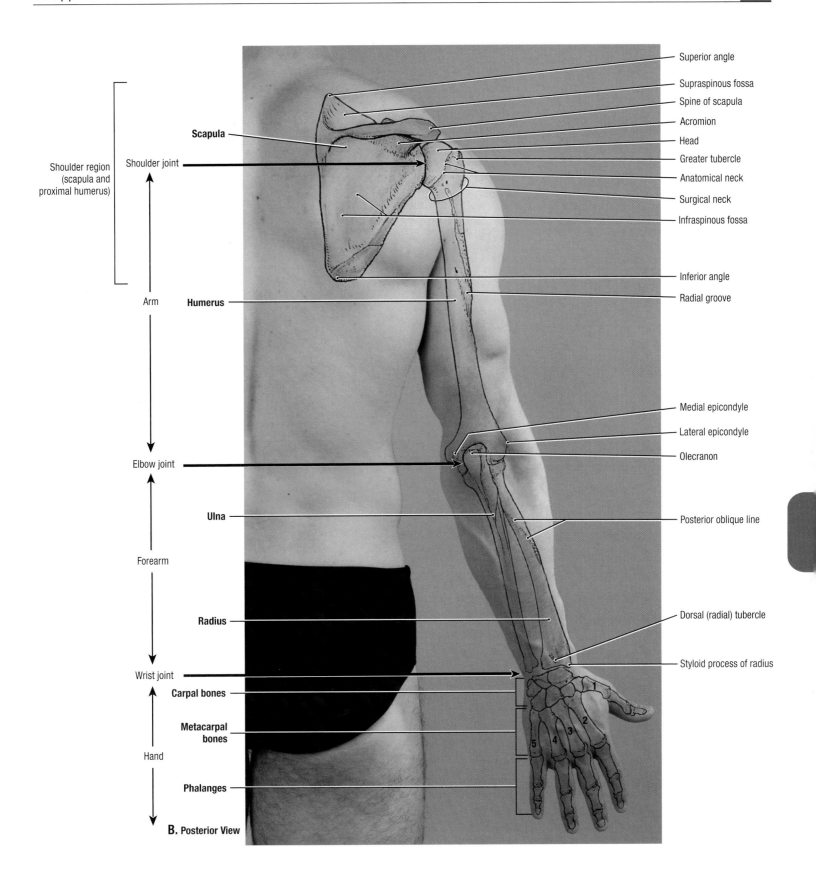

Shoulder region (scapula and proximal humerus)

Shoulder joint

Arm

Elbow joint

Forearm

Wrist joint

Hand

Scapula

Humerus

Ulna

Radius

Carpal bones

Metacarpal bones

Phalanges

B. Posterior View

Superior angle

Supraspinous fossa

Spine of scapula

Acromion

Head

Greater tubercle

Anatomical neck

Surgical neck

Infraspinous fossa

Inferior angle

Radial groove

Medial epicondyle

Lateral epicondyle

Olecranon

Posterior oblique line

Dorsal (radial) tubercle

Styloid process of radius

6.1   **Regions and bones of upper limb** *(continued)*

To identify features of bones see Figures 6.31, 6.59, 8.83, and Table 6.11.

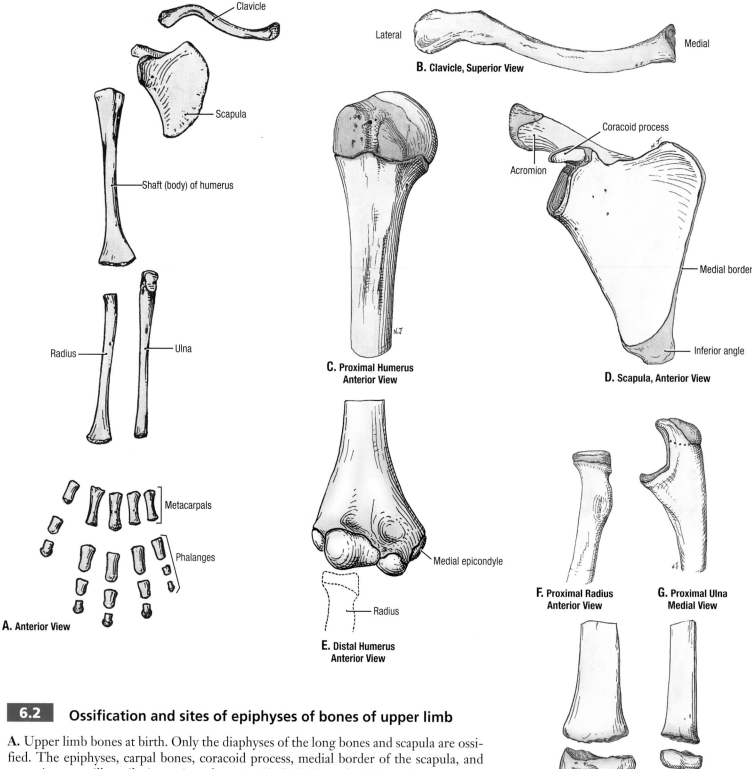

A. Anterior View

B. Clavicle, Superior View

C. Proximal Humerus Anterior View

D. Scapula, Anterior View

E. Distal Humerus Anterior View

F. Proximal Radius Anterior View

G. Proximal Ulna Medial View

H. Distal Radius Anterior View

I. Distal Ulna Anterior View

6.2   Ossification and sites of epiphyses of bones of upper limb

A. Upper limb bones at birth. Only the diaphyses of the long bones and scapula are ossified. The epiphyses, carpal bones, coracoid process, medial border of the scapula, and acromion are still cartilaginous (not shown in **A**). **B–I.** Sites of epiphyses *(darker orange regions)*.

- The ends of the long bones are ossified by the formation of one or more secondary centers of ossification; these epiphyses develop from birth to approximately 20 years of age in the clavicle, humerus, radius, ulna, metacarpals, and phalanges.
- When the epiphysis and shaft (body) fuse, active bone growth stops. Sometimes, as in the proximal humerus, many epiphyses fuse to form a single mass that later fuses with the diaphysis, i.e., three centers of ossification develop (for the head, greater tubercle, and lesser tubercle) and fuse into a single mass by the 7th year and to the diaphysis by the 24th year.
- The epiphysis of the clavicle is the last of the long bone epiphyses to fuse (by 31 years), but the clavicle is one of the first bones to begin ossification. The acromial epiphysis of the scapula may persist into adult life.

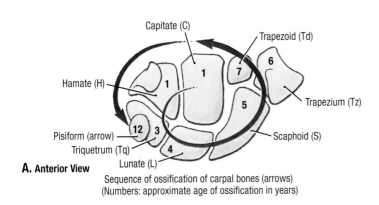

A. Anterior View

Sequence of ossification of carpal bones (arrows)
(Numbers: approximate age of ossification in years)

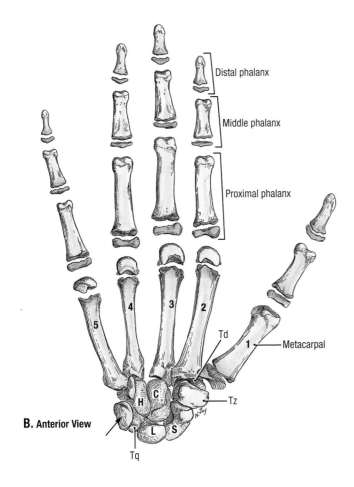

B. Anterior View

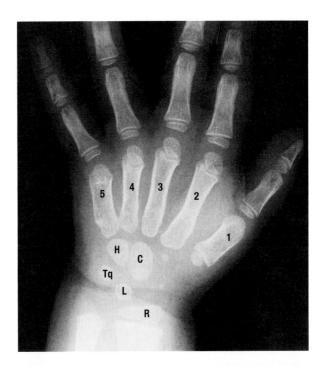

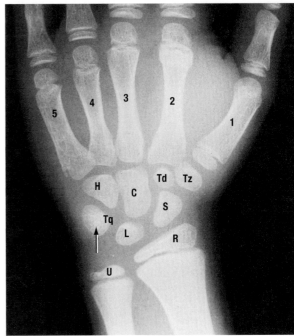

C. Anteroposterior View

6.3	**Epiphyses of hand and wrist**

A. Sequence of ossification of carpal bones. **B.** Ossification of bones of hand. Note that the phalanges have a single proximal epiphysis and metacarpals 2, 3, 4, and 5 have single distal epiphyses. The 1st metacarpal behaves as a phalanx by having a proximal epiphysis. Short-lived epiphyses may appear at the other ends of metacarpals 1 and/or 2. There are individual and gender differences in sequence and timing. **C.** Radiographs of stages of ossification of wrist and hand. *Top*, a 2½-year-old child; the lunate is ossifying, and the distal radial epiphysis *(R)* is present *(C*, capitate; *H*, hamate; *Tq*, triquetrum; *L*, lunate). *Bottom*, an 11-year-old child. All carpal bones are ossified *(S*, scaphoid; *Td*, trapezoid; *Tz*, trapezium; *arrowhead*, pisiform), and the distal epiphysis of the ulna *(U)* has ossified.

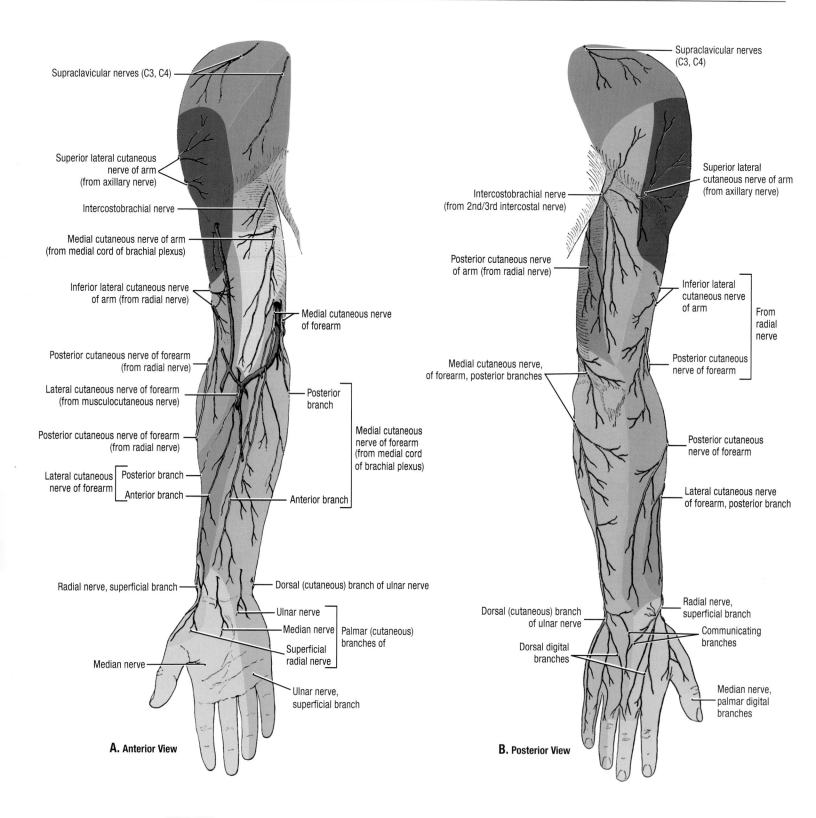

Supraclavicular nerves (C3, C4)

Superior lateral cutaneous nerve of arm (from axillary nerve)

Intercostobrachial nerve

Medial cutaneous nerve of arm (from medial cord of brachial plexus)

Inferior lateral cutaneous nerve of arm (from radial nerve)

Posterior cutaneous nerve of forearm (from radial nerve)

Lateral cutaneous nerve of forearm (from musculocutaneous nerve)

Posterior cutaneous nerve of forearm (from radial nerve)

Lateral cutaneous nerve of forearm — Posterior branch / Anterior branch

Radial nerve, superficial branch

Median nerve

Medial cutaneous nerve of forearm

Posterior branch

Medial cutaneous nerve of forearm (from medial cord of brachial plexus)

Anterior branch

Dorsal (cutaneous) branch of ulnar nerve

Ulnar nerve

Median nerve — Palmar (cutaneous) branches of

Superficial radial nerve

Ulnar nerve, superficial branch

A. Anterior View

Supraclavicular nerves (C3, C4)

Intercostobrachial nerve (from 2nd/3rd intercostal nerve)

Superior lateral cutaneous nerve of arm (from axillary nerve)

Posterior cutaneous nerve of arm (from radial nerve)

Inferior lateral cutaneous nerve of arm — From radial nerve

Posterior cutaneous nerve of forearm

Medial cutaneous nerve, of forearm, posterior branches

Posterior cutaneous nerve of forearm

Lateral cutaneous nerve of forearm, posterior branch

Dorsal (cutaneous) branch of ulnar nerve

Radial nerve, superficial branch

Dorsal digital branches

Communicating branches

Median nerve, palmar digital branches

B. Posterior View

6.4   Cutaneous nerves of upper limb

Cutaneous nerves in the subcutaneous tissue supply the skin of the upper limb. Of the five terminal branches of the brachial plexus (median, ulnar, radial, musculocutaneous, and axillary nerves), the first three usually contribute cutaneous branches to the hand. Five cutaneous nerves are derived from the posterior cord of the brachial plexus. One of these, the superior lateral cutaneous nerve of the arm, is a branch of the axillary nerve; the others are branches of the radial nerve. Two cutaneous nerves arise directly as branches from the medial cord—the medial cutaneous nerves of the arm and forearm.

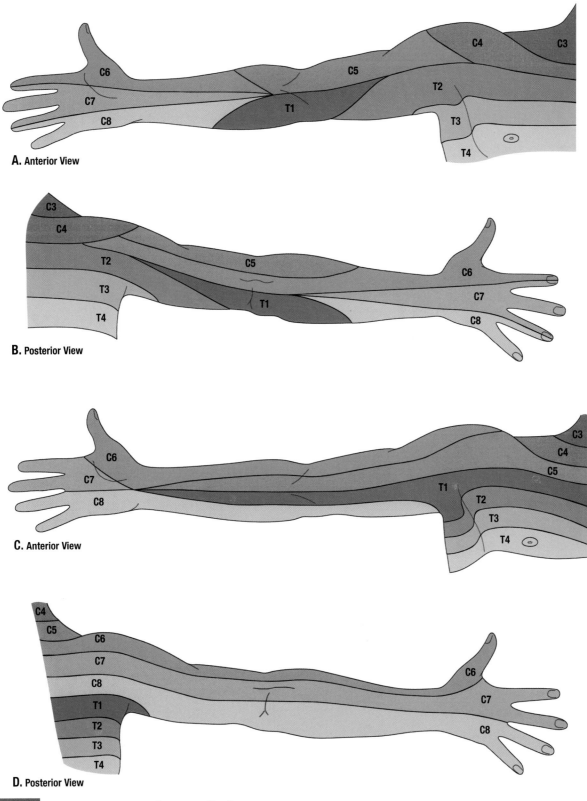

A. Anterior View

B. Posterior View

C. Anterior View

D. Posterior View

6.5   Dermatomes of upper limb

The area of skin supplied by (cutaneous branches derived from a single cord segment) a spinal nerve is called a *dermatome*. Adjacent dermatomes may overlap except at the axial line, which is the line of junction of dermatomes supplied from discontinuous spinal levels. Two alternate dermatome maps are commonly used. **A.** and **B.** are based on Fender FA. Foerster's scheme of the dermatomes. Arch Neurol Psychiatr 1939;41:688. **C.** and **D.** are based on Keegan JJ and Garrett FD. The segmental distribution of the cutaneous nerves in the limbs of man. Anat Rec 1948;102:409.

Preaxial

Postaxial

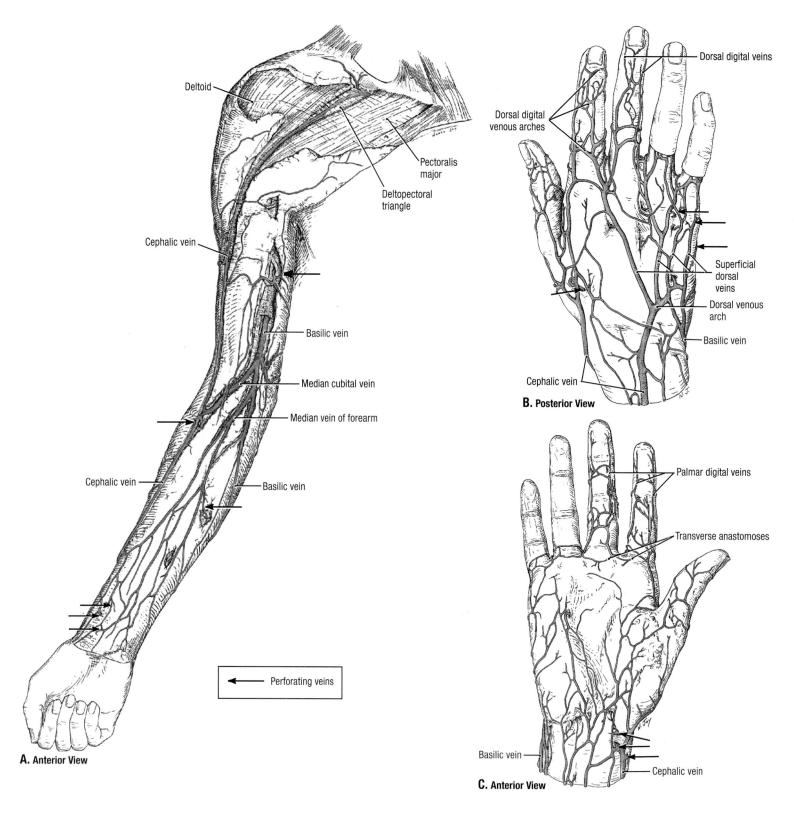

A. **Anterior View**

B. **Posterior View**

C. **Anterior View**

Deltoid

Pectoralis major

Deltopectoral triangle

Cephalic vein

Basilic vein

Median cubital vein

Median vein of forearm

Cephalic vein

Basilic vein

Dorsal digital veins

Dorsal digital venous arches

Superficial dorsal veins

Dorsal venous arch

Basilic vein

Cephalic vein

Palmar digital veins

Transverse anastomoses

Basilic vein

Cephalic vein

⟵ Perforating veins

6.6 **Superficial venous drainage of upper limb**

A. Forearm, arm, and pectoral region. **B.** Dorsal surface of hand. **C.** Palmar surface of hand. The *arrows* indicate where perforating veins penetrate the deep fascia. Blood is continuously shunted from these superficial veins in the subcutaneous tissue to deep veins via the perforating veins.

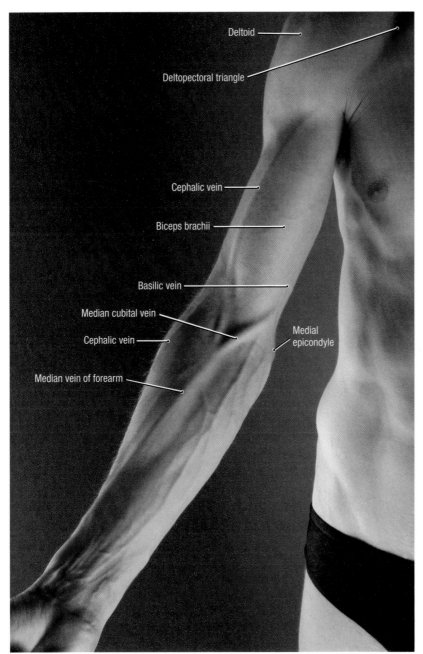

D. Anterior View

E. Posterior View

6.6   **Superficial venous drainage of upper limb** *(continued)*

D. Surface anatomy of veins of forearm and arm. **E.** Surface anatomy of veins on the dorsal surface of hand.

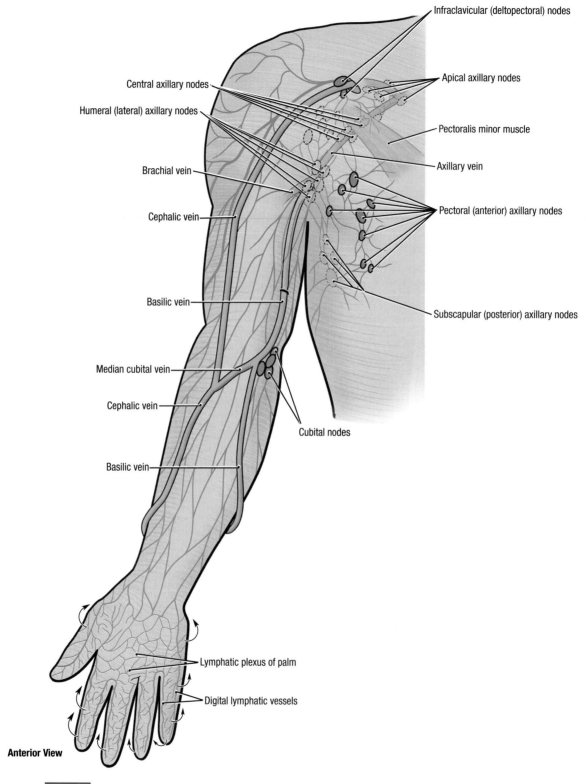

Infraclavicular (deltopectoral) nodes

Central axillary nodes

Apical axillary nodes

Humeral (lateral) axillary nodes

Pectoralis minor muscle

Brachial vein

Axillary vein

Cephalic vein

Pectoral (anterior) axillary nodes

Basilic vein

Subscapular (posterior) axillary nodes

Median cubital vein

Cephalic vein

Cubital nodes

Basilic vein

Lymphatic plexus of palm

Digital lymphatic vessels

Anterior View

6.7 Superficial lymphatic drainage of upper limb

Superficial lymphatic vessels arise from lymphatic plexuses in the digits, palm, and dorsum of the hand and ascend with the superficial veins of the upper limb. The superficial lymphatic vessels ascend through the forearm and arm, converging toward the cephalic and especially to the basilic vein to reach the axillary lymph nodes. Some lymph passes through the cubital nodes at the elbow and the deltopectoral nodes at the shoulder. Deep lymphatic vessels accompany the neurovascular bundles of the upper limb and end primarily in the humeral and central axillary lymph nodes.

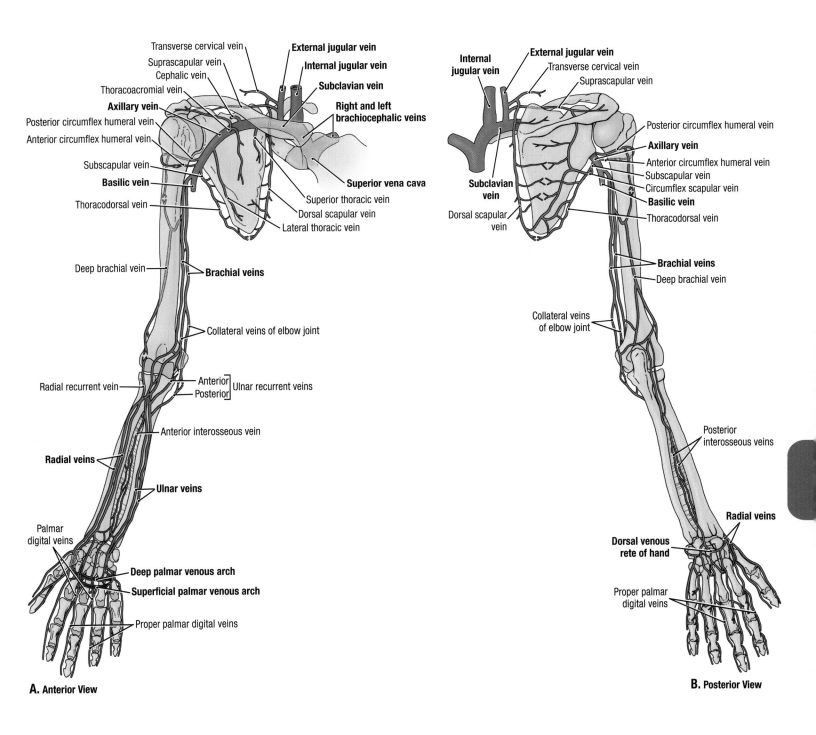

Transverse cervical vein
Suprascapular vein
Cephalic vein
Thoracoacromial vein
Axillary vein
Posterior circumflex humeral vein
Anterior circumflex humeral vein
Subscapular vein
Basilic vein
Thoracodorsal vein

External jugular vein
Internal jugular vein
Subclavian vein
Right and left brachiocephalic veins
Superior vena cava
Superior thoracic vein
Dorsal scapular vein
Lateral thoracic vein

Deep brachial vein
Brachial veins

Collateral veins of elbow joint

Radial recurrent vein
Anterior
Posterior } Ulnar recurrent veins

Anterior interosseous vein

Radial veins

Ulnar veins

Palmar digital veins

Deep palmar venous arch
Superficial palmar venous arch
Proper palmar digital veins

A. Anterior View

Internal jugular vein
External jugular vein
Transverse cervical vein
Suprascapular vein
Posterior circumflex humeral vein
Axillary vein
Anterior circumflex humeral vein
Subscapular vein
Circumflex scapular vein
Basilic vein
Thoracodorsal vein

Subclavian vein
Dorsal scapular vein

Brachial veins
Deep brachial vein

Collateral veins of elbow joint

Posterior interosseous veins

Radial veins

Dorsal venous rete of hand

Proper palmar digital veins

B. Posterior View

6.8 **Overview of the deep veins of the upper limb**

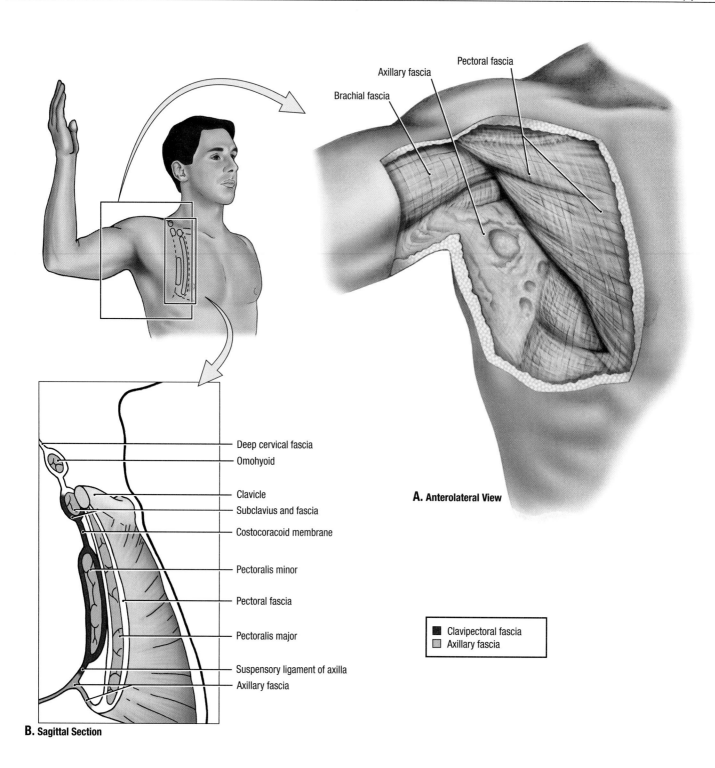

Pectoral fascia

Axillary fascia

Brachial fascia

A. Anterolateral View

Deep cervical fascia
Omohyoid

Clavicle
Subclavius and fascia

Costocoracoid membrane

Pectoralis minor

Pectoral fascia

Pectoralis major

Suspensory ligament of axilla
Axillary fascia

B. Sagittal Section

■ Clavipectoral fascia
□ Axillary fascia

6.9 Deep fascia of upper limb—axillary and clavipectoral fascia

A. Axillary fascia. The axillary fascia forms the floor of the axillary fossa. The axillary fascia is continuous with the pectoral fascia covering the pectoralis major muscle and the brachial fascia of the arm. **B.** Clavipectoral fascia. The clavipectoral fascia extends from the axillary fascia to enclose the pectoralis minor and subclavius muscles and then attaches to the clavicle. The part of the clavipectoral fascia superior to the pectoralis minor is the costocoracoid membrane and the part of the clavipectoral fascia inferior to the pectoralis minor is the suspensory ligament of the axilla. The suspensory ligament of the axilla supports the axillary fascia and pulls it and the skin inferior to it superiorly when the arm is abducted, forming the axilla or "armpit."

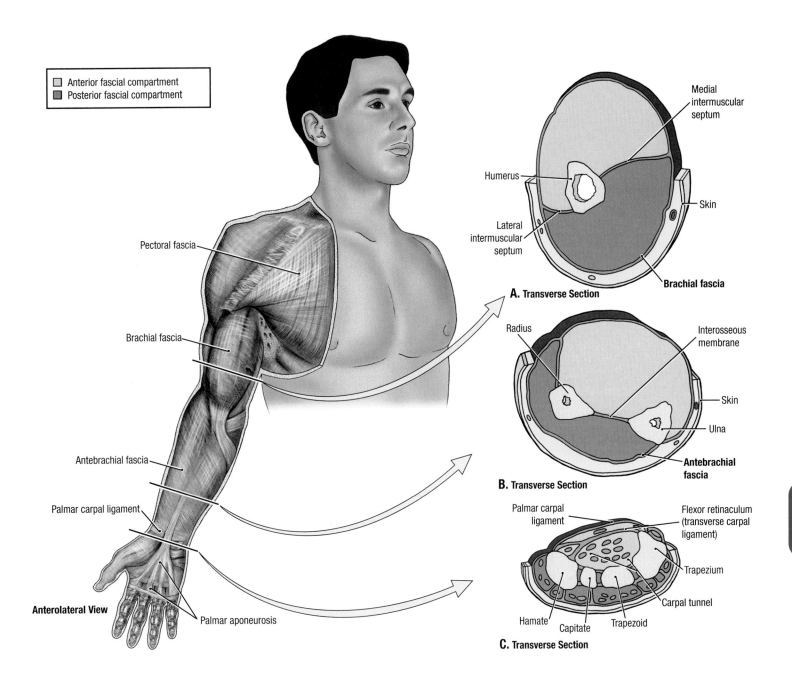

Anterolateral View

A. Transverse Section

B. Transverse Section

C. Transverse Section

6.10   Deep fascia of upper limb—brachial and antebrachial fascia

A. Brachial fascia. The brachial fascia is the deep fascia of the arm and is continuous superiorly with the pectoral and axillary layers of fascia. Medial and lateral intermuscular septa extend from the deep aspect of the brachial fascia to the humerus, dividing the arm into anterior and posterior fascial compartments. **B.** Antebrachial fascia. The antebrachial fascia surrounds the forearm and is continuous with the brachial fascia and deep fascia of the hand. The interosseous membrane separates the forearm into anterior and posterior fascial compartments. Distally the fascia thickens to form the palmar carpal ligament, which is continuous with the flexor retinaculum and dorsally with the extensor expansion. The deep fascia of the hand is continuous with the antebrachial fascia, and on the palmar surface of the hand it thickens to form the palmar aponeurosis. **C.** Flexor retinaculum (transverse carpal ligament). The flexor retinaculum extends between the medial and lateral carpal bones to form an osseofibrous tunnel, the carpal tunnel.

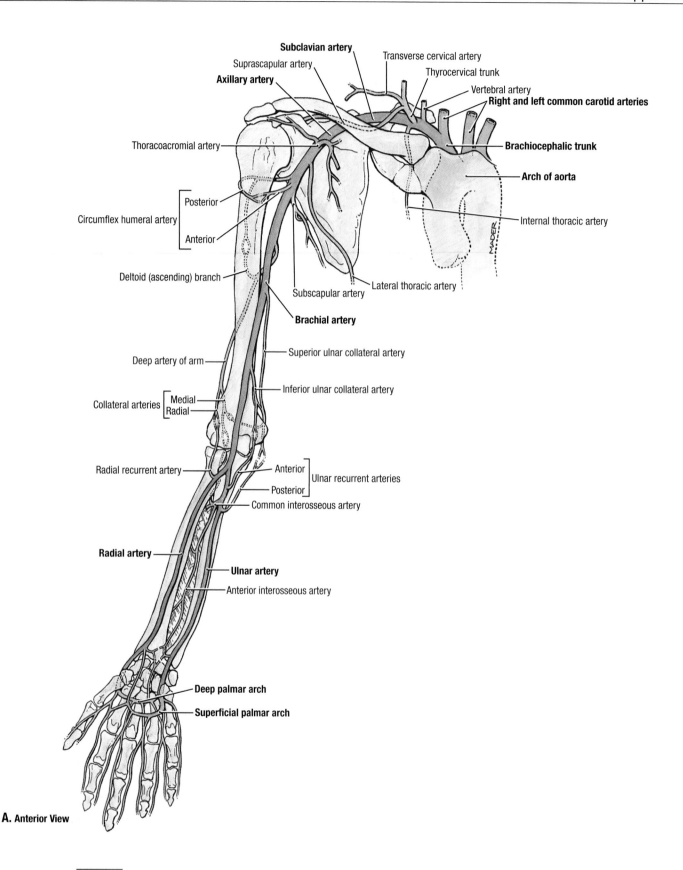

A. Anterior View

6.11 Arteries and arterial anastomoses of upper limb

A. Overview. **B.** Scapular anastomoses. **C.** Anastomoses of the elbow. **D.** Anastomoses of the hand. Joints receive blood from articular arteries that arise from vessels around joints. The arteries often anastomose or communicate to form networks to ensure blood supply distal to the joint throughout the range of movement.

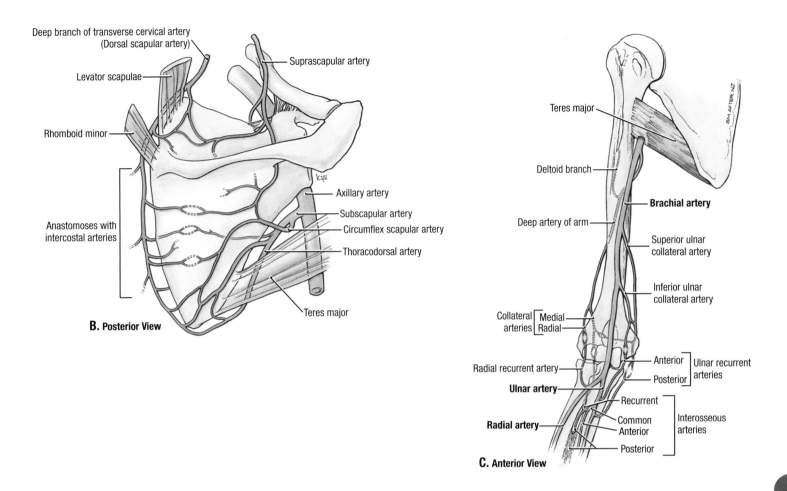

B. Posterior View

C. Anterior View

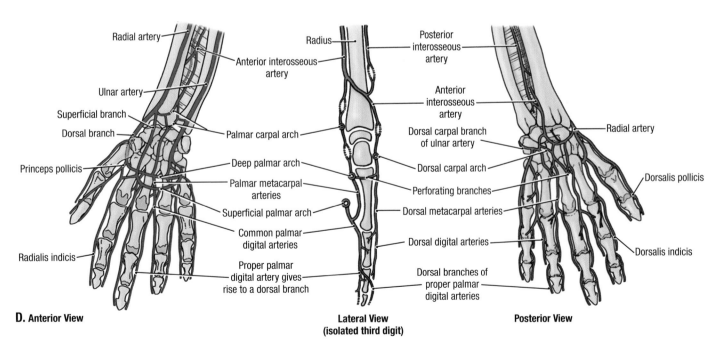

D. Anterior View

Lateral View
(isolated third digit)

Posterior View

6.11   **Arteries and arterial anastomoses of upper limb** *(continued)*

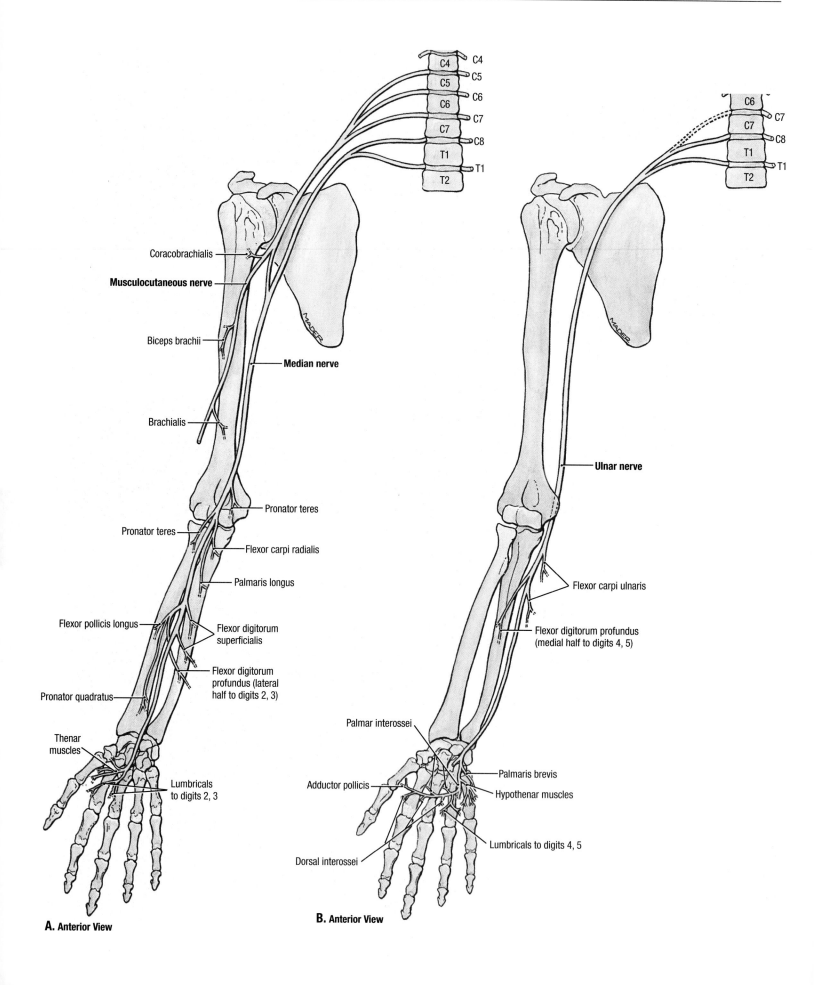

C4
C4
C5
C5
C6
C6
C7
C7
C7
C8
C8
T1
T1
T1
T2

C6
C6
C7
C7
C7
C8
C8
T1
T1
T1
T2

Coracobrachialis

Musculocutaneous nerve

Biceps brachii

Median nerve

Brachialis

Ulnar nerve

Pronator teres

Pronator teres

Flexor carpi radialis

Flexor carpi ulnaris

Palmaris longus

Flexor pollicis longus

Flexor digitorum superficialis

Flexor digitorum profundus (medial half to digits 4, 5)

Flexor digitorum profundus (lateral half to digits 2, 3)

Pronator quadratus

Thenar muscles

Lumbricals to digits 2, 3

Palmar interossei

Palmaris brevis

Adductor pollicis

Hypothenar muscles

Lumbricals to digits 4, 5

Dorsal interossei

A. Anterior View

B. Anterior View

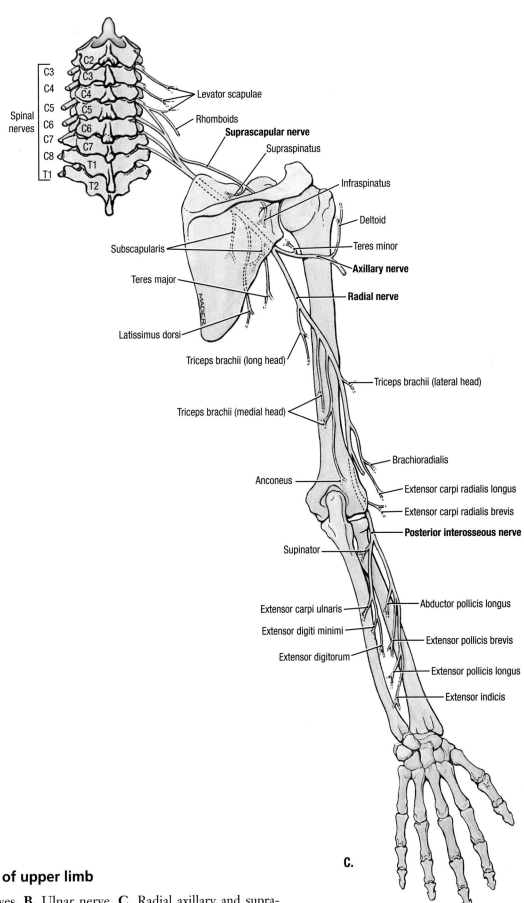

6.12   **Innervation of muscles of upper limb**

A. Median and musculocutaneous nerves. **B.** Ulnar nerve. **C.** Radial axillary and supra-scapular nerve.

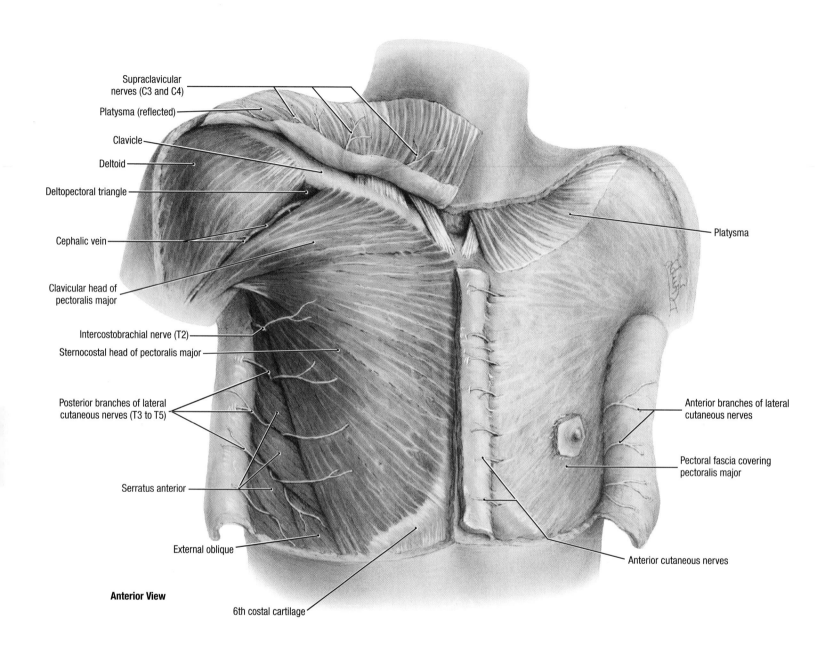

Supraclavicular nerves (C3 and C4)

Platysma (reflected)

Clavicle

Deltoid

Deltopectoral triangle

Cephalic vein

Clavicular head of pectoralis major

Intercostobrachial nerve (T2)

Sternocostal head of pectoralis major

Posterior branches of lateral cutaneous nerves (T3 to T5)

Serratus anterior

External oblique

Anterior View

6th costal cartilage

Platysma

Anterior branches of lateral cutaneous nerves

Pectoral fascia covering pectoralis major

Anterior cutaneous nerves

6.13   Superficial dissection, male pectoral region

- The platysma muscle, which usually descends to the 2nd or 3rd rib, is cut short on the right side, and together with the supraclavicular nerves, is reflected on the left side.
- The exposed intermuscular bony strip of the clavicle is subcutaneous and subplatysmal.
- The cephalic vein passes deeply to join the axillary vein in the deltopectoral triangle.
- The cutaneous innervation of the pectoral region by the supraclavicular nerves (C3 and C4) and upper thoracic nerves (T2 to T6); the brachial plexus (C5, C6, C7, C8, and T1) does not supply cutaneous branches to the pectoral region.

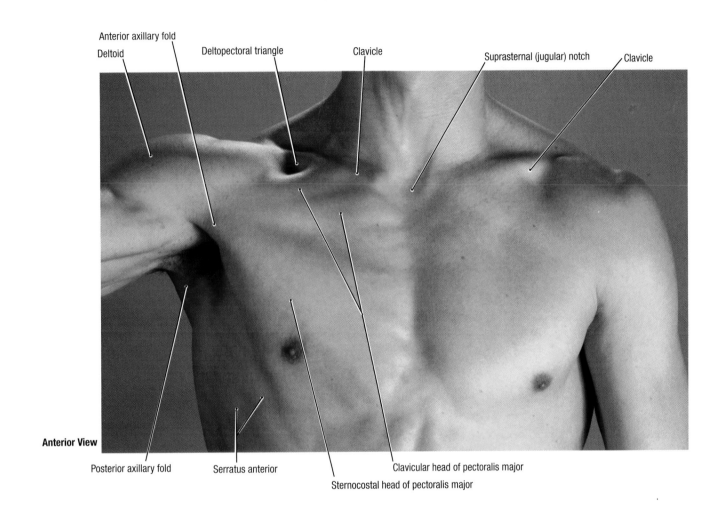

Anterior axillary fold

Deltoid

Deltopectoral triangle

Clavicle

Suprasternal (jugular) notch

Clavicle

Anterior View

Posterior axillary fold

Serratus anterior

Clavicular head of pectoralis major

Sternocostal head of pectoralis major

6.14   **Surface anatomy, male pectoral region**

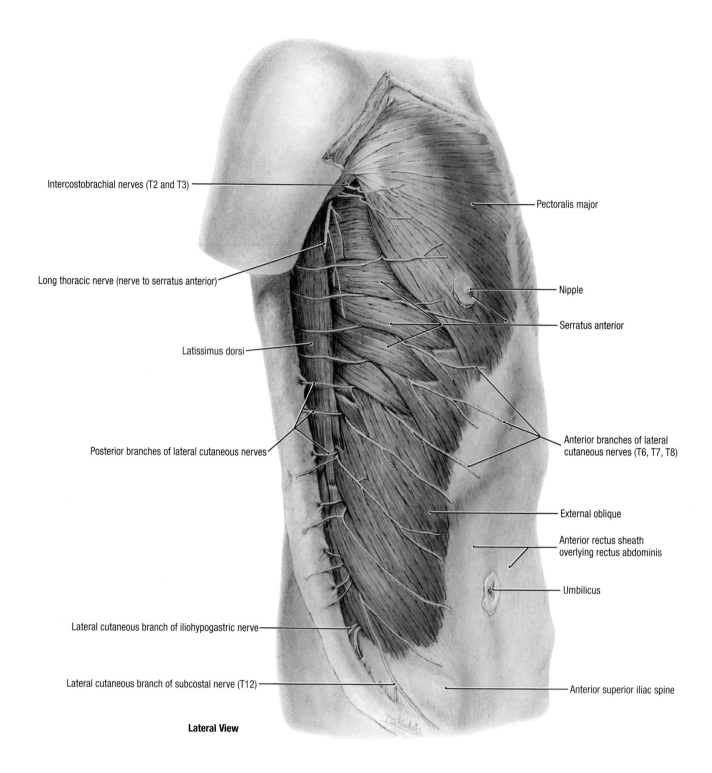

Intercostobrachial nerves (T2 and T3)

Long thoracic nerve (nerve to serratus anterior)

Latissimus dorsi

Posterior branches of lateral cutaneous nerves

Lateral cutaneous branch of iliohypogastric nerve

Lateral cutaneous branch of subcostal nerve (T12)

Pectoralis major

Nipple

Serratus anterior

Anterior branches of lateral cutaneous nerves (T6, T7, T8)

External oblique

Anterior rectus sheath overlying rectus abdominis

Umbilicus

Anterior superior iliac spine

Lateral View

6.15   **Superficial dissection of trunk**

- The slips of the serratus anterior interdigitate with the external oblique.
- The long thoracic nerve (nerve to serratus anterior) lies on the lateral (superficial) aspect of the serratus anterior; this nerve is vulnerable to damage from stab wounds and during surgery (e.g., radical mastectomy).
- The anterior and posterior branches of the lateral cutaneous nerves are dissected.

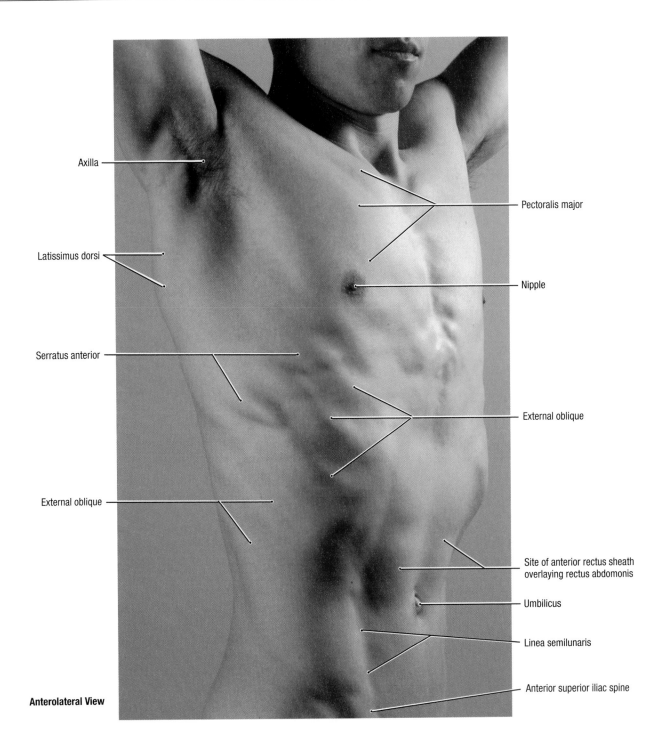

Axilla

Latissimus dorsi

Serratus anterior

External oblique

Anterolateral View

Pectoralis major

Nipple

External oblique

Site of anterior rectus sheath overlaying rectus abdomonis

Umbilicus

Linea semilunaris

Anterior superior iliac spine

6.16 **Surface anatomy of anterolateral aspect of the trunk**

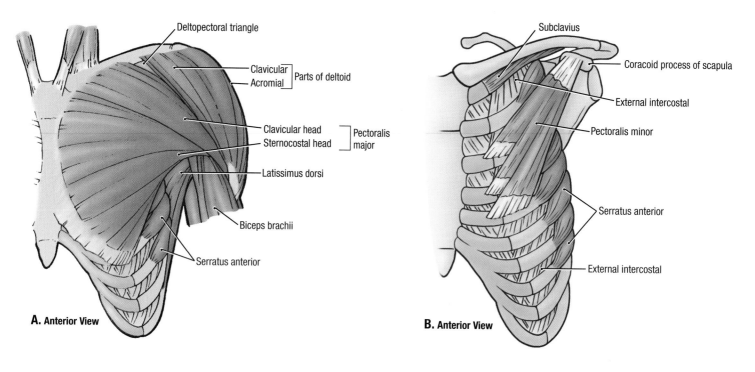

A. Anterior View

Deltopectoral triangle

Clavicular ⎤ Parts of deltoid
Acromial ⎦

Clavicular head ⎤ Pectoralis
Sternocostal head ⎦ major

Latissimus dorsi

Biceps brachii

Serratus anterior

Subclavius

Coracoid process of scapula

External intercostal

Pectoralis minor

Serratus anterior

External intercostal

B. Anterior View

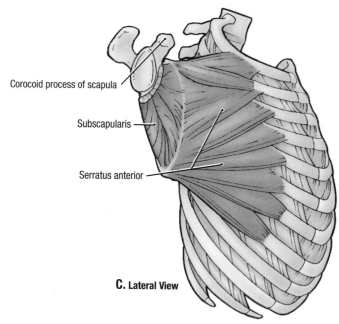

Corocoid process of scapula

Subscapularis

Serratus anterior

C. Lateral View

6.17   Pectoralis major and minor and serratus anterior

A. Pectoralis major. **B.** Pectoralis minor. **C.** Serratus anterior.

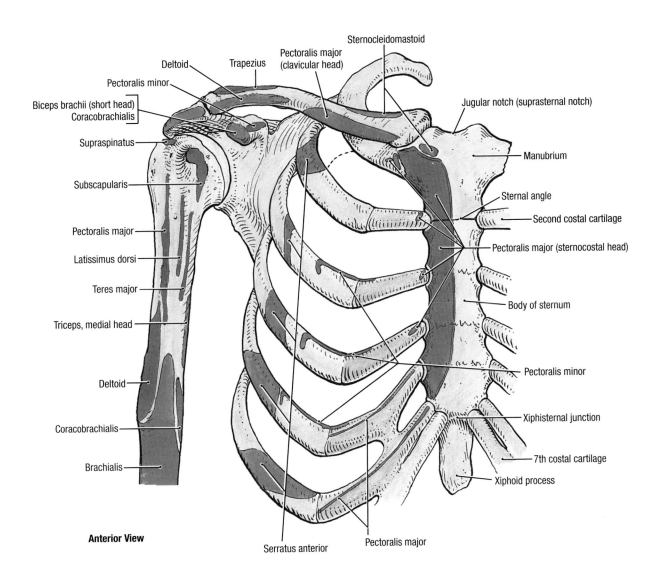

Deltoid
Pectoralis minor
Biceps brachii (short head)
Coracobrachialis
Supraspinatus
Subscapularis
Pectoralis major
Latissimus dorsi
Teres major
Triceps, medial head
Deltoid
Coracobrachialis
Brachialis

Trapezius
Pectoralis major (clavicular head)
Sternocleidomastoid

Jugular notch (suprasternal notch)
Manubrium
Sternal angle
Second costal cartilage
Pectoralis major (sternocostal head)
Body of sternum
Pectoralis minor
Xiphisternal junction
7th costal cartilage
Xiphoid process

Anterior View

Serratus anterior
Pectoralis major

TABLE 6.1   ANTERIOR THORACOAPPENDICULAR MUSCLES

MUSCLE	PROXIMAL ATTACHMENT (RED)	DISTAL ATTACHMENT (BLUE)	INNERVATION[a]	MAIN ACTIONS
Pectoralis major	Clavicular head: anterior surface of medial half of clavicle. Sternocostal head: anterior surface of sternum, superior six costal cartilages, and aponeurosis of external oblique muscle	Crest of greater tubercle of humerus (lateral lip)	Lateral and medial pectoral nerves; clavicular head (C5 and **C6**), sternocostal head (**C7**, **C8**, and T1)	Adducts and medially rotates humerus; draws scapula anteriorly and inferiorly. Acting alone: clavicular head flexes humerus and sternocostal head extends it from the flexed position
Pectoralis minor	3rd to 5th ribs near their costal cartilages	Medial border and superior surface of coracoid process of scapula	Medial pectoral nerve (C8 and T1)	Stabilizes scapula by drawing it inferiorly and anteriorly against thoracic wall
Subclavius	Junction of 1st rib and its costal cartilage	Inferior surface of middle third of clavicle	Nerve to subclavius (**C5** and C6)	Anchors and depresses clavicle
Serratus anterior	External surfaces of lateral parts of 1st to 8th ribs	Anterior surface of medial border of scapula	Long thoracic nerve (C5, **C6**, and C7)	Protracts scapula and holds it against thoracic wall; rotates scapula

[a]Numbers indicate spinal cord segmental innervation (e.g., C5 and C6 indicate that nerves supplying the clavicular head of pectoralis major are derived from 5th and 6th cervical segments of spinal cord). **Boldface** numbers indicate the main segmental innervation. Damage to these segments, or to motor nerve roots arising from them, results in paralysis of muscles concerned.

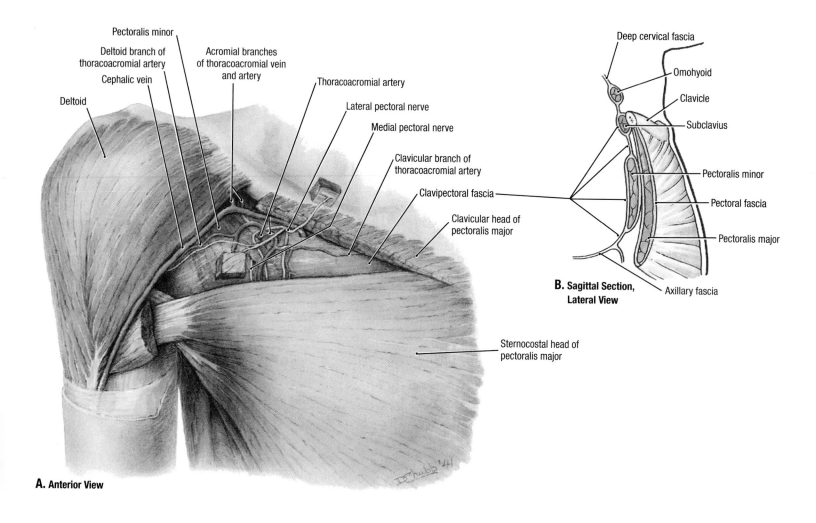

A. Anterior View

6.18 **Anterior wall of axilla and clavipectoral fascia**

A. Anterior wall of axilla. The clavicular head of the pectoralis major is excised, except for two cubes of muscle that remain to identify the branches of the lateral pectoral nerve. **B.** Clavipectoral fascia.

- The part of the clavipectoral fascia superior to the pectoralis minor (the costocoracoid membrane) is pierced by the cephalic vein, the lateral pectoral nerve, and the thoracoacromial vessels.
- The pectoralis minor and clavipectoral fascia are pierced by the medial pectoral nerve.
- Observe the trilaminar insertion of the pectoralis major from deep to superficial: inferior part of the sternocostal head, superior part of the sternocostal head, and clavicular head.

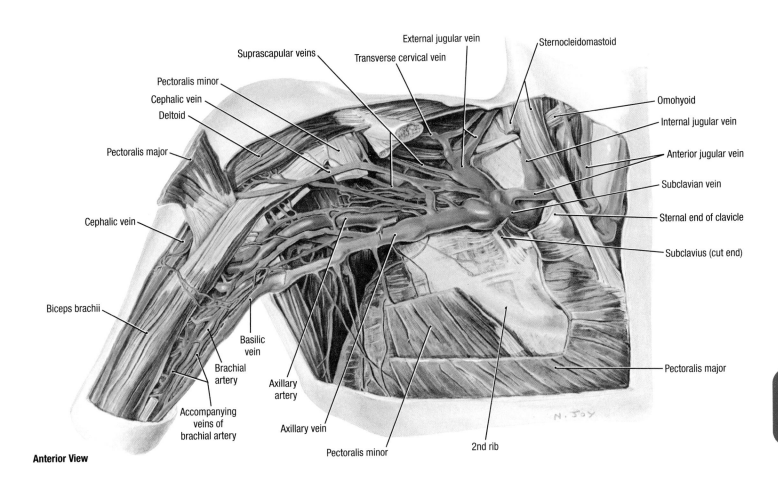

Anterior View

6.19   **Veins of axilla**

- The basilic vein joins the accompanying veins of the brachial artery to become the axillary vein at the inferior border of teres major, the axillary vein becomes the subclavian vein at the lateral border of the 1st rib, and the subclavian joins the internal jugular to become the brachiocephalic vein posterior to the sternal end of the clavicle.
- Numerous valves (enlargements in the vein) are shown; note three in the axillary vein and one in the subclavian vein, which is the last valve before the heart.
- The cephalic vein in this specimen bifurcates to end in the axillary and external jugular veins.
- Observe the three suprascapular veins: one entering the axillary vein, and two entering the external jugular vein.

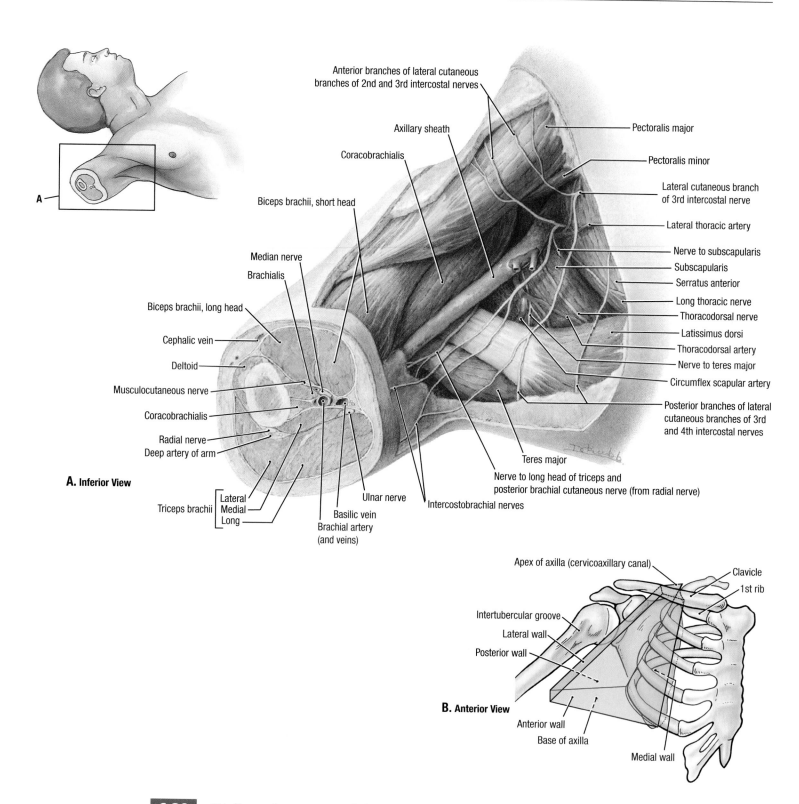

Anterior branches of lateral cutaneous branches of 2nd and 3rd intercostal nerves

Axillary sheath

Coracobrachialis

Biceps brachii, short head

Median nerve

Brachialis

Biceps brachii, long head

Cephalic vein

Deltoid

Musculocutaneous nerve

Coracobrachialis

Radial nerve
Deep artery of arm

A. Inferior View

Triceps brachii {Lateral, Medial, Long}

Ulnar nerve

Basilic vein
Brachial artery (and veins)

Intercostobrachial nerves

Nerve to long head of triceps and posterior brachial cutaneous nerve (from radial nerve)

Teres major

Pectoralis major

Pectoralis minor

Lateral cutaneous branch of 3rd intercostal nerve

Lateral thoracic artery

Nerve to subscapularis

Subscapularis

Serratus anterior

Long thoracic nerve

Thoracodorsal nerve

Latissimus dorsi

Thoracodorsal artery

Nerve to teres major

Circumflex scapular artery

Posterior branches of lateral cutaneous branches of 3rd and 4th intercostal nerves

Apex of axilla (cervicoaxillary canal)

Clavicle

1st rib

Intertubercular groove

Lateral wall

Posterior wall

B. Anterior View

Anterior wall

Base of axilla

Medial wall

6.20   Walls and contents of the axilla

A. Dissection. **B.** Location of axilla, schematic diagram.
- The walls of the axilla are the anterior wall (formed by the pectoralis major, pectoralis minor, and subclavius muscles), the posterior wall (formed by subscapularis, latissimus dorsi, and teres major muscles), the medial wall (formed by the serratus anterior muscle), and the lateral wall (formed by the intertubercular [bicipital] groove of the humerus), concealed by the biceps and coracobrachialis muscles.
- The axillary sheath surrounds the nerves and vessels (neurovascular bundle) of the upper limb.

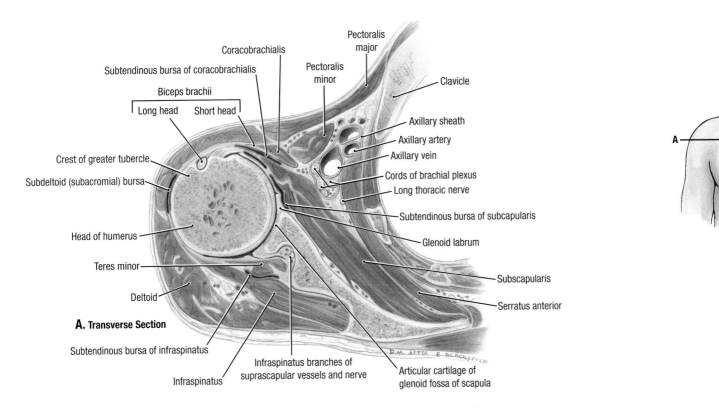

A. Transverse Section

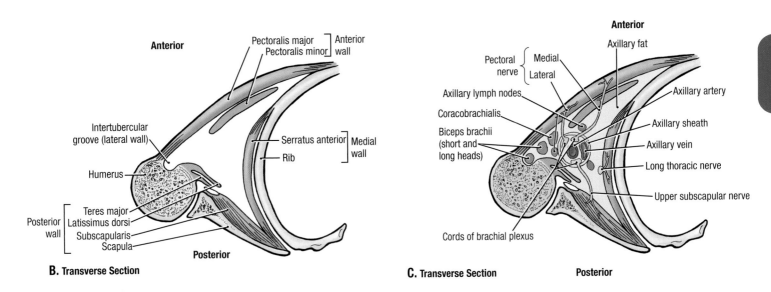

B. Transverse Section

C. Transverse Section

6.21   **Transverse sections through the shoulder joint and axilla**

A. Detailed section. **B.** Walls of axilla, schematic illustration. **C.** Walls and contents of axilla, schematic illustration.

In **A,** notice that

- The intertubercular (bicipital) groove containing the tendon of the long head of the biceps brachii muscle is directed anteriorly; the short head of the biceps muscle and the coracobrachialis and pectoralis minor muscles are sectioned just inferior to their attachments to the coracoid process.
- The fibrous capsule of the shoulder joint is thin posteriorly and partly fused with the tendon of infraspinatus; it is thicker anteriorly.

- The small glenoid cavity is deepened by the glenoid labrum.
- Bursae include the subdeltoid (subacromial) bursa, between the deltoid and greater tubercle; the subtendinous bursa of subscapularis, between the subscapularis tendon and scapula; and subtendinous bursa of coracobrachialis, between the coracobrachialis and subscapularis.
- The axillary sheath encloses the axillary artery and vein and the three cords of the brachial plexus to form a neurovascular bundle, surrounded by axillary fat.

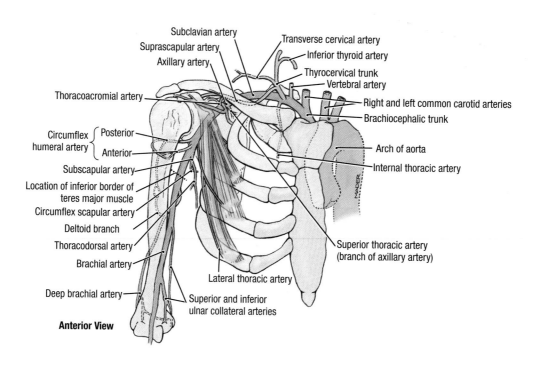

Anterior View

TABLE 6.2   ARTERIES OF THE PROXIMAL UPPER LIMB

ARTERY	ORIGIN	COURSE
Suprascapular	Thyrocervical trunk	Passes inferolaterally over anterior scalene muscle and phrenic nerve, crosses subclavian artery and brachial plexus, and runs laterally posterior and parallel to clavicle; it then passes to posterior aspect of scapula and supplies supraspinatus and infraspinatus muscles
Superior thoracic	Only branch of first part of axillary artery[a]	Runs anteromedially along superior border of pectoralis minor and then passes between it and pectoralis major to thoracic wall; helps to supply 1st and 2nd intercostal spaces and superior part of serratus anterior
Thoracoacromial	Second part of axillary artery[b]	Curls around superomedial border of pectoralis minor, pierces clavipectoral fascia, and divides into four branches
Lateral thoracic	Second part of axillary artery	Descends along axillary border of pectoralis minor and follows it onto thoracic wall
Subscapular	Third part of axillary artery[c]	Descends along lateral border of subscapularis and axillary border of scapula to its inferior angle, where it passes onto thoracic wall
Circumflex scapular	Subscapular artery	Curves around lateral border of scapula and enters infraspinous fossa
Thoracodorsal	Subscapular artery	Continues course of subscapular artery and accompanies thoracodorsal nerve to latissimus dorsi
Anterior and posterior circumflex humeral	Third part of axillary artery	These arteries anastomose to form a circle around surgical neck of humerus; larger posterior circumflex humeral artery passes through quadrangular space with axillary nerve
Deep brachial	Brachial artery near its origin	Accompanies radial nerve through radial groove in humerus and takes part in anastomosis around elbow joint
Ulnar collateral (superior and inferior)	Superior ulnar collateral artery arises from brachial artery near middle of arm; inferior ulnar collateral artery arises from brachial artery just superior to elbow	Superior ulnar collateral artery accompanies ulnar never to posterior aspect of elbow; inferior ulnar collateral artery divides into anterior and posterior branches; both ulnar collateral arteries take part in anastomosis around elbow joint

[a]First part of the axillary artery is located between the lateral border of the 1st rib and the medial border of pectoralis minor.
[b]Second part of the axillary artery lies posterior to pectoralis minor.
[c]Third part of the axillary artery extends from the lateral border of pectoralis minor to the inferior border of teres major, where it becomes the brachial artery.

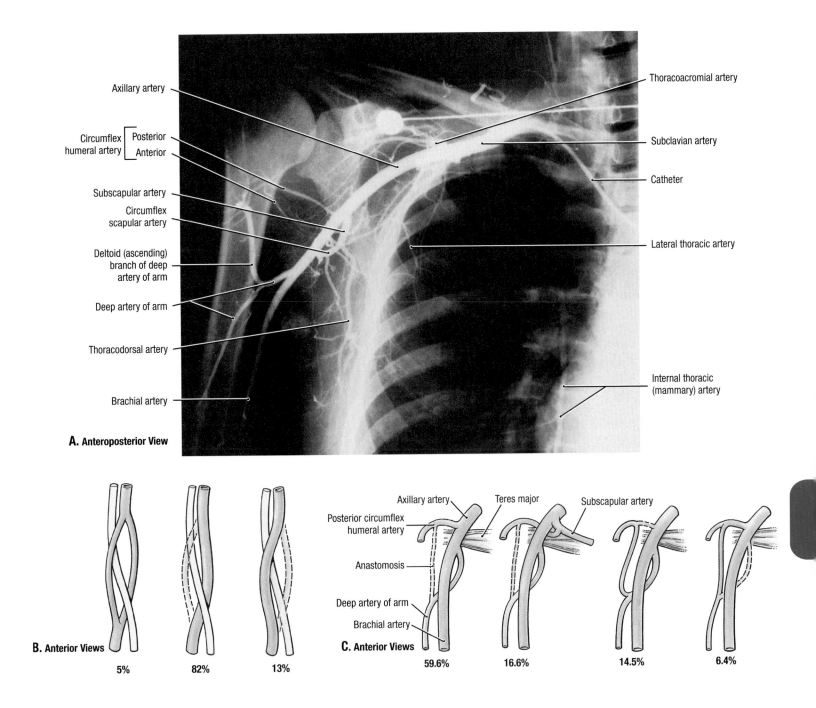

A. Anteroposterior View

Axillary artery

Circumflex humeral artery — Posterior / Anterior

Subscapular artery

Circumflex scapular artery

Deltoid (ascending) branch of deep artery of arm

Deep artery of arm

Thoracodorsal artery

Brachial artery

Thoracoacromial artery

Subclavian artery

Catheter

Lateral thoracic artery

Internal thoracic (mammary) artery

B. Anterior Views

5%   82%   13%

C. Anterior Views

Axillary artery   Teres major   Subscapular artery

Posterior circumflex humeral artery

Anastomosis

Deep artery of arm

Brachial artery

59.6%   16.6%   14.5%   6.4%

6.22   **Axillary arteriogram and arterial anomalies**

A. Axillary arteriogram. **B.** Relationship of median nerve and brachial artery. The variable relationship of these two structures can be explained developmentally. In a study of 307 limbs, both primitive brachial arteries persisted in 5%, the posterior in 82%, and the anterior in 13%. **C.** Variations of the posterior circumflex humeral artery and deep artery of arm. Four variations in origin of the posterior humeral circumflex artery and deep artery of arm are shown; in 2.9%, the arteries were otherwise irregular. Percentages are based on 235 specimens.

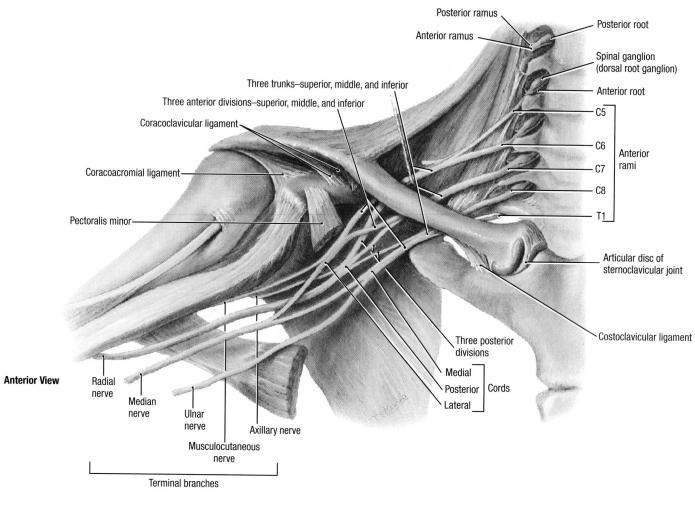

Posterior ramus

Anterior ramus

Posterior root

Spinal ganglion (dorsal root ganglion)

Anterior root

Three trunks—superior, middle, and inferior

Three anterior divisions—superior, middle, and inferior

Coracoclavicular ligament

C5

C6

C7

Anterior rami

C8

Coracoacromial ligament

T1

Pectoralis minor

Articular disc of sternoclavicular joint

Three posterior divisions

Costoclavicular ligament

Medial

Posterior   Cords

Lateral

Anterior View

Radial nerve

Median nerve

Ulnar nerve

Axillary nerve

Musculocutaneous nerve

Terminal branches

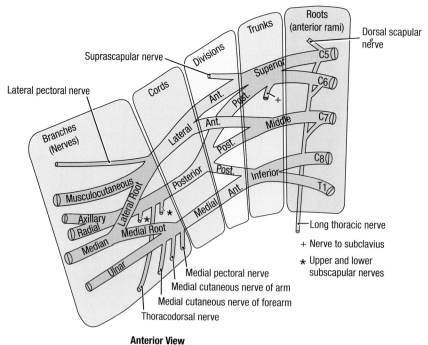

Roots (anterior rami)

Trunks

Divisions

Dorsal scapular nerve

Suprascapular nerve

Cords

C5

Superior

Lateral pectoral nerve

Branches (Nerves)

Ant.

Post.

+

C6

Ant.

Middle

C7

Lateral

Post.

Musculocutaneous

Post.

C8

Posterior

Inferior

Axillary

Radial

Lateral Root

Medial

Ant.

T1

Median

Medial Root

★   ★

Long thoracic nerve

Ulnar

+ Nerve to subclavius

★ Upper and lower subscapular nerves

Medial pectoral nerve

Medial cutaneous nerve of arm

Medial cutaneous nerve of forearm

Thoracodorsal nerve

Anterior View

TABLE 6.3 BRANCHES OF BRACHIAL PLEXUS

NERVE	ORIGIN	COURSE	DISTRIBUTION (INNERVATION)
Supraclavicular branches			
Dorsal scapular	Anterior ramus of C5 with a frequent contribution from C4	Pierces scalenus medius, descends deep to levator scapulae, and enters deep surface of rhomboids	Rhomboids and occasionally supplies levator scapulae
Long thoracic	Anterior rami of C5–C7	Descends posterior to C8 and T1 rami and passes distally on external surface of serratus anterior	Serratus anterior
Nerve to subclavius	Superior trunk receiving fibers from C5 and C6 and often C4	Descends posterior to clavicle and anterior to brachial plexus and subclavian artery	Subclavius and sternoclavicular joint
Suprascapular	Superior trunk receiving fibers from C5 and C6 and often C4	Passes laterally across posterior triangle of neck, through suprascapular notch under superior transverse scapular ligament	Supraspinatus, infraspinatus, and shoulder joint
Infraclavicular branches			
Lateral pectoral	Lateral cord receiving fibers from C5–C7	Pierces clavipectoral fascia to reach deep surface of pectoral muscles	Primarily pectoralis major but sends a loop to medial pectoral nerve that innervates pectoralis minor
Musculocutaneous	Lateral cord receiving fibers from C5–C7	Enters deep surface of coracobrachialis and descends between biceps brachii and brachialis	Coracobrachialis, biceps brachii, and brachialis; continues as lateral cutaneous nerve of forearm
Median	Lateral root is a continuation of lateral cord, receiving fibers from C6 and C7; medial root is a continuation of medial cord receiving fibers from C8 and T1	Lateral root joins medial root to form median nerve lateral to axillary artery	Flexor muscles in forearm (except flexor carpi ulnaris, ulnar half of flexor digitorum profundus, and five hand muscles) and skin of palm and $3\frac{1}{2}$ digits lateral to a line bisecting fourth digit and the dorsum of the distal halves of these digits
Medial pectoral	Medial cord receiving fibers from C8 and T1	Passes between axillary artery and vein and enters deep surface of pectoralis minor	Pectoralis minor and part of pectoralis major
Medial cutaneous nerve of arm	Medial cord receiving fibers from C8 and T1	Runs along the medial side of axillary vein and communicates with intercostobrachial nerve	Skin on medial side of arm
Medial cutaneous nerve of forearm	Medial cord receiving fibers from C8 and T1	Runs between axillary artery and vein	Skin over medial side of forearm
Ulnar	A terminal branch of medial cord receiving fibers from C8 and T1 and often C7	Passes down medial aspect of arm and runs posterior to medial epicondyle to enter forearm	Innervates one and one-half flexor muscles in forearm, most small muscles in hand, and skin of hand medial to a line bisecting fourth digit (ring finger) anteriorly and posteriorly
Upper subscapular	Branch of posterior cord receiving fibers from C5 and C6	Passes posteriorly and enters subscapularis	Superior portion of subscapularis
Thoracodorsal	Branch of posterior cord receiving fibers from C6–C8	Arises between upper and lower subscapular nerves and runs inferolaterally to latissimus dorsi	Latissimus dorsi
Lower subscapular	Branch of posterior cord receiving fibers from C5 and C6	Passes inferolaterally, deep to subscapular artery and vein, to subscapularis and teres major	Inferior portion of subscapularis and teres major
Axillary	Terminal branch of posterior cord receiving fibers from C5 and C6	Passes to posterior aspect of arm through quadrangular space[a] in company with posterior circumflex humeral artery and then winds around surgical neck of humerus; gives rise to lateral cutaneous nerve of arm	Teres minor and deltoid, shoulder joint, and skin over inferior part of deltoid
Radial	Terminal branch of posterior cord receiving fibers from C5–C8 and C11	Descends posterior to axillary artery; enters radial groove with deep brachial artery to pass between long and medial heads of triceps	Triceps brachii, anconeus, brachioradialis, and extensor muscles of forearm; supplies skin on posterior aspect of arm and forearm through posterior cutaneous nerves of arm and forearm

[a]Quadrangular space is bounded superiorly by subscapularis and teres minor, inferiorly by teres major, medially by long head of triceps, and laterally by humerus.

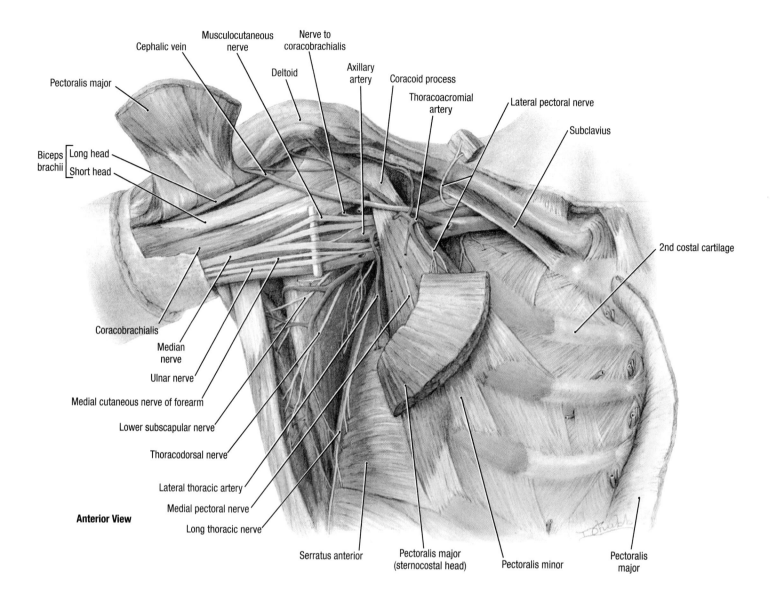

Anterior View

6.23 Structures of axilla

- The pectoralis major muscle is reflected, and the clavipectoral fascia is removed.
- The subclavius and pectoralis minor are the two deep muscles of the anterior wall.
- The axillary artery passes posterior to the pectoralis minor muscle, a fingerbreadth from the tip of the coracoid process.
- The axillary vein lies medial to the axillary artery.
- The median nerve, followed proximally, leads by its lateral root to the lateral cord and musculocutaneous nerve and by its medial root to the medial cord and ulnar nerve. These four nerves and the medial cutaneous nerve of the forearm are derived from the anterior division of the plexus and are raised on a *stick*. The lateral root of the median nerve may occur as several strands.
- The cube of muscle superior to the clavicle is cut from the clavicular head of the pectoralis major muscle.

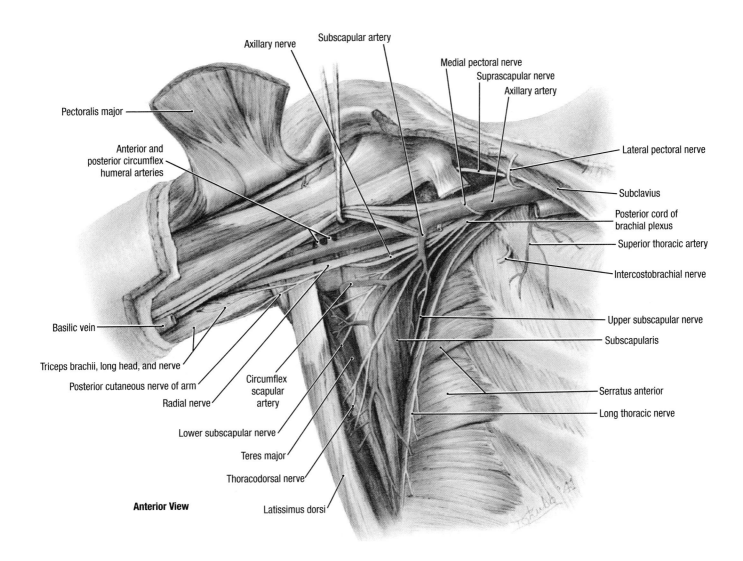

Anterior View

6.24 Posterior and medial walls of axilla

- The pectoralis minor muscle is excised, the lateral and medial cords of the brachial plexus are retracted, and the axillary vein is removed.
- The posterior cord and its two terminal branches (the radial and axillary nerves) lie posterior to the axillary artery.
- The nerve to the serratus anterior lies on the serratus anterior muscle; superiorly, some fat may intervene.
- The suprascapular nerve passes toward the base of the coracoid process.
- The subscapular artery, the largest branch of the axillary artery, in this specimen arises more proximally than usual; typically it arises at the inferior border of subscapularis.

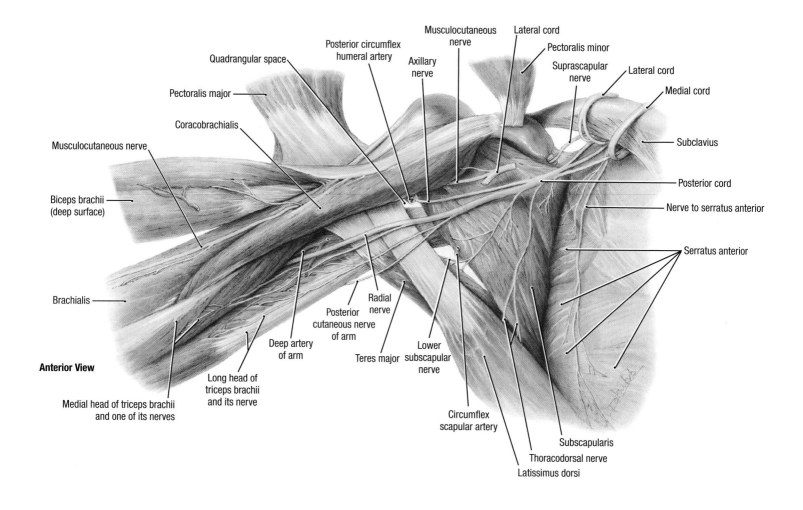

6.25 **Posterior wall of axilla, musculocutaneous nerve, and posterior cord**

- The pectoralis major and minor muscles are reflected laterally, the lateral and medial cords of the brachial plexus are reflected superiorly, and the arteries, veins, and median and ulnar nerves are removed.
- Coracobrachialis arises with the short head of the biceps brachii muscle from the tip of the coracoid process and attaches halfway down the medial aspect of the humerus.
- The musculocutaneous nerve pierces the coracobrachialis muscle and supplies it, the biceps, and the brachialis before becoming the lateral cutaneous nerve of the forearm.
- The posterior cord of the plexus is formed by the union of the three posterior divisions; it supplies the three muscles of the posterior wall of the axilla and then bifurcates into the radial and axillary nerves.
- In the axilla, the radial nerve gives off the nerve to the long head of the triceps brachii muscle and a cutaneous branch; in this specimen, it also gives off a branch to the medial head of the triceps. It then enters the radial groove of the humerus with the deep artery of the arm.
- The axillary nerve passes through the quadrangular space with the posterior circumflex humeral artery. The borders of the quadrangular space are superiorly, the lateral border of the scapula; inferiorly, the teres major; laterally, the humerus; and medially, the long head of triceps brachii. The circumflex scapular artery traverses the triangular space.

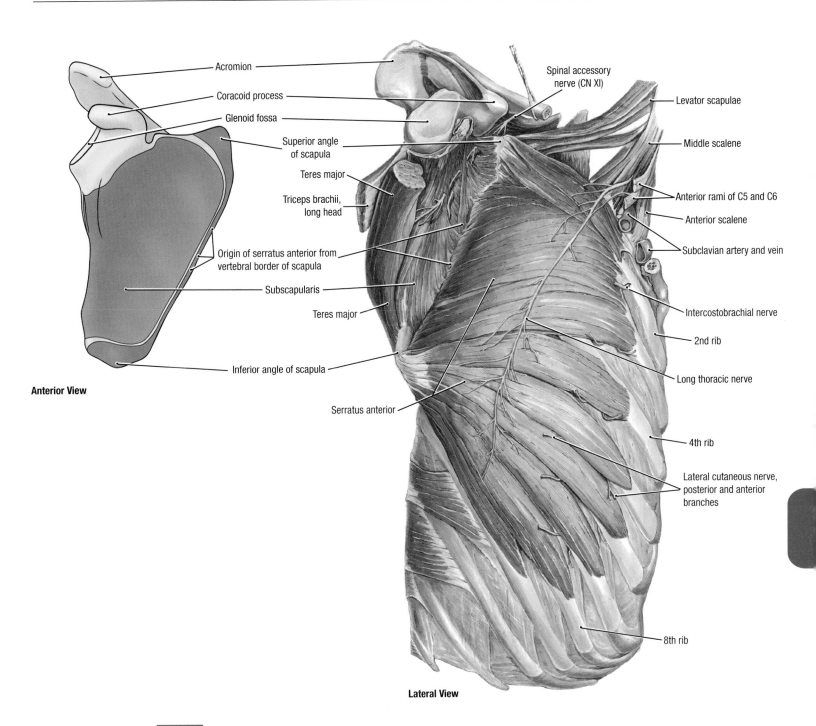

Anterior View

Lateral View

6.26   **Serratus anterior and subscapularis**

Serratus anterior. Scapular attachments of subscapularis *(red)* and serratus anterior *(blue)* *(left)* and dissection *(right)*.

- The serratus anterior muscle, which forms the medial wall of the axilla, has a fleshy belly extending from the superior eight (here nine) ribs in the midclavicular line to the medial border of the scapula. The fibers from the 1st rib and the arch between the 1st and 2nd ribs converge on the superior angle of the scapula; those from the 2nd and 3rd ribs diverge to spread thinly along the medial border; and the remainder (from the 4th to 9th ribs), which form the bulk of the muscle, converge on the inferior angle via a tendinous insertion.
- The long thoracic nerve to serratus anterior arises from spinal nerves C5, C6, and C7 and courses the whole length of the muscle; the fibers from C5 and C6 pierce the middle scalene and appear lateral to the brachial plexus; and those from C7 descend posterior to the plexus.
- The brachial plexus and subclavian artery emerge between the anterior and middle scalene muscles; the subclavian vein is separated from the artery by the anterior scalene.

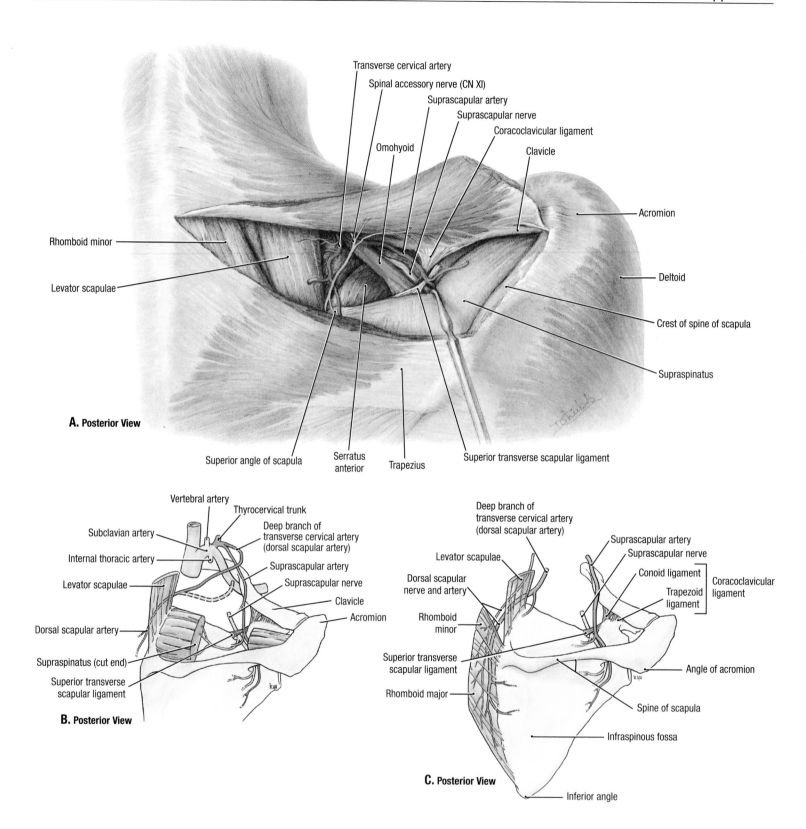

A. Posterior View

Transverse cervical artery
Spinal accessory nerve (CN XI)
Suprascapular artery
Suprascapular nerve
Coracoclavicular ligament
Clavicle
Omohyoid
Acromion
Rhomboid minor
Deltoid
Levator scapulae
Crest of spine of scapula
Supraspinatus
Superior angle of scapula
Serratus anterior
Trapezius
Superior transverse scapular ligament

B. Posterior View

Vertebral artery
Thyrocervical trunk
Subclavian artery
Deep branch of transverse cervical artery (dorsal scapular artery)
Internal thoracic artery
Suprascapular artery
Levator scapulae
Suprascapular nerve
Clavicle
Acromion
Dorsal scapular artery
Supraspinatus (cut end)
Superior transverse scapular ligament

C. Posterior View

Deep branch of transverse cervical artery (dorsal scapular artery)
Levator scapulae
Suprascapular artery
Suprascapular nerve
Dorsal scapular nerve and artery
Conoid ligament
Coracoclavicular ligament
Trapezoid ligament
Rhomboid minor
Superior transverse scapular ligament
Angle of acromion
Rhomboid major
Spine of scapula
Infraspinous fossa
Inferior angle

6.27   Suprascapular region

A. Dissection. At the level of the superior angle of the scapula, the transverse part of the trapezius muscle is separated, and the incision is carried laterally along the crest of the spine of the scapula. **B** and **C.** Suprascapular and dorsal scapular arteries. Note that the suprascapular artery runs posterior to the clavicle before crossing superior to the superior transverse scapular ligament and that the suprascapular nerve crosses inferior to the ligament.

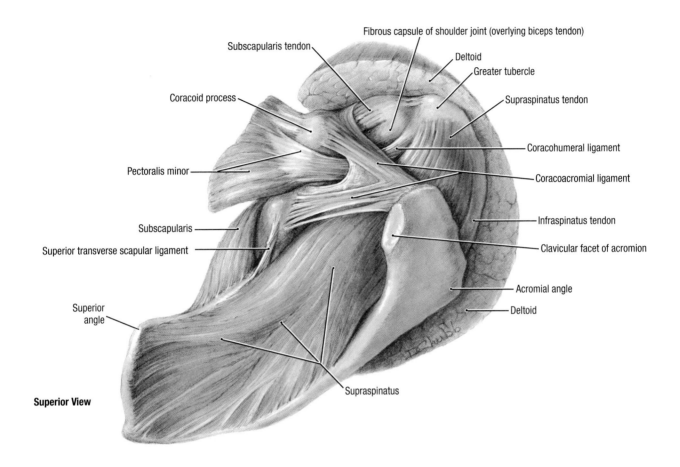

Superior View

6.28   Supraspinatus

- The clavicular facet on the acromion is small, oval, and obliquely set.
- The triangular coracoacromial ligament arches from the lateral border and base of the coracoid process to the acromion between the articular facet and tip.
- Part of the pectoralis minor tendon divides the coracoacromial ligament into two parts and continues as the anterior part of the coracohumeral ligament to the greater tubercle of the humerus.
- The supraspinatus muscle passes inferior to the coracoacromial arch and then lies between the deltoid muscle superiorly and the capsule of the shoulder joint inferiorly; the supraspinatus muscle and the acromial part of the deltoid muscle are the abductors of the joint. Abduction must be initiated by the supraspinatus muscle. The deltoid muscle becomes fully effective as an abductor following the initial 15° of abduction.

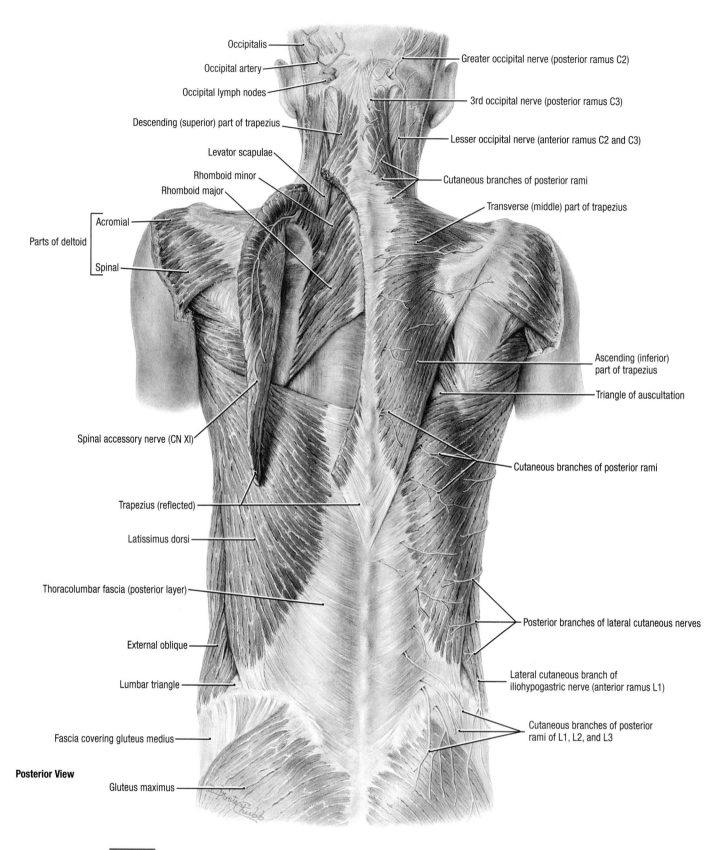

Occipitalis

Occipital artery

Occipital lymph nodes

Descending (superior) part of trapezius

Levator scapulae

Rhomboid minor

Rhomboid major

Acromial

Parts of deltoid

Spinal

Greater occipital nerve (posterior ramus C2)

3rd occipital nerve (posterior ramus C3)

Lesser occipital nerve (anterior ramus C2 and C3)

Cutaneous branches of posterior rami

Transverse (middle) part of trapezius

Ascending (inferior) part of trapezius

Triangle of auscultation

Spinal accessory nerve (CN XI)

Cutaneous branches of posterior rami

Trapezius (reflected)

Latissimus dorsi

Thoracolumbar fascia (posterior layer)

External oblique

Lumbar triangle

Fascia covering gluteus medius

Posterior View

Gluteus maximus

Posterior branches of lateral cutaneous nerves

Lateral cutaneous branch of iliohypogastric nerve (anterior ramus L1)

Cutaneous branches of posterior rami of L1, L2, and L3

6.29 **Cutaneous nerves of back and first two muscle layers**

The trapezius muscle is cut and reflected on the left side. The superficial or first muscle layer consists of the trapezius and latissimus dorsi muscles, and the second layer of the levator scapulae and rhomboids. Observe the cutaneous branches of the posterior rami.

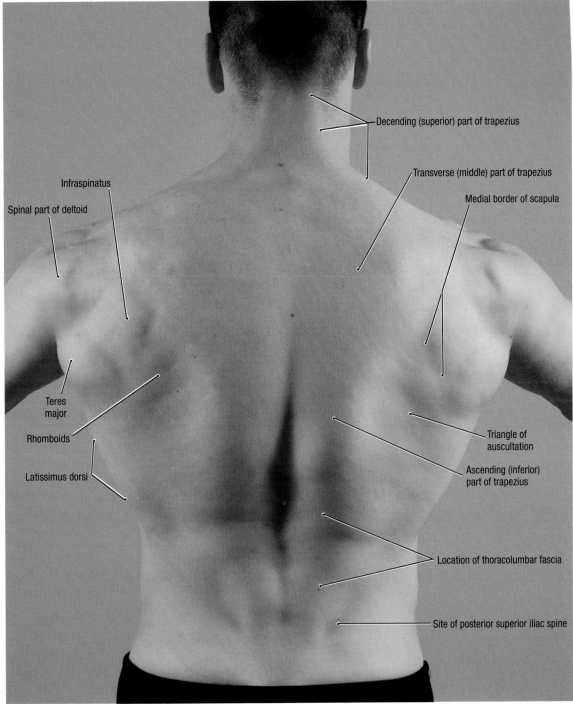

Posterior View

6.30   **Surface anatomy of superficial back**

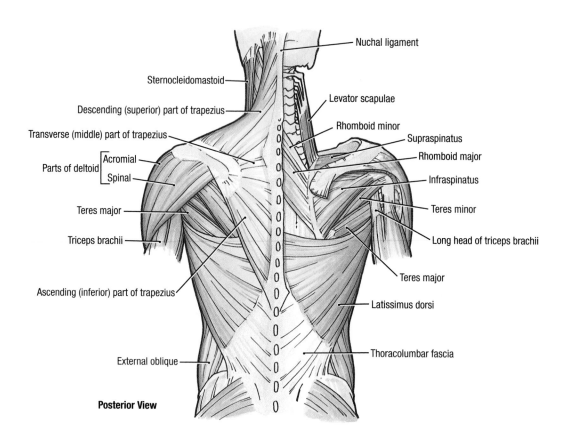

Nuchal ligament

Sternocleidomastoid

Descending (superior) part of trapezius

Transverse (middle) part of trapezius

Parts of deltoid { Acromial / Spinal }

Teres major

Triceps brachii

Ascending (inferior) part of trapezius

External oblique

Posterior View

Levator scapulae

Rhomboid minor

Supraspinatus

Rhomboid major

Infraspinatus

Teres minor

Long head of triceps brachii

Teres major

Latissimus dorsi

Thoracolumbar fascia

TABLE 6.4   SUPERFICIAL BACK MUSCLES

MUSCLE	PROXIMAL ATTACHMENT	DISTAL ATTACHMENT	INNERVATION	MAIN ACTIONS
Trapezius	Medial third of superior nuchal line; external occipital protuberance, nuchal ligament, and spinous processes of C7–T12	Lateral third of clavicle, acromion, and spine of scapula	Spinal accessory nerve (CN XI) and cervical nerves (C3 and C4)	Elevates, retracts, and rotates scapula; descending part elevates, transverse part retracts, and ascending part depresses scapula; descending and ascending part act together in superior rotation of scapula
Latissimus dorsi	Spinous processes of inferior six thoracic vertebrae, thoracolumbar fascia, iliac crest, and inferior three or four ribs	Floor of intertubercular (bicipital) groove of humerus	Thoracodorsal nerves (C6, C7, and C8)	Extends, adducts, and medially rotates humerus; raises body toward arms during climbing
Levator scapulae	Posterior tubercles of transverse processes of C1–C4 vertebrae	Superior part of medial border of scapula	Dorsal scapular (C5) and cervical (C3 and C4) nerves	Elevates scapula and tilts its glenoid cavity inferiorly by rotating scapula
Rhomboid minor and major	*Minor:* nuchal ligament and spinous processes of C7 and T1 vertebrae *Major:* spinous processes of T2–T5 vertebrae	Medial border of scapula from level of spine to inferior angle	Dorsal scapular nerve (C4 and **C5**)	Retracts scapula and rotates it to depress glenoid cavity; fixes scapula to thoracic wall
Deltoid	Lateral third of clavicle, acromion, and spine of scapula	Deltoid tuberosity of humerus	Axillary nerve (C5 and C6)	*Clavicular (anterior) part:* flexes and medially rotates arm *Aromial (middle) part:* abducts arm *Spinal (posterior) part:* extends and laterally rotates arm

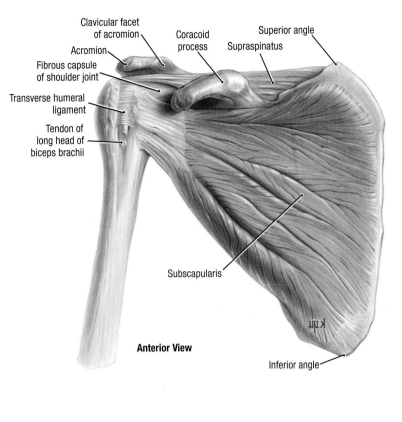

Clavicular facet
of acromion

Coracoid
process

Superior angle

Supraspinatus

Acromion

Fibrous capsule
of shoulder joint

Transverse humeral
ligament

Tendon of
long head of
biceps brachii

Subscapularis

Anterior View

Inferior angle

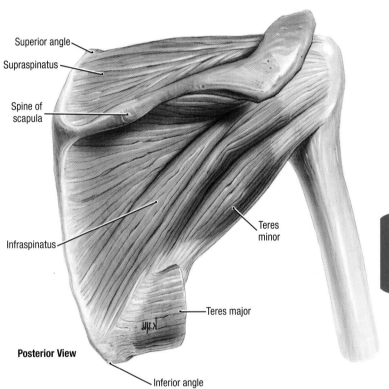

Superior angle

Supraspinatus

Spine of
scapula

Infraspinatus

Teres
minor

Teres major

Posterior View

Inferior angle

TABLE 6.5   ROTATOR CUFF MUSCLES

MUSCLE	PROXIMAL ATTACHMENT	DISTAL ATTACHMENT	INNERVATION	MAIN ACTIONS
Supraspinatus	Supraspinous fossa of scapula	Superior facet on greater tubercle of humerus	Suprascapular nerve (C4, **C5**, and C6)	Helps deltoid to abduct arm and acts with rotator cuff muscles[a]
Infraspinatus	Infraspinous fossa of scapula	Middle facet on greater tubercle of humerus	Suprascapular nerve (**C5** and C6)	Laterally rotate arm; help to hold humeral head in glenoid cavity of scapula
Teres minor	Superior part of lateral border of scapula	Inferior facet on greater tubercle of humerus	Axillary nerve (**C5** and C6)	
Teres major	Dorsal surface of inferior angle of scapula	Crest of lesser tubercle (medial lip) of humerus	Lower subscapular nerve (**C6** and C7)	Adducts and medially rotates arm
Subscapularis	Subscapular fossa	Lesser tubercle of humerus	Upper and lower subscapular nerves (C5, **C6**, and C7)	Medially rotates arm and adducts it; helps to hold humeral head in glenoid cavity

[a]Collectively, the supraspinatus, infraspinatus, teres minor, and subscapularis muscles are referred to as the rotator cuff muscles. Their prime function during all movements the of shoulder joint is to hold the head of the humerus in the glenoid cavity of scapula.

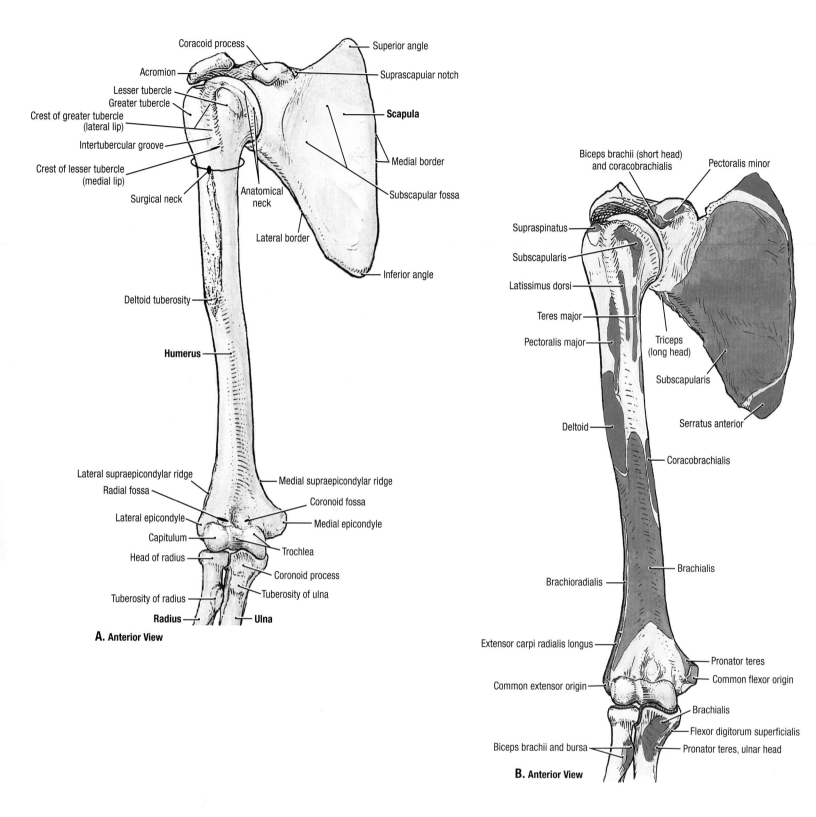

6.31 **Bones of proximal upper limb**

A. Bony features, anterior aspect. **B.** Muscle attachment sites, anterior aspect.

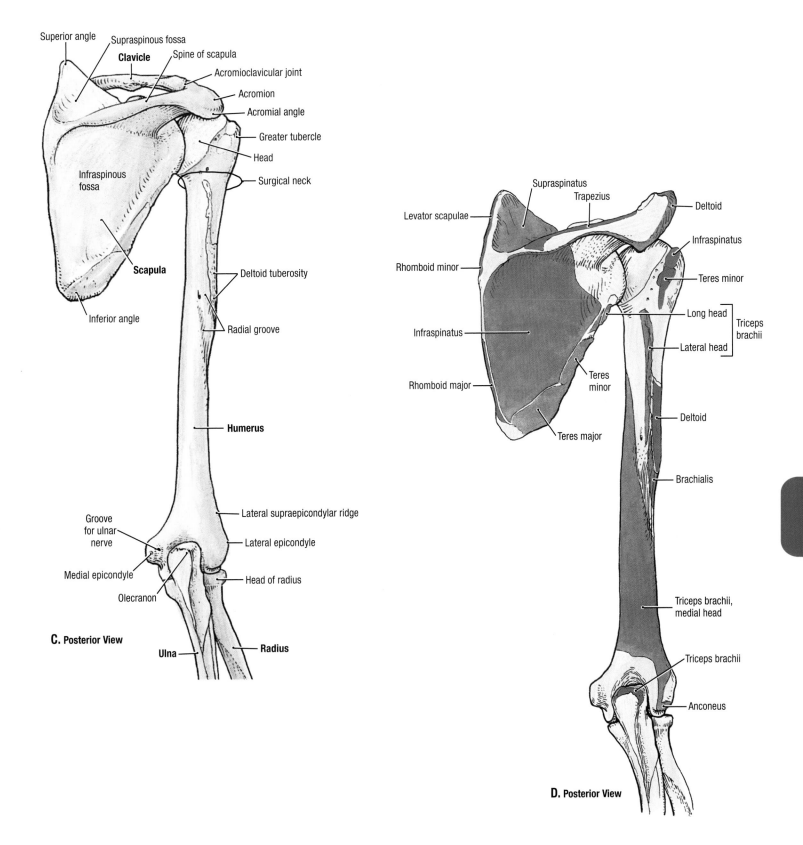

C. Posterior View

Superior angle
Supraspinous fossa
Clavicle
Spine of scapula
Acromioclavicular joint
Acromion
Acromial angle
Greater tubercle
Head
Surgical neck
Infraspinous fossa
Scapula
Deltoid tuberosity
Inferior angle
Radial groove
Humerus
Lateral supraepicondylar ridge
Groove for ulnar nerve
Lateral epicondyle
Medial epicondyle
Olecranon
Head of radius
Ulna
Radius

D. Posterior View

Levator scapulae
Supraspinatus
Trapezius
Deltoid
Infraspinatus
Rhomboid minor
Teres minor
Long head
Lateral head
Triceps brachii
Infraspinatus
Rhomboid major
Teres minor
Teres major
Deltoid
Brachialis
Triceps brachii, medial head
Triceps brachii
Anconeus

6.31 **Bones of proximal upper limb (continued)**

C. Bony features, posterior aspect. **D.** Muscle attachment sites, posterior aspect.

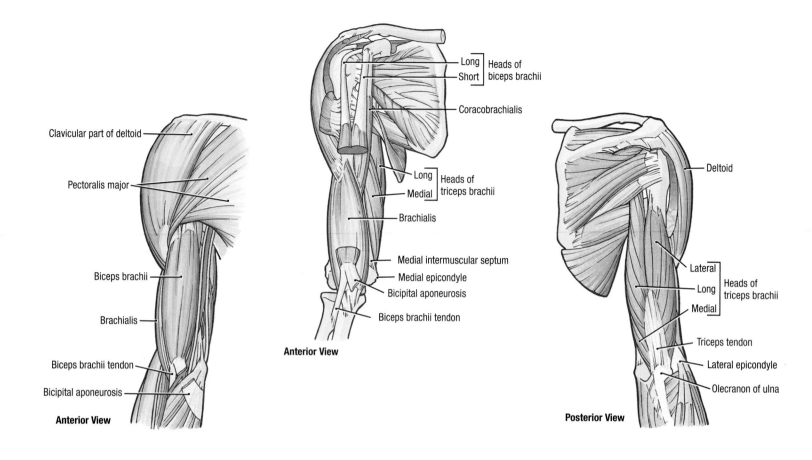

Clavicular part of deltoid

Pectoralis major

Biceps brachii

Brachialis

Biceps brachii tendon

Bicipital aponeurosis

Anterior View

Long ⎤ Heads of
Short ⎦ biceps brachii

Coracobrachialis

Long ⎤ Heads of
Medial ⎦ triceps brachii

Brachialis

Medial intermuscular septum

Medial epicondyle

Bicipital aponeurosis

Biceps brachii tendon

Anterior View

Deltoid

Lateral ⎤
Long ⎬ Heads of
Medial ⎦ triceps brachii

Triceps tendon

Lateral epicondyle

Olecranon of ulna

Posterior View

TABLE 6.6   ARM MUSCLES

MUSCLE	PROXIMAL ATTACHMENT	DISTAL ATTACHMENT	INNERVATION	MAIN ACTIONS
Biceps brachii	*Short head:* tip of coracoid process of scapula *Long head:* supraglenoid tubercle of scapula	Tuberosity of radius and fascia of forearm through bicipital aponeurosis	Musculocutaneous nerve (C5 and **C6**)	Supinates forearm and, when supine, flexes forearm
Brachialis	Distal half of anterior surface of humerus	Coronoid process and tuberosity of ulna		Flexes forearm in all positions
Coracobrachialis	Tip of coracoid process of scapula	Middle third of medial surface of humerus	Musculocutaneous nerve (C5, **C6**, and C7)	Helps to flex and adduct arm
Triceps brachii	*Long head:* infraglenoid tubercle of scapula *Lateral head:* posterior surface of humerus, superior to radial groove *Medial head:* posterior surface of humerus, inferior to radial groove	Proximal end of olecranon of ulna and fascia of forearm	Radial nerve (C6, **C7**, and **C8**)	Extends the forearm; it is chief extensor of forearm; long head steadies head of abducted humerus
Anconeus	Lateral epicondyle of humerus	Lateral surface of olecranon and superior part of posterior surface of ulna	Radial nerve (C7, C8, and T1)	Assists triceps in extending forearm; stabilizes elbow joint; abducts ulna during pronation

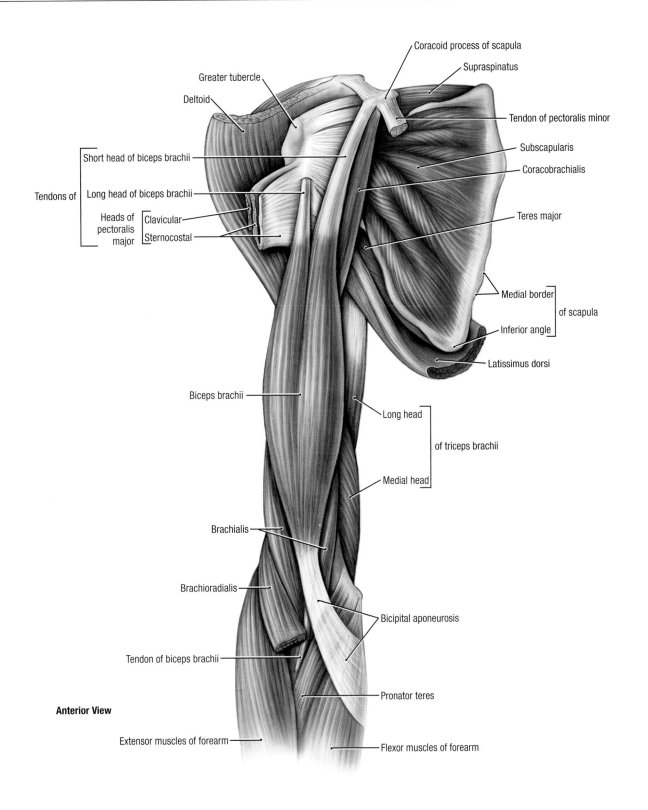

6.32 **Muscles of anterior aspect of arm—I**

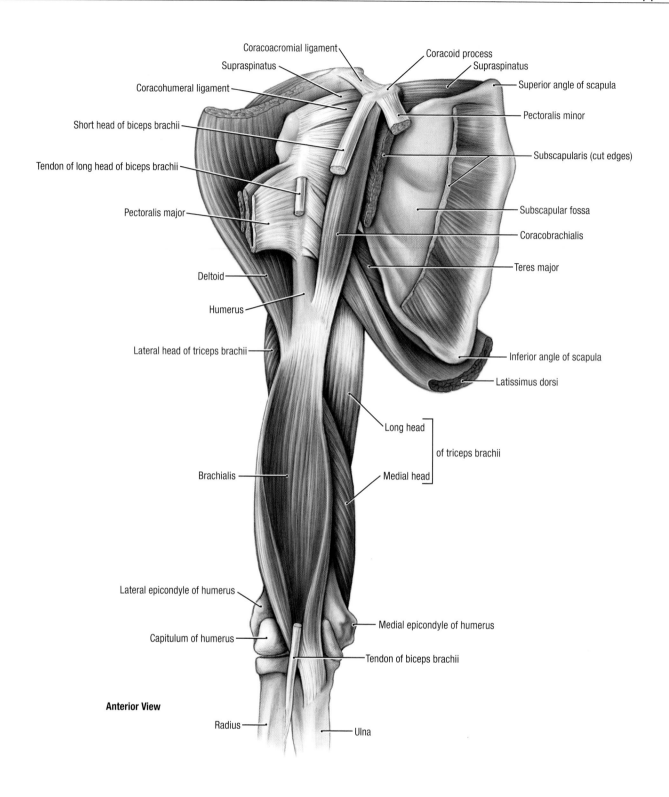

Anterior View

6.33 Muscles of anterior aspect of arm—II

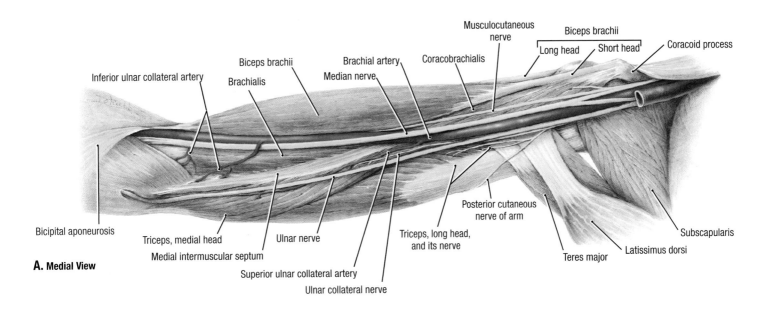

A. Medial View

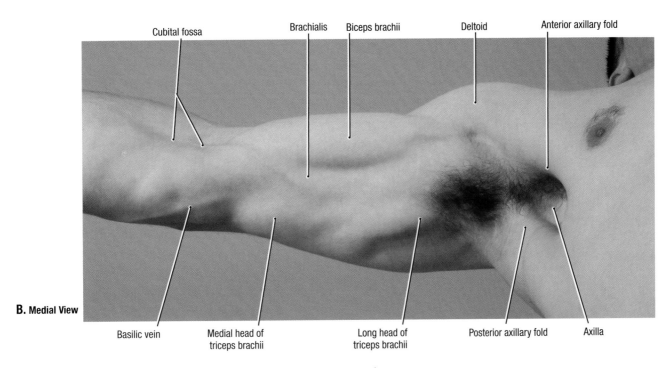

B. Medial View

6.34   Medial aspect of arm

A. Dissection. B. Surface anatomy.

- Three muscles, the biceps, brachialis, and coracobrachialis, lie in the anterior compartment of the arm; the triceps brachii lies in the posterior compartment. The medial intermuscular septum separates these two muscle groups in the distal two thirds of the arm.
- The axillary artery passes a fingerbreadth inferior to the tip of the coracoid process and courses posterior to the coracobrachialis. At the inferior border of the teres major, the axillary artery changes names to become the *brachial artery* and continues distally on the anterior aspect of brachialis.
- The median nerve lies adjacent to the axillary and brachial arteries and then crosses the artery from lateral to medial.

- Proximally, the ulnar nerve is adjacent to the medial side of the artery, passes posterior to the medial intermuscular septum, and descends on the medial head of triceps to pass posterior to the medial epicondyle; here, the ulnar nerve is palpable.
- The superior ulnar collateral artery and ulnar collateral branch of the radial nerve (to medial head of the triceps) accompany the ulnar nerve in the arm.
- The musculocutaneous nerve supplies and often perforates the coracobrachialis muscle, follows the lateral side of the brachial artery, and passes deeply between the biceps and brachialis.

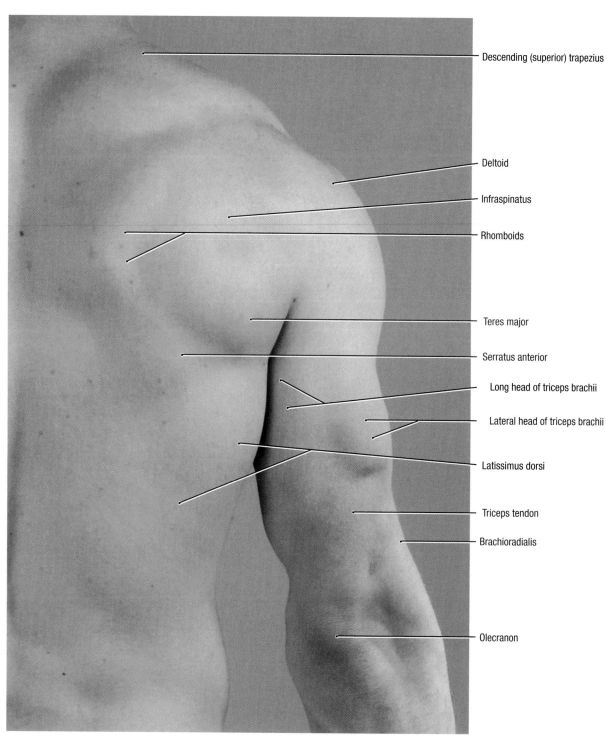

Descending (superior) trapezius

Deltoid

Infraspinatus

Rhomboids

Teres major

Serratus anterior

Long head of triceps brachii

Lateral head of triceps brachii

Latissimus dorsi

Triceps tendon

Brachioradialis

Olecranon

Posterior View

6.35 **Surface anatomy of the scapular region and posterior aspect of the arm**

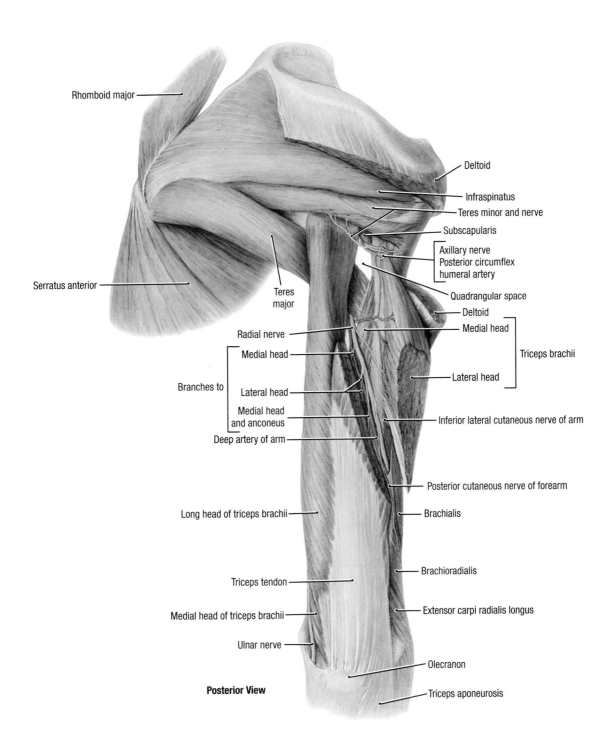

Rhomboid major

Deltoid

Infraspinatus

Teres minor and nerve

Subscapularis

Axillary nerve
Posterior circumflex
humeral artery

Serratus anterior

Teres
major

Quadrangular space

Deltoid

Radial nerve

Medial head

Triceps brachii

Medial head

Branches to

Lateral head

Lateral head

Medial head
and anconeus

Inferior lateral cutaneous nerve of arm

Deep artery of arm

Posterior cutaneous nerve of forearm

Long head of triceps brachii

Brachialis

Triceps tendon

Brachioradialis

Medial head of triceps brachii

Extensor carpi radialis longus

Ulnar nerve

Olecranon

Posterior View

Triceps aponeurosis

6.36   Triceps brachii and related nerves

- The long head is the most medial head of the triceps and attaches to the infraglenoid tubercle of the scapula. The lateral head is divided and reflected laterally to reveal its attachment to the humerus. The medial head is attached to the deep surface of the triceps tendon, which attaches to the olecranon and antebrachial fascia.
- The radial nerve passes between the proximal attachments of the long and medial heads of the triceps brachii muscle.
- The axillary nerve, passing through the quadrangular space, supplies the deltoid and teres minor muscles.
- The ulnar nerve follows the medial border of the triceps.

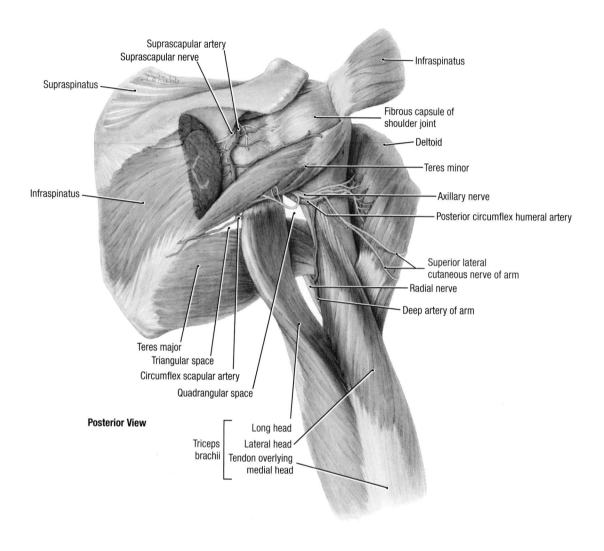

Suprascapular artery
Suprascapular nerve
Supraspinatus
Infraspinatus
Infraspinatus
Fibrous capsule of shoulder joint
Deltoid
Teres minor
Axillary nerve
Posterior circumflex humeral artery
Superior lateral cutaneous nerve of arm
Radial nerve
Deep artery of arm
Teres major
Triangular space
Circumflex scapular artery
Quadrangular space
Triceps brachii { Long head, Lateral head, Tendon overlying medial head }

Posterior View

6.37 **Dorsal scapular and subdeltoid regions**

- The infraspinatus muscle, aided by the teres minor and posterior fibers of the deltoid, rotates the humerus laterally.
- The long head of the triceps muscle passes between the teres minor (a lateral rotator) and teres major (a medial rotator) muscles.
- The long head of the triceps muscle separates the quadrangular space from the triangular space.
- Regarding the distribution of the suprascapular and axillary nerves, each comes from C5 and C6; each supplies two muscles—the suprascapular nerve innervates the supraspinatus and infraspinatus and the axillary nerve innervates the teres minor and deltoid muscles. Both nerves supply the shoulder joint, but only the axillary nerve has a cutaneous branch.

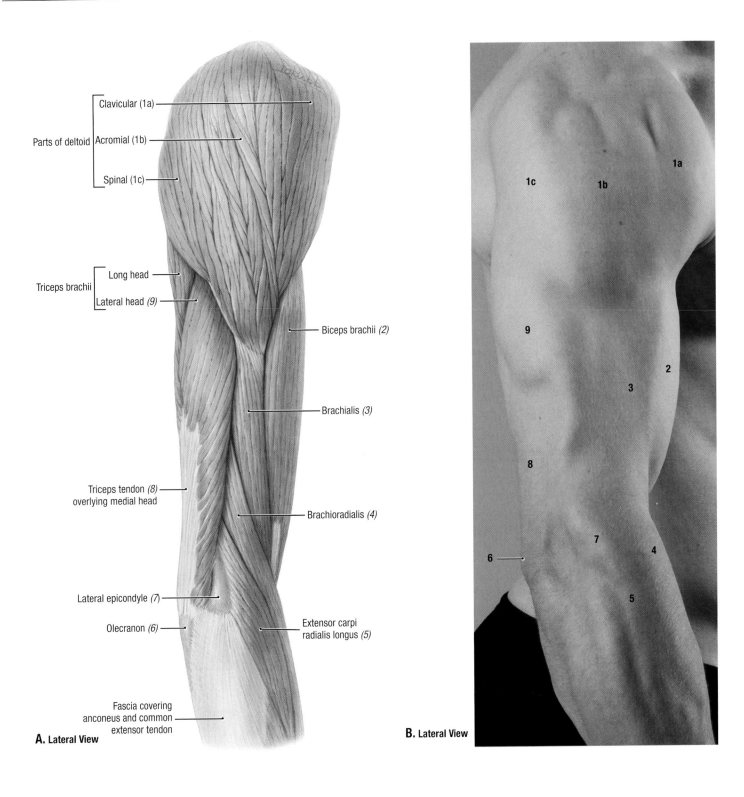

A. Lateral View

Parts of deltoid
- Clavicular (1a)
- Acromial (1b)
- Spinal (1c)

Triceps brachii
- Long head
- Lateral head (9)

Biceps brachii (2)

Brachialis (3)

Triceps tendon (8) overlying medial head

Brachioradialis (4)

Lateral epicondyle (7)

Olecranon (6)

Extensor carpi radialis longus (5)

Fascia covering anconeus and common extensor tendon

B. Lateral View

6.38 Lateral aspect of arm

A. Dissection (*numbers* in parentheses refer to structures in **B**). **B.** Surface anatomy.

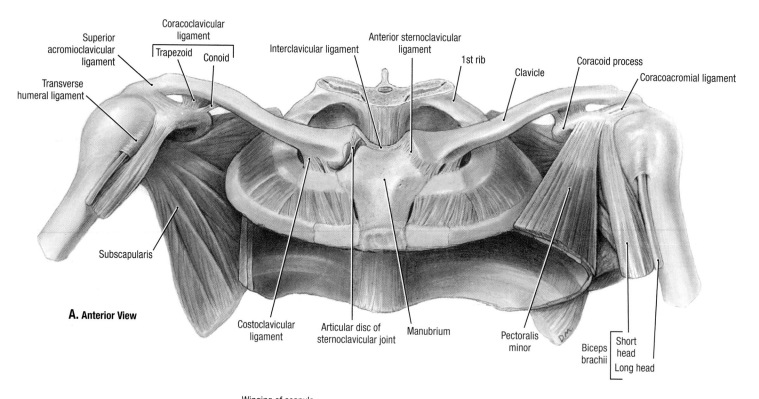

Superior acromioclavicular ligament

Coracoclavicular ligament

Trapezoid

Conoid

Transverse humeral ligament

Interclavicular ligament

Anterior sternoclavicular ligament

1st rib

Clavicle

Coracoid process

Coracoacromial ligament

Subscapularis

A. Anterior View

Costoclavicular ligament

Articular disc of sternoclavicular joint

Manubrium

Pectoralis minor

Biceps brachii

Short head

Long head

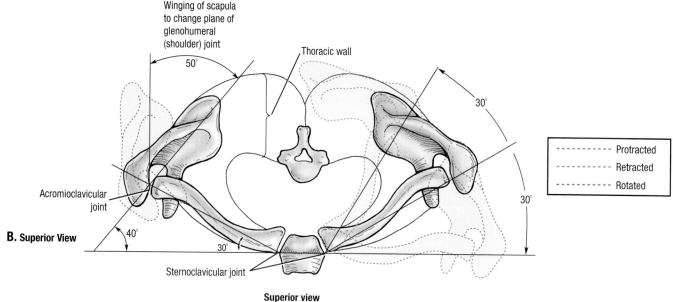

Winging of scapula to change plane of glenohumeral (shoulder) joint

Thoracic wall

50°

30°

Acromioclavicular joint

30°

30°

- - - - Protracted
- - - - Retracted
- - - - Rotated

B. Superior View

40°

30°

Sternoclavicular joint

Superior view

6.39 Pectoral girdle

A. Dissection. **B.** Clavicular movements at the sternoclavicular and acromioclavicular joints during rotation, protraction, and retraction of the scapula on the thoracic wall *(left side)* and winging of the scapula *(right side)*.

- The pectoral (shoulder) girdle consists of the sternoclavicular, acromioclavicular, and shoulder (glenohumeral) joints; the mobility of the clavicle is essential to the movement of the upper limb.
- The sternoclavicular joint is the only joint connecting the upper limb (appendicular skeleton) to the trunk (axial skeleton). The articular disc of the sternoclavicular joint divides the joint

cavity into two parts and attaches superiorly to the clavicle and inferiorly to the first costal cartilage; the disc resists superior and medial displacement of the clavicle.

- In **B,** note that when the serratus anterior is paralyzed because of injury to the long thoracic nerve, the medial border of the scapula moves laterally and posteriorly away from the thoracic wall, giving the scapula the appearance of a wing. The arm cannot be abducted beyond the horizontal position because the serratus anterior cannot rotate the glenoid cavity superiorly to allow complete abduction of the arm.

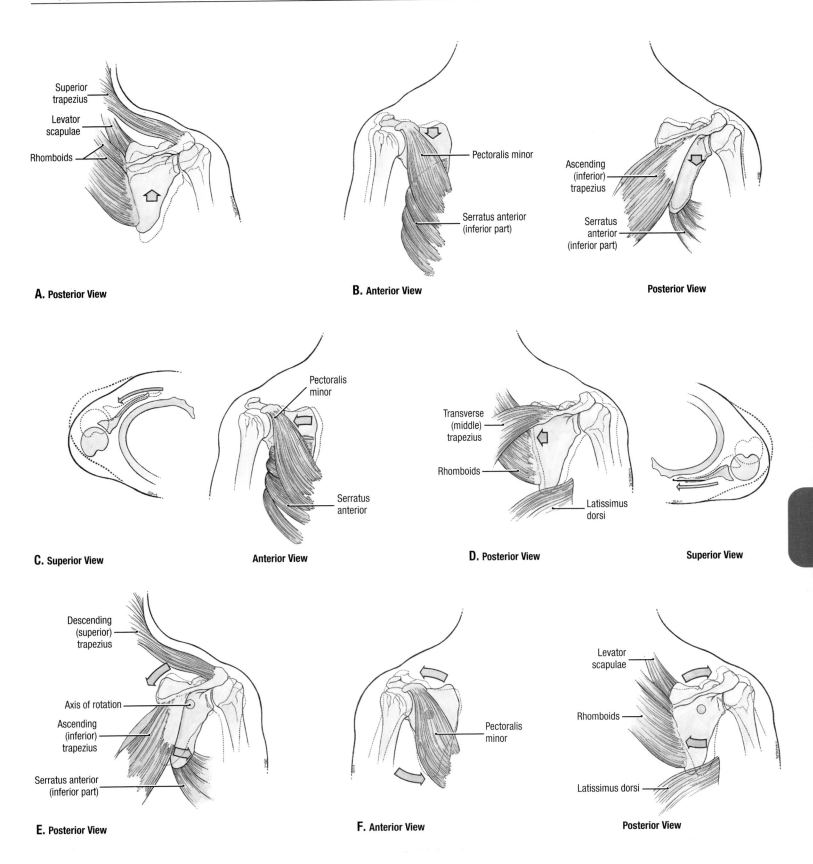

A. Posterior View

B. Anterior View   Posterior View

C. Superior View   Anterior View

D. Posterior View   Superior View

E. Posterior View

F. Anterior View   Posterior View

6.40   **Scapular movements**

A. Elevation. **B.** Depression. **C.** Protraction. **D.** Retraction. **E.** Elevation with superior (upward) rotation of glenoid fossa. **F.** Depression with inferior (downward) rotation of glenoid fossa. The *dotted outlines* represent the starting position for each movement.

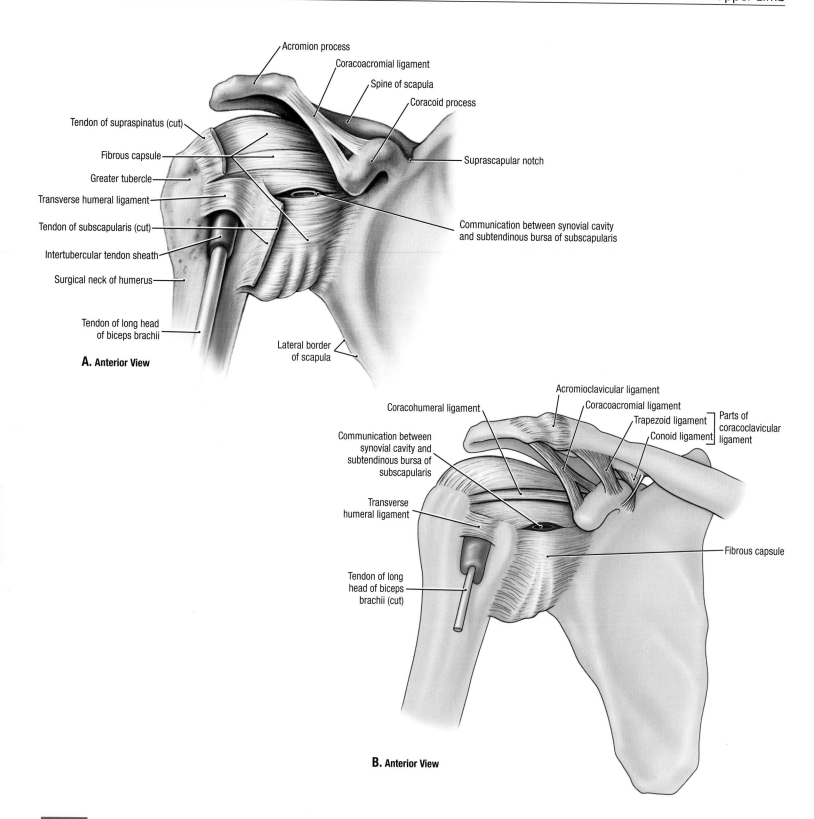

A. Anterior View

Acromion process
Coracoacromial ligament
Spine of scapula
Coracoid process
Tendon of supraspinatus (cut)
Fibrous capsule
Greater tubercle
Transverse humeral ligament
Tendon of subscapularis (cut)
Intertubercular tendon sheath
Surgical neck of humerus
Tendon of long head of biceps brachii
Suprascapular notch
Communication between synovial cavity and subtendinous bursa of subscapularis
Lateral border of scapula

B. Anterior View

Coracohumeral ligament
Acromioclavicular ligament
Coracoacromial ligament
Trapezoid ligament
Conoid ligament
Parts of coracoclavicular ligament
Communication between synovial cavity and subtendinous bursa of subscapularis
Transverse humeral ligament
Tendon of long head of biceps brachii (cut)
Fibrous capsule

6.41 Anterior aspect of shoulder joint

A. Dissection. **B.** Schematic illustration.
- The loose fibrous capsule surrounds the joint and is attached medially to the margin of the glenoid cavity and laterally to the anatomical neck of the humerus.
- The strong coracoclavicular ligament, consisting of the conoid and trapezoid ligaments, provides stability to the acromioclav-icular joint and prevents the scapula from being driven medially and the acromion from being driven inferior to the clavicle.
- The coracoacromial ligament, together with the acromion and coracoid process, forms the coracoacromial arch; the arch prevents superior displacement of the head of the humerus.

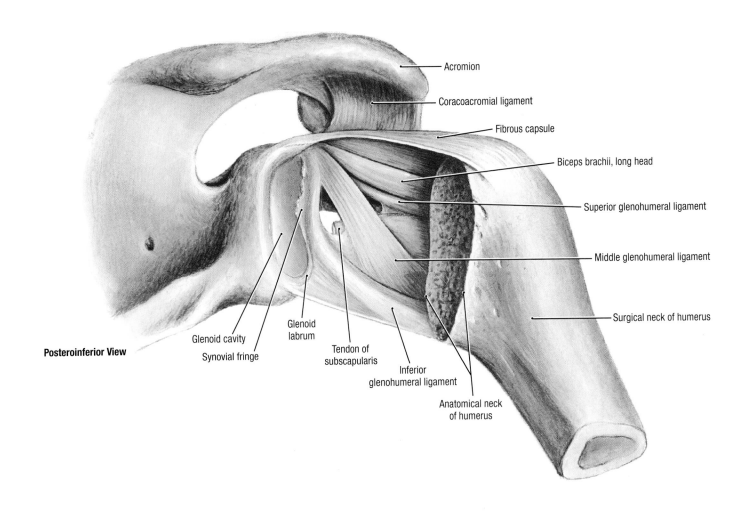

Acromion

Coracoacromial ligament

Fibrous capsule

Biceps brachii, long head

Superior glenohumeral ligament

Middle glenohumeral ligament

Surgical neck of humerus

Posteroinferior View

Glenoid cavity

Synovial fringe

Glenoid labrum

Tendon of subscapularis

Inferior glenohumeral ligament

Anatomical neck of humerus

6.42   **Posterior aspect of the interior of shoulder joint**

- The joint is exposed from the posterior aspect by cutting away the posteroinferior part of the capsule and sawing off the head of the humerus.
- The glenohumeral ligaments are visible from within the joint, but are not easily seen externally.
- The glenohumeral ligaments and tendon of the long head of biceps brachii muscle converge on the supraglenoid tubercle.
- The slender superior glenohumeral ligament lies parallel to the tendon of the long head of biceps brachii. The middle ligament is free medially because the subscapularis bursa communicates with the joint cavity superior and inferior to this ligament in this individual; often there is only a single site of communication.

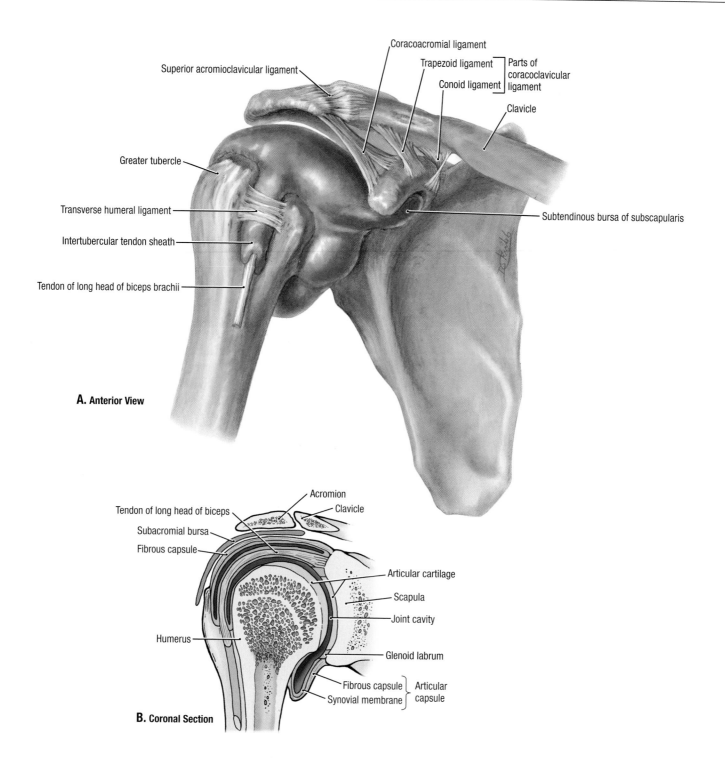

Coracoacromial ligament

Trapezoid ligament — Parts of coracoclavicular ligament

Conoid ligament

Superior acromioclavicular ligament

Clavicle

Greater tubercle

Transverse humeral ligament

Subtendinous bursa of subscapularis

Intertubercular tendon sheath

Tendon of long head of biceps brachii

A. Anterior View

Acromion

Tendon of long head of biceps

Clavicle

Subacromial bursa

Fibrous capsule

Articular cartilage

Scapula

Joint cavity

Humerus

Glenoid labrum

Fibrous capsule ⎱ Articular
Synovial membrane ⎰ capsule

B. Coronal Section

6.43 **Synovial capsule of shoulder joint and ligaments at the lateral end of the clavicle**

A. Dissection. **B.** Schematic coronal section.
The capsule has two prolongations: (1) where it forms a synovial sheath for the tendon of the long head of the biceps muscle in its osseofibrous tunnel and (2) inferior to the coracoid process, where it forms a bursa between the subscapularis tendon and margin of the glenoid cavity—the subtendinous bursa of the subscapularis.

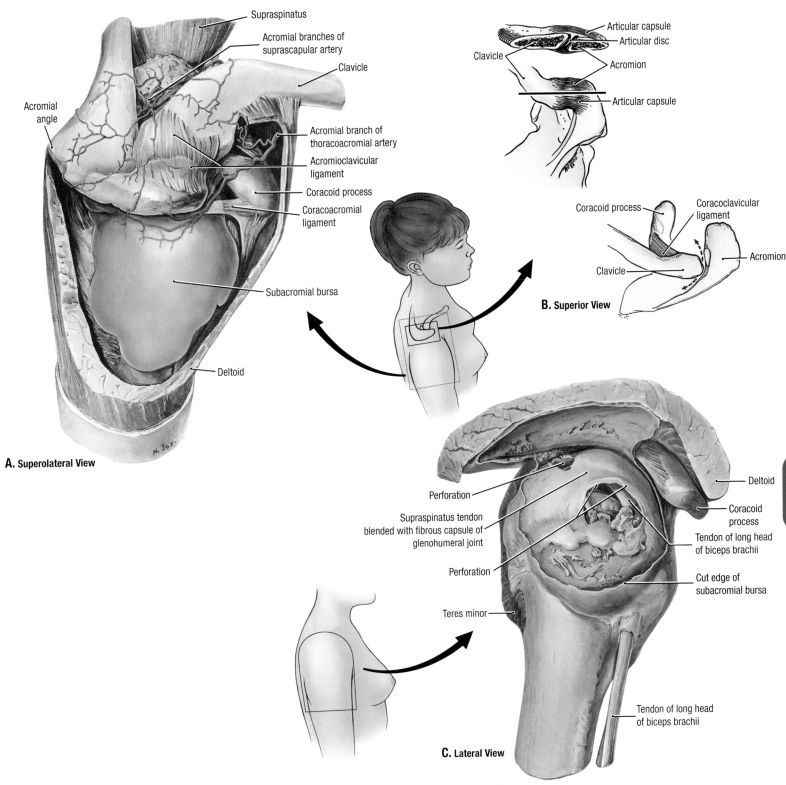

A. Superolateral View

B. Superior View

C. Lateral View

6.44   **Lateral aspect of subacromial bursa and acromioclavicular joint**

A. Subacromial bursa. The bursa has been injected with purple latex. **B.** Acromioclavicular joint. **C.** Attrition of supraspinatus tendon. As a result of wearing away of the supraspinatus tendon and underlying capsule, the subacromial bursa and shoulder joint come into communication. The intracapsular part of the tendon of the long head of biceps muscle becomes frayed, leaving it adherent to the intertubercular groove. Of 95 dissecting room subjects, none of the 18 younger than 50 years of age had a perforation, but 4 of the 19 who were 50 to 60 years and 23 of the 57 older than 60 years had perforations. The perforation was bilateral in 11 subjects and unilateral in 14.

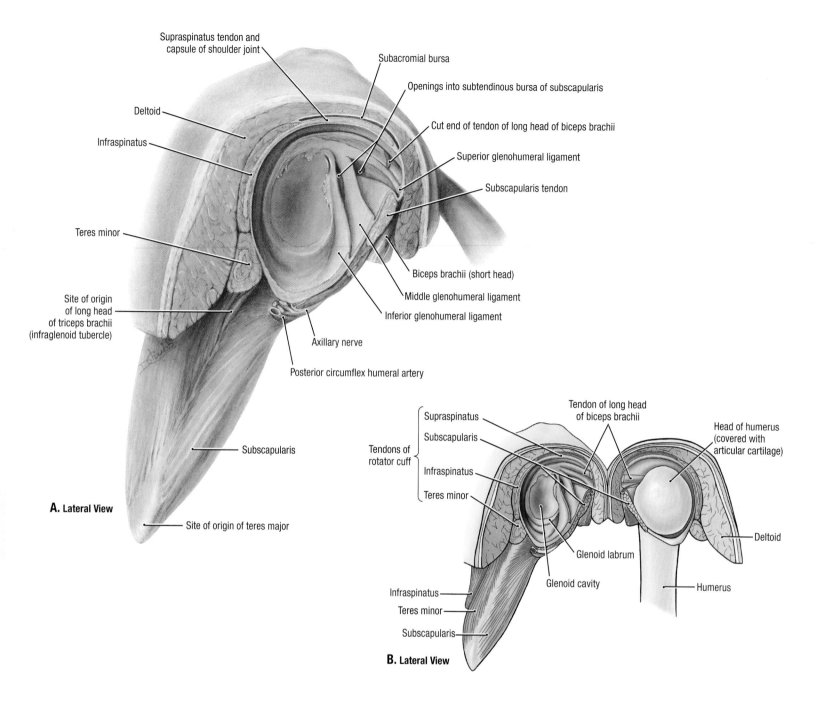

Supraspinatus tendon and capsule of shoulder joint

Subacromial bursa

Openings into subtendinous bursa of subscapularis

Deltoid

Infraspinatus

Cut end of tendon of long head of biceps brachii

Superior glenohumeral ligament

Subscapularis tendon

Teres minor

Biceps brachii (short head)

Middle glenohumeral ligament

Inferior glenohumeral ligament

Site of origin of long head of triceps brachii (infraglenoid tubercle)

Axillary nerve

Posterior circumflex humeral artery

Subscapularis

A. Lateral View

Site of origin of teres major

Supraspinatus

Subscapularis

Tendons of rotator cuff

Infraspinatus

Teres minor

Tendon of long head of biceps brachii

Head of humerus (covered with articular cartilage)

Deltoid

Glenoid labrum

Glenoid cavity

Humerus

Infraspinatus

Teres minor

Subscapularis

B. Lateral View

6.45   Lateral aspect of glenohumeral ligaments and rotator cuff

A. Dissection. **B.** Schematic illustration.

- The fibrous capsule of the joint is thickened anteriorly by the three glenohumeral ligaments, which converge from the humerus to attach with the tendon of the long head of biceps brachii muscle to the supraglenoid tubercle.
- The subacromial bursa is between the acromion and deltoid superiorly and the tendon of supraspinatus inferiorly.
- The four short rotator cuff muscles (supraspinatus, infraspinatus, teres minor, and subscapularis) cross the joint, blend with the capsule, and help hold the head of the humerus in the glenoid fossa.
- The axillary nerve and posterior circumflex humeral artery are in contact with the capsule inferiorly.
- The subscapularis bursa opens superior and often inferior to the middle glenohumeral ligament.

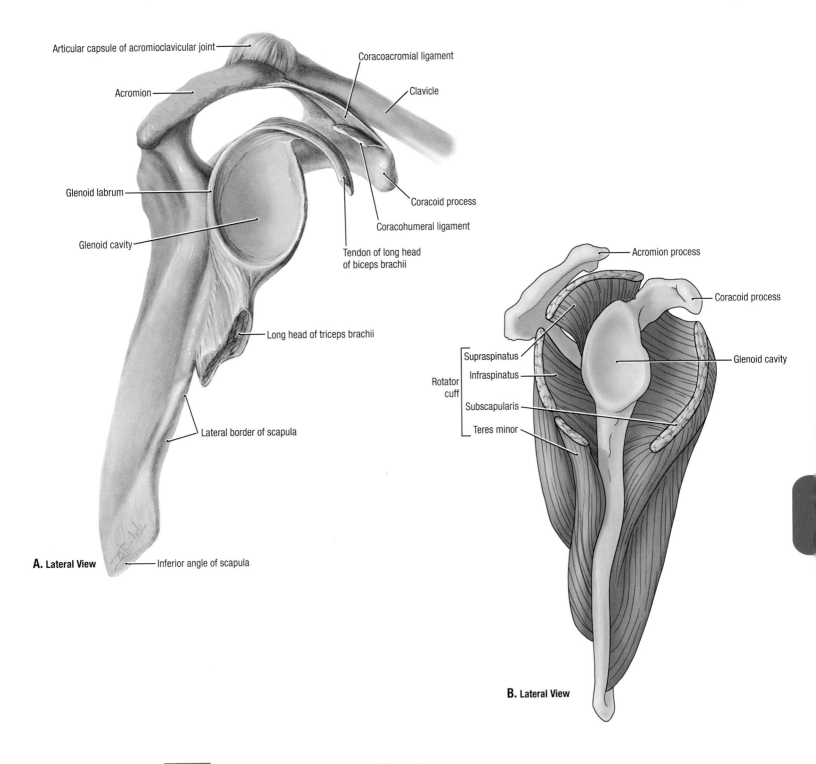

Articular capsule of acromioclavicular joint

Coracoacromial ligament

Acromion

Clavicle

Glenoid labrum

Coracoid process

Glenoid cavity

Coracohumeral ligament

Tendon of long head of biceps brachii

Long head of triceps brachii

Lateral border of scapula

A. Lateral View   Inferior angle of scapula

Acromion process

Coracoid process

Supraspinatus

Glenoid cavity

Rotator cuff

Infraspinatus

Subscapularis

Teres minor

B. Lateral View

6.46   Relationships of glenoid cavity

A. Dissection. **B.** Diagram of the muscles of the rotator cuff as they are related to the scapula and glenoid cavity.

- The coracoacromial arch (coracoid process, coracoacromial ligament, and acromion) prevents superior displacement of the head of the humerus.
- The long head of the triceps brachii muscle arises just inferior to the glenoid cavity.
- The long head of the biceps brachii muscle arises just superior to the glenoid cavity; distally it curves anterior to the front of the head of the humerus.
- The main function of the musculotendinous rotator cuff is to hold the large head of the humerus in the smaller and shallow glenoid cavity of the scapula, both during the relaxed state (by tonic contraction) and during active abduction.

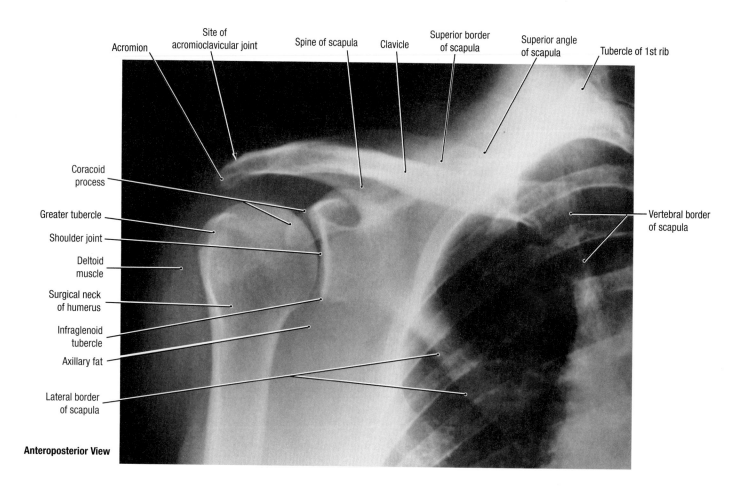

Acromion

Site of acromioclavicular joint

Spine of scapula

Clavicle

Superior border of scapula

Superior angle of scapula

Tubercle of 1st rib

Coracoid process

Greater tubercle

Shoulder joint

Deltoid muscle

Surgical neck of humerus

Infraglenoid tubercle

Axillary fat

Lateral border of scapula

Vertebral border of scapula

Anteroposterior View

6.47 **Radiograph of shoulder**

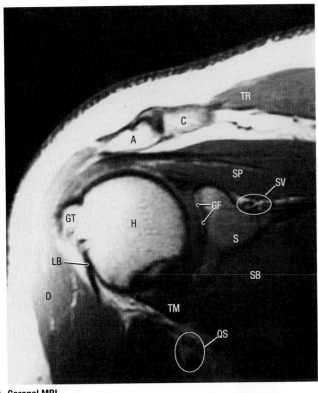

A. Coronal MRI

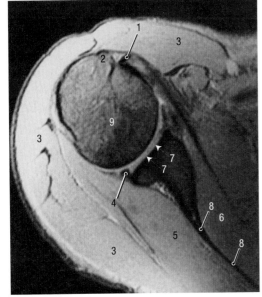

B. Transverse MRI

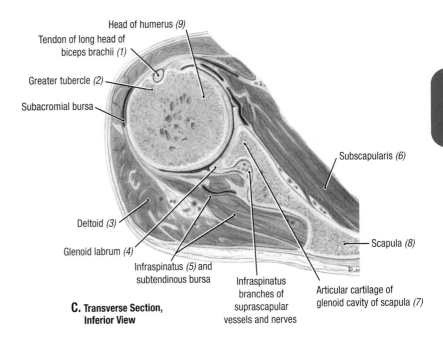

Head of humerus (9)
Tendon of long head of biceps brachii (1)
Greater tubercle (2)
Subacromial bursa
Subscapularis (6)
Deltoid (3)
Glenoid labrum (4)
Infraspinatus (5) and subtendinous bursa
Infraspinatus branches of suprascapular vessels and nerves
Articular cartilage of glenoid cavity of scapula (7)
Scapula (8)

C. Transverse Section, Inferior View

> **6.48** **Transverse section and MRI through shoulder joint and axilla**

A. Coronal MRI. *A,* acromion; *C,* clavicle; *D,* deltoid; *GF,* glenoid cavity; *GT,* crest of greater tubercle; *H,* head of humerus; *LB,* long head of the biceps brachii; *QS,* quadrangular space; *S,* scapula; *SB,* subscapularis; *SP,* supraspinatus; *SV,* suprascapular vessels and nerve; *TM,* teres minor; *TR,* trapezius. **B.** Transverse MRI. **C.** Transverse section (*numbers* in **C** refer to structures labeled in **B**).

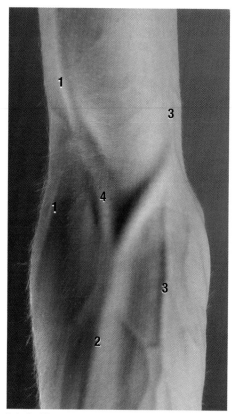

A. Anterior View

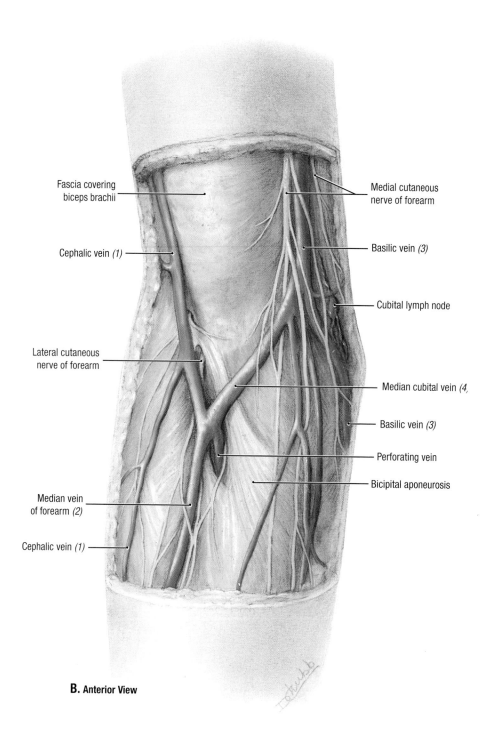

Fascia covering biceps brachii

Medial cutaneous nerve of forearm

Cephalic vein *(1)*

Basilic vein *(3)*

Cubital lymph node

Lateral cutaneous nerve of forearm

Median cubital vein *(4,*

Basilic vein *(3)*

Perforating vein

Bicipital aponeurosis

Median vein of forearm *(2)*

Cephalic vein *(1)*

B. Anterior View

6.49   Cubital fossa

A. Surface anatomy. **B.** Superficial dissection (*numbers* in parentheses refer to structures in **A**).

- The cubital fossa is the triangular space inferior to the elbow crease.
- In the forearm, the superficial veins (cephalic, median, basilic, and their connecting veins) make a variable, M-shaped pattern.
- The cephalic and basilic veins occupy the bicipital grooves, one on each side of the biceps brachii. In the lateral bicipital groove, the lateral cutaneous nerve of the forearm appears just superior to the elbow crease; in the medial bicipital groove, the medial cutaneous nerve of the forearm becomes cutaneous at approximately the midpoint of the arm.

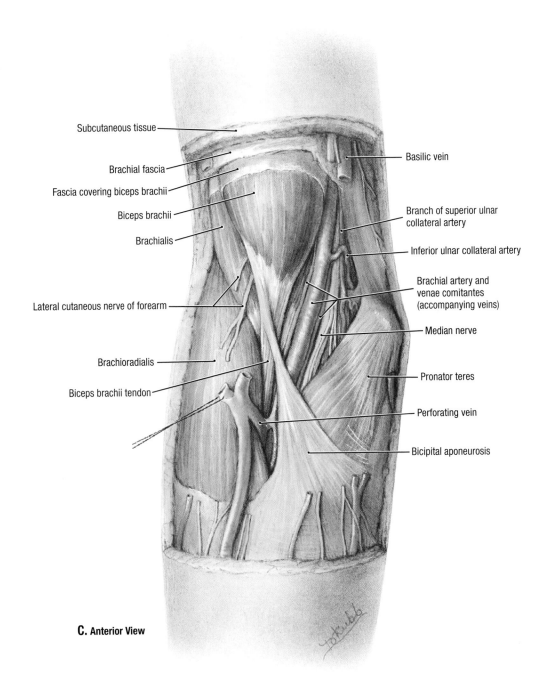

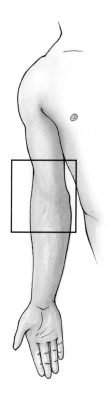

Subcutaneous tissue

Brachial fascia

Fascia covering biceps brachii

Biceps brachii

Brachialis

Lateral cutaneous nerve of forearm

Brachioradialis

Biceps brachii tendon

Basilic vein

Branch of superior ulnar collateral artery

Inferior ulnar collateral artery

Brachial artery and venae comitantes (accompanying veins)

Median nerve

Pronator teres

Perforating vein

Bicipital aponeurosis

C. Anterior View

6.49 **Cubital fossa (continued)**

C. Boundaries and contents of the cubital fossa.

- The cubital fossa is bound laterally by the brachioradialis and medially by the pronator teres. The apex is where these two muscles meet distally.
- The three chief contents of the cubital fossa are the biceps brachii tendon, brachial artery, and median nerve.
- The biceps brachii tendon, on approaching its insertion, rotates through 90°, and the bicipital aponeurosis extends medially from the tendon proximally.

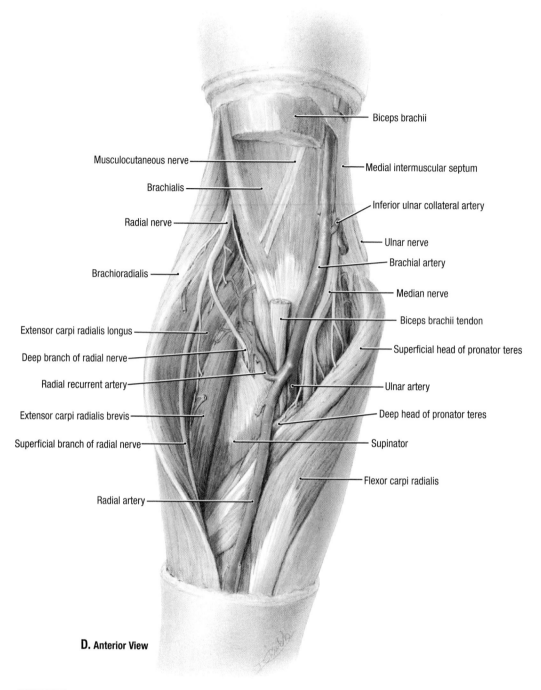

Biceps brachii

Musculocutaneous nerve

Medial intermuscular septum

Brachialis

Inferior ulnar collateral artery

Radial nerve

Ulnar nerve

Brachial artery

Brachioradialis

Median nerve

Extensor carpi radialis longus

Biceps brachii tendon

Deep branch of radial nerve

Superficial head of pronator teres

Radial recurrent artery

Ulnar artery

Extensor carpi radialis brevis

Deep head of pronator teres

Superficial branch of radial nerve

Supinator

Flexor carpi radialis

Radial artery

D. Anterior View

6.49　**Cubital fossa (continued)**

D. Floor of the cubital fossa.
- Part of the biceps brachii muscle is excised, and the cubital fossa is opened widely, exposing the alis and supinator muscles in the floor of the fossa.
- The deep branch of the radial nerve pierces the supinator.
- The brachial artery lies between the biceps tendon and median nerve and divides into two branches, the ulnar and radial arteries.
- The median nerve supplies the flexor muscles. With the exception of the twig to the deep head of pronator teres, its motor branches arise from its medial side.
- The radial nerve supplies the extensor muscles. With the exception of the twig to brachioradialis, its motor branches arise from its lateral side. In this specimen, the radial nerve has been displaced laterally, so even its lateral branches appear to run medially.

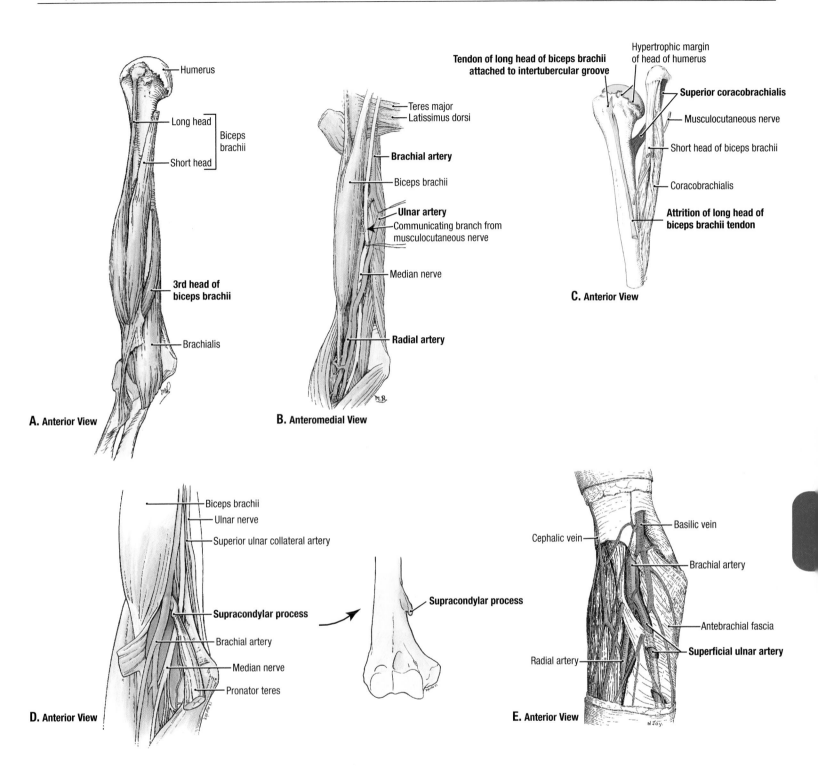

A. Anterior View

Humerus
Long head
Short head
} Biceps brachii
3rd head of biceps brachii
Brachialis

B. Anteromedial View

Tendon of long head of biceps brachii attached to intertubercular groove
Teres major
Latissimus dorsi
Brachial artery
Biceps brachii
Ulnar artery
Communicating branch from musculocutaneous nerve
Median nerve
Radial artery

C. Anterior View

Hypertrophic margin of head of humerus
Superior coracobrachialis
Musculocutaneous nerve
Short head of biceps brachii
Coracobrachialis
Attrition of long head of biceps brachii tendon

D. Anterior View

Biceps brachii
Ulnar nerve
Superior ulnar collateral artery
Supracondylar process
Brachial artery
Median nerve
Pronator teres

Supracondylar process

E. Anterior View

Cephalic vein
Basilic vein
Brachial artery
Antebrachial fascia
Superficial ulnar artery
Radial artery

6.50   Anomalies

A. Third head of biceps brachii. In this case, there is also attrition of the biceps tendon. **B.** Anomalous division of brachial artery. In this case, the median nerve passes between the radial and ulnar arteries, which arise high in the arm. The musculocutaneous and median nerves commonly communicate, as shown here. **C.** Attrition of the tendon of the long head of biceps brachii and superior coracobrachialis. **D.** Supracondylar process of humerus. A fibrous band joins this supracondylar process to the medial epicondyle; the median nerve passes through the foramen, and the brachial artery may go with it. This may be a cause of entrapment. **E.** Superficial ulnar artery.

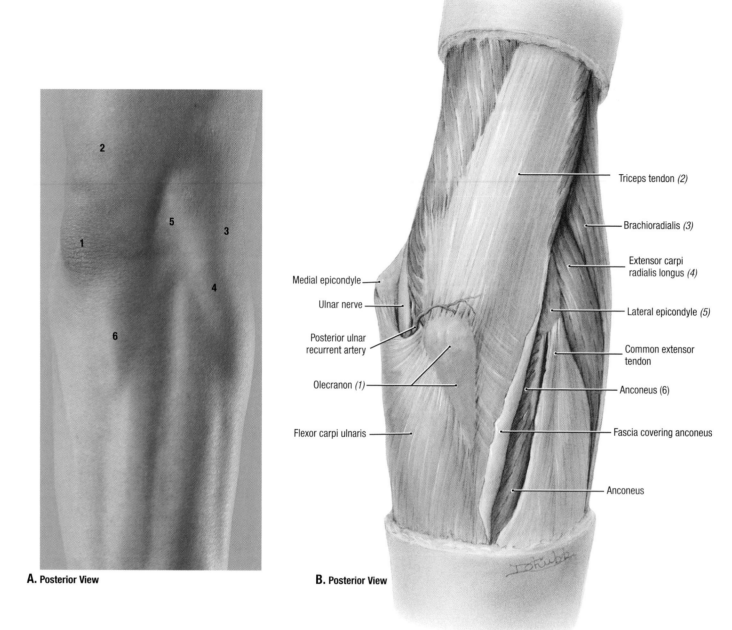

A. Posterior View

B. Posterior View

Triceps tendon (2)

Brachioradialis (3)

Extensor carpi
radialis longus (4)

Medial epicondyle

Ulnar nerve

Lateral epicondyle (5)

Common extensor
tendon

Posterior ulnar
recurrent artery

Olecranon (1)

Anconeus (6)

Fascia covering anconeus

Flexor carpi ulnaris

Anconeus

6.51 **Posterior aspect of elbow**

A. *Surface anatomy.* **B.** *Superficial dissection (numbers in parentheses refer to structures in* **A***).*

- The triceps brachii is inserted into the superior surface of the olecranon and, through the deep fascia covering the anconeus, into the lateral border of olecranon;
- The posterior surfaces of the medial epicondyle, lateral epicondyle, and olecranon are subcutaneous and palpable.
- The ulnar nerve, also palpable, runs subfascially posterior to the medial epicondyle; distal to this point, it disappears deep to the two heads of the flexor carpi ulnaris.
- The flexor carpi ulnaris has two heads; one arises from the common flexor tendon, and the other from the medial border of the olecranon and posterior border of the shaft of the ulna.

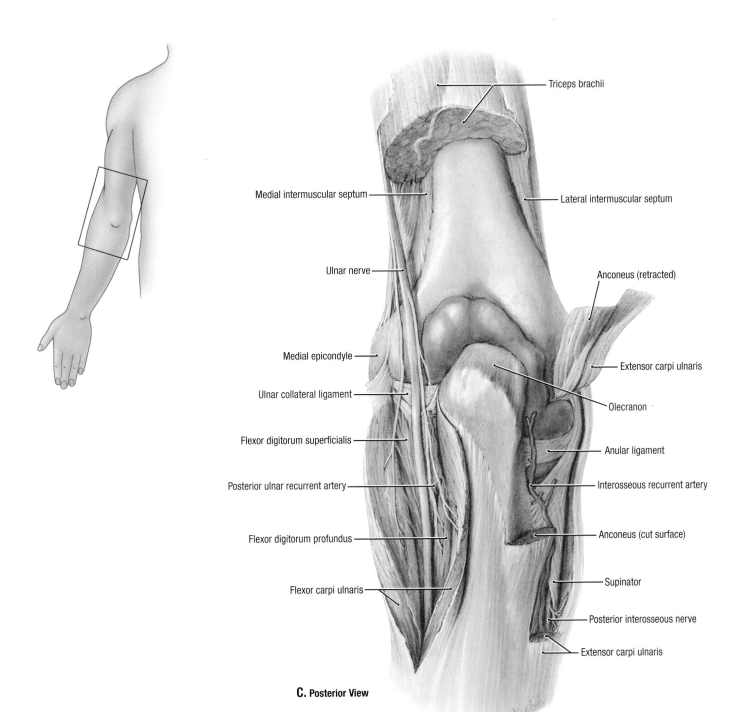

C. Posterior View

6.51   **Posterior aspect of elbow (continued)**

C. Deep dissection. The distal portion of the triceps brachii muscle was removed.

- The ulnar nerve descends subfascially within the posterior compartment of the arm passing posterior to the medial epicondyle. Next it passes posterior to the ulnar collateral ligament of the elbow joint and then between the flexor carpi ulnaris and flexor digitorum profundus muscles.
- The ulnar nerve supplies the flexor carpi ulnaris muscle, half of the flexor digitorum profundus muscle, and the elbow joint.
- The posterior interosseous nerve (continuation of the deep branch of the radial nerve) pierces the supinator inferior to the head of the radius.

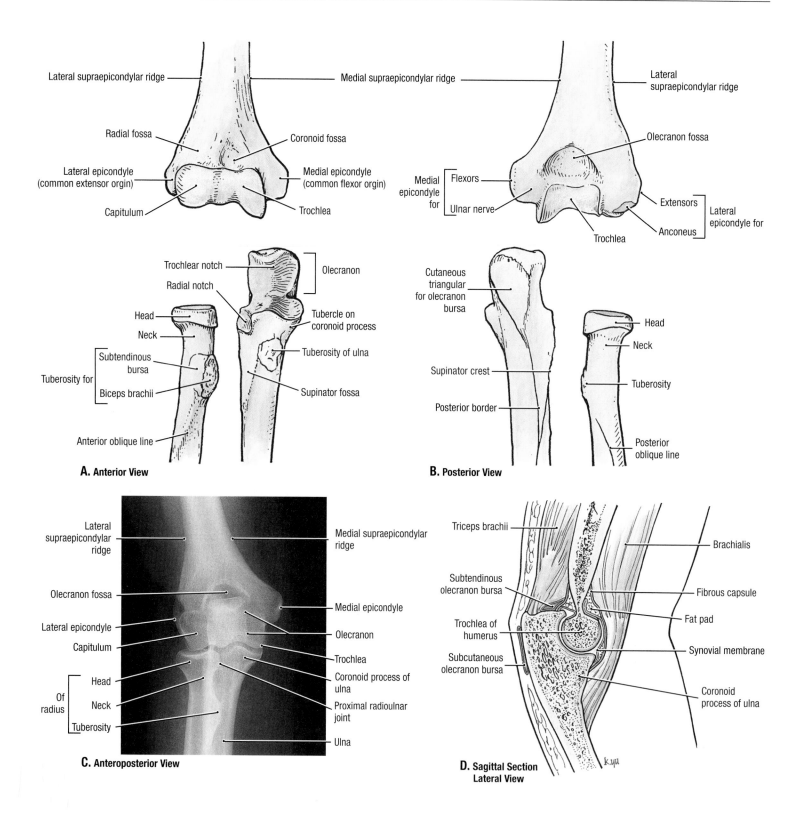

A. Anterior View

Lateral supraepicondylar ridge
Medial supraepicondylar ridge
Radial fossa
Coronoid fossa
Lateral epicondyle (common extensor orgin)
Medial epicondyle (common flexor orgin)
Capitulum
Trochlea
Trochlear notch
Olecranon
Radial notch
Head
Tubercle on coronoid process
Neck
Tuberosity of ulna
Subtendinous bursa
Tuberosity for
Biceps brachii
Supinator fossa
Anterior oblique line

B. Posterior View

Lateral supraepicondylar ridge
Olecranon fossa
Medial epicondyle for
Flexors
Ulnar nerve
Extensors
Anconeus
Lateral epicondyle for
Trochlea
Cutaneous triangular for olecranon bursa
Head
Neck
Supinator crest
Tuberosity
Posterior border
Posterior oblique line

C. Anteroposterior View

Lateral supraepicondylar ridge
Medial supraepicondylar ridge
Olecranon fossa
Medial epicondyle
Lateral epicondyle
Olecranon
Capitulum
Trochlea
Head
Coronoid process of ulna
Of radius
Neck
Proximal radioulnar joint
Tuberosity
Ulna

D. Sagittal Section Lateral View

Triceps brachii
Brachialis
Subtendinous olecranon bursa
Fibrous capsule
Fat pad
Trochlea of humerus
Subcutaneous olecranon bursa
Synovial membrane
Coronoid process of ulna

6.52   Bones and imaging of elbow region

A. Anterior aspect of bones and bony features. **B.** Posterior aspect of bones and bony features. **C.** Radiograph of elbow joint. **D.** Section of humeroulnar joint.

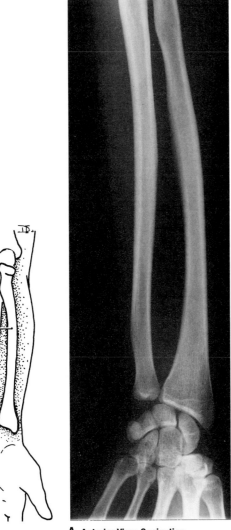

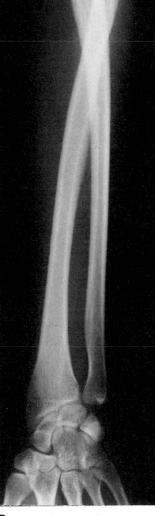
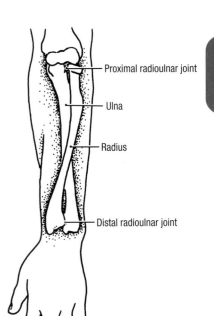

A. Anterior View, Supination   **B. Anterior View, Pronation**

6.53   **Superior and inferior radioulnar joints**

A. Radiograph of forearm in supination. **B.** Anterior radiograph of forearm in pronation.
Note that the radius crosses the ulna when the forearm is pronated.

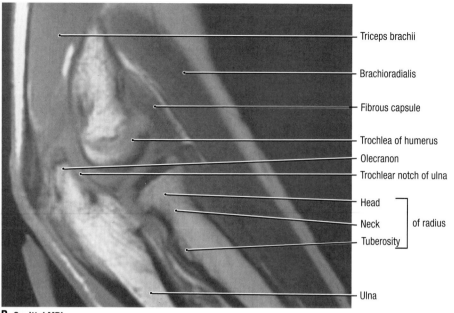

A. Medial View

B. Sagittal MRI

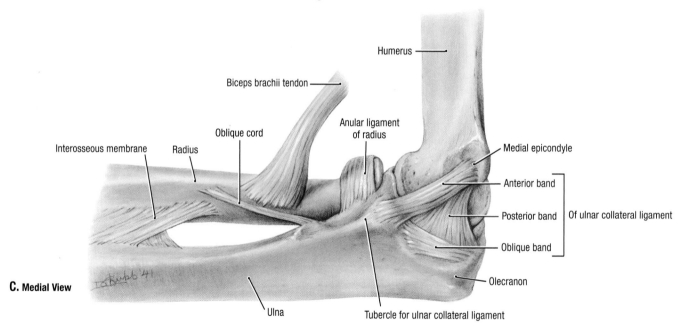

C. Medial View

6.54 **Medial aspect of elbow**

A. Bony features. **B.** Sagittal MRI through elbow joint. **C.** Ulnar (medial) collateral ligament. The anterior band is a strong, round cord that is taut when the elbow joint is extended. The posterior band is a weak fan that is taut in flexion of the joint. The oblique fibers deepen the socket for the trochlea of the humerus.

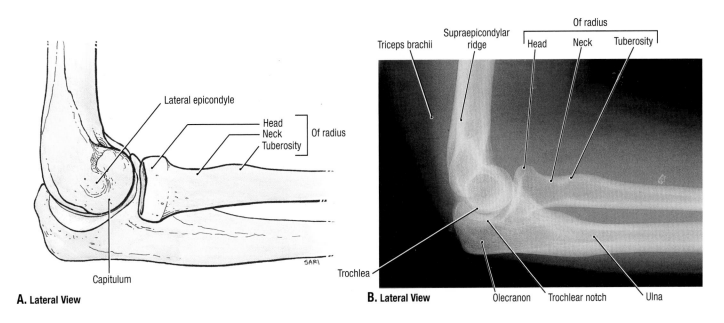

A. Lateral View

- Lateral epicondyle
- Head
- Neck — Of radius
- Tuberosity
- Capitulum

B. Lateral View

- Triceps brachii
- Supraepicondylar ridge
- Of radius
  - Head
  - Neck
  - Tuberosity
- Trochlea
- Olecranon
- Trochlear notch
- Ulna

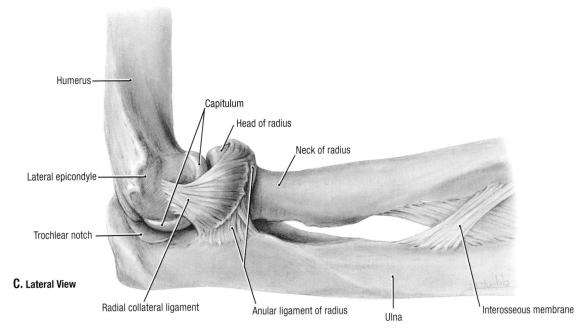

C. Lateral View

- Humerus
- Capitulum
- Head of radius
- Neck of radius
- Lateral epicondyle
- Trochlear notch
- Radial collateral ligament
- Anular ligament of radius
- Ulna
- Interosseous membrane

6.55   **Lateral aspect of elbow**

A. Bony features. **B.** Lateral radiograph. **C.** Radial (lateral) collateral ligament. The fan-shaped lateral ligament is attached to the anular ligament of the radius, but the superficial fibers of the lateral ligament continue onto the radius.

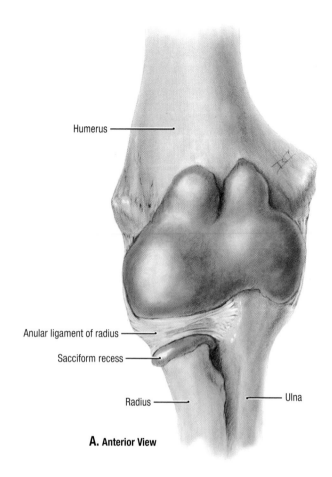

Humerus

Anular ligament of radius

Sacciform recess

Radius ——————— Ulna

A. Anterior View

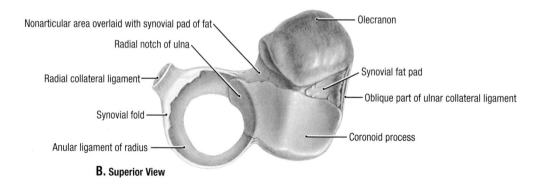

Nonarticular area overlaid with synovial pad of fat

Radial notch of ulna

Radial collateral ligament

Synovial fold

Anular ligament of radius

Olecranon

Synovial fat pad

Oblique part of ulnar collateral ligament

Coronoid process

B. Superior View

6.56 Articular cavity of elbow joint

A. Synovial membrane of elbow and proximal radioulnar joints. The cavity of the elbow was injected with wax. The fibrous capsule was removed, and the synovial membrane remains. **B.** Socket for head of radius and trochlea of humerus. The anular ligament secures the head of the radius to the radial notch of the ulna and with it forms a cup-shaped socket (i.e., wide superiorly, narrow inferiorly). The anular ligament is bound to the humerus by the radial collateral ligament of the elbow. A common childhood injury is displacement of the head of the radius after traction on a pronated forearm. Part of the anular ligament becomes trapped between the radial head and the capitulum.

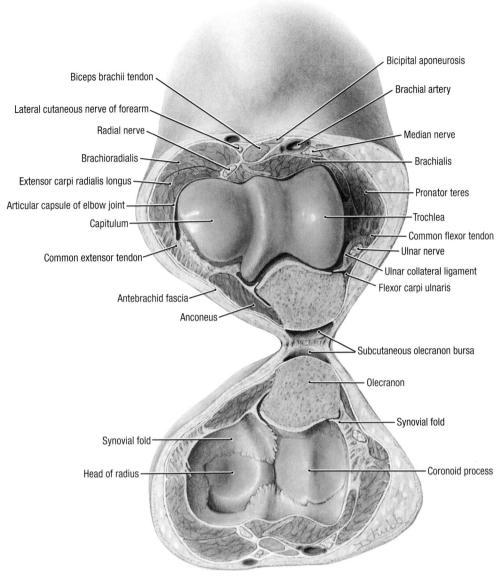

Transverse Section

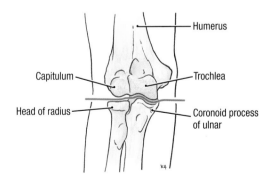

6.57   Transverse section through elbow joint

- The radial nerve is in contact with the joint capsule, the ulnar nerve is in contact with the ulnar collateral ligament, and the median nerve is separated from the joint capsule by the brachialis muscle.
- Synovial folds containing fat overlie the periphery of the head of the radius and the nonarticular indentations on the trochlear notch of the ulna.
- The subcutaneous olecranon bursa lies between the olecranon process and the skin.

TABLE 6.7   ARTERIES OF FOREARM

ARTERY

Radial
Origin:
In cubital fossa, as smaller terminal division of brachial artery
Course/Distribution:
Runs distally under cover of brachioradialis, lateral to flexor carpi radialis, defining boundary between flexor and extensor compartments and supplying the radial aspect of both; after giving rise to a superficial branch near the radiocarpal joint, it transverses anatomical snuff box to pass between heads of 1st dorsal interosseous muscle, forming the deep palmar arch (with deep branch of ulnar artery)

Ulnar
Origin:
In cubital fossa, as larger terminal division of brachial artery
Course/Distribution:
Passes distally between 2nd and 3rd layers of flexor muscles, supplying ulnar aspect of flexor compartment; passes superficial to flexor retinaculum at wrist, continuing as the superficial palmar arch (with superficial branch of radial) after its deep palmar branch joins the deep palmar arch

Radial recurrent
Origin:
In cubital fossa, as 1st (lateral) branch of radial artery
Course/Distribution:
Courses proximally, superficial to supinator, passing between brachioradialis and brachialis to anastomose with radial collateral artery

Anterior and posterior ulnar recurrent
Origin:
In and immediately distal to cubital fossa, as 1st and 2nd medial branches of ulnar artery
Course/Distribution:
Course proximally to anastomose with the inferior and superior ulnar collateral arteries, respectively, forming collateral pathways anterior and posterior to the medial epicondyle of the humerus

Common interosseous
Origin:
Immediately distal to the cubital fossa, as 1st lateral branch of ulnar artery
Course/Distribution:
Terminates almost immediately, dividing into anterior and posterior interosseous arteries

Anterior and posterior interosseous
Origin:
Distal to radial tubercle, as terminal branches of common interosseous
Course/Distribution:
Pass to opposite sides of interosseous membrane; anterior artery runs on interosseous membrane; posterior artery runs between superficial and deep layers of extensor muscles as primary artery of compartment

Interosseous recurrent
Origin:
Initial part of posterior interosseous artery
Course/Distribution:
Courses proximally between lateral epicondyle and olecranon, deep to anconeus, to anastomose with middle collateral artery

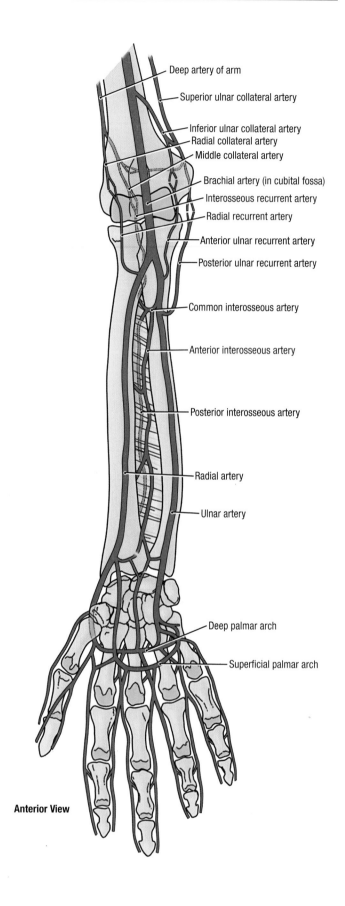

Deep artery of arm
Superior ulnar collateral artery
Inferior ulnar collateral artery
Radial collateral artery
Middle collateral artery
Brachial artery (in cubital fossa)
Interosseous recurrent artery
Radial recurrent artery
Anterior ulnar recurrent artery
Posterior ulnar recurrent artery
Common interosseous artery
Anterior interosseous artery
Posterior interosseous artery
Radial artery
Ulnar artery
Deep palmar arch
Superficial palmar arch

Anterior View

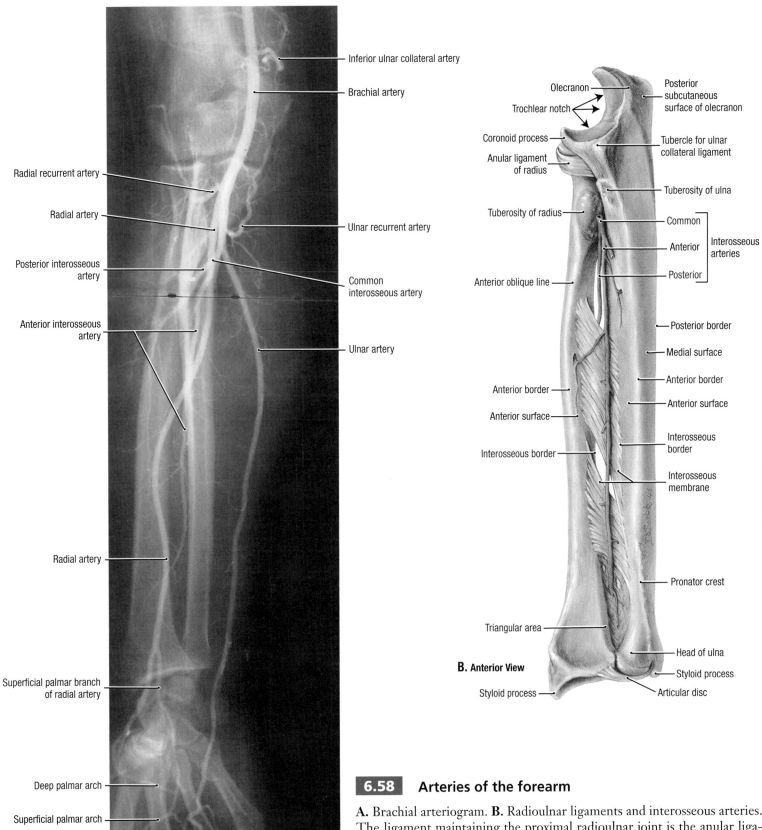

A. Anteroposterior View

Inferior ulnar collateral artery

Brachial artery

Radial recurrent artery

Radial artery

Ulnar recurrent artery

Posterior interosseous artery

Common interosseous artery

Anterior interosseous artery

Ulnar artery

Radial artery

Superficial palmar branch of radial artery

Deep palmar arch

Superficial palmar arch

Olecranon

Trochlear notch

Posterior subcutaneous surface of olecranon

Coronoid process

Anular ligament of radius

Tubercle for ulnar collateral ligament

Tuberosity of ulna

Tuberosity of radius

Common

Anterior — Interosseous arteries

Posterior

Anterior oblique line

Posterior border

Medial surface

Anterior border

Anterior surface

Anterior border

Anterior surface

Interosseous border

Interosseous border

Interosseous membrane

Pronator crest

Triangular area

Head of ulna

Styloid process

Styloid process

Articular disc

B. Anterior View

6.58   Arteries of the forearm

A. Brachial arteriogram. **B.** Radioulnar ligaments and interosseous arteries. The ligament maintaining the proximal radioulnar joint is the anular ligament, that for the distal joint is the articular disc, and that for the middle joint is the interosseous membrane. The interosseous membrane is attached to the interosseous borders of the radius and ulna, but it also spreads onto their surfaces.

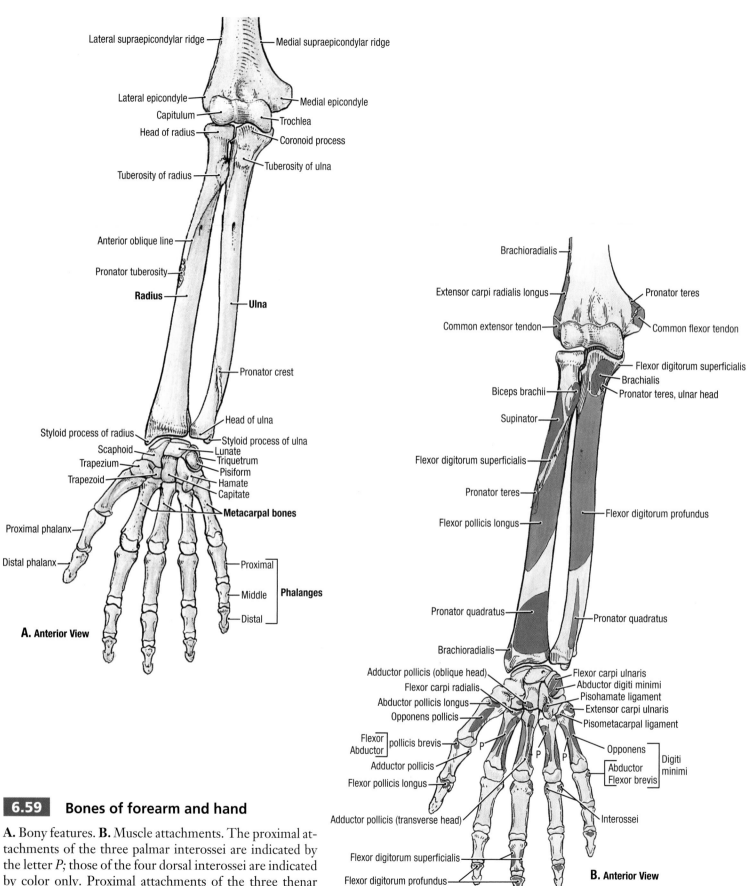

A. Anterior View

B. Anterior View

6.59 Bones of forearm and hand

A. Bony features. **B.** Muscle attachments. The proximal attachments of the three palmar interossei are indicated by the letter *P;* those of the four dorsal interossei are indicated by color only. Proximal attachments of the three thenar and two of the hypothenar muscles are not shown.

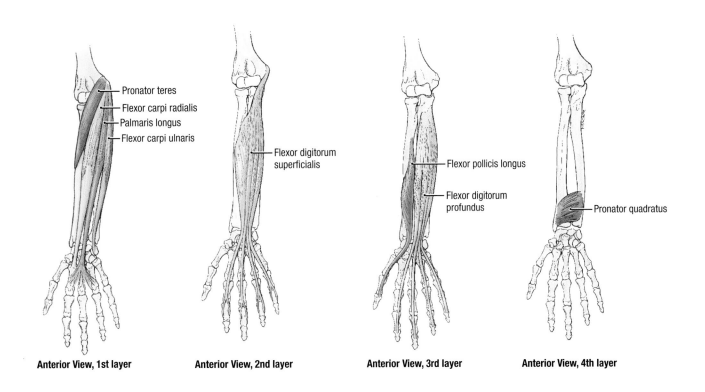

Anterior View, 1st layer   **Anterior View, 2nd layer**   **Anterior View, 3rd layer**   **Anterior View, 4th layer**

TABLE 6.8   MUSCLES ON THE ANTERIOR SURFACE OF THE FOREARM

MUSCLE	PROXIMAL ATTACHMENT	DISTAL ATTACHMENT	INNERVATION	MAIN ACTIONS
Pronator teres	Medial epicondyle of humerus and coronoid process of ulna	Middle of lateral surface of radius (pronator tuberosity)	Median nerve (C6 and **C7**)	Pronates forearm and flexes elbow
Flexor carpi radialis	Medial epicondyle of humerus	Base of 2nd metacarpal		Flexes wrist and abducts hand
Palmaris longus	Medial epicondyle of humerus	Distal half of flexor retinaculum and palmar aponeurosis	Median nerve (C7 and **C8**)	Flexes wrist and tightens palmar aponeurosis
Flexor carpi ulnaris	*Humeral head:* medial epicondyle of humerus; *Ulnar head:* olecranon and posterior border of ulna	Pisiform, hook of hamate, and 5th metacarpal	Ulnar nerve (C7 and **C8**)	Flexes wrist and adducts hand
Flexor digitorum superficialis	*Humeroulnar head:* medial epicondyle of humerus, ulnar collateral ligament, and coronoid process of ulna *Radial head:* superior half of anterior border of radius	Bodies of middle phalanges of medial four digits	Median nerve (C7, **C8**, and T1)	Flexes PIPs of medial four digits; acting more strongly, it flexes MCPs and hand
Flexor digitorum profundus	Proximal three quarters of medial and anterior surfaces of ulna and interosseous membrane	Bases of distal phalanges of medial four digits	*Medial part:* ulnar nerve (**C8** and T1) *Lateral part:* median nerve (**C8** and T1)	Flexes DIPs of medial four digits; assists with flexion of wrist
Flexor pollicis longus	Anterior surface of radius and adjacent interosseous membrane	Base of distal phalanx of thumb	Anterior interosseous nerve from median (**C8** and T1)	Flexes phalanges of 1st digit (thumb)
Pronator quadratus	Distal fourth of anterior surface of ulna	Distal fourth of anterior surface of radius		Pronates forearm; deep fibers bind radius and ulna together

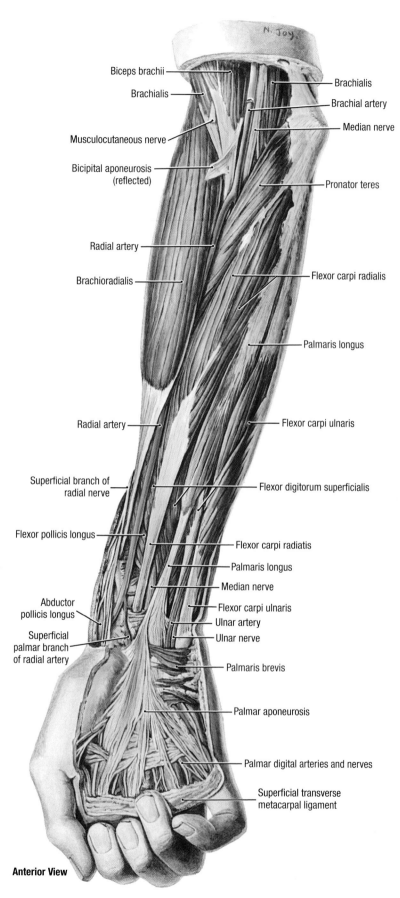

Anterior View

6.60   Superficial muscles of the forearm and palmar aponeurosis

- At the elbow, the brachial artery lies between the biceps tendon and median nerve. It then bifurcates into the radial and ulnar arteries.
- At the wrist, the radial artery is lateral to the flexor carpi radialis tendon, and the ulnar artery is lateral to flexor carpi ulnaris tendon.
- In the forearm, the radial artery lies between the flexor and extensor compartments. The muscles lateral to the artery are supplied by the radial nerve, and those medial to it by the median and ulnar nerves; thus, no motor nerve crosses the radial artery.
- The lateral group of muscles, represented by the brachioradialis muscle, slightly overlaps the radial artery, which is otherwise superficial.
- The four superficial muscles (pronator teres, flexor carpi radialis, palmaris longus, and flexor carpi ulnaris) all attach proximally to the medial epicondyle.
- The palmaris longus muscle, in this specimen, has an anomalous distal belly; this muscle usually has a small belly at the common flexor origin and a long tendon that is continued into the palm as the palmar aponeurosis. The palmaris longus is absent in approximately 14% of limbs.

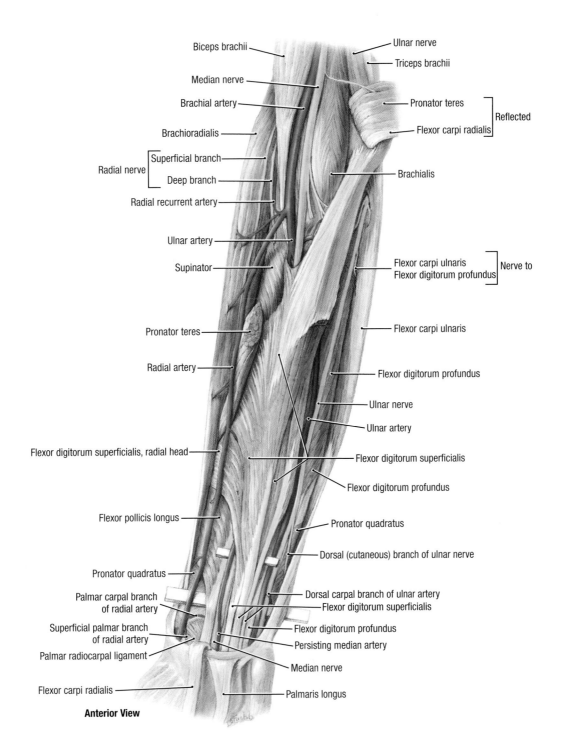

Biceps brachii — Ulnar nerve — Triceps brachii — Median nerve — Pronator teres — Brachial artery — Flexor carpi radialis — Reflected — Brachioradialis — Brachialis — Radial nerve — Superficial branch — Deep branch — Radial recurrent artery — Ulnar artery — Flexor carpi ulnaris — Flexor digitorum profundus — Nerve to — Supinator — Pronator teres — Flexor carpi ulnaris — Radial artery — Flexor digitorum profundus — Ulnar nerve — Ulnar artery — Flexor digitorum superficialis, radial head — Flexor digitorum superficialis — Flexor digitorum profundus — Flexor pollicis longus — Pronator quadratus — Dorsal (cutaneous) branch of ulnar nerve — Pronator quadratus — Dorsal carpal branch of ulnar artery — Flexor digitorum superficialis — Palmar carpal branch of radial artery — Superficial palmar branch of radial artery — Flexor digitorum profundus — Persisting median artery — Palmar radiocarpal ligament — Median nerve — Flexor carpi radialis — Palmaris longus

Anterior View

6.61   Flexor digitorum superficialis and related structures

- Note the attachment of the flexor digitorum superficialis muscle to the humerus, ulna, and radius.
- The ulnar artery passes obliquely posterior to the flexor digitorum superficialis; at the medial border of the muscle, the ulnar artery joins the ulnar nerve.
- The ulnar nerve descends vertically near the medial border of the superficialis; it is exposed by splitting the septum between the superficialis and flexor carpi ulnaris muscles.
- The median nerve descends vertically posterior to the superficialis and appears distally at its lateral border.
- The median artery in this specimen persists as an embryologic remnant.

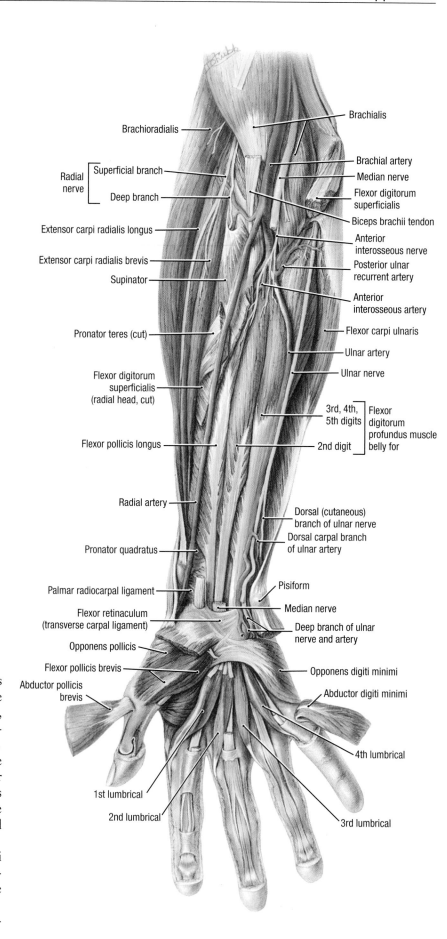

Brachioradialis

Brachialis

Radial nerve
- Superficial branch
- Deep branch

Brachial artery

Median nerve

Flexor digitorum superficialis

Biceps brachii tendon

Extensor carpi radialis longus

Anterior interosseous nerve

Extensor carpi radialis brevis

Posterior ulnar recurrent artery

Supinator

Anterior interosseous artery

Flexor carpi ulnaris

Pronator teres (cut)

Ulnar artery

Ulnar nerve

Flexor digitorum superficialis (radial head, cut)

3rd, 4th, 5th digits | Flexor digitorum profundus muscle belly for

Flexor pollicis longus

2nd digit

Radial artery

Dorsal (cutaneous) branch of ulnar nerve

Dorsal carpal branch of ulnar artery

Pronator quadratus

Palmar radiocarpal ligament

Pisiform

Median nerve

Flexor retinaculum (transverse carpal ligament)

Deep branch of ulnar nerve and artery

Opponens pollicis

Flexor pollicis brevis

Opponens digiti minimi

Abductor pollicis brevis

Abductor digiti minimi

4th lumbrical

1st lumbrical

2nd lumbrical

3rd lumbrical

6.62   **Deep flexors of the digits and related structures**

- The two deep digital flexor muscles, flexor pollicis longus and flexor digitorum profundus, arise from the flexor aspects of the radius, interosseous membrane, and ulna between the origin of flexor digitorum superficialis proximally and pronator quadratus distally.
- The ulnar nerve enters the forearm posterior to the medial epicondyle, then descends between the flexor digitorum profundus and flexor carpi ulnaris and is joined by the ulnar artery. At the wrist the ulnar nerve and artery pass anterior to the flexor retinaculum and lateral to the pisiform to enter the palm.
- At the elbow, the ulnar nerve supplies the flexor carpi ulnaris and the medial half of the flexor digitorum profundus muscles; superior to the wrist, it gives off the dorsal (cutaneous) branch.
- The four lumbricals (L1–L4) arise from the flexor digitorum profundus tendons.

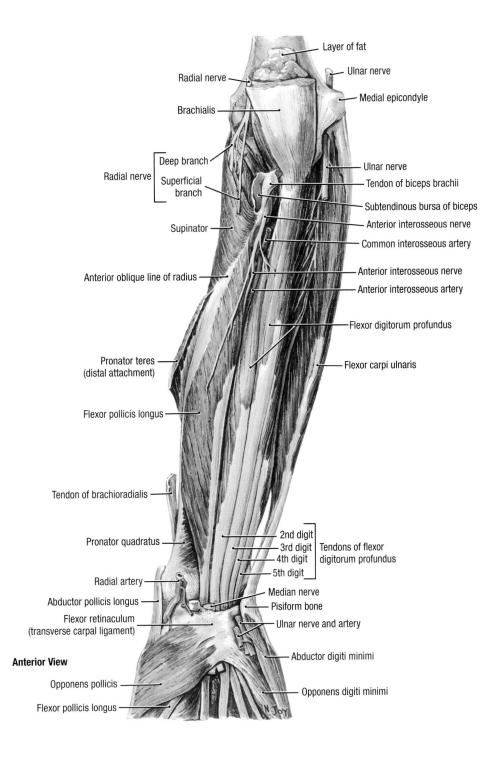

6.63   Deep flexors of the digits and supinator

- The five tendons of the deep digital flexors (flexor pollicis longus and flexor digitorum profundus) lie side by side as they enter the carpal tunnel.
- The biceps brachii muscle attaches to the medial aspect of the radius; hence, it can supinate the forearm, whereas the pronator teres muscle, by attaching to the lateral surface, can pronate the forearm.
- The deep branch of the radial nerve pierces and innervates the supinator muscle.
- The anterior interosseous nerve and artery disappear between the flexor pollicis longus and flexor digitorum profundus muscles to lie on the interosseous membrane.

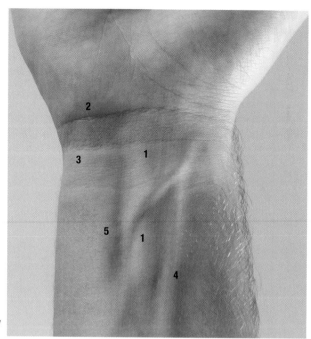

A. Anterior View

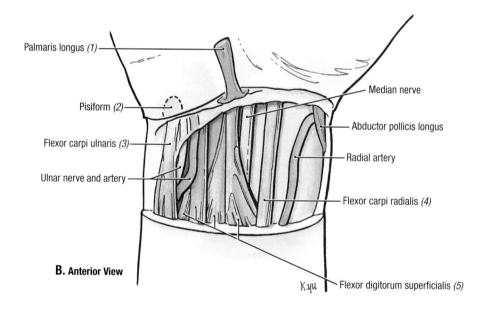

Palmaris longus *(1)*

Pisiform *(2)*

Flexor carpi ulnaris *(3)*

Ulnar nerve and artery

Median nerve

Abductor pollicis longus

Radial artery

Flexor carpi radialis *(4)*

B. Anterior View

K.yu

Flexor digitorum superficialis *(5)*

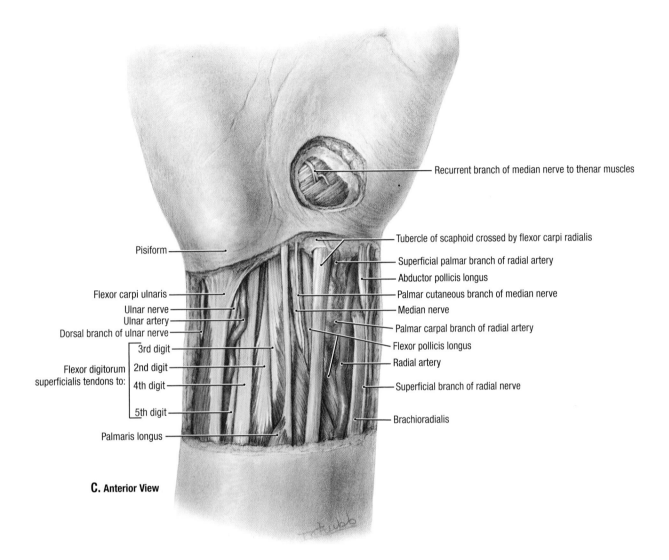

C. Anterior View

Recurrent branch of median nerve to thenar muscles

Tubercle of scaphoid crossed by flexor carpi radialis

Superficial palmar branch of radial artery

Abductor pollicis longus

Palmar cutaneous branch of median nerve

Median nerve

Palmar carpal branch of radial artery

Flexor pollicis longus

Radial artery

Superficial branch of radial nerve

Brachioradialis

Pisiform

Flexor carpi ulnaris

Ulnar nerve

Ulnar artery

Dorsal branch of ulnar nerve

Flexor digitorum superficialis tendons to: 3rd digit / 2nd digit / 4th digit / 5th digit

Palmaris longus

6.64   **Structures of anterior aspect of wrist**

A. Surface anatomy. **B.** Schematic illustration. **C.** Dissection.
In **C**:
- The distal skin incision follows the transverse skin crease at the wrist. The incision crosses the pisiform, to which the flexor carpi ulnaris muscle attaches, and the tubercle of the scaphoid, to which the tendon of flexor carpi radialis muscle is a guide.
- The palmaris longus tendon bisects the transverse skin crease; deep to its lateral margin is the median nerve.
- The radial artery passes deep to the abductor pollicis longus muscle.
- The flexor digitorum superficialis tendons to the 3rd and 4th digits become anterior to those of the 2nd and 5th digits.
- The recurrent branch of the median nerve to the thenar muscles lies within a circle whose center is 2.5 to 4 cm distal to the tubercle of the scaphoid.

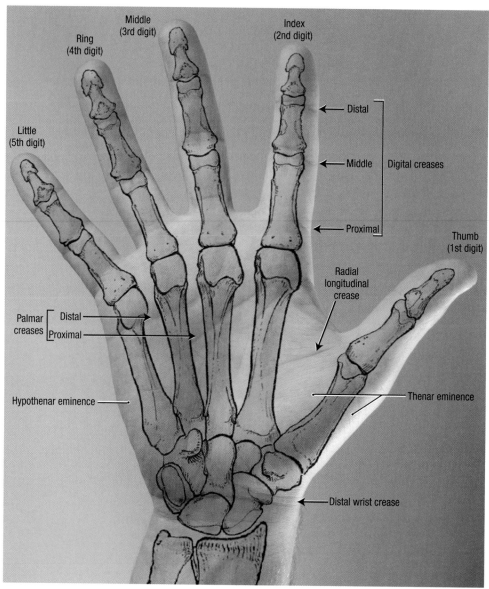

Anterior View

6.65   **Skin creases of wrist and hand**

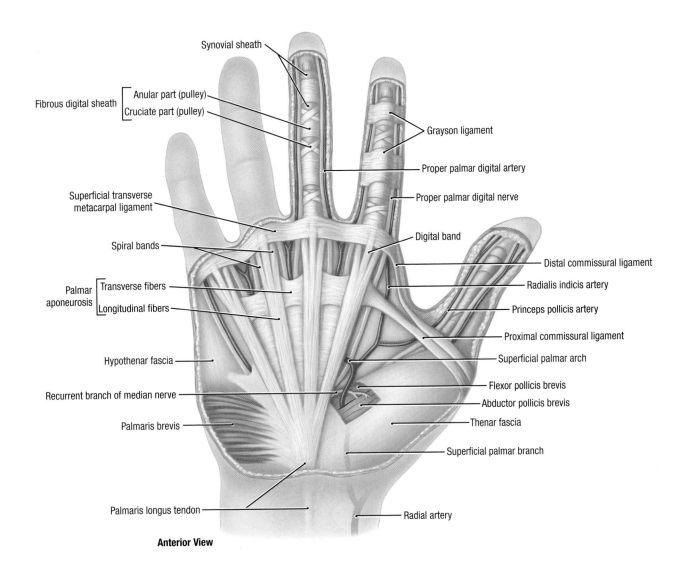

Synovial sheath

Fibrous digital sheath
- Anular part (pulley)
- Cruciate part (pulley)

Grayson ligament

Proper palmar digital artery

Superficial transverse metacarpal ligament

Proper palmar digital nerve

Digital band

Spiral bands

Distal commissural ligament

Palmar aponeurosis
- Transverse fibers
- Longitudinal fibers

Radialis indicis artery

Princeps pollicis artery

Proximal commissural ligament

Hypothenar fascia

Superficial palmar arch

Recurrent branch of median nerve

Flexor pollicis brevis

Abductor pollicis brevis

Palmaris brevis

Thenar fascia

Superficial palmar branch

Palmaris longus tendon

Radial artery

Anterior View

6.66   **Palmar aponeurosis**

- The palmar fascia is thin over the thenar and hypothenar eminences, but thick centrally, where it forms the palmar aponeurosis, and in the digits, where it forms the fibrous digital sheaths.
- The distal end (base) of the palmar aponeurosis divides into four digital bands.

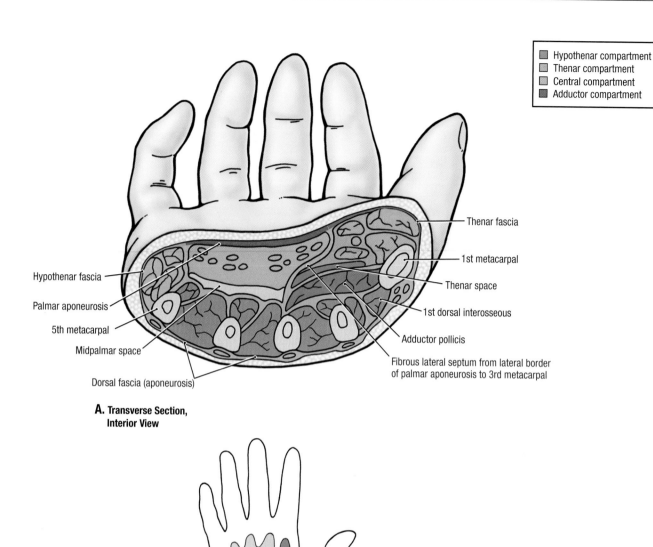

☐ Hypothenar compartment
☐ Thenar compartment
☐ Central compartment
☐ Adductor compartment

Thenar fascia

1st metacarpal

Thenar space

1st dorsal interosseous

Adductor pollicis

Fibrous lateral septum from lateral border
of palmar aponeurosis to 3rd metacarpal

Hypothenar fascia

Palmar aponeurosis

5th metacarpal

Midpalmar space

Dorsal fascia (aponeurosis)

A. Transverse Section,
Interior View

Thenar space

Midpalmar space

B. Anterior View

6.67   Compartments, spaces, and fascia of the palm

A. Transverse section through the middle of the palm showing the fascial compartments of the hand. **B.** Fascial spaces of palm.

- The midpalmar space lies posterior to the central compartment, is bounded medially by the hypothenar compartment, and is related distally to the synovial sheath of the third, fourth, and fifth digits.
- The thenar space lies posterior to the thenar compartment and is related distally to the synovial sheath of the index finger.
- The midpalmar and thenar spaces are separated by a septum that passes from the palmar aponeurosis to the third metacarpal.

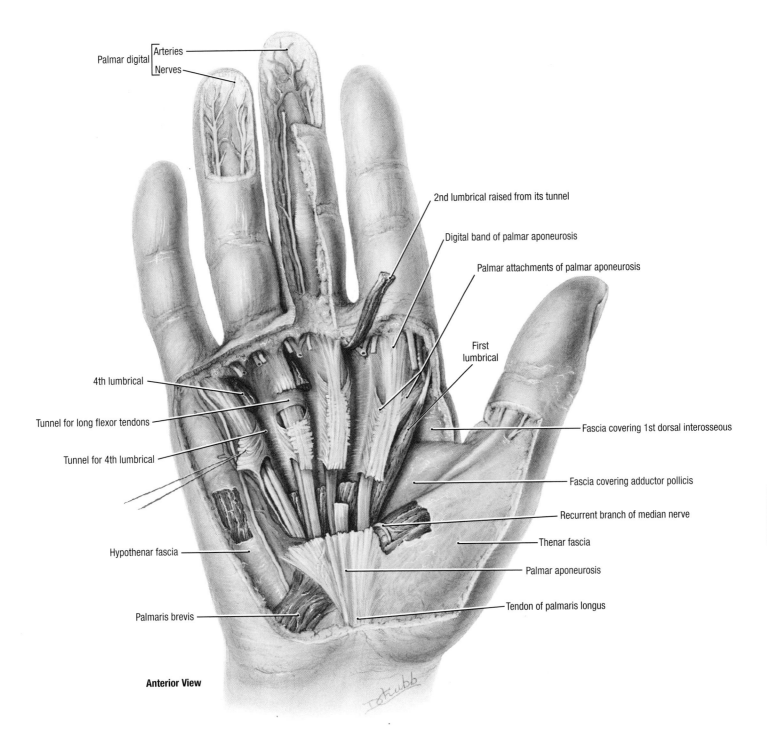

Palmar digital ⌈ Arteries
　　　　　　 ⌊ Nerves

2nd lumbrical raised from its tunnel

Digital band of palmar aponeurosis

Palmar attachments of palmar aponeurosis

First lumbrical

4th lumbrical

Tunnel for long flexor tendons

Fascia covering 1st dorsal interosseous

Tunnel for 4th lumbrical

Fascia covering adductor pollicis

Recurrent branch of median nerve

Hypothenar fascia

Thenar fascia

Palmar aponeurosis

Palmaris brevis

Tendon of palmaris longus

Anterior View

6.68　Attachments of palmar aponeurosis, digital vessels, and nerves

- From the palmar aponeurosis, longitudinal fibers (digital bands) enter the fingers; the other fibers, forming an extensive fibroareolar septa, pass posteriorly to the palmar ligaments (see Fig. 6.74) and, more proximally, to the fascia covering the interossei. Thus, two sets of tunnels exist in the distal half of the palm: (1) tunnels for long flexor tendons and (2) tunnels for lumbricals, digital vessels, and digital nerves.
- In the dissected middle finger, note the absence of fat deep to the skin creases of the fingers.

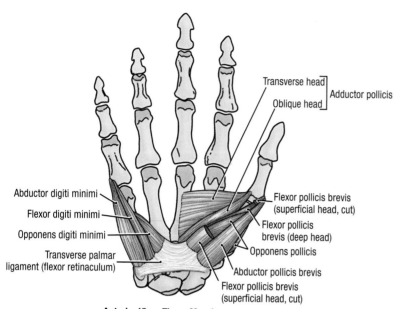

Transverse head ⎤
Oblique head ⎦ Adductor pollicis

Abductor digiti minimi

Flexor digiti minimi

Opponens digiti minimi

Transverse palmar
ligament (flexor retinaculum)

Flexor pollicis brevis
(superficial head, cut)

Flexor pollicis
brevis (deep head)

Opponens pollicis

Abductor pollicis brevis

Flexor pollicis brevis
(superficial head, cut)

**Anterior View, Thena Muscles,
Hypothenar Muscles, and
Adductor Pollicis**

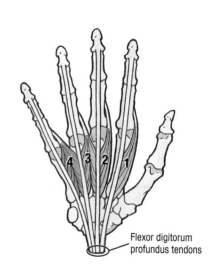

4 3 2 1

Flexor digitorum
profundus tendons

**Anterior View, Lumbricals
(1–4)**

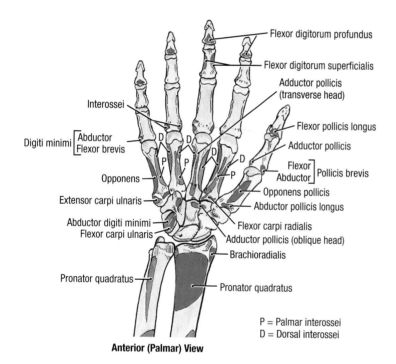

Flexor digitorum profundus

Flexor digitorum superficialis

Adductor pollicis
(transverse head)

Interossei

Digiti minimi ⎡ Abductor
          ⎣ Flexor brevis

D   D

D

D

Flexor pollicis longus

Adductor pollicis

Flexor ⎤
Abductor ⎦ Pollicis brevis

Opponens

P   P

P

P

Opponens pollicis

Extensor carpi ulnaris

Abductor pollicis longus

Abductor digiti minimi
Flexor carpi ulnaris

Flexor carpi radialis

Adductor pollicis (oblique head)

Brachioradialis

Pronator quadratus

Pronator quadratus

P = Palmar interossei
D = Dorsal interossei

Anterior (Palmar) View

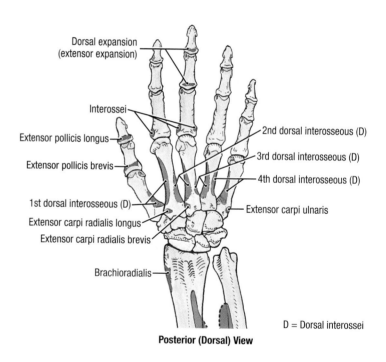

Dorsal expansion
(extensor expansion)

Interossei

Extensor pollicis longus

Extensor pollicis brevis

1st dorsal interosseous (D)

Extensor carpi radialis longus

Extensor carpi radialis brevis

2nd dorsal interosseous (D)

3rd dorsal interosseous (D)

4th dorsal interosseous (D)

Extensor carpi ulnaris

Brachioradialis

D = Dorsal interossei

Posterior (Dorsal) View

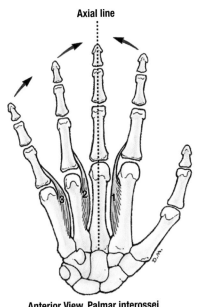

Anterior View, Palmar interossei
(Adduction)

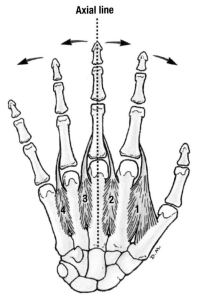

Anterior View, Dorsal interossei
(Abduction)

TABLE 6.9   MUSCLES OF HAND

MUSCLE	PROXIMAL ATTACHMENT	DISTAL ATTACHMENT	INNERVATION	MAIN ACTIONS
Abductor pollicis brevis	Flexor retinaculum and tubercles of scaphoid and trapezium	Lateral side of base of proximal phalanx of thumb	Recurrent branch of median nerve (**C8** and T1)	Abducts thumb and helps oppose it
Flexor pollicis brevis	Flexor retinaculum (transverse carpal ligament) and tubercle of trapezium			Flexes thumb
Opponens pollicis		Lateral side of first metacarpal		Opposes thumb toward center of palm and rotates it medially
Adductor pollicis	*Oblique head:* bases of second and third metacarpals, capitate, and adjacent carpal bones *Transverse head:* anterior surface of body of third metacarpal	Medial side of base of proximal phalanx of thumb	Deep branch of ulnar nerve (C8 and **T1**)	Adducts thumb toward middle digit
Abductor digiti minimi	Pisiform	Medial side of base of proximal phalanx of digit 5	Deep branch of ulnar nerve (C8 and T1)	Abducts digit 5
Flexor digiti minimi brevis	Hook of hamate and flexor retinaculum (transverse carpal ligament)			Flexes proximal phalanx of digit 5
Opponens digiti minimi		Medial border of fifth metacarpal		Draws fifth metacarpal anteriorly and rotates it, bringing digit 5 into opposition with thumb
Lumbricals 1 and 2	Lateral two tendons of flexor digitorum profundus	Lateral sides of extensor expansions of digits 2–5	Median nerve (C8 and **T1**)	Flex digits at metacarpophalangeal joints and extend interphalangeal joints
Lumbricals 3 and 4	Medial three tendons of flexor digitorum profundus		Deep branch of ulnar nerve (C8 and **T1**)	
Dorsal interossei 1–4	Adjacent sides of two metacarpals (D)	Extensor expansions and bases of proximal phalanges of digits 2–4	Deep branch of ulnar nerve (C8 and **T1**)	Abduct digits 2–5 and assist lumbricals
Palmar interossei 1–3	Palmar surfaces of second, fourth, and fifth metacarpals (P)	Extensor expansions of digits and bases of proximal phalanges of digits 2, 4, and 5		Adduct digits 2, 4, and 5 and assist lumbricals

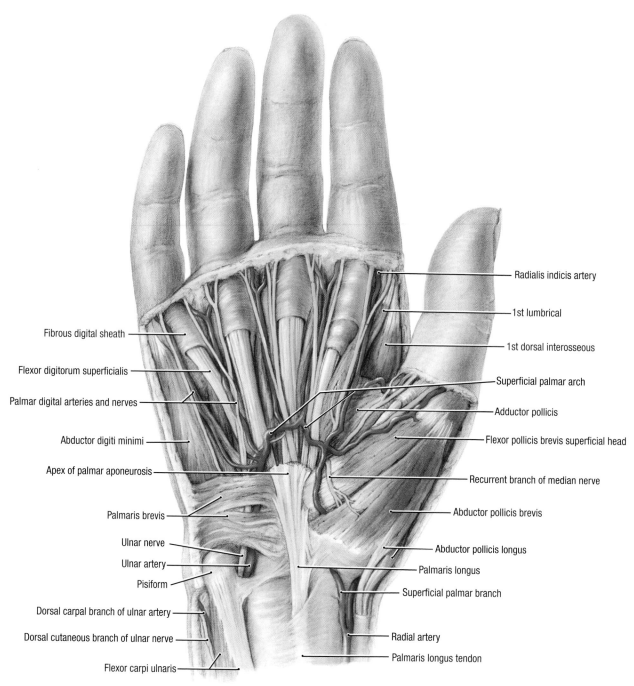

Radialis indicis artery

1st lumbrical

1st dorsal interosseous

Superficial palmar arch

Adductor pollicis

Flexor pollicis brevis superficial head

Recurrent branch of median nerve

Abductor pollicis brevis

Abductor pollicis longus

Palmaris longus

Superficial palmar branch

Radial artery

Palmaris longus tendon

Fibrous digital sheath

Flexor digitorum superficialis

Palmar digital arteries and nerves

Abductor digiti minimi

Apex of palmar aponeurosis

Palmaris brevis

Ulnar nerve

Ulnar artery

Pisiform

Dorsal carpal branch of ulnar artery

Dorsal cutaneous branch of ulnar nerve

Flexor carpi ulnaris

Anterior View

6.69 **Superficial dissection of palm**

- The skin, superficial fascia, palmar aponeurosis, and thenar and hypothenar fasciae have been removed.
- The superficial palmar arch, located immediately deep to the fasciae, is formed by the ulnar artery and completed by the superficial palmar branch of the radial artery.
- The four lumbricals lie posterior to the digital vessels and nerves. The lumbricals arise from the lateral sides of the flexor digitorum profundus tendons and are inserted into the lateral sides of the dorsal expansions of the corresponding digits. The medial two lumbricals are bipennate and also arise from the medial sides of adjacent flexor digitorum profundus tendons.
- In the digits, a (proper) digital artery and nerve lie on each side of the fibrous digital sheath.
- Note the canal (Guyon) through which the ulnar vessels and nerve pass medial to the pisiform.

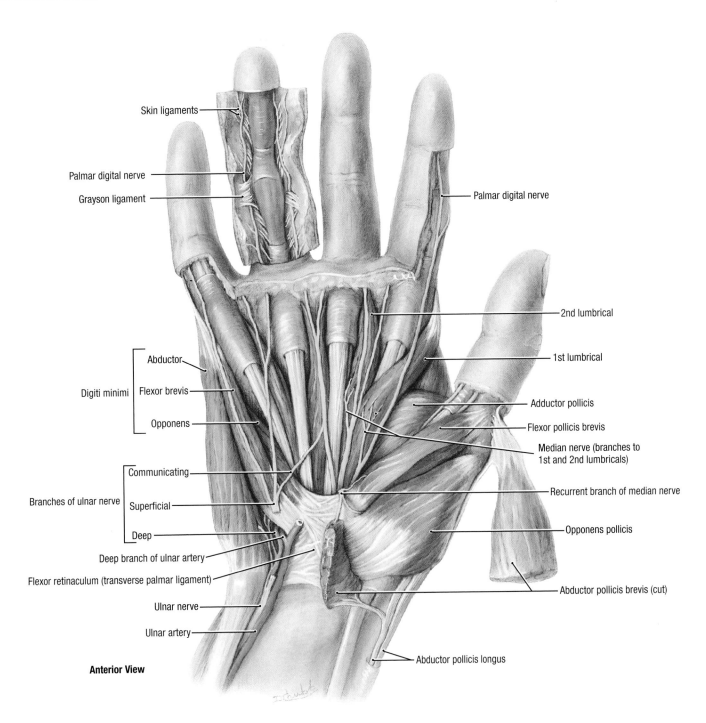

Skin ligaments

Palmar digital nerve

Grayson ligament

Palmar digital nerve

2nd lumbrical

1st lumbrical

Abductor

Digiti minimi — Flexor brevis

Opponens

Adductor pollicis

Flexor pollicis brevis

Median nerve (branches to 1st and 2nd lumbricals)

Communicating

Branches of ulnar nerve — Superficial

Deep

Recurrent branch of median nerve

Deep branch of ulnar artery

Flexor retinaculum (transverse palmar ligament)

Opponens pollicis

Ulnar nerve

Ulnar artery

Abductor pollicis brevis (cut)

Abductor pollicis longus

Anterior View

6.70   Superficial dissection of palm, ulnar, and median nerves

- The three thenar muscles (flexor pollicis brevis, abductor pollicis brevis, and opponens pollicis) and the three hypothenar muscles (flexor digiti minimi, abductor digiti minimi, and opponens digiti minimi) attach to the transverse carpal ligament (flexor retinaculum) and the four marginal carpal bones (scaphoid, trapezium, pisiform and hamate), which are united by the ligament.
- The median nerve is distributed to five muscles (three thenar and lateral two lumbrical muscles) and provides cutaneous branches to three digits.
- The ulnar nerve supplies all other short (intrinsic) muscles in the hand (three hypothenar, two lumbricals, interossei, and adductor pollicis) and provides cutaneous branches to the medial one and a half digits; the communicating branch joins the ulnar and median nerves.
- The recurrent branch (motor branch) of the median nerve arises from the lateral side of the nerve at the distal border of the transverse carpal ligament (flexor retinaculum) and innervates the thenar muscles. The recurrent branch lies superficially and can easily be severed.

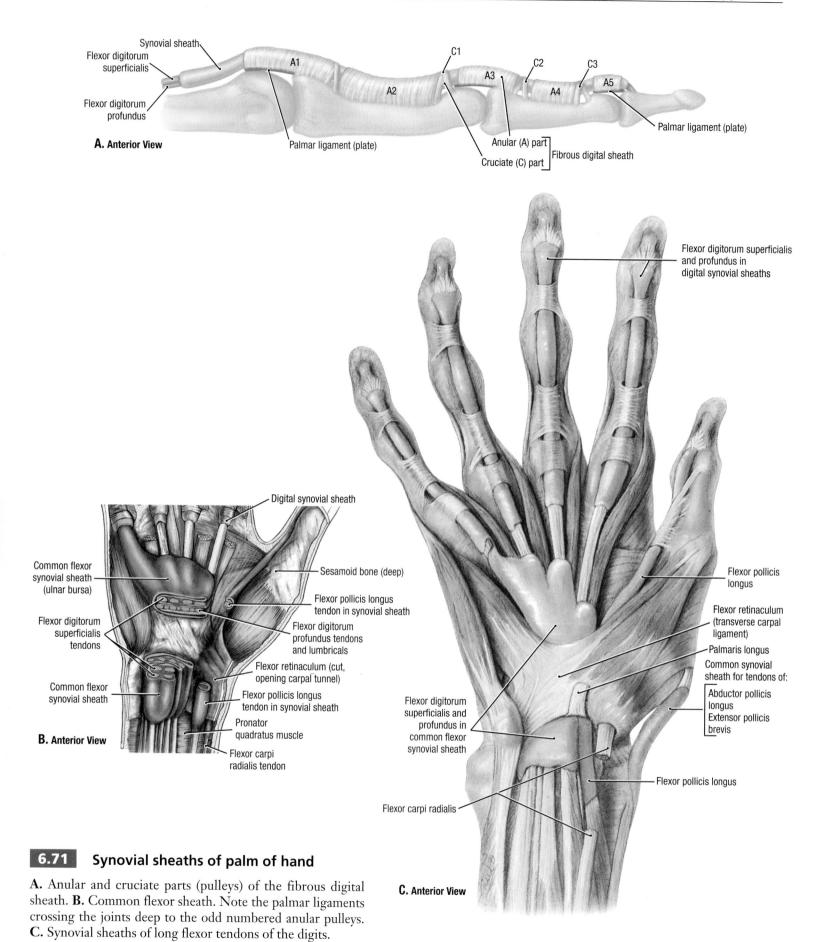

A. Anterior View

B. Anterior View

C. Anterior View

6.71 **Synovial sheaths of palm of hand**

A. Anular and cruciate parts (pulleys) of the fibrous digital sheath. **B.** Common flexor sheath. Note the palmar ligaments crossing the joints deep to the odd numbered anular pulleys. **C.** Synovial sheaths of long flexor tendons of the digits.

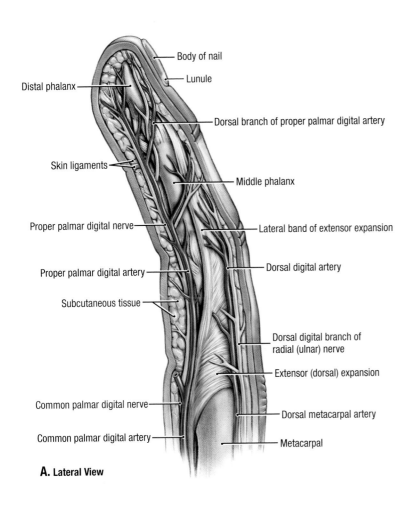

A. Lateral View

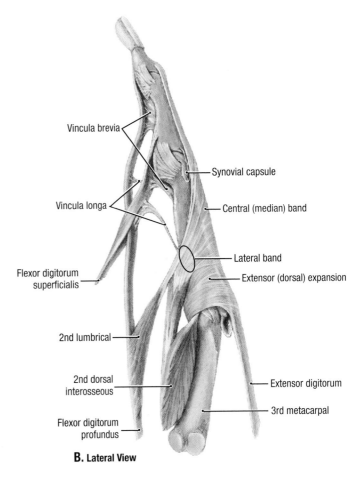

B. Lateral View

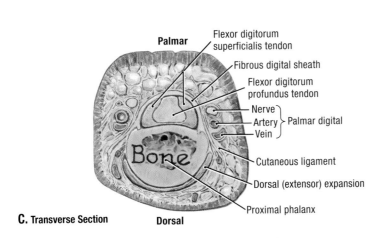

C. Transverse Section

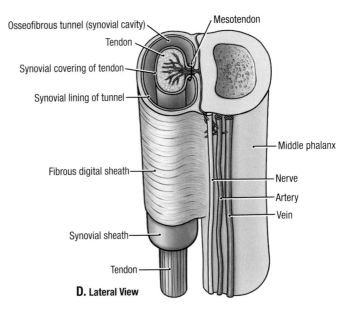

D. Lateral View

6.72   **Digital tendons, vessels, and nerves**

A. Digital vessels and nerves. **B.** Extensor expansion of the 3rd (middle) digit. **C.** Transverse section through the proximal phalanx. **D.** Osseofibrous tunnel and synovial tendon sheath.

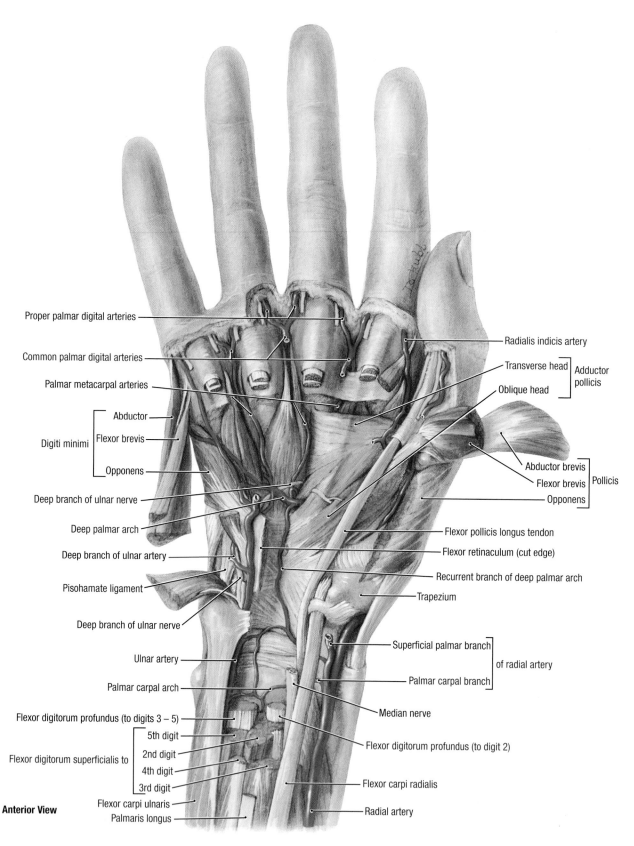

Proper palmar digital arteries

Common palmar digital arteries

Palmar metacarpal arteries

Abductor
Flexor brevis — Digiti minimi
Opponens

Deep branch of ulnar nerve

Deep palmar arch

Deep branch of ulnar artery

Pisohamate ligament

Deep branch of ulnar nerve

Ulnar artery

Palmar carpal arch

Flexor digitorum profundus (to digits 3 – 5)

Flexor digitorum superficialis to
5th digit
2nd digit
4th digit
3rd digit

Flexor carpi ulnaris

Anterior View

Palmaris longus

Radialis indicis artery

Transverse head
Oblique head — Adductor pollicis

Abductor brevis
Flexor brevis — Pollicis
Opponens

Flexor pollicis longus tendon

Flexor retinaculum (cut edge)

Recurrent branch of deep palmar arch

Trapezium

Superficial palmar branch
of radial artery
Palmar carpal branch

Median nerve

Flexor digitorum profundus (to digit 2)

Flexor carpi radialis

Radial artery

6.73 **Deep dissection of palm**

- The deep branch of the ulnar artery joins the radial artery to form the deep palmar arch.
- The transverse and oblique heads of the adductor pollicis muscle attach to the medial side of the proximal phalanx of the

thumb; there is usually a sesamoid bone contained in the tendon of insertion.
- The opponens pollicis attaches along the length of the first metacarpal.

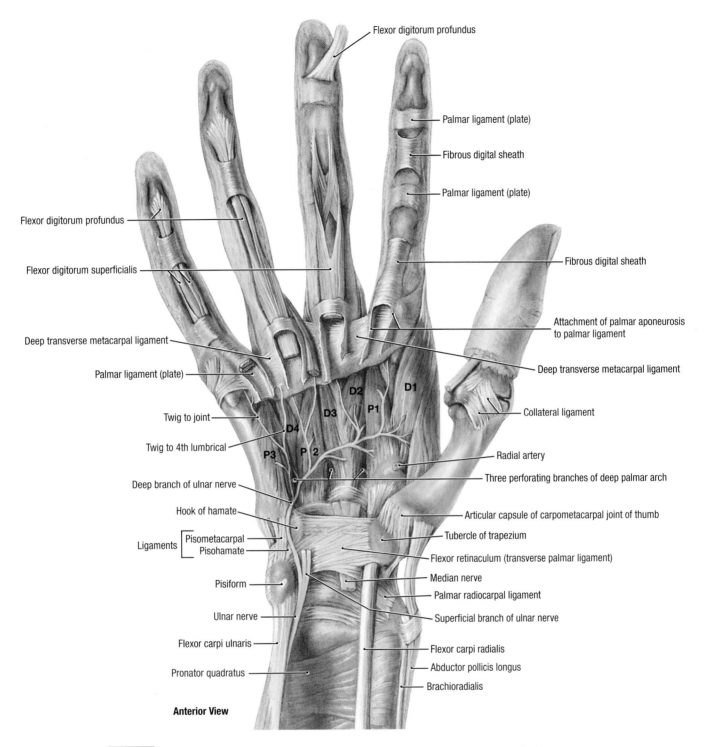

Flexor digitorum profundus

Palmar ligament (plate)

Fibrous digital sheath

Palmar ligament (plate)

Flexor digitorum profundus

Flexor digitorum superficialis

Fibrous digital sheath

Deep transverse metacarpal ligament

Attachment of palmar aponeurosis to palmar ligament

Palmar ligament (plate)

Deep transverse metacarpal ligament

D2   **D1**

Twig to joint

D3 **P1**

Collateral ligament

D4

Twig to 4th lumbrical

P3 **P 2**

Radial artery

Three perforating branches of deep palmar arch

Deep branch of ulnar nerve

Hook of hamate

Articular capsule of carpometacarpal joint of thumb

Ligaments ⎡ Pisometacarpal
       ⎣ Pisohamate

Tubercle of trapezium

Flexor retinaculum (transverse palmar ligament)

Median nerve

Pisiform

Palmar radiocarpal ligament

Ulnar nerve

Superficial branch of ulnar nerve

Flexor carpi ulnaris

Flexor carpi radialis

Pronator quadratus

Abductor pollicis longus

Brachioradialis

Anterior View

| **6.74** | **Deep dissection of palm and digits with deep branch of ulnar nerve** |

- The flexor carpi ulnaris muscle continues beyond the pisiform as the pisohamate and the pisometacarpal ligament.
- Three unipennate palmar *(P1–3)* and four bipennate dorsal *(D1–4)* interosseous muscles are illustrated; the palmar interossei adduct the fingers, and the dorsal interossei abduct the fingers in relation to the axial line, an imaginary line drawn through the long axis of the 3rd digit (see Table 6.9).
- The deep transverse metacarpal ligaments unite the palmar ligaments; the lumbricals pass anterior to the deep transverse metacarpal ligament, and the interossei pass posterior to the ligament.

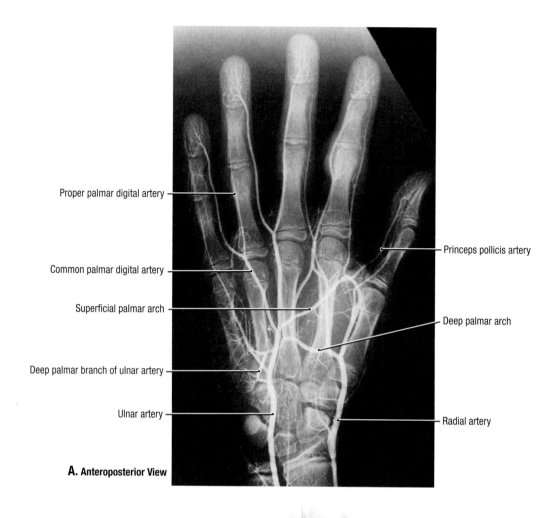

Proper palmar digital artery

Common palmar digital artery

Superficial palmar arch

Deep palmar branch of ulnar artery

Ulnar artery

Princeps pollicis artery

Deep palmar arch

Radial artery

A. Anteroposterior View

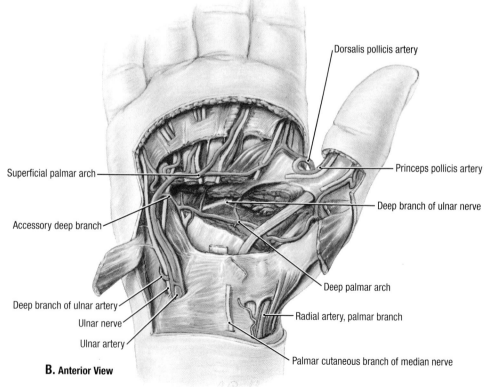

Dorsalis pollicis artery

Superficial palmar arch

Accessory deep branch

Deep branch of ulnar artery

Ulnar nerve

Ulnar artery

Princeps pollicis artery

Deep branch of ulnar nerve

Deep palmar arch

Radial artery, palmar branch

Palmar cutaneous branch of median nerve

B. Anterior View

6.75 **Arterial supply of hand**

A. Arteriogram of the hand. **B.** Dissection of palmar arterial arches. The superficial palmar arch is usually completed by the superficial palmar branch of the radial artery, but in this specimen the dorsalis pollicis artery completes the arch.

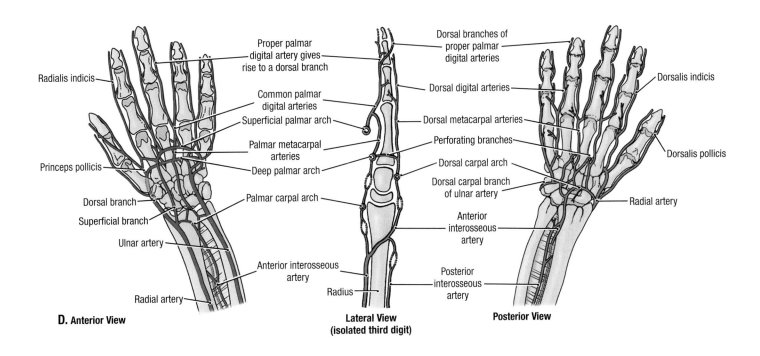

Radialis indicis

Proper palmar digital artery gives rise to a dorsal branch

Common palmar digital arteries

Superficial palmar arch

Palmar metacarpal arteries

Princeps pollicis

Deep palmar arch

Dorsal branch

Palmar carpal arch

Superficial branch

Ulnar artery

Anterior interosseous artery

Radial artery

D. Anterior View

Dorsal branches of proper palmar digital arteries

Dorsal digital arteries

Dorsal metacarpal arteries

Perforating branches

Dorsal carpal arch

Dorsal carpal branch of ulnar artery

Anterior interosseous artery

Posterior interosseous artery

Radius

Lateral View (isolated third digit)

Dorsalis indicis

Dorsalis pollicis

Radial artery

Posterior View

TABLE 6.10   ARTERIES OF HAND

ARTERY	ORIGIN	COURSE
Superficial palmar arch	Direct continuation of ulnar artery; arch is completed on lateral side by superficial branch of radial artery or another of its branches	Curves laterally deep to palmar aponeurosis and superficial to long flexor tendons; curve of arch lies across palm at level of distal border of extended thumb
Deep palmar arch	Direct continuation of radial artery arch is completed on medial side by deep branch of ulnar artery	Curves medially, deep to long flexor tendons and is in contact with bases of metacarpals
Common palmar digitals	Superficial palmar arch	Pass directly on lumbricals to webbings of digits
Proper palmar digitals	Common palmar digital arteries	Run along sides of digits 2–5
Princeps pollicis	Radial artery as it turns into palm	Descends on palmar aspect of first metacarpal and divides at the base of proximal phalanx into two branches that run along sides of thumb
Radialis indicis	Radial artery, but may arise from princeps pollicis artery	Passes along lateral side of index finger to its distal end
Dorsal carpal arch	Radial and ulnar arteries	Arches within fascia on dorsum of hand

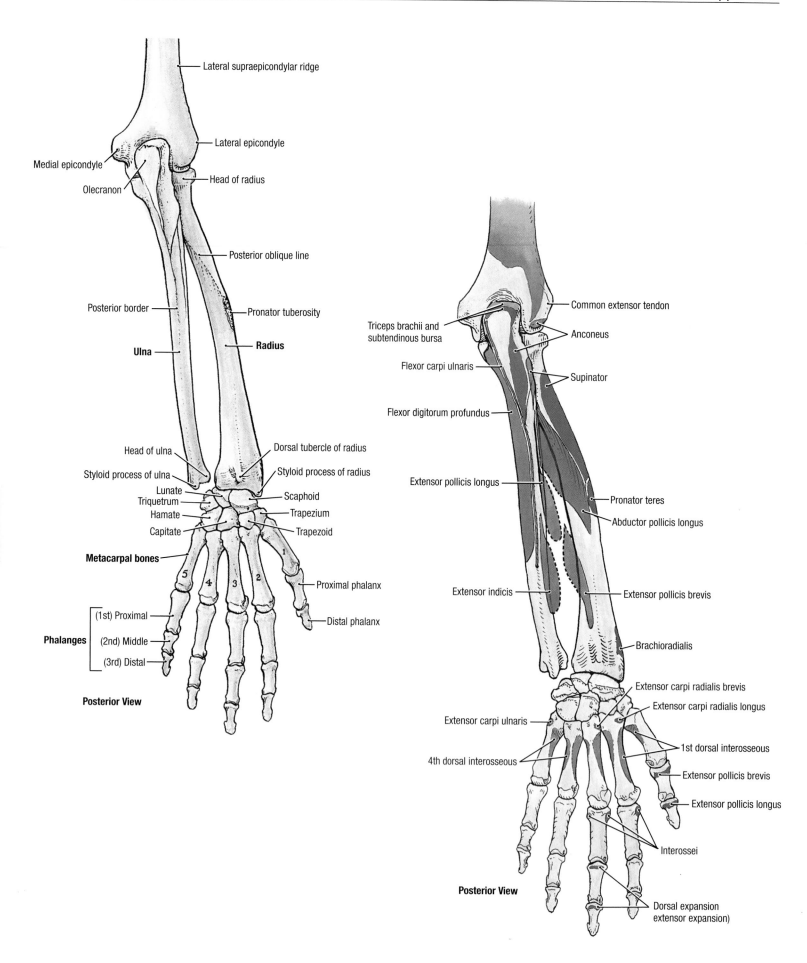

Lateral supraepicondylar ridge

Lateral epicondyle

Medial epicondyle

Olecranon

Head of radius

Posterior oblique line

Posterior border

Pronator tuberosity

Ulna

Radius

Head of ulna

Dorsal tubercle of radius

Styloid process of ulna

Styloid process of radius

Lunate

Scaphoid

Triquetrum

Hamate

Trapezium

Capitate

Trapezoid

Metacarpal bones

Proximal phalanx

Distal phalanx

(1st) Proximal

Phalanges

(2nd) Middle

(3rd) Distal

Posterior View

Triceps brachii and subtendinous bursa

Common extensor tendon

Anconeus

Flexor carpi ulnaris

Supinator

Flexor digitorum profundus

Extensor pollicis longus

Pronator teres

Abductor pollicis longus

Extensor indicis

Extensor pollicis brevis

Brachioradialis

Extensor carpi radialis brevis

Extensor carpi radialis longus

Extensor carpi ulnaris

1st dorsal interosseous

4th dorsal interosseous

Extensor pollicis brevis

Extensor pollicis longus

Interossei

Posterior View

Dorsal expansion extensor expansion)

TABLE 6.11   MUSCLES ON THE POSTERIOR SURFACE OF THE FOREARM

MUSCLE	PROXIMAL ATTACHMENT	DISTAL ATTACHMENT	INNERVATION	MAIN ACTIONS
Brachioradialis	Proximal two thirds of lateral supraepicondylar ridge of humerus	Lateral surface of distal end of radius	Radial nerve (C5, **C6**, and C7)	Flexes forearm
Extensor carpi radialis longus	Lateral supraepicondylar ridge of humerus	Base of second metacarpal bone	Radial nerve (C6 and C7)	Extend and abduct hand at wrist joint
Extensor carpi radialis brevis		Base of third metacarpal bone	Deep branch of radial nerve (**C7** and C8)	
Extensor digitorum	Lateral epicondyle of humerus	Extensor expansions of medial four digits	Posterior interosseous nerve (C7 and C8), a branch of the radial nerve	Extends medial four digits at metacarpophalangeal joints; extends hand at wrist joint
Extensor digiti minimi		Extensor expansion of fifth digit		Extends fifth digit at metacarpophalangeal and interphalangeal joints
Extensor carpi ulnaris	Lateral epicondyle of humerus and posterior border of ulna	Base of fifth metacarpal bone		Extends and adducts hand at wrist joint
Anconeus	Lateral epicondyle humerus	Lateral surface of olecranon and superior part of posterior surface of ulna	Radial nerve (C7, C8, and T1)	Assists triceps in extending elbow joint; stabilizes elbow joint; abducts ulna during pronation
Supinator	Lateral epicondyle of humerus, radial collateral and anular ligaments, supinator fossa, and crest of ulna	Lateral, posterior, and anterior surfaces of proximal third of radius	Deep branch of radial nerve (C5 and **C6**)	Supinates forearm, i.e., rotates radius to turn palm anteriorly
Abductor pollicis longus	Posterior surfaces of ulna, radius, and interosseous membrane	Base of first metacarpal bone	Posterior interosseous nerve (C7 and **C8**)	Abducts thumb and extends it at carpometacarpal joint
Extensor pollicis brevis	Posterior surface of radius and interosseous membrane	Base of proximal phalanx of thumb		Extends proximal phalanx of thumb at metacarpophalangeal joint
Extensor pollicis longus	Posterior surface of middle third of ulna and interosseous membrane	Base of distal phalanx of thumb		Extends distal phalanx of thumb at metacarpophalangeal and interphalangeal joints
Extensor indicis	Posterior surface of ulna and interosseous membrane	Extensor expansion of second digit		Extends second digit and helps to extend hand

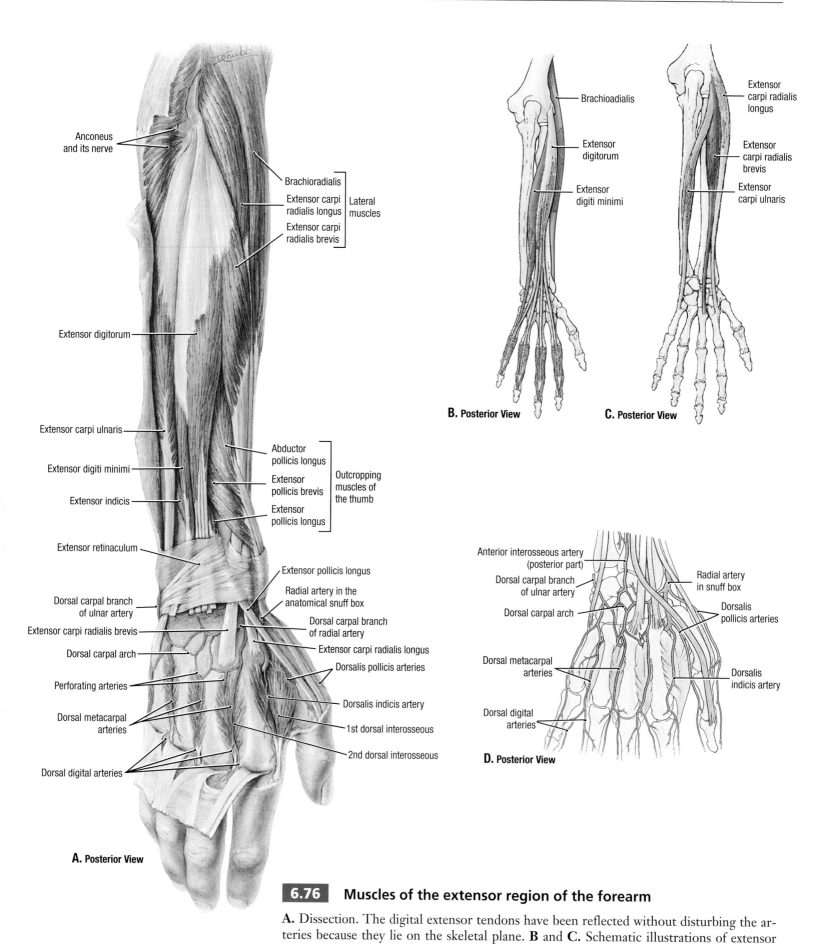

Anconeus and its nerve

Brachioradialis
Extensor carpi radialis longus } Lateral muscles
Extensor carpi radialis brevis

Extensor digitorum

Extensor carpi ulnaris

Extensor digiti minimi

Extensor indicis

Abductor pollicis longus
Extensor pollicis brevis } Outcropping muscles of the thumb
Extensor pollicis longus

Extensor retinaculum

Dorsal carpal branch of ulnar artery

Extensor carpi radialis brevis

Dorsal carpal arch

Perforating arteries

Dorsal metacarpal arteries

Dorsal digital arteries

Extensor pollicis longus
Radial artery in the anatomical snuff box
Dorsal carpal branch of radial artery
Extensor carpi radialis longus
Dorsalis pollicis arteries
Dorsalis indicis artery
1st dorsal interosseous
2nd dorsal interosseous

A. Posterior View

Brachioadialis
Extensor digitorum
Extensor digiti minimi

B. Posterior View

Extensor carpi radialis longus
Extensor carpi radialis brevis
Extensor carpi ulnaris

C. Posterior View

Anterior interosseous artery (posterior part)
Dorsal carpal branch of ulnar artery
Dorsal carpal arch

Dorsal metacarpal arteries

Dorsal digital arteries

Radial artery in snuff box
Dorsalis pollicis arteries
Dorsalis indicis artery

D. Posterior View

6.76 **Muscles of the extensor region of the forearm**

A. Dissection. The digital extensor tendons have been reflected without disturbing the arteries because they lie on the skeletal plane. **B** and **C.** Schematic illustrations of extensor muscles. **D.** Arteries on dorsum of hand.

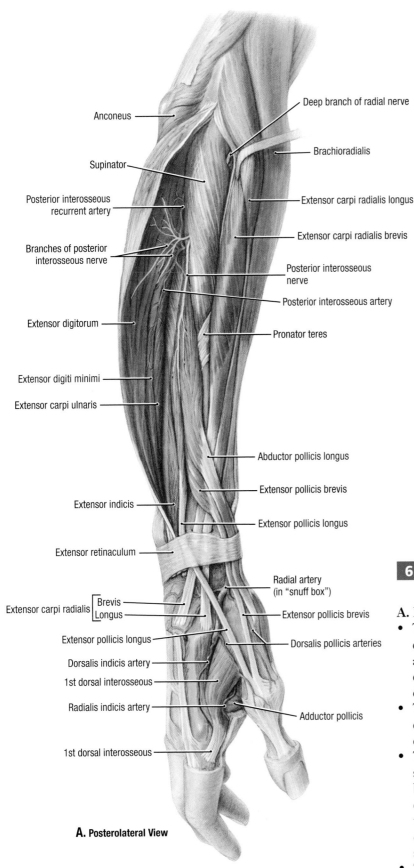

Anconeus

Deep branch of radial nerve

Supinator

Brachioradialis

Posterior interosseous recurrent artery

Extensor carpi radialis longus

Extensor carpi radialis brevis

Branches of posterior interosseous nerve

Posterior interosseous nerve

Posterior interosseous artery

Extensor digitorum

Pronator teres

Extensor digiti minimi

Extensor carpi ulnaris

Abductor pollicis longus

Extensor pollicis brevis

Extensor indicis

Extensor pollicis longus

Extensor retinaculum

Radial artery (in "snuff box")

Extensor carpi radialis ⌈ Brevis

⌊ Longus

Extensor pollicis brevis

Extensor pollicis longus

Dorsalis pollicis arteries

Dorsalis indicis artery

1st dorsal interosseous

Radialis indicis artery

Adductor pollicis

1st dorsal interosseous

A. Posterolateral View

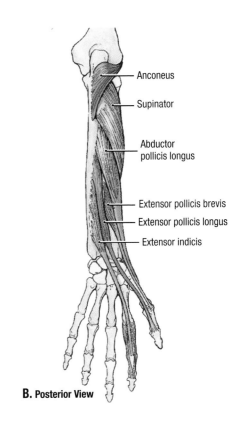

Anconeus

Supinator

Abductor pollicis longus

Extensor pollicis brevis

Extensor pollicis longus

Extensor indicis

B. Posterior View

6.77   **Deep structures on extensor aspect of forearm**

A. Dissection. **B.** Schematic illustration.

- Three "outcropping" muscles of the thumb (abductor pollicis longus, extensor pollicis brevis, and extensor pollicis longus) emerge between the extensor carpi radialis brevis and the extensor digitorum.
- The plane from which the thumb muscles emerge has been opened proximally to the lateral epicondyle, exposing the supinator.
- The laterally retracted brachioradialis and extensor carpi radialis longus and brevis are innervated by the deep branch of the radial nerve; the other extensor muscles are supplied by the posterior interosseous nerve, which is a continuation of the deep branch of the radial nerve that pierced the supinator.
- The tendons of the three outcropping muscles of the thumb pass to the bases of the three long bones of the thumb: the metacarpal, proximal phalanx, and distal phalanx.

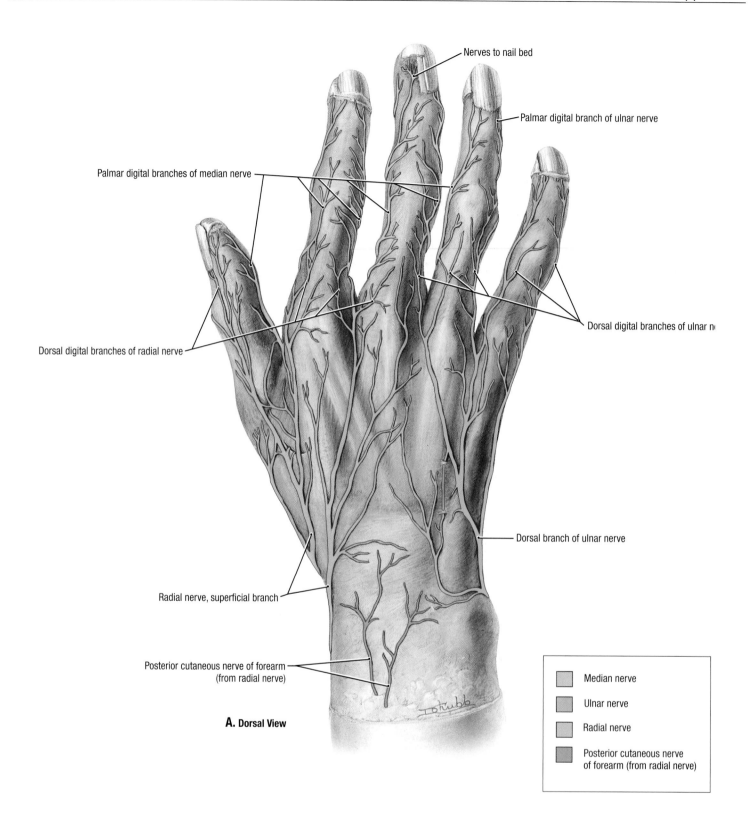

Nerves to nail bed

Palmar digital branch of ulnar nerve

Palmar digital branches of median nerve

Dorsal digital branches of ulnar n[

Dorsal digital branches of radial nerve

Dorsal branch of ulnar nerve

Radial nerve, superficial branch

Posterior cutaneous nerve of forearm
(from radial nerve)

A. Dorsal View

	Median nerve
	Ulnar nerve
	Radial nerve
	Posterior cutaneous nerve of forearm (from radial nerve)

6.78 **Cutaneous innervation of hand**

A. Dissection of nerves of dorsum of hand.

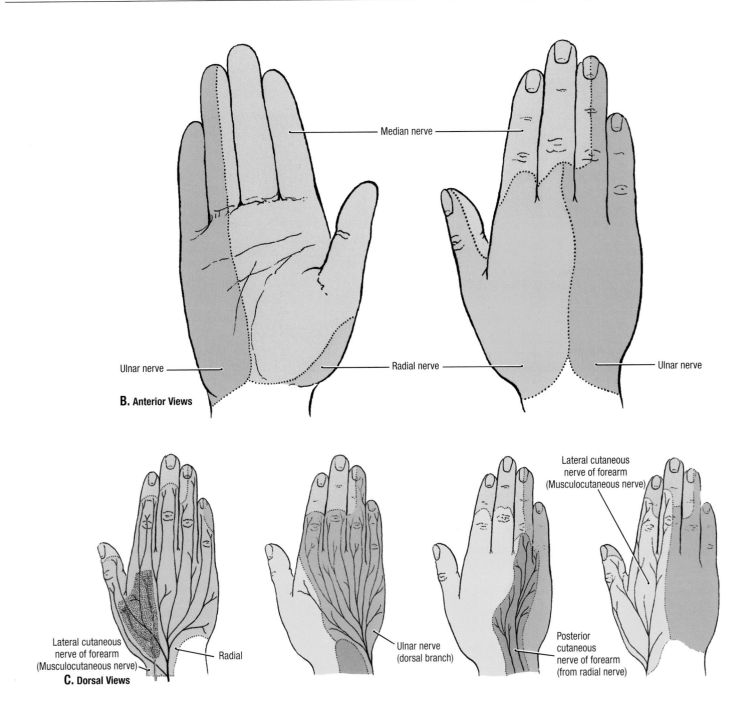

Median nerve

Ulnar nerve   Radial nerve   Ulnar nerve

B. Anterior Views

Lateral cutaneous
nerve of forearm
(Musculocutaneous nerve)

Lateral cutaneous
nerve of forearm
(Musculocutaneous nerve)   Radial

C. Dorsal Views

Ulnar nerve
(dorsal branch)

Posterior
cutaneous
nerve of forearm
(from radial nerve)

6.78   **Cutaneous innervation of hand** *(continued)*

B. Distribution of the cutaneous nerves to the palm and dorsum of the hand, schematic illustration. **C.** Variations in pattern of cutaneous nerves in dorsum of hand.

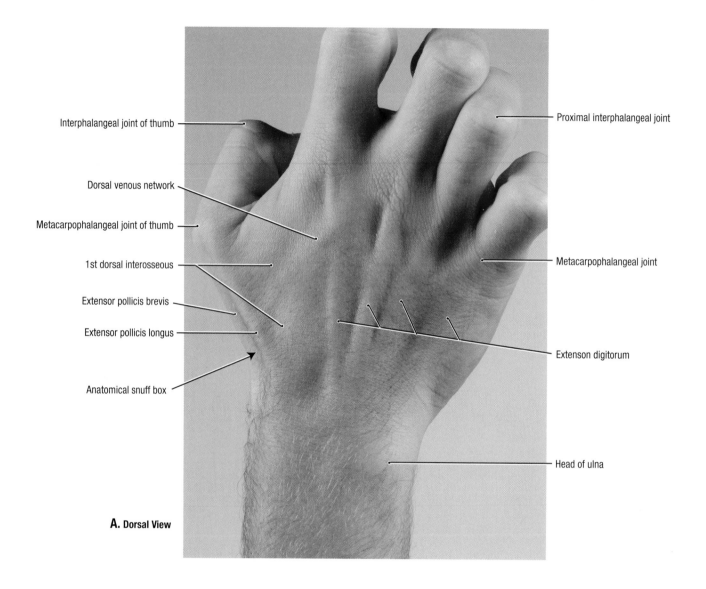

Interphalangeal joint of thumb

Dorsal venous network

Metacarpophalangeal joint of thumb

1st dorsal interosseous

Extensor pollicis brevis

Extensor pollicis longus

Anatomical snuff box

Proximal interphalangeal joint

Metacarpophalangeal joint

Extenson digitorum

Head of ulna

A. Dorsal View

6.79   **Dorsum of hand**

A. Surface anatomy. The interphalangeal joints are flexed, and the metacarpophalangeal joints are hyperextended to demonstrate the extensor digitorum tendons. **B.** Synovial sheaths. **C.** Transverse section of distal forearm (*numbers* refer to structures labeled in **B**).

- Six sheaths occupy the six osseofibrous tunnels deep to the extensor retinaculum. They contain nine tendons: tendons for the thumb in sheaths 1 and 3, tendons for the extensors of the wrist in sheaths 2 and 6, and tendons for the extensors of the wrist and fingers in sheaths 4 and 5.
- The tendon of the extensor pollicis longus hooks around the dorsal tubercle of radius to pass obliquely across the tendons of the extensor carpi radialis longus and brevis to the thumb.

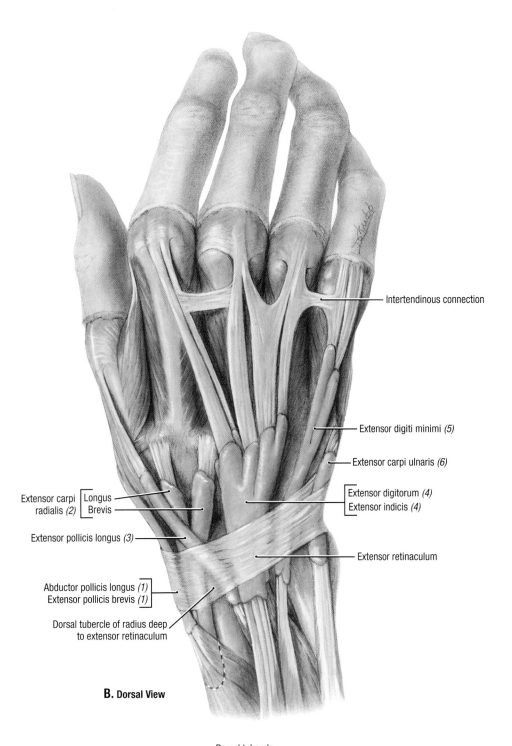

Intertendinous connection

Extensor digiti minimi *(5)*

Extensor carpi ulnaris *(6)*

Extensor carpi | Longus
radialis *(2)* | Brevis

Extensor digitorum *(4)*
Extensor indicis *(4)*

Extensor pollicis longus *(3)*

Extensor retinaculum

Abductor pollicis longus *(1)*
Extensor pollicis brevis *(1)*

Dorsal tubercle of radius deep
to extensor retinaculum

B. Dorsal View

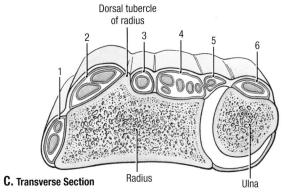

Dorsal tubercle
of radius

C. Transverse Section

Radius

Ulna

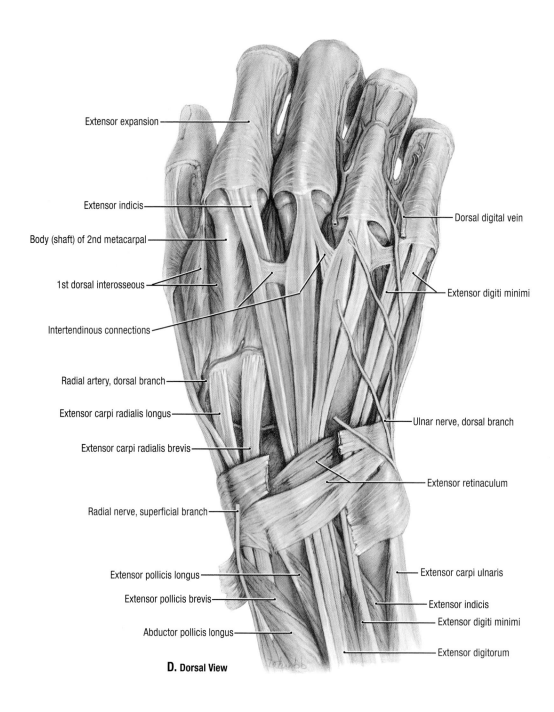

Extensor expansion

Extensor indicis

Body (shaft) of 2nd metacarpal

1st dorsal interosseous

Intertendinous connections

Radial artery, dorsal branch

Extensor carpi radialis longus

Extensor carpi radialis brevis

Radial nerve, superficial branch

Extensor pollicis longus

Extensor pollicis brevis

Abductor pollicis longus

Dorsal digital vein

Extensor digiti minimi

Ulnar nerve, dorsal branch

Extensor retinaculum

Extensor carpi ulnaris

Extensor indicis

Extensor digiti minimi

Extensor digitorum

D. Dorsal View

6.79   **Dorsum of hand** *(continued)*

D. Tendons on dorsum of hand and extensor retinaculum.
- The deep fascia is thickened to form the extensor retinaculum; the retinaculum stretches obliquely from one ridge on the radius to another; medially, it passes distal to the ulna to attach to the pisiform and triquetrum bones.
- Proximal to the knuckles, intertendinous connections extend between the tendons of the digital extensors and, thereby, restrict the independent action of the fingers.
- The body (shaft) of the 2nd metacarpal is not covered with an extensor tendon.

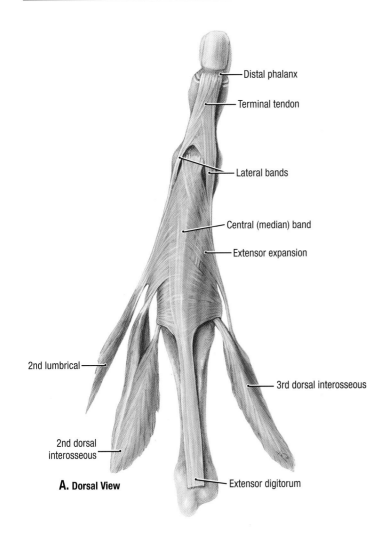

A. Dorsal View

Labels: Distal phalanx, Terminal tendon, Lateral bands, Central (median) band, Extensor expansion, 2nd lumbrical, 3rd dorsal interosseous, 2nd dorsal interosseous, Extensor digitorum

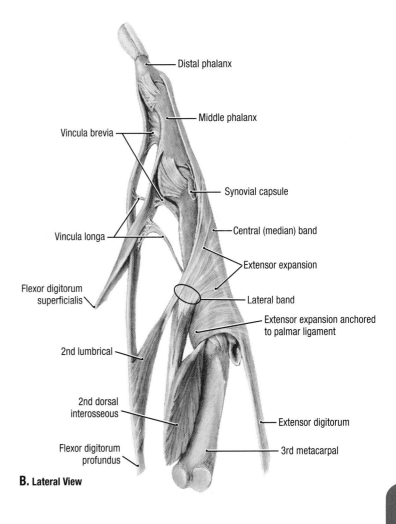

B. Lateral View

Labels: Distal phalanx, Middle phalanx, Vincula brevia, Synovial capsule, Vincula longa, Central (median) band, Extensor expansion, Flexor digitorum superficialis, Lateral band, Extensor expansion anchored to palmar ligament, 2nd lumbrical, 2nd dorsal interosseous, Extensor digitorum, Flexor digitorum profundus, 3rd metacarpal

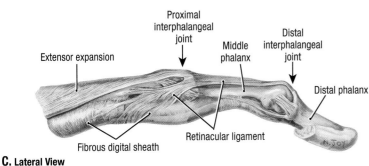

C. Lateral View

Labels: Extensor expansion, Proximal interphalangeal joint, Middle phalanx, Distal interphalangeal joint, Distal phalanx, Fibrous digital sheath, Retinacular ligament

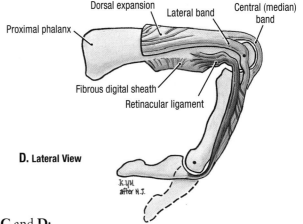

D. Lateral View

Labels: Dorsal expansion, Lateral band, Central (median) band, Proximal phalanx, Fibrous digital sheath, Retinacular ligament

6.80   Extensor (dorsal) expansion of 3rd digit

A. Dorsal aspect. **B.** Lateral aspect. **C.** Retinacular ligaments of extended digit. **D.** Retinacular ligaments of flexed digit.

In **A** and **B:**

- The interossei are partly inserted into the bases of the proximal phalanges and partly into the extensor expansion.
- The 2nd lumbrical inserts into the radial side of the expansion.
- The hood covering the head of the metacarpal is attached to the palmar ligament, preventing "bowstringing" of the extensor tendon and expansion.
- In **B,** note that contraction of the muscles attaching to the lateral band will produce flexion of the metacarpophalangeal joint and extension of the interphalangeal joints.

In **C** and **D:**

- The retinacular ligament is a fibrous band that runs from the proximal phalanx and fibrous digital sheath obliquely across the middle phalanx and two interphalangeal joints to join the dorsal expansion, and then to the distal phalanx.
- On flexion of the distal interphalangeal joint, the retinacular ligament becomes taut and pulls the proximal joint into flexion; on extension of the proximal joint, the distal joint is pulled by the ligament into nearly complete extension.

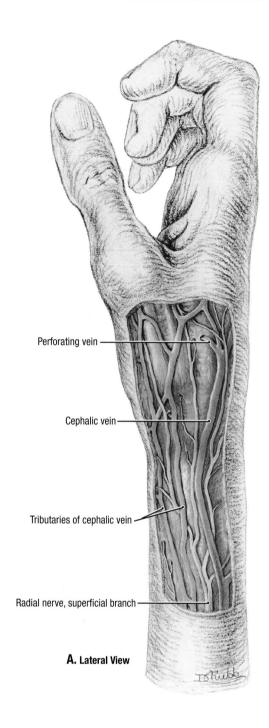

Perforating vein

Cephalic vein

Tributaries of cephalic vein

Radial nerve, superficial branch

A. Lateral View

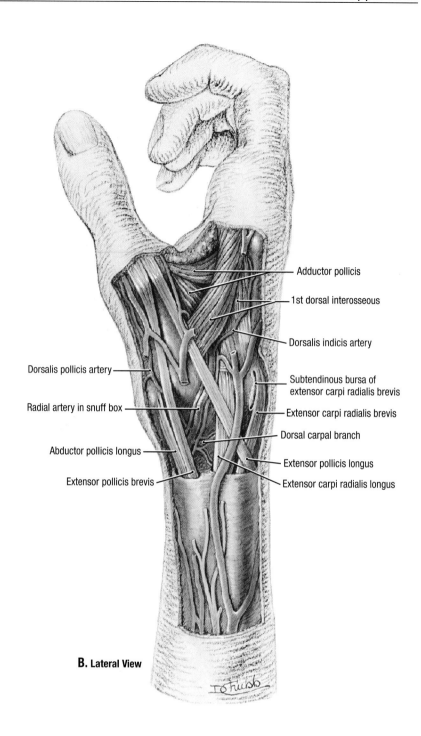

Adductor pollicis

1st dorsal interosseous

Dorsalis indicis artery

Dorsalis pollicis artery

Subtendinous bursa of extensor carpi radialis brevis

Radial artery in snuff box

Extensor carpi radialis brevis

Dorsal carpal branch

Abductor pollicis longus

Extensor pollicis longus

Extensor pollicis brevis

Extensor carpi radialis longus

B. Lateral View

6.81   Lateral aspect of wrist and hand

A. Anatomical snuff box—I. **B.** Anatomical snuff box—II.
In **A**:
- The depression at the base of the thumb, the "anatomical snuff box," retains its name from an archaic habit.
- Note the superficial veins, including the cephalic vein and/or its tributaries, and cutaneous nerves crossing the snuff box.
- Perforating veins and articular nerves pierce the deep fascia.
In **B**:
- Three long tendons of the thumb form the boundaries of the

snuff box; the extensor pollicis longus forms the medial boundary and the abductor pollicis longus and extensor pollicis brevis the lateral boundary.
- The radial artery and its communicating veins cross the floor of the snuff box and travel between the two heads of the 1st dorsal interosseous.
- The adductor pollicis and 1st dorsal interosseous are supplied by the ulnar nerve.

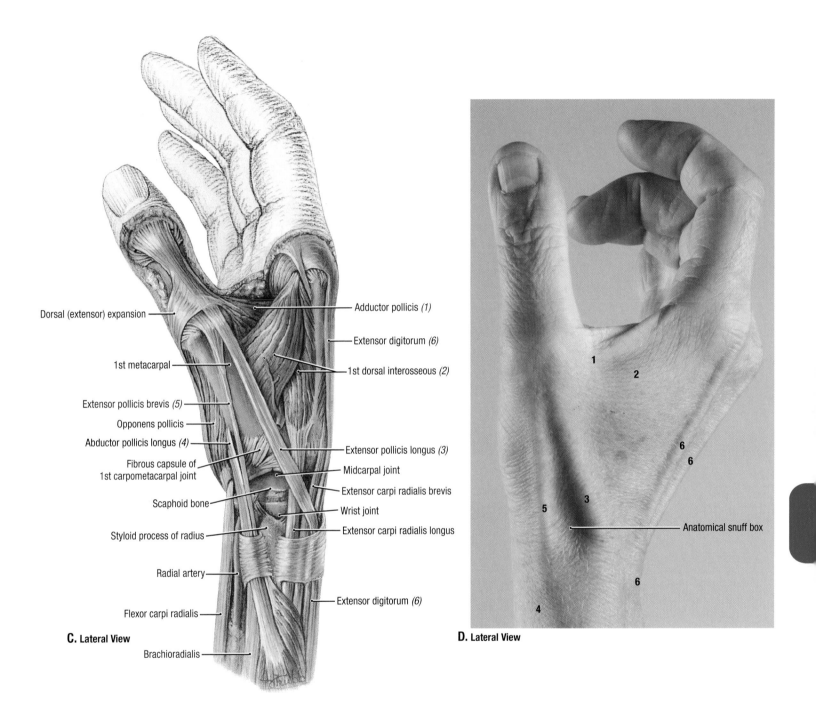

C. Lateral View

Dorsal (extensor) expansion

Adductor pollicis *(1)*

Extensor digitorum *(6)*

1st metacarpal

1st dorsal interosseous *(2)*

Extensor pollicis brevis *(5)*

Opponens pollicis

Abductor pollicis longus *(4)*

Extensor pollicis longus *(3)*

Fibrous capsule of 1st carpometacarpal joint

Midcarpal joint

Scaphoid bone

Extensor carpi radialis brevis

Wrist joint

Styloid process of radius

Extensor carpi radialis longus

Radial artery

Extensor digitorum *(6)*

Flexor carpi radialis

Brachioradialis

D. Lateral View

Anatomical snuff box

6.81   **Lateral aspect of wrist and hand (*continued*)**

C. Anatomical snuff box—III. **D.** Surface anatomy.

In **C:**

- Note the scaphoid bone, the wrist joint proximal to the scaphoid, and the midcarpal joint distal to it.
- The abductor pollicis brevis and adductor pollicis muscles are partly inserted into the dorsal (extensor) expansion.

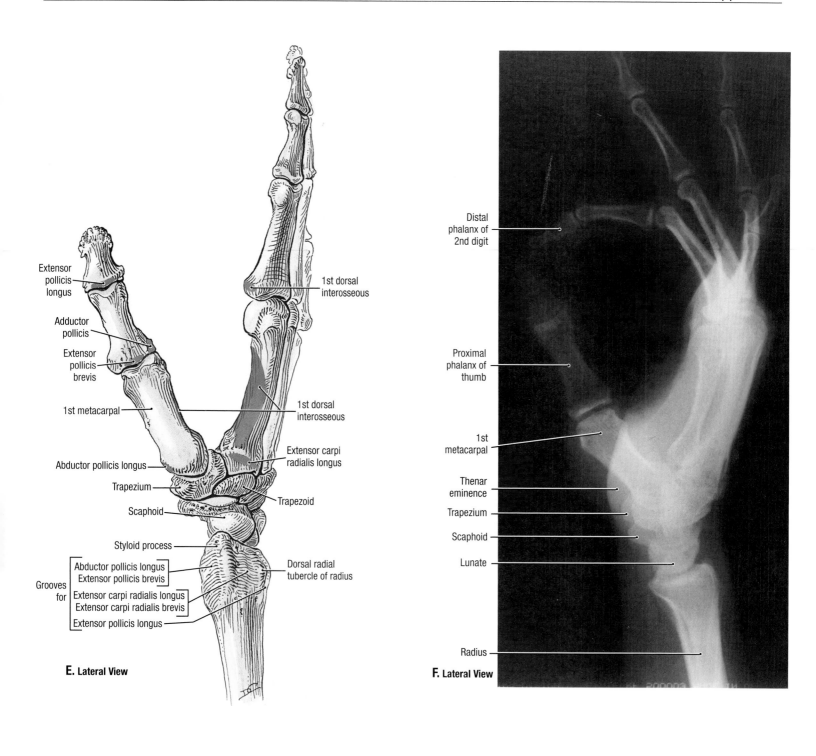

Extensor pollicis longus

Adductor pollicis

Extensor pollicis brevis

1st metacarpal

Abductor pollicis longus

Trapezium

Scaphoid

Styloid process

Grooves for
- Abductor pollicis longus
- Extensor pollicis brevis
- Extensor carpi radialis longus
- Extensor carpi radialis brevis
- Extensor pollicis longus

1st dorsal interosseous

1st dorsal interosseous

Extensor carpi radialis longus

Trapezoid

Dorsal radial tubercle of radius

E. Lateral View

Distal phalanx of 2nd digit

Proximal phalanx of thumb

1st metacarpal

Thenar eminence

Trapezium

Scaphoid

Lunate

Radius

F. Lateral View

6.81   **Lateral aspect of wrist and hand** *(continued)*

E. Bony hand showing muscle attachments. **F.** Radiograph.

In **E:**

- Note the attachments of muscles to bone (red and dark blue areas). The articular surfaces are pale blue.
- The anatomical snuff box is limited proximally by the styloid process of the radius and distally by the base of the 1st metacarpal; aspects of the two lateral bones of the carpus (scaphoid and trapezium) form the floor of the snuff box.

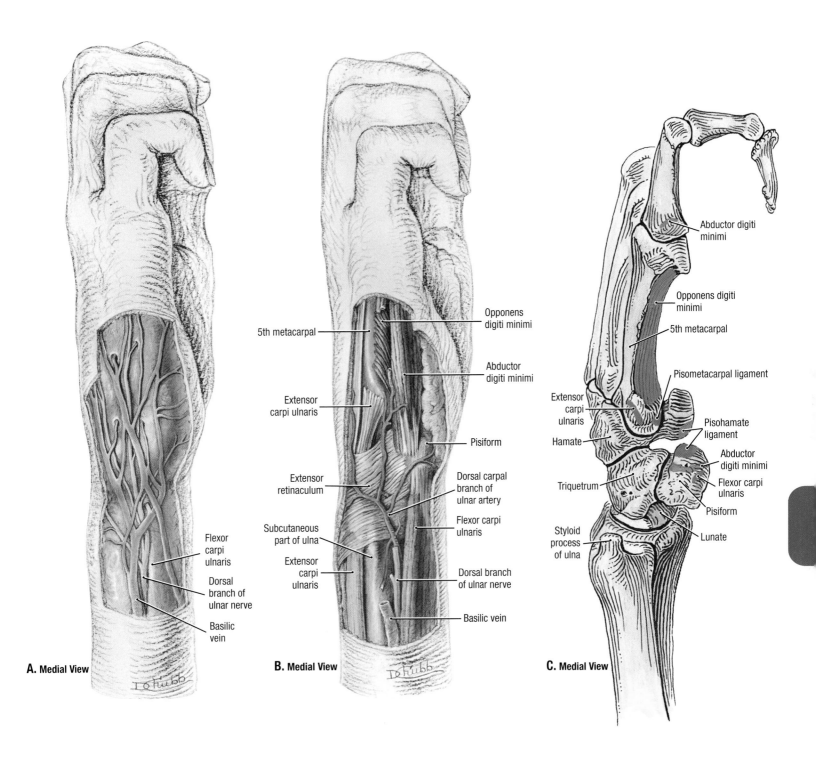

Opponens digiti minimi

5th metacarpal

Abductor digiti minimi

Extensor carpi ulnaris

Pisiform

Extensor retinaculum

Dorsal carpal branch of ulnar artery

Flexor carpi ulnaris

Subcutaneous part of ulna

Extensor carpi ulnaris

Dorsal branch of ulnar nerve

Basilic vein

Flexor carpi ulnaris

Dorsal branch of ulnar nerve

Basilic vein

Abductor digiti minimi

Opponens digiti minimi

5th metacarpal

Pisometacarpal ligament

Extensor carpi ulnaris

Pisohamate ligament

Hamate

Abductor digiti minimi

Flexor carpi ulnaris

Triquetrum

Pisiform

Styloid process of ulna

Lunate

A. Medial View   **B. Medial View**   **C. Medial View**

6.82 **Medial aspect of wrist and hand**

A. Superficial dissection. **B.** Deep dissection. **C.** Bony hand showing muscle attachments. Note that the extensor carpi ulnaris is inserted directly into the base of the fifth metacarpal, but the flexor carpi ulnaris inserts indirectly to the base of the fifth metacarpal and the hook of the hamate through the pisiform and pisohamate and pisometacarpal ligaments.

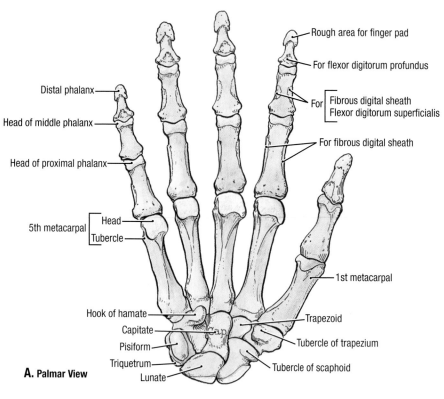

Distal phalanx

Head of middle phalanx

Head of proximal phalanx

5th metacarpal — Head / Tubercle

Hook of hamate
Capitate
Pisiform
Triquetrum
Lunate

A. Palmar View

Rough area for finger pad

For flexor digitorum profundus

For Fibrous digital sheath / Flexor digitorum superficialis

For fibrous digital sheath

1st metacarpal

Trapezoid
Tubercle of trapezium
Tubercle of scaphoid

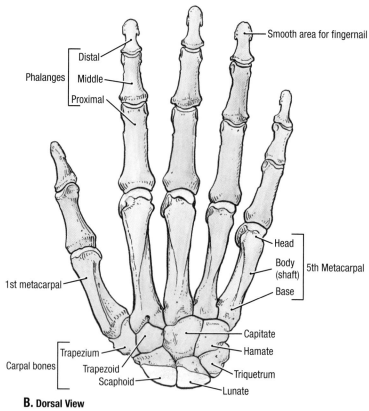

Smooth area for fingernail

Phalanges — Distal / Middle / Proximal

1st metacarpal

Head / Body (shaft) / Base — 5th Metacarpal

Capitate
Hamate
Triquetrum
Lunate

Carpal bones — Trapezium / Trapezoid / Scaphoid

B. Dorsal View

6.83 **Bones of hand**

A. Palmar view. **B.** Dorsal view.
The eight carpal bones form two rows: in the distal row, the hamate, capitate, trapezoid, and trapezium; the trapezium forming a saddle-shaped joint with the 1st metacarpal; in the proximal row, the scaphoid, lunate, and pisiform; the pisiform superimposed on the triquetrum.

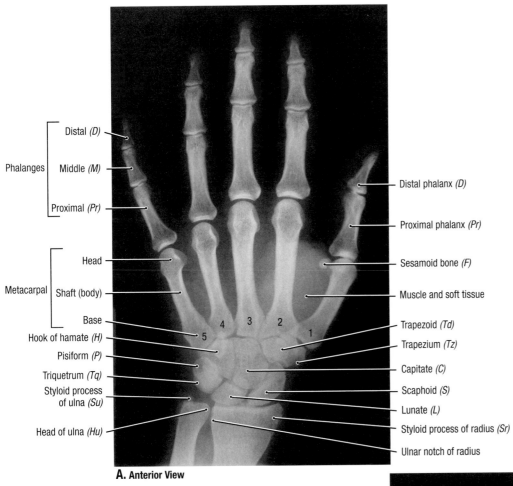

A. Anterior View

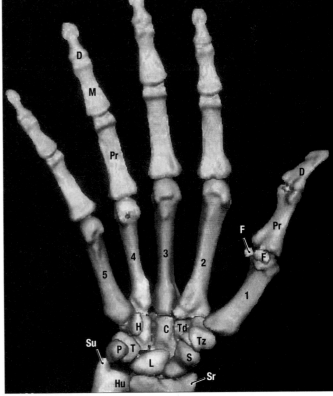

B. Anterior View

6.84   **Imaging of bones of wrist and hand**

A. Radiograph. **B.** Three-dimensional computer-generated image of wrist and hand (*letters* correspond to structures labeled in **A**).

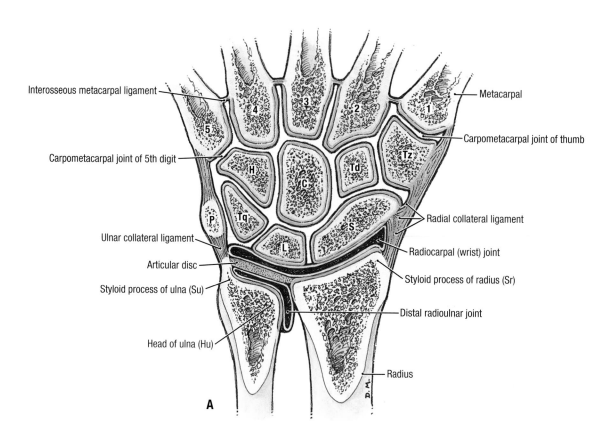

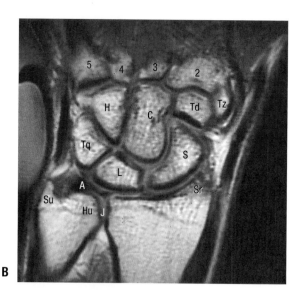

6.85   **Coronal section of wrist**

A. Schematic illustration. **B.** Coronal MRI. *A*, articular disc; *J*, distal radioulnar joint (*letters* correspond to structures labeled in **A** and Figure 6.84A).

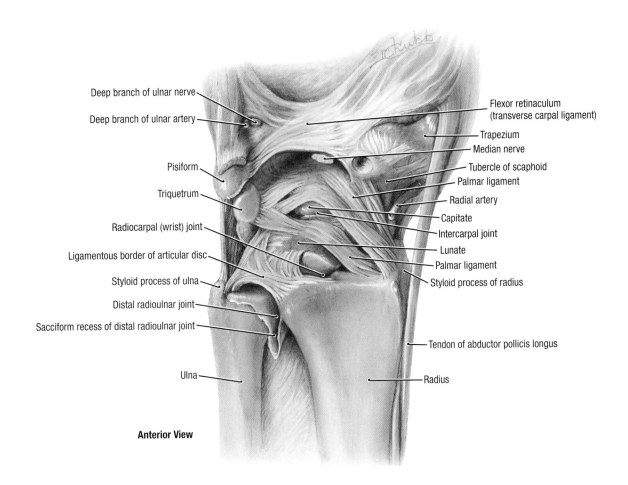

Deep branch of ulnar nerve

Deep branch of ulnar artery

Pisiform

Triquetrum

Radiocarpal (wrist) joint

Ligamentous border of articular disc

Styloid process of ulna

Distal radioulnar joint

Sacciform recess of distal radioulnar joint

Ulna

Flexor retinaculum (transverse carpal ligament)

Trapezium

Median nerve

Tubercle of scaphoid

Palmar ligament

Radial artery

Capitate

Intercarpal joint

Lunate

Palmar ligament

Styloid process of radius

Tendon of abductor pollicis longus

Radius

Anterior View

6.86   **Ligaments of distal radioulnar, radiocarpal, and intercarpal joints**

The hand is forcibly extended. Observe the anterior (palmar) ligaments passing from the radius to the two rows of carpal bones; they are strong and directed, so that the hand moves with the radius during supination.

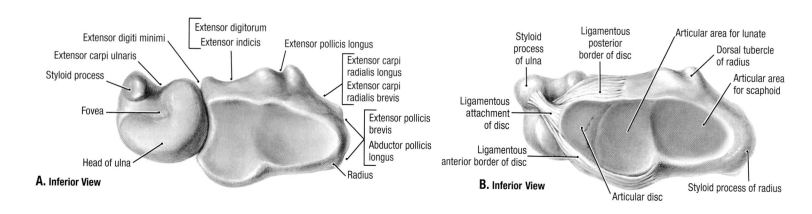

A. Inferior View

Extensor digiti minimi
Extensor carpi ulnaris
Styloid process
Fovea
Head of ulna
Extensor digitorum
Extensor indicis
Extensor pollicis longus
Extensor carpi radialis longus
Extensor carpi radialis brevis
Extensor pollicis brevis
Abductor pollicis longus
Radius

B. Inferior View

Styloid process of ulna
Ligamentous posterior border of disc
Articular area for lunate
Dorsal tubercle of radius
Articular area for scaphoid
Ligamentous attachment of disc
Ligamentous anterior border of disc
Articular disc
Styloid process of radius

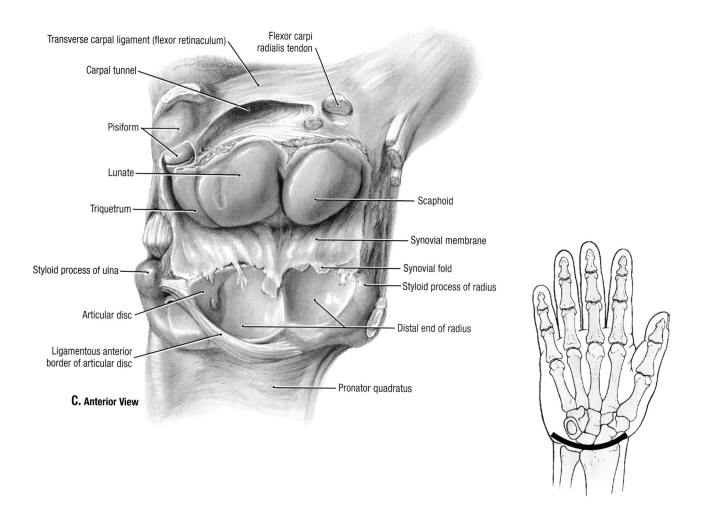

C. Anterior View

Transverse carpal ligament (flexor retinaculum)
Carpal tunnel
Pisiform
Lunate
Triquetrum
Styloid process of ulna
Articular disc
Ligamentous anterior border of articular disc
Flexor carpi radialis tendon
Scaphoid
Synovial membrane
Synovial fold
Styloid process of radius
Distal end of radius
Pronator quadratus

6.87 **Radiocarpal (wrist) joint**

A. Distal ends of radius and ulna showing grooves for tendons on the posterior aspects. **B.** Articular disc. The articular disc unites the distal ends of the radius and ulna; it is fibrocartilaginous at the triangular area between the head of the ulna and the lunate bone, but ligamentous and pliable elsewhere. The cartilaginous part commonly has a fissure or perforation, as shown here. **C.** Articular surface of the radiocarpal joint, which is opened anteriorly. The lunate articulates with the radius and articular disc; only during adduction of the wrist does the triquetrum come into articulation with the disc. The perforation in the disc and the associated roughened surface of the lunate are a common occurrence.

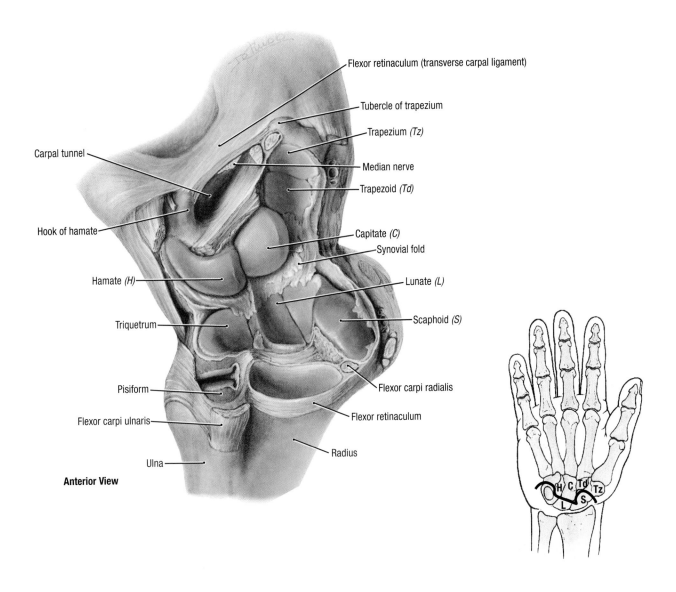

Flexor retinaculum (transverse carpal ligament)
Tubercle of trapezium
Trapezium *(Tz)*
Median nerve
Trapezoid *(Td)*
Capitate *(C)*
Synovial fold
Lunate *(L)*
Scaphoid *(S)*
Flexor carpi radialis
Flexor retinaculum
Radius

Carpal tunnel
Hook of hamate
Hamate *(H)*
Triquetrum
Pisiform
Flexor carpi ulnaris
Ulna

Anterior View

6.88   **Articular surfaces of midcarpal (transverse carpal) joint, opened anteriorly**

- The transverse carpal ligament (flexor retinaculum) is cut; the proximal part of the ligament, which spans from the pisiform to the scaphoid, is relatively weak; the distal part, which passes from the hook of the hamate to the tubercle of the trapezium, is strong.
- Observe the sinuous surfaces of the opposed bones: the trapezium and trapezoid together form a concave, oval surface for the scaphoid, and the capitate and hamate together form a convex surface for the scaphoid, lunate, and triquetrum.

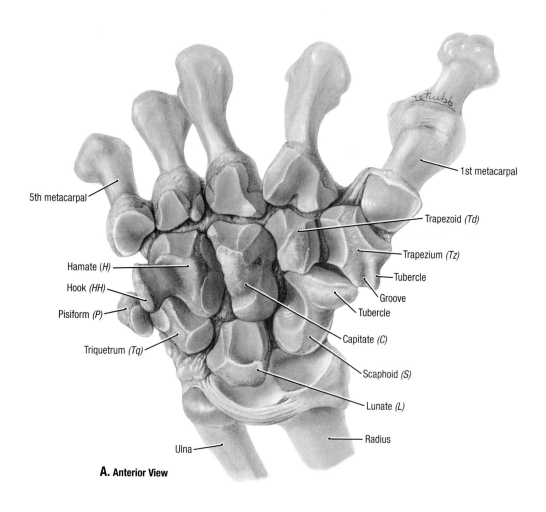

A. Anterior View

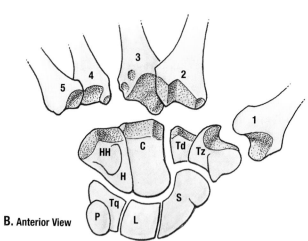

B. Anterior View

6.89 **Carpal bones and bases of metacarpals**

A. Open intercarpal and carpometacarpal joints. The dorsal ligaments remain intact, and all the joints have been hyperextended, permitting study of articular facets. **B.** Diagram of the articular surfaces of the carpometacarpal joints (*letters* refer to structures labeled in **A**).

• The capitate articulates with three metacarpals (2nd, 3rd, and 4th).

• The 2nd metacarpal articulates with three carpals (trapezium, trapezoid, and capitate).

• The 2nd and 3rd carpometacarpal joints are practically immobile; the 1st is saddle shaped, and the 4th and 5th are hinge shaped synovial joints.

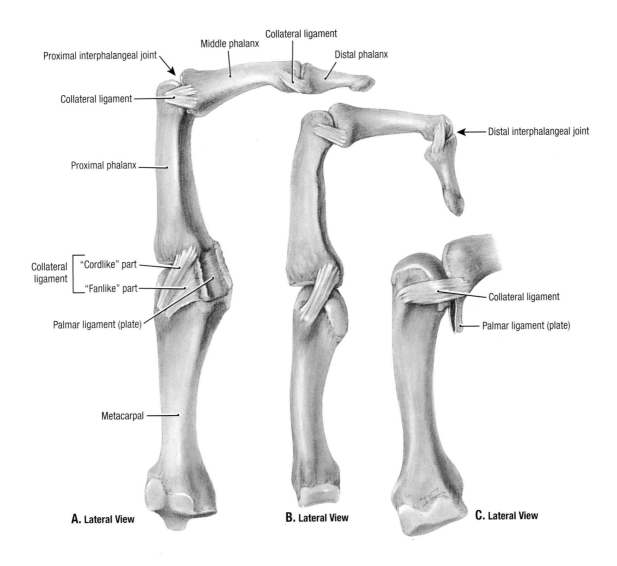

A. Lateral View   **B.** Lateral View   **C.** Lateral View

6.90   Collateral ligaments of the metacarpophalangeal and interphalangeal joints

A. Extended metacarpophalangeal and distal interphalangeal joints. **B.** Flexed interphalangeal joints. **C.** Flexed metacarpophalangeal joint.

- A fibrocartilaginous plate, the palmar ligament, hangs from the base of the proximal phalanx, is fixed to the head of the metacarpal by the weaker, fanlike part of the collateral ligament **(A)**, and moves like a visor across the metacarpal head **(C)**.
- The extremely strong, cordlike parts of the collateral ligaments of this joint **(A and B)** are eccentrically attached to the metacarpal heads; they are slack during extension and taut during flexion **(C)**, so the fingers cannot be spread (abducted) unless the hand is open; the interphalangeal joints have similar ligaments.

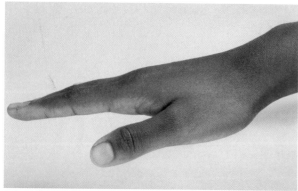

A. Lateral View

B. Lateral View

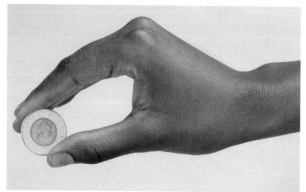

C. Lateral View

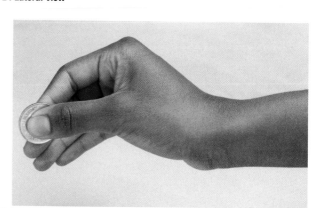

D. Lateral View

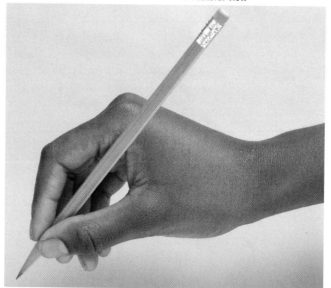

E. Lateral View

6.91 **Grasp and pinch**

A. The extended hand. **B.** Power grip (cylindrical grip) to hold cylinder. When grasping an object, the metacarpophalangeal and interphalangeal joints are flexed, but the radiocarpal and transverse carpal joints are extended. Without wrist extension the grip is weak and insecure. **C.** Tip pinch. **D.** Lateral (key) pinch. **E.** Tripod pinch. The wrist and fingers are held firmly by the long flexor and extensor tendons, while the intrinsic hand muscles perform fine movements of the digits.

F. Anterior View

G. Anterior View

H. Anterior View

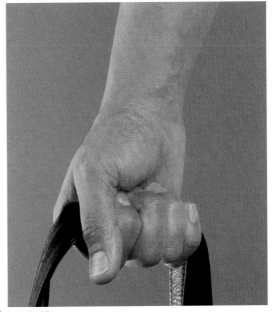

J. Anterior View

I. Lateral View

6.91 Grasp and pinch (continued)

F. Loosely held rod. **G.** Firmly gripped rod (power grip). The second and third carpometacarpal joints are rigid and stable, but the fourth and fifth are hinge joints permitting flexion and extension. **H.** Power grip (centralized grip) to hold screwdriver. **I.** Disc grasp (power grip) to unscrew lid of jar. **J.** Hook grasp. This grasp involves primarily the long flexors of the fingers, which are flexed to a varying degree depending on the size of the object.

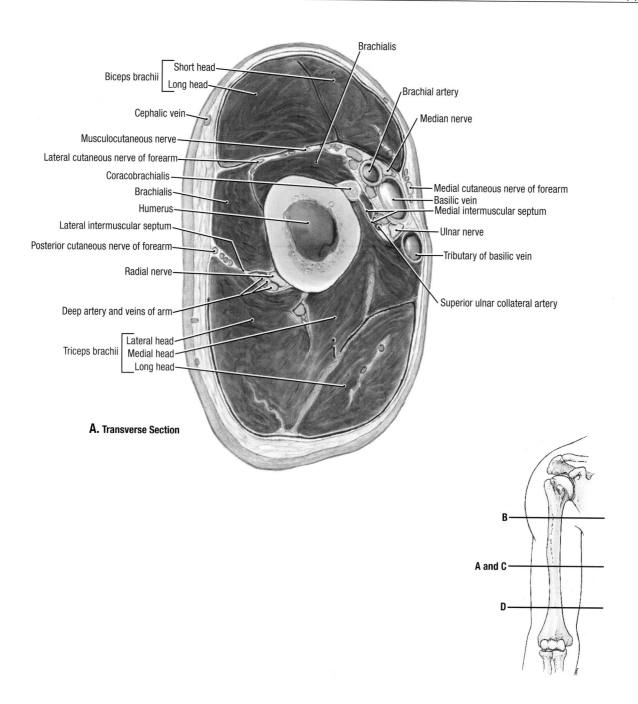

A. Transverse Section

Biceps brachii — Short head, Long head
Cephalic vein
Musculocutaneous nerve
Lateral cutaneous nerve of forearm
Coracobrachialis
Brachialis
Humerus
Lateral intermuscular septum
Posterior cutaneous nerve of forearm
Radial nerve
Deep artery and veins of arm
Triceps brachii — Lateral head, Medial head, Long head

Brachialis
Brachial artery
Median nerve
Medial cutaneous nerve of forearm
Basilic vein
Medial intermuscular septum
Ulnar nerve
Tributary of basilic vein
Superior ulnar collateral artery

B
A and C
D

6.92   **Transverse section and transverse (axial) MRIs of the arm**

A. Transverse section through arm

- The body (shaft) of the humerus is nearly circular, and its cortex is thickest at this level.
- Note the three heads (lateral, medial, and long) of the triceps muscle in the posterior compartment of the arm.
- The radial nerve and deep artery and veins of arm lie in contact with the radial groove of the humerus.
- The musculocutaneous nerve lies in the plane between the biceps and brachialis muscles.
- The median nerve crosses to the medial side of the brachial artery and its venae communicantes, the ulnar nerve passes posteriorly onto the medial side of the triceps muscle, and the basilic vein (appearing here as two vessels) has pierced the deep fascia.

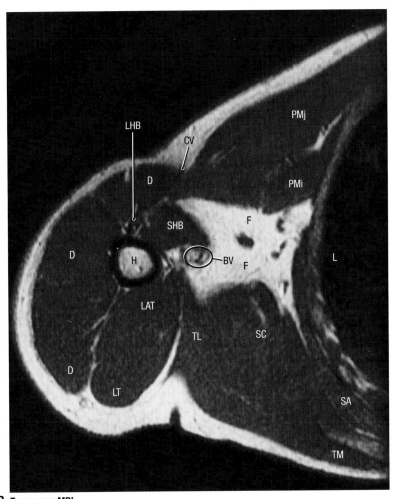

B. Transverse MRI

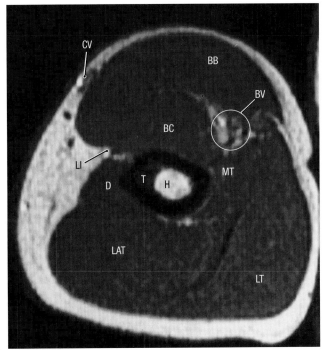

C. Transverse MRI

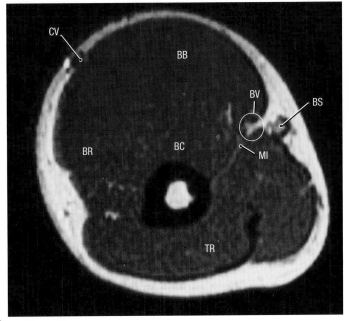

D. Transverse MRI

BB	Biceps brachii	LT	Long head of triceps brachii
BC	Brachialis	MI	Medial intermuscular septum
BR	Brachioradialis	MT	Medial head of triceps brachii
BS	Basilic vein	PMi	Pectoralis minor
BV	Brachial vessels and nerves	PMj	Pectoralis major
CV	Cephalic vein	SA	Serratus anterior
D	Deltoid	SC	Subscapularis
F	Fat in axilla	SHB	Short head of biceps brachii
H	Humerus	T	Deltoid tuberosity
L	Lung	TL	Teres major and latissimus dorsi
LAT	Lateral head of triceps brachii	TM	Teres minor
LHB	Long head of biceps brachii	TR	Triceps brachii
LI	Lateral intermuscular septum		

6.92   **Transverse section and transverse (axial) MRIs of the arm** *(continued)*

B. Transverse MRI through the proximal arm. **C.** Transverse MRI though the middle of the arm. **D.** Transverse MRI through the distal arm.

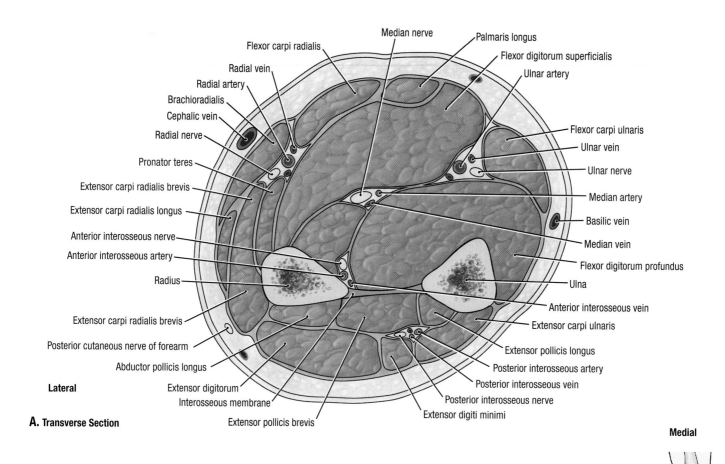

Median nerve

Flexor carpi radialis

Palmaris longus

Radial vein

Flexor digitorum superficialis

Radial artery

Ulnar artery

Brachioradialis

Cephalic vein

Radial nerve

Flexor carpi ulnaris

Pronator teres

Ulnar vein

Extensor carpi radialis brevis

Ulnar nerve

Extensor carpi radialis longus

Median artery

Anterior interosseous nerve

Basilic vein

Anterior interosseous artery

Median vein

Radius

Flexor digitorum profundus

Extensor carpi radialis brevis

Ulna

Posterior cutaneous nerve of forearm

Anterior interosseous vein

Abductor pollicis longus

Extensor carpi ulnaris

Extensor pollicis longus

Lateral

Posterior interosseous artery

Extensor digitorum

Posterior interosseous vein

Interosseous membrane

Posterior interosseous nerve

A. Transverse Section

Extensor pollicis brevis

Extensor digiti minimi

Medial

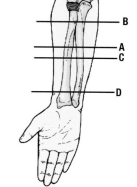

6.93   **Transverse section and transverse (axial) MRIs of forearm**

A. Transverse section. **B.** Transverse MRI through the proximal forearm. **C.** Transverse MRI through the middle forearm. **D.** Transverse MRI through the distal forearm.

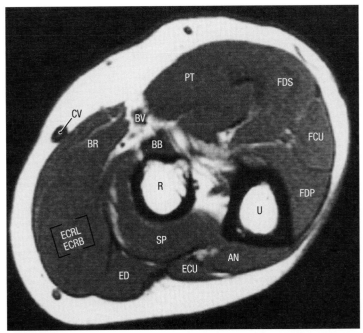

B. Transverse MRI

AN	Anconeus
APL	Abductor pollicis longus
AV	Anterior interosseous vessels and nerve
BB	Biceps brachii
BR	Brachioradialis
BV	Brachial vessels
CV	Cephalic vein
ECRB	Extensor carpi radialis brevis
ECRL	Extensor carpi radialis longus
ECU	Extensor carpi ulnaris
ED	Extensor digitorum
EPB	Extensor pollicis brevis
EPL	Extensor pollicis longus
FCR	Flexor carpi radialis
FCU	Flexor carpi ulnaris
FDP	Flexor digitorum profundus
FDS	Flexor digitorum superficialis
FPL	Flexor pollicis longus
INT	Interosseous membrane
PQ	Pronator quadratus
PT	Pronator teres
R	Radius
RV	Radial vessels
SP	Supinator
U	Ulna
UV	Ulnar vessels and nerve

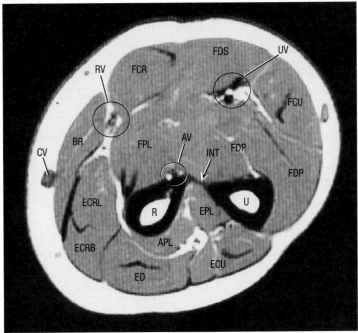

C. Transverse MRI

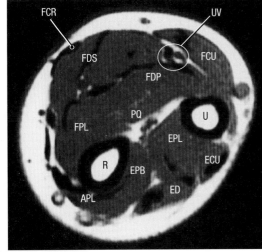

D. Transverse MRI

6.93   Transverse section and transverse (axial) MRIs of forearm *(continued)*

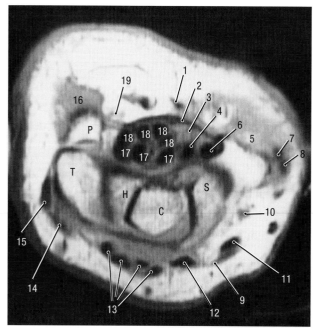

A. Transverse MRI

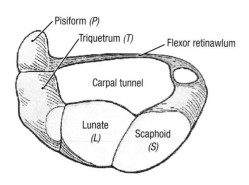

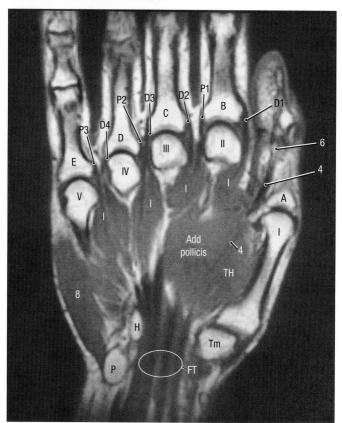

B. Coronal MRI

6.94   Transverse (axial) section and MRIs through the carpal tunnel

A. Transverse MRI through the proximal carpal tunnel (*numbers* and *letters* in MRIs refer to structures in **D**). **B.** Coronal MRI of wrist and hand showing the course of the long flexor tendons in the carpal tunnel (*numbers* and *letters* in MRIs refer to structures in **D**). *FT*, long flexor tendons in carpal tunnel; *TH*, thenar muscles; *P*, pisiform; *H*, hook of hamate; *Tm*, trapezium; *I*, interossei, *A–E*, Proximal phalanges.

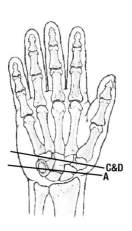

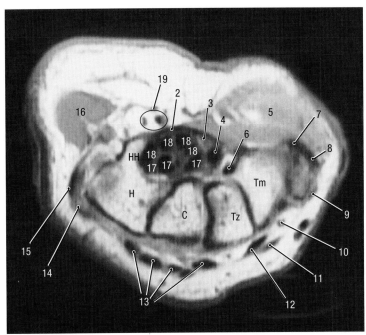

C. Transverse MRI

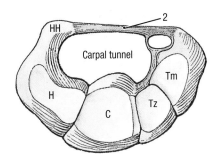

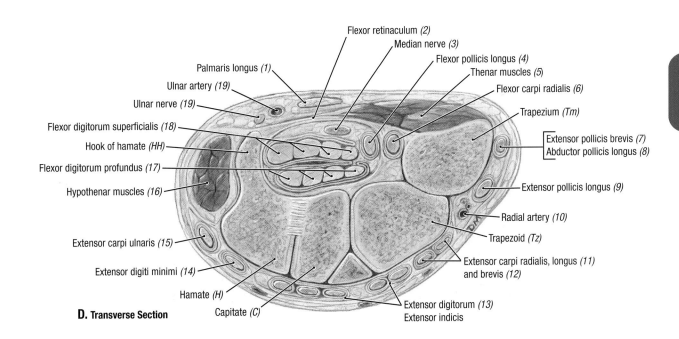

D. Transverse Section

Palmaris longus (1)
Flexor retinaculum (2)
Median nerve (3)
Flexor pollicis longus (4)
Thenar muscles (5)
Flexor carpi radialis (6)
Trapezium (Tm)
Extensor pollicis brevis (7)
Abductor pollicis longus (8)
Extensor pollicis longus (9)
Radial artery (10)
Trapezoid (Tz)
Extensor carpi radialis, longus (11) and brevis (12)
Extensor digitorum (13) Extensor indicis
Ulnar artery (19)
Ulnar nerve (19)
Flexor digitorum superficialis (18)
Hook of hamate (HH)
Flexor digitorum profundus (17)
Hypothenar muscles (16)
Extensor carpi ulnaris (15)
Extensor digiti minimi (14)
Hamate (H)
Capitate (C)

6.94   Transverse (axial) section and MRIs through the carpal tunnel (*continued*)

C. Transverse MRI through the distal carpal tunnel (*numbers* and *letters* in MRIs refer to structures in **D**). **D.** Transverse section of carpal tunnel through the distal row of carpal bones.

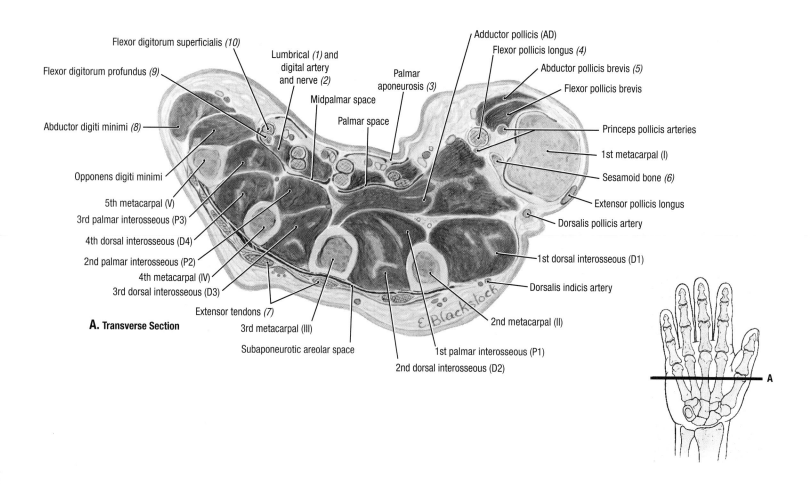

A. Transverse Section

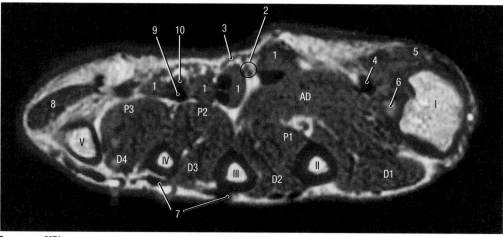

B. Transverse MRI

6.95 **Transverse section and MRI through palm (metacarpals)**

A. Transverse section through the adductor pollicis. *D1–D4*, dorsal interossei; *P1–P3*, palmar interossei. **B.** Axial (transverse) MRI through the adductor pollicis.

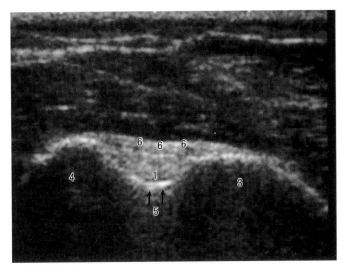

A. Transverse Scan

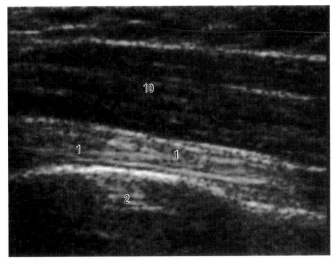

B. Sagittal Scan

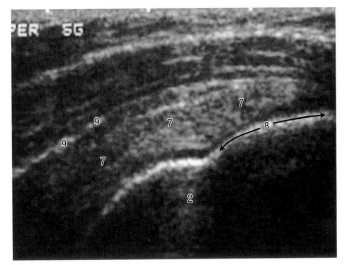

C. Coronal Oblique Scan

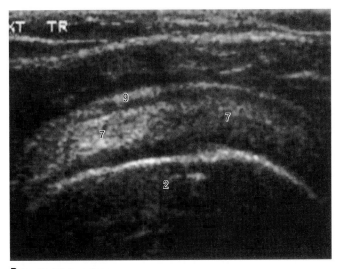

D. Sagittal Oblique Scan

1	Tendon of long head of biceps brachii
2	Humerus
3	Lesser tubercle of humerus
4	Greater tubercle of humerus
5	Bicipital groove of humerus
6	Transverse humeral ligament
7	Supraspinatus
8	Attachment site of supraspinatus to greater tubercle of humerus
9	Subacromial bursal fat
10	Deltoid

6.96 Ultrasound scans of shoulder

A. Transverse (axial) scan. **B.** Sagittal scan. **C.** Coronal oblique scan along the supraspinatus. **D.** Sagittal oblique scan.

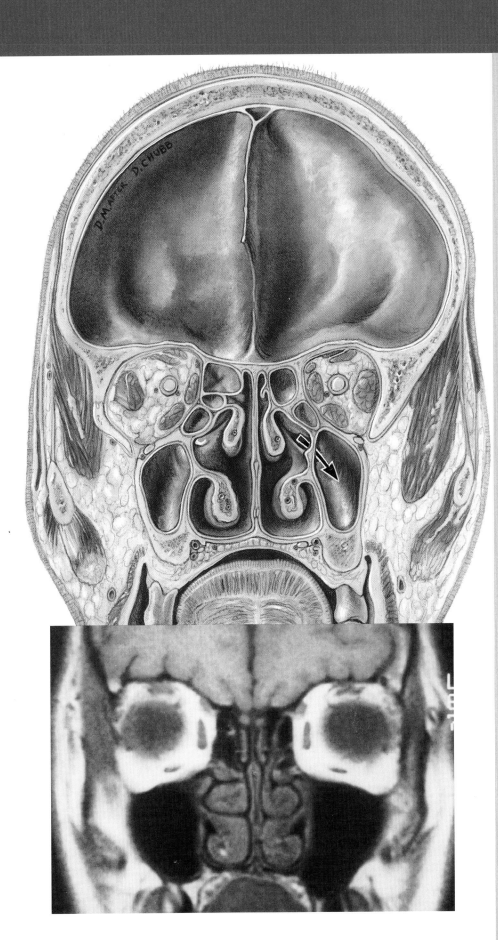

- Cranium **588**
- Face and Scalp **600**
- Circulation and Innervation of Cranial Cavity **606**
- Meninges, Meningeal Spaces **610**
- Cranial Base and Cranial Nerves **614**
- Blood Supply of Brain **622**
- Overview of Brain and Ventricles **626**
- Telencephalon (Cerebrum) and Diencephalon **629**
- Brainstem and Cerebellum **635**
- Anatomical Sections of Brain **638**
- Orbit and Eyeball **640**
- Parotid Region **652**
- Temporal Region and Infratemporal Fossa **654**
- Temporomandibular Joint **662**
- Tongue **666**
- Palate **672**
- Teeth **675**
- Nose, Paranasal Sinuses, and Pterygopalatine Fossa **680**
- Ear **696**
- Lymphatic Drainage of Head **710**
- Imaging of Head: Coronal, Axial, Sagittal Scans **711**

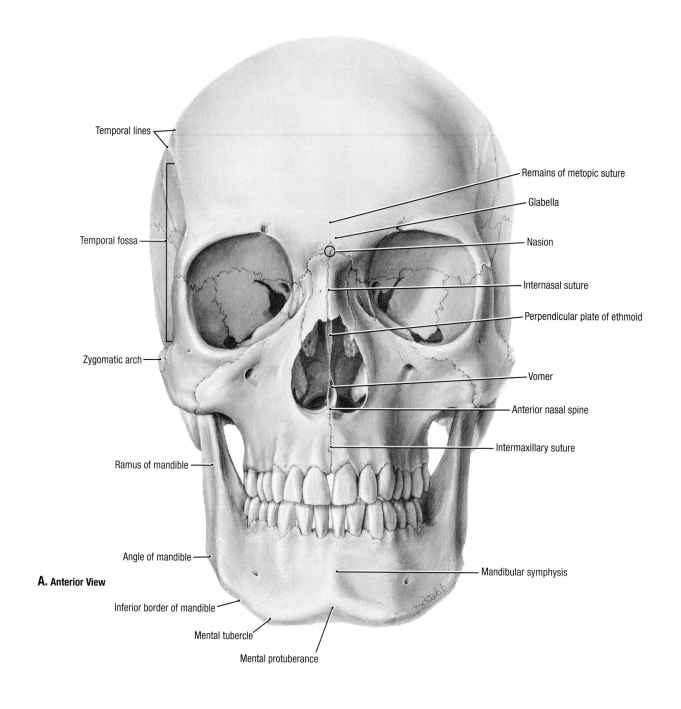

Temporal lines

Temporal fossa

Zygomatic arch

Ramus of mandible

Angle of mandible

A. Anterior View

Inferior border of mandible

Mental tubercle

Mental protuberance

Remains of metopic suture

Glabella

Nasion

Internasal suture

Perpendicular plate of ethmoid

Vomer

Anterior nasal spine

Intermaxillary suture

Mandibular symphysis

7.1   **Cranium, facial (frontal) aspect**

A. Bony cranium. **B.** Diagram of cranium. The bones forming the cranium are color coded.
For the orbital cavity, see also Figure 7.36A.

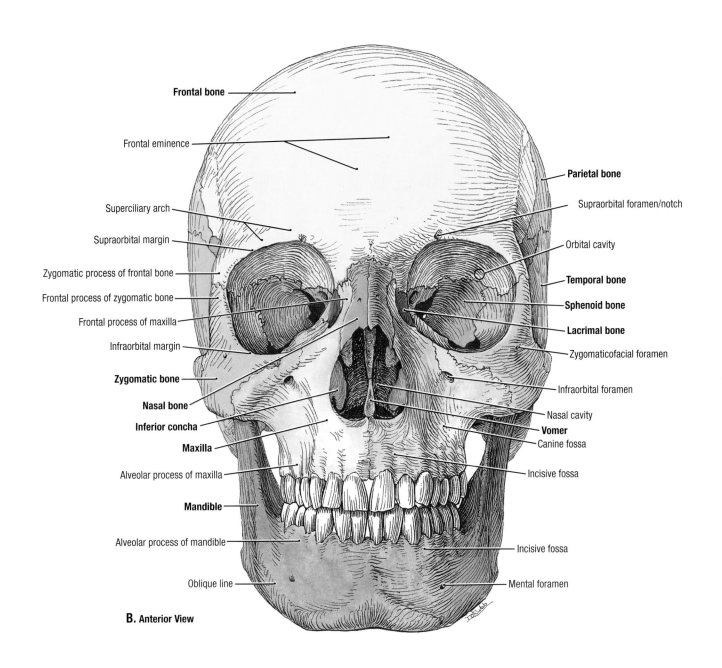

Frontal bone

Frontal eminence

Superciliary arch

Supraorbital margin

Zygomatic process of frontal bone

Frontal process of zygomatic bone

Frontal process of maxilla

Infraorbital margin

Zygomatic bone

Nasal bone

Inferior concha

Maxilla

Alveolar process of maxilla

Mandible

Alveolar process of mandible

Oblique line

Parietal bone

Supraorbital foramen/notch

Orbital cavity

Temporal bone

Sphenoid bone

Lacrimal bone

Zygomaticofacial foramen

Infraorbital foramen

Nasal cavity

Vomer

Canine fossa

Incisive fossa

Incisive fossa

Mental foramen

B. Anterior View

7.1   **Cranium, facial (frontal) aspect** *(continued)*

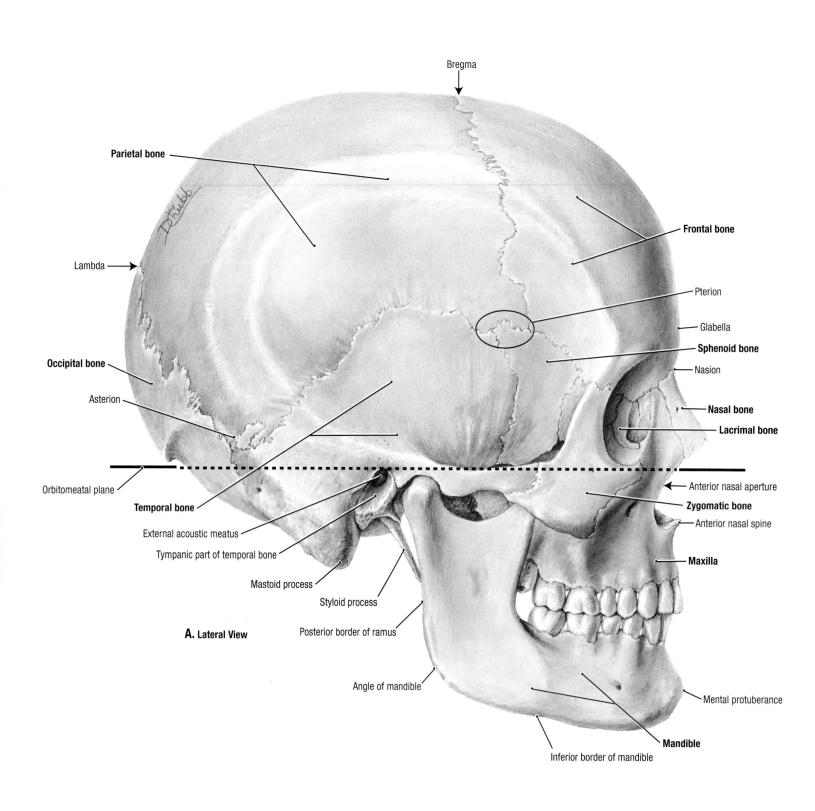

Bregma

Parietal bone

Lambda

Occipital bone

Asterion

Frontal bone

Pterion

Glabella

Sphenoid bone

Nasion

Nasal bone

Lacrimal bone

Orbitomeatal plane

Anterior nasal aperture

Zygomatic bone

Anterior nasal spine

Temporal bone

Maxilla

External acoustic meatus

Tympanic part of temporal bone

Mastoid process

Styloid process

Posterior border of ramus

A. Lateral View

Angle of mandible

Mental protuberance

Mandible

Inferior border of mandible

7.2   **Cranium, lateral aspect**

A. Bony cranium. **B.** Diagram of cranium.

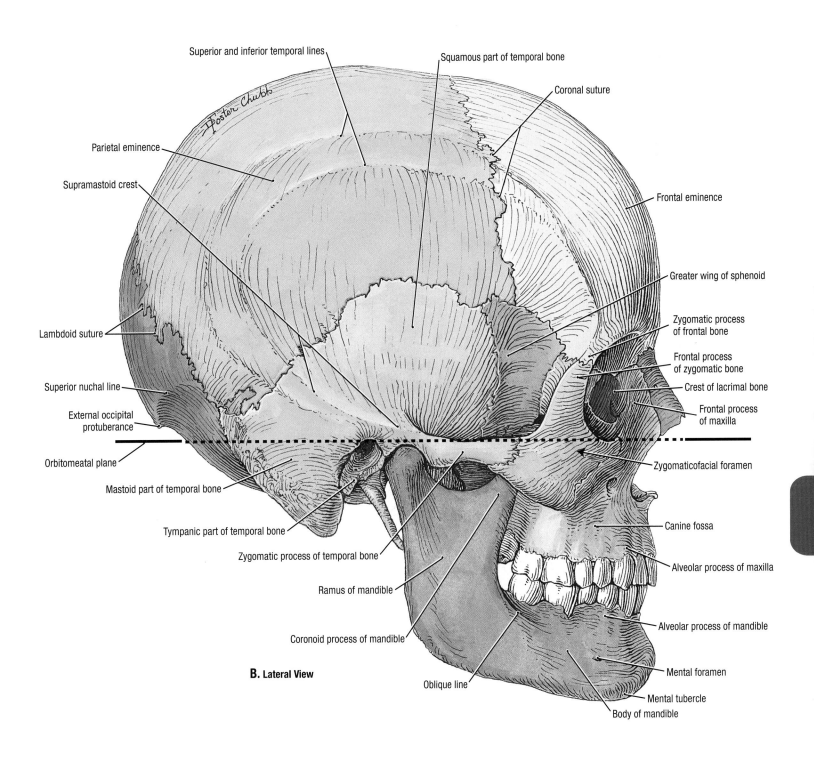

Superior and inferior temporal lines

Squamous part of temporal bone

Coronal suture

Parietal eminence

Supramastoid crest

Frontal eminence

Lambdoid suture

Greater wing of sphenoid

Zygomatic process of frontal bone

Frontal process of zygomatic bone

Crest of lacrimal bone

Superior nuchal line

Frontal process of maxilla

External occipital protuberance

Orbitomeatal plane

Zygomaticofacial foramen

Mastoid part of temporal bone

Tympanic part of temporal bone

Canine fossa

Zygomatic process of temporal bone

Alveolar process of maxilla

Ramus of mandible

Alveolar process of mandible

Coronoid process of mandible

Mental foramen

B. Lateral View

Oblique line

Mental tubercle

Body of mandible

7.2   **Cranium, lateral aspect** *(continued)*

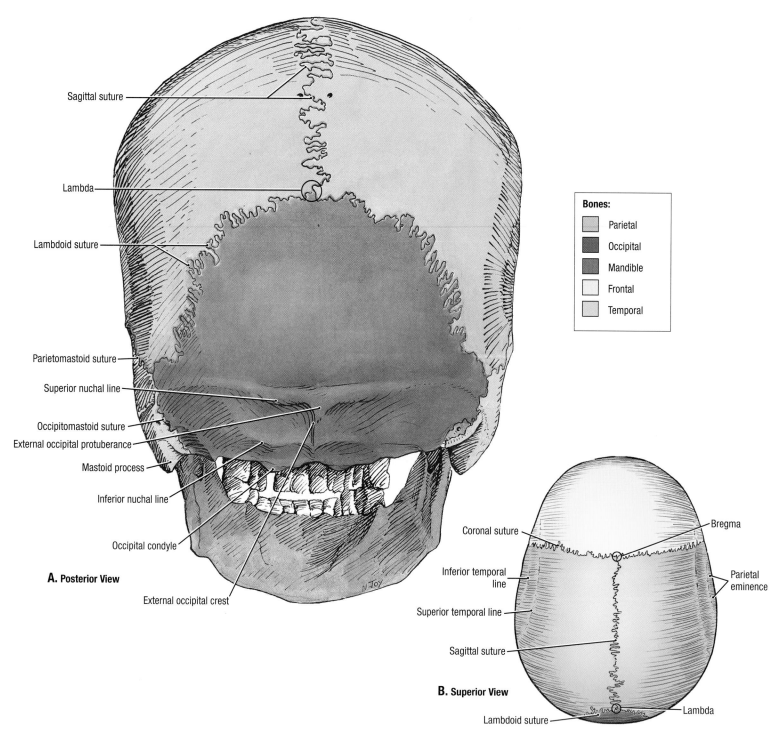

Sagittal suture

Lambda

Lambdoid suture

Parietomastoid suture

Superior nuchal line

Occipitomastoid suture

External occipital protuberance

Mastoid process

Inferior nuchal line

Occipital condyle

A. Posterior View

External occipital crest

Bones:

Parietal

Occipital

Mandible

Frontal

Temporal

Coronal suture

Bregma

Inferior temporal line

Parietal eminence

Superior temporal line

Sagittal suture

Lambda

B. Superior View

Lambdoid suture

7.3 **Neurocranium, occipital aspect, calvaria, and anterior part of posterior cranial fossa**

A. Bones and bony features of the posterior aspect of the neurocranium. **B.** Bones of neurocranium (calvaria or skullcap), superior (vertical) aspect.

Note in **A:**

• This surface of the cranium is convex and includes parts of the parietal, occipital, and temporal (mastoid part) bones; near the center is the lambda, located at the junction of the sagittal and lambdoid sutures.

• The superior nuchal line curves laterally from the external occipital protuberance to the mastoid process.

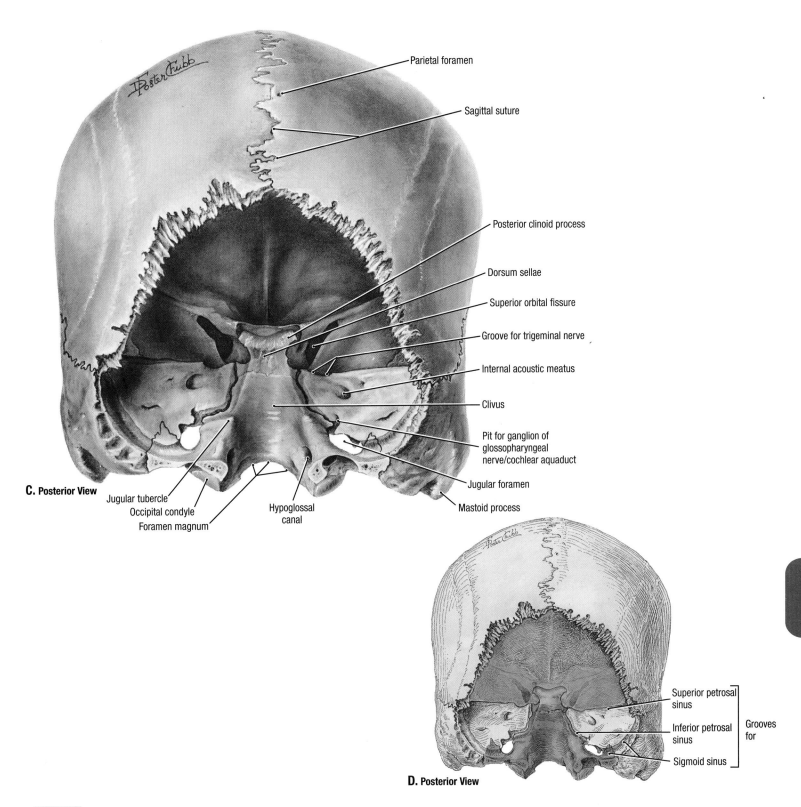

C. Posterior View

- Parietal foramen
- Sagittal suture
- Posterior clinoid process
- Dorsum sellae
- Superior orbital fissure
- Groove for trigeminal nerve
- Internal acoustic meatus
- Clivus
- Pit for ganglion of glossopharyngeal nerve/cochlear aquaduct
- Jugular foramen
- Mastoid process
- Jugular tubercle
- Occipital condyle
- Foramen magnum
- Hypoglossal canal

D. Posterior View

- Superior petrosal sinus
- Inferior petrosal sinus
- Sigmoid sinus
- Grooves for

7.3   **Neurocranium, occipital aspect, calvaria and anterior part of posterior cranial fossa (continued)**

C and **D.** Neurocranium after removal of occipital squama.
C. Bony features of anterior half of posterior cranial fossa. **D.** Bones and grooves for sinuses in posterior cranial fossa.

- The dorsum sellae is a quadrangular plate of bone that projects from the body of the sphenoid; the posterior clinoid processes compose its superolateral corners.

- The clivus is the slope descending from the dorsum sellae to the foramen magnum; it is formed by the body of the sphenoid and the basilar part of the occipital bone.

- The grooves for the sigmoid sinus and inferior petrosal sinus lead inferiorly to the jugular foramen.

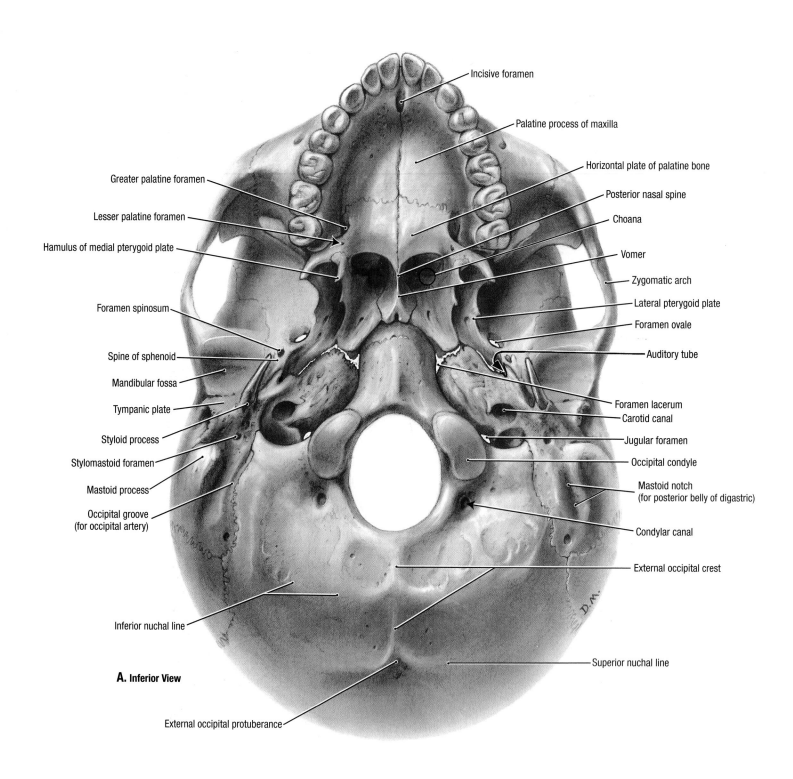

Incisive foramen

Palatine process of maxilla

Greater palatine foramen

Horizontal plate of palatine bone

Lesser palatine foramen

Posterior nasal spine

Hamulus of medial pterygoid plate

Choana

Vomer

Zygomatic arch

Foramen spinosum

Lateral pterygoid plate

Foramen ovale

Spine of sphenoid

Auditory tube

Mandibular fossa

Tympanic plate

Foramen lacerum

Carotid canal

Styloid process

Jugular foramen

Stylomastoid foramen

Occipital condyle

Mastoid process

Mastoid notch
(for posterior belly of digastric)

Occipital groove
(for occipital artery)

Condylar canal

External occipital crest

Inferior nuchal line

Superior nuchal line

A. Inferior View

External occipital protuberance

7.4   **Cranium, inferior aspect**

A. Bony cranium. **B.** Diagram of cranium with bones color coded.

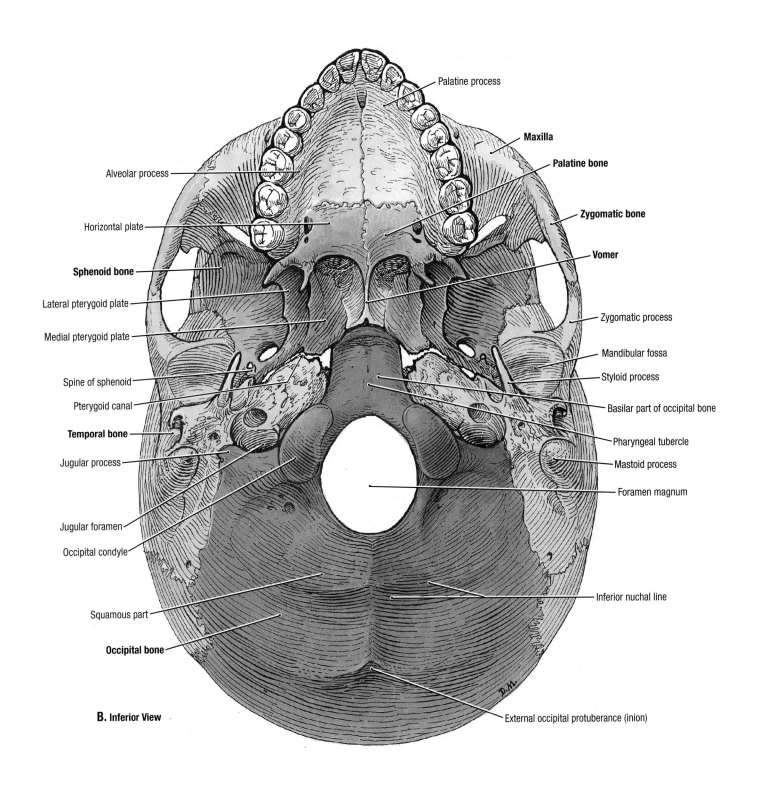

Palatine process

Maxilla

Palatine bone

Zygomatic bone

Vomer

Alveolar process

Horizontal plate

Sphenoid bone

Lateral pterygoid plate

Medial pterygoid plate

Zygomatic process

Mandibular fossa

Styloid process

Basilar part of occipital bone

Spine of sphenoid

Pterygoid canal

Temporal bone

Jugular process

Pharyngeal tubercle

Mastoid process

Foramen magnum

Jugular foramen

Occipital condyle

Squamous part

Inferior nuchal line

Occipital bone

External occipital protuberance (inion)

B. Inferior View

7.4   **Cranium, inferior aspect** *(continued)*

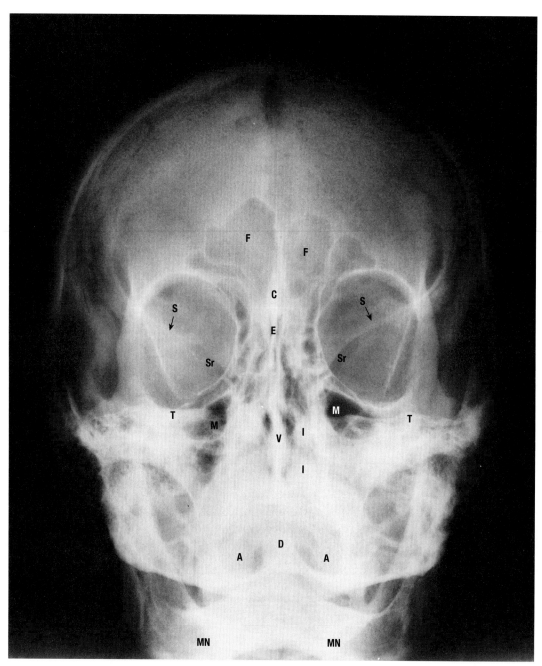

A. Anteroposterior View

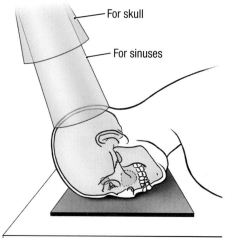

For skull

For sinuses

7.5   Radiographs of the cranium

A. Posteroanterior (Caldwell) radiograph. Observe in **A:**

- The labeled features include the superior orbital fissure *(Sr)*, lesser wing of the sphenoid *(S)*, superior surface of the petrous part of the temporal bone *(T)*, crista galli *(C)*, frontal sinus *(F)*, mandible *(MN)*, and maxillary sinus *(M)*.
- The nasal septum is formed by the perpendicular plate of the ethmoid *(E)* and the vomer *(V)*; note the inferior and middle conchae *(I)* of the lateral wall of the nose.
- Superimposed on the facial skeleton is the dens *(D)* and lateral masses of the atlas *(A)*.

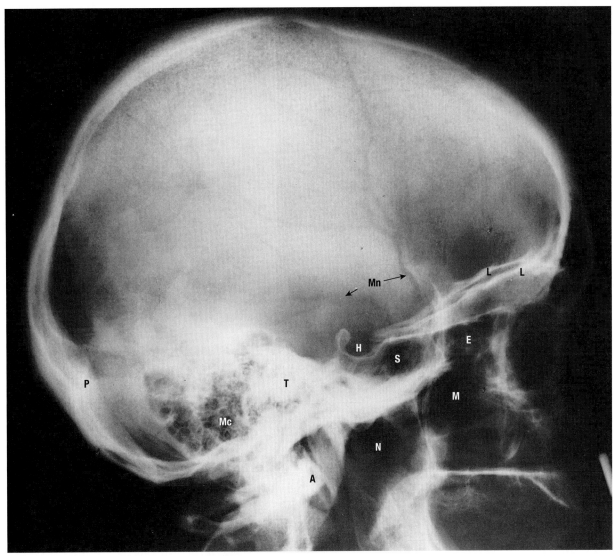

B. Lateral View

7.5   **Radiographs of the cranium** *(continued)*

B. Lateral radiograph of the cranium.
- The labeled features include the paranasal sinuses: ethmoidal *(E)*, sphenoidal *(S)*, and maxillary *(M)*, and the hypophyseal fossa *(H)* for the pituitary gland, the petrous part of the temporal bone *(T)*, mastoid cells *(Mc)*, grooves for the branches of the middle meningeal vessels *(Mn)*, arch of the atlas *(A)*, internal occipital protuberance *(P)* and the nasopharynx *(N)*.
- The right and left orbital plates of the frontal bone are not superimposed; thus, the floor of the anterior cranial fossa appears as two lines *(L)*.

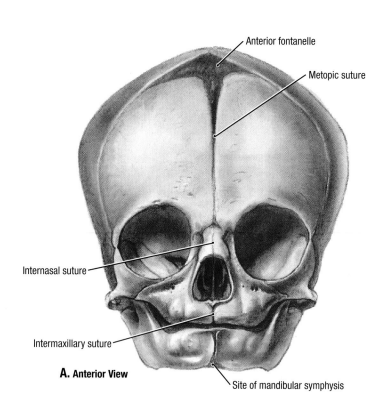

Anterior fontanelle

Metopic suture

Internasal suture

Intermaxillary suture

A. Anterior View

Site of mandibular symphysis

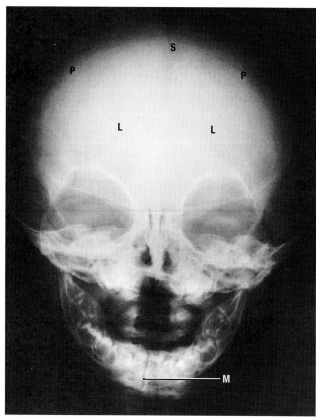

B. Anteroposterior View

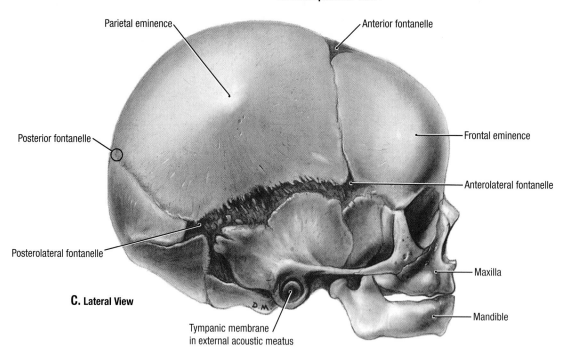

Parietal eminence

Anterior fontanelle

Posterior fontanelle

Frontal eminence

Anterolateral fontanelle

Posterolateral fontanelle

Maxilla

C. Lateral View

Mandible

Tympanic membrane
in external acoustic meatus

7.6   Child's cranium

A. Cranium at birth, anterior aspect. **B.** Radiograph of 6½-month-old child. **C.** Cranium at birth, lateral aspect.

- The maxilla and mandible are small; most of the ramus of the mandible lies at the level of the body. The mandibular symphysis (symphysis menti), which closes during the second year,

and the metopic suture, which closes during the sixth year, are still open (unfused).

- The orbital cavities are large, but the face is small; the facial skeleton forming only one eighth of the whole cranium, while in the adult, it forms one third.

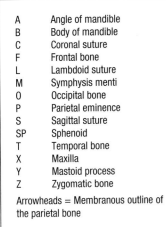

A	Angle of mandible
B	Body of mandible
C	Coronal suture
F	Frontal bone
L	Lambdoid suture
M	Symphysis menti
O	Occipital bone
P	Parietal eminence
S	Sagittal suture
SP	Sphenoid
T	Temporal bone
X	Maxilla
Y	Mastoid process
Z	Zygomatic bone

Arrowheads = Membranous outline of the parietal bone

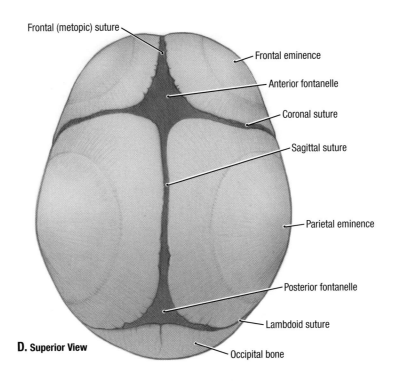

D. Superior View

Frontal (metopic) suture
Frontal eminence
Anterior fontanelle
Coronal suture
Sagittal suture
Parietal eminence
Posterior fontanelle
Lambdoid suture
Occipital bone

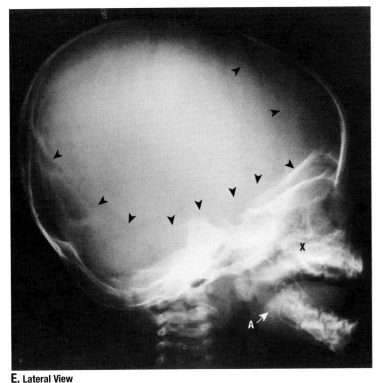

E. Lateral View

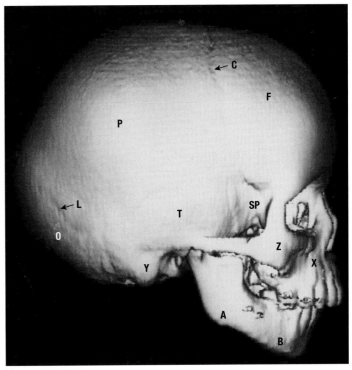

F. Lateral View

7.6   Child's cranium *(continued)*

D. Cranium at birth, superior aspect. **E.** Radiograph of 6½-month-old child. **F.** Three-dimensional computer-generated images of 3-year-old child's cranium.

- The eminence of the parietal bone is a rounded cone. Ossification, which starts at the eminences, has not yet reached the four angles of the parietal bone; accordingly, these regions are membranous, and the membrane is blended with the pericranium externally and the dura mater internally to form the fontanelles. The fontanelles are usually closed by the second year; there is no mastoid process until the second year.

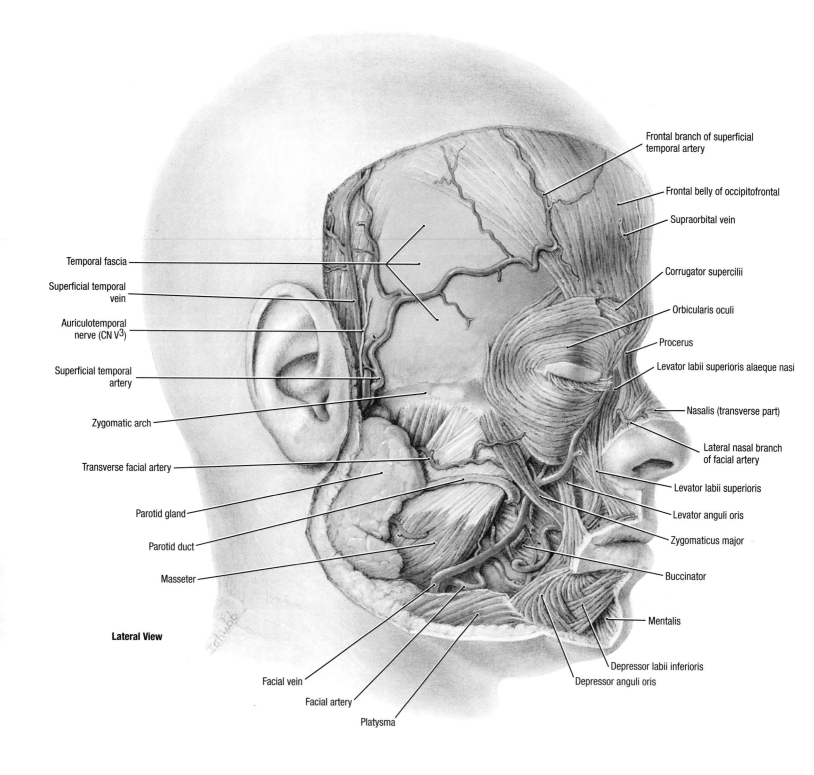

Frontal branch of superficial temporal artery

Frontal belly of occipitofrontal

Supraorbital vein

Corrugator supercilii

Orbicularis oculi

Procerus

Levator labii superioris alaeque nasi

Nasalis (transverse part)

Lateral nasal branch of facial artery

Levator labii superioris

Levator anguli oris

Zygomaticus major

Buccinator

Mentalis

Depressor labii inferioris

Depressor anguli oris

Temporal fascia

Superficial temporal vein

Auriculotemporal nerve (CN V^3)

Superficial temporal artery

Zygomatic arch

Transverse facial artery

Parotid gland

Parotid duct

Masseter

Lateral View

Facial vein

Facial artery

Platysma

7.7 Muscles of facial expression and arteries of the face

- The muscles of facial expression are the superficial sphincters and dilators of the openings of the head; all are supplied by the facial nerve (CN VII). The masseter and temporalis (the latter covered here by temporal fascia) are muscles of mastication that are innervated by the trigeminal nerve.
- A pulse can be taken along the course of the facial artery, where it winds around the inferior border of the mandible; it can also be felt at the superficial temporal artery, where the vessel crosses the zygomatic arch anterior to the ear.

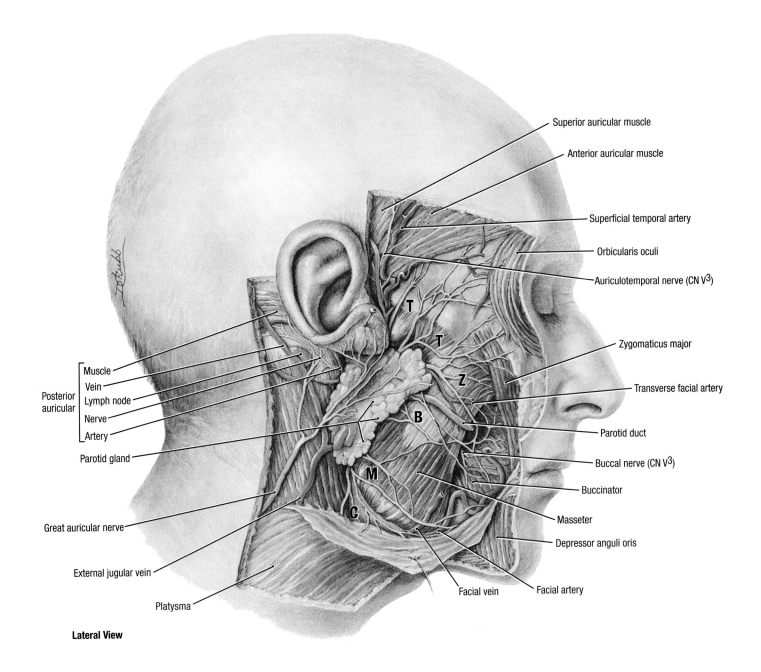

Superior auricular muscle

Anterior auricular muscle

Superficial temporal artery

Orbicularis oculi

Auriculotemporal nerve (CN V³)

Zygomaticus major

Transverse facial artery

Parotid duct

Buccal nerve (CN V³)

Buccinator

Masseter

Depressor anguli oris

Facial vein   Facial artery

Platysma

External jugular vein

Great auricular nerve

Parotid gland

Posterior auricular
- Muscle
- Vein
- Lymph node
- Nerve
- Artery

T

T

Z

B

M

C

Lateral View

7.8 **Relationships of the branches of the facial nerve and vessels to the parotid gland and duct**

- The parotid duct extends across the masseter muscle just inferior to the zygomatic arch; the duct turns medially to pierce the buccinator.
- The facial nerve (CN VII) innervates the muscles of facial expression; it forms a plexus within the parotid gland, the branches of which radiate over the face, anastomosing with each other and the branches of the trigeminal nerve. After emerging from the stylomastoid foramen, the main stem of the facial nerve has posterior auricular, digastric, and stylohyoid branches; the parotid plexus gives rise to temporal *(T)*, zygomatic *(Z)*, buccal *(B)*, marginal mandibular *(M)*, cervical *(C)*, and posterior auricular branches. During parotidectomy (surgical excision of the gland), identification, dissection, and preservation of the branches of facial nerve are critical.

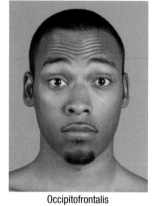

Occipitofrontalis

Corrugator supercilii

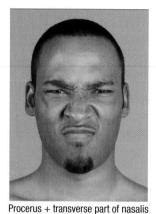

Procerus + transverse part of nasalis

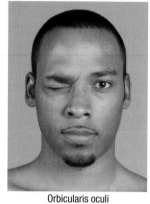

Orbicularis oculi

Lev. labii sup. alaeque nasi + alar part of nasalis

Buccinator + orbicularis oris

Zygomaticus major + minor

Risorius

Risorius + depressor labii inferioris

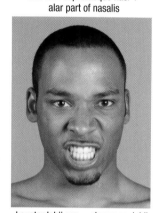

Levator labii sup. + depressor labii

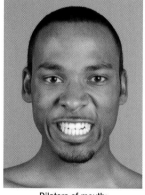

Dilators of mouth:
Risorius plus levator labii superioris + depressor labii inferioris

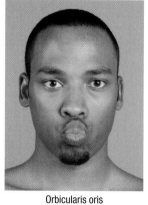

Orbicularis oris

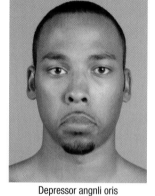

Depressor angnli oris

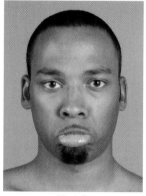

Mentalis

Platysma

D. Anterior Views

7.9   Muscles of facial expression

A. Orbicularis oculi. Note the palpebral *(P)* and orbital *(O)* parts of the orbicularis oculi muscle. The lacrimal portion (not shown) passes posterior to the lacrimal sac and helps facilitate the spread lacrimal secretions. **B.** Gentle closure of eyelid—palpebral part. **C.** Tight closure of eyelid—orbital part. **D.** Actions of selected muscles of facial expression.

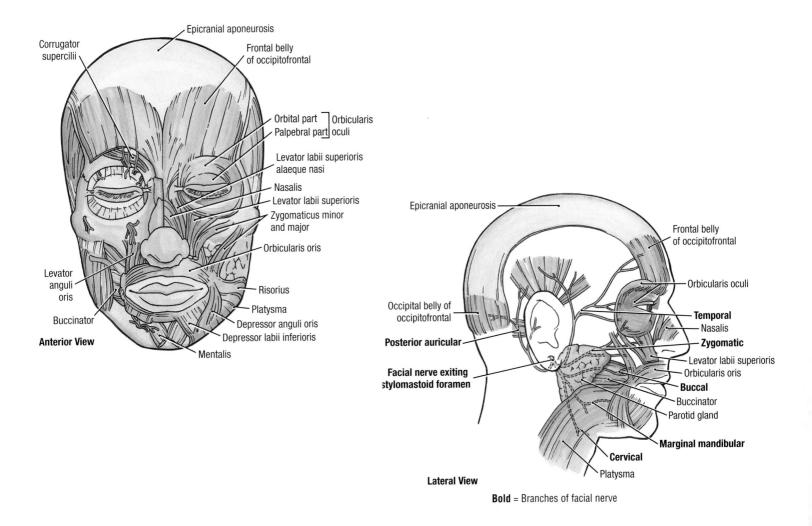

Anterior View

Lateral View

Bold = Branches of facial nerve

TABLE 7.1 MAIN MUSCLES OF FACIAL EXPRESSION

MUSCLE[a]	ORIGIN	INSERTION	ACTION(S)
Frontal belly of occipitofrontal	Epicranial aponeurosis	Skin of forehead	Elevates eyebrows and forehead
Orbicularis oculi	Medial orbital margin, medial palpebral ligament, and lacrimal bone	Skin around margin of orbit; tarsal plate	Closes eyelids
Nasalis	Superior part of canine ridge of maxilla	Nasal cartilages	Flares nostrils
Orbicularis oris	Some fibers arise near median plane of maxilla superiorly and mandible inferiorly; other fibers arise from deep surface of skin	Mucous membrane of lips	Compresses and protrudes lips (e.g., purses them during whistling, sucking, and kissing)
Levator labii superioris	Frontal process of maxilla and infraorbital region	Skin of upper lip and alar cartilage of nose	Elevates lip, dilates nostril, and raises angle of mouth
Platysma	Superficial fascia of deltoid and pectoral regions	Mandible, skin of cheek, angle of mouth, and orbicularis oris	Depresses mandible and tenses skin of lower face and neck
Mentalis	Incisive fossa of mandible	Skin of chin	Protrudes lower lip
Buccinator	Mandible, pterygomandibular raphe, and alveolar processes of maxilla and mandible	Angle of mouth	Presses cheek against molar teeth to keep food between teeth; expels air from oral cavity as occurs when playing a wind instrument

[a] All of these muscles are supplied by the facial nerve (CN VII).

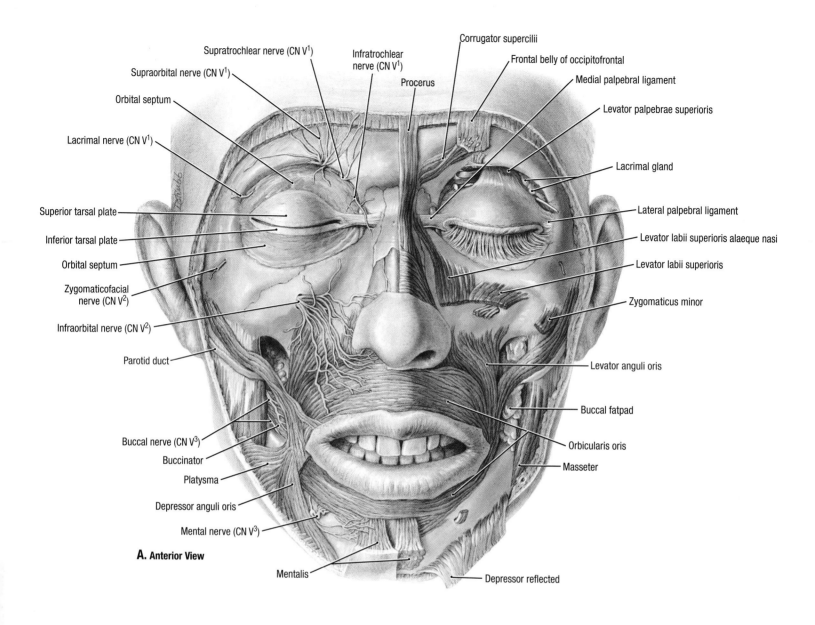

Supratrochlear nerve (CN V¹)

Supraorbital nerve (CN V¹)

Orbital septum

Lacrimal nerve (CN V¹)

Infratrochlear nerve (CN V¹)

Procerus

Corrugator supercilii

Frontal belly of occipitofrontal

Medial palpebral ligament

Levator palpebrae superioris

Lacrimal gland

Superior tarsal plate

Inferior tarsal plate

Orbital septum

Zygomaticofacial nerve (CN V²)

Infraorbital nerve (CN V²)

Parotid duct

Lateral palpebral ligament

Levator labii superioris alaeque nasi

Levator labii superioris

Zygomaticus minor

Levator anguli oris

Buccal fatpad

Buccal nerve (CN V³)

Buccinator

Platysma

Depressor anguli oris

Mental nerve (CN V³)

Orbicularis oris

Masseter

A. Anterior View

Mentalis

Depressor reflected

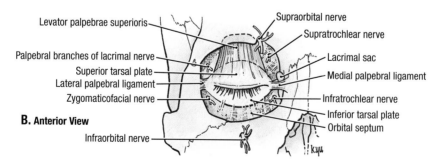

Levator palpebrae superioris

Palpebral branches of lacrimal nerve

Superior tarsal plate

Lateral palpebral ligament

Zygomaticofacial nerve

B. Anterior View

Infraorbital nerve

Supraorbital nerve

Supratrochlear nerve

Lacrimal sac

Medial palpebral ligament

Infratrochlear nerve

Inferior tarsal plate

Orbital septum

7.10 **Cutaneous branches of trigeminal nerve, muscles of facial expression, and eyelid**

A. Dissection of face. **B.** Orbital septum and eyelid.

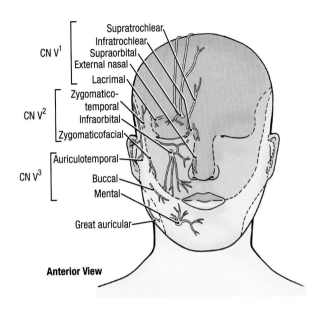

Anterior View

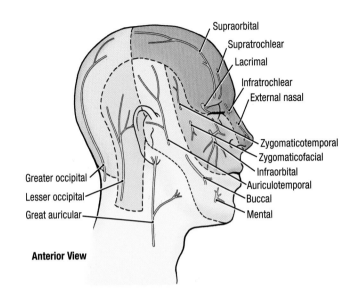

Anterior View

TABLE 7.2   NERVES OF FACE AND SCALP

NERVE	ORIGIN	COURSE	DISTRIBUTION
Frontal	Ophthalmic nerve (CN V^1)	Crosses orbit on superior aspect of levator palpebrae superioris; divides into supraorbital and supratrochlear branches	Skin of forehead, scalp, upper eyelid, and nose; conjunctiva of upper lid and mucosa of frontal sinus
Supraorbital	Continuation of frontal nerve (CN V^1)	Emerges through supraorbital notch, or foramen, and breaks up into small branches	Mucous membrane of frontal sinus and conjunctiva (lining) of upper eyelid; skin of forehead as far as vertex
Supratrochlear	Frontal nerve (CN V^1)	Passes superiorly on medial of supraorbital nerve and divides into two or more branches	Skin in middle of forehead to hairline
Infratrochlear	Nasociliary nerve (CN V^1)	Follows medial wall of orbit to upper eyelid	Skin and conjunctiva (lining) of upper eyelid
Lacrimal	Ophthalmic nerve (CN V^1)	Passes through palpebral fascia of upper eyelid near lateral angle (canthus) of eye	Lacrimal gland and small area of skin and conjunctiva of lateral part of upper eyelid
External nasal	Anterior ethmoidal nerve (CN V^1)	Runs in nasal cavity and emerges on face between nasal bone and lateral nasal cartilage	Skin on dorsum of nose, including tip of nose
Zygomatic	Maxillary nerve (CN V^2)	Arises in floor of orbit, divides into zygomatico-facial and zygomaticotemporal nerves, which traverse foramina of same name	Skin over zygomatic arch and anterior temporal region; carries postsynaptic parasympathetic fibers from pterygopalatine ganglion to lacrimal nerve
Infraorbital	Terminal branch of maxillary nerve (CN V^2)	Runs in floor of orbit and emerges at infraorbital foramen	Skin of cheek, lower lid, lateral side of nose and inferior septum and upper lip, upper premolar incisors and canine teeth; mucosa of maxillary sinus and upper lip
Auriculotemporal	Mandibular nerve (CN V^3)	From posterior division of CN V^3, it passes between neck of mandible and external acoustic meatus to accompany superficial temporal artery	Skin anterior to ear and posterior temporal region, tragus and part of helix of auricle, and roof of exterior acoustic meatus and upper tympanic membrane
Buccal	Mandibular nerve (CN V^3)	From the anterior division of CN V^3 in infratemporal fossa, it passes anteriorly to reach cheek	Skin and mucosa of cheek, buccal gingiva adjacent to 2nd and 3rd molar teeth
Mental	Terminal branch of inferior alveolar nerve (CN V^3)	Emerges from mandibular canal at mental foramen	Skin of chin and lower lip and mucosa of lower lip

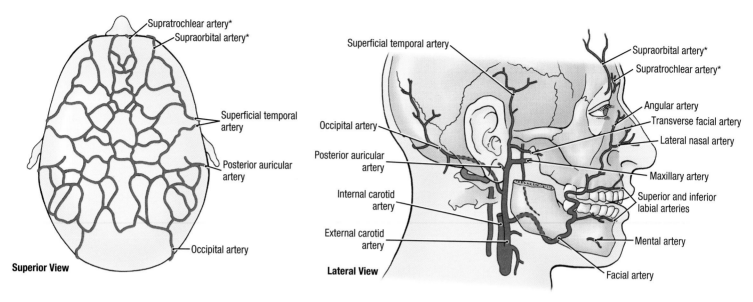

Superior View

Lateral View

*Source= internal carotid artery (ophthalmic artery); all other labeled
arteries are from external carotid

TABLE 7.3 ARTERIES OF FACE AND SCALP

ARTERY	ORIGIN	COURSE	DISTRIBUTION
Facial	External carotid artery	Ascends deep to submandibular gland, winds around inferior border of mandible and enters face	Muscles of facial expression and face
Inferior labial	Facial artery near angle of mouth	Runs medially in lower lip	Lower lip and chin
Superior labial		Runs medially in upper lip	Upper lip and ala (side) and septum of nose
Lateral nasal	Facial artery as it ascends alongside nose	Passes to ala of nose	Skin on ala and dorsum of nose
Angular	Terminal branch of facial artery	Passes to medial angle (canthus) of eye	Superior part of cheek and lower eyelid
Occipital	External carotid artery	Passes medial to posterior belly of digastric and mastoid process; accompanies occipital nerve in occipital region	Scalp of back of head, as far as vertex
Posterior auricular		Passes posteriorly, deep to parotid, along styloid process between mastoid and ear	Scalp posterior to auricle and auricle
Superficial temporal	Smaller terminal branch of external carotid artery	Ascends anterior to ear to temporal region and ends in scalp	Facial muscles and skin of frontal and temporal regions
Transverse facial	Superficial temporal artery within parotid gland	Crosses face superficial to masseter and inferior to zygomatic arch	Parotid gland and duct, muscles and skin of face
Mental	Terminal branch of inferior alveolar artery	Emerges from mental foramen and passes to chin	Facial muscles and skin of chin
*Supraorbital	Terminal branch of ophthalmic artery, a branch of internal carotid artery	Passes superiorly from supraorbital foramen	Muscles and skin of forehead and scalp
*Supratrochlear		Passes superiorly from supratrochlear notch	Muscles and skin of scalp

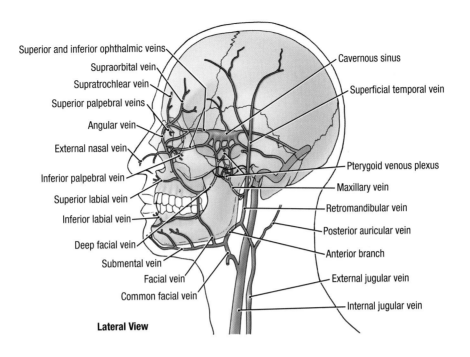

Superior and inferior ophthalmic veins
Supraorbital vein
Supratrochlear vein
Superior palpebral veins
Angular vein
External nasal vein
Inferior palpebral vein
Superior labial vein
Inferior labial vein
Deep facial vein
Submental vein
Facial vein
Common facial vein

Cavernous sinus
Superficial temporal vein
Pterygoid venous plexus
Maxillary vein
Retromandibular vein
Posterior auricular vein
Anterior branch
External jugular vein
Internal jugular vein

Lateral View

TABLE 7.4 VEINS OF FACE

VEIN	ORIGIN	COURSE	TERMINATION	AREA DRAINED
Supratrochlear	Begins from a venous plexus on the forehead and scalp, through which it communicates with the frontal branch of the superficial temporal vein, its contralateral partner, and the supraorbital vein	Descends near the midline of the forehead to the root of the nose where it joins the supraorbital vein	Angular vein at the root of the nose	Anterior part of scalp and forehead
Supraorbital	Begins in the forehead by anastomosing with a frontal tributary of the superficial temporal vein	Passes medially superior to the orbit and joins the supratrochlear vein; a branch passes through the supraorbital notch and joins with the superior ophthalmic vein		
Angular	Begins at root of nose by union of supratrochlear and supraorbital veins	Descends obliquely along the root and side of the nose to the inferior margin of the orbit	Becomes the facial vein at the inferior margin of the orbit	In addition to above, drains upper and lower lids and conjunctiva; may receive drainage from cavernous sinus
Facial	Continuation of angular vein past inferior margin of orbit	Descends along lateral border of the nose, receiving external nasal and inferior palpebral veins, then obliquely across face to mandible; receives anterior division of retromandibular vein, after which it is sometimes called the common facial vein	Internal jugular vein opposite or inferior to the level of the hyoid bone	Anterior scalp and forehead, eyelids, external nose, and anterior cheek, lips, chin, and submandibular gland
Deep facial	Pterygoid venous plexus	Runs anteriorly on maxilla above buccinator and deep to masseter, emerging medial to anterior border of masseter onto face	Enters posterior aspect of facial vein	Infratemporal fossa (most areas supplied by maxillary artery)
Superficial temporal	Begins from a widespread plexus of veins on the side of the scalp and along the zygomatic arch	Its frontal and parietal tributaries unite anterior to the auricle; it crosses the temporal root of the zygomatic arch to pass from the temporal region and enters the substance of the parotid gland	Joins the maxillary vein posterior to the neck of the mandible to form the retromandibular vein	Side of the scalp, superficial aspect of the temporal muscle, and external ear
Retromandibular	Formed anterior to the ear by the union of the superficial temporal and maxillary veins	Runs posterior and deep to the ramus of the mandible through the substance of the parotid gland; communicates at its inferior end with the facial vein	Unites with the posterior auricular vein to form the external jugular vein	Parotid gland and masseter muscle

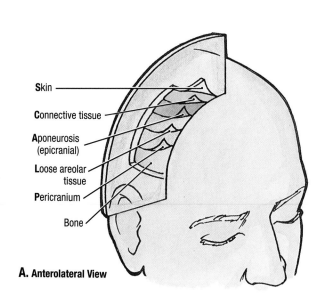

A. Anterolateral View

Skin

Connective tissue

Aponeurosis (epicranial)

Loose areolar tissue

Pericranium

Bone

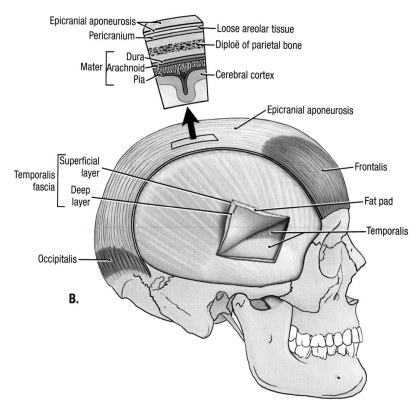

Epicranial aponeurosis

Pericranium

Loose areolar tissue

Diploë of parietal bone

Dura

Mater Arachnoid

Pia

Cerebral cortex

Epicranial aponeurosis

Temporalis fascia — Superficial layer / Deep layer

Frontalis

Fat pad

Temporalis

Occipitalis

B.

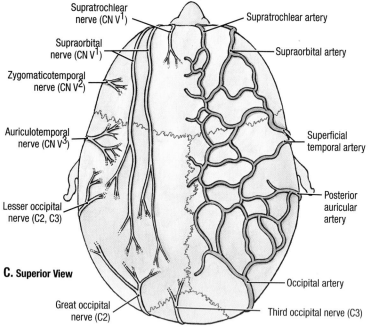

Supratrochlear nerve (CN V¹)

Supraorbital nerve (CN V¹)

Zygomaticotemporal nerve (CN V²)

Auriculotemporal nerve (CN V³)

Lesser occipital nerve (C2, C3)

Great occipital nerve (C2)

Supratrochlear artery

Supraorbital artery

Superficial temporal artery

Posterior auricular artery

Occipital artery

Third occipital nerve (C3)

C. Superior View

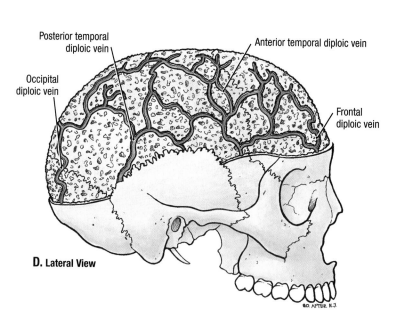

Posterior temporal diploic vein

Occipital diploic vein

Anterior temporal diploic vein

Frontal diploic vein

D. Lateral View

7.11   **Branches of facial nerve, muscles of facial expression, and scalp**

A. Layers of scalp. **B.** Occipitofrontalis and temporal muscles and fascia. **C.** Sensory nerves and arteries of the scalp. **D.** Diploic veins. The outer layer of the compact bone of the cranium has been filed away, exposing the channels for the diploic veins in the cancellous bone that composes the diploë.

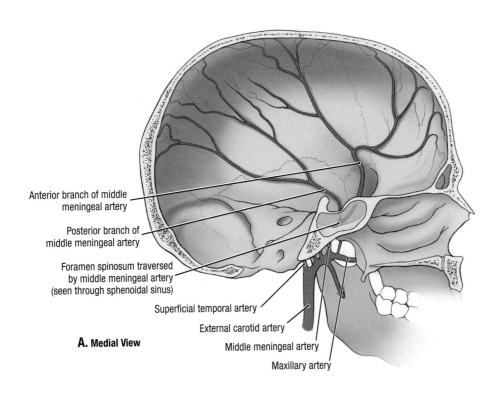

Anterior branch of middle
meningeal artery

Posterior branch of
middle meningeal artery

Foramen spinosum traversed
by middle meningeal artery
(seen through sphenoidal sinus)

Superficial temporal artery

External carotid artery

Middle meningeal artery

Maxillary artery

A. Medial View

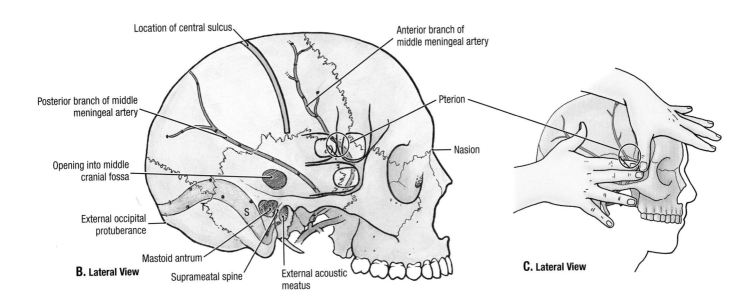

Location of central sulcus

Anterior branch of
middle meningeal artery

Posterior branch of middle
meningeal artery

Pterion

Opening into middle
cranial fossa

Nasion

External occipital
protuberance

Mastoid antrum

Suprameatal spine

External acoustic
meatus

B. Lateral View

C. Lateral View

7.12   Middle meningeal artery and pterion

A. Course of the middle meningeal artery in the cranium. **B.** Surface projections of internal features of the neurocranium. **C.** Locating the pterion. The pterion is located two fingers breadth superior to the zygomatic arch and one thumb breadth posterior to the frontal process of the zygomatic bone (approximately 4 cm superior to the midpoint of the zygomatic arch); the anterior branch of the middle meningeal artery crosses the pterion, a site at which the artery is frequently torn resulting in an extradural (epidural) hematoma.

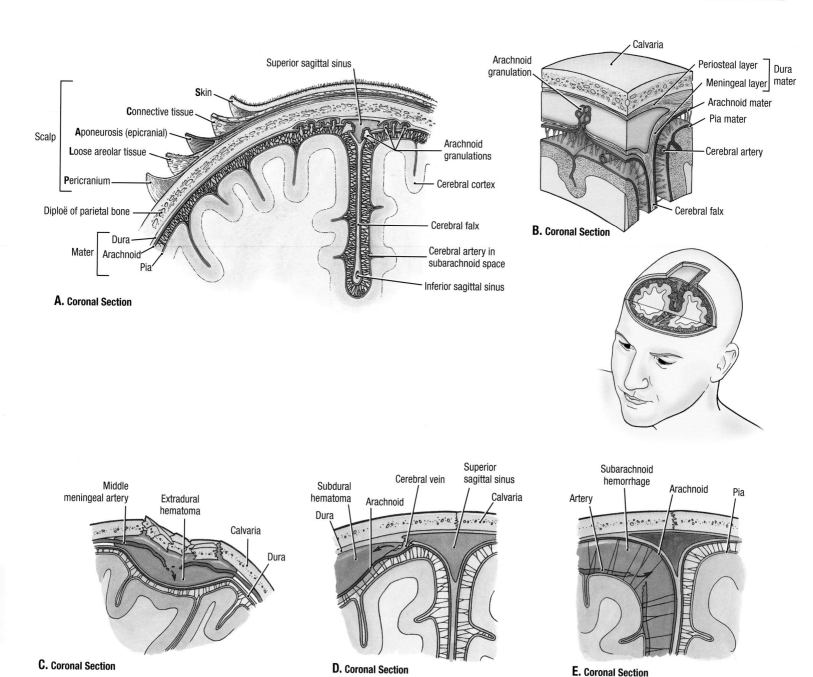

A. Coronal Section

B. Coronal Section

C. Coronal Section

D. Coronal Section

E. Coronal Section

7.13 **Layers of the scalp and meninges**

A. Scalp, cranium, and meninges. **B.** Meninges and their relationship to the calvaria. The three meningeal spaces include the extradural (epidural) space between the cranial bones and dura, which is a potential space normally (it becomes a real space pathologically if blood accumulates in it); the similarly potential subdural space between the dura and arachnoid; and the subarachnoid space, the normal realized space between the arachnoid and pia, which contains cerebrospinal fluid (CSF). **C.** Extradural (epidural) hematomas result from bleeding from a torn middle meningeal artery. **D.** Subdural hematomas commonly result from tearing of a cerebral vein as it enters the superior sagittal sinus. **E.** Subarachnoid hemorrhage results from bleeding within the subarachnoid space, e.g., from rupture of an aneurysm.

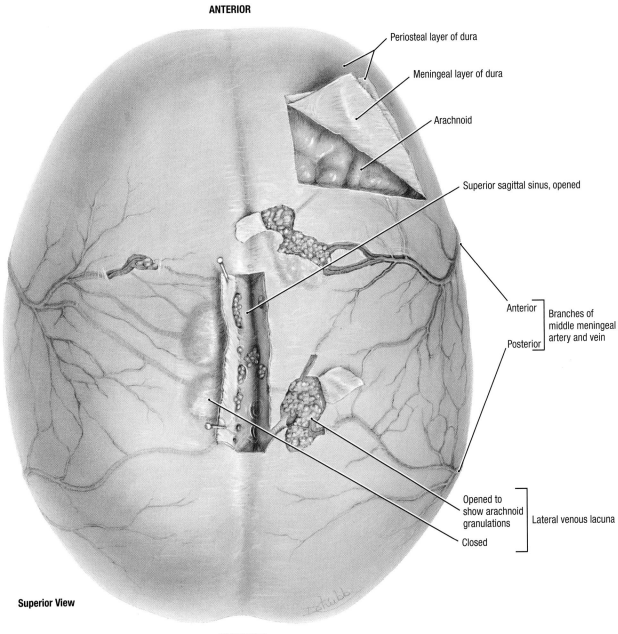

ANTERIOR

Periosteal layer of dura

Meningeal layer of dura

Arachnoid

Superior sagittal sinus, opened

Anterior ⎱ Branches of
Posterior ⎰ middle meningeal artery and vein

Opened to show arachnoid granulations ⎱ Lateral venous lacuna
Closed ⎰

Superior View

POSTERIOR

7.14   External surface of dura mater and arachnoid granulations

- The calvaria is removed. In the median plane, the thick roof of the superior sagittal sinus is partly pinned aside, and laterally, the thin roofs of two lateral lacunae are reflected.
- The middle meningeal artery lies in a venous channel (middle meningeal vein), which enlarges superiorly into a lateral lacunae. A channel or channels drain the lateral lacuna into the superior sagittal sinus.
- Arachnoid granulations in the lacunae are responsible for absorption of CSF from the subarachnoid space into the venous system.

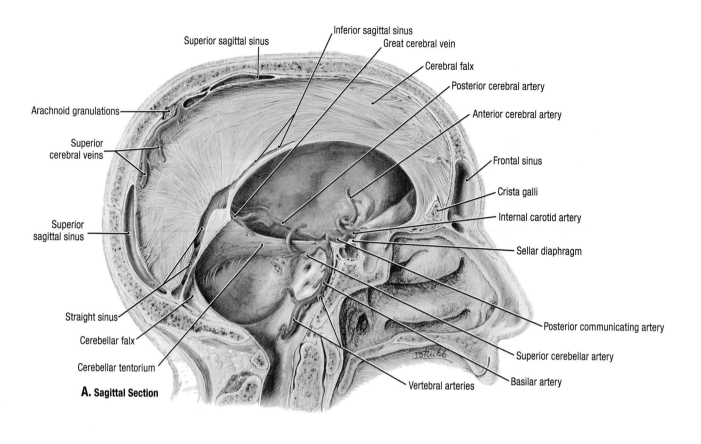

A. Sagittal Section

- Superior sagittal sinus
- Inferior sagittal sinus
- Great cerebral vein
- Cerebral falx
- Posterior cerebral artery
- Anterior cerebral artery
- Arachnoid granulations
- Superior cerebral veins
- Frontal sinus
- Crista galli
- Internal carotid artery
- Superior sagittal sinus
- Sellar diaphragm
- Straight sinus
- Posterior communicating artery
- Cerebellar falx
- Superior cerebellar artery
- Cerebellar tentorium
- Basilar artery
- Vertebral arteries

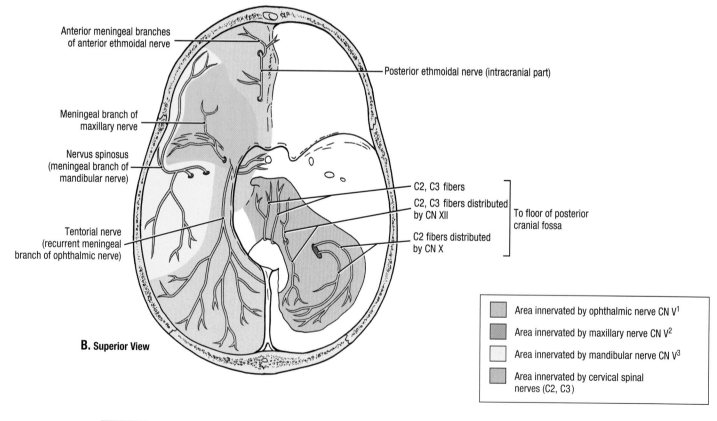

B. Superior View

- Anterior meningeal branches of anterior ethmoidal nerve
- Posterior ethmoidal nerve (intracranial part)
- Meningeal branch of maxillary nerve
- Nervus spinosus (meningeal branch of mandibular nerve)
- C2, C3 fibers
- C2, C3 fibers distributed by CN XII
- To floor of posterior cranial fossa
- Tentorial nerve (recurrent meningeal branch of ophthalmic nerve)
- C2 fibers distributed by CN X

- ▨ Area innervated by ophthalmic nerve CN V^1
- ▨ Area innervated by maxillary nerve CN V^2
- ▢ Area innervated by mandibular nerve CN V^3
- ▨ Area innervated by cervical spinal nerves (C2, C3)

7.15   Dura mater

A. Reflections of the dura mater: falx cerebri, falx cerebelli, tentorium cerebelli, and diaphragma sellae. **B.** Innervation of the dura of the cranial base.

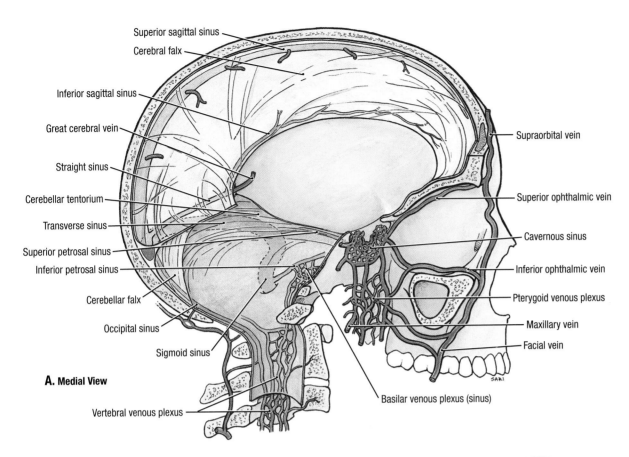

A. Medial View

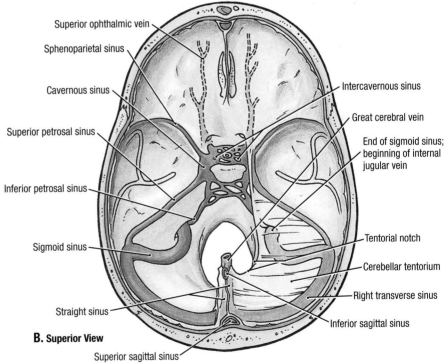

B. Superior View

7.16   **Venous sinuses of the dura mater**

A. Schematic of head, sagittally sectioned. **B.** Venous sinuses of the cranial fossae.

- The superior sagittal sinus is at the superior border of the falx cerebri, and the inferior sagittal sinus is in its free border. The great cerebral vein joins the inferior sagittal sinus to form the straight sinus, which runs obliquely in the junction between the falx cerebri and tentorium cerebelli. The occipital sinus is in the attached border of the falx cerebelli.
- The superior sagittal sinus usually becomes the right transverse sinus, right sigmoid sinus, and right internal jugular vein; the straight sinus similarly drains into the left transverse sinus, which then continues into the left sigmoid sinus and left internal jugular vein.
- The cavernous sinus communicates with the veins of the face through the ophthalmic veins and pterygoid plexus of veins and with the beginning and end of the sigmoid sinus through the superior and inferior petrosal sinuses.

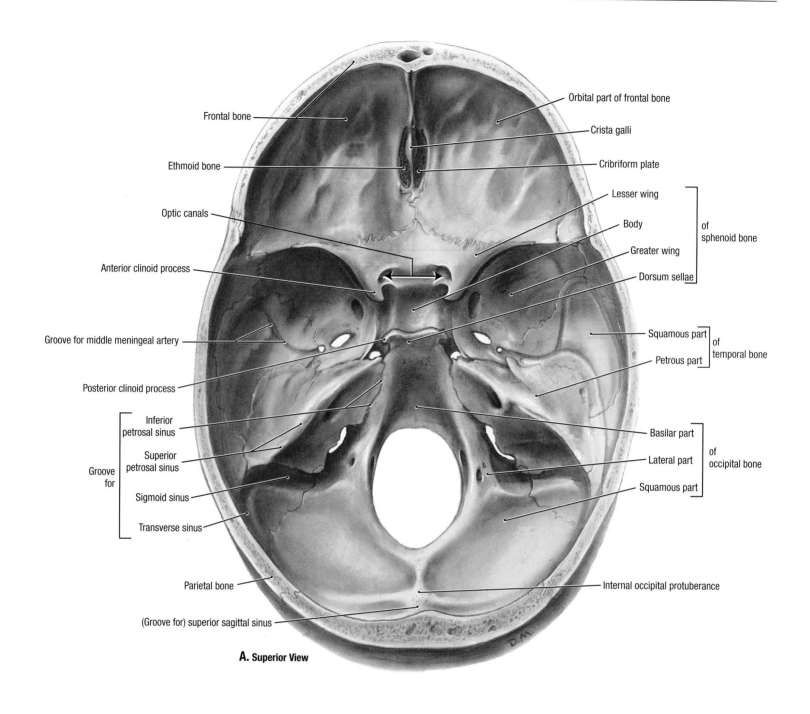

Frontal bone

Ethmoid bone

Optic canals

Anterior clinoid process

Groove for middle meningeal artery

Posterior clinoid process

Groove for —
Inferior petrosal sinus

Superior petrosal sinus

Sigmoid sinus

Transverse sinus

Parietal bone

(Groove for) superior sagittal sinus

Orbital part of frontal bone

Crista galli

Cribriform plate

Lesser wing

Body

Greater wing

Dorsum sellae

of sphenoid bone

Squamous part

Petrous part

of temporal bone

Basilar part

Lateral part

Squamous part

of occipital bone

Internal occipital protuberance

A. Superior View

7.17   Interior of the cranial base

A. Bony cranial base. **B.** Diagrammatic cranial base with bones color coded.
In **A**:

- Three bones contribute to the anterior cranial fossa: the orbital part of the frontal bone, the cribriform plate of the ethmoid, and the lesser wing of the sphenoid.
- The four parts of the occipital bone are the basilar, right and left lateral, and squamous.

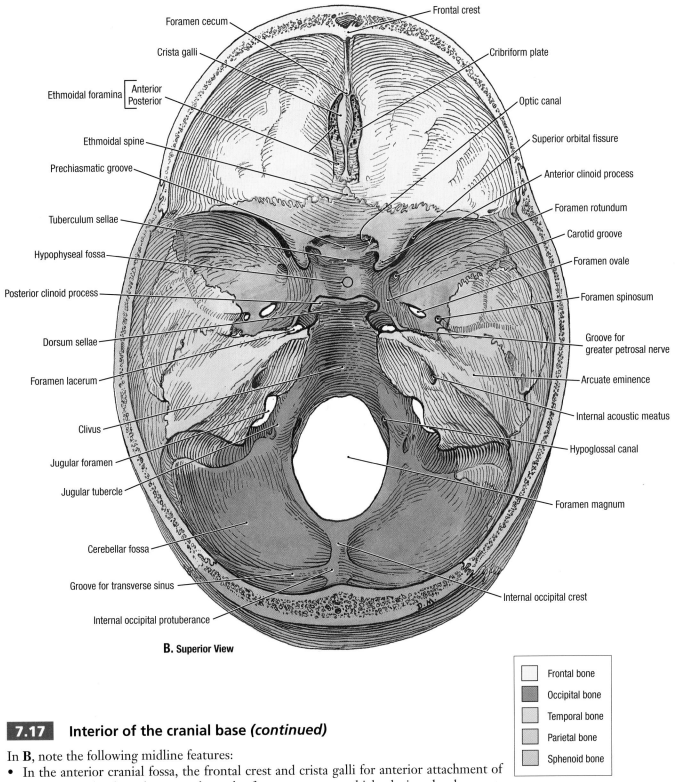

Foramen cecum

Crista galli

Ethmoidal foramina { Anterior / Posterior

Ethmoidal spine

Prechiasmatic groove

Tuberculum sellae

Hypophyseal fossa

Posterior clinoid process

Dorsum sellae

Foramen lacerum

Clivus

Jugular foramen

Jugular tubercle

Cerebellar fossa

Groove for transverse sinus

Internal occipital protuberance

Frontal crest

Cribriform plate

Optic canal

Superior orbital fissure

Anterior clinoid process

Foramen rotundum

Carotid groove

Foramen ovale

Foramen spinosum

Groove for greater petrosal nerve

Arcuate eminence

Internal acoustic meatus

Hypoglossal canal

Foramen magnum

Internal occipital crest

B. Superior View

Frontal bone

Occipital bone

Temporal bone

Parietal bone

Sphenoid bone

7.17   Interior of the cranial base (continued)

In **B**, note the following midline features:

- In the anterior cranial fossa, the frontal crest and crista galli for anterior attachment of the falx cerebri have between them the foramen cecum, which, during development, transmits a vein connecting the superior sagittal sinus with the veins of the frontal sinus and root of the nose.
- In the middle cranial fossa, the tuberculum sellae, hypophyseal fossa, and dorsum sellae constitute the sella turcica (L. Turkish saddle).
- In the posterior cranial fossa, note the clivus, foramen magnum, internal occipital crest for attachment of the falx cerebelli, and the internal occipital protuberance, from which the grooves for the transverse sinuses course laterally.

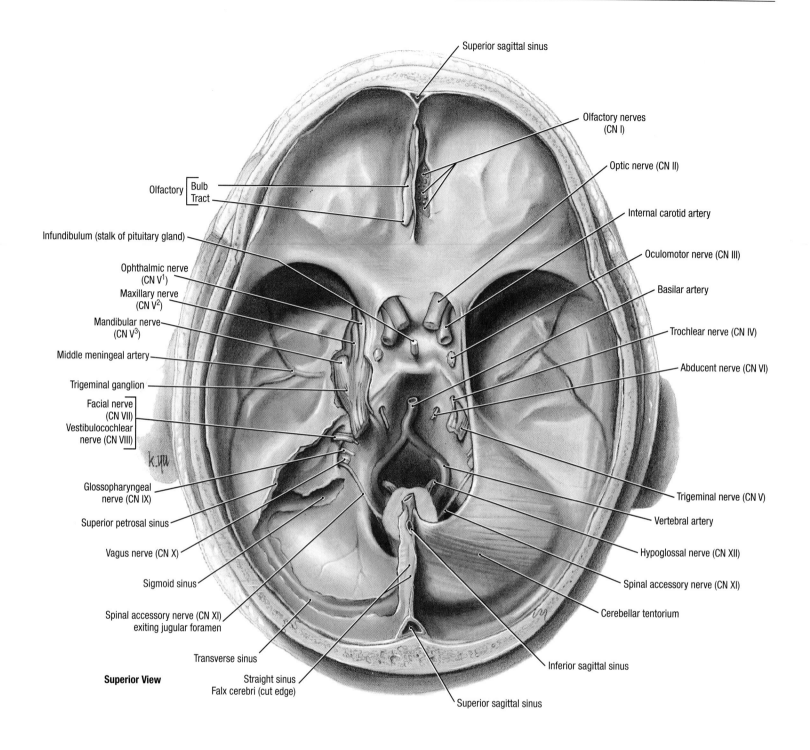

Superior sagittal sinus

Olfactory nerves (CN I)

Optic nerve (CN II)

Internal carotid artery

Oculomotor nerve (CN III)

Basilar artery

Trochlear nerve (CN IV)

Abducent nerve (CN VI)

Trigeminal nerve (CN V)

Vertebral artery

Hypoglossal nerve (CN XII)

Spinal accessory nerve (CN XI)

Cerebellar tentorium

Inferior sagittal sinus

Superior sagittal sinus

Olfactory { Bulb / Tract

Infundibulum (stalk of pituitary gland)

Ophthalmic nerve (CN V¹)

Maxillary nerve (CN V²)

Mandibular nerve (CN V³)

Middle meningeal artery

Trigeminal ganglion

Facial nerve (CN VII)

Vestibulocochlear nerve (CN VIII)

Glossopharyngeal nerve (CN IX)

Superior petrosal sinus

Vagus nerve (CN X)

Sigmoid sinus

Spinal accessory nerve (CN XI) exiting jugular foramen

Transverse sinus

Superior View

Straight sinus

Falx cerebri (cut edge)

Superior sagittal sinus

7.18 **Nerves and vessels of the interior of the base of the cranium**

- On the left of the specimen, the dura mater roofing the trigeminal cave is cut away to expose the trigeminal nerve and its three branches and the sigmoid sinus. The cerebellar tentorium is removed to reveal the transverse and superior petrosal sinuses.
- The frontal lobes of the cerebrum are located in the anterior cranial fossa, the temporal lobes in the middle cranial fossa, and the brainstem and cerebellum in the posterior cranial fossa; the occipital lobes rest on the tentorium.
- Note the location of the 12 cranial nerves and the internal carotid, vertebral, basilar, and middle meningeal arteries.

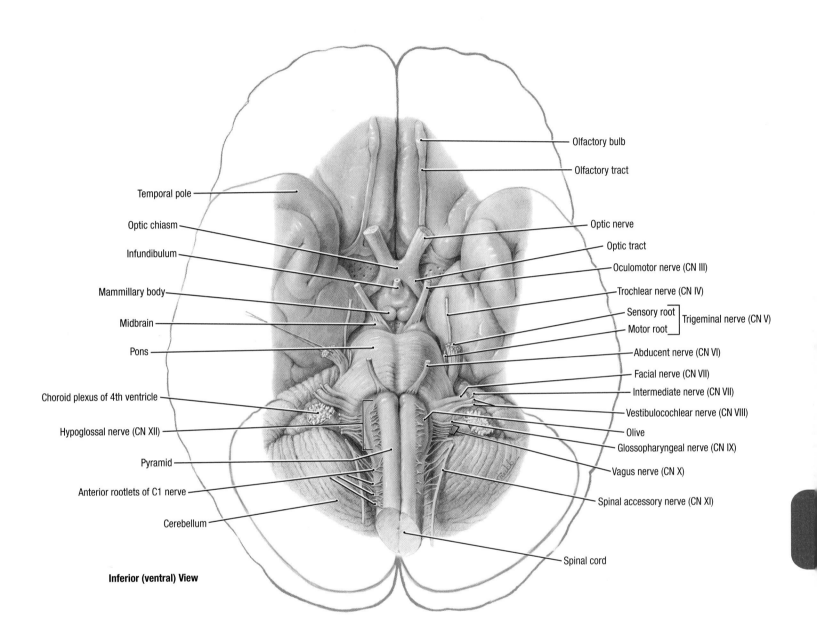

Olfactory bulb

Olfactory tract

Temporal pole

Optic chiasm

Infundibulum

Mammillary body

Midbrain

Pons

Choroid plexus of 4th ventricle

Hypoglossal nerve (CN XII)

Pyramid

Anterior rootlets of C1 nerve

Cerebellum

Optic nerve

Optic tract

Oculomotor nerve (CN III)

Trochlear nerve (CN IV)

Sensory root ⎤
Motor root ⎦ Trigeminal nerve (CN V)

Abducent nerve (CN VI)

Facial nerve (CN VII)

Intermediate nerve (CN VII)

Vestibulocochlear nerve (CN VIII)

Olive

Glossopharyngeal nerve (CN IX)

Vagus nerve (CN X)

Spinal accessory nerve (CN XI)

Spinal cord

Inferior (ventral) View

7.19   **Base of brain and superficial origins of cranial nerves**

The olfactory nerves that end in the olfactory bulbs are not shown. Figure 9.1 is an enlargement of this figure.

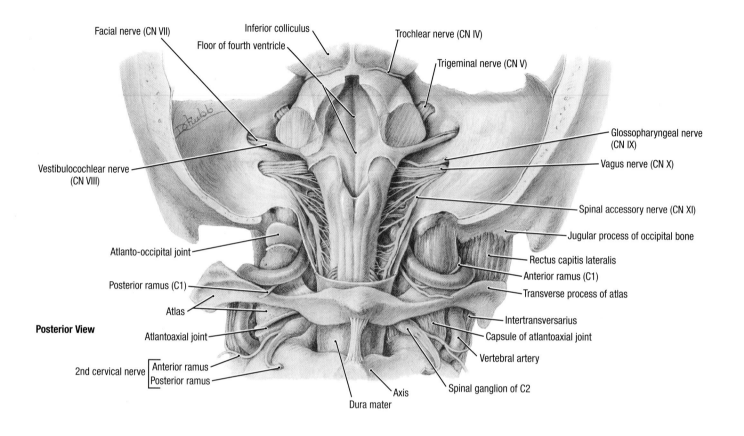

Facial nerve (CN VII)

Inferior colliculus

Floor of fourth ventricle

Trochlear nerve (CN IV)

Trigeminal nerve (CN V)

Glossopharyngeal nerve (CN IX)

Vestibulocochlear nerve (CN VIII)

Vagus nerve (CN X)

Spinal accessory nerve (CN XI)

Jugular process of occipital bone

Atlanto-occipital joint

Rectus capitis lateralis

Anterior ramus (C1)

Posterior ramus (C1)

Transverse process of atlas

Atlas

Intertransversarius

Posterior View

Capsule of atlantoaxial joint

Atlantoaxial joint

Vertebral artery

2nd cervical nerve ⎰ Anterior ramus
⎱ Posterior ramus

Spinal ganglion of C2

Axis

Dura mater

7.20   Posterior exposure of cranial nerve

- The trochlear nerves (CN IV) are unique in arising from the dorsal aspect of the mid-brain, just inferior to the inferior colliculi; the sensory and motor roots of the trigeminal nerves (CN V) pass anterolaterally to enter the mouths of the trigeminal caves; the facial (CN VII) and vestibulocochlear (CN VIII) nerves course laterally to the internal acoustic meatus; the glossopharyngeal nerves (CN IX) pierce the dura mater separately but pass with the vagus (CN X) and spinal accessory (CN XI) nerves through the jugular foramina; and the rootlets of the spinal accessory nerves arise asymmetrically from opposite sides of the medulla and spinal cord.
- The vertebral arteries are elevated from their grooves on the posterior arch of the atlas.
- The posterior ramus of the 1st cervical nerve (suboccipital nerve) passes between the vertebral artery and posterior arch of the atlas; when present, its anterior ramus curves around the atlanto-occipital joint.
- The 2nd cervical nerve has a large spinal ganglion and posterior ramus (greater occipital nerve); its anterior ramus is relatively small.

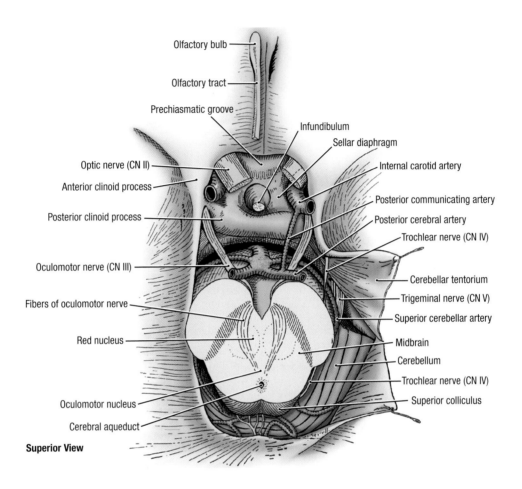

Olfactory bulb

Olfactory tract

Prechiasmatic groove

Infundibulum

Sellar diaphragm

Optic nerve (CN II)

Internal carotid artery

Anterior clinoid process

Posterior communicating artery

Posterior clinoid process

Posterior cerebral artery

Trochlear nerve (CN IV)

Oculomotor nerve (CN III)

Cerebellar tentorium

Fibers of oculomotor nerve

Trigeminal nerve (CN V)

Superior cerebellar artery

Red nucleus

Midbrain

Cerebellum

Trochlear nerve (CN IV)

Oculomotor nucleus

Superior colliculus

Cerebral aqueduct

Superior View

7.21   Tentorial notch

- The forebrain has been removed by cutting through the midbrain, revealing the tentorial notch through which the brainstem extends from the posterior into the middle cranial fossa.
- On the right side of the specimen, the tentorium cerebelli is divided and reflected. The trochlear nerve (CN IV) passes around the midbrain under the free edge of the tentorium cerebelli; the roots of the trigeminal nerve (CN V) enter the mouth of the trigeminal cave.
- There is a circular opening in the diaphragma sellae for the infundibulum, the stalk of the pituitary gland.
- The oculomotor nerve (CN III) passes laterally around the posterior clinoid process and then passes between the posterior cerebral and superior cerebellar arteries.

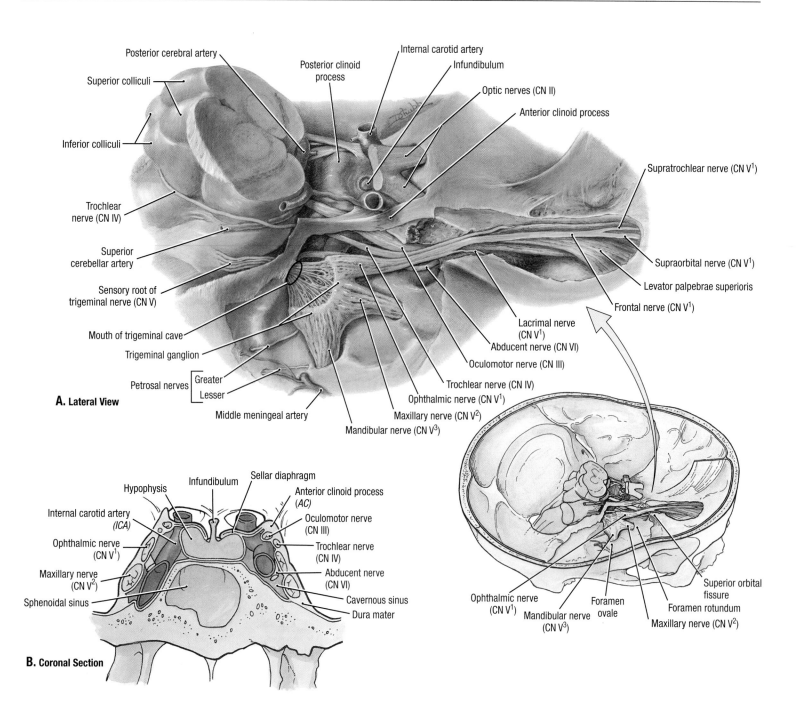

A. Lateral View

Posterior cerebral artery
Superior colliculi
Inferior colliculi
Trochlear nerve (CN IV)
Superior cerebellar artery
Sensory root of trigeminal nerve (CN V)
Mouth of trigeminal cave
Trigeminal ganglion
Petrosal nerves Greater / Lesser
Middle meningeal artery
Mandibular nerve (CN V³)
Maxillary nerve (CN V²)
Ophthalmic nerve (CN V¹)
Trochlear nerve (CN IV)
Oculomotor nerve (CN III)
Abducent nerve (CN VI)
Lacrimal nerve (CN V¹)
Frontal nerve (CN V¹)
Levator palpebrae superioris
Supraorbital nerve (CN V¹)
Supratrochlear nerve (CN V¹)
Anterior clinoid process
Optic nerves (CN II)
Infundibulum
Internal carotid artery
Posterior clinoid process

B. Coronal Section

Hypophysis
Internal carotid artery (ICA)
Ophthalmic nerve (CN V¹)
Maxillary nerve (CN V²)
Sphenoidal sinus
Infundibulum
Sellar diaphragm
Anterior clinoid process (AC)
Oculomotor nerve (CN III)
Trochlear nerve (CN IV)
Abducent nerve (CN VI)
Cavernous sinus
Dura mater

Ophthalmic nerve (CN V¹)
Mandibular nerve (CN V³)
Foramen ovale
Foramen rotundum
Superior orbital fissure
Maxillary nerve (CN V²)

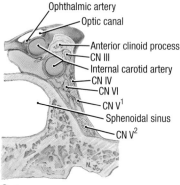

C. Coronal Section

Ophthalmic artery
Optic canal
Anterior clinoid process
CN III
Internal carotid artery
CN IV
CN VI
CN V¹
Sphenoidal sinus
CN V²

7.22 Nerves and vessels of middle cranial fossa—I

A. Superficial dissection and superficial dissection in situ in the cranial base. The tentorium cerebelli is cut away to reveal the courses of the trochlear nerve and roots of the trigeminal nerve in the posterior cranial fossa. The dura is largely removed from the middle cranial fossa. The roof of the orbit is partly removed. **B** and **C.** Coronal sections through the cavernous sinus.

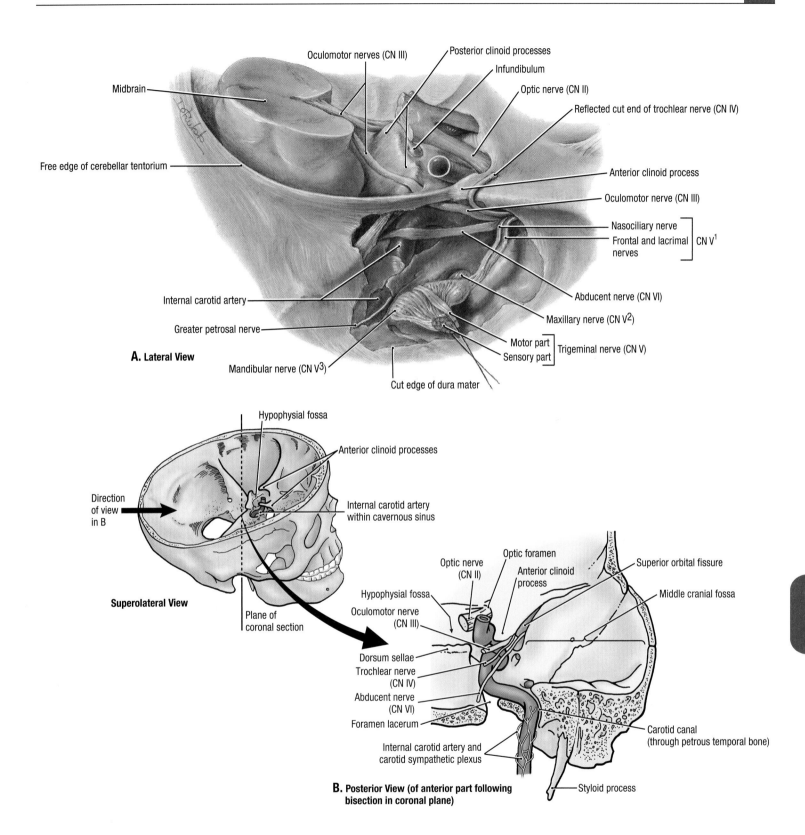

Oculomotor nerves (CN III)

Posterior clinoid processes

Infundibulum

Optic nerve (CN II)

Midbrain

Reflected cut end of trochlear nerve (CN IV)

Free edge of cerebellar tentorium

Anterior clinoid process

Oculomotor nerve (CN III)

Nasociliary nerve

Frontal and lacrimal nerves } CN V^1

Abducent nerve (CN VI)

Internal carotid artery

Maxillary nerve (CN V^2)

Greater petrosal nerve

Motor part
Sensory part } Trigeminal nerve (CN V)

A. Lateral View

Mandibular nerve (CN V^3)

Cut edge of dura mater

Hypophysial fossa

Anterior clinoid processes

Direction
of view
in B

Internal carotid artery
within cavernous sinus

Superolateral View

Plane of
coronal section

Optic nerve
(CN II)

Optic foramen

Anterior clinoid
process

Superior orbital fissure

Middle cranial fossa

Hypophysial fossa

Oculomotor nerve
(CN III)

Dorsum sellae

Trochlear nerve
(CN IV)

Abducent nerve
(CN VI)

Foramen lacerum

Internal carotid artery and
carotid sympathetic plexus

Carotid canal
(through petrous temporal bone)

**B. Posterior View (of anterior part following
bisection in coronal plane)**

Styloid process

7.23   Nerves and vessels of middle cranial fossa—II

A. Deep dissection. The roots of the trigeminal nerve are divided, withdrawn from the mouth of the trigeminal cave, and turned anteriorly. The trochlear nerve is reflected anteriorly. **B.** Course of the internal carotid artery.

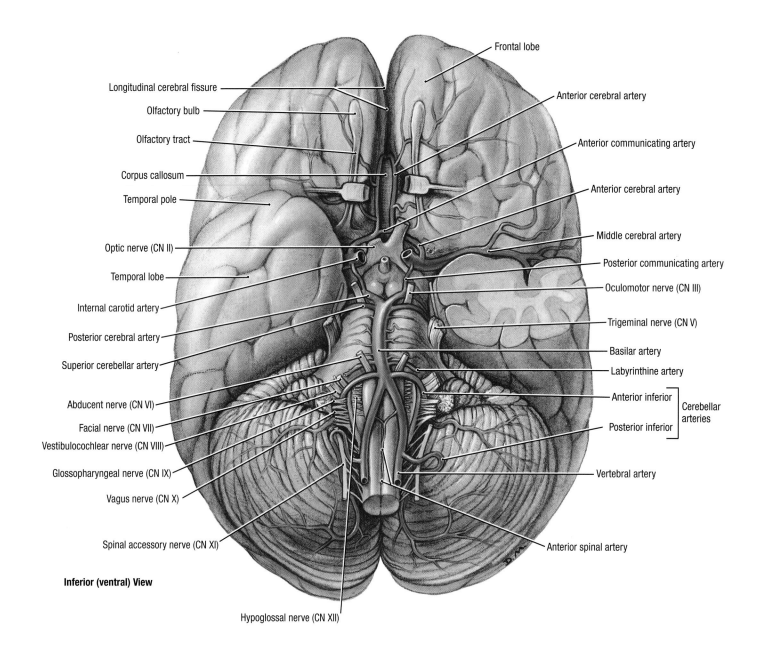

Frontal lobe

Longitudinal cerebral fissure

Olfactory bulb

Olfactory tract

Corpus callosum

Temporal pole

Optic nerve (CN II)

Temporal lobe

Internal carotid artery

Posterior cerebral artery

Superior cerebellar artery

Abducent nerve (CN VI)

Facial nerve (CN VII)

Vestibulocochlear nerve (CN VIII)

Glossopharyngeal nerve (CN IX)

Vagus nerve (CN X)

Spinal accessory nerve (CN XI)

Anterior cerebral artery

Anterior communicating artery

Anterior cerebral artery

Middle cerebral artery

Posterior communicating artery

Oculomotor nerve (CN III)

Trigeminal nerve (CN V)

Basilar artery

Labyrinthine artery

Anterior inferior

Posterior inferior

Cerebellar arteries

Vertebral artery

Anterior spinal artery

Inferior (ventral) View

Hypoglossal nerve (CN XII)

7.24 **Base of brain and cerebral arterial circle**

The left temporal pole is removed to enable visualization of the middle cerebral artery in the lateral fissure. The frontal lobes are separated to expose the anterior cerebral arteries and corpus callosum.

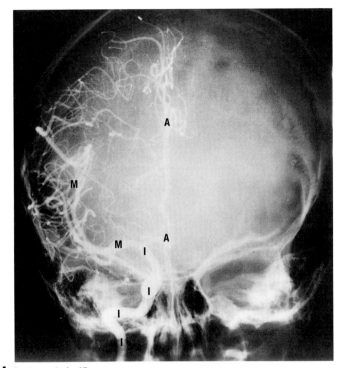

A. Posteroanterior View

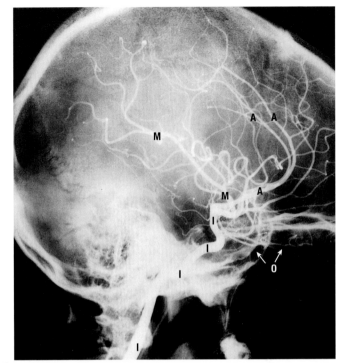

B. Lateral View

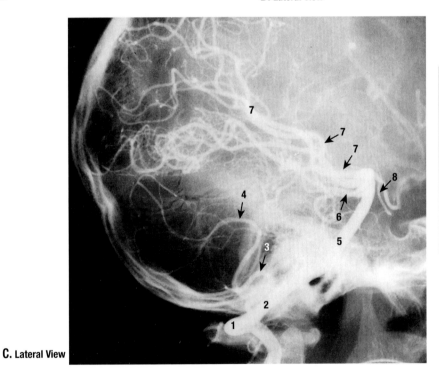

C. Lateral View

1	Vertebral artery on posterior arch of atlas
2	Vertebral artery entering skull through foramen magnum
3	Posterior inferior cerebellar artery
4	Anterior inferior cerebellar artery
5	Basilar artery
6	Superior cerebellar artery
7	Posterior cerebellar artery
8	Posterior communicating artery

7.25   Arteriograms

A and **B.** Carotid arteriogram. The four letter *Is* indicate the parts of the internal carotid artery: cervical, before entering the cranium; petrous, within the temporal bone; cavernous, within that venous sinus; and cerebral, within the cranial subarachnoid space. Note the anterior cerebral artery *(A)*, middle cerebral artery *(M)*, and opthalmic artery *(O)*. **C.** Vertebral arteriogram.

Anterior cerebral
(A2 segment)

Anterior communicating

Distal medial striate

Anteromedial central

Anterior cerebral
(A1 segment)

Anterolateral central
striate (lenticulostriate)

Ophthalmic

Internal carotid

Middle
cerebral

Hypophyseal

Anterior choroidal

Posterior communicating artery

Posteromedial central

Posterior cerebral

P2 P1 P1 P2

Posterolateral
central

Superior cerebellar

Pontine

Labyrinthine

Basilar

Anterior inferior
cerebellar

Vertebral

Posterior inferior
cerebellar

Anterior spinal

Inferior (Ventral) View

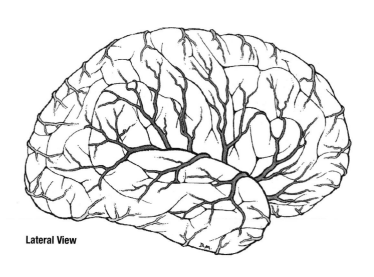

Lateral View

Blood is supplied to the cerebral hemispheres by the anterior (*dark green*), middle (*purple*), and posterior (*light green*) cerebral arteries.

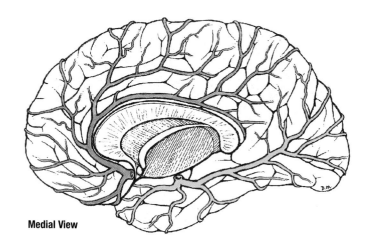

Medial View

Posterior cerebral (P1)	Anterior spinal	
Superior cerebellar	Posterior spinal	
Anterior cerebral	Long circumferential branches	
Middle cerebral		
Anterior inferior cerebellar	Short circumferential branches	Basilar
Posterior inferior cerebellar		
Vertebral	Paramedian branches	

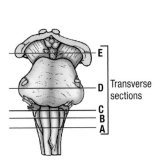

Transverse sections

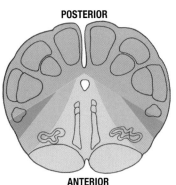

POSTERIOR

ANTERIOR

Medulla oblongata

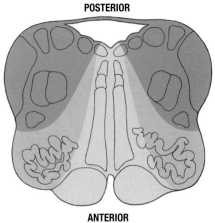

POSTERIOR

ANTERIOR

Medulla oblongata

POSTERIOR

ANTERIOR

Medulla oblongata

POSTERIOR

ANTERIOR

Pons

POSTERIOR

ANTERIOR

Midbrain

TABLE 7.5   ARTERIAL SUPPLY TO THE BRAIN

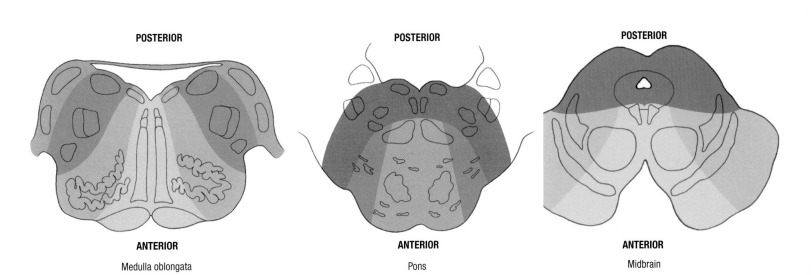

ARTERY	ORIGIN	DISTRIBUTION
Vertebral	Subclavian artery	Cranial meninges and cerebellum
Posterior inferior cerebellar	Vertebral artery	Posteroinferior aspect of cerebellum
Basilar	Formed by junction of vertebral arteries	Brainstem, cerebellum, and cerebrum
Pontine		Numerous branches to brainstem
Anterior inferior cerebellar	Basilar artery	Inferior aspect of cerebellum
Superior cerebellar		Superior aspect of cerebellum
Internal carotid	Common carotid artery at superior border of thyroid cartilage	Gives branches in cavernous sinus and provides supply to brain
Anterior cerebral	Internal carotid artery	Cerebral hemispheres, except for occipital lobes
Middle cerebral	Continuation of the internal carotid artery distal to anterior cerebral artery	Most of lateral surface of cerebral hemispheres
Posterior cerebral	Terminal branch of basilar artery	Inferior aspect of cerebral hemisphere and lobe
Anterior communicating	Anterior cerebral artery	Cerebral arterial circle
Posterior communicating	Posterior cerebral artery	

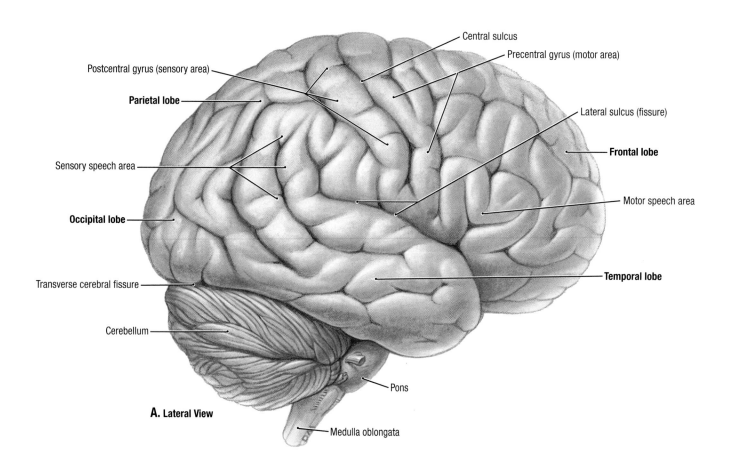

Central sulcus
Precentral gyrus (motor area)
Postcentral gyrus (sensory area)
Parietal lobe
Lateral sulcus (fissure)
Frontal lobe
Sensory speech area
Motor speech area
Occipital lobe
Temporal lobe
Transverse cerebral fissure
Cerebellum
Pons
A. Lateral View
Medulla oblongata

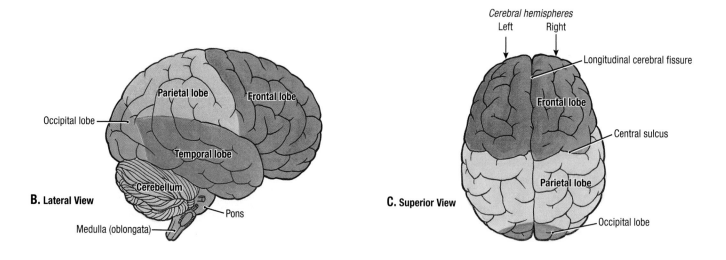

Parietal lobe
Frontal lobe
Occipital lobe
Temporal lobe
Cerebellum
B. Lateral View
Pons
Medulla (oblongata)

Cerebral hemispheres
Left   Right
Longitudinal cerebral fissure
Frontal lobe
Central sulcus
Parietal lobe
C. Superior View
Occipital lobe

7.26   Brain

A. Cerebrum, cerebellum, and brainstem, lateral aspect. **B.** Lobes of the cerebral hemispheres, lateral aspect. **C.** Lobes of the cerebral hemispheres, superior aspect.

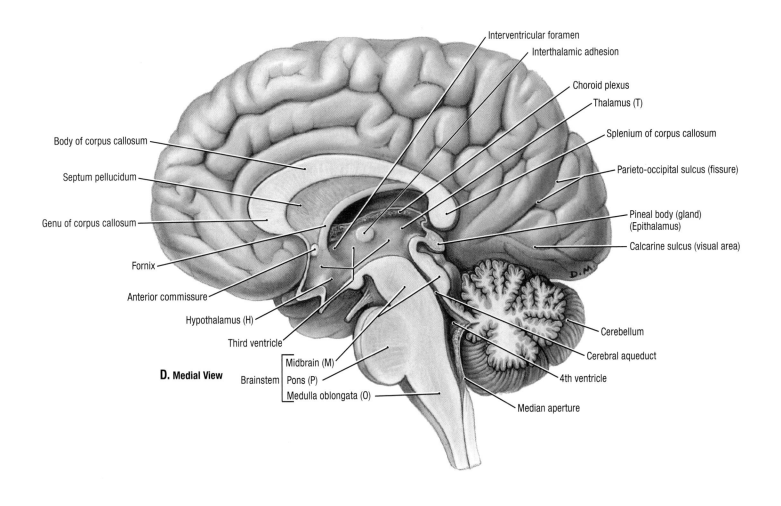

Interventricular foramen

Interthalamic adhesion

Choroid plexus

Thalamus (T)

Splenium of corpus callosum

Parieto-occipital sulcus (fissure)

Pineal body (gland)
(Epithalamus)

Calcarine sulcus (visual area)

Cerebellum

Cerebral aqueduct

4th ventricle

Median aperture

Body of corpus callosum

Septum pellucidum

Genu of corpus callosum

Fornix

Anterior commissure

Hypothalamus (H)

Third ventricle

D. Medial View

Brainstem { Midbrain (M)
Pons (P)
Medulla oblongata (O) }

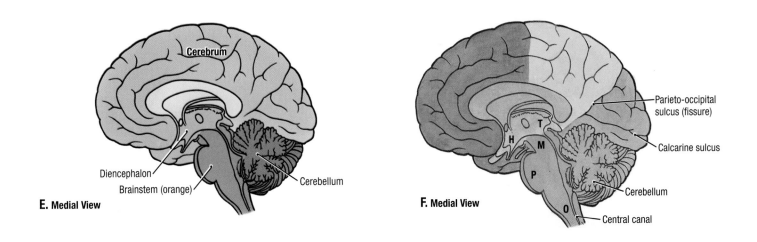

Cerebrum

Diencephalon

Brainstem (orange)

Cerebellum

E. Medial View

Parieto-occipital
sulcus (fissure)

Calcarine sulcus

Cerebellum

Central canal

F. Medial View

7.26   **Brain (continued)**

D. Cerebrum, cerebellum, and brainstem, median section. **E.** Parts of the cerebrum, median section. **F.** Lobes of the cerebral hemisphere, median section.

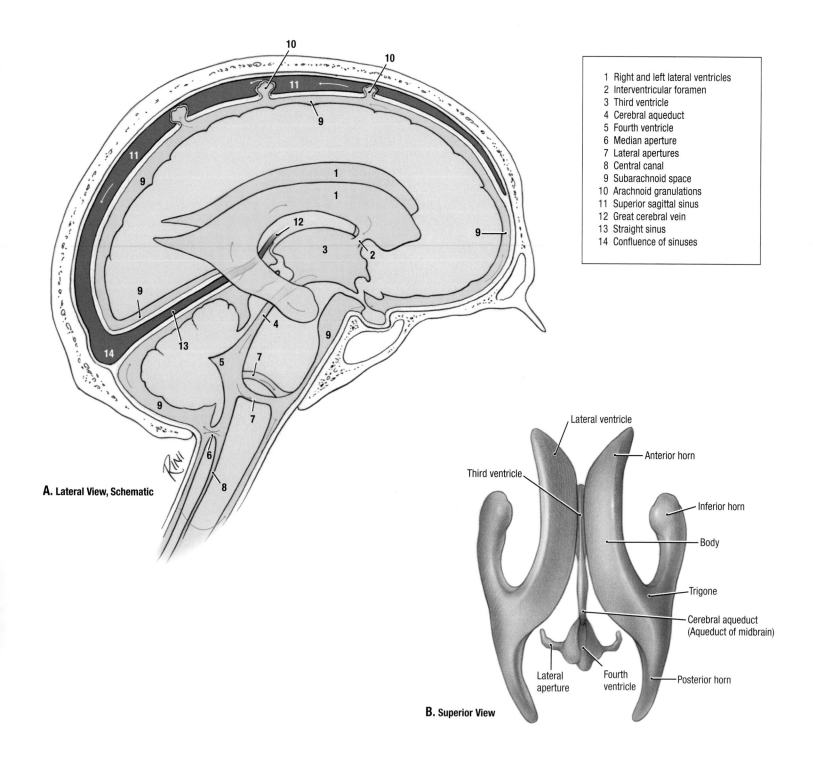

1 Right and left lateral ventricles
2 Interventricular foramen
3 Third ventricle
4 Cerebral aqueduct
5 Fourth ventricle
6 Median aperture
7 Lateral apertures
8 Central canal
9 Subarachnoid space
10 Arachnoid granulations
11 Superior sagittal sinus
12 Great cerebral vein
13 Straight sinus
14 Confluence of sinuses

A. Lateral View, Schematic

Lateral ventricle

Anterior horn

Third ventricle

Inferior horn

Body

Trigone

Cerebral aqueduct
(Aqueduct of midbrain)

Lateral
aperture

Fourth
ventricle

Posterior horn

B. Superior View

7.27   Ventricular system

A. Circulation of CSF. **B.** Ventricles: lateral, third, and fourth.

- The ventricular system consists of two lateral ventricles located in the cerebral hemispheres, a third ventricle located between the right and left halves of the diencephalon, and a fourth ventricle located in the posterior parts of the pons and medulla.
- CSF secreted by choroid plexus in the ventricles drains via the interventricular foramen from the lateral to the third ventricle, via the cerebral aqueduct from the third to the fourth ventricle, and via median and lateral apertures into the subarachnoid space. CSF is absorbed by arachnoid granulations into the venous sinuses (especially the superior sagittal sinus).

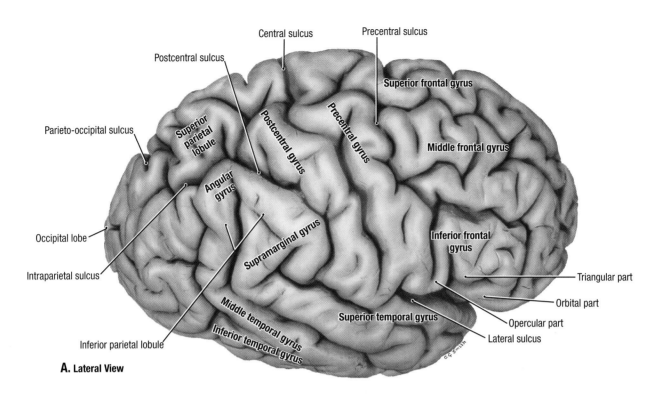

A. Lateral View

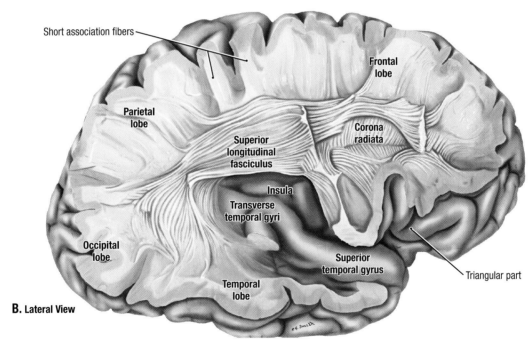

B. Lateral View

7.28   **Serial dissections of the lateral aspect of the cerebral hemisphere**

The dissections begin from the lateral surface of the cerebral hemisphere (**A**) and proceed sequentially medially (**B** to **F**).
A. Sulci and gyri of the lateral surface of one cerebral hemisphere. Each gyrus is a fold of cerebral cortex with a core of white matter. The furrows are called *sulci*. The pattern of sulci and gyri formed shortly before birth is recognizable in some adult brains, as shown in this specimen. Usually the expanding cortex acquires secondary foldings, which make identification of this basic pattern more difficult. **B.** Superior longitudinal fasciculus, transverse temporal gyri, and insula. The cortex and short association fiber bundles around the lateral fissure have been removed.

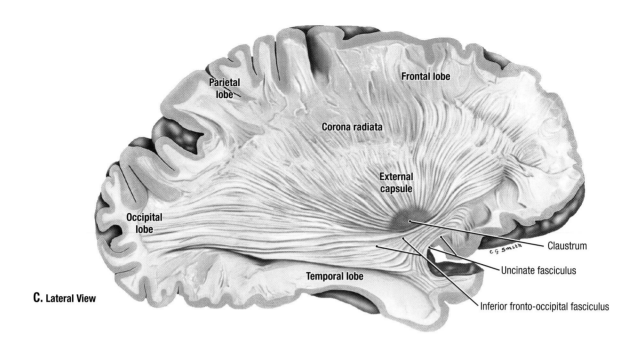

C. Lateral View

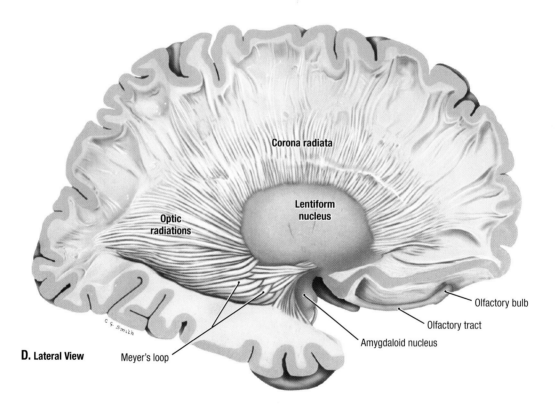

D. Lateral View

7.28 **Serial dissections of the lateral aspect of the cerebral hemisphere** *(continued)*

C. Uncinate and inferior fronto-occipital fasciculi and external capsule. The external capsule consists of projection fibers that pass between the claustrum laterally and the lentiform nucleus medially. **D.** Lentiform nucleus and corona radiata. The inferior longitudinal and uncinate fasciculi, claustrum, and external capsule have been removed. The fibers of the optic radiations convey impulses from the right half of the retina of each eye; the fibers extending closest to the temporal pole (Meyer's loop) carry impulses from the lower portion of each retina.

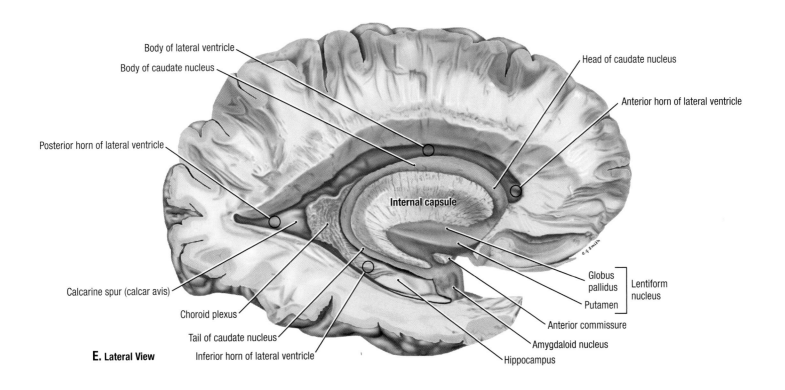

Body of lateral ventricle

Body of caudate nucleus

Posterior horn of lateral ventricle

Internal capsule

Head of caudate nucleus

Anterior horn of lateral ventricle

Globus pallidus ⎱ Lentiform nucleus
Putamen

Calcarine spur (calcar avis)

Choroid plexus

Tail of caudate nucleus

Inferior horn of lateral ventricle

Anterior commissure

Amygdaloid nucleus

Hippocampus

E. Lateral View

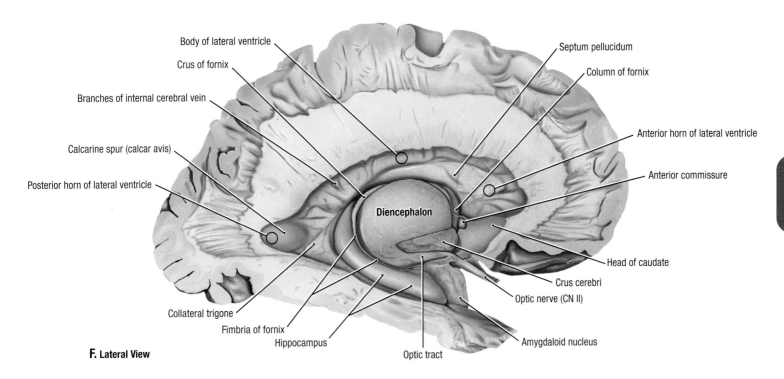

Body of lateral ventricle

Crus of fornix

Branches of internal cerebral vein

Calcarine spur (calcar avis)

Posterior horn of lateral ventricle

Diencephalon

Septum pellucidum

Column of fornix

Anterior horn of lateral ventricle

Anterior commissure

Head of caudate

Crus cerebri

Optic nerve (CN II)

Amygdaloid nucleus

Collateral trigone

Fimbria of fornix

Hippocampus

Optic tract

F. Lateral View

7.28   **Serial dissections of the lateral aspect of the cerebral hemisphere (continued)**

E. Caudate and amygdaloid nuclei and internal capsule. The lateral wall of the lateral ventricle, the marginal part of the internal capsule, the anterior commissure, and the superior part of the lentiform nucleus have been removed. **F.** Lateral ventricle, hippocampus, and diencephalon. The inferior parts of the lentiform nucleus, internal capsule, and caudate nucleus have been removed.

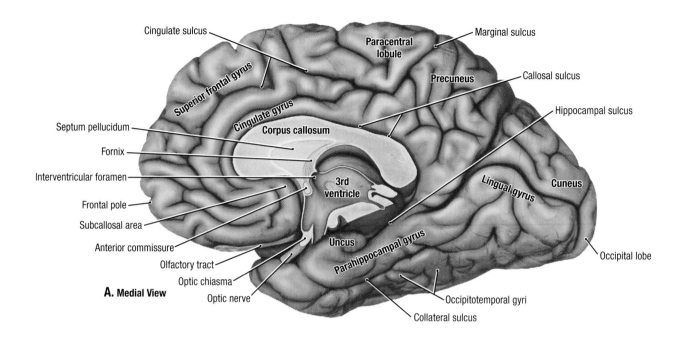

A. Medial View

Cingulate sulcus — Superior frontal gyrus — Cingulate gyrus — Corpus callosum — Septum pellucidum — Fornix — Interventricular foramen — Frontal pole — Subcallosal area — Anterior commissure — Olfactory tract — Optic chiasma — Optic nerve — 3rd ventricle — Uncus — Parahippocampal gyrus — Paracentral lobule — Precuneus — Marginal sulcus — Callosal sulcus — Hippocampal sulcus — Lingual gyrus — Cuneus — Occipital lobe — Occipitotemporal gyri — Collateral sulcus

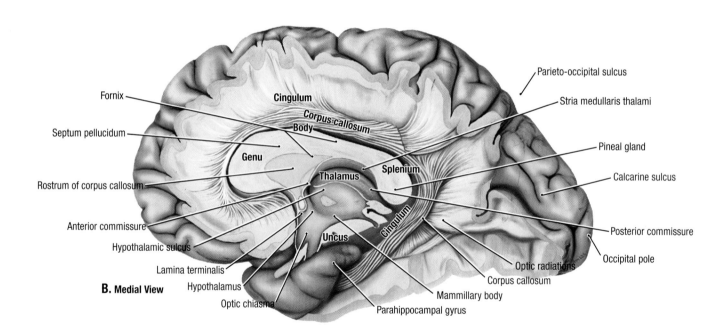

B. Medial View

Fornix — Septum pellucidum — Rostrum of corpus callosum — Anterior commissure — Hypothalamic sulcus — Lamina terminalis — Hypothalamus — Optic chiasma — Cingulum — Corpus callosum — Body — Genu — Thalamus — Splenium — Uncus — Cingulum — Parahippocampal gyrus — Optic chiasma — Mammillary body — Parieto-occipital sulcus — Stria medullaris thalami — Pineal gland — Calcarine sulcus — Posterior commissure — Occipital pole — Optic radiations — Corpus callosum

7.29   **Serial dissections of the medial aspect of cerebral hemisphere**

The dissections begin from the medial surface of the cerebral hemisphere (**A**) and proceed sequentially laterally (**B** to **D**).

A. Sulci and gyri of medial surface of cerebral hemisphere. The corpus callosum consists of the rostrum, genu, body, and splenium; the cingulate and parahippocampal gyri from the limbic lobe. **B.** Cingulum. The cortex and short association fibers were removed from the medial aspect of the hemisphere. The cingulum is a long association fiber bundle that lies in the core of the cingulate and parahippocampal gyri.

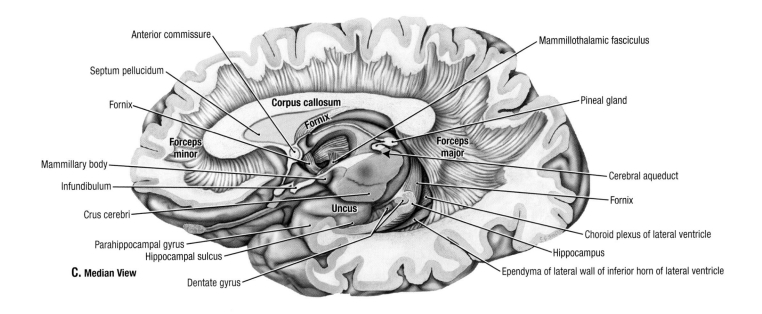

C. Median View

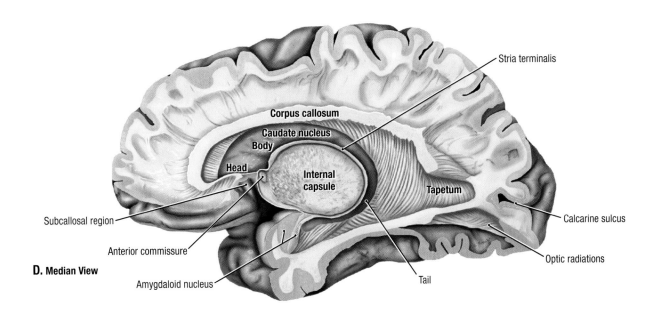

D. Median View

7.29 **Serial dissections of the medial aspect of cerebral hemisphere *(continued)***

C. Fornix, mamillothalamic tract, and forceps major and minor. The cingulum and a portion of the wall of the third ventricle have been removed. The fornix begins at the hippocampus and terminates in the mammillary body by passing anterior to the interventricular foramen and posterior to the anterior commissure. The mamillothalamic fasciculus emerges from the mammillary body and terminates in the anterior nucleus of the thalamus. **D.** Caudate nucleus and internal capsule. The diencephalon was removed, along with the ependyma of the lateral ventricle, except where it covers the caudate and amygdaloid nuclei.

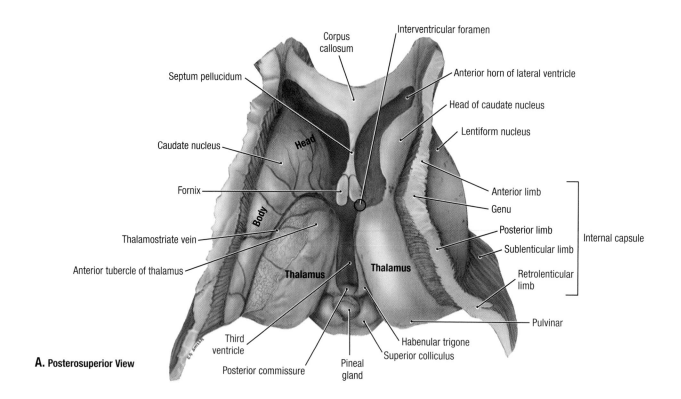

A. Posterosuperior View

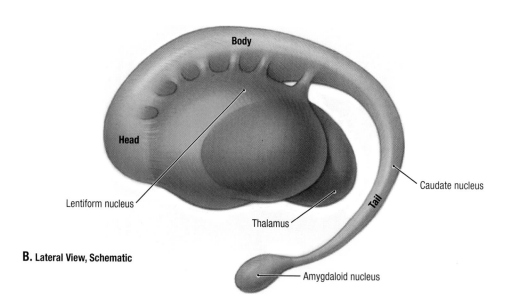

B. Lateral View, Schematic

7.30 Caudate and lentiform nuclei

A. Relationship to the lateral ventricles and internal capsule. The dorsal surface of the diencephalon has been exposed by dissecting away the two cerebral hemispheres, except the anterior part of the corpus callosum, the inferior part of the septum pellucidum, the internal capsule, and the caudate and lentiform nuclei. On the right side of the specimen, the thalamus, caudate, and lentiform nuclei have been cut horizontally at the level of the interventricular foramen. The parts of the internal capsule include the anterior, posterior, retrolenticular sublenticular limbs, and genu. **B.** Schematic illustration of nuclei.

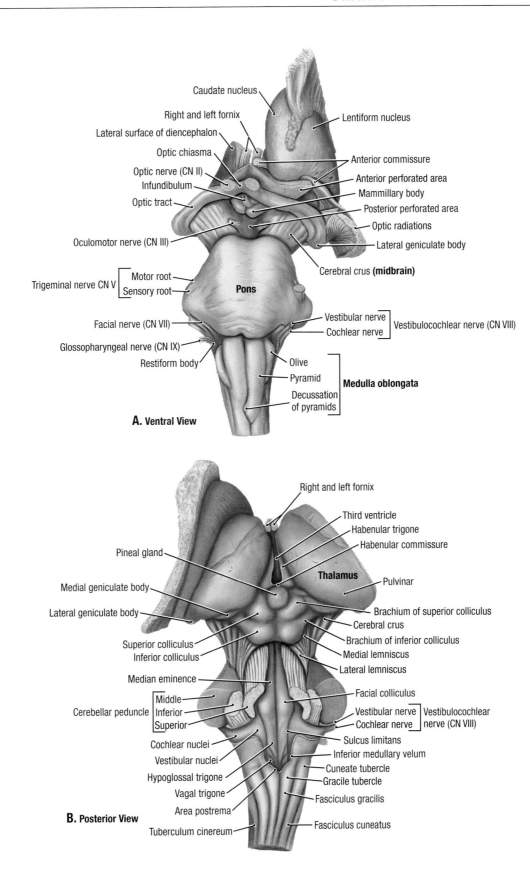

A. Ventral View

Caudate nucleus
Right and left fornix
Lateral surface of diencephalon
Optic chiasma
Optic nerve (CN II)
Infundibulum
Optic tract
Oculomotor nerve (CN III)
Trigeminal nerve CN V
Motor root
Sensory root
Facial nerve (CN VII)
Glossopharyngeal nerve (CN IX)
Restiform body

Lentiform nucleus
Anterior commissure
Anterior perforated area
Mammillary body
Posterior perforated area
Optic radiations
Lateral geniculate body
Cerebral crus **(midbrain)**
Pons
Vestibular nerve
Cochlear nerve
Vestibulocochlear nerve (CN VIII)
Olive
Pyramid
Decussation of pyramids
Medulla oblongata

B. Posterior View

Right and left fornix
Third ventricle
Habenular trigone
Habenular commissure
Thalamus
Pulvinar
Pineal gland
Medial geniculate body
Lateral geniculate body
Brachium of superior colliculus
Cerebral crus
Brachium of inferior colliculus
Medial lemniscus
Lateral lemniscus
Superior colliculus
Inferior colliculus
Median eminence
Middle
Cerebellar peduncle   Inferior
Superior
Facial colliculus
Vestibular nerve
Cochlear nerve
Vestibulocochlear nerve (CN VIII)
Cochlear nuclei
Vestibular nuclei
Hypoglossal trigone
Vagal trigone
Area postrema
Tuberculum cinereum
Sulcus limitans
Inferior medullary velum
Cuneate tubercle
Gracile tubercle
Fasciculus gracilis
Fasciculus cuneatus

7.31   **Brainstem**

A. Ventral aspect. **B.** Dorsal aspect.

The brainstem has been exposed by removing the cerebellum, all of the right cerebral hemisphere, and the major portion of the left hemisphere.

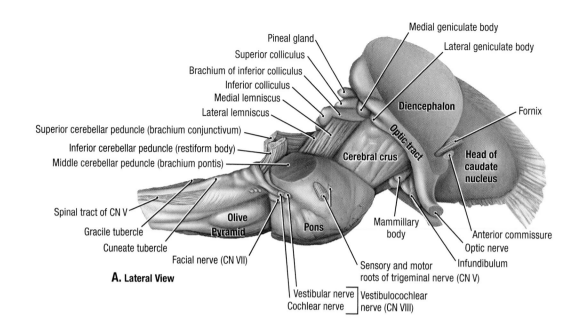

Medial geniculate body
Lateral geniculate body
Pineal gland
Superior colliculus
Brachium of inferior colliculus
Inferior colliculus
Medial lemniscus
Lateral lemniscus
Superior cerebellar peduncle (brachium conjunctivum)
Inferior cerebellar peduncle (restiform body)
Middle cerebellar peduncle (brachium pontis)
Diencephalon
Fornix
Optic tract
Cerebral crus
Head of caudate nucleus
Spinal tract of CN V
Olive
Pyramid
Pons
Gracile tubercle
Cuneate tubercle
Mammillary body
Anterior commissure
Facial nerve (CN VII)
Optic nerve
Infundibulum
Sensory and motor roots of trigeminal nerve (CN V)
A. Lateral View
Vestibular nerve
Cochlear nerve
Vestibulocochlear nerve (CN VIII)

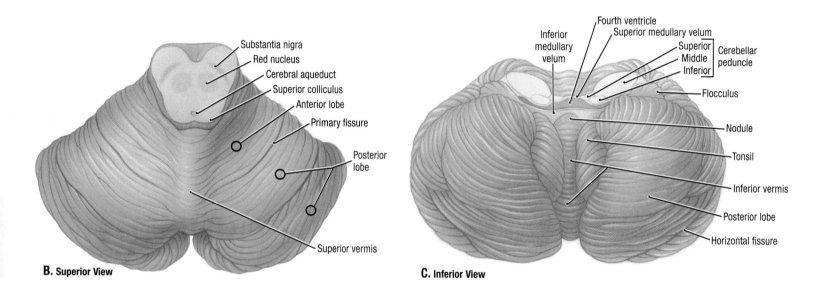

Substantia nigra
Red nucleus
Cerebral aqueduct
Superior colliculus
Anterior lobe
Primary fissure
Posterior lobe
Superior vermis
B. Superior View

Fourth ventricle
Inferior medullary velum
Superior medullary velum
Superior
Middle
Inferior
Cerebellar peduncle
Flocculus
Nodule
Tonsil
Inferior vermis
Posterior lobe
Horizontal fissure
C. Inferior View

7.32 **Brainstem and cerebellum**

A. Lateral aspect of the brainstem. The brainstem has been exposed by removing the cerebellum, all of the right cerebral hemisphere, and the major portion of the left hemisphere. **B.** Superior view of the cerebellum. The right and left cerebellar hemispheres are united by the superior vermis; the anterior and posterior lobes are separated by the primary fissure. **C.** Inferior view of cerebellum. The flocculonodular lobe, the oldest part of the cerebellum, consists of the flocculus and nodule; the cerebellar tonsils typically extend into the foramen magnum.

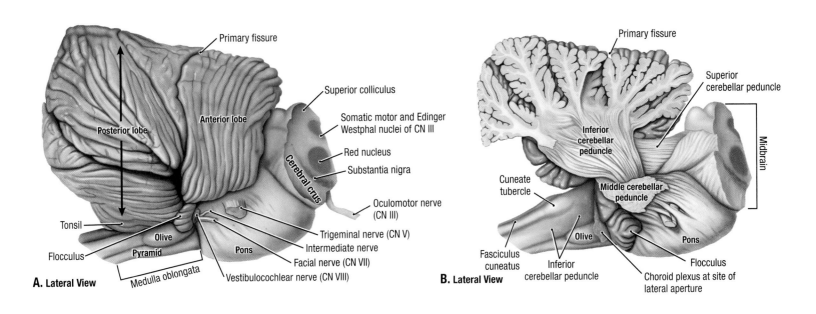

A. Lateral View

Primary fissure

Posterior lobe

Anterior lobe

Superior colliculus

Somatic motor and Edinger Westphal nuclei of CN III

Red nucleus

Substantia nigra

Cerebral crus

Oculomotor nerve (CN III)

Tonsil

Olive

Pyramid

Pons

Flocculus

Medulla oblongata

Trigeminal nerve (CN V)

Intermediate nerve

Facial nerve (CN VII)

Vestibulocochlear nerve (CN VIII)

B. Lateral View

Primary fissure

Superior cerebellar peduncle

Inferior cerebellar peduncle

Middle cerebellar peduncle

Midbrain

Cuneate tubercle

Olive

Pons

Fasciculus cuneatus

Inferior cerebellar peduncle

Flocculus

Choroid plexus at site of lateral aperture

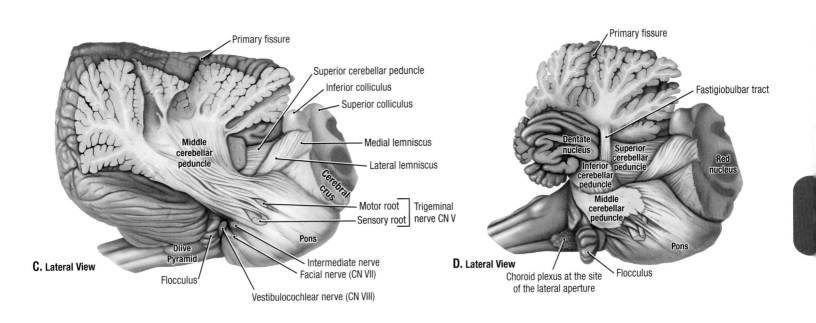

C. Lateral View

Primary fissure

Superior cerebellar peduncle

Inferior colliculus

Superior colliculus

Middle cerebellar peduncle

Medial lemniscus

Lateral lemniscus

Cerebral crus

Motor root ⎤
Sensory root ⎦ Trigeminal nerve CN V

Olive

Pyramid

Pons

Flocculus

Intermediate nerve

Facial nerve (CN VII)

Vestibulocochlear nerve (CN VIII)

D. Lateral View

Primary fissure

Fastigiobulbar tract

Dentate nucleus

Superior cerebellar peduncle

Inferior cerebellar peduncle

Red nucleus

Middle cerebellar peduncle

Pons

Choroid plexus at the site of the lateral aperture

Flocculus

7.33   Serial dissections of the cerebellum

The series begins with the lateral surface of the cerebellar hemispheres (**A**) and proceeds medially in sequence (**B** to **D**).
A. Cerebellum and brainstem. **B.** Inferior cerebellar peduncle. The fibers of the middle cerebellar peduncle were cut dorsal to the trigeminal nerve and peeled away to expose the fibers of the inferior cerebellar peduncle. **C.** Middle cerebellar peduncle. The fibers of the middle cerebellar peduncle were exposed by peeling away the lateral portion of the lobules of the cerebellar hemisphere. **D.** Superior cerebellar peduncle and dentate nucleus. The fibers of the inferior cerebellar peduncle were cut just dorsal to the previously sectioned middle cerebellar peduncle and peeled away until the gray matter of the dentate nucleus could be seen.

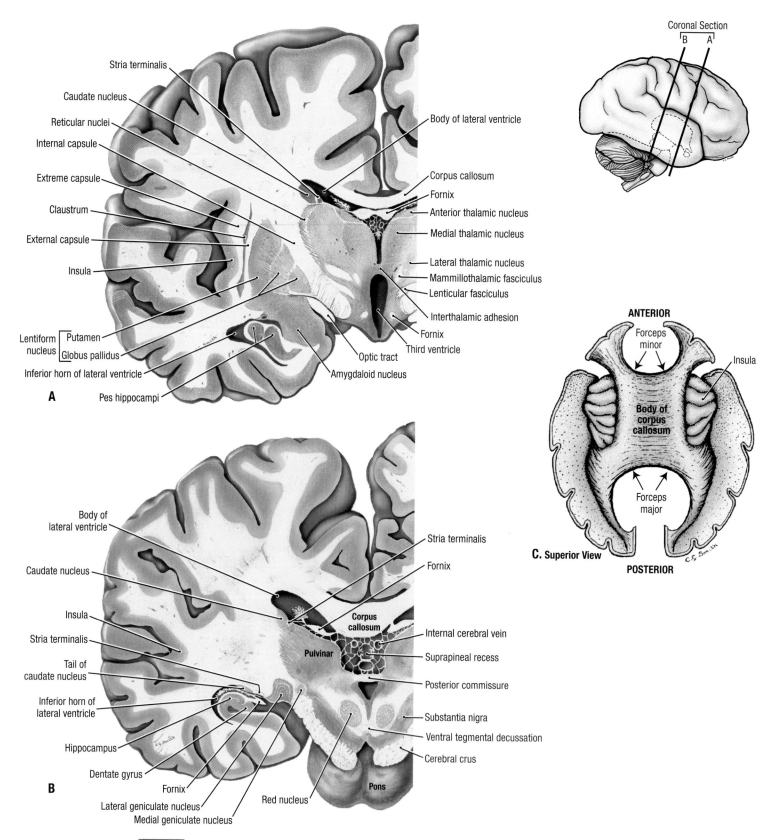

Coronal Section

ANTERIOR
Forceps
minor
Insula
Body of
corpus
callosum
Forceps
major
C. Superior View
POSTERIOR

7.34   **Coronal sections through the brain**

A. Level of middle of the diencephalon. **B.** Level of pulvinar and junction of the midbrain and diencephalon. **C.** Corpus callosum. The body of the corpus callosum connects the two cerebral hemispheres; the minor (frontal) forceps (at the genu of corpus callosum) connects the frontal lobes, and the major (occipital) forceps (at splenium) connects the occipital lobes.

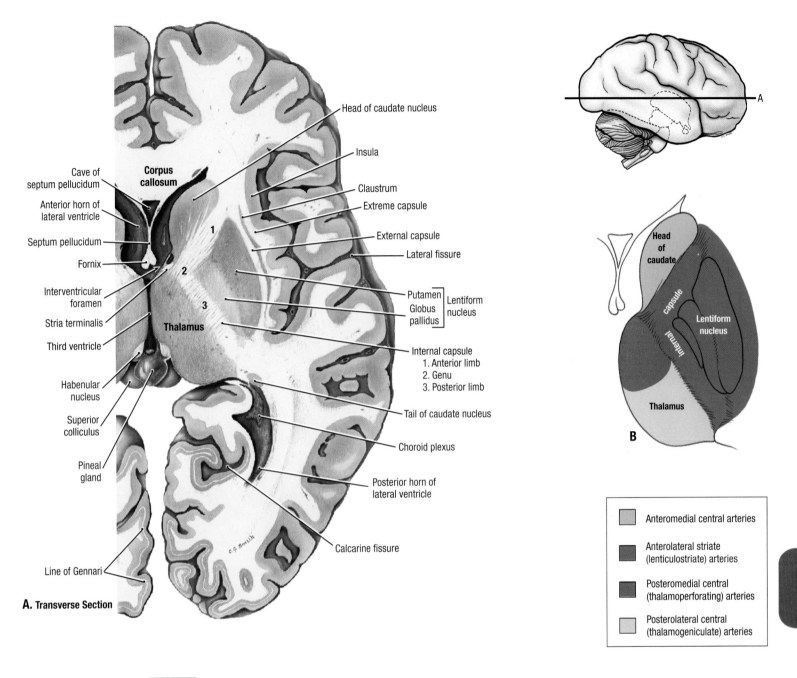

Head of caudate nucleus

Insula

Corpus callosum

Cave of septum pellucidum

Anterior horn of lateral ventricle

Septum pellucidum

Fornix

Interventricular foramen

Stria terminalis

Third ventricle

Habenular nucleus

Superior colliculus

Pineal gland

Line of Gennari

Claustrum

Extreme capsule

External capsule

Lateral fissure

Putamen — Lentiform
Globus pallidus — nucleus

Internal capsule
1. Anterior limb
2. Genu
3. Posterior limb

Tail of caudate nucleus

Choroid plexus

Posterior horn of lateral ventricle

Calcarine fissure

Thalamus

A. Transverse Section

Head of caudate

Internal capsule

Lentiform nucleus

Thalamus

B

Anteromedial central arteries

Anterolateral striate (lenticulostriate) arteries

Posteromedial central (thalamoperforating) arteries

Posterolateral central (thalamogeniculate) arteries

7.35   **Axial sections through the thalamus, caudate nucleus, and lentiform nucleus**

A. Relationships of the internal capsule. **B.** Blood supply of region.

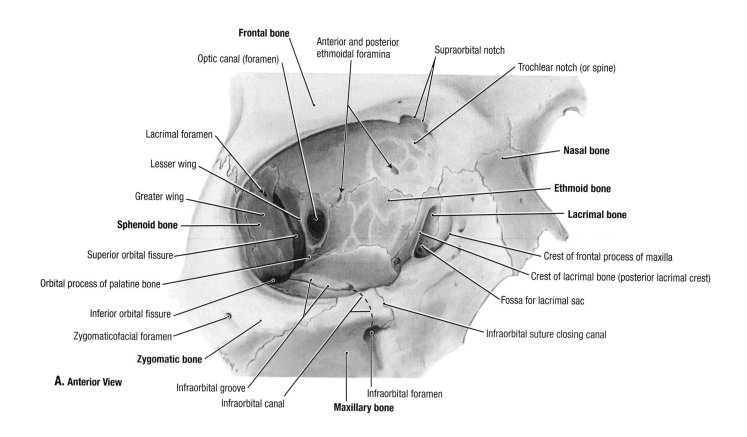

Frontal bone

Optic canal (foramen)

Anterior and posterior ethmoidal foramina

Supraorbital notch

Trochlear notch (or spine)

Lacrimal foramen

Lesser wing

Greater wing

Sphenoid bone

Superior orbital fissure

Orbital process of palatine bone

Inferior orbital fissure

Zygomaticofacial foramen

Zygomatic bone

A. Anterior View

Infraorbital groove

Infraorbital canal

Infraorbital foramen

Maxillary bone

Nasal bone

Ethmoid bone

Lacrimal bone

Crest of frontal process of maxilla

Crest of lacrimal bone (posterior lacrimal crest)

Fossa for lacrimal sac

Infraorbital suture closing canal

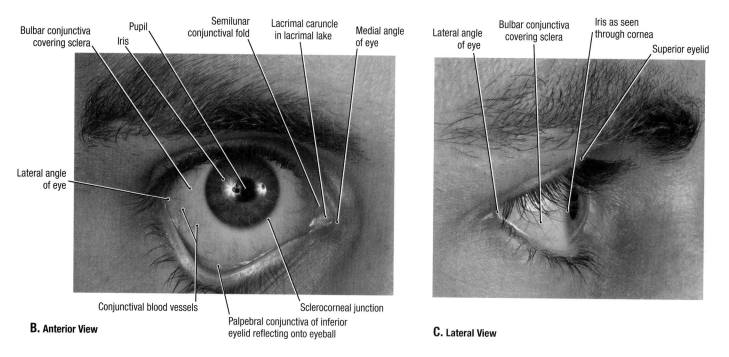

Bulbar conjunctiva covering sclera

Pupil

Iris

Semilunar conjunctival fold

Lacrimal caruncle in lacrimal lake

Medial angle of eye

Lateral angle of eye

Conjunctival blood vessels

Sclerocorneal junction

Palpebral conjunctiva of inferior eyelid reflecting onto eyeball

B. Anterior View

Lateral angle of eye

Bulbar conjunctiva covering sclera

Iris as seen through cornea

Superior eyelid

C. Lateral View

7.36 **Orbital cavity and surface anatomy of the eye**

A. Bones and features of the orbital cavity. **B** and **C.** Surface anatomy of the eye. In **B,** the inferior eyelid is everted to demonstrate the palpebral conjunctiva.

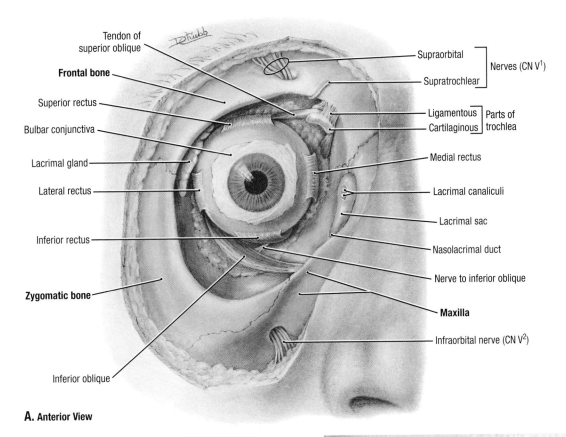

Tendon of superior oblique

Frontal bone

Superior rectus

Bulbar conjunctiva

Lacrimal gland

Lateral rectus

Inferior rectus

Zygomatic bone

Inferior oblique

Supraorbital

Supratrochlear

} Nerves (CN V¹)

Ligamentous } Parts of
Cartilaginous } trochlea

Medial rectus

Lacrimal canaliculi

Lacrimal sac

Nasolacrimal duct

Nerve to inferior oblique

Maxilla

Infraorbital nerve (CN V²)

A. Anterior View

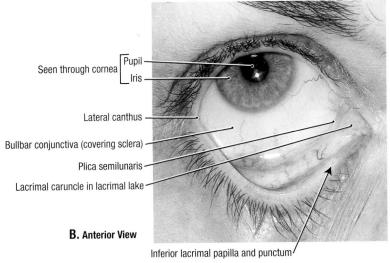

Seen through cornea { Pupil

Iris

Lateral canthus

Bullbar conjunctiva (covering sclera)

Plica semilunaris

Lacrimal caruncle in lacrimal lake

B. Anterior View

Inferior lacrimal papilla and punctum

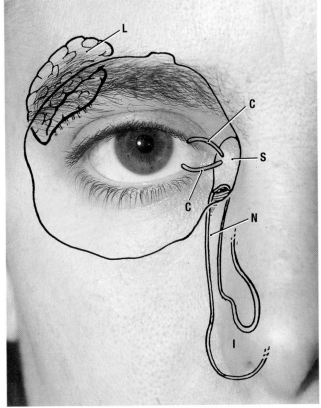

L

C

S

C

N

I

C. Anterior View

7.37　Eye and lacrimal apparatus

A. Anterior dissection of orbital cavity. The eyelids, orbital septum, levator palpebrae superioris, and some fat are removed. **B.** Surface features, with the inferior eyelid everted. **C.** Surface projection of lacrimal apparatus. Tears, secreted by the lacrimal gland (*L*) in the superolateral angle of the bony orbit, pass over the eyeball and enter the lacrimal lake at the medial angle of the eye; from here they drain through the lacrimal puncta and lacrimal canaliculi (*C*) to the lacrimal sac (*S*). The lacrimal sac drains into the nasolacrimal duct (*N*), which empties into the inferior meatus (*I*) of the nose.

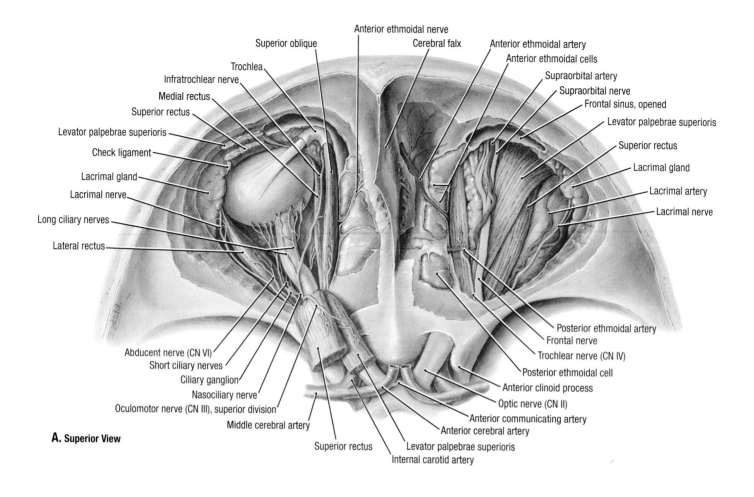

Anterior ethmoidal nerve
Superior oblique
Cerebral falx
Anterior ethmoidal artery
Trochlea
Anterior ethmoidal cells
Infratrochlear nerve
Supraorbital artery
Medial rectus
Supraorbital nerve
Superior rectus
Frontal sinus, opened
Levator palpebrae superioris
Levator palpebrae superioris
Check ligament
Superior rectus
Lacrimal gland
Lacrimal gland
Lacrimal nerve
Lacrimal artery
Long ciliary nerves
Lacrimal nerve
Lateral rectus

Posterior ethmoidal artery
Frontal nerve
Abducent nerve (CN VI)
Trochlear nerve (CN IV)
Short ciliary nerves
Posterior ethmoidal cell
Ciliary ganglion
Anterior clinoid process
Nasociliary nerve
Optic nerve (CN II)
Oculomotor nerve (CN III), superior division
Anterior communicating artery
Middle cerebral artery
Anterior cerebral artery
Superior rectus
Levator palpebrae superioris
Internal carotid artery

A. Superior View

7.38 Orbital cavity, superior approach

A. Superficial dissection.

On the right side of figure **A:**

- The orbital plate of the frontal bone is removed.
- The levator palpebrae superioris muscle lies superficial to the superior rectus muscle.
- The trochlear, frontal, and lacrimal nerves lie immediately inferior to the roof of the orbital cavity.

On the left side of figure **A:**

- The levator palpebrae and superior rectus muscles are reflected.
- The superior division of the oculomotor nerve (CN III) supplies the superior rectus and levator palpebrae muscles.
- The trochlear nerve (CN IV) lies on the medial side of the superior oblique muscle, and the abducent nerve (CN VI) on the medial side of the lateral rectus muscle.
- The lacrimal nerve runs superior to the lateral rectus muscle supplying sensory fibers to the conjunctiva and skin of the superior eyelid; it receives a communicating branch of the zygomaticotemporal nerve carrying secretory motor fibers from the pterygopalatine ganglion to the lacrimal gland.
- The ciliary ganglion, placed between the lateral rectus muscle and the optic nerve (CN II), gives rise to many short ciliary nerves; the nasociliary nerve gives rise to two long ciliary nerves that anastomose with each other and the short ciliary nerves.

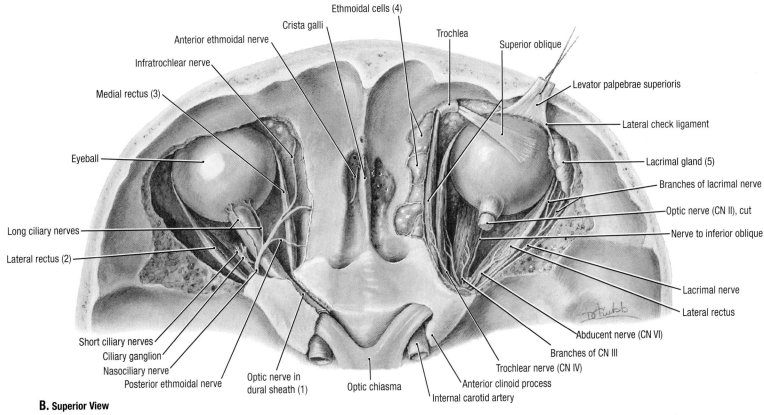

Ethmoidal cells (4)

Crista galli

Trochlea

Superior oblique

Anterior ethmoidal nerve

Infratrochlear nerve

Levator palpebrae superioris

Medial rectus (3)

Lateral check ligament

Eyeball

Lacrimal gland (5)

Branches of lacrimal nerve

Optic nerve (CN II), cut

Long ciliary nerves

Nerve to inferior oblique

Lateral rectus (2)

Lacrimal nerve

Lateral rectus

Short ciliary nerves

Abducent nerve (CN VI)

Ciliary ganglion

Branches of CN III

Nasociliary nerve

Trochlear nerve (CN IV)

Posterior ethmoidal nerve

Optic nerve in dural sheath (1)

Optic chiasma

Anterior clinoid process

Internal carotid artery

B. Superior View

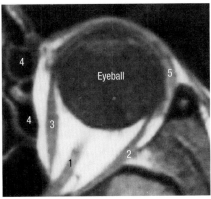

C. Axial MRI

7.38   Orbital cavity, superior approach *(continued)*

B. Deep dissection before *(right side)* and after *(left side)* section of the optic nerve. **C.** Transverse (axial) MRI of orbital cavity. (The *numbers* refer to structures labeled in **B**).

Observe on the right side of figure **B:**

- The eyeball occupies the anterior half of the orbital cavity.
- Nerves supplying the four recti (superior, medial, inferior, lateral) enter their ocular surfaces (the superior rectus is not shown).

Observe on the left of figure **B:**

- The ciliary ganglion lies posteriorly between the lateral rectus muscle and the sheath of the optic nerve.
- The nasociliary nerve sends a branch to the ciliary ganglion and crosses the optic nerve, where it gives off two long ciliary nerves (sensory to the eyeball and cornea) and the posterior ethmoidal nerve (to the sphenoidal sinus and posterior ethmoidal cells). The nasociliary nerve then divides into the anterior ethmoidal and infratrochlear nerves.

Frontal nerve
Lacrimal nerve
Short ciliary nerve
Dural sheath covering optic nerve
Superior rectus
Ciliary ganglion
Nasociliary nerve
Abducent nerve (CN VI)
Lateral rectus
Oculomotor nerve (CN III), inferior branch
Foramen rotundum
Maxillary nerve (V²)
Sphenopalatine artery
Maxillary nerve (V²)
Maxillary artery
Infraorbital nerve
Infraorbital artery
Zygomatic bone

A. Lateral View

Levator palpebrae superioris
Lacrimal gland
Conjunctival sac
Inferior rectus
Inferior oblique
Nerve to inferior oblique

7.39 **Lateral aspect of the orbit and structure of the eyelid**

A. Dissection.

• The ciliary ganglion receives sensory fibers from the nasociliary branches of VI, postsynaptic sympathetic fibers from the continuation of the internal carotid plexus extending along the ophthalmic artery, and presynaptic parasympathetic fibers from the inferior branch of the oculomotor nerve; only the latter synapse in the ganglion.

• Interruption of a cervical sympathetic trunk causes paralysis of the ipsilateral superior tarsal muscle, resulting in ptosis. This is part of Horner's syndrome, which also includes constriction of the pupil, posterior retraction of the eye, and redness, dryness, and increased temperature on the affected side of the face.

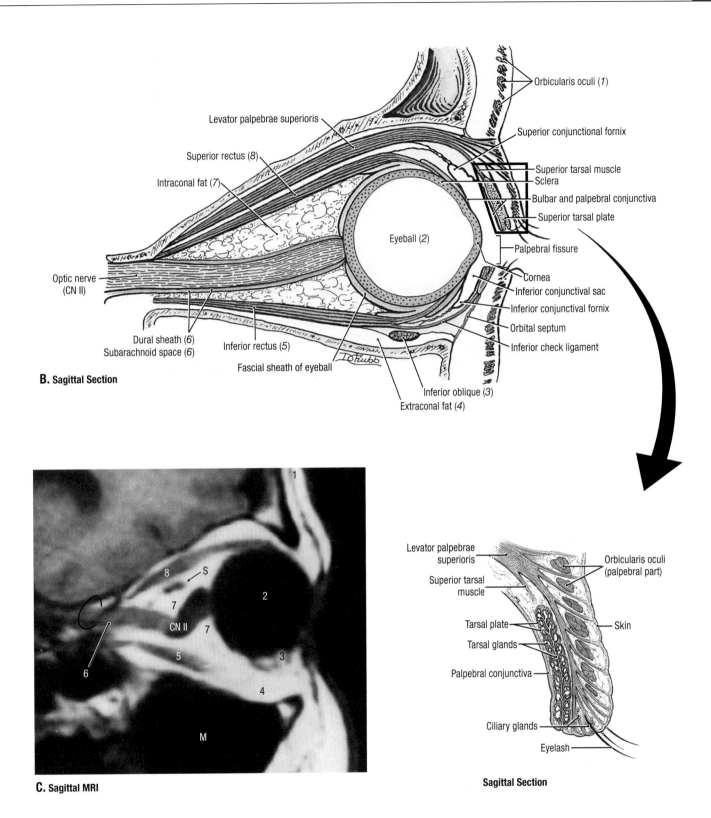

B. Sagittal Section

Orbicularis oculi (*1*)

Levator palpebrae superioris

Superior conjunctional fornix

Superior rectus (*8*)

Superior tarsal muscle

Sclera

Intraconal fat (*7*)

Bulbar and palpebral conjunctiva

Superior tarsal plate

Eyeball (*2*)

Palpebral fissure

Optic nerve (CN II)

Cornea

Inferior conjunctival sac

Inferior conjunctival fornix

Orbital septum

Inferior check ligament

Dural sheath (*6*)

Subarachnoid space (*6*)

Inferior rectus (*5*)

Fascial sheath of eyeball

Inferior oblique (*3*)

Extraconal fat (*4*)

C. Sagittal MRI

Levator palpebrae superioris

Orbicularis oculi (palpebral part)

Superior tarsal muscle

Tarsal plate

Skin

Tarsal glands

Palpebral conjunctiva

Ciliary glands

Eyelash

Sagittal Section

7.39   **Lateral aspect of the orbit and structure of the eyelid (*continued*)**

B. Sagittal section of orbit and eyelid. **C.** Sagittal MRI through the optic nerve. The *numbers* refer to structures labeled in **B**; *S*, superior ophthalmic vein; *M*, maxillary sinus; *circled*, optic foramen.

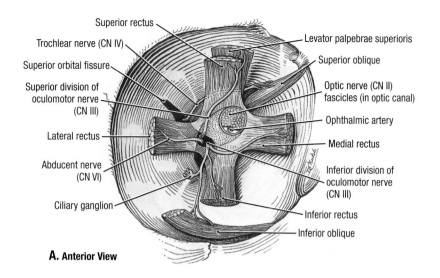

A. Anterior View

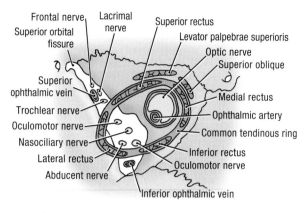

B. Anterior View

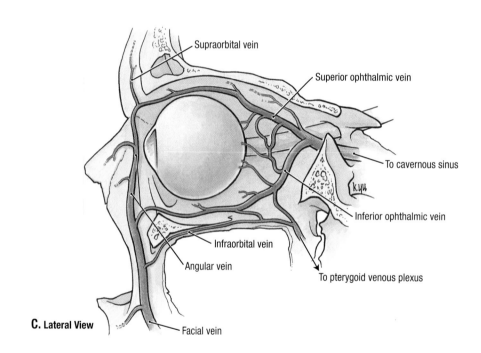

C. Lateral View

7.40 Nerves and veins of the orbit

A and **B.** Nerves of orbit in relation to the orbital fissures and the common tendinous ring. The common tendinous ring is formed by the origin of the four recti and encircles the dural sheath of the optic nerve, CN VI, and the superior and inferior branches of CN III; the nasociliary nerve also passes through this cuff. **C.** Ophthalmic veins. The superior and inferior ophthalmic veins receive the vorticose veins from the eyeball (not shown) and empty into the cavernous sinus posteriorly and the pterygoid plexus inferiorly and communicate with the facial and supraorbital veins anteriorly.

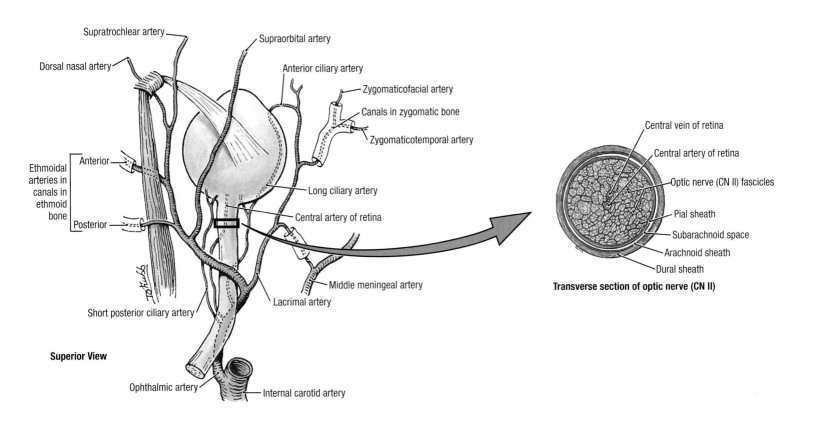

Superior View

Transverse section of optic nerve (CN II)

TABLE 7.6 ARTERIES OF ORBIT

ARTERY	ORIGIN	COURSE AND DISTRIBUTION
Ophthalmic	Internal carotid artery	Traverses optic foramen to reach orbital cavity
Central artery of retina	Ophthalmic artery	Runs in dural sheath of optic nerve and pierces nerve near eyeball; appears at center of optic disc; supplies optic retina (except cones and rods)
Supraorbital		Passes superiorly and posteriorly from supraorbital foramen to supply forehead and scalp
Supratrochlear		Passes from supraorbital margin to forehead and scalp
Lacrimal		Passes along superior border of lateral rectus muscle to supply lacrimal gland, conjunctiva, and eyelids
Dorsal nasal		Courses along dorsal aspect of nose and supplies its surface
Short posterior ciliary		Pierces sclera at periphery of optic nerve to supply choroid, which, in turn, supplies cones and rods of optic retina
Long posterior ciliary		Pierces sclera to supply ciliary body and iris
Posterior ethmoidal		Passes through posterior ethmoidal foramen to posterior ethmoidal cells
Anterior ethmoidal		Passes through anterior ethmoidal foramen to anterior cranial fossa; supplies anterior and middle ethmoidal cells, frontal sinus, nasal cavity, and skin on dorsum of nose
Anterior ciliary	Muscular (rectus) branches of ophthalmic artery	Pierces sclera at attachments of rectus muscles and forms network in iris and ciliary body
Infraorbital	Third part of maxillary artery	Passes along infraorbital groove and foramen to face

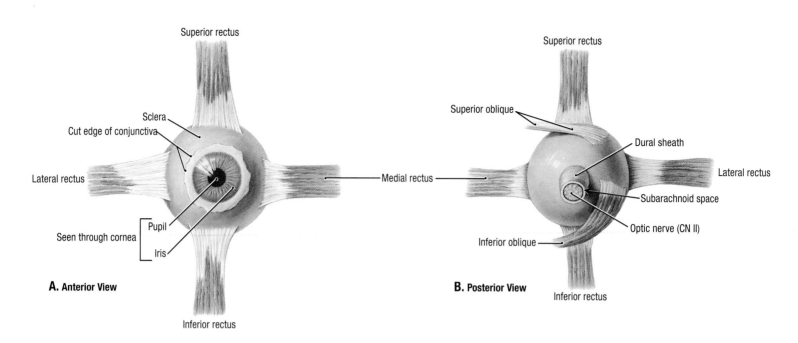

Superior rectus

Sclera
Cut edge of conjunctiva

Lateral rectus

Medial rectus

Pupil
Seen through cornea
Iris

A. Anterior View

Inferior rectus

Superior rectus

Superior oblique

Dural sheath

Lateral rectus

Subarachnoid space

Optic nerve (CN II)

Inferior oblique

B. Posterior View

Inferior rectus

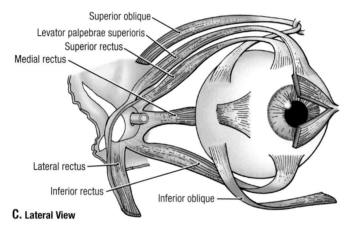

Superior oblique
Levator palpebrae superioris
Superior rectus
Medial rectus

Lateral rectus
Inferior rectus
Inferior oblique

C. Lateral View

TABLE 7.7 MUSCLES OF ORBIT

MUSCLE	ORIGIN	INSERTION	INNERVATION	MAIN ACTIONS
Levator palpebrae superioris	Lesser wing of sphenoid bone, superior and anterior to optic canal	Tarsal plate and skin of superior (upper) eyelid	Oculomotor nerve (CN III); deep layer (superior tarsal muscle) is supplied by sympathetic fibers	Elevates superior (upper) eyelid
Superior rectus	Common tendinous ring	Sclera just posterior to cornea	Oculomotor nerve (CN III)	Elevates, adducts, and rotates eyeball medially
Inferior rectus				Depresses, adducts, and rotates eyeball medially
Lateral rectus			Abducent nerve (CN VI)	Abducts eyeball
Medial rectus			Oculomotor nerve (CN III)	Adducts eyeball
Superior oblique	Body of sphenoid bone	Its tendon passes through a fibrous ring or trochlea, changes its direction, and inserts into sclera deep to superior rectus muscle	Trochlear nerve (CN IV)	Abducts, depresses, and medially rotates eyeball
Inferior oblique	Anterior part of floor of orbit	Sclera deep to lateral rectus muscle	Oculomotor nerve (CN III)	Abducts, elevates, and laterally rotates eyeball

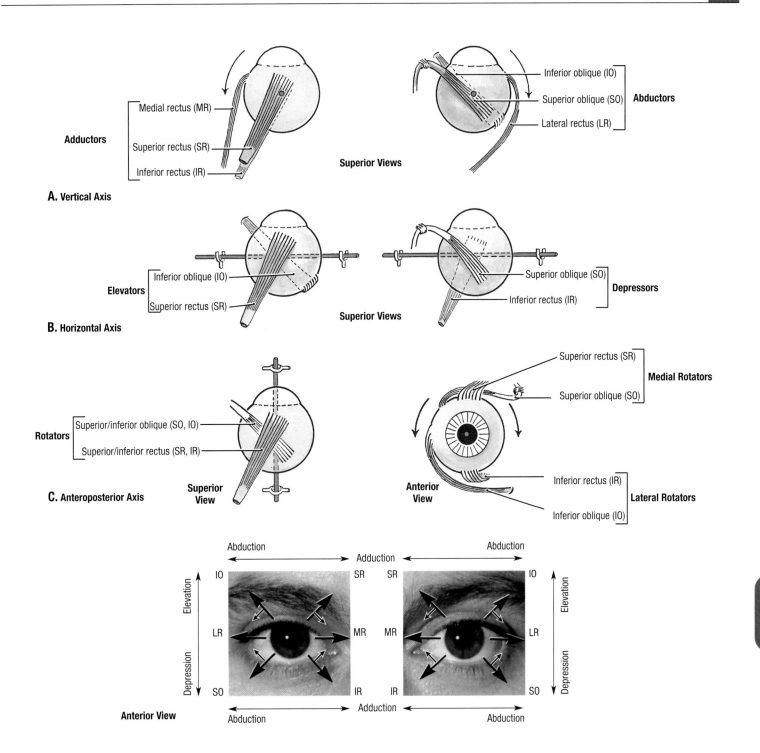

A. Vertical Axis

B. Horizontal Axis

C. Anteroposterior Axis

TABLE 7.8	ACTIONS OF MUSCLES OF THE ORBIT		
	ACTIONS		
MUSCLE	VERTICAL AXIS **(A)**	HORIZONTAL AXIS **(B)**	ANTEROPOSTERIOR AXIS **(C)**
Superior rectus	Elevates	Adducts	Rotates medially
Inferior rectus	Depresses	Adducts	Rotates laterally
Superior oblique	Depresses	Abducts	Rotates medially
Inferior oblique	Elevates	Abducts	Rotates laterally
Medial rectus	N/A	Adducts	N/A
Lateral rectus	N/A	Abducts	N/A

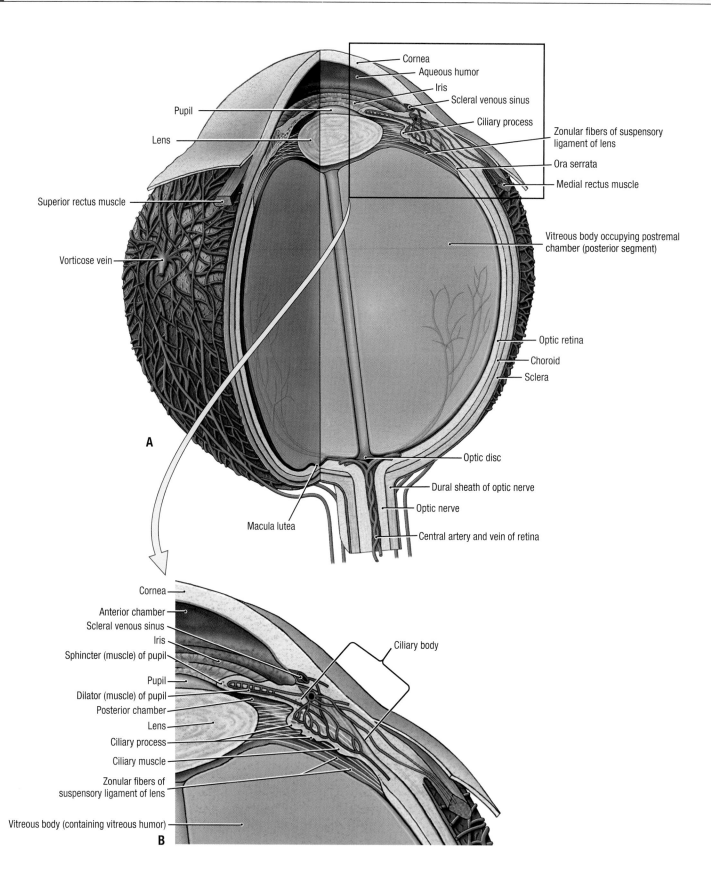

7.41 **Illustration of a dissected eyeball**

A. Parts of the eyeball. **B.** Ciliary region. The aqueous humor is produced by the ciliary processes and provides nutrients for the avascular cornea and lens; the aqueous humor drains into the scle-ral venous sinus (also called the *sinus venosus sclerae* or *canal of Schlemm*). If drainage of the aqueous humor is reduced significantly, pressure builds up in the chambers of the eye (glaucoma).

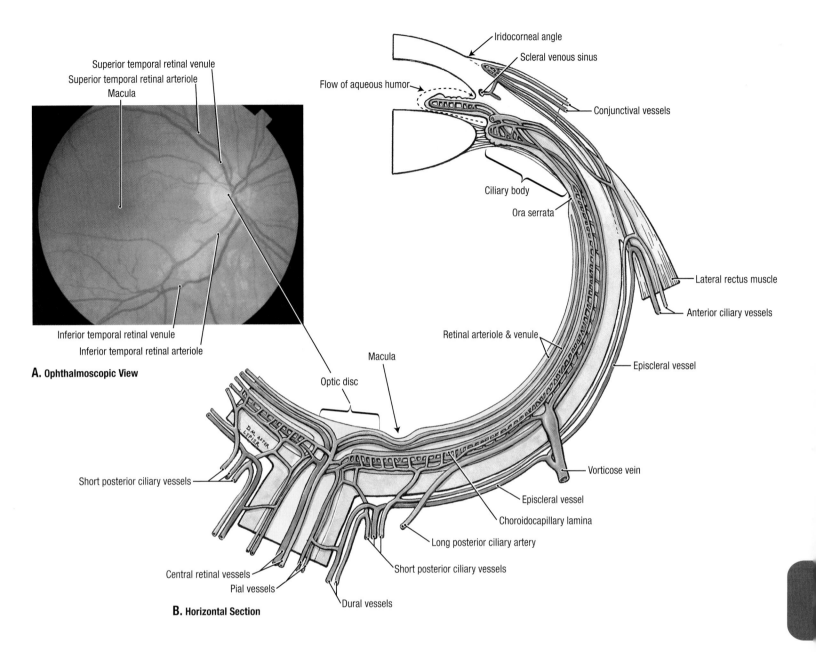

A. Ophthalmoscopic View

Superior temporal retinal venule
Superior temporal retinal arteriole
Macula

Inferior temporal retinal venule
Inferior temporal retinal arteriole

Iridocorneal angle
Scleral venous sinus
Flow of aqueous humor
Conjunctival vessels
Ciliary body
Ora serrata
Lateral rectus muscle
Anterior ciliary vessels
Retinal arteriole & venule
Episcleral vessel
Macula
Optic disc
Episcleral vessel
Vorticose vein
Choroidocapillary lamina
Long posterior ciliary artery
Short posterior ciliary vessels
Short posterior ciliary vessels
Central retinal vessels
Pial vessels
Dural vessels

D.M. AFTER
ESPER

B. Horizontal Section

7.42   Ocular fundus and blood supply to the eyeball

A. Right ocular fundus, ophthalmoscopic view. Retinal venules (wider) and retinal arterioles (narrower) radiate from the center of the oval optic disc, formed in relation to the entry of the optic nerve into the eyeball. The *round, dark area* lateral to the disc is the macula; branches of vessels extend to this area, but do not reach its center, the fovea centralis, a depressed spot that is the area of most acute vision. It is avascular but, like the rest of the outermost (cones and rods) layer of the retina, is nourished by the adjacent choriocapillaris. **B.** Blood supply to eyeball. The eyeball has three layers: (a) the external, fibrous layer is the sclera and cornea; (b) the middle, vascular layer is the choroid, ciliary body, and iris; and (c) the internal, neural layer or retina consists of a pigment cell layer and a neural layer. The central artery of the retina, a branch of the ophthalmic artery, is an end artery. Of the eight posterior ciliary arteries, six are short posterior ciliary arteries and supply the choroid, which in turn nourishes the outer, nonvascular layer of the retina. Two long posterior ciliary arteries, one on each side of the eyeball, run between the sclera and choroid to anastomose with the anterior ciliary arteries, which are derived from muscular branches. The choroid is drained by posterior ciliary veins, and four to five vorticose veins drain into the ophthalmic veins.

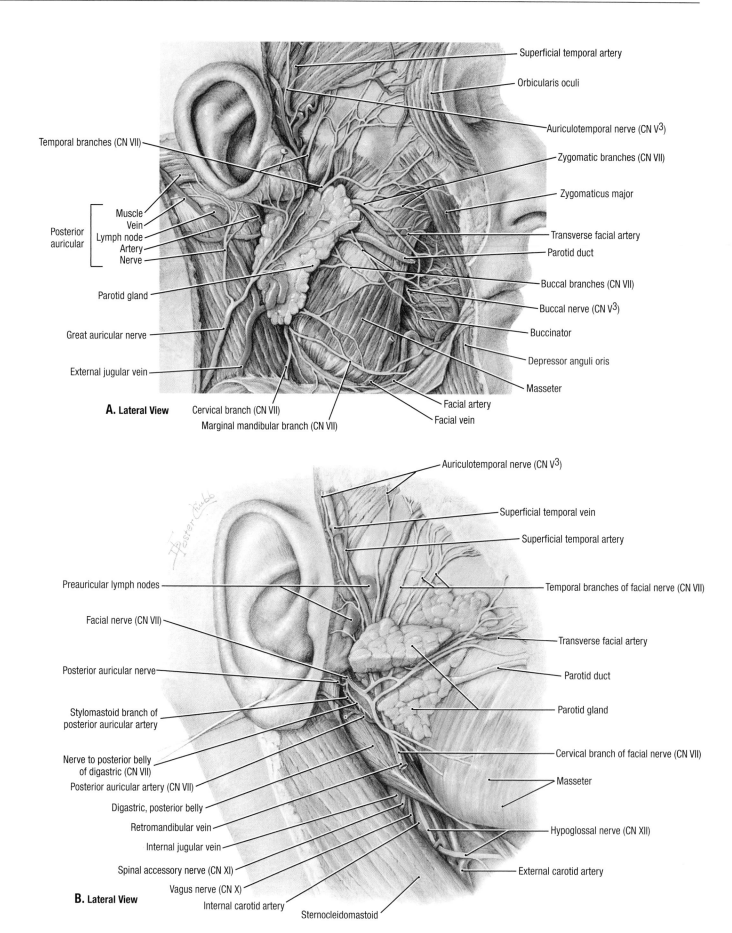

Superficial temporal artery

Orbicularis oculi

Auriculotemporal nerve (CN V³)

Zygomatic branches (CN VII)

Zygomaticus major

Transverse facial artery

Parotid duct

Buccal branches (CN VII)

Buccal nerve (CN V³)

Buccinator

Depressor anguli oris

Masseter

Facial artery

Facial vein

Temporal branches (CN VII)

Posterior auricular { Muscle, Vein, Lymph node, Artery, Nerve }

Parotid gland

Great auricular nerve

External jugular vein

A. Lateral View

Cervical branch (CN VII)

Marginal mandibular branch (CN VII)

Auriculotemporal nerve (CN V³)

Superficial temporal vein

Superficial temporal artery

Temporal branches of facial nerve (CN VII)

Transverse facial artery

Parotid duct

Parotid gland

Cervical branch of facial nerve (CN VII)

Masseter

Hypoglossal nerve (CN XII)

External carotid artery

Preauricular lymph nodes

Facial nerve (CN VII)

Posterior auricular nerve

Stylomastoid branch of posterior auricular artery

Nerve to posterior belly of digastric (CN VII)

Posterior auricular artery (CN VII)

Digastric, posterior belly

Retromandibular vein

Internal jugular vein

Spinal accessory nerve (CN XI)

Vagus nerve (CN X)

Internal carotid artery

Sternocleidomastoid

B. Lateral View

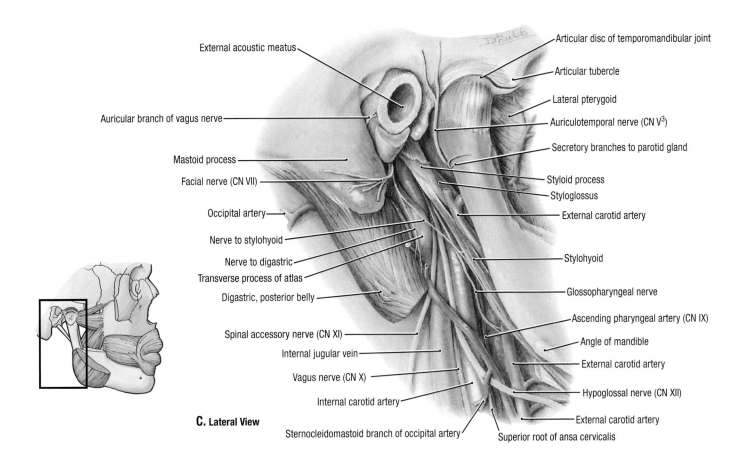

External acoustic meatus
Articular disc of temporomandibular joint
Articular tubercle
Lateral pterygoid
Auricular branch of vagus nerve
Auriculotemporal nerve (CN V³)
Secretory branches to parotid gland
Mastoid process
Facial nerve (CN VII)
Styloid process
Styloglossus
Occipital artery
External carotid artery
Nerve to stylohyoid
Nerve to digastric
Transverse process of atlas
Stylohyoid
Digastric, posterior belly
Glossopharyngeal nerve
Spinal accessory nerve (CN XI)
Ascending pharyngeal artery (CN IX)
Internal jugular vein
Angle of mandible
External carotid artery
Vagus nerve (CN X)
Hypoglossal nerve (CN XII)
Internal carotid artery
External carotid artery
C. Lateral View
Sternocleidomastoid branch of occipital artery
Superior root of ansa cervicalis

7.43   Parotid region

A and **B.** Relationships of the branches of the facial nerve and vessels to parotid gland and duct. **A.** Superficial dissection. **B.** Deep dissection with part of the gland removed. The facial nerve (CN VII) supplies motor innervation to the muscles of facial expression; it forms a plexus within the parotid gland and the branches of which radiate over the face, anastomosing with each other and the branches of the trigeminal nerve. During parotidectomy (surgical excision of the gland), identification, dissection, and preservation of the facial nerve are critical. **C.** Deep dissection following removal of the parotid gland. The facial nerve, posterior belly of the digastric muscle, and its nerve are retracted; the external carotid artery, stylohyoid muscle, and the nerve to the stylohyoid remain in situ. The internal jugular vein, internal carotid artery, and glossopharyngeal (CN IX), vagus (CN X), accessory (CN XI), and hypoglossal (CN XII) nerves cross anterior to the transverse process of the atlas and deep to the styloid process. Care must be taken during surgical procedures involving the temporomandibular joint to preserve the branches of the facial nerve that overlie the joint and the articular branches of the auriculotemporal nerve that enter the joint.

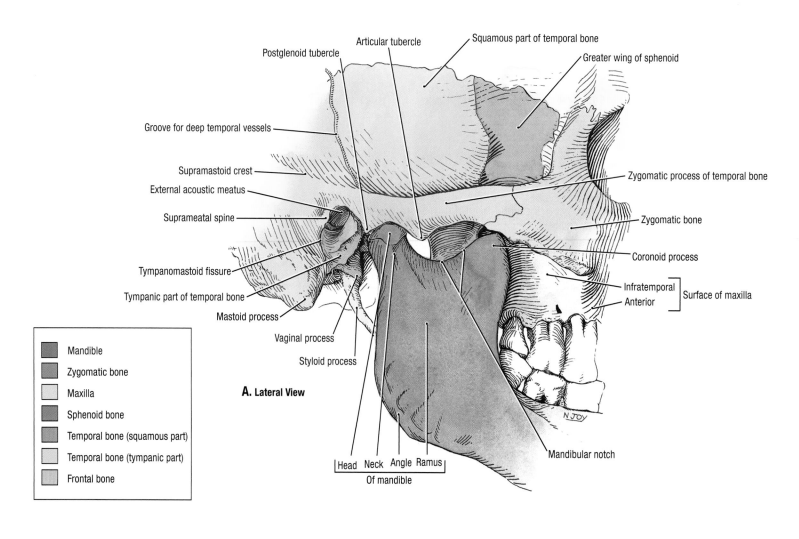

Articular tubercle
Postglenoid tubercle
Squamous part of temporal bone
Greater wing of sphenoid
Groove for deep temporal vessels
Supramastoid crest
External acoustic meatus
Suprameatal spine
Tympanomastoid fissure
Tympanic part of temporal bone
Mastoid process
Vaginal process
Styloid process
Zygomatic process of temporal bone
Zygomatic bone
Coronoid process
Infratemporal
Anterior
Surface of maxilla
Mandibular notch

Mandible
Zygomatic bone
Maxilla
Sphenoid bone
Temporal bone (squamous part)
Temporal bone (tympanic part)
Frontal bone

A. Lateral View

Head Neck Angle Ramus
Of mandible

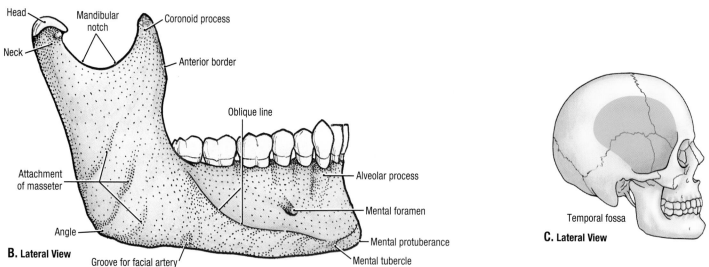

Head
Mandibular notch
Coronoid process
Neck
Anterior border
Oblique line
Attachment of masseter
Alveolar process
Mental foramen
Angle
Mental protuberance
B. Lateral View
Groove for facial artery
Mental tubercle

Temporal fossa
C. Lateral View

7.44 **Temporal and infratemporal fossa and mandible**

A. Bones and bony features. Note that the zygomatic process of the temporal bone is the boundary between the temporal fossa superiorly and the infratemporal fossa inferiorly. **B.** External surface of the mandible. **C.** Temporal fossa.

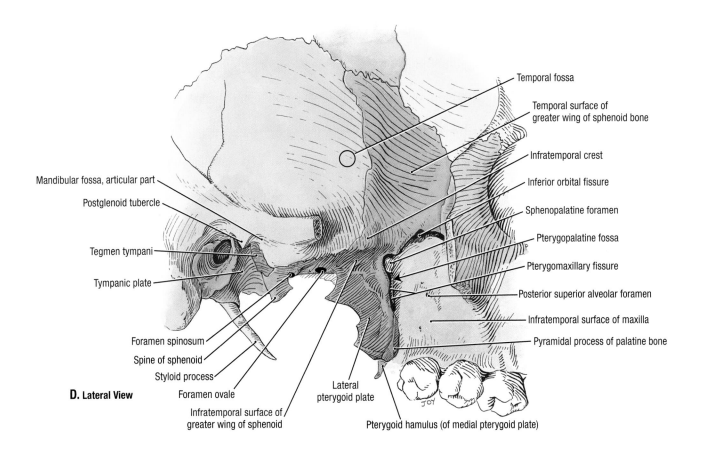

D. Lateral View

Temporal fossa
Temporal surface of greater wing of sphenoid bone
Infratemporal crest
Inferior orbital fissure
Sphenopalatine foramen
Pterygopalatine fossa
Pterygomaxillary fissure
Posterior superior alveolar foramen
Infratemporal surface of maxilla
Pyramidal process of palatine bone

Mandibular fossa, articular part
Postglenoid tubercle
Tegmen tympani
Tympanic plate
Foramen spinosum
Spine of sphenoid
Styloid process
Foramen ovale
Infratemporal surface of greater wing of sphenoid
Lateral pterygoid plate
Pterygoid hamulus (of medial pterygoid plate)

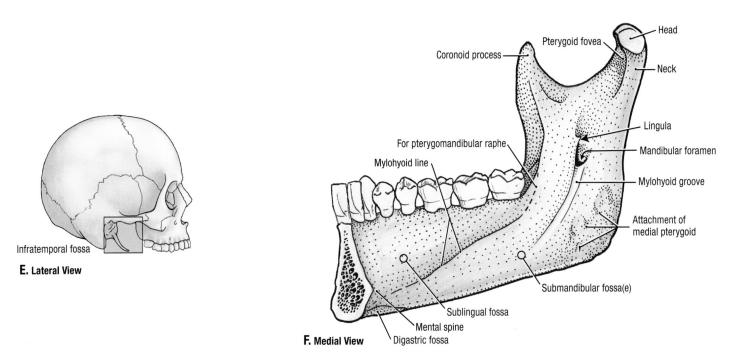

E. Lateral View

Infratemporal fossa

F. Medial View

Coronoid process
Pterygoid fovea
Head
Neck
Lingula
Mandibular foramen
Mylohyoid groove
Attachment of medial pterygoid
Submandibular fossa(e)

For pterygomandibular raphe
Mylohyoid line
Sublingual fossa
Mental spine
Digastric fossa

7.44 **Temporal and infratemporal fossa and mandible (continued)**

D. Bones and bony features of the infratemporal fossa. The mandible and part of the zygomatic arch have been removed. **E.** Infratemporal fossa. **F.** Internal surface of the mandible.

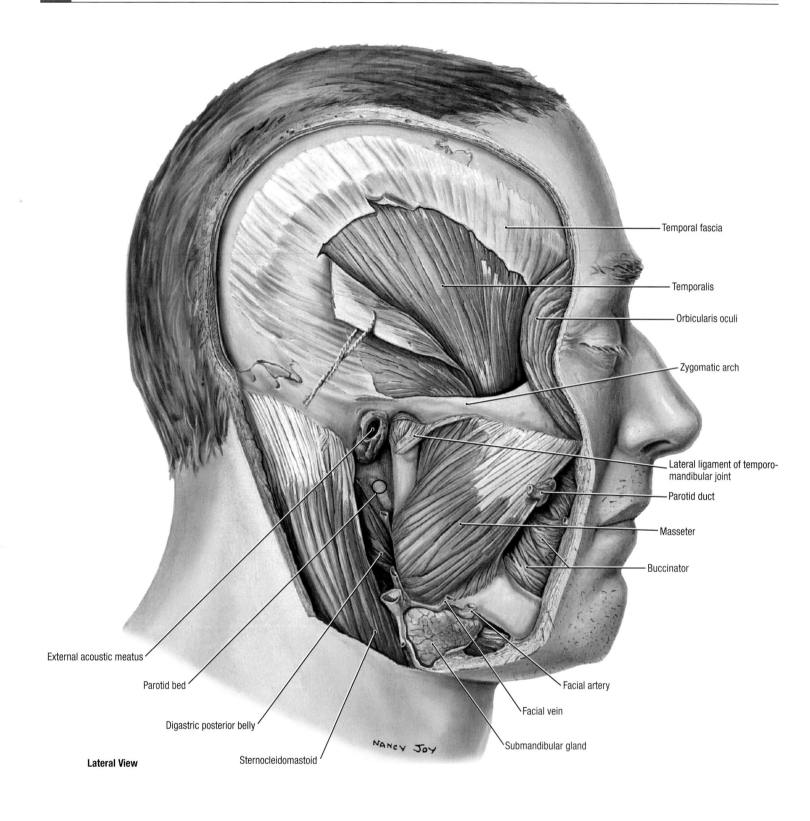

Temporal fascia

Temporalis

Orbicularis oculi

Zygomatic arch

Lateral ligament of temporo-mandibular joint

Parotid duct

Masseter

Buccinator

Facial artery

Facial vein

Submandibular gland

Sternocleidomastoid

Digastric posterior belly

Parotid bed

External acoustic meatus

NANCY JOY

Lateral View

7.45   **Temporalis and masseter—I**

- The temporalis and masseter muscles are supplied by the trigeminal nerve, and both elevate the mandible. The orbicularis oculi and buccinator muscles are supplied by the facial nerve.
- The sternocleidomastoid muscle is the chief flexor of the head and neck; it forms the lateral part of the posterior boundary of the parotid region.

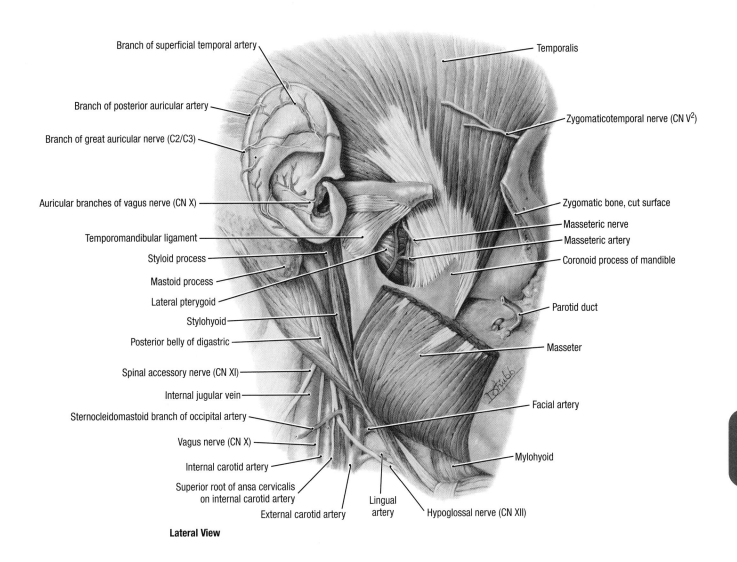

Branch of superficial temporal artery

Branch of posterior auricular artery

Branch of great auricular nerve (C2/C3)

Auricular branches of vagus nerve (CN X)

Temporomandibular ligament

Styloid process

Mastoid process

Lateral pterygoid

Stylohyoid

Posterior belly of digastric

Spinal accessory nerve (CN XI)

Internal jugular vein

Sternocleidomastoid branch of occipital artery

Vagus nerve (CN X)

Internal carotid artery

Superior root of ansa cervicalis on internal carotid artery

External carotid artery

Lingual artery

Hypoglossal nerve (CN XII)

Temporalis

Zygomaticotemporal nerve (CN V²)

Zygomatic bone, cut surface

Masseteric nerve

Masseteric artery

Coronoid process of mandible

Parotid duct

Masseter

Facial artery

Mylohyoid

Lateral View

7.46   **Temporalis and masseter—II**

- Part of the zygomatic arch and the masseter muscle have been removed to expose the attachment of the temporalis muscle to the coronoid process of the mandible.
- The internal jugular vein, internal carotid artery, and the vagus (CN X) are contained within a fascial carotid sheath (not shown). The external carotid artery and its lingual, facial, and occipital branches, and the spinal accessory (CN XI) and hypoglossal (CN XII) nerves pass deep to the posterior belly of the digastric muscle.

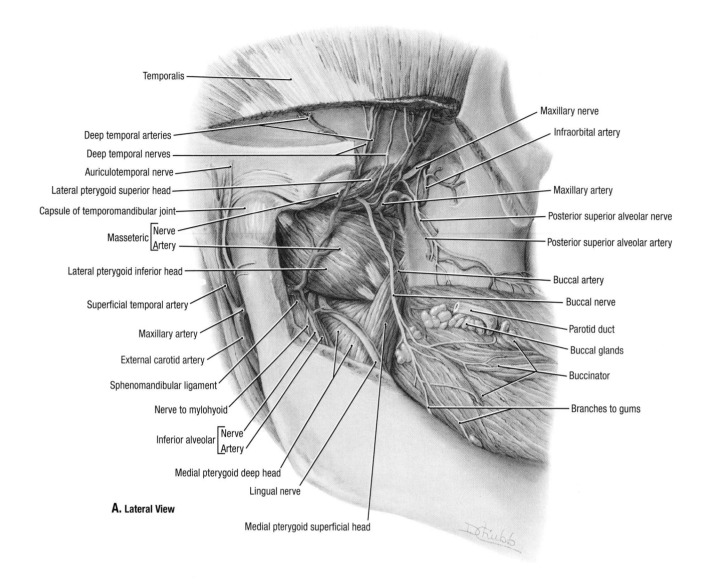

Temporalis

Deep temporal arteries

Deep temporal nerves

Auriculotemporal nerve

Lateral pterygoid superior head

Capsule of temporomandibular joint

Masseteric [Nerve / Artery]

Lateral pterygoid inferior head

Superficial temporal artery

Maxillary artery

External carotid artery

Sphenomandibular ligament

Nerve to mylohyoid

Inferior alveolar [Nerve / Artery]

Medial pterygoid deep head

Lingual nerve

A. Lateral View

Medial pterygoid superficial head

Maxillary nerve

Infraorbital artery

Maxillary artery

Posterior superior alveolar nerve

Posterior superior alveolar artery

Buccal artery

Buccal nerve

Parotid duct

Buccal glands

Buccinator

Branches to gums

7.47 Infratemporal region

A. Superficial dissection.

- The maxillary artery, the larger of two terminal branches of the external carotid, is divided into three parts relative to the lateral pterygoid muscle: the first or retromandibular part gives rise to branches that pass through bony foramina to enter the temporal bone, cranium, and mandible; the second part—in contact with the lateral pterygoid—supplies blood to the muscles of the region; and the third or pterygopalatine part gives rise to branches that accompany the branches of CN V^2 or the pterygopalatine ganglion arising within the pterygopalatine fossa (e.g., infraorbital and posterior superior alveolar arteries).
- The buccinator is pierced by the parotid duct, the ducts of the buccal glands, and sensory branches of the buccal nerve.
- The lateral pterygoid muscle arises by two heads (parts), one head from the roof, and the other head from the medial wall of the infratemporal fossa; both heads insert in relation to the temporomandibular joint—the superior head attaching primarily to the articular disc of the joint and the inferior head primarily to the anterior aspect of the neck of the mandible (pterygoid fovea).

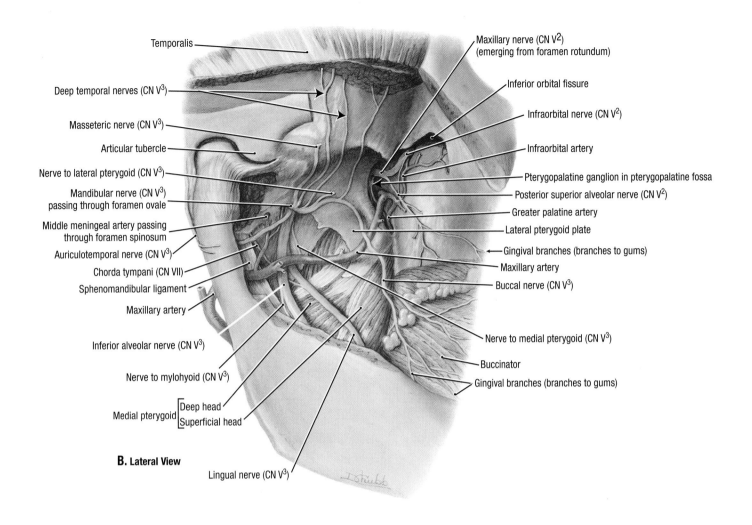

B. Lateral View

7.47   Infratemporal region (continued)

B. Deeper dissection.

- The lateral pterygoid muscle and most of the branches of the maxillary artery have been removed to expose the mandibular (CN V³) nerve entering the infratemporal fossa through the foramen ovale and the middle meningeal artery and vein (not shown) passing through the foramen spinosum.
- The medial pterygoid muscle arises from the medial surface of the lateral pterygoid plate; it has a small, superficial head that arises from the pyramidal process of the palatine bone.
- The inferior alveolar and lingual nerves descend on the medial pterygoid muscle. The inferior alveolar gives off the mylohyoid nerve to the mylohyoid muscle and anterior belly of the digastric muscle, and the lingual receives the chorda tympani, which carries secretory parasympathetic fibers and fibers of taste.
- Motor nerves arising from CN V³ supply the four muscles of mastication: the masseter, temporalis, and lateral and medial pterygoids. The buccal branch of the mandibular nerve is sensory; the buccal branch of the facial nerve is the motor supply to the buccinator muscle.

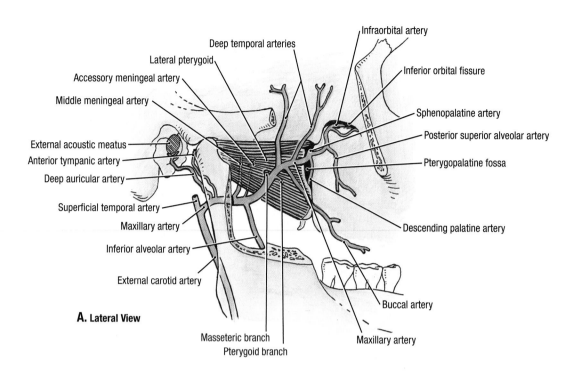

A. Lateral View

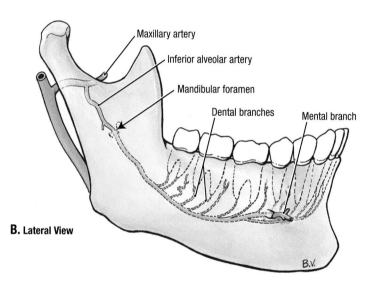

B. Lateral View

7.48 Branches of maxillary artery

A. Infratemporal region. **B.** Mandible.

- The maxillary artery arises at the neck of the mandible and is divided into three parts by the lateral pterygoid; it can pass medial or lateral to the lateral pterygoid.
- The branches of the first or retromandibular part pass through foramina or canals: the deep auricular to the external acoustic meatus, the anterior tympanic to the tympanic cavity, the middle and accessory meningeal to the cranial cavity, and the inferior alveolar to the mandible and teeth.
- The branches of the second part (directly related to the lateral pterygoid) supply muscles via the masseteric, deep temporal, pterygoid, and buccal branches.
- The branches of the third (pterygopalatine) part (posterior superior alveolar, infraorbital, descending palatine, and sphenopalatine arteries) arise immediately proximal to and within the pterygopalatine fossa.

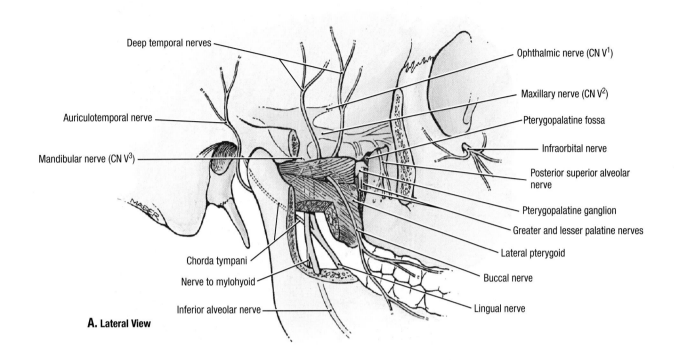

Deep temporal nerves

Ophthalmic nerve (CN V¹)

Auriculotemporal nerve

Maxillary nerve (CN V²)

Pterygopalatine fossa

Infraorbital nerve

Mandibular nerve (CN V³)

Posterior superior alveolar nerve

Pterygopalatine ganglion

Greater and lesser palatine nerves

Lateral pterygoid

Chorda tympani

Nerve to mylohyoid

Buccal nerve

Inferior alveolar nerve

Lingual nerve

A. Lateral View

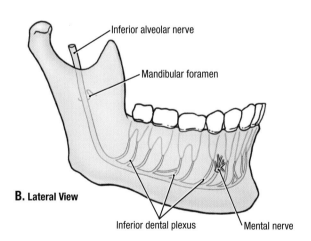

Inferior alveolar nerve

Mandibular foramen

B. Lateral View

Inferior dental plexus

Mental nerve

7.49 **Branches of maxillary nerve**

A. Infratemporal region and pterygopalatine fossa. Branches of the maxillary (CN V²) and mandibular (CN V³) nerves accompany branches from the three parts of the maxillary artery. **B.** Mandible.

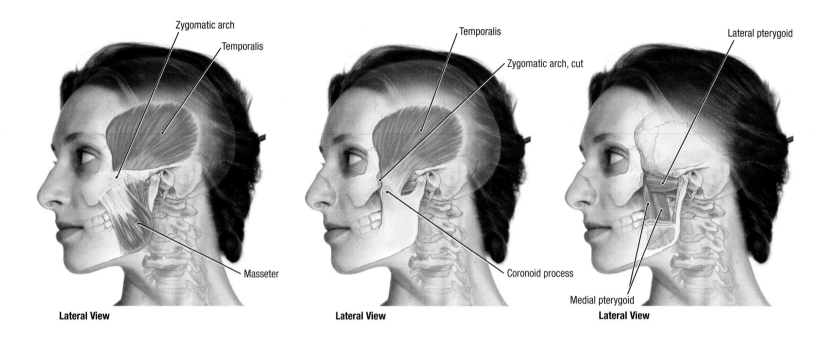

Zygomatic arch
Temporalis
Masseter
Lateral View

Temporalis
Zygomatic arch, cut
Coronoid process
Lateral View

Lateral pterygoid
Medial pterygoid
Lateral View

TABLE 7.9   MUSCLES ACTING ON THE TEMPOROMANDIBULAR JOINT

MUSCLE	ORIGIN	INSERTION	INNERVATION	MAIN ACTIONS
Temporalis	Floor of temporal fossa and deep surface of temporal fascia	Tip and medial surface of coronoid process and anterior border of ramus of mandible	Deep temporal branches of mandibular nerve (CN V^3)	Elevates mandible, closing jaws; its posterior fibers retrude mandible after protrusion
Masseter	Inferior border and medial surface of zygomatic arch	Lateral surface of ramus of mandible and its coronoid process	Mandibular nerve (CN V^3) through masseteric nerve that enters its deep surface	Elevates and protrudes mandible, thus closing jaws; deep fibers retrude it
Lateral pterygoid	*Superior head:* infratemporal surface and infratemporal crest of greater wing of sphenoid bone *Inferior head:* lateral surface of lateral pterygoid plate	Neck of mandible, articular disc, and capsule of temporomandibular joint	Mandibular nerve (CN V^3) through lateral pterygoid nerve from anterior trunk, which enters its deep surface	Acting together, they protrude mandible and depress chin; acting alone and alternately, they produce side-to-side movements of mandible
Medial pterygoid	*Deep head:* medial surface of lateral pterygoid plate and pyramidal process of palatine bone *Superficial head:* tuberosity of maxilla	Medial surface of ramus of mandible, inferior to mandibular foramen	Mandibular nerve (CN V^3) through medial pterygoid nerve	Helps elevate mandible, closing jaws; acting together, they help protrude mandible; acting alone, it protrudes side of jaw; acting alternately, they produce a grinding motion

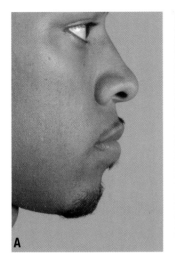
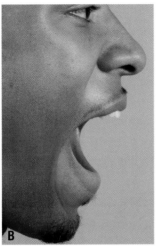
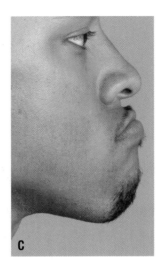
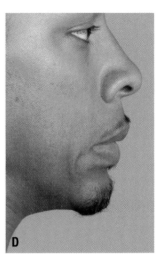

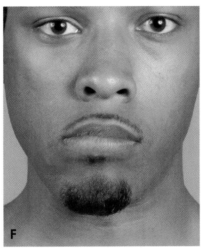
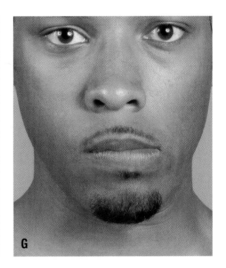

Anterior Views

TABLE 7.10   MOVEMENTS OF THE TEMPOROMANDIBULAR JOINT

MOVEMENTS	MUSCLE(S)
Elevation (close mouth) **(A)**	Temporal, masseter, and medial pterygoid
Depression (open mouth) **(B)**	Lateral pterygoid and suprahyoid and infrahyoid
Protrusion (protrude chin) **(C & E)**	Lateral pterygoid, masseter, and medial pterygoid
Retrusion (retrude chin) **(D)**	Temporal (posterior oblique and near horizontal fibers) and masseter
Lateral movements (grinding and chewing) **(F & G)**	Temporal of same side, pterygoids of opposite side, and masseter

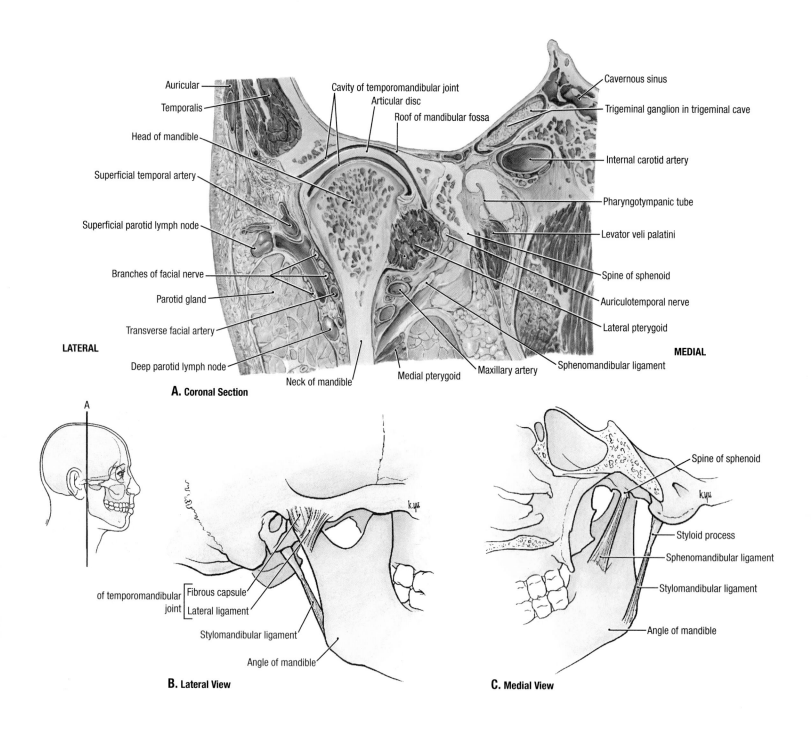

Auricular
Temporalis
Head of mandible
Superficial temporal artery
Superficial parotid lymph node
Branches of facial nerve
Parotid gland
Transverse facial artery
LATERAL
Deep parotid lymph node
Neck of mandible

Cavity of temporomandibular joint
Articular disc
Roof of mandibular fossa

Cavernous sinus
Trigeminal ganglion in trigeminal cave
Internal carotid artery
Pharyngotympanic tube
Levator veli palatini
Spine of sphenoid
Auriculotemporal nerve
Lateral pterygoid
MEDIAL
Sphenomandibular ligament
Maxillary artery
Medial pterygoid

A. **Coronal Section**

A

of temporomandibular
joint [Fibrous capsule
Lateral ligament]
Stylomandibular ligament
Angle of mandible

B. **Lateral View**

Spine of sphenoid
Styloid process
Sphenomandibular ligament
Stylomandibular ligament
Angle of mandible

C. **Medial View**

7.50   Temporomandibular joint

A. Coronal section. **B.** Temporomandibular joint and stylomandibular ligament. The fibrous capsule of the temporomandibular joint attaches to the margins of the articular area of the temporal bone and around the neck of the mandible; the lateral (temporomandibular) ligament strengthens the lateral aspect of the joint. **C.** Stylomandibular and sphenomandibular ligaments. The strong sphenomandibular ligament descends from near the spine of the sphenoid to the lingula of the mandible and is the "swinging hinge" by which the mandible is suspended; the weaker stylomandibular ligament is a thickened part of the parotid sheath that joins the styloid process to the angle of the mandible.

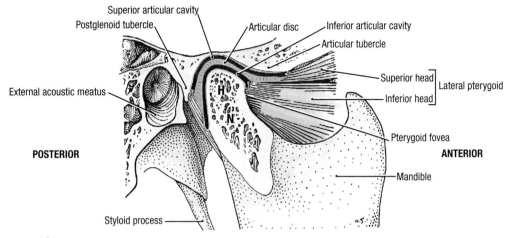

Superior articular cavity
Postglenoid tubercle
Articular disc
Inferior articular cavity
Articular tubercle
External acoustic meatus
Superior head ⎤
Inferior head ⎦ Lateral pterygoid
POSTERIOR
Pterygoid fovea
ANTERIOR
Mandible
Styloid process

A. Sagittal Section

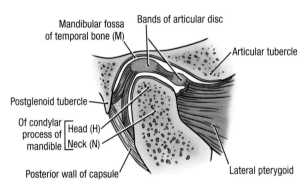

Mandibular fossa of temporal bone (M)
Bands of articular disc
Articular tubercle
Postglenoid tubercle
Of condylar process of mandible ⎡ Head (H)
⎣ Neck (N)
Posterior wall of capsule
Lateral pterygoid

B. Closed Mouth, Sagittal Section

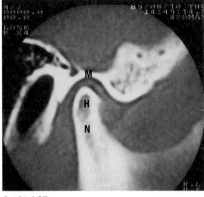

Sagittal CT

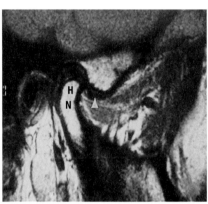

Sagittal MRI

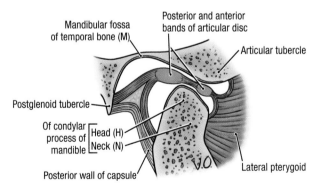

Mandibular fossa of temporal bone (M)
Posterior and anterior bands of articular disc
Articular tubercle
Postglenoid tubercle
Of condylar process of mandible ⎡ Head (H)
⎣ Neck (N)
Posterior wall of capsule
Lateral pterygoid

C. Open Mouth, Sagittal Section

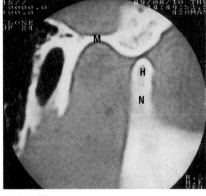

Sagittal CT

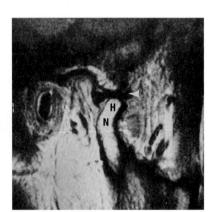

Sagittal MRI

7.51   Sectional anatomy of temporomandibular joint (TMJ)

A. TMJ and related structures, sagittal section **B.** Sagittal orientation figure, CT, and MRI—mouth closed. **C.** Sagittal orientation figure, CT, and MRI—mouth opened widely. The articular disc divides the articular cavity into superior and inferior compartments, each lined by a separate synovial membrane: one lines the fibrous capsule superior to the disc, and the other lines the capsule inferior to the disc. During yawning or taking large bites, excessive contraction of the lateral pterygoids can cause the head of the mandible to dislocate (pass anterior to the articular tubercle). In this position, the mouth remains wide open, and the person cannot close it without manual distraction.

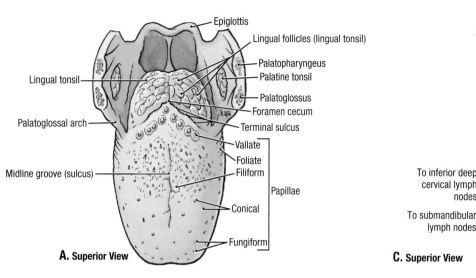

To superior deep cervical lymph nodes

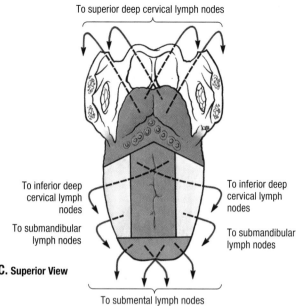

To inferior deep cervical lymph nodes

To submandibular lymph nodes

To inferior deep cervical lymph nodes

To submandibular lymph nodes

To submental lymph nodes

C. Superior View

Epiglottis

Lingual follicles (lingual tonsil)

Lingual tonsil

Palatopharyngeus

Palatine tonsil

Palatoglossus

Foramen cecum

Terminal sulcus

Palatoglossal arch

Vallate

Foliate

Filiform

Midline groove (sulcus)

Papillae

Conical

Fungiform

A. Superior View

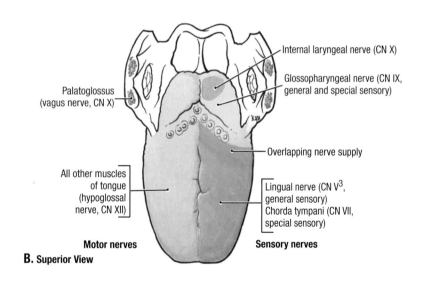

Internal laryngeal nerve (CN X)

Glossopharyngeal nerve (CN IX, general and special sensory)

Palatoglossus (vagus nerve, CN X)

Overlapping nerve supply

All other muscles of tongue (hypoglossal nerve, CN XII)

Lingual nerve (CN V^3, general sensory)
Chorda tympani (CN VII, special sensory)

Motor nerves

Sensory nerves

B. Superior View

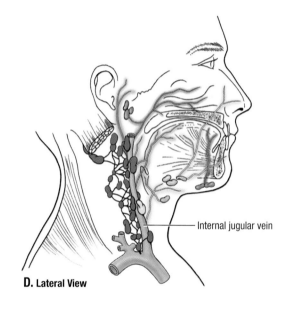

Internal jugular vein

D. Lateral View

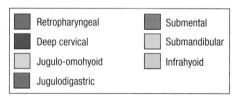

Retropharyngeal	Submental
Deep cervical	Submandibular
Jugulo-omohyoid	Infrahyoid
Jugulodigastric	

7.52   Tongue

A. Features of dorsum of the tongue. The foramen cecum is the upper end of the primitive thyroglossal duct; the arms of the V-shaped terminal sulcus diverge from the foramen, demarcating the posterior third of the tongue from the anterior two thirds. **B.** General sensory, special sensory (taste), and motor innervation of tongue. **C.** Lymphatic drainage of dorsum of tongue. **D.** Lymphatic drainage of tongue, mouth, nasal cavity, and nose.

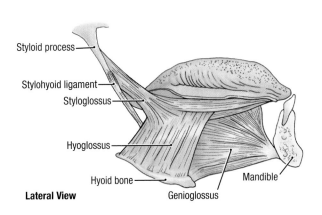

Lateral View

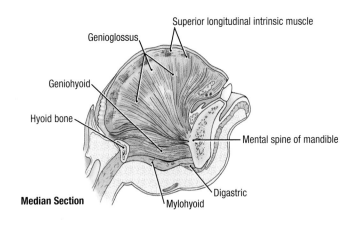

Median Section

TABLE 7.11   MUSCLES OF TONGUE

EXTRINSIC MUSCLES

MUSCLE	ORIGIN	INSERTION	INNERVATION	MAIN ACTION
Genioglossus	Superior part of mental spine of mandible	Dorsum of tongue and body of hyoid bone	Hypoglossal nerve (CN XII)	Depresses tongue; its posterior part pulls tongue anteriorly for protrusion[a]
Hyoglossus	Body and greater horn of hyoid bone	Side and inferior aspect of tongue		Depresses and retracts tongue
Styloglossus	Styloid process and stylohyoid ligament	Side and inferior aspect of tongue		Retracts tongue and draws it up to create a trough for swallowing
Palatoglossus	Palatine aponeurosis of soft palate	Side of tongue	Cranial root of CN XI via pharyngeal branch of CN X and pharyngeal plexus	Elevates posterior part of tongue

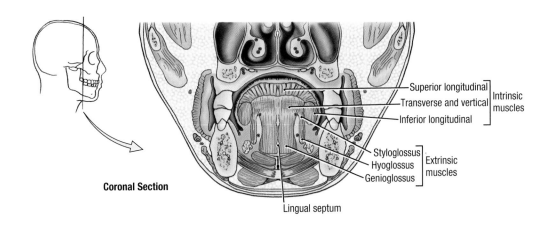

Coronal Section

INTRINSIC MUSCLES

MUSCLE	ORIGIN	INSERTION	INNERVATION	MAIN ACTION
Superior longitudinal	Submucous fibrous layer and median fibrous septum	Margins of tongue and mucous membrane	Hypoglossal nerve (CN XII)	Curls tip and sides of tongue superiorly and shortens tongue
Inferior longitudinal	Root of tongue and body of hyoid bone	Apex of tongue		Curls tip of tongue inferiorly and shortens tongue
Transverse	Median fibrous septum	Fibrous tissue at margins of tongue		Narrows and elongates the tongue[a]
Vertical	Superior surface of borders of tongue	Inferior surface of borders of the tongue		Flattens and broadens the tongue[a]

[a]Acts simultaneously to protrude tongue.

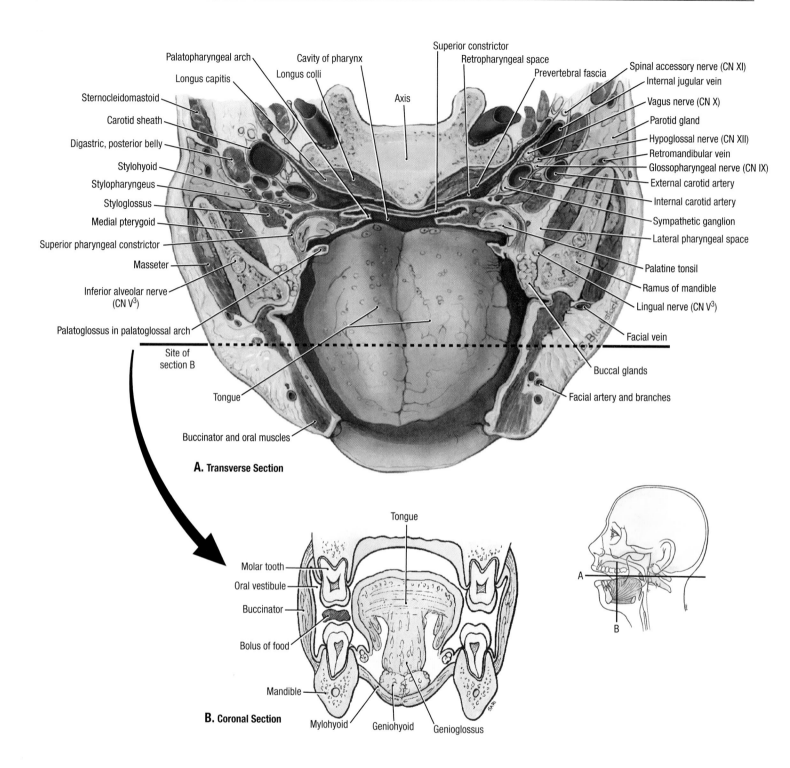

A. Transverse Section

B. Coronal Section

7.53 **Sections through mouth**

A. The viscerocranium has been sectioned at the C1 vertebral level, the plane of section passing through the oral fissure anteriorly. The retropharyngeal space (opened up in this specimen) allows the pharynx to contract and relax during swallowing; the retropharyngeal space is closed laterally at the carotid sheath and limited posteriorly by the prevertebral fascia. The beds of the parotid glands are also demonstrated. **B.** Schematic coronal section demonstrating how the tongue and buccinator (or, anteriorly, the orbicularis oris) work together to retain food between the teeth when chewing. The buccinator and superior part of the orbicularis oris are innervated by the buccal branch of the facial nerve (CN VII).

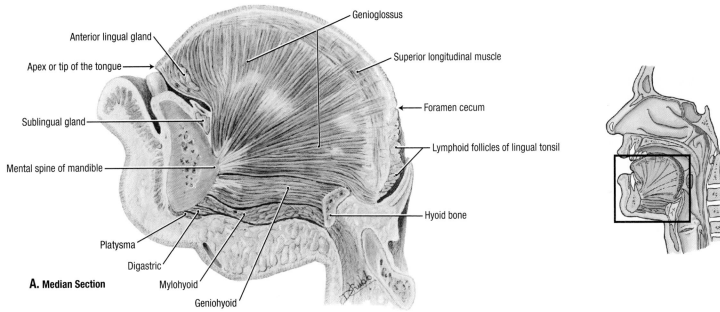

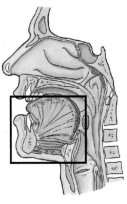

A. Median Section

Labels (clockwise from upper left):
Anterior lingual gland
Apex or tip of the tongue
Sublingual gland
Mental spine of mandible
Platysma
Digastric
Mylohyoid
Geniohyoid
Genioglossus
Superior longitudinal muscle
Foramen cecum
Lymphoid follicles of lingual tonsil
Hyoid bone

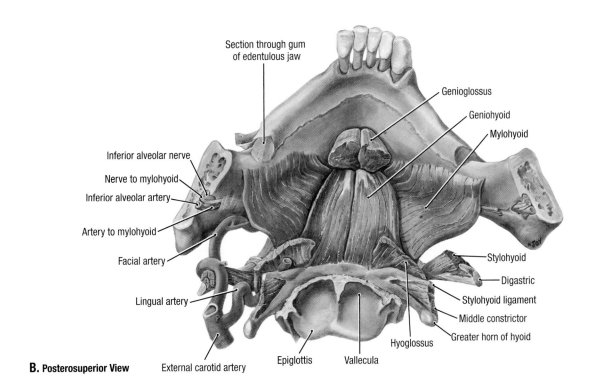

B. Posterosuperior View

Labels:
Section through gum of edentulous jaw
Inferior alveolar nerve
Nerve to mylohyoid
Inferior alveolar artery
Artery to mylohyoid
Facial artery
Lingual artery
External carotid artery
Epiglottis
Vallecula
Hyoglossus
Genioglossus
Geniohyoid
Mylohyoid
Stylohyoid
Digastric
Stylohyoid ligament
Middle constrictor
Greater horn of hyoid

7.54 Tongue and floor of mouth

A. Median section though the tongue and lower jaw. The tongue is composed mainly of muscle; the extrinsic muscles alter the position of the tongue, and intrinsic muscles alter its shape. The genioglossus is the extrinsic muscle apparent in this plane, and the superior longitudinal muscle is the intrinsic muscle. **B.** Muscles of the floor of the mouth viewed superiorly. The mylohyoid muscle extends between the two mylohyoid lines of the mandible. It has a thick, free posterior border and becomes thinner anteriorly.

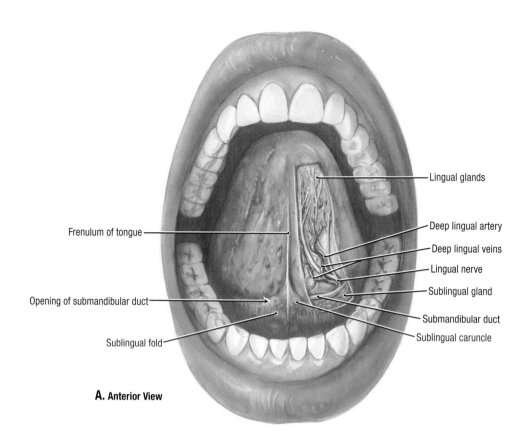

A. Anterior View

Lingual glands

Deep lingual artery

Deep lingual veins

Lingual nerve

Sublingual gland

Submandibular duct

Sublingual caruncle

Frenulum of tongue

Opening of submandibular duct

Sublingual fold

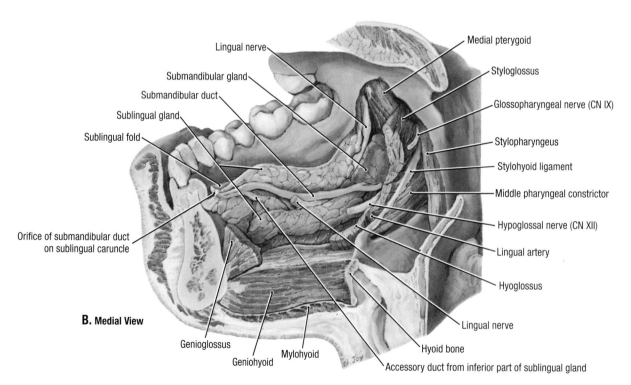

Lingual nerve

Submandibular gland

Submandibular duct

Sublingual gland

Sublingual fold

Orifice of submandibular duct on sublingual caruncle

Genioglossus

Geniohyoid

Mylohyoid

Hyoid bone

Accessory duct from inferior part of sublingual gland

Lingual nerve

Lingual artery

Hyoglossus

Hypoglossal nerve (CN XII)

Middle pharyngeal constrictor

Stylohyoid ligament

Stylopharyngeus

Glossopharyngeal nerve (CN IX)

Styloglossus

Medial pterygoid

B. Medial View

7.55 **Inferior surface of the tongue, sublingual and submandibular glands**

A. Inferior surface of tongue and floor of the mouth. The thin sublingual mucosa has been removed on the left side. **B.** Sublingual and submandibular glands. The tongue has been excised. The geniohyoid muscle (inferiorly), the middle constrictor (posteriorly), and the cut edge of the mucous membrane (superiorly) are intact. The genioglossus muscle (anteriorly), hyoglossus (inferiorly), and styloglossus (posteriorly) are divided.

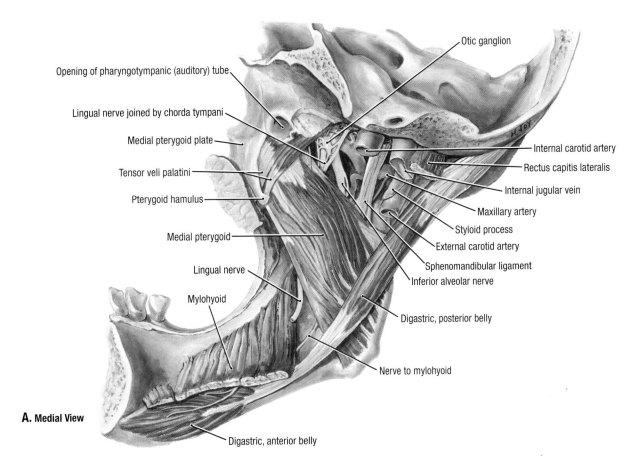

A. Medial View

Otic ganglion
Opening of pharyngotympanic (auditory) tube
Lingual nerve joined by chorda tympani
Medial pterygoid plate
Tensor veli palatini
Pterygoid hamulus
Medial pterygoid
Lingual nerve
Mylohyoid
Internal carotid artery
Rectus capitis lateralis
Internal jugular vein
Maxillary artery
Styloid process
External carotid artery
Sphenomandibular ligament
Inferior alveolar nerve
Digastric, posterior belly
Nerve to mylohyoid
Digastric, anterior belly

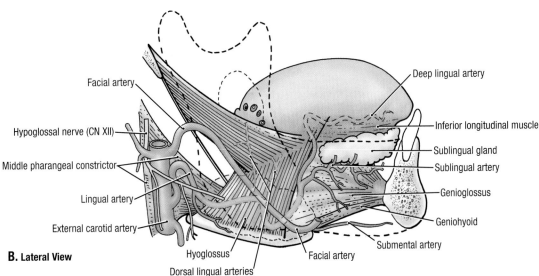

B. Lateral View

Facial artery
Hypoglossal nerve (CN XII)
Middle pharangeal constrictor
Lingual artery
External carotid artery
Hyoglossus
Dorsal lingual arteries
Deep lingual artery
Inferior longitudinal muscle
Sublingual gland
Sublingual artery
Genioglossus
Geniohyoid
Submental artery
Facial artery

7.56 **Muscles and vessels on the medial surface of the mandible and base of skull**

A. Structures related to the medial surface of mandible. The otic ganglion lies medial to the mandibular nerve (CN V³) and between the foramen ovale superiorly and the medial ptery-goid muscle inferiorly. The tensor veli palatini muscle usually lies directly medial to the gan-glion. **B.** Course and distribution of the lingual artery. The deep lingual artery supplies the body of the tongue, the dorsal lingual artery supplies the root of the tongue and palatine ton-sil, and the sublingual branch supplies the floor of the mouth.

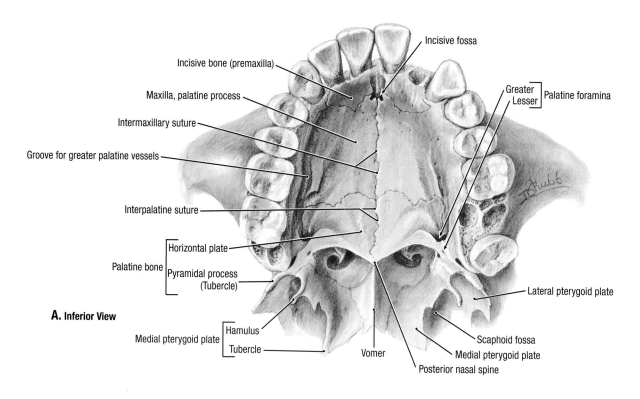

Incisive fossa

Incisive bone (premaxilla)

Maxilla, palatine process

Intermaxillary suture

Groove for greater palatine vessels

Greater
Lesser ⎫ Palatine foramina

Interpalatine suture

Palatine bone ⎰ Horizontal plate
⎱ Pyramidal process (Tubercle)

Lateral pterygoid plate

A. Inferior View

Medial pterygoid plate ⎰ Hamulus
⎱ Tubercle

Scaphoid fossa

Medial pterygoid plate

Vomer

Posterior nasal spine

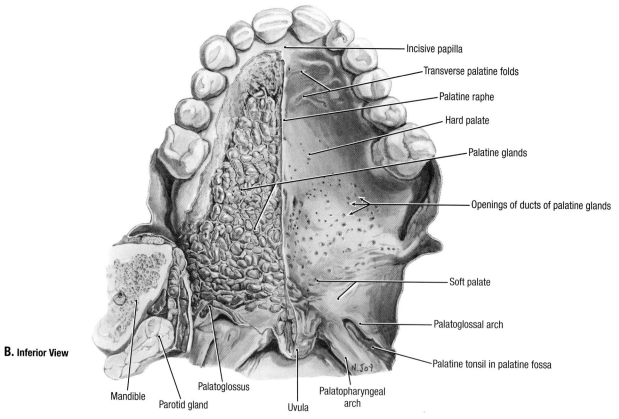

Incisive papilla

Transverse palatine folds

Palatine raphe

Hard palate

Palatine glands

Openings of ducts of palatine glands

Soft palate

Palatoglossal arch

Palatine tonsil in palatine fossa

B. Inferior View

Mandible

Parotid gland

Palatoglossus

Uvula

Palatopharyngeal arch

7.57 **Palate**

The palate consists of bony (hard palate), aponeurotic, and muscular (soft palate) parts. **A.** Bones of the hard palate. The palatine aponeurosis stretches between **B.** Mucous membrane and glands of palate. Note the abundant mucosal palatine glands on the right side and their openings in the intact mucosa on the left side.

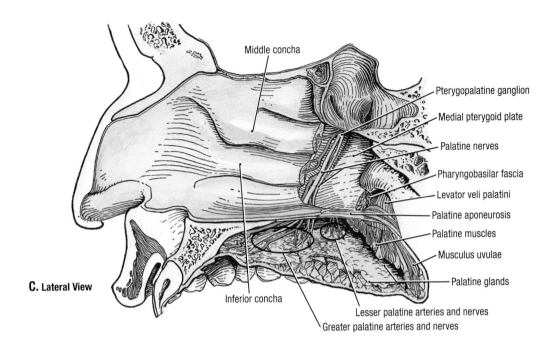

C. Lateral View

Middle concha
Pterygopalatine ganglion
Medial pterygoid plate
Palatine nerves
Pharyngobasilar fascia
Levator veli palatini
Palatine aponeurosis
Palatine muscles
Musculus uvulae
Palatine glands
Lesser palatine arteries and nerves
Greater palatine arteries and nerves
Inferior concha

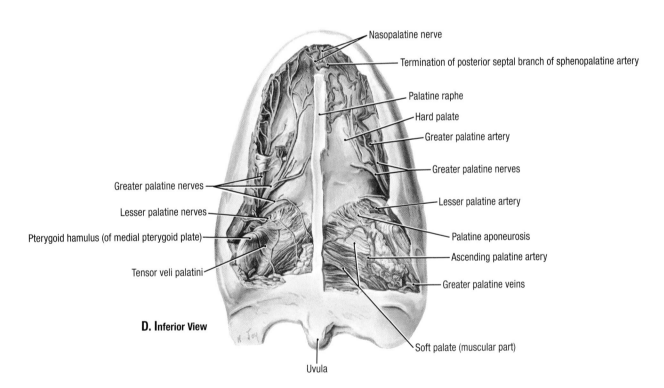

D. Inferior View

Nasopalatine nerve
Termination of posterior septal branch of sphenopalatine artery
Palatine raphe
Hard palate
Greater palatine artery
Greater palatine nerves
Lesser palatine artery
Palatine aponeurosis
Ascending palatine artery
Greater palatine veins
Soft palate (muscular part)
Uvula
Tensor veli palatini
Pterygoid hamulus (of medial pterygoid plate)
Lesser palatine nerves
Greater palatine nerves

7.57   **Palate** *(continued)*

C and **D.** Nerves and vessels of palate. **C.** The lateral wall of nasal cavity is shown. The posterior ends of the middle and inferior conchae are excised along with the mucoperiosteum; the thin, perpendicular plate of the palatine bone is removed to expose the palatine nerves and arteries. **D.** Dissection of an edentulous palate. The greater palatine nerve supplies the gingivae and hard palate, the nasopalatine nerve the incisive region, and the lesser palatine nerves the soft palate.

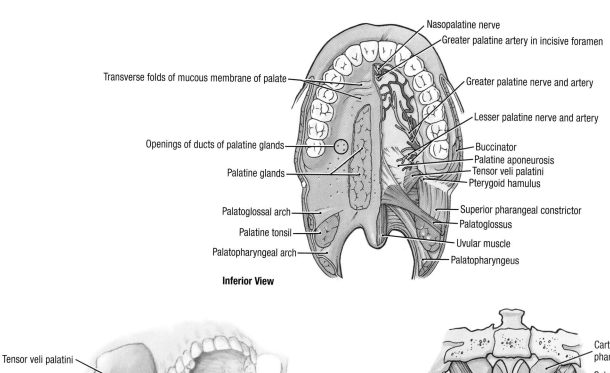

Inferior View

Anterolateral View

Posterior View

TABLE 7.12 MUSCLES OF SOFT PALATE

MUSCLE	SUPERIOR ATTACHMENT	INFERIOR ATTACHMENT	INNERVATION	MAIN ACTION(S)
Levator veli palatini	Cartilage of pharyngotympanic tube and petrous part of temporal bone	Palatine aponeurosis	Pharyngeal branch of vagus nerve through pharyngeal plexus	Elevates soft palate during swallowing and yawning
Tensor veli palatini	Scaphoid fossa of medial pterygoid plate, spine of sphenoid bone, and cartilage of pharyngotympanic tube		Medial pterygoid nerve (CN V^3) through otic ganglion	Tenses soft palate and opens mouth of pharyngotympanic tube during swallowing and yawning
Palatoglossus	Palatine aponeurosis	Side of tongue	Pharyngeal branch of vagus nerve (CN X) via the pharyngeal plexus	Elevates posterior part of tongue and draws soft palate onto tongue
Palatopharyngeus	Hard palate and palatine aponeurosis	Lateral wall of pharynx		Tenses soft palate and pulls walls of pharynx superiorly, anteriorly, and medially during swallowing
Musculus uvulae	Posterior nasal spine and palatine aponeurosis	Mucosa of uvula		Shortens uvula and pulls it superiorly

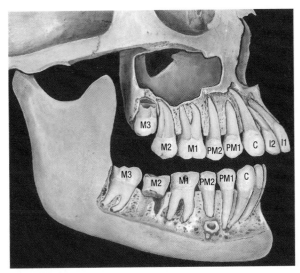

A. Lateral View

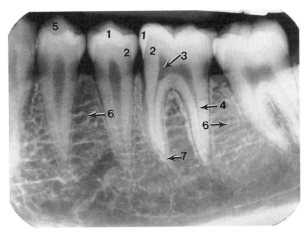

B. Lateral Radiograph

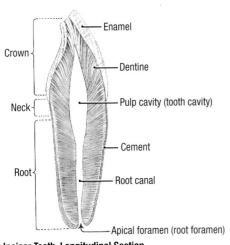

Incisor Tooth, Longitudinal Section

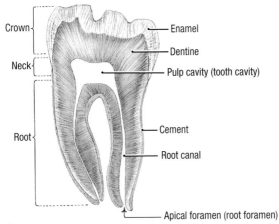

C. Molar Tooth, Longitudinal Section

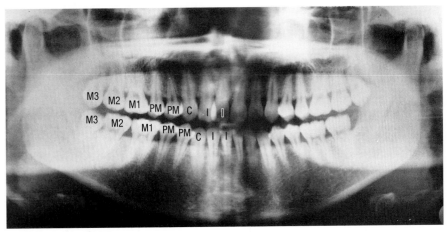

D. Pantomographic Radiograph

7.58   Permanent teeth—I

A. Teeth in situ with roots exposed. Incisors *(I1, I2)*, canine *(C1)*, premolars *(PM1, PM2)*, and molars *(M1, M2, M3)*. The roots of the 2nd lower molar have been removed. **B.** Lateral radiograph. *(1)* enamel, *(2)* dentin, *(3)* pulp chamber, *(4)* pulp canal, *(5)* buccal cusp, *(6)* alveolar bone, and *(7)* root apex. **C.** Longitudinal sections of incisor and molar teeth. **D.** Pantomographic radiograph of mandible and maxilla. The left lower third molar is not present.

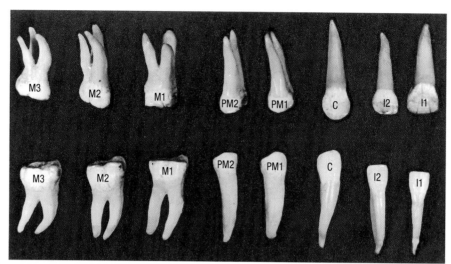

A. Vestibular View

Maxillary Teeth

Mandibular Teeth

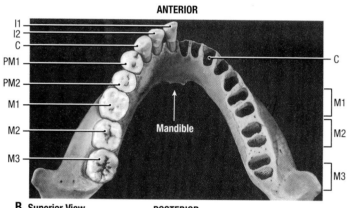

B. Superior View

C. Superior View

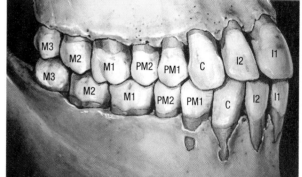

D. Anterolateral View

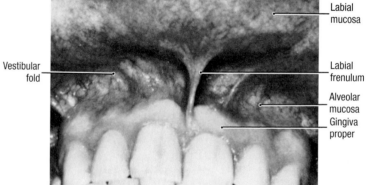

E. Anterior View

7.59 **Permanent teeth—II**

A. Removed teeth, displaying roots. There are 32 permanent teeth; 8 are on each side of each dental arch on the top (maxillary teeth) and bottom (mandibular teeth): 2 incisors *(I1–2)*, 1 canine *(C)*, 2 premolars *(PM1–2)*, and 3 molars *(M1–3)*. **B.** Permanent mandibular teeth and their sockets. **C.** Permanent maxillary teeth and their sockets. **D.** Teeth in occlusion. **E.** Vestibule and gingivae of the maxilla.

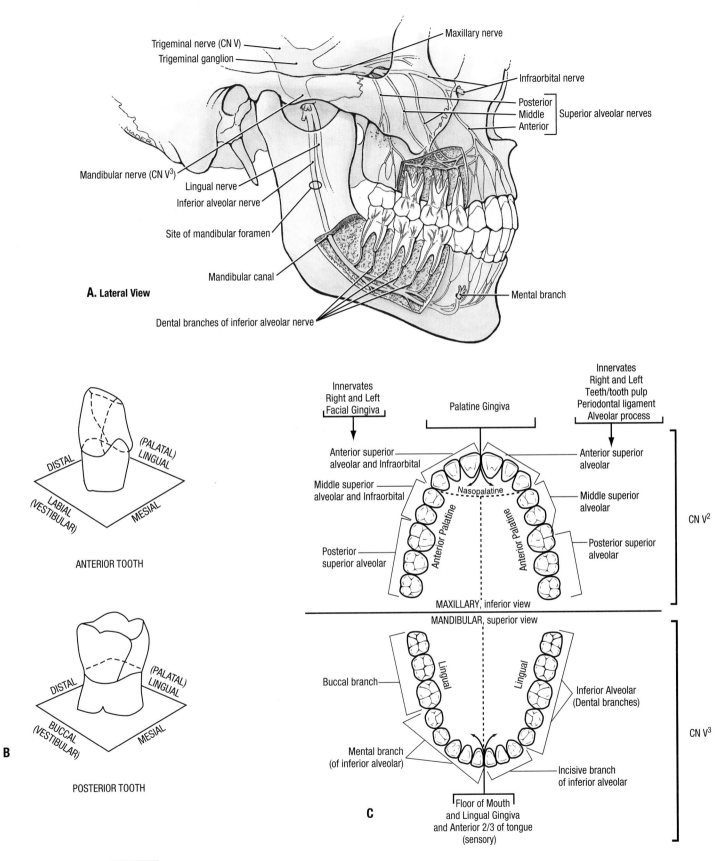

A. Lateral View

- Trigeminal nerve (CN V)
- Trigeminal ganglion
- Maxillary nerve
- Infraorbital nerve
- Posterior
- Middle — Superior alveolar nerves
- Anterior
- Mandibular nerve (CN V³)
- Lingual nerve
- Inferior alveolar nerve
- Site of mandibular foramen
- Mandibular canal
- Mental branch
- Dental branches of inferior alveolar nerve

ANTERIOR TOOTH

- DISTAL
- (PALATAL) LINGUAL
- LABIAL (VESTIBULAR)
- MESIAL

POSTERIOR TOOTH

- DISTAL
- (PALATAL) LINGUAL
- BUCCAL (VESTIBULAR)
- MESIAL

B

Innervates Right and Left Facial Gingiva

Palatine Gingiva

Innervates Right and Left Teeth/tooth pulp Periodontal ligament Alveolar process

- Anterior superior alveolar and Infraorbital
- Anterior superior alveolar
- Middle superior alveolar and Infraorbital
- Nasopalatine
- Middle superior alveolar
- Anterior Palatine
- Anterior Palatine
- Posterior superior alveolar
- Posterior superior alveolar

CN V²

MAXILLARY, inferior view

MANDIBULAR, superior view

- Buccal branch
- Lingual
- Lingual
- Inferior Alveolar (Dental branches)
- Mental branch (of inferior alveolar)
- Incisive branch of inferior alveolar
- Floor of Mouth and Lingual Gingiva and Anterior 2/3 of tongue (sensory)

CN V³

C

7.60   Innervation of teeth

A. Superior and inferior alveolar nerves. **B.** Surfaces of an incisor and molar tooth. **C.** Innervation of the mouth and teeth.

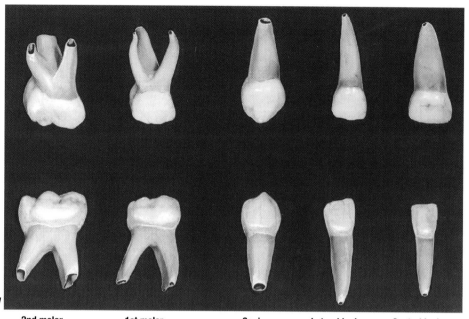

A. Vestibular View

| 2nd molar | 1st molar | Canine | Lateral incisor | Central incisor |

Maxillary teeth

Mandibular teeth

Inferior View of Maxillary Teeth

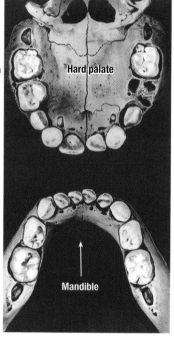

Hard palate

Mandible

B. Superior View of Mandibular Teeth

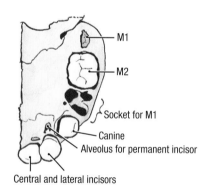

M1

M2

Socket for M1

Canine

Alveolus for permanent incisor

Central and lateral incisors

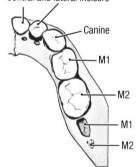

Canine

M1

M2

M1

M2

7.61 **Primary teeth**

A. Removed teeth. There are 20 primary (deciduous) teeth, 5 in each half of the mandible and 5 in each maxilla. They are named central incisor, lateral incisor, canine, 1st molar *(M1)*, and 2nd molar *(M2)*. Primary teeth differ from permanent teeth in that the primary teeth are smaller and whiter; the molars also have more bulbous crowns and more divergent roots. **B.** Teeth in situ, younger than 2 years of age. Permanent teeth are colored orange; the crowns of the unerupted 1st and 2nd permanent molars are partly visible.

TABLE 7.13 PRIMARY AND SECONDARY DENTITION

DECIDUOUS TEETH	MEDIAL INCISOR	LATERAL INCISOR	CANINE	FIRST MOLAR	SECOND MOLAR		
Eruption (months)[a]	6–8	8–10	16–20	12–16	20–24		
Shedding (years)	6–7	7–8	10–12	9–11	10–12		

[a] In some normal infants, the first teeth (medial incisors) may not erupt until 12 to 13 months of age.

Between 6 and 12 years, the primary teeth are replaced by permanent teeth *(orange)*. *I,* incisor; *PM,* premolar; *M,* molar teeth.

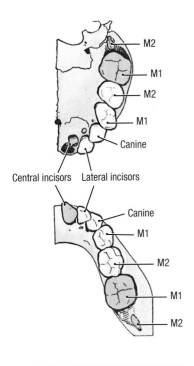

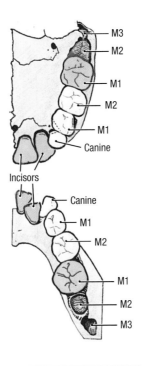

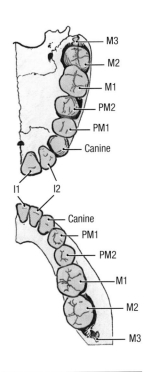

Age: 6–7 years

The 1st molars (6-year molars) have fully erupted, the primary central incisor has been shed, the lower central incisor is almost fully erupted, and the upper central incisor is descending into the vacated socket.

Age: 8 years

All of the permanent incisors have erupted; however, the lower lateral incisor is only partially erupted.

Age: 12 years

The primary teeth have been replaced by 20 permanent teeth, and the 1st and 2nd molars (12-year molars) have erupted; the canines, 2nd premolars, and 2nd molars (especially those in the upper jaw) have not erupted fully, nor have their bony sockets closed around them. By age 12, 28 permanent teeth are in evidence; the last 4 teeth, the 3rd molars, may erupt any time after this, or never.

PERMANENT TEETH	MEDIAL INCISOR	LATERAL INCISOR	CANINE	FIRST PREMOLAR	SECOND PREMOLAR	FIRST MOLAR	SECOND MOLAR	THIRD MOLAR
Eruption (years)	7–8	8–9	10–12	10–11	11–12	6–7	12	13–25

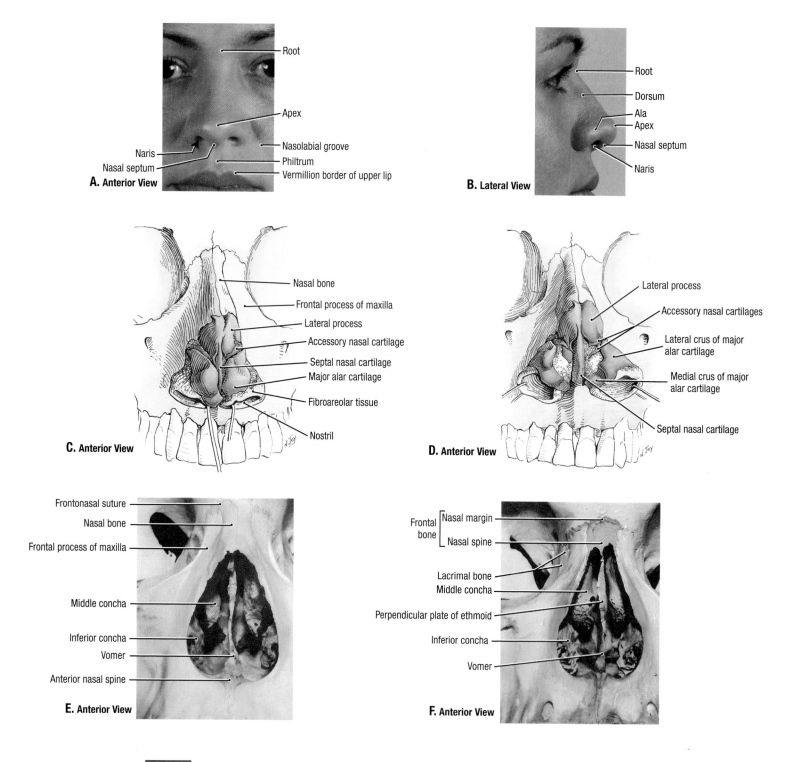

A. Anterior View
- Root
- Apex
- Nasolabial groove
- Naris
- Nasal septum
- Philtrum
- Vermillion border of upper lip

B. Lateral View
- Root
- Dorsum
- Ala
- Apex
- Nasal septum
- Naris

C. Anterior View
- Nasal bone
- Frontal process of maxilla
- Lateral process
- Accessory nasal cartilage
- Septal nasal cartilage
- Major alar cartilage
- Fibroareolar tissue
- Nostril

D. Anterior View
- Lateral process
- Accessory nasal cartilages
- Lateral crus of major alar cartilage
- Medial crus of major alar cartilage
- Septal nasal cartilage

E. Anterior View
- Frontonasal suture
- Nasal bone
- Frontal process of maxilla
- Middle concha
- Inferior concha
- Vomer
- Anterior nasal spine

F. Anterior View
- Frontal bone [Nasal margin / Nasal spine]
- Lacrimal bone
- Middle concha
- Perpendicular plate of ethmoid
- Inferior concha
- Vomer

7.62   Surface anatomy, cartilages, and bones of nose

A. Surface features of anterior aspect of nose. **B.** Surface features of lateral aspect of nose. **C.** Alar cartilages, with the septum pulled inferiorly. **D.** Alar cartilages, separated and retracted laterally. **E.** Lower conchae and bony septum seen through the piriform aperture. The margin of the piriform aperture is sharp and formed by the maxillae and nasal bones. **F.** Nasal bones removed. The areas of the frontal processes of the maxillae (*yellow*) and of the frontal bone (*blue*) that articulate with the nasal bones can be seen.

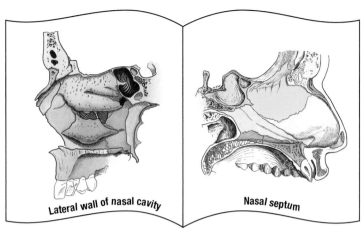

Right Nasal Cavity

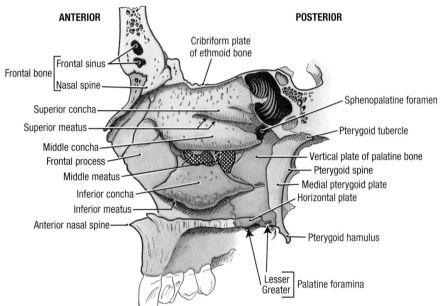

A. Medial View of Lateral Wall

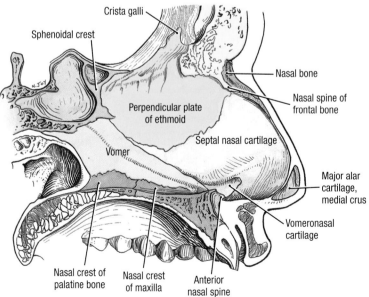

B. Lateral View of Nasal Septum

7.63   **Bones of the lateral wall and septum of the nose**

A. Lateral wall of nose. (Frontal, *peach*; nasal, *green*; maxilla, *purple*; lacrimal, *blue*; ethmoid, *bright yellow*; palatine, *red*; sphenoid, *gray*.) The superior and middle conchae are parts of the ethmoid bone, whereas the inferior concha is itself a bone. **B.** Nasal septum. (Nasal, *green*; frontal, *light yellow*; sphenoid, *dark gray*; ethmoid, *yellow*; palatine, *red*; maxilla, *purple*; vomer, *peach*; septal and alar cartilage, *light gray*).

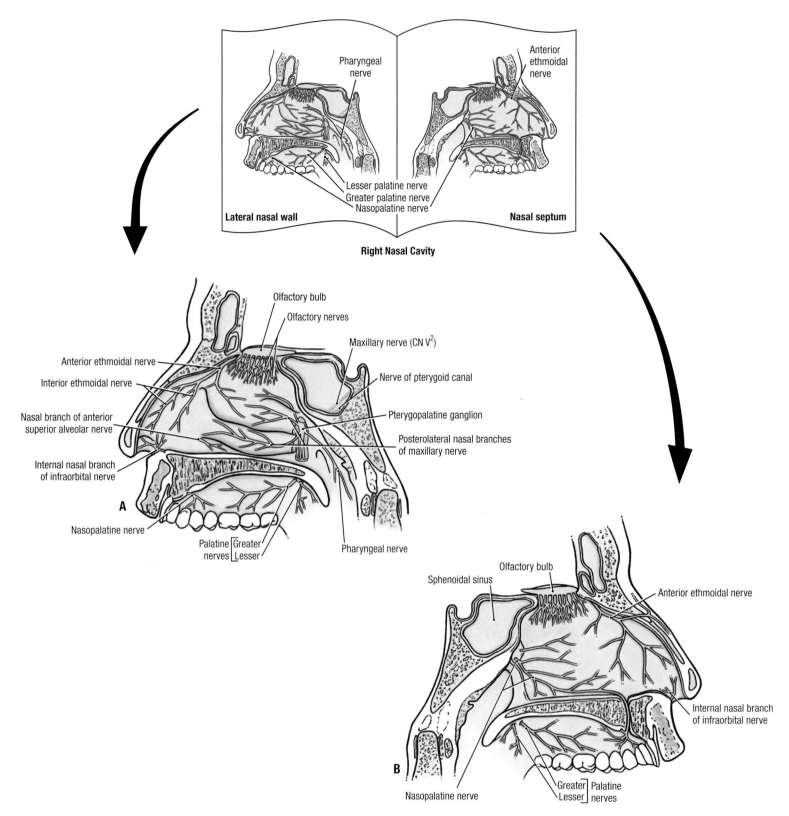

7.64   **Innervation of lateral wall and septum of the nose**

A. Lateral wall of nose. The olfactory neuroepithelium is in the superior part of the lateral and septal walls of the nasal cavity. The central processes of the olfactory neurosensory cells of each side form approximately 20 bundles that together form an olfactory nerve (CN I). **B.** Nasal septum. The nasopalatine nerve from the pterygopalatine ganglion supplies the posteroinferior septum, and the anterior ethmoidal nerve (branch of V1) supplies the anterosuperior septum.

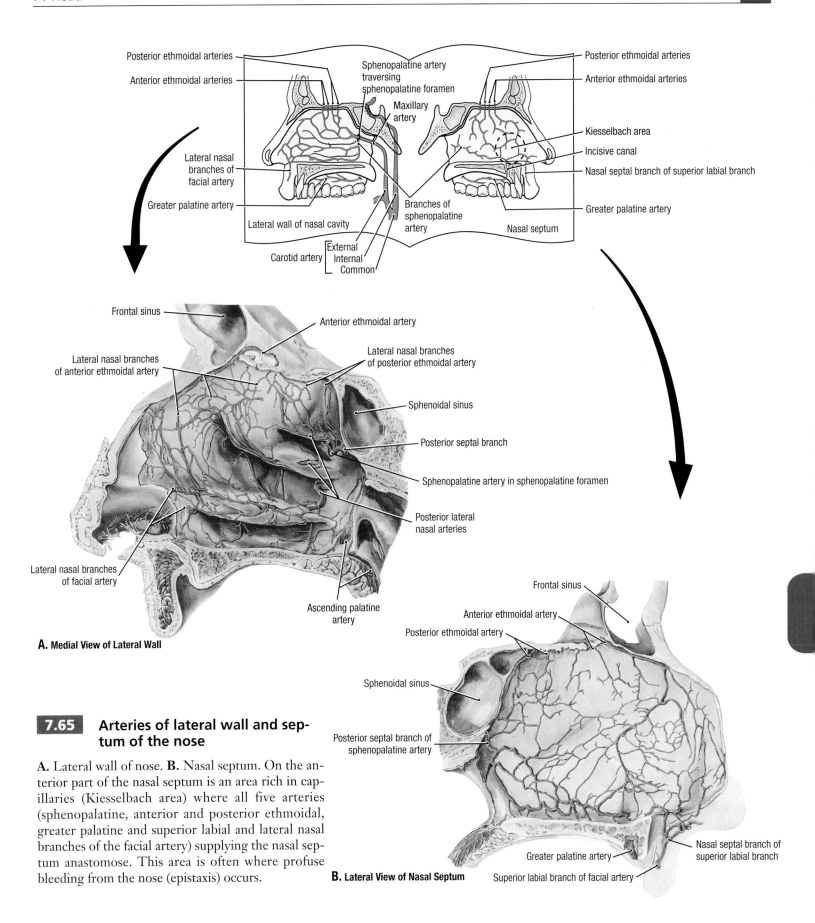

A. Medial View of Lateral Wall

B. Lateral View of Nasal Septum

7.65   **Arteries of lateral wall and septum of the nose**

A. Lateral wall of nose. **B.** Nasal septum. On the anterior part of the nasal septum is an area rich in capillaries (Kiesselbach area) where all five arteries (sphenopalatine, anterior and posterior ethmoidal, greater palatine and superior labial and lateral nasal branches of the facial artery) supplying the nasal septum anastomose. This area is often where profuse bleeding from the nose (epistaxis) occurs.

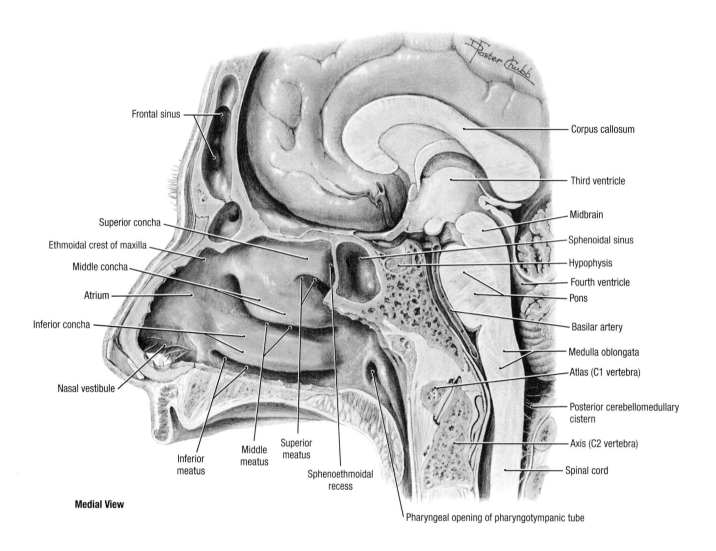

Frontal sinus

Corpus callosum

Third ventricle

Midbrain

Superior concha

Sphenoidal sinus

Ethmoidal crest of maxilla

Hypophysis

Middle concha

Fourth ventricle

Atrium

Pons

Inferior concha

Basilar artery

Medulla oblongata

Atlas (C1 vertebra)

Nasal vestibule

Posterior cerebellomedullary cistern

Axis (C2 vertebra)

Inferior meatus

Middle meatus

Superior meatus

Spinal cord

Sphenoethmoidal recess

Medial View

Pharyngeal opening of pharyngotympanic tube

7.66   **Right half of hemisected head demonstrating upper respiratory tract**

- The vestibule is superior to the nostril and anterior to the inferior meatus; hairs grow from its skin-lined surface. The atrium is superior to the vestibule and anterior to the middle meatus.
- The inferior and middle conchae curve inferiorly and medially from the lateral wall, dividing it into three nearly equal parts and covering the inferior and middle meatuses, respectively. The middle concha ends inferior to the sphenoidal sinus, and the inferior concha ends inferior to the middle concha, just anterior to the orifice of the auditory tube. The superior concha is small and anterior to the sphenoidal sinus.
- The roof comprises an anterior sloping part corresponding to the bridge of the nose; an intermediate horizontal part; a perpendicular part anterior to the sphenoidal sinus; and a curved part, inferior to the sinus, that is continuous with the roof of the nasopharynx.

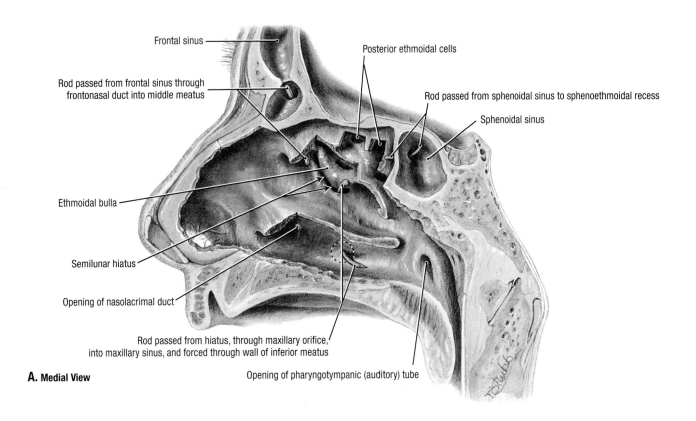

Frontal sinus

Rod passed from frontal sinus through
frontonasal duct into middle meatus

Posterior ethmoidal cells

Rod passed from sphenoidal sinus to sphenoethmoidal recess

Sphenoidal sinus

Ethmoidal bulla

Semilunar hiatus

Opening of nasolacrimal duct

Rod passed from hiatus, through maxillary orifice,
into maxillary sinus, and forced through wall of inferior meatus

Opening of pharyngotympanic (auditory) tube

A. Medial View

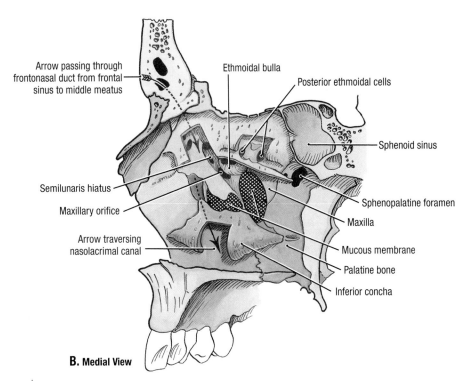

Arrow passing through
frontonasal duct from frontal
sinus to middle meatus

Ethmoidal bulla

Posterior ethmoidal cells

Sphenoid sinus

Semilunaris hiatus

Maxillary orifice

Sphenopalatine foramen

Maxilla

Arrow traversing
nasolacrimal canal

Mucous membrane

Palatine bone

Inferior concha

B. Medial View

7.67   **Communications through the lateral wall of the nasal cavity**

A. Dissection. Parts of the superior, middle, and inferior conchae are cut away. **B.** Bones. (Frontal, *light yellow*; nasal, *green*; maxilla, *purple*; lacrimal, *blue*; ethmoid, *bright yellow*; palatine, *red*; sphenoid, *gray*; inferior concha, *brown*.) Note one *arrow* passing from the frontal sinus through the frontonasal duct into the middle meatus and another *arrow* coming from the anteromedial orbit via the nasolacrimal canal.

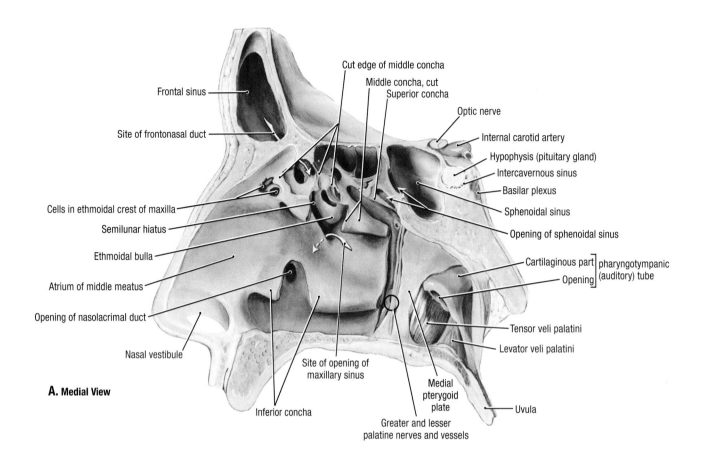

A. Medial View

Frontal sinus

Site of frontonasal duct

Cells in ethmoidal crest of maxilla

Semilunar hiatus

Ethmoidal bulla

Atrium of middle meatus

Opening of nasolacrimal duct

Nasal vestibule

Cut edge of middle concha

Middle concha, cut

Superior concha

Optic nerve

Internal carotid artery

Hypophysis (pituitary gland)

Intercavernous sinus

Basilar plexus

Sphenoidal sinus

Opening of sphenoidal sinus

Cartilaginous part ⎤ pharyngotympanic
Opening ⎦ (auditory) tube

Tensor veli palatini

Levator veli palatini

Uvula

Medial pterygoid plate

Greater and lesser palatine nerves and vessels

Site of opening of maxillary sinus

Inferior concha

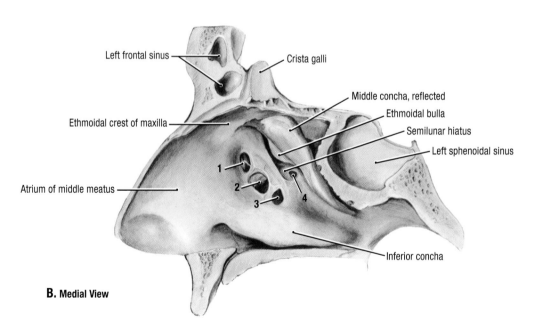

B. Medial View

Left frontal sinus

Ethmoidal crest of maxilla

Atrium of middle meatus

Crista galli

Middle concha, reflected

Ethmoidal bulla

Semilunar hiatus

Left sphenoidal sinus

Inferior concha

7.68 **Paranasal sinuses, openings, and palatine muscles in the lateral wall of the nasal cavity**

A. Dissection. Parts of the middle and inferior conchae and lateral wall of the nasal cavity are cut away to expose the nerves and vessels in the palatine canal and the extrinsic palatine muscles. **B.** Accessory maxillary orifices. In addition to the primary, or normal, ostium (not shown), there are four secondary, or acquired, ostia (numbered *1–4*); these result from breakdown of the membrane shown in *cross-hatching* in Figure 7.67B.

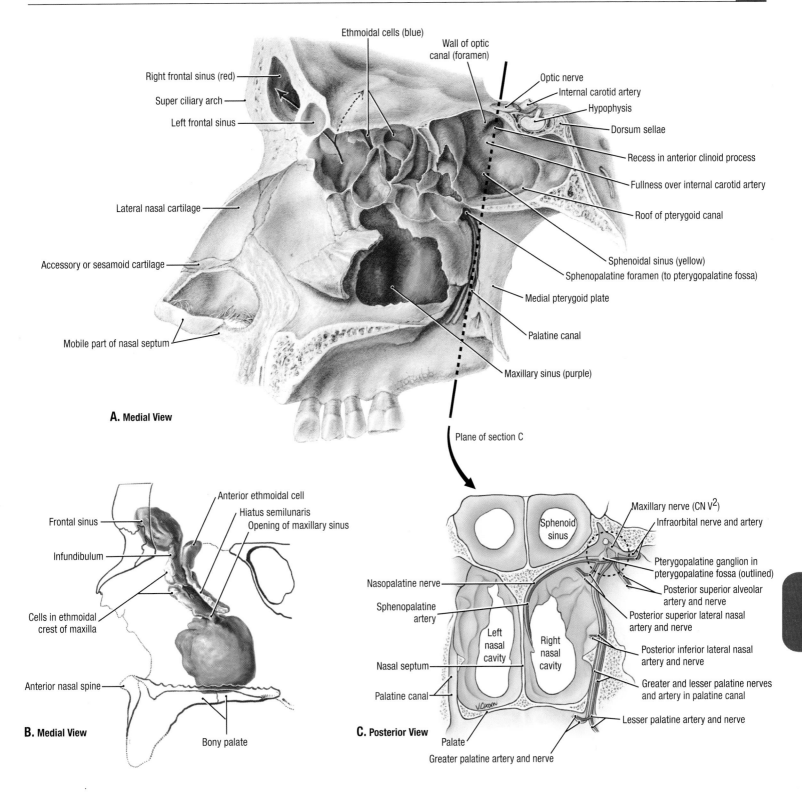

Ethmoidal cells (blue)

Wall of optic canal (foramen)

Right frontal sinus (red)

Super ciliary arch

Left frontal sinus

Optic nerve

Internal carotid artery

Hypophysis

Dorsum sellae

Recess in anterior clinoid process

Fullness over internal carotid artery

Roof of pterygoid canal

Lateral nasal cartilage

Accessory or sesamoid cartilage

Sphenoidal sinus (yellow)

Sphenopalatine foramen (to pterygopalatine fossa)

Medial pterygoid plate

Palatine canal

Mobile part of nasal septum

Maxillary sinus (purple)

A. Medial View

Plane of section C

Anterior ethmoidal cell

Hiatus semilunaris

Opening of maxillary sinus

Frontal sinus

Infundibulum

Cells in ethmoidal crest of maxilla

Anterior nasal spine

B. Medial View

Bony palate

Maxillary nerve (CN V²)

Infraorbital nerve and artery

Sphenoid sinus

Pterygopalatine ganglion in pterygopalatine fossa (outlined)

Posterior superior alveolar artery and nerve

Nasopalatine nerve

Sphenopalatine artery

Left nasal cavity

Right nasal cavity

Posterior superior lateral nasal artery and nerve

Posterior inferior lateral nasal artery and nerve

Nasal septum

Palatine canal

Greater and lesser palatine nerves and artery in palatine canal

Lesser palatine artery and nerve

C. Posterior View

Palate

Greater palatine artery and nerve

7.69   **Paranasal sinuses**

A. Opened sinuses, color coded. **B.** Cast of frontal and maxillary sinuses. **C.** Coronal section through nasal cavities, sphenoidal sinuses, and right pterygopalatine fossa, in the plane of the palatine canal, demonstrating the course of the nasopalatine and greater and lesser palatine nerves.

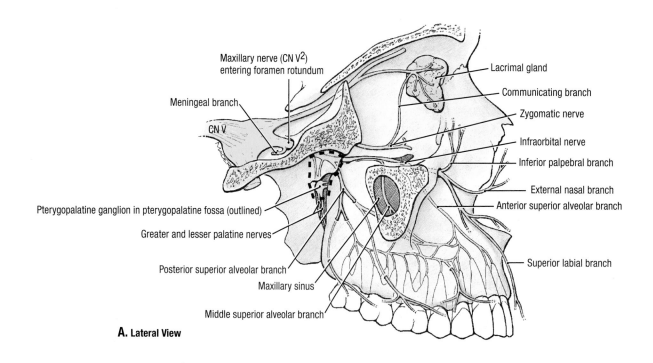

Maxillary nerve (CN V²) entering foramen rotundum

Meningeal branch

CN V

Lacrimal gland

Communicating branch

Zygomatic nerve

Infraorbital nerve

Inferior palpebral branch

External nasal branch

Anterior superior alveolar branch

Pterygopalatine ganglion in pterygopalatine fossa (outlined)

Greater and lesser palatine nerves

Posterior superior alveolar branch

Maxillary sinus

Superior labial branch

Middle superior alveolar branch

A. Lateral View

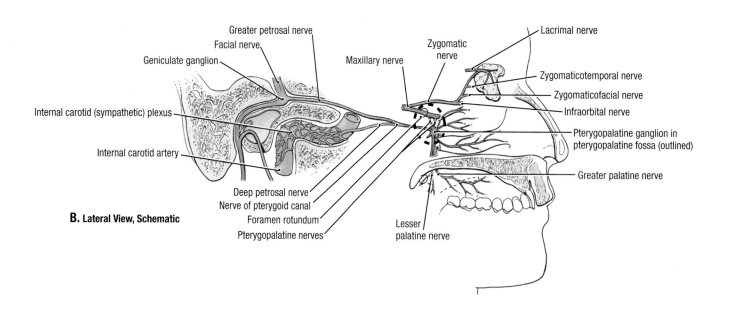

Greater petrosal nerve

Facial nerve

Geniculate ganglion

Maxillary nerve

Zygomatic nerve

Lacrimal nerve

Zygomaticotemporal nerve

Zygomaticofacial nerve

Infraorbital nerve

Internal carotid (sympathetic) plexus

Pterygopalatine ganglion in pterygopalatine fossa (outlined)

Internal carotid artery

Greater palatine nerve

Deep petrosal nerve

Nerve of pterygoid canal

Foramen rotundum

Pterygopalatine nerves

Lesser palatine nerve

B. Lateral View, Schematic

7.70 **Nerves of the pterygopalatine fossa**

A. Maxillary nerve and branches. **B.** Autonomic innervation of the lacrimal gland.

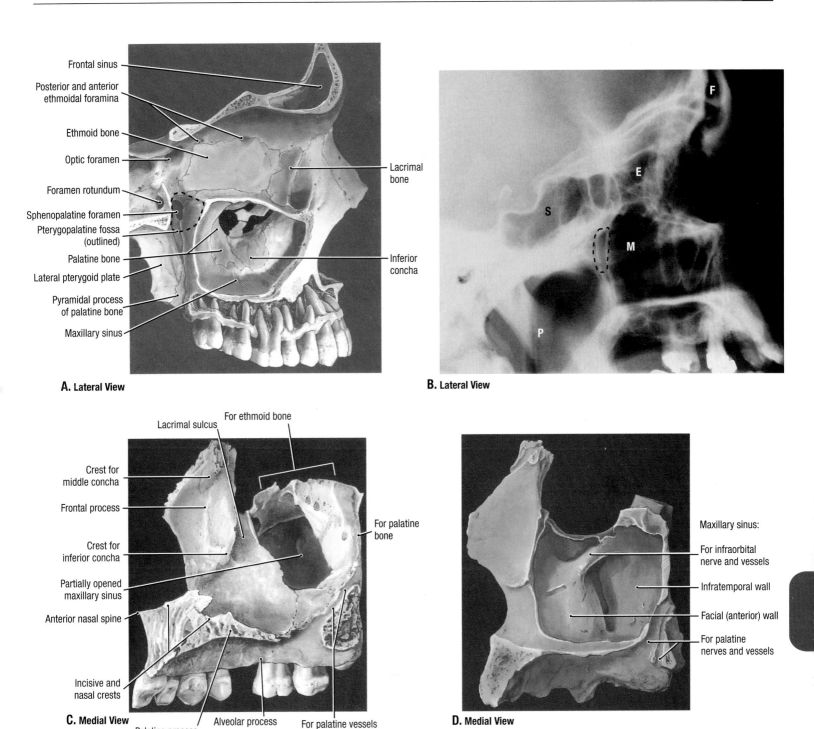

Frontal sinus

Posterior and anterior ethmoidal foramina

Ethmoid bone

Optic foramen

Foramen rotundum

Sphenopalatine foramen

Pterygopalatine fossa (outlined)

Palatine bone

Lateral pterygoid plate

Pyramidal process of palatine bone

Maxillary sinus

Lacrimal bone

Inferior concha

A. Lateral View

F

E

S

M

P

B. Lateral View

Lacrimal sulcus

For ethmoid bone

Crest for middle concha

Frontal process

Crest for inferior concha

Partially opened maxillary sinus

Anterior nasal spine

Incisive and nasal crests

For palatine bone

C. Medial View

Palatine process

Alveolar process

For palatine vessels and nerves

Maxillary sinus:

For infraorbital nerve and vessels

Infratemporal wall

Facial (anterior) wall

For palatine nerves and vessels

D. Medial View

7.71 Maxillary sinus

A. Medial half of the right viscerocranium following sagittal sectioning through the maxillary sinus. The inferior concha (*orange*) and palatine bone (*pink*) form part of the medial wall of the maxillary sinus. Note the ethmoid (*yellow*) and lacrimal (*blue*) bones of the medial wall of the orbital cavity and the sphenopalatine foramen opening into the nasal cavity from the pterygopalatine fossa. **B.** Imaging of paranasal sinuses. *F*, frontal sinus; *E*, ethmoidal sinus; *S*, sphenoidal sinus; *M*, maxillary sinus; *P*, pharynx; *dotted lines*, pterygopalatine fossa. **C.** Isolated maxilla with partially opened maxillary sinus. **D.** Isolated maxilla with opened maxillary sinus.

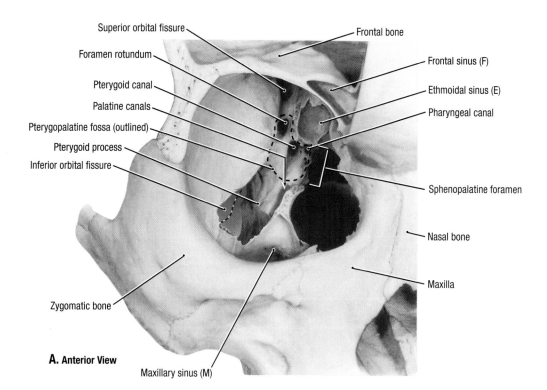

A. Anterior View

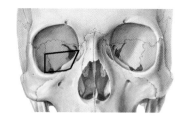

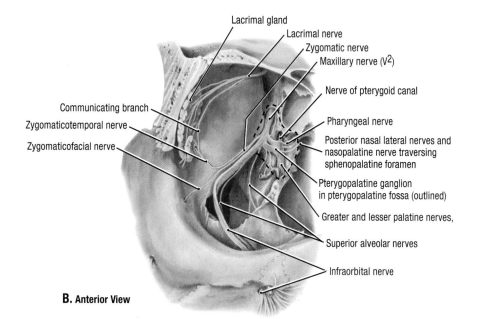

B. Anterior View

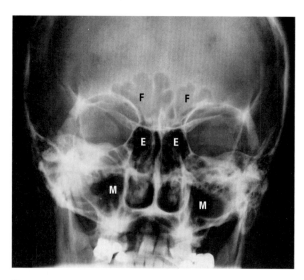

C. Anteroposterior View

7.72 **Pterygopalatine fossa, orbital approach**

A. Bones and foramina. **B.** Maxillary nerve. In **A** and **B**, the pterygopalatine fossa has been exposed through the floor of the orbit and maxillary sinus. **C.** Radiograph of cranium. Letters refer to sinuses labeled in **A**.

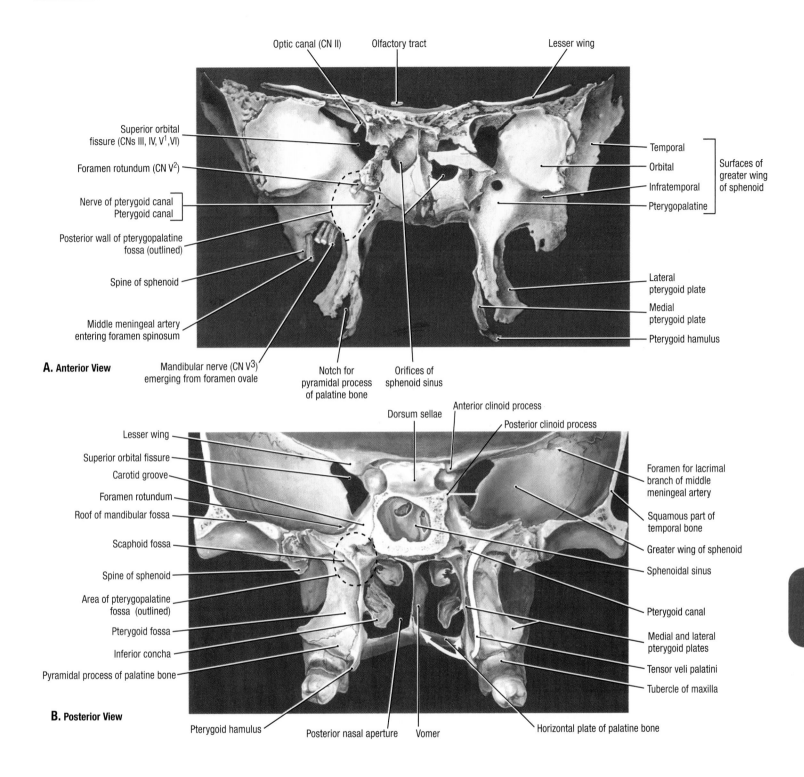

A. Anterior View

Optic canal (CN II) — Olfactory tract — Lesser wing

Superior orbital fissure (CNs III, IV, V¹, VI)

Foramen rotundum (CN V²)

Nerve of pterygoid canal
Pterygoid canal

Posterior wall of pterygopalatine fossa (outlined)

Spine of sphenoid

Middle meningeal artery entering foramen spinosum

Temporal
Orbital
Infratemporal
Pterygopalatine
— Surfaces of greater wing of sphenoid

Lateral pterygoid plate
Medial pterygoid plate
Pterygoid hamulus

Mandibular nerve (CN V³) emerging from foramen ovale

Notch for pyramidal process of palatine bone

Orifices of sphenoid sinus

B. Posterior View

Dorsum sellae — Anterior clinoid process
Posterior clinoid process

Lesser wing
Superior orbital fissure
Carotid groove
Foramen rotundum
Roof of mandibular fossa
Scaphoid fossa
Spine of sphenoid
Area of pterygopalatine fossa (outlined)
Pterygoid fossa
Inferior concha
Pyramidal process of palatine bone

Foramen for lacrimal branch of middle meningeal artery
Squamous part of temporal bone
Greater wing of sphenoid
Sphenoidal sinus
Pterygoid canal
Medial and lateral pterygoid plates
Tensor veli palatini
Tubercle of maxilla

Pterygoid hamulus   Posterior nasal aperture   Vomer   Horizontal plate of palatine bone

7.73   Sphenoid bone and sphenoidal sinus

A. Isolated sphenoid bone demonstrating external surfaces of the greater wing of the sphenoid. **B.** Anterior part of the cranium following coronal sectioning in the plane of the foramina lacerum, demonstrating the cranial surface of the greater wing and the interior of the sphenoidal sinus. The tensor veli palatini (*white arrow*) attaches to the spine of the sphenoid, scaphoid fossa, and cartilage of the auditory tube, and passes inferiorly to hook around the hamulus of the medial pterygoid plate before passing into the palatine aponeurosis.

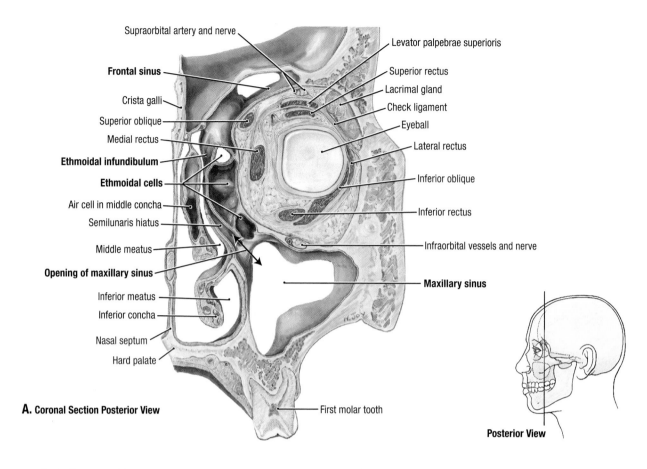

Supraorbital artery and nerve

Levator palpebrae superioris

Frontal sinus

Superior rectus

Crista galli

Lacrimal gland

Superior oblique

Check ligament

Medial rectus

Eyeball

Ethmoidal infundibulum

Lateral rectus

Ethmoidal cells

Inferior oblique

Air cell in middle concha

Semilunaris hiatus

Inferior rectus

Middle meatus

Infraorbital vessels and nerve

Opening of maxillary sinus

Maxillary sinus

Inferior meatus

Inferior concha

Nasal septum

Hard palate

A. Coronal Section Posterior View

First molar tooth

Posterior View

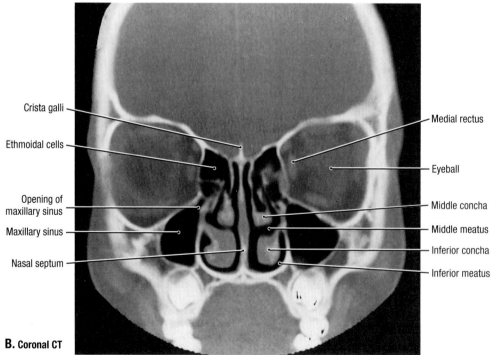

Crista galli

Medial rectus

Ethmoidal cells

Eyeball

Opening of maxillary sinus

Middle concha

Maxillary sinus

Middle meatus

Nasal septum

Inferior concha

Inferior meatus

B. Coronal CT

7.74 **Cross sectional anatomy of the head—I: orbit, paranasal sinuses, and nasal cavity**

A. Anatomical section of right side of the head. **B.** CT scan.

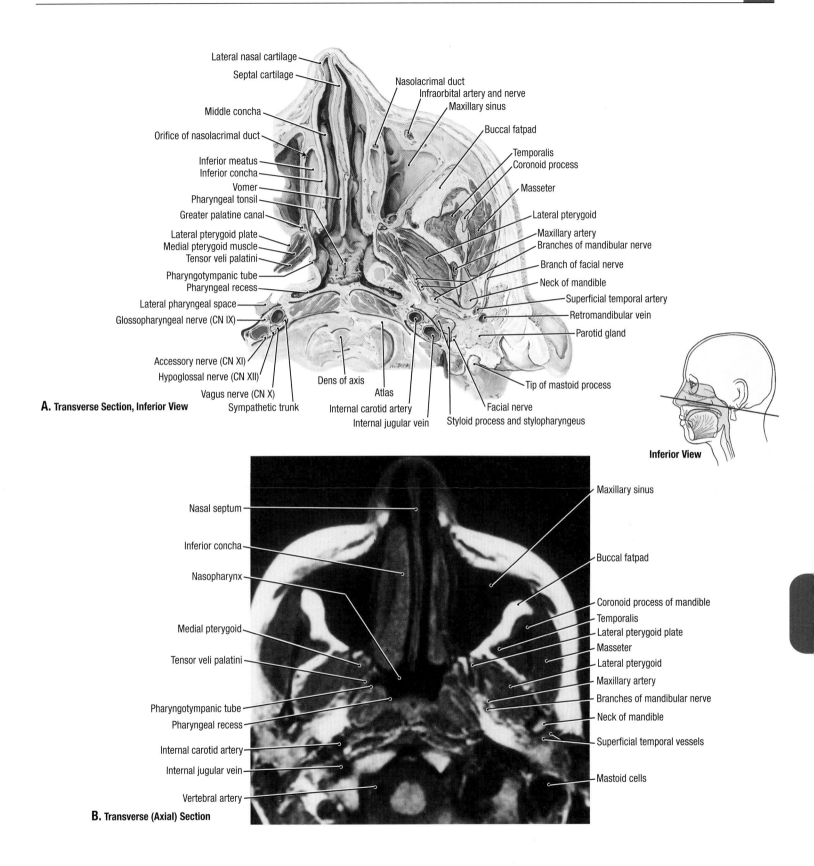

A. Transverse Section, Inferior View

Lateral nasal cartilage
Septal cartilage
Middle concha
Orifice of nasolacrimal duct
Inferior meatus
Inferior concha
Vomer
Pharyngeal tonsil
Greater palatine canal
Lateral pterygoid plate
Medial pterygoid muscle
Tensor veli palatini
Pharyngotympanic tube
Pharyngeal recess
Lateral pharyngeal space
Glossopharyngeal nerve (CN IX)
Accessory nerve (CN XI)
Hypoglossal nerve (CN XII)
Vagus nerve (CN X)
Sympathetic trunk

Nasolacrimal duct
Infraorbital artery and nerve
Maxillary sinus
Buccal fatpad
Temporalis
Coronoid process
Masseter
Lateral pterygoid
Maxillary artery
Branches of mandibular nerve
Branch of facial nerve
Neck of mandible
Superficial temporal artery
Retromandibular vein
Parotid gland
Tip of mastoid process

Dens of axis
Atlas
Internal carotid artery
Internal jugular vein
Facial nerve
Styloid process and stylopharyngeus

Inferior View

B. Transverse (Axial) Section

Nasal septum
Inferior concha
Nasopharynx
Medial pterygoid
Tensor veli palatini
Pharyngotympanic tube
Pharyngeal recess
Internal carotid artery
Internal jugular vein
Vertebral artery

Maxillary sinus
Buccal fatpad
Coronoid process of mandible
Temporalis
Lateral pterygoid plate
Masseter
Lateral pterygoid
Maxillary artery
Branches of mandibular nerve
Neck of mandible
Superficial temporal vessels
Mastoid cells

7.75   **Cross sectional anatomy of the head—II: nasal cavity and nasopharynx, maxillary sinus, and infratemporal fossa**

A. Anatomical section of left side of head. **B.** MRI scan.

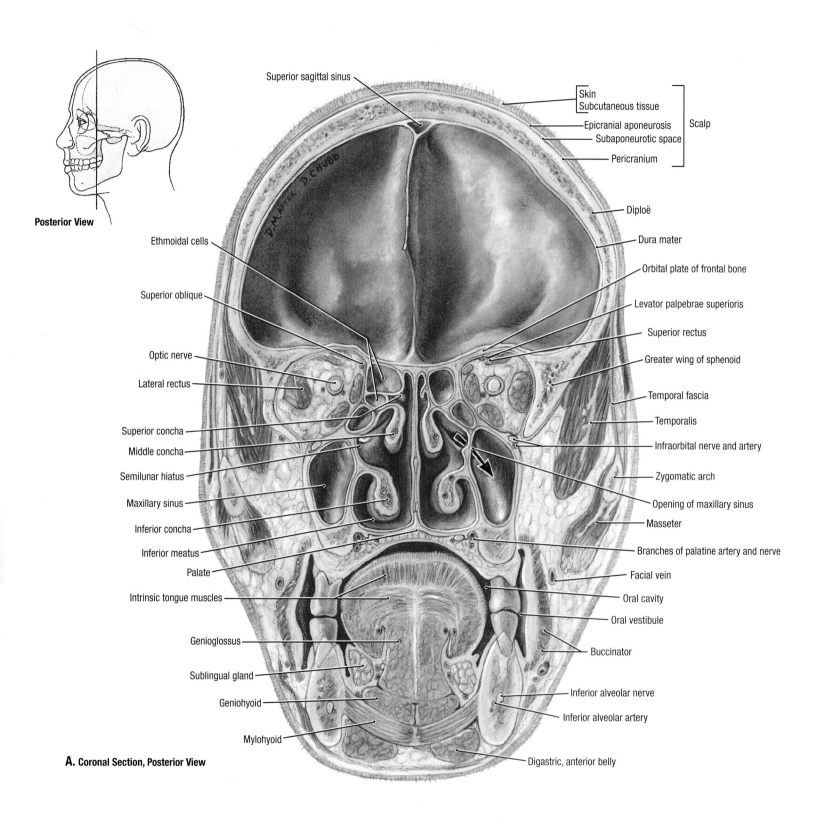

Posterior View

Superior sagittal sinus

Skin
Subcutaneous tissue
Epicranial aponeurosis — Scalp
Subaponeurotic space
Pericranium

Diploë

Dura mater

Orbital plate of frontal bone

Levator palpebrae superioris

Superior rectus

Greater wing of sphenoid

Temporal fascia

Temporalis

Infraorbital nerve and artery

Zygomatic arch

Opening of maxillary sinus

Masseter

Branches of palatine artery and nerve

Facial vein

Oral cavity

Oral vestibule

Buccinator

Inferior alveolar nerve

Inferior alveolar artery

Digastric, anterior belly

Ethmoidal cells

Superior oblique

Optic nerve

Lateral rectus

Superior concha

Middle concha

Semilunar hiatus

Maxillary sinus

Inferior concha

Inferior meatus

Palate

Intrinsic tongue muscles

Genioglossus

Sublingual gland

Geniohyoid

Mylohyoid

A. Coronal Section, Posterior View

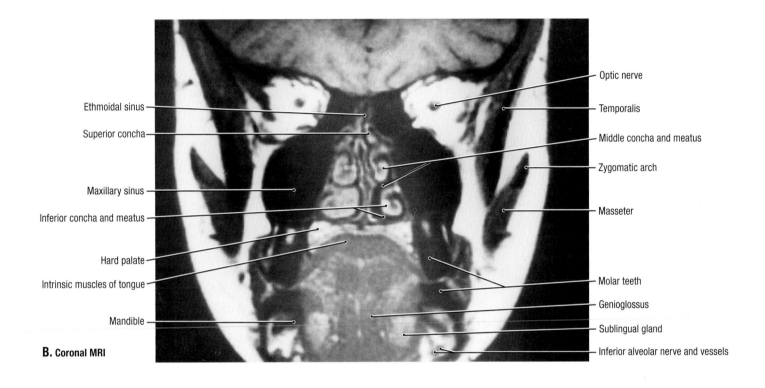

Ethmoidal sinus

Superior concha

Maxillary sinus

Inferior concha and meatus

Hard palate

Intrinsic muscles of tongue

Mandible

B. Coronal MRI

Optic nerve

Temporalis

Middle concha and meatus

Zygomatic arch

Masseter

Molar teeth

Genioglossus

Sublingual gland

Inferior alveolar nerve and vessels

7.76 **Cross sectional anatomy of the head—III: cranial, nasal/paranasal, and oral cavities**

A. Anatomical section through head. **B.** MRI of head.

- The thin orbital plate of the frontal bone forms a "roof" over the orbit and a "floor" for the anterior cranial fossa.
- The palate forms the floor of the nasal cavity and the roof of the oral cavity.
- The maxillary sinus forms the inferior part of the lateral wall of the nose; the middle concha covers the hiatus semilunaris into which the maxillary sinus opens *(arrow)*.
- The mylohyoid muscle, suspended between the right and left halves of the mandible, forms a diaphragm supporting the structures of the oral cavity.

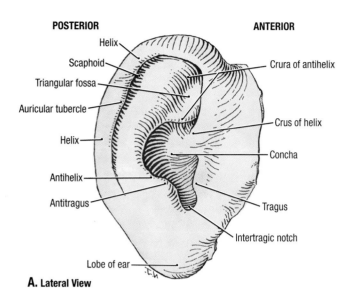

POSTERIOR   **ANTERIOR**

Helix
Scaphoid
Triangular fossa
Auricular tubercle
Helix
Antihelix
Antitragus

Crura of antihelix
Crus of helix
Concha
Tragus
Intertragic notch
Lobe of ear

A. Lateral View

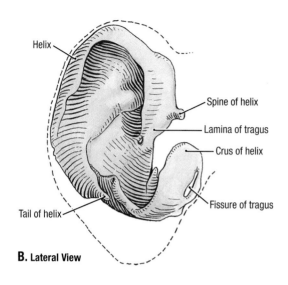

Helix

Spine of helix
Lamina of tragus
Crus of helix
Fissure of tragus
Tail of helix

B. Lateral View

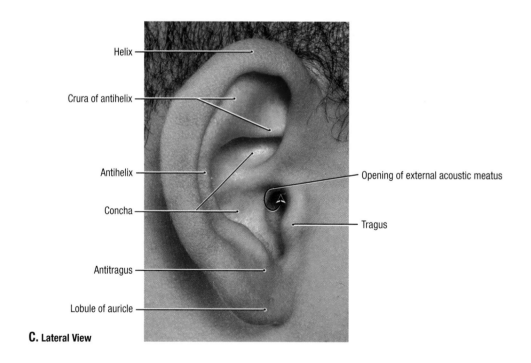

Helix
Crura of antihelix
Antihelix
Concha
Antitragus
Lobule of auricle

Opening of external acoustic meatus
Tragus

C. Lateral View

7.77   **Auricle**

A. Features of auricle. **B.** Cartilage of auricle. **C.** Surface anatomy of auricle.

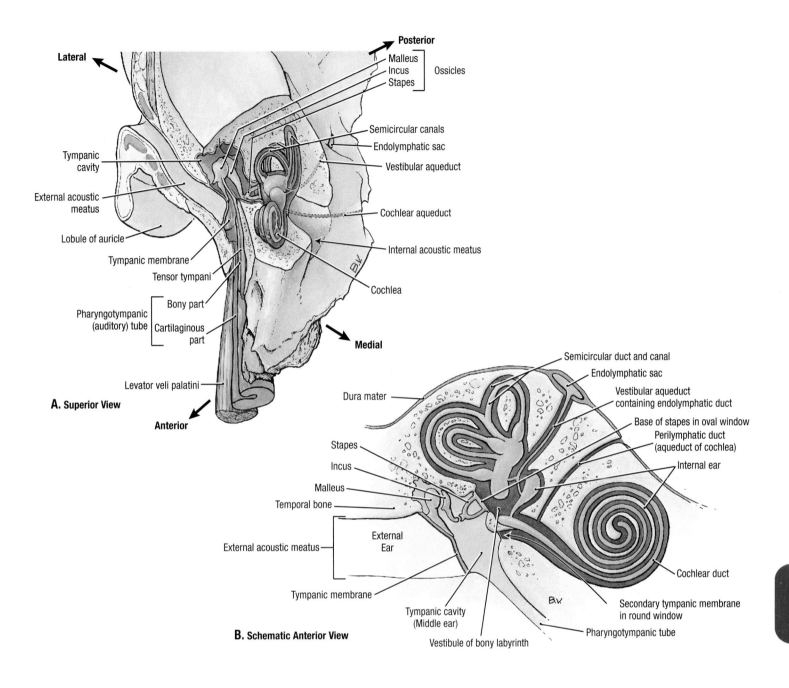

7.78   **External, middle, and internal ear—I: overviews**

A. Right temporal bone and auricle, sectioned in planes of (1) externa acoustic meatus and (2) pharyngotympanic tube. **B.** Schematic section of petrous temporal bone.

- The external ear comprises the auricle and external acoustic (auditory) meatus.
- The middle ear (tympanum) lies between the tympanic membrane and internal ear. Three ossicles extend from the lateral to the medial walls of the tympanum. Of these, the malleus is attached to the tympanic membrane. The stapes is attached by the anular ligament to the fenestra vestibuli (oval window), and the incus connects to the malleus and stapes. The pharyngotympanic tube, extending from the nasopharynx, opens into the anterior wall of the tympanic cavity.
- The membranous labyrinth comprises a closed system of membranous tubes and bulbs filled with fluid (endolymph) and bathed in surrounding fluid, called *perilymph* (*purple* in **B**), both membranous labyrinth and perilymph are contained within the bony labyrinth.

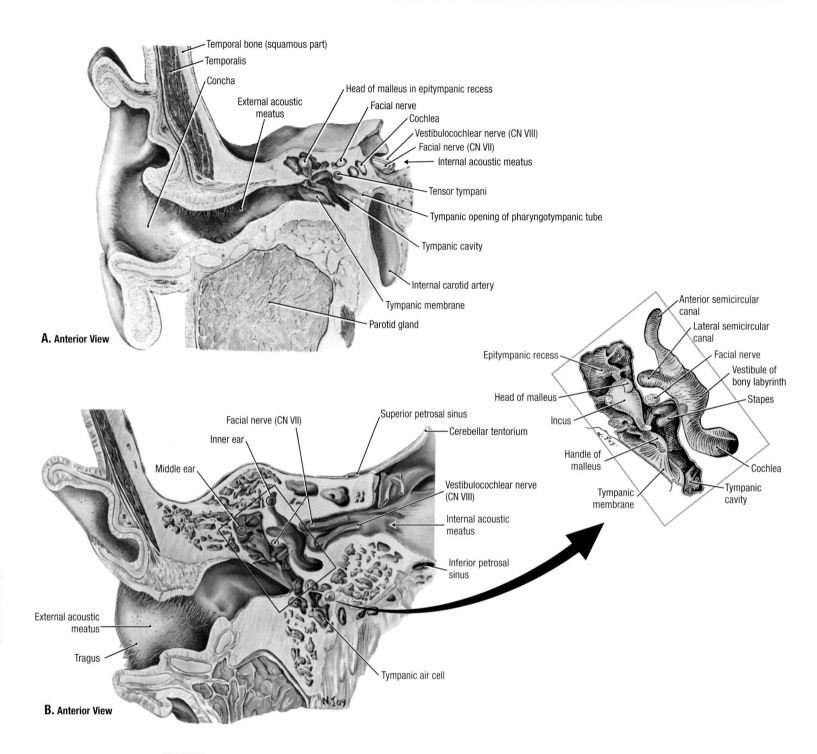

A. Anterior View

- Temporal bone (squamous part)
- Temporalis
- Concha
- External acoustic meatus
- Head of malleus in epitympanic recess
- Facial nerve
- Cochlea
- Vestibulocochlear nerve (CN VIII)
- Facial nerve (CN VII)
- Internal acoustic meatus
- Tensor tympani
- Tympanic opening of pharyngotympanic tube
- Tympanic cavity
- Internal carotid artery
- Tympanic membrane
- Parotid gland

B. Anterior View

- External acoustic meatus
- Tragus
- Middle ear
- Inner ear
- Facial nerve (CN VII)
- Superior petrosal sinus
- Cerebellar tentorium
- Vestibulocochlear nerve (CN VIII)
- Internal acoustic meatus
- Inferior petrosal sinus
- Tympanic air cell

- Anterior semicircular canal
- Lateral semicircular canal
- Facial nerve
- Vestibule of bony labyrinth
- Stapes
- Epitympanic recess
- Head of malleus
- Incus
- Handle of malleus
- Tympanic membrane
- Tympanic cavity
- Cochlea

7.79   External, middle, and internal ear—II: coronally sectioned

A. Posterior portion. **B.** Anterior portion. The inset (*outlined by the box*) is an enlargement of the structures of the middle and internal ear as they appear in **B.**

- The external acoustic meatus is about 3 cm long; half is cartilaginous and half is bony. It is narrowest at the isthmus, near the junction of the cartilaginous and bony parts.
- The external acoustic meatus is innervated by the auriculotemporal branch of the mandibular nerve (CN V^3) and the auricular branches of the vagus nerve (CN X); the middle ear is innervated by the glossopharyngeal nerve (CN IX).
- The cartilaginous part of the external acoustic meatus is lined with thick skin; the bony part is lined with thin epithelium that adheres to the periosteum and forms the outermost layer of the tympanic membrane.

POSTERIOR   ANTERIOR

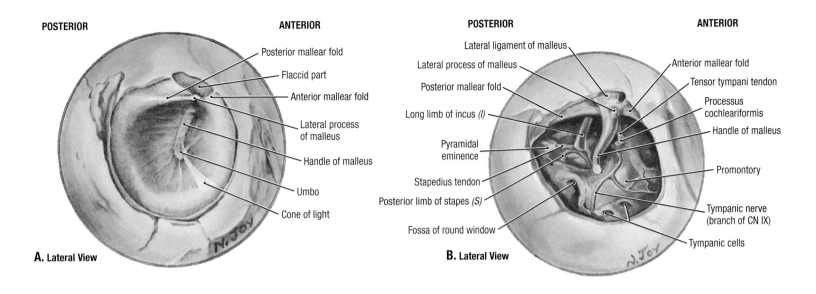

POSTERIOR   ANTERIOR

A. Lateral View

- Posterior mallear fold
- Flaccid part
- Anterior mallear fold
- Lateral process of malleus
- Handle of malleus
- Umbo
- Cone of light

B. Lateral View

- Lateral ligament of malleus
- Lateral process of malleus
- Posterior mallear fold
- Long limb of incus (I)
- Pyramidal eminence
- Stapedius tendon
- Posterior limb of stapes (S)
- Fossa of round window
- Anterior mallear fold
- Tensor tympani tendon
- Processus cochleariformis
- Handle of malleus
- Promontory
- Tympanic nerve (branch of CN IX)
- Tympanic cells

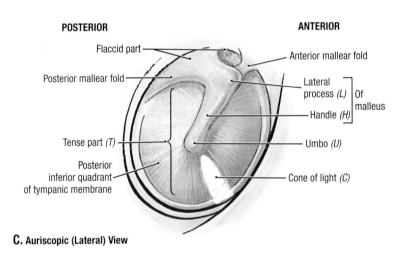

POSTERIOR   ANTERIOR

C. Auriscopic (Lateral) View

- Flaccid part
- Posterior mallear fold
- Anterior mallear fold
- Lateral process (L) } Of malleus
- Handle (H)
- Tense part (T)
- Posterior inferior quadrant of tympanic membrane
- Umbo (U)
- Cone of light (C)

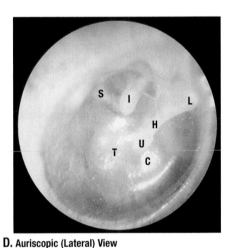

D. Auriscopic (Lateral) View

7.80   Tympanic membrane

A. External (lateral) surface of tympanic membrane. **B.** Tympanic membrane removed, demonstrating structures that lie medially. **C.** Diagram of auriscopic view of tympanic membrane. **D.** Auriscopic view of tympanic membrane.

- The oval tympanic membrane is a shallow cone deepest at the central apex, the umbo, where the membrane is attached to the tip of the handle of the malleus. The handle of the malleus is attached to the membrane along its entire length as it extends anterosuperiorly toward the periphery of the membrane.
- Superior to the lateral process of the malleus, the membrane is thin (the "flaccid part," or pars flaccida); the flaccid part lacks the radial and circular fibers present in the remainder of the membrane ("tense part," or pars tensa). The junction between the two parts is marked by anterior and posterior mallear folds.
- The lateral surface of the tympanic membrane is innervated by the auricular branch of the auriculotemporal nerve (CN V³) and the auricular branch of the vagus nerve (CN X); the medial surface is innervated by tympanic branches of CN IX.

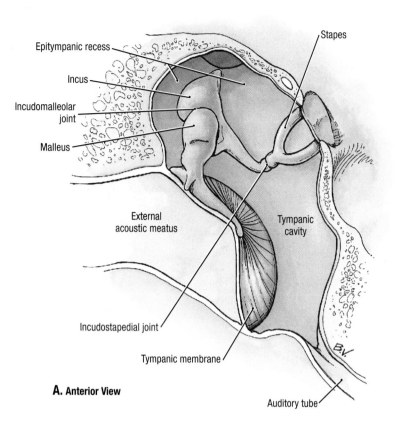

Epitympanic recess

Incus

Incudomalleolar joint

Malleus

External acoustic meatus

Incudostapedial joint

Tympanic membrane

A. Anterior View

Stapes

Tympanic cavity

Auditory tube

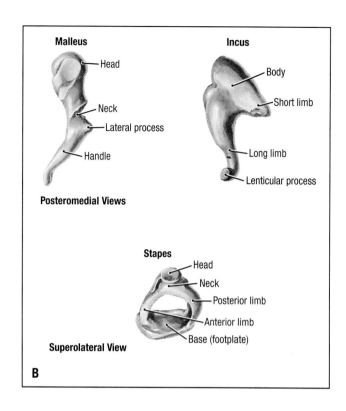

Malleus
Head
Neck
Lateral process
Handle
Posteromedial Views

Incus
Body
Short limb
Long limb
Lenticular process

Stapes
Head
Neck
Posterior limb
Anterior limb
Base (footplate)
Superolateral View

B

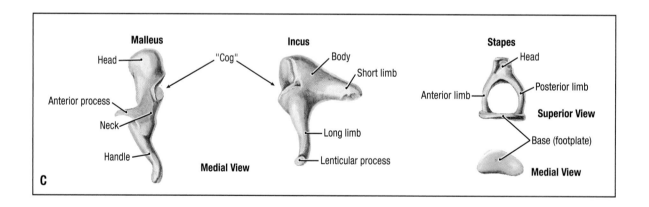

Malleus
Head
Anterior process
Neck
Handle
Medial View

"Cog"

Incus
Body
Short limb
Long limb
Lenticular process
Medial View

Stapes
Head
Anterior limb
Posterior limb
Superior View
Base (footplate)
Medial View

C

7.81 Ossicles of the middle ear

A. Ossicles in situ, as revealed by a coronal section of the temporal bone. **B** and **C.** Isolated ossicles.

- The head of the malleus and body and short process of the incus lie in the epitympanic recess, and the handle of the malleus is embedded in the tympanic membrane.
- The saddle-shaped articular surface of the head of the malleus and the reciprocally shaped articular surface of the body of the incus form the incudomalleolar synovial joint.
- A convex articular facet at the end of the long process of the incus articulates with the head of the stapes to compose the incudostapedial synovial joint.

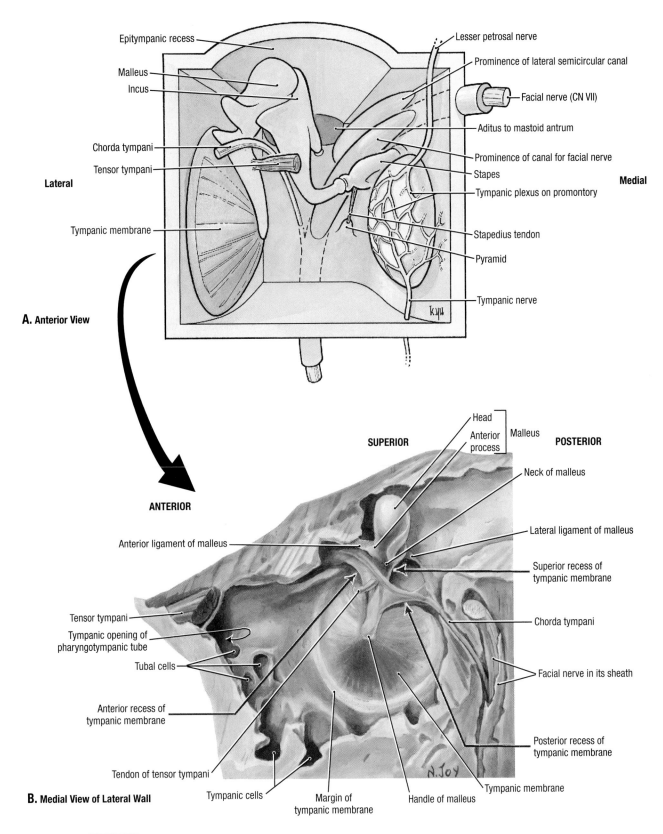

A. Anterior View

B. Medial View of Lateral Wall

7.82 Structures of the tympanic cavity

A. Schematic illustration of the tympanic cavity with the anterior wall removed. **B.** Lateral wall of the tympanic cavity. The facial nerve lies within the facial canal surrounded by a tough periosteal tube; the chorda tympani leaves the facial nerve and lies within two crescentic folds of mucous membrane, crossing the neck of the malleus superior to the tendon of tensor tympani and following the anterior process and anterior ligament of the malleus.

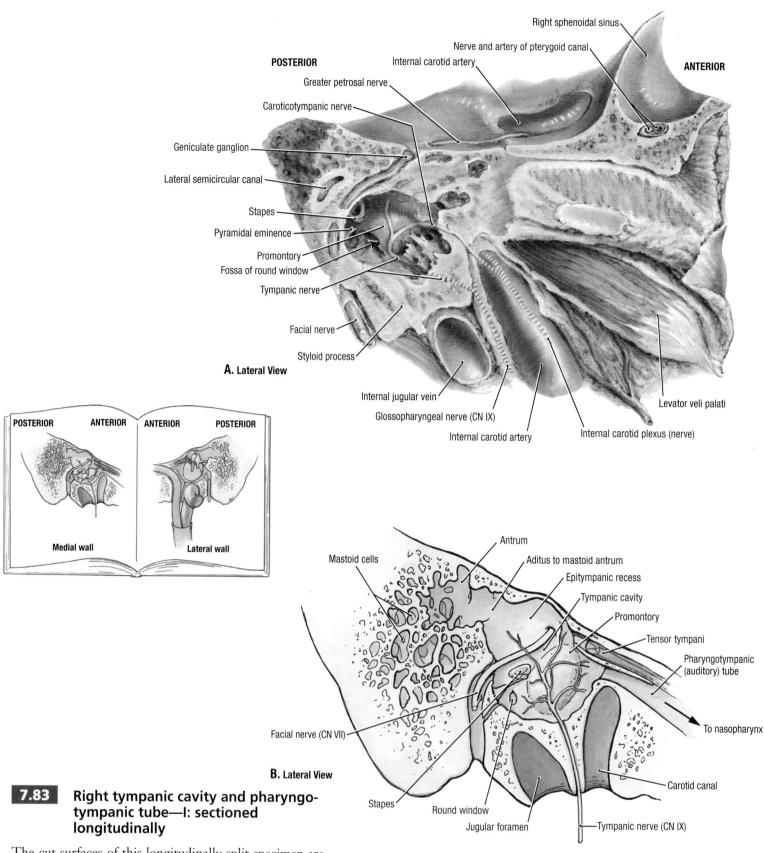

POSTERIOR

ANTERIOR

Right sphenoidal sinus

Nerve and artery of pterygoid canal

Internal carotid artery

Greater petrosal nerve

Caroticotympanic nerve

Geniculate ganglion

Lateral semicircular canal

Stapes

Pyramidal eminence

Promontory

Fossa of round window

Tympanic nerve

Facial nerve

Styloid process

A. Lateral View

Internal jugular vein

Glossopharyngeal nerve (CN IX)

Internal carotid artery

Levator veli palati

Internal carotid plexus (nerve)

POSTERIOR   ANTERIOR   ANTERIOR   POSTERIOR

Medial wall

Lateral wall

Mastoid cells

Antrum

Aditus to mastoid antrum

Epitympanic recess

Tympanic cavity

Promontory

Tensor tympani

Pharyngotympanic (auditory) tube

To nasopharynx

Facial nerve (CN VII)

Carotid canal

Stapes

Round window

Jugular foramen

Tympanic nerve (CN IX)

B. Lateral View

7.83   Right tympanic cavity and pharyngo-tympanic tube—I: sectioned longitudinally

The cut surfaces of this longitudinally split specimen are displayed as pages in a book.
A. Dissection of medial wall. **B.** Schematic illustration of medial wall.

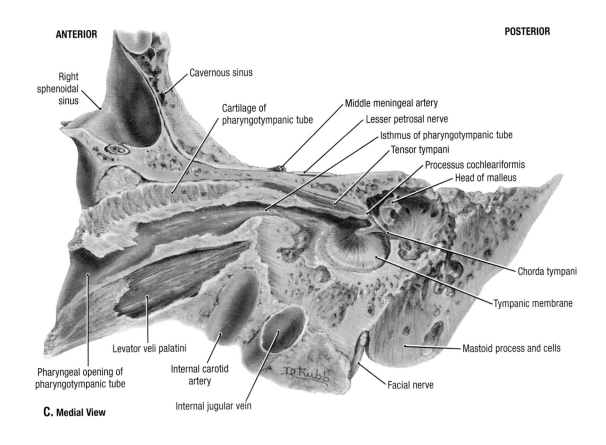

ANTERIOR

POSTERIOR

Right sphenoidal sinus

Cavernous sinus

Cartilage of pharyngotympanic tube

Middle meningeal artery

Lesser petrosal nerve

Isthmus of pharyngotympanic tube

Tensor tympani

Processus cochleariformis

Head of malleus

Chorda tympani

Tympanic membrane

Mastoid process and cells

Facial nerve

Internal jugular vein

Internal carotid artery

Levator veli palatini

Pharyngeal opening of pharyngotympanic tube

C. Medial View

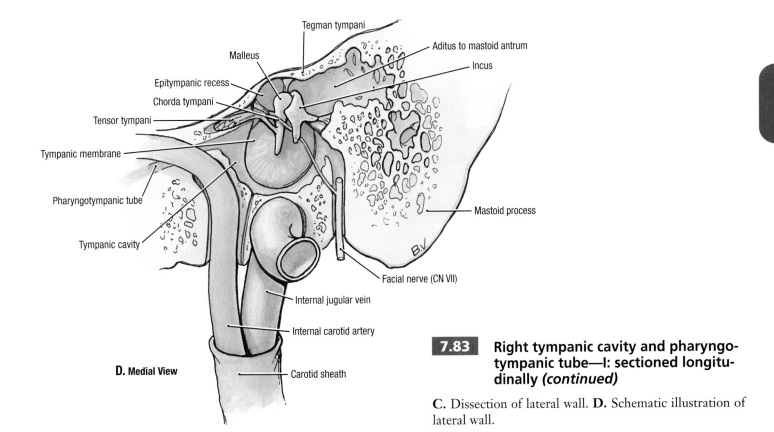

Tegman tympani

Malleus

Epitympanic recess

Chorda tympani

Tensor tympani

Tympanic membrane

Pharyngotympanic tube

Tympanic cavity

Aditus to mastoid antrum

Incus

Mastoid process

Facial nerve (CN VII)

Internal jugular vein

Internal carotid artery

Carotid sheath

D. Medial View

7.83   **Right tympanic cavity and pharyngotympanic tube—I: sectioned longitudinally** *(continued)*

C. Dissection of lateral wall. **D.** Schematic illustration of lateral wall.

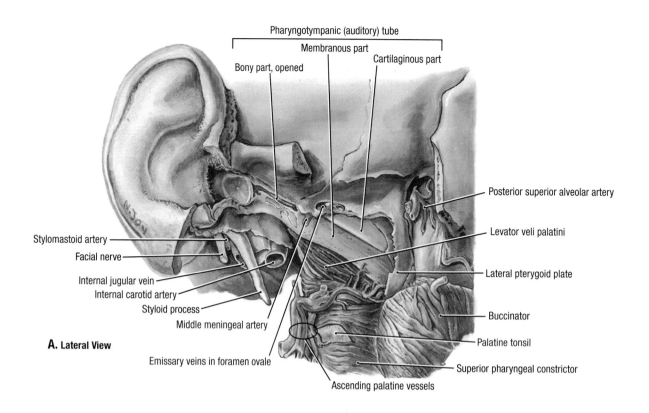

Pharyngotympanic (auditory) tube
Membranous part
Bony part, opened
Cartilaginous part

Posterior superior alveolar artery

Levator veli palatini

Stylomastoid artery

Facial nerve

Internal jugular vein

Internal carotid artery

Styloid process

Middle meningeal artery

Lateral pterygoid plate

Buccinator

Palatine tonsil

Superior pharyngeal constrictor

A. Lateral View

Emissary veins in foramen ovale

Ascending palatine vessels

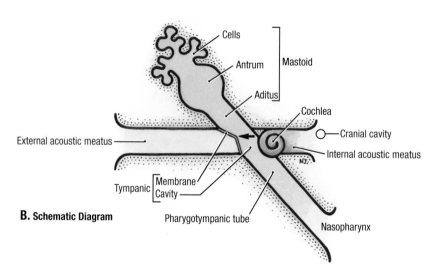

Cells

Antrum   Mastoid

Aditus

Cochlea

External acoustic meatus

Cranial cavity

Internal acoustic meatus

Tympanic [Membrane / Cavity]

B. Schematic Diagram

Pharygotympanic tube

Nasopharynx

7.84 **Right tympanic cavity and pharyngotympanic tube—II: relationships**

A. Dissection demonstrating lateral aspect of tube and structures located medially. **B.** Schematic illustration demonstrating relationship to internal and external acoustic meatuses. **C.** Dissection demonstrating medial (pharyngeal) aspect of tube and structures located laterally. **D.** Relationship of tympanic cavity to internal carotid artery, sigmoid sinus, and middle cranial fossa.

- The general direction of the pharyngotympanic tube is superior, posterior, and lateral from the nasopharynx to the tympanic cavity.
- The cartilaginous part of the tube rests throughout its length on the levator veli palatini muscle.
- The tegmen tympani forms the roof of the tympanic cavity and mastoid antrum.
- The internal carotid artery is the primary relationship of the anterior wall, the internal jugular vein is the primary relationship of the floor, and the facial nerve is the primary relationship of the posterior wall.

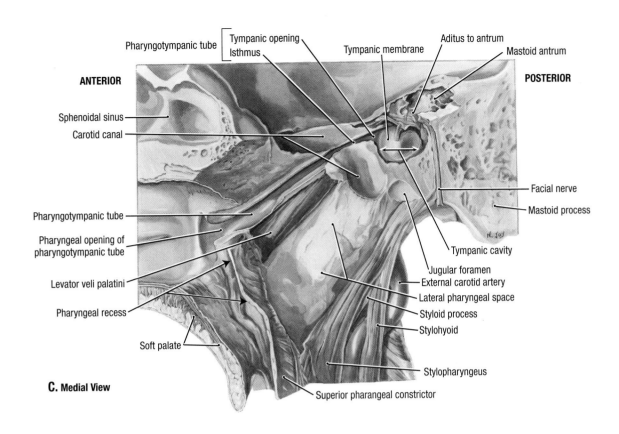

Pharyngotympanic tube

Tympanic opening
Isthmus

Tympanic membrane

Aditus to antrum

Mastoid antrum

ANTERIOR

POSTERIOR

Sphenoidal sinus

Carotid canal

Facial nerve

Mastoid process

Pharyngotympanic tube

Pharyngeal opening of
pharyngotympanic tube

Levator veli palatini

Pharyngeal recess

Tympanic cavity

Jugular foramen

External carotid artery

Lateral pharyngeal space

Styloid process

Stylohyoid

Soft palate

Stylopharyngeus

C. Medial View

Superior pharangeal constrictor

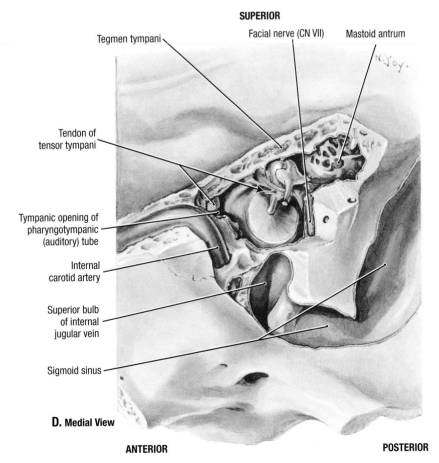

SUPERIOR

Tegmen tympani

Facial nerve (CN VII)

Mastoid antrum

Tendon of
tensor tympani

Tympanic opening of
pharyngotympanic
(auditory) tube

Internal
carotid artery

Superior bulb
of internal
jugular vein

Sigmoid sinus

D. Medial View

ANTERIOR

POSTERIOR

7.84 **Right tympanic cavity and pharyngotympanic tube—II: relationships** *(continued)*

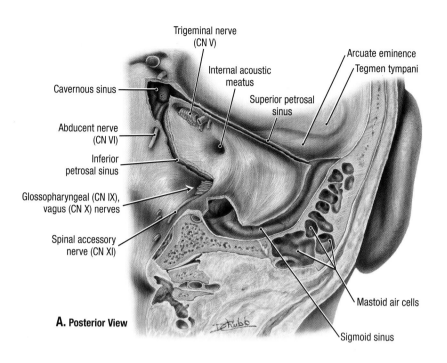

Trigeminal nerve (CN V)

Internal acoustic meatus

Arcuate eminence

Tegmen tympani

Superior petrosal sinus

Cavernous sinus

Abducent nerve (CN VI)

Inferior petrosal sinus

Glossopharyngeal (CN IX), vagus (CN X) nerves

Spinal accessory nerve (CN XI)

Mastoid air cells

Sigmoid sinus

A. Posterior View

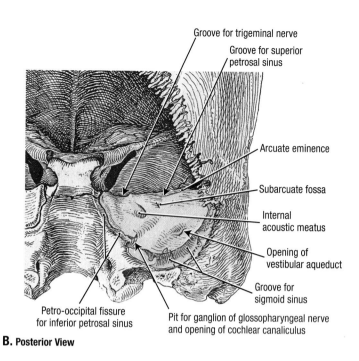

Groove for trigeminal nerve

Groove for superior petrosal sinus

Arcuate eminence

Subarcuate fossa

Internal acoustic meatus

Opening of vestibular aqueduct

Groove for sigmoid sinus

Petro-occipital fissure for inferior petrosal sinus

Pit for ganglion of glossopharyngeal nerve and opening of cochlear canaliculus

B. Posterior View

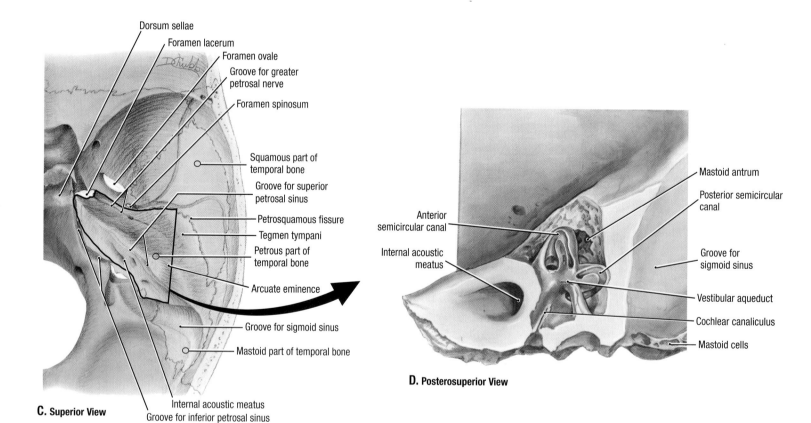

Dorsum sellae

Foramen lacerum

Foramen ovale

Groove for greater petrosal nerve

Foramen spinosum

Squamous part of temporal bone

Groove for superior petrosal sinus

Petrosquamous fissure

Tegmen tympani

Petrous part of temporal bone

Arcuate eminence

Groove for sigmoid sinus

Mastoid part of temporal bone

Internal acoustic meatus

Groove for inferior petrosal sinus

C. Superior View

Anterior semicircular canal

Internal acoustic meatus

Mastoid antrum

Posterior semicircular canal

Groove for sigmoid sinus

Vestibular aqueduct

Cochlear canaliculus

Mastoid cells

D. Posterosuperior View

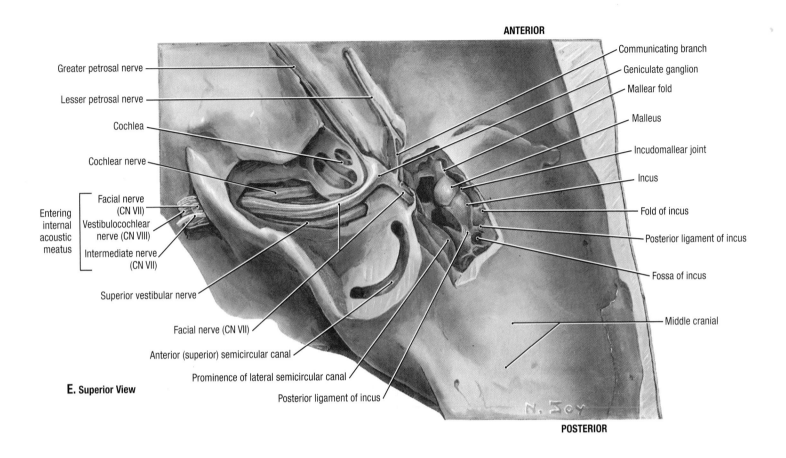

ANTERIOR

Greater petrosal nerve

Lesser petrosal nerve

Cochlea

Cochlear nerve

Entering internal acoustic meatus —
- Facial nerve (CN VII)
- Vestibulocochlear nerve (CN VIII)
- Intermediate nerve (CN VII)

Superior vestibular nerve

Facial nerve (CN VII)

Anterior (superior) semicircular canal

Prominence of lateral semicircular canal

Posterior ligament of incus

E. Superior View

Communicating branch

Geniculate ganglion

Mallear fold

Malleus

Incudomallear joint

Incus

Fold of incus

Posterior ligament of incus

Fossa of incus

Middle cranial

POSTERIOR

7.85   **Petrous temporal bone: external and internal relationships**

A. Venous sinuses and cranial nerves related to the petrous temporal bone. **B.** Petrous temporal bone in situ. The occipital squama has been removed. **C.** Features and parts of temporal bone. **D.** Semicircular canals and aqueducts in situ. **E.** Middle and inner ear in situ. The tegmen tympani has been removed to expose the middle ear, the arcuate eminence has been removed to expose the anterior semicircular canal, and the course of the facial and vestibulocochlear nerves through the internal acoustic meatus and internal ear is demonstrated.

- The aqueduct of the vestibule transmits the endolymphatic duct, which expands into a blind pouch, the endolymphatic sac; the sac is located deep to the dura mater on the posterior surface of the petrous part of the temporal bone and serves as a storage reservoir for excess endolymph formed by blood capillaries within the membranous labyrinth.
- The perilymphatic duct (within the canaliculus of the cochlea) opens at the base of the pyramidal pit for the glossopharyngeal ganglion; the duct runs from the scala vestibuli in the basal turn of the cochlea to an extension of the subarachnoid space around the glossopharyngeal, vagus, and accessory nerves (CN IX, X, and XI).
- At the geniculate ganglion, the facial nerve executes a sharp bend, called the *genu,* and then curves posteroinferiorly within the bony facial canal; the thin lateral wall of the facial canal separates the facial nerve from the tympanic cavity of the middle ear.

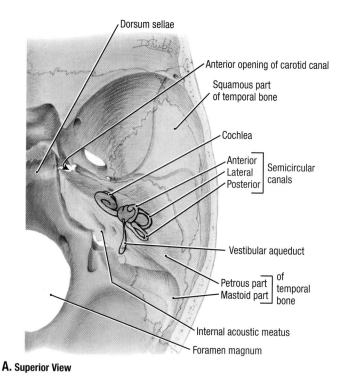

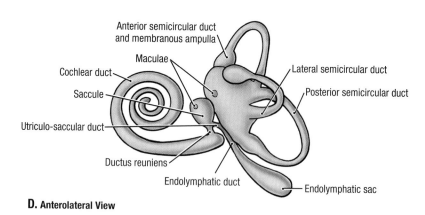

A. **Superior View**

Anterior (superior) semicircular canal and ampulla
Facial canal, opened (canal for facial nerve)
Lateral semicircular canal and ampulla
Posterior semicircular canal and ampulla
Cupula
Cochlea 2nd turn
1st turn
Oval window
Vestibule
Round window

B. **Anterolateral View**

Dorsum sellae
Anterior opening of carotid canal
Squamous part of temporal bone
Cochlea
Anterior
Lateral   Semicircular
Posterior   canals
Vestibular aqueduct
Petrous part   of
Mastoid part   temporal bone
Internal acoustic meatus
Foramen magnum

Anterior (superior) semicircular duct and membranous ampulla
Lateral semicircular duct
Common crus
Cochlear duct
Posterior semicircular duct
Utricle
Saccule
Endolymphatic sac
Ductus reuniens

C. **Anterolateral View**

Anterior semicircular duct and membranous ampulla
Maculae
Cochlear duct
Lateral semicircular duct
Saccule
Posterior semicircular duct
Utriculo-saccular duct
Ductus reuniens
Endolymphatic duct
Endolymphatic sac

D. **Anterolateral View**

7.86   **Bony and membranous labyrinths**

A. Location of bony labyrinth. **B.** Walls of left bony labyrinth (otic capsule). **C.** Left bony and membranous labyrinth. **D.** Left membranous labyrinth.

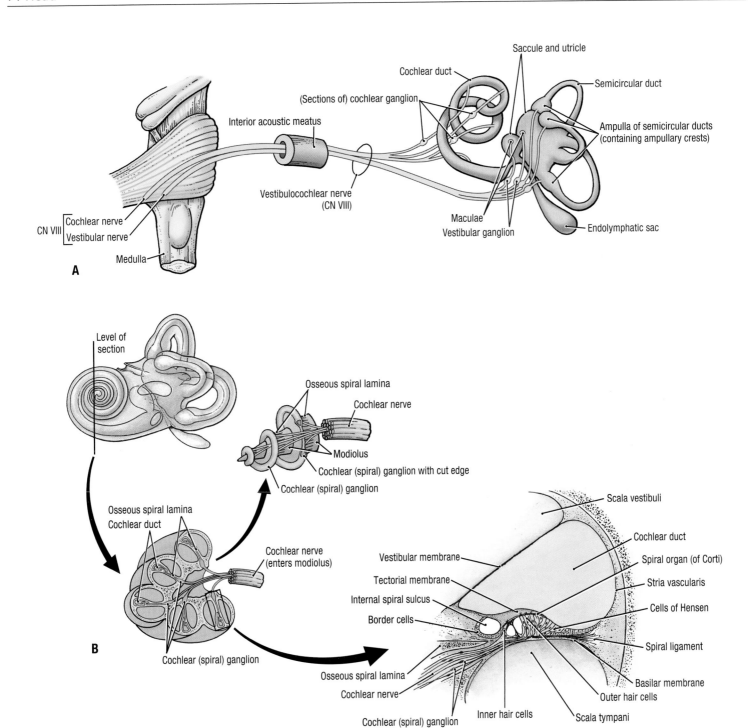

7.87   **Vestibulocochlear nerve and structure of cochlea**

A. Distribution of vestibulocochlear nerve (schematic). **B.** Structure of cochlea. The cochlea has been sectioned along the bony core of the cochlea (modiolus), the axis about which the cochlea winds. An isolated modiolus is shown after the turns of the cochlea are removed, leaving only the spiral lamina winding around it. The large drawing shows the details of the area enclosed in the rectangle, including a cross-section of the cochlear duct of the membranous labyrinth.

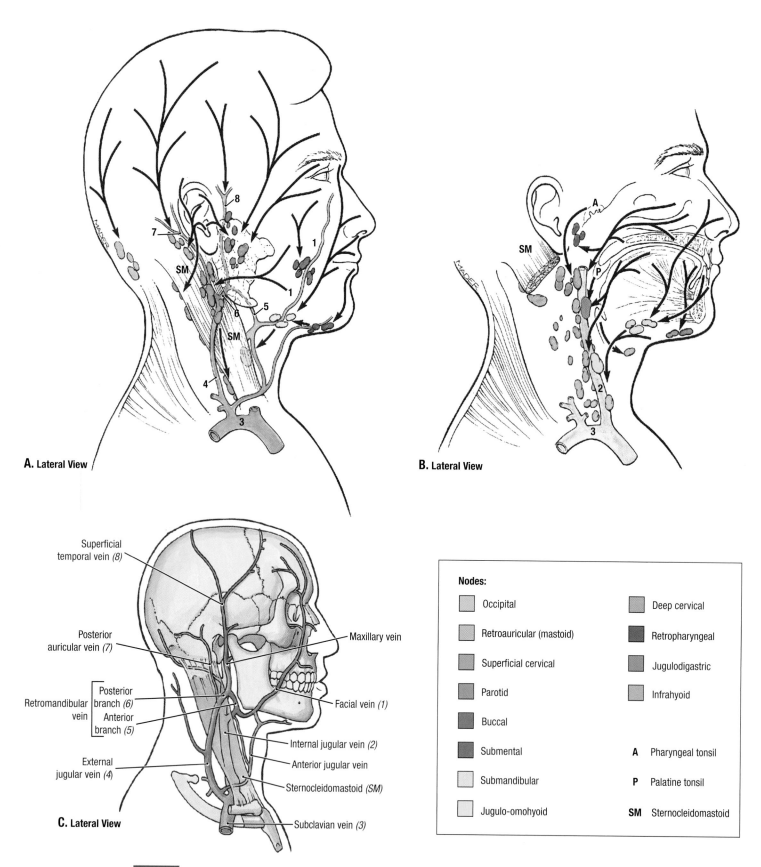

A. Lateral View

B. Lateral View

C. Lateral View

Superficial temporal vein (8)

Posterior auricular vein (7)

Retromandibular vein {
Posterior branch (6)
Anterior branch (5)

External jugular vein (4)

Maxillary vein

Facial vein (1)

Internal jugular vein (2)

Anterior jugular vein

Sternocleidomastoid (SM)

Subclavian vein (3)

Nodes:

Occipital

Retroauricular (mastoid)

Superficial cervical

Parotid

Buccal

Submental

Submandibular

Jugulo-omohyoid

Deep cervical

Retropharyngeal

Jugulodigastric

Infrahyoid

A Pharyngeal tonsil

P Palatine tonsil

SM Sternocleidomastoid

7.88 Lymphatic and venous drainage of the head and neck

A. Superficial lymphatic drainage. **B.** Deep lymphatic drainage. **C.** Venous drainage of head and neck.

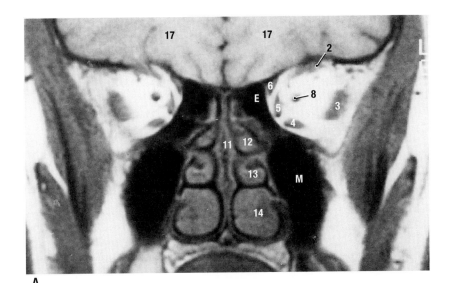

A

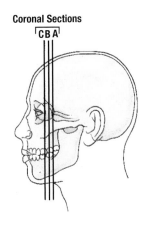

Coronal Sections

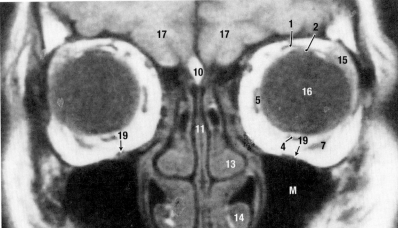

B

C

1	Levator palpebrae superioris
2	Superior rectus
3	Lateral rectus
4	Inferior rectus
5	Medial rectus
6	Superior oblique
7	Inferior oblique
8	Optic nerve
9	Olfactory bulb
10	Crista galli
11	Nasal septum
12	Superior concha
13	Middle concha
14	Inferior concha
15	Lacrimal gland
16	Eyeball
17	Frontal lobe
18	Tongue
19	Infraorbital vessels and nerve
M	Maxillary sinus
E	Ethmoidal sinus

7.89 **Coronal MRIs through the orbit**

See orientation drawing for sites of scans **A–C.**

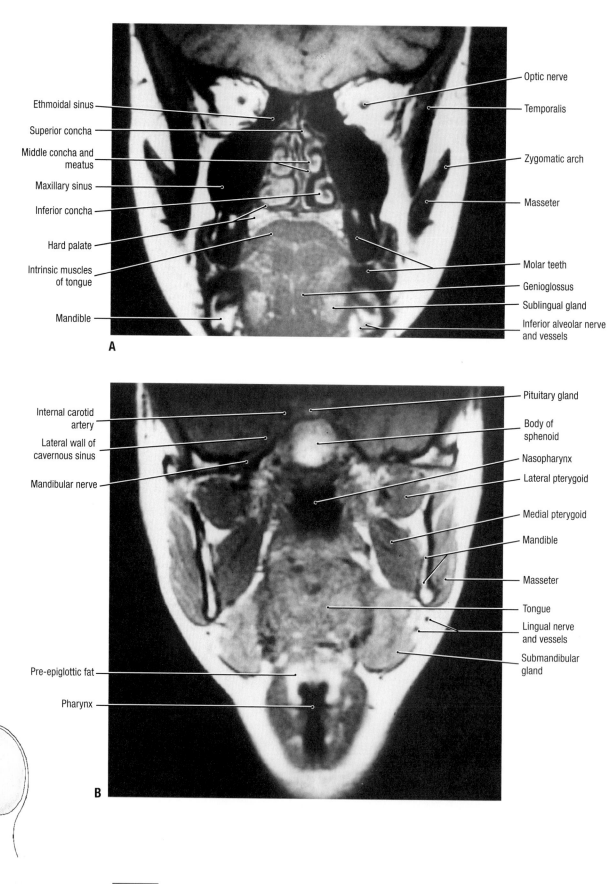

Ethmoidal sinus

Superior concha

Middle concha and meatus

Maxillary sinus

Inferior concha

Hard palate

Intrinsic muscles of tongue

Mandible

Optic nerve

Temporalis

Zygomatic arch

Masseter

Molar teeth

Genioglossus

Sublingual gland

Inferior alveolar nerve and vessels

A

Internal carotid artery

Lateral wall of cavernous sinus

Mandibular nerve

Pituitary gland

Body of sphenoid

Nasopharynx

Lateral pterygoid

Medial pterygoid

Mandible

Masseter

Tongue

Lingual nerve and vessels

Submandibular gland

Pre-epiglottic fat

Pharynx

B

Coronal Sections

A B

7.90 **Coronal MRIs of the head**

See orientation drawing for sites of scans **A** and **B**.

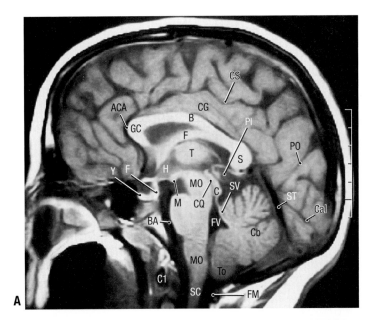

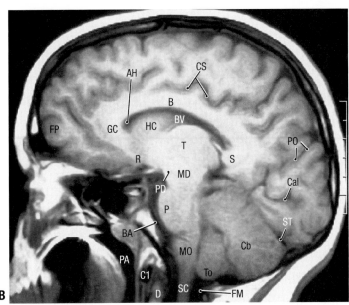

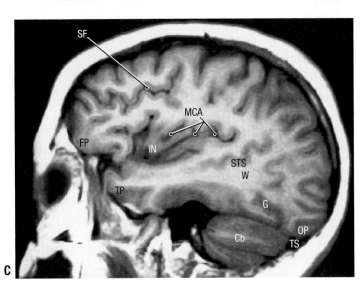

ACA	Anterior cerebral artery
AH	Anterior horn of lateral ventricle
B	Body of corpus callosum
BA	Basilar artery
BV	Body of lateral ventricle
C	Colliculi
C1	Anterior tubercle of atlas
Cal	Calcarine sulcus
Cb	Cerebellum
CG	Cingulate gyrus
CQ`	Cerebral aqueduct
CS	Cingulate sulcus
D	Dens (odontoid process)
F	Fornix
FM	Foramen magnum
FP	Frontal pole
FV	Fourth ventricle
G	Cerebral cortex (gray matter)
GC	Genu of corpus callosum
H	Hypothalamus
HC	Head of caudate nucleus
I	Infundibulum
IN	Insular cortex
M	Mammillary body
MCA	Middle cerebral artery
MD	Midbrain
MO	Medulla oblongata
OP	Occipital pole
P	Pons
PA	Pharynx
PD	Cerebral peduncle
PI	Pineal
PO	Parieto-occipital fissure
R	Rostrum of corpus callosum
S	Splenium of corpus callosum
SC	Spinal cord
SF	Superior frontal sulcus
ST	Straight sinus
STS	Superior temporal sulcus
SV	Superior medullary vellum
T	Thalamus
To	Cerebellar tonsil
TP	Temporal pole
TS	Transverse sinus
W	White matter
Y	Hypophysis

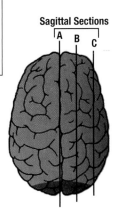

Sagittal Sections

7.91   **Sagittal MRIs of the brain (T1 weighted)**

See orientation drawing for sites of scans **A–C.**

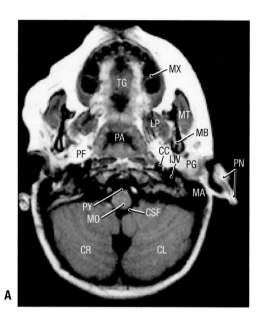

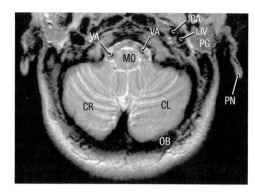

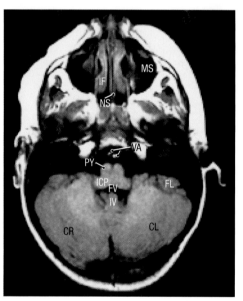

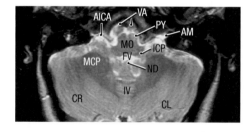

AICA	Anterior inferior cerebellar artery
AM	Internal acoustic meatus
BA	Basilar artery
CA	Cerebral aqueduct
CB	Ciliary body
CC	Common carotid artery
CI	Colliculi
CL	Left cerebellar hemisphere
CP	Cochlear perilymph
CR	Right cerebellar hemisphere
CSF	CSF in subarachnoid space
DS	Dorsum sellae
EB	Eyeball
F	CN VII and CN VIII
FC	Facial colliculus
FI	Fat in infratemporal fossa
FL	Flocculus
FV	Fourth ventricle
G	Gray matter
HF	Hypophyseal fossa
HP	Hippocampus
I	Infundibulum
IC	Interpeduncular cistern
ICA	Internal carotid artery
ICP	Inferior cerebellar peduncle
IF	Inferior concha
IH	Inferior horn (lateral ventricle)
IJV	Internal jugular vein
IP	Interpeduncular fossa

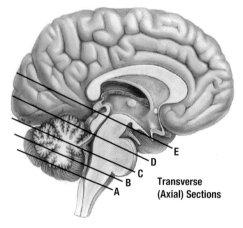

7.92 **Axial (transverse) MRIs through the brainstem**

See orientation drawing for sites of scans **A–E**. Images on left side of page are T1 weighted, and images on the right side are T2 weighted.

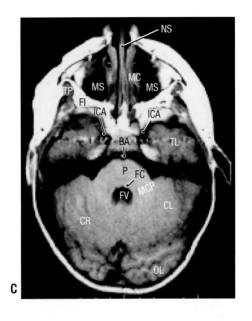

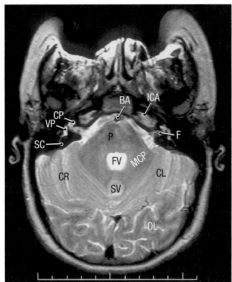

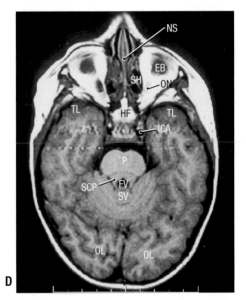

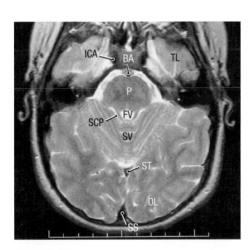

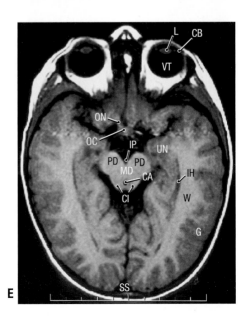

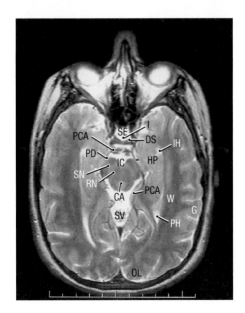

IV	Inferior vermis
L	Lens
LP	Lateral pterygoid
MA	Mastoid air cells
MB	Mandible
MC	Middle concha
MCP	Middle cerebellar peduncle
MD	Midbrain
MO	Medulla oblongata
MS	Maxillary sinus
MT	Masseter
MX	Maxilla
ND	Nodule of cerebellum
NS	Nasal septum
OB	Occipital bone
OC	Optic chiasm
OL	Occipital lobe
ON	Optic nerve (CNII)
P	Pons
PA	Pharynx
PCA	Posterior cerebral artery
PD	Cerebral crus
PF	Parapharyngeal fat
PG	Parotid gland
PH	Posterior horn (lateral ventricle)
PN	Pinna
PY	Pyramid
RN	Red nucleus
SC	Semicircular canals
SCP	Superior cerebellar peduncle
SE	Suprasellar cistern
SH	Superior concha
SN	Substantia nigra
SS	Superior sagittal sinus
ST	Straight sinus
SV	Superior vermis
TG	Tongue
TL	Temporal lobe
TP	Temporalis
UN	Uncus
VA	Vertebral artery
VP	Vestibular perilymph
VT	Vitreous body
W	White matter

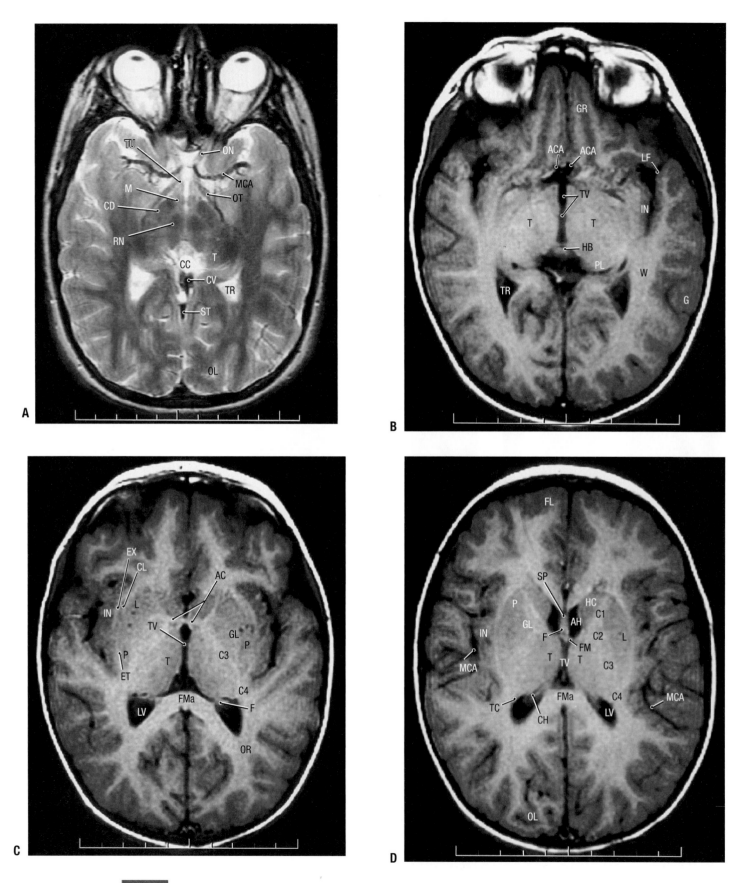

7.93 Axial (transverse) MRIs through the cerebral hemispheres

See orientation drawing for sites of scans **A–F. A** is T2 weighted, and **B–F** are T1 weighted.

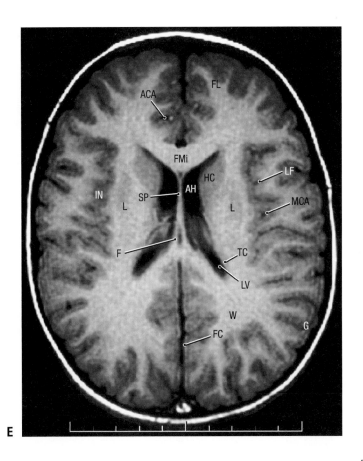

E

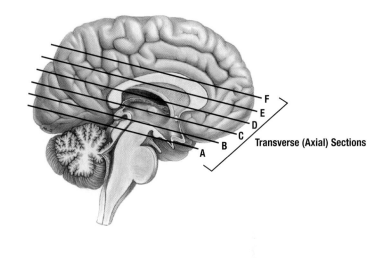

Transverse (Axial) Sections

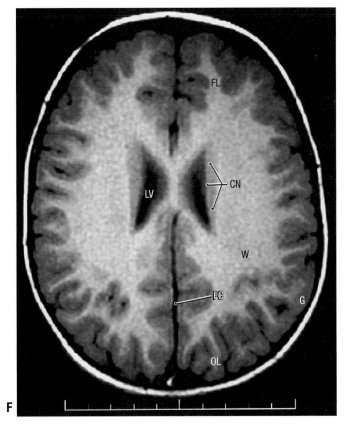

F

AC	Anterior commissure	GR	Gyrus rectus
ACA	Anterior cerebral artery	HB	Habenular commissure
AH	Anterior horn of lateral ventricle	HC	Head of caudate nucleus
C1	Anterior limb of internal capsule	IN	Insular cortex
C2	Genu of internal capsule	L	Lentiform nucleus
C3	Posterior limb of internal capsule	LF	Lateral fissure
C4	Retrolenticular limb of internal capsule	LV	Lateral ventricle
		M	Mammillary body
CC	Collicular cistern	MCA	Middle cerebral artery
CD	Cerebral peduncle	OL	Occipital lobe
CH	Choroid plexus	ON	Optic nerve
CL	Claustrum	OR	Optic radiations
CN	Caudate nucleus	OT	Optic tract
CV	Great cerebral vein	P	Putamen
ET	External capsule	PL	Pulvinar
EX	Extreme capsule	RN	Red nucleus
F	Fornix	SP	Septum pellucidum
FC	Falx cerebri	ST	Straight sinus
FL	Frontal lobe	T	Thalamus
FM	Interventricular foramen	TC	Tail of caudate nucleus
FMa	Forceps major	TR	Trigone of lateral ventricle
FMi	Forceps minor	TU	Tuber cinereum
G	Gray matter	TV	Third ventricle
GL	Globus pallidus	W	White matter

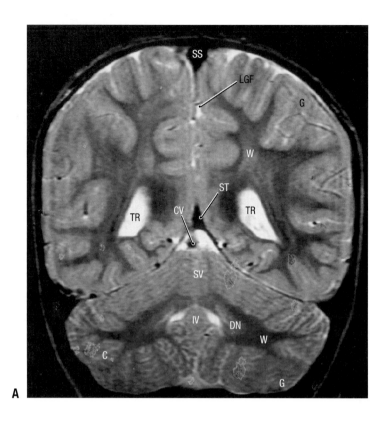

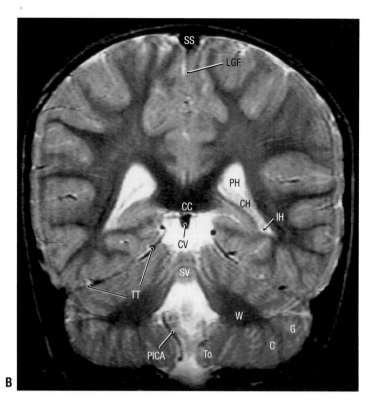

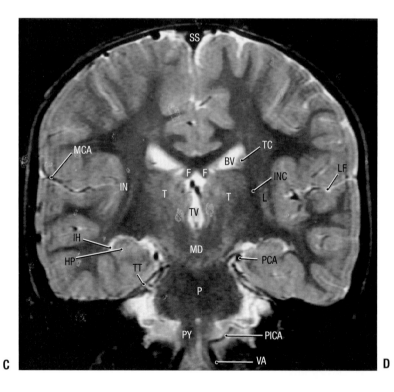

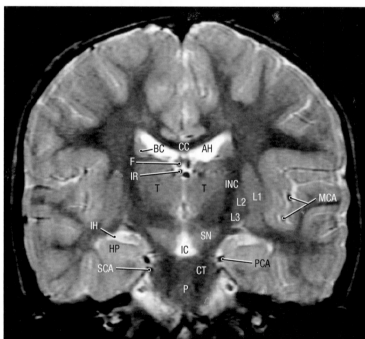

7.94   Coronal MRIs of brain (T2 weighted)

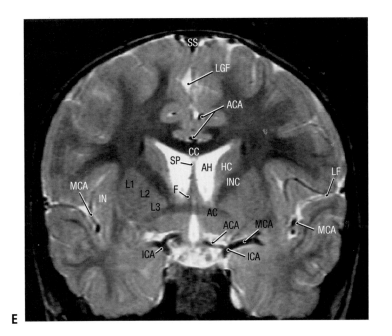

E

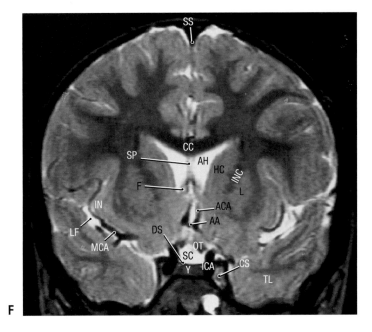

F

AA	Anterior communicating artery	L3	Internal (medial) segment of globus pallidus
AC	Anterior commissure		
ACA	Anterior cerebral artery	LF	Lateral fissure
AH	Anterior horn of lateral ventricle	LGF	Longitudinal fissure
BC	Body of caudate nucleus	MCA	Middle cerebral artery
BV	Body of lateral ventricle	MD	Midbrain
C	Cerebellum	OT	Optic tract
CC	Corpus callosum	P	Pons
CH	Choroid plexus	PCA	Posterior cerebral artery
CS	Cavernous sinus	PH	Posterior horn of lateral ventricle
CT	Corticospinal tract	PICA	Posterior inferior cerebellar artery
CV	Great cerebral vein		
DN	Dentate nucleus	PY	Pyramid
DS	Diaphragma sellae	S	Carotid siphon
F	Fornix	SC	Supracerebellar cistern
FV	Fourth ventricle	SCA	Superior cerebellar artery
G	Gray matter	SN	Substantia nigra
HC	Head of caudate nucleus	SP	Septum pellucidum
HP	Hippocampus	SS	Superior sagittal sinus
IC	Interpeduncular cistern	ST	Straight sinus
ICA	Internal carotid artery	SV	Superior vermis
IH	Inferior horn of lateral ventricle	T	Thalamus
IN	Insular cortex	TC	Tail of caudate nucleus
INC	Internal capsule	TL	Temporal lobe
IV	Inferior vermis	To	Cerebellar tonsil
IR	Intercerebral vein	TR	Trigone of lateral ventricle
L	Lentiform nucleus	TT	Tentorium cerebelli
L1	Putamen	TV	Third ventricle
L2	External (lateral) segment of globus pallidus	VA	Vertebral artery
		W	White matter
		Y	Hypophysis

Coronal Sections

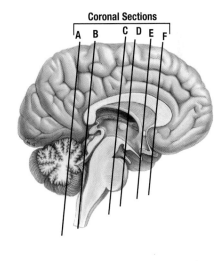

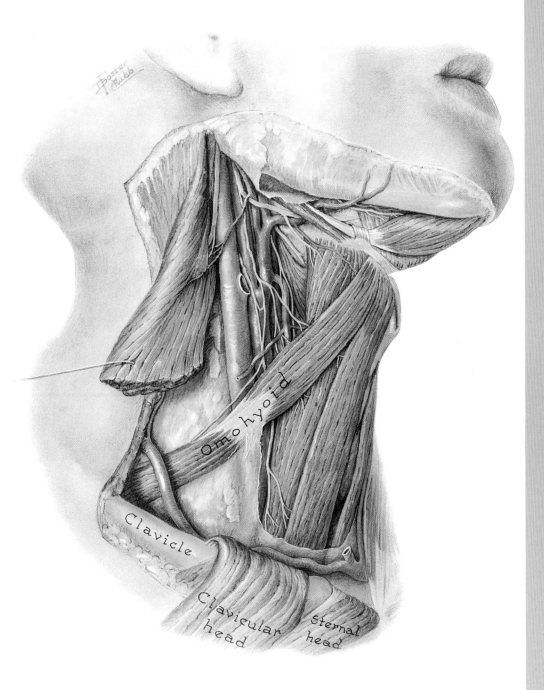

■ Cervical Fascia and Regions of
   the Neck **722**
■ Lateral Cervical Region
   (Posterior Triangle) **730**
■ Veins and Arteries of Neck **734**
■ Anterior Cervical Region
   (Anterior Triangle) **738**
■ Root of Neck **756**
■ Prevertebral Region **761**
■ External Cranial Base **764**
■ Pharynx **768**
■ Isthmus of Fauces and
   Tonsils **772**
■ Larynx **778**
■ Sectional Anatomy and
   Imaging: Neck **786**

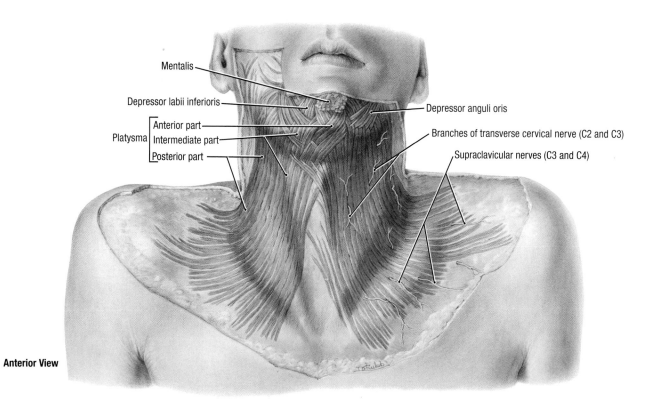

Mentalis

Depressor labii inferioris

Anterior part
Platysma — Intermediate part
Posterior part

Depressor anguli oris

Branches of transverse cervical nerve (C2 and C3)

Supraclavicular nerves (C3 and C4)

Anterior View

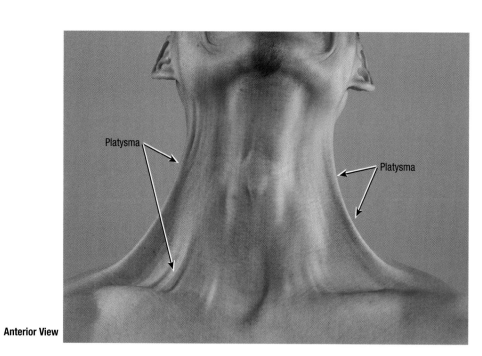

Platysma

Platysma

Anterior View

TABLE 8.1 PLATYSMA

MUSCLE	SUPERIOR ATTACHMENT	INFERIOR ATTACHMENT	INNERVATION	MAIN ACTION
Platysma	*Anterior part:* Fibers interlace with contralateral muscle *Intermediate part:* Fibers pass deep to depressors anguli oris and labii inferioris to attach to inferior border of mandible, lateral lower lip *Posterior part:* Skin/subcutaneous tissue of lower face lateral to mouth	Subcutaneous tissue overlying superior parts of pectoralis major (and sometimes anteromedial deltoid and/or trapezius muscle(s))	Cervical branch of facial nerve (CN VII)	Draws corner of mouth inferiorly and widens it as in expressions of sadness and fright; draws the skin of the neck superiorly, forming tense vertical and oblique ridges over the anterior neck

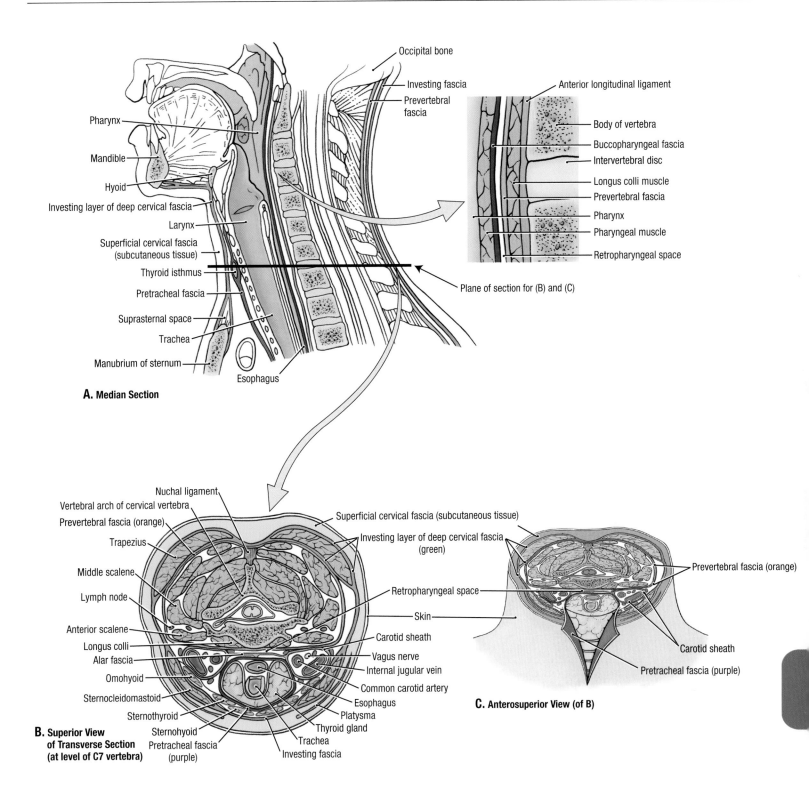

A. Median Section

Occipital bone
Investing fascia
Prevertebral fascia
Pharynx
Mandible
Hyoid
Investing layer of deep cervical fascia
Larynx
Superficial cervical fascia (subcutaneous tissue)
Thyroid isthmus
Pretracheal fascia
Suprasternal space
Trachea
Manubrium of sternum
Esophagus

Anterior longitudinal ligament
Body of vertebra
Buccopharyngeal fascia
Intervertebral disc
Longus colli muscle
Prevertebral fascia
Pharynx
Pharyngeal muscle
Retropharyngeal space
Plane of section for (B) and (C)

B. Superior View of Transverse Section (at level of C7 vertebra)

Nuchal ligament
Vertebral arch of cervical vertebra
Prevertebral fascia (orange)
Trapezius
Middle scalene
Lymph node
Anterior scalene
Longus colli
Alar fascia
Omohyoid
Sternocleidomastoid
Sternothyroid
Sternohyoid
Pretracheal fascia (purple)

Superficial cervical fascia (subcutaneous tissue)
Investing layer of deep cervical fascia (green)
Retropharyngeal space
Skin
Carotid sheath
Vagus nerve
Internal jugular vein
Common carotid artery
Esophagus
Platysma
Thyroid gland
Trachea
Investing fascia

C. Anterosuperior View (of B)

Prevertebral fascia (orange)
Carotid sheath
Pretracheal fascia (purple)

8.1   Cervical fascia

Sectional demonstrations of the fasciae of the neck. **A.** Fasciae of the neck are continuous inferiorly and superiorly with thoracic and cranial fasciae. The *inset* illustrates the fascia of the retropharyngeal region. **B.** The concentric layers of fascia are apparent in this transverse section of neck at the level indicated in **A. C.** Relationship of the main layers of deep cervical fascia and the carotid sheath. Midline access to the cervical viscera is possible with minimal disruption of tissues.

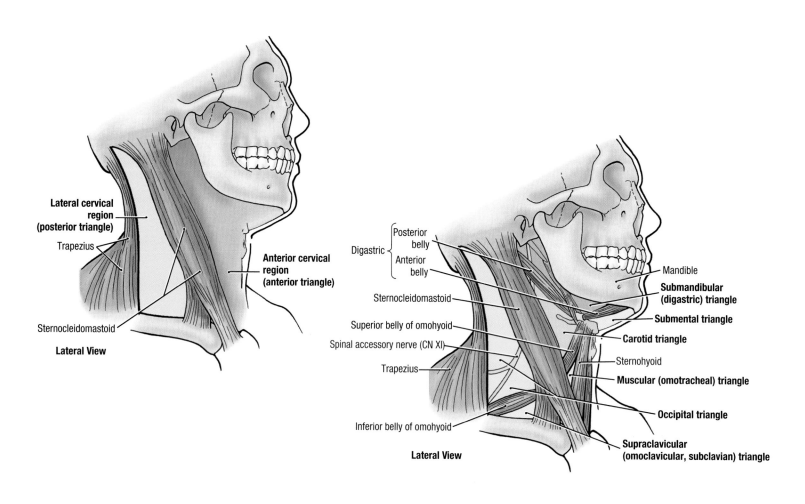

Lateral cervical region (posterior triangle)

Trapezius

Sternocleidomastoid

Lateral View

Anterior cervical region (anterior triangle)

Digastric — Posterior belly / Anterior belly

Sternocleidomastoid

Superior belly of omohyoid

Spinal accessory nerve (CN XI)

Trapezius

Inferior belly of omohyoid

Lateral View

Mandible

Submandibular (digastric) triangle

Submental triangle

Carotid triangle

Sternohyoid

Muscular (omotracheal) triangle

Occipital triangle

Supraclavicular (omoclavicular, subclavian) triangle

TABLE 8.2   CERVICAL TRIANGLES AND CONTENTS

LATERAL CERVICAL REGION (POSTERIOR TRIANGLE)	MAIN CONTENTS
Occipital triangle	Part of external jugular vein, posterior branches of cervical plexus of nerves, spinal accessory nerve, trunks of brachial plexus, transverse cervical artery, cervical lymph nodes
Supraclavicular (omoclavicular, subclavian) triangle	Subclavian artery (3rd part), part of subclavian vein (sometimes), suprascapular artery, supraclavicular lymph nodes
ANTERIOR CERVICAL REGION (ANTERIOR TRIANGLE)	**MAIN CONTENTS**
Submandibular (digastric) triangle	Submandibular gland almost fills triangle; submandibular lymph nodes, hypoglossal nerve, mylohyoid nerve, parts of facial artery and vein
Submental triangle	Submental lymph nodes, small veins that unite to form anterior jugular vein
Carotid triangle	Carotid sheath containing common carotid artery and its branches, internal jugular vein and its tributaries, and vagus nerve; external carotid artery and some of its branches; hypoglossal nerve and superior root of ansa cervicalis; spinal accessory nerve; thyroid, larynx, and pharynx; deep cervical lymph nodes; branches of cervical plexus
Muscular (omotracheal) triangle	Sternothyroid and sternohyoid muscles, thyroid and parathyroid glands

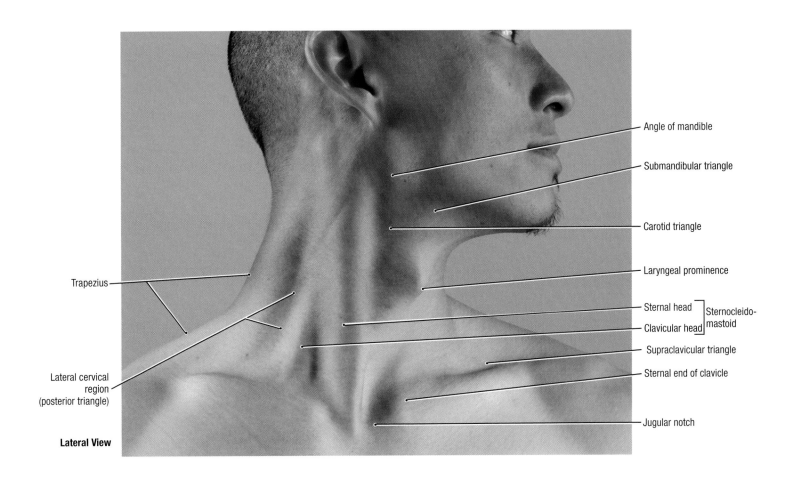

Trapezius

Lateral cervical
region
(posterior triangle)

Lateral View

Angle of mandible

Submandibular triangle

Carotid triangle

Laryngeal prominence

Sternal head ⎤ Sternocleido-
Clavicular head ⎦ mastoid

Supraclavicular triangle

Sternal end of clavicle

Jugular notch

TABLE 8.3   STERNOCLEIDOMASTOID AND TRAPEZIUS

MUSCLE	SUPERIOR ATTACHMENT	INFERIOR ATTACHMENT	INNERVATION	MAIN ACTION
Sternocleidomastoid	Lateral surface of mastoid process of temporal bone and lateral half of superior nuchal line	Sternal head: anterior surface of manubrium of sternum Clavicular head: superior surface of medial third of clavicle	Spinal accessory nerve (motor) and C2 and C3 nerves (pain and proprioception)	Tilts head to one side (i.e., laterally); flexes neck and rotates it so face is turned superiorly toward opposite side; acting together, the two muscles flex the neck so chin is thrust forward
Trapezius	Medial third of superior nuchal line, external occipital protuberance, nuchal ligament spinous processes of C7–T12 vertebrae	Lateral third of clavicle, acromion, and spine of scapula	Spinal accessory nerve (motor) and C3 and C4 nerves (pain and proprioception)	Elevates, retracts, and rotates scapula; descending fibers elevate the scapula, transverse fibers retract it, and ascending fibers depress it

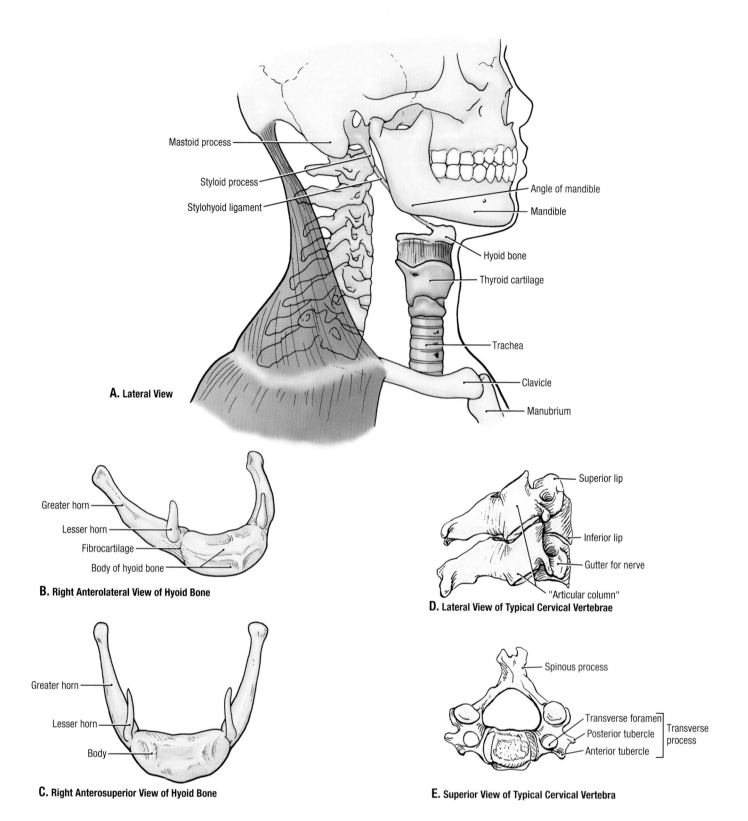

A. Lateral View

Mastoid process

Styloid process

Stylohyoid ligament

Angle of mandible

Mandible

Hyoid bone

Thyroid cartilage

Trachea

Clavicle

Manubrium

B. Right Anterolateral View of Hyoid Bone

Greater horn

Lesser horn

Fibrocartilage

Body of hyoid bone

C. Right Anterosuperior View of Hyoid Bone

Greater horn

Lesser horn

Body

D. Lateral View of Typical Cervical Vertebrae

Superior lip

Inferior lip

Gutter for nerve

"Articular column"

E. Superior View of Typical Cervical Vertebra

Spinous process

Transverse foramen

Posterior tubercle

Anterior tubercle

Transverse process

8.2 **Bones and cartilages of the neck**

A. Bony and cartilaginous landmarks. **B** and **C.** Features of hyoid bone. **D** and **E.** Features of typical cervical vertebrae.

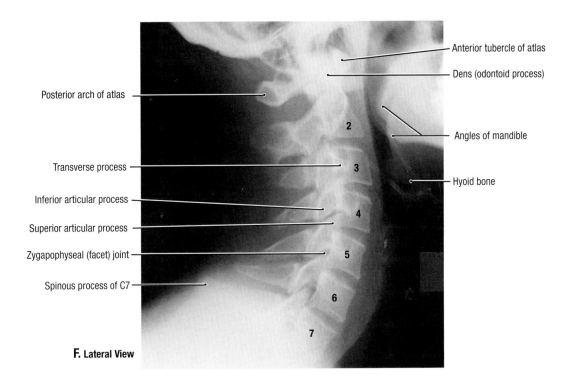

F. Lateral View

Anterior tubercle of atlas

Dens (odontoid process)

Angles of mandible

Hyoid bone

Posterior arch of atlas

Transverse process

Inferior articular process

Superior articular process

Zygapophyseal (facet) joint

Spinous process of C7

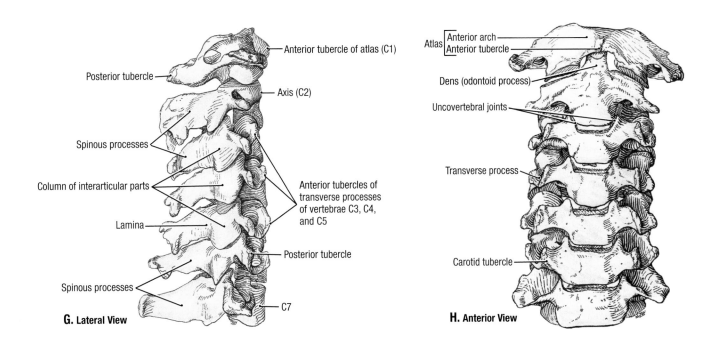

Anterior tubercle of atlas (C1)

Posterior tubercle

Axis (C2)

Spinous processes

Column of interarticular parts

Anterior tubercles of transverse processes of vertebrae C3, C4, and C5

Lamina

Posterior tubercle

Spinous processes

C7

G. Lateral View

Atlas   Anterior arch

Anterior tubercle

Dens (odontoid process)

Uncovertebral joints

Transverse process

Carotid tubercle

H. Anterior View

8.2   **Bones and cartilages of the neck *(continued)***

F. Radiograph of cervical vertebrae. Because the upper cervical vertebrae lie posterior to the upper and lower jaws and teeth, they are best seen radiographically in lateral views. **G** and **H.** Articulated cervical vertebrae.

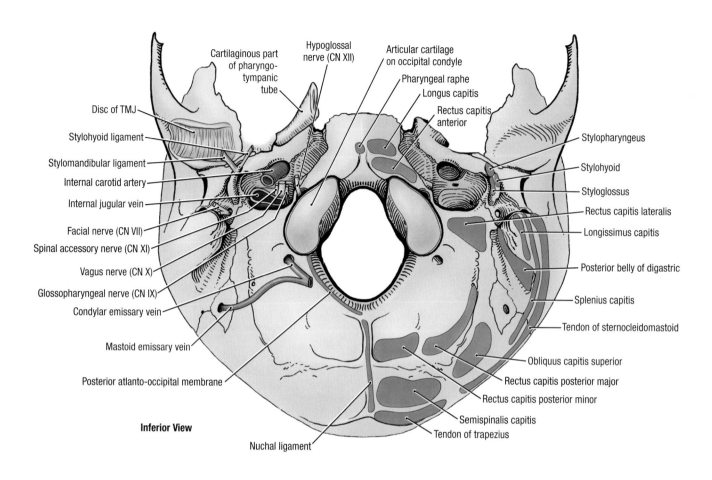

Disc of TMJ

Stylohyoid ligament

Stylomandibular ligament

Internal carotid artery

Internal jugular vein

Facial nerve (CN VII)

Spinal accessory nerve (CN XI)

Vagus nerve (CN X)

Glossopharyngeal nerve (CN IX)

Condylar emissary vein

Mastoid emissary vein

Posterior atlanto-occipital membrane

Inferior View

Cartilaginous part of pharyngo-tympanic tube

Hypoglossal nerve (CN XII)

Articular cartilage on occipital condyle

Pharyngeal raphe

Longus capitis

Rectus capitis anterior

Stylopharyngeus

Stylohyoid

Styloglossus

Rectus capitis lateralis

Longissimus capitis

Posterior belly of digastric

Splenius capitis

Tendon of sternocleidomastoid

Obliquus capitis superior

Rectus capitis posterior major

Rectus capitis posterior minor

Semispinalis capitis

Tendon of trapezius

Nuchal ligament

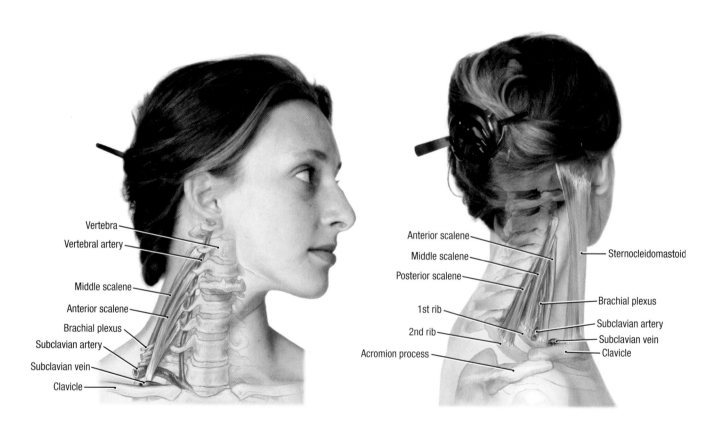

Vertebra

Vertebral artery

Middle scalene

Anterior scalene

Brachial plexus

Subclavian artery

Subclavian vein

Clavicle

Anterior scalene

Middle scalene

Posterior scalene

1st rib

2nd rib

Acromion process

Sternocleidomastoid

Brachial plexus

Subclavian artery

Subclavian vein

Clavicle

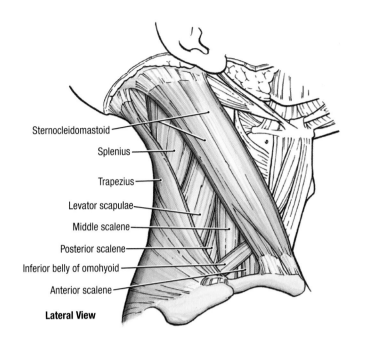

Sternocleidomastoid
Splenius
Trapezius
Levator scapulae
Middle scalene
Posterior scalene
Inferior belly of omohyoid
Anterior scalene

Lateral View

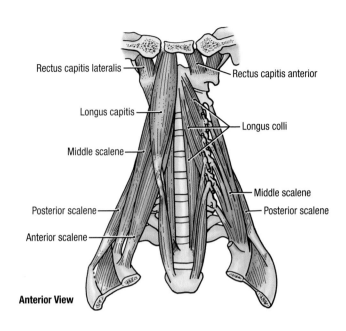

Rectus capitis lateralis
Longus capitis
Middle scalene
Posterior scalene
Anterior scalene
Rectus capitis anterior
Longus colli
Middle scalene
Posterior scalene

Anterior View

TABLE 8.4   PREVERTEBRAL MUSCLES

MUSCLE	SUPERIOR ATTACHMENT	INFERIOR ATTACHMENT	INNERVATION	MAIN ACTION
Anterior Muscles				
Longus colli	Anterior tubercle of C1 vertebra (atlas); bodies of C1–C3 and transverse processes of C3–C6 vertebrae	Bodies of C5–T3 vertebrae, transverse process of C3–C5 vertebrae	Anterior rami of C2–C6 spinal nerves	*Flexes neck with rotation (torsion) to opposite side if acting unilaterally
Longus capitis	Basilar part of occipital bone	Anterior tubercles of C3–C6 transverse processes	Anterior rami of C1–C3 spinal nerves	**Flexes head
Rectus capitis anterior	Base of skull, just anterior to occipital condyle	Anterior surface of lateral mass of C1 vertebra (atlas)	Branches from loop between C1 and C2 spinal nerves	
Rectus capitis lateralis	Jugular process of occipital bone	Transverse process of C1 vertebra (atlas)		Flexes head and helps to stabilize it
Lateral Muscles				
Splenius capitis	Inferior half of nuchal ligament and spinous processes of superior six thoracic vertebrae	Lateral aspect of mastoid process and lateral third of superior nuchal line	Posterior rami of middle cervical spinal nerves	***Laterally flexes and rotates head and neck to same side; acting bilaterally, they extend head and neck
Levator scapulae	Posterior tubercles of transverse processes of C1–C4 vertebrae	Superior part of medial border of scapula	Dorsal scapular nerve C5 and cervical spinal nerves C3 and C4	Elevates scapula and tilts its glenoid cavity inferiorly by rotating scapula
Posterior scalene	Posterior tubercles of transverse processes of C4–C6 vertebrae	External border of 2nd rib	Anterior rami of cervical spinal nerves C7 and C8	Flexes neck laterally; elevates 2nd rib during forced inspiration
Middle scalene	Posterior tubercles of transverse processes of C2–C7 vertebrae	Superior surface of 1st rib, posterior to groove for subclavian artery	Anterior rami of cervical spinal nerves	Flexes neck laterally; elevates 1st rib during forced inspiration
Anterior scalene	Transverse processes of C4–C6 vertebrae	1st rib	Cervical spinal nerves C4, C5, and C6	Elevates 1st rib; laterally flexes and rotates neck

*Flexion of neck: anterior (or lateral if so stated) bending of cervical vertebrae C2–C7.

**Flexion of head: anterior (or lateral if so stated) bending of head relative to vertebral column of atlanto-occipital joints.

***Rotation of head occurs at atlantoaxial joints.

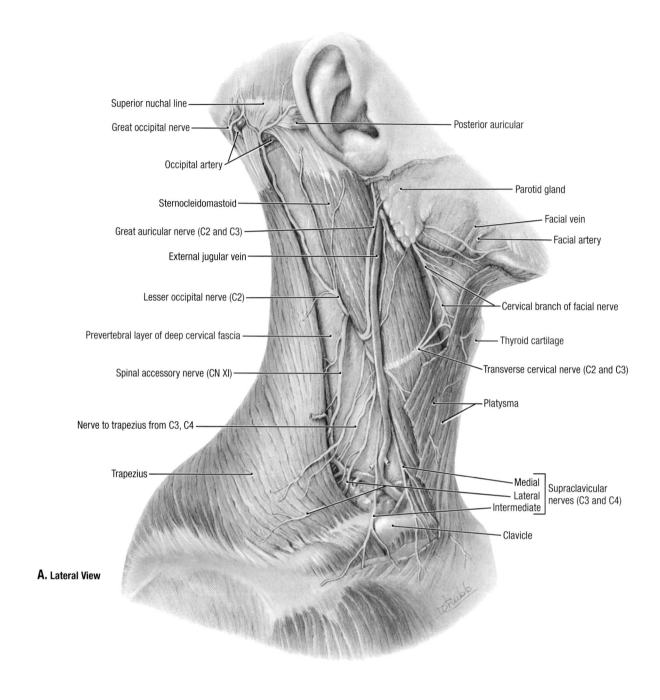

Superior nuchal line

Great occipital nerve

Occipital artery

Sternocleidomastoid

Great auricular nerve (C2 and C3)

External jugular vein

Lesser occipital nerve (C2)

Prevertebral layer of deep cervical fascia

Spinal accessory nerve (CN XI)

Nerve to trapezius from C3, C4

Trapezius

Posterior auricular

Parotid gland

Facial vein

Facial artery

Cervical branch of facial nerve

Thyroid cartilage

Transverse cervical nerve (C2 and C3)

Platysma

Medial

Lateral — Supraclavicular nerves (C3 and C4)

Intermediate

Clavicle

A. Lateral View

8.3 Lateral cervical region—I

A. Subcutaneous fat, the part of the platysma muscle overlying the inferior part of the lateral cervical region, and the investing layer of deep cervical fascia have all been removed.

- The external jugular vein descends vertically across the sternocleidomastoid muscle, from posterior to the angle of the mandible to the posterior border of the muscle, where it pierces the prevertebral layer of deep cervical fascia superior to the clavicle.
- The spinal accessory nerve (CN XI), the only motor nerve superficial to the prevertebral layer of deep cervical fascia, disappears deep to the anterior border of the trapezius approximately 2 fingers' breadth superior to the clavicle.

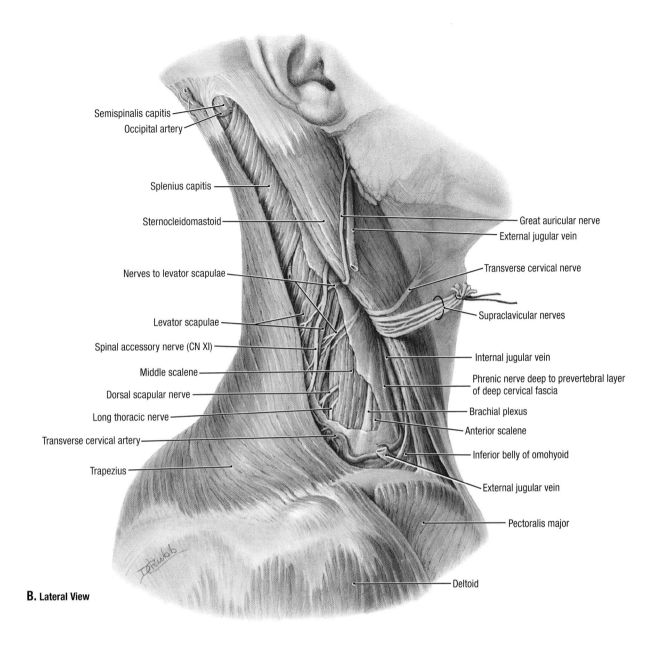

Semispinalis capitis

Occipital artery

Splenius capitis

Sternocleidomastoid

Nerves to levator scapulae

Levator scapulae

Spinal accessory nerve (CN XI)

Middle scalene

Dorsal scapular nerve

Long thoracic nerve

Transverse cervical artery

Trapezius

Great auricular nerve

External jugular vein

Transverse cervical nerve

Supraclavicular nerves

Internal jugular vein

Phrenic nerve deep to prevertebral layer of deep cervical fascia

Brachial plexus

Anterior scalene

Inferior belly of omohyoid

External jugular vein

Pectoralis major

Deltoid

B. Lateral View

8.3   Lateral cervical region—II

B. The prevertebral layer of deep cervical fascia has been partially removed, and the motor nerves and most of the floor of the triangle are exposed.

- The muscles that form the floor of the superior part of the triangle are the semispinalis capitis, splenius capitis, and levator scapulae.
- The spinal accessory nerve supplies both the sternocleidomastoid and trapezius muscles; between them, it courses along the levator scapulae muscle but is separated from it by the prevertebral layer of deep cervical fascia.
- The phrenic nerve supplies the diaphragm (C3, C4, C5) and is located deep to the prevertebral layer of deep cervical fascia and on the anterior surface of the anterior scalene; if severed, paralysis of the ipsilateral half of the diaphragm results.

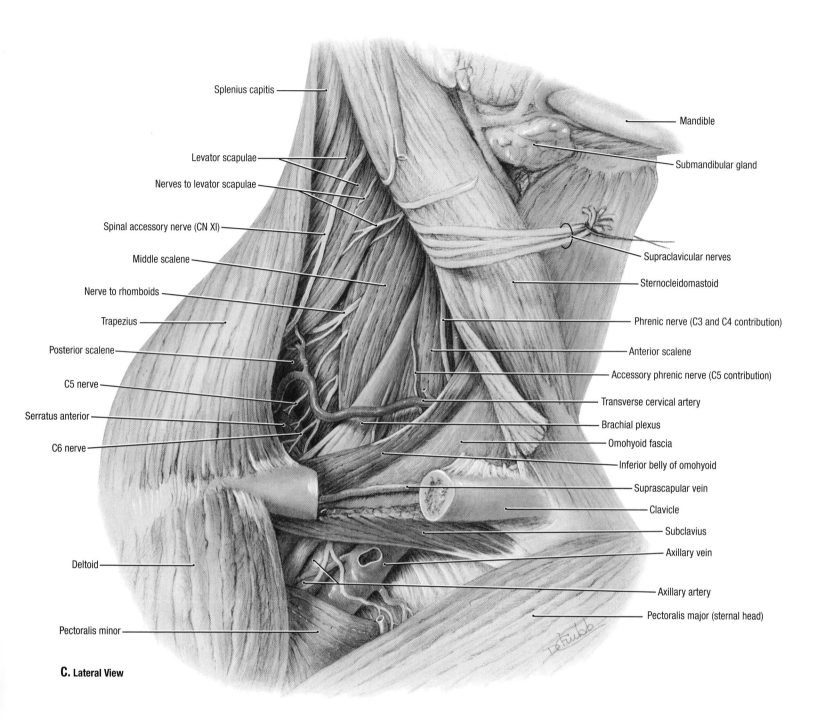

Splenius capitis

Mandible

Levator scapulae

Submandibular gland

Nerves to levator scapulae

Spinal accessory nerve (CN XI)

Supraclavicular nerves

Middle scalene

Sternocleidomastoid

Nerve to rhomboids

Phrenic nerve (C3 and C4 contribution)

Trapezius

Anterior scalene

Posterior scalene

Accessory phrenic nerve (C5 contribution)

C5 nerve

Transverse cervical artery

Serratus anterior

Brachial plexus

C6 nerve

Omohyoid fascia

Inferior belly of omohyoid

Suprascapular vein

Clavicle

Subclavius

Axillary vein

Deltoid

Axillary artery

Pectoralis major (sternal head)

Pectoralis minor

C. Lateral View

8.3   Lateral cervical region—III

C. The clavicular head of the pectoralis major muscle and part of the clavicle have been removed.

- The anterior, middle, and posterior scalenes and the serratus anterior form the floor of the inferior part of the region.
- The omohyoid fascia (a thickened part of the muscular part of the pretracheal fascia) lies between the omohyoid and subclavius muscles and forms a pulley for the intermediate tendon of the omohyoid.
- The brachial plexus emerges between the anterior and middle scalene muscles.

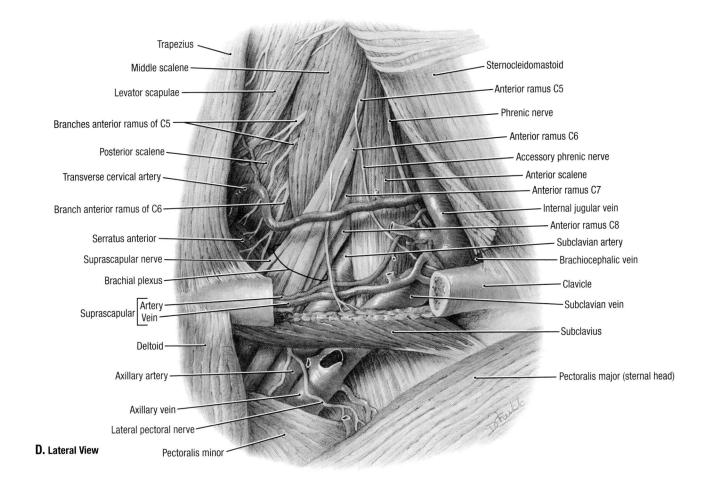

Trapezius

Middle scalene

Levator scapulae

Branches anterior ramus of C5

Posterior scalene

Transverse cervical artery

Branch anterior ramus of C6

Serratus anterior

Suprascapular nerve

Brachial plexus

Suprascapular { Artery / Vein

Deltoid

Axillary artery

Axillary vein

Lateral pectoral nerve

D. Lateral View   Pectoralis minor

Sternocleidomastoid

Anterior ramus C5

Phrenic nerve

Anterior ramus C6

Accessory phrenic nerve

Anterior scalene

Anterior ramus C7

Internal jugular vein

Anterior ramus C8

Subclavian artery

Brachiocephalic vein

Clavicle

Subclavian vein

Subclavius

Pectoralis major (sternal head)

8.3   Lateral cervical region—IV

D. The omohyoid muscle and fascia have been removed, and the brachial plexus and sub-clavian vessels are exposed.

- The anterior scalene lies between the subclavian artery and vein.
- The anterior rami of C5–T1 form the brachial plexus; the anterior ramus of T1 lies posterior to the subclavian artery; the brachial plexus and subclavian artery emerge between the middle and anterior scalene muscles.
- The subclavian vein is separated from the subclavian artery by the anterior scalene muscle; the subclavian vein is often used for insertion of a central venous catheter and, therefore, the relationships of the subclavian vein to the sternocleidomastoid muscle, clavicle, 1st rib, and cupula of the pleura are of clinical importance.

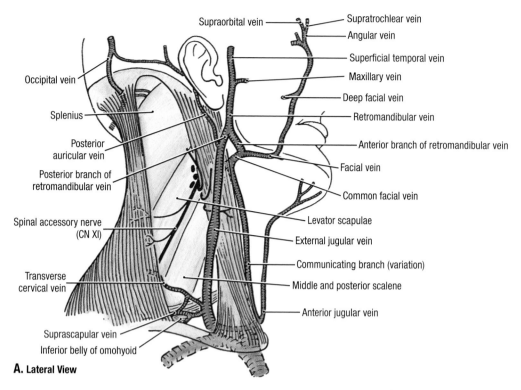

A. Lateral View

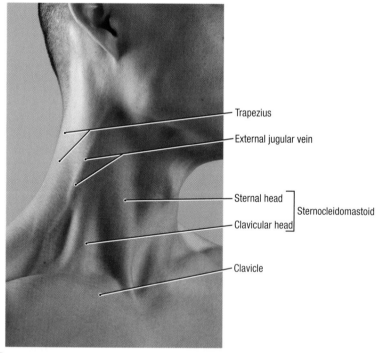

B. Lateral View

8.4   **Superficial veins of the neck**

A. Schematic illustration of superficial veins of the neck. The superficial temporal and maxillary veins merge to form the retromandibular vein. The posterior division of the retromandibular vein unites with the posterior auricular vein to form the external jugular vein. The facial vein receives the anterior division of the retromandibular vein, forming the common facial vein that empties into the internal jugular vein. **B.** Surface anatomy of the external jugular vein and the muscles bounding the lateral cervical region (posterior triangle) of the neck.

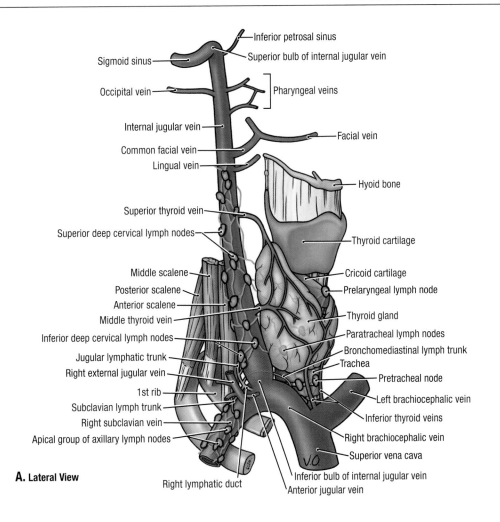

Inferior petrosal sinus
Sigmoid sinus
Superior bulb of internal jugular vein
Occipital vein
Pharyngeal veins
Internal jugular vein
Facial vein
Common facial vein
Lingual vein
Hyoid bone
Superior thyroid vein
Superior deep cervical lymph nodes
Thyroid cartilage
Middle scalene
Cricoid cartilage
Posterior scalene
Prelaryngeal lymph node
Anterior scalene
Middle thyroid vein
Thyroid gland
Inferior deep cervical lymph nodes
Paratracheal lymph nodes
Jugular lymphatic trunk
Bronchomediastinal lymph trunk
Right external jugular vein
Trachea
1st rib
Pretracheal node
Subclavian lymph trunk
Left brachiocephalic vein
Right subclavian vein
Inferior thyroid veins
Apical group of axillary lymph nodes
Right brachiocephalic vein
Superior vena cava
A. Lateral View
Right lymphatic duct
Inferior bulb of internal jugular vein
Anterior jugular vein

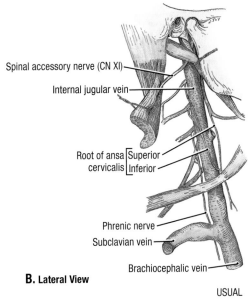

Spinal accessory nerve (CN XI)
Internal jugular vein
Root of ansa cervicalis [Superior / Inferior]
Phrenic nerve
Subclavian vein
Brachiocephalic vein
B. Lateral View
USUAL

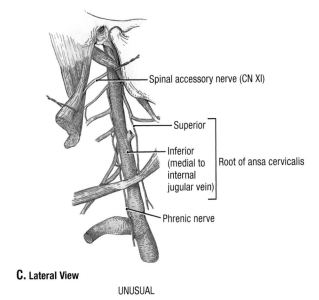

Spinal accessory nerve (CN XI)
Superior
Inferior (medial to internal jugular vein)
Root of ansa cervicalis
Phrenic nerve
C. Lateral View
UNUSUAL

8.5   Deep veins of the neck

A. Internal jugular vein, tributaries, and related structures. **B** and **C.** Variable relationships of the spinal accessory nerve (CN XI), phrenic nerve, and ansa cervicalis to the internal jugular and subclavian veins. **B.** Usual relationships. **C.** Unusual (less common) relationships. Studies in Dr. Grant's laboratory showed the spinal accessory nerve crossing anterior to the internal jugular vein in 70% of 188 specimens and crossing posterior to it in 30%. The phrenic nerve typically passes posterior to the site of union of the subclavian, internal jugular, and brachiocephalic veins, but occasionally, the branch from C5 or, less commonly, the entire phrenic nerve, passes anterior to the subclavian vein. More than half of the time, the inferior root of the ansa cervicalis passes superficial to the internal jugular vein.

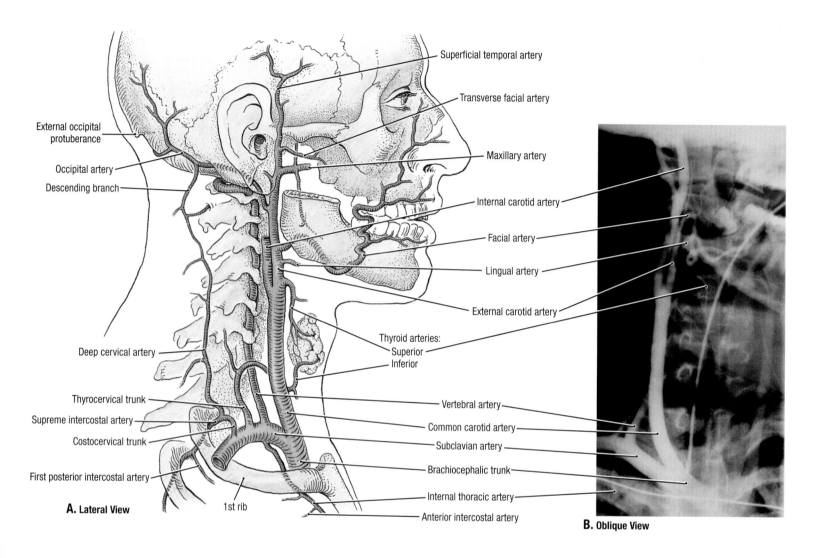

Superficial temporal artery

Transverse facial artery

External occipital protuberance

Occipital artery

Descending branch

Maxillary artery

Internal carotid artery

Facial artery

Lingual artery

External carotid artery

Thyroid arteries:
Superior
Inferior

Deep cervical artery

Thyrocervical trunk

Supreme intercostal artery

Costocervical trunk

First posterior intercostal artery

Vertebral artery

Common carotid artery

Subclavian artery

Brachiocephalic trunk

Internal thoracic artery

Anterior intercostal artery

A. Lateral View

1st rib

B. Oblique View

8.6 **Arteries of the neck**

A. Overview. **B.** Carotid arteriogram.
- The common carotid and subclavian arteries supply blood to the head.
- The common carotid artery terminates at the C4 vertebral level as the internal and external carotid arteries. In general, the internal carotid supplies structures inside the head, and the external carotid supplies the exterior. However, the ophthalmic artery (from the internal carotid) sends supraorbital and supratrochlear arteries to the forehead/anterior scalp.
- The vertebral artery, a branch of the subclavian artery, enters the foramen magnum and contributes to the cerebral arterial circle (of Willis).
- The deep cervical artery (from the costocervical trunk) anastomoses with the descending branch of the occipital artery and branches of the vertebral artery; thus it provides a potential collateral route by which blood might reach the head.

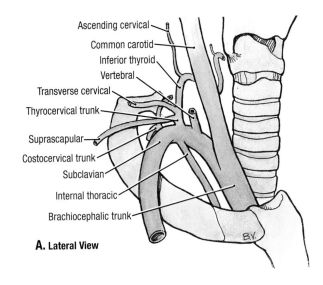

A. Lateral View

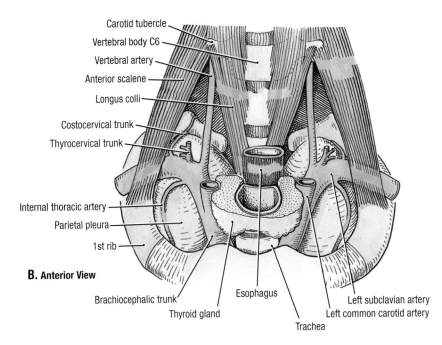

B. Anterior View

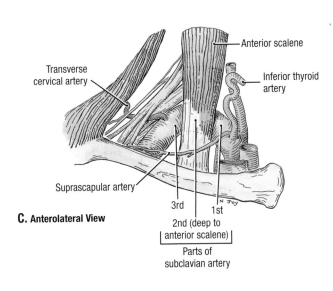

C. Anterolateral View

8.7   **Subclavian artery**

A. The subclavian artery and its branches. **B.** Course of the inferior part of vertebral arteries. The vertebral artery typically arises from the posterosuperior aspect of the subclavian artery and enters the transverse foramen of the C6 vertebra, but it may enter a more superior transverse foramen. **C.** Parts of subclavian artery. The anterior scalene divides the subclavian artery into three parts. Some branches from the second or third part (most commonly, the dorsal scapular and, less frequently, the transverse cervical and suprascapular arteries pass posterolaterally through the brachial plexus).

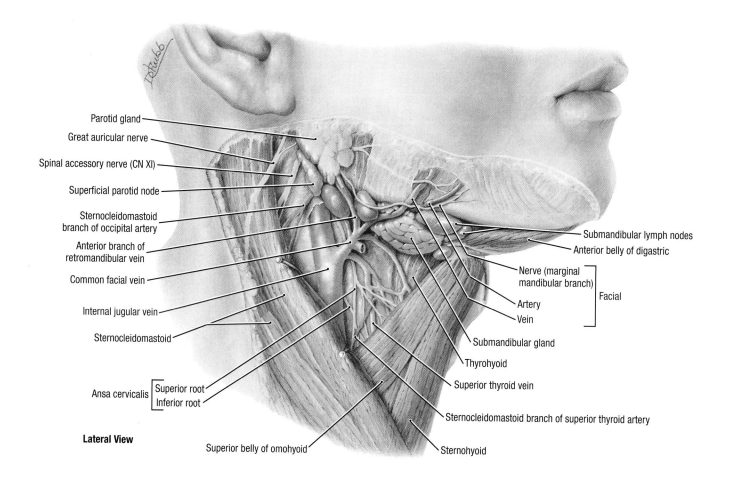

Parotid gland

Great auricular nerve

Spinal accessory nerve (CN XI)

Superficial parotid node

Sternocleidomastoid branch of occipital artery

Anterior branch of retromandibular vein

Common facial vein

Internal jugular vein

Sternocleidomastoid

Ansa cervicalis — Superior root / Inferior root

Lateral View

Superior belly of omohyoid

Submandibular lymph nodes

Anterior belly of digastric

Nerve (marginal mandibular branch) — Facial

Artery — Facial

Vein — Facial

Submandibular gland

Thyrohyoid

Superior thyroid vein

Sternocleidomastoid branch of superior thyroid artery

Sternohyoid

8.8   Anterior cervical region—I

The skin, subcutaneous tissue (with platysma), and the investing layer of deep cervical fascia (including the sheaths of the parotid and submandibular glands) have been removed.

- The spinal accessory nerve (CN XI) enters the deep surface of the sternocleidomastoid muscle and is joined along its anterior border by the sternocleidomastoid branch of the occipital artery.
- The (common) facial vein joins the internal jugular vein near the level of the hyoid bone; here, the common facial vein is joined by several other veins.
- The sternocleidomastoid branch of the superior thyroid artery descends near the superior border of the omohyoid muscle.
- The retromandibular and facial veins run superficial to the submandibular gland.
- The submandibular lymph nodes lie deep to the investing layer of deep cervical fascia in the submandibular triangle; some of the nodes lie in the submandibular gland.

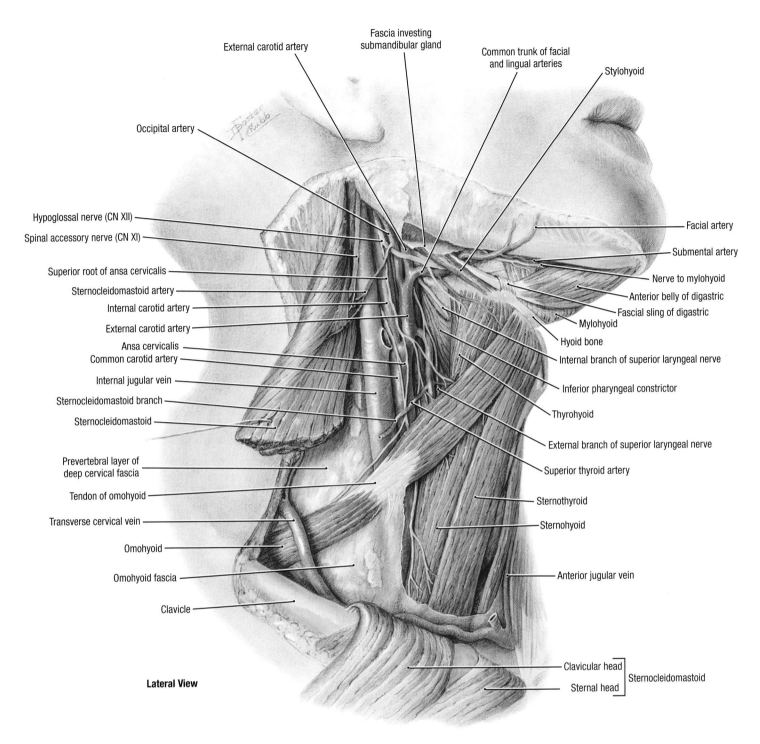

External carotid artery

Fascia investing
submandibular gland

Common trunk of facial
and lingual arteries

Stylohyoid

Occipital artery

Hypoglossal nerve (CN XII)

Spinal accessory nerve (CN XI)

Superior root of ansa cervicalis

Sternocleidomastoid artery

Internal carotid artery

External carotid artery

Ansa cervicalis

Common carotid artery

Internal jugular vein

Sternocleidomastoid branch

Sternocleidomastoid

Prevertebral layer of
deep cervical fascia

Tendon of omohyoid

Transverse cervical vein

Omohyoid

Omohyoid fascia

Clavicle

Lateral View

Facial artery

Submental artery

Nerve to mylohyoid

Anterior belly of digastric

Fascial sling of digastric

Mylohyoid

Hyoid bone

Internal branch of superior laryngeal nerve

Inferior pharyngeal constrictor

Thyrohyoid

External branch of superior laryngeal nerve

Superior thyroid artery

Sternothyroid

Sternohyoid

Anterior jugular vein

Clavicular head ⎫
                ⎬ Sternocleidomastoid
Sternal head   ⎭

8.9   Anterior cervical region—II

- The tendon of the digastric muscle is connected to the hyoid bone by a fascial sling derived from the muscular part of the pretracheal layer of deep cervical fascia; the tendon of the omohyoid muscle is similarly tethered to the clavicle.
- In this specimen, the facial and lingual arteries arise from a common trunk and pass deep to the stylohyoid and digastric muscles.
- The hypoglossal nerve (CN XII) appears from deep to the posterior belly of the digastric in contact with, or slightly medial to, the spinal accessory nerve (CN XI). The hypoglossal nerve crosses the internal and external carotid arteries and gives off two branches, the superior root of the ansa cervicalis and the nerve to the thyrohyoid muscle, before passing anteriorly between the mylohyoid and hyoglossus muscles.

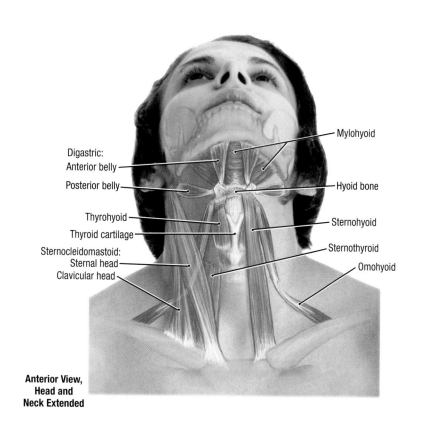

Anterior View, Head and Neck Extended

Digastric:
Anterior belly
Posterior belly
Thyrohyoid
Thyroid cartilage
Sternocleidomastoid:
Sternal head
Clavicular head

Mylohyoid
Hyoid bone
Sternohyoid
Sternothyroid
Omohyoid

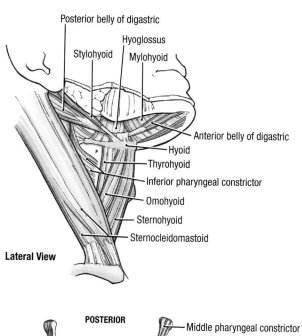

Lateral View

Posterior belly of digastric
Stylohyoid
Hyoglossus
Mylohyoid
Anterior belly of digastric
Hyoid
Thyrohyoid
Inferior pharyngeal constrictor
Omohyoid
Sternohyoid
Sternocleidomastoid

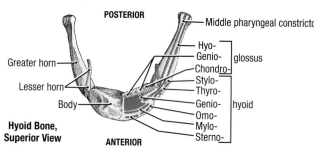

Hyoid Bone, Superior View

POSTERIOR
Middle pharyngeal constrictor
Greater horn
Lesser horn
Body
Hyo-
Genio- glossus
Chondro-
Stylo-
Thyro-
Genio- hyoid
Omo-
Mylo-
Sterno-
ANTERIOR

TABLE 8.5   SUPRAHYOID AND INFRAHYOID MUSCLES

MUSCLE	ORIGIN	INSERTION	INNERVATION	MAIN ACTIONS
Suprahyoid Muscles				
Mylohyoid	Mylohyoid line of mandible	Raphe and body of hyoid bone	Nerve to mylohyoid, a branch of inferior alveolar nerve of CN V³	Elevates hyoid bone, floor of mouth and tongue during swallowing and speaking
Geniohyoid	Inferior mental spine mandible	Body of hyoid bone	C1 via the hypoglossal nerve (CN XII)	Pulls hyoid bone anterosuperiorly, shortens floor of mouth, and widens pharynx
Stylohyoid	Styloid process of temporal bone		Cervical branch of facial nerve (CN VII)	Elevates and retracts hyoid bone, thereby elongating floor of mouth
Digastric	Anterior belly: digastric fossa of mandible Posterior belly: mastoid notch of temporal bone	Intermediate tendon to body and greater horn of hyoid bone	Anterior belly: nerve to mylohyoid, a branch of inferior alveolar nerve (CN V³) Posterior belly: facial nerve (CN VII)	Elevates hyoid bone and steadies it during swallowing and speaking; depresses mandible against resistance
Infrahyoid Muscles				
Sternohyoid	Manubrium of sternum and medial end of clavicle	Body of hyoid bone	C1–C3 by a branch of ansa cervicalis	Depresses hyoid bone after it has been elevated during swallowing
Omohyoid	Superior border of scapula near suprascapular notch	Inferior border of hyoid bone		Depresses, retracts, and steadies hyoid bone
Sternothyroid	Posterior surface of manubrium of sternum	Oblique line of thyroid cartilage	C2 and C3 by a branch of ansa cervicalis	Depresses hyoid bone and larynx
Thyrohyoid	Oblique line of thyroid cartilage	Inferior border of body and greater horn of hyoid bone	C1 via hypoglossal nerve (CN XII)	Depresses hyoid bone and elevates larynx

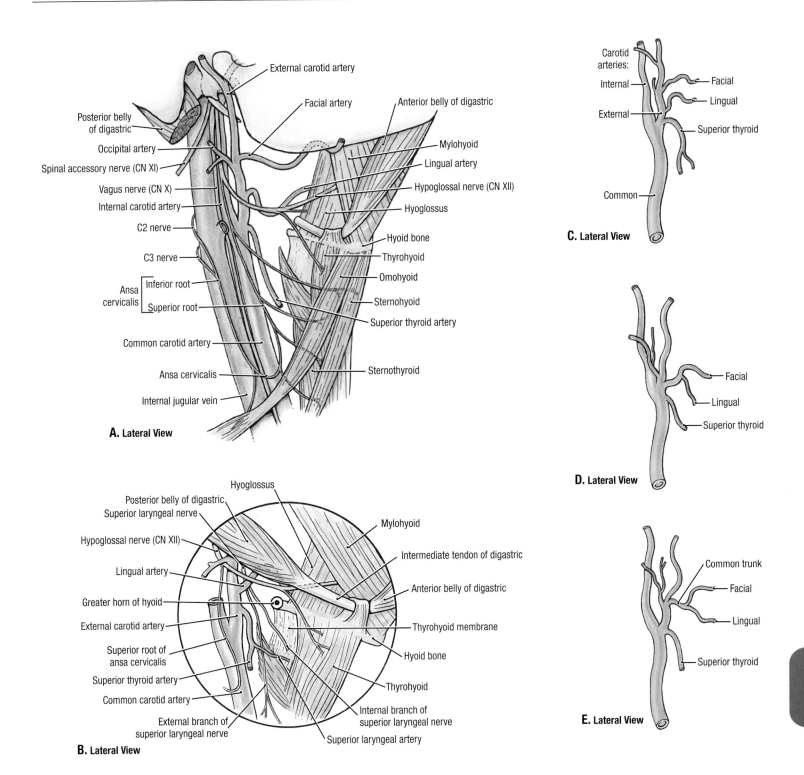

A. Lateral View

- External carotid artery
- Posterior belly of digastric
- Occipital artery
- Spinal accessory nerve (CN XI)
- Vagus nerve (CN X)
- Internal carotid artery
- C2 nerve
- C3 nerve
- Ansa cervicalis — Inferior root / Superior root
- Common carotid artery
- Ansa cervicalis
- Internal jugular vein
- Facial artery
- Anterior belly of digastric
- Mylohyoid
- Lingual artery
- Hypoglossal nerve (CN XII)
- Hyoglossus
- Hyoid bone
- Thyrohyoid
- Omohyoid
- Sternohyoid
- Superior thyroid artery
- Sternothyroid

B. Lateral View

- Hyoglossus
- Posterior belly of digastric
- Superior laryngeal nerve
- Hypoglossal nerve (CN XII)
- Lingual artery
- Greater horn of hyoid
- External carotid artery
- Superior root of ansa cervicalis
- Superior thyroid artery
- Common carotid artery
- External branch of superior laryngeal nerve
- Mylohyoid
- Intermediate tendon of digastric
- Anterior belly of digastric
- Thyrohyoid membrane
- Hyoid bone
- Thyrohyoid
- Internal branch of superior laryngeal nerve
- Superior laryngeal artery

C. Lateral View

- Carotid arteries:
- Internal
- External
- Common
- Facial
- Lingual
- Superior thyroid

D. Lateral View

- Facial
- Lingual
- Superior thyroid

E. Lateral View

- Common trunk
- Facial
- Lingual
- Superior thyroid

8.10   Relationships of nerves and vessels in the anterior triangle of the neck

A. Ansa cervicalis and strap muscles. **B.** Hypoglossal nerve and internal and external branches of superior laryngeal nerve. The tip of the greater horn of the hyoid bone, indicated with a *circle* is the reference point for many structures. **C–E.** Variation in the origin of the lingual artery as studied by Dr. Grant in 211 specimens. In 80%, the superior thyroid, lingual, and facial arteries arose separately **(C)**; in 20%, the lingual and facial arteries arose from a common stem inferiorly **(D)** or high on the external carotid artery **(E).** In one specimen, the superior thyroid and lingual arteries arose from a common stem.

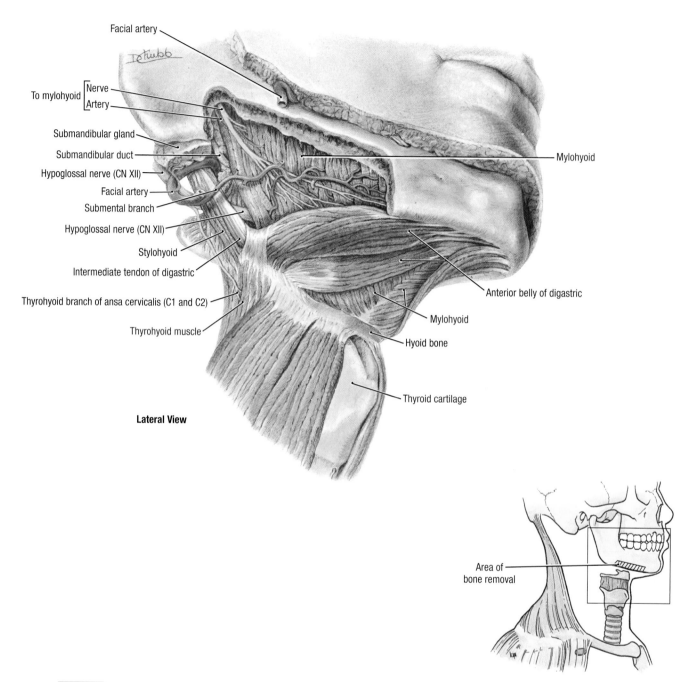

Facial artery

To mylohyoid [Nerve / Artery]

Submandibular gland

Submandibular duct

Hypoglossal nerve (CN XII)

Facial artery

Submental branch

Hypoglossal nerve (CN XII)

Stylohyoid

Intermediate tendon of digastric

Thyrohyoid branch of ansa cervicalis (C1 and C2)

Thyrohyoid muscle

Mylohyoid

Anterior belly of digastric

Mylohyoid

Hyoid bone

Thyroid cartilage

Lateral View

Area of bone removal

8.11   Suprahyoid region—I

Structures overlying the mandible and a portion of the body of the mandible have been removed.

- The anterior region of the neck extends from the mandible superiorly to the sternum inferiorly; it is divided by the hyoid bone into suprahyoid and infrahyoid parts.
- The stylohyoid and posterior belly of the digastric muscle form the posterior border of the submandibular triangle; the facial artery passes superficial to these muscles.
- The anterior belly of the digastric muscle forms the anterior border of the submandibular triangle. In this specimen, the anterior belly has an additional origin from the hyoid bone; the mylohyoid muscle forms the medial wall of the triangle and has a thick, free posterior border.
- The nerve to the mylohyoid, which supplies the mylohyoid muscle and anterior belly of the digastric muscle, is accompanied by the mylohyoid branch of the inferior alveolar artery posteriorly and the submental branch of the facial artery anteriorly.

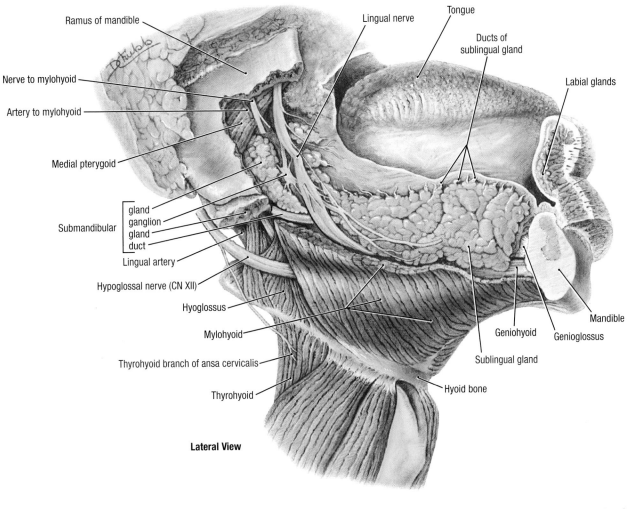

Ramus of mandible

Nerve to mylohyoid

Artery to mylohyoid

Medial pterygoid

Submandibular — gland / ganglion / gland / duct

Lingual artery

Hypoglossal nerve (CN XII)

Hyoglossus

Mylohyoid

Thyrohyoid branch of ansa cervicalis

Thyrohyoid

Lingual nerve

Tongue

Ducts of sublingual gland

Labial glands

Mandible

Geniohyoid

Genioglossus

Sublingual gland

Hyoid bone

Lateral View

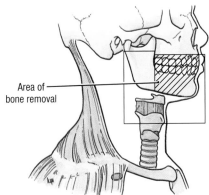

Area of bone removal

8.12 **Suprahyoid region—II**

The body and adjacent portion of the ramus of the mandible have been removed.

- The sublingual salivary gland lies posterior to the mandible and is in contact with the deep part of the submandibular gland posteriorly.
- Numerous fine ducts pass from the superior border of the sublingual gland to open on the sublingual fold of the overlying mucosa.
- Anteriorly and inferiorly, the nerve and artery to the mylohyoid (cut) and the lingual nerve course between the medial pterygoid muscle and ramus of the mandible.
- The lingual nerve lies between the sublingual gland and the deep part of the submandibular gland; the submandibular ganglion is suspended from this nerve.

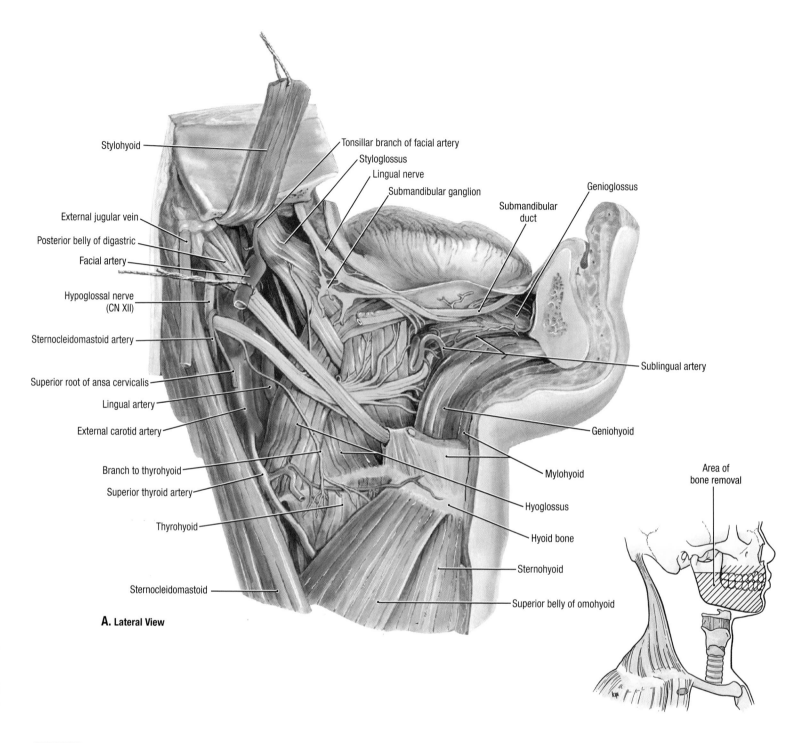

Stylohyoid

Tonsillar branch of facial artery

Styloglossus

Lingual nerve

Submandibular ganglion

Genioglossus

Submandibular duct

External jugular vein

Posterior belly of digastric

Facial artery

Hypoglossal nerve (CN XII)

Sternocleidomastoid artery

Superior root of ansa cervicalis

Lingual artery

External carotid artery

Branch to thyrohyoid

Superior thyroid artery

Thyrohyoid

Sternocleidomastoid

A. Lateral View

Sublingual artery

Geniohyoid

Mylohyoid

Hyoglossus

Hyoid bone

Sternohyoid

Superior belly of omohyoid

Area of bone removal

8.13   Suprahyoid region—III

A. All of the right half of the mandible, except the superior part of the ramus, has been removed. The stylohyoid muscle is reflected superiorly, and the posterior belly of the digastric muscle is left in situ.

- The hyoglossus muscle ascends from the greater horn and body of the hyoid bone to the side of the tongue.
- The styloglossus muscle is crossed by the tonsillar branch of the facial artery posterosuperiorly and interdigitates with bundles of the hyoglossus muscle inferiorly.
- The hypoglossal nerve supplies all of the muscles of the tongue,

both extrinsic and intrinsic, except the palatoglossus (a palatine muscle, innervated by CN X).

- The submandibular duct runs anteriorly in contact with the hyoglossus and genioglossus muscles to its opening on the side of the lingual frenulum.
- The lingual nerve is in contact with the mandible posteriorly, looping inferior to the submandibular duct and ending in the tongue. The submandibular ganglion is suspended from the lingual nerve; twigs leave the nerve to supply the mucous membrane.

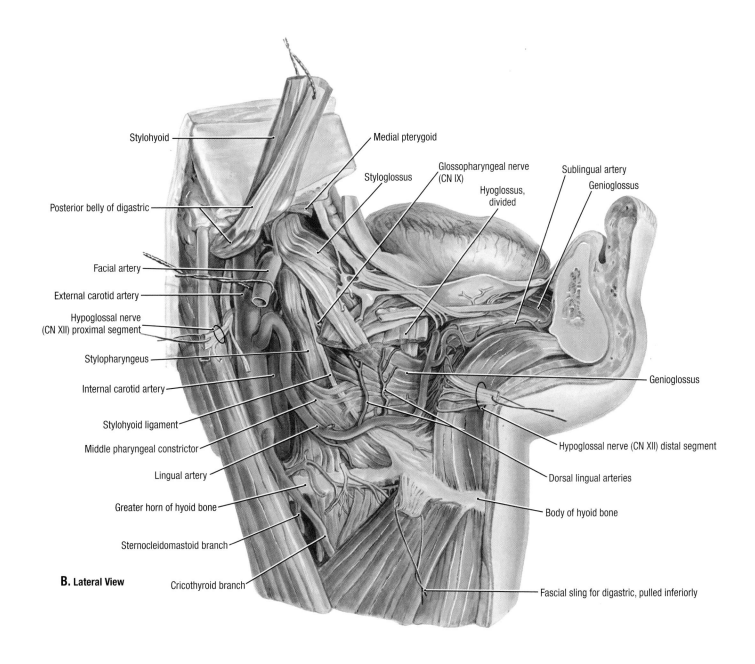

Stylohyoid

Medial pterygoid

Styloglossus

Glossopharyngeal nerve (CN IX)

Hyoglossus, divided

Sublingual artery

Genioglossus

Posterior belly of digastric

Facial artery

External carotid artery

Hypoglossal nerve (CN XII) proximal segment

Stylopharyngeus

Internal carotid artery

Stylohyoid ligament

Middle pharyngeal constrictor

Lingual artery

Greater horn of hyoid bone

Sternocleidomastoid branch

Genioglossus

Hypoglossal nerve (CN XII) distal segment

Dorsal lingual arteries

Body of hyoid bone

B. Lateral View

Cricothyroid branch

Fascial sling for digastric, pulled inferiorly

8.13   **Suprahyoid region—IV**

B. The stylohyoid and posterior belly of the digastric muscle are reflected superiorly, the hypoglossal nerve is divided, and the hyoglossus muscle is mostly removed.

- The lingual artery passes deep to the hyoglossus muscle, close to the greater horn of the hyoid, and then passes lateral to the middle pharyngeal constrictor muscle, stylohyoid ligament, and genioglossus muscle and turns into the tongue as the deep lingual arteries.
- The branches of the lingual artery include (a) muscular branches, (b) dorsal lingual arteries that supply the tonsil bed, and (c) the sublingual branch that supplies the sublingual gland and anterior part of the floor of the mouth.

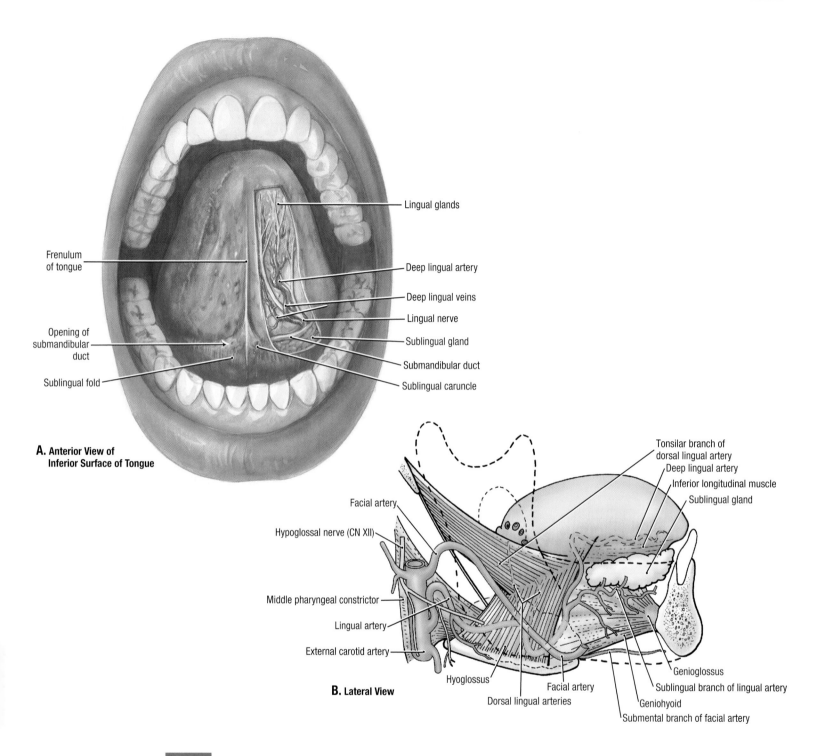

A. Anterior View of Inferior Surface of Tongue

Lingual glands

Frenulum of tongue

Deep lingual artery

Deep lingual veins

Lingual nerve

Opening of submandibular duct

Sublingual gland

Submandibular duct

Sublingual fold

Sublingual caruncle

Tonsilar branch of dorsal lingual artery

Deep lingual artery

Inferior longitudinal muscle

Sublingual gland

Facial artery

Hypoglossal nerve (CN XII)

Middle pharyngeal constrictor

Lingual artery

External carotid artery

Hyoglossus

Dorsal lingual arteries

Facial artery

Genioglossus

Sublingual branch of lingual artery

Geniohyoid

Submental branch of facial artery

B. Lateral View

8.14 Floor of mouth with tongue elevated and lingual artery

A. Inferior surface of the tongue and floor of the mouth. **B.** Course of the lingual artery.

In **A:**

• The inferior (sublingual) surface of the tongue is covered by a mucous membrane through which the underlying deep lingual veins can be seen.

• The sublingual caruncle, a papilla on each side of the frenulum, marks the location of the opening of the submandibular duct.

In **B:**

• The dorsal lingual arteries supply the root of the tongue and palatine tonsil, the deep lingual artery supplies the body of the tongue, and the sublingual branch supplies the floor of the mouth.

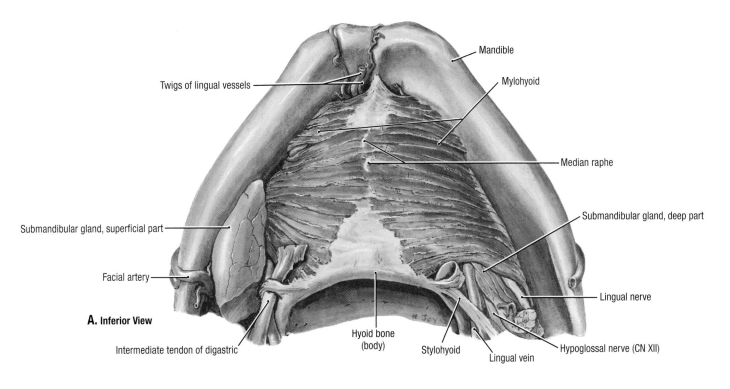

Twigs of lingual vessels

Mandible

Mylohyoid

Median raphe

Submandibular gland, deep part

Submandibular gland, superficial part

Facial artery

Lingual nerve

A. Inferior View

Intermediate tendon of digastric

Hyoid bone (body)

Stylohyoid

Lingual vein

Hypoglossal nerve (CN XII)

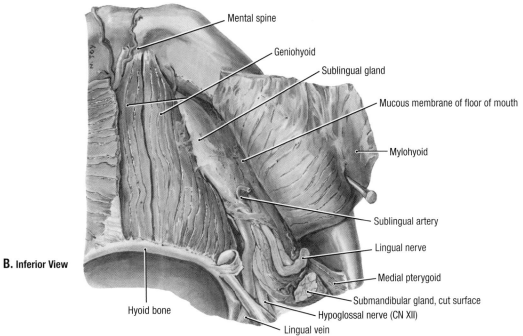

Mental spine

Geniohyoid

Sublingual gland

Mucous membrane of floor of mouth

Mylohyoid

Sublingual artery

Lingual nerve

Medial pterygoid

Submandibular gland, cut surface

Hypoglossal nerve (CN XII)

Lingual vein

B. Inferior View

Hyoid bone

8.15   Submental triangle

The anterior bellies of the digastric muscle have been removed. **A.** Mylohyoid muscles. The right and left mylohyoid muscles, which together form the floor of the mouth, attach to the mylohyoid line of the mandible, the median raphe, and the hyoid bone. Note also that the submandibular gland "grasps" the posterior border of the mylohyoid muscle, passing both superficial (mainly) and deep to it. **B.** Geniohyoid muscles. The left mylohyoid muscle and part of the right are reflected. The geniohyoid muscle extends from the mental spine of the mandible to the body of the hyoid bone.

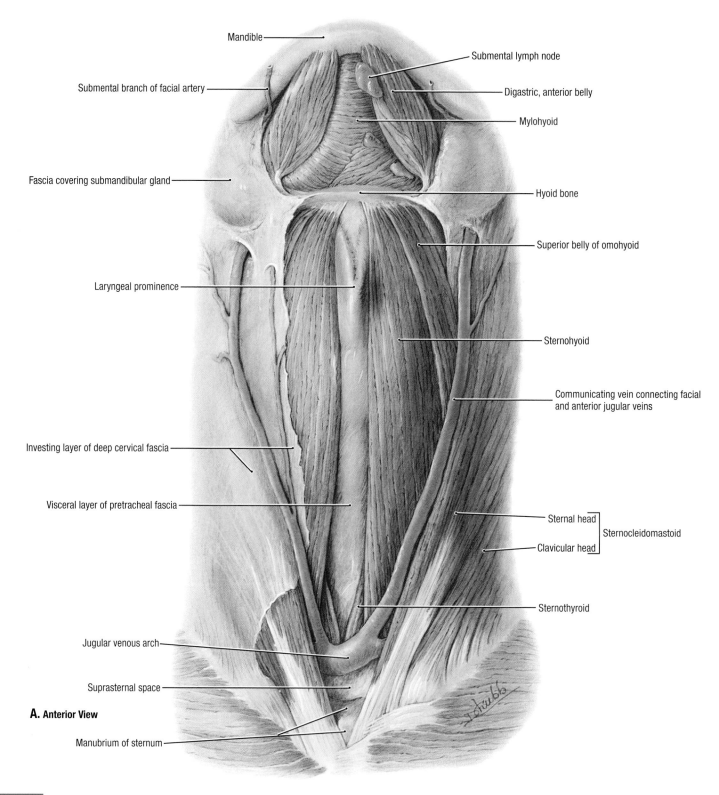

Mandible

Submental lymph node

Submental branch of facial artery

Digastric, anterior belly

Mylohyoid

Fascia covering submandibular gland

Hyoid bone

Superior belly of omohyoid

Laryngeal prominence

Sternohyoid

Communicating vein connecting facial and anterior jugular veins

Investing layer of deep cervical fascia

Visceral layer of pretracheal fascia

Sternal head ⎫
Clavicular head ⎭ Sternocleidomastoid

Sternothyroid

Jugular venous arch

Suprasternal space

A. Anterior View

Manubrium of sternum

8.16 Anterior cervical region—I

A. Infrahyoid part and submental triangle (suprahyoid part): superficial dissection. The investing and much of the muscular part of the pretracheal layers of deep cervical fascia have been removed.

- The anterior bellies of the digastric muscles form the sides of the suprahyoid part, or submental triangle (floor of the mouth); the hyoid bone forms the base of the triangle, and the mylohy-

oid muscles are its floor; some submental lymph nodes are its contents.

- The infrahyoid part is shaped like an elongated diamond and is bounded by the sternohyoid muscle superiorly and sternothyroid muscle inferiorly.

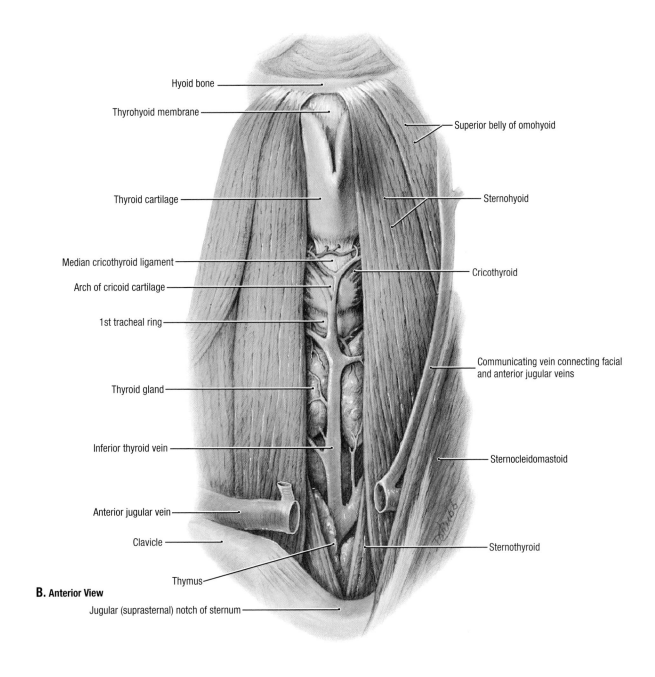

Hyoid bone

Thyrohyoid membrane

Superior belly of omohyoid

Thyroid cartilage

Sternohyoid

Median cricothyroid ligament

Cricothyroid

Arch of cricoid cartilage

1st tracheal ring

Communicating vein connecting facial
and anterior jugular veins

Thyroid gland

Inferior thyroid vein

Sternocleidomastoid

Anterior jugular vein

Clavicle

Sternothyroid

Thymus

B. Anterior View

Jugular (suprasternal) notch of sternum

| **8.16** | **Anterior cervical region—II** |

B. Infrahyoid part: muscular layer. The pretracheal fascia, right anterior jugular vein, and jugular venous arch have been removed.

- An enlarged thymus projects superiorly from the thorax.
- The two superficial depressors of the larynx ("strap muscles") are the omohyoid (only the superior belly of which is seen here) and sternohyoid.

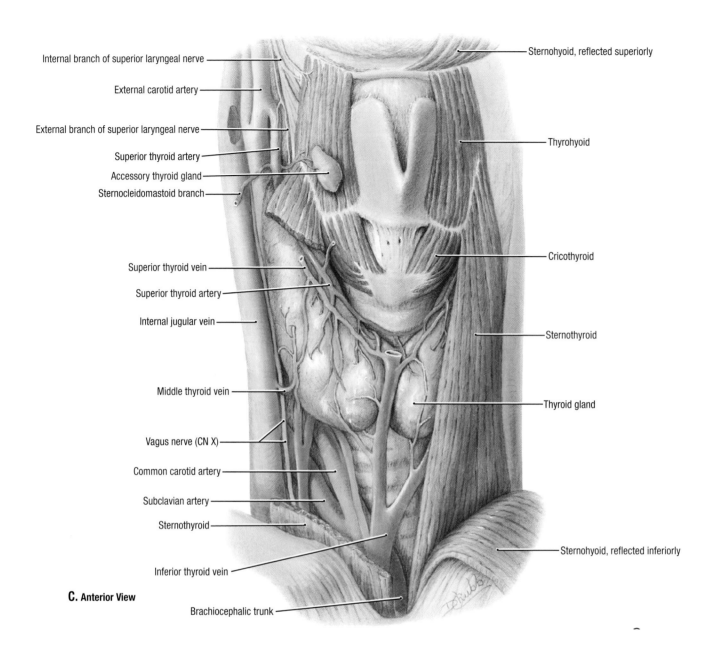

Internal branch of superior laryngeal nerve

External carotid artery

External branch of superior laryngeal nerve

Superior thyroid artery

Accessory thyroid gland

Sternocleidomastoid branch

Superior thyroid vein

Superior thyroid artery

Internal jugular vein

Middle thyroid vein

Vagus nerve (CN X)

Common carotid artery

Subclavian artery

Sternothyroid

Inferior thyroid vein

C. Anterior View

Brachiocephalic trunk

Sternohyoid, reflected superiorly

Thyrohyoid

Cricothyroid

Sternothyroid

Thyroid gland

Sternohyoid, reflected inferiorly

8.16 **Anterior cervical region—III**

C. Infrahyoid part: endocrine layer. On the left side of the specimen, the sternohyoid and omohyoid muscles are reflected, exposing the sternothyroid and the thyrohyoid muscles; on the right side of the specimen, the sternothyroid muscle is largely excised. **D.** Schematic illustration of the arterial supply of the thyroid gland. The superior and inferior thyroid arteries supply the thyroid gland. In about 10% of people a thyroid ima artery arises from the brachiocephalic trunk or arch of the aorta and continues to the isthmus of the thyroid gland.

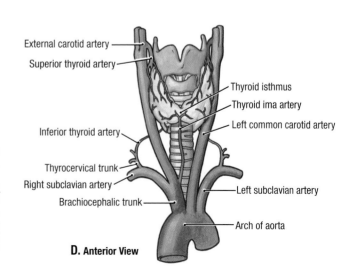

External carotid artery

Superior thyroid artery

Thyroid isthmus

Thyroid ima artery

Left common carotid artery

Inferior thyroid artery

Thyrocervical trunk

Right subclavian artery

Brachiocephalic trunk

Left subclavian artery

Arch of aorta

D. Anterior View

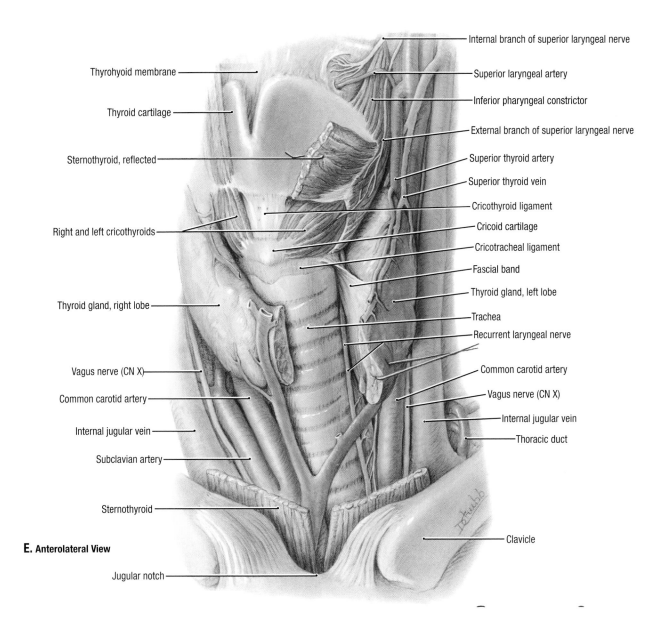

Internal branch of superior laryngeal nerve

Thyrohyoid membrane

Superior laryngeal artery

Inferior pharyngeal constrictor

Thyroid cartilage

External branch of superior laryngeal nerve

Superior thyroid artery

Sternothyroid, reflected

Superior thyroid vein

Cricothyroid ligament

Right and left cricothyroids

Cricoid cartilage

Cricotracheal ligament

Fascial band

Thyroid gland, left lobe

Thyroid gland, right lobe

Trachea

Recurrent laryngeal nerve

Vagus nerve (CN X)

Common carotid artery

Common carotid artery

Vagus nerve (CN X)

Internal jugular vein

Internal jugular vein

Thoracic duct

Subclavian artery

Sternothyroid

Clavicle

E. Anterolateral View

Jugular notch

8.16   Anterior cervical region—IV

E. Infrahyoid part: respiratory layer. The isthmus of the thyroid gland is divided, and the left lobe is retracted. The left recurrent laryngeal nerve ascends on the lateral aspect of the trachea between the trachea and esophagus. The internal branch of the superior laryngeal nerve runs along the superior border of the inferior pharyngeal constrictor muscle and pierces the thyrohyoid membrane. The external branch of the superior laryngeal nerve lies adjacent to the inferior pharyngeal constrictor muscle and supplies its lower portion; it continues to run along the anterior border of the superior thyroid artery, passing deep to the superior attachment of the sternothyroid muscle, and then supplies the cricothyroid muscle. **F.** Schematic illustration of the venous drainage of the thyroid gland. Except for the superior thyroid veins, the thyroid veins are not paired with arteries of corresponding names.

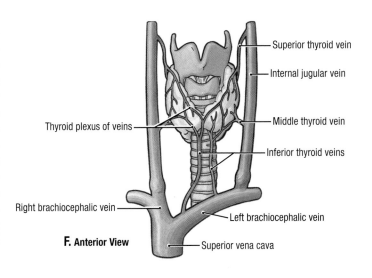

Superior thyroid vein

Internal jugular vein

Thyroid plexus of veins

Middle thyroid vein

Inferior thyroid veins

Right brachiocephalic vein

Left brachiocephalic vein

F. Anterior View

Superior vena cava

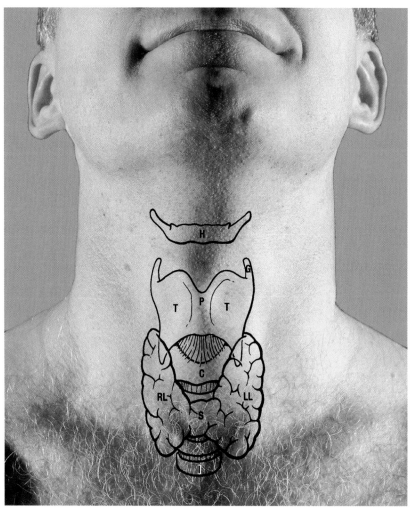

A. Anterior View

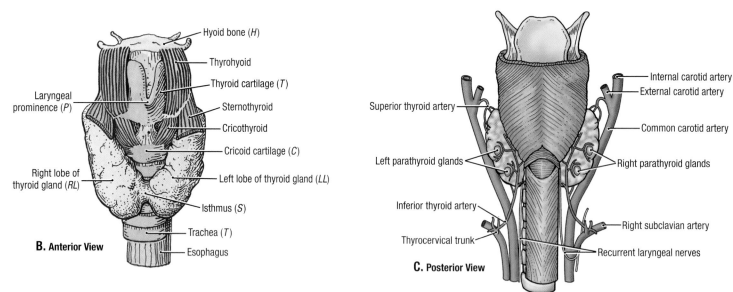

Hyoid bone (*H*)

Thyrohyoid

Laryngeal prominence (*P*)

Thyroid cartilage (*T*)

Sternothyroid

Cricothyroid

Right lobe of thyroid gland (*RL*)

Cricoid cartilage (*C*)

Left lobe of thyroid gland (*LL*)

Isthmus (*S*)

Trachea (*T*)

Esophagus

B. Anterior View

Superior thyroid artery

Internal carotid artery

External carotid artery

Common carotid artery

Left parathyroid glands

Right parathyroid glands

Inferior thyroid artery

Right subclavian artery

Thyrocervical trunk

Recurrent laryngeal nerves

C. Posterior View

8.17 **Thyroid and parathyroid glands**

A. Surface anatomy of hyoid bone, laryngeal cartilages, and thyroid gland. **B.** Parts of the thyroid gland and its relationships. **C.** Parathyroid glands, demonstrating their primary blood supply from the inferior thyroid artery.

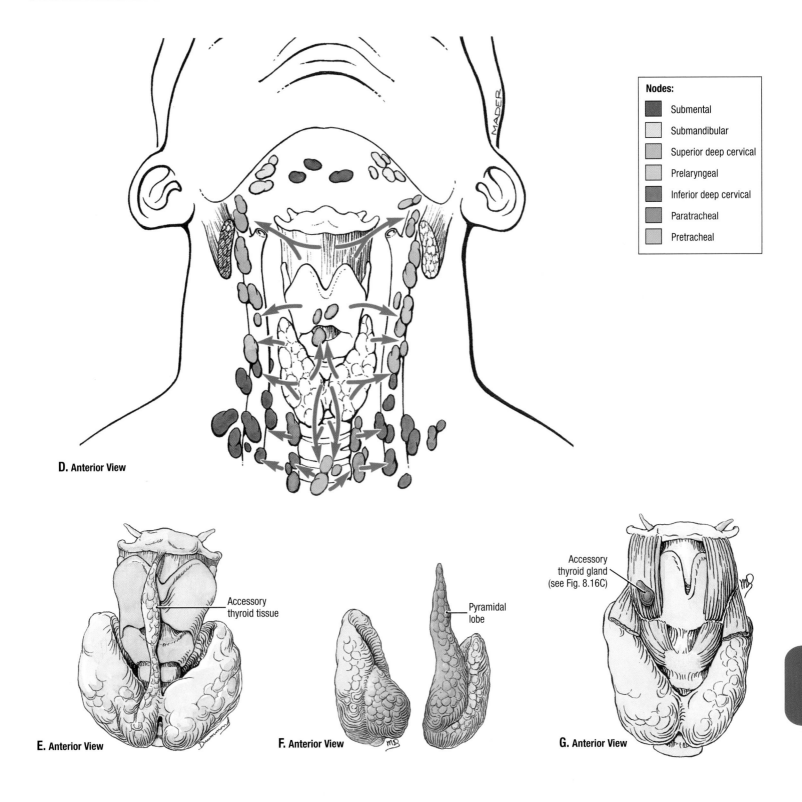

Nodes:
- Submental
- Submandibular
- Superior deep cervical
- Prelaryngeal
- Inferior deep cervical
- Paratracheal
- Pretracheal

D. Anterior View

Accessory thyroid tissue

E. Anterior View

Pyramidal lobe

F. Anterior View

Accessory thyroid gland (see Fig. 8.16C)

G. Anterior View

8.17   **Thyroid and parathyroid glands** *(continued)*

D. Lymphatic drainage of the thyroid gland, larynx, and trachea. **E.** Accessory thyroid tissue along the course of the thyroglossal duct. **F.** Approximately 50% of glands have a pyramidal lobe that extends from near the isthmus to or toward the hyoid bone; the isthmus is occasionally absent, in which case the gland is in two parts. **G.** An accessory thyroid gland can occur between the suprahyoid region and arch of the aorta (see Fig. 8.16C).

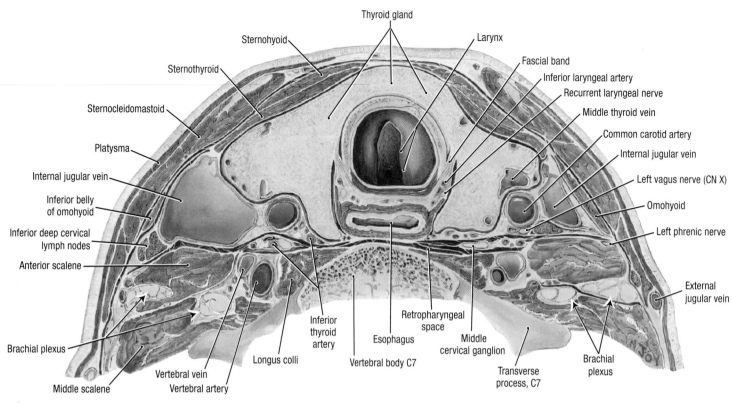

A. Transverse Section, Inferior View

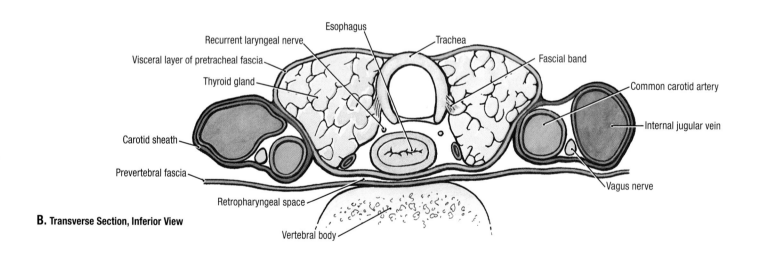

B. Transverse Section, Inferior View

8.18 **Sectional anatomy of the visceral compartment of the neck**

A. Anatomical section of neck at the level of the isthmus of the thyroid gland (C7 vertebral level). The thyroid gland is asymmetrically enlarged and lies anterior to most of the carotid sheath and its contents. **B.** Diagrammatic section demonstrating the visceral layer of the pretracheal fascia and the carotid sheath.

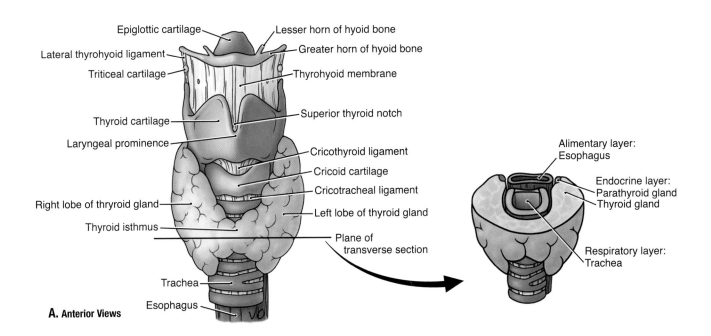

A. Anterior Views

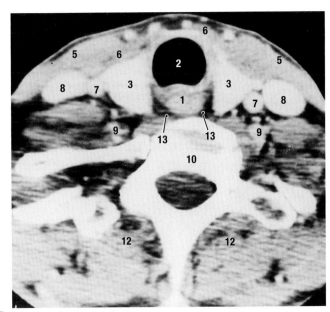

B. Transverse Section, Inferior View

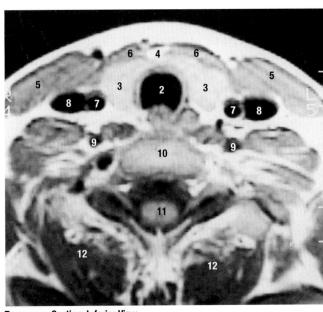

C. Transverse Section, Inferior View

8.19   Imaging of the thyroid gland

A. Orientation figure demonstrating the relations of thyroid gland and the approximate level and plane of the sections. **B** and **C.** MRI of the anterior neck. *1*, Esophagus; *2*, trachea; *3*, lobes of thyroid gland; *4*, thyroid isthmus; *5*, sternocleidomastoid; *6*, strap muscles; *7*, common carotid artery; *8*, internal jugular vein; *9*, vertebral artery; *10*, vertebral body; *11*, spinal cord in cerebrospinal fluid; *12*, deep muscles of back; *13*, retropharyngeal space.

Arrow = direction
of retraction

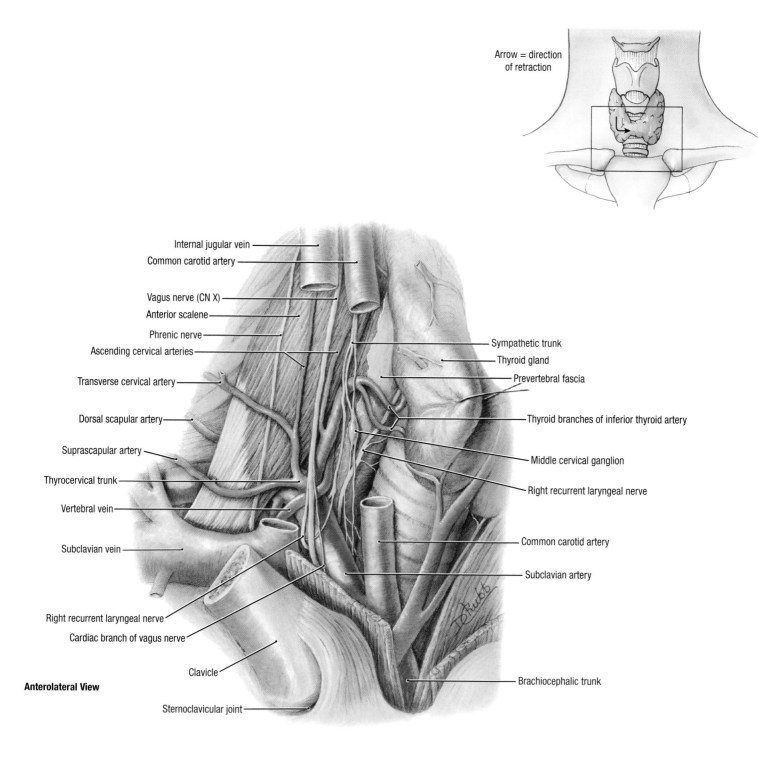

Internal jugular vein

Common carotid artery

Vagus nerve (CN X)

Anterior scalene

Phrenic nerve

Ascending cervical arteries

Transverse cervical artery

Dorsal scapular artery

Suprascapular artery

Thyrocervical trunk

Vertebral vein

Subclavian vein

Right recurrent laryngeal nerve

Cardiac branch of vagus nerve

Clavicle

Anterolateral View

Sternoclavicular joint

Sympathetic trunk

Thyroid gland

Prevertebral fascia

Thyroid branches of inferior thyroid artery

Middle cervical ganglion

Right recurrent laryngeal nerve

Common carotid artery

Subclavian artery

Brachiocephalic trunk

8.20 **Root of the neck—I**

Dissection of the right side of the root of the neck. The clavicle is cut, sections of the common carotid artery and internal jugular vein are removed, and the right lobe of the thyroid gland is retracted. The right vagus nerve crosses the first part of the subclavian artery and gives off an inferior cardiac branch and the right recurrent laryngeal nerve. The right recurrent laryngeal nerve loops inferior to the subclavian artery and passes posterior to the common carotid artery on its way to the posterolateral aspect of the trachea.

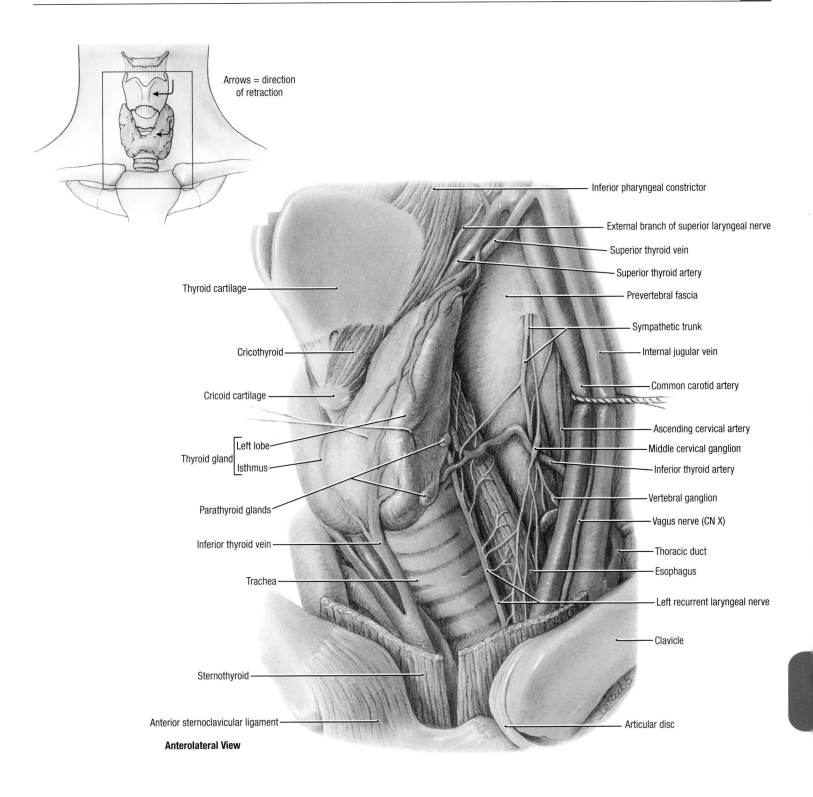

Arrows = direction of retraction

Inferior pharyngeal constrictor

External branch of superior laryngeal nerve

Superior thyroid vein

Superior thyroid artery

Thyroid cartilage

Prevertebral fascia

Sympathetic trunk

Cricothyroid

Internal jugular vein

Common carotid artery

Cricoid cartilage

Ascending cervical artery

Left lobe

Middle cervical ganglion

Thyroid gland

Inferior thyroid artery

Isthmus

Vertebral ganglion

Parathyroid glands

Vagus nerve (CN X)

Inferior thyroid vein

Thoracic duct

Esophagus

Trachea

Left recurrent laryngeal nerve

Clavicle

Sternothyroid

Anterior sternoclavicular ligament

Articular disc

Anterolateral View

8.21 Root of the neck—II

Dissection of the left side of the root of the neck. The three structures contained in the carotid sheath (internal jugular vein, common carotid artery, and vagus nerve) are retracted. The left recurrent laryngeal nerve ascends on the lateral aspect of the trachea, just anterior to the recess between the trachea and esophagus. The recurrent laryngeal nerves are vulnerable to injury during thyroidectomy and other surgeries in the anterior cervical region of the neck. Because the terminal branch of this nerve, the inferior laryngeal nerve, innervates the muscles moving the vocal folds, injury to the nerve results in paralysis of the vocal folds.

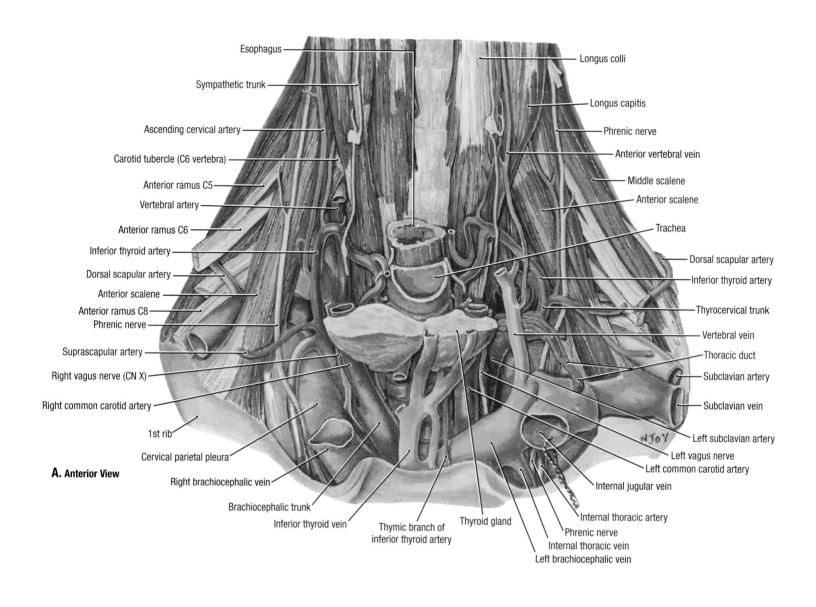

A. Anterior View

Esophagus

Sympathetic trunk

Ascending cervical artery

Carotid tubercle (C6 vertebra)

Anterior ramus C5

Vertebral artery

Anterior ramus C6

Inferior thyroid artery

Dorsal scapular artery

Anterior scalene

Anterior ramus C8

Phrenic nerve

Suprascapular artery

Right vagus nerve (CN X)

Right common carotid artery

1st rib

Cervical parietal pleura

Right brachiocephalic vein

Brachiocephalic trunk

Inferior thyroid vein

Thymic branch of inferior thyroid artery

Thyroid gland

Longus colli

Longus capitis

Phrenic nerve

Anterior vertebral vein

Middle scalene

Anterior scalene

Trachea

Dorsal scapular artery

Inferior thyroid artery

Thyrocervical trunk

Vertebral vein

Thoracic duct

Subclavian artery

Subclavian vein

Left subclavian artery

Left vagus nerve

Left common carotid artery

Internal jugular vein

Internal thoracic artery

Phrenic nerve

Internal thoracic vein

Left brachiocephalic vein

8.22　Root of the neck—III

A. Deep anterior dissection. **B.** Schematic illustration of termination of the thoracic duct. The thoracic duct arches laterally in the neck, passing posterior to the carotid sheath and anterior to the vertebral artery, thyrocervical trunk, and subclavian arteries; it enters the angle formed by the junction of the left subclavian and internal jugular veins to form the left brachiocephalic vein (the left venous angle).

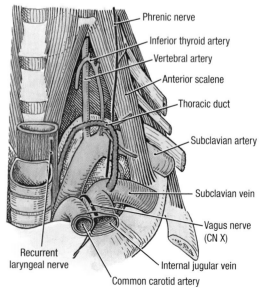

Phrenic nerve

Inferior thyroid artery

Vertebral artery

Anterior scalene

Thoracic duct

Subclavian artery

Subclavian vein

Vagus nerve (CN X)

Internal jugular vein

Recurrent laryngeal nerve

Common carotid artery

B. Anterior View

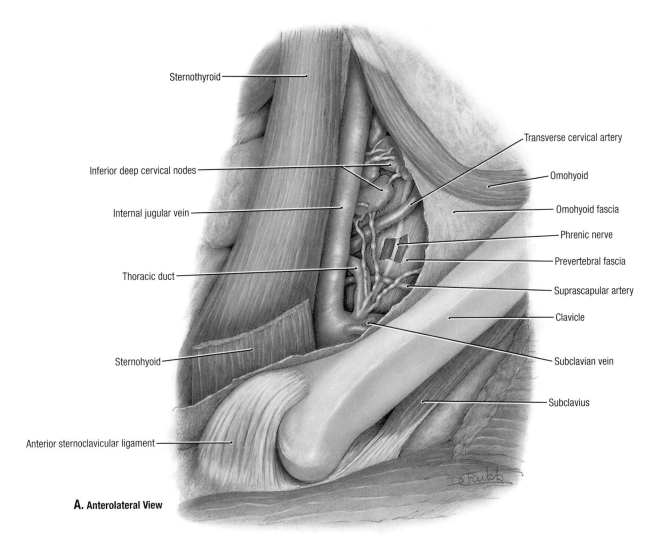

Sternothyroid

Inferior deep cervical nodes

Internal jugular vein

Thoracic duct

Sternohyoid

Anterior sternoclavicular ligament

Transverse cervical artery

Omohyoid

Omohyoid fascia

Phrenic nerve

Prevertebral fascia

Suprascapular artery

Clavicle

Subclavian vein

Subclavius

A. Anterolateral View

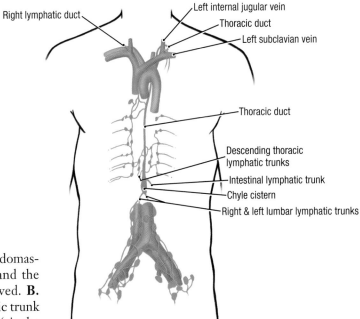

Right lymphatic duct

Left internal jugular vein

Thoracic duct

Left subclavian vein

Thoracic duct

Descending thoracic lymphatic trunks

Intestinal lymphatic trunk

Chyle cistern

Right & left lumbar lymphatic trunks

B. Anterior View

8.23   Root of the neck—IV

A. Dissection of termination of the thoracic duct. The sternocleidomastoid muscle is removed, the sternohyoid muscle is resected, and the omohyoid portion of the pretracheal fascia is partially removed. **B.** Overview of lymphatic drainage of the trunk. The right lymphatic trunk and thoracic duct also receive drainage from the head and neck (via the jugular lymphatic trunk) and the upper limb (via the subclavian trunk). Some or all of the trunks may enter the venous angles independently.

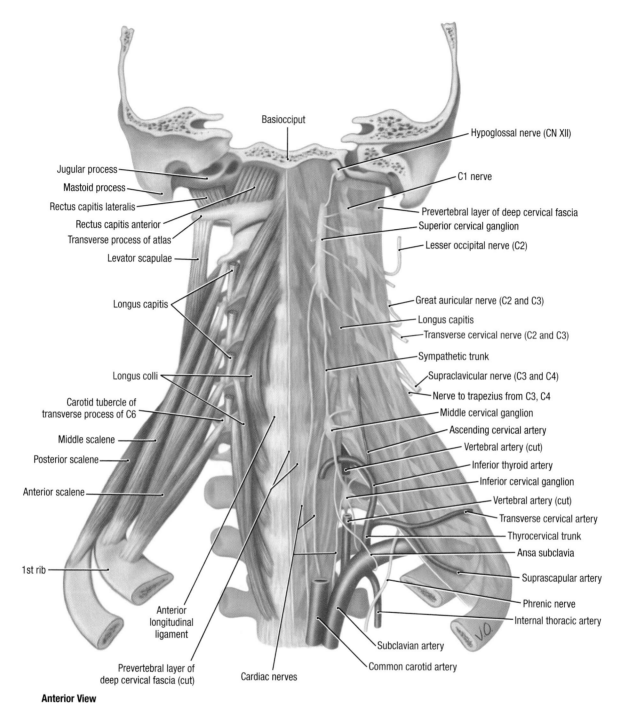

Basiocciput
Hypoglossal nerve (CN XII)
Jugular process
Mastoid process
Rectus capitis lateralis
Rectus capitis anterior
Transverse process of atlas
Levator scapulae
Longus capitis
Longus colli
Carotid tubercle of
transverse process of C6
Middle scalene
Posterior scalene
Anterior scalene
1st rib
Anterior
longitudinal
ligament
Prevertebral layer of
deep cervical fascia (cut)
Cardiac nerves
C1 nerve
Prevertebral layer of deep cervical fascia
Superior cervical ganglion
Lesser occipital nerve (C2)
Great auricular nerve (C2 and C3)
Longus capitis
Transverse cervical nerve (C2 and C3)
Sympathetic trunk
Supraclavicular nerve (C3 and C4)
Nerve to trapezius from C3, C4
Middle cervical ganglion
Ascending cervical artery
Vertebral artery (cut)
Inferior thyroid artery
Inferior cervical ganglion
Vertebral artery (cut)
Transverse cervical artery
Thyrocervical trunk
Ansa subclavia
Suprascapular artery
Phrenic nerve
Internal thoracic artery
Subclavian artery
Common carotid artery

Anterior View

TABLE 8.6 PREVERTEBRAL MUSCLES

MUSCLE	SUPERIOR ATTACHMENT	INFERIOR ATTACHMENT	INNERVATION	MAIN ACTION
Longus colli	Anterior tubercle of C1 vertebra (atlas)	Body of T3 vertebra with attachments to bodies of C1–C3 and transverse processes of C3–C6 vertebrae	Anterior rami of C2–C6 spinal nerves	Flexes cervical vertebrae
Longus capitis	Basilar part of occipital bone	Anterior tubercles of C3–C6 transverse processes	Anterior rami of C2 and C3 spinal nerves	Flexes cervical vertebrae and atlanto-occipital joint
Rectus capitis anterior	Base of skull, just anterior to occipital condyle	Anterior surface of lateral mass of C1 vertebra (atlas)	Branches from C1 and C2 spinal nerves	Flexes atlanto-occipital joint
Rectus capitis lateralis	Jugular process of occipital bone	Transverse process of C1 (atlas) vertebra		Flexes atlanto-occipital joint and helps to stabilize the head

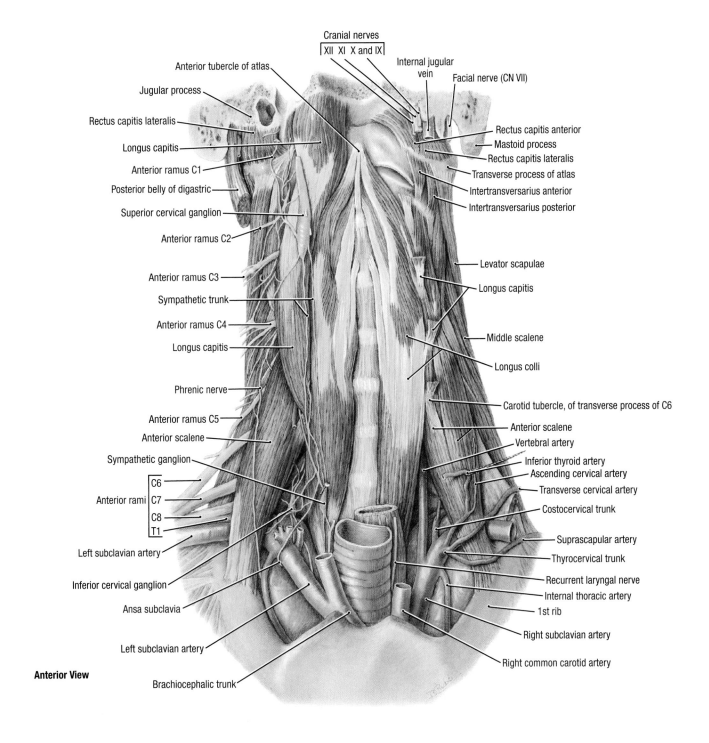

Cranial nerves
XII XI X and IX

Anterior tubercle of atlas

Jugular process

Rectus capitis lateralis

Longus capitis

Anterior ramus C1

Posterior belly of digastric

Superior cervical ganglion

Anterior ramus C2

Anterior ramus C3

Sympathetic trunk

Anterior ramus C4

Longus capitis

Phrenic nerve

Anterior ramus C5

Anterior scalene

Sympathetic ganglion

Anterior rami C6 C7 C8 T1

Left subclavian artery

Inferior cervical ganglion

Ansa subclavia

Left subclavian artery

Anterior View

Brachiocephalic trunk

Internal jugular vein

Facial nerve (CN VII)

Rectus capitis anterior

Mastoid process

Rectus capitis lateralis

Transverse process of atlas

Intertransversarius anterior

Intertransversarius posterior

Levator scapulae

Longus capitis

Middle scalene

Longus colli

Carotid tubercle, of transverse process of C6

Anterior scalene

Vertebral artery

Inferior thyroid artery

Ascending cervical artery

Transverse cervical artery

Costocervical trunk

Suprascapular artery

Thyrocervical trunk

Recurrent laryngal nerve

Internal thoracic artery

1st rib

Right subclavian artery

Right common carotid artery

8.24   **Prevertebral region**

The prevertebral fascia and the left longus capitis muscle are removed. The cervical plexus arises from anterior rami of C1, C2, C3, and C4 and the brachial plexus from anterior rami of C5, C6, C7, C8, and T1; the plexuses emerge in a coronal plane that passes anterior to the levator scapulae and middle scalene muscles.

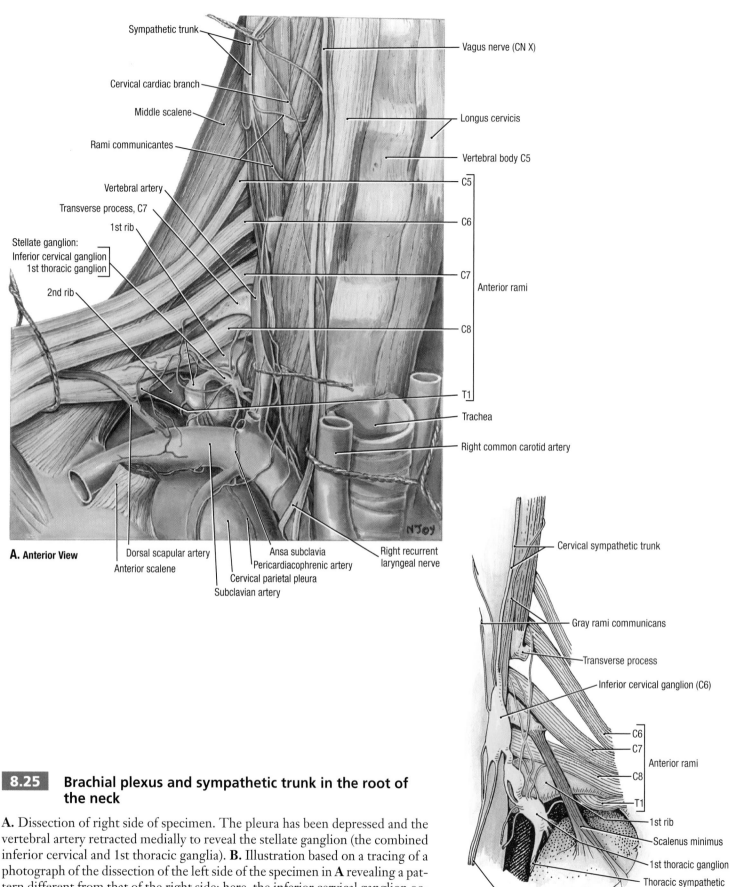

Sympathetic trunk

Cervical cardiac branch

Middle scalene

Rami communicantes

Vertebral artery

Transverse process, C7

1st rib

Stellate ganglion:
Inferior cervical ganglion
1st thoracic ganglion

2nd rib

Vagus nerve (CN X)

Longus cervicis

Vertebral body C5

C5
C6
C7
C8
T1

Anterior rami

Trachea

Right common carotid artery

A. Anterior View

Dorsal scapular artery
Anterior scalene
                    Ansa subclavia
                    Pericardiacophrenic artery
                    Cervical parietal pleura
Subclavian artery

Right recurrent
laryngeal nerve

Cervical sympathetic trunk

Gray rami communicans

Transverse process

Inferior cervical ganglion (C6)

C6
C7
C8
T1

Anterior rami

1st rib

Scalenus minimus

1st thoracic ganglion

Thoracic sympathetic
trunk

Cardiac and vascular branches

B. Anterior View

8.25   Brachial plexus and sympathetic trunk in the root of the neck

A. Dissection of right side of specimen. The pleura has been depressed and the vertebral artery retracted medially to reveal the stellate ganglion (the combined inferior cervical and 1st thoracic ganglia). **B.** Illustration based on a tracing of a photograph of the dissection of the left side of the specimen in **A** revealing a pattern different from that of the right side; here, the inferior cervical ganglion occupies its more common position between the transverse process of C7 and the 1st rib; the T1 ganglion contacts the 1st rib anteriorly and inferiorly.

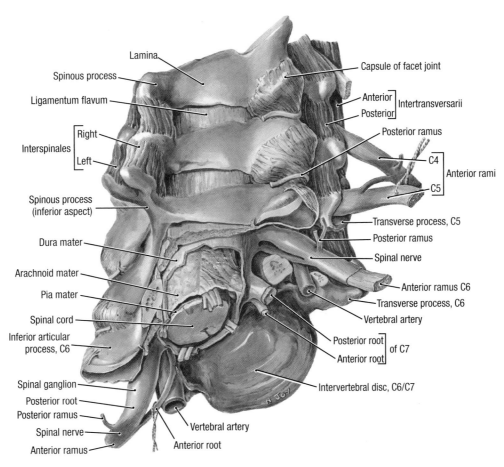

A. Posteroinferior View

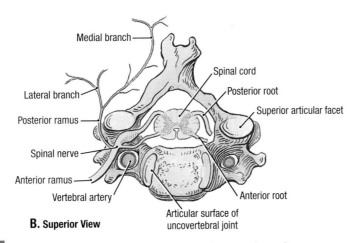

B. Superior View

8.26 **Lower cervical vertebrae and associated structures and nerves**

A. Relationship to cervical spinal cord, spinal nerves, and coverings. The anterior and posterior roots, each in a separate dural sheath, unite beyond the spinal ganglion to form a spinal nerve that immediately divides into a small posterior and large anterior ramus. The roots pass anterior to the zygapophysial joints and unite as they exit the intervertebral foramina and pass posterior to the vertebral artery. The posterior ramus curves dorsally around the superior articular process, and the anterior ramus rests on the transverse process, which is grooved to support it. **B.** Observing this diagram, one can appreciate the vulnerability of the vertebral artery, spinal cord, and nerve roots to arthritic expansion from articular processes and the vertebral body, particularly the lateral edge of the superior surface of the body, the uncovertebral joint (joint of Luschka).

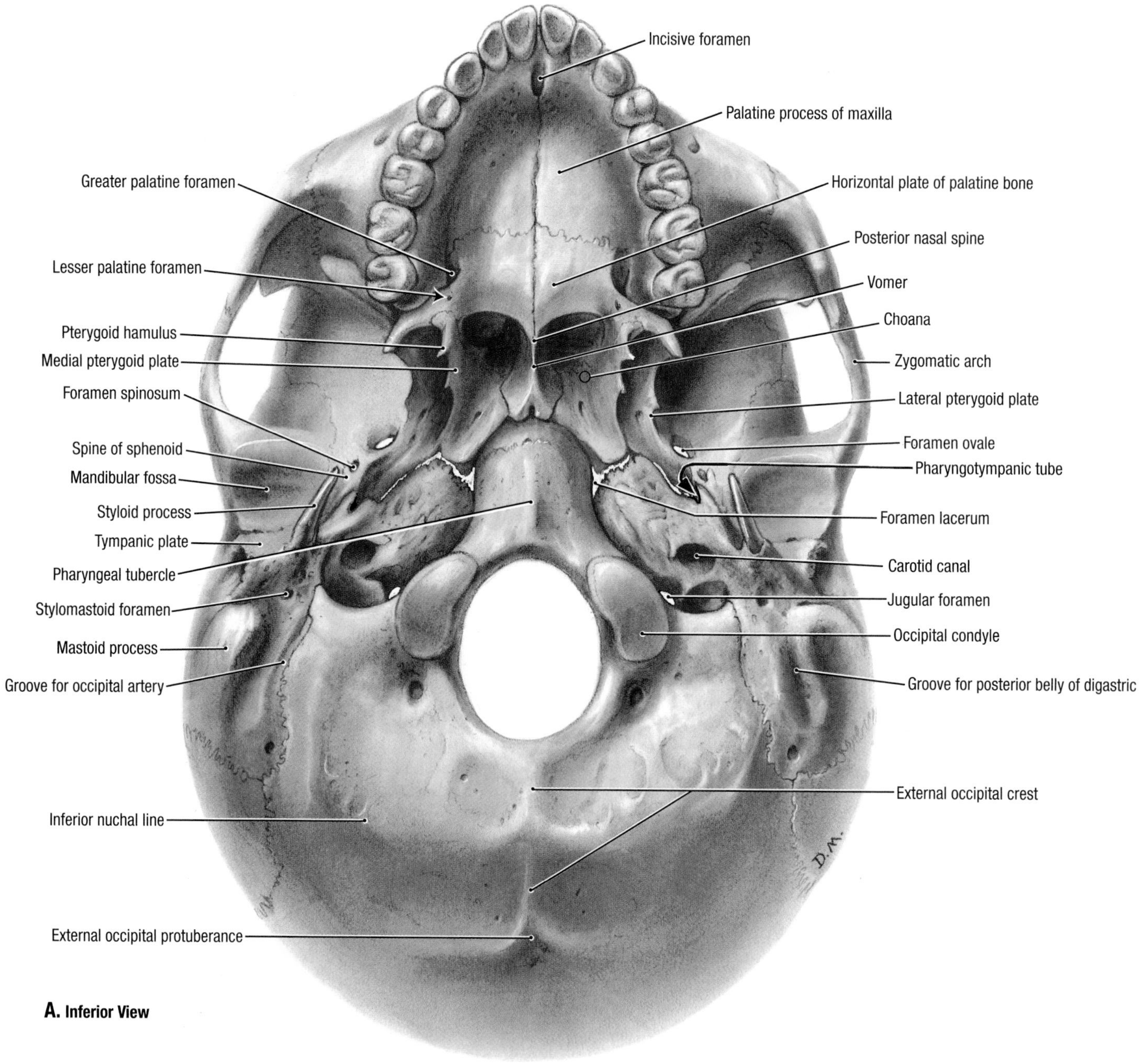

Incisive foramen

Palatine process of maxilla

Greater palatine foramen

Horizontal plate of palatine bone

Posterior nasal spine

Lesser palatine foramen

Vomer

Choana

Pterygoid hamulus

Zygomatic arch

Medial pterygoid plate

Lateral pterygoid plate

Foramen spinosum

Foramen ovale

Spine of sphenoid

Pharyngotympanic tube

Mandibular fossa

Styloid process

Foramen lacerum

Tympanic plate

Pharyngeal tubercle

Carotid canal

Stylomastoid foramen

Jugular foramen

Mastoid process

Occipital condyle

Groove for occipital artery

Groove for posterior belly of digastric

Inferior nuchal line

External occipital crest

External occipital protuberance

A. Inferior View

8.27 **External cranial base**

A. Features of the external aspect of the cranial base.

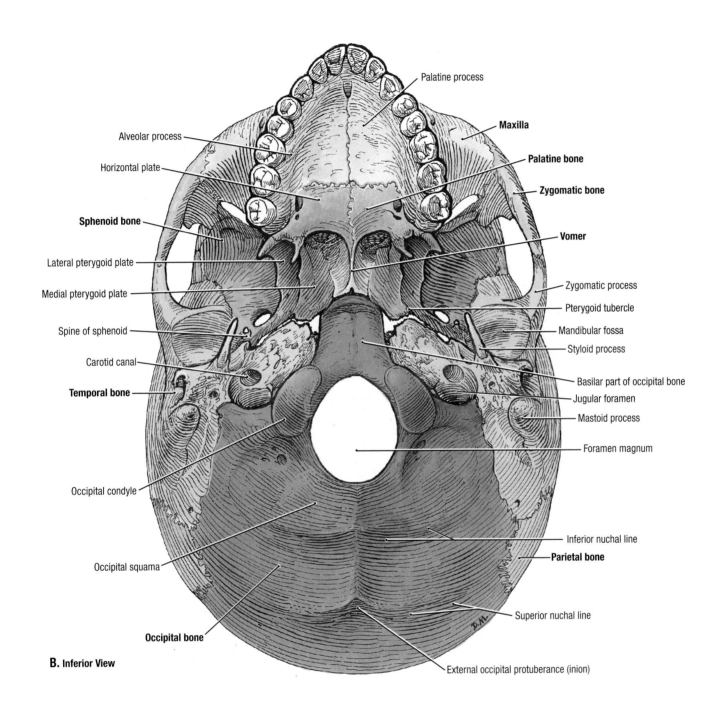

Palatine process

Alveolar process

Horizontal plate

Sphenoid bone

Lateral pterygoid plate

Medial pterygoid plate

Spine of sphenoid

Carotid canal

Temporal bone

Occipital condyle

Occipital squama

Occipital bone

Maxilla

Palatine bone

Zygomatic bone

Vomer

Zygomatic process

Pterygoid tubercle

Mandibular fossa

Styloid process

Basilar part of occipital bone

Jugular foramen

Mastoid process

Foramen magnum

Inferior nuchal line

Parietal bone

Superior nuchal line

External occipital protuberance (inion)

B. Inferior View

8.27   **External cranial base (continued)**

B. Diagram of external cranial base with bones color coded.

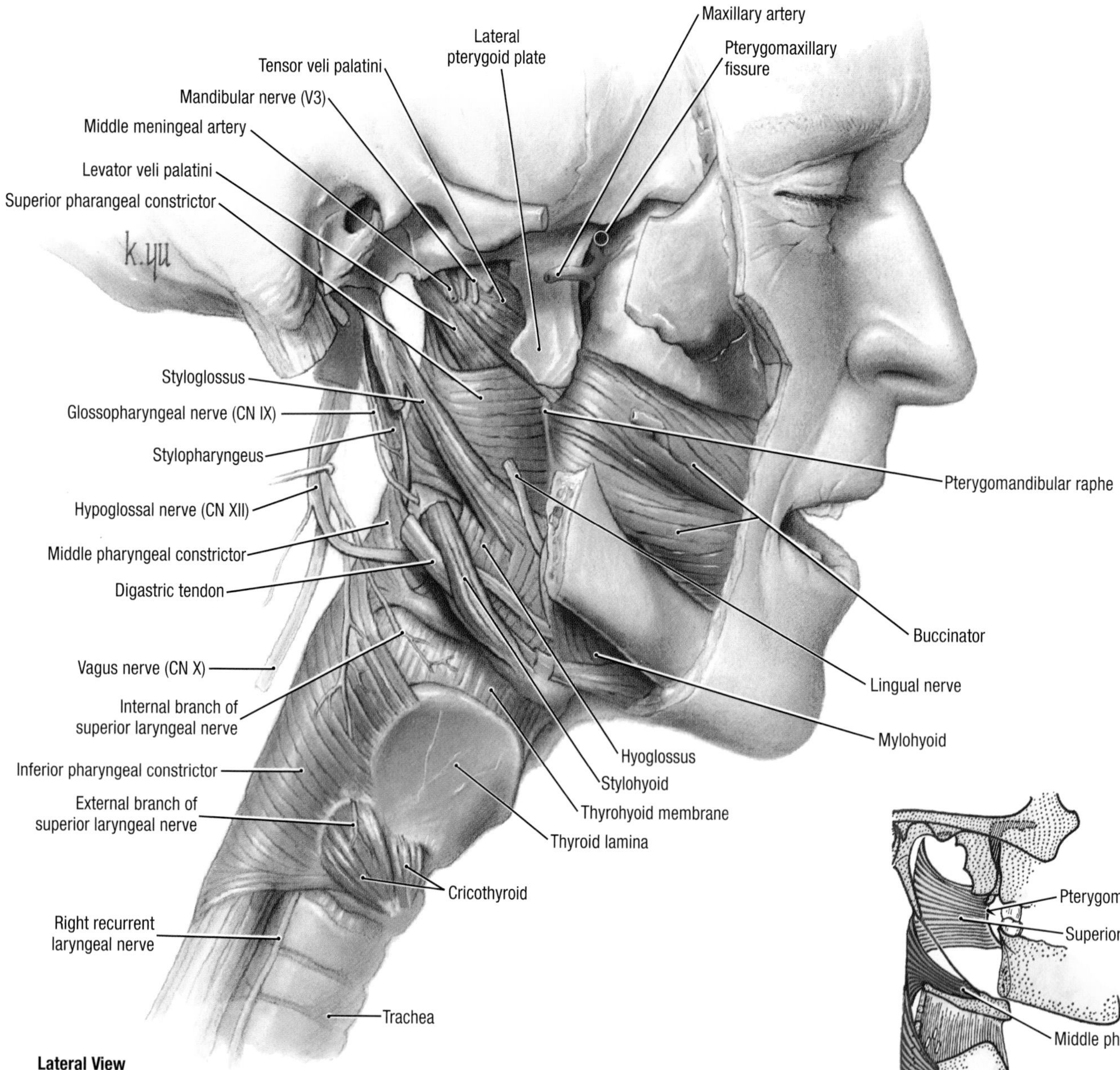

Lateral View

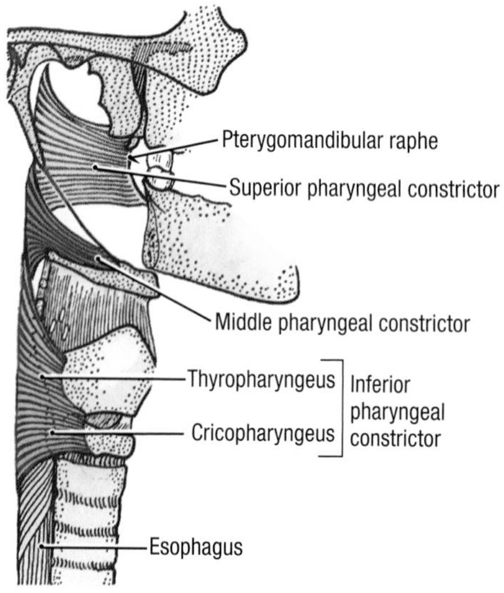

Maxillary artery

Lateral pterygoid plate

Pterygomaxillary fissure

Tensor veli palatini

Mandibular nerve (V3)

Middle meningeal artery

Levator veli palatini

Superior pharangeal constrictor

Styloglossus

Glossopharyngeal nerve (CN IX)

Stylopharyngeus

Hypoglossal nerve (CN XII)

Middle pharyngeal constrictor

Digastric tendon

Vagus nerve (CN X)

Internal branch of superior laryngeal nerve

Inferior pharyngeal constrictor

External branch of superior laryngeal nerve

Right recurrent laryngeal nerve

Trachea

Pterygomandibular raphe

Buccinator

Lingual nerve

Mylohyoid

Hyoglossus

Stylohyoid

Thyrohyoid membrane

Thyroid lamina

Cricothyroid

Lateral View

Pterygomandibular raphe

Superior pharyngeal constrictor

Middle pharyngeal constrictor

Thyropharyngeus

Cricopharyngeus

Inferior pharyngeal constrictor

Esophagus

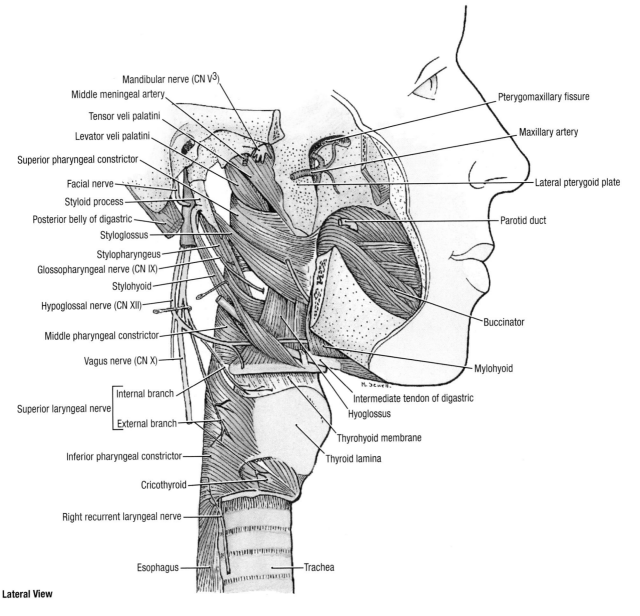

Lateral View

Labels (top to bottom, left side):
Mandibular nerve (CN V³)
Middle meningeal artery
Tensor veli palatini
Levator veli palatini
Superior pharyngeal constrictor
Facial nerve
Styloid process
Posterior belly of digastric
Styloglossus
Stylopharyngeus
Glossopharyngeal nerve (CN IX)
Stylohyoid
Hypoglossal nerve (CN XII)
Middle pharyngeal constrictor
Vagus nerve (CN X)
Superior laryngeal nerve — Internal branch / External branch
Inferior pharyngeal constrictor
Cricothyroid
Right recurrent laryngeal nerve
Esophagus

Labels (right side):
Pterygomaxillary fissure
Maxillary artery
Lateral pterygoid plate
Parotid duct
Buccinator
Mylohyoid
Intermediate tendon of digastric
Hyoglossus
Thyrohyoid membrane
Thyroid lamina
Trachea

TABLE 8.7 MUSCLES OF PHARYNX

MUSCLE	ORIGIN	INSERTION	INNERVATION	MAIN ACTION(S)
Superior pharyngeal constrictor	Pterygoid hamulus, pterygomandibular raphe, posterior end of mylohyoid line of mandible, and side of tongue	Pharyngeal raphe	Pharyngeal and superior laryngeal branches of vagus (CN X) through pharyngeal plexus	Constrict wall of pharynx during swallowing
Middle pharyngeal constrictor	Stylohyoid ligament and superior (greater) and inferior (lesser) horns of hyoid bone			
Inferior pharyngeal constrictor Thyropharyngeus	Oblique line of thyroid cartilage			
Cricopharyngeus	Side of cricoid cartilage	Contralateral side of cricoid cartilage	Pharyngeal and superior laryngeal branches of vagus (CN X) through pharyngeal plexus + external laryngeal plexus	Serves as superior esophageal sphincter
Palatopharyngeus	Hard plate and palatine aponeurosis	Posterior border of lamina of thyroid cartilage and side of pharynx and esophagus	Pharyngeal and superior laryngeal branches of vagus (CN X) through pharyngeal plexus	Elevate pharynx and larynx during swallowing and speaking
Salpingopharyngeus	Cartilaginous part of pharyngotympanic tube	Blends with palatopharyngeus		
Stylopharyngeus	Styloid process of temporal bone	Posterior and superior borders of thyroid cartilage with palatopharyngeus	Glossopharyngeal nerve (CN IX)	

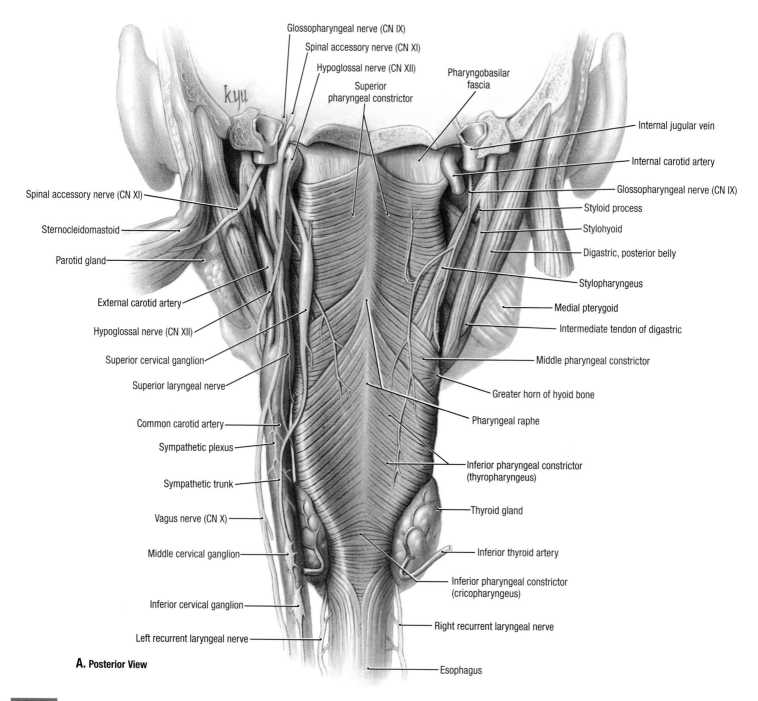

Glossopharyngeal nerve (CN IX)
Spinal accessory nerve (CN XI)
Hypoglossal nerve (CN XII)
Superior pharyngeal constrictor
Pharyngobasilar fascia
Internal jugular vein
Internal carotid artery
Glossopharyngeal nerve (CN IX)
Styloid process
Stylohyoid
Digastric, posterior belly
Stylopharyngeus
Medial pterygoid
Intermediate tendon of digastric
Middle pharyngeal constrictor
Greater horn of hyoid bone
Pharyngeal raphe
Inferior pharyngeal constrictor (thyropharyngeus)
Thyroid gland
Inferior thyroid artery
Inferior pharyngeal constrictor (cricopharyngeus)
Right recurrent laryngeal nerve
Esophagus

Spinal accessory nerve (CN XI)
Sternocleidomastoid
Parotid gland
External carotid artery
Hypoglossal nerve (CN XII)
Superior cervical ganglion
Superior laryngeal nerve
Common carotid artery
Sympathetic plexus
Sympathetic trunk
Vagus nerve (CN X)
Middle cervical ganglion
Inferior cervical ganglion
Left recurrent laryngeal nerve

A. Posterior View

8.28 External pharynx—I

A. Dissection. A large wedge of occipital bone (including the foramen magnum) and the articulated cervical vertebrae have been separated from the remainder (anterior portion) of the head and cervical viscera at the retropharyngeal space and removed.

- The pharynx is a unique portion of the alimentary tract, having a circular layer of muscle externally and a longitudinal layer internally.
- The circular layer of the pharynx consists of the three pharyngeal constrictor muscles (superior, middle and inferior), which overlap one another; their posterior aspect is flat and even slightly concave.
- On the right side of the specimen, the stylopharyngeus muscle

and glossopharyngeal nerve (IX) pass from the medial side of the styloid process anteromedially through the interval between the superior and middle pharyngeal constrictor muscles to become part of the internal longitudinal layer. The stylohyoid muscle passes from the lateral side of the styloid process anterolaterally and splits on its way to the hyoid bone to accommodate passage of the intermediate tendon of the digastric.

- Pharyngeal branches of the glossopharyngeal nerve (CN IX, shown on the right) and the vagus nerve (CN X, shown on the left) form the pharyngeal plexus, which provides most of the pharyngeal innervation. The glossopharyngeal nerve supplies the sensory component, while the vagus supplies motor innervation.

Cranial nerves

IX   XII   X IX XI

Jugular bulb

Styloid process

Pharyngobasilar fascia

Facial nerve (CN VII)

Stylohyoid

Parotid gland

Posterior belly of digastric

Posterior belly of digastric (cut)

Glossopharyngeal nerve (CN IX)

Stylopharyngeus

Superior cervical ganglion

Stylopharyngeus

Pharyngeal ⎤ Branches
Superior laryngeal ⎦ of CN X

Superior pharyngeal constrictor

Hypoglossal nerve (CN XII)

Ascending pharyngeal artery

Spinal accessory nerve (CN XI)

Submandibular gland

Sternocleidomastoid

Internal jugular vein

Carotid   ⎡ External
arteries ⎢ Internal
          ⎣ Common

Vagus nerve (CN X)

Common carotid artery

Sympathetic trunk

Inferior pharyngeal constrictor

Thyroid gland

Sheath of thyroid gland

Parathyroid gland

Inferior thyroid artery

Parathyroid ⎡ Superior
glands   ⎣ Inferior

Right recurrent laryngeal nerve

Left recurrent laryngeal nerve

Esophagus

Paratracheal lymph nodes

B. Posterior View

M. Sewell.

8.28   External pharynx—II

B. Illustration of a dissection similar to **A.** The sympathetic trunk (including the superior cervical ganglion), which normally lies posterior to the internal carotid artery, has been retracted medially.

- The pharyngobasilar fascia, between the superior pharyngeal constrictor muscle and the base of the skull, attaches the pharynx to the occipital bone and forms the wall of the non-collapsible pharyngeal recesses.
- As they exit the jugular foramen, CN IX lies anterior to CN X, and CN XI; CN XII, exiting the hypoglossal canal, lies medially.

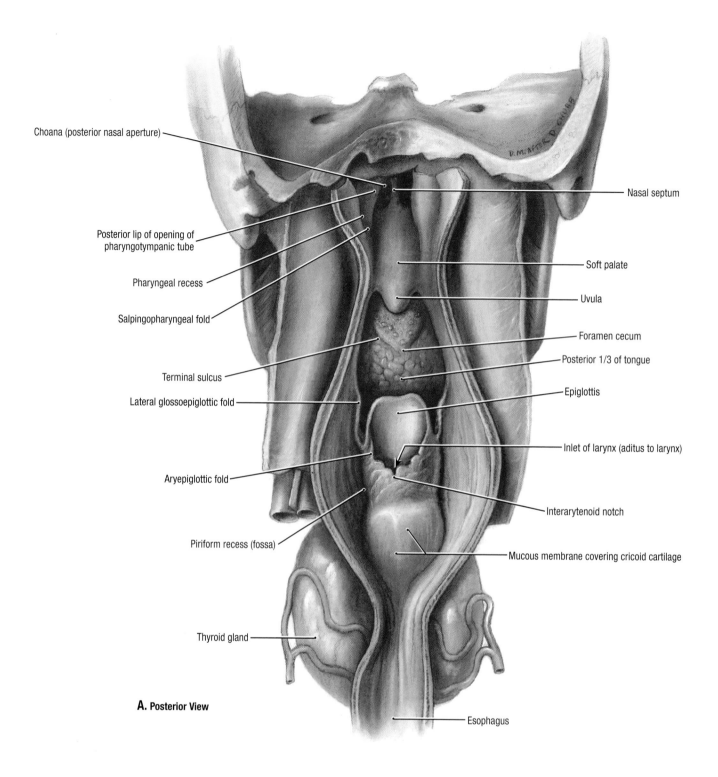

Choana (posterior nasal aperture)

Posterior lip of opening of
pharyngotympanic tube

Pharyngeal recess

Salpingopharyngeal fold

Terminal sulcus

Lateral glossoepiglottic fold

Aryepiglottic fold

Piriform recess (fossa)

Thyroid gland

Nasal septum

Soft palate

Uvula

Foramen cecum

Posterior 1/3 of tongue

Epiglottis

Inlet of larynx (aditus to larynx)

Interarytenoid notch

Mucous membrane covering cricoid cartilage

Esophagus

A. Posterior View

8.29 **Internal pharynx—I**

A. Dissection. The posterior wall of the pharynx has been split in the midline and the halves
retracted laterally to reveal the internal aspect of the anterior wall of the pharynx, occupied
by the nasal, oral, and laryngeal orifices. The pharynx has three parts (nasal, oral, and laryngeal) and extends from the base of the cranium to the inferior border of the cricoid cartilage,
where it narrows as it becomes the esophagus.

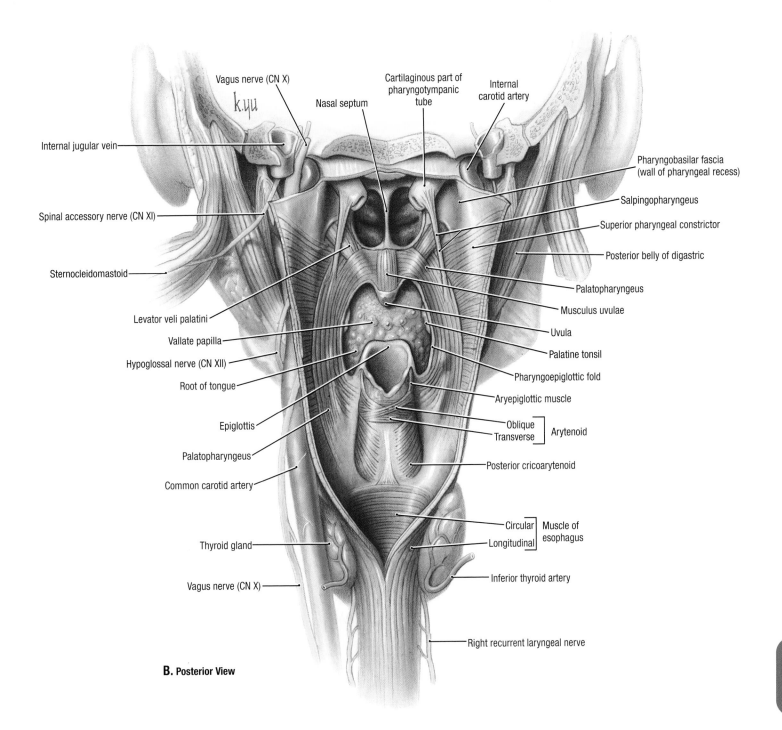

Vagus nerve (CN X)

Nasal septum

Cartilaginous part of pharyngotympanic tube

Internal carotid artery

Internal jugular vein

Pharyngobasilar fascia (wall of pharyngeal recess)

Spinal accessory nerve (CN XI)

Salpingopharyngeus

Superior pharyngeal constrictor

Posterior belly of digastric

Sternocleidomastoid

Palatopharyngeus

Musculus uvulae

Levator veli palatini

Uvula

Vallate papilla

Palatine tonsil

Hypoglossal nerve (CN XII)

Pharyngoepiglottic fold

Root of tongue

Aryepiglottic muscle

Epiglottis

Oblique
Transverse } Arytenoid

Palatopharyngeus

Posterior cricoarytenoid

Common carotid artery

Circular
Longitudinal } Muscle of esophagus

Thyroid gland

Inferior thyroid artery

Vagus nerve (CN X)

Right recurrent laryngeal nerve

B. Posterior View

8.29   Internal pharynx—II

B. Illustration. The posterior wall of the pharynx has been split in the midline and reflected laterally as in **A;** then, the mucous membrane was removed to expose the underlying musculature. The nasal part (nasopharynx) lies superior to the level of the soft palate and is continuous anteriorly, through the choanae, with the nasal cavities. The oral part (oropharynx) lies between the levels of the soft palate and larynx and communicates anteriorly with the oral cavity. The laryngeal part (laryngopharynx) lies posterior to the larynx and communicates with the cavity of the larynx through the aditus.

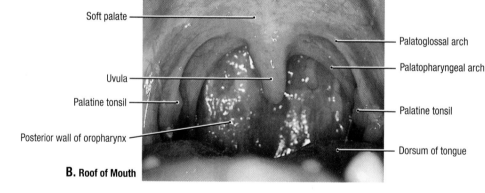

A. Roof of Mouth

Palatine glands
Greater palatine artery
Greater palatine nerve
Aponeurosis of tensor veli palatini
Lesser palatine artery
Lesser palatine nerve
Palatopharyngeus
Palatoglossus
Buccinator
Pterygomandibular raphe
Superior pharyngeal constrictor
Palatine tonsil
Musculus uvulae

Hard palate
Soft palate
Palatoglossal arch
Palatopharyngeal arch
Palatine tonsil
Uvula

Soft palate
Uvula
Palatine tonsil
Posterior wall of oropharynx
Palatoglossal arch
Palatopharyngeal arch
Palatine tonsil
Dorsum of tongue

B. Roof of Mouth

8.30   Isthmus of the fauces (oropharyngeal isthmus)—I

A. Oral cavity and isthmus demonstrating the sinus (fossa or bed) of the tonsils. **B.** Tonsillar sinuses (with palatine tonsils in situ) and oropharynx.

- The fauces (throat), the passage from the mouth to the pharynx, is bounded superiorly by the soft palate, inferiorly by the root (base) of the tongue, and laterally by the palatoglossal and palatopharyngeal arches.
- The palatine tonsils are located between the palatoglossal and palatopharyngeal arches, formed by mucosa overlying the similarly named muscles; the arches form the boundaries, and the superior pharyngeal constrictor the floor, of the tonsillar sinuses.
- The soft palate consists of an aponeurotic part (palatine aponeurosis), formed by the expanded tendon of the tensor veli palatini, and a muscular part, consisting of the tensor and levator (not shown) veli palatini, palatoglossus, palatopharyngeus, and musculus uvulae.
- The greater and lesser palatine nerves and vessels emerging from the greater and lesser palatine foramina, respectively, supply the hard and soft palates.

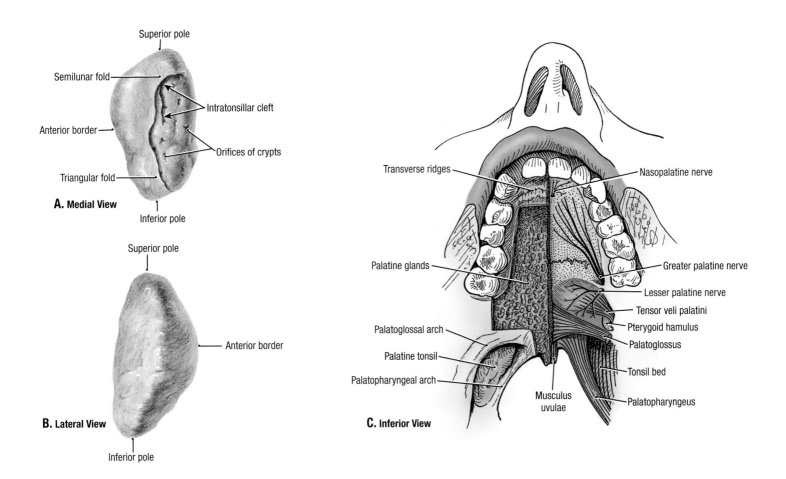

A. Medial View
- Superior pole
- Semilunar fold
- Intratonsillar cleft
- Anterior border
- Orifices of crypts
- Triangular fold
- Inferior pole

B. Lateral View
- Superior pole
- Anterior border
- Inferior pole

C. Inferior View
- Transverse ridges
- Nasopalatine nerve
- Palatine glands
- Greater palatine nerve
- Lesser palatine nerve
- Tensor veli palatini
- Pterygoid hamulus
- Palatoglossal arch
- Palatoglossus
- Palatine tonsil
- Tonsil bed
- Palatopharyngeal arch
- Musculus uvulae
- Palatopharyngeus

8.31   **Palatine tonsil and tonsillar sinus (fossa or bed)**

A and **B.** Isolated palatine tonsil.
- The fibrous capsule forms the lateral, or attached, surface of the tonsil; in removing the tonsil, the loose areolar tissue lying between the capsule and the thin pharyngobasilar fascia that forms the immediate bed of the tonsil, was easily separated.
- The capsule extends around the anterior border and slightly over the medial surface as a thin, free fold, covered with mucous membrane on both surfaces.
- On the medial, or free, surface, the orifices of the crypts extend right through the tonsil to the capsule; the intratonsillar cleft extends toward the superior pole.

C. Right side: Palatine tonsil in situ and glands of palatine mucosa. Left side: Palatine mucosa and tonsils removed demonstrating palatine nerves and muscles, including those underlying folds and wall of the tonsillar sinus.

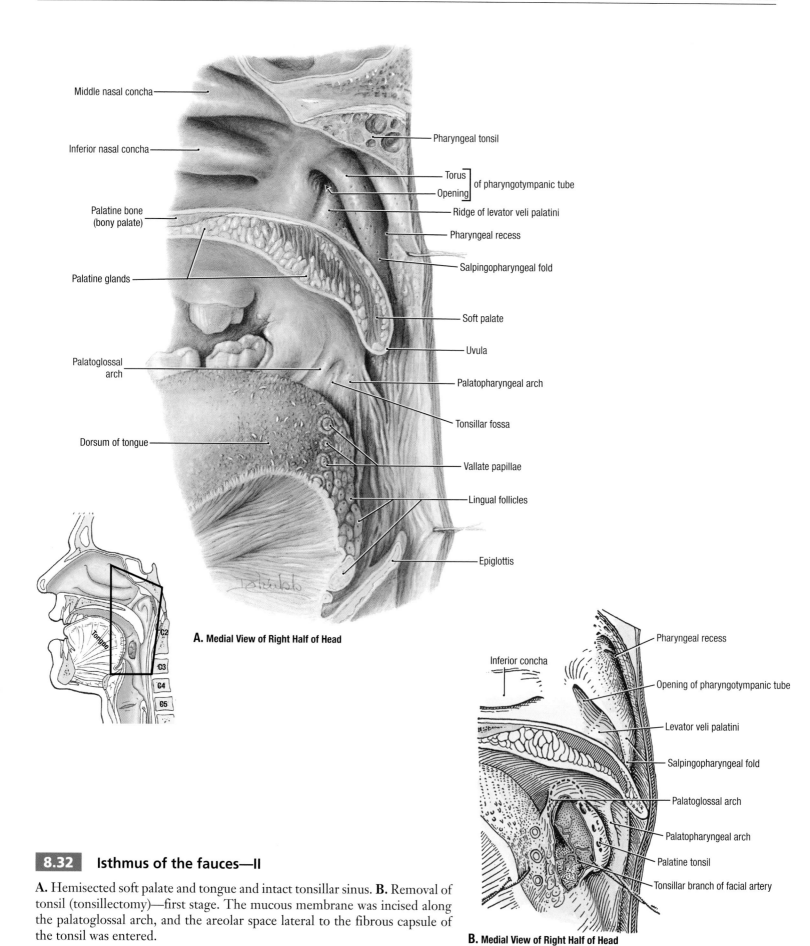

Middle nasal concha

Inferior nasal concha

Palatine bone
(bony palate)

Palatine glands

Palatoglossal
arch

Dorsum of tongue

Pharyngeal tonsil

Torus ⎤
        ⎬ of pharyngotympanic tube
Opening ⎦

Ridge of levator veli palatini

Pharyngeal recess

Salpingopharyngeal fold

Soft palate

Uvula

Palatopharyngeal arch

Tonsillar fossa

Vallate papillae

Lingual follicles

Epiglottis

A. Medial View of Right Half of Head

Inferior concha

Pharyngeal recess

Opening of pharyngotympanic tube

Levator veli palatini

Salpingopharyngeal fold

Palatoglossal arch

Palatopharyngeal arch

Palatine tonsil

Tonsillar branch of facial artery

B. Medial View of Right Half of Head

8.32 **Isthmus of the fauces—II**

A. Hemisected soft palate and tongue and intact tonsillar sinus. **B.** Removal of tonsil (tonsillectomy)—first stage. The mucous membrane was incised along the palatoglossal arch, and the areolar space lateral to the fibrous capsule of the tonsil was entered.

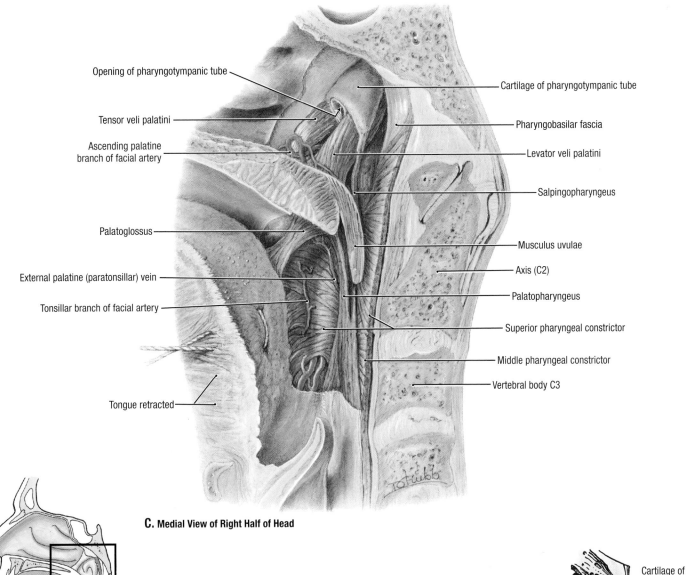

Opening of pharyngotympanic tube

Tensor veli palatini

Ascending palatine
branch of facial artery

Palatoglossus

External palatine (paratonsillar) vein

Tonsillar branch of facial artery

Tongue retracted

Cartilage of pharyngotympanic tube

Pharyngobasilar fascia

Levator veli palatini

Salpingopharyngeus

Musculus uvulae

Axis (C2)

Palatopharyngeus

Superior pharyngeal constrictor

Middle pharyngeal constrictor

Vertebral body C3

C. Medial View of Right Half of Head

8.32   Isthmus of the fauces—III

C. Muscles underlying tonsillar sinus and wall of nasopharynx. The palatine and pharyngeal tonsils and mucous membrane have been removed. The pharyngobasilar fascia, which attaches the pharynx to the basilar part of the occipital bone was also removed, except at the superior, arched border of the superior pharyngeal constrictor. **D.** Removal of tonsil—second stage. The anterior border of the tonsil was freed, and the superior part, which extends into the soft palate, was shelled out. The mucous membrane was cut along the palatopharyngeal arch.

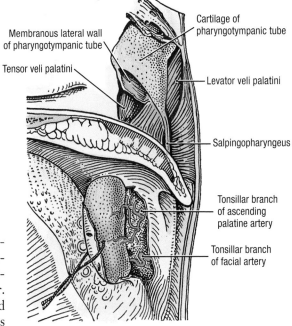

Membranous lateral wall
of pharyngotympanic tube

Tensor veli palatini

Cartilage of
pharyngotympanic tube

Levator veli palatini

Salpingopharyngeus

Tonsillar branch
of ascending
palatine artery

Tonsillar branch
of facial artery

D. Medial View of Right Half of Head

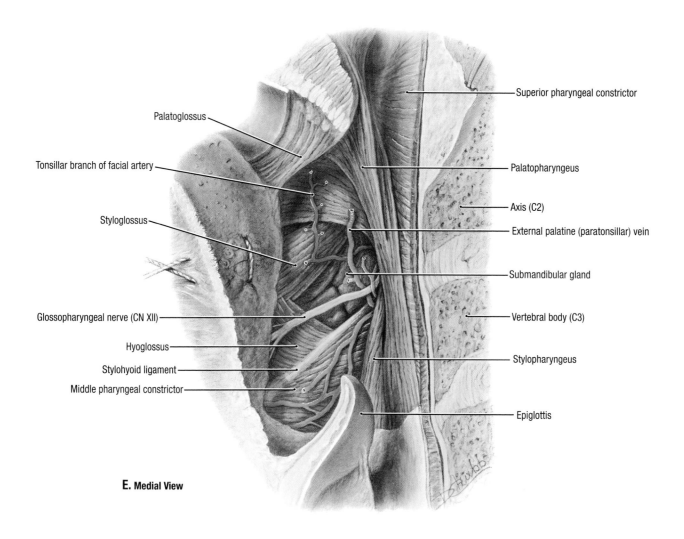

Palatoglossus

Tonsillar branch of facial artery

Styloglossus

Glossopharyngeal nerve (CN XII)

Hyoglossus

Stylohyoid ligament

Middle pharyngeal constrictor

Superior pharyngeal constrictor

Palatopharyngeus

Axis (C2)

External palatine (paratonsillar) vein

Submandibular gland

Vertebral body (C3)

Stylopharyngeus

Epiglottis

E. Medial View

8.32 Isthmus of the fauces—IV

E. Neurovascular structures of tonsillar sinus and longitudinal muscles of the pharynx.

- In this deeper dissection, the tongue was pulled anteriorly, and the inferior part of the origin of the superior pharyngeal constrictor muscle was cut away.
- The glossopharyngeal nerve passes to the posterior one third of the tongue and lies anterior to the stylopharyngeus muscle.
- The tonsillar branch of the facial artery sends a branch (cut short here) to accompany the glossopharyngeal nerve to the tongue; the submandibular gland is seen lateral to the artery and paratonsillar vein.
- The palatopharyngeus and stylopharyngeus muscles form the longitudinal muscle "coat" of the pharynx; the pharyngeal constrictor muscles form the circular coat; the stylopharyngeus muscle descends along the anterior border of the palatopharyngeus muscle.
- The styloglossus muscle passes to the anterior two thirds of the tongue, where it interdigitates with the hyoglossus muscle.

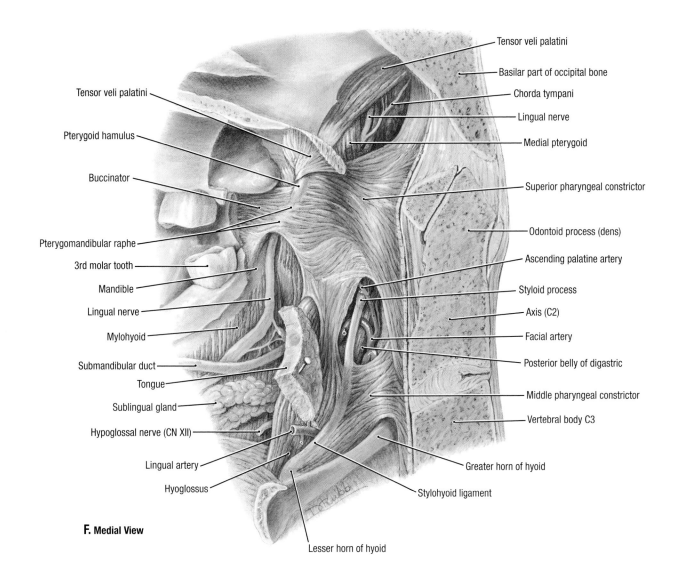

F. Medial View

8.32　Isthmus of the fauces—V

F. Circular muscles of the pharynx.

- The superior pharyngeal constrictor muscle arises from (a) the pterygomandibular raphe, which unites it to the buccinator muscle; (b) the bones at each end of the raphe (the hamulus of the medial pterygoid plate superiorly and the mandible inferiorly); and (c) the root (posterior part) of the tongue.
- The middle pharyngeal constrictor muscle arises from the angle formed by the greater and lesser horns of the hyoid bone and from the stylohyoid ligament; in this specimen, the styloid process is long and, therefore, a lateral relation of the tonsil.
- The facial artery arches superior to the posterior belly of the digastric muscle, and the lingual artery arches just inferior to it.
- The lingual nerve is joined by the chorda tympani, disappears at the posterior border of the medial pterygoid muscle, and reappears at the anterior border to follow the mandible.
- The tendon of the tensor veli palatini muscle hooks around the pterygoid hamulus.

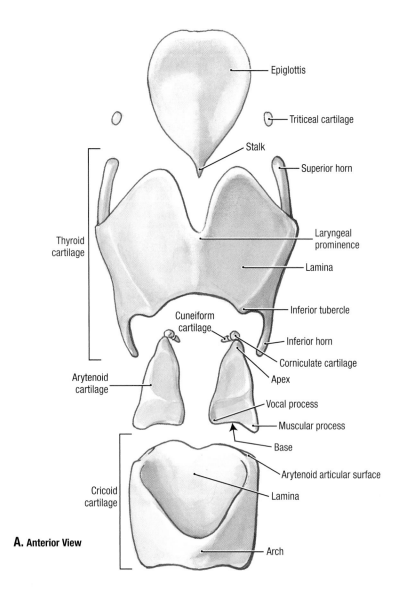

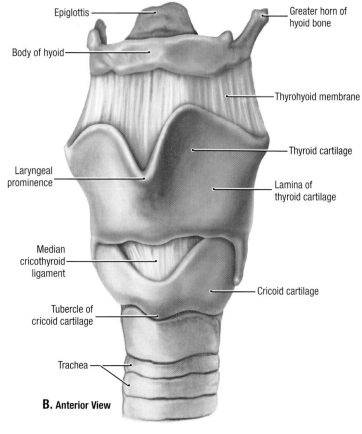

A. Anterior View

B. Anterior View

8.33 **Cartilages of the laryngeal skeleton—I**

A. Cartilages disarticulated and separated. **B.** Articulated laryngeal skeleton.
The larynx extends vertically from the tip of the epiglottis to the inferior border of the cricoid cartilage. The hyoid bone is generally not regarded as part of the larynx.

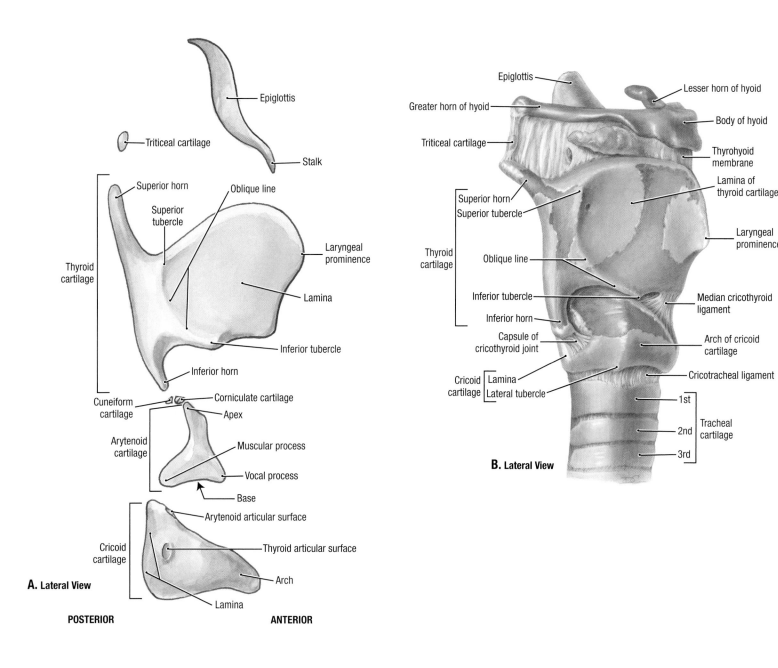

8.34 **Cartilages of the laryngeal skeleton—II**

A. Cartilages disarticulated and separated. **B.** Articulated laryngeal skeleton.
- The lamina of the thyroid cartilage projects anteriorly to form the laryngeal prominence.
- The cricoid cartilage has an arch anteriorly and a lamina posteriorly; the superior border of the arch is inclined, and the inferior border projects anteriorly beyond the trachea.
- The thyrohyoid membrane attaches the entire length of the superior border of the thyroid lamina to the inferior border of the body and greater horn of the hyoid bone; it is thickened posteriorly to form the thyrohyoid ligament and is pierced by the internal branch of the superior laryngeal nerve and vessels.

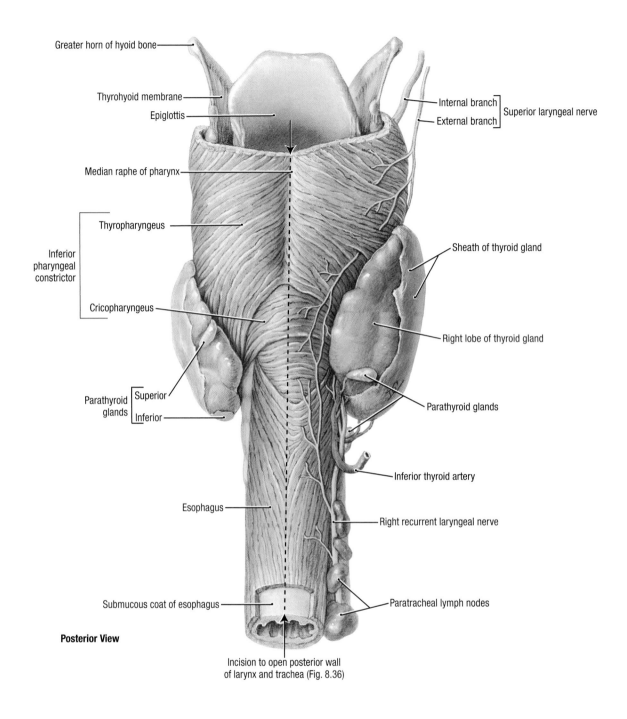

Greater horn of hyoid bone

Thyrohyoid membrane

Epiglottis

Median raphe of pharynx

Thyropharyngeus

Inferior pharyngeal constrictor

Cricopharyngeus

Parathyroid glands — Superior / Inferior

Esophagus

Submucous coat of esophagus

Internal branch — Superior laryngeal nerve
External branch

Sheath of thyroid gland

Right lobe of thyroid gland

Parathyroid glands

Inferior thyroid artery

Right recurrent laryngeal nerve

Paratracheal lymph nodes

Posterior View

Incision to open posterior wall of larynx and trachea (Fig. 8.36)

8.35 Posterior approach to the larynx—I

External larynx.
- The superior parathyroid gland lies in a crevice on the posterior border of the lobes of the thyroid gland. On the right side of the specimen, both parathyroid glands are low.
- The internal branch of the superior laryngeal nerve innervates the mucous membrane superior to the vocal folds, and the external laryngeal branch supplies the inferior pharyngeal constrictor and cricothyroid muscles.
- The recurrent laryngeal nerve supplies the esophagus, trachea, and inferior pharyngeal constrictor muscle and then continues into the larynx. At the larynx, it supplies sensory innervation to the area inferior to the vocal folds (cords) and motor innervation to all of the intrinsic muscles of the larynx, except the cricothyroid.

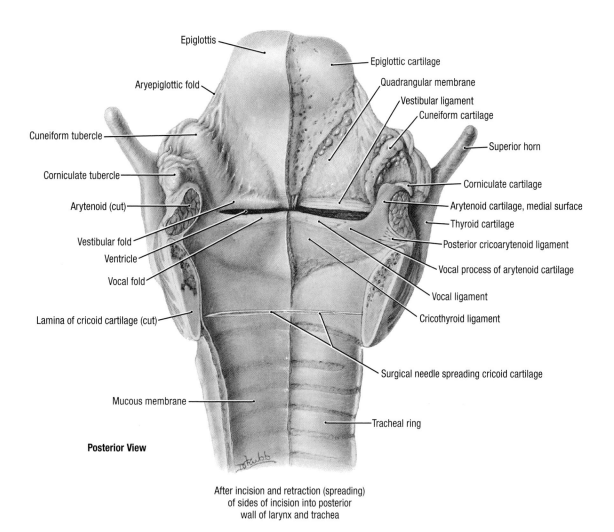

Epiglottis

Epiglottic cartilage

Aryepiglottic fold

Quadrangular membrane

Vestibular ligament

Cuneiform cartilage

Cuneiform tubercle

Superior horn

Corniculate tubercle

Corniculate cartilage

Arytenoid (cut)

Arytenoid cartilage, medial surface

Thyroid cartilage

Vestibular fold

Posterior cricoarytenoid ligament

Ventricle

Vocal process of arytenoid cartilage

Vocal fold

Vocal ligament

Lamina of cricoid cartilage (cut)

Cricothyroid ligament

Surgical needle spreading cricoid cartilage

Mucous membrane

Tracheal ring

Posterior View

After incision and retraction (spreading)
of sides of incision into posterior
wall of larynx and trachea

8.36   Posterior approach to the larynx—II

Interior of larynx.
- The posterior wall of the larynx was split in the median plane (see Figure 8.35), and the two sides held apart. On the left side of the specimen, the mucous membrane, which is the innermost coat of the larynx, is intact; on the right side of the specimen, the mucous and submucous coats were peeled off, and the next coat, consisting of cartilages, ligaments, and fibroelastic membrane, was uncovered.
- The three compartments of the larynx are (a) the superior compartment of the vestibule, superior to the level of the vestibular folds (false cords); (b) the middle, between the levels of the vestibular and vocal folds; and (c) the inferior, or infraglottic, cavity, inferior to the level of the vocal folds.
- The mucous membrane is smooth and adherent over the epiglottic cartilage and vocal ligaments and loose and wrinkled about the arytenoid cartilages.
- The quadrangular membrane is thickened inferiorly to form the vestibular ligament; the inferior part, the cricothyroid ligament (conus elasticus), begins inferiorly as the strong median cricothyroid ligament and ends superiorly as the vocal ligament. Between the vocal and vestibular ligaments, the membrane, lined with mucous membrane, is evaginated to form the wall of the ventricle.
- The cuneiform cartilage, composed of elastic cartilage and glands, is attached to the arytenoid cartilage near the posterior end of the vestibular ligament.

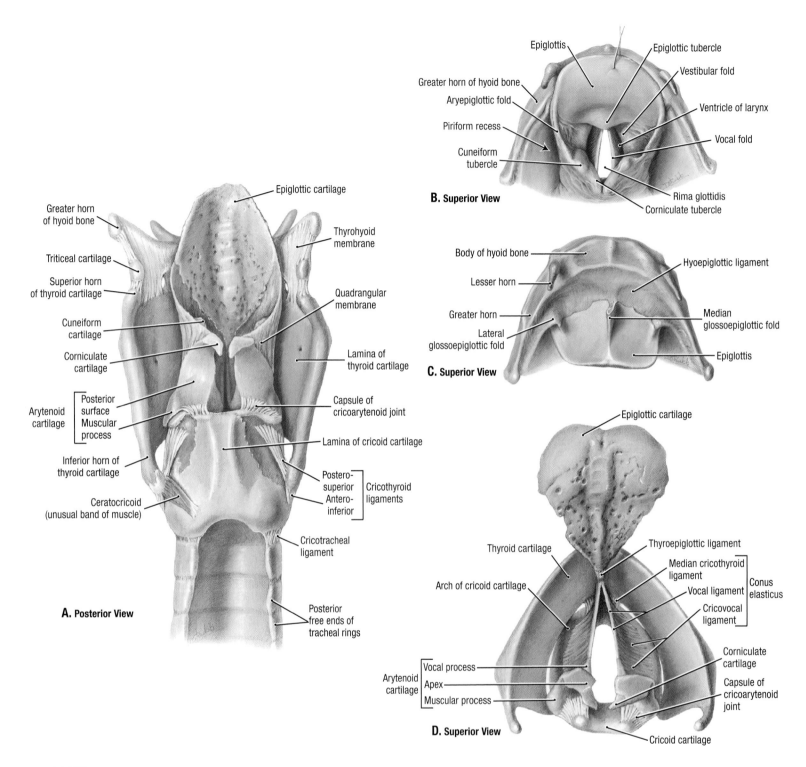

8.37 Joints, ligaments, and folds of the larynx

A. Posterior aspect of skeleton of larynx. **B.** Structures forming the vestibule of the larynx. **C.** Epiglottis and hyoepiglottic ligament. **D.** Conus elasticus and rima glottidis.

- The inlet, or aditus, to the larynx is bounded anteriorly by the epiglottis; posteriorly by the arytenoid cartilages, the corniculate cartilages that cap them, and the interarytenoid fold that unites them; and on each side by the aryepiglottic fold, which contains the superior end of the cuneiform cartilage.

- The rima glottidis is the aperture between the vocal folds. During normal respiration, it is narrow and wedge shaped; during forced respiration, it is wide. Variations in the tension and length of the vocal folds, in the width of the rima glottidis, and in the intensity of the expiratory effort produce changes in the pitch of the voice.

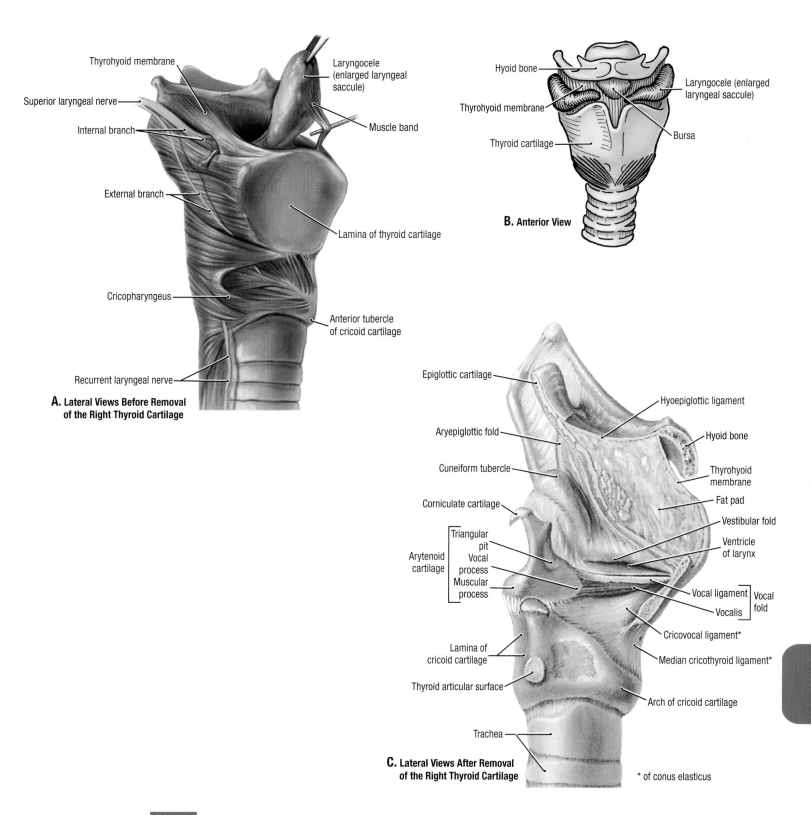

A. Lateral Views Before Removal of the Right Thyroid Cartilage

- Thyrohyoid membrane
- Superior laryngeal nerve
- Internal branch
- External branch
- Cricopharyngeus
- Recurrent laryngeal nerve
- Laryngocele (enlarged laryngeal saccule)
- Muscle band
- Lamina of thyroid cartilage
- Anterior tubercle of cricoid cartilage

B. Anterior View

- Hyoid bone
- Thyrohyoid membrane
- Thyroid cartilage
- Laryngocele (enlarged laryngeal saccule)
- Bursa

C. Lateral Views After Removal of the Right Thyroid Cartilage

- Epiglottic cartilage
- Aryepiglottic fold
- Cuneiform tubercle
- Corniculate cartilage
- Arytenoid cartilage
  - Triangular pit
  - Vocal process
  - Muscular process
- Lamina of cricoid cartilage
- Thyroid articular surface
- Trachea
- Hyoepiglottic ligament
- Hyoid bone
- Thyrohyoid membrane
- Fat pad
- Vestibular fold
- Ventricle of larynx
- Vocal ligament — Vocal fold
- Vocalis
- Cricovocal ligament*
- Median cricothyroid ligament*
- Arch of cricoid cartilage

* of conus elasticus

8.38   Ventricle and saccule of the larynx

A and **B.** Laryngocele. The laryngocele (enlarged laryngeal saccule) projects through the thyrohyoid membrane and communicates with the larynx through the ventricle. This air sac can form a bulge in the neck, especially on coughing. **C.** Interior of the larynx superior to the vocal folds. The larynx was sectioned near the median plane to reveal the interior of its left side. Inferior to this level, the right side of the intact larynx was dissected. The thyrohyoid membrane is intact; there is no laryngocele.

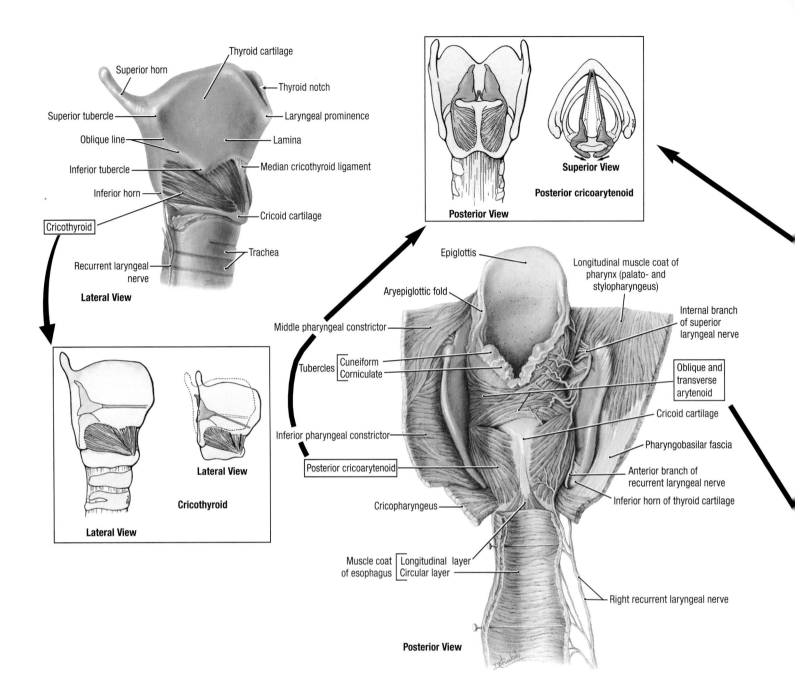

Lateral View

Cricothyroid

Lateral View

Cricothyroid

Lateral View

Posterior View

Superior View

Posterior cricoarytenoid

Posterior View

TABLE 8.8 MUSCLES OF LARYNX

MUSCLE	ORIGIN	INSERTION	INNERVATION	MAIN ACTION(S)
Cricothyroid	Anterolateral part of cricoid cartilage	Inferior margin and inferior horn of thyroid cartilage	External branch of laryngeal nerve	Tenses vocal fold
Posterior cricoarytenoid	Posterior surface of laminae of cricoid cartilage	Muscular process of arytenoid cartilage	Recurrent laryngeal nerve	Abducts vocal fold
Lateral cricoarytenoid	Arch of cricoid cartilage	Muscular process of arytenoid cartilage		Adducts vocal fold
Thyroarytenoid[a]	Posterior surface of thyroid cartilage	Muscular process of arytenoid process		Relaxes vocal fold
Transverse and oblique arytenoids[b]	One arytenoid cartilage	Opposite arytenoid cartilage		Close inlet of larynx by approximating arytenoid cartilages
Vocalis[c]	Angle between laminae of thyroid cartilage	Vocal ligament, between origin and vocal process of arytenoid cartilage		Alters vocal fold during phonation

[a] Superior fibers of the thyroarytenoid muscle pass into the aryepiglottic fold, and some of them reach the epiglottic cartilage. These fibers constitute the thyroepiglottic muscle, which widens inlet of larynx.

[b] Some fibers of the oblique arytenoid muscle continue as the aryepiglottic muscle.

[c] This slender muscular slip is derived from inferior deeper fibers of the thyroarytenoid muscle.

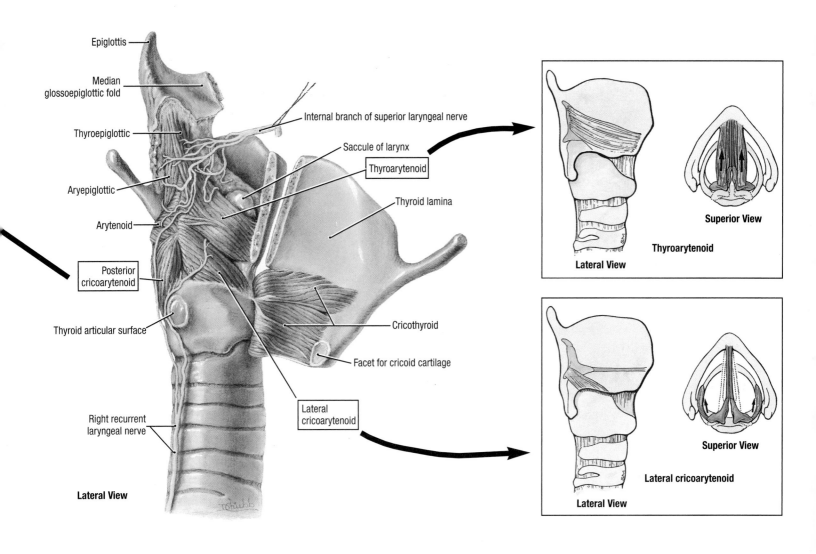

Epiglottis

Median glossoepiglottic fold

Thyroepiglottic

Aryepiglottic

Arytenoid

Posterior cricoarytenoid

Thyroid articular surface

Right recurrent laryngeal nerve

Lateral View

Internal branch of superior laryngeal nerve

Saccule of larynx

Thyroarytenoid

Thyroid lamina

Cricothyroid

Facet for cricoid cartilage

Lateral cricoarytenoid

Superior View

Thyroarytenoid

Lateral View

Superior View

Lateral cricoarytenoid

Lateral View

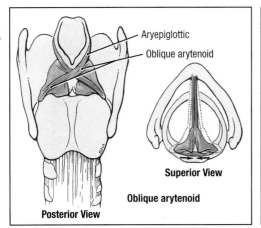

Aryepiglottic

Oblique arytenoid

Superior View

Oblique arytenoid

Posterior View

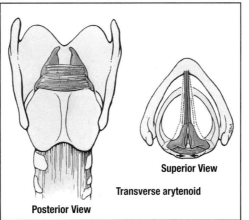

Superior View

Transverse arytenoid

Posterior View

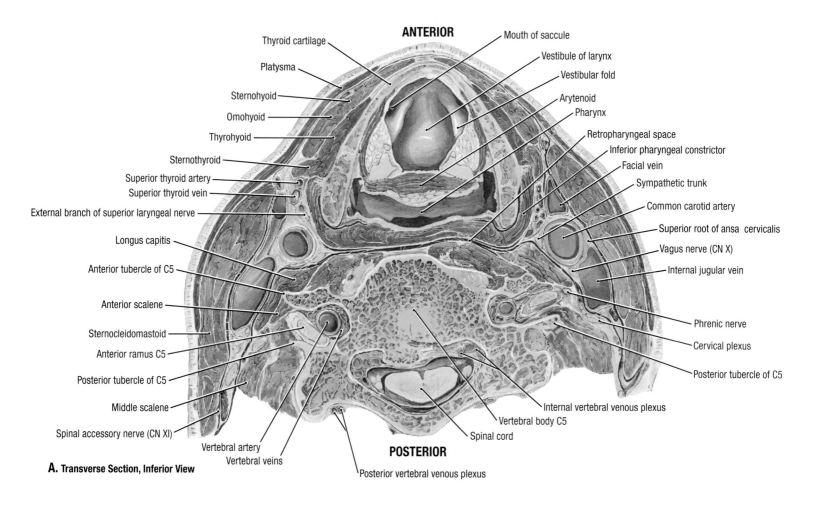

ANTERIOR

Thyroid cartilage
Platysma
Sternohyoid
Omohyoid
Thyrohyoid
Sternothyroid
Superior thyroid artery
Superior thyroid vein
External branch of superior laryngeal nerve
Longus capitis
Anterior tubercle of C5
Anterior scalene
Sternocleidomastoid
Anterior ramus C5
Posterior tubercle of C5
Middle scalene
Spinal accessory nerve (CN XI)

Mouth of saccule
Vestibule of larynx
Vestibular fold
Arytenoid
Pharynx
Retropharyngeal space
Inferior pharyngeal constrictor
Facial vein
Sympathetic trunk
Common carotid artery
Superior root of ansa cervicalis
Vagus nerve (CN X)
Internal jugular vein
Phrenic nerve
Cervical plexus
Posterior tubercle of C5

Internal vertebral venous plexus
Vertebral body C5
Spinal cord
Vertebral artery
Vertebral veins
Posterior vertebral venous plexus

POSTERIOR

A. Transverse Section, Inferior View

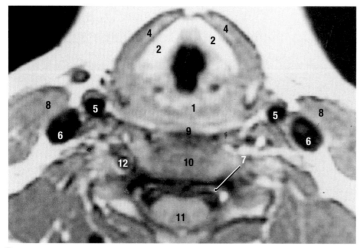

B. Transverse (axial) MRI

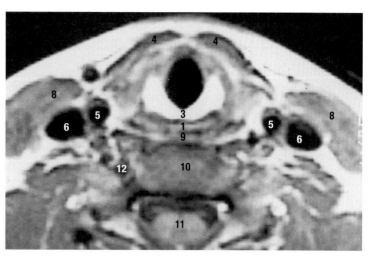

C. Transverse (axial) MRI

8.39   **Sectional anatomy of the neck—I**

A. Transverse anatomical section of neck through middle of larynx. **B.** Transverse (axial) MRI through thyroid cartilage. **C.** Transverse (axial) MRI through cricoid cartilage. For **B** and **C:** *1*, pharynx; *2*, thyroid cartilage; *3*, lamina of cricoid cartilage; *4*, infrahyoid (strap) muscles; *5*, common carotid artery; *6*, internal jugular vein; *7*, anterior root; *8*, sternocleidomastoid; *9*, inferior pharyngeal constrictor; *10*, vertebral body; *11*, spinal cord in cerebrospinal fluid; *12*, vertebral artery.

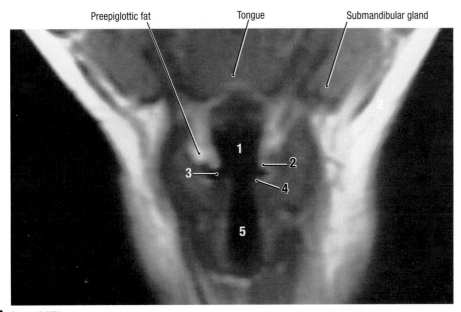

Preepiglottic fat Tongue Submandibular gland

A. Coronal MRI

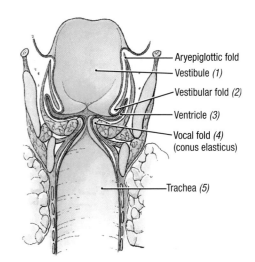

Aryepiglottic fold
Vestibule *(1)*
Vestibular fold *(2)*
Ventricle *(3)*
Vocal fold *(4)* (conus elasticus)
Trachea *(5)*

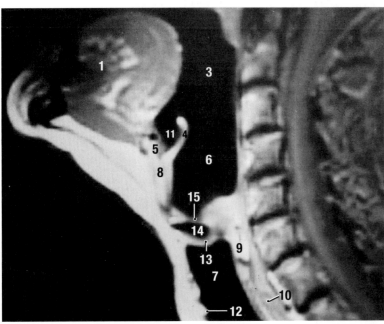

B. Sagittal MRI

Key for B

1	Genioglossus	9	Cricoid cartilage
2	Tongue	10	Esophagus
3	Oropharynx	11	Vallecula
4	Epiglottis	12	Tracheal ring
5	Hyoid bone	13	Vocal fold
6	Laryngopharynx	14	Ventricle
7	Trachea	15	Vestibular fold
8	Thyroid cartilage		

8.40 **Sectional anatomy of the neck—II**

A. MRI *(left)* and orientation figure *(right)* through larynx and trachea. Numbers in parentheses on diagram refer to numbered structures on MRI. **B.** MRI through the oropharynx, larynx, and trachea.

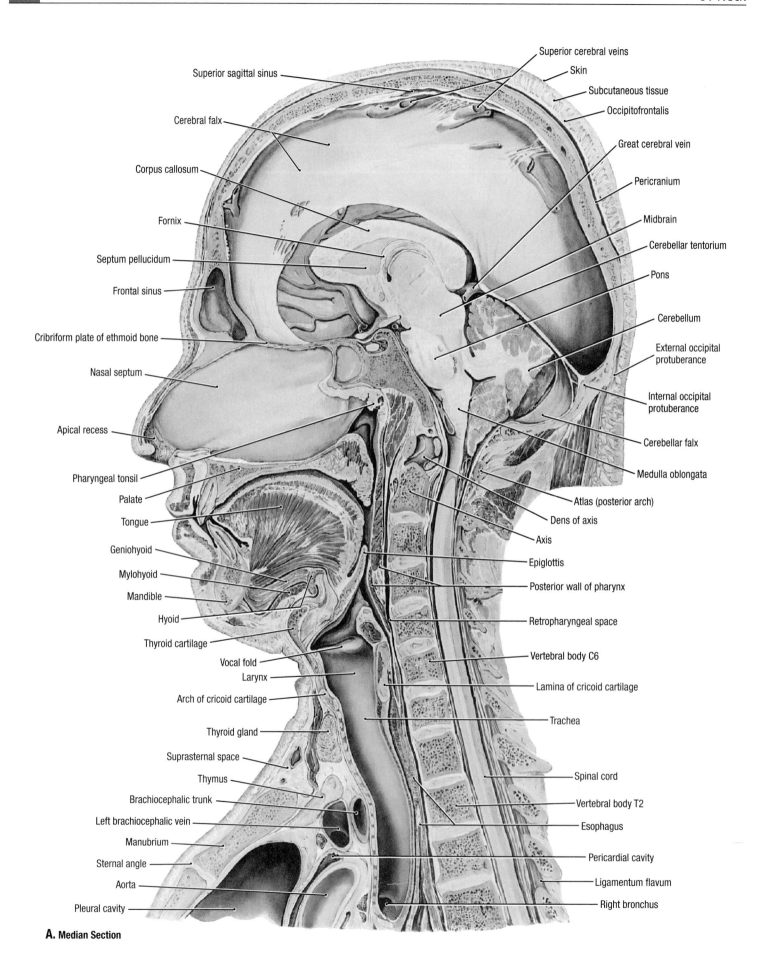

Superior cerebral veins

Superior sagittal sinus

Skin

Subcutaneous tissue

Cerebral falx

Occipitofrontalis

Great cerebral vein

Corpus callosum

Pericranium

Fornix

Midbrain

Septum pellucidum

Cerebellar tentorium

Frontal sinus

Pons

Cribriform plate of ethmoid bone

Cerebellum

External occipital protuberance

Nasal septum

Internal occipital protuberance

Apical recess

Cerebellar falx

Pharyngeal tonsil

Medulla oblongata

Palate

Atlas (posterior arch)

Tongue

Dens of axis

Geniohyoid

Axis

Mylohyoid

Epiglottis

Mandible

Posterior wall of pharynx

Hyoid

Retropharyngeal space

Thyroid cartilage

Vertebral body C6

Vocal fold

Lamina of cricoid cartilage

Larynx

Arch of cricoid cartilage

Trachea

Thyroid gland

Suprasternal space

Spinal cord

Thymus

Vertebral body T2

Brachiocephalic trunk

Left brachiocephalic vein

Esophagus

Manubrium

Sternal angle

Pericardial cavity

Aorta

Ligamentum flavum

Pleural cavity

Right bronchus

A. Median Section

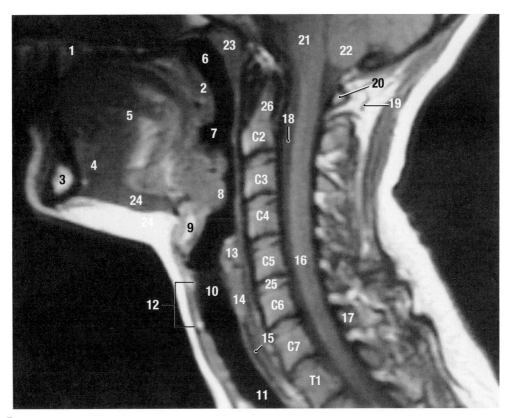

B. Sagittal MRI

1	Hard palate	14	Cricoid cartilage
2	Soft palate	15	Esophagus
3	Mandible	16	Spinal cord
4	Genioglossus	17	Spinous process
5	Tongue	18	Cerebrospinal fluid in
6	Nasopharynx		subarachnoid space
7	Oropharynx	19	Nuchal ligament
8	Epiglottis	20	Posterior arch of atlas
9	Hyoid bone	21	Medulla oblongata
10	Vestibule	22	Tonsil of cerebellum
11	Trachea	23	Pharyngeal tonsil (adenoid)
12	Thyroid cartilage	24	Geniohyoid
13	Arytenoid cartilage	25	Intervertebral disc
		26	Dens of axis

8.41   **Sectional anatomy of the neck—III**

A. Median anatomical section of head and neck. **B.** Corresponding median MRI of inferior head and superior neck.

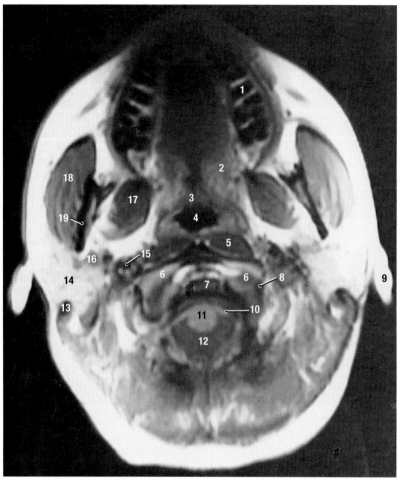

A. Transverse MRI

1 Tooth	23 Genioglossus
2 Palatine tonsil	24 Buccal fat
3 Uvula	25 Submandibular gland
4 Oropharynx	26 Intrinsic muscles of tongue
5 Longus capitis and colli	27 Vertebral body
6 Lateral mass of atlas	28 Lamina of vertebra
7 Dens (odontoid process)	29 Semispinalis cervicis
8 Vertebral artery	30 Semispinalis capitis
9 Auricle	31 Splenius capitis
10 Anterior rootlet	32 Trapezius
11 Spinal cord	33 Sternocleidomastoid
12 Cerebrospinal fluid in subarachnoid space	34 Internal jugular vein
13 Mastoid process	35 Bifurcation of common carotid artery
14 Parotid gland	36 Levator scapulae
15 Internal carotid artery	37 External jugular vein
16 Retromandibular vein	38 Common carotid artery
17 Medial pterygoid	39 Rima glottidis
18 Masseter	40 Vocal fold
19 Ramus of mandible	41 Thyrohyoid
20 Body of mandible	42 Thyroid cartilage
21 Mylohyoid	43 Sublingual gland
22 Hyoglossus	

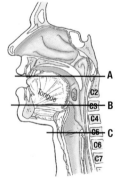

8.42 **Sectional anatomy of the neck—IV**

A–C. Transverse (axial) MRIs of neck. The line diagram indicates the location of the section of the MRIs.

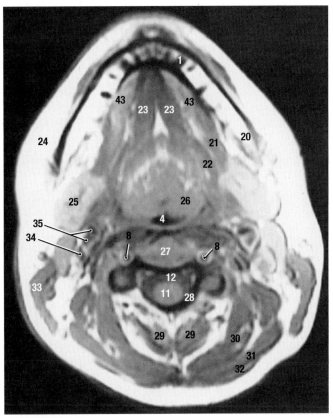

B. Transverse MRI

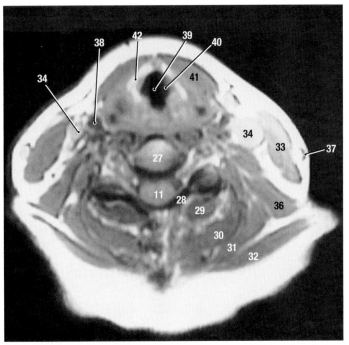

C. Transverse MRI

CHAPTER 9

CRANIAL NERVES

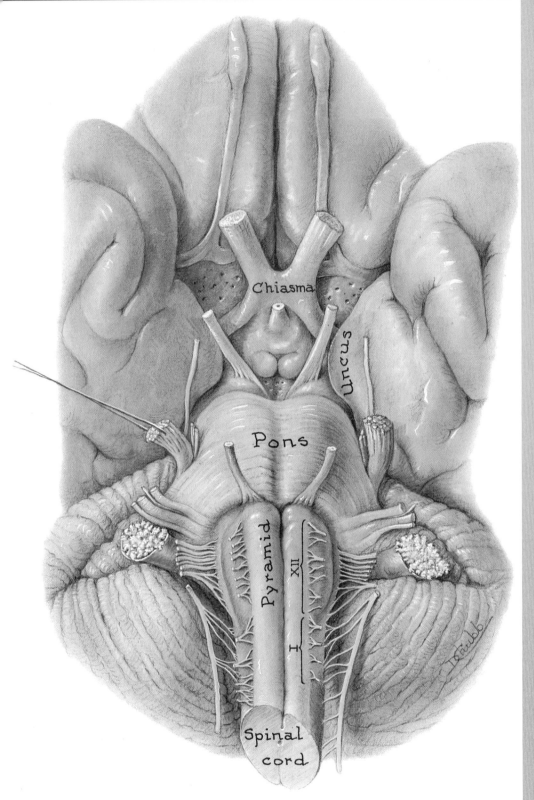

- Overview of Cranial Nerves **794**
- Cranial Nerve Nuclei **798**
- Cranial Nerve I: Olfactory **800**
- Cranial Nerve II: Optic **801**
- Cranial Nerves III, IV, and VI: Oculomotor, Trochlear, and Abducent **803**
- Cranial Nerve V: Trigeminal **806**
- Cranial Nerve VII: Facial **812**
- Cranial Nerve VIII: Vestibulocochlear **814**
- Cranial Nerve IX: Glossopharyngeal **816**
- Cranial Nerve X: Vagus **818**
- Cranial Nerve XI: Spinal Accessory **820**
- Cranial Nerve XII: Hypoglossal **821**
- Summary of Autonomic Ganglia of Head **822**
- Summary of Cranial Nerve Lesions **823**
- Sectional Imaging of Cranial Nerves: Transverse and Coronal Scans **824**

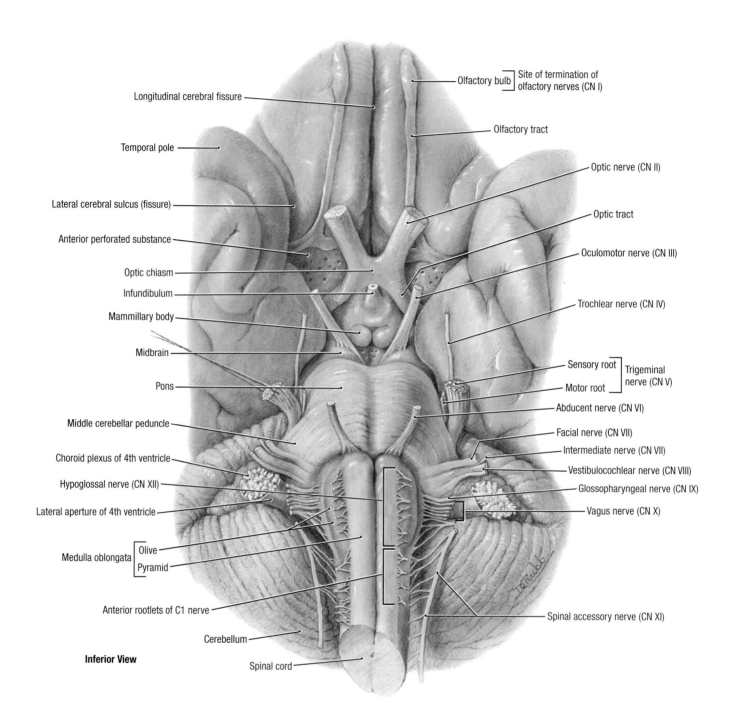

Longitudinal cerebral fissure

Temporal pole

Lateral cerebral sulcus (fissure)

Anterior perforated substance

Optic chiasm

Infundibulum

Mammillary body

Midbrain

Pons

Middle cerebellar peduncle

Choroid plexus of 4th ventricle

Hypoglossal nerve (CN XII)

Lateral aperture of 4th ventricle

Medulla oblongata — Olive / Pyramid

Anterior rootlets of C1 nerve

Cerebellum

Inferior View

Spinal cord

Olfactory bulb — Site of termination of olfactory nerves (CN I)

Olfactory tract

Optic nerve (CN II)

Optic tract

Oculomotor nerve (CN III)

Trochlear nerve (CN IV)

Sensory root / Motor root — Trigeminal nerve (CN V)

Abducent nerve (CN VI)

Facial nerve (CN VII)

Intermediate nerve (CN VII)

Vestibulocochlear nerve (CN VIII)

Glossopharyngeal nerve (CN IX)

Vagus nerve (CN X)

Spinal accessory nerve (CN XI)

9.1 Base of brain: superficial origins of cranial nerves

Cranial nerves are nerves that exit from the cranial cavity through openings in the cranium. There are 12 pairs of cranial nerves that are named and numbered in rostrocaudal sequence to their attachment to the brain and superior spinal cord. The olfactory nerves (CN I, not shown) end in the olfactory bulb. The entire origin of the spinal accessory nerve (CN XI) from the spinal cord is not included here; it extends inferiorly as far as the C6 spinal cord segment.

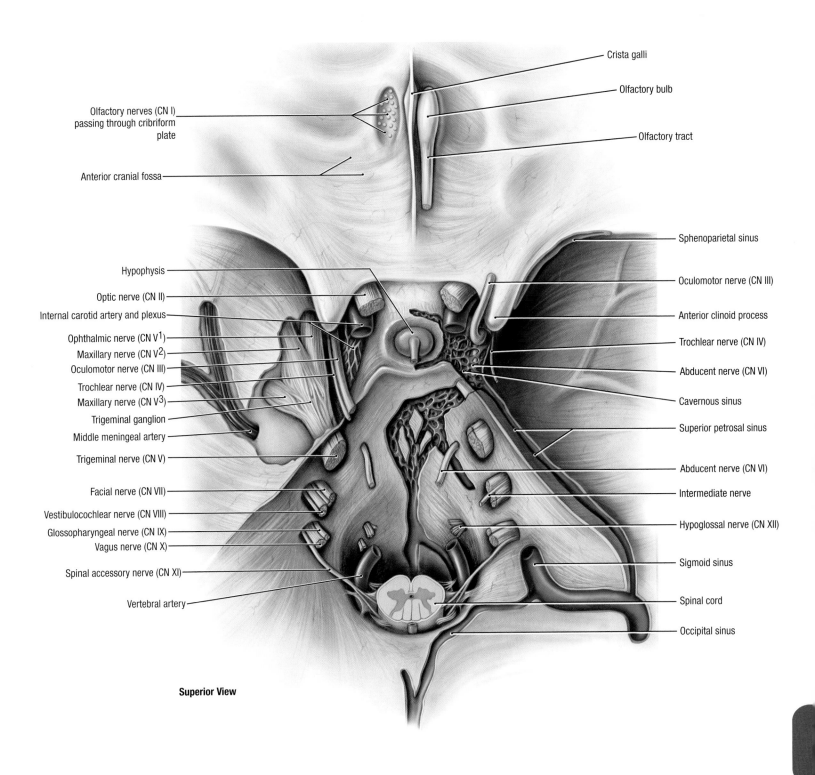

Olfactory nerves (CN I) passing through cribriform plate

Anterior cranial fossa

Hypophysis

Optic nerve (CN II)

Internal carotid artery and plexus

Ophthalmic nerve (CN V^1)

Maxillary nerve (CN V^2)

Oculomotor nerve (CN III)

Trochlear nerve (CN IV)

Maxillary nerve (CN V^3)

Trigeminal ganglion

Middle meningeal artery

Trigeminal nerve (CN V)

Facial nerve (CN VII)

Vestibulocochlear nerve (CN VIII)

Glossopharyngeal nerve (CN IX)

Vagus nerve (CN X)

Spinal accessory nerve (CN XI)

Vertebral artery

Crista galli

Olfactory bulb

Olfactory tract

Sphenoparietal sinus

Oculomotor nerve (CN III)

Anterior clinoid process

Trochlear nerve (CN IV)

Abducent nerve (CN VI)

Cavernous sinus

Superior petrosal sinus

Abducent nerve (CN VI)

Intermediate nerve

Hypoglossal nerve (CN XII)

Sigmoid sinus

Spinal cord

Occipital sinus

Superior View

9.2 **Cranial nerves and vessels of the interior of base of skull**

The venous sinuses have been opened on the right side. The ophthalmic division of the trigeminal nerve (CN V^1), and the trochlear (CN IV) and oculomotor (CN III) nerves have been dissected from the lateral wall of the cavernous sinus.

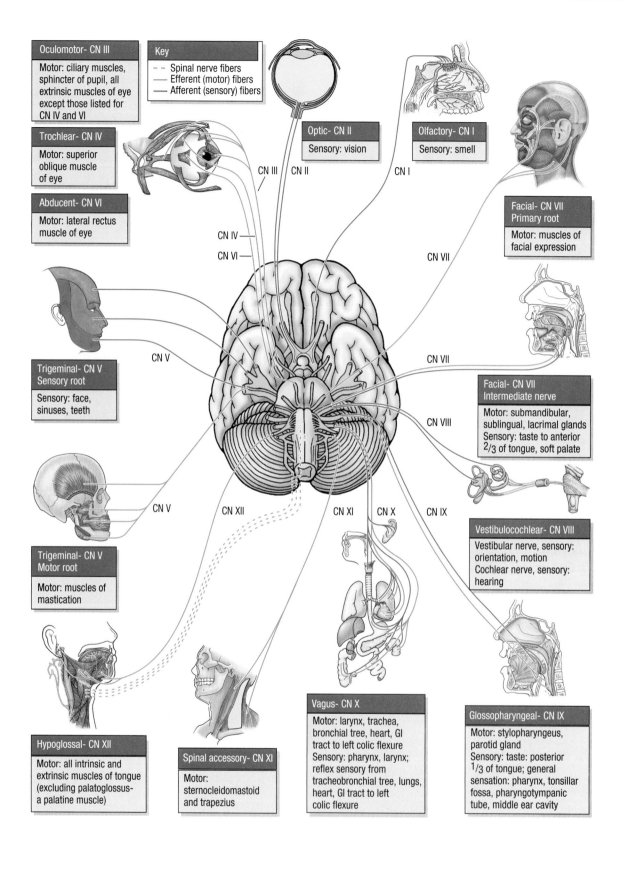

Oculomotor- CN III

Motor: ciliary muscles, sphincter of pupil, all extrinsic muscles of eye except those listed for CN IV and VI

Trochlear- CN IV

Motor: superior oblique muscle of eye

Abducent- CN VI

Motor: lateral rectus muscle of eye

Trigeminal- CN V
Sensory root

Sensory: face, sinuses, teeth

Trigeminal- CN V
Motor root

Motor: muscles of mastication

Hypoglossal- CN XII

Motor: all intrinsic and extrinsic muscles of tongue (excluding palatoglossus- a palatine muscle)

Spinal accessory- CN XI

Motor: sternocleidomastoid and trapezius

Key
-- Spinal nerve fibers
— Efferent (motor) fibers
— Afferent (sensory) fibers

Optic- CN II

Sensory: vision

Olfactory- CN I

Sensory: smell

Facial- CN VII
Primary root

Motor: muscles of facial expression

Facial- CN VII
Intermediate nerve

Motor: submandibular, sublingual, lacrimal glands
Sensory: taste to anterior 2/3 of tongue, soft palate

Vestibulocochlear- CN VIII

Vestibular nerve, sensory: orientation, motion
Cochlear nerve, sensory: hearing

Vagus- CN X

Motor: larynx, trachea, bronchial tree, heart, GI tract to left colic flexure
Sensory: pharynx, larynx; reflex sensory from tracheobronchial tree, lungs, heart, GI tract to left colic flexure

Glossopharyngeal- CN IX

Motor: stylopharyngeus, parotid gland
Sensory: taste: posterior 1/3 of tongue; general sensation: pharynx, tonsillar fossa, pharyngotympanic tube, middle ear cavity

CN III CN II CN I

CN IV

CN VI

CN V

CN V

CN XII CN XI CN X CN IX

CN VII

CN VII

CN VIII

TABLE 9.1   SUMMARY OF CRANIAL NERVES

NERVE	COMPONENTS	LOCATION OF NERVE CELL BODIES	CRANIAL EXIT	MAIN ACTION
Olfactory (CN I)	Special sensory	Olfactory epithelium (olfactory cells)	Foramina in cribriform plate of ethmoid bone	Smell from nasal mucosa of roof of each nasal cavity and superior sides of nasal septum and superior concha
Optic (CN II)	Special sensory	Retina (ganglion cells)	Optic canal	Vision from retina
Oculomotor (CN III)	Somatic motor	Midbrain	Superior orbital fissure	Motor to superior, inferior, and medial rectus, inferior oblique, and levator palpebrae superioris muscles; raises upper eyelid; turns eyeball superiorly, inferiorly, and medially
	Visceral motor	Presynaptic: midbrain; postsynaptic: ciliary ganglion		Parasympathetic innervation to sphincter pupillae and ciliary muscle; constricts pupil and accommodates lens of eye
Trochlear (CN IV)	Somatic motor	Midbrain		Motor to superior oblique that assists in turning eye inferolaterally
Trigeminal (CN V) Ophthalmic division (CN V¹)	General sensory	Trigeminal ganglion	Superior orbital fissure	Sensation from cornea, skin of forehead, scalp, eyelids, nose, and mucosa of nasal cavity and paranasal sinuses
Maxillary division (CN V²)	General sensory	Trigeminal ganglion	Foramen rotundum	Sensation from skin of face over maxilla including upper lip, maxillary teeth, mucosa of nose, maxillary sinuses, and palate
Mandibular division (CN V³)	Branchial motor	Pons	Foramen ovale	Motor to muscles of mastication, mylohyoid, anterior belly of digastric, tensor veli palatini, and tensor tympani
	General sensory	Trigeminal ganglion		Sensation from the skin over mandible, including lower lip and side of head, mandibular teeth, temporomandibular joint, and mucosa of mouth and anterior two thirds of the tongue
Abducent (CN VI)	Somatic motor	Pons	Superior orbital fissure	Motor to lateral rectus that turns eye laterally
Facial (CN VII)	Branchial motor	Pons	Internal acoustic meatus, facial canal, and stylomastoid foramen	Motor to muscle of facial expression and scalp; also supplies stapedius of middle ear, stylohyoid, and posterior belly of digastric
	Special sensory	Geniculate ganglion		Taste from anterior two thirds of tongue, floor of mouth, and palate
	General sensory			Sensation from skin of external acoustic meatus
	Visceral motor	Presynaptic: pons; postsynaptic: pterygopalatine ganglion and submandibular ganglion		Parasympathetic innervation to submandibular and sublingual salivary glands, lacrimal gland, and glands of nose and palate
Vestibulocochlear (CN VIII) Vestibular	Special sensory	Vestibular ganglion	Internal acoustic meatus	Vestibular sensation from semicircular ducts, utricle, and saccule related to position and movement of head
Cochlear	Special sensory	Spiral ganglion		Hearing from spiral organ
Glossopharyngeal (CN IX)	Branchial motor	Medulla	Jugular foramen	Motor to stylopharyngeus that assists with swallowing
	Visceral motor	Presynaptic: medulla; postsynaptic: otic ganglion		Parasympathetic innervation to parotid gland
	Visceral sensory	Inferior ganglion		Visceral sensation from parotid gland, carotid body and sinus, pharynx, and middle ear
	Special sensory	Superior ganglion		Taste from posterior third of tongue
	General sensory	Inferior ganglion		Cutaneous sensation from external ear
Vagus (CN X)	Branchial motor	Medulla		Motor to constrictor muscles of pharynx, intrinsic muscles of larynx, and muscles of palate except tensor veli palatini, and striated muscle in superior two thirds of esophagus
	Visceral motor	Presynaptic: medulla; postsynaptic: neurons in, on, or near viscera		Parasympathetic innervation to smooth muscle of trachea, bronchi, digestive tract, and cardiac muscle of heart
	Special sensory	Inferior ganglion		Visceral sensation from base of tongue, pharynx, larynx, trachea, bronchi, heart, esophagus, stomach, and intestine
	General sensory	Superior ganglion		Sensation from auricle, external acoustic meatus, and dura mater of posterior cranial fossa
	Somatic motor	Medulla		Motor to striated muscles of soft palate, pharynx via fibers that join CN X; larynx
Spinal accessory nerve (CN XI)	Somatic motor	Cervical spinal cord		Motor to sternocleidomastoid and trapezius
Hypoglossal (CN XII)	Somatic motor	Medulla	Hypoglossal canal	Motor to muscles of tongue (except palatoglossus)

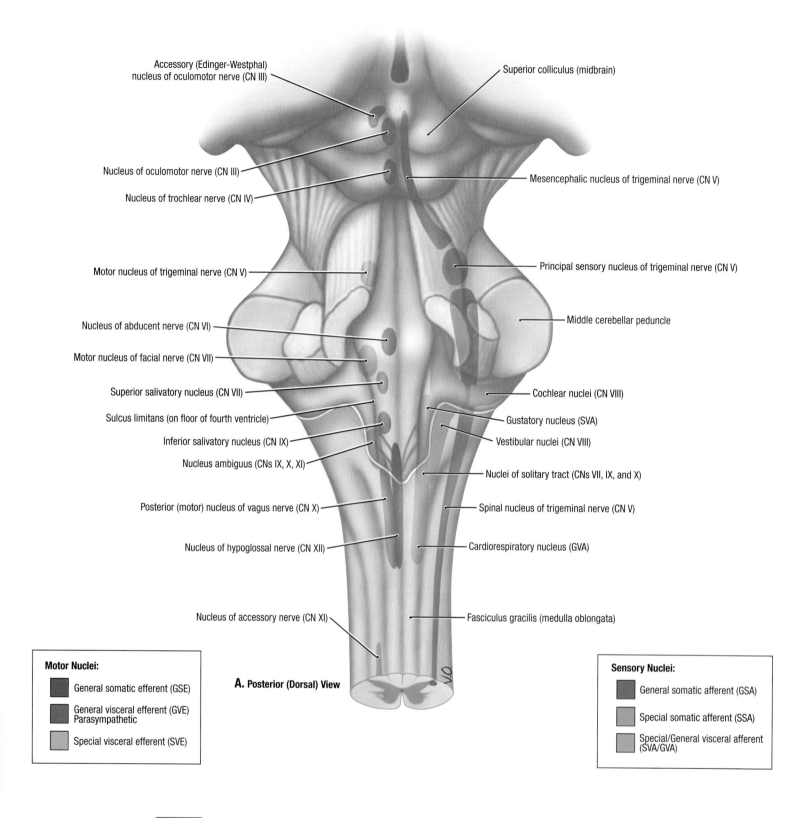

Accessory (Edinger-Westphal) nucleus of oculomotor nerve (CN III)

Superior colliculus (midbrain)

Nucleus of oculomotor nerve (CN III)

Mesencephalic nucleus of trigeminal nerve (CN V)

Nucleus of trochlear nerve (CN IV)

Motor nucleus of trigeminal nerve (CN V)

Principal sensory nucleus of trigeminal nerve (CN V)

Middle cerebellar peduncle

Nucleus of abducent nerve (CN VI)

Motor nucleus of facial nerve (CN VII)

Superior salivatory nucleus (CN VII)

Cochlear nuclei (CN VIII)

Sulcus limitans (on floor of fourth ventricle)

Gustatory nucleus (SVA)

Inferior salivatory nucleus (CN IX)

Vestibular nuclei (CN VIII)

Nucleus ambiguus (CNs IX, X, XI)

Nuclei of solitary tract (CNs VII, IX, and X)

Posterior (motor) nucleus of vagus nerve (CN X)

Spinal nucleus of trigeminal nerve (CN V)

Nucleus of hypoglossal nerve (CN XII)

Cardiorespiratory nucleus (GVA)

Nucleus of accessory nerve (CN XI)

Fasciculus gracilis (medulla oblongata)

A. Posterior (Dorsal) View

Motor Nuclei:

General somatic efferent (GSE)

General visceral efferent (GVE) Parasympathetic

Special visceral efferent (SVE)

Sensory Nuclei:

General somatic afferent (GSA)

Special somatic afferent (SSA)

Special/General visceral afferent (SVA/GVA)

9.3   Cranial nerve nuclei

The fibers of the cranial nerves are connected to nuclei (groups of nerve cell bodies in the central nervous system), in which afferent (sensory) fibers terminate and from which efferent (motor) fibers originate. Nuclei of common functional types (motor, sensory, parasympathetic, and special sensory nuclei) have a generally columnar placement within the brain stem, with the sulcus limitans demarcating motor and sensory columns.

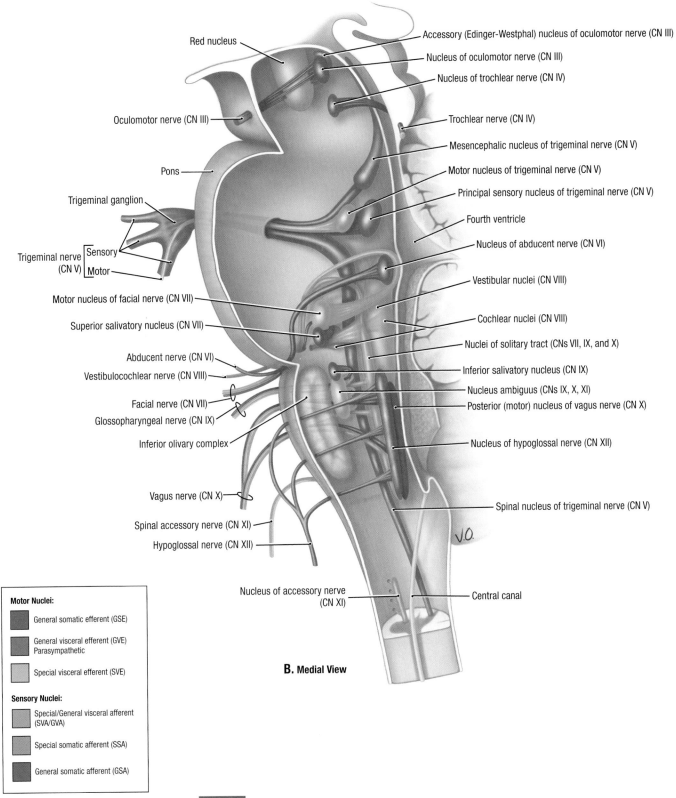

Red nucleus

Accessory (Edinger-Westphal) nucleus of oculomotor nerve (CN III)

Nucleus of oculomotor nerve (CN III)

Nucleus of trochlear nerve (CN IV)

Oculomotor nerve (CN III)

Trochlear nerve (CN IV)

Mesencephalic nucleus of trigeminal nerve (CN V)

Pons

Motor nucleus of trigeminal nerve (CN V)

Principal sensory nucleus of trigeminal nerve (CN V)

Trigeminal ganglion

Fourth ventricle

Trigeminal nerve (CN V) Sensory / Motor

Nucleus of abducent nerve (CN VI)

Vestibular nuclei (CN VIII)

Motor nucleus of facial nerve (CN VII)

Cochlear nuclei (CN VIII)

Superior salivatory nucleus (CN VII)

Nuclei of solitary tract (CNs VII, IX, and X)

Abducent nerve (CN VI)

Inferior salivatory nucleus (CN IX)

Vestibulocochlear nerve (CN VIII)

Nucleus ambiguus (CNs IX, X, XI)

Facial nerve (CN VII)

Posterior (motor) nucleus of vagus nerve (CN X)

Glossopharyngeal nerve (CN IX)

Inferior olivary complex

Nucleus of hypoglossal nerve (CN XII)

Vagus nerve (CN X)

Spinal nucleus of trigeminal nerve (CN V)

Spinal accessory nerve (CN XI)

Hypoglossal nerve (CN XII)

Nucleus of accessory nerve (CN XI)

Central canal

B. Medial View

Motor Nuclei:

General somatic efferent (GSE)

General visceral efferent (GVE) Parasympathetic

Special visceral efferent (SVE)

Sensory Nuclei:

Special/General visceral afferent (SVA/GVA)

Special somatic afferent (SSA)

General somatic afferent (GSA)

9.3   Cranial nerve nuclei *(continued)*

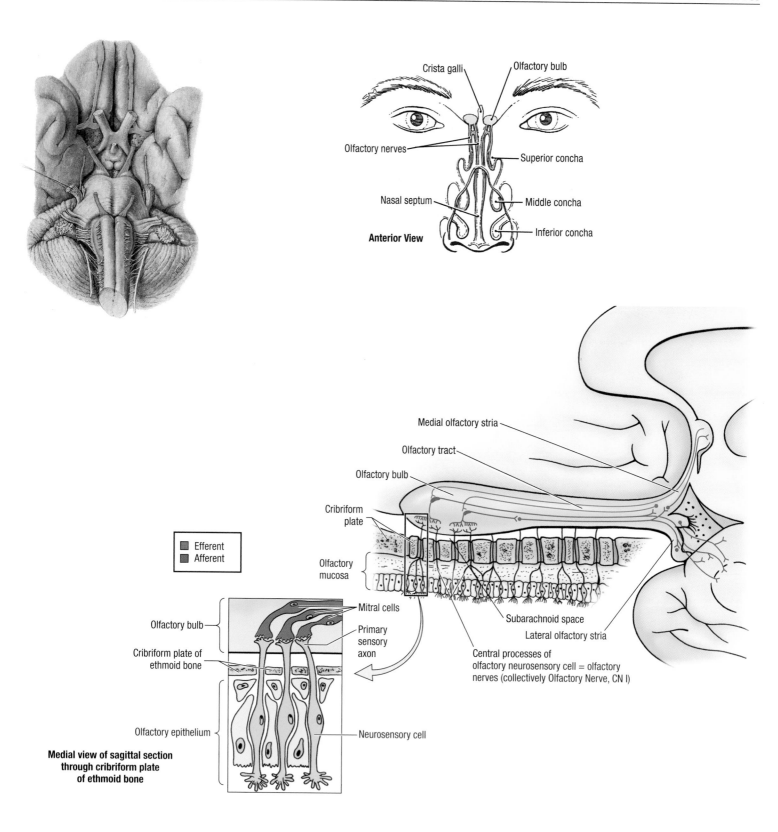

Crista galli

Olfactory bulb

Olfactory nerves

Superior concha

Nasal septum

Middle concha

Inferior concha

Anterior View

Medial olfactory stria

Olfactory tract

Olfactory bulb

Cribriform plate

Olfactory mucosa

Mitral cells

Primary sensory axon

Olfactory bulb

Cribriform plate of ethmoid bone

Olfactory epithelium

Medial view of sagittal section through cribriform plate of ethmoid bone

Neurosensory cell

■ Efferent
■ Afferent

Subarachnoid space

Lateral olfactory stria

Central processes of olfactory neurosensory cell = olfactory nerves (collectively Olfactory Nerve, CN I)

TABLE 9.2   OLFACTORY NERVE (CN I)

NERVE	FUNCTIONAL COMPONENTS	CELLS OF ORIGIN/TERMINATION	CRANIAL EXIT	DISTRIBUTION AND FUNCTIONS
Olfactory	Special sensory	Olfactory epithelium (olfactory cells/olfactory bulb)	Foramina in cribriform plate of ethmoid bone	Smell from nasal mucosa of roof and superior sides of nasal septum and superior concha of each nasal cavity

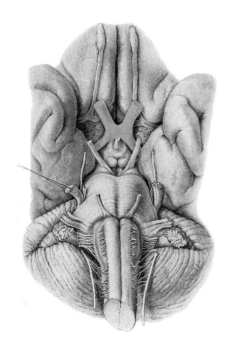

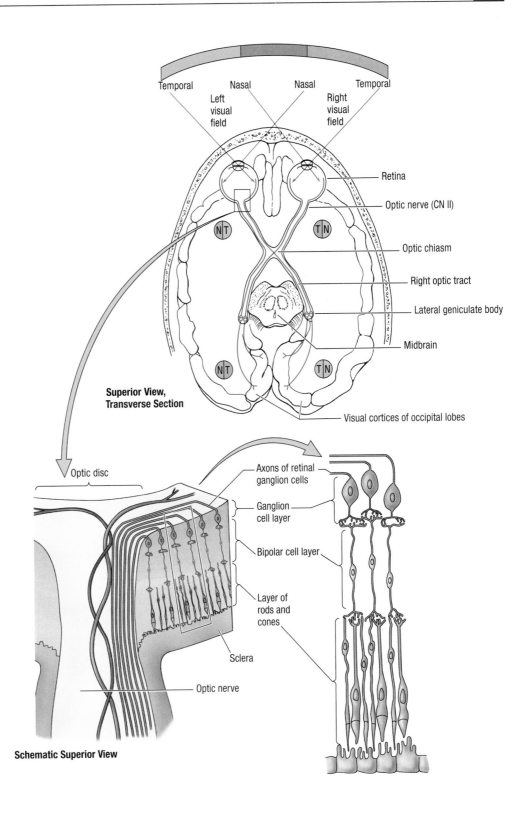

Temporal   Nasal   Nasal   Temporal

Left visual field

Right visual field

Retina

Optic nerve (CN II)

Optic chiasm

Right optic tract

Lateral geniculate body

Midbrain

Superior View, Transverse Section

Visual cortices of occipital lobes

Optic disc

Axons of retinal ganglion cells

Ganglion cell layer

Bipolar cell layer

Layer of rods and cones

Sclera

Optic nerve

Schematic Superior View

TABLE 9.3   OPTIC NERVE (CN II)

NERVE	FUNCTIONAL COMPONENTS	CELLS OF ORIGIN/TERMINATION	CRANIAL EXIT	DISTRIBUTION AND FUNCTIONS
Optic	Special sensory	Retina (ganglion cells)/lateral geniculate body (nucleus)	Optic canal	Vision from retina

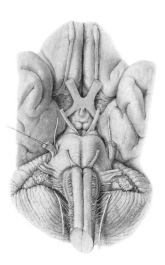

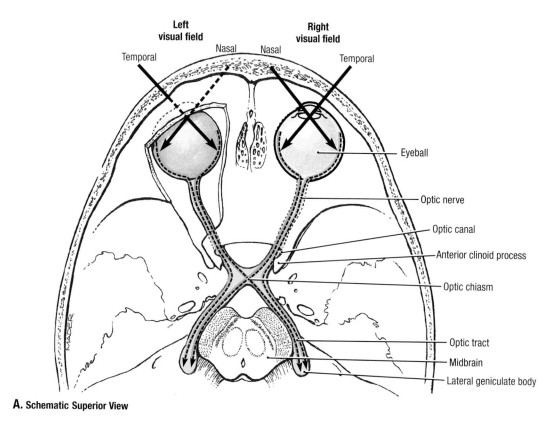

Left visual field

Right visual field

Temporal — Nasal — Nasal — Temporal

Eyeball

Optic nerve

Optic canal

Anterior clinoid process

Optic chiasm

Optic tract

Midbrain

Lateral geniculate body

A. Schematic Superior View

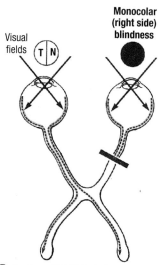

B. Section of Right Optic Nerve

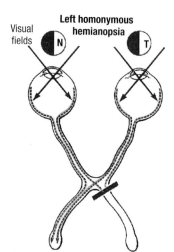

C. Section of Right Optic Tract

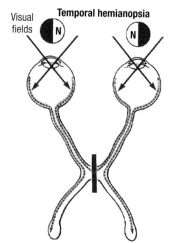

D. Section of Optic Chiasm

9.4 Visual pathway

A. The visual pathway in situ. **B–D.** Schematic illustrations of lesions of the visual pathway. A lesion of the right optic nerve *(top)* would result in blindness of the right eye (monocular blindness), a lesion of the right optic tract *(middle)* would eliminate vision from the left visual fields of both eyes (homonymous hemianopia), and a lesion of the optic chiasm *(bottom)* would reduce peripheral vision (bitemporal hemianopia or tunnel vision). The pituitary gland lies just posterior to the optic chiasma; expansion of this gland by a tumor would put pressure on the fibers that cross here.

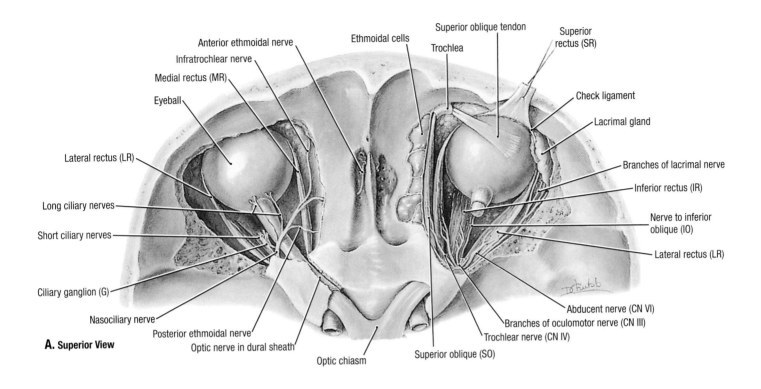

A. Superior View

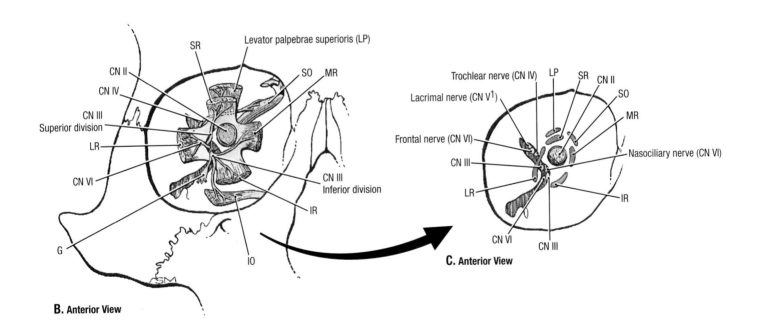

B. Anterior View

C. Anterior View

| 9.5 | Overview of muscles and nerves of orbit |

A. Orbital cavities, dissected from a superior approach. **B.** Structures of apex of orbit. **C.** Relationship of muscle attachments and nerves at apex of orbit.

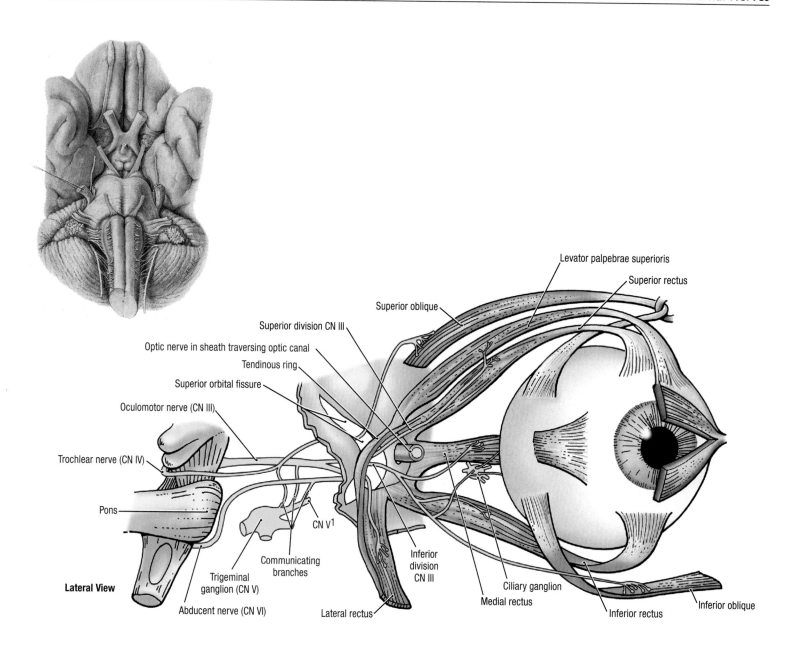

TABLE 9.4 OCULOMOTOR (CN III), TROCHLEAR (CN IV), AND ABDUCENT (CN VI) NERVES[a]

NERVE	FUNCTIONAL COMPONENTS	CELLS OF ORIGIN/TERMINATION	CRANIAL EXIT	DISTRIBUTION AND FUNCTIONS
Oculomotor	Somatic motor	Oculomotor nucleus	Superior orbital fissure	Motor to superior, inferior, and medial recti, inferior oblique, and levator palpebrae superioris muscles; raises upper eyelid; turns eyeball superiorly, inferiorly, and medially
	Visceral motor	Presynaptic: midbrain (Edinger-Westphal nucleus); postsynaptic: ciliary ganglion		Parasympathetic innervation to sphincter pupillae and ciliary muscle; constricts pupil and accommodates lens of eyeball
Trochlear	Somatic motor	Trochlear nucleus		Motor to superior oblique that assists in turning eyeball inferolaterally
Abducent	Somatic motor	Abducent nucleus		Motor to lateral rectus that turns eyeball laterally

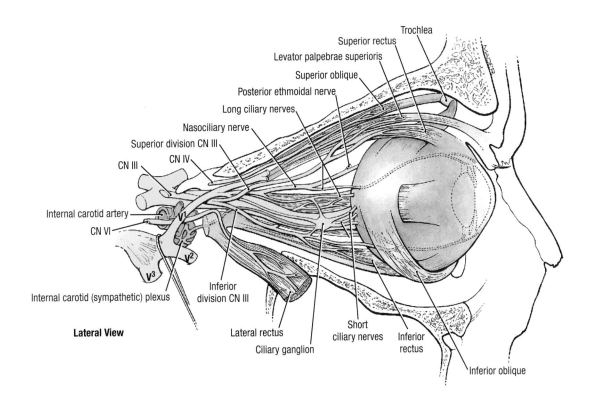

Trochlea
Superior rectus
Levator palpebrae superioris
Superior oblique
Posterior ethmoidal nerve
Long ciliary nerves
Nasociliary nerve
Superior division CN III
CN IV
CN III
Internal carotid artery
CN VI
V1
V2
V3
Internal carotid (sympathetic) plexus
Inferior division CN III
Lateral rectus
Ciliary ganglion
Short ciliary nerves
Inferior rectus
Inferior oblique

Lateral View

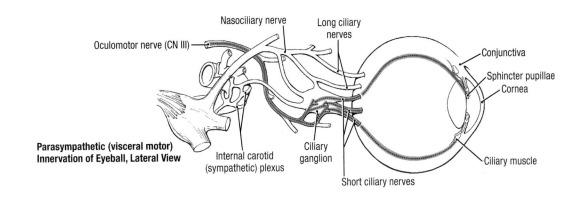

Nasociliary nerve
Long ciliary nerves
Oculomotor nerve (CN III)
Conjunctiva
Sphincter pupillae
Cornea
Ciliary muscle
Ciliary ganglion
Short ciliary nerves
Internal carotid (sympathetic) plexus

Parasympathetic (visceral motor) Innervation of Eyeball, Lateral View

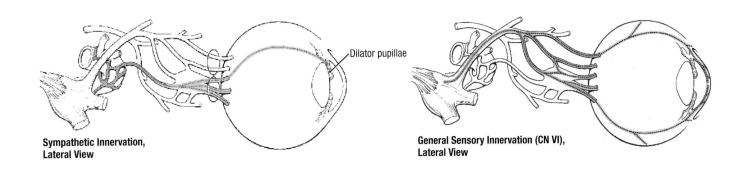

Dilator pupillae

Sympathetic Innervation, Lateral View

General Sensory Innervation (CN VI), Lateral View

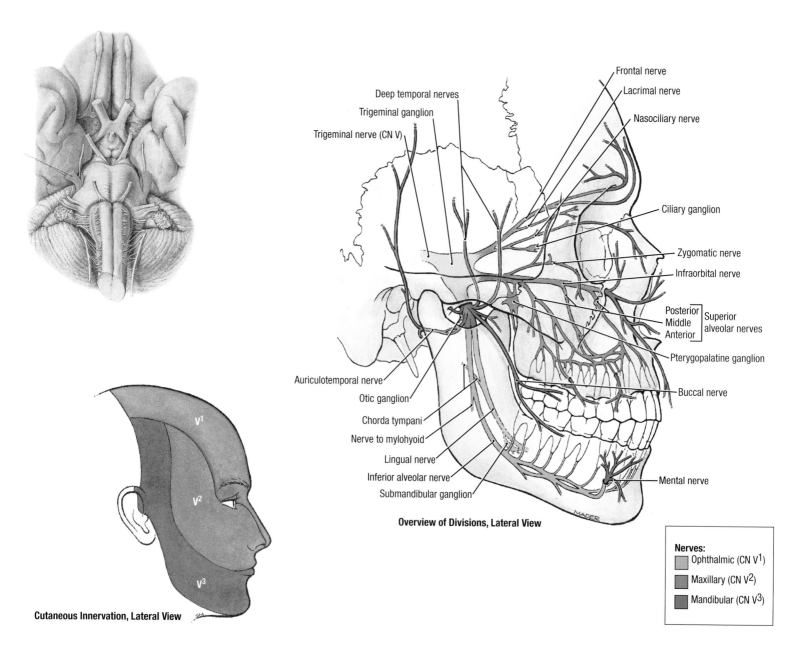

Overview of Divisions, Lateral View

Cutaneous Innervation, Lateral View

Nerves:
Ophthalmic (CN V^1)
Maxillary (CN V^2)
Mandibular (CN V^3)

TABLE 9.5   TRIGEMINAL NERVE (CN V)

NERVE DIVISION	FUNCTIONAL COMPONENTS	CELLS OF ORIGIN/TERMINATION	CRANIAL EXIT	DISTRIBUTION AND FUNCTIONS
Ophthalmic division (CN V^1)	General sensory	Trigeminal ganglion/spinal, principal and mesencephalic nucleus of CN V	Superior orbital fissure	Sensation from cornea, skin of forehead, scalp, eyelids, nose, and mucosa of nasal cavity and paranasal sinuses
Maxillary division (CN V^2)			Foramen rotundum	Sensation from skin of face over maxilla including upper lip, maxillary teeth, mucosa of nose, maxillary sinuses, and palate
Mandibular division (CN V^3)			Foramen ovale	Sensation from the skin over mandible, including lower lip and side of head, mandibular teeth, temporomandibular joint, and mucosa of mouth and anterior two thirds of tongue
	Branchial motor	Trigeminal motor nucleus		Motor to muscles of mastication, mylohyoid, anterior belly of digastric, tensor veli palatini, and tensor tympani

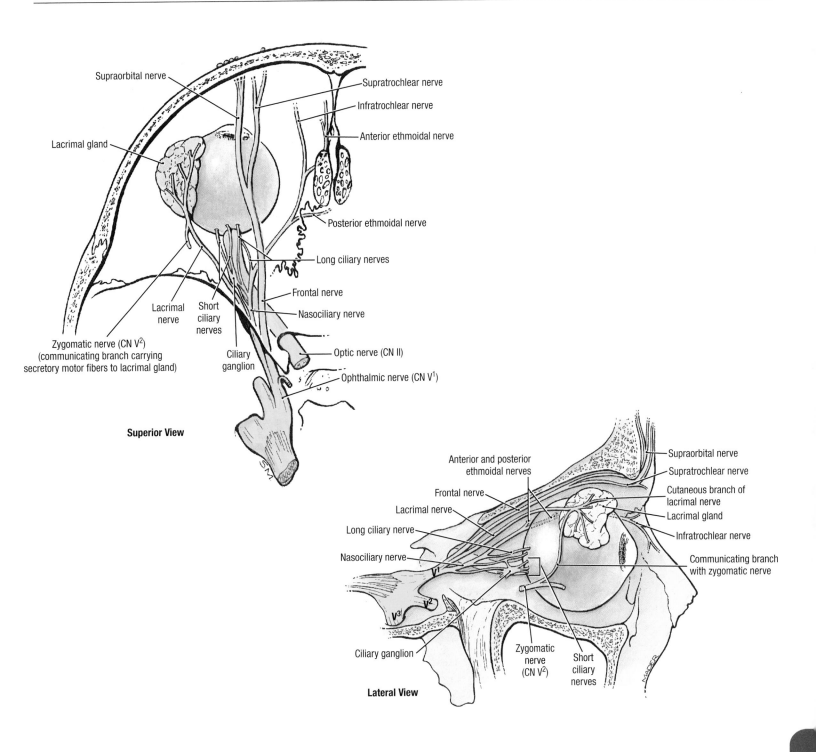

Superior View

Lateral View

TABLE 9.6   BRANCHES OF OPHTHALMIC NERVE (CN V¹)

FUNCTION	BRANCHES

The ophthalmic nerve is a sensory nerve passing through the superior orbital fissure that supplies the eyeball and conjunctiva, lacrimal gland and sac, nasal mucosa, frontal sinus, external nose, upper eyelid, forehead, scalp, and central dura mater of anterior cranial fossa	Lacrimal nerve
	
	Frontal nerve
	  Supraorbital nerve
	  Supratrochlear nerve
	
	Nasociliary nerve
	  Short ciliary nerves
	  Long ciliary nerves
	  Infratrochlear nerve
	  Anterior and posterior ethmoidal nerves

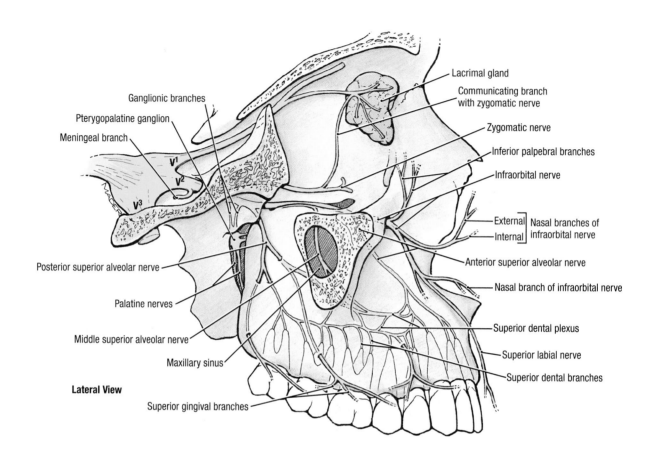

Ganglionic branches

Pterygopalatine ganglion

Meningeal branch

V¹
V²
V³

Posterior superior alveolar nerve

Palatine nerves

Middle superior alveolar nerve

Maxillary sinus

Lateral View

Superior gingival branches

Lacrimal gland

Communicating branch with zygomatic nerve

Zygomatic nerve

Inferior palpebral branches

Infraorbital nerve

External ⎤ Nasal branches of
Internal ⎦ infraorbital nerve

Anterior superior alveolar nerve

Nasal branch of infraorbital nerve

Superior dental plexus

Superior labial nerve

Superior dental branches

TABLE 9.7 BRANCHES OF MAXILLARY NERVE (CN V²)

FUNCTION	BRANCHES
The maxillary nerve is a sensory nerve passing through the foramen rotundum that supplies sensation to the face, upper teeth and gums, mucous membrane of the nasal cavity, palate and roof of the pharynx, maxillary, ethmoidal, and sphenoidal sinuses, and secretory fibers from the pterygopalatine ganglion, which pass with the zygomatic and lacrimal nerves to the lacrimal gland.	Meningeal branch

	Zygomatic nerve
	  Zygomaticofacial nerve
	  Zygomaticotemporal nerve

	Posterior superior alveolar nerves

	Infraorbital nerve
	  Anterior and middle superior alveolar nerves
	  Superior labial branches
	  Inferior palpebral branches
	  External and internal nasal branches

	Greater palatine nerve Posterior inferior lateral nasal branches
	Lesser palatine nerve Posterior superior lateral nasal branches
	Nasopalatine nerve
	Pharyngeal nerve

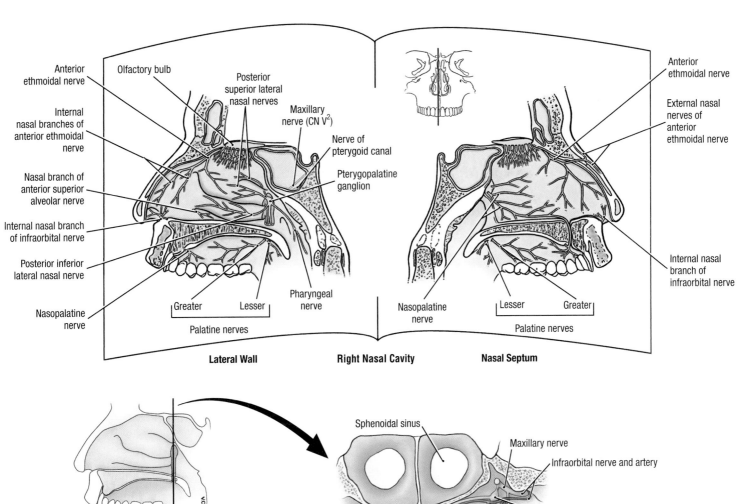

Lateral Wall   **Right Nasal Cavity**   **Nasal Septum**

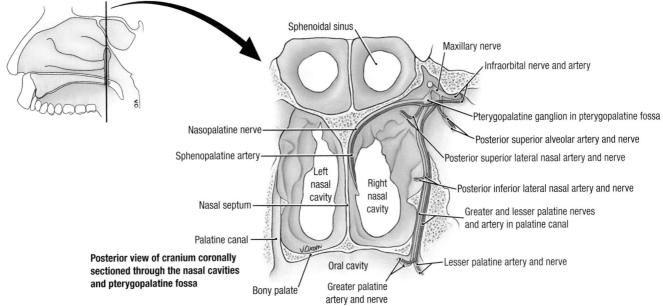

Posterior view of cranium coronally sectioned through the nasal cavities and pterygopalatine fossa

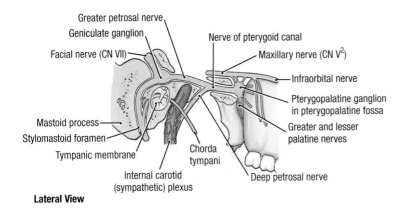

Lateral View

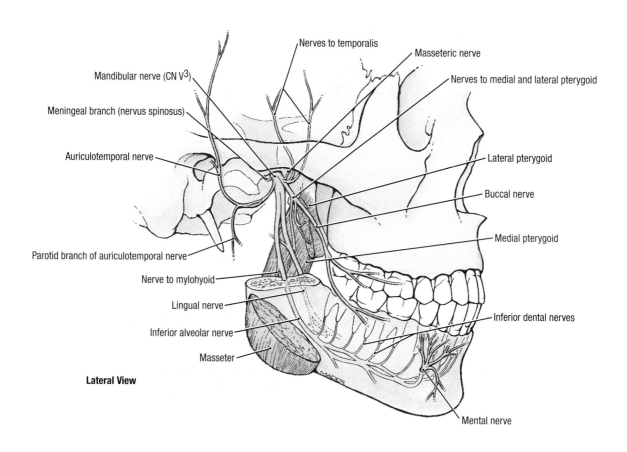

Nerves to temporalis

Masseteric nerve

Mandibular nerve (CN V³)

Nerves to medial and lateral pterygoid

Meningeal branch (nervus spinosus)

Auriculotemporal nerve

Lateral pterygoid

Buccal nerve

Medial pterygoid

Parotid branch of auriculotemporal nerve

Nerve to mylohyoid

Lingual nerve

Inferior dental nerves

Inferior alveolar nerve

Masseter

Lateral View

Mental nerve

TABLE 9.8   **BRANCHES OF MANDIBULAR NERVE (CN V³)**

FUNCTION	BRANCHES
The mandibular nerve is a sensory and motor nerve passing through the foramen ovale. General sensory branches (GSA) supply the lower teeth, gums, lip, auricle, external acoustic meatus, outer surface of tympanic membrane, cheek, anterior two thirds of tongue, and floor of mouth. CN V³ also conveys secretory fibers from the otic ganglion to the parotid gland. Taste to the anterior two thirds of the tongue and secretory motor fibers to the submandibular ganglion are conveyed by the chorda tympani. Postsynaptic fibers from the submandibular ganglion terminate in the submandibular and sublingual glands	Meningeal branch
	Buccal nerve
	Auriculotemporal nerve

	Inferior alveolar nerve
	  Inferior dental nerves
	  Mental nerve
	  Incisive nerve

	Lingual nerve
Motor branches (SVE) supply the muscles of mastication and other muscles derived from the embryonic branchial arches.	Masseter
	Temporalis
	Medial and lateral pterygoids
	Tensor veli palatini
	Mylohyoid
	Anterior belly of digastric
	Tensor tympani

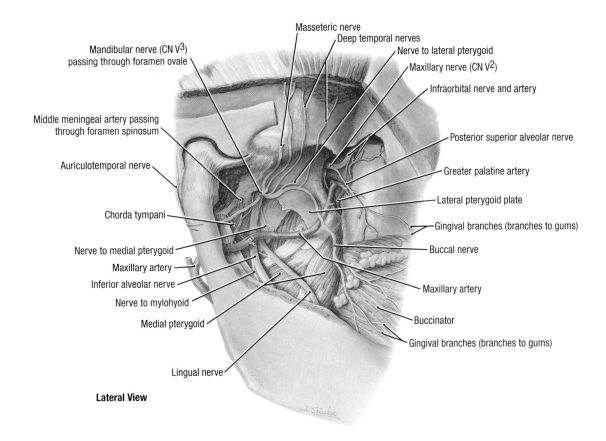

Masseteric nerve

Deep temporal nerves

Mandibular nerve (CN V3)
passing through foramen ovale

Nerve to lateral pterygoid

Maxillary nerve (CN V2)

Infraorbital nerve and artery

Middle meningeal artery passing
through foramen spinosum

Posterior superior alveolar nerve

Auriculotemporal nerve

Greater palatine artery

Lateral pterygoid plate

Chorda tympani

Gingival branches (branches to gums)

Nerve to medial pterygoid

Buccal nerve

Maxillary artery

Inferior alveolar nerve

Maxillary artery

Nerve to mylohyoid

Buccinator

Medial pterygoid

Gingival branches (branches to gums)

Lingual nerve

Lateral View

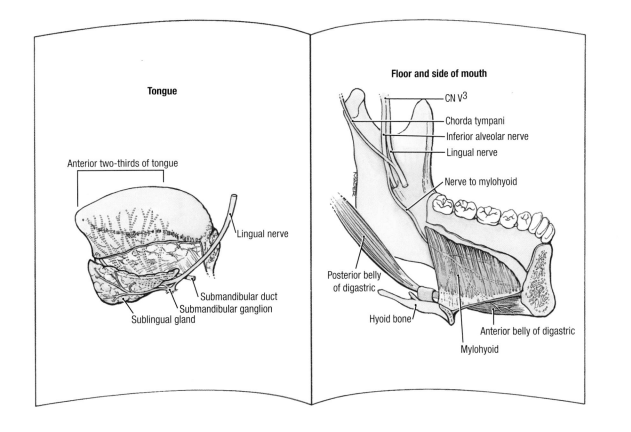

Tongue

Anterior two-thirds of tongue

Lingual nerve

Submandibular duct
Submandibular ganglion
Sublingual gland

Floor and side of mouth

CN V3

Chorda tympani
Inferior alveolar nerve
Lingual nerve

Nerve to mylohyoid

Posterior belly
of digastric

Hyoid bone

Anterior belly of digastric

Mylohyoid

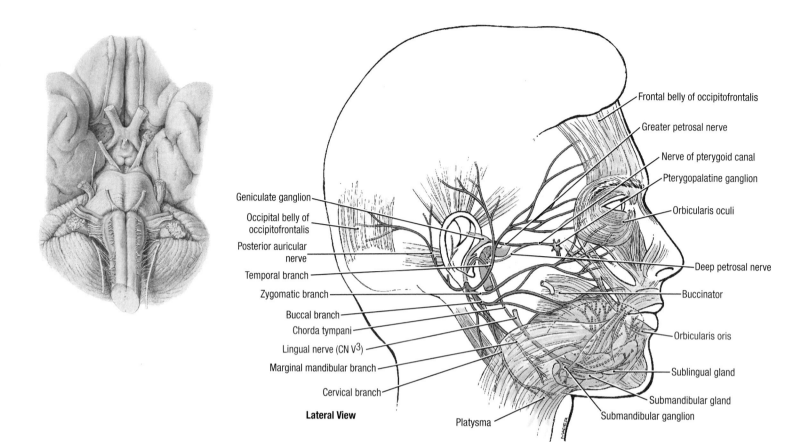

Frontal belly of occipitofrontalis
Greater petrosal nerve
Nerve of pterygoid canal
Pterygopalatine ganglion
Orbicularis oculi
Deep petrosal nerve
Buccinator
Orbicularis oris
Sublingual gland
Submandibular gland
Submandibular ganglion

Geniculate ganglion
Occipital belly of occipitofrontalis
Posterior auricular nerve
Temporal branch
Zygomatic branch
Buccal branch
Chorda tympani
Lingual nerve (CN V³)
Marginal mandibular branch
Cervical branch

Lateral View

Platysma

TABLE 9.9   FACIAL NERVE (CN VII), INCLUDING MOTOR ROOT AND NERVUS INTERMEDIUS[a]

NERVE BRANCH	FUNCTIONAL COMPONENTS	CELLS OF ORIGIN/TERMINATION	CRANIAL EXIT	DISTRIBUTION AND FUNCTIONS
Temporal, zygomatic, buccal, mandibular, cervical, posterior auricular nerves, nerve to posterior belly of digastric, nerve to stylohyoid, nerve to stapedius	Branchial motor	Facial motor nucleus	Stylomastoid foramen	Motor to muscles of facial expression and scalp; also supplies stapedius of middle ear, stylohyoid, and posterior belly of digastric
Intermediate nerve through chorda tympani	Special sensory	Geniculate ganglion/ solitary nucleus	Internal acoustc meatus/ facial canal/ petrotympanic fissure	Taste from anterior two thirds of tongue, floor of mouth, and palate
Intermediate nerve	General sensory	Geniculate ganglion/ spinal trigeminal nucleus	Internal acoustic meatus	Sensation from skin of external acoustic meatus
Intermediate nerve through greater petrosal nerve	Visceral sensory	Solitary nucleus	Internal acoustc meatus/ facial canal/foramen for greater petrosal nerve	Mucous membranes of nasopharynx and palate
Greater petrosal nerve[1] Chorda tympani[2]	Visceral motor	*Presynaptic:* superior salivatory nucleus *Postsynaptic:* pterygopalatine ganglion[1] and submandibular ganglion[2]	Internal acoustc meatus/ facial canal/foramen for greater petrosal nerve[1], petrotympanic fissure[2]	Parasympathetic innervation to lacrimal gland and glands of the nose and palate[1]; submandibular and sublingual salivary glands[2]

[a] See also Table 9.15.

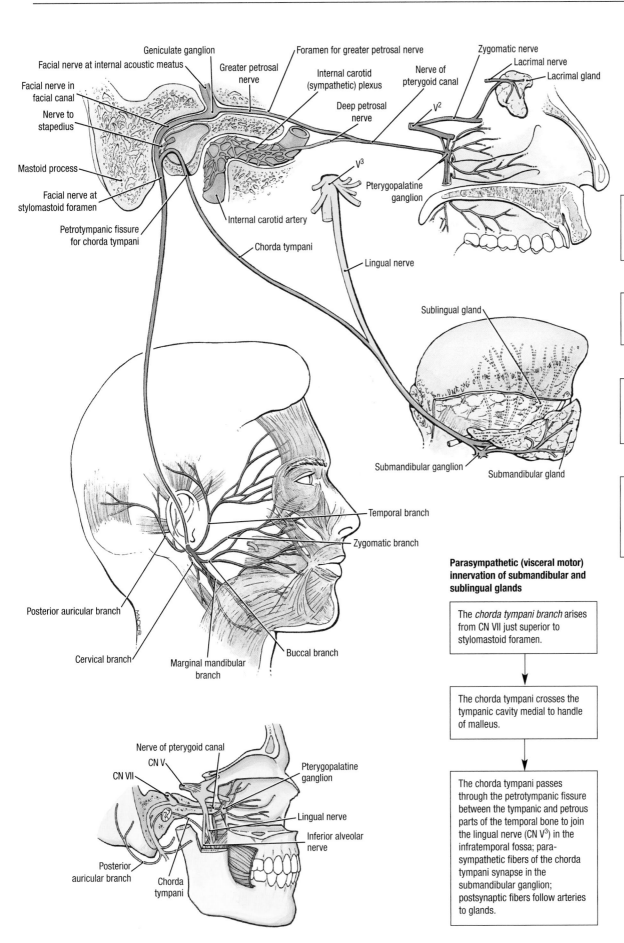

Geniculate ganglion

Facial nerve at internal acoustic meatus

Facial nerve in facial canal

Nerve to stapedius

Mastoid process

Facial nerve at stylomastoid foramen

Petrotympanic fissure for chorda tympani

Greater petrosal nerve

Foramen for greater petrosal nerve

Internal carotid (sympathetic) plexus

Deep petrosal nerve

Internal carotid artery

Chorda tympani

Zygomatic nerve

Lacrimal nerve

Lacrimal gland

Nerve of pterygoid canal

V^2

V^3

Pterygopalatine ganglion

Lingual nerve

Sublingual gland

Submandibular ganglion

Submandibular gland

Temporal branch

Zygomatic branch

Posterior auricular branch

Cervical branch

Marginal mandibular branch

Buccal branch

Nerve of pterygoid canal

CN V

CN VII

Posterior auricular branch

Chorda tympani

Pterygopalatine ganglion

Lingual nerve

Inferior alveolar nerve

Parasympathetic (visceral motor) innervation of lacrimal gland

Greater petrosal nerve arises from CN VII at the geniculate ganglion and emerges from the superior surface of the petrous part of the temporal bone to enter the middle cranial fossa.

Greater petrosal nerve joins the *deep petrosal nerve* (sympathetic) at the foramen lacerum to form the nerve of the pterygoid canal.

Nerve of the pterygoid canal travels through the pterygoid canal and enters the pterygopalantine fossa.

Parasympathetic fibers from the nerve of pterygoid canal in the pterygopalantine fossa synapse in the *pterygopalantine ganglion*.

Postsynaptic parasympathetic fibers from this ganglion innervate the *lacrimal gland* via the zygomatic nerve of CN V^2 and the lacrimal nerve CN V^1.

Parasympathetic (visceral motor) innervation of submandibular and sublingual glands

The *chorda tympani branch* arises from CN VII just superior to stylomastoid foramen.

The chorda tympani crosses the tympanic cavity medial to handle of malleus.

The chorda tympani passes through the petrotympanic fissure between the tympanic and petrous parts of the temporal bone to join the lingual nerve (CN V^3) in the infratemporal fossa; parasympathetic fibers of the chorda tympani synapse in the submandibular ganglion; postsynaptic fibers follow arteries to glands.

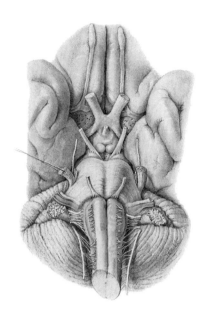

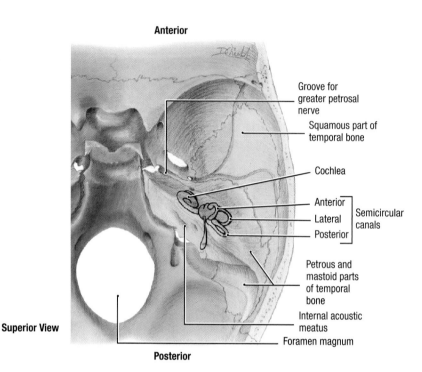

Anterior

Groove for greater petrosal nerve

Squamous part of temporal bone

Cochlea

Anterior

Lateral — Semicircular canals

Posterior

Petrous and mastoid parts of temporal bone

Internal acoustic meatus

Foramen magnum

Superior View

Posterior

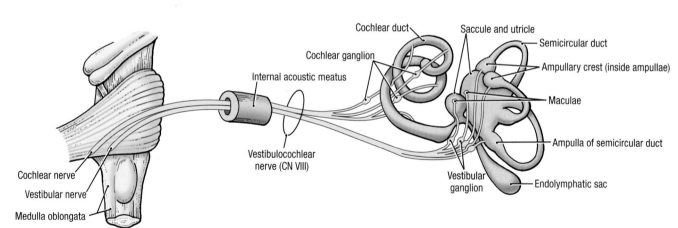

Cochlear duct

Cochlear ganglion

Saccule and utricle

Semicircular duct

Ampullary crest (inside ampullae)

Internal acoustic meatus

Maculae

Vestibulocochlear nerve (CN VIII)

Ampulla of semicircular duct

Cochlear nerve

Vestibular nerve

Vestibular ganglion

Endolymphatic sac

Medulla oblongata

TABLE 9.10   VESTIBULOCOCHLEAR NERVE (CN VIII)

PART OF VESTIBULOCOCHLEAR NERVE	FUNCTIONAL COMPONENTS	CELLS OF ORIGIN/TERMINATION	CRANIAL EXIT	DISTRIBUTION AND FUNCTIONS
Vestibular nerve	Special sensory	Vestibular ganglion/ vestibular nuclei	Internal acoustic meatus	Vestibular sensation from semicircular ducts, utricle, and saccule related to position and movement of head
Cochlear nerve	Special sensory	Spiral ganglion/ cochlear nuclei	Internal acoustic meatus	Hearing from spiral organ

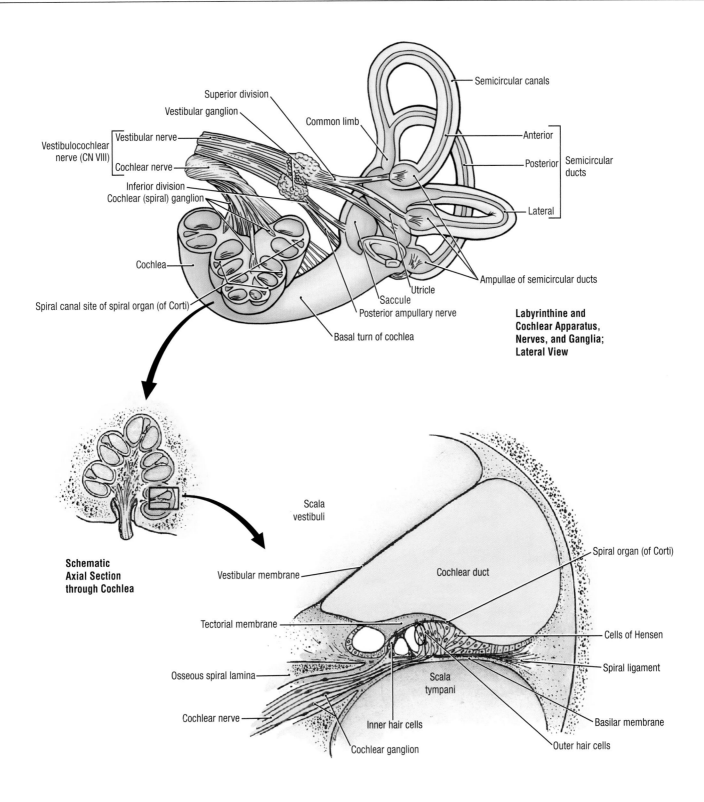

Labyrinthine and Cochlear Apparatus, Nerves, and Ganglia; Lateral View

Schematic Axial Section through Cochlea

Observe in the lower diagram:
- The cochlear duct is a spiral tube fixed to the internal and external walls of the cochlear canal by the spiral ligament.
- The triangular cochlear duct lies between the osseous spiral lamina and the external wall of the cochlear canal.
- The roof of the cochlear duct is formed by the vestibular membrane, and the floor by the basilar membrane and osseous spiral lamina.

- The receptor of auditory stimuli is the spiral organ (of Corti), situated on the basilar membrane; it is overlaid by the gelatinous tectorial membrane.
- The spiral organ contains hair cells that respond to vibrations induced in the endolymph by sound waves.
- The fibers of the cochlear nerve are axons of neurons in the spiral ganglion; the peripheral processes enter the spiral organ (of Corti).

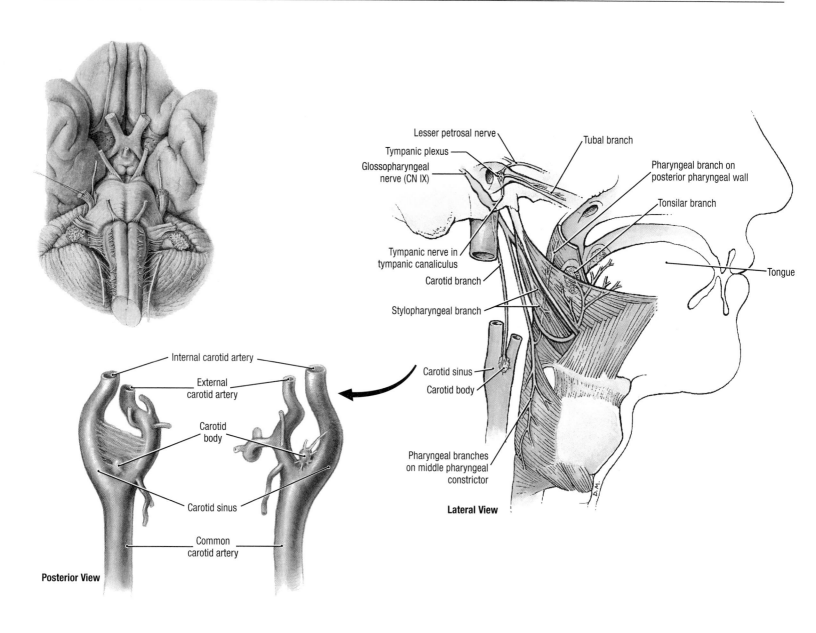

Lesser petrosal nerve
Tympanic plexus
Glossopharyngeal nerve (CN IX)
Tubal branch
Pharyngeal branch on posterior pharyngeal wall
Tonsilar branch
Tympanic nerve in tympanic canaliculus
Carotid branch
Stylopharyngeal branch
Tongue
Carotid sinus
Carotid body
Pharyngeal branches on middle pharyngeal constrictor
Lateral View

Internal carotid artery
External carotid artery
Carotid body
Carotid sinus
Common carotid artery
Posterior View

TABLE 9.11   GLOSSOPHARYNGEAL NERVE (CN IX)[a]

NERVE FUNCTIONS	FUNCTIONAL COMPONENTS	CELLS OF ORIGIN/TERMINATION	CRANIAL EXIT	DISTRIBUTION AND FUNCTIONS
Glossopharyngeal	Branchial motor	Nucleus ambiguus		Motor to stylopharyngeus that assists with swallowing
	Visceral motor	Presynaptic: inferior salivatory nucleus Postsynaptic: otic ganglion		Parasympathetic innervation to parotid gland
	Visceral sensory	Solitary nucleus, spinal trigeminal nucleus/ inferior ganglion	Jugular foramen	Visceral sensation from parotid gland, carotid body and carotid sinus, pharynx, and middle ear
	Special sensory	Solitary nucleus/inferior ganglion		Taste from posterior third of tongue
	General sensory	Spinal trigeminal nucleus/ superior ganglion		Cutaneous sensation from external ear

[a] See also Table 9.15.

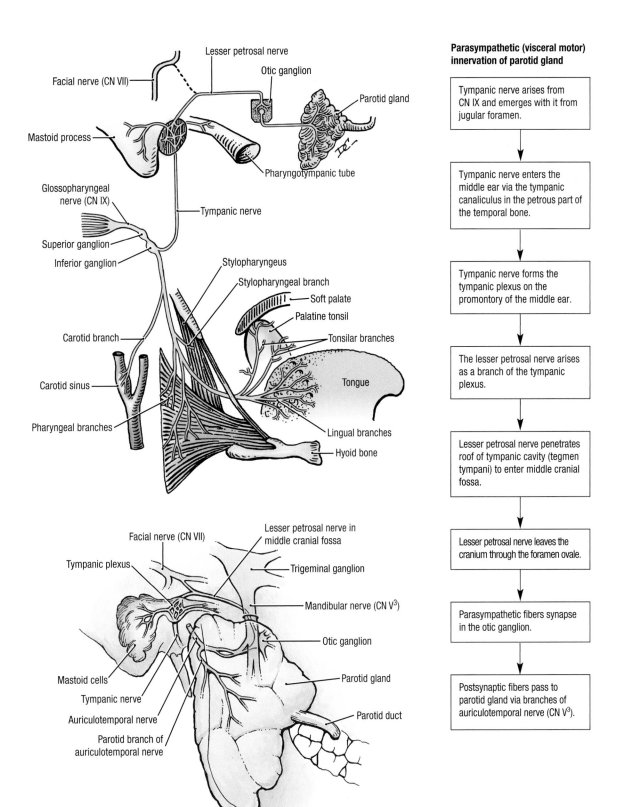

Lesser petrosal nerve

Otic ganglion

Facial nerve (CN VII)

Parotid gland

Mastoid process

Pharyngotympanic tube

Glossopharyngeal nerve (CN IX)

Tympanic nerve

Superior ganglion

Inferior ganglion

Stylopharyngeus

Stylopharyngeal branch

Soft palate

Palatine tonsil

Tonsilar branches

Tongue

Carotid branch

Carotid sinus

Lingual branches

Hyoid bone

Pharyngeal branches

Parasympathetic (visceral motor) innervation of parotid gland

Tympanic nerve arises from CN IX and emerges with it from jugular foramen.

↓

Tympanic nerve enters the middle ear via the tympanic canaliculus in the petrous part of the temporal bone.

↓

Tympanic nerve forms the tympanic plexus on the promontory of the middle ear.

↓

The lesser petrosal nerve arises as a branch of the tympanic plexus.

↓

Lesser petrosal nerve penetrates roof of tympanic cavity (tegmen tympani) to enter middle cranial fossa.

↓

Lesser petrosal nerve leaves the cranium through the foramen ovale.

↓

Parasympathetic fibers synapse in the otic ganglion.

↓

Postsynaptic fibers pass to parotid gland via branches of auriculotemporal nerve (CN V^3).

Facial nerve (CN VII)

Lesser petrosal nerve in middle cranial fossa

Tympanic plexus

Trigeminal ganglion

Mandibular nerve (CN V^3)

Otic ganglion

Parotid gland

Mastoid cells

Tympanic nerve

Auriculotemporal nerve

Parotid duct

Parotid branch of auriculotemporal nerve

Lateral View

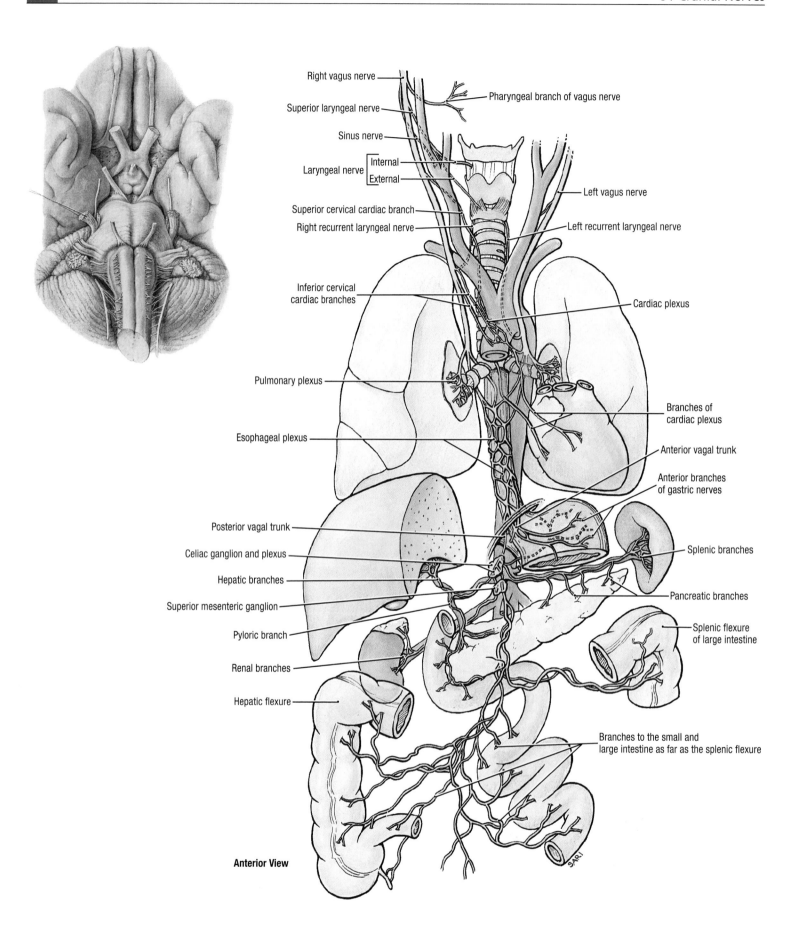

Right vagus nerve

Superior laryngeal nerve

Sinus nerve

Laryngeal nerve — Internal / External

Superior cervical cardiac branch

Right recurrent laryngeal nerve

Inferior cervical cardiac branches

Pulmonary plexus

Esophageal plexus

Posterior vagal trunk

Celiac ganglion and plexus

Hepatic branches

Superior mesenteric ganglion

Pyloric branch

Renal branches

Hepatic flexure

Pharyngeal branch of vagus nerve

Left vagus nerve

Left recurrent laryngeal nerve

Cardiac plexus

Branches of cardiac plexus

Anterior vagal trunk

Anterior branches of gastric nerves

Splenic branches

Pancreatic branches

Splenic flexure of large intestine

Branches to the small and large intestine as far as the splenic flexure

Anterior View

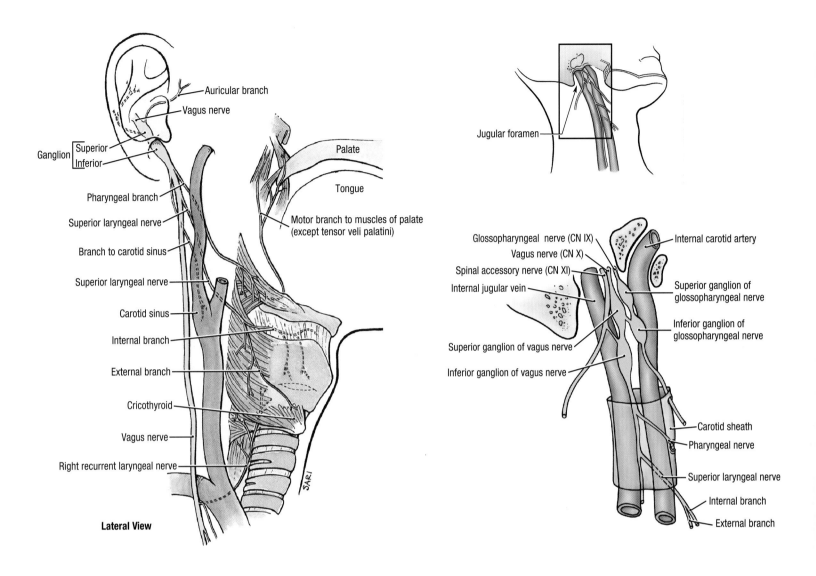

Lateral View

TABLE 9.12   VAGUS NERVE (CN X)

NERVE FUNCTIONS	FUNCTIONAL COMPONENTS	CELLS OF ORIGIN/TERMINATION	CRANIAL EXIT	DISTRIBUTION AND FUNCTIONS
Vagus	Branchial motor	Nucleus ambiguus		Motor to constrictor muscles of pharynx, intrinsic muscles of larynx, and muscles of palate, except tensor veli palatini, and striated muscle in superior two thirds of esophagus
	Visceral motor	*Presynaptic*: dorsal vagal nucleus *Postsynaptic*: neurons in, on, or near viscera	Jugular foramen	Parasympathetic innervation to smooth muscle of trachea, bronchi, digestive tract, and cardiac muscle
	Visceral sensory	Solitary nucleus, spinal trigeminal nucleus/inferior ganglion		Visceral sensation from base of tongue, pharynx, larynx, trachea, bronchi, heart, esophagus, stomach, and intestine
	Special sensory	Solitary nucleus/inferior ganglion		Taste from epiglottis and palate
	General sensory	Spinal trigeminal nucleus/superior or inferior ganglion		Sensation from auricle, external acoustic meatus, and dura mater of posterior cranial fossa

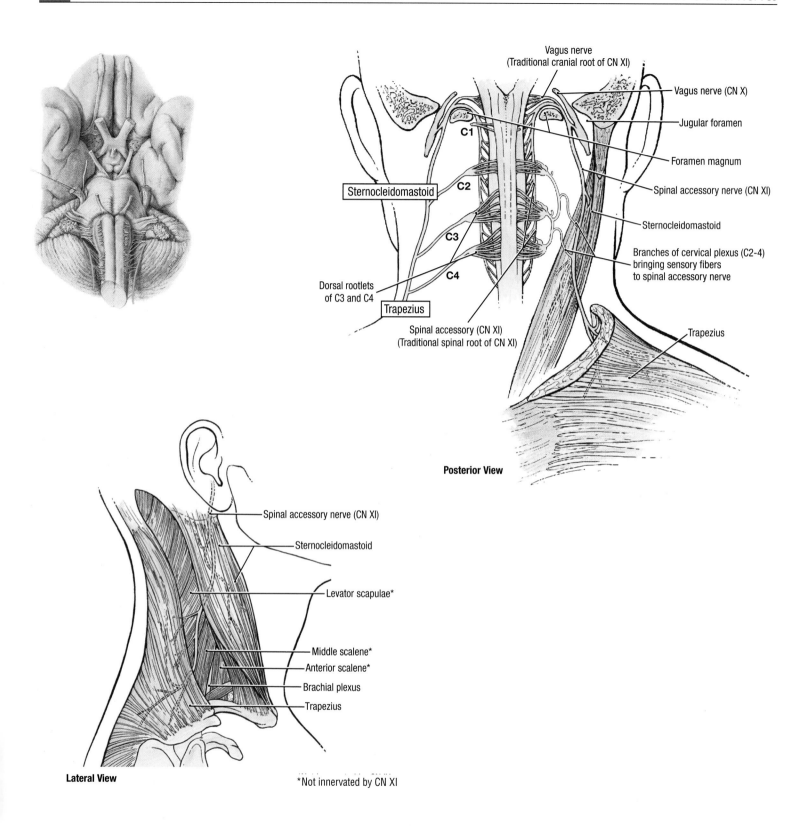

Vagus nerve
(Traditional cranial root of CN XI)

Vagus nerve (CN X)

Jugular foramen

Foramen magnum

Spinal accessory nerve (CN XI)

Sternocleidomastoid

C1

Sternocleidomastoid

C2

C3

Branches of cervical plexus (C2-4)
bringing sensory fibers
to spinal accessory nerve

C4

Dorsal rootlets
of C3 and C4

Trapezius

Trapezius

Spinal accessory (CN XI)
(Traditional spinal root of CN XI)

Posterior View

Spinal accessory nerve (CN XI)

Sternocleidomastoid

Levator scapulae*

Middle scalene*

Anterior scalene*

Brachial plexus

Trapezius

Lateral View   *Not innervated by CN XI

TABLE 9.13   SPINAL ACCESSORY NERVE (CN XI)

NERVE	FUNCTIONAL COMPONENTS	CELLS OF ORIGIN/TERMINATION	CRANIAL EXIT	DISTRIBUTION AND FUNCTIONS
Spinal accessory	Somatic motor	Accessory nucleus of spinal cord	Jugular foramen	Motor to sternocleidomastoid and trapezius

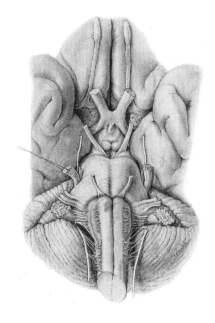

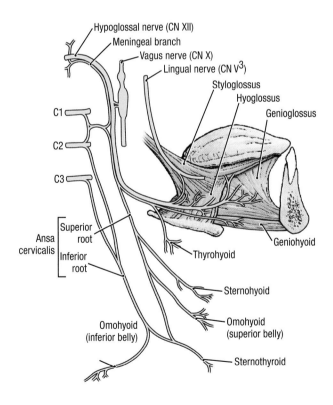

Hypoglossal nerve (CN XII)
Meningeal branch
Vagus nerve (CN X)
Lingual nerve (CN V³)
Styloglossus
Hyoglossus
Genioglossus

C1
C2
C3

Ansa cervicalis
Superior root
Inferior root

Thyrohyoid
Geniohyoid
Omohyoid (inferior belly)
Omohyoid (superior belly)
Sternothyroid
Sternohyoid

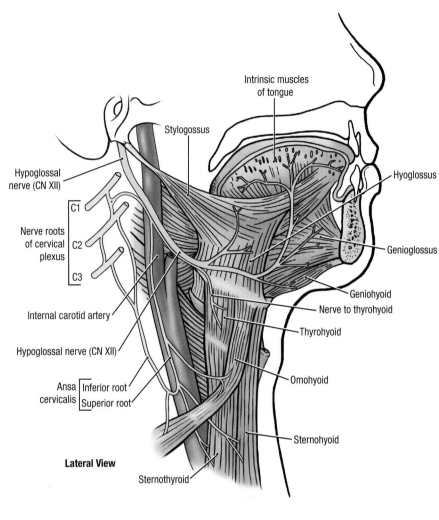

Intrinsic muscles of tongue
Styloglossus
Hypoglossal nerve (CN XII)
Nerve roots of cervical plexus
C1
C2
C3
Internal carotid artery
Hypoglossal nerve (CN XII)
Ansa cervicalis
Inferior root
Superior root
Hyoglossus
Genioglossus
Geniohyoid
Nerve to thyrohyoid
Thyrohyoid
Omohyoid
Sternohyoid
Sternothyroid

Lateral View

TABLE 9.14   HYPOGLOSSAL NERVE (CN XII)

NERVE	FUNCTIONAL COMPONENTS	CELLS OF ORIGIN/TERMINATION	CRANIAL EXIT	DISTRIBUTION AND FUNCTIONS
Hypoglossal	Somatic motor	Hypoglossal nucleus	Hypoglossal canal	Motor to muscles of tongue (except palatoglossus)

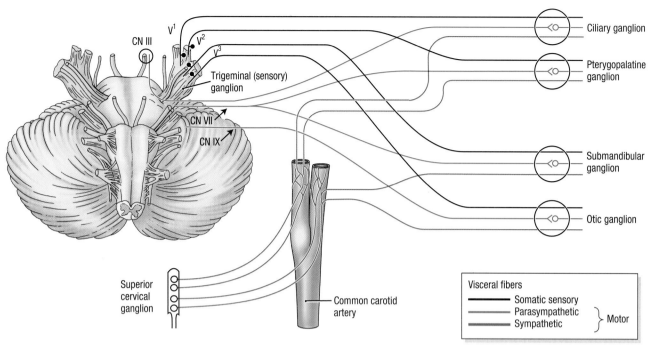

TABLE 9.15 AUTONOMIC GANGLIA OF THE HEAD

GANGLION	LOCATION	PARASYMPATHETIC ROOT (NUCLEUS OF ORIGIN)[a]	SYMPATHETIC ROOT	MAIN DISTRIBUTION
Ciliary	Located between optic nerve and lateral rectus, close to apex of orbit	Inferior branch of oculomotor nerve (Edinger-Westphal nucleus)	Branch from internal carotid plexus in cavernous sinus	Parasympathetic postsynaptic fibers from ciliary ganglion pass to ciliary muscle and sphincter pupillae of iris; sympathetic postsynaptic fibers from superior cervical ganglion pass to dilator pupillae and blood vessels of eye
Pterygopalatine	Located in pterygopalatine fossa, where it is attached by pterygopalatine branches of maxillary nerve; located just anterior to opening of pterygoid canal and inferior to CN V^2	Greater petrosal nerve from facial nerve (superior salivatory nucleus)	Deep petrosal nerve, a branch of internal carotid plexus that is continuation of postsynaptic fibers of cervical sympathetic trunk; fibers from superior cervical ganglion pass through pterygopalatine ganglion and enter branches of CN V^2	Parasympathetic postsynaptic fibers from pterygopalatine ganglion innervate lacrimal gland through zygomatic branch of CN V^2; sympathetic postsynaptic fibers from superior cervical ganglion accompany those branches of pterygopalatine nerve that are distributed to the nasal cavity, palate, and superior part of the pharynx
Otic	Located between tensor veli palatini and mandibular nerve; lies inferior to foramen ovale sphenoid bone	Tympanic nerve from glossopharyngeal nerve; from tympanic plexus, tympanic nerve continues as lesser petrosal nerve (inferior salivatory nucleus)	Fibers from superior cervical ganglion travel via plexus on middle meningeal artery	Parasympathetic postsynaptic fibers from otic ganglion are distributed to parotid gland through auriculotemporal nerve (branch of CN V^3); sympathetic postsynaptic fibers from superior cervical ganglion pass to parotid gland and supply its blood vessels
Submandibular	Suspended from lingual nerve by two short roots; lies on surface of hyoglossus muscle inferior to submandibular duct	Parasympathetic fibers join facial nerve and leave it in its chorda tympani branch, which unites with lingual nerve (superior salivatory nucleus)	Sympathetic fibers from superior cervical ganglion travel via the plexus on facial artery	Postsynaptic parasympathetic fibers from submandibular ganglion are distributed to the sublingual and submandibular glands; sympathetic fibers supply sublingual and submandibular glands and appear to be secretomotor

[a] For location of nuclei, see Figure 9.3.

TABLE 9.16   SUMMARY OF CRANIAL NERVES

NERVE	LESION TYPE AND/OR SITE	ABNORMAL FINDINGS
CN I	Fracture of cribriform plate	Anosmia (loss of smell); cerebrospinal fluid (CSF) rhinorrhea (leakage of CSF through nose)
CN II	Direct trauma to orbit or eyeball; fracture involving optic canal	Loss of pupillary constriction
	Pressure on optic pathway; laceration or intracerebral clot in temporal, parietal, or occipital lobes of brain	Visual field defects
	Increased CSF pressure	Swelling of optic disc (papilledema)
CN III	Pressure from herniating uncus on nerve; fracture involving cavernous sinus; aneurysms	Dilated pupil, ptosis, eye turns down and out; pupillary reflex on the side of the lesion will be lost
CN IV	Stretching of nerve during its course around brainstem; fracture of orbit	Inability to look down when the eye is adducted
CN V	Injury to terminal branches (particularly CN V^2) in roof of maxillary sinus; pathologic processes (tumors, aneurysms, infections) affecting trigeminal nerve	Loss of pain and touch sensations/paraesthesia on face; loss of corneal reflex (blinking when cornea touched); paralysis of muscles of mastication; deviation of mandible to side of lesion when mouth is opened
CN VI	Base of brain or fracture involving cavernous sinus or orbit	Eye does not move laterally; diplopia on lateral gaze
CN VII	Laceration or contusion in parotid region	Paralysis of facial muscles; eye remains open; angle of mouth droops; forehead does not wrinkle
	Fracture of temporal bone	As above, plus associated involvement of cochlear nerve and chorda tympani; dry cornea and loss of taste on anterior two thirds of tongue
	Intracranial hematoma ("stroke")	Weakness (paralysis) of lower facial muscles contralateral to the lesion, upper facial muscles are not affected because they are bilaterally innervated
CN VIII	Tumor of nerve	Progressive unilateral hearing loss; tinnitus (noises in ear); vertigo (loss of balance)
CN IX[a]	Brainstem lesion or deep laceration of neck	Loss of taste on posterior third of tongue; loss of sensation on affected side of soft palate; loss of gag reflex on affected side
CN X	Brainstem lesion or deep laceration of neck	Sagging of soft palate; deviation of uvula to unaffected side; hoarseness owing to paralysis of vocal fold; difficulty in swallowing and speaking
CN XI	Laceration of neck	Paralysis of sternocleidomastoid and superior fibers of trapezius; drooping of shoulder
CN XII	Neck laceration; basal skull fractures	Protruded tongue deviates toward affected side; moderate dysarthria (disturbance of articulation)

[a] Isolated lesions of CN IX are uncommon; usually, CN IX, X, and XI are involved together as they pass through the jugular foramen.

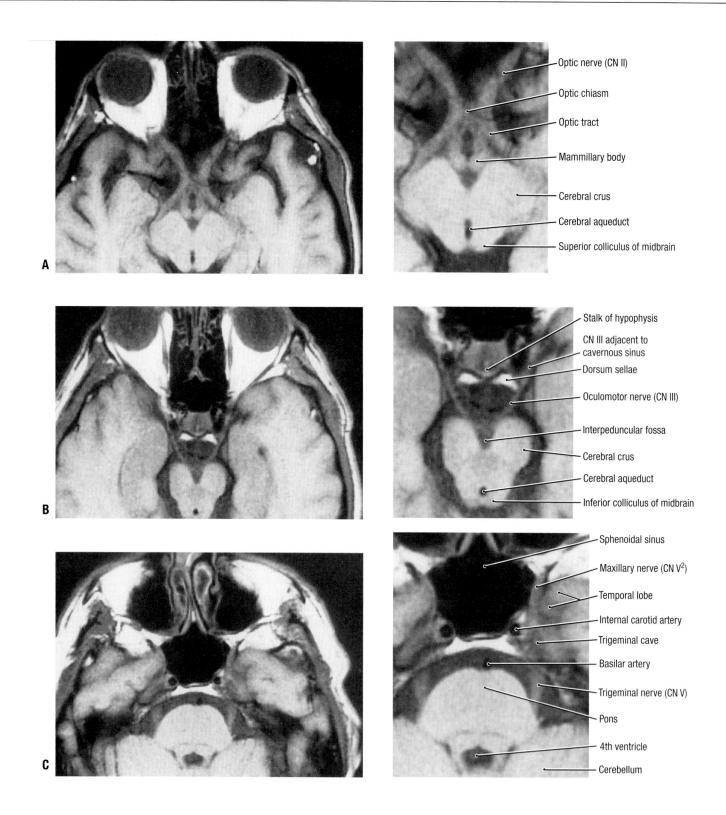

Optic nerve (CN II)

Optic chiasm

Optic tract

Mammillary body

Cerebral crus

Cerebral aqueduct

Superior colliculus of midbrain

Stalk of hypophysis

CN III adjacent to cavernous sinus

Dorsum sellae

Oculomotor nerve (CN III)

Interpeduncular fossa

Cerebral crus

Cerebral aqueduct

Inferior colliculus of midbrain

Sphenoidal sinus

Maxillary nerve (CN V^2)

Temporal lobe

Internal carotid artery

Trigeminal cave

Basilar artery

Trigeminal nerve (CN V)

Pons

4th ventricle

Cerebellum

9.6   **Transverse MRIs through head, showing cranial nerves**

A. Optic neve (CN II). **B.** Oculomotor nerve (CN III). **C.** Trigeminal nerve (CN V).

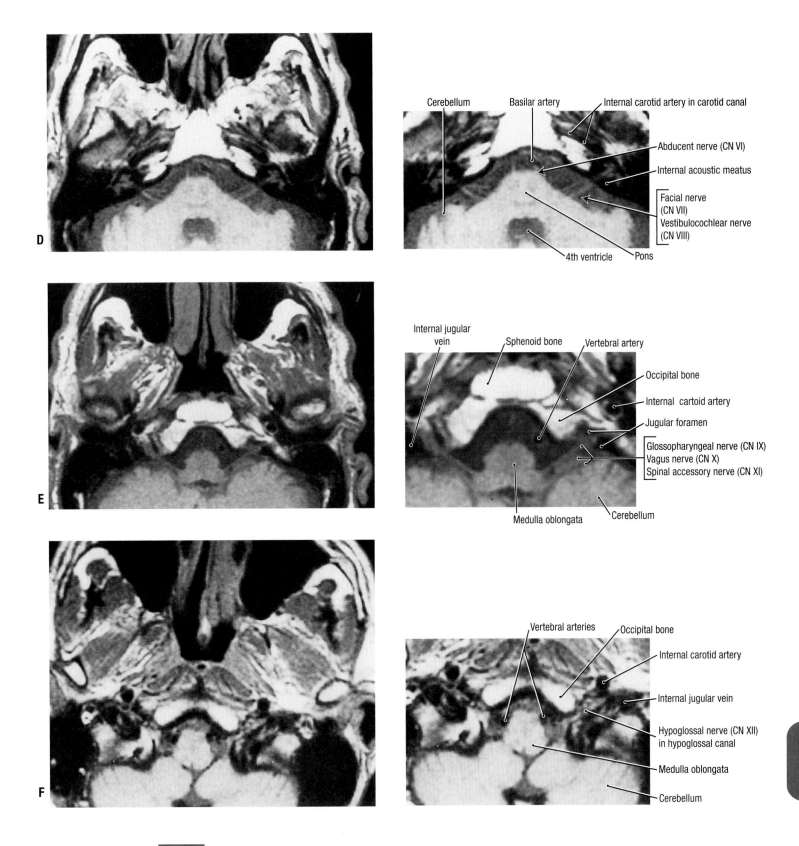

9.6   **Transverse MRIs through head, showing cranial nerves** *(continued)*

D. Abducent (CN VI), facial (CN VII), and vestibulocochlear (CN VIII) nerves. **E.** Glossopharyngeal (CN IX), vagus (CN X), and spinal accessory (CN XI) nerves. **F.** Hypoglossal nerve (CN XII).

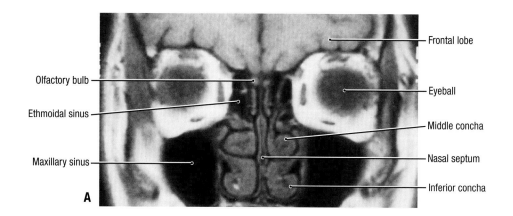

Frontal lobe

Olfactory bulb

Eyeball

Ethmoidal sinus

Middle concha

Nasal septum

Maxillary sinus

Inferior concha

A

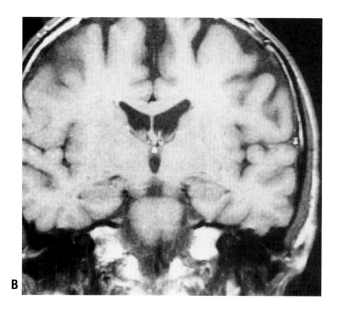

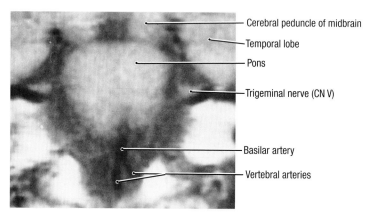

Cerebral peduncle of midbrain

Temporal lobe

Pons

Trigeminal nerve (CN V)

Basilar artery

Vertebral arteries

B

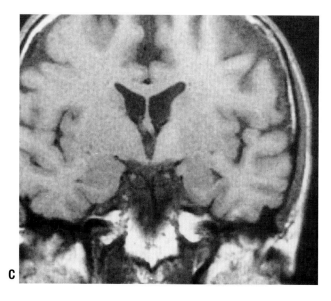

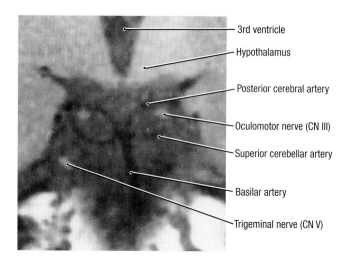

3rd ventricle

Hypothalamus

Posterior cerebral artery

Oculomotor nerve (CN III)

Superior cerebellar artery

Basilar artery

Trigeminal nerve (CN V)

C

9.7   **Coronal MRIs through head, showing cranial nerves**

A. Olfactory bulb. **B.** Trigeminal (CN V) nerve. **C.** Oculomotor (CN III) and trigeminal (CN V) nerves.

Abdomen
  autonomic nerve supply of, 168, 170
  magnetic resonance angiogram of, 180–181
  magnetic resonance imaging of, 176–179
  ultrasound of, 180–181
  vagus nerves in, 171
  viscera of, 92
Abdominal wall
  anterior, 97–98, 106
  anterolateral, 112
  posterior, 155, 161–163, 175
Acetabulum, 185, 268–269, 353, 378
Acromion, 460, 511, 513, 515
Ampulla
  hepatopancreatic, 130
  lateral, 708
  phrenic, 124
  posterior, 708
  rectum, 214, 224
Anastomoses
  knee, 398
  upper limb, 470–471
Anatomical snuff box, 560, 564
Angiogram
  ascending aorta, 88
  descending aorta, 88
  heart, 88
  pelvic, 216
  pulmonary, 41
  renal, 158
Angle
  acromial, 493, 498, 513
  costal, 13
  of eye, 640
  inferolateral, 293
  infrasternal, 10
  iridocorneal, 651
  sternal, 10, 12, 285, 479, 788
  subpubic, 185, 190–191
Ankle
  articular surfaces of, 439
  joint cavity of, 432
  ligaments of, 435
  magnetic resonance imaging, 436–437
  medial aspect of, 438
  posterior aspect of, 434
  synovial sheaths of, 412–413
  tendons of, 412–413
Ansa cervicalis, 735, 738, 741, 743, 786, 821
Ansa subclavia, 70, 762
Anterior longitudinal ligament, 15, 17
Antihelix, 696
Antitragus, 696
Antrum
  mastoid, 609, 701, 703
  pyloric, 121, 124, 177, 179
Anulus fibrosus, 300–304, 310
Anus, 153, 217, 243–244, 258, 260–261, 264–269
Aorta
  abdominal, 72, 117, 167

ascending, 41, 46, 49, 51, 68
  magnetic resonance angiogram of, 88
  magnetic resonance imaging of, 83, 85–86
  radiograph of, 27
bifurcation of, 179
branches of, 166
descending, 63, 68–69, 80
  computed tomography of, 6
  groove for, 35
  magnetic resonance angiogram of, 88
  magnetic resonance imaging of, 83–85, 87
general images of, 50, 57, 63, 71, 111, 133, 198, 326
magnetic resonance imaging of, 336
thoracic, 67, 72–73
  descending, 74, 80, 113
ultrasound of, 180–181
Aortic arch node, 43
Aperture
  lateral, 628
  median, 627–628
  nasal, 590
Aponeurosis
  bicipital, 500–502, 518, 529
  dorsal, 408
  epicranial, 316, 608, 610, 694
  of external oblique, 94, 97, 100, 102, 353
  gastrocnemius, 453
  of internal oblique, 107
  palatine, 673–674
  palmar, 534, 541, 543–545, 584
  plantar, 424, 428t, 436–437
  transversus abdominal, 308–310, 312
  triceps, 505
Appendices epiploicae, 132, 136, 140–141
Appendix
  blood supply to, 135
  general images of, 93, 120, 132, 135, 224, 232, 238
  referred pain from, 169t
Aqueduct
  cerebral, 593, 619, 627–628, 633, 713–714, 824
  cochlear, 697
  vestibular, 697, 706
Aqueous humor, 650
Arachnoid, 610
Arachnoid granulations, 610, 612, 628
Arch
  aortic
    branches of, 68
    general anatomy of, 27, 39, 47, 49, 52, 66–67, 69, 72, 76, 79
    groove for, 35
    magnetic resonance imaging of, 85, 87–88
  carpal
    dorsal, 553t
  ciliary, 687
  iliopectineal, 352

lumbocostal, 75
neural, 278
palatoglossal, 666, 672, 674, 772, 774
palatopharyngeal, 674, 772–774
palmar
  deep, 467, 470–471, 550, 552, 553t
  superficial, 464–465, 467, 470–471, 541, 546, 550, 552, 553t
pancreaticoduodenal
  anterior, 127
  posterior, 127, 129
plantar, 422
superciliary, 589
venous
  dorsal, 342–343, 347, 560
  dorsal digital, 464–465
  jugular, 748
  plantar, 346–347
vertebral, 16, 277, 301
zygomatic, 588, 594, 600, 656, 694, 712, 764–765
Areola, 4
Arm
  lateral aspect of, 507
  medial aspect of, 503
  muscles of, 501–502
Arteries
  alveolar, 658, 660, 669, 704–706
  anal, 245
  angular, 606t
  appendicular, 135–136
  arcuate, 346, 407, 409, 425
  auricular
    deep, 660
    posterior, 601, 606t, 608, 652, 657
  axillary, 8, 22, 470–471, 481, 483, 732–733
  basilar, 321, 616, 622, 625t, 684, 714, 824–826
  brachial, 470, 481–482, 484t, 519, 521, 529, 536, 578–579
  brachiocephalic, 64–65
  bronchial, 35, 67
    left, 72
    right, 73
  buccal, 658, 660
  carotid
    common, 48, 319, 714, 723, 736, 741, 750, 756, 790–791, 816
      left, 46, 49, 64–65, 68–69, 72, 79–80, 82, 85, 87, 470
      right, 26, 68–69, 82, 470
    external, 653, 658, 660, 668, 736, 739, 741, 744, 750, 816
    internal, 612, 616, 619–621, 625t, 642, 652–653, 663, 668, 671, 686, 693, 702, 704–706, 712, 728, 736, 739, 745, 771, 790–791, 805, 816, 819, 825
  celiac, 75–76, 80, 122, 176, 178–179
  central, of retina, 647t, 650
  cerebellar

Arteries—*continued*
anterior, 622–624, 716–720
anterior inferior, 625t, 714
inferior, 622, 624
posterior inferior, 625t, 715
superior, 612, 619, 624, 625t, 826
cerebral, 610
anterior, 612, 622, 624, 625t, 713
middle, 622, 624, 625t, 713, 719
posterior, 612, 619–622, 624, 625t
cervical
ascending, 329, 756–757, 761
deep, 317, 736
transverse, 470, 731–732, 756, 759, 761
ciliary
anterior, 647t
long posterior, 647t, 651
short posterior, 647t
circumflex femoral
lateral, 216, 346, 355, 375, 382–383
medial, 216, 346, 355, 370–371, 375,
382–383
circumflex humoral
anterior, 470, 484t
posterior, 470, 484t, 506, 514
circumflex iliac
deep, 102–103, 105–106, 175, 198, 212,
234, 352, 354
superficial, 97, 350, 354
circumflex scapular, 471, 482, 484t, 489–490,
506
colic
left, 138–139, 141
middle, 117, 119, 126–127, 136–137,
140–141
right, 126, 136–137
common carotid, 48, 319, 714, 723, 736, 741,
750, 756, 790–791, 816
left, 46, 49, 64–65, 68–69, 72, 79–80, 82,
85, 87, 470
right, 26, 68–69, 82, 470
common femoral, 216
common hepatic, 118–119, 122
common iliac, 167, 178, 214–216, 234, 346
left, 170, 198, 221, 232, 301
right, 111, 138, 219, 221
communicating
anterior, 622, 624, 625t, 642
posterior, 612, 619, 622, 625t
coronary, 52
atrioventricular nodal branch of, 52
circumflex branch of, 55
interventricular branch of
anterior, 52, 54
posterior, 52, 54
lateral diagonal of, 52
left, 46, 50, 52, 54–55, 59, 61, 83
marginal branch of, 52
right, 46–47, 49–50, 52, 54–55, 61
sinuatrial nodal branch of, 52
cremasteric, 103, 106, 111, 198
cystic, 122, 148–151
deep, 506
digital
common palmar, 471, 546, 549
dorsal, 409, 425
proper palmar, 541
dorsal, of penis, 108

ductus deferens, 111
epigastric, 97, 102
inferior, 98, 103, 106, 109, 132, 140, 198,
205, 212, 224, 232, 234, 346, 352
superficial, 350, 353
superior, 98
esophageal, 73
ethmoidal
anterior, 642, 647t, 683
posterior, 642, 647t, 683
facial, 319, 606t, 652, 656, 668, 730, 736, 738,
742
lateral nasal branch of, 600
submental branch of, 748
superior labial branch of, 683
tonsillar branch of, 744, 775
transverse, 600–601, 606t, 652, 663, 736
femoral, 103, 111–112, 195, 210, 240,
263–265, 268–269, 346, 352–355, 375,
381–382, 422, 452–454
fibular, 422, 423t
gastric
left, 73, 117, 119, 123, 126–127, 154, 171,
180–181
posterior, 122, 127
right, 119, 121–123, 127, 149, 171
short, 121, 127
gastroduodenal, 122–123, 125–126, 129, 149,
166, 171, 180–181
gastro-omental
left, 114–115, 118–119, 121–123, 127
right, 114, 118–119, 122–123, 126–127
genicular
descending, 346, 399
lateral inferior, 375, 386, 391, 398
lateral superior, 346, 375, 386, 391, 398
medial inferior, 375, 390, 398–399, 422
medial superior, 346, 375, 386, 390,
398–399, 422
gluteal
inferior, 194, 212, 215–216, 220, 234, 235t,
240, 346, 370–371, 375
superior, 212, 215–216, 220, 234, 235t, 346,
370–371
gonadal, 235t
hepatic, 129, 143, 176, 180–181
common, 122, 127, 149, 166, 171, 176
proper, 117, 122–123, 126–127
ileocolic, 126, 135–137, 140–141
iliac
external, 103, 155, 167–168, 192–193, 204,
212, 214–216, 228, 232, 234, 241,
264–269, 352, 361, 382
internal, 167–168, 192–193, 198, 212, 213t,
214–216, 220, 228, 232, 234, 235t,
241, 264–265, 270
iliolumbar, 212, 213t, 220, 234, 235t
indicis
dorsalis, 557, 564
radialis, 541, 546, 550
infraorbital, 644, 647t, 658–660, 687,
693–694, 811
intercostal
anterior, 20, 22–23, 326, 580, 736
general anatomy of, 79
posterior, 20, 73, 80, 736
superior, 73
interlobar, 158

interosseous
anterior, 470, 530–531, 536–537
common, 470, 530–531
posterior, 530–531, 557
interventricular
anterior, 46–47, 83
posterior, 47
jejunal, 127
labial
inferior, 606t
superior, 606t, 683
labyrinthine, 622
lacrimal, 642, 647, 647t
laryngeal
inferior, 754
superior, 741, 751
lingual, 669–671, 736, 741, 744–746, 777
lumbar, 167, 329
malleolar
lateral, 346, 407, 409
medial, 346, 407, 409
marginal, 137–138
left, 46
right, 46, 55
masseteric, 657
maxillary, 609, 658–660, 663, 736, 765, 811
medullary
anterior segmental, 329
posterior segmental, 329
meningeal
accessory, 660
middle, 609–611, 620–621, 647, 659, 703,
765, 811
mental, 606t
mesenteric
inferior, 117, 128, 138, 140–141, 155,
166–168, 170
superior, 76, 116–117, 119, 125–127, 129,
136–137, 141, 154–155, 166, 168,
170–171, 177–178, 180–181
metacarpal, dorsal, 471, 549
metatarsal
dorsal, 407, 409
plantar, 346
nasal, 683
dorsal, 647, 647t
lateral, 606t
obturator, 193–194, 212, 213t, 216, 234, 235t,
240, 346, 383
occipital, 307, 316–319, 606t, 608, 653, 657,
730–731, 739
ophthalmic, 646, 647t
ovarian, 167, 225, 228–229, 233, 235t, 241
palatine
descending, 660
greater, 659, 673–674, 683, 687, 772, 809
lesser, 673–674, 687, 772, 809
palmar digital, 543, 549–550, 552
common, 471, 553t
proper, 471, 549–550, 552, 553t
pancreatic
dorsal, 127
pancreaticoduodenal, 141
anterior inferior, 126
anterior superior, 126–127
superior, 122
penis, 254

perforating, 409
    first, 371, 375
    of foot, 425
    fourth, 375
    second, 371, 375
    third, 375
pericardiacophrenic, 78, 762
perineal, 192
phrenic, 155, 171, 174–175
plantar
    deep, 346, 409, 425
    lateral, 346, 418, 421–422, 423t, 436–437
    medial, 346, 418, 422, 423t, 436–437
pollicis
    dorsalis, 552, 553t, 584
    princeps, 552, 553t, 584
pontine, 625t
popliteal, 346, 367, 375, 384, 386, 398,
        419–421, 423t
pudendal, 201, 212, 234
    external, 346, 350, 353
    internal, 204, 210, 212, 213t, 215–216, 220,
        234, 235t, 261, 361, 370–371
pulmonary, 34–35, 41
    left, 26–27, 41, 46, 49, 52, 79, 83, 88
    right, 41, 47, 66, 83, 85, 88
radial, 471, 530–531, 535, 538–539, 541, 546,
        550–552, 565, 580
    dorsal branch of, 562
    superficial branch of, 562
radialis indicis, 541, 546, 550
rectal, 138–139, 192, 201, 212, 213t, 215, 234,
        235t, 241
recurrent
    radial, 470–471, 520, 530–531, 535
    ulnar, 470–471, 530–531, 536
renal, 158, 167, 177
    accessory, 156
    left, 166, 170, 178, 180–181, 219
    right, 129, 166, 171, 179–181, 219
sacral
    lateral, 212, 213t, 216, 220, 234, 235t
    median, 194, 220, 235t
    middle, 216
saphenous, 355
scapular, 492, 756, 758, 762
scrotal
    posterior, 245
segmental
    apical, 158–159
    inferior, 158–159
    posterior, 158–159
septal
    posterior nasal, 683
sigmoid, 138–139, 141
sphenopalatine, 644, 660, 683, 687, 809
spinal
    anterior, 329, 622
    posterior, 329
splenic, 115, 119, 122–123, 126–127, 156, 166,
        177–181
sternocleidomastoid, 739, 744
stylomastoid, 704–706
subclavian, 319, 470, 491–492, 728, 736–737,
        750–751, 756, 758, 762
    general anatomy of, 8, 23, 26, 70, 78–79
    groove for, 35
    left, 46–47, 49, 64–66, 68–69, 72, 79, 85,
        88, 737, 758, 761
    right, 25, 66, 68–69, 82, 761

sublingual, 744, 747
submental, 671
subscapular, 470–471, 482, 484t, 489–490
supraduodenal, 122, 126
supraorbital, 608, 647t, 692
suprarenal, 158, 167
suprascapular, 470, 484t, 506, 513, 737, 756,
        759, 761
supratrochlear, 608, 647t
tarsal
    lateral, 346, 407, 409, 425
    medial, 346, 407, 409
temporal
    frontal branch of, 600
    superficial, 600–601, 606t, 608–609, 652,
        657, 660, 693, 736
testicular, 107–108, 111, 128–129, 141,
        154–155, 167, 171, 198, 213t, 214–215,
        235t, 250
of thigh, 375
thoracic
    internal
        general anatomy of, 8, 20, 22–23, 48,
            64–65, 68, 70–72, 78–79, 470, 492,
            736–737, 761
        medial mammary branches of, 8
        perforating branches of, 8
    lateral, 8, 470, 484t
        lateral mammary branches of, 8
    long, 482, 489–490
    superior, 484t
thoracoacromial, 470, 479–480, 484t, 513
thoracodorsal, 471, 484t
thyroid, 736, 739, 741, 744
    inferior, 73, 754, 756, 758, 768–769, 780
    superior, 750–751, 757, 768–769, 786
tibial
    anterior, 346, 375, 398, 406–407, 409,
        419–422, 423t, 425
    anterior recurrent, 398–399, 407
    posterior, 346, 375, 398, 418, 421–422,
        423t, 425, 438, 445
tympanic, 660
ulnar, 470, 530–531, 534–535, 539, 546–547,
        552, 583
    deep branch of, 571
    dorsal branch of, 567
    dorsal carpal branch of, 567
ulnar collateral
    inferior, 470–471, 484t, 519, 531
    superior, 470–471, 484t, 531, 578–579
ulnar recurrent, 522–523
umbilical, 112, 140, 213t, 215, 234, 235t
uterine, 213t, 228–229, 232–234, 235t, 239,
        241
vaginal, 213t, 228, 233–234, 235t, 241
vertebral, 72, 82, 285, 318–321, 327, 329, 334,
        336, 470, 612, 616, 622, 624, 625t,
        728, 736–737, 754, 763, 786, 790–791
vesical, 216
    inferior, 212, 213t, 214–215
    superior, 212, 213t, 215, 234, 235t, 241
zygomaticotemporal, 647
Arteriograms
    carotid, 623
    celiac, 123
    coronary, 54
    inferior mesenteric, 139

popliteal, 422
    superior mesenteric, 137
    vertebral, 623
Arterioles
    retinal, 651
Articular column, 282–283
Asterion, 590
Atlas, 282–285, 320, 334, 597
    anterior tubercle of, 727
    posterior arch of, 727
    transverse process of, 618
Atrioventricular bundle, 62
Atrium
    computed tomography of, 6
    left, 47, 50, 58, 60, 62, 83–85, 88
    radiograph of, 27
    right, 26–27, 46–47, 49–50, 56, 60, 83–85, 88
        computed tomography of, 6
        general images of, 26–27, 46–47, 49–50, 56
Auricle
    of ear, 696
    left, 46–47, 49, 58, 60, 86
    right, 49, 60
Auscultation
    of heart, 44
    triangle of, 495
Axilla, 2
    anterior wall of, 480
    apex of, 482
    contents of, 482
    general images of, 477
    medial wall of, 480
    posterior wall of, 480, 489–490
    structures of, 488
    transverse sections through, 483
    veins of, 481
    walls of, 482
Axis, 282–285, 316, 334, 684, 788

Back
    surface anatomy of, 306, 495
Barium enema, 134
Barium swallow, 124
Benign prostatic hyperplasia, 211
Bile duct, 117, 126, 128, 145
Bladder
    female, 268–270
    general images of, 92–93, 112, 116–117, 132,
        158, 175, 195, 197, 207, 210, 214–215,
        239
    innervation of, 221
    magnetic resonance imaging of, 268–270
    male, 209, 221, 252, 264–265
    trigone of, 209, 228, 232
    ultrasound scan of, 211
Body
    anococcygeal, 224, 248, 263
    carotid, 816
    of corpus callosum, 627
    geniculate
        lateral, 635–636, 801
        medial, 635–636
    mammillary, 617, 632–633, 635–636, 716–720,
        794, 824
    perineal, 245, 259
    pineal, 627
    restiform, 635
    vertebral, 14–15, 76, 277, 290, 305, 335–336,
        737, 790–791
    vitreous, 650, 715

Brain
  arterial supply to, 625t
  lobes of, 626
    frontal, 626, 630
    occipital, 626, 630, 632
    parietal, 626, 630
    temporal, 626, 630
Brainstem, 627, 635–636
Breast
  anatomy of, 5
  anterior view of, 5
  anterolateral view of, 7
  arterial supply, 8
  axillary tail of, 4
  bed of, 7
  computed tomography of, 6
  lymphatic drainage of, 9
  sagittal view of, 5
Bregma, 592
Bronchogram
  of tracheobronchial tree, 39–40
Bronchopulmonary segments, 36–37
Bronchus, 28, 35
  anterior segmental, 41
  apical segmental, 39, 41
  apicoposterior segmental, 39–40
  basal, 40
  eparterial, 35
  inferior lobar
    left, 38–39
    right, 39, 72
  inferior segmental, 41
  left, 40, 72
  lingular, 41
  main
    right, 33, 38
  middle lobar, 38–39, 41, 72
  posterior segmental, 41
  right, 40, 788
  segmental, 36–37, 39
  superior lobar
    left, 38–39
    right, 38–39, 72
  superior segmental, 41
Bulb
  of duodenum, 179
  jugular, 769
  olfactory, 616–617, 619, 622, 630, 682, 711,
    794, 800, 826
  of penis, 197, 206, 214, 248, 254
Bulla
  ethmoidal, 685
Bundle branch
  left, 62
  right, 62
Bursa
  anserine, 397t
  gastrocnemius, 397t
  iliopectineal, 353
  infrapatellar, 403
  infraspinatus, 517
  infratendinous, 517
  obturator, 381
  olecranon, 524
  omental, 115–119
  popliteus, 387, 397t
  prepatellar, 392, 403

  retromammary, 5
  subacromial, 512–513, 517
  subdeltoid, 483
  subtendinous, 370, 483, 512, 537
  suprapatellar, 397t, 402–403
  triceps brachii, 554
  trochanteric, 370

Calcaneus, 338–339, 411, 419–421, 432, 434,
    445–447, 451
Calcarine spur, 631
Calices
  major, 158
  minor, 158
  renal, 157–158
Calvaria, 610
Canal
  adductor, 354
  anal, 199–201, 224, 255
    innervation of, 202
  carotid, 594, 621, 702, 764–765
  central, 628
  cervical, 233
  femoral, 352
  hypoglossal, 593, 615
  infraorbital, 640
  inguinal
    of female, 104–105
    of male, 102t
  mandibular, 677
  optic, 614–615, 640, 804
  palatine, 687, 690, 693
  pharyngeal, 690
  pterygoid, 595, 682, 687, 690–691, 702, 809,
    813
  pudendal, 221, 245
  root, 675
  sacral, 184, 192, 293, 295, 297
  semicircular, 697–698, 702, 706–708, 715,
    814–815
  vertebral, 310, 320, 334
Canines, 678
Canthus
  lateral, 641
Capitate, 461, 469, 532, 554, 568–569, 571, 573
Capitulum, 498–499, 527, 529, 532
Capsule
  articular, 320, 513
  of atlantoaxial joint, 618
  external, 639, 716–720
  extreme, 638–639, 716–720
  fibrous, 383
    of elbow, 524, 526
    of kidneys, 157
    of shoulder, 510
  hip joint, 373, 383
  internal, 633–634, 638–639, 716–720
  joint, 383, 392
  knee joint, 403
  synovial, 512, 549, 563
Cardiac impression, 35
Carotid siphon, 719
Carpal tunnel, 469, 572, 582–583
Cartilage
  alar, 680–681
  articular, 383, 392, 512
  arytenoid, 781–783, 787, 789

  corniculate, 781
  costal, 3, 10, 12, 22, 24, 48, 97–98, 176, 474,
    479, 488
  cricoid, 735, 749, 755, 757, 770, 778–779, 783,
    787–789
  cuneiform, 781
  epiglottic, 755, 781, 783
  nasal
    accessory, 680
    alar, 680–681
    lateral, 687
    septal, 680
    sesamoid, 687
  sesamoid, 687
  thyroid, 726, 730, 735, 742, 749, 757,
    778–779, 783, 787, 789–791
  triradiate, 379
  triticeal, 755
  vomeronasal, 681
Caruncle
  lacrimal, 640–641
  sublingual, 670, 746
Cauda equina, 301, 325, 330, 336
Cave, trigeminal, 824
Cavity
  articular, 665
  glenoid, 515
  medullary, 382
  nasal, 28, 120, 589
  oral, 120
  orbital, 589, 640, 642–643
  pericardial, 53, 788
  peritoneal, 116–117, 197
  pleural, 28–29
    roof of, 70
  pulp, 675
  tympanic, 697–698, 700–706
Cecum, 92–93, 117, 120, 132, 135–136, 140–141
Celiac arteriogram, 123
Cells
  border, 709
  ethmoidal
    anterior, 687, 692, 694, 803
    posterior, 685
  ganglion, 801
  hair, 815
  of Hensen, 709, 815
  mastoid, 597, 702
  mitral, 800
  tubal, 701
  tympanic, 699, 701
Cement, 675
Central band, 563
Cerebellum, 335, 617, 626, 637, 788
  tonsil of, 335, 789
Cerebral hemispheres, 626, 629
Cerebrospinal fluid, 274, 325, 334–335, 628,
    714, 790–791
Cervix, 224, 227, 233
Chiasm
  optic, 617, 632, 635, 643, 794, 801, 803
Choana, 594, 764–765, 770
Chorda tympani, 659, 661, 701, 703, 777, 806,
    811, 813
Chordae tendineae, 58
Choroid, 650
Cistern
  chyle, 72, 81, 111
  collicular, 716–720

interpeduncular, 714
lumbar, 301, 325
suprasellar, 715
Cisterna magna, 684
Claustrum, 630, 638–639
Clavicle
general images of, 2–4, 8, 10, 24, 39, 92, 283,
    460, 468, 474, 492, 515, 751, 759
radiograph of, 27
sternal end of, 481, 725
Cleft
intergluteal, 306
pudendal, 104–105
Clitoris, 257
angle of, 258
body of, 243, 258
crus of, 228, 258, 261, 263
dorsal artery of, 258, 261
dorsal nerve of, 242, 258, 261
dorsal vein of, 242
frenulum of, 260
glans of, 258, 260
prepuce of, 260–261
Clivus, 335, 593, 615
Coccyx, 184–185, 190, 193, 210, 214, 224, 239,
    248, 270, 274, 276, 292–293, 298
Cochlea, 697, 707–709
Colliculus
facial, 635
inferior, 618, 620–621, 635–636, 824
seminal, 209
superior, 619–621, 634–637, 639
Colon
ascending, 92–93, 117, 120, 125, 132, 134,
    136–137, 140, 152, 174, 176, 178
barium enema of, 134
descending, 92–93, 117, 120, 125, 132–134,
    139–141, 176, 178
referred pain to, 169t
sigmoid, 93, 120, 133–134, 138, 140–141, 152,
    198, 228, 264–267, 270
transverse, 92, 114–115, 118, 120, 132–134,
    137–138, 141, 177, 179
Commissure
anterior, 627, 631, 635–636, 716–720
habenular, 635
posterior, 632, 634, 638
Compressor urethrae, 247
Computed tomography
of breast, 6
of intervertebral disc, 301
of sacroiliac joint, 295
of thorax, 88–89
Concha
of ear, 696
inferior, 335, 589, 673, 680–681, 684–686,
    689, 691, 693, 695, 711–712, 714, 800,
    826
middle, 673, 680–681, 686, 711–712, 800, 826
nasal, 774
superior, 681, 695, 711–712, 715, 800
Condyle
femoral
    lateral, 362–363, 394, 400
    medial, 362–363, 392, 402
occipital, 592–595, 764–765
tibial
    lateral, 363, 400
    medial, 363, 400

Cone(s)
of eye, 801
medullary, 322, 325, 336
Conjunctiva
bulbar, 641
palpebral, 640, 645
Conus arteriosus, 57
Conus elasticus, 782
Cornea, 641, 645, 650
Corona radiata, 629–630
Coronary circulation, 55
Corpus callosum, 638, 684, 788
body of, 713
genu of, 713
rostrum of, 632, 713
splenium of, 627, 713
Cortex
cerebral, 610
insular, 713
renal, 157
Costovertebral articulations, 14
ligaments of, 15
Cranial nerves
abducent (CN VI), 616–617, 620–622, 642,
    644, 794, 796, 797t, 803–804, 823, 825
facial (CN VII), 616–617, 622, 635, 637, 652,
    698, 728, 738, 761, 794
    branches of, 608, 812
    cervical branch of, 730
    primary root of, 796, 797t, 803, 823, 825
    temporal branches of, 652, 812
glossopharyngeal (CN IX), 327, 593, 616–618,
    622, 635, 653, 666, 670, 706, 728, 765,
    776, 794, 796, 797t, 803, 816–817,
    823, 825
hypoglossal (CN XII), 327, 616–617, 652–653,
    670–671, 693, 739, 741–742, 745, 747,
    766, 769, 777, 796, 797t, 803, 821,
    823, 825
nuclei of, 798–799
oculomotor (CN III), 616–617, 619–622, 635,
    637, 642, 644, 646, 794, 796, 797t,
    803–804, 823, 825
olfactory (CN I), 616, 682, 796, 797t, 800,
    803, 823, 825
optic (CN II), 619, 622, 632, 635, 642–643,
    646, 650, 686, 694–695, 711–712, 794,
    796, 797t, 801, 803, 823–825
overview of, 794
posterior exposure of, 618
spinal accessory (CN XI), 307, 322, 327, 491,
    616–618, 622, 652, 657, 730, 732, 739,
    769, 794, 796, 797t, 803, 820, 823, 825
summary of, 823
trigeminal (V), 616, 618, 622, 635, 637, 826
    groove for, 593
    mandibular division of, 806, 810
    maxillary division of, 806, 808–809
    motor root of, 617, 635, 794, 796, 797t,
        803, 823, 825
    ophthalmic division of, 806–807
    sensory root of, 617, 620–621, 635, 794,
        796, 797t, 803, 823, 825
trochlear (CN IV), 616–621, 642, 794, 796,
    797t, 803–804, 823, 825
vagus (CN X), 48, 67, 616–618, 622, 652–653,
    728, 750–751, 756–757, 762, 769, 771,
    786, 794, 796, 797t, 803, 819, 823, 825

auricular branches of, 657
left, 25, 42, 49, 63, 65–66, 79, 758, 818
right, 25, 42, 65–66, 70, 75, 78, 758, 818
vestibulocochlear (CN VIII), 616–618, 622,
    635, 637, 698, 709, 794, 796, 797t,
    803, 814, 823, 825
Cranium
of child, 598–599
frontal aspect of, 588–589
inferior aspect of, 594
radiograph of, 596–597
Crest
ethmoidal, of maxilla, 684
frontal, 615
iliac, 175, 184–186, 222, 291–292, 298, 308,
    345, 363, 372–373, 379
infratemporal, 655
intertrochanteric, 291, 363
of lacrimal bone, 591, 640
nasal
    of maxilla, 681, 689
    of palatine bone, 681
obturator, 378
occipital
    external, 592, 594, 764–765
    internal, 615
pronator, 532
pubic, 100, 362
sacral, 292–293
sphenoidal, 681
supramastoid, 591, 654
supraventricular, 57
urethral, 209
Crista galli, 596, 612, 614–615, 643, 681, 686,
    692, 711, 800
Crus
cerebral, 631, 633, 635, 638, 824
of clitoris, 228, 258, 261, 263
of diaphragm, 166
of penis, 243, 248, 254, 264–265
Cuboid, 339, 433, 445–446
Cuneiforms, 405
distal, 427
first, 426, 447
lateral, 433, 444
medial, 451
middle, 426–427, 433, 444, 446
second, 426
third, 426
Cuneus, 632
Cusp
aortic valve, 55, 59, 84, 88
pulmonary valve, 60
septal, 57

Dartos tunic, 109
Decussation
ventral tegmental, 638
Deltoid, 2, 4
Dens
of atlas, 285
of axis, 284, 320, 727
of cervical vertebrae, 282–283, 334–335
Dentin, 675
Dermatomes
of breast, 7
of lower limb, 341
of neck, 331

Dermatomes—*continued*
 of trunk, 95, 331
 of upper limb, 463
Diaphragm
 central tendon of, 24, 71
 crura of, 26, 72, 75, 177, 180–181, 335
 dome of, 27, 39, 48, 86, 92–93, 115
 general anatomy of, 23–25, 49, 71, 81, 112, 156, 166
 radiograph of, 124
 sellar, 612, 619
 surface markings of, 45
Diencephalon, 627, 631
Digestive system, 120
Dilator pupillae, 805
Disc
 articular, 572, 653, 665, 757
 intervertebral, 274, 285, 300–302, 310, 324, 334–336, 723, 763
 optic, 650–651
Diverticulum
 ileal, 135
Duct
 bile, 117, 126, 128, 130, 145, 150, 180–181
 cochlear, 697, 708–709, 814
 cystic, 128–129, 148–149, 151
 efferent, 206
 frontonasal, 686
 hepatic
   accessory, 151
   common, 128, 130, 148, 150–151, 176
   left, 149–151
   right, 149–151
 lactiferous, 4–6
 lymphatic, 43
 nasolacrimal, 641, 685–686, 693
 pancreatic, 125, 130, 150
   accessory, 130–131
   development of, 131
   main, 130
   ventral, 150
 parotid, 600–601, 604, 652, 656–657, 766, 817
 semicircular, 708–709, 814–815
 submandibular, 670, 742–744, 746
 thoracic, 29, 43, 67, 71–72, 74, 77, 79–81, 113, 172, 326, 751, 757–758
Ductus deferens, 103, 107–108, 111–112, 155, 198, 204, 206, 208, 210, 214–215, 250
Ductus reuniens, 708
Duodenal cap, 124
Duodenum
 ascending part of, 125t
 blood supply to, 127
 bulb of, 179
 descending part of, 125t
 general images of, 92, 114, 116, 118, 126, 128, 140, 151, 154
 horizontal part of, 125t
 papilla of, 125
 referred pain to, 169t
 superior part of, 125t
Dura mater, 304, 313, 324–328

Ear
 auricle of, 696
 external, 697–698
 inner, 697–698
 middle, 697–698

 ossicles of, 697, 700
 pharyngotympanic tube, 28, 663, 671, 684–686, 693, 697, 700, 702–706, 764–765, 774–775
 semicircular canals of, 697–698, 702, 706–708, 715, 814–815
 tympanic cavity, 697–698, 700–706
Ejaculatory duct, 206, 208, 211, 214
Eminence
 arcuate, 615, 706
 frontal, 589, 591, 598–599
 iliopubic, 184, 190, 292, 353, 362
 intercondylar, 400, 402
 median, 635
 parietal, 591–592, 598–599
 pyramidal, 699, 702
Enamel, 675
Endometrium, 230, 270–271
Endoscopic retrograde cholangiography and pancreatography, 130, 151
Epicardium, 58
Epicondyle
 lateral, 362, 400, 498–499, 507, 522–524, 532
 medial, 362, 394, 498–499, 522–524, 532, 537
Epicondylus medialis, 388
Epididymis, 107–108, 110–111, 206, 214, 250
Epiglottis, 28, 669, 674, 770, 774, 776, 780, 787–788
Epiphyses
 of hand, 461
 upper limb bones, 460–461
 of wrist, 461
Esophageal varices, 153
Esophagus
 abdominal part of, 120
 cervical part of, 120
 computed tomography of, 6
 general anatomy of, 29, 34–35, 63–66, 70, 72, 75, 79, 81, 113, 176, 737, 758
 magnetic resonance imaging of, 335
 radiograph of, 124
 thoracic part of, 120
Ethmoid bone, 614, 640, 681
Expansion
 dorsal, 565
 extensor, 408–409, 562–563
Expiration, 24
External pudendal artery, 97
Eyelid
 structure of, 644
 superior, 640

Fabella, 449
Facet, 393
 articular
   inferior, 288, 300, 335
     magnetic resonance imaging of, 336
   superior, 277, 285, 288, 293, 334
Falx
 cerebellar, 612–613, 788
 cerebral, 610
Fascia
 alar, 723
 antebrachial, 469, 529
 axillary, 468
 brachial, 468–469
 buccopharyngeal, 723
 cervical, 723

 deep, 468, 730
 superficial, 723
 clavipectoral, 480
 cribriform, 345
 crural, 345
 deep, 410
 endopelvic, 238–239
 gluteal, 307–308, 311
 hypothenar, 541
 iliac, 241
 iliacus, 352
 investing, 723
 obturator, 192–193, 201, 241, 248
 omohyoid, 732, 759
 palmar, 542
 pectineal, 353–354
 pectoral, 3–4, 468
 pelvic, 241
   visceral, 208
 penis, 251
 perineal
   superficial, 245
 pharyngobasilar, 673, 768–769, 771, 775
 plantar, 424
 popliteus, 386, 390, 418
 presacral, 241
 pretracheal, 723
 prevertebral, 723, 756–757, 759
 prevesical, 241
 psoas, 155, 163, 192, 215, 241, 310, 352
 renal, 156, 162–163
 retrovesical, 197
 spermatic, 198, 250
   external, 106–108, 350, 353
   internal, 106–107, 109
 temporal, 600, 694
 temporalis, 608
 thenar, 541–542
 thoracolumbar, 161, 163, 307–308, 310, 312, 495
 transversalis, 102, 105, 112, 163
 upper limb, 468–469
 vesical, 241
Fascia lata, 100, 106, 255, 260, 263, 345, 352, 355, 357, 388, 452–454
Fascia transversalis, 102–103, 107
Fasciculus
 fronto-occipital, 630
 lenticular, 638
 mammillothalamic, 638
 superior longitudinal, 629
 uncinate, 630
Fasciculus cuneatus, 635, 637
Fasciculus gracilis, 635
Fat
 epidural, 336
 extraconal, 645
 intraconal, 645
 intrarenal, 180–181
 parapharyngeal, 715
 pararenal, 162–163
 perirenal, 156, 162–163, 177
 pre-epiglottic, 712
 prefemoral, 402
 retropubic, 224
 subperitoneal, 156
 suprapatellar, 402

Fat pad
  buccal, 604, 693, 790–791
  infrapatellar, 392, 403
  synovial, 528
Femur, 339, 396–397, 401
  condyle of
    lateral, 394
  head of, 186, 210, 270, 362, 376, 381
    blood supply to, 382
    ligament of, 383
  neck of, 240, 291, 377
  shaft of, 382
  transverse section of, 450
Fibers
  intercrural, 97
  postganglionic parasympathetic, 813
  postsynaptic parasympathetic, 221
Fibula, 338–339, 345, 397, 411, 415
  head of, 363, 394, 401, 407, 415
  neck of, 415
  transverse section of, 450
Fissure
  cerebral
    longitudinal, 622, 794
    transverse, 626
  lateral, 639
  oblique, 25, 35
  orbital
    inferior, 640, 655, 659, 690
    superior, 593, 596, 615, 621, 640, 690–691
  palpebral, 645
  primary, 637
  pterygomaxillary, 655, 765–766
  tympanomastoid, 654
Flexure
  colic
    left, 120
    right, 118, 132, 140
  hepatic, 134, 818
  splenic, 177, 335, 818
Flocculus, 637, 714
Folds
  alar, 392
  aryepiglottic, 770, 781–783
  axillary, 2
    anterior, 2, 475
    posterior, 2, 475, 503
  conjunctival, 640
  gastric, 121, 124
  gastropancreatic, 118
  glossoepiglottic, 770, 782, 785
  ileocecal, 135, 232
  mallear, 699
  palatine, 672
  pharyngoepiglottic, 771
  rectal, 199
  rectouterine, 224–225, 238–239
  sacrogenital, 198
  salpingopharyngeal, 770, 774
  semilunar, 133
  sublingual, 670, 746
  synovial, 301, 392, 444, 451, 528–529,
      572–573
  umbilical
    lateral, 112, 204, 225
    median, 112, 204
  vestibular, 676, 781–783, 786–787
  vocal, 781–783, 786–788

Follicles, 270–271, 774
Fontanelles
  anterior, 598–599
  anterolateral, 598
  posterior, 598
  posterolateral, 598
Foot, 338
  arteries of, 425
  bones of, 426–427
  development of, 451
  dorsal artery of, 423t
  dorsum of, 408–409, 444
  muscles of, 410
  sole of, 424, 446–447
Foramen
  alveolar, 655
  costotransverse, 278
  ethmoidal, 615
    anterior, 640, 689
    posterior, 640, 689
  incisive, 594, 764–765
  infraorbital, 589, 640
  interventricular, 627–628, 632, 634
  intervertebral, 274, 300, 336
  jugular, 593–595, 615, 702, 764–765, 820, 825
  mandibular, 655, 660, 677
  mental, 589, 591, 654
  nutrient, 383
  obturator, 184, 190, 291, 362, 379
  omental, 114, 116–117
  optic, 689
  palatine
    greater, 594, 672, 681, 764–765
    lesser, 594, 672, 681, 764–765
  parietal, 593
  sacral, 186, 188, 290, 293, 295
  sciatic
    greater, 188, 192, 298–299, 372–373
    lesser, 188, 192, 299
  sphenopalatine, 655, 681, 683, 687, 690
  sternal, 16
  stylomastoid, 594, 809
  supraorbital, 589
  transverse, 16, 278, 284, 321, 334–335
  vertebral, 277, 285
  zygomaticofacial, 589, 591, 640
Foramen cecum, 615, 666, 669, 770
Foramen lacerum, 594, 615, 621, 764–765
Foramen magnum, 335, 593, 595, 615, 713, 814
Foramen ovale, 58, 594, 615, 620–621, 655,
      704–706, 764–765
Foramen rotundum, 615, 644, 688–691
Foramen spinosum, 594, 609, 615, 655, 764–765
Forceps major, 633, 716–720
Forceps minor, 638, 716–720
Forearm
  arteries of, 530–531
  bones of, 532
  muscles of, 532, 533t, 534, 555t, 556–557
Fornix, 631, 633, 636, 638, 713, 788
  left, 635
  posterior, 224
  right, 635
  superior conjunctival, 645
Fossa
  acetabular
    anatomy of, 291, 378–380
    blood supply, 383

  canine, 589, 591
  cerebellar, 615
  cranial, 592
    middle, 609, 621
    nerves of, 620–621
    posterior, 612
    vessels of, 620–621
  cubital, 518–520
  digastric, 655
  gallbladder, 148–149
  glenoid, 483, 491, 511
  hypophyseal, 597, 615
  iliac, 175, 184–186, 292, 362
  incisive, 589, 672
  of incus, 707
  infraspinous, 499
  infratemporal, 654
  inguinal
    lateral, 112, 205
    medial, 112, 205
  intercondylar, 363
  interpeduncular, 824
  intrabulbar, 197
  ischioanal, 112, 204, 245, 248, 255, 257,
      259–260, 263, 268–269
  ischiorectal, 240
  lacrimal sac, 640
  mandibular, 594–595, 655, 665, 691, 764–765
  navicular, 196
  oval, 56
  pararectal, 198, 225
  paravesical, 198, 205, 225
  popliteal, 384–386, 403
  pterygopalatine, 655, 659, 688, 690–691
  radial, 524
  of round window, 699
  scaphoid, 672
  subarcuate, 706
  sublingual, 655
  subscapular, 498–499
  supraspinatus, 498
  supravesical, 112, 205
  temporal, 588, 654–655
  tonsillar, 774
  triangular, 696
Fovea
  pterygoid, 655, 665
Frenulum
  labial, 676
  of tongue, 670, 746
Frontal bone, 589–590
  nasal margin of, 680
  nasal spine of, 680
  orbital part of, 614, 694
  zygomatic process of, 591

Galactogram, 6
Gallbladder
  body of, 151
  fundus of, 92, 113, 151
  general images of, 114–115, 118, 120, 124,
      128, 130, 142–143, 148
  neck of, 151
  referred pain from, 169t
  venous drainage of, 148
Ganglion
  aorticorenal, 168, 170
  celiac, 75, 155, 168, 171, 174, 818

Ganglion—*continued*
   cervical, 757, 762, 768–769
   cervicothoracic, 75
   ciliary, 642–644, 803–806, 822
   cochlear, 709, 814
   geniculate, 688, 702, 707, 809, 812
   mesenteric
      inferior, 168, 170, 221
      superior, 168, 170
   otic, 671, 806, 817, 822
   parasympathetic, 221
   pterygopalatine, 659, 661, 673, 682, 688, 690,
      806, 809, 813, 822
   spinal, 17, 300, 319, 322, 324, 328, 618, 763
   stellate, 75
   submandibular, 743–744, 806, 812, 822
   sympathetic, 78–80, 165, 168, 170, 174,
      219–220, 237, 239, 328, 761
   thoracic, 762
   trigeminal, 616, 620–621, 663, 677
   vertebral, 757
   vestibular, 814
Gas
   in stomach, 39
Gemellus
   inferior, 367, 368t, 370–371, 373
   superior, 210, 367, 368t, 370–371, 373, 381
Genu
   of corpus callosum, 627
Gingiva proper, 676
Glabella, 588, 590
Glands
   buccal, 658, 668
   bulbourethral, 204, 206, 214, 242
   ciliary, 645
   labial, 743
   lacrimal, 604, 642–643, 688, 711, 803,
      808–809, 813
   lingual, 670, 746
   mammary, 5
   palatine, 672–673, 773
   parathyroid, 753, 757, 769, 780
   parotid, 600, 652, 730, 768–769, 790–791, 817
   pineal, 632–633, 639
   pituitary, 712
   prostate, 112, 207–208, 249
      peripheral zone of, 211
      transition zone of, 211
      ultrasound scan of, 211
   sublingual, 669–670, 694–695, 712, 746, 777,
      790–791, 812–813
   submandibular, 712, 732, 738, 742–743,
      747–748, 776, 812
   suprarenal, 155–156, 174
      left, 93, 125, 170–171, 176
      magnetic resonance imaging of, 335
      right, 93, 117, 125, 156, 171, 177
   thyroid, 723, 737, 750–751, 753, 755, 757
Globus pallidus, 631, 716–720
Grasping, 576–577
Groove
   atrioventricular, 46
   carotid, 615, 691
   costal, 13
   facial artery, 654
   greater petrosal nerve, 615, 814
   infraorbital, 640
   intertubular, 498–499
   mylohyoid, 655

   obturatory, 184
   occipital, 594
   petrosal sinus
      inferior, 614
      superior, 614
   prechiasmatic, 615, 619
   radial, 498
   sagittal sinus
      superior, 614
   sigmoid sinus, 614, 706
   transverse sinus, 614–615
   trigeminal nerve, 593, 706
Gyrus
   angular, 629
   cingulate, 713
   dentate, 633, 638
   frontal
      inferior, 629
      middle, 629
      superior, 629
   lingual, 632
   occipitotemporal, 632
   parahippocampal, 632–633
   postcentral, 626
   precentral, 629
   supramarginal, 629
   temporal
      inferior, 629
      superior, 629
Gyrus rectus, 716–720

Hamate, 461, 469, 532, 554, 567–569, 573–574
Hand
   arterial supply of, 552, 553t
   bones of, 532, 568–569
   cutaneous innervation of, 559
   dorsum of, 560–562
   lateral aspect of, 564–566
   medial aspect of, 567
   skin creases of, 540
Haustra, 132–134
Heart, 46–47, 50
   apex of, 26–28, 46, 58–59, 92
   auscultation of, 44
   conduction system of, 62
   crux of, 52, 62
   magnetic resonance angiogram of, 88
   sternocostal surface of, 49
   surface markings of, 44–45
Helix, 696
Hematoma
   subdural, 610
Hemorrhage
   subarachnoid, 610
Hiatus
   aortic, 166
   esophageal, 166, 171
   semilunar, 685–687, 692, 694
   urogenital, 193
Hip bone, 292, 338, 379
Hippocampus, 631, 638
Homonymous hemianopsia, 802
Hook of hamate, 568–569, 573
Humerus
   capitulum of, 502
   head of, 483, 517
   lymph nodes of, 9
   neck of, 511
   proximal, 460

   shaft of, 460
   trochlea of, 526
Hyaline plate, 304
Hyoid bone, 283, 669, 726–727, 735, 740,
      744–745, 748, 755, 768–769, 778–779,
      783, 787, 789, 817
Hypophysis, 620–621, 684, 686, 713, 824
Hypothalamus, 627, 826
Hysterosalpingogram, pelvic, 230–231

Ileum
   distal, 132
   general images of, 92, 120, 133, 136, 140, 232
   proximal, 132
   referred pain to, 169t
Ilium, 185, 192, 268–270, 291, 298, 339,
      372–373
Incisors, 676, 678
Incus, 697–698, 707
Infundibulum, 616–617, 619, 633, 635, 687, 713
   ethmoidal, 692
   pulmonary, 84
Inguinal hernia, 109
Inguinal region
   of female, 104–105
   of male, 100–103
Inspiration, 24
Insula, 638–639
Intertendinous connection, 561
Intestine, 268–269
Iris, 640, 650
Ischium, 185, 192, 197, 210, 263, 339, 373
   body of, 292, 379
   ramus of, 254, 292, 379

Jejunum
   general images of, 92, 120, 126, 136
   proximal, 132
   referred pain to, 169t
Joints
   acromioclavicular, 498, 515
   atlantoaxial
      capsule of, 618
      lateral, 285
      ligaments of, 320
      median, 285
   atlanto-occipital, 320, 618
   calcaneocuboid, 412, 442t
   carpometacarpal, 551, 570
   costochondral, 10
   costotransverse, 14–15
   craniovertebral, 321
   cubonavicular, 443
   cuneonavicular, 442t, 443
   elbow
      articular cavity of, 528
      bones of, 524
      imaging of, 524–525
      lateral aspect of, 527
      medial aspect of, 526
      posterior aspect of, 522–523
      transverse section of, 529
   facet, 763
   hip
      magnetic resonance imaging of, 381
      radiograph of, 380
   incudomalleal, 700, 707
   incudostapedial, 700
   intercarpal, 571

interchondral, 12
intermetatarsal, 442t
interphalangeal, 442t
    distal, 563, 575
    posterior, 560
    proximal, 563, 575
    of thumb, 560
manubriosternal, 10, 12, 48
metacarpophalangeal, 560, 575
metatarsophalangeal, 442t, 448
radiocarpal, 572
radioulnar
    distal, 525, 570–571
    proximal, 525
sacrococcygeal, 185
sacroiliac, 184–185, 190–191, 290–291, 294
shoulder
    anterior aspect of, 510
    magnetic resonance imaging of, 517
    posterior aspect of, 511
    radiograph of, 516
    synovial capsule of, 512
    transverse sections through, 483
sternoclavicular, 12, 82, 486, 756
sternocostal, 12
subtalar, 412, 442t
talocalcaneonavicular, 441
tarsal, 441
tarsometatarsal, 442t, 443
temporomandibular
    articular disc of, 653
    movements of, 663
    sectional anatomy of, 665
tibiofibular
    distal, 406
    proximal, 394, 406
uncovertebral, 282–283, 334, 727
xiphisternal, 10, 12, 48
zygapophysial, 283, 290, 300–302, 334, 336,
        727
Junction
    duodenojejunal, 117, 140
    esophagogastric, 121
    gastroesophageal, 179–181
    ileocecal, 120, 137
    xiphisternal, 479

Kidneys, 118
    anomalies of, 160
    calices of, 157
    cortex of, 157, 180–181
    ectopic pelvic, 160
    external features of, 157
    horseshoe, 160
    internal features of, 158
    left, 125, 154, 156, 178, 335
    lymphatic drainage of, 172–173
    medulla of, 157
    referred pain to, 169t
    right, 93, 125, 128, 174, 177, 335
    segments of, 159
    structure of, 157
Knee
    anastomoses, 398
    anterior aspect of, 388–389
    articular muscle of, 396
    articular surfaces of, 393
    distended, 396–397

lateral aspect of, 391
ligaments of, 393–394
magnetic resonance imaging of, 400–403
medial aspect of, 390
open view of, 392
radiograph of, 402
Kyphosis
    sacral, 275
    thoracic, 275

Labium majus, 104–105, 224, 233, 257, 268–269
Labium minus, 224, 233, 257
Labrum
    acetabular, 100, 376, 380
    glenoid, 483, 512, 515
Labyrinth
    membranous, 708
Lacrimal bone, 589–590, 640, 680, 689
Lacuna
    lateral venous, 611
Lamina, 334
    choroidocapillary, 651
    osseous spiral, 815
    thyroid, 765–766, 785
Large intestine
    colon
        ascending, 92–93, 117, 120, 125, 132, 134,
            136–137, 140, 152, 174, 176, 178
        barium enema of, 134
        descending, 92–93, 117, 120, 125, 132–134,
            139–141, 176, 178
        referred pain to, 169t
        sigmoid, 93, 120, 133–134, 138, 140–141,
            152, 198, 228, 264–267, 270
        transverse, 92, 114–115, 118, 120, 132–134,
            137–138, 141, 177, 179
    lymphatic drainage of, 172–173
Laryngocele, 783
Laryngopharynx, 335, 787
Larynx, 28, 120, 754, 781–785
Leg
    anterior compartment of, 404, 406
    arteries of, 407
    crural region of, 338
    magnetic resonance imaging of, 453–455
    muscles of, 410–411
    nerves of, 407
    posterior compartment of, 415–421
Lemniscus
    lateral, 635–637
    medial, 635–637
Ligaments
    acetabular, 378, 380, 383
    acromioclavicular, 508, 510, 513
    alar, 321
    anococcygeal, 240
    anterior longitudinal, 186
    anular, 526, 528
    arcuate
        general images of, 164, 166
        lateral, 166
        medial, 166, 174
        median, 166–167
    atlantoaxial, 321
    bifurcate, 433, 435, 444
    bladder, 208
    broad, of uterus, 224–225, 227–229, 238,
        270–271

calcaneocuboid, 435, 444–445
calcaneofibular, 412, 426, 433, 435, 438, 444
calcaneonavicular, 431, 445
    lateral, 445
    plantar, 445–447
calcaneotibial, 434
carpal
    palmar, 469
    transverse, 469
cervical, 241
check, 642–643, 645, 803
collateral
    fibular, 387, 391, 393–396, 399, 401
    radial, 527, 570
    tibial, 387, 390, 392–393, 395–396
    ulnar, 526, 529, 570
commissural
    distal, 541
    proximal, 541
conoid, 512
coracoacromial, 493, 508, 510–512, 515
coracoclavicular, 492, 508
coracohumeral, 493, 502, 510, 515
coronary, 117, 142–143, 154, 390, 393, 395,
        399
costoclavicular, 486
costotransverse, 15, 17
    lateral, 14, 18, 313
    posterior, 313
    superior, 313
costovertebral articulations, 15
costoxiphoid
    anterior, 12
cricothyroid, 751, 755, 778–779, 783
cricotracheal, 751, 755
cricovocal, 782–783
cruciate
    anterior, 393–395, 401, 403
    posterior, 393–395, 397, 401, 403
cruciform, 320–321
cubonavicular, 447
cuneonavicular, 433
cutaneous, 549
deltoid, 421, 433, 445, 447
denticulate, 322–323, 325, 327–328
falciform, 113–114, 117–118, 128, 142–144,
        154, 176, 180–181
fundiform, 101
gastrocolic, 119
gastrosplenic, 115, 117, 121, 156
glenohumeral, 514
    inferior, 514
        lateral, 511, 514
        middle, 511, 514
        superior, 514
    Grayson, 541
    hepatoduodenal, 121
    humeral
        transverse, 508, 512
    hyoepiglottic, 782–783
    iliofemoral, 186, 376–378, 381
    iliolumbar, 186, 188, 298–299
    inguinal, 96, 100–101, 103–105, 345, 350, 352,
        354
    intermetatarsal, 433, 447
    interspinous, 301, 304, 310
    intra-articular, 12, 15
    ischiofemoral, 188, 377

Ligaments—*continued*
  lacunar, 100, 192, 352–354, 381
  longitudinal
      anterior, 298, 302, 304, 320, 326, 723
        posterior, 302–304, 313
  lumbar, 298
  lumbocostal, 312
  meniscofemoral
      posterior, 394
  metacarpal, 541, 551, 570
  nuchal, 306, 308, 314, 317, 723, 728
  palmar, 548, 551, 571, 575
  palmar radiocarpal, 535–536, 551
  palpebral
      lateral, 604
      medial, 604
  patellar, 356–357, 359, 362, 390–391, 393,
          395–396, 399, 402–403, 406–407
  pectineal, 192, 352–353
  pelvic, 298
  perineal
      transverse, 254
  phrenicocolic, 115
  pisiform, 532
  pisohamate, 550–551, 567
  pisometacarpal, 532, 551, 567
  plantar, 431, 446
  popliteal
      oblique, 367
  pubic, 192
      inferior, 224
  pubofemoral, 186
  puboprostatic, 196, 207, 214, 249
  pubovesical, 241
  pulmonary, 34–35
  radiate, 15
  round
      of liver, 113, 128, 142–145
      of uterus, 104–105, 117, 224–225, 227–229,
          232–233, 238, 241, 257, 260, 268–269
  sacrococcygeal, 186
      posterior, 299
      ventral, 298
  sacroiliac, 186, 188, 192, 294
      anterior, 294, 298
      computed tomography of, 295
      interosseous, 294–295
      posterior, 294, 299, 372
  sacrospinous, 186, 188, 192, 210, 294,
          298–299, 373
  sacrotuberous, 186, 188, 192, 248, 294,
          298–299, 370–373, 377
  sphenomandibular, 658–659, 663, 671
  spiral, 815
  splenorenal, 117–118, 123, 154, 156
  sternoclavicular, 508, 757
  stylohyoid, 669, 726, 728, 777
  stylomandibular, 663, 728
  supraspinous, 188, 285, 299, 304
  suspensory, 4–6
      of axilla, 468
      of clitoris, 260
      of lens, 650
      of ovary, 224–225, 227, 229, 233, 238–239,
          241
      of penis, 107–108
  talocalcanean, 432–433, 435–437, 440, 444
  talofibular
      anterior, 444, 412, 433, 435
      posterior, 433, 436–438

  talonavicular
      dorsal, 433, 435, 444
  tarsometatarsal, 433, 446–447
  temporomandibular, 657
  thyroepiglottic, 782
  thyrohyoid, 755
  tibiofibular
      anterior inferior, 412, 432–433, 435, 444
      interosseous, 411, 436–437
      posterior, 433, 438
      transverse, 439
  transverse, 285
  trapezoid, 512
  triangular, 115, 117–119, 142–144, 154
  umbilical, 198, 204–205, 212, 224–225, 234,
          241
  uterosacral, 238, 241
  vestibular, 781
  vocal, 781–782
Ligamentum arteriosum, 41, 46–47, 49, 51,
          65–67, 79
Ligamentum flavum, 300–302, 304, 310, 763,
          788
Ligamentum patellae, 388
Ligamentum venosum, 143, 148
Line
  arcuate, 112, 205, 292
  gluteal
      anterior, 363, 372
      inferior, 363, 372, 379
      posterior, 363, 372
  intertrochanteric, 362, 376
  joint
      anterior, 295
      posterior, 295
  mylohyoid, 655
  nuchal
      inferior, 592, 594–595, 764–765
      superior, 314, 316, 592, 594–595
  oblique, 589, 591, 654
      posterior, 554
  pectinate, 199
  soleal, 363
  supracondylar
      lateral, 363
      medial, 363
  temporal, 588, 591
Linea alba, 2, 96–98, 100–101
Linea aspera, 363
Linea semilunaris, 96, 477
Lingula, 26, 35, 41, 655
Liver
  anterior surface of, 142
  caudate lobe of, 128, 143, 154, 176
  computed tomography of, 6
  general images of, 92–93, 120, 335
  inferior surface of, 143
  lobes of, 113, 115, 128, 142, 146t, 176
  lymphatic drainage of, 172–173
  posterior surface of, 143
  quadrate lobe of, 143, 149
  referred pain to, 169t
  round ligament of, 113
  segments of, 147
  superior surface of, 142
  veins of, 145
Lobes
  of brain
      frontal, 622, 626, 630
      occipital, 626, 630

  parietal, 626, 630
  temporal, 626, 630, 824, 826
  of liver, 113, 115, 128, 142, 146t, 176
Longissimus, 18
Lordosis
  cervical, 275
  lumbar, 275
Lower limb
  arteries of, 346
  bones of, 338
  cutaneous nerves of, 340
  dermatomes of, 341
  development of, 339
  fascia of, 345
  fascial compartments of, 345
  lymphatic drainage of, 344
  nerves of, 348, 349t
  veins of
      deep, 347
      superficial, 342–343
Lunate, 461, 532, 554, 568–569, 571–572, 574,
          582
Lung
  apex of, 25, 32, 39
  cardiac notch of, 26, 32
  computed tomography of, 6
  coronal section of, 28
  costal surface of, 25
  extent of, 30–31
  horizontal fissure of, 25
  inferior lobe of, 25
  innervation of, 42
  left, 25, 63, 66, 92
      bronchi of, 36–38
      cardiac notch of, 92
      lobes of, 32–33, 41
      magnetic resonance imaging of, 84–87, 335
      mediastinal surface of, 35
  lobes of, 25, 28, 32
  lymphatic drainage of, 43
  mediastinal surface of, 25, 34–35
  right, 6, 25, 63, 66, 92–93
      bronchi of, 36–38
      lobes of, 32–33, 41, 81
      magnetic resonance imaging of, 83–85, 87,
          335
      mediastinal surface of, 34
  root of, 25, 48
  superior lobe of, 25
  surface markings of, 45
  topography of, 26
Lymph nodes
  aortic arch, 43
  apical, 9
  appendicular, 173
  auricular, 601, 652
  axillary
      apical, 466
      central, 466
      humeral, 466
      posterior, 466
  bronchopulmonary, 35, 43
  buccal, 710
  celiac, 172–173
  central, 9
  cervical, 666
  cubital, 466, 518
  cystic, 173

gastric, 172
gastroomental, 172
hepatic, 173
humeral, 9
ileocolic, 173
iliac, 111
    common, 218, 237, 253, 262
    external, 218, 237, 253, 262
    internal, 218, 237, 253, 262
infraclavicular, 9, 466
infrahyoid, 666, 710
inguinal, 103, 203, 351
    deep, 218, 237, 253, 262, 344, 352, 710
    superficial, 218, 237, 253, 262, 344, 350,
        710
intermediate colic, 173
interpectoral, 9
jugulodigastric, 666
jugulo-omohyoid, 666, 710
lateral aortic, 173
lumbar, 174, 203, 237, 253, 262
mediastinal, 74
mesenteric, 172, 203
    inferior, 173, 218, 237, 253, 262
    superior, 173
occipital, 307, 710
pancreaticoduodenal, 172
paracolic, 141, 173
pararectal, 203
parasternal, 9, 22
paratracheal, 43, 735, 769, 780
parotid, 663, 710, 738
pectoral, 9
pectoralis, 466
phrenic, 173
popliteal, 344
preauricular, 652
prelaryngeal, 735
pretracheal, 735
pulmonary, 43, 74
pyloric, 172
retroauricular, 710
retropharyngeal, 666, 710
sacral, 218, 253, 262
splenic, 172
submandibular, 666, 710, 738
submental, 666, 710, 748
subscapular, 9
supraclavicular, 9
tracheobronchial, 43
Lymphangiogram, 351
Lymphatic system
    of breast, 9
    of kidneys, 172–173
    of large intestine, 172–173
    of liver, 172–173
    of lower limb, 344
    of lung, 43
    of lungs, 43
    of penis, 253
    of rectum, 203
    of scrotum, 253
    of spleen, 172
    of stomach, 172
    of testis, 111
    of upper limb, 466
    of urethra, 253
    of uterine tube, 237

of uterus, 237
of vagina, 237
Macula lutea, 650
Maculae, 814
Magnetic resonance imaging
    of abdomen, 176–179
    of arm, 579
    of carpal tunnel, 582–583
    of cervical spine, 334, 336
    of cranial nerves, 824–825
    of hip joint, 381
    of knee, 401–403
    of lumbar spine, 336
    of metacarpals, 584
    of palm, 584
    of patellofemoral joint, 400
    of pelvis
        female, 268–271
        male, 264–267
    of perineum
        female, 268–271
        male, 264–267
    of spinal cord, 335
    of thoracic spine, 335
    of thoracic vertebrae, 286
    of thorax, 82, 87
    of vertebral column, 274
    of wrist, 570
Malleolus
    lateral
        of ankle, 342, 405, 408, 415, 433, 435–437,
            439
        of knee, 404, 407
    medial
        of ankle, 338, 342–343, 405, 408, 415, 418,
            421, 433, 438–439, 446, 455
        of knee, 404
Malleus, 697–699, 701
Mammogram, 6
Mandible
    alveolar process of, 589
    angle of, 283, 588, 590, 654, 663, 725–727
    body of, 591, 790–791
    in child, 598
    head of, 654
    inferior border of, 588, 590
    neck of, 663, 693
    ramus of, 588, 591, 654, 790–791
Manubrium, 10, 40, 82, 479
Margin
    acetabular, 378
    costal, 10, 24
    infraorbital, 589
    nasal, 680
    supraorbital, 589
Mater
    arachnoid, 322, 325–328, 610, 763
    dura, 610–612, 618, 621, 694
    pia, 326, 610, 763
Maxilla, 589–590, 595
    alveolar process of, 589
    anterior surface of, 654
    in child, 598
    frontal process of, 591
    infratemporal surface of, 654
Meatus
    acoustic
        external, 590, 653, 656, 660, 665, 697–698,
            700
        internal, 593, 615, 698, 706–708, 825

auditory, 714
    inferior, 681, 684, 692–695
    middle, 681, 684, 686, 692, 694
    superior, 681, 684, 692, 694
Mediastinum, 78
    anterior, 29
    left side of, 79
    middle, 29
    posterior, 29, 75, 80
    subdivisions of, 29
    superior, 64–67, 70, 75
    topography of, 26
Medulla oblongata, 334–335, 626, 635, 684, 715,
        719, 788, 794, 825
Membrane
    atlanto-occipital
        anterior, 320
        posterior, 318, 320, 728
    basilar, 709
    costocoracoid, 468
    intercostal, 17–18, 22
    interosseous, 406–407, 433, 453, 469,
        526–527, 580
    obturator, 186, 192, 240, 353, 378, 383
    perineal, 197, 245, 248, 258, 260, 268–270
    quadrangular, 781
    synovial, 380, 383, 392, 399, 403, 524, 572
    tectorial, 321, 815
    thyrohyoid, 741, 749, 751, 755, 765–766,
        778–780, 783
    tympanic, 598, 697–699, 703, 809
    vestibular, 815
Meninges, 327
Meniscus
    lateral, 391, 393–396, 399, 401
    medial, 359, 390, 392–395, 399, 401
Mesoappendix, 135, 232, 238
Mesocolon
    sigmoid, 117, 133, 140, 198, 225, 228, 232
    transverse, 115–118, 141, 144
Mesoesophagus, 71
Mesometrium, 229
Mesosalpinx, 227, 229
Mesovarium, 227, 229
Metacarpals, 460
    fifth, 542, 567, 574
    first, 542, 568–569, 574, 584
    magnetic resonance imaging of, 584
    second, 562
Metatarsals, 339, 405, 421, 426–427, 431,
        446–447
    fifth, 447
    first, 433, 447
Meyer's loop, 630
Midbrain, 617, 619, 684, 714, 788, 801
Molars, 676, 678, 695
Mons pubis, 260
Mucosa
    alveolar, 676
    labial, 676
Muscles
    abductor digiti minimi, 412, 428t, 436–437,
        445, 532, 536–537, 544, 545t, 547,
        550, 567, 584
    abductor hallucis, 421, 427, 428t, 429, 430t,
        431, 438, 447
    adductor brevis, 210, 348, 359, 360t, 362–364,
        378, 452–454, 550

Muscles—*continued*

   adductor hallucis obliquus, 447

   adductor longus, 210, 348, 354–357, 359, 360t, 361–362, 364, 375, 378, 382

   adductor magnus, 255, 263–269, 348, 355, 359, 360t, 361–364, 366, 370–371, 375, 377–378, 387, 390, 398–399, 452–454

   adductor pollicis, 472, 532, 543, 545t, 550, 564–566, 584

      brevis, 541, 545t, 546–547

      longus, 473, 534, 538–539, 546–547, 551, 554, 555t, 556–557, 564, 566, 571, 580

   anconeus, 473, 499, 500t, 522–523, 554, 555t, 556–557, 581

   articular, 348

   aryepiglottic, 771

   arytenoids, 784t

   auricular

      anterior, 601

      posterior, 601, 652, 730

      superior, 601

   biceps brachii, 472, 478–479, 481, 483, 488, 500t, 502, 507–508, 510–511, 514–515, 517, 519–520, 534–535, 578–579, 581

      anomalies of, 521

      long head of, 22

      short head of, 22

   biceps femoris, 348, 362–363, 365–367, 369t, 370–371, 378, 384–386, 388, 393, 396–397, 403, 407, 416–417, 419–421, 452–454

   brachialis, 472, 479, 482, 498–499, 500t, 501–502, 505, 507, 519–520, 529, 534–536, 578–579

   brachioradialis, 473, 498–499, 501–502, 504–505, 507, 520, 522–523, 526, 529, 532, 534–536, 539, 551, 555t, 556–557, 581

   buccinator, 600–601, 603t, 604, 652, 656, 658, 694, 704–706, 765, 811–812

   bulbospongiosus, 196–197, 204, 243, 245, 246t, 258, 260

   coccygeus, 192–194, 196, 220, 242, 249, 259, 361

   coracobrachialis, 22, 472, 479, 482–483, 498–499, 500t, 501–502, 578–579

   corpus cavernosum, 214, 243, 245, 251–252, 254, 264–267

   corpus spongiosum, 206, 214, 243, 245, 251–252, 254, 266–267

   corrugator supercili, 600, 604

   cremaster, 101, 105–106, 109

   cricoarytenoid

      lateral, 784t

      posterior, 784t

   cricopharyngeus, 780

   cricothyroid, 749–751, 757, 765–766, 784t, 819

   dartos, 109

   deltoid, 307–308, 464–465, 473–474, 478, 494, 496t, 507, 514, 731–733

      spinal part of, 306

   depressor anguli oris, 600–601, 604, 652, 722

   depressor labii inferioris, 600, 722

   detrusor, 209

   digastric, 316, 652–653, 656, 669, 671, 694, 724, 728, 740t, 742, 745, 748, 761

   erector spine, 163, 306, 309–312, 314, 315t

extensor carpi radialis

      brevis, 473, 536, 544, 554, 555t, 556–557, 561–562, 565, 580–581

      longus, 473, 498–499, 505, 507, 520, 522–523, 536, 544, 554, 555t, 556–557, 561–562, 565, 580–581

   extensor carpi ulnaris, 473, 532, 554, 555t, 556–557, 561–562, 564, 567, 572, 580 581, 583

   extensor digiti minimi, 473, 555t, 556–557, 561–562, 580, 583

   extensor digitorum, 473, 549, 555t, 556–557, 560–562, 565, 572, 581, 583

      brevis, 348, 407–408, 410, 412–413

      longus, 348, 404, 405t, 406, 408, 410, 412–413, 455

   extensor hallucis

      brevis, 404–405, 410, 413

      longus, 348, 404–405, 405t, 408, 410, 413, 453

   extensor indicis, 473, 554, 555t, 556–557, 561–562, 572, 583

   extensor pollicis

      brevis, 473, 544, 548, 554, 555t, 556–557, 561–562, 564, 580–581, 583

      longus, 473, 544, 555t, 556–557, 560–562, 564–566, 581, 583–584

   extensor retinaculum, 561–562, 567

   facial expression, 600, 602, 603t

   fibularis, 405t

   fibularis brevis, 348, 407, 410–411, 413, 416, 433–434, 436–437, 444, 453, 455

   fibularis longus, 348, 405, 407, 410–411, 413, 416, 431, 434, 436–437, 444, 447, 453

   fibularis tertius, 348, 405, 408, 410, 412

   flexor carpi

      radialis, 472, 520, 532, 533t, 534–535, 538, 565, 573, 581, 583

      ulnaris, 522–523, 529, 532, 533t, 534–539, 546, 550–551, 567, 573, 580–581

   flexor digiti minimi, 430t, 544, 545t, 547, 550

   flexor digitorum

      brevis, 348, 428t, 429, 431, 436–437, 548

      longus, 348, 414t, 415, 417–421, 429, 432, 438, 453

      profundus, 472, 532, 533t, 535–537, 550–551, 554, 563, 583

      superficialis, 472, 532, 533t, 535–536, 538–539, 546, 548–551, 580, 584

   flexor hallucis

      brevis, 348, 429, 430t, 431, 447

      longus, 348, 414t, 415, 417–421, 424, 427, 429, 431–432, 434, 445–447, 453

   flexor pollicis

      brevis, 541, 544, 546–547, 550, 584

      longus, 472, 532, 533t, 534–535, 539, 548, 581, 583–584

   flexor retinaculum, 469, 547–548, 550–551, 571, 573, 582–583

   frontalis, 608

   gastrocnemius, 348, 361, 363, 365–367, 384–387, 390–391, 396–397, 403–404, 410, 414t, 415–417, 455

   genioglossus, 667t, 669, 694–695, 712, 743, 745, 787, 789–791, 821

   geniohyoid, 669, 694, 740t, 743–744, 747, 788–789, 821

   gluteus maximus, 192, 197, 210, 240, 245, 248,

255, 259, 263–265, 268–269, 306–309, 311, 345, 361, 365–367, 368t, 370–371, 381, 452–454, 494

   gluteus medius, 192, 264–265, 306, 363, 365–367, 368t, 370–371, 377, 381, 494

   gluteus minimus, 264–265, 359, 362–363, 366–367, 368t, 370–371, 376, 378

   gracilis, 348, 354–355, 357, 359, 360t, 362, 371, 384–386, 390, 405, 419–421, 452–455

   hyoglossus, 667t, 671, 743, 745, 765, 777, 790–791, 821

   hypothenar, 472, 583

   iliacus, 112, 155, 164t, 175, 204, 240, 264–267, 295, 348, 352, 355, 357, 358t, 364

   iliococcygeus, 193–194, 242, 248–249

   iliocostalis, 18

      cervicis, 309, 314

      lumborum, 309, 314

      thoracis, 309, 314

   iliopsoas, 184, 198, 255, 264–265, 268–269, 352–355, 358t, 376–377, 381–382

   infraspinatus, 473, 483, 497t, 504, 506, 514–515

   intercostal, 17, 156

      external, 19, 21, 24, 81, 311, 313, 322, 478

      innermost, 19, 21, 81

      internal, 19, 21, 23–24, 81

   interossei

      dorsal, 472, 544, 545t, 546, 549, 554, 557, 560, 565–566, 584

      palmar, 472, 545t, 546, 584

   interosseous

      dorsal, 431t

      first dorsal, 408

      plantar, 431t

   interspinales, 315t, 316

   interspinalis, 310, 318

   intertransversarius, 310, 312, 314, 315t, 319

      anterior, 761, 763

      posterior, 761, 763

   ischiocavernosus, 243, 245, 246t, 260

   latissimus dorsi, 4, 7, 71, 94, 163, 306–308, 310, 473, 476–477, 479, 482, 489–490, 496t, 498–499, 501–502

   levator anguli oris, 600, 604

   levator ani, 112, 193–194, 196, 204, 208, 214, 220, 233, 240–241, 245, 255, 258, 264–269

   levator costarum, 18, 21, 311–314, 315t

   levator labii superioris, 600, 603t

   levator labii superioris alaeque nasi, 600

   levator palpebrae superioris, 642–644, 646, 648t, 692, 694, 711, 803–804

   levator scapulae, 307–309, 318–319, 471, 473, 491, 496t, 729t, 731–733, 761, 790–791, 820

   levator veli palatini, 663, 673, 674t, 686, 703–706, 765–766, 771, 775

   longissimus

      capitis, 311, 316–319, 728

      lumborum, 314

      thoracis, 309–310, 314

   longus capitis, 318, 668, 729t, 758, 760t, 761

   longus cervicis, 762

   longus colli, 67, 78, 318, 668, 723, 729t, 737, 754, 758, 760t

   lumbricals, 421, 429t, 472, 536, 545t, 546, 563, 584

masseter, 600–601, 604, 652, 657, 662t, 694–695, 712, 715, 790–791
mentalis, 600, 603t, 604, 722
multifidus, 310–312, 314
mylohyoid, 657, 669, 671, 694, 739, 740t, 741–744, 747–748, 765–766, 788, 790–791
nasalis, 603t
oblique
   external, 2–3, 19, 22, 24, 71, 94–95, 97, 101, 112–113, 155, 162–163, 306–308, 310, 312, 476–477, 494
   inferior, 316–318, 648t–649t
   internal, 19, 24, 98, 101–103, 106–107, 112–113, 155, 161–163, 308, 310, 364
   superior, 316–318, 648t
obliquus capitis
   superior, 728
obturator externus, 197, 210, 228, 240, 255, 259, 264–270, 348, 360t, 362, 373, 375–376, 378
obturator internus, 112, 192–193t, 201, 204, 210, 220, 240, 263–270, 361, 367, 368t, 373, 376, 381
occipitalis, 307, 316
occipitofrontalis, 603t, 788, 812
omohyoid, 468, 480, 731, 734, 738, 740t, 741, 744, 749, 759, 786, 821
opponens digiti minimi, 536–537, 544, 545t, 550, 567, 584
opponens pollicis, 536–537, 545t, 547, 550, 565
orbicularis oculi, 600–601, 603t, 645, 652, 656, 812
orbicularis oris, 604, 812
palatine, 673
palatoglossus, 666, 667t, 668, 672, 674t, 772–773, 776
palatopharyngeus, 666, 674t, 766t, 771–773, 775–776
palmaris
   brevis, 472, 543, 546
   longus, 472, 533t, 534–535, 538, 543, 546, 548, 550, 580, 583
papillary, 58
   anterior, 57, 59, 61–62
   magnetic resonance imaging of, 84
   posterior, 57, 59, 61
   septal, 57
pectinate, 56
pectineus, 210, 263–270, 348, 353–355, 357, 359, 360t, 362, 364, 376, 378, 381–382
pectoralis major
   clavicular head of, 2–3, 475, 478, 480
   general images of, 22, 82, 94, 98, 464–465, 468, 474, 478, 479t, 481–482, 488–490, 502, 579, 731
   sternocostal head of, 2–3, 475, 478, 480, 732
pectoralis minor, 22, 466, 478–479, 479t, 482–483, 488, 498–499, 508, 579, 732–733
perineal
   transverse, 246t, 247, 249, 260–261
pes anserinus, 390, 396, 419–421
pharyngeal, 723
piriformis, 192–193, 195, 220, 264–265,

268–269, 361, 367, 368t, 371–372, 376
plantaris, 348, 363, 366–367, 385–387, 414t, 415, 453
platysma, 474, 600–601, 603t, 604, 669, 722, 730, 812
popliteus, 348, 363, 367, 375, 385–386, 396–397, 403, 414t, 415, 417–418, 455
princeps pollicis, 471
procerus, 600
pronator quadratus, 472, 532, 533t, 536–537, 544, 548, 551, 572, 581
pronator teres, 472, 501–502, 519–520, 529, 533t, 534–535, 554, 581
psoas, 125, 140–141, 163, 165, 177–178, 204, 264–267, 290, 295, 310, 335, 348, 355, 375
psoas major, 164t, 166, 175, 228, 240, 301, 322, 357, 358t, 359, 377
pterygoid
   lateral, 653, 657, 662t, 665, 712, 810
   medial, 659, 662t, 663, 712, 743, 745, 747, 777, 810
pubococcygeus, 192–195, 242, 248
puborectalis, 193–194, 196–197, 199, 210, 249, 263–265
pubovaginalis, 195, 259
quadratus femoris, 220, 255, 268–269, 358t, 363, 367, 368t, 370–372, 377, 396, 402
quadratus lumborum, 129, 155, 163, 164t, 165–166, 175, 177, 222, 310, 312
quadratus plantae, 429t, 436–438
radialis indicis, 471
rectus
   inferior, 641, 644, 646, 648t–649t, 692, 711, 804–805
   lateral, 641–644, 646, 648t–649t, 651, 803, 805
   medial, 641, 643, 646, 648t–649t, 650, 711, 803–804
   superior, 642, 646, 649t, 650, 692, 711, 803, 805
rectus abdominis, 2, 19, 22, 24, 96–98, 112–113, 132, 177, 197, 204–205, 210, 241, 264–265, 270
rectus capitis
   anterior, 318, 729t, 760t
   lateralis, 318, 671, 728, 729t, 760t, 761
   posterior
      major, 316–319, 728
      minor, 317–318, 728
rectus capitis lateralis, 618
rectus femoris, 255, 263–265, 348, 354–357, 358t, 359, 361–365, 376–378, 381, 388, 452–454
rhomboid major, 307–308, 473, 494–495, 496t, 499, 504
rhomboid minor, 307–308, 471, 473, 492, 494–495, 496t, 499, 504
rotator cuff, 497t
rotatores
   brevis, 313
   longus, 313
salpingopharyngeus, 674, 766t, 771, 775
sartorius, 255, 264–265, 348, 352, 354–357, 358t, 359, 361–364, 381, 384, 388, 393, 405, 416, 419–421, 452–454
scalene, 334
   anterior, 22–24, 70, 78, 80, 491, 723, 728,

729t, 731–733, 735, 737, 754, 758, 761–762, 786, 820
   middle, 22, 24, 80, 491, 723, 728, 729t, 731–733, 735, 754, 758, 761, 786, 820
   posterior, 22, 24, 308, 728, 729t, 732–733, 735, 758
scalenus
   medius, 319
   minimus, 70
semimembranosus, 348, 361, 363, 366–367, 369t, 370–371, 378, 384–387, 390, 397, 402, 416–417, 452–454
semispinalis, 18
   capitis, 308–309, 311, 314, 316–319, 728, 731, 790–791
   cervicis, 314, 316–318, 790–791
   thoracis, 314
semitendinosus, 348, 361, 366, 369t, 370–371, 378, 385–386, 390, 402, 416, 452–453
serratus anterior, 2, 4, 7, 22, 94–95, 97–98, 308, 474–478, 479t, 482–483, 488, 491–492, 504–505, 579, 732
serratus posterior
   inferior, 21, 71, 161–162, 308
   superior, 21, 24, 308
soleus, 348, 363, 367, 384–387, 404, 410, 414t, 415–418, 453, 455
spinalis, 309
   thoracis, 314
splenius, 308, 315t, 318
   capitis, 314, 317, 319, 728, 729t, 731–732, 790–791
   cervicis, 309, 311, 314, 319
sternocleidomastoid, 22, 308–309, 318–319, 334, 481, 652, 656, 723, 725, 731, 733, 738–739, 748–749, 754, 790–791, 820
sternohyoid, 22–23, 723, 739, 740t, 741, 744, 748, 750, 754, 759, 786, 821
sternothyroid, 22–23, 723, 738–739, 741, 748–751, 754, 759
styloglossus, 653, 667t, 668, 728, 745, 765–766, 776
stylohyoid, 653, 668, 728, 739, 740t, 742, 747, 765, 786
stylopharyngeus, 668, 670, 693, 728, 766, 766t, 769, 776, 817
subclavius, 22, 468, 479t, 489–490, 732–733, 759
subscapularis, 473, 478–479, 482–483, 489–491, 497t, 498–499, 501–502, 508, 514–515, 517, 579
supinator, 473, 520, 532, 537, 554, 555t, 556–557, 581
supraspinatus, 473, 479, 492–493, 497t, 498–499, 506, 513, 515
suspensory, 125
tarsal, 645
temporalis, 608, 656, 659, 662t, 663, 694–695
tensor of fascia lata, 240, 358, 367, 381
tensor tympani, 697–698, 701–703
tensor veli palatini, 671, 673, 674t, 686, 691, 697, 765–766, 773, 777
teres major, 306, 308, 471, 473, 479, 482, 489–490, 497t, 501–502, 504, 579
teres minor, 473, 483, 498–499, 506, 513–514, 579
thenar, 472

Muscles—*continued*
  thyroarytenoid, 784t
  thyrohyoid, 738–739, 740t, 741–743, 750, 786,
     790–791
  tibialis
    anterior, 348, 405, 405t, 406, 408, 413, 421,
      427, 433
    posterior, 348, 414t, 416, 418, 421, 427,
      431–432, 436–438, 445–446, 453, 455
  transverse abdominal, 19, 23–24, 96, 102–103,
     107, 113, 155, 163, 178
  transverse thoracic, 21–23
  transversospinal, 315t
  trapezius
    ascending, 306–307
    descending, 307, 495, 504
    general images of, 308, 317–319, 479, 492,
      496t, 723, 725, 730, 733, 790–791, 820
    inferior, 494
    transverse, 306–307, 495
  triceps brachii, 473, 479, 482, 499, 500t,
     501–505, 507, 515, 526, 554, 578–579
  uvulae, 673, 674t, 771–772
  uvular, 674
  vastus intermedius, 255, 264–265, 348, 356,
     358t, 359, 362–364, 396, 452–454
  vastus lateralis, 255, 263, 348, 355–356, 358t,
     359, 362–365, 376–377, 388, 391,
     452–454
  vastus medialis, 348, 356–357, 358t, 359,
     361–362, 364, 367, 388, 390, 396, 399,
     402, 416, 452–454
  zygomaticus
    major, 600–601, 652
    minor, 604
Myelogram, 325
Myocardium, 58
Myometrium, 230, 268–271
Myotome, 332

Nail bed, 558
Nasal bone, 590, 640, 690
Nasion, 588, 590, 609
Nasopharynx, 335, 597, 789
Navicular, 405, 421, 433, 445–446
Neck
  arteries of, 736
  bones of, 726
  cartilages of, 726
  veins of
    deep, 735
    superficial, 734
Nerves
  abducent, 616–617, 620–622, 642, 644, 794,
     796, 797t, 803–804, 823, 825
  abductor hallucis, 421
  accessory, 307, 319, 322, 327, 492, 668, 728
  alveolar
    inferior, 658–659, 668, 671, 677, 694–695,
      712, 806, 810
    superior, 677, 682, 688, 806, 808–809, 811
  anal
    inferior, 202, 219, 221, 245, 253
  auricular
    great, 601, 657, 730–731
    posterior, 601, 652, 812
  auriculotemporal, 600–601, 605t, 608, 659,
     661, 663, 806

axillary, 333, 473, 486, 487t, 506, 514
buccal, 601, 604, 605t, 652, 658–659, 661,
  806, 811
calcaneal, 424
caroticotympanic, 702
cavernous, 253
cervical, 618, 722
ciliary, 642, 805
clunial, 374t
coccygeal, 220
cochlear, 707
cranial (*see* Cranial nerves)
cutaneous, 219
  anterior, 3
    femoral, 222
  inferior lateral, 462, 505
  lateral, 164–165, 307, 474, 476, 494, 505,
    518, 529, 559
    anterior branch of, 3, 7, 94
    posterior branch of, a, 3, 7
  medial, 462, 518, 578–579
  perforating, 223
  posterior, 462, 558
  superior lateral, 462
  thigh, 257
dental, 810
ethmoidal
  anterior, 612, 643, 682, 803, 807
  inferior, 682
  posterior, 612, 643, 803, 807
facial, 616–617, 622, 635, 637, 652, 698, 728,
  738, 761, 794
  branches of, 608, 812
  cervical branch of, 730
  primary root of, 796, 797t, 803, 823, 825
  temporal branches of, 652, 812
femoral, 112, 155, 164–165, 175, 204, 210,
  222, 240, 268–269, 333, 348, 349t,
  352–354, 364
femoral cutaneous, 222–223
  anterior, 340, 354
  lateral, 340, 352, 354
  posterior, 340, 370–371, 381
fibular
  common, 333, 348, 349t, 366, 384–386,
    406–407, 410, 416, 418
  deep, 333, 340, 348, 349t, 406–407
  superficial, 333, 340, 348, 349t, 407, 410,
    453
frontal, 605t, 620–621, 642, 646, 806–807
gastric, 818
genitofemoral, 103–106, 108, 155, 164–165,
  198, 222, 352
  femoral branch of, 222, 340, 350
  genital branch of, 222, 340
glossopharyngeal, 327, 593, 616–618, 622,
  635, 653, 666, 670, 706, 728, 765, 776,
  794, 796, 797t, 803, 816–817, 823, 825
gluteal
  inferior, 223, 371, 374t
  superior, 223, 371, 374t
hypogastric, 168, 170, 202
  left, 219, 239, 253
  right, 219, 221, 237, 239
hypoglossal, 327, 616–617, 652–653, 670–671,
  693, 739, 741–742, 745, 747, 766, 769,
  777, 796, 797t, 803, 821, 823, 825
iliohypogastric, 94, 98, 101, 106, 162,
  164–165, 222, 340, 364, 476, 494

ilioinguinal, 97–98, 101, 104–106, 108,
  164–165, 222, 250, 257, 340, 350, 364
infraorbital, 604, 605t, 641, 644, 659, 661,
  677, 682, 687–688, 693–694, 806,
  808–809, 811
infratrochlear, 605t, 803, 807
intercostal, 322, 326, 482
  anterior ramus of, 20
  collateral branches of, 17–18
  first, 22, 70
  general anatomy of, 17–19, 78–81
  lateral cutaneous branch of, 19
  ninth, 19
  posterior ramus of, 20
  second, 23
intercostobrachial, 3, 7, 94, 462, 474, 476,
  482, 489–490
intermediate, 321, 617, 637, 794, 796, 797t,
  803, 812, 823, 825
interosseous, 333
  anterior, 537, 580
  posterior, 473, 557
labial
  posterior, 257
lacrimal, 604, 605t, 620–621, 642–644, 690,
  806, 813
laryngeal
  internal, 666
  recurrent, 751, 754
    left, 42, 49, 65–67, 75, 79, 757, 768–769,
     818
    right, 25, 42, 66, 70, 762, 765, 768–769,
     771, 780, 785, 818
  superior, 741, 750–751, 757, 765–766, 783
lingual, 658–659, 661, 668, 677, 712, 747, 777,
  810
long thoracic, 487t, 488
mandibular, 616, 621, 661, 691, 712, 765–766,
  806, 810
masseteric, 657–658, 810–811
maxillary, 616, 620–621, 644, 658–659, 661,
  677, 682, 687–688, 690, 806, 808–809,
  811, 824
  meningeal branch of, 612
medial antebrachial cutaneous, 487t
medial brachial cutaneous, 487t
median, 333, 472, 486, 487t, 520, 534,
  538–539, 546–547, 550, 571
meningeal
  recurrent, 300
mental, 604, 605t, 661, 806, 810
musculocutaneous, 333, 472, 482, 486, 487t,
  490, 520, 534
mylohyoid, 658
nasal
  external, 605t
nasociliary, 621, 642–643, 646, 803, 805–807
nasopalatine, 673, 682, 687, 773
obturator
  anterior, 354, 376
  branches of, 222
  cutaneous branches of, 340, 364
  general images of, 112, 164–165, 175, 192,
    194, 215, 220, 222, 240, 333, 348, 349t
  posterior, 376
occipital
  greater, 307, 316, 318–319, 494, 608, 730
  lesser, 307, 494, 608, 730
  third, 307, 319, 608

oculomotor, 616–617, 619–622, 635, 637, 642, 644, 646, 794, 796, 797t, 803–804, 823, 825

olfactory, 616, 682, 796, 797t, 800, 803, 823, 825

ophthalmic, 616, 620–621, 661, 806–807

optic, 619, 622, 632, 635, 642–643, 646, 650, 686, 694–695, 711–712, 794, 796, 797t, 801, 803, 823–825

palatine
    greater, 674, 682, 686, 688, 690, 772–773, 809
    lesser, 674, 682, 686, 688, 690, 772–773, 809

palmar digital, 543, 546–547, 549

pectoral
    lateral, 479, 487t, 488, 733
    medial, 479, 487t, 488–490

perineal, 192, 219, 221, 253

petrosal
    deep, 812–813
    greater, 620–621, 701–702, 707, 809, 812–813
    lesser, 620–621, 701, 703, 707, 816

pharyngeal, 653, 682, 690

phrenic
    general anatomy of, 25, 48, 65, 70, 78, 731–732, 735, 754, 756, 758–759, 761
    left, 42, 71, 73, 79
    right, 42, 71

plantar
    lateral, 333, 340, 348, 430, 445
    medial, 333, 340, 348, 418, 430

pterygopalatine, 688

pudendal, 168, 202, 204, 219–223, 240, 253, 261, 370, 374t, 381

radial, 333, 462, 473, 486, 487t, 489–490, 535, 537, 539, 557, 559, 578–579
    dorsal digital branch of, 558
    palmar digital branch of, 558
    superficial branch of, 558, 564

rectal, 192

saphenous, 333, 340, 349t, 354, 364
    infrapatellar branch of, 340

scapular, dorsal, 487t, 731

sciatic, 164–165, 168, 175, 204, 210, 214, 219, 221–223, 240, 255, 263, 348, 349t, 366, 370–372, 374t, 381, 452–454

scrotal
    posterior, 219, 221, 245, 253

spinal, 300, 322, 326, 328, 336, 473

spinal accessory, 307, 322, 327, 491, 616–618, 622, 652, 657, 730, 732, 739, 769, 794, 796, 797t, 803, 820, 823, 825

splanchnic
    general anatomy of, 17, 71, 75
    greater, 75, 78–81, 168, 170
    least, 75, 168, 170
    lesser, 75, 80, 168, 170
    lumbar, 168, 170, 174, 202, 221, 237, 253
    pelvic, 196, 202, 215, 219–221, 223, 237, 253
    sacral, 202, 237, 253

subcostal, 94–95, 162, 164–165, 340, 476

subscapular, 487t, 488

supraclavicular, 3, 462, 474, 722, 730–732

supraorbital, 604, 605t–606t, 608, 620–621, 641–642, 692, 807

suprascapular, 487t, 489–490, 492, 506

supratrochlear, 604, 605t–606t, 608, 620–621, 641, 807

sural, 340, 349t, 412

sural cutaneous
    lateral, 340, 384
    medial, 340, 384, 416
    posterior, 340

temporal, 658

tentorial, 612

thoracic
    anterior ramus of, 17
    long, 94, 476, 482–483, 731
    posterior ramus of, 17–18, 313
    twelfth, 106

thoracodorsal, 482, 487t, 488

tibial, 333, 340, 348, 349t, 366, 384–386, 416, 438

trigeminal, 616, 618, 622, 635, 637, 826
    groove for, 593
    mandibular division of, 806, 810
    maxillary division of, 806, 808–809
    motor root of, 617, 635, 794, 796, 797t, 803, 823, 825
    ophthalmic division of, 806–807
    sensory root of, 617, 620–621, 635, 794, 796, 797t, 803, 823, 825

trochlear, 616–621, 642, 794, 796, 797t, 803–804, 823, 825

tympanic, 699, 701

ulnar, 487t, 505, 520, 534–535, 537, 539, 546–547, 552, 559, 580, 583
    deep branch of, 571
    dorsal branch of, 462, 558
    dorsal digital branch of, 558
    general images of, 333, 462, 472
    palmar digital branch of, 558

vagus, 48, 67, 616–618, 622, 652–653, 728, 750–751, 756–757, 762, 769, 771, 786, 794, 796, 797t, 803, 819, 823, 825
    auricular branches of, 657
    left, 25, 42, 49, 63, 65–66, 79, 758, 818
    right, 25, 42, 65–66, 70, 75, 78, 758, 818

vestibular, 707

vestibulocochlear, 616–618, 622, 635, 637, 698, 709, 794, 796, 797t, 803, 814, 823, 825

zygomatic, 605t, 688, 690, 806–809, 813

zygomaticofacial, 604, 688, 690

zygomaticotemporal, 608, 657, 688, 690

Nervus spinosus, 612, 810

Nipple, 5–7

Node
    atrioventricular, 62
    lymph (see Lymph nodes)
    sinuatrial, 52, 62
    tracheobronchial, 43

Nose
    bones of, 680
    cartilage of, 680
    surface anatomy of, 680

Nostril, 680

Notch
    acetabular, 379
    cardiac, 26, 32, 35
    clavicular, 12
    costal, 12
    interarytenoid, 770

intercondylar, 394

jugular, 12, 479, 725, 751

mandibular, 654

mastoid, 594

radial, 528

sacral
    superior, 293

sacrococcygeal, 293

sciatic
    greater, 184, 221, 292, 363, 379
    lesser, 184, 192, 292, 363, 379

supraorbital, 589, 640

suprascapular, 510

suprasternal, 2, 12, 475, 479

tentorial, 613, 619

thyroid, 755

trochlear, 526–527, 640

ulnar, 569

vertebral
    inferior, 277, 300
    superior, 277, 300

Nucleus
    amygdaloid, 631, 633, 638
    caudate, 631, 634–636, 638–639, 713, 716–720
    dentate, 637
    geniculate
        lateral, 638
        medial, 638
    habenular, 639
    lentiform, 630–631, 634–635, 638–639, 716–720
    oculomotor, 619
    red, 619, 636, 638, 715
    reticular, 638
    thalamic
        anterior, 638
        lateral, 638
        medial, 638

Nucleus pulposus, 302, 304, 310

Occipital bone, 320, 335, 590, 595, 599, 614, 715, 723, 825
    jugular process of, 618

Olecranon, 504–505, 522–523, 554

Olive, 617, 635, 794

Omentum
    greater, 113–116, 121, 132–133, 140
    lesser, 114–115, 117, 121, 143–144, 171

Orbicular zone, 377, 380

Organ of Corti, 709, 815

Orifices
    atrioventricular
        left, 59
        right, 56
    ileocecal, 135
    ureteric, 209, 228, 233
    urethral
        external, 214, 221, 251
        female, 258–260
        internal, 209
    vaginal, 258–260

Oropharynx, 335, 787, 789–791

Os trigonum, 449

Ossicles, 697, 700

Ostium
    abdominal, 227
    external, 233
    internal, 233
    uterine, 233

Ovary, 224, 227, 230, 270
  left, 117, 271
  magnetic resonance imaging of, 268–269

Pain
  referred, 169
Palate, 28
  hard, 672–673, 695, 772, 789
  soft, 672–673, 674t, 772, 789
Palatine bone, 595, 640, 672, 774
Palm
  deep dissection of, 550
  lymphatic plexus of, 466
  superficial dissection of, 546–547
  synovial sheath of, 548
Pancreas, 92–93, 116, 118
  blood supply to, 127
  body of, 126, 177
  general images of, 125–126
  head of, 126, 129, 177, 179
  magnetic resonance imaging of, 177
  tail of, 115, 126, 154, 156, 179
  uncinate process of, 177, 180–181
Papilla
  incisive, 672
  lacrimal
    inferior, 641
  major duodenal, 125
  minor duodenal, 125
  vallate, 771, 774
Parietal bone, 589–590, 610
Patella, 343, 345, 356, 388, 392, 400
  bipartite, 449
Pectoral girdle, 508
Pectoral region, 2–4
  female, 4
  lymph nodes of, 9
  male, 2–3
  superficial dissection of, 474
  surface anatomy of, 477
  trunk, 476–477
Pedicle
  of cervical vertebrae, 283
Peduncles
  cerebellar, 635–637, 714, 716–720, 824
Pelvic diaphragm, 194
Pelvic wall, 214–215
Pelvis
  anatomical position of, 184
  anterior, 204
  anterior view of, 187
  bones of, 185
  bony, 184
  divisions of, 185
  female, 187, 190, 224
    autonomic nerves of, 239
    hysterosalpingogram of, 230–231
    magnetic resonance imaging of, 268–270
    ultrasound of, 230–231, 271
  floor of, 194–195
  ligaments of, 186, 188
  magnetic resonance imaging of, 266–267
  male, 187, 190, 197–198, 206–207, 210
    innervation of, 219
    lymphatic drainage of, 218
    veins of, 217
  posterior view of, 189

renal
  autonomic nerve supply of, 170
  general images of, 139, 158, 177
Penis, 250–251
  artery of, 254
  body of, 244, 250, 252
  bulb of, 197, 206, 214, 248, 254
  corona of, 244, 250–251
  crus of, 243, 248, 254, 264–265
  dorsal nerve of, 219, 221, 242, 250–251, 253–254
  dorsal vein of, 108, 214, 242, 250–251, 254–255
  fascia of, 251
  glans, 196, 214, 244, 250–252
  innervation of, 253
  lymphatic drainage of, 253
  root of, 244, 250, 252, 254–255, 264–265
  suspensory ligament of, 107
Perforated substance, anterior, 794
Pericardium, 49
  fibrous, 25, 29, 48, 51, 53, 63
  magnetic resonance imaging of, 84
  parietal layer of, 25
  serous, 25, 50–51, 53, 63
Pericranium, 694, 788
Perilymph, 714
Perineum
  female, 242–243, 256–259, 262–263
  magnetic resonance imaging of, 266–267
  male, 221, 242–245, 248–249
  muscles of, 246t
  posterior view of, 204
Peristaltic wave, 124
Peritoneum
  female pelvis, 225, 238
  general images of, 112, 128, 208, 224
  parietal, 116–117, 132, 156
  of posterior abdomen, 140
  visceral, 116
Perpendicular plate of ethmoid, 588, 596, 680
Phalanges
  of foot, 339
  of hand, 460, 554
  third, 563
Phalanx
  distal
    of foot, 405, 426, 451
    of hand, 461, 532, 554, 563, 566, 568–569
  middle
    of foot, 405, 426
    of hand, 461, 532, 554, 563, 566, 568–569
  proximal
    of foot, 405, 426
    of hand, 461, 532, 549, 554, 563, 566, 568–569
Pharyngeal constrictor
  inferior, 751, 757, 766t, 768–769, 780, 786
  middle, 745, 766t, 768–769, 775–776
  superior, 674, 704–706, 766t, 768–769, 772, 775–776
Pharyngotympanic tube, 28, 663, 671, 684–686, 693, 697, 700, 702–706, 764–765, 774–775
Pharynx, 28, 120, 723
Pinching, 576–577
Pinna, 715

Pisiform, 461, 532, 536, 538–539, 567, 572, 574, 582
Plane
  orbitomeatal, 590–591
Plate
  cribriform, 614–615, 681, 800
  epiphyseal, 451
  horizontal, 594–595, 681
  pterygoid
    lateral, 594–595, 655, 659, 672, 689, 691, 704–706, 764–765, 811
    medial, 594–595, 672–673, 681, 686, 691, 764–765
  tarsal
    inferior, 604
    superior, 604
  tympanic, 594, 764–765
  vertical, of palatine bone, 681
Pleura, 28
  cervical, 28, 65, 67
  costal, 28, 71, 78–79
  cupula of, 64–65
  diaphragmatic, 28, 71
  extent of, 30–31
  mediastinal, 28, 78–79
  parietal, 28–29, 76, 81, 322, 737, 758, 762
  visceral, 28–29
Plexus
  aortic, 75
  brachial, 70, 483, 731–732, 754, 762
    branches of, 487t
    general images of, 22, 70, 80
  cardiac, 66, 818
  carotid, 688, 702, 805, 809
  celiac, 818
  cervical, 786
  choroid, 627, 631, 633, 637, 639, 716–720, 794
  coccygeal, 220
  dental, 808–809
  esophageal, 42, 63, 75
  external
    anterior, 305
    posterior, 305
  hypogastric, 198
    inferior, 168, 170, 202, 219, 221, 237, 239, 253
    superior, 141, 170, 202, 219, 221, 232, 237, 253
  intermesenteric, 168, 170, 237
  lumbar, 164–165
  lumbosacral, 175, 222
  lymphatic, 43
    of palm, 466
    subpleural, 43
  ovarian, 66
  pampiniform, 107–108, 250
  phrenic, 155, 171
  prostatic, 219, 253
  pulmonary, 42, 49, 65–66
  rectal, 219
  renal, 168, 170
  sacral, 220
  subareolar lymphatic, 9
  testicular, 66
  tympanic, 701, 817
  ureteric, 170

uterovaginal, 237, 239
venous
  basilar, 613
  pampiniform, 214
  pial, 329
  prostatic, 209–211, 217
  pterygoid, 613, 646
  rectal, 199, 201, 217
  vertebral, 305, 326, 329, 613, 786
  vesical, 209, 217, 219
vesical, 253
Plica semilunaris, 641
Pole
  frontal, 632, 713
  of kidneys, 157–158
  occipital, 632, 713
  temporal, 617, 622, 794
  transverse, 713
Pons, 617, 626, 637, 684, 713, 715, 788, 794, 804
Popliteal region, 338
Porta hepatis, 114, 143, 148–149
Portacaval system, 153
Portal triad, 117, 128–129, 145–146, 154
Pouch
  hepatorenal, 128
  rectouterine, 116, 224–226, 238–240
  rectovesical, 198, 205, 214–215
  retrovesical, 197
  vesical
    transverse, 225
  vesicouterine, 224–226, 270
Prepuce
  of clitoris, 260–261
  of penis, 196, 251
Process
  accessory
    of lumbar vertebrae, 288
    of thoracic vertebrae, 286, 288
  acromion, 510
  alveolar
    of mandible, 589, 591, 595, 654
    of maxilla, 589, 591, 689
  articular
    inferior, 290, 297, 727
    superior, 290, 297, 300, 302, 727
  ciliary, 650
  clinoid
    anterior, 614, 619–621, 642–643, 687
    posterior, 593, 614–615, 619–621
  coracoid, 27, 460, 478, 491, 493, 498–499,
      508, 513, 515, 662
  coronoid, 524, 528, 532, 591, 654, 657
  frontal
    of maxilla, 589, 640, 680
    of zygomatic bone, 589
  jugular, 595, 618
  lateral, 680
  mammillary
    of lumbar vertebrae, 288
    of thoracic vertebrae, 286
  mastoid, 334–335, 590, 592–595, 653, 693,
      703, 726, 761, 790–791, 813, 817
  odontoid, 777
  palatine, 594–595, 672, 764–765
  spinous, 177, 274, 277, 285, 305, 311, 317,
      334, 763
  styloid, 532, 566, 570, 590, 595, 657, 671, 693,
      726, 766, 768–769, 777

supracondylar, 521
transverse, 727
  coccyx, 293
  thoracic vertebrae, 286, 334
  vertebral, 277, 284
uncinate, 282–283
vaginal, 654
xiphoid, 10, 12, 29, 175, 177, 285, 479
zygomatic
  of frontal bone, 589, 591
  of temporal bone, 591
Processus cochleariformis, 699, 703
Processus vaginalis, 225
Prominence
  laryngeal, 725, 748, 755, 778–779
Prostatic utricle, 214
Protuberance
  mental, 588, 590, 654
  occipital
    external, 314, 316–317, 335, 590–591,
        594–595, 736, 764–765, 788
    internal, 614–615, 788
Pterion, 590, 609
Pterygoid hamulus, 674, 691, 773, 777
Pubic arch, 185, 190–191
Pubic bone, 220, 228, 232, 266–269
Pubis
  body of, 184, 291–292, 362
  pecten, 184, 190–191, 195, 292, 362
  ramus of
    inferior, 291, 362, 379
    superior, 184, 190, 268–269, 291–292, 362
Pulvinar, 634–635
Pupil, 640
Putamen, 631
Pyramid
  renal, 157–158
Pyramids, 617, 635, 715
  decussation of, 635

Radiations
  optic, 630, 632–633
Radiographs
  of ankle, 427
  of atlas, 285
  of axis, 285
  of biliary passages, 150
  of cervical spine, 280, 283
  of chest, 27
  of cranium, 596
  of epiphyses, 339
  of esophagus, 124
  of foot, 427
  of hip joint, 380
  of inferior thoracic spine, 290
  of knee, 400, 402
  of lumbar spine, 281, 289–290
  of pelvis, 291
  of small intestine, 124
  of stomach, 124
Radius, 460
  distal, 460
  dorsal tubercle of, 554
  head of, 498–499, 524, 527, 529
  neck of, 524, 526–527
  proximal, 460
  styloid process of, 532, 554, 572

tuberosity of, 498–499, 524, 527
ulnar notch of, 569
Ramus
  anterior, 324, 761, 763
  gray, 221
  ischiopubic, 190–191, 362
  posterior, 763
Ramus communicans, 17, 78–81, 202, 219–220,
    222, 322, 326, 328
Raphe
  median, 747
  palatine, 673
  perineal, 244
  pterygomandibular, 655, 765, 777
  scrotal, 244
Recess
  apical, 788
  costodiaphragmatic, 25, 28, 71, 76, 86,
      113–115, 155–156
  costomediastinal, 71
  epitympanic, 700–701, 703
  ileocecal
    inferior, 135
    superior, 135
  pharyngeal, 693, 770, 774
  piriform, 770, 782
  sphenoethmoidal, 684
  splenic, 156
  suprapineal, 638
Rectum
  ampulla of, 214, 224
  general images of, 93, 116–117, 120, 193–194,
      198, 200, 207, 210, 232, 234, 238–239,
      249, 263
  innervation of, 202
  lymphatic drainage of, 203
  magnetic resonance imaging of, 264–267
  ultrasound scan of, 211
Referred pain, 169
Renal column, 157–158
Renal medulla, 157
Renal papilla, 157–158
Respiration
  muscles of, 24
Respiratory system, 28
Retina, 801
Retinaculum, 380
  extensor
    inferior, 404, 408, 412–413
    superior, 407–408
  fibular
    inferior, 407
    superior, 412, 416
  flexor, 418–421
  patellar, 396
  synovial, 382
Ribs
  angle of, 308
  anomalies of, 16
  atypical, 13
  bicipital, 16
  bifid, 16
  cervical, 16
  costovertebral articulations, 14
  eighth, 10–11, 13, 18, 308, 491
  eleventh, 10–11, 13, 156, 161–162, 175
  facets of, 13

Ribs—*continued*
  fifth, 10–11
  first, 10–11, 13, 23–24, 26–27, 34–35, 64–65, 482, 508, 737, 762
  floating, 11
  fourth, 10–11, 22, 26
  head of, 13–14, 17, 84, 336
  inferior facet of, 13
  lumbar, 312
  neck of, 13, 313
  ninth, 10–11, 19, 156
  second, 10–11, 13, 23–24, 481, 762
  seventh, 10–11, 18
  shaft of, 13
  sixth, 10–11, 13, 26
  superior facet of, 13
  tenth, 10–11, 26, 155, 308–309
  third, 10–11
  tubercle of, 13–14, 313
  twelfth, 13, 156, 158, 166, 175, 290
  typical, 13
Ridge
  supracondylar
    lateral, 499, 524, 532, 554
    medial, 524, 532
Rima glottidis, 782, 790–791
Ring
  femoral, 195, 352–354
  inguinal
    deep, 106, 112, 241
    superficial, 101, 104–105, 108, 250, 353, 355
  tendinous, 804
  tracheal, 749, 781, 787
Rods, 801
Rugae, 121

Sac
  conjunctival, 644
  dural, 285, 324–325, 336
  endolymphatic, 697, 708–709, 814
  lacrimal, 640–641
  pericardial, 48, 50–51, 78–79, 113
Saccule, 708, 814
Sacral promontory, 184–185
Sacrum
  ala of, 184–185, 190, 195, 290–291, 293, 295
  apex of, 293
  body of, 184, 190, 293
  cornua of, 292–293
  general images of, 158, 184–185, 215, 239, 274, 276, 292, 361
  lateral mass of, 295
  magnetic resonance imaging of, 266–267, 270
  promontory of, 191
Scala tympani, 815
Scala vestibuli, 709
Scalp, 610, 694
Scaphoid, 461, 532, 554, 565–566, 573, 582
Scapula
  coracoid process of, 478, 501–502
  general images of, 10–11, 93, 460
  inferior angle of, 11, 491, 502
  lateral border of, 515
  medial border of, 495, 498–499
  movements of, 509
  spine of, 11

  superior angle of, 492, 498–499
  surface anatomy of, 504
Sclera, 801
Scrotum, 244, 250
  innervation of, 253
  lymphatic drainage of, 253
Sellae
  dorsum, 593, 615, 687, 708, 714, 824
  tuberculum, 615
Seminal vesicle, 112, 196, 204–206, 208, 210–211, 214, 252, 264–267
Seminiferous tubule, 110
Septomarginal trabecula, 57, 61–62
Septum
  intermuscular
    general images of, 371, 386, 391, 421, 452–454
    lateral, 469, 578–579
    medial, 469, 578–579
  interventricular, 57–59, 62, 84
  nasal, 596, 687, 711, 715, 770, 788, 800
  orbital, 604, 645
  rectovesical, 208, 214
Septum pellucidum, 627, 631, 633, 639, 716–720, 788
Sesamoid bone, 420, 429, 447, 449, 548, 569, 584
Sheath
  axillary, 482–483
  carotid, 703, 723, 754, 819
  dural, 645, 803
  femoral, 350, 352–353
  fibrous digital, 541, 551, 563
  hypogastric, 241
  rectus, 112
    anterior, 476–477
  synovial, 541, 548
Sinus
  aortic, 59
  basilar, 686
  carotid, 816
  cavernous, 613, 646, 663, 703
  confluence of, 628
  coronary, 47, 50, 53, 56
    opening of, 62
    valve of, 56
  ethmoidal, 597, 690, 711, 826
  frontal, 596–597, 612, 642, 681, 683–685, 687, 689–690, 788
  intercavernous, 613, 686
  lactiferous, 4–5
  maxillary, 596–597, 686–690, 711, 715, 808–809, 826
  occipital, 613
  pericardial, 51
    transverse, 60
  petrosal
    inferior, 593, 613, 698, 735
    superior, 593, 613, 698
  prostatic, 209
  renal, 157
  sagittal
    inferior, 610, 612–613, 616
    superior, 610–613, 616, 628, 694, 788
  scleral venous, 650
  sigmoid, 321, 593, 613, 616, 735
  sphenoidal, 594, 597, 683–686, 691, 702, 824

  sphenoparietal, 613
  straight, 612, 713, 716–720
  tonsillar, 773
  transverse
    right, 613
Sinus venarum, 56
Sinus venosus sclerae, 651
Small intestine
  general images of, 92–93, 116, 335
  interior of, 132
  magnetic resonance imaging of, 179
  mesentery of, 126, 133, 140, 144
  radiograph of, 124
  referred pain to, 169t
Somatic nervous system, 333
Space
  epidural, 274
  intercostal
    contents of, 20
    eighth, 10
    inferior, 19
    ninth, 11
    vertebral ends of, 17–18
  midpalmar, 542, 584
  palmar, 542, 584
  pharyngeal, 668
  rectorectal, 241
  rectovaginal, 241
  retropharyngeal, 668, 723, 754, 786, 788
  retropubic, 204, 207, 214, 224
  subaponeurotic, 584, 694
  subarachnoid, 301, 325, 330, 610, 628, 800
  suprasternal, 723, 748, 788
  thenar, 542
  triangular, 506
  vesicocervical, 241
Speech area, 626
Spermatic cord, 97–98, 102, 108–110, 210, 214, 250, 255, 264–265, 351, 381
Sphenoid bone, 589–590
  body of, 614
  dorsum sellae of, 614
  greater wing of, 591, 614, 640, 654–655, 691
  lesser wing of, 596, 614, 640, 654, 691
  sinus of, 594
Sphincter
  anal, 197, 199, 201–202, 214
    external, 243, 245, 246t, 248–249
    internal, 248–249
  pyloric, 121
  urethral, 197, 211
    external, 221, 242, 246t, 247
    internal, 221, 270
  urethrovaginal, 242, 247, 259
Spinal cord, 617, 684, 713, 789, 794
  blood vessels of, 329
  cervical region of, 323
  general images of, 83, 177, 274, 285, 313, 322–323, 326–327
  gray matter of, 328
  horn of
    anterior, 323
    posterior, 323
  lumbar region of, 323
  magnetic resonance imaging of, 335
  rootlets of, 326–327
  thoracic region of, 323
  white matter of, 328

Spine
 ethmoidal, 615
 iliac
  anterior inferior, 184–186, 190, 291–292,
   353, 362, 376, 379, 383
  anterior superior, 92, 94, 97, 100, 103–104,
   175, 184–186, 190, 291–292, 338,
   352–354, 357, 362, 364, 376, 378–379,
   382, 476–477
  posterior inferior, 184, 292, 363, 372, 379
  posterior superior, 184, 188, 292, 299, 306,
   312, 363, 371, 379, 495
  ischial, 184–185, 191–193, 195, 221, 264–265,
   291, 294, 298–299, 363, 372
 mental, 655, 747
 nasal, 680
  anterior, 588, 590, 680, 689
  posterior, 594, 672, 764–765
 suprameatal, 609, 654
Spleen
 blood supply to, 127
 general images of, 92–93, 115, 119, 335
 lymphatic drainage of, 172
 referred pain to, 169t
 visceral surface of, 123
Spondylolisthesis, 297
Spondylolysis, 297
Stapes, 697–698, 702
Sternum
 anomalies of, 16
 body of, 10, 12, 175, 285, 479
 general images of, 2, 12, 23, 29, 70
 magnetic resonance imaging of, 83–84
 manubrium of, 24, 723
 ossification of, 16
 pericardial sac in relation to, 48
 transverse ridge of, 12
Stomach, 92, 114, 119, 121
 angular incisure of, 124
 fundus of, 121, 124
 gas in, 39
 greater curvature of, 115, 124
 lymphatic drainage of, 172
 magnetic resonance imaging of, 335
 muscular layer of, 121
 pylorus of, 92, 115, 121, 124, 126, 130
 radiograph of, 124
 referred pain to, 169t
Stria
 olfactory, 800
Stria vascularis, 709
Subarachnoid hemorrhage, 610
Subcutaneous tissue, 3
Subdural hematoma, 610
Substantia nigra, 636, 638, 715
Sulcus
 calcarine, 627, 633
 callosal, 632
 central, 609, 626, 629
 cerebral, 794
 cingulate, 632, 713
 collateral, 632
 frontal, 713
 hippocampal, 632–633
 hypothalamic, 632
 interatrial, 50
 intraparietal, 629

 lacrimal, 689
 lateral, 626, 629
 marginal, 632
 parieto-occipital, 627, 629, 632
 postcentral, 629
 precentral, 629
Suprascapular region, 492
Sustentaculum tali, 421, 426–427, 429, 434–437,
 445–447
Suture
 coronal, 591, 599
 frontonasal, 680
 infraorbital, 640
 intermaxillary, 588, 598, 672
 internasal, 588, 598
 interpalatine, 672
 lambda, 590, 592
 lambdoid, 591–592, 599
 metopic, 588, 598
 occipitomastoid, 592
 parietomastoid, 592
 sagittal, 592–593, 599
Symphysis
 mandibular, 588, 598
 pubic, 100, 116, 184–186, 190–195, 197, 206,
  215, 224, 232, 238, 241, 249, 254,
  258–259, 263–267, 270, 338, 362, 382
Synostosis, 296

Talus, 339, 405, 411, 426, 432–433
Teeth
 canines, 678
 eruption of, 679
 incisors, 676, 678
 innervation of, 677
 mandibular, 676, 678
 maxillary, 676, 678
 molars, 676, 678
 permanent, 675
 primary, 678
 secondary to, 679
Tegmen tympani, 655
Temporal bone, 589–590, 595, 814
 mastoid part of, 591, 706
 petrous part of, 596, 614, 706–707
 squamous part of, 591, 614, 706, 708
 tympanic part of, 590, 654
 zygomatic process of, 654
Tendinous cords, 57, 59, 61
Tendon
 biceps brachii, 519, 521, 529, 536–537
 biceps femoris, 391, 410
 calcaneal, 410, 412, 419–421, 433, 437–438,
  444
 common extensor, 554
 common flexor, 529
 conjoint, 106
 digastric, 765
 extensor digitorum
  brevis, 404
  longus, 404
 extensor hallucis longus, 404
 flexor carpi radialis, 572
 flexor digitorum profundus, 549
 flexor digitorum superficialis, 549
 flexor pollicis longus, 550
 infraspinatus, 493

 obturator internus, 371, 377
 palmaris longus, 541, 546
 plantaris, 385
 popliteus, 391, 393, 395
 psoas minor, 174, 352
 quadriceps, 393, 403
 stapedius, 699, 701
 subscapularis, 514
 terminal, 563
 tibialis anterior muscle, 404
 triceps, 500, 507, 522–523
Tenia coli, 132–133, 135–136, 140, 198
Tentorium
 cerebellar, 612–613, 616, 619, 698, 719, 788
Terminal filum, 197, 322, 324–325, 330
Terminalis
 crista, 56
 lamina, 632
 linea, 186
 stria, 633, 638–639
 sulcus, 49
Testis, 107–108, 110, 197, 214
 appendix of, 110
 blood supply to, 111
 lymphatic drainage of, 111
Thalamus, 627, 716–720
Thigh
 anteromedial aspect of, 364
 blood supply, 375
 bones of, 362
 femoral region of, 338
 lateral aspect of, 365
 magnetic resonance imaging of, 452–454
 muscles of
  anterior, 357, 358t
  medial, 359, 360t, 361
 posterior aspect of, 366, 375
 surface anatomy of, 356
Thoracic wall
 anterior, 23
 external, 22
Thorax
 bony, 10–11
 contents of, 25
 magnetic resonance imaging of, 82
Thymus, 64–65, 749
Tibia, 338–339, 345, 359, 363, 394, 401, 405,
 411, 450
Tongue, 28, 120, 335, 666, 667t, 711, 774, 816
Tonsil, 636, 666
 lingual, 666, 669
 palatine, 666, 672, 674, 704–706, 772–773,
  817
 pharyngeal, 774, 789
Trabeculae carneae, 57–59
Trachea
 arterial supply, 73
 general anatomy of, 25, 39–40, 48, 64–65, 67,
  70, 72, 80, 82, 120, 726, 754, 757, 765,
  787–788
 magnetic resonance imaging of, 87, 285, 335
Tracheobronchial tree, 39–40
Tract
 corticospinal, 719
 fastigiobulbar, 637
 iliopubic, 112
 iliotibial, 345, 354, 357, 359, 362, 365, 367,
  381, 391–392, 395, 404–405, 407, 410

Tract—*continued*
  mammillothalamic, 633
  olfactory, 616–617, 619, 622, 630, 632, 794,
    800
  optic, 617, 631, 635, 794, 801, 824
Tragus, 696, 698
Trapezium, 461, 469, 532, 550–551, 554, 566,
  568–569, 571, 573–574, 582
Trapezoid, 461, 469, 532, 554, 566, 568–569,
  573, 583
Triangle
  carotid, 724–725
  cervical, 724
  deltopectoral, 2–3, 464–465, 474–475, 478
  femoral, 354–355
  lumbar, 308, 494
  muscular, 724
  occipital, 724
  submandibular, 724–725
  submental, 724, 747
  supraclavicular, 724–725
  vertebrocostal, 166
Trigone
  habenular, 634
  hypoglossal, 635
  vagal, 635
Triquetrum, 461, 532, 554, 567, 572
Trochanter
  greater, 291, 338–339, 362–363, 371–372,
    376–377, 381, 383
  lesser, 291, 338–339, 362–363, 372–373,
    376–377
Trochlea, 526, 532, 642
  fibular, 426
Trunk
  brachiocephalic, 46–49, 68, 72, 78, 80, 470,
    736–737, 750, 756, 758, 788
  bronchomediastinal, 9, 43, 735
  celiac, 116, 119, 125, 127, 155, 166–168, 170
  costocervical, 72, 736–737
  dermatomes of, 95
  general anatomy of, 94
  lumbosacral, 164–165, 192, 195, 219–222,
    253, 266–267, 270
  lymphatic
    bronchomediastinal, 9, 74
    jugular, 9, 74
    subclavian, 9, 43, 74
  pulmonary, 26–27, 41, 46, 49–52, 57, 59, 63,
    66, 83, 86–87
  sympathetic, 17, 70–71, 75, 78–81, 164, 168,
    170, 174, 219–222, 237, 239, 253, 322,
    756–757, 762, 786
  thyrocervical, 68, 319, 470, 736–737, 758
  vagal, 171, 818
Tubercle
  adductor, 354, 362–363, 367, 415
  articular, 653, 659, 665
  of atlas, 282–285, 316–317, 320, 727
  auricular, 696
  carotid, 282–284, 727, 737, 761
  corniculate, 781–782
  cuneate, 635–637
  cuneiform, 781–783
  epiglottic, 782
  Gerdy, 405
  gracile, 635–636

greater, 483, 493, 498–499, 501–502
  iliac, 345
  jugular, 327, 593, 615
  lesser, 493, 498–499
  medial, 415, 434, 436–437, 439
  mental, 588, 591, 654
  pharyngeal, 595
  posterior, 727
  postglenoid, 654–655, 665
  pterygoid, 681
  pubic, 100, 102–105, 184–185, 190–191, 291,
    353–354, 376, 379
  radial, 566
  of rib, 13–14, 313
  ribs, 13–14
  of scaphoid, 568–569
  of trapezium, 551, 568–569, 573
Tuberculum cinereum, 635, 717
Tuberosity
  calcaneal, 424, 426, 429
  cuboid, 426
  deltoid, 498
  gluteal, 363
  ischial, 184, 188, 192, 240, 248–249, 255, 259,
    264–265, 268–269, 291–292, 338, 345,
    363, 367, 372–373, 378–379
  pronator, 554
  radial, 526
  sacral, 293
  tibial, 404–405, 407
  ulna, 532
Tunica albuginea, 110
Tunica vaginalis, 107, 109–111, 197

Ulna, 460, 571
  distal, 460
  head of, 554, 560, 569–570
  proximal, 460
  styloid process of, 554, 569
Ultrasound
  of abdomen, 180–181
  of hepatic veins, 145
  of pelvis
    female, 230–231
    male, 211
Umbilicus, 94–95, 112, 153, 476
Umbo, 699
Uncus, 632, 715
Upper limb
  anastomoses of, 470–471
  arteries of, 470–472
  bones of, 458–460, 497t
  cutaneous nerves of, 462
  deep fascia of, 468–469
  deep veins of, 467
  dermatomes of, 463
  lymphatic drainage of, 466
  muscles of
    innervation of, 472–473
  regions of, 458–459
  venous drainage of, 464–465
Urachus, 252
Ureter, 125, 139, 176, 252
  anomalies of, 160
  bifid, 160
  female, 232, 242
  general images of, 93, 112, 128–129, 141, 154,
    174, 198, 204, 206, 228, 239

male, 242
  referred pain from, 169t
  retrocaval, 160
Urethra, 116, 194, 206, 234
  female, 263, 270
  innervation of, 221, 253
  intermediate, 252
  lymphatic drainage of, 253
  magnetic resonance imaging of, 270
  male, 255, 266–267
  spongy, 209, 214, 254
  ultrasound scan of, 211
Urinary bladder
  female, 268–270
  general images of, 92–93, 112, 116–117, 132,
    158, 175, 195, 197, 207, 210, 214–215,
    239
  innervation of, 221
  magnetic resonance imaging of, 268–270
  male, 209, 221, 252, 264–265
  trigone of, 209, 228, 232
  ultrasound scan of, 211
Uterine tube, 117, 224–225, 227–228, 230, 233,
  238
  lymphatic drainage of, 237
Uterus, 116–117, 227
  adnexa of, 227
  body of, 270
  cervix of, 270
  fundus of, 228, 230, 270
  lymphatic drainage of, 237
  magnetic resonance imaging of, 268–270
  round ligament of, 104–105, 117
Utricle, 708, 814–815
Uvula, 672, 686, 770, 772, 790–791

Vagina, 116, 195, 224, 233–234, 263
  fornix of, 230, 233, 238
  lymphatic drainage of, 237
  magnetic resonance imaging of, 268–270
  vestibule of, 257, 261, 268–269
Vaginal wall, 258–259
Vallecula, 669, 787
Valve
  anal, 199
  aortic, 55, 59–60, 84, 88
  atrioventricular, 61
  heart, 61 (*see also specific valve*)
    auscultation of, 44
    surface markings of, 44
  mitral, 58–59
  pulmonary, 57, 60–61
  tricuspid, 57
Vas deferens, 211
Vasa recta, 129, 136
Vasa recta duodeni, 127
Veins
  angular, 607t, 734
  appendicular, 152
  auricular
    great, 652
    posterior, 317, 601, 652, 710, 734
  axillary, 22, 467, 481, 483, 732–733
  azygos, 76–77, 153, 174, 176, 217, 326
    arch of, 47, 49, 66–67
    general anatomy of, 29, 71, 73–74, 78,
      80–81
    magnetic resonance imaging of, 83–84, 88

basilic, 464–467, 481–482, 489–490, 503, 518–519, 567, 578–579
basivertebral, 304–305, 329
brachial, 466–467
brachiocephalic, 23, 733
    left, 46, 48, 64–65, 74, 76, 78–79, 85, 467, 735, 758, 788
    right, 46, 48–49, 64–65, 70, 74, 76, 78, 82, 87, 467
cardiac, 53
    anterior, 46, 49
    great, 46–47, 49–50, 53, 58
    middle, 47, 50, 53
    small, 46–47, 50, 53
cephalic, 3–4, 464–466, 474, 479, 481, 564, 578–580
cerebral
    great, 612–613, 716–720, 788
    internal, 631, 638
    superior, 612, 788
cervical
    deep, 316–317
    transverse, 467, 481, 731, 734, 739
circumflex femoral
    lateral, 347
    medial, 347, 355
circumflex humoral
    anterior, 467
    posterior, 467
circumflex iliac
    deep, 102, 105, 198, 217
    superficial, 97, 342
circumflex subscapular, 467
colic
    left, 152
    middle, 117, 119, 152
    right, 126, 152, 178
collateral
    of elbow joint, 467
cremasteric, 102
cubital
    medial, 464–465, 518
cystic, 148, 152
digital
    dorsal, 342, 562
    of hand, 464–465
    palmar, 464–465, 467
diploic
    anterior temporal, 608
    frontal, 608
    occipital, 608
    posterior temporal, 608
dorsal
    of penis, 108
emissary
    condylar, 728
    mastoid, 728
epigastric, 153
    inferior, 98, 103, 205, 217
    superficial, 342, 350
esophageal, 153, 217
facial, 600–601, 607t, 613, 646, 652, 656, 668, 694, 710, 730, 734–735, 738, 786
femoral, 103, 112, 195, 210, 240, 263–269, 342, 347, 352–353, 381, 452–454
femoral cutaneous
    lateral, 342
    medial, 342

fibular, 343, 347
    common, 347
gastric
    left, 119, 126, 149, 152–153, 217
    right, 119, 121, 148–149, 152
    short, 121, 152
gastrocolic, 119
gastro-omental, 121, 152
genicular
    descending, 347
gluteal
    inferior, 217, 347
    superior, 217, 264–265, 347
great saphenous, 97–98, 106
hemiazygos, 76–77, 81, 176, 326, 335
    accessory, 76–77
    magnetic resonance imaging of, 84
hepatic, 145–146, 155, 176, 178–179
ileal, 152
ileocolic, 126, 152
iliac
    common, 77, 214, 217, 232, 266–267
    external, 103, 155, 192, 198, 204, 214–215, 217, 241, 352, 361
    internal, 214, 217, 241, 264–265, 270, 352
iliolumbar, 77
infraorbital, 646
intercerebral, 716–717
intercostal, 326
    anterior, 22–23
    general anatomy of, 79
    left superior, 65, 74, 76–77, 79
    posterior, 76–77, 80
    superior, 70
interosseous, 467, 580
intervertebral, 305, 329
jejunal, 152
jugular
    anterior, 481, 710, 734–735, 739, 749
    external, 467, 481, 601, 652, 730–731, 734, 744, 790–791
    internal, 48, 319, 467, 481, 652, 666, 668, 702, 710, 714, 723, 728, 731, 738–739, 750, 754, 756, 768–769, 771, 819
        left, 9, 43, 82
        right, 26, 43
    of knee
        lateral inferior, 347
        lateral superior, 347
        medial inferior, 347
lingual, 670, 735, 746
lumbar, 77, 167
marginal, 53
maxillary, 613, 710, 734
median
    of axilla, 482
    of forearm, 464–465, 580
mesenteric
    inferior, 117, 119, 125–126, 141, 152–153, 176, 217
    superior, 119, 125–126, 141, 152, 154, 177–178, 180–181
oblique, 53
obturator, 193–194, 215, 217, 240
occipital, 317, 734
ophthalmic
    inferior, 613
    superior, 613, 646

ovarian, 225
palatine, 673, 775–776
pancreatic, 152
pancreaticoduodenal, 152
    posterior superior, 148
paraumbilical, 112, 153, 217
perforating, 518, 564
perivaginal, 268–270
pharyngeal, 735
plantar, 347
popliteal, 343–344, 367, 384, 386, 417, 419–421
portal, 117, 125–126, 128, 143, 148–149, 152–153, 178, 180–181
pudendal, 201, 210
    external, 342, 350
    internal, 217, 261, 347
pulmonary, 34
    inferior, 78–79
    left, 33, 41, 46–47, 51, 58, 62–63, 66
    right, 33, 46–47, 50–51, 58, 60, 63, 66, 83, 88
radial, 467, 580
radicular
    anterior segmental, 329
    posterior segmental, 329
rectal, 201, 217
    inferior, 153, 217
    middle, 153, 217
    superior, 152–153, 217
recurrent
    radial, 467
    ulnar, 467
renal, 76–77, 119, 154, 177–181
retromandibular, 607t, 652, 668, 693, 710, 734, 790–791
retroperitoneal, 153
saphenous
    great, 97–98, 106, 263, 342–343, 345, 350–354, 364
    small, 342, 344, 384, 412
scapular, dorsal, 467
sigmoid, 152
spinal, anterior, 329
splenic, 115, 119, 126, 152–153, 156, 177–181
subclavian, 467, 481, 728, 733, 735, 756, 758–759
    general anatomy of, 22, 26, 48, 78, 491
    left, 43, 79, 82
    right, 43, 735
subcostal, 77
supraorbital, 600, 607t, 613, 646
suprascapular, 467, 481, 732, 734
supratrochlear, 607t, 734
temporal
    superficial, 607t, 652, 734
testicular, 119, 128, 141, 154–155, 198, 215
thalamostriate, 634
thoracic
    inferior, 750
    internal, 22–23, 64–65, 70, 78–79, 758
    lateral, 467
    superior, 467, 750
thoracoacromial, 480
thoracodorsal, 467
thyroid, 735, 738
    inferior, 64–65, 758

Veins—*continued*
    middle, 750
    superior, 751, 757, 786
  tibial
    anterior, 419–421
  ulnar, 467, 580
  ventricular
    left posterior, 47, 53
  vertebral, 754, 756, 786
  vesical
    inferior, 217
  vorticose, 650–651
Velum
  inferior medullary, 636
  superior medullary, 636
Vena cavae
  inferior
    computed tomography of, 6
    general images of, 46–47, 50–51, 56, 63, 71,
      74, 76–78, 80, 113, 128–129, 143, 146,
      152, 174, 179, 198, 228, 352
    groove for, 34
    magnetic resonance imaging of, 85, 87
    radiograph of, 27
  superior
    general anatomy of, 41, 47–51, 56–58, 60,
      62–63, 65, 70, 76, 467, 735
    groove for, 34
    magnetic resonance imaging of, 85, 87–88
    radiograph of, 27

Ventricles
  of brain, 617–618, 627–628
  fourth, 617–618, 627–628, 684, 714, 794, 824
  of heart
    left, 6, 27, 46–47, 49–50, 58–60, 62, 85
    right, 46, 49–50, 57, 59, 62, 84, 86, 88
  lateral, 628, 631, 633–634, 638, 713, 716–720
  third, 627–628, 632, 634, 684, 716–720, 826
Vermis
  inferior, 636, 714
  superior, 636
Vertebrae
  anatomy of, 277
  anomalies of, 296
  cervical, 282–284, 726, 763
    homologous parts of, 278
    movements of, 280
    overview of, 274, 276
    radiographs of, 280
  costovertebral articulations, 14
  homologous parts of, 278
  lumbar, 288–290, 324–325
    homologous parts of, 278
    movements of, 281
    overview of, 274, 276
    radiograph of, 290
    radiographs of, 281
  magnetic resonance imaging of, 334
  sacral, 324
    homologous parts of, 278

    spondylolisthesis of, 297
    spondylolysis of, 297
  thoracic, 82–84
    homologous parts of, 278
    overview of, 274, 276, 286–287
    radiograph of, 290
Vertebral column
  curvatures of, 275
  lumbar region of, 304
  overview of, 274
Vestibule, 228, 684, 694, 708
Vincula brevia, 549, 563
Vincula longa, 549, 563
Visual pathways, 802
Vomer, 588–589, 594–595, 672, 680, 691, 693,
    764–765

Window
  oval, 708
  round, 699, 702
Wrist
  anterior aspect of, 539
  epiphyses of, 461
  lateral aspect of, 564–566
  medial aspect of, 567
  skin creases of, 540

Zygomatic bone, 589, 595, 640, 644, 690